AA002262

2006 5th International Power Electronics and Motion Control Conference

Shanghai, China
13-16 August 2006

Volume 3 of 4

IEEE Catalog Number: 06EX1405
ISBN: 1-4244-0448-7

Copyright © 2006 by The Institute of Electrical and Electronics Engineers, Inc.
All Rights Reserved

Copyright and Reprint Permissions: Abstracting is permitted with credit to the source. Libraries are permitted to photocopy beyond the limit of U.S. copyright law for private use of patrons those articles in this volume that carry a code at the bottom of the first page, provided the per-copy fee indicated in the code is paid through Copyright Clearance Center, 222 Rosewood Drive, Danvers, MA 01923.

For other copying, reprint or republications permission, write to IEEE Copyrights Manager, IEEE Operations Center, 445 Hoes Lane, Piscataway, New Jersey USA 08854. All rights reserved.

IEEE Catalog Number: 06EX1405

ISBN: 1-4244-0448-7

Library of Congress: 2006925601

Additional Copies of This Publication Are Available from:

IEEE Service Center
445 Hoes Lane
Piscataway, NJ 08854
IEEE Service Center
445 Hoes Lane
Piscataway, NJ 08854
Phone: (800) 678-IEEE
            (732) 981-1393
Fax: (732) 981-9667
E-mail: customer-service@ieee.org

2006 5th International Power Electronics and Motion Control Conference

Shanghai, China
13-16 August 2006

IEEE Catalog Number: CFP06792-POD
ISBN: 978-1-42440-448-3

Table of Contents

Design Challenges For Distributed Power Systems 1
Fred C. Lee, Ming Xu, Shuo Wang, Bing Lu

A Smarter Grid for Improving System Reliability and Asset Utilization 16
D. Divan, H. Johal

Medium-Voltage Power Conversion Systems in the Next Generation 23
Hirofumi Akagi, Shigenori Inoue

Modern Electrical Drives: Design and Future Trends 31
R. W. De Doncker

Power Semiconductors development trends 39
L. Lorenz

Power Electronics in Wind Turbine Systems 46
F. Blaabjerg, Z. Chen, R. Teodorescu, F. Iov

Sustainable Energy and Mobility, and Challenges to Power Electronics 57
C.C.Chan

Wind farms with increased transient stability margin provided by a STATCOM 63
Marta Molinas, Jon Are Suul, Tore Undeland

A New Super Junction LDMOS with N+-Floating Layer 70
Baoxing Duan, Bo Zhang, Zhaoji Li

**Unified Power Flow Controller: Comparison of Two Advanced Control Schemes and Performance
Analysis for Power Flow Control** 74
Liu Liming, Zhu Pengcheng, Kang Yong, Chen Jian

A New Analytical Model for the Surface Electrical Field Distribution of Double RESURF LDMOS 79
Qi Li, Zhaoji Li

A Novel Centralized HID Ballast System with Power-Bus 83
Xiaodong Lu, Bo Yang, Jiande Wu, Xiangning He

Research on a Novel Structure of SiGeC/Si Heterojunction Power Diodes 88
Liu Jing, Gao Yong, Ma Li

Gate driving of high power IGBT by wireless transmission 92
Stéphane Bréhaut, François Costa

The Characteristics of Thyristor Controlled Reactance Series Compensation by Adjustable Coupling 97
Guo-rong Zhu, Min-zu Li, Yong Kang

Dual-Side Cooled Novel IPM and Improved Capability of Inverter for Elevated-Temperature Operations 102
Jie (Jay) Chang, Changming Liao

An Improved Current-Doubler with Coupled Inductors 108
T.-F. Wu, C.-T. Tsai, W.-C. Lin, Y.-M. Chen

Monolithic Integration of Trench Power JFET with Schottky Diode 113
Yang Gao, Jie Chen, Alex. Q. Huang

Sequential Color LED Back-Light Driving System for LCD Panels 117
C.-C. Chen, C.-Y. Wu, P.-C. Lu, Y.-M. Chen, T.-F. Wu

**Development of Large Capacity Programmable Harmonic Current Generator Based on Three-phase-
four-wire Configuration** 122
LIU Tao, ZHUO Fang, CHEN Bo, ZHAI Xi, WANG Zhao-an

A Universal Digital Platform and Software Library for Power Electronic Systems Integration 127
Haibing HU, Tianjun Jin, Wenxi YAO, Zhengyu LU, Zhaoming Qian

Table of Contents

Unipolar SiC Devices - Latest Achievements on the Way to a New Generation of High Voltage Power Semiconductors 132
Peter Friedrichs

Implementation of GA-trained GRNN for Intelligent Fast Charger for Ni-Cd Batteries 137
Panom Petchjatuporn, Noppadol Khaehintung, Khamron Sunat, Phaophak Sirisuk, Wiwat Kiranon

Modelling and Analysis of a Novel Transformer with Ability to Suppress Conducted interference 142
Zongxiang Chen, Pengsheng Ye, Junmin Pan

An Observer-Based Three-Phase Current Reconstruction using DC Link Measurement in PMAC Motors 147
Li Ying, Nesimi Ertugrul

Experiment Research of Chaotic PWM Suppressing EMI in Converter 152
R. Yang, B. Zhang , F.Li, J.J. Jiang

Emitter Size Effect in 4H-SiC BJT 157
Yan Gao, Alex Q. Huang, Sumi Krishnaswami, Anant K. Agarwal, Charles Scozzie

PSIM and SIMULINK Co-simulation for Threelevel Adjustable Speed Drive Systems 161
Zhang Yongchang, Zhao Zhengming, Baihua, Yuan Liqiang, Zhang Haitao

Three-Phase Z-Source AC-AC Converter for Motor Drives 166
Xu-Peng Fang

Construction and Application of Macro Model for ZVS Resonant Mode Controller MC34067 171
Wei Chen, Yilei Gu, Zhengyu Lu, Zhaoming Qian

Optimum Design of Hollow Conductor in Stator Winding for Large Evaporative Hydro-generator 175
Z. Wen , L. Ruan, G. Gu

Rotor Suspension Principle and Decoupling Control for Self-bearing Induction Motors 179
Tengchao Zhang, Huangqiu Zhu, Yuxin Sun

Field Oriented Control of Linear Induction Motor Considering Attraction Force & End-Effects 184
Jianqiang Liu, Fei Lin, Zhongping Yang, Trillion Q. Zheng

Series Resonant High Frequency Link Sine-wave Inverter System Modeling using Sampled Data 189
Jin Xiaoyi, Dong Wei, Sun Xiaofeng, Wu Weiyang

Maximal Power Point Tracking under Speed-Mode Control for Wind Energy Generation System with Doubly Fed Introduction Generator 194
Y. Zhao, X. D. Zou, Y. N. Xu, Y. Kang, J. Chen

Effective Mobility in Nano-Scaled n-MOSFETs 199
Yue-Hua Dai, Jun-Ning Chen, Dao-Ming Ke, Jia-E Sun

Investigation on the Factors Affecting Inrush Current of Transformers Based on Finite Element Modeling 204
M. Reza Feyzi, M. B. B. Sharifian

An Improved Support Vector Machine Method for Harmonic and Inter-harmonic Detecting 209
Ma Li, Liu Kaipei, Lei Xiao

A Common Mode and Differential Mode Integrated EMI Filter 214
Liu Nan, Yang Yugang

Electromagnetism Model and Characteristic Simulation of Novel Claw Pole Generator with Permanent Magnet Outer Rotor 219
Fengge Zhang, Haijun Bai, Shifu Zhang, Hans Pert Gruenberger, Eugen Nolle

An Improved Adaptive Filter for Voltage and Current Reference Extraction 224
A. Abedini, A. Nasiri

Simulation Analysis on Current SVM Algorithm of Matrix Rectifier 229
Xi-jun Yang, Peng-sheng Ye, Xiang Liu, Xing-hua Yang, Jian-quan Wang, Luan-guo Zhang

Table of Contents

Study of Measurement Approach of Loop Gain of Converter 236
Weiping Zhang, Yunpeng Chen, Yuanchao Liu, Dongyan Zhang, Zheng Meng

A Stand-Alone Hybrid Generation System Combining Solar Photovoltaic and Wind Turbine with Simple Maximum Power Point Tracking Control 242
Nabil A. Ahmed, Masafumi Miyatake

Design Optimization of Industrial Motor Drive Power Stage Using Genetic Algorithms 249
F. Wang, W. Shen, D. Boroyevich, S. Ragon, V. Stefanovic, M. Arpilliere

FEM Based Simulation of a Permanent Magnet Synchronous Motor Performance Characteristics 254
L. Petkovska, G. Cvetkovski

Analytical Modeling of Semiconductor Losses in Matrix Converters 259
Bingsen Wang, Giri Venkataramanan

Nonlinear Robust Sliding Mode Control for PM Linear Synchronous Motors 267
Xi Zhang, Junmin Pan

Dynamic Analysis of PWM Switching DC-DC Converters 272
Liu Jian, Wang Yuanbin

A Novel LLC Resonant Converter Topology: Voltage Stresses of All Components in Secondary Side Being Half of Output Voltage 276
Yilei Gu, Zhengyu Lu, Zhaoming Qian

On the hybrid automaton models and control synthesis of a single inductor, double output boost converter 281
Sreekumar C, Vivek Agarwal

Complex Intermittency in Voltage-Mode Controlled Buck Converter 286
Zheng-Ping Li, Yu-Fei Zhou, Jun-Ning Chen

Dual Mode Control Multiphase DC/DC Converter for CPU Power 291
Li-Wei Lin, Chung-Hsing Chang, Huang-Jen Chiu, Shann-Chyi Mou

An Analog Implementation of Pulse-Width-Modulation Based Sliding Mode Controller for DC-DC Boost Converters 296
Siew-Chong Tan, Y. M. Lai, Chi K. Tse

Low Cost Electronic Ballast with Buck Converter as PFC Stage 301
Li Xiangrong, Xu Dianguo, Zhang Xiangjun

A New Converter Architecture for Future Generations of Microprocessors 306
Dodi Garinto

A Combined ZVS Converter with Naturally Sharing Input-Current and High Voltage Gain 311
Linbing Wang, Bo Yang

Matrix Coefficient Polynomial Description Model of DC-DC Converters Based on Switched Linear Systems 316
Yongping Zhang, Bo Zhang, Zongbo Hu, Dongyuan Qiu, Guiping Du

Development of DC-DC Multiple Converter based on Push-pull Forward Topology accomplished 321
Weihao Hu, Yunqing Pei, Zhaoan Wang

Voltage Fed and Current Fed Full Bridge Converter for the Use in Three Phase Grid Connected Fuel Cell Systems 325
M. Mohr, F.-W. Fuchs

Small-Signal Modeling of Asymmetrical Half Bridge Flyback Converter 332
Tso-Min Chen, Chern-Lin Chen

A DSP Based Controller for High Power Dual-Phase DC-DC Converters 337
Xin Guo, Xuhui Wen, Ermin Qiao

Table of Contents

Effective Load Resistance; A New Method to Evaluate DC/DC converters Efficiency 342
Alan Elbanhawy

Calculation of Power Loss in Output Diode of a Flyback Switching DC-DC Converter 346
Jiaxin Chen, Jianguo Zhu, Youguang Guo

A Multiple Output Forward Converter Adopting Weighted Time-Sharing Control and Switch-Linear Hybrid Scheme 351
Xiaodong Liu, Songqin Hu, Sizhou Sun

A Novel Soft-Switching PWM Full-Bridge DC/DC Converter with DC Busline Series Switch-Parallel Capacitor Edge Resonant Snubber Assisted by High-Frequency Transformer Leakage Inductor 356
Khairy Fathy, Toshimitsu Doi, Keiki Morimoto, Hyun Woo Lee, Mutsuo Nakaoka

High-Efficiency Cascode Forward Converter of Low Power PEMFC System 361
Jiann-Fuh Chen, Wei-Shih Liu, Ray-Lee Lin, Tsorng-Juu Liang, Ching-Hsiung Liu

Control of Bifurcation by Fuzzy Logic Controller for Current-mode Boost Converters 368
Noppadol Khaehintung, Phaophak Sirisuk, Anantawat Kunakorn

An Improved Three-Level Soft-Switching DC/DC Converter 373
Z. L. Lou, Z. S. Wang

A Novel Soft Switching Bidirectional DC/DC Converter and Design Consideration 378
Ma Gang, Qu Wenlong, Liu Yuanyuan

State-Variable Description and Analysis of a DC-Rail ZVT Inverter Feeding a Permanent Magnet Synchronous Motor 382
Ming Zhengfeng, Zhong Yanru

Analysis, Simulations and Experiments Of A Novel ZVS -ZCS Inverter With Pulse Current Feedback Transformer Auxiliary Commutation 386
Yaogang, Mahamnad Mansoor Khan, Chenchen

A Novel Eddy-Current Based Far-Infrared Rays Radiant Planner Heater using High-Frequency ZVT-PWM Inverter 392
Hisayuki Sugimura, Bishwajit Saha, Hideki Omori, Hyun Woo Lee, Mutsuo Nakaoka

3 Phases-3 Devices AC Voltage Regulator With Quasi-Zero Switching 397
Qianzhi Zhou, Wenhua Hu, Bin Wu

Study on Power Decoupling Control of Three Phase Voltage Source PWM Rectifiers 401
Wang Jiuhe, Yin Hongren, Zhang Jinlong, Li Huade

A Fully Digital Controlled 3KW, Single-Stage Power Factor Correction Converter Based on Full-Bridge Topology 406
HANG Li-jun, YANG Yue-feng, SU Bin, LU Zheng-yu, QIAN Zhao-ming

A New ZVT Power Factor Corrected Three-Phase AC-AC Converter with Single-Phase HF Link 411
T. H. Abdelhamid, A. Sabzali

Simple Bridge-Type AC/DC Converters with Natural Input-Current-Shaper 417
Hsing-Fu Liu, Chih-Yu Wu, Chin Sun, Lon-Kou Chang

Rough Controlling TSC for Reactive Current Compensation in Traction Substations 423
Hongsheng Su, Qunzhan Li

A Digitally Controlled 4-kW Single-Phase Bridgeless PFC Circuit for Air Conditioner Motor Drive Applications 428
Yong Li, Toshio Takahashi

Optimized Electrical Design for Single Phase PFC Active IPEM 433
Qiaoliang Chen, Xu Yang, Zhao-an Wang

Table of Contents

A Novel Topology of APFC with On-Line Half-Bridge UPS Controlled by DSP 438
Xuejun Ma, Xuezhi Hu, Hongxia Wu, XuWu Chen

Nonlinear Current Control of Single-Phase PFC Suitable for Mixed-Signal IC Implementation 442
Min Chen, Anu Mathew, Jian Sun

A Novel Detection Method for Three-Phase Reactive Current 449
Zong Ming, Wang Fengxiang, Hua Funian, Sun Yidan

**Selective Harmonic Controlling for Three-Level High Power Active Front End Converter with Low
Switching Frequency** 453
Hui Zhang, Kaipei Liu,

A Unity Power Factor Three-Phase Buck Type SVPWM Rectifier Based on Direct Phase Control Scheme 458
LI Yabin, Li Heming, Peng Yonglong

3-Phase Current-Source SMES-UPS Based on TFSC and its Control Strategies Control Strategies 463
WANG Fu-sheng, LI Hong-mei

**A novel control scheme of 230kA DC power source using thyristor, Phase-shifting rectifier transformer
and On-load tap changer** 468
Qiao Shutong, Jiang Jianguo, Zuo Dongsheng, Wu Xiaojie

Research on Control Method of Double-Mode Inverter with Grid-Connection and Stand-Alone 473
Herong Gu, Zilong Yang, Deyu Wang, Weiyang Wu

Power and Energy Management of a Dual- Energy Source Electric Vehicle - Policy Implementation Issues 478
P.C.K. Luk, L.C. Rosario

**Study on Non Contact Automatic on-Load Voltage Regulating Distributing Transformer Based on Solid
State Relay** 484
Zhao-Yulin, Dong-Shoutian, Li-Jiahui, Yao-Xin, Zheng-Na, Liu-Xueli

The Principle of a Novel Arc-suppression Coil and its Implementation 489
Cheng Lu, Chen Qiaofu, Zhang Yu, Zhang Changzheng

Grid Connection to Stand Alone Transitions of Slip Ring Induction Generator During Grid Faults 494
G. Iwanski, W. Koczara

System Control of Power Electronics Interfaced Distribution Generation Units 499
D. Feng, Z. Chen

**Test Loadability of Power Systems using A Networked Power Electronic Devices Control and
Measurement System** 505
Sheng Yang, Venkataramana Ajjarapu, Bo Zhang

**Test-Bed of Doubly Fed Induction Generator for Variable-Speed Constant-Frequency Wind Power
Generation** 510
S. Y. Yang, X. Zhang, C. W. Zhang, R. X. Cao

Control strategy of Hybrid sources for Transport applications using supercapacitors and batteries 515
M.B. Camara, H. Gualous, F. Gustin, A. Berthon

Wind Generator Stabilization With Doubly-Fed Asynchronous Machine 520
Li Wu, Zhixin Wang

**Design Consideration of a Novel Digital Bidirectional Constant Current Source Used in Hybrid Electric
Vehicle** 526
Qingbo Hu, Zhengyu Lü

A Single-Phase Grid-Connected Inverter System With Zero Steady-State Error 532
Guo Xiaoqiang, Zhao Qinglin, Wu Weiyang

DC Transformer with Line Frequency Ripple Cancellation 537
Sen Dou, Wilson Wu, Annabelle Pratt, Pavan Kumar

vii

Table of Contents

A Novel PWM Method for Stacked Flying Capacitor Inverter 542
Gangui Yan, Gang Mu, Yafeng Huang, Wenhua Liu

Study on a New Method of Voltage-Source Induction Heating Load-Matched 549
Li Jin-gang, Zhong Yan-ru, Zhao Miao

An Alternating-master-salve Parallel Control Research for Single Phase Paralleled Inverters Based on CAN Bus 554
Zhang Chunjiang, Chen Guitao, Guo Zhongnan, Wu Weiyang

Analysis and Design of a Novel Dual Secondary Winding and Dual Power Bridge High Frequency Link Inverter 559
Zhang Zhe, Zhang Chunjiang, Wu Weiyang, Gu Herong, Shen Hong

Reduction of Common Mode EMI in a Full-Bridge Converter through Automatic Tuning of Gating Signals 564
Kai Zhang, Yunbin Zhou, Yonggao Zhang, Yong Kang

Phase Multilevel Inverter Fault Diagnosis and Tolerant Control Technique 569
Wang Baocheng, Wang Jie, Sun Xiaofeng Wu Junjuan, Wu Weiyang

Microcontroller-Based Single Phase Inverter Using a New Switching Strategy 574
K. Meghriche, O. Mansouri, A. Cherifi

Study of Stability Regions in Parallel Connected Boost Converters 580
Yuehui Huang, Chi. K. Tse

A Novel Analysis and Design Method for Integrated Magnetics 585
Zheng Feng, Weihao Hu, Pei Yun-qing

Investigation on the Space Vector PWM for Large Power Three-Level DC-Link Voltage Source Inverter Equipped with IGCTs 589
Wang Chengsheng, Li Chongjian, Li Yaohua, Zhao Xiaotan

Status and Opportunities of Photovoltaic Inverters in Grid-Tied and Micro-Grid Systems 593
Xiaoming Yuan, Yingqi Zhang

Adaptive Neuro-Fuzzy Control with Fuzzy Supervisory Learning Algorithm for Speed Regulation of 4-Switch Inverter Brushless DC Machines 597
A. Halvaei Niasar, H. Moghbelli, A. Vahedi

Combined Modulation and Harmonic Suppression 602
Cheng Weibin, Zhong Yanru, Jin Shun

Application Research of Maximum Wind-energy Tracing Controller Based Adaptive Control Strategy in WECS 607
Changhong Shao, Xiangjun Chen, Zhonghua Liang

Research on Synchrodrive Control Technology for Wind Turbine Adjustable-Pitch System Based on Adaptive decoupling Control 612
Hongche Guo, Qingding Guo

Limit-Trajectory Single- and Two-Mode Overmodulation Technology 617
Shun Jin, Yan-ru Zhong

Multiphase Permanent Magnet Motor Drive System Based on A Novel Multiphase SVPWM 622
Shan Xue, Xuhui Wen, Zhao Feng

Novel Random-Harmonic Elimination PWM Technique for Single-Switch Three-Phase AC-DC Buck Converter 627
Guang-Hui Tan, Wenchuan Ma, Yanchao Ji, Hongxiang Yu, Wancai Xu

viii

Table of Contents

FPGA Based Multichannel PWM Pulse Generator for Multi-modular Converters or Multilevel Converters 632
Liqiao Wang, Weiyang Wu

Cascaded Multilevel Converters with Non-Integer or Dynamically Changing DC Voltage Ratios 637
Shuai Lu, Keith A. Corzine

Practical Thermal Design Considerations for IPEM-based Converter 642
Qiaoliang Chen, Xu Yang, Zhao-an Wang

Realization of an FPGA-Based Space-Vector PWM Controller 647
Zhou Yuan, Xu Fei-peng, Zhou Zhao-yong

Chaotifying Control of Permanent Magnet Synchronous Motor 652
Hai Peng Ren, Chong Zhao Han

Analysis of PMLSM Direct Thrust Control System Based on Sliding Mode Variable Structure 657
Junyou Yang, Guofeng He, Jiefan Cui

Carrier-based Pulse Width Modulation for Three-Level Inverters: Neutral Point Potential and Output Voltage Distortion 662
Jang-Hwan Kim, Seung-Ki Sul

AC Current Sensorless Control of Three-Phase Three-Wire PWM rectifiers under the Unbalanced Source Voltage 669
Jia-peng Xu, Yu-peng Tang

Waveform Library Control of Converter 674
Xiaofeng Sun, Bin Wang, Meng Lingjie, Weiyang Wu

d-model Adaptive Algorithm Based on Plant-Parameterization 679
Zhao Feng, Liu Weiguo

Dynamics and Control of Electronic Cascaded Systems 684
Wen Wei, Xu Haiping, Wen Xuhui, Shi Wenqing

The Controlling Strategy for Electronic Ballast of HID Lamps 688
Weiping Zhang, Xiaohan Guan, Xusen Zhao, Hongtao Li, Zhengang Liu

Voltage Spectra of Three-Level Inverters with Three-Phase Modulation 693
S. Halász, I. Varjasi

Design of Motion Control System Used for Filter Rod Production Machine 699
Yang Qingyu, Ge Sibo, Ye Kesong, Shi Ren

Magnetic Pole Identification for PMSM at Zero Speed Based on Space Vector PWM 703
Jiangang Hu, Longya Xu, Jingbo Liu

Study on Stagewise Control of Connecting DFIG to the Grid 708
Xueguang Zhang, Dianguo Xu, Yongqiang Lang, Hongfei Ma

Generalized Control Approach for Active Power Filters 713
Xiaoyu Wang, Jinjun Liu, Chang Yuan, Zhaoan Wang

Novel Circuit Configuration for Hybrid Reactive Power Compensator 718
H.L Jou, J.C Wu, J.J. Yang, W.P. Hsu

Shunt Active Power Filter with Sample Time Staggered Space Vector Modulation Based Cascade Multilevel Converters 724
Liqiao Wang, Weiyang Wu

Shunt Active Power Filter Synthesizing Resistive Loads by Means of Adaptive Inverse Control 729
Wu Yanfeng, Wu Zhengguo, Li Hua, Li Hui

Table of Contents

Single Neutral Element Self-Adaptive PID Controller Used In SVC 734
Zeng Guang, Ke Min-qian, Su Yan-min, Fu Qi-gang

A Novel Shunt Single-Phase Active Power Filter for High Voltage Application 739
Zhang Changzheng, Chen Qiaofu, Zhao Youbin, Chen Yuda, Cheng Lu

Three-phase Active Power Filter Based on Space Vector and One-cycle Control 744
Wang Yong, Shen Songhua, Guan Miao

Implementation of a Shunt-Series Compensator for Nonlinear and Voltage Sensitive Load 748
Bor-Ren Lin, Chien-Lan Huang

Three-Phase Active Filter using a Single-Phase STATCOM Structure with Asymmetrical Dead-band Control 753
Seyyed Hossein Hosseini, Mehran Sabahi

Mitigation of Voltage Sag Using Adaptive Neural Network with Dynamic Voltage Restorer 759
M. R. Banaei, S. H. Hosseini, M. Darkalee Khajee

Mitigation of Current Harmonic Using Adaptive Neural Network with Active Power Line Conditioner 764
M. R. Banaei, S. H. Hosseini

A direct control strategy for UPQC in three-phase four-wire system 769
Tan Zhili, Li Xun, Chen Jian, Kang Yong, Duan Shanxu

Three-Phase Harmonic Selective Active Filter Using Multiple Adaptive Feed Forward Cancellation Method 774
Lewei Qian, David Cartes, Qiang Zhang

Reactive Power Compensation in Distribution Networks with STATCOM by Fuzzy Logic Theory Application 779
Seyyed Hossein Hosseini, Reza Rahnavard, Yousef Ebrahimi

A Distributed Fuel Cell Based Generation and Compensation System to Improve Power Quality 784
Haimin Tao, Jorge L. Duarte, Marcel A. M. Hendrix

Parallel Control of Three-Phase Three-Wire Shunt Active Power Filters 789
Xueliang Wei, Ke Dai, Xin Fang, Pan Geng, Fang Luo, Yong Kang

Study and Design of Noninductive Bus bar for high power switching converter 794
Zhiling Qiu, Hongyan Zhang, Guozhu Chen

A New Minimum Torque-ripple and Sensorless Control Scheme of BLDC Motors Based on RBF Networks 798
Juan Wang, Hongwei Liu, Yuran Zhu, Bo Cui, Huijuan Duan

Improved Modelling and Calculation on Electromagnetic Transient of Power Transformer 802
Chen Zhe, Wen Yuanfang, Lu Guojun

The Simulation and the Experimental Research of the Stator Bars' Evaporative Cooling System in the Three Gorges' Hydrogenerator 808
Ruan Lin, Gu Guobiao, Tian Xindong, Yuan JiaYi

An Investigation of Multi-phase Transverse Flux Permanent Magnet Machine 813
G.Q. Bao, J.K.Wang, D.Zhang, J.Z. Jiang

Suspension Principle and Digital Control for Bearingless Permanent Magnet Slice Motors 817
Huangqiu Zhu, Liang Fang

The effect of parameter variations on the performance of indirect vector controlled induction motor drive 821
A. Shiri, A. Vahedi, A. Shoulaie

Magnetic Field Analysis and Performance Calculation for New Type of Claw Pole Motor with Permanent Magnet Outer Rotor 826
Fengge Zhang, Shifu Zhang, Haijun Bai, Eugen Nolle, Hans Pert Gruenberger

x

Table of Contents

Performance Analysis of a PM Claw Pole SMC Motor with Brushless DC Control Scheme 831
Youguang Guo, Jianguo Zhu, Jiaxin Chen, Jianxun Jin

Solving Induction Motor Equivalent Circuit using Numerical Methods for an In-Service and Nonintrusive Motor Efficiency Estimation Method 836
Bin Lu, Wei Qiao, Thomas G. Habetler, Ronald G. Harley

Fault Investigation of X-by-wire Permanent Magnet Synchronous Machine 842
L. Feng, A. Binder, A. Rentschler, A. Paweletz, D. Guenther

PLC-Based Speed Control of DC Motor 847
Ashraf Salah El Din Zein El Din

H8 Control of Adjustable-Pitch Wind Turbine Adjustable-Pitch System 853
Hongche Guo, Qingding Guo

The Motion Control Algorithm based on Quaternion Rotation for a Permanent Magnet Spherical Stepper Motor 857
Qun-jing Wang, Kun Xia

Research on Restraining Thrust Force Ripple for Permanent Magnet Linear Synchronous Motor 862
Cui Jiefan, Wu Hui, Sun Qing, Zhang Yi, Zhao Lijun

Using Recurrent Fuzzy Wavelet Neural Network to Control AC Servo System 866
Yan Tang, Wei Sun, Yaonan Wang, Xiaohua Zhai

new topology of multi - level - converter for harmonic reduction 870
Frank Grundmann, Jian Xie

PWM Based Sensing and Control of Magnetic Bearings 875
Zhuliang Yeic, Flalph Vansencc

Position Sensorless Direct Torque Control of Synchronous Reluctance with Permanent Magnet Motor 880
Jiang Dong, Zhao Zhengming, Duan Yao, Guo Wei

Counter-Rotating Permanent Magnet Brushless DC Motor for Underwater Propulsion 885
Jianqi Qiu, Cenwei Shi, Mengjia Jin, Ruiguang Lin

A Special Flux-weakening Control Scheme of PMSM - Incorporating and Adaptive to Wide-Range Speed Regulation 890
Song Chi, Longya Xu

Model-based Disturbance Attenuation for Linear Motor Servo System 896
Guiqiu Liu, Qingding Guo

A Fuzzy-Wavelet-Network-Based Position Control for PMSM 899
Wang Jun, Peng Hong, Xia Ling

Stability Analysis of Magnetic Bearing with Resonance Circuit 903
Zong Ming, Wang Fengxiang, Sun Yidan, Wang Jiqiang

Flux-Weakening Characteristics of Trapezoidal Back-EMF Machines in Brushless DC and AC Modes 908
Z.Q. Zhu, J.X. Shen, D. Howe

A Cost Effective Sensorless Control Method for Permanent Magnet Synchronous Motors Based on Average Terminal Voltage 913
Cheng-Hu Chen, Wei-Chih Tai, Ming-Yang Cheng

DSP-based Discrete-Time Reaching Law Control of Switched Reluctance Motor 918
Ge Baoming, Zhao Nan

Digital Control System on Bearingless Permanent Magnet-type Synchronous Motors 923
Jianming Deng, Huangqiu Zhu, Yang Zhou

xi

Table of Contents

Practical Issues in Sensorless Control of PM Brushless Machines Using Third-Harmonic Back-EMF 928
J.X. Shen, Z.Q. Zhu, D. Howe

Switched Reluctance Motors Drive for the Electrical Traction in Shearer 933
H. Chen

Research on Three-level Inverter of Six-phase Synchronous Motor 937
Yao Wenxi, Hu Haibing, Lu Zhengyu, Xu Haijie

Doubly-Salient Permanent-Magnet Machine with Skewed Rotor and Six-State Commutating Mode 942
Yongbin Li, Chris Mi

Sensorless Control and PMSM Drive System for Compressor Applications 947
Dongsheng Li, Takahiro Suzuki, Kiyoshi Sakamoto, Yasuo Notohara, Tsunehiro Endo, Chikara Tanaka, Tatsuo Ando

Analysis and Experimental Study of Slot Effect in Synchronous Reluctance Permanent Magnet Motors 952
Wei Guo, Zhengming Zhao,Yingchao Zhang

A New BLDC Motor Drives Method Based on BUCK Converter for Torque Ripple Reduction 958
Zhang Xiaofeng, Lu Zhengyu

Performance Investigation of a Fault-Tolerant Brushless Permanent Magnet AC Motor Drive 962
Jingwei Zhu, Nesimi Ertugrul, Wen Liang Soong

Current sensorless integral variable structure controller of synchronous reluctance motor 967
Huann-Keng Chiang, Chien-An Chen, Bor-Ren Lin, Kai-Sheng Hsu

An Improved Sliding Mode Observer for Speed Sensorless Vector Control Drive of PMSM 972
K. Paponpen, M. Konghirun

Analysis of an AC fed direct converter for a switched reluctance machine in aerospace applications 977
S. J. Forrest, J. Wang, G. W. Jewell, C. M. Johnson, S.D. Calverley

Direct Torque Control of an Interior Permanent Magnet Synchronous Machine fed by a Direct AC-AC Converter 983
D. Xiao, M. F. Rahman

A Novel Modular Permanent Magnet Drive System Design 989
Wen Ouyang, Nicholas Lemberg, Ruoping Yao, T.A.Lipo

Research on Digital Control Systems for Large Power AC-DC-AC Converters with Synchronous Motor Load 995
Xiaotan Zhao, Chongjian Li, Weihui Sheng, Yaohua Li

About the Prediction of Undesired Higher Current and Torque Harmonics of Inverter Driven Motors with Numerical Methods 999
C. Grabner

A Method of Stator Voltage Error Compensation in MRAS Sensorless Vector Control of Induction Motor 1006
Wen Xuhui, Chen Guilan, Han Li

Systematic Design of Fuzzy Logic Based Hybrid On-Line Minimum Input Power Search Control Strategy for Efficiency Optimization of IM 1012
Zhang Liwei, Liu Jun, Wen Xuhui, Trillion Q. Zheng

Research on an AC Variable-frequency Power Dynamometer Based on PWM Rectifier and Fuzzy Direct Torque Control 1017
Jia-qiang Yang, Jin Huang

Characteristic Research of Bearing Currents in Inverter-Motor Drive Systems 1023
Xing Shancheng, Wu Zhengguo

Research on a New Motor Drive Control System for Electric Transit Bus 1027
SHAO Gui-xin, ZHANG Cheng-ning

xii

Table of Contents

New Micro-Drive Series For Induction Motors & Survey of Market Trends..1032
Henrik Rosendal Andersen, Ruimin Tan, Zhang Hui

Robust Backstepping Control of Induction Motor Drives Using Artificial Neural Networks......................1038
J. Soltani, R. Yazdanpanah

Robust Nonlinear Control of Linear Induction Motor taking into account the Primary End Effects.................1043
J. Soltani, M.A. Abbasian

A Novel Adaptive Scheme for Stator Resistance Estimation in Sensorless Induction Motor Drives.................1049
Han Li, Wen Xuhui, Chen Guilan

Ripple-Free Sampling of Current Signals in Drives with Carrier-based PWM Patterns.................................1054
Haihui Lu, Qiang Yin, Russel J. Kerkman, Thomas A. Nondahl

Study of Speed Sensorless Control Methodology for Single Inverter Parallel Connected Dual Induction Motors Based on the Dynamic Model...1061
Shi Wei, Wang Ruxi, Wang Yue, He Yanhui, Wang Zhaoan, Liu Jinjun

ADC architecture with direct binary output for digital controllers of high-frequency SMPS.................1066
Tao Zhou, Jianping Xu

Analysis and Evaluation of a High-Voltage AC Amplifier for Electrostatic Suspension.................................1071
F. T. Han, Q. P. Wu, K. Liu, Z. Y. Gao

Design and Development of a 50kW Z-Source Inverter for Fuel Cell Vehicles...1076
Miaosen Shen, Alan Joseph, Yi Huang, Fang Z. Peng, Zhaoming Qian

Identification and improvement of stray coupling effect in an L-C-L common mode EMI filter.................1081
Junping He, Wei Chen, Jianguo Jiang

High Step-up Converter Associated with Soft-Switching Circuit with Partial Energy Processing for Livestock Stunning Applications...1086
S. -Y. Tseng, S.-H. Tseng, J. -Z. Shiang

A Computationally Intelligent Methodologies and Sliding Mode Control Based Traction control System for in-wheel driven EV...1091
Ming Zhengfeng, NI Guangzheng

A Low-Cost Gate Driver Design Using Bootstrap Capacitors for Multilevel MOSFET Inverters.................1096
J. J. Graczkowski, K. L. Neff, X. Kou

An Effective Method to Suppress Resonance in Input LC Filter of a PWM Current-Source Rectifier.................1101
Y.W. Li, B. Wu, N. Zargari, J. Wiseman, D. Xu

Topological and Modulation Design of Three-Level Z-Source Inverters...1107
P. C. Loh, F. Gao F. Blaabjerg

Investigation of Power Supplies for a Piezoelectric Brake Actuator in Aircrafts.................................1112
Rongyuan Li, Norbert Fröhleke, Hermann Wetzel, Joachim Böcker

A Line Power-Supply for LED Lighting using Piezoelectric Transformers in Class-E Topology.................1117
F.E. Bisogno, S. Nittayarumphong, M. Radecker, A. V. Carazo, R. N. do Prado

Integrating Large Wind Farms into Weak Power Grids with Long Transmission Lines.................................1122
Richard Piwko, Nicholas Miller, Juan Sanchez-Gasca, Xiaoming Yuan, Renchang Dai, James Lyons

Turn-on Condition and Characteristics of Highpower Semiconductor Switch RSD.................................1129
Y. M. Zhou, Y. H. Yu, H. G. Chen, L. Liang

The analysis and simulation of power circuits for high voltage converter.................................1133
S. I. Volskiy, Y. Y. Skorokhod, V. V. Shergin

A novel IGCT-based Half-controlled Bridge Type Fault Current Limiter.................................1138
Wanmin Fei, Yanli Zhang

xiii

Table of Contents

Influence of Proton Irradiation dose on the Performance of Local Lifetime Controlled Power Diode with Proximity Gettering of Platinum 1143
B.D. Han, D.Q. Hu, S.S. Xie, Y.P. Jia, B.W. Kang

IMPLEMENTATION OF A HIGHER QUALITY DC POWER CONVERTER 1148
Barsoum, N.N., YII, M.L.

Design of a Digital Programmable Control IC for Single-Phase Controlled Rectifiers 1154
Ming-Fa Tsai, Fu-Jing Ke, Ying-De Lin, Jui-Kum Wang

Feasibility Study of AlGaN/GaN HEMT for Multimegahertz DC/DC Converter Applications 1159
Yang Gao, Alex Q. Huang

The Mechanism Analysis of IGBT Module Invalidation 1162
Xu Aide, Fan Yinhai, Wang Xinxin, Liu Yuanyuan

A New Injection Efficiency Controlled GTO 1167
Wang Cailin, Gao Yong, Zhang Ruliang

Implementation and Analysis of 3-phase Voltage Sourced Regenerative Rectifier 1171
Rui Chen, Qiongxuan Ge, Shijie Li

Design and Implementation of Electronic Ballast for Fluorescent Lamps with Low Lighting Flicker 1178
Yang-Sheng Lin, Chun-An Cheng, Jiann-Fuh Chen, Tsorng-Juu Liang, Wei-Shih Liu

A Floating-point Coprocessor Configured by a FPGA in a Digital Platform Based on Fixed-point DSP for Power Electronics 1183
Haibing HU, Tianjun Jin, Xianmiao Zhang, Zhengyu LU, Zhaoming Qian

An Analytical Model for 4H-SiC Super-Junction Devices 1188
L.C. Yu, K. Sheng

Architecture Implementation of Class-D Amplifiers Using Digital-Controlled Multiphase-Interleaved PWM Technique 1192
Yu-Tzung Lin, Chi-Yang Lee, Ying-Yu Tzou,

Integrated IC-like Thyristor-based Switching Structure for Pulse Current Generation to Electronic Ignition 1198
C. L. Zhang, K. S. Jeon, C. H. Ahn, J. D. Park, E. D. Kim, Na Zhi, Yong Gao

A Wide Bandwidth Current Probe Based on Rogowski Coil and Hall Sensor 1202
Dong Li, Guiyou Chen

Voltage Dip Detection Based on an Efficient Least Squares Algorithm for D-STATCOM Application 1207
Thip Manmek, Chathura P. Mudannayake, Colin Grantham

Optimal Design and Analysis on Bearingless Permanent Magnet-type Synchronous Motors Using Finite Element Method 1213
Chang Jiang , Huangqiu Zhu, Zhenyue Huang

The Restrain of Harmonic Circulating Currents between Parallel Inverters 1218
Yu Zhang, Shanxu Duan, Yong Kang, Jian Chen

Simulation of Permanent Magnet Synchronous Motor with Dual Closed Loop by Time-Stepping Finite Element Model 1223
Xinhua Liu, Jianzhong Jiang, Yu Gong, Ye Ding

Online Dynamic Parameter Estimation of Transformer Equivalent Circuit 1228
M. Reza Feyzi, Mehran Sabahi

Worst-Case Tolerance Analysis for a Power Electronic System by Modified Genetic Algorithms 1233
Toshiji Kato, Kaoru Inoue, Kazuya Nishimae

The Reduction of Force Ripples of PMLSM Using Field Oriented Control Method 1238
Yu-wu Zhu, Kun-seok Jung, Yun-hyun Cho

xiv

Table of Contents

Analysis and Design of Signal Stage AC/DC Converter with Resonant Model PFC 1243
Weiping Zhang, Liangrui Lin, Dongyan Zhang, Xusen Zhao

Low Frequency Model for the Metal Halide Lamp 1248
Weiping Zhang, Yuanchao Liu, Xiaoqiang Zhang, Hongtao Li, Wenji Liu

**H8 Robust Controller Based on Local Feedback Recurrent Neural Network for Permanent Magnet
Linear Synchronous Motor** 1253
Junyou Yang, Naiguang Fa, Ruijuan Chen

Parameter Estimate Modeling of Electronic Transformer 1258
Jiaju Wu, Hidehiko Sugimoto, Changkun Wang

Analysis and Design of Boost DC-DC Converters for Intrinsic Safety 1267
Shu-Lin Liu, Jian Liu, Hong Mao

Modeling and Fuzzy Logic with Integrator Control for the ZVZCS PWM DC/DC Converter 1273
Shen Hong, Wan Jianru, Yang Xiaobo, Wu Weiyang, Wang Xiaohuan

ZVS DC-DC Converter with Parallel-Connected Current Doubler Rectifier 1278
Bor-Ren Lin, Shuh-Chuan Tsay, Chun-Sheng Yang, Chien-Lan Huang

Study on the Dynamical Model and Analytical Method for DC-DC Switching Converter 1283
Li-Li Wang, Yu-Fei Zhou, Jun-Ning Chen

A Novel Topology Family of Single-stage Parallel Mode Uninterruptible AC/DC Converter with PFC 1288
Xuejun Ma, Hongxia Wu, Congsheng Huang, Xuwen Huang

**Analysis and Design of an Automatic-Current-Sharing Control Based on Average-Current Mode for
Parallel Boost Converters** 1293
Wenxun Xiao, Bo Zhang, Dongyuan Qiu

A Novel Digital Charge Control for DC-DC Converters 1298
Shi Wenqing, Xu Haiping, Wen Xuhui, Wen Wei

**An Asymmetrical Switched Capacitor and Lossless Inductor Quasi-Resonant Snubber-Assisted ZCS-
PWM DC-DC Converter with High frequency Link** 1302
Khairy Fathy, Keiki Morimoto, Toshimitsu Doi, Hyun Woo Lee, Mutsuo Nakaoka

**A Divided Voltage Half-Bridge High Frequency Soft-Switching PWM DC-DC Converter with High and
Low Side DC Rail Active Edge Resonant Snubbers** 1307
Khairy Fathy, Keiki Morimoto, Toshimitsu Doi, Hiroyuki Ogiwara, Hyun Woo Lee, Mutsuo Nakaoka

Dynamic Analysis of a Current Source Inductively Coupled Power Transfer System 1312
Wenqi Zhou, Hao Ma

A New Topology of Capacitor-Clamp Cascade Multilevel Converters 1318
Anees Abu Sneineh, Ming-Yan Wang, Kai Tian

Evaluation of Semiconductor Losses in Cryogenic DC-DC Converters 1323
C. Jia, A. J. Forsyth

**Design and Performance Evaluation of a 10-kW Interleaved Boost Converter for a Fuel Cell Electric
Vehicle** 1328
G. Calderon-Lopez, A. J. Forsyth, D. R. Nuttall

**Analysis of Abnormal Phenomenon in Common-Source-type Forward Converter with Self-driven
Synchronous Rectifier** 1333
Kentaro Fukushima, Takayoshi Hashimoto, Tamotsu Ninomiya, Takeshi Segawa

Power Quality Conditioning in Distributed Generation Systems 1338
R.K. Járdán, I. Nagy

xv

Table of Contents

Active Clamp Forward Converter Combined with Dither Voltage Generator for Poultry Stunning Applications..1343
S. -Y. Tseng, H.-T. Wen, H.-H. Chang, J. -S. Kuo

A Novel Zero-Voltage Switching Resonant Pole Inverter 1348
Sanbo Pan, Junmin Pan

Analysis of Three-Level ZVS PWM Inverter for Induction Heating Applications 1353
A. Jangwanitlert, J. Songboonkaew, W. Thammasiriroj, J.C. Balda

Dual Duty Cycle Controlled Voltage Source Soft-Switching High Frequency Inverter with AC Load Side Reverse Blocking Switched Resonant Capacitor 1358
Khairy Fathy, Ju-Sung Kang, Hiroyuki Ogiwara, Bin Eiuo, Hideki Omori, Hyun Woo Lee, Mutsuo Nakaoka

A Switched-Capacitor Lossless Inductor ZCS Snubber-Assisted Series Load Resonant High Frequency Inverter with Dual Mode Pulse Modulation Scheme..1363
Khairy Fathy, Takaaki Okude, Hideki Omori, Hyun Woo Lee, Mutsuo Nakaoka

Topologies of Switch-Linear Hybrid Power Conversion & Special Operation States....................1368
Lu-sheng Ge, Qian-zhi Zhou, Wu bin

Single Reverse Blocking Switch Type Pulse Density Modulation Controlled ZVS Inverter with Boost Transformer for Dielectric Barrier Discharge Lamp Dimmer..1372
Hisayuki Sugimura, Bishwajit Saha, Hideki Omori, Hyun-Woo Lee, Mutsuo Nakaoka

PDM Controlled Series Load Resonant Soft Switching High Frequency Inverter for Induction Heated Toner Fixing Outer Roller with Inner Cylindrical Working Coil Stator 1377
Hisayuki Sugimura, Hideki Omori, Hyun Woo Lee, Mutsuo Nakaoka

Zero-Voltage and Zero-Current Switching Two-Transformer Full-Bridge Converter Using the Output-Voltage-Doubler 1382
H.K. Yoon, E.S. Choi, S.K. Han, G.W. Moon, M.J. Youn

A Single-stage Boost-Flyback PFC Converter 1387
Zhao Qinglin, Wen Yi, Wu Weiyang, Chen Zhe

Control Bifurcation in PFC Boost Converter under Peak Current-Mode Control....................1392
Yi-Jing Ke, Yu-Fei Zhou, Jun-Ning Chen

Analysis and Design of One-Cycle-Controlled Dual-Boost Power Factor Corrector 1397
Yue-feng Yao, Yuan-rui Chen

A Novel Single-phase Buck PFC Converter Based on One-cycle Control................................1401
Chen Bing, Xie Yun-Xiang, Huang Feng, Chen Jiang-Hui

Modeling and Simulation of Three Phase High Power Factor PWM Rectifier factor correction. 1406
Yu Fang, Yong Xie, Yan Xing

Effect of the Ripple Current on Power Factor of CRM Boost APFC 1412
A. Abramovitz

Simulated Study of Three-Phase Single-Switch PFC Converter with Harmonic Injected PWM by MATLAB..1416
Zhanlong Li, Yupeng Tang

A Simple Digital Controller for Constant Instantaneous Input Power type Three-Phase Boost Rectifier under Unbalanced System..1421
Jin Ai-Juan, Li Hang-Tian, Li Shao-Long

An Improved and Digital Current Control Strategy for One Cycle Control Based Three-Phase Boost Rectifier under Unbalanced System..1426
Li Shao-Long, Jin Ai-Juan, Li Hang-Tian

Table of Contents

Control Method for Power Quality Compensation Based on Levenberg-Marquardt Optimized BP Neural Networks 1431
Zhou Ming, Wan Jian-Ru, Wei Zhi-Qiang, Cui Jian

A Nonlinear Method for Hybrid Electromagnetic Suspension 1436
Junwei Cui, Jianhui Wang

New topology of multi - level - converter for harmonic reduction 1442
Frank Grundmann, Jian Xie

Model Reference Adaptive Control based on Neural Network for Electrode System in Electric Arc Furnace 1447
Zhang Shi-feng, Zhang Shao-De, Li Kun, Zheng Xiao

STATCOM ETO Failure Analysis 1450
Zhong Du, Bin Chen, Chong Han, Zhaoning Yang, Wenchao Song, Subhashish Bhattacharya, Alex Q. Huang

Modeling and Control of Three-phase Voltage Source PWM Rectifier 1454
Yao Chen, Xin Min Jin

Mitigation of Electric Arc Furnace Voltage Flicker Using Static Synchronous Compensator 1458
Y.F. Wang, J.G. Jiang, L.S. Ge, X.J. Yang

Design of Distributed FACTS Controller and Considerations for Transient Characteristics 1463
Gaidi Ning, Shijie He, Yue Wang, Lei Yao, Zhaoan Wang

A Wind-Power Generation System Having a Function of Suppressing Line Voltage Deviation 1468
Y. Nakayama, S. Fukuda, M. Futami, M. Ichinose, S. Ohara, H. Kita

A Novel Active Islanding Detection Method of Grid-connected Photovoltaic Inverters Based on Current-Disturbing 1473
Zhang Chunjiang, Liu Wei, San Guocheng, Wu Weiyang

Grid Connection of Doubly-Fed Induction Generators in Wind Energy Conversion System 1477
Ahmed G. Abo-Khalil, Dong-Choon Lee, Se-Hyun Lee

Active and Reactive Power Control of DFIG for Wind Energy Conversion under Unbalanced Grid Voltage 1482
Jeong-Ik Jang, Young-Sin Kim, Dong-Choon Lee

A BASIC STUDY OF FUZZY-LOGIC-BASED POWER SYSTEM STABILIZATION WITH DOUBLY-FED ASYNCHRONOUS MACHINE 1487
Li Wu, Zhixin Wang

Quantitative Analysis on Different Modes of Energy Optimal Control for Series Power Quality Controllers 1492
Huang Xinming, Liu Jinjun, Zhang Hui

Resonance inverter power system for improving plasma sterilization effect 1497
Y.M Kim, J.Y Kim, M. C Jo, S.H Lee, S.P Mun, H.W Lee, S.K Kwon, K.Y Suh

Generic optimization for SMPS design with Smart Scan and Genetic Algorithm 1502
Heidi H.T. Yeung, N. K. Poon, Stephen L. Lai

Novel Single-Stage Isolated Buck-Boost Inverter Based on Improved SPWM Control Method 1507
Guang-Hui Tan, Fanpeng Zeng, Yanchao Ji, Xi Chen, Hua Wang

On the Effects of Voltage Loop in Paralleled Converters Under Master-Slave Current Sharing 1512
Yuehui Huang, Chi K. Tse

Improved Control for Parallel Inverter with Current-Sharing Control Scheme 1517
Zhao Qinglin, Chen Zhongying, Wu Weiyang

A Novel Digital Controlled battery charger for High power UPS application 1522
Fang Luo, Yong Kang, Shan Xu Duan, Xueliang Wei

xvii

Table of Contents

A Novel High Input Power Factor Single-Stage Single-Phase AC/AC Converter 1527
Chien-Ming Wang, Chien-Yeh Ho, Maoh-Chin Jiag

Research on the Power Sharing of the Parallel Inverters without Control Interconnection Basing on
Droop Characteristic 1532
Kan Jiarong, Xie Shaojun

Analysis and Design of Repetitive controlled Inverter System with High Dynamic Performance........................... 1537
Mingzhu Li, Zhongyi He, Yan Xing

Study on a large-volume high-performance programmable voltage disturbance source........................... 1542
Zhan Qizhi, Zhuo Fang, Dong Wenjuan, Wang Zhao'an

1 KW Dual Interleaved Boost Converter for Low Voltage Applications.. 1546
Heinz van der Broeck, Ibrahim Tezcan

Control of Multilevel Flying Capacitor Inverters for High Performance.. 1551
L. Zhang, S. J. Watkins, Duan Qi Chang

Analysis of Harmonics in Input Line Current for Matrix Converter based on Double Input Line-toline
Voltages... 1557
Guo Yougui, Deng Wenlang, Zhu Jianlin

Research on Neutral-point Balancing Control for Three-level NPC Inverter Based on Correlation between
Carrier-based PWM and SVPWM 1560
Wenxiang Song, Guocheng Chen, Xiaoyu Ding, Mantang Shu

Instantaneous Voltage Regulated Seamless Transfer Control Strategy for Utility-interconnected Fuel cell
Inverters with an LCL-filter.. 1566
Guoqiao Shen, Dehong Xu, Xiaoming Yuan

An Anti-windup Design Method for Internal Model Control Based on H8 Optimization 1571
Hou Yansong, Li Hua

Study on Pwm Control Strategy of Photovoltaic Grid-connected Generation System........................... 1576
Shi-cheng Zheng, Pei-zhen Wang, Lu-sheng Ge

Robust Sliding Model Control for Regenerative Braking of Electric Vehicle.. 1581
Min Ye, Zhifeng Bai, Binggang. Cao

A Self-adaptive Fuzzy Control Scheme of High Frequency Link SPWM Inverters.. 1585
Herong Gu, Deyu Wan, Weiyang Wu

Using Automatic Frequency Shifting Techniques for LLC-SRC Output Voltage Regulation 1590
Kuo-Kai Shyu, Ching-Ming Lai, Ko-Wen Jwo, Ming-Ho Pan, Chung-Ping Ku

Design and Test of Novel Programmable Digital Three Phases SPWM Chip... 1595
Yang Yuan, Gao Yong, Chen Lijie

An Improved Performance of Five-Leg Inverter in Two Induction Motor Drives.. 1598
Ryuji Omata, Kazuo Oka, Atsushi Furuya, Shuji Matsumoto, Yusuke Nozawa, Kouki Matsuse

Adaptive Three Dimensional Space Vector Modulation in abc Coordinates for Three Phase Four Wire
Split Capacitor Converter... 1603
Xiao-bo Yang, Wei-yang Wu, Hong Shen

Inverters Parallel Operation Based on CAN 1608
Yong Wu, Xianglong Jiang, Jinbang Xu, Qingyi Wang, Shuyun Wan

EMI Reduction Method for a Single-Phase PWM Inverter by Suppressing Common-Mode Currents with
Complementary Switching... 1613
Toshiji Kato, Kaoru Inoue, Koji Akimasa

xviii

Table of Contents

Analysis and Design of a Novel Dual Secondary Winding and Dual Power Bridge High Frequency Link Inverter 1618
Zhang Zhe, Zhang Chunjiang, Wu Weiyang, Gu Herong, Shen Hong

Research of Complex Fuzzy Control on-off Magnetism Team Motor Speed-Adjusting System 1623
Zhao Ming-fu, Chen Yan, Zhang Zhi-yuan, Dong Chun, DongYu

A New BLDC Motor Drives Method Based on BUCK Converter for Torque Ripple Reduction 1626
Zhang Xiaofeng, Lu Zhengyu

Design of Wind Turbine Generator Control System 1630
Chen Guiyou, Zhou Li, Sun Tongjing, Wang Zhongmin

Non-touching Intelligent Control System of Water Inteneraring Equipment Based on Sodion Exchange 1634
Chen Guiyou, Zhang Qingfan, Zhou Li, Luo Donghua

Investigation of Hybrid Modeling and Control for DC-DC Converters 1637
Hao Ma, Feng Qi, Wenqi Zhou

Effect of Peak Current Mode Control on Transient Response for VRM Application 1641
Seiya Abe, Tamotsu Ninomiya

Modulations for Voltage Source Rectification and Voltage Source Inversion Using Direct Space Vector Approach 1646
Keping You, M. F. Rahman

Synchronization of Voltage Waveforms in Basic Topologies of Dual Inverter-Fed Motor Drives 1651
V. Oleschuk, F. Profumo, A. Tenconi, R. Bojoi, A.M. Stankovic

Research on Fast Magnetic Valve Controllable Reactor 1657
Zhang Jian-wen, Cai Xu

Study and comparison of fault tolerant shunt threephase active filter topologies 1663
H. El Brouji, P. Poure, S. Saadate

Application of GA-BP in Fault Diagnosis of Power Circuit of SVC 1669
Zeng Guang, Xi Yu-fan, Su Yan-min, Zhang Jing-Gang

The Optimization-Sliding Mode Control For Three-Phase Three-Wire DSP-based Active Power Filter 1674
Zhou Wei-ping, Liu Da-ming, Wu Zheng-guo, Xia Li, and Yang Xuan-fang

Three-Phase DVR using a Single-Phase Structure with Combined Hysteresis/ Dead-band Control 1679
Seyyed Hossein Hosseini, Mehran Sabahi

Harmonic Detection Based on the TLS Estimation Algorithm 1684
Liu Kaipei, Zhang Junmin

Control Strategy Study of Hybrid Active Power Filter 1689
Jia Zhang, Guohong Zeng

Novel Harmonic Free Single Phase Variable Inductor Based on Active Power Filter Strategy 1693
Mu Xianmin, Wang Jianze, Ji Yanchao, Wei Xiaoxia, Fu Xiangyun

A Multi-Output Series Resonant Inverter with Asymmetrical Voltage-Cancellation Control for Induction-Heating Cooking Appliances 1697
S.H. Hosseini, A. Yazdanpanah Goharrizi, E. Karimi

Capacitor Voltage Control in a Cascaded Multilevel Inverter as a Static Var Generator 1703
M. Li, J. N. Chiasson, L. M. Tolbert

DC-link Pumping-up Voltage Suppression of a Series Active Voltage Regulator With Phase Shift Control 1708
G. C. Xiao, Z. L. Hu, C. H. Nan, Z. A. Wang

The Fuzzy Soft-startup Controller of Active Power Filter 1713
He Na, Wu Jian, Xu Dianguo

Table of Contents

A Novel Control Method for DSTATCOM Using Artificial Neural Network..1718
Yang Xiao-ping, Zhong Yan-ru, Wang Yan

A Detailed Analysis of Unexpected DC-side Voltage Boost in Series Power Quality Controllers 1722
Yuan Chang, Liu Jinjun, Wang Xiaoyu, Wang Zhaoan

Comparative Analysis of Popular Control Schemes for Parallel Active Power Filter and Experimental Verification..1726
Xiaoyu Wang, Jinjun Liu, Chang Yuan, Zhaoan Wang

Accurate Modeling of the Three Phase Induction Motor Including Saturation Effects................................1731
E. V. N. Souza, S. R. Naidu

A study on the reliability evaluation of driving parts for note handling units 1736
Joo Han Kim, Jung Kee Chung, Ha Kyeong Sung, Se Hyun Rhyu

Analysis on Toothless Permanent Magnet Machine with Halbach Array...1741
Xu Yanliang, Feng Kaijie

Improvement in Reliability of Doubly Salient Permanent Magnet Motor Drive..1746
Wenxiang Zhao, Ming Cheng, Xiaoyong Zhu, Wei Hua, Jianzhong Zhang

A New Approach of Modeling the Saturated Induction and Synchronous Salient Pole Machines 1751
A. Câmpeanu, M. Badica

Inductance characteristics of 3-phase fluxswitching permanent magnet machine with doubly-salient structure 1758
Wei Hua, Cheng Ming

Performance Index Evaluations of a Micro Axialflux Switched-reluctance Motor.......................................1763
Cheng-Tsung Liu, Yen-Ming Chen, Da-Chen Pang

Study of Variable Frequency Operation of Induction Generator for Wind Power..1768
Noriyuki Kimura, Mitsuhiro Hirao, Toshimitsu Morizane, Katsunori Taniguchi

Optimal Power Control Strategy of Maximizing Wind Energy Tracking and Conversion for VSCF Doubly Fed Induction Generator System 1773
H. Li, Z. Chen, John K. Pedersen

Design and Evaluation of a Dual Mechanical Port Machine and System 1779
Longya Xu, Yuan Zhang

Characteristic Analysis on Overhang Effect in Axial Flux PM Synchronous Motors with Slotted Winding 1784
WonYoung Jo, YunHyun Cho, YonDo Chun, DaeHyun Koo

Design and Analysis of a Double-Stator Cup-Rotor Directly Driven Permanent Magnet Wind Power Generator 1788
Dong Zhang, Shuangxia Niu, K. T. Chau, J. Z. Jiang, Yu Gong

Feasibility Analysis of Accelerometer Configuration of Non-gyro Micro Inertial Measurement Unit....................1793
Ding Mingli, Zhou Qingdong, Wang Qi, Wang Changhong

Design of Fractional-Order a PI Controller with two modes...1797
Wen Li, Yoichi Hori

Sliding Mode Robust Tracking Control Based on Learning Feedforward Compensation for High Precision Linear Servo System 1802
Zhu Guoxin, Guo Qingding, Zhao Ximei

Application of Fuzzy Self-learning Sliding Mode Variable Structure Control in Linear AC Servo System..............1806
Qing Hu, Shuo Jie, Dongmei Yu

Dynamics Research of Robot Manipulator 1811
Zhibing Shu, Caizhong Yan, Hairong Zhang

Table of Contents

Advanced Angle Control Schemes for Stator Hybrid Excited Doubly Salient Motor Drive 1815
Xiaoyong Zhu, Ming Cheng, Wenxiang Zhao, Wenguang Li

A Design Method of Reconfigurable Controller for AC Position Servo Systems 1820
Wu Qinmu, Qin Yi, Li Yesong

Position Sensorless Control of PMSM Based on a Novel Sliding Mode Observer over Wide Speed Range 1825
Song Chi, Student Member, Longya Xu,

Design of Motion Control System Used for Filter Rod Production Machine 1832
Yang Qingyu, Ge Sibo, Ye Kesong, Shi Ren

Analysis and Implementation of Sensorless Position Detection in a Permanent Magnet Generator 1836
Sebastian Rosado, Xiangfei Ma, Fred Wang, Jerry Francis, Dushan Boroyevich

Torque-Speed Characteristics of Interior-Magnet Machines in Brushless AC and DC Modes, with Particular Reference to Their Flux-Weakening Performance 1841
Y. F. Shi, Z. Q. Zhu, D. Howe

H8 Robust Control for Dual Linear Motors Servo System 1846
Zhao Ximei, Guo Qingding

Research on Linear Motor Driving System Based on Wavelet Transform 1849
Cui Jiefan, Zhao Lijun, Wang Hemin, Wan Junzhu, Jiang Lili

Study on Rotor Position Detection Error in Sensorless BLDC Motor Drives 1853
Li Qiang, Wang Ruixia

A New Scheme to Direct Torque Control of Interior Permanent Magnet Synchronous Machine Drives for Constant Inverter Switching Frequency and Low Torque Ripple 1858
Jun Zhang, M. Faz Rahman, Colin Grantham

A Modified Direct Toque Control for Interior Permanent Magnet Synchronous Motor Drive Without a Speed Sensor 1863
Yanping Xu, Yanru Zhong, Hui Yang

Direct Torque Control for Interior Permanent Magnet Synchronous Motors Using Matrix Converters 1867
D. Xiao, M. F. Rahman

A Neural Network Based Initial Position Detection Method To Permanent Magnet Synchronous Machines 1872
Mengjia Jin, P.C.K Luk, Jianqi Qiu, Cenwei Shi, Ruiguang Lin

A New Recurrent Fuzzy Neural Network Sliding Mode Position Controller Based on Vector Control of PMLSM Using SVM 1877
Junyou Yang, Ruijuan Chen, Naiguang Fa

DSP Implementation of Rotor Position Detection Method for Hybrid Stepper Motors 1882
M. Bendjedia, Y. Ait-Amirat, B. Walther, A. Berthon

An In-Wheel Switched Reluctance Motor for Electric Vehicles 1887
P.C.K. Luk, P. Jinupun

Speed Sensorless Vector Control of Induction Motor Based on Full-Order Flux Observer 1892
Shanshan Wu, Yongdong Li, Zedong Zheng

A Parameter Identification Method for General Inverter-fed Induction Motor Drive 1896
Xiaochun Jiang, Geng Yang, Yunfei Wang

Indirect Rotor Field Orientation Vector Control for Induction Motor Drives in the Absence of Current Sensors 1901
Z. S. WANG, S. L. HO

A Robust Adaptive Sliding-Mode Controller for Slip Power Recovery Induction Machine Drives 1906
J.Soltani, A. Farrokh Payam

Table of Contents

Identification of the Rotor Time Constant in Induction Machines without Speed Sensor...........................1912
M. Li, J.N. Chiasson, M. Bodson, L.M. Tolbert

Adaptive Control of Doubly Fed Field-Oriented Induction Machine Based On Recursive Least Squares Method Taking the Iron Loss Into account...1917
N. R. Abjadi, J. Askari, J. Soltani

Analysis and Design of PDM Converter with High Frequency Link for HEV Drive System.......................1922
Ma Xianmin

A Multi-Directional Power Converter for a Hybrid Renewable Energy Distributed Generation System with Battery Storage..1926
Mei Qiang, Wu Wei-Yang, Xu Zhen-lin

Four-bridge Multilevel Converters Based on Hybrid-clamped Techniques..1931
Xiaofeng Wang, Yan Deng, Xiangning He

Standardization of Input/Output Impedance Specifications of Buck Converters Based on the System Integration Concept..1936
Tao Wu, Xinbo Ruan

Research on The Magnetic Integration in Three-Level ZCS Quasi-Resonant Buck Converter..................1942
Jiang Ying, Xiang Hui-jie, Yang Yu-gang, Liu Nan

Decoupling Control of Magnetically Levitated Induction Motor with Inverse System Theory 1947
Yang Zhou, Huangqiu Zhu, Tianbo Li

Fault Detection and Accommodation for Nonlinear Systems Using Fuzzy Neural Networks 1952
H. Xue, J.G. Jiang

A Novel Constant Power Control of High Frequency Electronic Ballast Applying the PLL Technique for a Metal Halide Lamp...1957
Chang-Hua Lin, Chung-Lun Ou, Tien-Shuo Liu, Ken-Chuan Hsu

The Voltage Stability Research of Ship Electric Power System 1962
Fanyinhai Zhaomin

Parasitic Gate Resistance and Switching Performance...1967
Alan Elbanhawy

PWM Rectifier with DC Reverse-Blocking Diode for High-Reliability Generating Apparatus and Its Application to Gas Heat Pump System...1971
Akio Toba, Toshihiro Maeda, Kouetsu Fujita, Tomohiko Kato

A Novel Stator Section Crossing Method of Long Stator Linear Synchronous Motor for Maglev Vehicles..............1976
Qian Zhang, Fei Lin, Xiaojie You, Trillion Q. Zheng

Common Mode Current Suppression in Full-Bridge Converter Based on Simulated Annealing Algorithm 1981
Yonggao Zhang, Kai Zhang, Yunbin Zhou, Yong Kang

Summary of Distance Measurement Based on Vision in Localization Technology 1986
Handong Zhang, Gang Wang, Yuwan Cen

The studies of Single-phase Inverter Fault Diagnosis Based on D-S Evidential Theory and Fuzzy Logical Theory...1991
Wang Baocheng, Li Danhe, Sun Xiaofeng, Wu Weiyang

A Novel Single-Stage High-Power-Factor Electronic Ballast with Symmetrical Half-Bridge Topology 1995
Chien-Ming Wang, Chien-Yeh Ho

Smoothed-Power Output Supply System for Battery of Stand-alone Renewable Power System Using EDLC..2000
Y. Jia, R. Shibata, N. Yamamura, M. Ashida

xxii

Table of Contents

Supercapacitors characterization for hybrid vehicle applications 2005
F. Rafik, H. Gualous, R. Gallay, A. Crausaz, A. Berthon

Power Transfer Maximization and Di/Dt Based Extremum Tracking for a Swing Engine Based Portable Power System 2010
Satish Rajagopalan, Deepak M. Divan, Ronald G. Harley, J. Rhett Mayor

3D FEA of the Stator of the Linear Magnetic Flux Compression Generator 2015
Yanjie Cao, Chengxue Wang

The Effect of Current Control Strategies on Power Consumption of a Magnetically Levitated Turbomolecular Pump 2018
A.E. Hartavi, R.N. Tuncay, M.N. Sahinkaya

Direct Torque Control of an Interior Permanent Magnet Synchronous Machine fed by a Direct AC-AC Converter 2023
D. Xiao, M. F. Rahman

Control of Distributed Power Systems 2029
Z. Chen, Y. Hu, F. Blaaberg

Characteristic Research of Bearing Currents in Inverter-Motor Drive Systems

Xing Shancheng, Wu Zhengguo

Navy university of engineering, Wuhan, China

xingshancheng@163.com

Abstract—To gain better velocity modulation function and reduce the switch loss, the PWM voltage usually possesses very high frequency and dv/dt in inverter fed induction motor system. The common mode coupling current in the common mode coupling circuit will come into being inside the motor when this style PWM voltage acts on the winding of induction motor. The common mode coupling current component which crossed the motor bearing will give rise to damage to the bearing. This paper gives a measure method for the bearing currents of induction motor, divides the two kinds of bearing currents, conduction mode bearing current and discharge mode bearing current by experiment, analyzes the characteristic of the two kinds of bearing currents and their relation to common mode voltage, common mode current and shaft voltage. The experiment result shows that the conduction mode bearing currents appear chiefly at low motor speeds and is characterized by its occurrences in synchronization with the edges of common mode voltage pulse. Its value is small and the damage to the bearings is negligible. The discharge mode bearing current appears chiefly at high motor speeds. Its occurrences are not necessarily in synchronization with the edges of the common mode voltage pulses. Contrarily, the largest peak current usually occurs sometime during the interval of the highest level of common mode voltage. Its value, contrary to the conduction mode bearing current, is large and it is the main factor caused the bearings to be damaged.

Keywords- common mode coupling; common mode voltage; common mode current; bearing current

I. INTRODUCTION

Despite a lot of benefits brought by the extensive application of inverter fed induction motor system, there are backside effects which we must take into account. The common mode voltage and the parasitic coupling effect under high frequency condition will exist in the system unavoidably. The common mode coupling effect can give rise to common mode currents. Part of the common mode currents can pass through the motor bearing and cause damage to the motor bearing. The service life of motor bearings is shortened and the service life of motor is shortened accordingly[1-6]. So it is very important to research the motor bearing currents in inverter fed induction motor system. This paper gives a measure technique for the bearing currents of induction motor; divides the two kinds of bearing currents, conduction mode bearing current and discharge mode bearing current by experiment, analyzes the characteristic of the two kinds of bearing currents and their relation to common mode voltage, common mode current and shaft voltage.

II. MEASUREMENT METHOD OF BEARING CURRENT

A special modification to an ordinary induction motor has been made for experimental verification of the parasitic coupling and common mode voltage effects inside a squirrel cage induction motor. Such a modification has also made it possible to perform a direct measurement of true bearing, shaft voltages and currents. Fig. 1 depicts an idealized representation of this machine.

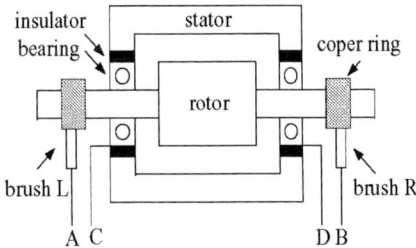

Fig. 1 Modified induction motor for bearing current measurement

In this modified motor. Insulation materials are inserted between the outer races of bearings and the stator case, which ensures electrical isolation between the stator and the rotor. Brushes are used to connect wires A and B to the left and right shaft ends, respectively, of the motor, while wires C and D are connected, respectively, to the outer races of the left and right bearings. It is shown in Fig. 2. The test motor will simulated a realistic machine without insulated bearings. Hence, almost all bearing currents will now flow into wires C and D, providing a means to measure the true bearing currents. In addition, the grounding terminal G_M on the motor stator case is used for ground connection.

1-4244-0448-7/06/$25.00 ©2006 IEEE

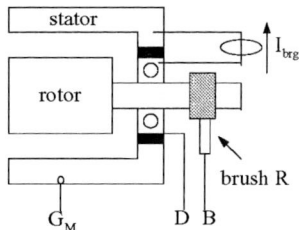

Fig. 2 Measurement method of bearing currents

III. VERIFICATION OF THE BEARING CURRENTS

An important experiment is to verify the existence of bearing currents caused by PWM inverter in normal field applications of inverter-motor systems. A setup similar to practical installation of inverter–motor system is desirable. The experimental inverter-motor system connection is shown in Fig. 3. The motor three-phase winding input terminals are denoted by "a", "b" and "c". The motor winding neutral point is also available which is indicated by "N" in the figure. An isolation transformer T is used to decouple the system from the earth. The platform upon which the motor and the inverter are situated is also insulated. The bearing wire D is connected to the motor stator to simulate a realistic motor without insulated bearings. The three-phase cable which connects the inverter output to the motor input is contained inside a metal conduit, while the two ends of the conduit are connected, respectively, to the inverter and motor case grounding terminals, "G" and "G_M". The conduit is then grounded to the earth. Such a setup is most often encountered as is recommended by drive manufactures except that the isolation transformer is usually not required. It is noted that the use of a three-phase utility isolation transformer is to reduce other possible coupling paths from the inverter negative DC bus to the earth ground. Thus, the measured bearing current amplitude will be somewhat conservative.

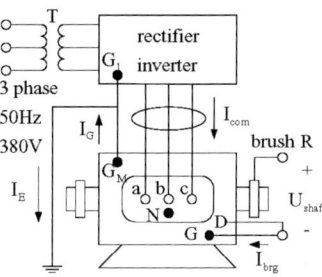

Fig. 3 Setup for bearing current measurement

In the test system shown in Fig. 3, the bearing current I_{brg} is measured by attaching a current probe to wire D, while the shaft voltage U_{shaft} is measured by connecting a voltage probe between brush B and motor stator case. To facilitate an analysis of the experimental results and to avoid other types of bearing currents, only one bearing is connected in the test system. When both bearings are connected in the system, the amplitude of bearing currents may change depending on whether both bearings have the same conductivity. However, the characteristics of bearing currents were seen to remain almost the same as those obtained from the above test setup.

A. Measuement of Voltage Induced Bearing Current

With above setup, the bearing current I_{brg}, the grounding current I_G in conduit, and the motor stator winding neutral point voltage V_N are plotted in Fig. 4. In Figs. 4a and 4b, the motor neutral point voltage V_N steps up and down which can be used as a reference for recognizing the inverter PWM switching. For each step change in V_N or PWM switching, there is a current pulse produced in the grounding cable. The grounding current I_G is thus dv/dt dependent and remains consistent at different motor or shaft speeds. However, the bearing current I_{brg} in Fig. 4a is totally different from that in Fig. 4b. It is noticed that at extreme low motor speeds or low inverter output frequencies, for example, 2Hz, the bearing current appears as short pulses which are dv/dt dependent. The peak value of the pulses is about 50mA, and the duty cycle is very small. While at high motor speeds or high inverter output frequencies, such as 35Hz, the bearing current becomes dominated by some large spikes with a peak amplitude of 200-300mA. An interesting observation is that the current spikes are neither dv/dt dependent nor with a regular frequency of occurrence.

The bearing current spikes are present in a wide rang of motor speeds, while the dv/dt dependent bearing current is observable only at very low motor speeds. For the current spikes, their amplitude and frequency of occurrence were observed to vary with motor speed, motor temperature and other factors. At very high speeds, for example, 60Hz inverter output frequency, the rate of generation of current spikes was observed to decrease.

The change of bearing current characteristics with motor speeds and other conditions has been found to be related to the electric behavior of rotating bearings. At different operating conditions, a bearing may exhibit randomly good conductivity, no-conductivity, or values between. Depending on the conductivity, there are two different types of bearing currents which we have termed the conduction mode bearing current and the discharge mode bearing current.

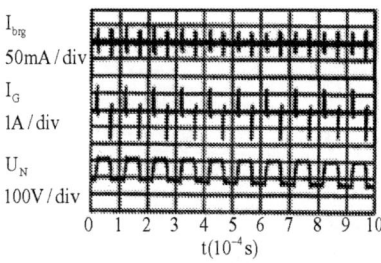

a) Bearing current with motor running at low speed

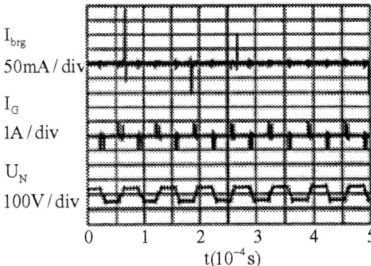

b) Bearing current with motor running at high speed

Fig. 4 Different bearings current at different motor speeds

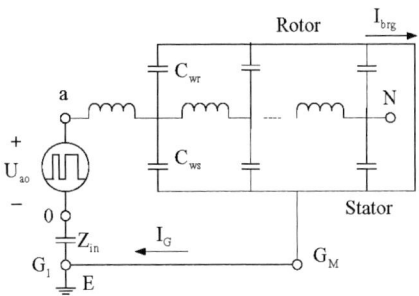

b) Explanation of conduction mode bearing current

Fig. 5 Conduction mode bearing current

B. Conduction Mode Current

The conduction mode bearing current is referred to as the current in bearings which exhibit continuously good conductivity for a certain period of time. This type of bearing current exists typically at low motor shaft speeds when the mechanical contact of bearing balls with races is good and the internal impedances of bearings are small.

By including the shaft voltage across the bearings and repeating the experiment, a measurement resulting from a low shaft speed is shown in Fig. 5a. The short-circuit behavior of the bearings is readily confirmed because the shaft voltage across the bearings is almost equal to zero. Due to its low impedance, a bearing can be approximated by a short-circuit and thus the air-gap capacitor of the motor does not appear in the circuit. The per phase bearing current model can be simplified to Fig. 5b. Based on the model, it is easily understood that whenever the inverter switches, there will be a current pulse in the bearings, which is very similar to the current pulses in the grounding cable.

By correlating to motor neutral common mode voltage, V_N, which shows all PWM switching states in the inverter, the relationship between PWM switching and the bearing current as well as the common mode current in the cable G can be clearly identified. It is noted that the conduction mode bearing current is characterized by its occurrences in synchronization with the edges of common mode voltage pulse.

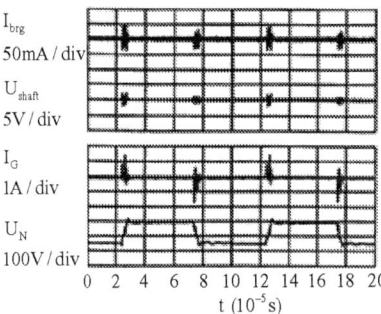

a) Measured conduction mode bearing current

C. Discharge Mode Bearing Current

As is well known in the study of classical bearing current problems, when a bearing rotates at high speeds, a thin film of lubricant usually build up between the bearing balls and the races. Thus, the bearing balls become floating in lubricant when rotating at high speeds. The thickness of the film is normally 2-3 micron. Due to this mechanism, rotating bearings can behave like an open circuit, or high impedance. However, once a bearing becomes open-circuited, it could return to a low impedance state if the contact between the balls and the races regains control or if the lubricant film breaks down due to the presence of electric fields caused by the shaft voltage. The same phenomenon is also observed in our experiment. Actually, the discharge mode current can be explained by such bearing behavior.

As is shown in Fig. 6a, the measurement of Fig. 4b has been repeated with the shaft voltage V_{shaft} included. It is seen that the current spikes usually appear after the shaft voltage is present for a short period of time. The presence of shaft voltage indicates a high impedance state of the bearing, which is also evident from the observation that the dv/dt related conduction mode bearing current becomes almost zero. Whenever, a current spike occurs, the shaft voltage immediately drops to zero. This implied that a sudden short-circuit happens inside the bearing.

The experimental results can also be explained by referring to the circuit model shown in Fig. 6b. During the period when the bearings become open-circuited, or lose conductivity, the coupling current to the rotor can not return through the bearings to the stator ground. Instead, this current charges the air-gap capacitor, and all energy is temporarily stored in the air-gap capacitor. That is why the dv/dt related bearing current becomes not observable during this period. The motor shaft voltage appears as a result of charge accumulation in the air-gap capacitor. Since the shaft voltage or the air-gap capacitor voltage is applied across a thin film of bearing lubricant, the electric field inside the bearing housing may become very intensive. When the field potential reaches a certain level, the film may break down and thus produce a sudden short-circuit path and all the stored energy is damped to the bearing. The amplitude of current spike is obviously dependent on the amplitude of shaft voltage and the impedance of the short circuit path. Usually, the spikes could be as high as several amperes depending on the motor and inverter rating, switching schemes, and grounding system configurations.

The type of bearing current which is produced due to charge-up of air-gap capacitor by common mode voltages and then discharge of air-gap capacitor into bearings will be called the discharge mode bearing current in this paper. It can be seen that occurrences of the discharge mode current are not necessarily in synchronization with the edges of the common mode voltage pulses. On the contrary, the largest peak current usually occurs sometime during the interval of the highest level of common mode voltage at the motor neutral point.

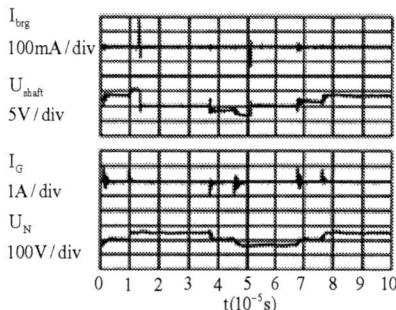

a) Measured discharge mode bearing current

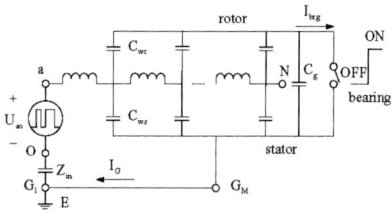

b) Explanation of discharge mode bearing current

Fig. 6 Discharge mode bearing current

IV. RESULTS

Bearing current generation depends on the bearing electrical behavior. In general, the electrical properties of a rotating bearing are more complicated than expected. At different operating conditions, a bearing may exhibit randomly good conductivity, no conductivity, or values between. Depending on bearing conductivity, bearing currents can be classed as the conduction mode and the discharge mode.

The conduction mode bearing current is dv/dt dependent which occurs only when a bearing becomes short-circuit or low impedance. The discharge mode bearing current is caused by storage of coupling currents in the air-gap capacitor which produces a shaft voltage

across the bearings. The shaft voltage may force the lubricant films inside bearings to breakdown and thus produce a discharge current spike through the bearings.

Based on our observations, the conduction mode or the continuous short-circuit inside bearing actually seldom happens under normal operating conditions. It appears only when a motor runs nears zero shaft speed. This is reasonable because the continuous short-circuit can only result from the continuous ball and race contact which is impossible at high speeds.

In almost all other operating condition, one only sees the discharge mode bearing current. As the discharge mode bearing current has much higher amplitude, it is this component that should be suspected to be a major contribution to bearing damage. Finally, as was observed, bearing currents increase as coupling increases. Therefore, bearing currents will become more serious in drive systems with good grounding and low impedance between the motor and inverter cases. Also, the bearing current problem can be expected to increase as the size of the machine increases with rating faster than the bearing surface area.

APPENDIX

Test motor and inverter specifications

Induction motor: volts: 220; hp: 8; rpm: 1165

Bearings: frt. 206sff. ext.207sff

Inverter: input: 380V, 50Hz; output: 220V, 18A, 2-50Hz; hp: 5; dc bus: 320V

Modulation: 15kHz sine-triangle PWM

REFERENCES

[1] Kerszenbaum, "Shaft Currents in Electric Machines Fed by Solid State Drives", *IEEE Conf. Proc. Industrial and Commercial Power System Tech. Conf.*, 1992, pp. 71-79.

[2] J. Erdman, R. J. Kerkman, D. Schlegel., "Effect of PWM Inverters on AC Motor Bearing Currents and Shaft Voltages". *IEEE Proceedings of the 10th Annual Applied Power Electronics Conference and Exposition (APEC)*, 1995, Dallas, Vol. 1, pp. 24-33

[3] S. Chen, T. A. Lipo and D. Fitzgerald, "Source of Induction Motor Bearing Currents Caused by PWM Inverters," *IEEE Transactions on Energy Conversion*, Vol. 11, No. 1, March 1996

[4] DoyleFBusse, "Bearing currents and their relationship to PWM drives", *IEEE Trans. Power Electronics*, 1997, 12(2): 243-252.

[5] J. Alan Lawson, "Motor bearing fluting." *Conference Record of Pulp and Paper Industry Technical Conference, IEEE Industrial Application Society*, 1993, pp. 32-35

[6] Hugh Boyanton, "Bearing damage due to electrical discharge." *Publication by Shaft Grounding Systems Inc., Albony, OR.*

2006 5th International Power Electronics and Motion Control Conference

Research on a New Motor Drive Control System for Electric Transit Bus

SHAO Gui-xin, ZHANG Cheng-ning

School of Mechanics/Vehicle and Transportation Engineering, Beijing Institute of Technology, Beijing 100081, China

Abstract—To eliminate the deficiencies in the conventional drive control system of Direct Current (DC) motor for electric vehicle, we design a new Permanent Magnetic Direct Current (PMDC) motor with the enhanced exciting windings ,namely, enhanced magnetism motor and develop its drive control system for electric transit bus. It can obtain enhanced torque with enhancing magnetism in low speed and enhanced speed with weakening magnetism in high speed automatically. On the basis of data and graph of experiments, we analyze its control theory and characteristics. Contrasting the enhanced magnetism motor to conventional DC motor with exciting windings in series, the enhanced magnetism motor as great advantages on characteristics of control way, configuration of controller and regenerative braking. Heat assessing experiment shows that switched loss is main part of power loss when IGBT switched in high frequency and it will affect the controller characteristics. Experiments of vehicle running show that the drive control system's anti-jamming ability is strong and the regulated performance is fast and smooth, and it's characteristic of enhanced torque with enhancing magnetism in low speed and enhanced speed with weakening magnetism in high speed can meet the electric transit bus's demand of power characteristics very well.

Keywords: electric transit bus, enhanced magnetism motor, weakening magnetism and regulating speed system, switched loss, peak voltage

I. INTRODUCTION

Because energy crisis and natural disaster caused by the atmosphere pollution become more and more serious, electric vehicles, that have advantages of no pollution, zero emission, energy resources saving, low noise, of great help to alleviate pollution of cities and to increase utilization rate of energy resource, have become one of the advanced developing directions of automotive industry. Electric vehicles have been investigated in China or abroad. Electric transit bus, as a new way in municipal transportation, becomes more and more important.

Motor drive control system is the most important part of electric vehicle and its configuration and characteristics decide the electric vehicle performance. For municipal transportation, electric transit bus should have following characteristics of wide regulating speed range for buses run in high speed, large start-up torque for good acceleration, wide high efficiency range and good regenerative braking to increase utilization rate of energy resource.

Sponsored by the National 863 project（2003AA501800）

Analyzing conventional DC motor systems, they have following disadvantages. Separate excited motor can not meet the demand of power characteristics when buses start or accelerate, including permanent magnetic motor. Although series-wound motor hasn't above disadvantage, however, it has complex configuration and instability while bus regenerative braking. The deficiencies in DC motor with extra and series exciting windings are its inherent controller is too complex and lacks reliability.

In order to avoid above disadvantages and base on the theory of DC motor with compound exciting windings, we design a new Permanent Magnetic Direct Current (PMDC) motor with the enhanced exciting windings and develop its drive control system for electric transit bus. In the new system, enhanced exciting windings are connected in the continued-current loop and a bran-new theory of weakening magnetism and regulating speed automatically comes into beings. It can obtain enhanced torque with enhancing magnetism in low speed and enhanced speed with weakening magnetism in high speed automatically. It can realize regenerative braking because it can run in double quadrant. It adopts PWM control in high frequency and runs in low noise.

II. WEAKENING MAGNETISM TO REGULATE SPEED THEORY

The schematic of the enhanced magnetism motor drive control system for electric transit bus is shown in Fig. 1. The main function of the motor controller is to control drive torque and regenerative torque by receiving singles of ACCEL from the acceleration pedal, of BRAKE from brake pedal and of motor-armature current feedback from

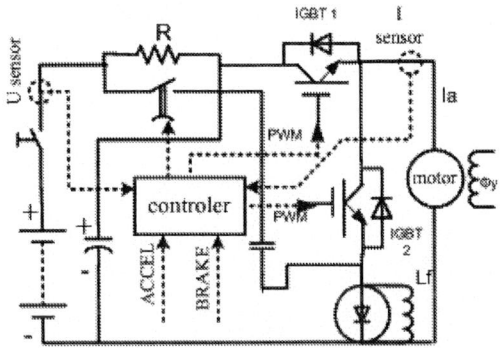

Figure 1. The schematic of control system

power circuit.

While electric transit bus starts up, accelerates or runs in high speed, BRAKE single is locked and IGBT2 is cut-off. Controller receives ACCEL single from acceleration pedal and sends high frequency PWM single to control IGBT1 to switch in high frequency. Match motor-armature current to the ACCEL single in closed loop by regulating the duty of PWM single. While IGBT1 is closed, the work loop is battery pack, IGBT1, motor and battery pack; and no current goes through the enhanced exciting windings connected in the continued-current loop. While IGBT1 is cut-off, the work loop is motor, the enhanced exciting windings, the reversed protecting diode inside the IGBT2 module and motor; current goes through the enhanced exciting windings. During that time, the continued-current functions are not only to protect circuit but also to enhance magnetism to regulate speed.

The newly weakening magnetism and regulating speed system applies to BJD6100-EV bus. Its Mathematical model is shown as below.

$$T = C_m(\phi_y + \phi_e) \times i_a \tag{1}$$

$$n = (V_b \times D + r_a \times i_a)/(C_c(\phi_y + \phi_e)) \tag{2}$$

$$\phi_e = l_f \times i_f \tag{3}$$

$$i_f = i_a \times (1 - D) \tag{4}$$

$$v_e = C_e \times (\phi_y + \phi_e) \times n \tag{5}$$

Where, C_m is torque constant of motor, C_e is back electromotive force constant of motor, ϕ_y is permanent magnetism, ϕ_e is enhanced magnetism, r_a is motor-armature resistance, i_a is motor-armature current, l_f is inductance of continued-current and enhanced exciting windings, i_f is current of continued-current and enhanced exciting windings, v_e is back electromotive force of motor-armature and D is the duty of PWM single.

Base on above equations, we can regulate i_a, i_f and ϕ_e by controlling the ACEEL to set motor-armature current desired value and the duty of PWM by executing adjusting arithmetic. Increasing torque by enhancing magnetism to meet demand when electric transit bus starts up and increasing speed by weakening magnetism to let electric transit bus run in high speed.

While electric transit bus is braking, ACCEL single blanks off and IGBT1 module is always cut-off. Controller receives BRAKE single from brake pedal and sends high frequency PWM to IGBT2 module to make IGBT2 module switched in high frequency. The duty of PWM single controls the motor-armature current and makes it equal to the desired value set by BRAKE single. No current goes through the continued-current loop because there is a reversed diode parallel connected with the enhanced exciting windings. There is no enhanced magnetism all the time so that back electromotive force of motor-armature is stable. Because IGBT2 module switched in high frequency, back electromotive force of motor-armature will hoik. When back electromotive force of motor-armature exceeds the battery pack voltage, battery pack will be charging so that regenerative braking takes place.

III. ANALYSIS OF QUIESCENT CHARACTERISTICS OF THE ENHANCED MAGNETISM MOTOR

A. Analysis of Inherent Acceleration Characteristics of the Newly Motor System

While electric transit bus is running in low speed, e.g. starting up, we know that back electromotive force of motor-armature is low from equation (5) and the motor-armature current can match the desired value set by ACCEL when the duty of PWM is small. From equations (1), (3) and (4), we know that great current goes through the enhanced exciting windings during most time and the effect of enhanced magnetism is so great that the motor system has great torque and electric transit bus has good power characteristics. Back electromotive force is rising as motor speed increase and the duty of PWM must increase little by little to let the motor-armature current equal to the desired current value set by ACCEL. From equations (2), (3) and (4), we know that the time when current goes through the enhanced exciting windings decreases little by little and the enhanced magnetism is weakened so that the newly motor runs in higher speed at constant power.

Base on above analysis, we know above process of regulating speed is executing automatically and the only thing we do is to set the desired motor-armature current value by ACCEL single through acceleration pedal. Contrast with conventional permanent magnetism DC motor, the weakening magnetism and regulating speed system has much superiority on characteristics of regulating speed, torque, peak power and efficiency.

The dynamics characteristics of one electric vehicle are decided by its drive control system. In order to obtain the working performances of the newly system, we do following experiments. Set desired armature current by regulating ACCEL signal to obtain the relation between torque and speed and the results are shown in Fig. 2. While motor runs at constant speed, torque is increasing as motor-armature current is rising. While motor runs as motor-armature current is constant, torque is increasing as motor speed is rising. Namely, while motor runs in low speed, torque is enhanced by enhanced magnetism; while

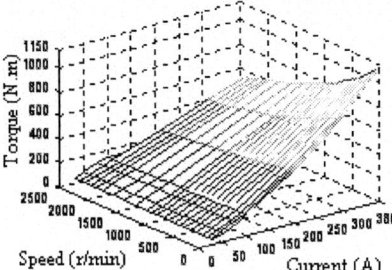

Figure 2. Torque versus speed and current with continued-current

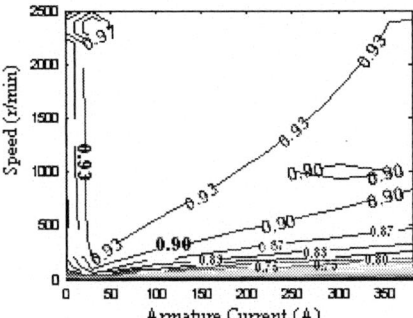

Figure 3. Efficiency map of the enhanced magnetism motor

speed is rising and enhanced magnetism is weakening, torque is decreasing little by little to realize that motor can run at large constant power in wide range. Fig. 3 shows the efficiency map relative to armature currents and speeds and the system energy efficiency is larger than 85% within 88.2% working range.

B. Characteristics Contrast between the Pattern with Enhanced Exciting Windings and That of Without

When enhanced exciting windings is connected in continued-current loop, the new motor system can regulate magnetism under the control of PWM single so that it has the characteristic of large torque with enhanced magnetism in low speed and large wide constant power range with weakened magnetism in high speed. When there is no enhanced exciting winding connected in continued-current loop, the enhanced magnetism motor is equivalent to a separate excited permanent motor. Here, we can calculate its characteristics by the known mathematics model and experiments.

Torque comparison between with enhanced exciting windings and without is shown in Fig.4. It is clear that the torque of the enhanced magnetism motor is much larger than that of the other's in low speed and the torque of the enhanced magnetism motor is decreasing as speed is rising and enhanced magnetism is weakening. At last, the torque of the new system is equivalent to that of the DC motor with extra exciting windings. During the whole process, the part of permanent magnetism is constant so

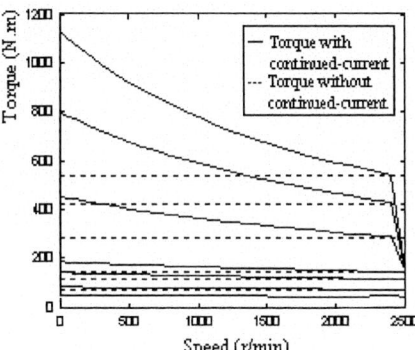

Figure 4. Torque contrast between with continued-current and without

that the torque can meet the load torque still when the enhanced magnetism motor runs in high speed.

Contrast the peak torque and peak power between the enhanced magnetism motor and the equivalent DC motor and show results in Fig.5 and Table Ⅰ. When the enhanced windings are connected in the continued-current loop, external characteristic of the enhanced magnetism motor is similar to that of engine and can meet the torque and power demands of the electric transit bus very well. The torque and energy efficiency of the enhanced magnetism motor are much great than those of the equivalent DC motor with extra exciting windings when they are at the same speed.

C. Characteristics Contrast between the Enhanced Magnetism Motor and Conventional Sries-wind Motor

Base on the weakening magnetism and regulating speed theory, we know that the process of weakening magnetism is executing automatically under the control of the duty of PWM single while electric transit bus runs and no extra parts such as chopper and contactor are needed to control current value. Namely, the configuration of the new control system is not only succinct but also reliable.

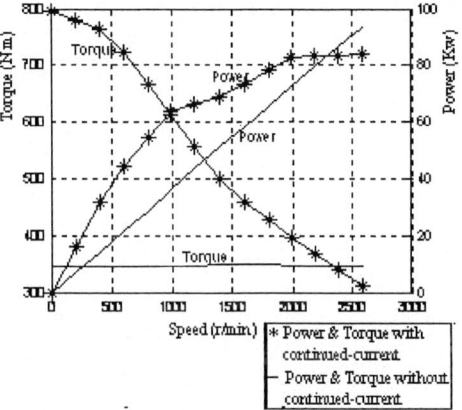

Figure 5. Peak power and peak torque contrast between two patterns

TABLE I. DATA OF TORQUE AND POWER CONTRAST BETWEEN TWO PATTERNS, WHILE MOTOR-ARMATURE CURRENT IS 300A

Speed (r/min)	Motor without continued-current			Motor with continued-current		
	Torque (N.m)	Power out (Kw)	η	Torque (N.m)	Power out (Kw)	η
0	345	0	0	796.5	0	0
487	346	17.515	0.576	755	38.54	0.778
980	344	35.24	0.684	617	63.58	0.877
1472	348	53.77	0.843	480	69.73	0.886
1976	347	71.85	0.856	400	82.88	0.891
2624	344	94.29	0.893	306.8	83.77	0.915

While electric transit bus runs in high speed, the enhanced magnetism is weakened as the duty of PWM decreases while the permanent magnetism is constant. However, the conventional series-wind DC motor has to decrease its magnetism and torque to enhance speed. Contrast both, the enhanced magnetism motor can meet both demands of torque and power much better than the latter when vehicle runs in high speed.

While electric transit bus brakes, the enhanced magnetism motor is equivalent to DC motor with extra exciting windings. The permanent magnetism is stable with constant direction when bus regenerative braking is happening. However the conventional DC motor with series exciting windings has to change the exciting current direction and be connected with assistant circuit made of contactor or power diode to make the exciting magnetism stable.

Contrast between both, in one word, the enhanced magnetism motor system is much better reliable, stable and economic.

IV. ANALYSIS OF CHARACTERISTICS OF CONTROLLER

A. Characteristics of Controller Power Loss

Insulated Gate Bipolar Transistor (IGBT) is a new type electric power semiconductor controlled by magnetism field and can be self-closed. Connected with high speed of MOSFET and low resistance of bipolar semiconductor together, IGBT has the characteristics of high input resistance, low power loss under control of voltage, supporting large voltage and current.

Power loss of IGBT includes breakover loss, open state loss, switched loss and drive grid loss. Among above losses, breakover loss and switched loss is the primary. Beakover loss is decided by saturated voltage between Collector and Emitter of IGBT and laden current. Switched loss consists of turn-on loss and turnoff loss. At the moment when IGBT is turned on and the voltage of IGBT still equals to battery pack voltage, the diode can't come back from breakover to reversed cut-off immediately but a large current with large peak value goes through it before IGBT can endure a large reverse voltage. In fact, the current that goes through IGBT at that moment is the sum of the reverse comeback current and laden current. At the moment when IGBT is turned off and the voltage of IGBT rises to equal to battery pack voltage, the

laden current turns to go through diode from IGBT. Base on above analysis, switched loss of IGBT is large. In the new drive control system, IGBT module switched in high frequency under the control of PWM single, the voltage of IGBT between Collector and Emitter when IGBT cut-off is 384 volt and the current goes through circuit is 300A so that the heat from power loss of IGBT is large enough to influent controller characteristics.

Some investigations on power loss have shown that switched loss increases linearly to switched frequency. When switched frequency is larger than 5 kHz, the main loss is switched loss. Detailed trend is shown in Fig. 6.

When parameters of system capability, the saturated voltage and current while IGBT is closed are set, the breakover loss increases linearly to the duty of PWM single without influence of the frequency of PWM. IGBT switched loss is the absolute integral of product of instant voltage and instant current of IGBT when it switches. Although IGBT can turn on or turn off within several microseconds, instant large voltage and current make switched loss increase greatly. In the new drive control system, the switched frequency is 16 kHz so that it is necessary to assess the switched loss and its influence to the new controller.

Base on the theory that IGBT breakover loss is constant when motor runs in the same power, same motor speed and same motor-armature, we design experiments as below. Initialize the motor runs at the same speed, same input/output power and same motor-armature current, and then set the switched frequencies 16 kHz and 8 kHz orderly and respectively. Record the duration of experiments and the final temperatures when temperature of controller doesn't vary any more. The results are shown in Tab. II.

The duration of the temperature variety is 8 minutes and the final temperature is 79 ℃ when the switched frequency is 16 kHz. The other duration is 30 minutes and the final temperature is 65.9℃.Both meet the schedule switched of 5 minutes, but it is clear the switched loss of the latter is much less and it is necessary that to connect to capacitance in parallel to reduce switch loss.

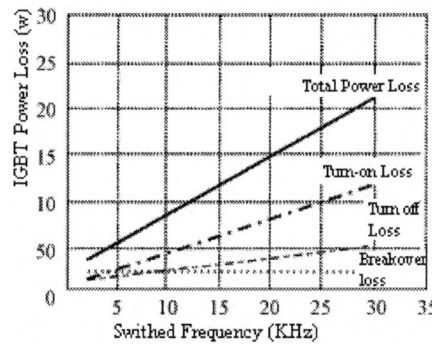

Figure 6. Power loss versus switched frequency.

TABLE II. SWITCHED LOSS ESTIMATION EXPERIMENTS

Parameters / Freq.(kHz)	Speed (r/min)	Input Power (kW)	Output Power (kW)	Motor-armature Current (A)	Balanceable Temperature (℃)	Duration of time (minute)
16	1975	93.1	82.88	300	79	8
8	1978	92.3	81.73	300	65.9	30

Ambient temperature: 24.5℃, Relative humidity: 26%

B. Characteristics of IGBT Peak Voltage

The peak voltage between Collector and Emitter of IGBT is the main cause to damage IGBT and it is caused by stray inductance of snubber circuit and the comeback process of diode in snubber circuit when IGBT turns off. Fig. 7 shows the peak voltage variation at various motor speeds and various motor-armature currents. At the same curve, the motor speed is constant. At various motor speeds, the peak voltage has the same variation trend relative to motor-armature current: peak voltage rises linearly to motor-armature current. When motor-armature current is constant, the peak voltage doesn't change if we change the motor speed.

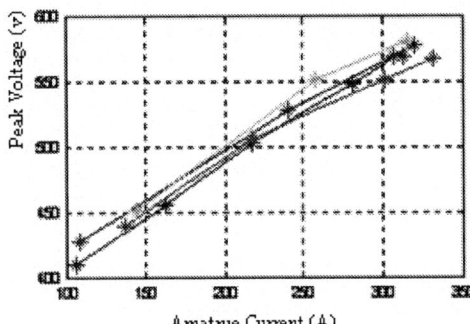

Figure 7. IGBT peak voltage variation relative to motor-armature current

V. EXPERIMENTS OF THE ELECTRIC TRANSIT BUS RUNNING

The electric transit bus equipped with the enhanced magnetism motor system is BJD6110-EV and is battery-operated. The rated storage battery voltage is 384 volts and battery pack can furnish 9100 Watt-hours. Its gross mass with no load is 12100 Kilograms and it equips with a retarder and no transmission. The experiments of the electric transit bus running show that maximum velocity is 75 Km/h, the maximum gradient is great than 20%, and the duration of accelerating (0~60 Km/h) is less than 55 seconds. The regenerative braking is of great prominence.

VI. CONCLUSIONS

The enhanced magnetism motor and its drive control system for electric transit bus can eliminate the deficiencies of the conventional DC motor drive control systems for electric vehicles. It not only can obtain enhanced torque with enhancing magnetism in low speed and enhanced speed with weakening magnetism in high speed automatically, but also its configuration is simple and its energy efficiency is high.

The process of weakening magnetism and regulating speed is executing automatically so that it meet electric transit bus power demands very well.

Contrasting to conventional series-wind DC motor, the new one has great advantages on characteristics of control way, configuration of controller and regenerative braking.

Heat assessing experiment shows that switched loss is main part of power loss when IGBT switched in high frequency and it will affect the controller temperature and performance. Experiments of vehicle running show that

the drive control system's anti-jamming ability is strong and the regulation performance is fast and smooth, and it's characteristic of enhanced torque with enhancing magnetism in low speed and enhanced speed with weakening magnetism in high speed can meet the electric transit bus's demand of power characteristics very well.

REFERENCES

[1] SUN Feng-chun, ZHANG Cheng-ning, ZHU Jia-guang. Electric Vehicle- the Important Vehicles in 21st Century[M]. Beijing: Beijing Institute of Technology press, 1997. (in chinese)

[2] SUN Feng-chun, ZHANG Cheng-ning, LI Xue-yong, etal. Electric vehicle traction motor with performance of automatic weakening magnet and regulating speed[P]. China Patent:CN 99107924,1999.

[3] CM Delco Remy. "Propulsion Systems for Electric Vehicles".

[4] ZHANG Cheng-ning, SUN Feng-chun. Analysis of Control Characteristics for Drive System of Electric City Bus[J]. TRANSACTIONS OF CHINA ELECTROTECHNIAL SOCIETY, Vol. 19, No. 1, Jan. 2004.

[5] ZHANG Cheng-ning, SUN Feng-chun, YU Xiao-jiang. Characteristics Contrast of Electric Vehicles on Convergence Drive[M]. Journal of China Ordnance—Tank, Armored Car and Engine Fascicule, Vol. 4,1995.

[6] MING Zheng-feng, NI Guang-zheng, HUANG Xiao-dong, ZHOU Wen-yun, ZHONG Yan-ru, TONG Jian-li. Analysis of Transducer on Resonance Transition Soft Switching Technology (1).

[7] YU Zhi-sheng. Vehicle Theory[M]. Beijing: China Machine Press,1996.

[8] WANG Ying, WANG Jing. Electric Drive Systems Summarization of Electric Vehicles[J]. Electric Engineering, 1998 (4): 8~12.

2006 5th International Power Electronics and Motion Control Conference

New Micro-Drive Series For Induction Motors & Survey of Market Trends

Henrik Rosendal Andersen, Ruimin Tan and Zhang Hui

Danfoss DD-MC, R&D Dep., Beijing, China

Abstract—Ultimo 2006 Danfoss launches the first inhouse-developed micro drive series for the global market. The drive is developed by the Danfoss R&D office in Beijing and manufactured using Danfoss production facilities close to Shanghai. The scope of this paper is to compare the new product with the state-of-the-art of the market.

Keywords - micro drives; low-cost; strip-down concept; state-of-the-art; performance parameters for comparison

I. INTRODUCTION

Danfoss Drives A/S has until now focused on producing high-performance standard motor drives for the global market. Results are the VLT®5000 and VLT®2800 series. In 2004 the concept of tomorrow was introduced by the VLT®AutomationDrive FC 300 series shown in fig. 1. FC 300 includes FC 302 and FC 301 covering a power range up to megawatts. "High performance" relates to every aspect of the list: Mechanical concept, thermal rating, EMC, harmonic line performance, robustness & lifetime, galvanic isolation principle, user interface and control performance.

Now it is time to launch the first inhouse-developed micro-drive series, referred to as VLT® Micro Drive FC 51. This is achieved by a Danfoss R&D office in Beijing and Danfoss production facilities in the Shanghai area. Most people perceive "micro" as referring to size. But to be successful the term is related to cost also. Hence, a micro drive has to be based on a strip-down concept. The minimum required hardware and software to make a motor spin should be integrated only, along with a suitable level of application hardware and software. Needs for upgrading the performance have to be purchased separately or by advising customers to buy FC 300 etc.

II. STATE-OF-THE-ART IN MICRO DRIVES

The state-of-the-art in micro drives is highlighted by analyzing various products in the segment in terms of the performance indicators in section I. The analysis assumes the PWM-modulated voltage-source B6 inverter with a diode-rectifier-based front-end. This is the choice for general-purpose industrial standard drives.

A. Mechanical Concept

The bookstyle shape used for FC 300 in fig. 1 is rare in micro drives. Most micro drives exhibit a cubic shape.

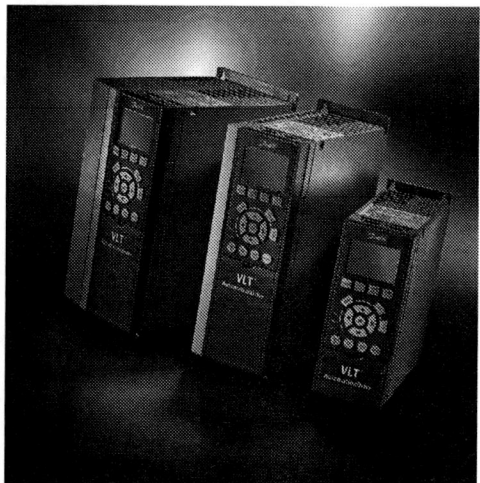

Figure 1. FC 300 from Danfoss Drives A/S up to 7.5kW.

The heatsink is placed in the back of the drive in the depth dimension, with the fins sticking out in the back. A power module is mounted on the heatsink. The power PCB for the module gives a wall between the heatsink and the rest of the electronics. The electronics are surrounded by a plastic cover and the power PCB. Typically, more PCBs are stacked on top of the power PCB requiring several interconnections by flat cables, pin connectors etc. The plastic cover may be open in the top, bottom and the 2 sides of the drive to cool the electronics. Hence, the tightness of enclosures in the micro market is normally IP20 as a maximum. An advantage of the cubic style is the thermal separation of the heatsink and the electronics ensured by the power PCB. This makes the perspective of avoiding a fan more likely to succeed. Many micro drives up to 0.75kW are without a fan. Also, quasi-cold-plate and through-the-back-panel constructions are favored by the cubic style. A drawback is the holes in the plastic cover for cooling the electronics allowing pollution to enter. True side-by-side mounting in cabinets can be problematic for the same reason. Another drawback is the connections between the stacked PCBs. Such connections are a root cause for quality related issues. Finally, a cubic shape means that, less drives can be mounted side-by-side. On the other hand the bookstyle shape means a deeper and higher unit.

1-4244-0448-7/06/$25.00 ©2006 IEEE

B. Thermal Rating

The thermal rating of a drive is given by the rated ambient temperature and the loss capacity. This translates into voltage and current capacity, switching frequency and motor-cable length. Micro drives are rated for an ambient temperature of 40°C. Operation is maintained to 50°C with derating of the output power and sometimes extended opening of the plastic cover. Side-by-side mounting often requires derating. 200V micro drives are specified as 200 to 240V units. 400V drives are specified as 380 to 480V units. The input-current rating varies. The input current depends on the line impedance. It is a measure of how stiff a line a given drive can be connected to. Some drives can be connected to a 250kVA transformer only. Other drives are compatible with 500kVA transformers without requiring line reactors. At rated load, the simulation in fig. 2 shows the line current versus line inductance of a 230V/0.75kW drive with a single- (1Φ) and 3-phase (3Φ) front-end. A 3Φ 400V/0.75kW drive is considered also. A 500kVA transformer corresponds to less than 0.1% in fig. 2. L% is given by (1). f_{line} is the line frequency, L_{line} is the line inductance pr. phase, $U_{line,ph}$ is the line phase voltage, P_{shaft} is the rated motor shaft power, η_{tot} is system efficiency set to 0.8 in this paper for convenience.

$$L\% = 100 \cdot \frac{2\pi \cdot f_{line} \cdot L_{line}}{z_{base}}, \; z_{base} = \frac{\eta_{tot} \cdot U_{line,ph}^2}{P_{shaft}} \qquad (1)$$

The rated output current of micro drives is given by standards like UL508C and commercial motors etc. The transient capacity is 150% current for 60s. Rated current at rated temperature is achieved at a 4kHz switching frequency. The motor-cable-length specification of most micro drives is in the range of 50m. This means full cable reflections [1] and full cable-related switching loss.

C. EMC

Demands for EMC vary from one region to another. In Europe EMC is a must. Industrial drives must comply with at least EN55011, class A2. In USA and Asia lower demands are observed. Regardless of regulations, a virtual minimum demand exists for drive manufacturers always. Customers complain, if a purchased drive interferes with existing equipment. According to micro drives on the market the trend seems to be a minimum level of onboard EMC performance. Different variants for different markets are observed also. This may be American/Asian variants with virtually no EMC-filter followed by a European variant with an integrated EN55011, class A1 filter suited for a short motor cable (≈10m). For 1Φ 200V drives and at a minimum motor-cable length (<5m), the filter may comply with EN55011, class B also. Generally, the trend is to design low-leakage filters compatible with residual current devices (RCDs) and IT (isolated terra) grids. In terms of immunity all micro drives are designed to sustain burst and surge transients, as required for obtaining a CE approval.

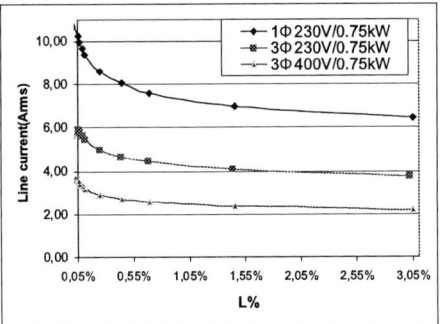

Figure 2. Line current versus line impedance.

D. Harmonic Line Performance

The trend is to rely on passive solutions for enabling the diode-rectifier-based front-end to comply with European line-harmonics standards like IEC1000-3-2 and IEC1000-3-12. All manufacturers agree that, means for reduction of line harmonics cannot be integrated in a micro drive. A DC-inductor as in FC 300 is not compatible mechanically and thermally. The best offer is extra power terminals for connecting an external DC-choke. The alternative is an external AC-choke, which does not require extra terminals. The penalty of the AC-choke is the extra commutation voltage drop at 3Φ rectification [2]. A stripped-down micro drive without inductors draws a heavy line current (see fig. 2). The ratio of the line-to-motor current is over 250% for 1Φ drives. For 3Φ drives the ratio is over 150%, whereas a FC 300 draws less than 100%. For the customer this means extra installation costs to be considered. At 3Φ rectification another issue, when choosing a micro drive without inductors, is the unbalance performance defined by IEC1000-24. A FC 300 will handle a 3% unbalance on the line voltages. A micro drive will not handle more than 1 to 2% at rated load without a substantial reduction of the lifetime. Further, the question on voltage distortion and commutation notches defined by IEC1000-24 and IEC146-1 should be considered. Without inductors this is difficult to handle. Finally, a drive without inductance can never be as robust against bad line transients as defined by VDE160. In terms of shaft performance, the missing inductor means that, the DC-link voltage ripple is typically increased compared to a FC 300. This gives more torque ripple in the motor at high-speed and high-torque operation.

E. Robustness & Lifetime

The robustness of a drive is related to the concept, protection features and the main power components. All micro drives are overcurrent, under/overvoltage and overtemperature protected. Also, shorts between the motor terminals are managed. The earth-fault protection varies. Some micro drives are protected during start-up only. Others are protected during operation also. The trend in power-module technology is small-sized IGBT modules without a base plate. The junction-to-heatsink thermal impedance is relatively higher for these modules meaning that, some micro drives may not be suited for constant-torque, low-speed applications. The DC-link capacitors of a micro drive determine the lifetime. The trend is to employ 400V electrolytic capacitors, some rated for 85ºC and others for 105ºC. The relative DC-link capacitance C% for more existing micro drives is given in fig. 3 at 1Φ 230V/0.75kW. In average, a 6.8% capacitor is used. Almost a factor 2 in C% is observed. A small 8.5% capacitor is an optimistic design at a stiff line, even at low-ESR capacitors and optimum cooling. A 5% capacitor should not give any problems, maybe apart from cost and size. Using a 5% capacitor could indicate a poor thermal management of the drive. The relative capacitance is defined by (2). C_{dctot} is the total DC-link capacitance.

$$C\% = 100 \cdot \frac{X_c}{z_{base}}, \quad X_c = \frac{1}{2\pi \cdot f_{line} \cdot C_{dctot}} \qquad (2)$$

Depending on the current ripple and the internal drive temperature, a 6 to 7% capacitor gives a lifetime of 5 years. The current ripple in the capacitor depends on the line impedance as may be deduced from fig. 2.

F. Galvanic Isolation Principle

The trend is to use the negative DC-link bus as reference for the basic part of the control circuitry. Hence, the IGBT gate-driver circuitry can be based on low-cost principles, and simple resistive sensing principles can be used for temperature, voltage and current. Galvanic isolation is required for the I/O customer interface. Typically, low-cost optocouplers are used, since the communication speed and the number of I/O channels are low in a micro drive.

G. User Interface

Micro drives employ simple local control panels (LCPs) for viewing and programming parameters. Besides a display and keypads, a potentiometer for easy operation is often employed. Some LCPs are detachable giving a userfriendly installation, operation, reconfiguration and maintenance. In this case a mounting cable is often supplied to enable the LCP to be mounted on the front of a cabinet. The detachable LCP allows the customer to purchase a drive without one. Still, many micro drives come with a built-in LCP. This gives the lowest cost, if a LCP is needed. Digital and analog I/O channels and a serial communication interface are observed for existing micro drives. Usually, 3 to 5 digital inputs and a voltage/current input are offered. Some employ an output relay and an analogue voltage/current output also. A serial interface such as RS232 or RS485 is preferred due to cost and reliability. Thus, it is easy to get a parameter setup from a PC or copy settings from other drives. And different drives can be connected together and controlled by a PC.

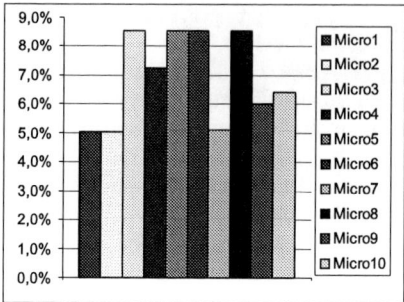

Figure 3. C% of 1Φ 200V/0.75kW micro drives.

H. Control Performance

A low-cost micro controller is the typical choice for a micro drive making high-performance vector control difficult. Some parties may use the term "flux-vector" performance as offered by FC 302 having a high-performance DSP onboard. By a look at the processor, motor parameters and the current-sensing principle it turns out that, "flux-vector" obviously means scalar control with some steady-state slip compensation. Hence, more micro drives exhibit a good steady-state open-loop speed control performance. A few micro drives exhibit a good dynamic control performance also. To achieve this, the 3 motor currents are sensed by a shunt in each low-side inverter leg. A penalty of this sensing principle is the need for an open emitter in the low-side IGBTs. This means more module pins and a more expensive and potentially larger power module. Also the principle requires an advanced sampling technique at a high output voltage, unless workarounds are applied. A workaround may be to omit overmodulation in the PWM generation giving a reduced output voltage and to apply an asymmetric modulation strategy (120PWM, [3,4]) at high speed. For reference, space-vector modulation [5] is the normal PWM strategy in micro drives as in most other drives.

III. THE DANFOSS MICRO DRIVE

VLT® Micro Drive FC 51 covers a power range from 0.18 to 7.5kW. The same mechanics are used for 200 and 400V units to ensure a simple concept. Mechanically, a dedicated 200V series is not offered. This means that, UL demands for creapage & clearance are determined by 400V. This may give a drawback on the physical size of the 200V drive, but the advantage is a solid family look. FC 51 comes in 3 frames as shown in fig. 4. To date, the frames are the smallest from Danfoss. All are aligned with market standards. The smallest frame is rated up to 0.75kW in 200 and 400V. The middle frame is rated up to 1.5kW in 200V and 2.2kW in 400V. The largest frame is rated for 3.7kW in 200V and 7.5kW in 400V. All drives in the middle and largest frame employ a built-in brake chopper. Up to 0.75kW enough braking torque can be obtained by advanced motor control using the motor for

1034

dumping the braking energy. Alternatively, the customer may use the available DC-link power terminals. A frame-3 drive may act as a brake-chopper module for several frame-1 drives installed side-by-side in a cabinet. The basic power circuit in the 400V version of FC 51 is shown in fig. 5. Available power terminals are marked with a dot. The terminals are all placed in the bottom of the drive giving a single-entry construction.

Figure 4.   Micro-Drive FC 51 series from Danfoss.

A.   Mechanical Concept

FC 51 is bookstyle-shaped giving a clear family look. The heatsink is placed in the side of the drive with the fins sticking out in the side instead of in the back. In fig. 4 this is not visible due to the plastic cover. A simple built-up, with a minimum of connections between different PCB's, is ensured. Loose electrical wires are avoided similar to FC 300. One of the strongest features of the drive is the tightness of the enclosure, setting a new standard. FC 51 comes as IP20 as standard. NEMA1 is offered as standard also without an additional cover for the top of the drive. Electronic components cannot be seen through holes in the plastic cover, and air stream is not allowed to enter the electronics. Therefore, the drive is robust in polluted environments, which are often seen in Asian applications believed to be an important target for the product. As all electronics are exposed to a relatively large heatsink surface compared to a cubic construction, the mechanical concept can only be obtained by applying a fan. If not, the physical dimensions of the drive go too large relative to market standards.

B.   Thermal Rating

FC 51 follows the state-of-the-art in terms of the voltage rating and ambient temperature. Derating at side-by-side mounting is not required due to the mechanical concept. The output current of a drive determines the internal temperature rise. Hence, output current and size are coupled. Therefore, FC 51 is designed for standard motors. Some American and Asian motors require a high current-to-power ratio. Compared to modern motors over 10% more current may be required. For such low-efficient motors Danfoss recommends to buy a one-size-up drive. Fig. 6 compares the FC 51 output-current rating to the average of several existing 200V micro drives. FC 51 is competitive and complies with UL508C. Compared to a standard ABB motor a suited rating is offered (see fig. 6). Transiently, 150% current can be drawn for 60s. Another loss-producing factor is the motor cable. A long motor cable increases the inverter loss and impacts the rectifier-side also. If a customer requires a low-cost micro drive, then this customer must consider the installation cost also. Therefore, it does not make sense to design a micro drive for an expensive 150m motor cable. This customer should buy FC 300. A typical cable length for a micro drive is expected to be 10m, eventhough FC 51 will operate at a 50m motor cable.

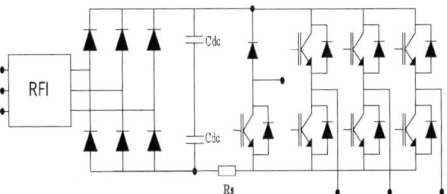

Figure 5.   Power circuit of FC 51.

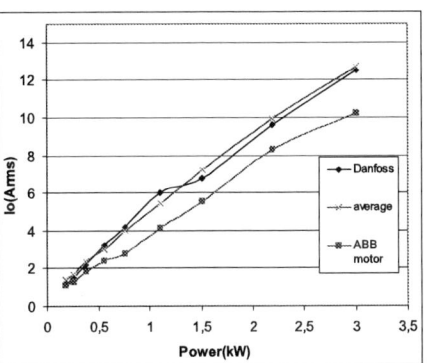

Figure 6.   Output current level of FC 51.

C.   EMC

FC 51 complies with EN55011 A1 up to a 15m motor cable. Longer cable lengths saturate the built-in common-mode choke, due to the higher leakage current. A measurement verifying the compliance is shown in fig. 7 for a 1Φ 200V/0.75kW drive. The 1Φ drive complies with class B also, at a shorter cable length. The EMC performance is ensured, while maintaining a low leakage current. Hence, FC 51 is RCD compatible and suited for IT-lines. Finally, the drive is immune to surge and burst transients. Additional EMC performance is obtained by ordering a separate power option, which supports a 50m shielded motor cable.

D. Harmonic Line Performance

FC 51 does not offer integrated harmonic performance. Hence, the drive violates IEC1000-3-2 and IEC1000-3-12. The compatibility is ensured by a separate power option including AC-inductors. The passive solution is a solid choice for a 3Φ drive. The test result in table I verifies that, a 1Φ 200V/0.37kW drive complies with IEC1000-3-2 also, if a 2% AC-choke is applied. In the test the drive draws 649W on the 230V line at a 3.7A line current. The THD of the current is 82%. The power factor is 0.75. The compliance comes with a 3% reduced average DC-link voltage as drawback. The advantage is that, the customer gets a robust and low-cost harmonic solution. Danfoss will not offer 0.55kW drives, and 0.75kW drives draw more than 1kW input power, which is the limit in IEC1000-3-2. Hence, the compliance is ensured by a passive solution.

E. Robustness & Lifetime

FC 51 has adopted protection features from FC 301 in terms of software and hardware. FC 51 tolerates coupling on the output phases, FC 51 tolerates to be ramped up and –down, while controlling a motor with a large load of inertia, resulting in repeated hammering into the current and voltage protections. FC 51 tolerates short-circuit conditions on the output. For earth-fault protection a principle from the FC 301 (patent pending) is reused. This means that, the earth-fault protection of FC 51 is good, eventhough cost is saved relative to the FC 301. The IGBT module of FC 51 is without a base plate due to cost and size. But the drive is suited for constant-torque/low-speed applications still. The relative DC-link capacitance of FC 51 is 40% for 3Φ 400V units, 18% for 3Φ 200V units and 6% for 1Φ 200V units, which is in the high-end of the segment according to fig. 3. This is achieved by a good thermal management of the capacitors. FC 51 tolerates continuous full load operation at a low line impedance. The drive may be coupled to a 500kVA transformer without inserting line reactors. Hence, it is a true industrial compliant drive suited for large industrial plants having a private transformer installation. In this operating condition FC 51 exhibits a lifetime of 5 years. A measured DC-link voltage ripple is shown in fig. 8. A 3Φ 400V/0.75kW drive is powering a 1.1kW motor operated at 50Hz and loaded with 5Nm. Channel 1 shows the DC-link voltage at 20V/div (ch. 1 offset = 500V). Channel 2 shows the line current at 5A/div. The RMS current is 2.8A at the 0.15% line inductance of the Beijing grid. Channel 3 shows the motor current at 2A/div. The RMS current is 2.1A. A 50m motor cable was used explaining the high-frequency ripple on the motor current. The peak-to-peak voltage ripple is around 40V relative to an average DC-link voltage of 580V. This equals a relative ripple of 7%, which modulates the motor current slightly (see fig. 8) giving a 300Hz torque ripple.

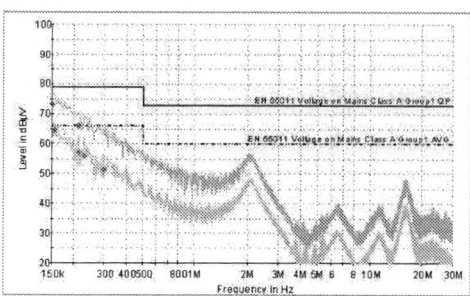

Figure 7.   EMC test of 1Φ 200V/0.75kW drive.

TABLE I.
IEC1000-3-2 TEST OF 1Φ 230V/0.37KW DRIVE

Harmonic NO.	Test result [A]	Limit [A]
3	2.1	2.3
5	1.04	1.14
7	0.29	0.77
11	0.13	0.33
13	0.08	0.21
17	0.05	0.13

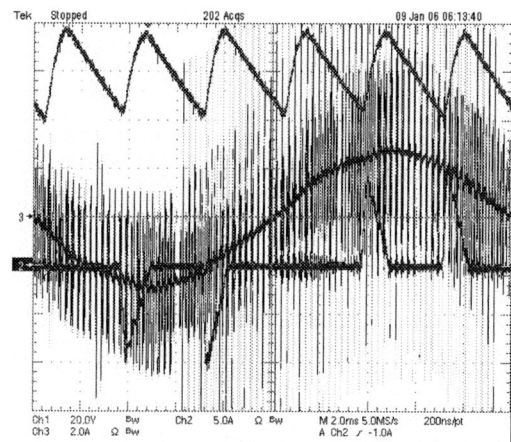

Figure 8.   Voltage ripple of 400V/0.75kW drive.

F. Galvanic Isolation Principle

The negative DC-link bus is the reference for the electronics. Optocouplers are used for the I/O channels to isolate the customer from the power voltages.

G. User Interface

To achieve the best man-machine interaction (MMI) in the market, FC 51 has adopted a simplified version of the FC 300 LCP, which won the IF Design Award in 2004 for novel design, user-friendliness and multi-functionality. Fig. 4 shows the LCP. The hot-plugging property of the detachable LCP makes the parameter setup easy. The LCP can be offered with a potentiometer also to accommodate the Asian market. Parameters are arranged in menus giving a structure known from mobile phones. Both parameter and value are displayed simultaneously, which is rare in the segment. 2 analog voltage/current inputs, 4 digital inputs (NPN/PNP) and 1 analog current/voltage output are designed in FC 51 along with an output relay. A RS-485 interface with protection is built in for communication with other drives or a PC.

H. Control Performance

FC 51 employs the voltage-vector-control principle VVC+ [6], giving it a shaft performance better than the VLT®2800 and almost as good as the VLT®5000 and FC 300. FC 51 is space-vector-modulated at a 4kHz rated switching frequency. The shaft performance of the drive is achieved by a single current-sensing shunt in the DC-link (see fig. 5, R_s). Green's principle for reconstructing the phase currents from the DC-link current is applied [7,8]. Compared to 3 shunts in the low-side inverter legs, this gives less circuitry, and means that simpler power modules are usable. Hence, the customer will achieve a high shaft performance at the lowest cost. Measured steady-state torque curves are shown in fig. 9 for a 400V/0.75kW drive powering a 3Φ, 4-pole, 400V, 1.1kW induction motor. The torque is varied from 0 to 7.5Nm, and the speed is varied from 30rpm to 1500rpm. The speed is almost independent on torque. The fine performance is achieved at ultra-low speed also. A nominal torque step of 5Nm is applied in fig. 10 showing that, FC 51 regains the steady-state speed level in 300ms. This is very competitive at a moment of inertia of around 2 times the value of the induction motor. The dynamic command means that, the motor is able to generate a starting torque of more than 150%, when a 150% motor current is drawn.

IV. CONCLUSION

A new micro-drive series from Danfoss was presented in this paper and compared to existing products on the market. The drive is developed and produced in China. State-of-the-art equipment is offered by Danfoss with emphasis on the mechanical concept, the MMI and the shaft performance. The future will show, whether Danfoss will be able to dominate the micro segment by offering customers the quality known from other Danfoss products.

REFERENCES

[1] P. Enjeti et al, "The Effect of Long Motor Cables on PWM Inverter Fed AC Motor Drive Systems", APEC'95.

[2] Kjeld Thorborg, "Power Electronics", 1988, ISBN 0-13-686577-1

[3] H.R. Andersen, "Motor Drives for Variable-Speed Compressors", 1996, ISBN 87-89179-13-7 (bd. 1-3).

[4] 0. Ojo, "The Generalized Discontinuous PWM Scheme for Three-Phase Voltage Source Inverters", IEEE Trans. On Indust. Elec., VOL. 51, NO. 6, Dec. 2004.

[5] Paul Thøgersen et al, "Stator Flux Oriented Asynchronous Vector Modulation for AC-Drives, PESC'90.

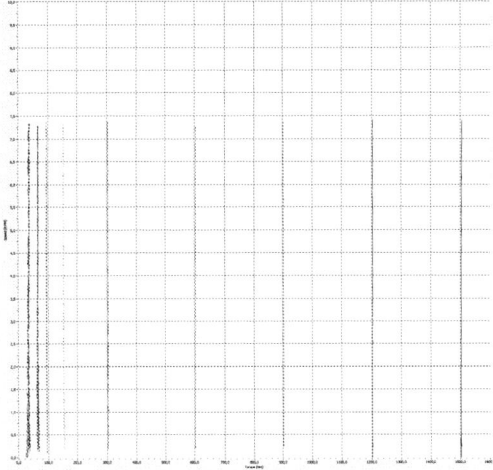

Figure 9.   Steady-state speed vs. torque.

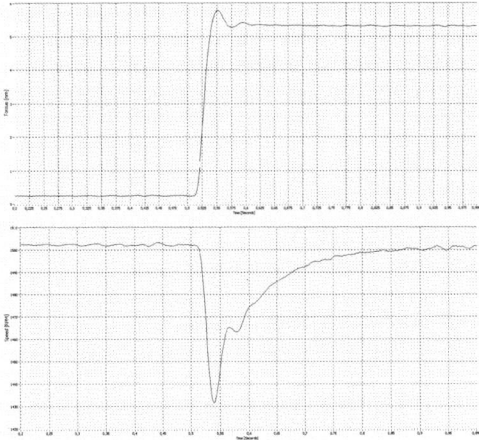

Figure 10. Torque-step performance.

[6] Paul Thøgersen et al, "New High Performance Vector Controlled AC-Drive with Automatic Energy Optimizer, EPE'95.

[7] Frede Blaabjerg et al, "Single Current Sensor Technique in the DC-Link and the Ultimate Solution", IAS'96.

[8] S. Nielsen et al, US-patent 5,969,958.

2006 5th International Power Electronics and Motion Control Conference

Robust Backstepping Control of Induction Motor Drives Using Artificial Neural Networks

J. Soltani[*], R. Yazdanpanah[**]

Electrical and Computer Engineering, Isfshan University of Technology, Isfahan, Iran

[*] j1234sm@cc.iut.ac.ir
[**] rezayazdanpanah@gmail.com

Abstract– **In this paper, using the three-phase Induction Motor(IM) fifth order model in a stationary two axis reference frame whit stator current and rotor flux as state variables, a conventional backstepping controller is designed first for speed and rotor flux control of an IM drive. Then in order to make the control system stable and robust against the parameter uncertainties as well as the unknown load torque, in the next stage the backstepping controller is combined with an Artificial Neural Network (ANN). It will be shown that the proposed composite controller is capable of compensating the parameters variations and rejecting the external load torque disturbance. The overall system stability is proved by Lyapunov theory. It is also shown that the method of ANN training, guarantees the boundedness of errors and ANN weighs. The validity and effectiveness of the controller is verified by computer simulation.**

Keywords- backstepping; induction motor; ANN; nonlinear systems; robust

I. INTRODUCTION

In the recent two decades, nonlinear control methods such as input-output feedback linearization and Sliding-Mode (SM) control have been applied to the IM drive. Specially in these years, in the field of adaptive and robust control, there has been a tremendous amount of activity on a special control scheme known as "backstepping" [1],[2],[3]. A major problem of backstepping control approach is that certain function must be "linear in the unknown system parameters" and some very tedious analysis is needed to determine a "regression matrix" [3]. It must be noted that in adaptive backstepping control, problem of finding regression matrix is more difficult in comparison with conventional backstepping method.

To overcome the above problem, in [4], a combination of the backstepping control with ANN has been proposed. According to this method, in the process of backstepping controller design, two ANN are used to estimate a nonlinear function. Therefore no need to find the regression matrix for on-line estimation of unknown parameters.

In [4], using the ANN, the theory of robust backstepping control has been presented for strictly feedback nonlinear systems. This method has been applied to a single arm robot[5] and to a rotor-flux Field Oriented Control (FOC) IM drive[6].

One may note that the FOC methods are in fact a type of partial feedback linearization control technique in which the zero dynamic stability can not be proved. As a result, it is not guaranteed that system model would be robust to parameters variation. In addition in these control methods, the field orientation can be achieved only in the system steady state conditions.

To overcome the above problems, in this paper, using the fifth order model of IM in a fixed stator reference frame, based on control theory described in [4], a composite nonlinear controller is designed that makes the IM drive system control robust and stable against the parameter uncertainties and external load torque. In this control approach, a two level SVPWM inverter feeds the IM drive.

II. IM MODEL

The IM fifth order model in fixed two axis reference frame with rotor fluxes and stator currents as state variables [7] is given as

$$\frac{d\omega}{dt} = \frac{3n_p M}{2JL_r}(\psi_{ra}i_{sb} - \psi_{rb}i_{sa}) - \frac{T_l}{J} \tag{1}$$

$$\frac{d\psi_{ra}}{dt} = -\frac{R_r}{L_r}\psi_{ra} - n_p\omega\psi_{rb} + \frac{R_r}{L_r}Mi_{sa} \tag{2}$$

$$\frac{d\psi_{rb}}{dt} = -\frac{R_r}{L_r}\psi_{rb} + n_p\omega\psi_{ra} + \frac{R_r}{L_r}Mi_{sb} \tag{3}$$

$$\frac{di_{sa}}{dt} = \frac{MR_r}{\sigma L_s L_r^2}\psi_{ra} + \frac{n_p M}{\sigma L_s L_r}\omega\psi_{rb}$$
$$- \left\{\frac{M^2 R_r + L_r^2 R_s}{\sigma L_s L_r^2}\right\}i_{sa} + \frac{1}{\sigma L_s}u_{sa} \tag{4}$$

$$\frac{di_{sb}}{dt} = \frac{MR_r}{\sigma L_s L_r^2}\psi_{rb} - \frac{n_p M}{\sigma L_s L_r}\omega\psi_{ra}$$
$$- \left\{\frac{M^2 R_r + L_r^2 R_s}{\sigma L_s L_r^2}\right\}i_{sb} + \frac{1}{\sigma L_s}u_{sb} \tag{5}$$

where $i_{sa}, i_{sb}, \psi_{ra}, \psi_{rb}, u_{sa}, u_{sb}$ are the stator currents, rotor fluxes and stator voltages, respectively. Subscripts a, b indicate a vector components in the fixed stator reference frame. Subscripts r, s indicate rotor and stator components. ω is rotor angular mechanical speed and $\sigma = 1 - M^2/(L_s L_r)$.

L_s, L_r are per-phase stator and rotor spatial inductances, respectively. M is per phase magnetizing inductance. n_p is number of pole pairs. R_s, R_r are stator and rotor resistances, respectively.

III. ROBUST BACKSTEPPING CONTROL

A. ANN Basics

Define W as the collection of ANN weighs, then the net output is [4]

$$y = W^T \phi(x) \qquad (6)$$

Let S be a compact simply connected set of $\mathbb{R}^n$, with map $f: S \to \mathbb{R}^n$, define $C^m(s)$ the functional space such that f is continuous. A general nonlinear function $f(x) \in C^m(S)$, $x(t) \in S$ can be approximated by a neural network as

$$f(x) = W^T \phi(x) + \varepsilon(x) \qquad (7)$$

with $\varepsilon(x)$ a ANN functional reconstruction error vector and $\phi(x)$ is sigmoid activation function.

B. Robust Backstepping Control of IM Using ANN

Using the well known fifth order IM model in a stator two axis reference frame where the rotor fluxes and stator currents are assumed as state variables [7], the robust nonlinear controller is designed in the following way.

Dividing the above IM model into two nonlinear subsystems, where i_{sa}, i_{sb} are the outputs for the first subsystem which are simultaneously assumed the fictitious inputs of the second sub-system.

Assume that:

Assumption 1: The reference trajectories ω^r and ψ_r^r are differentiable and bounded.

Assumption 2: The load torque is an unknown constant and resistances, inductances and moment of inertia are unknown and bounded.

In the first step of the controller design, i_{sa}, i_{sb} are assumed as fictitious controls for the second sub-system. The main objective is to obtain these controls so that the desired rotor speed and rotor flux amplitude signals are perfectly tracked in spite of machine parameters and external load torque uncertainties. Considering ω^r and ψ_r^r as references for ω and ψ_r, tracking error equations are

$$e_1 = \omega - \omega^r$$
$$e_2 = \psi_{ra}^2 + \psi_{rb}^2 - \psi_r^{r2} = \psi_r^2 - \psi_r^{r2} \qquad (8)$$

Then

$$\dot{D}_1 e = F_1 + G_1 i \ , \ F_1 = \begin{bmatrix} \dfrac{-T_l L_r}{M} - J\dfrac{L_r}{M}\dot{\omega}^r \\ -\dfrac{2}{M}(\psi_r^2) - 2\dfrac{L_r}{R_r M}\psi_r^r \dot{\psi}_r^r \end{bmatrix}$$

$$G_1 = \begin{bmatrix} -\dfrac{3n_p}{2}\psi_{rb} & \dfrac{3n_p}{2}\psi_{ra} \\ 2\psi_{ra} & 2\psi_{rb} \end{bmatrix}, D_1 = \begin{bmatrix} J\dfrac{L_r}{M} & 0 \\ 0 & \dfrac{L_r}{R_r M} \end{bmatrix} \qquad (9)$$

It is clear that G_1 is known and invertible. By treating $\bar{i}$ as a fictitious input, a controller for the ideal $\bar{i}$ is designed as

$$\bar{i} = G_1^{-1}[-\hat{F}_1 - K_1 e] \qquad , K_1 > 0 \qquad (10)$$

where K_1 a design parameter and $\hat{F}_1$ the estimate of F_1 which will be estimated in the next section with a two layer ANN. Substituting (10) into (9) gives

$$\dot{D}_1 e = F_1 - \hat{F}_1 - K_1 e + G_1 \eta \ , \quad \eta = i - \bar{i} \qquad (11)$$

In the second step, the control $u(u_{sa}, u_{sb})$ are obtained in such a way that η in equation (11), becomes as small as possible. Differentiating η with respect to time, yields

$$D_2 \dot{\eta} = F_2 + G_2 u \qquad (12)$$

where

$$u = \begin{bmatrix} u_{sa} \\ u_{sb} \end{bmatrix}, G_2 = \begin{bmatrix} 1 & 0 \\ 0 & 1 \end{bmatrix}, D_2 = \sigma L_s \begin{bmatrix} 1 & 0 \\ 0 & 1 \end{bmatrix}$$

$$F_2 = \dots + D_2 \{ G_1^{-1} (\dot{\hat{F}}_1 + K_1 e) + G_1^{-1} \hat{F}_1 \qquad (13)$$
$$+ G_1^{-1} K_1 D_1^{-1} (F_1 - \hat{F}_1 - K_1 e + G_1 \eta) \}$$

To make η as small as possible, the following control is chosen

$$u = G_2^{-1}[-\hat{F}_2 - K_2 \eta - G_1^T e] \qquad (14)$$

In (14), $\hat{F}_2$ is an estimate of F_2 that like the first step, a two layer ANN is used to estimate it. In addition a term $-G_1^T e$ is added in (14) which is necessary to cancel the effect of $G_1 \eta$ in (11).

Combining (12) and (14), gives

$$D_2 \dot{\eta} = F_2 - \hat{F}_2 - K_2 \eta - G_1^T e \qquad (15)$$

C. F_1, F_2 Approximation Using ANN

In this section, functions F_1, F_2 are approximated by two two-layer ANN. In adaptive backstepping control, it is assumed that functions F_1, F_2 are linear in term of known regression matrices, however in ANN method, there is no limitation for these functions. Using ANNs approximation property, F_1, F_2 as outputs of two two-layer ANN with constant weights W_i, is assumed to be as follows

$$F_1 = W_1^T \phi_1 + \varepsilon_1 \quad , \|\varepsilon_1\| < \varepsilon_{1N} = cte$$
$$F_2 = W_2^T \phi_2 + \varepsilon_2 \quad , \|\varepsilon_2\| < \varepsilon_{2N} = cte \tag{16}$$

where ϕ_1, ϕ_2 provide suitable basis functions. From (16), one can find that net reconstruction error $\varepsilon_i(x)$ is bounded by a known constant ε_{iN}.

Assumption 3: The ideal weighs are bounded by known positive values so that

$$\|W_1\|_F \le W_{1M} \quad , \|W_2\|_F \le W_{2M} \tag{17}$$

Or equivalently:

$$\|Z\|_F \le Z_M \quad , \ Z = diag\{W_1, W_2\} \tag{18}$$

The actual inputs to ANN1 are $\psi_r, \omega_r, \dot{\psi}_r^r, \dot{\psi}_r^r$ and actual inputs to ANN2 are $\omega, \omega^r, \dot{\omega}^r, \psi_r, \dot{\psi}_r^r, \psi_{ra}, \psi_{rb}$, $\ddot{\psi}_r^r, i_{sa}, i_{sb}, e_1, e_2$.

On line ANN approximation of F_1 is

$$\hat{F}_1 = \hat{W}_1^T \phi_1 \tag{19}$$

Then error dynamic equation of (11) becomes

$$D_1 \dot{e} = \widetilde{W}_1^T \phi_1 - K_1 e + G_1 \eta + \varepsilon_1 \tag{20}$$

where $\widetilde{W}_1 = W_1 - \hat{W}_1$. Similarly, approximation of F_2 is assumes as

$$\hat{F}_2 = \hat{W}_2^T \phi_2 \tag{21}$$

Then error dynamic (15) will be

$$D_2 \dot{\eta} = \widetilde{W}_2^T \phi_2 - K_2 \eta - G_1^T e + \varepsilon_2 \tag{22}$$

Note that there is a term $G_1 \eta$ in (20) and a term $-G_1^T e$ in (22). This means there are couplings between the error dynamics (20) and (22).

D. Updating ANNs Weights

In this part, the stability of proposed controller, is proved based on Lyapunov stability theory. This analysis shows that tracking errors and updated weighs are Uniformly Ultimately Bounded (UUB).

Theory : Let the desired trajectories ω^r, ψ_r^r be bounded. Take the control input (14) with weigh updates be provided by

$$\dot{\hat{W}}_1 = \Gamma_1 \phi_1 e^T - k_\omega \Gamma_1 \|\zeta\| \hat{W}_1$$
$$\dot{\hat{W}}_2 = \Gamma_2 \phi_2 e^T - k_\omega \Gamma_2 \|\zeta\| \hat{W}_2 \tag{23}$$

with any constant matrices $\Gamma_1 = \Gamma_1^T > 0, \Gamma_2 = \Gamma_2^T > 0$ and scalar positive constant k_ω. Then the errors

$\eta(t), e(t)$ are UUB. ANN updated weights are bounded. The errors $\eta(t), e(t)$ can be kept as small as desired by increasing gains K_i. Proof of this theory can be find in [4].

Note 1: Small tracking error bounds may be achieved by selecting large control gain K. The parameter k_ω offers a design tradeoff between the relative eventual magnitudes of $\|\zeta\|$ and $\|\widetilde{Z}\|_F$, a smaller k_ω yields a smaller $\|\zeta\|$ and a larger $\|\widetilde{Z}\|_F$, and vice versa.

Note 2 : If $\hat{W}_i(0)$ are taken as zeroes the linear proportional control term $-K\zeta$ stabilizes the system on an interim basis.

IV. SYSTEM SIMULATION

Based on proposed control strategy described in previous section, the block diagram of IM drive control is shown in Fig. 1.

A C^{++} computer program was developed for system simulation. In this program, the nonlinear equations are solved based on static forth order Range-Kutta method. The proposed control method, is tested for a three-phase IM with parameters shown in Table (1). In this simulation, the controller gains are obtained by trial and error method which are given as

$$K_1 = diag\{1525, 1550\} \ , K_2 = diag\{5000, 1550\}$$
$$k_\omega = 1 \quad , \Gamma_i = 10I$$

Table 1 : IM PARAMETERS

Stator resistance	$R_s = 0.18\Omega$
Rotor resistance	$R_r = 0.15\Omega$
Rotor nominal flux linkage	$\psi_r^r = 1.3 Wb.turns$
Number of pole pairs	$n_p = 1$
Stator inductance	$L_s = 0.0699H$
Rotor inductance	$L_r = 0.0699H$
Mutual inductance	$M = 0.068H$
Nominal rotor speed	$\omega^r = 220 rad/s$
Moment of inertia	$J = .0586 kgm^2$

Simulation results shown in Fig. 2 , are obtained in the case of an exponential reference flux rising up from zero to $1.3W.T$ at $t = 0$s, down to $0.8W.T$ at $t = 3$s with a time constant of $\tau = 0.05s$, an exponential reference speed from zero to $220 rad/s$ at $t = 0.3$s, rising up to $350 rad/s$ at $t = 3$s with a time constant of $\tau = 0.1s$, a step load torque disturbance from zero to $40N.m.$ at $t = 2$s and motor electromechanical parameters assumed to be twice their nominal values at $t = 1$s. In addition, the

1040

steady state tracking errors e_2, e_1 are also shown in this figure.

Fig. 3 shows the simulation results obtained for an exponential reference flux rising up from zero to $1.3W.t$ at $t = 0s$ and an exponential reference speed rising up from zero to $220rad/s$ at $t = 0.3s$, a load torque profile which is also shown in Fig.3 and motor electromechanical parameters assumed to be twice their nominal values at $t = 0s$.

The IM rotor flux control is obtained for an exponential reference speed rising up from zero to $220rad/s$ at $t = 0.3s$ and an exponential flux reference from zero to

$1.3W.t$ at $t = 0s$, down to $0.8W.t$ at $t = 2s$ and rising up to $1.3W.t$ at $t = 3.5s$, a step up load torque from zero to $40N.m$ at $t = 1s$ is shown in Fig. 4. In addition the IM speed control is obtained for an exponential reference flux rising up from zero to $1.3W.t$ at $t = 0s$ and an exponential reference speed from zero to $220rad/s$ at $t = 0.3s$, down to $-220rad/s$ at $t = 2s$, rising up to $220rad/s$ at $t = 3.5s$, a step load torque from zero to $40N.m.$ at $t = 1s$ is shown in Fig. 5. In flux and speed control performance, motor electromechanical parameters assumed to be twice their nominal values at $t = 0s$.

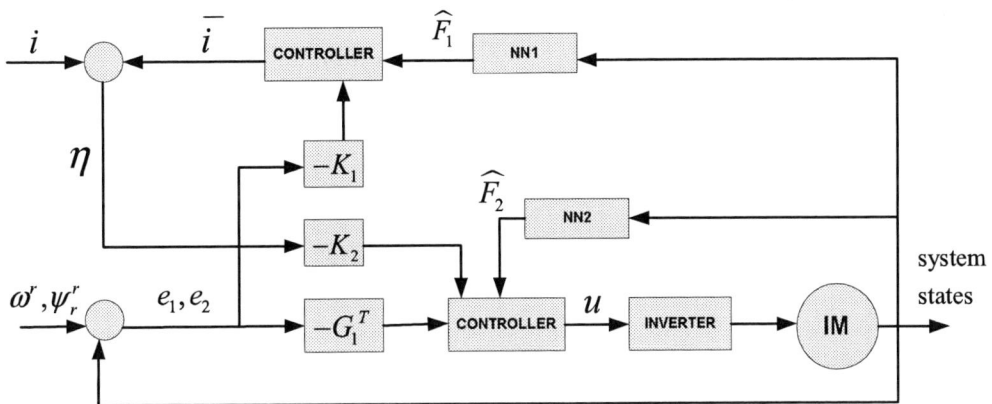

Fig. 1 : The overall block diagram of IM drive control

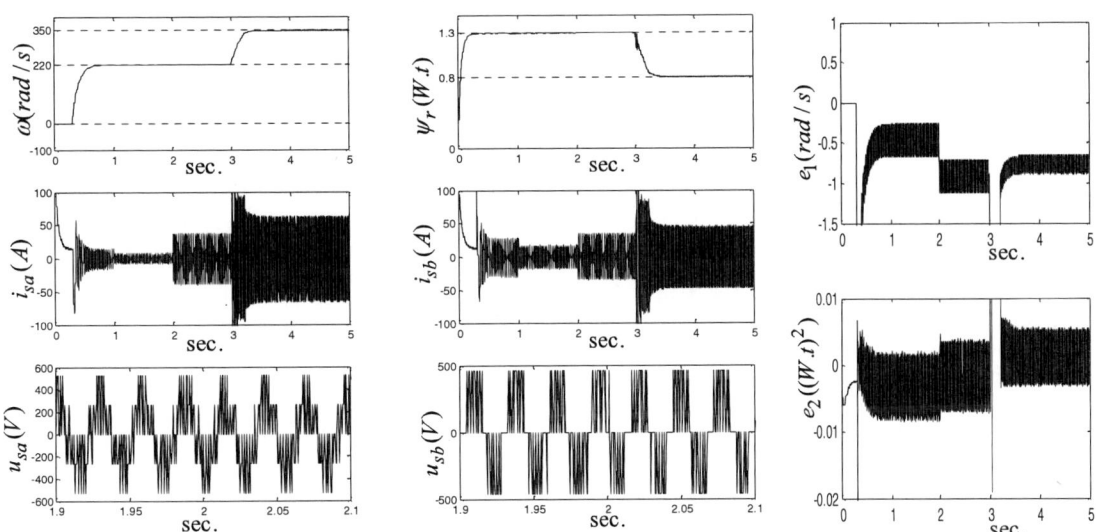

Fig. 2 : IM performance using robust backstepping controller

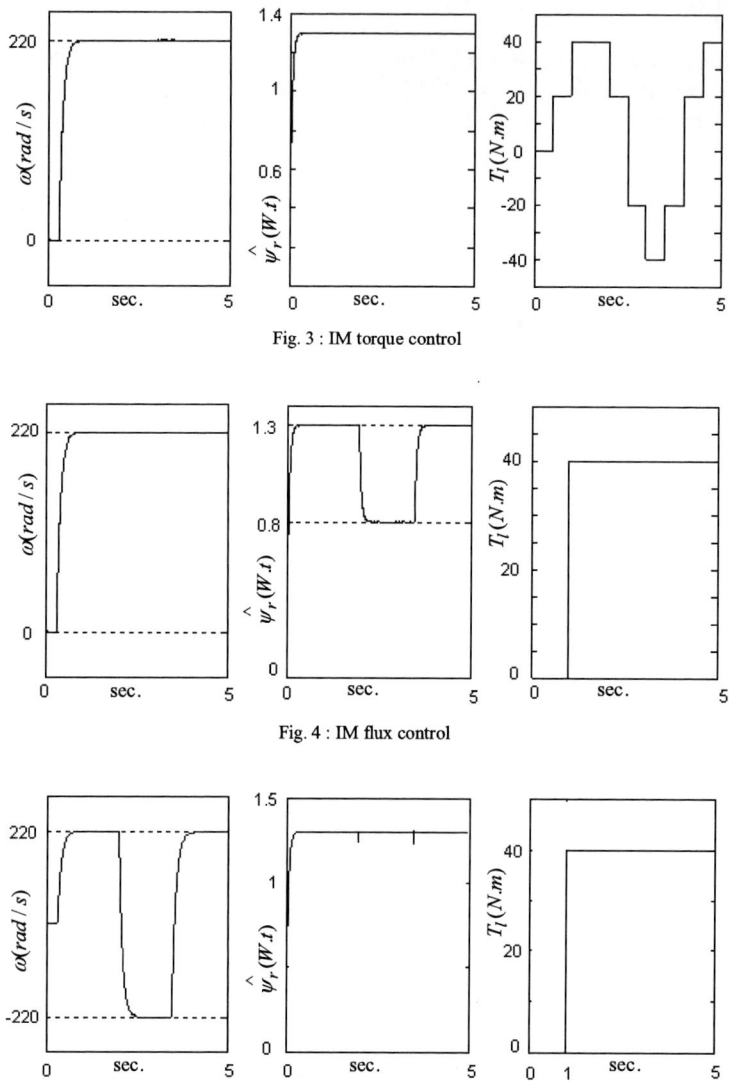

Fig. 3 : IM torque control

Fig. 4 : IM flux control

Fig. 5 : IM speed control

CONCLUSIONS

In this paper, a composite nonlinear controller has been proposed for the IM rotor flux and speed tracking control. The nonlinear controller is designed based on the IM fifth order model in a fixed two axis reference frame, combining the backstepping control and ANN. The overall stability of this controller is proved by Lyapunov theory. Computer simulation results obtained, confirm the effectiveness and validity of the proposed controller. These results also confirm that the drive system control is robust and stable against the parameters uncertainties and unknown load torque disturbance.

REFERENCES

[1] I. Kanellakopoulos, P. V. Kokotovic and A. S. Morse, "Systematic Design of Adaptive Controllers for Feedback Linearizable Systems," *IEEE Trans. Automat. Contr.*, vol. 36, pp. 1241–1253, 1991.

[2] P. V. Kokotovic, "Bode lecture: The joy of feedback," *IEEE Contr. Syst. Mag.*, No. 3, pp. 7–17, June 1992.

[3] M. Krstic, I. Kannellakopoulos, and P. Kokotovic, *Nonlinear and Adaptive Control Design*, Wiley and Sons Inc., New York, 1995.

[4] C. M. Kwan, F. L. Lewis, "Robust Backstepping Control of Nonlinear Systems Using Neural Networks," *IEEE Trans. Systems, Man and Cybernetics*, vol. 30, No. 6, Nov. 2000.

[5] O. Kuljaca, N. Swamy, F. L. Lewis and Ch. M. Kwan "Design and Implementation of Industrial Neural Network Controller Using Backstepping," *IEEE Trans. Industrial Electronics*, vol. 50, No. 1, Feb. 2003.

[6] C. M. Kwan, F. L. Lewis, "Robust Backstepping Control of Induction Motors Using Neural Networks," *IEEE Trans. Neural Networks*, vol. 11, No. 5, Sep. 2000.

[7] P. C. Krause, *Analysis of Electric Machinery*, McGraw-Hill Inc., 1986.

2006 5th International Power Electronics and Motion Control Conference

Robust Nonlinear Control of Linear Induction Motor taking into account the Primary End Effects

J. Soltani, M.A. Abbasian
Isfahan University of Technology,Isfahan,Iran
j1234sm@cc.iut.ac.ir, a_abbasian@ec.iut.ac.ir

Abstract—This paper presents a robust nonlinear controller for speed and flux control of primary type Linear Induction Motor (LIM) Drive. The proposed controller is designed based on combination of input-output feedback linearization and Sliding-Mode (SM) control taking into account the LIM end effects. The stability of proposed composite controller is proved by Lyaponuv theory. The effectiveness and validity of the nonlinear controller system is supported by computer simulation. Simulation results obtained, confirm that the drive system controller is robust an stable against the parameters variations and external unknown load force.

Keywords- linear induction motor, nonlinear control , sliding-mode, end effect

I. INTRODUCTION

The Linear Induction Motor (LIM) has many excellent performance features such as high-starting thrust force, alleviation of gear between motor and the motion devices, reduction of mechanical losses and the size of motion devices, high speed operation, silence, and so on [1]. Because of above the advantages, the primary type LIM, shown in Fig. 1, has been used widely in the field of industrial processes and transportation applications. The driving principles of the LIM are similar to the traditional Rotary Induction Motor (RIM), but its control characteristics are more complicated than the RIM, and the motor parameters are time-varying due to the change of operating conditions, such as speed mover, temperature, and configuration of rail [2].

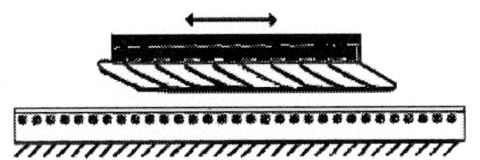

Figure. 1 Linear Induction Motor, Primary Type

In a LIM, the primary winding corresponds to the stator winding of a rotary induction motor (RIM), while the secondary corresponds to the rotor. There are some characteristics differences between the RIM and LIM. The main difference is that the primary of the LIM has a finite length, and therefore, there is a fringing field at both ends of the primary. The infinitely long secondary enters the air-gap field, carries the magnetic flux along with it, and makes, the distribution of the electromagnetic quantities nonuniform, resulting in considerable electric and force losses [3]. The losses, as well as the flux-profile attenuation, become severer as the speed increases. Such a phenomena is called 'end effect' of LIM.

An accurate equivalent circuit model is indispensable for high performance control for LIM drive. Most of the existing models of a LIM depend on field theory [4,5]. Hence, they can not be directly applied for the nonlinear control and in the most of the present studies, the RIM model is used in order to control LIM [6-8]. so, because of the end effect, they can not be valid in high speed operation of LIM. In [6], an adaptive controller is used in order to control LIM at low speeds. But the controller is robust only against the mechanical parameter uncertainties. In [7], an indirect field oriented control is used to control LIM at low speed which is not robust in respect to all machine parameter variations. In [8], a robust nonlinear controller has been proposed for a primary type LIM witch is based on combination of SM control and input-output feedback linearization. Using the RIM model, the drive system of [8] is robust to parameters variation only at low speed operation. Using the control method presented in [8], the main objective of this paper is to introduce a robust nonlinear controller for LIM which will be robust and stable subject to the parameter uncertainties and external unknown load force at low and high speed operation.

II. NONLINEAR DECOUPLED CONTROL

Using the fifth order model of LIM, in a two axis reference frame, attached to the secondary [9]:

$$v_{dx} = R_x i_{dx} + \frac{d\lambda_{dx}}{dt} - \frac{n_p\pi}{h}V_e\lambda_{qx} \qquad (1)$$

1-4244-0448-7/06/$25.00 ©2006 IEEE

1043

$$v_{qx} = R_x i_{qx} + \frac{d\lambda_{qx}}{dt} + \frac{n_p \pi}{h} V_e \lambda_{dx} \qquad (2)$$

$$v_{dy} = R_y i_{dy} + \frac{d\lambda_{dy}}{dt} = 0 \qquad (3)$$

$$v_{qy} = R_y i_{qy} + \frac{d\lambda_{qy}}{dt} = 0 \qquad (4)$$

where subscripts "x" and "y" denote the primary and secondary values; subscripts "d" and "q" denote the d-axis and q-axis values; and v , i , λ and R denote the voltage, current, flux and resistance values, respectively.

The d:q-axis flux linkage of the primary and secondary are given by:

$$\lambda_{dx} = L_{lx} i_{dx} + L_m (1 - f(Q))(i_{dx} + i_{dy}) \qquad (5)$$

$$\lambda_{qx} = L_{lx} i_{qx} + L_m (1 - f(Q))(i_{qx} + i_{qy}) \qquad (6)$$

$$\lambda_{dy} = L_{ly} i_{dy} + L_m (1 - f(Q))(i_{dx} + i_{dy}) \qquad (7)$$

$$\lambda_{qy} = L_{ly} i_{qy} + L_m (1 - f(Q))(i_{qx} + i_{qy}) \qquad (8)$$

with:

$$f(Q) = \frac{(1 - e^{-Q})}{Q} \qquad (9)$$

$$Q = \frac{2 n_p h R_y}{(L_m + L_{lr})|V_e|} \qquad (10)$$

where $L_x = L_{lx} + L_m$ is the primary inductance, $L_y = L_{ly} + L_m$ is the secondary inductance, L_{lx} is the primary leakage inductance, L_{ly} is the secondary leakage inductance and L_m is the magnetizing inductance at zero speed.

The secondary flux amplitude is defined as follows:

$$\lambda_y = \sqrt{(\lambda_{dy})^2 + (\lambda_{qy})^2} \qquad (11)$$

using (3) and (4) the following can be obtained:

$$i_{dy} = -\frac{\dot{\lambda}_{dy}}{R_y} \qquad (12)$$

$$i_{qy} = -\frac{\dot{\lambda}_{qy}}{R_y} \qquad (13)$$

Replacing (12) and (13) respectively into (7) and (8), the following equations can be derived:

$$\lambda_{qy} = f_1 i_{qx} + f_2 \lambda_{qy} \qquad (14)$$

$$\lambda_{dy} = f_1 i_{dx} + f_2 \lambda_{dy} \qquad (15)$$

where:

$$f_1 = \frac{L_m (1 - f(Q))}{\frac{L_m}{R_y}(1 - f(Q)) + \frac{L_{ly}}{R_y}} \qquad (16)$$

$$f_2 = \frac{1}{\frac{L_m}{R_y}(1 - f(Q)) + \frac{L_{ly}}{R_y}} \qquad (17)$$

Using Equations (11), (14) and (15), the derivative of λ_y with respect to time (t), is derived as:

$$\dot{\lambda}_y = f_2 \lambda_y + \frac{f_1 \lambda_{dy}}{\lambda_y} i_{dx} + \frac{f_1 \lambda_{qy}}{\lambda_y} i_{qx} \qquad (18)$$

In addition, the trust force is given by:

$$F_e = K_f (\lambda_{dy} i_{qx} - \lambda_{qy} i_{dx}) = M \dot{V}_e + D V_e + F_l \qquad (19)$$

$$f_3 = \frac{3}{2} \frac{n_p \pi}{h} \frac{L_m (1 - f(Q))}{L_{ly} + L_m (1 - f(Q))} \qquad (20)$$

where V_e is the secondary velocity, n_p is the pair of poles, h is the pole pitch, F_e is the electromagnetic force, F_l is the external force disturbance, M is the total mass of moving element and D is the viscous friction coefficient .

From (19), the LIM motion dynamics is expressed by:

$$\dot{V}_e = -\frac{D}{M} V_e + \frac{f_3}{M}(\lambda_{dy} i_{qx} - \lambda_{qy} i_{dx}) \qquad (21)$$

In Equations (18) and (21), i_{dx} and i_{qx} are the control inputs, where λ_y and V_e are assumed to be the system outputs. Thus, the LIM dynamic is a coupled system. Since there is no direct relationship between the outputs and inputs, hence it is difficult to design the control inputs i_{dx} and i_{qx} so that the system outputs λ_y and V_e can track the desired trajectories accurately. To solve this problem, the nonlinear state-feedback theory is used to derive the system controller laws which is capable of decoupling the LIM speed and its secondary flux. From (18) and (21), one can obtain that:

$$\begin{pmatrix} \dot{\lambda}_y \\ \dot{V}_e \end{pmatrix} = \begin{pmatrix} f_2 \lambda_y \\ -\dfrac{D V_e + F_l}{M} \end{pmatrix} + \begin{pmatrix} \dfrac{f_1 \lambda_{dy}}{\lambda_y} & \dfrac{f_1 \lambda_{qy}}{\lambda_y} \\ -\dfrac{f_3 \lambda_{qy}}{M} & \dfrac{f_3 \lambda_{dy}}{M} \end{pmatrix} \begin{pmatrix} i_{dx} \\ i_{qx} \end{pmatrix}$$

$$\equiv b + A \begin{pmatrix} i_{dx} \\ i_{qx} \end{pmatrix} \qquad (22)$$

where:

$$A = \begin{pmatrix} \dfrac{f_1 \lambda_{dy}}{\lambda_y} & \dfrac{f_1 \lambda_{qy}}{\lambda_y} \\[3mm] -\dfrac{f_3 \lambda_{qy}}{M} & \dfrac{f_3 \lambda_{dy}}{M} \end{pmatrix} \qquad (23)$$

$$b = \begin{pmatrix} f_2 \lambda_y \\[2mm] -\dfrac{DV_e + F_l}{M} \end{pmatrix} \qquad (24)$$

Based on input-output feedback linearization [8]:

$$\begin{pmatrix} i_{dx} \\ i_{qx} \end{pmatrix} = A^{-1}\left(-b + \begin{pmatrix} I_\lambda \\ I_v \end{pmatrix}\right) \qquad (25)$$

where I_λ and I_v are the new control inputs.

If the uncertainties occur (i.e. the parameters of the system are deviated from the nominal value or an external force is added into the system) the ideal feedback linearisation design shown in (25) cannot guarantee the desired performance and the stability of the controlled system may be destroyed. To ensure the stability of the controlled system, despite the existence of the uncertain system dynamics, a newly designed sliding-mode feedback linearisation control system, that comprises a sliding-mode flux controller and a sliding-mode velocity controller, is proposed as:

$$\begin{pmatrix} i_{dx} \\ i_{qx} \end{pmatrix} = \bar{A}^{-1}\left(-\bar{b} + \begin{pmatrix} I_\lambda^a \\ I_v^a \end{pmatrix}\right) \qquad (26)$$

where I_λ^a and I_v^a are the new actual inputs; $\bar{A}$ and $\bar{b}$ can be obtained from (23) and (24), using the nominal parameter values without external force disturbance. Consider the system parameter variations and external force disturbance, the relationship between the inputs and outputs of the dynamic model can be rewritten [8]:

$$\begin{pmatrix} \dot{\lambda}_y \\ \dot{V}_e \end{pmatrix} = (\bar{b} + \Delta b) + (\bar{A} + \Delta A)\begin{pmatrix} i_{dx} \\ i_{qx} \end{pmatrix}$$

$$\equiv \bar{b} + \bar{A}\begin{pmatrix} i_{dx} \\ i_{qx} \end{pmatrix} + \begin{pmatrix} \psi_\lambda \\ \psi_v \end{pmatrix} \qquad (27)$$

where the lumped uncertainty vector $[\psi_\lambda \; \psi_v]^T = \Delta b + \Delta A [i_{dx} \; i_{qx}]^T$, in which ΔA and Δb denote the uncertainties introduced by mechanical and electrical parameters. The lumped uncertainties ψ_λ and ψ_v are assumed to be bounded, i.e. $|\psi_\lambda| \le \eta_\lambda$ and $|\psi_v| \le \eta_v$. Substituting (26) into (27), the decoupled dynamic model can be given as:

$$\begin{pmatrix} \dot{\lambda}_y \\ \dot{V}_e \end{pmatrix} = \begin{pmatrix} I_\lambda^a + \psi_\lambda \\ I_v^a + \psi_v \end{pmatrix} \qquad (28)$$

III. SLIDING-MODE FLUX AND SPEED CONTROLLERS

Assuming the following SM switching surfaces:

$$\begin{cases} S_\lambda(t) = e_\lambda(t) + \gamma \displaystyle\int_0^t e_\lambda(\tau)d\tau \\[2mm] S_v(t) = e_v(t) + \rho \displaystyle\int_0^t e_v(\tau)d\tau \end{cases} \qquad (29)$$

where γ and ρ are positive constants; $e_\lambda = \lambda_y - \lambda_y^*$ and $e_v = V_e - V_e^*$ represent the flux and speed tracking error and λ_y^* and V_e^* are the secondary flux command and speed command, respectively. Differentiating (29) with respect to time (t) and using equitation (28), yields:

$$\begin{cases} \dot{S}_\lambda(t) = \psi_\lambda + I_\lambda^a(t) - \dot{\lambda}_y^* + \gamma e_\lambda(t) \\[2mm] \dot{S}_v(t) = \psi_v + I_v^a(t) - \dot{V}_e^* + \rho e_v(t) \end{cases} \qquad (30)$$

Using (30), Based on Lyapanouv theory, the SM controllers can be obtained as:

$$\begin{cases} I_\lambda^a = -\gamma e_\lambda(t) + \dot{\lambda}_y^* - \hat{\eta}_\lambda(t)\,\mathrm{sgn}(S_\lambda(t)) \\[2mm] I_v^a = -\rho e_v(t) + \dot{V}_e^* - \hat{\eta}_v(t)\,\mathrm{sgn}(S_v(t)) \end{cases} \qquad (31)$$

$$\begin{cases} \dot{\hat{\eta}}_\lambda(t) = \alpha|S_\lambda(t)| \\[2mm] \dot{\hat{\eta}}_v(t) = \beta|S_v(t)| \end{cases} \qquad (32)$$

where $\hat{\eta}_\lambda$ and $\hat{\eta}_v$ are the estimated values of η_λ and η_λ; α and β are positive constants, and $\mathrm{sgn}(\cdot)$ is sign function. Define a Lyapunov function as:

$$V_\lambda(t) = \frac{1}{2}S_\lambda^2(t) + \frac{1}{2}S_v^2(t) + \frac{1}{2\alpha}\tilde{\eta}_\lambda^2(t) + \frac{1}{2\beta}\tilde{\eta}_v^2(t) \quad (33)$$

where $\tilde{\eta}_\lambda(t) = \hat{\eta}_\lambda(t) - \eta_\lambda$ and $\tilde{\eta}_v(t) = \hat{\eta}_v(t) - \eta_v$. Derivativing (33) with respect to the time (t), linking (30), (31) and (32), it is resulted that:

$$\begin{aligned} \dot{V}(t) &= S_\lambda(t)\dot{S}_\lambda(t) + S_v(t)\dot{S}_v(t) + \frac{1}{\alpha}\tilde{\eta}_\lambda(t)\dot{\tilde{\eta}}_\lambda(t) + \frac{1}{\beta}\tilde{\eta}_v(t)\dot{\tilde{\eta}}_v(t) \\ &= S_\lambda(t)\psi_\lambda - \eta_\lambda|S_\lambda(t)| + S_v(t)\psi_v - \eta_v|S_v(t)| \\ &\le |S_\lambda(t)||\psi_\lambda| - \eta_\lambda|S_\lambda(t)| + |S_v(t)||\psi_v| - \eta_v|S_v(t)| \\ &= -|S_\lambda(t)|(\eta_\lambda - |\psi_\lambda|) - |S_v(t)|(\eta_v - |\psi_v|) \le 0 \end{aligned} \qquad (34)$$

Defining the following function:

$$P(t) \equiv |S_\lambda(t)|(\eta_\lambda - |\psi_\lambda|) + |S_v(t)|(\eta_v - |\psi_v|) \le \dot{V}(t) \quad (35)$$

then:

$$\int_0^t P(\tau)d\tau \le V(0) - V(t) \qquad (36)$$

Since $V(0)$ is bounded and also $V(t)$ is nonincreasing and bounded, therefore:

$$\lim_{t \to \infty} \int_{0}^{t} P(\tau)dt < \infty \qquad (37)$$

In addition, $P_\lambda(t)$ is bounded. Therefore from Barbalat's lemma, we have:

$$\lim_{t \to \infty} P(t) = \circ \qquad (38)$$

It means that $S_\lambda(t) \to 0$ and $S_v(t) \to 0$ as $t \to \infty$. As a result, the proposed SM flux and speed controllers are asymptotically stable, even if system uncertainties exist. Moreover, the flux-tracking errors e_λ and e_v will converge to zero since $S_\lambda(t) = 0$ and $S_v(t) = 0$ as $t \to \infty$.

IV. System Simulation

The overall block diagram of the proposed control system is shown in Fig. 2 A C^{++} computer program was developed to model this system on P.C.

In this program, a static runge-kutta fourth order method is used to solve the system nonlinear equations. The effectiveness and validity of the proposed approach is tested for a three-phase 5 KW, 380 V, two poles,60 Hz type.

Simulation results shown in Fig. 3 are obtained in the condition of an exponentional speed reference from 0 to 1 m/sec rising up to 2 m/sec at t=1sec with a time constant of $\tau_v = .1$ sec, an exponentional reference flux signal from zero to 0.2W-t at t=0 with $\tau_f = .1$ sec, a step up load force from zero to 5Nm at t=0, stepped up to 10Nm at t=.5sec, stepped down to zero at t=1.5sec and $R_s = 2R_{sn}, R_r = 2R_{rn}, D = 2D_n, M = 2M_n, L_m = 2L_{mn}$, where R_s, R_r, D, M and L_m respectively are the stator resistance, the rotor resistances, the viscous friction coefficient, the total mass of moving element and magnetizing inductance at zero speed. Note that the subscript n shows the nominal parameters. In addition, the parameters of the proposed control system are given as follows:

$$\alpha = 3, \beta = 3, \rho = 200, \gamma = 50 \qquad (39)$$

Fig. 4 shows the drive system performance in the condition of an sinusoidal speed reference with amplitude of 1 m/sec and frequency of 1 Hz, an exponentional reference flux signal from zero to 0.2W-t at t=0 with $\tau_f = .1$ sec, a step up load force from zero to 5Nm at t=0, stepped up to 10Nm at t=.5sec, stepped down to zero at t=1.5sec, and machine parameters are chosen as: $R_s = R_{sn}, R_r = 1.5R_{rn}, D = 1.5D_n, M = 2M_n, L_m = 1.5L_{mn}$.

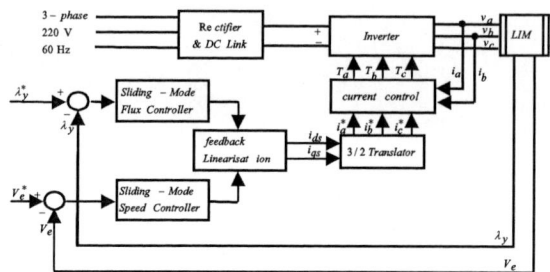

Figure 2. Simulated block diagram

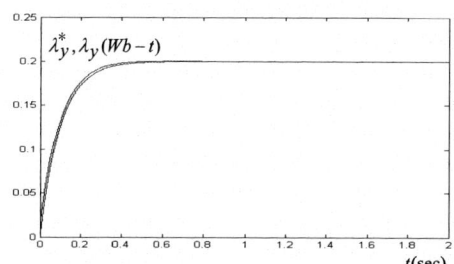

Figure 3a. Secondary flux tracking responses

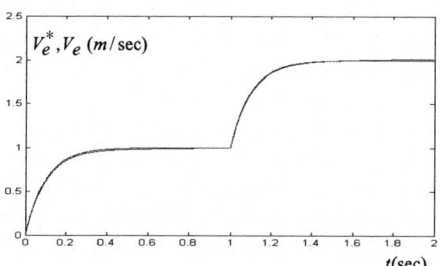

Figure 3b. Mover speed responses

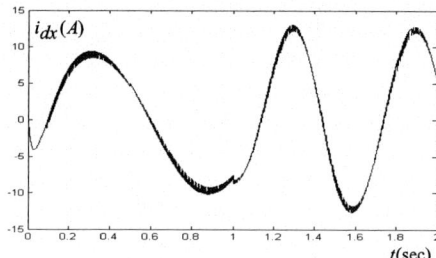

Figure 3c. d-axis current

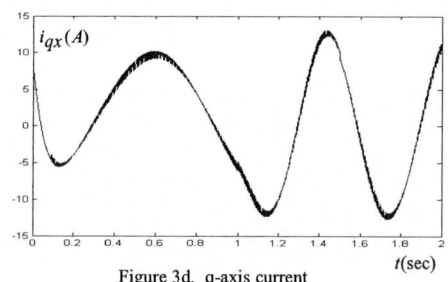

Figure 3d. q-axis current

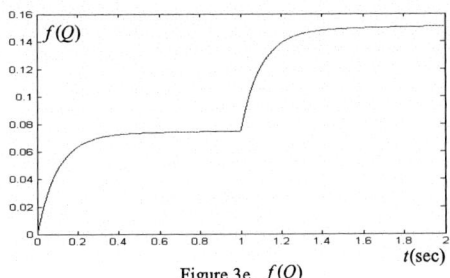

Figure 3e. $f(Q)$

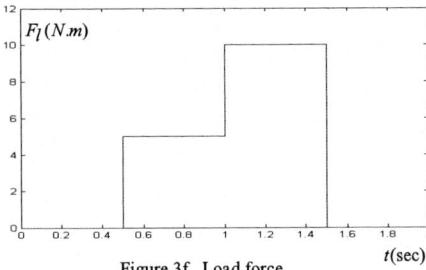

Figure 3f. Load force

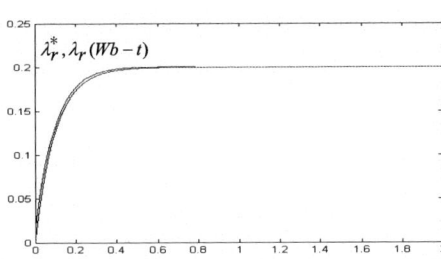

Figure 4a. Secondary flux tracking responses

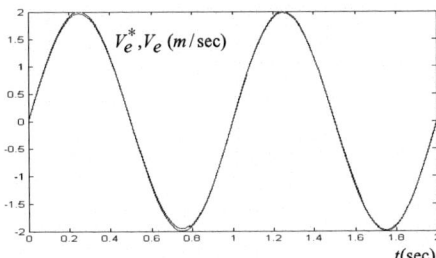

Figure 4b. Mover speed tracking responses

Figure 4c. d-axis current

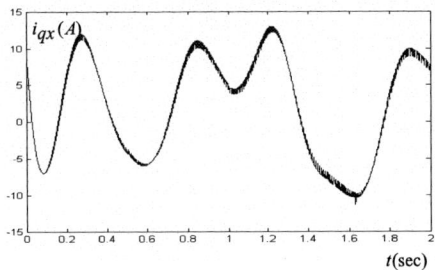

Figure 4d. d-axis current

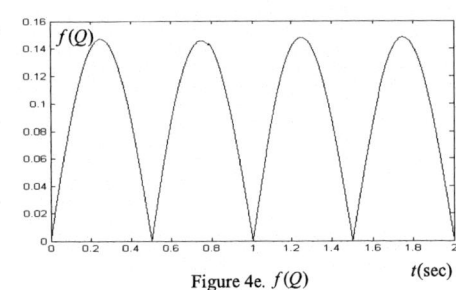

Figure 4e. $f(Q)$

V. CONCLUSIONS

A robust nonlinear controller has been proposed for a primary type LIM with taking into account the primary end effects. The nonlinear controller is designed based on combination of SM control and input-output feedback linearisation. The system control laws are obtained based on Lyaponuv Theory. The drive system performance is verified by computer simulation. Simulation results obtained, show that the drive system is capable of compensation the parameters variations and uncertainties that usually exist in the unknown load force disturbances.

REFERENCES

[1] I. Takahashi, and Y. Ide, "Decoupling control of thrust and attractive force of LIM using a space vector control inverter," *IEEE Trans. Ind. Appl.*, vol.29, No. 1, pp.161-167, 1993.

[2] G. H. Abdou and S.A. Sherif, "Theoretical and experimental design of LIM in automated manufacturing systems," *IEEE Trans. Ind. Applicant..*, Vol. 27, pp. 286-293, Mar/Apr 1991.

[3] Jan Jamali, "End Effect in Linear Induction and Rotating Electrical Machines," *IEEE Transaction on Energy Conversion*, Vol. 18, No. 3, September 2003.

[4] T.A. Nondahl and D.W. Novotny, "Three-phase pole-by-pole model of a linear induction machine," *Proc. IEE*, Vol. 127, Pt. B, No. 2, pp. 68-82, 1980.

[5] J.F. Gieras, G.E. Dawson and A. R. Eastham, "A New Longitudinal end effect factor for Linear Induction Motors," *IEEE. Trans. On Magnetics*, Vol. EC-2, No. 1, pp. 152-159, 1987.

[6] F.-J. Lin and C.-C. Lee, "Adaptive backstepping sliding mode control for linear induction motor drive to track periodic refrences," *IEE Proc.-Electr. Power Appl.*, Vol. 147, No. 6, November 2000.

[7] F.-J. Lin and P.-H. Shen, "Adaptive backstepping sliding mode control for linear induction motor," *IEE Proc.-Electr. Power Appl.,* Vol. 149, No. 3, May 2002.

[8] R.-J. Wai and W.-K. Liu, "Nonlinear decoupled control for linear induction motor servo-drive using the sliding-mode technique," *IEE Proc.-Control Theory Appl.,* Vol. 148, No. 3, May 2001.

[9] G. Kang and K. Nam, "Field-oriented control scheme for linear induction with the end effect," *IEE Proc.-Electr. Power Appl.,* Vol. 152, No. 6, November 2005.

2006 5th International Power Electronics and Motion Control Conference

A Novel Adaptive Scheme for Stator Resistance Estimation in Sensorless Induction Motor Drives

Han Li*†, Wen Xuhui*, Chen Guilan*

*Institute of Electrical Engineering Chinese Academy of Science, Beijing, P. R. China 100080
†Graduate School of Chinese Academy of Science, Beijing, P. R. China 100080

Abstract—This paper presents a new adaptive scheme for online estimation of stator resistance in speed-sensorless induction motor drives. The method is based on the adaptive control theory of Model Reference Adaptive System(MRAS) approach with Luenberger observer. And the stability of the observer with stator resistance estimation in sensorless vector control of induction motors is proved by the Lyapunov's theorem. Furthermore, the method has been compared with the classical adaptive scheme based on the integration algorithm. And in fact, the proposed method is a generalization of the integration adaptive algorithm. Finally the feasibility of the scheme is verified by simulation.

Keywords-adaptive scheme; Luenberger observer; induction motor; sensorless; stator resistance

Symbols List

variables and parameters:

$u_{1\alpha}, u_{1\beta}$ stator voltage α, β component in stator reference frame.

$i_{1\alpha}, i_{1\beta}$ stator current α, β component in stator reference frame.

$\psi_{2\alpha}, \psi_{2\beta}$ rotor flux α, β component in stator reference frame.

R_1, R_2 stator and rotor resistance.

L_1, L_2 stator and rotor self-inductance.

L_m mutual inductance.

σ leakage coefficient. $\sigma = 1 - \dfrac{L_m^2}{L_1 L_2}$.

τ_2 rotor time constant. $\tau_2 = \dfrac{L_2}{R_2}$.

ω_r motor angular velocity.

superscripts:

1   stator variables

2   rotor variables

∧   estimated quantities.

I. INTRODUCTION

Vector control is already the industrial standard for high-performance induction motor drives. In these applications, a rotational transducer such as a shaft encoder is used. However, due to the cost and fragility of mechanical speed sensors and to the difficulty of installing that kind of sensor in many applications, speed-sensorless systems are preferred for the next generation of commercial drives[1][2].

The stator resistance changes dramatically with temperature, especially at low and very low speeds with heavy loads. The stator resistance variation has a great influence on the speed estimation at the low speed region since the actual flux deviates from their set values due to stator resistance variations. Hence the estimation of stator resistance is vital in sensorless speed control applications especially at low and very low speed regions[3]. Several methods have been proposed to on-line estimate the stator resistance of induction motors in speed sensorless vector control drives. In [4], an adaptive integration algorithm based on MRAS approach has been proposed to tune the stator resistance from electrical measurements. Another proposal[5] use different model-based approach, but no stability proof is given and only limited experimental results are presented.

In this paper, to ensure robustness and accuracy of speed estimation in spite of stator resistance variations, a new adaptive scheme for stator resistance estimation is proposed. This method is based on adaptive control theory of MRAS using Luenberger observer[6], and the stability of the observer is proved by the Lyapunov's theorem. In fact, the proposed method is a generalization of the classical integration adaptive algorithm in [4]. Furthermore, a direct field-oriented drive based on this new stator resistance estimator and the speed observer in [7] is built. The system has been tested by numerical simulation and the feasibility of the adaptive scheme is verified.

II. INDUCTION MOTOR MODEL

A standard smooth-air-gap model for the induction motor can be described by following state equations in the stator reference frame.

$$\frac{d}{dt}\mathbf{x} = \mathbf{A}\mathbf{x} + \mathbf{B}\mathbf{u} \qquad (1)$$

$$\mathbf{y} = \mathbf{C}\mathbf{x} \qquad (2)$$

where

$\mathbf{u} = \begin{bmatrix} u_{1\alpha} & u_{1\beta} \end{bmatrix}^T$ is the stator voltage vector.

$\mathbf{x} = \begin{bmatrix} i_{1\alpha} & i_{1\beta} & \psi_{2\alpha} & \psi_{2\beta} \end{bmatrix}^T$ is the motor states containing stator current and rotor flux vectors.

1-4244-0448-7/06/$25.00 ©2006 IEEE

$\mathbf{y} = \begin{bmatrix} i_{1\alpha} & i_{1\beta} \end{bmatrix}^T$ is the system output of induction motor

$\mathbf{A} = \begin{bmatrix} \mathbf{A}_{11} & \mathbf{A}_{12} \\ \mathbf{A}_{21} & \mathbf{A}_{22} \end{bmatrix}$ is the system matrix of induction

motor

$\mathbf{B} = \begin{bmatrix} \mathbf{B}_1 & \mathbf{O}_2 \end{bmatrix}^T$ is the control matrix of induction motor

$\mathbf{C} = \begin{bmatrix} \mathbf{I}_2 & \mathbf{O}_2 \end{bmatrix}$ is the output matrix of induction motor

and

$$\mathbf{A}_{11} = -\left(\frac{R_1}{\sigma L_1} + \frac{1-\sigma}{\sigma \tau_2} \right) \mathbf{I}_2$$

$$\mathbf{A}_{12} = \frac{L_m}{\sigma L_1 L_2} \left\{ \left(\frac{1}{\tau_2} \right) \mathbf{I}_2 - \omega_r \mathbf{J}_2 \right\}$$

$$\mathbf{A}_{21} = \left(\frac{L_m}{\tau_2} \right) \mathbf{I}_2$$

$$\mathbf{A}_{22} = -\left(\frac{1}{\tau_2} \right) \mathbf{I}_2 + \omega_r \mathbf{J}_2$$

$$\mathbf{B}_1 = \left(\frac{1}{\sigma L_1} \right) \mathbf{I}_2$$

$$\mathbf{I}_2 = \begin{bmatrix} 1 & 0 \\ 0 & 1 \end{bmatrix} \quad \mathbf{J}_2 = \begin{bmatrix} 0 & -1 \\ 1 & 0 \end{bmatrix} \quad \mathbf{O}_2 = \begin{bmatrix} 0 & 0 \\ 0 & 0 \end{bmatrix}$$

III. THE LUENBERGER OBSERVER

The Luenberger observer (Fig.1) which estimates the stator current and the rotor flux in stator reference frame is written by following equations.

$$\frac{d\hat{\mathbf{x}}}{dt} = \hat{\mathbf{A}}\hat{\mathbf{x}} + \mathbf{B}\mathbf{u} + \mathbf{G}(\hat{\mathbf{y}} - \mathbf{y}) = \hat{\mathbf{A}}\hat{\mathbf{x}} + \mathbf{B}\mathbf{u} + \mathbf{G}(\mathbf{C}\hat{\mathbf{x}} - \mathbf{y})$$

$$= (\hat{\mathbf{A}} + \mathbf{GC})\hat{\mathbf{x}} + \mathbf{B}\mathbf{u} - \mathbf{G}\mathbf{y} = \mathbf{A}_L \hat{\mathbf{x}} + \mathbf{B}_L \mathbf{u}_L$$

(3)

$$\hat{\mathbf{y}} = \mathbf{C}\hat{\mathbf{x}} \tag{4}$$

where

$\hat{\mathbf{x}} = \begin{bmatrix} \hat{i}_{1\alpha} & \hat{i}_{1\beta} & \hat{\psi}_{2\alpha} & \hat{\psi}_{2\beta} \end{bmatrix}^T$ is the estimated states of induction motor.

$\mathbf{A}_L = \hat{\mathbf{A}} + \mathbf{GC}$ is system matrix of the observer.

$\mathbf{B}_L = \begin{bmatrix} \mathbf{B} & -\mathbf{G} \end{bmatrix}$ is the control matrix of the observer.

$\mathbf{u}_L = \begin{bmatrix} \mathbf{u} & \mathbf{y} \end{bmatrix}$ is the input matrix of the observer.

$\mathbf{G}$ is the observer gain matrix.

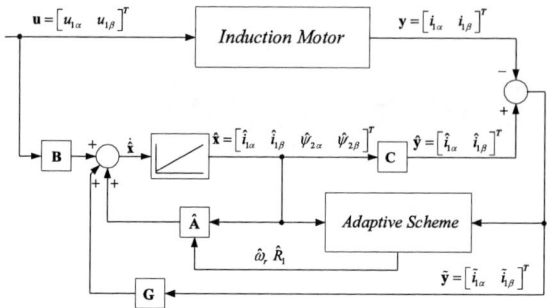

Fig.1 Block Diagram of MRAS with Speed and Stator Resistance Adaptive Scheme

The observer gain matrix is decided so that the Luenberger observer can be stable. There are many methods[6] to configure the gain matrix. The most frequently used method is based on the fact that the observer eigenvalues are chosen in such a way that they are proportional to the motor eigenvalues[2]. An induction motor itself is stable, so the Luenberger observer is also stable in usual operation.

IV. THE PROPOSED ADAPTIVE SCHEME

A. The Model Reference Adaptive System Approach

The proposed adaptive scheme is based on the Model Reference Adaptive System approach as shown in Fig.1. The induction motor(1) is considered as the Reference Model(RM) while the Luenberger observer(3) that contains the unknown stator resistance and rotor speed is regarded as the Adjustable Model(AM). By comparing the outputs of the RM and of the AM, the rotor speed and the stator resistance which vary with the motor temperature, are derived from the adaptive scheme.

From (1) and (3) the estimation error of the stator current and rotor flux between the RM and the AM is described by the following equation.

$$\frac{d}{dt}\tilde{\mathbf{x}} = (\mathbf{A} + \mathbf{GC})\tilde{\mathbf{x}} + \tilde{\mathbf{A}}\hat{\mathbf{x}} \tag{5}$$

where

$$\tilde{\mathbf{x}} = \hat{\mathbf{x}} - \mathbf{x} = \begin{bmatrix} \tilde{i}_{1\alpha} & \tilde{i}_{1\beta} & \tilde{\psi}_{2\alpha} & \tilde{\psi}_{2\beta} \end{bmatrix}^T$$

$$= \begin{bmatrix} \hat{i}_{1\alpha} - i_{1\alpha} & \hat{i}_{1\beta} - i_{1\beta} & \hat{\psi}_{2\alpha} - \psi_{2\alpha} & \hat{\psi}_{2\beta} - \psi_{2\beta} \end{bmatrix}^T$$

is the estimation error vector between motor and Luenberger observer.

$\tilde{\mathbf{A}}$ is the error matrix caused by the stator resistance and speed variation.

Let $\tilde{R}_1 = \hat{R}_1 - R_1, \tilde{\omega}_r = \hat{\omega}_r - \omega_r$

$\tilde{\mathbf{A}}$ can be expressed by the following equation.

$$\tilde{\mathbf{A}} = \hat{\mathbf{A}} - \mathbf{A} = \begin{bmatrix} \hat{\mathbf{A}}_{11} - \mathbf{A}_{11} & \hat{\mathbf{A}}_{12} - \mathbf{A}_{12} \\ \hat{\mathbf{A}}_{21} - \mathbf{A}_{21} & \hat{\mathbf{A}}_{22} - \mathbf{A}_{22} \end{bmatrix} = \begin{bmatrix} \tilde{\mathbf{A}}_{11} & \tilde{\mathbf{A}}_{12} \\ \tilde{\mathbf{A}}_{21} & \tilde{\mathbf{A}}_{22} \end{bmatrix}$$

where:

$$\tilde{\mathbf{A}}_{11} = -\frac{\tilde{R}_1}{\sigma L_1}\mathbf{I}_2$$

$$\tilde{\mathbf{A}}_{12} = \frac{-\tilde{\omega}_r L_m}{\sigma L_1 L_2}\mathbf{J}_2$$

$$\tilde{\mathbf{A}}_{21} = 0$$

$$\tilde{\mathbf{A}}_{22} = \tilde{\omega}_r \mathbf{J}_2$$

In order to derive the adaptive scheme, Lyapunov stability theorem is utilized. This theorem gives a sufficient condition for the uniform asymptotic stability of a non-linear system by using a Lyapunov function V. The Lyapunov function must be continuous, positive definite and differentiable. The sufficient condition for the uniform asymptotic stability is that the time derivative of the Lyapunov function is negative definite.

The following Lyapunov function is defined as a candidate:

$$V = \tilde{\mathbf{x}}^T \tilde{\mathbf{x}} + \frac{\tilde{R}_1^2}{\lambda_1 \sigma L_1} + \frac{L_m \tilde{\omega}_r^2}{\lambda_2 \sigma L_1 L_2} \qquad (6)$$

where, λ_1 and λ_2 are positive constants.

The time derivative of V becomes

$$\frac{d}{dt}V = \left[\frac{d}{dt}\tilde{\mathbf{x}}^T\right]\tilde{\mathbf{x}} + \tilde{\mathbf{x}}^T\left[\frac{d}{dt}\tilde{\mathbf{x}}\right]$$
$$+ \frac{2\tilde{R}_1}{\lambda_1 \sigma L_1}\frac{d\tilde{R}_1}{dt} + \frac{2L_m\tilde{\omega}_r}{\lambda_2 \sigma L_1 L_2}\frac{d}{dt}\tilde{\omega}_r \qquad (7)$$

By substituting (5) into (7), After some calculation and finally the derivative of the Lyapunov function can be expressed as follows:

$$\frac{d}{dt}V = \tilde{\mathbf{x}}^T\left[(\mathbf{A}+\mathbf{GC})^T + (\mathbf{A}+\mathbf{GC})\right]\tilde{\mathbf{x}}$$
$$- \frac{2\tilde{R}_1}{\sigma L_1}\left(\tilde{i}_{1\alpha}\hat{i}_{1\alpha} + \tilde{i}_{1\beta}\hat{i}_{1\beta}\right) - \frac{2\tilde{\omega}_r L_m}{\sigma L_1 L_2}\left(\tilde{i}_{1\beta}\hat{\psi}_{2\alpha} - \tilde{i}_{1\alpha}\hat{\psi}_{2\beta}\right)$$
$$+ 2\tilde{\omega}_r\tilde{\psi}_{2\beta}\hat{\psi}_{2\alpha} - 2\tilde{\omega}_r\tilde{\psi}_{2\alpha}\hat{\psi}_{2\beta}$$
$$+ \frac{2\tilde{R}_1}{\lambda_1 \sigma L_1}\frac{d\tilde{R}_1}{dt} + \frac{2L_m\tilde{\omega}_r}{\lambda_2 \sigma L_1 L_2}\frac{d}{dt}\tilde{\omega}_r \qquad (8)$$

In rotor field oriented control of the induction motor, $2\tilde{\omega}_r\tilde{\psi}_{2\beta}\hat{\psi}_{2\alpha} - 2\tilde{\omega}_r\tilde{\psi}_{2\alpha}\hat{\psi}_{2\beta}$ can be neglected as the rotor flux errors $\tilde{\psi}_{2\alpha}$ $\tilde{\psi}_{2\beta}$ are supposed to be very small. Thus the derivative of the Lyapunov function can be simplified as follows:

$$\frac{d}{dt}V = \tilde{\mathbf{x}}^T\left[(\mathbf{A}+\mathbf{GC})^T + (\mathbf{A}+\mathbf{GC})\right]\tilde{\mathbf{x}}$$
$$+ \frac{2\tilde{R}_1}{\lambda_1 \sigma L_1}\frac{d\tilde{R}_1}{dt} - \frac{2\tilde{R}_1}{\sigma L_1}\left(\tilde{i}_{1\alpha}\hat{i}_{1\alpha} + \tilde{i}_{1\beta}\hat{i}_{1\beta}\right) \qquad (9)$$
$$+ \frac{2L_m\tilde{\omega}_r}{\lambda_2 \sigma L_1 L_2}\frac{d}{dt}\tilde{\omega}_r - \frac{2\tilde{\omega}_r L_m}{\sigma L_1 L_2}\left(\tilde{i}_{1\beta}\hat{\psi}_{2\alpha} - \tilde{i}_{1\alpha}\hat{\psi}_{2\beta}\right)$$

The terms $\frac{d}{dt}V$ in this equation must be negative, in order to satisfy the stability criteria of Lyapunov. The first term $\tilde{\mathbf{x}}^T\left[(\mathbf{A}+\mathbf{GC})^T + (\mathbf{A}+\mathbf{GC})\right]\tilde{\mathbf{x}}$ is always negative, because of the imposed eigenvalues of the Luenberger observer. The other terms can be set to a value less than zero [7] as shown in the following expressions to fulfill Lyapunov criteria.

$$\frac{2\tilde{R}_1}{\lambda_1 \sigma L_1}\frac{d\tilde{R}_1}{dt} - \frac{2\tilde{R}_1}{\sigma L_1}\left(\tilde{i}_{1\alpha}\hat{i}_{1\alpha} + \tilde{i}_{1\beta}\hat{i}_{1\beta}\right) = -\frac{2K_1}{\sigma L_1}\tilde{R}_1^2 \quad (10)$$

$$\frac{2L_m\tilde{\omega}_r}{\lambda_2 \sigma L_1 L_2}\frac{d}{dt}\tilde{\omega}_r - \frac{2\tilde{\omega}_r L_m}{\sigma L_1 L_2}\left(\tilde{i}_{1\beta}\hat{\psi}_{2\alpha} - \tilde{i}_{1\alpha}\hat{\psi}_{2\beta}\right) = -\frac{2L_m K_2}{\sigma L_1 L_2}\tilde{\omega}_r^2$$
$$(11)$$

where K_1 and K_2 are positive constants.

B. The Proposed Stator Resistance Estimation Algorithm

Equation (10) can be used to estimated the stator resistance and with some mathematical calculation, the following differential equation can be obtained:

$$\frac{d\tilde{R}_1}{dt} + \lambda_1 K_1 \tilde{R}_1 = \lambda_1\left(\tilde{i}_{1\alpha}\hat{i}_{1\alpha} + \tilde{i}_{1\beta}\hat{i}_{1\beta}\right) \qquad (12)$$

By using the Laplace transform, the equation (12) becomes the following continuous transfer function.

$$\tilde{R}_1 = \frac{\dfrac{1}{K_1}}{\dfrac{1}{\lambda_1 K_1}s + 1}\left(\tilde{i}_{1\alpha}\hat{i}_{1\alpha} + \tilde{i}_{1\beta}\hat{i}_{1\beta}\right) \qquad (13)$$

Equation (13) represents a first order system with time constant $\frac{1}{\lambda_1 K_1}$ and gain $\frac{1}{K_1}$, whose input is $\left(\tilde{i}_{1\alpha}\hat{i}_{1\alpha} + \tilde{i}_{1\beta}\hat{i}_{1\beta}\right)$ and output is $\tilde{R}_1$.

The discrete form of (13) with sample time T using Euler method and the discrete transfer function are expressed by the following equations:

$$\tilde{R}_1(k+1) = \tilde{R}_1(k)(1 - T\lambda_1 K_1) + T\lambda_1\left(\tilde{i}_{1\alpha}\hat{i}_{1\alpha} + \tilde{i}_{1\beta}\hat{i}_{1\beta}\right)$$
$$(14)$$

$$\tilde{R}_1 = \frac{b_1}{z - a_1}\left(\tilde{i}_{1\alpha}\hat{i}_{1\alpha} + \tilde{i}_{1\beta}\hat{i}_{1\beta}\right) \qquad (15)$$

where $a_1 = (1 - T\lambda_1 K_1)$, $b_1 = T\lambda_1$ and k is the sample instant at time kT.

Assuming that the real value of stator resistance at the instant k is equal to the estimated value at the instant (k-1).

$$R_1(k) = \hat{R}_1(k-1) \qquad (16)$$

The estimated stator resistance at the instant k can be iteratively calculated from the following equation.

$$\hat{R}_1(k) = \tilde{R}_1(k) + R_1(k) = \tilde{R}_1(k) + \hat{R}_1(k-1) \quad (17)$$

This is equivalent with a discrete integral operation. The adaptive estimation of stator resistance can be described as Fig 2 from (15) and (17).

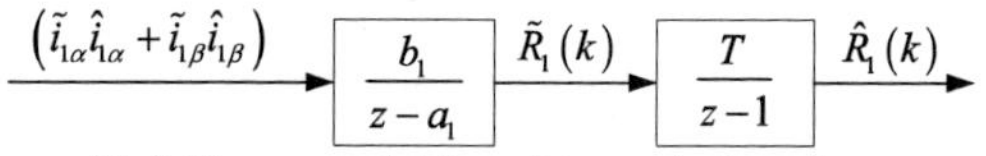

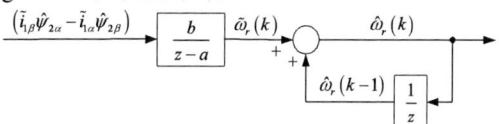

Fig 2. The proposed stator resistance adaptive scheme

C. The Speed Estimation Algorithm

Equation(11) can be used to estimated the rotor speed. The result is the same as proposed in [7]. The speed estimation algorithm is illustrated in detail as the following.

A differential equation can be obtained from (11).

$$\frac{d}{dt}\tilde{\omega}_r + \lambda_2 K_2 \tilde{\omega}_r = \lambda_2\left(\tilde{i}_{1\beta}\hat{\psi}_{2\alpha} - \tilde{i}_{1\alpha}\hat{\psi}_{2\beta}\right) \quad (18)$$

By using the Laplace transform, the (18) becomes the following continuous tranfer function.

$$\tilde{\omega}_r = \frac{\dfrac{1}{K_2}}{\dfrac{1}{\lambda_2 K_2}\mathbf{s}+1}\left(\tilde{i}_{1\beta}\hat{\psi}_{2\alpha} - \tilde{i}_{1\alpha}\hat{\psi}_{2\beta}\right) \quad (19)$$

Equation (19) represents a first order system with time constant $\dfrac{1}{\lambda_2 K_2}$ and gain $\dfrac{1}{K_2}$, whose input is $\left(\tilde{i}_{1\beta}\hat{\psi}_{2\alpha} - \tilde{i}_{1\alpha}\hat{\psi}_{2\beta}\right)$ and output is $\tilde{\omega}_r$. The discrete form of (19) with sample time T using Euler method and the discrete transfer function are expressed by the following equations:

$$\tilde{\omega}_r(k+1) = \tilde{\omega}_r(k)\left(1-T\lambda_2 K_2\right) + T\lambda_2\left(\tilde{i}_{1\beta}\hat{\psi}_{2\alpha} - \tilde{i}_{1\alpha}\hat{\psi}_{2\beta}\right) \quad (20)$$

$$\tilde{\omega}_r = \frac{b}{z-a}\left(\tilde{i}_{1\beta}\hat{\psi}_{2\alpha} - \tilde{i}_{1\alpha}\hat{\psi}_{2\beta}\right) \quad (21)$$

where $a = \left(1-T\lambda_2 K_2\right)$, $b=T\lambda_2$

Assuming that the real value of rotor speed at the instant k is equal to the estimated value at the instant $(k-1)$.

$$\omega_r(k) = \hat{\omega}_r(k-1) \quad (22)$$

The estimated rotor speed at the instant k can be iteratively calculated from the following equation.

$$\hat{\omega}_r(k) = \tilde{\omega}_r(k) + \omega_r(k) = \tilde{\omega}_r(k) + \hat{\omega}_r(k-1) \quad (23)$$

This is equivalent with a discrete integral operation. The adaptive estimation of rotor speed can be described as Fig.3 from (21) and (23).

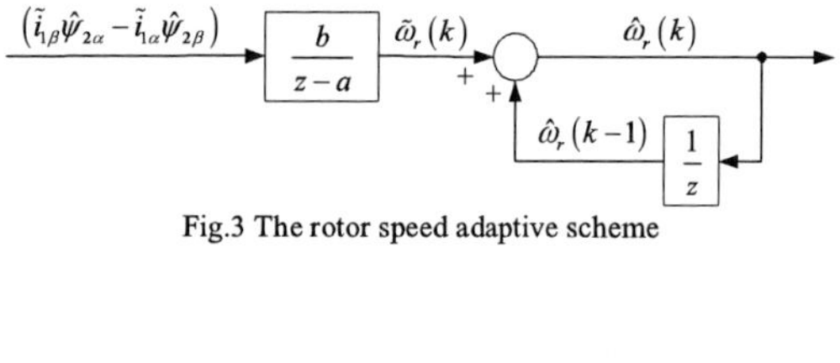

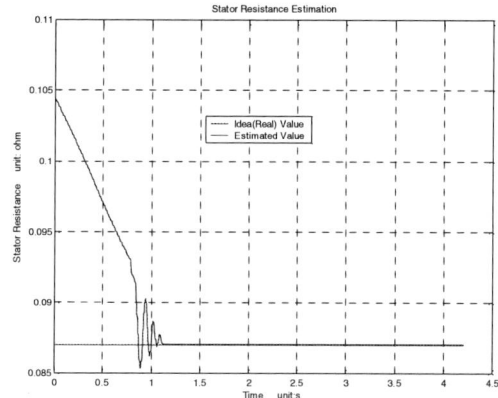

Fig.3 The rotor speed adaptive scheme

D. Comparing With The Classical Adaptive Integration Scheme

However, if the constants K_1 and K_2 are set to zero, (10) and (11) can be simplified to obtain the classical adaptive integration scheme proposed in [4]. The classical adaptation mechanism is described as the below.

$$\frac{d\tilde{R}_1}{dt} = \lambda_1\left(\tilde{i}_{1\alpha}\hat{i}_{1\alpha} + \tilde{i}_{1\beta}\hat{i}_{1\beta}\right) \quad (24)$$

$$\hat{\omega}_r = \lambda_{2p}\left(\tilde{i}_{1\beta}\hat{\psi}_{2\alpha} - \tilde{i}_{1\alpha}\hat{\psi}_{2\beta}\right) + \lambda_{2i}\int\left(\tilde{i}_{1\beta}\hat{\psi}_{2\alpha} - \tilde{i}_{1\alpha}\hat{\psi}_{2\beta}\right)dt \quad (25)$$

V. SIMULATION RESULTS AND DISCUSSION

To verified the proposed adaptive scheme, a full discrete SIMULINK model is built. The induction motor rated quantities and parameters are shown in Appendix. The power electronics devices including the induction motor, the inverter and voltage/current sensors are discreted by Powergui in MATLAB with a sample time 1us. The vector control, the Luenberger observer and the adaptive scheme are programmed with a 100us sample time. The time constants of the first order system in (13) and (19) are set to 400us and 200 us. This benefits the estimated values with less noise. Fig.4 and Fig.5 show the simulation result of stator resistance estimation. Fig.6 shows the simulation result of rotor speed estimation and measurement. The initial values of the estimated stator resistance is set to 1.2 and 0.8 times as much as the actual resistance value.

There is a very little difference between the idea(real) value and the estimated value because the power electronics devices run at 1us sample time while the Luenberger observer with the adaptive scheme runs at 100us sample time. In addition, the purpose of the Luenberger observer is to approach the induction motor, so when the state of Luenberger observer is very close to the state of induction motor, the adaptive scheme is halted to keep the current estimated values. This also contributes to the little difference. A little difference between the idea(real) and estimated value always exists. The classical integration adaptive scheme also has the similar results.

Fig.4 The initial values of the estimated stator resistance is 1.2 times as the actual value

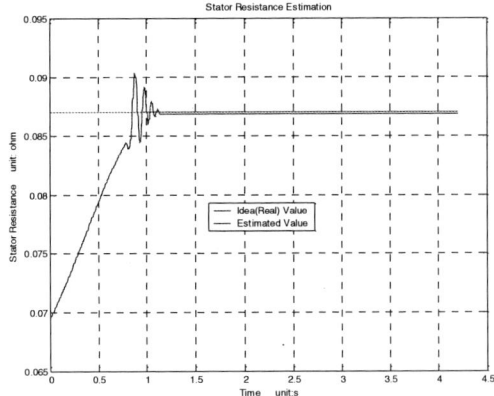

Fig.5 The initial values of the estimated stator resistance is 0.8 times as the actual value

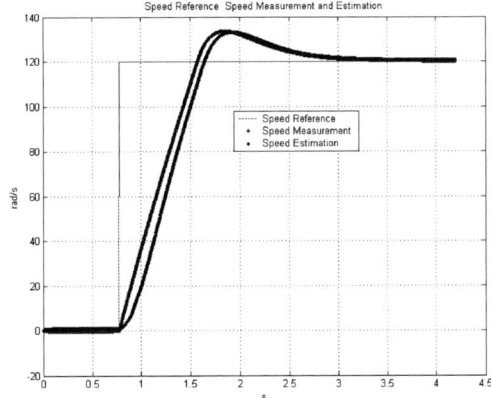

Fig.6 The estimated rotor speed vs. the real rotor speed

VI. CONCLUSION

In this paper, a new adaptive scheme for on-line stator resistance estimation in sensorless vector control of induction motors is proposed. This method is based on the adaptive control theory of Model Reference Adaptive System approach with Luenberger observer. And the stability of the observer with the proposed stator resistance adaptive estimation algorithm is proved by the Lyapunov's theorem. The derived adaptive scheme is the combination of a first order system and an integrator. The first order system acts as a low pass filter that the estimated parameters are less noisy as expected. Compared with the classical adaptive scheme based on the integration algorithm, the proposed method is a generalization of the classical mechanism. The validity of the proposed method has been verified by simulation. The real stator resistance

can be estimated by the proposed adaptive scheme, so the influence of the stator variations on the flux estimation and field oriented control of the induction motor can be reduced.

APPENDIX

The induction motor rated quantities and parameters are:

Rated Output Power	50	HP
Rated Torque	200	Nm
Rated Voltage	460	V
Rated Frequency	60	Hz
Rated Flux	0.98	Vs
Poles	4	
Rated Speed	1800	rpm
Max Speed	9000	rpm
Inertia	1.662	kg m^2
Stator Resistance	0.087	Ω
Rotor Resistance	0.228	Ω
Stator Leakage Inductance	0.8	mH
Rotor Leakage Inductance	0.8	mH
Main Inductance	34.7	mH

REFERENCES

[1]. Guidi,G; Umida,H.; "A novel stator resistance estimation method for speed-sensorless induction motor drives," Industry Applications, IEEE Transactions on Volume 36, Issue 6, Nov.-Dec. 2000 Page(s):1619 - 1627

[2]. Kubota,H.; Matsuse,K.; "Speed sensorless field-oriented control of induction motor with rotor resistance adaptation," Industry Applications, IEEE Transactions on Volume 30, Issue 5, Sept.-Oct. 1994 Page(s):1219 – 1224

[3]. Kojabadi,H.M.; Chang,L.; Doraiswami,R.; "A novel adaptive observer for very fast estimation of stator resistance in sensorless induction motor drives," Power Electronics Specialist Conference, 2003. PESC '03. 2003 IEEE 34th Annual Volume 3, 15-19 June 2003 Page(s):1455 - 1459 vol.3

[4]. Kubota,H.; Matsuse,K.; Nakano,T.; "DSP-based speed adaptive flux observer of induction motor," Industry Applications, IEEE Transactions on Volume 29, Issue 2, March-April 1993 Page(s):344 - 348

[5]. Byeong-Seok,Lee; Krishnan,R.; "Adaptive stator resistance compensator for high performance direct torque controlled induction motor drives," Industry Applications Conference, 1998. Thirty-Third IAS Annual Meeting. The 1998 IEEE Volume 1, 12-15 Oct. 1998 Page(s):423 - 430 vol.1

[6]. Peter Vas; "Sensorless Vector and Direct Torque Control," Oxford University Press 1998

[7]. Griva,G.; Profumo,F.; Bojoi,R.; Bostan,V.; Cuius,M.; Ilas,C.; "General adaptation law for MRAS high performance sensorless induction motor drives," Power Electronics Specialists Conference, 2001. PESC. 2001 IEEE 32nd Annual Volume 2, 17-21 June 2001 Page(s):1197 - 1202 vol.2

2006 5th International Power Electronics and Motion Control Conference

Ripple-Free Sampling of Current Signals in Drives with Carrier-based PWM Patterns

Haihui Lu [*], Qiang Yin [*], Russel J. Kerkman [**], Thomas A. Nondahl [***]

[*] Rockwell Automation Research (Shanghai) CO., LTD. Shanghai, 200233, CHINA
[**] Standard Drives, Rockwell Automation, Mequon, WI 53029, USA
[***] Advanced Technology Laboratories, Rockwell Automation, Milwaukee, WI 53204, USA

hlu@ra.rockwell.com, qyin@ra.rockwell.com, rjkerkman@ra.rockwell.com, tanondahl@ra.rockwell.com

Abstract —**This paper investigated the possibility of detecting the instantaneous fundamental component of current signals in electrical drives by sampling the current at the carrier's peak/valley (midpoints of zero vectors V0 and V7) in carrier-based PWM pulse patterns. PWM pattern that does not employ zero vectors or employs only one zero vector was also investigated. The observation can be extended to any carrier-based PWM patterns in which the time sequencing of the selected vectors are symmetrical in each carrier period. The conclusion was verified from both simulation and experiment.**

Keywords - **sampling; current; discontinuous; drives; pulse width modulation.**

1 INTRODUCTION

The control board of modern electrical drive is usually comprised of a high-speed digital processor (DSP) with a certain number of analog and digital inputs and outputs. The fast progress made in DSP eliminates the long-standing limitation on processing speed. However, the interaction between the DSP and the rest of the drive still remains a critical issue. One of these is the conversion of the analog signals into digital signals and their introduction into the computing unit. Sampling of the motor currents is especially important for drive performance since any detection error in the current feedback will pass along the inner current control loop. However, the current sampling is troublesome because PWM modulation results in rich ripple in the motor currents. Attempts to solve them have only been reported in a few references [1], [2].

The most common approach to deal with the current ripple is to add a low-pass filter in the signal conditioning circuit. That approach introduces phase delay which can not be handled easily with control. Theoretical background and a practical solution for sampling of average value of motor currents (locally averaged over a sampling interval) were provided in [1]. The current was first converted into a pulse

train by a voltage-to-frequency converter, with the frequency of the pulse train proportional to the current. The pulse train was then counted /integrated by a counter. The counter was then read every sampling interval. When the sampling interval is multiples of the PWM carrier period, the differentiated counter values represented the locally averaged value and was ripple-free. It takes advantage of the assumption that the mean value of the harmonic current component for every PWM carrier period may be equal to zero.

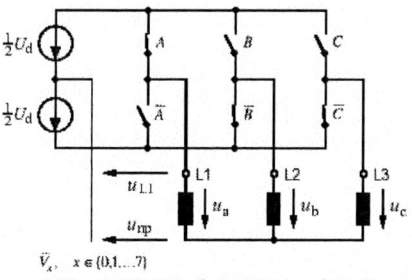

$\vec{V}_x$, $x \in \{0,1,...7\}$
$\{ABC\}$, $A,B,C \in \{1 \text{ (conducting)}, 0 \text{ (non conducting)}\}$

(a)

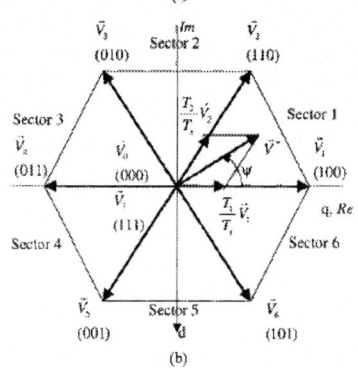

(b)

Fig.1. (a) Three-phase inverter with 3 bridges (b) The definition of switching-state vectors in complex plane

1-4244-0448-7/06/$25.00 ©2006 IEEE   1054

The authors of [1] are skeptical about sampling the instantaneous fundamental value of current due to the harmonics ripple in the sampled current. On the other hand, [7] suggested that it was possible to detect the fundamental component of current by sampling the current at the mid points of zero space vectors in a PWM pulse pattern. Authors of [2] proposed a method of detecting the instantaneous value by sampling the current at the midpoints of zero space vectors of SPWM pulse pattern. The instantaneous value sampled at these points was equal to the fundamental component.

This paper investigates the idea of [2] on more carrier-based PWM pulse patterns. The carrier-based PWM patterns compare a high-frequency triangle carrier with three reference signals and creating gating pulses for the power switches in the drive shown in Fig.1 (a). The switching-state vectors are defined by a combination of conducting/non-conducting of the power switches. Altogether there are 8 combinations shown in Fig.1 (b). There are 6 active states $\vec{V}_1, \vec{V}_2, \cdots \vec{V}_6$ and 2 zero states $\vec{V}_0, \vec{V}_7$. Different zero-sequence component can be subtracted from the three reference signals to form different PWM patterns. For example, if the zero-sequence component is comprised of successive mid values of three reference signals, classical SVPWM is achieved. There are many versions of zero-sequence component, resulting in many carrier-based PWM patterns [4] [5] [6], such as DPWM0, DPWM1, DPWM2, DPWM3, DPWMMAX, DPWMMIN, THIPWM1/6 and THIPWM1/4. The common feature of these PWM patterns are the selected vectors are symmetrically distributed in each carrier period. Recently, some works [3] are done to modify the carrier-based SVPWM patterns in order to reduce the common mode voltage. By those approaches, zero vectors are replaced with selected active vectors. The methods are named NSVM1, NSVM2, and NSVM3. To avoid a lengthy paper, only SVPWM, DPWM0 and NSVM3 will be investigated in this paper. Theoretical analysis, simulation and experimental tests all concluded that the fundamental component can be detected at the carrier's peak/valley no matter it is zero vector or not.

2 THEORETICAL ANALYSIS OF RIPPLE CURRENT

2.1 Ripple Current in Complex Plane

To analyze the characteristics of the ripple current under carrier-based PWM patterns, it is assumed that the frequency of the carrier $f_s = 1/T_s$ is much higher than the frequency of the reference signals and that the reference signals are constant over each carrier period T_s. The three-phase load component in Fig.1 (a) consists of resistance, inductance and electromotive force corresponding to each phase. Since the voltage drop on the resistance is trivial compared with that of the other two for an induction motor load, the resistance is neglected similar to [5] [6] [8]. Then

the equivalent circuit of an induction motor is shown in Fig.2.

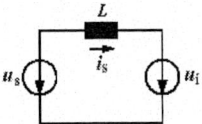

Fig.2. Equivalent Circuit of Induction Motor

Under these assumptions, the voltages and currents can be separated on the "ripple" components, which change over the carrier period T_s while the "fundamental" components remain constant over the same period. To investigate the trajectory of the ripple current $\tilde{I}$ in the complex plane in fig.1 (b), the constant fundamental values are removed from the differential equation. The differential equation with only the ripple components (designated with~) is considered.

$$\vec{\tilde{V}} = L\frac{d\vec{\tilde{I}}}{dt}, \qquad 0 \le t \le T_s \qquad (1)$$

where $\vec{\tilde{V}}$ is the space vector of the ripple voltage equal to the difference between the actual voltage vector and the reference voltage $\vec{V}$. Take sector 1 of Fig.1 (b) as an example. The change of current during each individual time segment follows from (1)

$$\frac{\Delta\vec{\tilde{I}}_{V_1}}{\Delta T_{V_1}} = (\vec{V}_1 - \vec{V}^*)/L \qquad (2)$$

$$\frac{\Delta\vec{\tilde{I}}_{V_2}}{\Delta T_{V_2}} = (\vec{V}_2 - \vec{V}^*)/L \qquad (3)$$

$$\frac{\Delta\vec{\tilde{I}}_{V_0}}{\Delta T_{V_0}} = -\vec{V}^*/L \qquad (4)$$

$$\frac{\Delta\vec{\tilde{I}}_{V_7}}{\Delta T_{V_7}} = -\vec{V}^*/L \qquad (5)$$

2.2 Ripple Current Trajectory of SVPWM

Take sector 1 as an example. The reference voltage $\vec{V}^*$ is synthesized with active vectors $\vec{V}_1, \vec{V}_2$ and zero vectors $\vec{V}_0, \vec{V}_7$. The timing of each space vector is calculated as in [6]. Sequence of the vectors is shown in Fig.3. Note that the time distribution of $\vec{V}_0$ and $\vec{V}_7$ are equal. The space vectors of the ripple voltages are shown in Fig.4. They are, $\vec{V}_1 - \vec{V}^*$, $\vec{V}_2 - \vec{V}^*$ and $-\vec{V}^*$ separately. The ripple current trajectory is determined by (1). The trajectory of ripple current for SVPWM in sector 1 is illustrated in Fig.5. The trajectory in solid line corresponds to the first half of the carrier period while that in dot-dashed line corresponds to the second half of the carrier period. Note that the trajectory crosses zero at instants t_0, t_4, t_8, which are the peak and valley of the triangle carrier signal.

This confirms that current sampled at the peak/valley of the triangle carrier represents the fundamental component and is ripple free. When the investigation extends to other sectors, the same observation can be got.

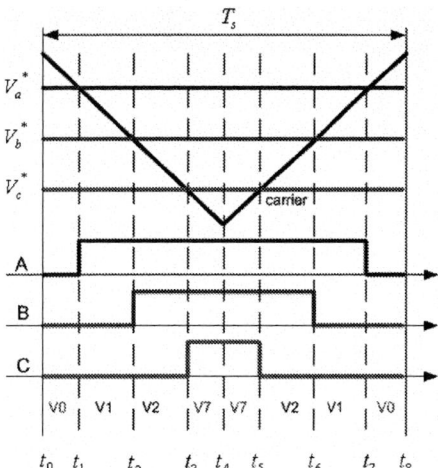

Fig.3 Timing of Space Vectors – SVPWM

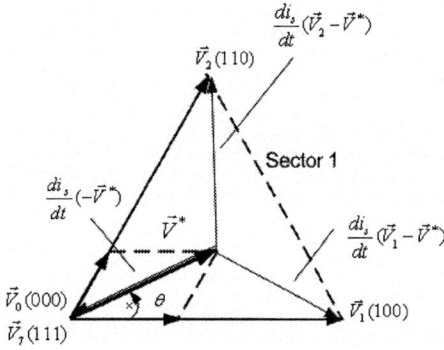

Fig.4. Space Vectors of Ripple Voltage – SVPWM

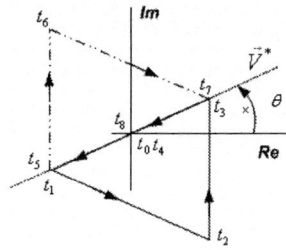

Fig.5. Trajectories of Ripple Current – SVPWM

1.1 Ripple Current Trajectory of DPWM0

DPWM0 [4] differs from SVPWM in that only two phases commutate in each sector. So it is also called two-phase modulation. Still take sector 1 for illustration. Fig.6 shows its timing sequence. There is only one zero vector applied in each carrier period.

The space vectors of the ripple voltages are shown in Fig.7. They are $\vec{V}_1 - \vec{V}^*$, $\vec{V}_2 - \vec{V}^*$ and $-\vec{V}^*$ separately. The ripple current trajectory is determined by (1) and is shown in Fig.8. Similar to the trajectory of SVPWM, the solid line corresponds to the first half of the carrier period and the dot-dashed line corresponds to the second half of the carrier period. The ripple trajectory crosses zero at instants t_0, t_3, t_6. Again these are the peak and valley time point of the carrier. The same observation can be got for all sectors. So current sampled at the peak/valley of the carrier represents the fundamental component.

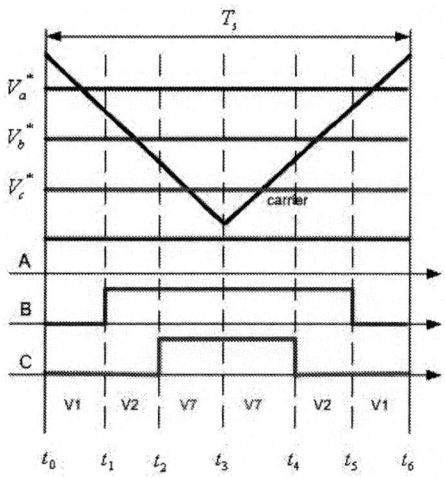

Fig.6. Timing of Space Vectors – DPWM0

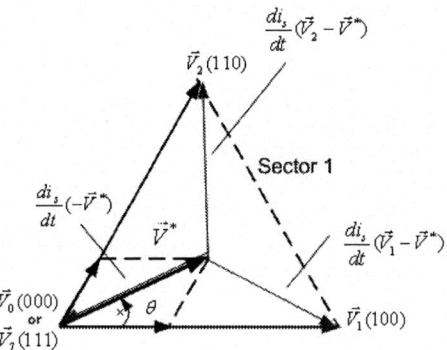

Fig.7. Space Vectors of Ripple Voltage – DPWM0

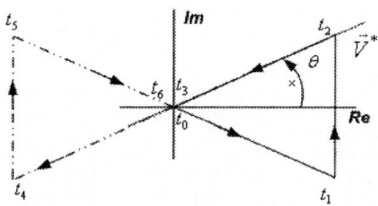

Fig.8. Trajectories of Ripple Current – DPWM0

1.2 Ripple Current Trajectory of NSVM3

NSVM3 [3] differs from SVPWM in that zero vectors are replaced by active vectors that are adjacent to vectors employed for synthesizing the reference vector. Its timing sequence for sector 1 is shown in Fig.9.

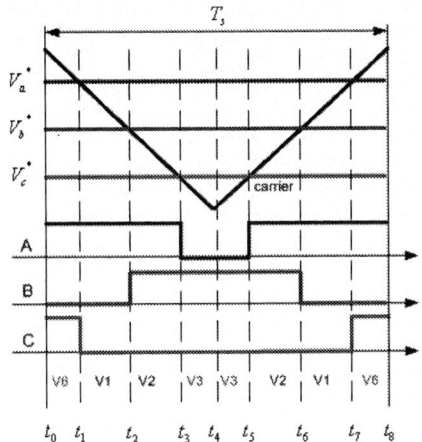

Fig.9. Timing of Space Vectors –NSVM3

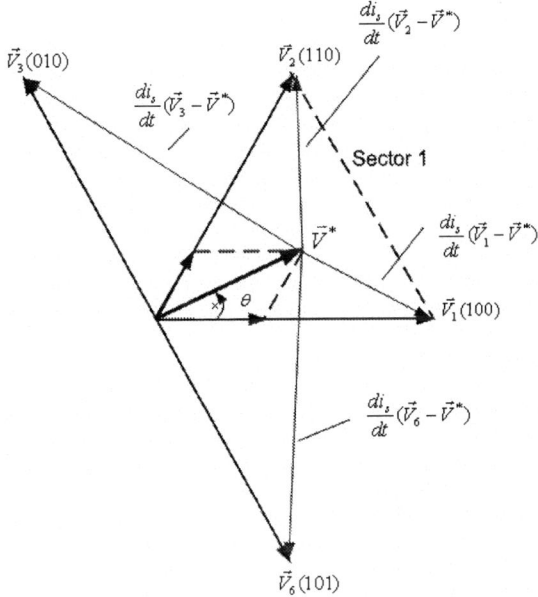

Fig.10. Space Vectors of Ripple Voltage – NSVM3

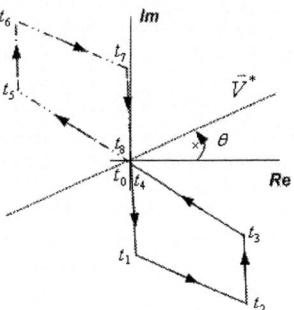

Fig.11. Trajectories of Ripple Current – NSVM3

The space vectors of the ripple voltages are shown in Fig.10. They are $\vec{V}_1 - \vec{V}^*$, $\vec{V}_2 - \vec{V}^*$, $\vec{V}_3 - \vec{V}^*$ and $\vec{V}_6 - \vec{V}^*$ separately. The ripple current trajectory is determined by (1) and is shown in Fig.11. Note the trajectory crosses zero at instants t_0, t_4, t_8, which are peak/valley of triangle carrier. Even though the zero vectors are replaced by non-zero vectors, current sampled at the peak/valley of the carrier still represents the fundamental component and is ripple free.

There is a word of caution here. When the dwelling time of zero vectors is very short, the current sampling at carrier's peak/valley is too close to the switching event. No matter whether the zero vectors are replaced by active vectors or not, the transient response of the current sensor coupled with the excited parasitics can yield erroneous data. Cable length, packaging and placement of the sensor all play a role. These technical issues [9] are beyond the scope of this paper.

2 SIMULATION RESULTS

To quantify the ripple of sampled current, a definition of Ripple Index K_{ripple} was introduced.

$$K_{ripple} = I_{ripple} / I_{fundamental} \qquad (6)$$

I_{ripple} is the RMS value of the ripple current (sampled current subtract fundamental component). $I_{fundamental}$ is the RMS value of the fundamental component. The selected 3 PWM patterns were simulated with RL load. Ripple index was calculated for 10 equally distributed sampling instants within each carrier period. The influence of switching frequency f_s and modulation index was also investigated. The definition of modulation index m is the same as in [6].

2.1 Ripple Index vs. Sampling Instant – SVPWM

The ripple current component with SVPWM is shown in Fig.12. Its correspondence with triangle carrier in time domain is shown in Fig.13. Note that the ripple current crosses zero at the peak/valley of the carrier, which is consistent with Fig.5. The calculated ripple index for different conditions is shown in Fig.14 and Fig.15.

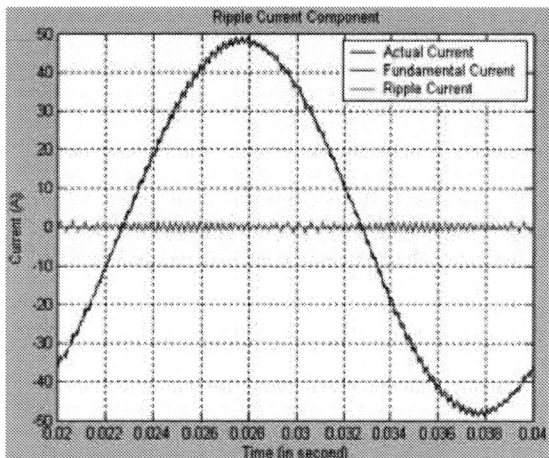

Fig.12. Ripple Current Component – SVPWM

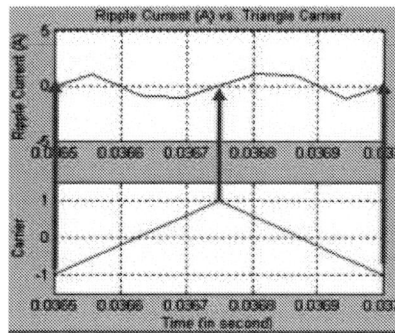

Fig.13. Ripple Current vs. Carrier – SVPWM

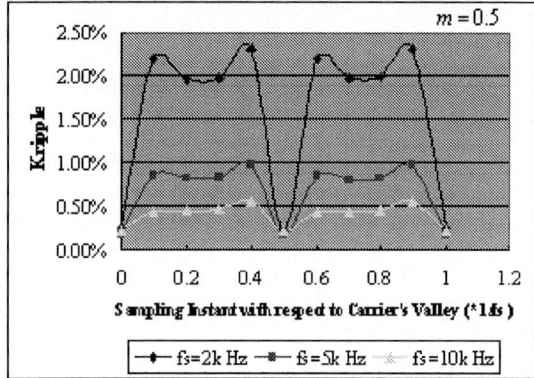

Fig.14. Kripple vs. Sampling Instant – SVPWM (a)

Generally, the higher the switching frequency is, the lower the ripple index is. The lower the modulation index is, the higher the ripple index is. The x-axis indicates the sampling instant, with 0 and 1 corresponding to carrier's valley and 0.5 corresponding to carrier's peak. It is easily

observed that the ripple index achieves minimum when sampling instant is chosen to be the carrier's peak/valley.

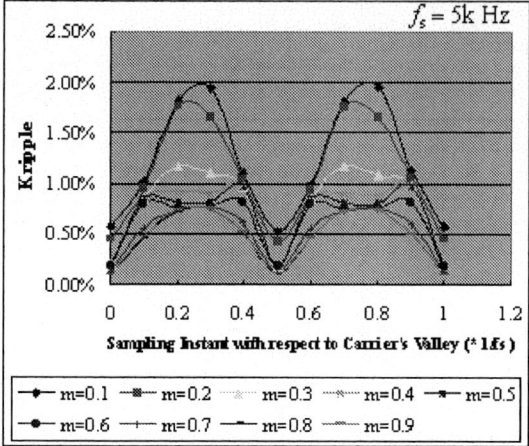

Fig.15. Kripple vs. Sampling Instant – SVPWM (b)

2.2 Ripple Index vs. Sampling Instant – DPWM0

The ripple current component with DPWM0 is shown in Fig.16. Its correspondence with triangle carrier in time domain is shown in Fig.17. Note that the ripple current crosses zero at the peak/valley of the carrier, which is consistent with Fig.8. The calculated ripple index for different conditions is shown in Fig.18 and Fig.19. It is easily observed that the ripple index achieves minimum when sampling is made at the peak/valley of the carrier signal no matter with the switching frequency and modulation index.

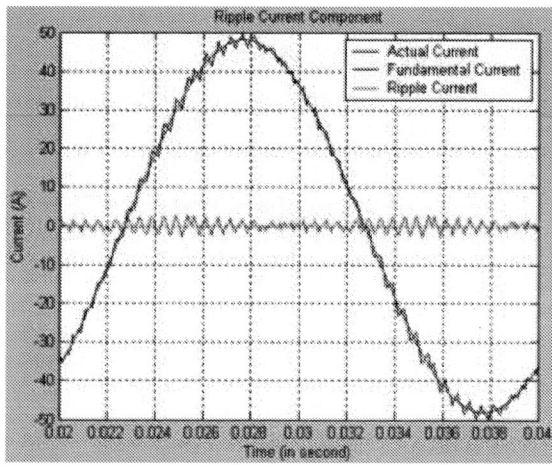

Fig.16. Ripple Current Component – DPWM0

1058

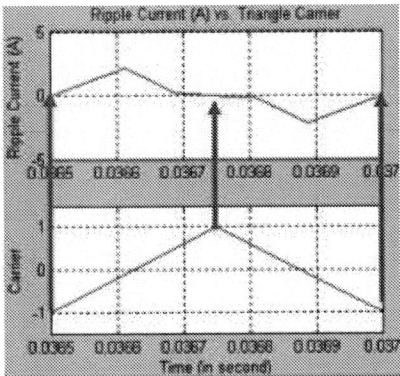

Fig.17. Ripple Current vs. Carrier – DPWM0

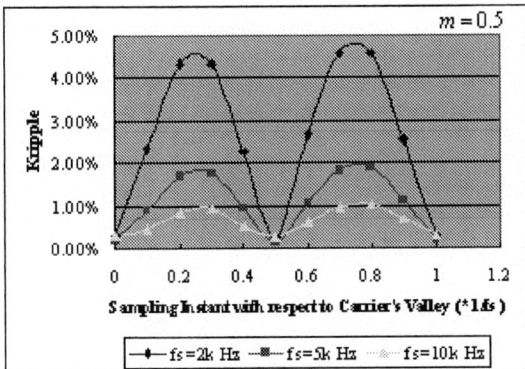

Fig.18. Kripple vs. Sampling Instant – DPWM0 (a)

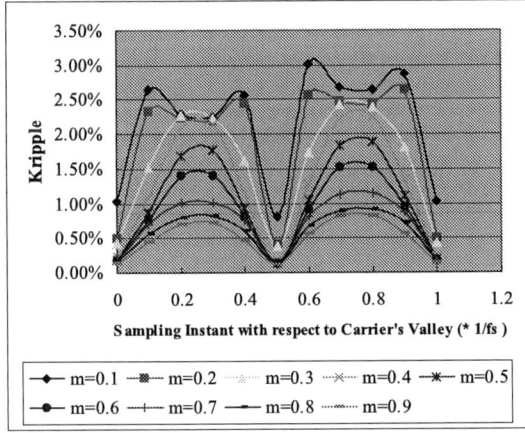

Fig.19. Kripple vs. Sampling Instant – DPWM0 (b)

2.3 Ripple Index vs. Sampling Instant – NSVM3

The ripple current component with NSVM3 is shown in Fig.20. Its correspondence with triangle carrier in time domain is shown in Fig.21. Note that the ripple current crosses zero at the peak/valley of the carrier, which is consistent with Fig.11. The calculated ripple index for different conditions is shown in Fig.22 and Fig.23. Again, it

is observed that the ripple index achieves minimum when sampling is made at the peak/valley of the carrier signal.

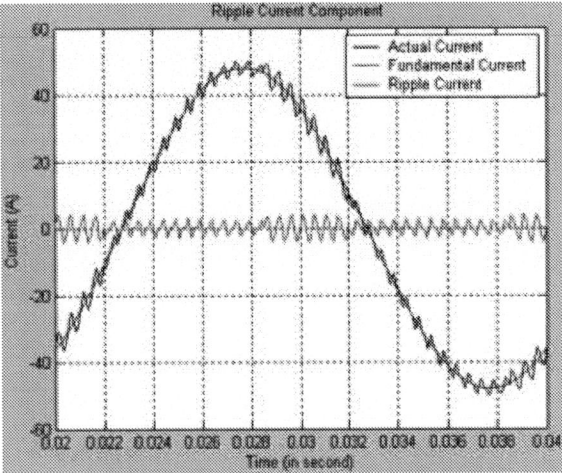

Fig.20. Ripple Current Component – NSVM3

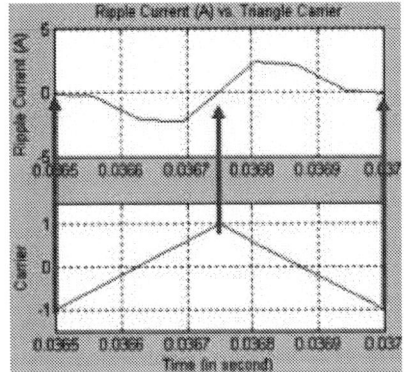

Fig.21. Ripple Current vs. Carrier – NSVM3

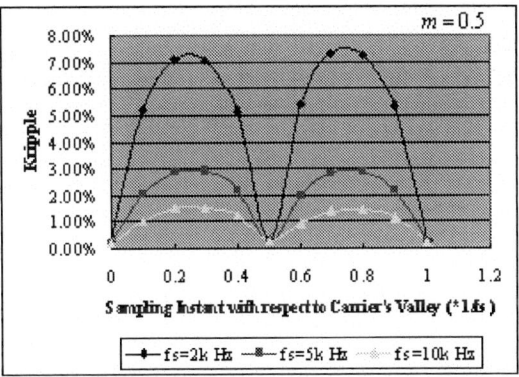

Fig.22. Kripple vs. Sampling Instant – NSVM3 (a)

1059

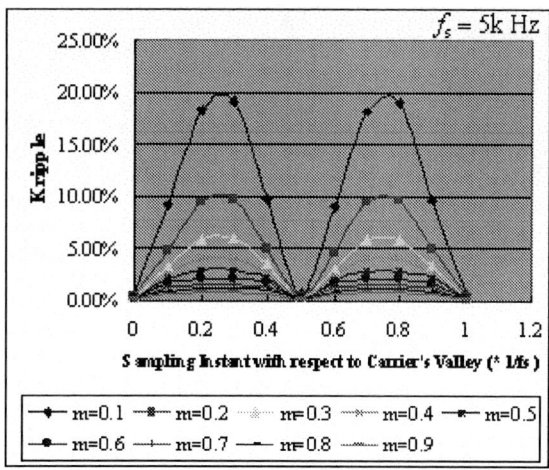

Fig.23. Kripple vs. Sampling Instant – NSVM3 (b)

3 EXPERIMENTAL RESULTS

This method of sampling instantaneous fundamental current component was experimentally verified. Some of the testing conditions are: (a) 50Hz reference voltage frequency; (b) 2 kHz switching frequency; (c) 3.5us dead-time; (d) R-L load (10 mH, 3.2 ohm). TI's DSP TMS320LF2407 EVM was used for control. The A/D conversion was started by the software timer's (carrier) underflow event (valley) and period event (peak) in turn. The sampled current was then sent out via D/A conversion. Fig.24 and Fig.25 illustrates the performance of this instantaneous current sampling technique for DPWM0 and NSVM3 separately. CH1 is the sampled current. CH3 is the actual current. CH4 indicates the valley and peak of the carrier. The sampled current does not change steep due to the settling time of D/A conversion.

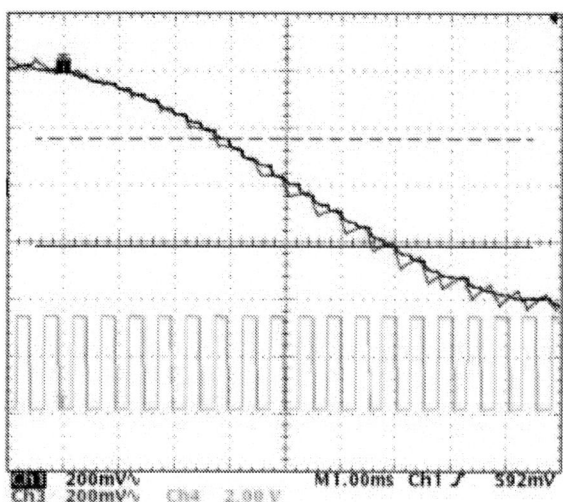

Fig.24. Instantaneous Ripple-Free Current Sampling – DPWM0

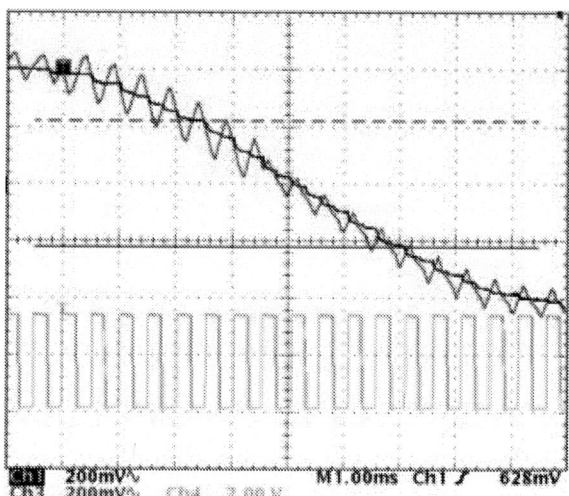

Fig.25. Instantaneous Ripple-Free Current Sampling – NSVM3

4 CONCLUSION

This paper investigated the possibility of detecting instantaneous fundamental current component for electrical AC drives. The conclusion was drawn from theoretical analysis, simulation study and experimental verification. The instantaneous fundamental current can be sampled at the peak/valley of the carrier signal for all symmetric carrier-based PWM patterns, no matter with the choice of zero vector selection.

5 REFERENCES

[1] T. Matsui, T. Okuyama, J. Takahashi, T. Sukegawa and K. Kamiyama, "A high accuracy current component detection method for fully digital vector-controlled PWM VSI-fed AC drives," IEEE. *Trans. Power Electronics, vol.5, no.1, January 1990*, pp. 62–68.

[2] V. Blasko, V. Kaura and W. Niewiadomski, "Sampling of discontinuous voltage and current signals in electrical drives: a system approach," *IEEE Trans. Industry Applications*, vol.34, no.5, Sept./Oct. 1998, pp.1123-1130.

[3] Yen-Shin Lai, Fu-San Shyu, "Optimal common-mode voltage reduction PWM Technique for induction motor drives with considering the dead-time effects for inverter control" , IEEE IAS Annual Meeting, 2003. (0-7803-7883-0/03)

[4] A.M. Hava, R.J. Kerkman, and T.A. Lipo, "Simple analytical and graphical methods for carrier-based PWM-VSI drives," IEEE. *Trans. Power Electronics, vol.14, no.1, January 1999*, pp. 49–61.

[5] V. Blasko, "Analysis of a Hybrid PWM based on modified space vector and triangle-comparison methods," *IEEE Trans. Industry Applications*, vol.33, no.3, May/June. 1997, pp.756-764.

[6] J.Holtz, "Pulswidth modulation for electric power conversion, " Proceedings of the IEEE, vol.82, no.8, Aug.1994, pp.1194-1214

[7] F. Bauer and G. Heinle, "Eliminating the harmonics from measured current values in PWM drives," EPE conf. pp. 343-348, Italy,1991

[8] Q.Yin, R.J. Kerkman, T.A. Nondahl and H.H. Lu, "Analytical investigation of the switching frequency harmonic characteristic for common mode reduction modulator, " IEEE IAS conf. CD 2005

[9] J.Pankau, D. Leggate, D.W. Schlegel, R.J. Kerkman, and G.L. Skibiniski, "High-Frequency Modeling of Current Sensors, " *IEEE Trans. Industry Applications*, vol.35, no.6, Nov./Dec. 1999, pp.1374-1382.

2006 5th International Power Electronics and Motion Control Conference

Study of Speed Sensorless Control Methodology for Single Inverter Parallel Connected Dual Induction Motors Based on the Dynamic Model

Shi Wei   Wang Ruxi   Wang Yue   He Yanhui   Wang Zhaoan   Liu Jinjun

Xi'an Jiaotong University Xi'an 710049 CHINA

Email: xjtuhour@gmail.com   yuewang@mail.xjtu.edu.cn

Abstract---In this paper, a complete close-loop speed sensorless dual-motor drive system was designed based on the dynamic model and an adaptive observer which is based on MRAC was applied to each induction motor, in which stator fluxs, rotor fluxs and motor speeds were estimated. The validity and effectiveness of this proposed method were confirmed through different balanced and unbalanced load tests of simulations.

Keywords-dual motors;dynamic model;speed sensorless

I. INTRODUCTION

Traditionally, one inverter drives only one single motor in most of the applications of vector control systems. However, the systems in which one inverter drives multiple induction motors connected in parallel, have been the subject of intensive research recently due to their potential applications in electric railway traction systems[1] and their obvious excellence of low cost, compactness and lightness.

Speed sensorless systems could be more economic..In most of the multiple motors drive systems, it is common to attach the speed sensor to only one motor properly which is chose from all the motors drived. When the unbalance of load and current arise for some reason, the drive system could be unstable.

Many field oriented control schemes have been proposed to realize vector control on multi-motor drive systems [2-6]. In this paper, we designed a complete close loop system based on this dynamic model. To perform sensorless control, we adapt flux observer to estimate stator flux,and rotor flux, and motor speed. The validity and effectiveness of this proposed method were confirmed through simulations.

II. ADAPTIVE FLUX OBSERVER

A. Description of Induction Motor

An induction motor can be described by the following state equation in the stationary reference frame:

$$\frac{d}{dt}\begin{bmatrix} \varphi_s \\ \varphi_r \end{bmatrix} = \begin{bmatrix} A_{11} & A_{12} \\ A_{21} & A_{22} \end{bmatrix}\begin{bmatrix} \varphi_s \\ \varphi_r \end{bmatrix} + Bv_s = Ax + Bv_s . \quad (1)$$

$$i_s = Cx . \quad (2)$$

Where

$\varphi_s = \begin{bmatrix} \varphi_{ds} & \varphi_{qs} \end{bmatrix}^T$:Stator flux.

$\varphi_r = \begin{bmatrix} \varphi_{dr} & \varphi_{qr} \end{bmatrix}^T$:Rotor flux.

$i_s = \begin{bmatrix} i_{ds} & i_{qs} \end{bmatrix}^T$:Stator curret.

$v_s = \begin{bmatrix} v_{ds} & v_{qs} \end{bmatrix}^T$:Stator voltage.

$A_{11} = -R_s L_r \sigma I$.

$A_{12} = R_s L_r \sigma I$.

$A_{21} = R_r L_m \sigma I$.

$A_{22} = -R_r L_s \sigma I + \omega_r J$.

$$B = \begin{bmatrix} 1 & 0 & 0 & 0 \\ 0 & 1 & 0 & 0 \\ 0 & 0 & 1 & 0 \\ 0 & 0 & 0 & 1 \end{bmatrix} .$$

$$C = \sigma \begin{bmatrix} L_r & 0 & -L_m & 0 \\ 0 & L_r & 0 & -L_m \end{bmatrix} .$$

$$I = \begin{bmatrix} 1 & 0 \\ 0 & 1 \end{bmatrix} .$$

$$J = \begin{bmatrix} 0 & -1 \\ 1 & 0 \end{bmatrix} .$$

R_s , R_r :Stator and rotor resistance.

L_s , L_r : Stator and rotor self-inductance.

M :Mutul inductance.

1-4244-0448-7/06/$25.00 ©2006 IEEE

σ :Leakage coefficient , $\sigma = 1/\left(L_s L_r - M^2\right)$.

τ_r :Rotor time constant, $\tau_r = L_r / R_r$.

ω_r :Motor angular velocity.

B. Stator and Rotor Flux Observer

The adaptive flux observer,which estimates the stator and rotor flux simultaneously,is written by the following equation:

$$\frac{d}{dt}\hat{x} = \hat{A}\hat{x} + Bv_s + G\left(\hat{i}_s - i_s\right) \quad . \tag{3}$$

Where $\wedge$ means the estimated values and G is the observer gain matrix ,which is decided so that (3) can be stable.In this paper , G is calculated by the following equation so that the observer poles are proportional to those of the induction motor(proportional constant k >0):

$$G = \begin{bmatrix} g_1 & g_2 & g_3 & g_4 \\ -g_2 & g_1 & -g_4 & g_3 \end{bmatrix}^T \quad . \tag{4}$$

$$g_1 = \left(k-1\right)R_s \tag{5}$$

$$g_2 = 0 \quad . \tag{6}$$

$$g_3 = \left(L_s R_r + R_s L_r k^2 - k R_s L_r - k L_s R_r\right)/L_m. \tag{7}$$

$$g_4 = \left(k-1\right)\omega_r /\left(L_m \sigma\right). \tag{8}$$

C. Adaptive Scheme for Rotor speed

In order to eliminate rotational transducers, the motor speed is estimated by the following adaptive scheme .See Fig. 1. This is derived by using the Lyapunov's stability theorem[7]:

$$\hat{\omega}_r = \left(K_p + K_i / s\right)\left(i_{ds}\hat{i}_{qs} - i_{qs}\hat{i}_{ds}\right). \tag{9}$$

K_p, K_i :arbitrary positive gain.

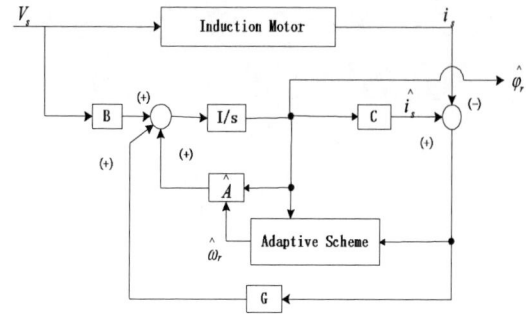

Fig.1. Block diagram of adaptive flux observer

III. DYNAMIC MODEL OF PARALELL CONNECTED INDUCTION MOTORS

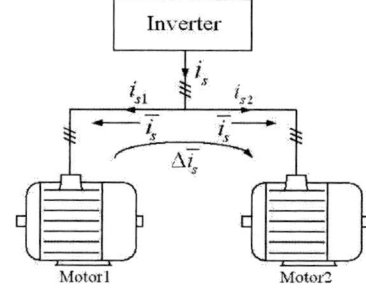

Fig. 2.   Current model of dual induction motors

Fig.2. shows the structure of the dynamic model. The stator current i_s divides into two parts: i_{s1} and i_{s2} , which are respectively the current in motor 1 and motor 2, and can both be represented as the sum of $\bar{i}_s$ and $\Delta\bar{i}_s$.

Where $\bar{i}_s$ is the current that flows equally in both motors and $\Delta\bar{i}_s$ is the average circulating current that flows around the two motors. They are expressed as a function of i_{s1} and i_{s2} :

$$\bar{i}_s = \frac{i_{s1} + i_{s2}}{2} \tag{10}$$

$$\Delta\bar{i}_s = \frac{i_{s2} - i_{s1}}{2} \tag{11}$$

In synchronous reference frame, the rotor flux and the electromagnetic torque of the motors are given by:

$$\frac{d}{dt}\phi_r^e + \left(S_r + j\left(\omega_e - \omega_r\right)\right)\phi_r^e = S' \cdot i_s^e \tag{12}$$

$$T_{em} = \frac{3PL_m}{L_r}\left(\phi_r^e \times i_s^e\right) \tag{13}$$

If the equation (12) is applied to the two motors, then the average rotor flux could be controlled as:

$$\bar{i}_{ds\_fc}^{e*} = \frac{\bar{S}_r \bar{\phi}_{dr}^{e*} + \Delta \bar{S}_r \Delta \bar{\phi}_{dr}^{e} + \Delta \bar{\omega}_r \Delta \bar{\phi}_{qr}^{e} - \Delta \bar{S}' \Delta \bar{i}_{ds}^{e}}{\bar{S}'} \tag{14}$$

$$\bar{i}_{qs}^{e*} = \frac{\left(\dfrac{\bar{T}^*}{\dfrac{3}{4}P\bar{L}'_m}\right) + \Delta \bar{\phi}_{qr}^{e}\Delta \bar{i}_{ds}^{e} - \Delta \bar{\phi}_{dr}^{e}\Delta \bar{i}_{qs}^{e}}{\bar{\phi}_{dr}^{e}} \tag{15}$$

$$\bar{i}_{ds\_tc}^{e*} = \frac{\left(\dfrac{-\Delta \bar{T}^*}{\dfrac{3}{4}P\bar{L}'_m}\right) + \Delta \bar{\phi}_{dr}^{e}\bar{i}_{qs}^{e*} + \bar{\phi}_{dr}^{e}\Delta \bar{i}_{qs}^{e}}{\Delta \bar{\phi}_{qr}^{e}} \tag{16}$$

To the rotor flux, with an analogous method one can control the average electromagnetic torque with the q-axis stator current and control the average differential electromagnetic torque with the d-axis stator current as Eq.15 and Eq.16

Noteworthy is that the d-axis stator current component could be used as $\bar{i}_{ds\_fc}^{e*}$ (Eq.14) when one tries to control the average rotor flux $\bar{\phi}_{dr}^{e*}$, while the d-axis stator current component could be used as $\bar{i}_{ds\_tc}^{e*}$ when one tries to control the average differential torque $\Delta\bar{T}^*$ (Eq.16). Meanwhile, one should note that $\Delta\bar{\phi}_{qr}^{e}$ in Eq. (16) must be a non-zero variable for the differential torque control. Furthermore, the transition between differential torque control and flux control should begin with the following equation (Where K is a constant to adjust the speed of this transition):

$$\bar{i}_{ds}^{e*} = \left(\bar{i}_{ds\_fc}^{e*} - \bar{i}_{ds\_tc}^{e*}\right)\exp\left(-K\left|\Delta\bar{\phi}_{qr}^{e}\right|^2\right) + \bar{i}_{ds\_tc}^{e*} \tag{17}$$

During the period of unbalanced load to balanced load, $\Delta\bar{\phi}_{qr}^{e}$ will have little vibration around the value of zero before it converged to zero. The vibration will lead to a

quick choice between the $\bar{i}_{ds\_fc}^{e*}$ and $\bar{i}_{ds\_tc}^{e*}$. And equation (17) this paper chose will reduce this influence[8].

IV. SYSTEM CONFIGURATION OF PROPOSED METHOD

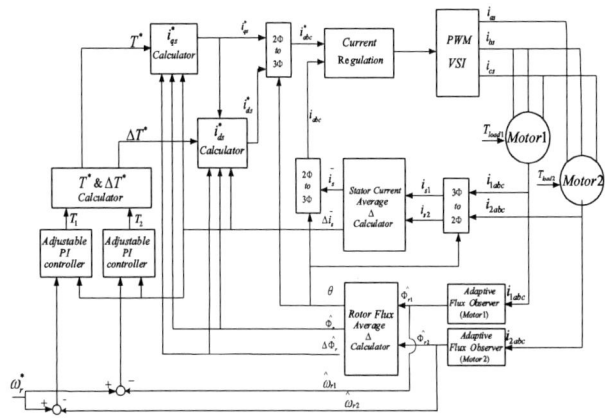

Fig. 3. System configuration of proposed method

The system configuration of proposed method is shown in fig 3. Because these two motors are parallel connected and fed by one inverter, their synchronous speed and stator voltage are exactly the same. When their loads are unbalanced, the speeds of the two motors will be different. It means, we can not eliminate the speed difference between these two motors when the running conditions of the two motors are unbalanced. Note that we want to control the average and differential torque of the two motors during the unbalanced process. Then the conventional PI controller will not suitable in this system. We add two adjustable PI controllers to this system. When the loads are balanced, this PI controllerhas both the proportional component and integral component. If the loads are unbalanced, the effects of the integral component will be eliminated. This could be achieved by the following equation:

$$K_i^* = K_i \exp(-K_1\left|\Delta\bar{\phi}_{qr}^{e}\right|) \tag{18}$$

K_i^* is the integral coefficient of the adjustable PI controller. K_1 is a constant to control the transition speed.

Table 1 Specifications of tested induction motors

Parameter	Value
Output	37.3 kW
Pole Pairs	2
Voltage	460 V
Current	80 A

Motor Speed	1710 r/min
Mutual Inductance	34.7 mH
Stator Resistance	0.087 Ω
Rotor Resistance	0.228 Ω
Stator Self Inductance	35.5 mH
Rotor Self Inductance	35.5 mH

The relevant parameters of the motors are listed in table1. Both motors have the same specifications.

V. RESULTS AND DISCUSSION

A. Speed Step Change Test

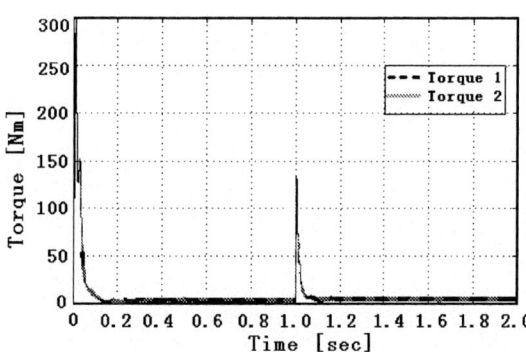

(a)Torque response

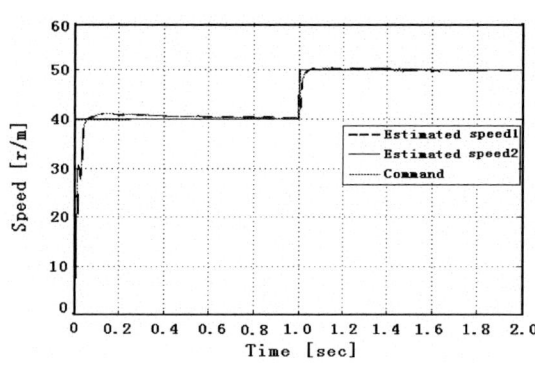

(b) Motor speed

Fig.4. Speed step change test (Simulation results)

Fig.4 shows the simulation results on the condition that the speed command changes from 400 to 500 rpm at 1s. Neither motor has a load. From Fig.4 (b) we can see that the speeds of two motors can follow the command quickly and accurately. Fig.4 means that when the parameters and motor speed of each motor are the same, the two induction motors can be treated as one induction motor.

B. Unbalanced-Load Test of proposed method

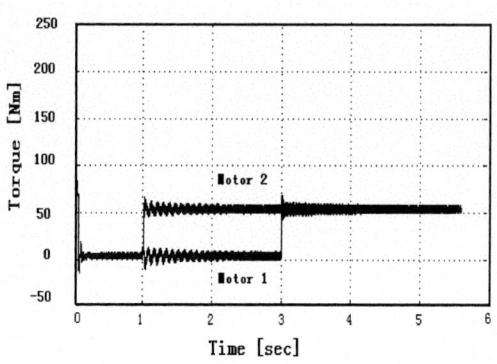

(a)

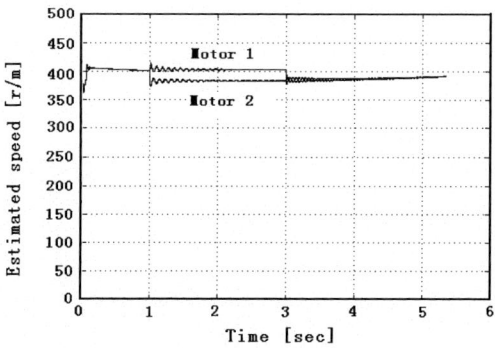

(b)

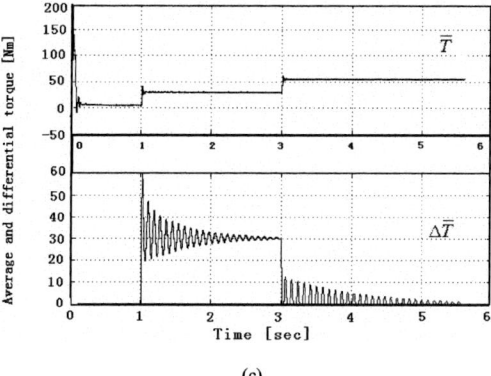

(c)

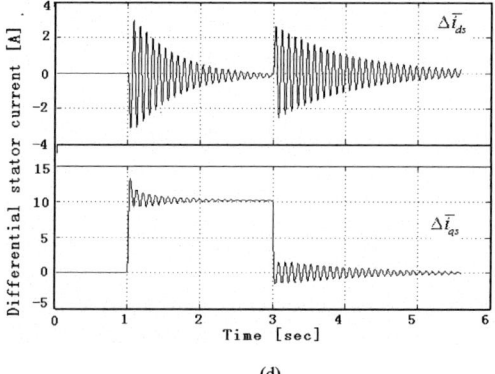

(d)

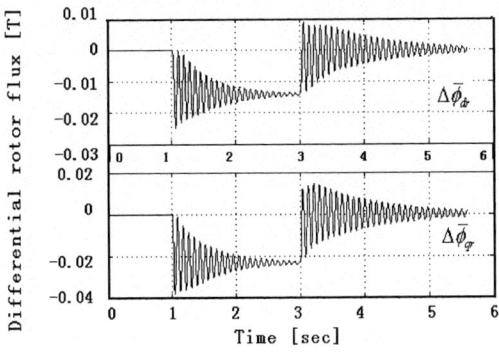

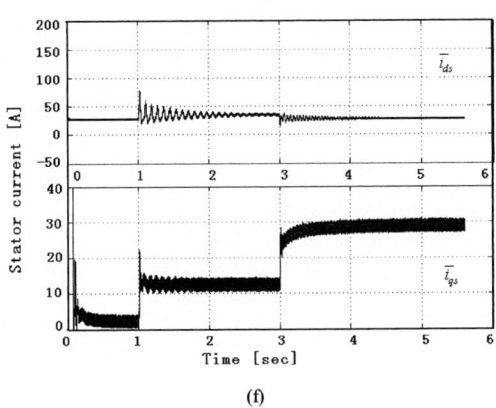

(f)

Fig 5. unbalanced-load test of the proposed method (Simulation results)

     (a) Torque response

     (b) Speed change

     (c) Calculated average and differential torque.

     (d) Differential stator current

     (e) Differential rotor flux

     (f) Stator current.

In the Unbalanced load test, the speed command is set to 400r/min. At the very beginning (time 0s), neither motor has a load. Motor2 is given a 50Nm load after 1s and Motor1 is given a 50Nm after 3s. The torque response follows the load for each motor in (a).

The speed changes of the two motors are shown in (b).The speed of motor2 decreases during the unbalanced process while the speed of motor1 almost remains the same. When the two motor's loads are balanced, both the speeds of two motors become the same and match the speed command. The calculated average and differential torques which were shown in Fig5 as $\overline{T}^{*}$ and $\Delta\overline{T}$ are calculated correctly and quickly in (c). Fig 5 (d) shows the circulating stator currents flow. The differential rotor flux and the stator current are shown in (e) and (f).

VI. CONCLUSION

In this paper, a method of speed sensorless vector control for dual induction motors connected in parallel was proposed. This proposed method had a good dynamic characteristic on torque response which was thought to be important in railway traction system. The validity and effectiveness of this proposed method were confirmed through different unbalanced load tests of simulations .

REFERENCES

[1]THOMAS M. JAHNS, et al . Recent Advances in Power Electronics Technology for Industrial and Traction Machine Drives. PROCEEDINGS OF THE IEEE, VOL. 89, NO. 6, JUNE 2001

[2]Patrick M. Kelecy, et al. Control Methodology for Single Inverter, Parallel Connected Dual Induction Motor Drives for Electric Vehicles. IEEE. 1994.

[3]K.Matsuse,Y.Kouno, H. Kawai, and S. Yokomizo, A speed-sensorless vector control method of parallel-connected dual induction motor fed by a single inverter. IEEE,2002

[4]Hirotoshi Kawai, Yusuke Kouno, and Kouki Matsuse. Characteristics of Speed Sensorless Vector Controlled Dual Induction Motor Drive Connected in Parallel Fed by A Single Inverter. IEEE,2002

[5]Yasushi Matsumoto, et al. A Stator-Flux-Based Vector Control Method for Parallel-Connected Multiple Induction motors Fed by A Single Inverter. IEEE,1998

[6]Yasushi Matsumoto, et al. A Novel Vector Control of Single-Inverter Multiple-Induction-motors Drives for Shinkansen Traction System. IEEE,2001

[7]Wang Jian, "Study on direct torque control system of asynchronous motor without speed sensor."ELECTRIC DRIVE FOR LOCOMOTIVES.No2,2005.

[8]Wang Ruxi, "Study of Control Methodology for Single Inverter Parallel Connected Dual Induction Motors Based on the Dynamic Model." IEEE-PESC Conference Records 2006.

2006 5th International Power Electronics and Motion Control Conference

ADC architecture with direct binary output for digital controllers of high-frequency SMPS

Tao ZHOU and Jianping XU

Department of Electrical Engineering, Southwest Jiaotong University, Chengdu, China

Abstract—**Without adopting encoder to convert "thermometer code" into binary code, analog-to-digital converter (ADC) architecture with direct binary code output is proposed for digital controllers of high-frequency switching mode power supply (SMPS). It can provide large conversion ranges with required precision and reduce ADC circuit complexity. The output can be compressed into a much shorter addressing character in logarithmic law and lead to reduction of memory size in the look-up table of the digital compensator.**

Keywords-digital controller; ADC; delay-line; flash; power converter; switching mode power supply (SMPS)

I. INTRODUCTION

Digital controllers for pulse-width modulation (PWM) SMPS have become more and more popular due to their low power consumption, immunity to analog component variations, ability to interface with digital systems, flexibility to implement sophisticated control schemes, and potentially faster design process [1-3]. The digital controller as shown in Fig.1 consists of three key building blocks: analog-to-digital converter (ADC), digital regulator and digital pulse-width modulator (DPWM). It serves as the feedback controller of power converters to regulate the controlled quantity to match the reference over input voltage, load currents, temperature and other parameters variations. For example, for a digital controlled voltage-mode PWM power converter, the sensed output voltage is sampled and compared to the reference voltage by ADC to produce digital error signal, which is passed to the digital compensator containing look-up tables [4]. These two building blocks together perform the function as the compensating error amplifier in analog controllers. The output of the digital compensator is the input to the DPWM, which produces a constant-frequency variable duty-ratio signal to control the on-off states of the switching power transistors.

Therefore, ADC plays a very important role in digital controllers. The conversion characteristics of ADC should satisfy voltage regulation requirements, i.e.: (i) in steady state, the dc output voltage must be equal to the reference voltage with some allowed tolerance, and the analog equivalent of the least significant bit (LSB) must not be greater than this tolerance; (ii) in transient state (including load or input voltage transients), the output voltage must always stay in the conversion range around

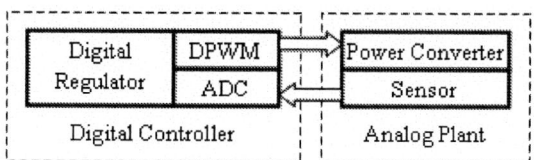

Figure 1.   Block diagram of digital controlled power converters

reference, and this conversion range is determined by the analog equivalent of the most significant bit (MSB). To satisfy the requirement of extremely fast responses, especially for high-frequency power converters, ADC topologies with low latency are desirable, since delays correspond to phase shift that may degrade the loop response. To meet these requirements, conventional high-speed, high-resolution ADCs consume much power, need large chip area and require external precision analog components. Therefore, flash ADCs [5] and delay-line ADCs [6-8] have been proposed respectively, which are introduced in Section II. As an improvement, Section III proposes an implementation technique with direct binary code output by exponential quantization and tap selection, the output of which can be compressed into a much shorter addressing character of the desired information stored in look-up tables [4]. Section IV presents the simulation result of exponential quantization method.

II. FLASH ADC AND DELAY-LINE ADC

A. Flash ADC Architecture

A windowed single stage flash topology reported in [5] is shown in Fig.2. It converts the digital reference (D_{ref}) into analog reference voltage (V_{ref}) by a digital-to-analog converter (DAC). An offset network with uniform steps (V_q) is installed with a number of comparators to make quantization bins around V_{ref}. The sensed output voltage (V_o) is fed into the other input of the comparators and is compared with V_{ref} to generate the digital representation (D_e) of the error signal in "thermometer code", which is then converted by an encoder into binary code output (D_o) to serve as addressing character of the desired information stored in the look-up table. According to V_{ref}, DAC does not need to change very fast. Such kind of ADC doesn't need multiple comparisons (like pipeline ADCs) or digital filtering (like $\Sigma \Delta$ ADCs), so it can perform quite fast conversion. However, a high-resolution flash ADC that covers the full range between ground and V_{in} will

1-4244-0448-7/06/$25.00 ©2006 IEEE

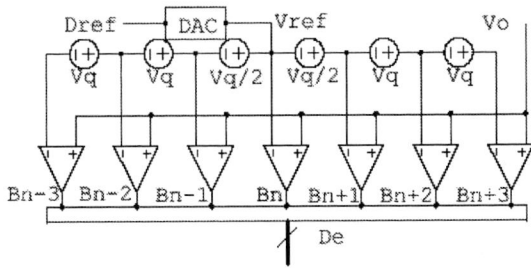

Figure 2.   Block diagram single-stage flash ADC

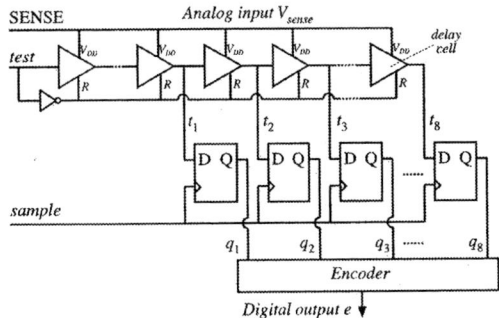

Figure 3.   Basic delay-line ADC configuration

demand excessive power and silicon area, as well as complicated logic circuits in the encoder. Therefore, it can only be realized in a small window around V_{ref}.

B. Delay-line ADC Architecture

Based on the principle that the propagation delay of a logic gate in a standard CMOS process is approximately inversely proportional to the supply voltage in certain condition, delay-line ADC configuration as shown in Fig. 3 has been proposed for the controller of high-frequency power converters [6]. The basic delay-line configuration as shown in Fig.3 is formed by a string of delay cells (consisting of logic gates) supplied from the sensed analog voltage as the input of the ADC. To perform a conversion, at the beginning of a conversion period, a test pulse propagates through the delay line supplied by the analog input voltage of the ADC. After a fixed time interval, the taps are sampled through the sampling unit consisting of a string of D-type flip-flops. Compared to the reference word, the sampled result serves as the digital representation of the error signal in "thermometer code", and is then converted by encoder into binary code, to serve as addressing character of corresponding information stored in the look-up table. However, in practice, the reference voltage determined by the reference word cannot be precisely controlled due to process and temperature variations. Therefore, calibration of reference is required and several techniques with digital subtraction are proposed, such as using the standard bandgap techniques [6], using two matched delay lines configuration [7] and using differential delay line configuration [8]. The basic delay-line configuration does not require any precision analog components and can be implemented using standard logic gates. However, in maintaining the required regulation range, if the resolution bits increases, the sampled taps and the encoder complexity will increase exponentially. Thus a high-precision delay-line ADC can only cover a small regulation range with a few taps around reference position.

From above discussion we can know that Flash ADCs perform a direct analog-to-digital conversion. We can adjust the resolution and the conversion range by modifying the step value and the number of steps in the offset network. While delay-line ADCs perform an indirect analog-to-digital conversion via time domain, the tap delay and the number of taps determine the effective LSB resolution and the conversion range. However, both architectures are combinational circuits, and they have some characteristics in common. They both obtain the digital error signal firstly in "thermometer code", and then need an encoder to convert it into desired binary code. They are both only suitable for a small regulation range, due to logic circuit complexity, power consumption and silicon area while regulating with high resolution in large ranges. In the following section, we will propose a new implementation technique to overcome these problems.

III. ADC WITH DIRECT BINARY CODE OUTPUT

To obtain larger regulation ranges with high precision, implementation techniques (consisting of quantization technique and tap selection technique) with direct binary code output are proposed for both flash ADCs and delay-line ADCs, which can still be compressed into a much shorter addressing character in logarithmic law. We call them modified ADCs comparing to conventional flash ADCs and delay-line ADCs. For the modified delay-line ADCs, the calibration methods with digital subtraction proposed in [6-8] cannot be used any more, and a new calibration technique is specially proposed, which is also universal to general cases of delay-line ADC.

A. Tap Selection

To obtain direct binary code output, the taps should be taken with 2-based exponentially increasing division, as shown in Fig. 4. It means that the controller can perform coarse regulations while V_o is far away from V_{ref} and perform fine regulations while V_o is close to V_{ref}. Therefore, with required precision, much fewer taps are needed to cover a required regulation range to make larger conversion ranges possible. For each sampling time, the right taps should be chosen by a selector to directly give the digital representation (D_e) of the error signal in binary code. We have also proposed a selector configuration as shown in Fig. 4, of which the selecting rule is that the sign bit (D_s) takes the value of the

1067

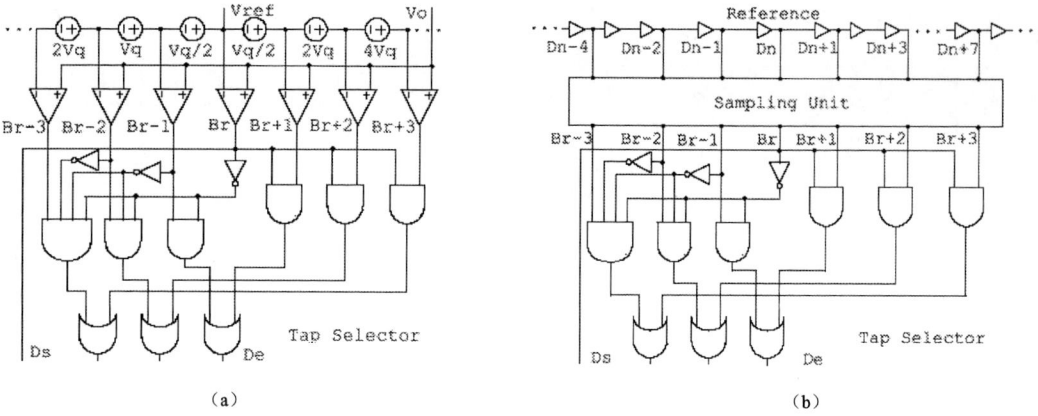

Figure 4.   Flash ADC (a) and delay-line ADC (b) with tap selector

reference tap (B_r), and:

(i) when V_o is larger than or equal to V_{ref}, $B_r =1$, the error signal is marked as positive and the tap selector chooses the taps on the left side of B_r, from B_{r+1} as LSB;

(ii) when V_o is smaller than V_{ref}, $B_r=0$, the error signal is marked as negative and the tap selector chooses the taps on the right side of B_r, from B_{r-1} as LSB until meeting the first '1' in B_{r-i}, and takes the rest tap outputs, from B_{r-i-1}, as '0'.

To make the rule obvious, Table I shows the corresponding relationship among tap outputs (from B_1 to B_7, with B_4 as reference tap), the sign bit (D_s) and the selector output error signal (D_e) for a range from $-4V_q$ to $+7V_q$, where V_q is the equivalent voltage precision represented by the least step in flash ADCs and by one delay in delay-line ADCs.

To illustrate that the modified ADC with direct binary code output can enormously reduce logic circuit complexity, the numbers of required logic gates for different regulation ranges are given in Table II. We can see the possibility of enlarging regulation range by adding only a few logic gates, including multiple-input AND, 2-input OR and NOT. For example, if the precision is required as V_q and the regulation range (V_o-V_{ref}) is from $-4V_q$ to $+7V_q$, the modified ADCs need only 7 taps, 6 ANDs, 3 ORs, and 3 NOTs as shown in Fig. 4, but conventional ADCs need 12 taps and a relatively complicated encoder. To double the regulation range, the modified ADCs need add only 2 taps, 2 ANDs, 1 OR and 1 NOT. However, conventional ADCs have to double the existed resources. When the regulation range covers from $-512V_q$ to $+1023V_q$, the modified ADCs need 20 taps, 20 ANDs, 10 ORs and 10 NOTs, which is still feasible, but conventional ADCs need 1536 taps and extremely complicated encoder, which seems impossible to realize.

B.   Logarithmic Compression

Since the division of the taps increases in 2-based exponential law, such kind of binary code output can be compressed into a much shorter addressing character in 2-based logarithmic law as shown in Table I:

$$D_c=\log_2(D_e+1), \qquad \text{if } D_s=1 \qquad (1)$$

$$D_c=\log_2(D_e)+1, \qquad \text{if } D_s=0 \qquad (2)$$

where D_s is the sign bit, D_e is the selector output error signal and D_c is the compressed addressing character. The compression can be quickly realized in digital computation by adder and shift instructions in the digital signal processing core. And the compressing process is the same process of addressing the information stored in the look-up table. Shorter addressing characters correspond to less stored information and lead to reduction of required memory size. The conversion range versus required addressing character bit number characteristic is shown in Fig. 5 for both conventional ADCs and the modified ADCs. We can see that a

TABLE I.   DATA CONVERSION RELATIONSHIP

B_1	B_2	B_3	B_4	B_5	B_6	B_7	D_s	D_e	D_c
1	1	1	1	1	1	1	1	111	11
1	1	1	1	1	1	0	1	011	10
1	1	1	1	1	0	0	1	001	01
1	1	1	1	0	0	0	1	000	00
1	1	1	0	0	0	0	0	001	01
1	1	0	0	0	0	0	0	010	10
1	0	0	0	0	0	0	0	100	11

TABLE II.   REQUIRED RESOURCES

(V_o-V_{ref}) / V_q	AND	OR	NOT	b_o	b_c
-4 ~ +7	6	3	3	4	3
-64 ~ +127	14	7	7	8	4
-128 ~ +255	16	8	8	9	5
-256 ~ +511	18	9	9	10	5
-512 ~ +1023	20	10	10	11	5
-16384 ~ +32767	30	15	15	16	5

1068

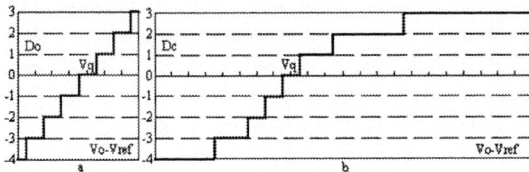

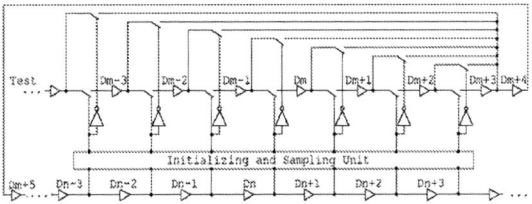

Figure 5. Addressing character versus error voltage
(a) Conventional ADCs; (b) New ADCs

same-length addressing character can cover a much larger conversion range in modified ADCs than in conventional ADCs. For example, as shown in Table II, with a regulation range from $-512V_q$ to $+1023V_q$, a 5-bit ($b_c=5$) addressing character (consisting of D_s and D_c) is enough to address among 21 consecutive memory elements, while the encoder output of conventional ADCs is a 11-bit ($b_o=11$) addressing character (D_o) to address among 1536 consecutive memory elements. Therefore, the modified ADCs without encoder lead to enormous reduction of logically processing resources and memory size, especially for wide regulation ranges.

C. Calibration for Delay-line ADC

In modified delay-line ADCs as shown in Fig. 4, the divisions among the taps are not uniform and the highest precision stays only with the reference taps B_r, of which the value is always taken from the reference position D_n. Therefore, the approaches of calibration using digital subtraction [6-8] will change the reference tap and lose required precision in new ADCs. Thus, a new technique using standard bandgap techniques with compensation line is proposed here. This method is also universal to general cases of delay-line ADC.

A compensation line, consisting of a number of delay cells connected in line and in parallel with some switches, is installed in the front part of the delay line as shown in Fig. 6. By controlling the on-off states of the switches to add or remove some delay cells on the delay line, the reference position can be adjusted and maintained at D_n. Two conversions including reference conversion and input conversion are performed in each conversion period. At the beginning of the reference conversion, the on-off states of the switches are initialized that the delay cells behind D_m on the compensation line are by-passed, and then the reference voltage is applied to the delay line. If the reference conversion is ideal, the reference position is right at D_n and the reference conversion result will be the same as the initial word and won't change the initial on-off states of the switches. However, the actual result can be different because of process and temperature variations. The idea of the compensation line is to use the reference conversion result as a feedback command, to (i) by-pass some more delay cells while the test signal propagates too slow and (ii) add in line some more delay cells while the test signal propagates too fast, for the purpose of maintaining the right reference position at D_n during the input conversion. The compensation line of the example shown in Fig 6 contains 7 delay cells from

Figure 6.   Calibration line implementation

D_{m-3} to D_{m+3}, of which D_{m+1}, D_{m+2} and D_{m+3} are by-passed by initialization. If test signal propagates too slow and reaches D_{n-1} at the sampling moment during reference conversion, one cell fewer than at D_n, with $D_n=0$, the compensation line will by-pass one delay cell (D_m), on the delay line and to maintain the reference position at D_n during next input conversion; if test signal propagates too fast and reaches D_{n+2} at the sampling moment during reference conversion, two cells more than at D_n, with $D_{n+1}=1$, $D_{n+2}=1$, the compensation line will add in line two more delay cells (D_{m+1} and D_{m+2}) and maintain the reference position at D_n during next input conversion. And then during the input conversion, the conversion result will be the precisely calibrated value of the error signal in binary code, as mentioned above in Subsection A. If desired, the reference conversion for the purpose of calibration does not have to be performed in every conversion period.

Compared to the calibration techniques with digital subtraction, the error signal can be output directly from the selector without being passed to a computational unit. And the implementation of the compensation line needs just some more delay cells and a number of switches, which consume fewer resources than the computational unit.

IV.   SIMULATION RESULT OF QUANTIZATION METHOD

Although some detailed information may be lost during coarse regulations, a controller with modified ADC can still achieve the same required performance as conventional ADCs with uniform division. Simulation in Matlab/Simulink environment is performed for both uniform quantization method and 2-based exponential quantization method of output voltage controlled by the same controller, of which the transfer function of the controller is described as:

$$G_c(s) = \frac{14.3s^2 + 651400s + 7.2 \times 10^9}{s^2 + 125600s} \qquad (3)$$

for a buck converter with 5V input voltage, 1.6V output voltage, L=1uH, C=1620uF, 0.004ohm equivalent serial resistor of filter capacitor, 0.5 ADC gain and 250 kHz switching frequency, the simulations results are given in Figure 7, which indicate that the transient responses to load variation from 16A to 32A with both uniform quantization method and exponential quantization method need about 10 switching periods (equivalent to 40 us).

1069

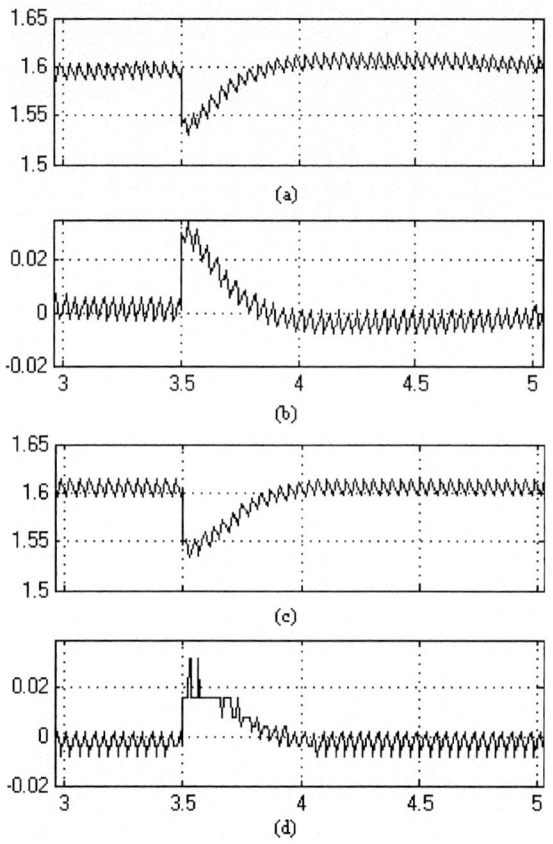

Figure 7. Output voltage V_o (V) versus time (10^{-4}s) during transient response to load variation from 16A to 32A. (a) output voltage with uniform quantization; (b)error signal with uniform quantization; (c) output voltage with exponential quantization; (d) error signal with

V. CONCLUSION

Modified ADC architecture with direct binary code output is proposed for both flash ADCs and delay-line ADCs in digital controller of SMPS. Compared to conventional architectures, it can provide large regulation ranges with high precision because of taking taps with exponentially varied division, reduce logic circuit complexity by adopting a simple tap selector instead of the complicated encoder, and lead to reduction of memory size of look-up tables in the digital compensator by compressing the output into a short addressing character. A new technique of calibration with compensation line is proposed for modified delay-line ADCs, which is also universal to general cases of delay-line ADC.

REFERENCE

[1] T. W. Martin, and S. S. Ang, "Digital control for switching converters," Industrial Electronics, 1995. ISIE '95., Proceedings of the IEEE International Symposium on Volume 2, pp.480 - 484 vol.2, 10-14 July 1995

[2] P. Vallittu, T. Suntio, and S. J. Ovaska, "Digital control of power supplies-opportunities and constraints," Industrial Electronics Society, 1998. IECON '98. Proceedings of the 24th Annual Conference of the IEEE Volume 1, pp.562 - 567 vol.1, 31 Aug.-4 Sept. 1998

[3] Y. Duan, and H. Jin, "Digital controller design for switchmode power converters," Applied Power Electronics Conference and Exposition, 1999. APEC '99. Fourteenth Annual Volume 2, pp.967 - 973 vol.2, 14-18 March 1999

[4] A. Prodic, D. Maksimovic, "Design of a digital PID regulator based on look-up tables for control of high-frequency DC-DC converters," Computers in Power Electronics, 2002. Proceedings. 2002 IEEE Workshop pp.18 – 22, on 3-4 June 2002 [5] Jinwen Xiao, A. V. Peterchev and S. R. Sanders, "Architecture and IC implementation of a digital VRM controller," Power Electronics Specialists Conference, 2001. PESC. 2001 IEEE 32nd Annual Volume 1, pp.38 - 47 vol. 1, 17-21 June 2001

[6] B. J. Patella, A. Prodic, A. Zirger, and D. Maksimovic, "High-frequency digital PWM controller IC for DC-DC converters," Power Electronics, IEEE Transactions on Volume 18, Issue 1, Part 2, pp.438 - 446, Jan. 2003

[7] D. Maksimovic, R. Zane, and R. Erickson, "Impact of digital control in power electronics," Power Semiconductor Devices and ICs, 2004. Proceedings. ISPSD '04. The 16th International Symposium pp.13 – 22 on 24-27 May 2004

[8] H. Peng, and D. Maksimovic, "Digital current-mode controller for DC-DC converters," Applied Power Electronics Conference and Exposition, 2005. APEC 2005. Twentieth Annual IEEE Volume 2, pp.899 – 9056, 10 March 2005

2006 5th International Power Electronics and Motion Control Conference

Analysis and Evaluation of a High-Voltage AC Amplifier for Electrostatic Suspension

F. T. Han, Q. P. Wu, K. Liu and Z. Y. Gao

Tsinghua University / Department of Precision Instruments and Mechanology, Beijing, China

Abstract—**High-voltage amplifiers with an output voltage typically in the kilovolts range are widely used for electrostatic force generation in electrostatic suspension. This paper describes a high-voltage ac amplifier using a nanocrystalline soft magnetic alloy-based output transformer in an effort to provide fast dynamic response, low power loss, and excellent thermal stability. An analysis and an equivalent circuit model of the high-voltage transformer are presented for performance evaluation. Simulated and measured results are provided to confirm the validity of the transformer model. The excellent characteristics of the high-voltage ac amplifier were proved experimentally for high-performance operation of electrostatic suspension.**

Keywords--High-voltage amplifier; nanocrystalline soft magnetic core; transformer model; electrostatic suspension

I. INTRODUCTION

Electrostatic suspension offers the advantage of directly levitating a wide variety of materials without any direct mechanical contact. The suspended object is usually supported by strong electric fields between one or more pairs of stator electrodes and the suspended object [1-3]. As a result, high voltage levels, typically in the kilovolts range, are required to achieve electrostatic suspension. Generally, each pair of electrodes is energized by controllable voltages from high-voltage amplifiers, supplying the electrodes with electric charge.

High-voltage ac amplifiers which are widely used for generation of electrostatic forces necessarily use a high-ratio output transformer to step up the usual supply voltage (20-30V) to the kilovolt suspension output level [1,4]. Such magnetically stable core materials as high quality permalloy or thin plate silicon steel are usually used. However, both permalloy and advanced silicon steel are not favorable for operating frequency higher than about 2kHz due to their loss characteristics [4].

An alternative to the ac amplifier is the dc amplifier which behaves an ideal solution for electrostatic suspension [2]. An AM-based dc amplifier operated at a carrier frequency of 30kHz has been introduced using an output transformer made of MnZn-ferrite core in recent years [5]. However, the transformer core made of power ferrites behaves poor temperature stability and not

suitable for such suspension subjected to a wide range of operating temperature.

The output transformer is the heart of the high-voltage ac amplifier. It occupies more volume than any other part, and a gyro suspension system uses six transformers to achieve three-degree-of-freedom support of a spherical rotor [2]. The miniaturization of the transformer is limited mainly by the saturation magnetic flux density (B_s) and the power loss of the core material. Recently, a new generation of core material has become available as an industrial product—the so called nanocrystalline soft magnetic alloys (NSMAs) which exhibit high B_s, high permeability, and low core losses [6]. These characteristics indicate that nanocrystalline alloys are suitable for the core material of high-voltage transformers. Another remarkable advantage of the nanocrystalline material compared to the other options is the fact that important parameters like hysteresis losses and permeability are almost constant within the usual working temperature of -40~120°C. In contrast to ferrites, there is no tendency to overheating due to a negative temperature coefficient even at temperature above 120°C. This helps to reduce the core size, improve the power efficiency and thermal stability for such suspension systems used in high-precision inertial sensors.

II. HIGH-VOLTAGE TRANSFORMER

The study here does not aim at optimizing the transformer design. Instead, it investigates the use of NSMA core-based transformers for high-voltage ac amplifiers and presents a general analysis on an accurate transformer model. Such a mathematical model can be applied to analyze subsequent high-voltage amplifiers.

A. Transformer Description

The transformer was designed with a NSMA-based C-C core pair and a high turns ratio. Table I summarizes the magnetic characteristics of the nanocrystalline alloy. The number of turns of the primary windings is 28 and that of the secondary windings is 5600. The secondary windings of the transformer has a center tap by which is used to connect with ground and provide a common reference for each electrode voltages.

B. Equivalent Model of High-Voltage Transformer

The authors would like to thank the National Natural Science Foundation of China for financial support through Project 50577036.

1-4244-0448-7/06/$25.00 ©2006 IEEE

TABLE I.
MATERIAL CHARACTERISTICS OF NSMA

Parameter	Value
Saturation magnetic flux density B_s	1.25 T
Coerectivity H_c	0.8 A/m
Magnetostriction λ_s	2×10^{-6}
Electrical resistibility ρ	1.3 mΩ·m
Core Loss (20kHz, 0.5T) W	25 W/kg
Maximum working temperature T_{max}	150 °C

The high-voltage transformer operates between widely different voltage levels and, thus, has a high turns ratio to provide 2000 volts from the 24-volt supply. It is bandwidth limited by its internal parameters such as leakage inductance, stray capacitance between the windings, and magnetizing inductance. The analysis of the transformer is most easily achieved by the use of an electrical equivalent circuit [7]. A lumped-element equivalent transformer model is illustrated in Fig.1, where:

r_1 --primary winding resistance;

r_2 --secondary winding resistance referred to the primary;

r_m --equivalent core loss;

L_s --leakage inductance referred to the primary;

L_m --magnetizing inductance;

C_m --secondary inter-winding capacitance referred to the primary;

n --turns ratio between primary and secondary windings.

It is shown by analysis that the transformer model includes two internal *RLC* resonant networks. In the parallel resonance condition, the leakage inductance and secondary winding resistance are treated as short circuits. The resonance is due to magnetizing inductance and stray capacitance of the transformer, and approximated by

$$G_p(s) = \frac{V_2(s)}{V_1(s)} = \frac{a \cdot n}{1 + Q_p \left(\dfrac{s}{\omega_p} + \dfrac{\omega_p}{s} \right)} \quad (1)$$

where $a = r_m / (r_1 + r_m)$, $\omega_p = 2\pi f_p = 1/\sqrt{L_m C_m}$, and $Q_p = ar_1 / \omega_p L_m = ar_1 C_m \omega_p$.

Since the series resonance frequency is much higher than that of the parallel resonance, the parallel network including magnetizing inductance and core loss is treated as open circuit. The simplest transfer function representing the series resonance is

$$G_s(s) = \frac{V_2(s)}{V_1(s)} = \frac{Q_s \omega_s}{s} \cdot \frac{n}{1 + Q_s \left(\dfrac{s}{\omega_s} + \dfrac{\omega_s}{s} \right)} \quad (2)$$

where $\omega_s = 2\pi f_s = 1/\sqrt{L_s C_m}$, $Q_s = L_s \omega_s /(r_1 + r_2)$.

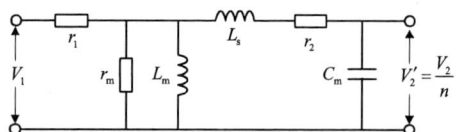

C. Simulation of Transformer Model

The transformer model parameters used for analysisthrough the paper are shown in Table II. It can be calculated that the transformer is operated at a parallel resonance with $f_p = 8.93$ kHz and $Q_p = 0.0216$ together at a series resonance with $f_s = 61.94$ kHz and $Q_s = 5.97$. It is readily seen that the bandwidth of parallel resonance is very wide due to a sufficiently low Q value while the series resonance qualifies a high-Q circuit. Since the high-voltage transformer will be tuned by a sinusoidal voltage at its parallel resonant frequency, the series resonance should be suppressed with an aim to attenuate the high-order harmonics of output voltage waveform and improve the dynamic performance of resulting high-voltage amplifiers.

TABLE II.
TRANSFORMER MODEL PARAMETERS

$r_1(\Omega)$	$r_2(\Omega)$	$r_m(\Omega)$	L_m(mH)	L_s(μH)	C_m(μF)	n
0.80	0.125	365	0.635	14.2	0.50	200

In order to suppress the series resonance, an external resistor labeled R_s, which is in series with the primary of high-voltage transformer, is utilized to produce desired resonance condition. The computed frequency responses for various R_s are shown in Fig.2. It is noted from Fig.2 that a larger value of R_s will produce an apparent change in the form of reduced series resonance and increased parallel resonance, as expected from (1) and (2). For a shunting resistor of $5\,\Omega$, the Q-value of the parallel resonance and the series resonance are 0.154 and 0.933, and for a resistor of $10\,\Omega$, 0.284 and 0.505, respectively. Consequently, R_s will be a critical parameter in design of high-voltage ac amplifiers.

II. ANALYSIS OF HIGH-VOLTAGE AMPLIFIER

A schematic diagram of the high-voltage ac amplifier is shown in Fig.3, where the input signal is labeled V_{in} and the output signal is labeled V_e. The carrier is a sinusoidal signal of a fixed frequency ω_c and a linear AM modulator is realized via a wide bandwidth, four-quadrant analog multiplier [5]. After modulation an output amplifier is followed and ac coupled through a

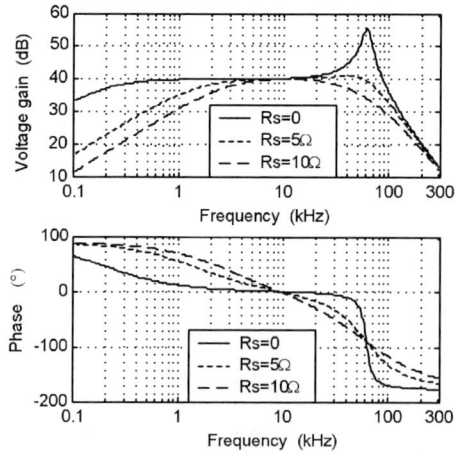

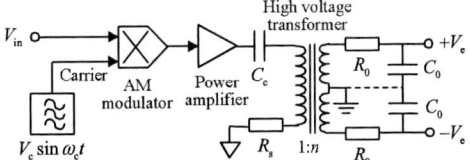

Fig.3. Schematic diagram of high-voltage ac amplifier.

capacitor C_c to drive subsequent step-up transformer.

There may be a pair of series resistors labeled R_0 added to the secondary terminals to limit current and provide isolation since the electrodes are also used for rotor position sensing. Note that the outputs of the amplifier, which are of the same magnitude and opposite polarity, will be applied on a pair of electrodes. The load of the high-voltage ac amplifier can be treated approximately as a capacitive load with a nominal value C_0 between a centered rotor and each electrode [3].

A. Modeling and Analysis of Output Circuit

As the NSMA core-based transformers are used in high-voltage ac amplifiers, the load condition as well as the output characteristics of the power amplifiers should be considered. By modifying Fig.3, the equivalent model of the output stage circuit referred to the primary side is shown in Fig.4, where $R_1 = R_s + r_1 + r_o$, $R_2 = 2R_0 / n^2$, $C_1 = C_m + n^2 C_w$, $C_2 = n^2 C_0 / 2$, r_o and C_w are output impedance of the power amplifiers and cable capacitance, respectively.

According to the circuit model shown in Fig.4, it is found by analysis that there are three resonance conditions at different frequency ranges. Considering the output V_{pa} from the power amplifier as an ac voltage source, another series resonance is introduced in the low-frequency range by the coupling capacitance and magnetizing inductance. The resulting resonance condition is similar to (2) and given by $\omega_{s1} = 2\pi f_{s1} = 1 / \sqrt{L_m C_C}$, $Q_{s1} = \omega_{s1} L_m / R_1$.

In the mid-frequency region, the equivalent model used for analysis is similar to those considered in the transformer model and the resulted parallel resonance condition is given by $\omega_p = 2\pi f_p = 1 / \sqrt{L_m (C_1 + C_2)}$, $Q_p = r_m R_1 / [\omega_p L_m (r_m + R_1)]$.

Whereas the equivalent model used for analysis in the high-frequency range is somewhat complex due to given load conditions. However, if the value of R_2 is smaller enough than that of R_1 , then the equivalent model used for analysis in the high-frequency range is similar to those considered in (2) and given by $\omega_{s2} = 2\pi f_{s2} = 1 / \sqrt{L_s (C_1 + C_2)}$, $Q_{s2} = \omega_{s2} L_s / R_1$.

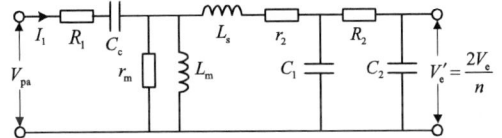

Fig.4. Equivalent model of output stage circuit.

Since the carrier frequency of these high-voltage amplifiers is usually set at a value equal to the frequency at parallel resonance, the two series resonance conditions should be suppressed sufficiently for desired dynamic response.

B. Frequency Response

The frequency characteristics for the equivalent circuit shown in Fig.4 can be expressed as $G(j\omega) = V_e(j\omega) / V_{pa}(j\omega)$. The computed frequency responses for various R_s are shown in Fig.5 for the following set of parameters: $r_o = 0.6\,\Omega$, $R_0 = 0$, $C_C = 47.5\,\mu F$, $C_1 = 0.75\,\mu F$, and $C_2 = 1.0\,\mu F$. Based on the indicated parameters, it can be calculated that three resonance frequencies in output stage circuit are $f_{s1} = 916\,Hz$, $f_p = 4.77\,kHz$, and $f_{s2} = 30.3\,kHz$ respectively. Comparison of Figs.2 and 5 shows that the load capacitance would lower the resonant frequency of the transformer circuits, as expected from above analysis, and thus reduce the maximum operating frequency of the high-voltage amplifiers. It is clear from Fig.5 that an apparent change can be produced in the form of reduced series resonance and improved parallel resonance by increasing the value of R_s . It was found that the output stage offers a bandwidth of $47.1\,kHz$, $16.5\,kHz$, and $8.09\,kHz$ for $R_s = 0$, $5\,\Omega$, and $10\,\Omega$, respectively.

For the test of frequency response, a pair of fixed capacitors was used to simulate above load conditions instead of actual electrode-rotor capacitances. The frequency responses of V_{PA} to V_e have been measured using an Agilent network analyzer. The measured resonance frequencies are found to be $f_{s1} = 910\,Hz$, $f_p = 4.75\,kHz$, and $f_{s2} = 30.5\,kHz$, respectively. There is an excellent agreement between the theoretically predicated curves and the experimentally measured frequency responses, as shown in Figs.5-6. It was observed that the measured bandwidths are $46.4\,kHz$, $16.7\,kHz$, and $8.17\,kHz$ for various R_s . Additionally, it was noted by analysis and experiments that there is little difference to output response at frequency near ω_p when the value of R_0 varies from 0 to $100\,k\Omega$.

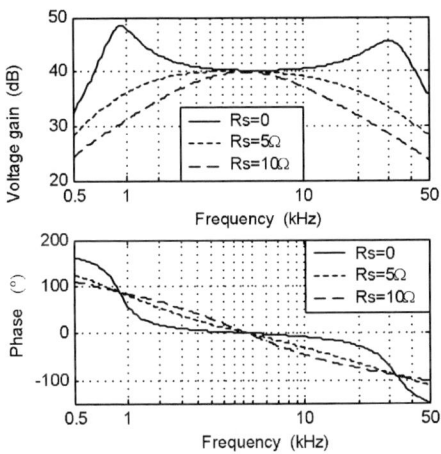

Fig.5. Simulated frequency responses of output stage circuit.

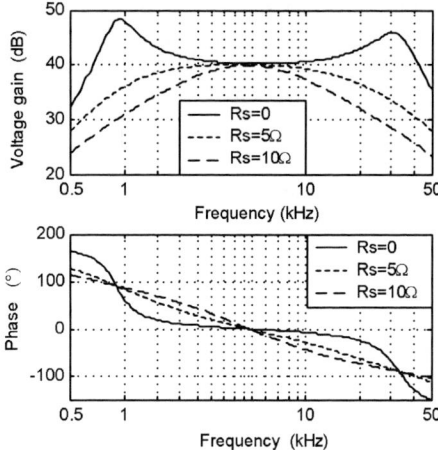

Fig.6. Experimental frequency responses of output stage circuit.

The bandwidth of the output stage should be chosen so that the modulated carrier signal is transferred unaffectedly. In the following analysis, R_s is set to 5Ω by trial and error in order to produce desired dynamic response and noise rejection. It is clear that the output stage offers a bandwidth which is much higher than the closed-loop bandwidth of gyro suspension systems with a typical value of 800Hz or less.

C. Impendence Characteristics

Great power savings can be made if the output stage is carefully engineered to operate at the highest possible impedance level. The load impedance of the power amplifying stage is actually a function of the signal frequency and defined as $Z(j\omega) = V_{pa}(j\omega)/I_1(j\omega)$, where I_1 is the current flowing in the primary windings.

Fig.7 shows the frequency dependence of the impedance for various values of R_0. It is clear from Fig.7 that a smaller value of the resistor R_0 can produce an apparent change in the form of increased load impedance near the carrier frequency. As can be seen from Fig.6, the frequency at which the maximum impedance magnitude occurs can be determined to be 4.77kHz, and the maximum impedance magnitudes are 329.1Ω, 189.3Ω, 142.4Ω for $R_0 = 0, 47k\Omega, 100k\Omega$, respectively.

In order to verify the predication, the impedance characteristics of the output stage circuit have also been measured and are shown in Fig.8. The measured peak

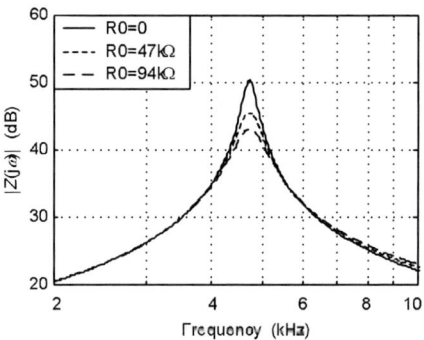

Fig.7. Simulated input impedance of output stage circuit.

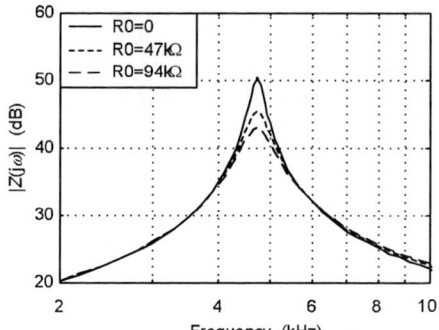

Fig.8. Experimental input impedance of output stage circuit.

impedances for various R_0 are 326.5Ω, 183.8Ω, and 139.4Ω, respectively. It is readily seen from Figs.5-8 that the presented model provides accurate predication on actual frequency characteristics of high-voltage transformer loop. A reasonable compromise between load impedance and noise reduction of position sensors was made with $R_0 = 47k\Omega$ in subsequent test.

IV. EXPERIMENTAL RESULTS AND DISSCUSSION

The transformer, operated at a carrier frequency of 4.75kHz, has been tested successfully in high-voltage ac amplifiers for electrostatic suspension application. For all of the experimental results reported in this section, parameter settings for the transformer are given in Table I and output stage parameters are the same as those in section III.

A. Output Voltage

Maximum output voltage was specified by requiring that an overload force in any direction can be produced for a centered rotor. Considering a possible gap-size variation of 10% due to the thermal environments to which the gyro housing will subjected, the required maximum output voltage was set at 2000 volts. The corresponding load capability with a centered rotor is more than $10g$ ($1g = 9.81 \text{m/s}^2$) in each direction [3].

The measured data indicates that the ac amplifier provides a peak output voltage ranging from 0 to 2020 volts and a mean voltage gain of 201.4 with a nonlinearity of less than 1.5%. In addition, the output voltage can be further increased by selecting a larger turns ratio of the high-voltage transformer for electrostatic suspension with higher electrode voltages.

B. Power Consumption

The consumption of power consists of the quiescent power P_0 used by the integrated power devices and the output circuit power necessary to drive the capacitive load. The measured power loss of the ac amplifier over the entire output range is plotted in Fig.9. The overall input power under steady-state conditions can be expressed approximately as follows:

$$P_a = P_0 + aV_0 \qquad (3)$$

Note that $\alpha = 0.76\,\text{W/kV}$, $P_0 = 0.61\,\text{W}$ and the power loss increases linearly with increasing output voltage until a maximum power of $2.13\,\text{W}$.

On the other hand, the dynamic power consumption of an ac amplifier [4], which incorporates a permalloy core-based transformer operated at a carrier frequency of $1.5\,\text{kHz}$ and the same load conditions, is about three times larger than the measured data in Fig.9. Furthermore, the dynamic power in Fig.9 is only half of those in AM-based dc amplifiers [5]. The reduction of the dissipated power is important not only in terms of energy consumption, but it allows for the reduction of the size of power devices and heat sinks with a reduction of the overall dimension of suspension electronics.

The power dissipation in high-voltage amplifiers can be further reduced by the introduction of a LR series network instead of R_0 to keep enough frequency separation from the position sensing circuit. The experimental results, shown in Fig.9 for comparison, indicate that a reduction of 39.4% on the dynamic power can be achieved by a resistor of $4.7\,\text{k}\Omega$ in series with an inductor of $10\,\text{mH}$.

C. Thermal Stability

It is of interest to test the temperature stability for output transformers made of nanocrystalline alloy core. Fig.10 shows the temperature dependence of frequency responses from V_{in} to V_{e} for the high-voltage ac amplifier shown in Fig.4. The resonant frequency of the transformer loop is found to exhibit much higher stability against temperature change in the temperature range of $20\,^{\circ}\text{C}$ to $120\,^{\circ}\text{C}$. The measured bandwidth are $13.8\,\text{kHz}$, $13.3\,\text{kHz}$, and $12.7\,\text{kHz}$ for operating temperature at $20\,^{\circ}\text{C}$, $70\,^{\circ}\text{C}$, and $120\,^{\circ}\text{C}$, respectively. Simulation results show that the difference between various experimental curves is mainly due to variation of the copper resistance of the transformer windings, which is equivalent to variation of R_1.

It is also observed from the experimental data that the presented ac amplifier yields a more stable performance within the usual operating temperature range of 20-$120\,^{\circ}\text{C}$. Both the power consumption and the output voltages of the ac amplifier is found to exhibit much higher stability against temperature change.

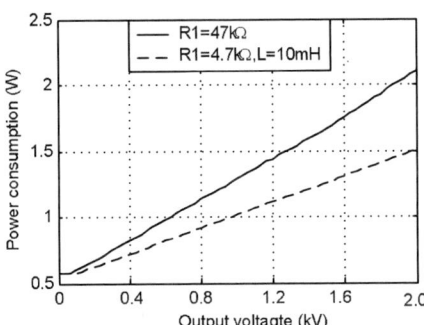

Fig.9. Power consumption vs. output voltage of the amplifier.

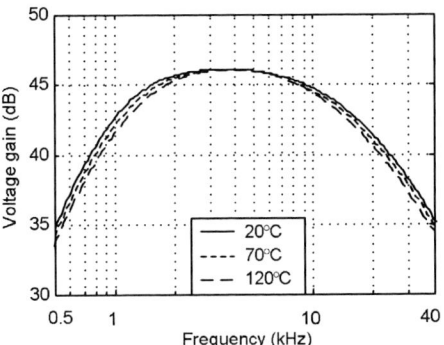

Fig.10. Temperature dependence of frequency response.

V. CONCLUSION

An equivalent model and analysis for high-voltage transformer loop is presented in this paper to well into its frequency characteristics. Analytical and experimental results of these transformers have confirmed the validity of the presented transformer model. The excellent characteristics of the high-voltage ac amplifier operated at a carrier frequency of $4.75\,\text{kHz}$ were proved experimentally. It was found that NSMA core-based transformers is significantly superior in terms of efficiency, operating frequency, and thermal stability. These results confirm the successful use of the NSMA core-based transformer for such electrostatic suspension systems requiring small size, low power consumption, wide temperature range, and fast dynamic response.

REFERENCES

[1] K. W. Exworthy, "Research in electrostatically supported vacuum gyroscopes—Volume III ESVG suspension research," Project Report, Honeywell Inc., Nov. 1968.

[2] C. H. Wu, "DC electrostatic gyro suspension system for the Gravity Probe B experiment," Ph.D. dissertation, Dept. Aeronaut. Astronaut., Stanford Univ., Standford, CV, 1994.

[3] F. T. Han, Z. Y. Gao, D. M. Li, and Y. L. Wang, "Nonlinear compensation of active electrostatic bearings supporting a spherical rotor," *Sens. Actuators A Phys.*, vol.119, pp.177-186, Mar. 2005.

[4] M. Yan, "Experimental research on electrostatic suspension circuits with 15kHz carrier frequency and integrated chips," Master's thesis (in Chinese), Dept. Precision Instrum. Mechanol., Tsinghua Univ., Beijing, China, 1991.

[5] F. T. Han, Z. Y. Gao, and Y. L. Wang, "Performance of a high-voltage DC amplifier for electrostatic levitation applications," *IEEE Trans. Ind. Electron.*, vol.50, pp.1253-1258, Dec. 2003.

[6] A. Makino, T. Hatanai, and Y. Naitoh, "Application of nanocrystalline soft magnetic Fe-M-B (M=Zr, Nb) Alloys NANOPERM," *IEEE Transactions on Magnetics*, vol.33, pp.3793-3798, Sept. 1997.

[7] S. Y. Hui, H. S. Chung, and S. C. Tang, "Coreless printed circuit board (PCB) transformers for power MOSFET/IGBT gate drive circuits," *IEEE Trans. Power Electron.*, vol.14, pp.422-430, May, 1999.

Design and Development of a 50kW Z-Source Inverter for Fuel Cell Vehicles

Miaosen Shen[1], Alan Joseph[1], Yi Huang[1], Fang Z. Peng[1,2], Zhaoming Qian[2]

1. ECE department, Michigan State University, USA, fzpeng@egr.msu.edu

2. College of electrical engineering, Zhejiang University, Hangzhou, China

ABSTRACT: **A detailed design process of the Z-source inverter is presented in this paper. A dc rail clamp circuit is used to reduce the overshoot of the device during turn off. The thermal and 3-D design process is gone through, and the loss calculation of the inverter is discussed, which is different from traditional PWM inverters. A 50 kW inverter for fuel cell vehicle is developed to demonstrate the validity of the design process. Experimental results confirmed the design process and demonstrated the high efficiency characteristic of the Z-source inverter.**

Keywords: Z-source inverter, Fuel cell vehicle, boost

I. INTRODUCTION

Fuel cells as an alternative power source provides much higher fuel efficiency than internal combustion engines (ICEs). The automotive industry started to push fuel cells into practical application and have already developed several types of fuel cell vehicles [1-3]. Unlike batteries, fuel cells have a unique polarization curve as shown in Fig.1. The output voltage of the fuel cell is highly dependent upon the load current and it drops quickly when the load current increases. Used to drive the traction motor and fed by the fuel cell, a power conditioner is one of the key technologies in fuel cell vehicles, and is also a significant portion of the cost. Traditionally, a voltage fed PWM inverter is used. With a PWM inverter alone, the output voltage of the inverter is lower than the fuel cell voltage and the current has to be high to output the required power, especially during high power operation when the fuel cell voltage is low. At the same time, the voltage rating of the inverter has to be high enough to sustain the no load fuel cell voltage. This results in oversized inverter and motor. A dc/dc boost converter is used to boost the fuel cell voltage before feeding to the PWM inverter in some cases [4]. This will reduce the requirement of the PWM inverter as well as the motor. However, the extra dc/dc converter increases the size and weight of the power conditioner and reduces the efficiency as well. A power conditioner configuration using the Z-source inverter is proposed in [5], which provides a single stage reliable solution. The associated relatively low cost and high efficiency make this topology very attractive. The design consideration of the Z-source inverter hasn't been fully discussed yet. This paper presents a detailed design process of the Z-source inverter using a 50kW inverter for fuel cell vehicle as an example, the circuit design, thermal and 3-D design will be presented together with testing results.

This material is based upon work partially supported by DoE FreedomCar Program via ORNL and Partially Supported by NSF under Grant No. 0424039.

II. CONFIGURATION, SPECIFICATIONS, AND BASIC OPERATION PRINCIPLES

By using an X shape LC network, the Z-source inverter as shown in Fig.2 can handle shoot through states when both switches in the same phase leg are turned on, furthermore, the output voltage can be boosted by intentionally inserting some shoot through states in the PWM [6,7]. Because of the symmetry of the network, the current through the two inductors and the voltage across the two capacitors are identical. From previous literatures, there are two basic operation modes as shown in Fig.3 and several relationships of the Z-source inverter as listed below:

$$V_C = \frac{1-D_0}{1-2D_0} V_{in}, \tag{1}$$

$$V_{PN} = \frac{1}{1-2D_0} V_{in}, \tag{2}$$

$$\hat{V}_o = \frac{M}{2} V_{PN}, \tag{3}$$

where V_c is the capacitor voltage, V_{PN} is the peak voltage across the PN, $\hat{V}_o$ is the peak output phase voltage, V_{in} is the input voltage, i.e. the fuel cell voltage, D_0 is the shoot through duty ratio, M is the modulation index.

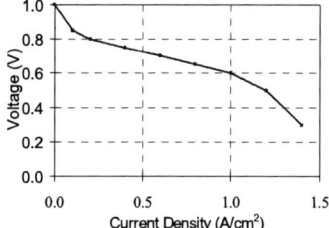

Fig.1. Fuel cell polarization curve

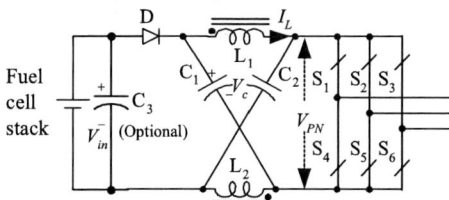

Fig. 2. Schematic of the Z-source inverter system.

Coolant inlet temperature of 75°C with a flow rate of 1.5 gallon/minute is the temperature requirement. Fig.1 shows the fuel cell polarization curve for the fuel cell the system is designed for, the open load voltage is 420 V, the maximum output current is 200 A at peak power of 50 kW and 250 V. The target maximum output power of the inverter is 50 kW. The components to be designed or selected include the two inductors, L_1 & L_2; the three capacitors, C_1, C_2, and C_3; the inverter switches; and the diode, D.

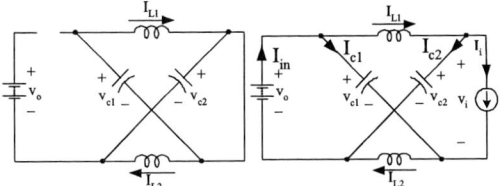

(a) Shoot through mode   (b) Non shoot through mode
Fig.3. Operation modes of the Z-source inverter

III. CIRCUIT PARAMETER DESIGN

Under traditional PWM operation, when there is no shoot-through, the capacitor voltage is always equal to the input voltage; therefore, there is no voltage across the inductor and only a pure dc current going through the inductors. The purpose of the inductors is to limit the current ripple through the devices during boost mode with shoot-through. During shoot-though as shown in Fig.3 (a), the inductor current increases linearly, and the voltage across the inductor is equal to the voltage across the capacitor; during non-shoot-through modes (six active modes and the two traditional zero modes) as shown in Fig.3 (b), the inductor current decreases linearly and the voltage across the inductor is the difference between the input voltage and the capacitor voltage. The average current through the inductor equals to that through the diode, which is

$$\bar{I}_L = \frac{P}{V_{in}}, \qquad (4)$$

where P is the total power and V_{in} is the input voltage.

The average inductor current at 50-kW and 250-V (fuel cell voltage at output power of 50-kW) input is

$$I_L = \frac{50000}{250} = 200 \text{ A}. \qquad (5)$$

Before designing the parameters, a control method has to be selected. As discussed in literature [7], the maximum boost method gives the highest boost ratio and lowest voltage stress, but due to the output frequency associated current ripple through the inductors, it is only suitable for high frequency operation or constant speed operation. For fuel cell vehicle, which is a variable speed drive system, maximum constant boost control [7] is chosen. The maximum current through the inductor occurs when the maximum shoot-through happens, which causes maximum ripple current. In our design, 30% (60% peak to peak) current ripple through the inductors during maximum power operation was chosen. Therefore, the allowed ripple current is 120-A, and the maximum current through the inductor is 260-A. Considering the auxiliary power consumption by automotive accessory loads, the fuel cell output voltage will always be less than the open circuit voltage, 420-V, therefore, a 600-V device was selected and 400-V is designed to be the maximum operating voltage across the switches. The maximum shoot-through duty cycle can be calculated by selecting the PN voltage to be 400-V:

$$\frac{1}{1-2D_0} = \frac{400}{250}. \qquad (6)$$
$$D_0 = 0.1875$$

For a switching frequency of 10 kHz, the shoot-through time per cycle is 18.75 μs. The capacitor voltage during that condition is

$$V_c = 250 * \frac{1-D_0}{1-2D_0} = 325V. \qquad (7)$$

To keep the current ripple less than 120-A, the inductance must be no less than

$$\frac{18.75 * 325}{120} = 50.8 \ \mu H. \qquad (8)$$

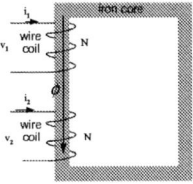

Fig. 4. Coupled inductors.

To minimize the size and weight of the inductors, the two inductors are built together on one core, as shown in Fig. 4. For a single coil on one core, the flux through the core is

$$\phi = PNi, \qquad (9)$$

where P is a constant related to the core material and dimension, N is the number of turns of the coil, and i is the current through the coil. The inductance of the coil is

$$L = \frac{N\phi}{i} = PN^2. \qquad (10)$$

For the two inductors in the Z-source inverter, because of the symmetry of the circuit, the current through the inductors is always exactly the same. For two coils on one core with exactly the same current, i, the flux through the core is

$$\phi = 2PNi. \qquad (11)$$

The resulted inductance of each coil when supplying exactly the same current to the two coils is

$$L = \frac{N\phi}{i} = 2PN^2. \qquad (12)$$

The inductance of each coil is doubled. Therefore, equivalently, we only need to build two coils with 25.4 μH/260-A peak each on one core. A Metglas AMCC_250 core was selected to reduce the loss.

The purpose of the capacitor in the Z-source network is to absorb the current ripple and maintain a fairly constant voltage so as to keep the output voltage sinusoidal. During shoot-through, the capacitor charges the inductors, and the current through the capacitor equals to the current through the inductor. Therefore, the voltage ripple across the capacitor can be roughly calculated by

$$\Delta V_C = \frac{I_{av} T_0}{C}, \qquad (13)$$

where I_{av} is the average current through the inductor, T_0 is the shoot-through period per switching cycle, and C is the capacitance of the capacitor. To limit the capacitor voltage ripple to 3% at peak power, the required capacitance is

$$C = \frac{200*18.75\mu}{325*3\%} = 384.6\mu F . \qquad (14)$$

Another function of the capacitor is to absorb the ripple current. The current through the capacitor can be calculated for a given operation condition. For induction machines, the power factor at high power is usually fairly high, so 0.9 was used for the calculation at 50kW, which gives 111 A rms current at peak power [8]. Electronic Concepts UL31 500-V/200-uF film capacitors were selected, with two connected in parallel to form one conceptual capacitor in the Z-source inverter. The fuel cell is a double-layer capacitor by itself, so theoretically no capacitor is needed in parallel with it. However, to minimize the high-frequency current path, one UL31 was used in parallel with the fuel cell (C_3).

The semiconductor devices are selected based on the current through them and the maximum voltage across them. The maximum voltage across the switches and the diode are 400-V. The peak current through the switches occurred at the peak power of 50-kW. From [8], the maximum current through the switches can be calculated based on the following equation,

$$I_{s\,max} = \frac{1}{2}I_{load\,max} + \frac{2}{3}I_{L\,max} . \qquad (15)$$

Using Eq. (15), the maximum current through the switches is 272-A. The average current through the diode, D equals the average current through the inductor, which is 200-A. The peak current through the diode is twice the inductor current during traditional zero states, therefore, the peak current through the diode is 520 A. The following devices were selected, considering the high temperature requirement: a 600-V/600-A six-pack IPM PM600CLA060 for the inverter bridge, and two 600-V/600-A diode QRS0660T30s in parallel for the input diode.

IV. DC RAIL CLAMP CIRCUIT DESIGN

Because of shoot through state, there can not be any dc capacitors right across the PN of the inverter bridge, otherwise huge loss will occur during shoot through. The high frequency loop of the Z-source inverter is highlighted with darkened lines in Fig.5. As can be seen from Fig.5, the loop length is much longer than a traditional voltage source inverter where a decoupling capacitor is placed right across the PN of the inverter bridge. In order to reduce the overshoot of the devices, a dc rail clamp circuit is developed as shown in Fig.6.

Two capacitors, C_4 and C_5, and one diode, D_1 connected in series is connected right across PN of the inverter bridge. Another two diodes, D_2 and D_3, in series are connected to the main power circuit from the clamp circuit forming a discharge loop for C_4. From Fig.7 (a), when the current to the inverter, I_i, has a step change, the dc rail clamping circuit provides an extra absorbing path for the extra current maintained by the parasitic inductance of the main bus-bar, thus helping to reduce the overshoot voltage across the device. Fig.7 (b) shows the two paths for discharging the two capacitors,

C_4 and C_5, in the clamping circuit. From Fig.7 (b), C_4 can be discharged through C_2, C_3, D_2, and D_3, C_5 can be discharged to C_1. The series connection of D_1, D_2, and D_3 are in parallel with the main diode D. The forward voltage drop of the main diode D is relatively higher than the low current ones. The reason to put D_2 and D_3 in series is to make sure that majority of the steady state input current goes through the main diode D instead of D_1, D_2, and D_3. The capacitors used in the clamp circuit are 1uF ceramic capacitors, and the diodes are TO220 package diodes.

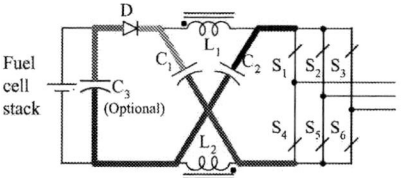

Fig.5. High frequency loop of the Z-source inverter

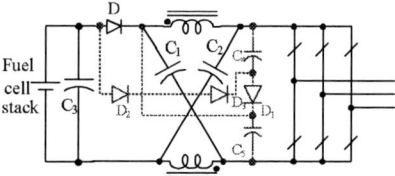

Fig.6. DC rail clamp circuit

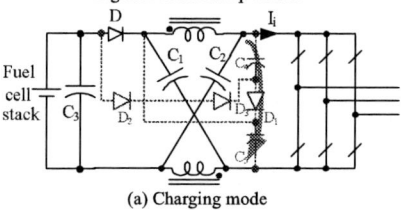

(a) Charging mode

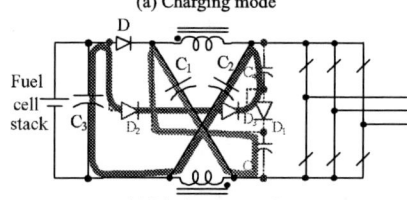

(b) Discharging mode

Fig.7. Operation of the dc rail clamp circuit

V. THERMAL AND 3-D DESIGN

Thermal design is very crucial for inverter design to choose a proper heat sink. With the extra shoot through states, the switching loss and the conduction loss of the switches are different from traditional PWM inverters, also, different PWM scheme can result in different losses. For this inverter, modified space vector PWM control is adopted, with which the switching states in one cycle is shown in Fig.8 (this method has the same effect as the maximum constant boost control in [7]). Basically, one of the zero states in traditional SVPWM is replaced by shoot through state and the duration of the shoot through state is determined by the requirement of voltage boost and it is always a constant for steady state operation. During the shoot through state, all six switches are turned on. The loss calculation method for traditional PWM inverters can be found in [9] and the expression of switching loss of each IGBT for a 3-phase inverter is

1078

cited below:

$$P_{sw} = (E_{swon} + E_{swoff})f_{sw}\frac{1}{2\pi}\int_0^\pi \sin x dx \,, \qquad (16)$$

where E_{swon} and E_{swoff} are the turn on and turn off energy loss of the IGBT at peak current, f_{sw} is the switching frequency. For the Z-source inverter, assume the load voltage and phase a current is shown in Fig.9, where the load current is lagging by α. In the shaded area, where the output voltage of phase a is the maximum among the three phases, all switching actions in phase a are turned into shoot through switching. Therefore the modified switching loss calculation has to be applied. There are two parts of switching losses: traditional switching (switching actions between traditional states) and shoot through switching (switching states between shoot through state and traditional states). The switching loss of traditional switching can be calculated by (17).

$$P_{swt} = (E_{swon} + E_{swoff})f_{sw}\frac{1}{2\pi}(\int_0^\pi \sin x dx - \frac{1}{2}\int_{\frac{\pi}{6}-\alpha}^{\frac{5\pi}{6}-\alpha}|\sin x|dx) \,, \qquad (17)$$

where E_{swon} and E_{swoff} are the turn on and turn off energy loss of the IGBT at peak current. During shoot through state, the current from the dc side is $2I_L$[8], where I_L is the inductor current. Assuming that the current is evenly distributed in three phase legs, the average switching current of shoot through state is $2I_L/3$. In each cycle, there are 3 shoot through switching, thus the shoot through switching loss of each IGBT is:

$$P_{sws} = \frac{1}{2}f_{sw}(E_{swons} + E_{swoffs}) \,, \qquad (18)$$

where E_{swons} and E_{swoffs} are the turn on and turn off energy loss corresponding to switching current of $\frac{2I_L}{3}$.

The reverse recovery loss of the free wheeling diodes is reduced because some of the turn off states of the diodes turn into shoot through turn off, so the reduction can be calculated in a way similar as in (17).

Fig.8. Switching states sequence in one cycle

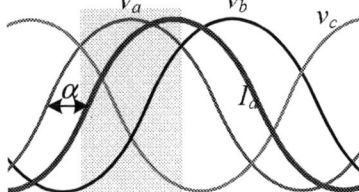

Fig.9. Load voltage and current

The conduction loss of the IGBTs and the diodes are also changed because of the shoot through states. Assuming the shoot through duty ratio is D_0 and the corresponding conduction losses of IGBTs and diodes for traditional PWM inverter under the same load current are P_{onIGBT} and $P_{ondiode}$ respectively, the conduction losses of the Z-source inverter during traditional states are:

$$P_{onIGBTtraditional} = (1 - D_0)P_{onIGBT} \qquad (19)$$

$$P_{ondiode\,traditional} = (1 - D_0)P_{ondiode} \qquad (20)$$

Assuming that the inductor current is high enough so that all IGBTs are on during shoot through state, the average current through the IGBTs during shoot through is $\frac{2I_L}{3}$, the conduction loss of IGBTs during shoot through is

$$P_{onIGBTshoothrough} = D_0 V_{CE(sat)} * \frac{2}{3}I_L \,, \qquad (21)$$

where $V_{CE(sat)}$ is the saturation voltage of the IGBT. From all above discussion, total loss of the inverter bridge can be calculated. The calculated loss of the inverter at 50kW is 1.35kW, the temperature rise of the IPM junction to the heat sink can be calculated from the thermal resistance of the IPM. To limit the junction temperature of the IPM below 125°C at 75°C inlet coolant temperature, a heat sink with thermal resistance of 0.01°C/W is chosen. The 3-D design of the inverter is shown in Fig.10. The final dimension of the inverter is 11"*12"*5.5". The final assembly is shown in Fig.11. As can be seen from the assembly, there is still a great potential to reduce the size of the inverter by using customized capacitors.

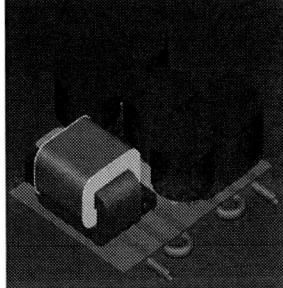

Fig.10. 3-D design

Fig.11. Final assembly

VI. TESTING RESULTS

Fig.12 (a) and (b) show the experimental results at 50 kW operation with 250 V input voltage supplying an induction motor. As can be seen from the experimental results, the inverter operates in boost mode with shoot-through. The PN voltage across the device is boosted to around 380-V, thus increasing the output voltage. The modulation index of the PWM for 50-kW conditions is 0.957. The output voltage of the inverter is 218 V line to line, while the obtainable output voltage for a traditional inverter at 250-V input is 153-V with a modulation index of 1. It was successfully demonstrated that the Z-source inverter can greatly boost the output voltage as desired. Also, the inductor current and capacitor voltage are as predicted with average of 200-A and 30% current ripple. The motor current is pure sinusoidal, which confirms that the Z-source inverter will produce very low harmonics. The voltage across the switch at 250-A load current is shown in Fig.13. As can be seen from Fig.13, the overshoot of the switch is clamped to around 15% with the dc rail clamp circuit.

The inverter efficiency is measured for different load condition as shown in Fig.14. The operating point with a black cross is under boost condition, the other points are under tradition PWM operation without any shoot through. As can be seen from Fig.14, the inverter achieves high efficiency. Also, the loss at 50kW is very close to the loss we calculated in the thermal design, which verifies the loss calculation method.

VII. CONCLUSION

A detailed design process of the Z-source inverter for fuel cell vehicles is provided. A dc rail clamp circuit is presented which helps reduce the voltage overshoot. Thermal design and 3-D design method is reviewed. Experimental results are presented to confirm the validity of the design process. The inverter presented in this paper doesn't include a battery, which is necessary for fuel cell vehicles. The battery can be connected directly or through a dc/dc converter in parallel with one of the capacitors in the Z-source network. However, the design process of the inverter is still the same.

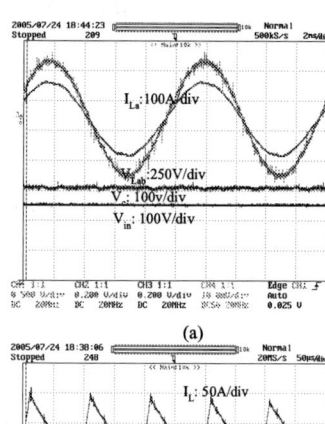

(a)

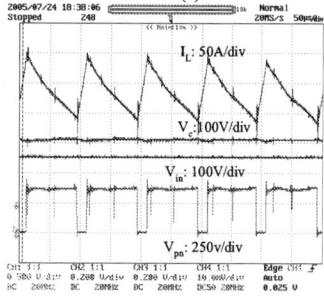

(b)

I_{La} : load line current; V_{Lab}: output line to line voltage after the monitoring LC filter; V_c: capacitor voltage; V_{in}: input voltage; I_L: inductor current; V_c: capcitor voltage; V_{in}: input voltage; V_{pn}: PN voltage

Fig.12 Experimental result at 50kW

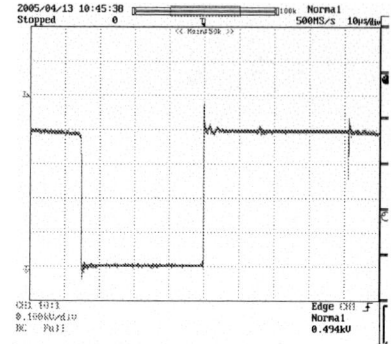

Fig.13. Switching voltage at 250-A load current

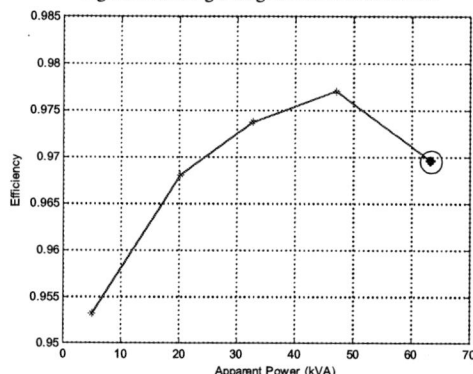

Fig. 14. Measured efficiencies: The operating point in the circle is under boost mode, other operating points are under traditional PWM operation

REFERENCES:

[1] Tadaichi Matsumoto; Nobuo Watanabe; Hiroshi Sugiura; Tetsuhiro Ishikawa; "Development of Fuel-Cell Hybrid Vehicle," Fuel cell power for transportation 2002 conference, SAE 2002 World congress, March 2000, Ref: 2002-01-0096

[2] B. D'Souza, H. Rawlins, J. Machuca, C. Larson, M. Shuck, B. Shaffer, T. Maxwell, M. Parten, D. Vines, and J. Jones, "Texas Tech University Developes Fuel Cell Powered Hybrid Electric Vehicle for FutureCar Challenge 1998," SAE 1999, Ref: 1999-01-0612.

[3] J. Adams, W. Yang, K. Oglesby, and K. Osborne, "The Development of Ford's P2000 Fuel Cell Vehicle," SAE 2000, Ref: 2000-01-1061.

[4] K. Rajashekara, "Power Conversion and Control Strategies for Fuel Cell Vehicles," in Proc. Industrial Electronics Society, 2003, vol. 3, pp. 2865- 2870.

[5] Kent Holland, Miaosen Shen, Fang Z. Peng, "Z-source inverter control for traction drive of fuel cell - battery hybrid vehicles" record of IEEE Industry Applications Conference, 2005 Volume 3, 2-6 Oct., 2005 pp.1651 – 1656

[6] F. Z. Peng, "Z-Source Inverter," IEEE Transactions on Industry Applications, vol. 39, No. 2, pp. 504-510, March/April 2003.

[7] Miaosen Shen, Jin Wang, Alan Joseph, Fang Z. Peng, Leon M. Tolbert, and Donald J. Adams, "Maximum constant boost control of the Z source inverter." in Proc. IEEE IAS'04, 2004, p.142.

[8] Miaosen Shen, Alan Joseph, Jin Wang ,Fang Z. Peng, and Donald J. Adams, "Comparison of Traditional Inverters and Z-Source Inverter for Fuel Cell Vehicles," in Proc. IEEE Power Electronics in Transportation, October 2004, Novi, MI, p. 125.

[9] "General Considerations: IGBT & IPM modules", Powerex Application Notes, A10-A27.

2006 5th International Power Electronics and Motion Control Conference

Identification and improvement of stray coupling effect in an L-C-L common mode EMI filter

Junping He[*], Wei Chen[**], Jianguo Jiang[***]

[*] Shenzhen graduate school, Harbin Institute of Technology /Shenzhen, P.R. China
[**] Delta Electronics (Shanghai) CO.LTD /Shanghai, P.R. China
[***] Tsinghua University /Beijing, P.R. China
Email: hejunping@tsinghua.org.cn

Abstract—**the high frequency attenuation performance of an EMI filter is often deteriorated greatly by the stray electromagnetic field existing in interior or exterior of a filter. In this paper, the stray electromagnetic coupling factors in an EMI common mode filter are investigated by different experiments after theory analysis. A novel metal insulated grids method is proposed to separate electric and magnetic field effect between two CM chokes. The parasitic parameter influence of passive components on insertion loss is also analyzed and compared with stray coupling effect. In the end, a novel equivalent paralleled capacitance decrease design is introduced and validated by experiment. Some coupling restrain methods are introduced in this paper too.**
Keywords—*stray electromagnetic coupling; parasitic parameter ; EMI filter; insertion loss; equivalent parallel capacitance*

I. INTRODUCTION

The high dv/dt and di/dt in semiconductor switch device in power electronic converter easily tend to produce strong electromagnetic interference emission. The power line EMI filter is a basic and effective control method to decrease conducted emission and radiation emission [1, 2]. However actual high frequency performance of an EMI filter is often far less than expected performance, which bases on ideal design parameter. Besides some familiar factors, such as equivalent EMI source impedance of power converter, parasitic parameters of an inductor and a capacitor [3,4,5], tiny stray electromagnetic coupling existing at inside and outside of filter influence high frequency insertion loss too [6,7,8,9]. Especially for on-board EMI filter, which is made up of discrete components and there is no metal shield outside, its performance can be deteriorated heavily by stray coupling effect. Paper [6] analyzed magnetic coupling mechanism between an EMI filter and PFC main circuit and studied coupling influence on differential mode conducted EMI emission. But the paper did not study stray electromagnetic coupling effect at interior of a filter. Paper [7],[8] investigated stray magnetic coupling effect and its influence existing in internal of an differential mode EMI filter in detail. However above two researches analyzed magnetic coupling and its influence on DM emission only, careful study on electric coupling and its effect on emission is still a virgin in EMI filter by now. In fact, DM conducted emission of a power converter generally is more easily

controlled than CM conducted emission and CM emission debug work occupies more time and expend. So it is necessary and useful to investigate stray electric coupling at inside of an EMI filter and its effect on CM high frequency attenuation performance.

Aiming at above question, this paper carefully analyzes and measures stray electromagnetic couplings and their effect using an L-C-L type CM EMI filter as an example. The possible stray electromagnetic couplings among components and interconnect and theirs influence are analyzed firstly. Then their electromagnetic nature and effect of stray coupling are identified and confirmed by experiments. The influence of high frequency parasitic parameter of passive component on filter performance is compared with that of stray coupling. In the end, an equivalent paralleled capacitance decrease design of CM choke is introduced and validated by experiment.

II. CM FILTER & ITS STRAY COUPLINGS ANALYSIS

A. L-C-L CM EMI filter and its test layout

Figure1 shows the power line EMI filter studied. This filter consists of two symmetrical winding separated toroid CM chokes, labeled as CM_1 and CM_2, two C_y capacitors and two C_x capacitors. Their design values are listed as shown. The whole filter is without metal shield shell. According to CM insertion loss test rules, the two C_x capacitors are short-circuited during test, they can be ignored in following study. Then the EMI filter is a typical L- C-L structure if seen according to CM paths. The filter test layout and principle of CM insertion loss performance are shown as Figure 2. All passive components and interconnection conductors are arranged along a straight line. The GND is a low impedance copper

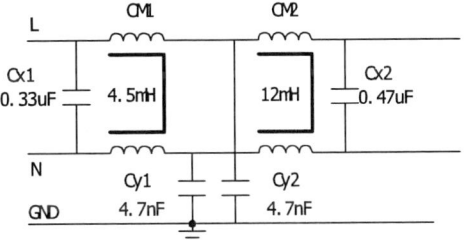

Figure1. EMI filter circuit

1-4244-0448-7/06/$25.00 ©2006 IEEE   1081

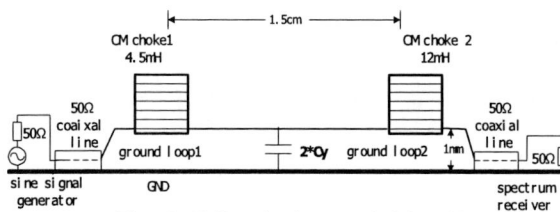

Figure2. CM insertion loss test principle and layout

board. R&S ESVN30 instrument is used to test filter attenuation characteristic.

Figure 3 shows both measured insertion loss curve and calculated insertion loss curve that is based on ideal design capacitance and inductance value. It is clear to see that there is a great difference above 1 MHz between these two curves. It can be reasonably judged that this may be caused by stray electromagnetic coupling effects or parasitic parameters of passive components. Then the stray coupling in filter is investigated firstly in the paper.

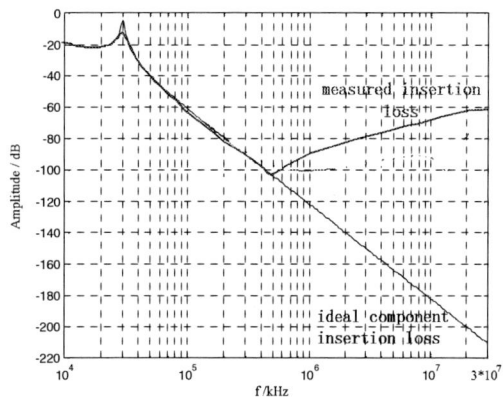

Figure3. Insertion loss curves measured and calculated ideally

B. Stray electromagnetic couplings inside of CM EMI filter

Strictly speaking, there must be some tiny stray electric and magnetic couplings among CM chokes, Cy capacitors and interconnection conductors in the filter. These tiny stray couplings can produce filter performance degraded. In conducted EMI frequency band (150kHz to 30MHz), the size of EMI filter is far less than $1/2\pi$ of the minimum electromagnetic wavelength, then the stray electromagnetic couplings can be presented using lumped mutual capacitance and mutual inductance. However these coupling capacitors and inductors are so small that their values are difficult to calculated and measured for an actual discrete EMI filter. Further more, the number of stray couplings will increase directly with ratio of $2C_N^2$. N is the number of passive components and current loops. These makes it is harder to build an accurate coupling filter model. So this paper adopts both theory analysis and experiment methods to identify the main stray coupling factors in the L-C-L CM filter.

The CM chokes, C_y capacitor line and current loops are clearly shown in Figure4. Those possible stray electric couplings and magnetic couplings in filter are also labeled

using dot lines. There are five possible couplings and their detail information is listed in below.

M_1---stray magnetic coupling between CM_1 and CM_2

C_1---stray electric coupling between CM_1 and CM_2

M_2---stray magnetic coupling between CM_1 and ground loop 2

M_3---stray magnetic coupling between CM_2 and ground loop 1

M_4---stray magnetic coupling between ground loop1 and ground loop2

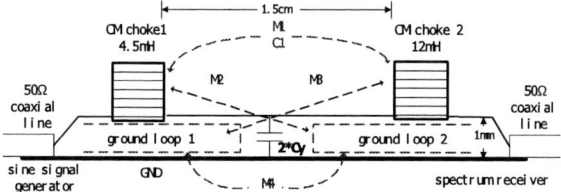

Figure4. Possible stray couplings in CM filter

It should be pointed out that mutual inductance M_4 between ground loop1 and loop2 is not exact according to strict mutual inductance definition for there is a common branch C_y between them. This situation can be solved using partial inductance concept and method. Because C_y branch is vertical to the interconnect conductor between chokes and is short, the mutual inductance between branches is zero. Then current branches can be decoupled. So the difference of mutual inductance between loops and branches is ignored in this paper.

C. Stray electromagnetic coupling influence analysis

Above five possible stray couplings are many and they can be simplified after careful analysis. The CM choke is separated windings toroid structure and the two windings are symmetrical in position and turn number. The stray fluxes around CM choke are also symmetrical when CM current flows. The dot lines in figure 5 show these stray fluxes. It can be seen clearly that the stray fluxes of choke are symmetrical to ground loops. The net fluxes pass through ground loops are zero. So mutual inductance M_2 and M_3 are zero and can be ignored.

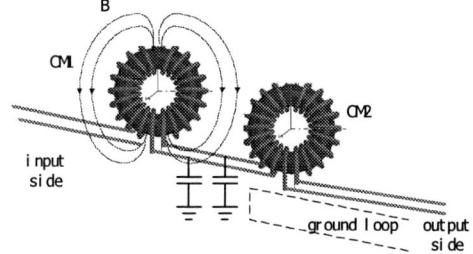

Figure5. CM chokes stray fluxes and ground loop layout

After above analysis, only stray coupling C_1, M_1 and M_4 are left. Their influences on insertion loss are difficult to predicted using calculation and have to be investigated using experiments.

1082

III. STRAY ELECTROMAGNETIC COUPLINGS EXPERIMENT IDENTIFICATION IN L-C-L CM FILTER

In order to find out coupling influence and effective countermeasure accurately, it is necessary to identification the roles of stray couplings C_1, M_1 and M_4 on filter insertion loss characteristic.

A. Electric coupling identification between CM chokes

The stray electromagnetic coupling between CM chokes can be observed visibly by changing their distance. But this method can't distinguish the respective role of C_1 and M_1 clearly for both are decreased in the same time. In conducted frequency band, usual depth ($\geq$0.25mm) metal board such as copper and iron is high shield effectiveness on both electric and magnetic field ($\geq$20dB), so this method can't distinguish electric coupling and magnetic coupling too. In order to identify the main stray coupling and its influence accurately, a novel insulated metal grid method is proposed and validated. During experiment procedure, the ground loop height is decreased to 1mm above GND to avoid the magnetic coupling influence of M4.

The structure of insulated metal grid is shown in figure 6. It is made up of a two parts. One is a horizontal grid array. The other is a vertical grid array. All of the metal conductors are thin insulated copper wires and they are welded at a single point. The size of each grid mesh is 2mm*2mm and the wire diameter is 0.1mm. This insulated metal grid has a special shielding effectiveness, which blocks electric field effectively and passes flux field smoothly. Because insulated grid mesh can't produce eddy by induction. There is no opposite eddy flux to cancel

external flux, so the magnetic field shielding effectiveness of the grid is close to zero. When the grid single point is grounded, these grid meshes shield electric field effectively. Figure 7 shows the grid shielding effectiveness curves on electric field and magnetic field measured. It is clearly shown that the electric field shield effectiveness is far higher than that on magnetic field and the maxim magnetic shield effectiveness is only 3dB in whole conducted frequency band. So electric or magnetic field can be distinguished by insulated metal grid method. As a comparison, 0.25mm copper board shield effectiveness on electric field and magnetic field are measured too. Test results shows that both electric and magnetic shield effectiveness are higher than 20dB in conducted frequency band and can't identify electric field and magnetic field.

A large area insulated metal grid is inserted between CM_1 and CM_2 and grounded using shortest wide conductor as shown as dot line in figure 8. Figure 9 shows CM filter insertion loss measured before and after inserting insulated grid. These curves show that the high frequency attenuation performance is improved greatly after adding electric field shield. So C_1 play a significant influence. In order to test magnetic coupling M_1 influence, a same area copper board with 0.25 mm depth is pasted behind insulated grid tightly. This means an electromagnetic shield is inserted. Figure 12 shows the insertion loss measured. It is clear that only a small change happen compared with electric shield only. So M_1 influence is little. These experiments proves that the stray electric coupling C_1 between CM chokes is a key coupling factor in this L-C-L CM filter.

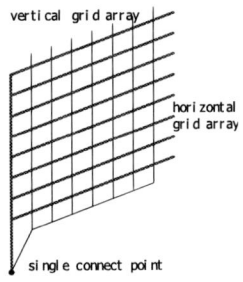

Figure6. Insulated metal grid structure

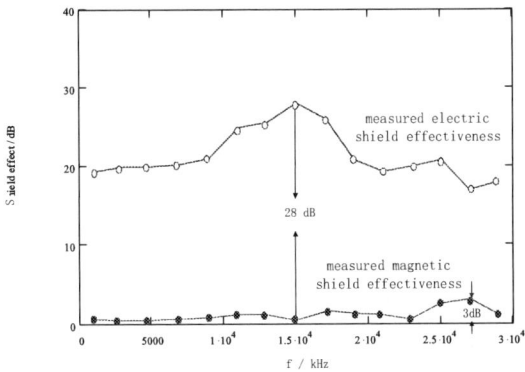

Figure7. Grid's electric and magnetic shielding effectiveness measured

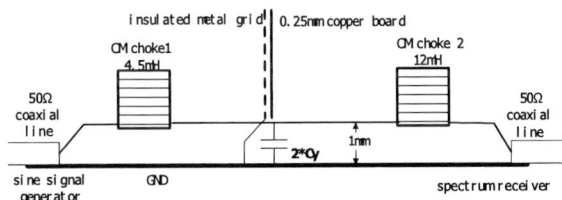

Figure8. CM insertion loss test layout with shield

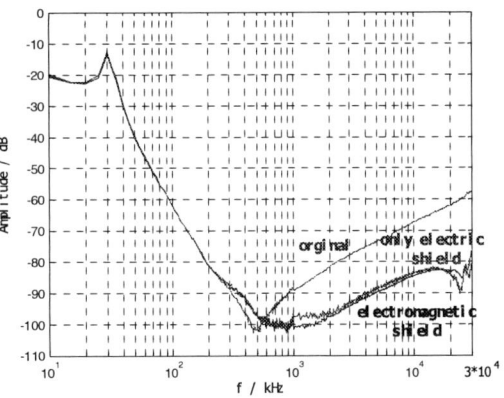

Figure9. CM insertion loss curves measured

B. Magnetic coupling identification between ground loops

In order to identify magnetic coupling effect between ground loops accurately, the stray electromagnetic coupling influence of CM chokes should be decreased as low as possible. So above subsection electromagnetic shield is adopted in this experiment, the M_4 is changed enough by increasing ground loop height from 1mm to 8mm as shown in Figure 10. Figure 11 shows insertion loss curves measured at 1mm and 8mm height respectively. It is clearly shown that the difference is only about 1-2dB. So magnetic coupling M_4 isn't the main influence factor in this filter.

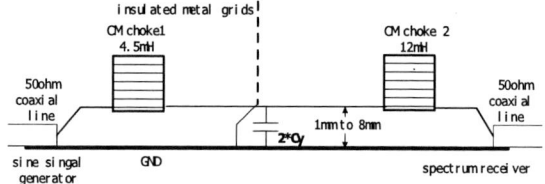

Figure10. Magnetic coupling effect test layout

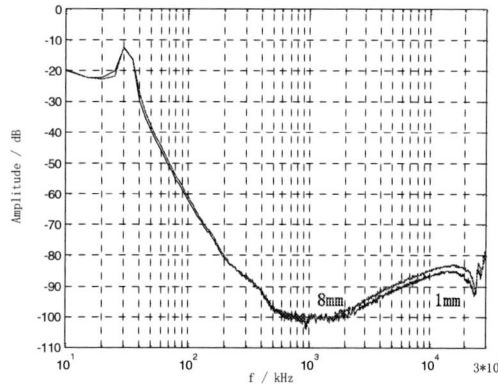

Figure11. CM insertion loss curves measured

According to above test results and analyses, stray electromagnetic coupling influences filter attenuation performance indeed. The high frequency insertion loss of filter can be improved by control stray coupling in filter. Some methods, such as enough distance between components, small current loop area, adapt shield and better filter layout, are effective if used correctly.

IV. PASSIVE COMPONENT PARASITIC PARAMETER INFLUENCE AND IMPROVEMENT

A. Components parasitic parameter influence

Although the CM attenuation performance improves significant after decreasing stray coupling as low as possible, the measured high frequency insertion loss is still far less than ideal filter which is made up of ideal choke and capacitor. So the influence of passive component parasitic parameter is further investigated in this section.

High frequency parasitic parameter of CM chokes, interconnection conductors and C_y capacitor can be measured by using impedance analyzer HP4294. Figure 12 shows the high frequency CM filter model including all parasitic parameters and insertion loss test circuit. Those

parasitic parameter values of all passive components are listed too. The filter insertion loss including parasitic parameter can be calculated according to this model.

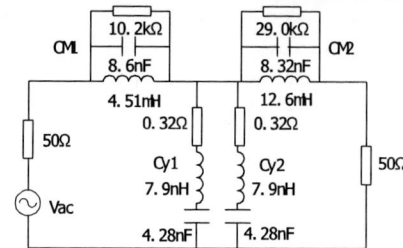

Figure12. Insertion loss test circuit including parasitic parameter effect

Figure 13 shows some insertion loss curves under different conditions, such as original layout, using electromagnetic shield method only, considering parasitic parameter effect only and ideal components. It is quite valuable to pay attention to two facts.

(1) Filter attenuation performance improve greatly if stray coupling effect is controlled as low as possible in a filter. Although there still is 2-10dB difference to the insertion loss predicted based on parasitic parameter filter model, their tendency and value are similar in general. This figure also hints that there is a limit for decreasing stray coupling technique, which is decided by component parasitic parameters.

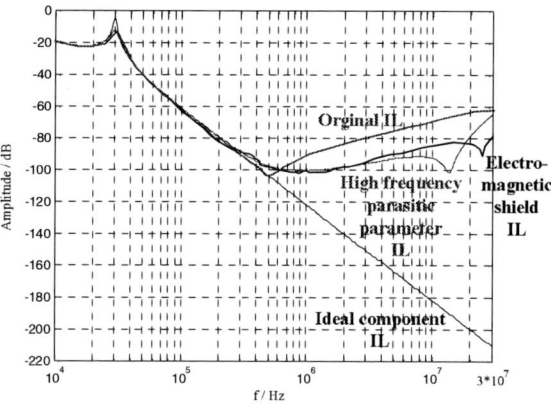

Figure13. L-C-L CM filter insertion loss under different conditions

(2) After decreasing stray coupling effect in a filter, component parasitic parameters become the main factor that influences high frequency performance.

B. CM choke equivalent parallel capacitance decrease

It is well known that equivalent parallel capacitance (EPC) is the main parasitic parameter of an inductor and equivalent series inductance (ESL) is the main parasitic parameter of a capacitor. Because an inductor is usually designed and made by power converter manufactory self, it is desirable to design and adopt a small EPC inductor in EMI filter.

Figure 14 shows a general CM choke structure and its EPC components. The choke consists of a toroid core and two winding. It is clear that CM choke EPC is made up of two parts. One is caused by the capacitance among turns, which is labeled as C_{tt}. The other part is caused by the capacitance among turn and toroid core, which is labeled as C_{tc}. There are some well known techniques to decrease EPC, such as better winding arrangement. A novel EPC decrease technique is proposed in the bellow. If the choke core is grounded, then C_{tc} will not only change into C_y and but also EPC also is decreased in the same time. So this will helpful for filter performance improvement.

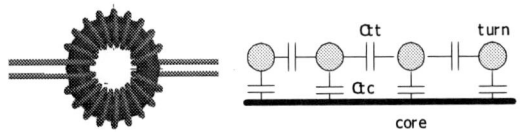

a. CM choke structure b. parasitic capacitance

Figure14. CM choke structure and its EPC components

Figure 15 shows the novel EPC decrease structure proposed. In order to ground a non-metal magnetic core, a thin copper strip is pasted tightly in inside surface and outside surface of toroid. It should be paid attention that the copper strip is open at ends to avoid shorten current during above procedure. Then it is easy to ground the core by a copper strip lead. Figure 16 shows the factual effect of the novel structure small EPC CM choke. The high frequency insertion loss increase about 5dB from 5MHz to 25MHz when choke core is grounded. This measured result validates the novel design although insertion loss is still less than expected. This error phenomenon needs further research.

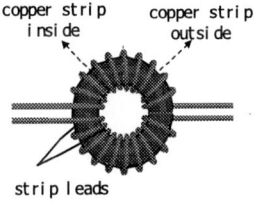

Figure15. Novel choke structure with EPC decrease

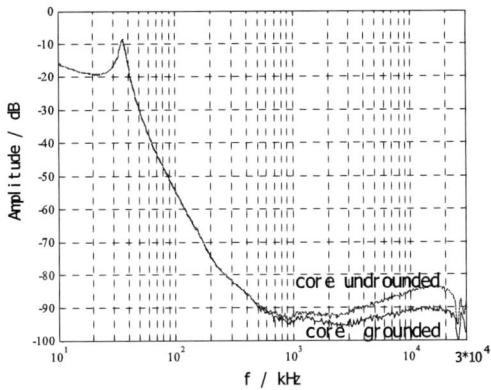

Figure16. Novel structure choke effectiveness

V. CONCLUSION

(1) The internal stray couplings greatly deteriorate high frequency attenuation performance in an L-C-L CM EMI filter. The high frequency insertion loss can be improved by decreasing stray coupling carefully.

(2) A novel insulated metal grid is proposed and validated，which permits flux pass smoothly and blocks electric field effectively. The stray electric coupling between CM chokes is proved the key coupling factor in L-C-L CM filter by using insulated metal grid method.

(3) The filter high frequency performance can be improved further more by decreasing component parasitic parameter after decreasing stray couplings in filter interior.

(4) The equivalent parallel capacitance of CM choke can be decreased by grounding core and it is good for the performance improvement of a filter.

ACKNOWLSDGEMENT

The authors would like to thank Delta Electric Company for financial support and providing experiment instruments.

REFERENCES

[1]J.D. Van Wyk, Fred .C Lee. *Power Electronics Technology—Status and Future.* PESC Record-IEEE Annual Power Electronics Specialists Conference. 1999. 3-12

[2]Tihanyi. *Electromagnetic Compatibility in Power Electronics.* New York. IEEE Press, 1995

[3]Dongbing Zhang, Dan Y. Chen, Mark J. Nave, et al. *Measurement of noise source impedance of off-line converters.* IEEE Transactions on Power Electronics, 2000, Vol.15 No.5: 820-825

[4]D.H. Liu, J.G. Jiang. *High Frequency Characteristic Analysis of EMI filter in Switch Mode Power Supply.* PESC Record - IEEE Annual Power Electronics Specialists Conference. 2002. 2039-2043

[5] Timothy C. Neugebauer, Joshua W. Phinney, David J. Perreault. *Filters and components with inductance cancellation.* Proceedings of IEEE Industrial Applications Society Annual Meeting, 2002. 939-947

[6] Junping He，Wei Chen，Jiangguo Jiang. *Analysis on the EMI effect of stray magnetic field from main circuit of a PFC switched mode power supply*, Proceedings of the CSEE, Vol.25, No.14, 2005.151-157

[7] Theodore M zeeff, Todd H Hubing, Thomas P. Van, David Pommerenke. *Analysis of simple two-capacitor low-pass filters.* IEEE Transactions on EMC. 2003, Vol.45, No.4. 595-601

[8] Wang. Shuo, Lee. Fred .C, Chen Danyang, et. al. *Effects of parasitic parameters on EMI filter performance.* IEEE Transactions on Power Electronics, Vol.19, No.3, May, 2004. 869-877

[9]C.P.Wang, D.H.Liu, Jianguo Jiang. *Study of coupling effects among passive components used in power electronic devices.* Proceedings of 4th international power electronics and motion control conference. 2004. S16.1-59.

2006 5th International Power Electronics and Motion Control Conference

High Step-up Converter Associated with Soft-Switching Circuit
with Partial Energy Processing
for Livestock Stunning Applications

S. -Y. Tseng, S.-H. Tseng and J. -Z. Shiang

Department of Electrical Engineering

Chien Kuo Technology University

Changhua, Changhua City, Taiwan, China

E-mail: sytseng@cc.ctu.edu.tw

Tel: 886-4-7111111 ext. 3234

Fax: 886-4-7111111 ext. 3200

Abstract-This paper presents a high step-up converter associated with a full-bridge inverter for livestock stunning applications. The proposed converter adopts a two stage converter and uses partial energy processing to reduce voltage stress and switching loss and then, its switches are operated in a complementary manner to reduce component counts and circuit structure complexity. To further improve its conversion efficiency, the proposed soft-switching converter is formed by introducing coupled inductors to the high step-up converter, which can achieve zero-voltage switching at turn-on transition. Compared with the conventional two-stage step-up converter and that with the switch integration, the proposed converter can improve conversion efficiency of 6 ~ 8% over that with two-stage hard switching under full load condition, and can reduce cost of that with the switch integration. In this research, the output voltage waveforms generated from the converter and inverter are with frequency varying from 50 Hz to 800 Hz, amplitude varying from zero to its breakdown voltage, as high as 200 V, and duty ratio changing from 0.3 to 0.7. Performance measurements from a prototype have verified the feasibility of the overall system design.

I. INTRODUCTION

In the world, many developed countries are highly concerned about animal. In particular, livestock must be rendered unconscious and insensible to pain before they are exsanguinated [1]-[8]. In humane slaughter methods, there exist two major methods: carbon dioxide (CO_2) and manual electrical stunning methods [1]-[6]. Since the CO_2 method has more limitations and requires higher cost, the manual electrical stunning method has been used more popular.

Livestock with electrical stunning mainly induces an epileptiform seizure by the amount of current passing the brain. When livestock is induced an epileptiform seizure, its impulse signal propagation between neurons will be suppressed to cause an unconsciousness. The illustration and equivalent circuit of impulse signal propagations between neurons in the livestock are shown in Figs. 1 and 2, respectively. Their propagation process and operational principle are detailed in [9]. From the operational principle of the equivalent circuit, it can be found that the minimum current and voltage required for stunning a pig is about 1.3 A and 180 V, respectively, and they must sustain at least 3 s. To generate the specified electrical waveforms, line voltage

or battery voltage is used as an input voltage source. When line voltage boosted up with a low-frequency transformer is adopted, it is easy to cause ecchymosis in the carcasses of pigs, resulting in low meat quality. If a stunner system chops the dc voltage into square waveforms with power switches and boosts them through a low-frequency transformer, its output voltage regulation will depend on the battery voltage. Thus, it will easily causes a bone fractures in the carcasses of pigs, causing a low meat quality. Additionally, they also have a larger volume and size, and heavier weight. To solve above problems, a dc/dc converter with PWM control is adopted for livestock stunning applications.

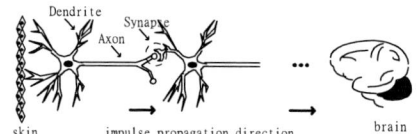

Fig. 1. Block diagram of procedure for impulse propagation between neurons.

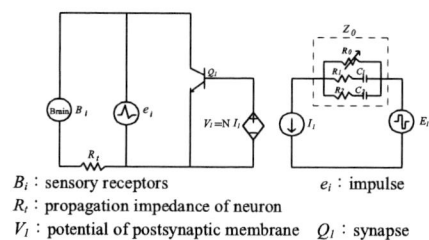

B_i : sensory receptors $\qquad$ e_i : impulse

R_t : propagation impedance of neuron

V_l : potential of postsynaptic membrane $\quad$ Q_l : synapse

I_l : current passing pig

Z_0 : equivalent impedance of pig $\qquad$ E_i : stunning voltage

Fig. 2. An equivalent circuit for describing the impulse signal propagation between neurons.

Since the stunner system in slaughterhouse is exposed to a higher humidity environment, it is required to operate under a safe condition. Based on the required condition, an independent voltage source, such as battery source, is more viable. To use battery as the source of a dc/dc converter, it usually needs a higher step-up voltage ratio. In dc/dc converter with single stage, a push-pull, half-bridge or full-bridge converter associated with a transformer and high

1-4244-0448-7/06/$25.00 ©2006 IEEE

turns ratio can be adopted [10]-[11]. Due to high turns ratio of transformer, there exist a large amount of leakage inductance in the converters, causing a higher spike voltage across switch and a low conversion efficiency. As a consequence, they need a larger capacity of soft-switching circuit or snubber to solve the problem, resulting in a high cost. To reduce component stress and cost, two boost converters cascade connection is adopted in the stunner system, as shown in Fig. 3. Although it can achieve high step-up voltage ratio, it will yield low conversion efficiency due to their input energy processed two times. To further improve conversion efficiency, a two-stage boost converter with the synchronous switches technique to integrate their active switches into a single stage can be adopted, achieving high conversion efficiency [12], as shown in Fig. 4. Since each switch in the conventional converter with the synchronous switch technique must afford the highest voltage and current stresses, there exists a large spike voltage or inrush current in the conventional one. To solve the above problems, it needs a snubber with larger power capacity, increasing its cost.

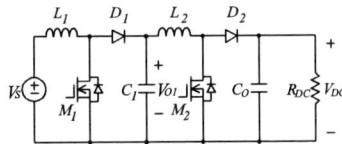

Fig. 3. Schematic diagram of the conventional two stages boost converter.

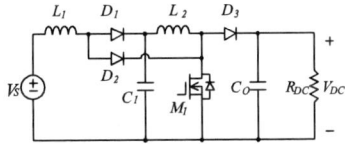

Fig. 4. Schematic diagram of the conventional boost converter with switch integration.

To trade off conversion efficiency and the cost of a stunner system, we propose a two stage boost converter with partial energy processing to improve its conversion efficiency [9], as shown in Fig. 5. Since the switches between the first and the secondary stages in the proposed converter are operated in complementary manner, it has a simpler circuit structure, a higher step-up voltage ratio, a higher conversion efficiency and lower component stress over the conventional two-stage converter. To further increase conversion efficiency, we introduce a pair of coupled inductors L_{f1} and L_{f2} into the high step-up converter to achieve the feature of zero-voltage switching (ZVS) at turn-on transition, as shown in Fig. 6. Therefore, the proposed stunner system can yield higher efficiency and reduce its weight, size and volume significantly.

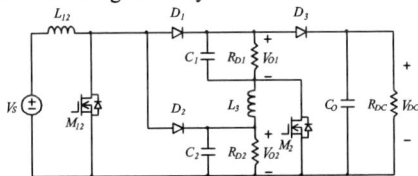

Fig. 5. Schematic diagram of the high step-up converter.

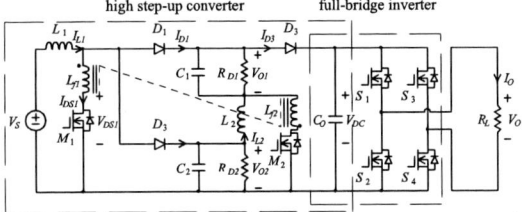

Fig. 6. Schematic diagram of the proposed high step-up converter associated with soft-switching circuit.

II. Design of the Proposed Converter

Due to page limitation, operational principle of the proposed converter does not describe in this paper. Its key waveforms are shown in Fig. 7. In this section, design of the proposed converter is presented, including determination of duty ratio D, inductors L_1 and L_2 and the coupled inductors L_{f1} and L_{f2}. In the following, they are derived briefly.

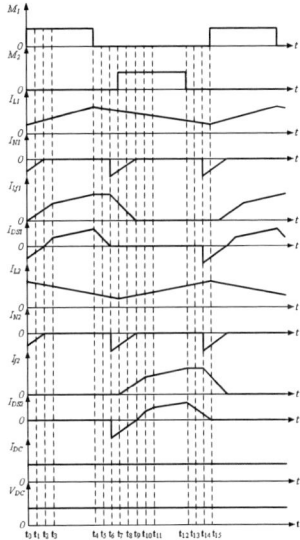

Fig. 7. Key waveforms of the proposed high step-up converter

A. Duty Ratio D

To determine duty ratio, we must first attain input to output voltage transfer ratio M. Since the coupled inductors L_{f1} and L_{f2} only helps switches M_1 and M_2 to achieve soft-switching feature, they do not affect transfer ratio M of the proposed two-stage boost converter. That is, transfer ratio M_1 (the first stage) and M_2 (the secondary stage) will be the same as the conventional one, in which their expression are derived in [9]. Thus, M_1 and M_2 can be respectively expressed as follows:

$$M_1 = \frac{V_{O1}}{V_S} = \frac{V_{O2}}{V_S} = \frac{1}{1-D_1}, \tag{1}$$

and

$$M_2 = \frac{V_{DC}}{V_{O2}} = \frac{V_{DC}}{V_{O2}} = \frac{2-D_2}{1-D_2}, \tag{2}$$

where D_1 is the duty ratio of the first stage and D_2 is the

be seen that the input put to output transfer ratio of the proposed overall system can be determined as

$$M = \frac{V_{DC}}{V_S} = \frac{(2 - D_2)}{(1 - D_1)(1 - D_2)}. \tag{3}$$

When $D_1 = D_2 = D$, (3) can be rewritten as

$$M = \frac{V_{DC}}{V_S} = \frac{2 - D}{(1 - D)^2}. \tag{4}$$

In practice, when switches M_1 and M_2 are operated complementarily, the overall transfer ratio M can be rewritten as

$$M = \frac{V_{DC}}{V_S} = \frac{1 + D}{D(1 - D)}. \tag{5}$$

To compare the transfer ratio among the conventional two-stage boost converter, the proposed converter with complementary operation and without that, Fig. 8 illustrates their input to output voltage transfer ratio. From Fig. 8, it can be seen that the proposed converter without complementary operation has the highest transfer ratio. In practically applications and according to the requirement of the input to output voltage transfer ration, we can have the proposed converter stay in a proper operational state. When an operational state of the proposed converter is specified, its duty ratios D_1 and D_2 can be determined.

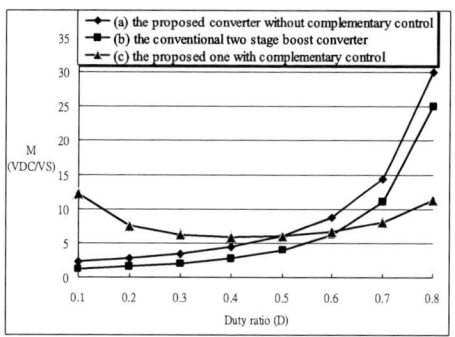

Fig. 8. Plot of input to output voltage transfer ratio M versus duty ratio D among the conventional two stages boost converter, the proposed one without complementary operation and that with complementary operation.

B. Inductors L_1 and L_2 Design

Since the proposed converter is operated in continuous conduction mode (CCM), inductors L_1 and L_2 must be greater than inductor L_{1B} and L_{2B} which are the boundary inductance of inductors L_1 and L_2 as it is operated in the boundary of CCM and discontinuous conduction mode (DCM), and their inductor currents I_{L1} and I_{L2} are shown in Fig. 9, we can derive the maximum inductor current $I_{L1(max)}$ as follows:

$$I_{L1(max)} = \frac{V_S}{L_1} DT_S. \tag{6}$$

Its average current $I_{L1(av)}$ can also be determined as

$$I_{L1(av)} = \frac{V_S}{2L_1} DT_S. \tag{7}$$

Assuming that input power P_i is equal to output power P_O, their relationship can be expressed as follows:

$$V_S I_{L1(av)} = V_{DC} I_{DC}, \tag{8}$$

where V_{DC} is the output dc-link voltage of the proposed

(8) can be rearranged and the boundary inductor L_{1B} can be expressed by

$$L_{1B} = \frac{(1 - D)^2 V_S}{2(2 - D) I_{DC}} DT_S. \tag{9}$$

From (9), it can be seen that when the proposed converter is operated in CCM, inductor L_1 must be greater than L_{1B}. In addition, the maximum value of inductor current $I_{L2(max)}$ shown in Fig. 10 can be also determined by

$$I_{L2(max)} = \frac{V_{O1}}{L_2} DT_S = \frac{V_S}{L_2(1 - D)} DT_S. \tag{10}$$

From the circuit of the proposed stunner system shown in Fig. 6, it can be found that the maximum of diode current $I_{D2(max)}$ is equal to that of inductor current $I_{L2(max)}$. In addition, the averaged current $I_{D2(av)}$ equals output current I_{DC}. Thus, the average current $I_{D2(av)}$ can be expressed as

$$I_{D2(av)} = I_{DC} = \frac{DV_S T_S}{2L_2}. \tag{11}$$

According to (11), the boundary value of inductor L_{2B} can be determined as

$$L_{2B} = \frac{DV_S T_S}{2I_{DC}}. \tag{12}$$

From (12), it can be also seen that inductor L_2 must be greater than L_{2B} to operate the proposed converter in CCM. From (9) and (12), it can be found that once the specifications of the proposed converter, duty ratio D and period T_S is specified, the boundary values of inductors L_{1B} and L_{2B} can be determined.

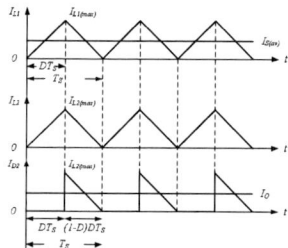

Fig. 9. Conceptual current waveforms of inductor currents I_{L1} and I_{L2} in the proposed converter operated in the boundary of CCM and DCM.

C. Coupled Inductors L_{f1} and L_{f2}

The coupled inductors L_{f1} and L_{f2} are used to achieve soft-switching feature. To achieve a ZVS feature, the energy stored in inductor L_{f1} must satisfy the following inequality:

$$\frac{1}{2} L_{f1} I^2_{DS1(max)} \geq \frac{1}{2} C_{M2} V^2_{DS2(off)}, \tag{13}$$

where $I_{DS1(max)}$ is the maximum value of switch current I_{DS1}, $V_{DS2(off)}$ $(= V_{DC} - V_{O1})$ is the voltage across switches M_2 during turn-off interval. Current $I_{DS1(max)}$ can be expressed as

$$I_{DS1(max)} = I_{DS1(0)} + \Delta I_{DS1(max)} = I_{DS1(0)} + \frac{DV_S T_S}{L_1}, \tag{14}$$

where $I_{DS1(0)}$ is the initial value of switch current I_{DS1}, in which the proposed converter is operated in CCM. Thus, the average input current $I_{i(av)}$ can be given as

$$I_{i(av)} = I_{DS1(0)} + \frac{DV_S T_S}{2L_1} \tag{15}$$

1088

From (15), $I_{DS1(0)}$ can be rewritten by

$$I_{DS1(0)} = I_{i(av)} - \frac{DV_S T_S}{2L_1}.$$ (16)

Substituting (16) in (14) results in

$$I_{DS1(max)} = I_{i(av)} + \frac{DV_S T_S}{2L_1}.$$ (17)

When switches M_1 and M_2 are operated complementarily, substituting (17) in (13) results in

$$L_{f1} \geq \frac{16 C_{M2} L_1^2 V_S^2}{D^2 (4L_1^2 I_{i(av)}^2 + 4DL_1 I_{i(av)} V_S T_S + D^2 V_S^2 T_S^2)}.$$ (18)

When inductor L_{f1} satisfies (18), the proposed two-stage boost converter can achieve ZVS feature. However, if inductor L_{f1} is too large, it will result in duty loss. From Fig. 8, it can be observed that the time interval of duty loss is between t_4 and t_8. Since time intervals t_{v4} (= $t_4 \sim t_5$) which charges the charges of capacitor C_{M1} form 0 to V_{O1}, and t_{v5} (= $t_5 \sim t_6$) which discharges the charges of capacitor C_{M2} from ($V_{DC} - V_{O1}$) to 0, they are much shorter than the time interval between t_6 and t_8. Thus, at time t_6, inductor current I_{Lf1} is approximately the maximum current $I_{DS1(max)}$, and at time t_6, switch current $I_{DS2(tv6)}$ can be expressed by

$$I_{DS(tv6)} = -\left(\frac{I_{i(av)}}{n} + \frac{DV_S T_S}{2nL_1}\right),$$ (19)

where n (= N_2/N_1) is turns ratio of the coupled inductors L_{f1} and L_{f2}. Since the maximum value of inductor current $I_{L2(max)}$ is equal to that of switch current $I_{DS2(max)}$, $I_{L2(max)}$ can be determined as

$$\begin{aligned} I_{L2(max)} &= I_{L2(0)} + \Delta I_{L2(max)} \\ &= I_{L2(0)} + \frac{V_{O1}(1-D)T_S}{L_2} \\ &= I_{L2(0)} + \frac{V_S T_S}{L_2}, \end{aligned}$$ (20)

where $I_{L2(0)}$ is the initial value of inductor current I_{L2}, in which the proposed converter is operated in CCM. According to the relationship of equations (15) ~ (17), the output average current $I_{O(av)}$ and the initial value $I_{L2(0)}$ can be respectively expressed as follows:

$$I_{O(av)} = I_{L2(0)} + \frac{V_S T_S}{2L_2},$$ (21)

and

$$I_{L2(0)} = I_{O(av)} - \frac{V_S T_S}{2L_2}.$$ (22)

Therefore, the current rate $\Delta I_{DS(tv68)}$ of change between t_6 and t_8 can be determined as

$$\begin{aligned} \Delta I_{DS(tv68)} &= -I_{DS2(tv6)} + I_{L2(0)} \\ &= \left(\frac{I_{i(av)}}{n} + \frac{DV_S T_S}{2nL_1}\right) + I_{O(av)} - \frac{V_S T_S}{2L_2}. \end{aligned}$$ (23)

Thus, the relationship between $\Delta I_{DS(tv68)}$ and the time interval t_{v68} of loss duty can be expressed by

$$\Delta I_{DS(tv68)} = \frac{(V_{DC} - V_{O1})t_{v68}}{L_{f2}} = \frac{2V_S t_{v68}}{DL_{f2}}.$$ (24)

From (24), it can be found that t_{v68} can be determined as follows:

$$t_{v68} = \frac{DL_f 2}{2V_S}\left(\frac{I_{i(av)}}{n} + \frac{DV_S T_S}{2nL_1}\right) + I_{O(av)} - \frac{V_S T_S}{2L_2}.$$ (25)

Once the maximum time $t_{v68(max)}$ is specified, L_{f2} can be satisfied the following inequality:

$$L_{f2} \geq \frac{2V_S t_{v68(max)}}{D\left(\frac{I_{i(av)}}{n} + \frac{DV_S T_S}{2nL_1}\right) + I_{O(av)} - \frac{V_S T_S}{2L_2}}.$$ (26)

Since L_{f1} is equal to L_{f2}/n^2, inductor L_{f1} can be expressed by

$$L_{f1} \geq \frac{2V_S t_{v68(max)}}{Dn^2\left(\frac{I_{i(av)}}{n} + \frac{DV_S T_S}{2nL_1}\right) + \frac{D(1-D)}{1+D}I_i - \frac{V_S T_S}{2L_2}}.$$ (27)

Therefore, to design the proposed converter operated in a proper operational condition, inductor L_{f1} must be satisfied (18) and (27) simultaneously. In practical, $t_{v68(max)}$ is indicated about $1/5 \sim 1/10$ $(1-D)T_S$.

III. Measured Results

To verify the performance of the proposed stunner, a prototype with the following specifications was implemented.

(A) Soft-switching high step-up converter
☐ input voltage V_S: DC 24 V,
☐ switching frequency f_{S1} : 50 kHz,
☐ output voltage V_O :DC 200 V, and
☐ maximum output current $I_{DC(max)}$: 2 A.
(B) Full-bridge inverter
☐ input voltage V_{dc}: DC 200 V,
☐ switching frequency f_{S2}: 400 Hz,
☐ duty ratio D_2: 0.5,
☐ output voltage V_O: ±200 V,
☐ maximum output current I_O: ±2 A, and
☐ maximum output power $P_{O(max)}$: 400 W.
According to the specifications, the components of high step-up converter are determined as follows:
☐ switch M_1: IRFP250, ☐ switch M_2: IRF264,
☐ diode $D_1 \sim D_3$: UF1606, ☐ inductor L_1: 70 µH,
☐ inductor L_2: 500 µH, ☐ inductor core of L_1: EE-42,
☐ inductor core of L_2: EE-35,
☐ output capacitor $C_1 \sim C_2$: 300 µF/ 250 V, and
☐ output capacitor C_3: 470 µF/ 400 V.
To generate ac voltage, the components of the full-bridge inverter are also determined as follows: switches $S_1 \sim S_4$: IRF840.

According to the designed values of components previously, and (18) and (27), plot of L_{f1} versus I_i is shown in Fig. 10. This figure shows the upper and lower bounds of qualified inductance for achieving a ZVS feature. Measured switch voltage V_{DS} and current I_{DS} waveforms of switch M_1 and M_2 respectively shown in Figs. 11(a) and (b) under 50% of the full load, illustrating ZVS features. According to the selected inductance shown in Fig. 10, efficiency comparison between the conventional and the proposed ones is plotted in Fig. 12. It can be seen that the efficiency of the proposed converter is higher than the conventional one and it is 88% under full load condition. In addition, the efficiency of overall stunner system is 84% under full load condition. Measured waveforms of output voltage V_O and output current I_O during pig stunning interval are shown in

Fig. 13, illustrating that pig stunning time duration is around 3 s and output current I_O is no greater than 2 A, with which good meat quality can be sustained.

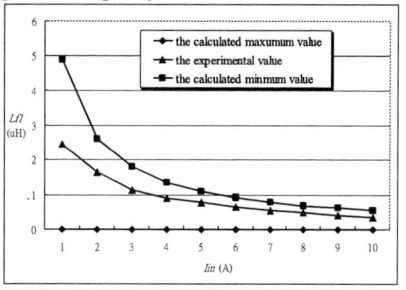

Fig. 10. Plot of inductance of the coupled inductors L_{f1} and L_{f2} versus input current I_i.

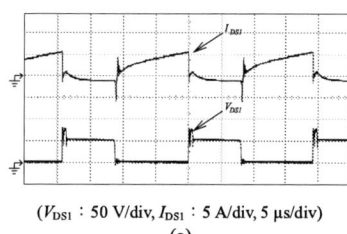

$(V_{DS1}:$ 50 V/div, $I_{DS1}:$ 5 A/div, 5 µs/div)

(a)

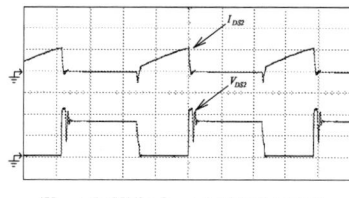

$(V_{DS2}:$ 50 V/div, $I_{DS2}:$ 5 A/div, 5 µs/div)

(b)

Fig. 11. Measured waveforms of voltage V_{DS} and current I_{DS} of (a) switch M_1 and (b) switch M_2 under 50% of the full load, illustrating ZVS features.

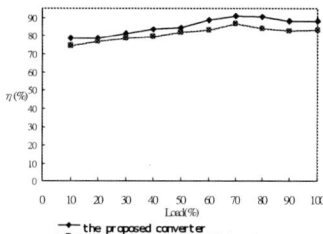

● the proposed converter
■ the conventional one with two stages

Fig. 12. Efficiency comparison between the proposed converter and the conventional one with two stages from light load to heavy load.

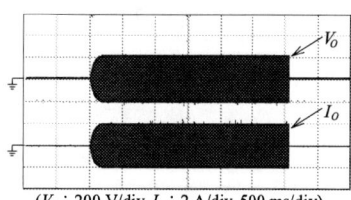

$(V_o:$ 200 V/div, $I_o:$ 2 A/div, 500 ms/div)

Fig. 13. Measured waveforms of output voltage V_O and output current I_O

IV. Conclusion

In this paper, coma mechanism of pig with electrical stunning has been briefly reviewed. Operational principle, steady-state analysis and design of the proposed soft-switching and a full-bridge inverter has been implemented to generate stunning electrical parameters, in which its current amplitude is ±1.8 A and its voltage amplitude is ±200 V. The proposed soft-switching two-stage boost converter with partial energy processing can achieve the efficiency around 88% under full load condition, and efficiency of overall stunner system is about 84%. From experimental process, it can be observed that the proposed stunner system can be effectively cause coma of pig. Additionally, from experimental results, it can also found that the proposed one can attain a good meat quality, while meet the regulation of animal welfare.

REFERENCES

[1] H. A. Channon, A. M. Payne and R. D. Warner, "Comparison of CO_2 Stunning with Manual Electrical Stunning (50Hz) of Pig on Carcass and Meat Quality," *Trans. on Meat Science*, 2002, pp.63—68.

[2] S. B. Wotton and M. O. Callaghan, "Electrical Stunning of Pigs: the Effect of Applied Voltage on Impedance to Current Flow and the Operation of a Fail-Safe Device," *Trans. on Meat Science*, 2002, pp. 203—208.

[3] E. Lambooij, *et al.*, "Some neural and behavioural aspects of electrical and mechanical stunning in ostriches," *Trans. on Meat Science*, 1999, pp. 339—345.

[4] E. Lambooij, *et al* l., "The Effects of Captive Bolt and Electrical Stunnin, and Restraining Methods on Broiler Meat Quality," *Trans. on Poultry Scien ce*, 1999, pp. 600—607.

[5] V. Sante, *et al.*, "Effect of Stunning Current Frequency on Carcass Downgrading and Meat Quality of Turkey," *Trans. on Poultry Science*, 2000, pp. 1208—1214.

[6] S.F. Bilgili, "Recent Advances in Electrical Stunning," *Trans. on Poultry Science*, 1999, pp. 282—286.

[7] B. Savenije, *et al.*, "Electrical Stunning and Exsanguination Decrease the Extracellular Volume in the Broiler Brain as Studied with Brain Impedance Recordings," *Trans. on Poultry Science*, 2000, pp. 1062-1066.

[8] W. D. McNeal, *et al.*, "Effects of Stunning and Decapitation on Broiler Activity During Bleeding, Blood Loss, Carcass, and Breast Meat Quality," *Trans. on Poultry Science*, 2003, pp. 163—168.

[9] S.-Y. Tseng, S.-H. Tseng and J. G. Huang, "High Step-up Converter with Partial Energy Processing for Livestock Stunning Applications," *Proceedings of the Applied Power Electronics Conference*, 2006, pp. 1537-1543.

[10] J.-C. Hung, *el at.*, "An Active-Clamp Push-Pull Converter for Battery Sourcing Applications," *Proceedings of the APEC*, Vol. 2, 2005, pp.1186 – 1192.

[11] H. Sakamoto, *el at.*, "A Self Oscillated Half Bridge Converter Using Impulse Resonant Soft-Switching," *Proceedings of the INTELEC*, 2002, pp.227 – 231.

[12] T.-F. Wu and T.-H. Yu, "Unified Approach to Developing Single-Stage Power Converters," *IEEE Trans. on Aerospace and Electronic Systems*, Vol. 34, 1998, pp.211 – 223.

A Computationally Intelligent Methodologies and Sliding Mode Control Based Traction control System for in-wheel driven EV

MING Zhengfeng*, NI guangzheng**
*Xidian University, Xi'an, China, 710071
**Zhejiang University, Hangzhou, China, 310027
E-mail: mzfxut@163.net

Abstract–A traction control system based on computationally intelligent methodologies and sliding mode control is developed in this paper. The basic idea of the proposed strategy is that the controller provided with computationally intelligent system is trained by using the sliding mode control. Therefore, the controller can guarantee that the movement of the controlled system can follow the ideal moving tracks, the drawback of control chattering occurred in the classical sliding mode control can be alleviated. Moreover, the robustness of systems is improved. The numerical results validate the proposed model and method.

Keywords–*Traction control (TC), Electrical Vehicle (EV), computationally intelligent methodologies, sliding mode control (SMC)*

I. INTRODUCTION

Nowadays, most high-ranking vehicles take Traction Control technique as standard or selected equipment. For EV, the torque controls of electromotor have more quickly responsibility and higher precision than the output torque of internal-combustion engine. Especially for in-wheel driven EV, torque of every driven wheel can be controlled individually. To realize high performance of Transaction Control, it provides an upstanding technique foundation. The advantages of Transaction Control on EV are as follows[4]:

(1) Because Transaction Control on EV is carried out through software platform, so the system cost is relative low and it's possible to realize Transaction Control with lower cost and higher performance.

(2) Because the response time of Transaction Control on EV is less than 10ms, so the dynamic performance is excellent. While traditional vehicles based on inner multi-mechanic system, for example, the response time of opening valve is commonly more than 200ms; addition to the delay of mechanic system, the actual response time is lower than expected.

(3) It's easy to design Transaction Control on EV because that the torque of electromotor is controlled by simply current controlling. But as to traditional vehicles,

unknown and strongly non-linearity exists not only in Transaction Control input but also in output torque of internal-combustion engine. So it's very difficult to create mathematics' model of controller.

Sliding Mode Control (SMC) is kind of changeable structural non-linearity control strategy. It's a kind of control method which makes system regularly move by changing controller structure under discipline of sliding mode surface according to the degree of system state departure from sliding mode surface. Its specialty lies on strongly robustness and applicability on handling with such unsure factors as model uncertainty and unknown disturbance. Corresponding the conflict between dynamic and stable performances is improved by simple control rules. In practice, pure SMC method has disadvantages as follows: Firstly, highly trembling while controller output, which is named, trembling problem, can lead to unexpected instability. Secondly, the feedback loops of SMC are easily affected by measure noise. Thirdly, SMC need a mass of system information to overcome the uncertainty of parameters. To alleviate those problems, some methods are proposed to improve original SMC. Those methods need perfect target mathematics' model and are based on primary equivalent control computing.

Computer Intelligent (CI) System is a brand new scientific method which is based on relatively mature theories—Neural Network、Ambiguous System and Inherit Arithmetic. CI has ability of adapting and handling under new circumstance which makes system can reason such as generalization, discovery, imagine and abstract. In another word, CI which includes practical self-adaptation conception, paradigm, arithmetic and implementation brings intelligent behaviors under complex and variable situations. It is the most important feature that CI directly handles with information from signal or digital layer while it need not create precise mathematic or logical model and doesn't rely on knowledge denotation. During the course of parameter adjusting in CI, error reverse broadcasting technique and Liebenberg Marquardt optimizing arithmetic are mostly popular. However, these methods are carefully used in noise situation and suddenly changed system. Difficulties can be effectively reduced by using changeable structural theory.

The fusion between CI and SMC aims at lessen problems in practical use for SMC. A Traction Control System for in-wheel driven EV is designed in this paper. The rate of parameter adjusting based on SMC is emphasized which not only guarantees that the movement of the controlled system can follow the ideal moving tracks, but also alleviates the drawback of control chattering occurred in the classical sliding mode control. Moreover, it enhances the adaptability of system uncertain parameters for the controller and improves the robustness of TC control system in essence.

II. MODELING OF ACCELERATOR FOR IN-WHEEL DRIVEN EV

A One Fourth Accelerator Model

In terms of studying in rules of vehicle control, most methods use simple wheel model and ignore so many factors like wind resistance force, scrolling resistance force and vertical load. According to [5], firstly this paper describes wheel accelerator model. One Fourth Accelerator Mode is as Figure1.

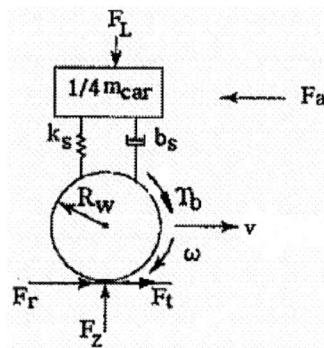

Figure1.   One Fourth Accelerator Mode

The movement equation and turning equation are:

$$m_{tot}\dot{v} = F_t - F_a$$

$$J_w\dot{\omega} = T_b - R_wF_t - R_wF_f$$

Where

$$F_t = \mu(\lambda)F_z \qquad \lambda = 1 - \frac{\omega R_w}{v}$$

$$m_{tot} = m_{tyre} + \frac{1}{4}m_{car} \quad F_a = \frac{1}{4}C_dA_fv^2$$

$$F_z = m_{tyre}g - \frac{m_{car}h_{cg}}{\ell}\dot{v} = m_{tyre}g - F_L$$

$$F_L = \frac{m_{car}h_{cg}}{\ell}\dot{v} \qquad m_{eq} = h_{cg}m_{car}/\ell$$

$$F_f = f_0 + 3.24f_s(K_{mpf}v)^{2.5}$$

The meanings of each parameter in Fig1 and equations above are

m_{car}—weight of vehicle，m_{tyre}—weight of tyre，m_{tot}—equivalent weight of one fourth wheel, k_s—elastic modulus of hanging spring, b_s—damp modulus of hanging damper, ℓ—length of vehicle bed, h_{cg}—height of bar center, T_b—torque of brake, F_t— friction force of ground, F_f—friction force of wheel scrolling, F_z—equivalent vertical load of vehicle, F_L—additional vertical load caused by inertia, F_a—wind resistance force while driving , R_w—wheel radius, ω —wheel angel speed , v—directly speed of vehicle, J_w—wheel scrolling inertia, F_z—wheel vertical load, F_a—wind resistance force, λ —slippage ratio, $\mu(\lambda)$—coherence modulus, function of λ , $K_{mph} = 1.2$.

B Wheel Model

Vehicle's movement relies on the force of wheels. The function

$$\mu(\lambda) = \frac{2\mu_p\lambda_p\lambda}{\lambda_p^2 + \lambda^2}$$

Is used to imitate the practical test curve by the wheel vendor. where, μ_p is peak value of coherence modulus , λ_p is μ_p relative slippage ratio value.

C Mechanic feature of EV

The sample in wheel driven EV adopts always magnetism no brush direct current outer rotor electronic motor system. Its mechanic feature under PWM adjusting pressure driven is:

$$T_b = k_t\frac{qU_N - k_en}{R_a}$$

where, k_t—torque constant of driven electronic motor, k_e—electromotive force constant of driven electronic motor, R_a—electronic motor resistance, q—ratio of space, U_N—inverse power voltage, battery pack voltage in this system, n—rotate speed of electronic motor, r/min。

III. DESIGN OF NEW TRANSACTION CONTROL SYSTEM OF IN-WHEEL DRIVEN EV

A System Description

For N non-linear system

$$x^{(r)} = f\left(x, \dot{x}, ..., x^{(r-1)}, t\right) + u$$

Where,

$$\mathbf{x} = \left[x, \dot{x}, ..., x^{(r-1)}\right]^T$$

and u are individually state vector and control input of system; f is a function evolving system unknown parameter vector. Expectative state vector's value and state deviation vector are defined:

$$\mathbf{x_d} = \left[x_d, \dot{x}_d, ..., \dot{x}_d^{(r-1)}\right]^T$$

$$\mathbf{e} = \mathbf{x} - \mathbf{x_d}$$

Basis on analysis before, Transaction Control System can be described as below equally

$$\dot{\lambda} = f(\dot{v}, \dot{\omega}, v, \omega, t)$$

$$\ddot{\lambda} = g(\dot{v}, \dot{\omega}, v, \omega, t) + c\dot{T}_b$$

Where

$$c = \frac{v}{J_w \omega^2 R_w} \qquad n = \frac{\omega}{2\pi} \cdot 60$$

$$\dot{T}_b = \frac{k_t}{R_a}(U_N \dot{q} - k_e \dot{n}) = \frac{k_t U_N}{R_a} \dot{q} - \frac{k_t k_e}{R_a} \cdot \frac{30}{\pi} \cdot \dot{\omega}$$

Therefore，equivalent model can be as follow:

$$\ddot{\lambda} = g(\dot{v}, \dot{\omega}, v, \omega, t) - \frac{k_t k_e}{R_a} \cdot \frac{30}{\pi} \cdot \dot{\omega} + c \frac{k_t U_N}{R_a} \dot{q}$$

$$= G(\dot{v}, \dot{\omega}, v, \omega, t) + u$$

Where, state vector

$$\mathbf{x} = [\lambda, \dot{\lambda}]^T$$

Control imputation

$$u = bu' \quad u'(t) = \dot{q} \quad b = \frac{k_t U_N c}{R_a}$$

B Selection of reference model

Basis on experience and type characteristic curve, reference slippage ratio is

$$x_{d1} = \lambda_d = \lambda_c + p_1 \cos(p_2 t)e^{-p_3 t}$$

Where $\quad p_1 > 0, p_2 > 0, p_3 > 0$

λ_c is input of reference model，stand for the best reference slippage ratio under different road station. Then

$$\dot{x}_{d1} = x_{d2} = \dot{\lambda}_d$$

$$= -p_1 p_3 \cos(p_2 t)e^{-p_3 t} - p_1 p_2 \sin(p_2 t)e^{-p_3 t}$$

$$\ddot{x}_{d1} = \ddot{\lambda}_d$$

$$= -2p_3 x_{d2} - (p_2^2 + p_3^2)x_{d1} + (p_2^2 + p_3^2)\lambda_c$$

$$= a_m x_{d2} + b_m x_{d1} + c_m \lambda_c$$

Where

$$a_m = 2p_3, b_m = -(p_2^2 + p_3^2), c_m = -b_m$$

C Design of SMC and CI Controller

The design aims at forcing system to move along the response as system anticipate which is produced by ideal control ratio u_d. Because that u_d is unknown, the key is to figure out compound signal iterated by CI Controller. Its primary idea is to train TC Controller using SMC, especially to design parameter adjusting ratio based on SMC.

In this paper, in terms of math，the expected and assumed move track is $\mathbf{x}_d(t)$ when $t \geq 0$，then u_d is the needed differential equation

$$x_d^{(r)} = f(\mathbf{x}_d, t) + u_d$$

If such a model and its control ratio exists, it means that when system ideal control ratio is known and initial condition is set as $\mathbf{x}(t=0) = \mathbf{x}_d(t=0)$, the system may move along expected track.. Based on the result of Lyapunov stability analisis, if control ratio above is applied to the system, the ideal control ratio is stable. As

$$\lim_{t \to \infty} u = u_d$$

Therefore, the method to figure out the difference between target value and controlled value by ideal control mode. The control error is defined as

$$s_c \overset{\Delta}{=} u - u_d$$

Then

$$u = u_d - \left(\Delta f + \Lambda_r^{-1} \left(\sum_{i=1}^{r-1} \Lambda_i e^{(i)} + \xi \operatorname{sgn}(s_p) \right) \right)$$

Where $\quad \Delta f = f(\mathbf{x}, t) - f(\mathbf{x}_d, t)$

If

$$\Delta f = -\Lambda_r^{-1} \left(\sum_{i=1}^{r-1} \Lambda_i e^{(i)} + \xi \operatorname{sgn}(s_p) \right)$$

then $u \equiv u_d$。When u_d is used in zero initial error system, $\mathbf{e}(t) \equiv 0$ is the result. Although u_d is not a computable variable，but here $s_c=0$, Ii is the target of designing controller. The following equation is the direct description:

$$\dot{s}_p = \Lambda_r \left(f + u - x_d^{(r)} + u_d - u_d \right) + \sum_{i=1}^{r-1} \Lambda_i e^{(i)}$$

$$= \Lambda_r (\Delta f + s_c) + \sum_{i=1}^{r-1} \Lambda_i e^{(i)}$$

So

$$s_c = \Lambda_r^{-1} \left(\dot{s}_p + \xi \operatorname{sgn}(s_p) \right)$$

In this paper, the idea of the design is to gain target SMC control ratio by using computable variable. Make sure that $s_c=0$ so that target control ratio can be compound.

The structure of controller is designed as

$$u = \boldsymbol{\varphi}^T \boldsymbol{\delta}$$

$$\boldsymbol{\delta} = [\mathbf{e}^T, 1]^T$$

This structure is fit for CI ADALINE(Adaptive Linear Elements), SFS(Standard Fuzzy Systems) and ANFIS(Adaptive Neuro-fuzzy Inference Systems) [3]。The aim of controller's parameter adjusting ratio is making the control error zero. The proper principle is that for any initial value, control error is going to zero in limited time. The boundary condition is defined as below:

$$\|\boldsymbol{\varphi}\| \leq B_\phi , \quad \|\boldsymbol{\delta}\| \leq B_\delta , \quad \|\dot{\boldsymbol{\delta}}\| \leq B_{\dot{\delta}} , \quad |u| \leq B_u ,$$
$$|u_d| \leq B_{u_d} , \quad |\dot{u}_d| \leq B_{\dot{u}_d} .$$

The control system Architecture is showed asFigure2

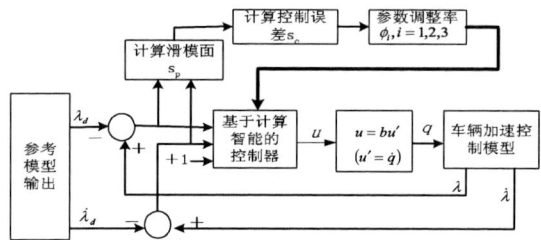

Figure2 Control System Architecture

Basis on 2-Lyapunov Function parameter adjusting ratio is designed as

$$\dot{\boldsymbol{\varphi}} = -K\left(\mu\mathbf{I} + p\boldsymbol{\delta}\boldsymbol{\delta}^T\right)^{-1}\boldsymbol{\delta}\,\mathrm{sgn}\left(s_c\right)$$

Selected probable Lyapunov Function is

$$V_A = \mu V_c + \rho\,\frac{1}{2}\left\|\frac{\partial V_c}{\partial \boldsymbol{\varphi}}\right\|^2$$

$$V_c\left(s_c\right) = \frac{s_c^2}{2}$$

Where, when

$$K > \left(\mu + \rho B_\delta^2\right)\left(B_\phi B_{\dot\delta} + B_{\dot u_d}\right) + \rho\left(B_u + B_{u_d}\right)B_\delta B_{\dot\delta}$$

It ensures that selected probable Lyapunov Function in equation is negative.

IV. SIMULATION RESULT AND ANALYSIS

When simulating , the primary constant parameter's value are : p_1=0.01; p_2=π; p_3=5; $\mu=1$; $\rho=1$; $\Lambda = [1,20]^T$; $\xi=1$; K=4100; ε=0.001 . One fourth model 's parameters' value are : m_{car}=370Kg; m_{tyre}= 40Kg; R_w=0.326m; J_w=13.7Kgm; ρ =0.00115; f_0=0.01; f_s=0.005; k_{mph}=1.2; ℓ =2.5m; A_f=2.04m^2; C_d=0.539; h=0.5; Parameters of electronic motor model are: k_t= 1.26, k_e=0.132, R_a=0.1, U_N=200V, T_{bmax}=4000Nm。

From Fig3 (g), it is indicated that system quickly comes into sliding mode when vehicle moves, while track converge into（0，0），all the control system is stable.

In Fig3 (a), the practical sliding ratio λ of controlled object is finally consistent with the output state of reference model λ_m .

From Fig3 (b), it is showed that when accelerating, speed v and wheel speed ωR_w are both reposefully increasing in the same way.

For output torque, Fig3(c) show that, in this paper, the problem of torque high frequency trembling existed in traditional SMC is well improved.

Fig3(d) (e) (f) are separately controller adjusting ratio ϕ_1, ϕ_2, ϕ_3 of relative simulations.

V. CONCLUSION

In this paper, a new method fused with SMC and CI is proposed which is use to design TC for in-wheel driven EV. SMC method is adopted to train controller based on CI. Control suggestion proposed is suitable for controlling of TC parameters' changeable non-linear system. Moreover, it not

only eliminates trembling problem existed in traditional SMC , but also enhance the robustness and stability of control. In the same time, controller needs not the parse formation of Function in the system. Through the study of simulation, it is proved in theory that the proposition can well improved adaptability for time changeable parameters of TC system and robustness for suddenly changed road surface. It can radically promote the stability of the whole system.

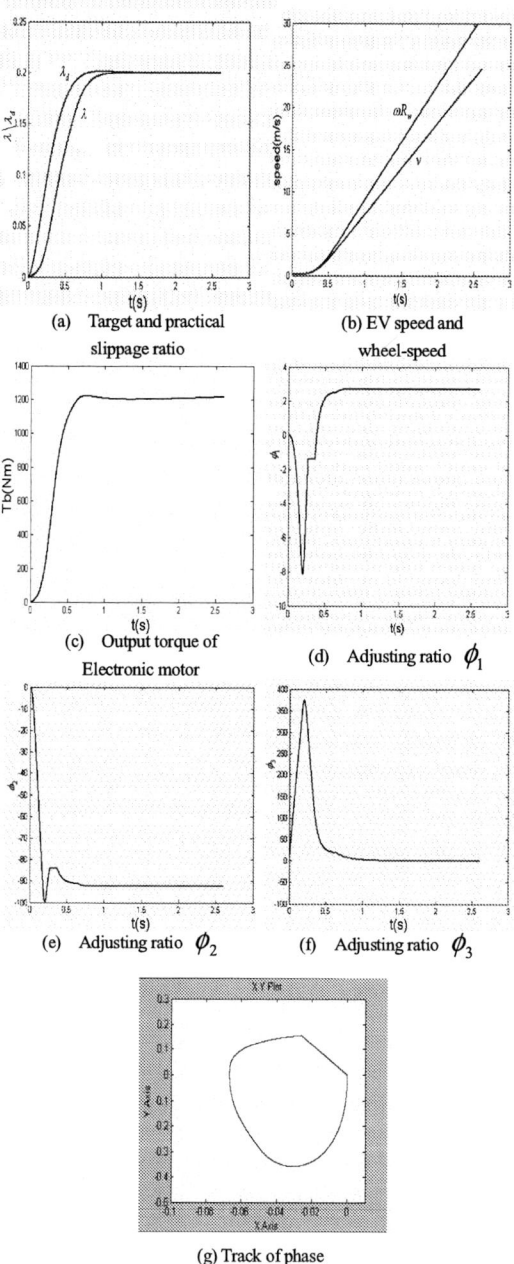

(a) Target and practical slippage ratio

(b) EV speed and wheel-speed

(c) Output torque of Electronic motor

(d) Adjusting ratio ϕ_1

(e) Adjusting ratio ϕ_2

(f) Adjusting ratio ϕ_3

(g) Track of phase

Figure4. Simulation Result of reference when target slippage ratio=0.15

REFERENCES

[1] Mehmet Önder Efe, Cem Ünsal, Okyay Kaynak and Xinghuo Yu, "Sliding mode Control of a Class of Uncertain Systems"[C], *Control Applications CCA2003, Proceedings of 2003 IEEE Conference on*, vol.1,pp.78-82,2003

[2] Okyay Kaynak, Kemalettin Erbatur and Meliksah Ertugrul, "The Fusion of Computationally Intelligent Methodologies and Sliding-Mode Control—A Survey"[J], *IEEE Trans. on Ind. Electronics*, vol.48, pp.4-17,Feb., 2001.

[3] Mehmet Önder Efe, Okyay Kaynak, "Variable Structure Systems Theory Based Training Strategies for Computationally Intelligent Systems"[C], *Industrial Electronics Society, IECON'01, The 27th Annual Conference of the IEEE Industrial Electronics Society*, vol.3, pp.1563-1576,2001.

[4] Y.Hori, Y. Toyoda, and Y. Tsuruoka, "Traction control of electric vehicle: Basic experimental results using the test EV, "UOT"," *IEEE Trans. Ind. Applicant.*, vol.34, pp.1131-1138, Sept./Oct. 1998.

[5] M.Milchik, "Vehicle Dynamics"[M], Beijing, People Communication publishing company, 1992.

[6] Qian Ming, "Sliding Mode Controller Design for ABS System"[D], Virginia, Virginia Polytechnic Institute and State University, pp.33 and pp.55, 1997.

[7] Zheng Limiao, "Car ABS and TC elements, circuit and repair"[M], Guangzhou, Guang Dong Science and Technology publishing company, 2000.

A Low-Cost Gate Driver Design Using Bootstrap Capacitors for Multilevel MOSFET Inverters

J. J. Graczkowski, K. L. Neff, *Student Member IEEE* and X. Kou, *Member IEEE*

University of Wisconsin – Platteville / Electrical Engineering Department, Platteville, WI, USA

Abstract — **Multilevel inverters require a large number of power semiconductors. The gate driver for each power semiconductor requires its own floating or isolated dc voltage source. Traditional methods on meeting the requirement are expensive and bulky. This paper presents a gate driver design for floating voltage source type multilevel inverters. Bootstrap capacitors are used to form the floating voltage sources in the design, which allows a single DC power supply to be used by all the gate drivers. Specially configured diode plus the proper charging cycles maintain adequate capacitor voltage levels. Such a simple and low cost solution allows user to utilize the easy accessible single-channel gate drivers for multilevel inverter applications without the extra cost on the bulky isolated dc power supplies for each gate driver. A prototype 3-cell 8-level floating voltage source inverter using the method illustrates the technique.**

Keywords- Multilevel Inverter; Bootstrap Capacitor; Pulse-Width Modulation; Gate Driver; MOSFET Inverter, Capacitor Voltage Balancing; Floating Voltage Source Inverter; Flying Capacitor Inverter.

I. INTRODUCTION

Multilevel inverters [1-5] are becoming popular in the modern drive systems, especially in medium voltage applications. Multilevel inverters improve the power converting quality by inserting more steps to the output voltages. A large number of switching elements is required and drive circuits are needed for each switch in order to efficiently change the state of the power semiconductor devices. The cost and complexity of multilevel inverters increase dramatically with the number of voltage levels. In addition to the increased number of switching elements, the drive circuits become more complicated when they cannot be referenced directly to the supply rails. In specific, if all of the power semiconductors could be referenced to ground, a single DC voltage source could provide power for all of the gate driver circuits. Unfortunately, multilevel inverter topologies typically have only one ground-referenced power switch in each phase. The remainder of the power devices are connected in a manner that each gate driver circuit shall be supplied by a source that is floating with respect to ground as well as the other supplies. This condition translates into the need for a floating voltage source for each gate driver circuit.

Floating sources have traditionally been provided either by separate isolation transformer circuits for each driver stage or by bootstrap capacitors. Isolation transformer circuits offer robust performance independent of inverter switching conditions. However, they also carry penalties in inverter efficiency, component size and cost. Capacitor power supplies are small and inexpensive, but require frequent recharging. Often known as the "bootstrap", capacitors have been successfully used as a power source for drivers for many years in 2-level inverters. However, its application in multilevel inverter is still a challenge. Some gate driver solutions have been proposed to work with multilevel inverters, which include separate floating voltage supplies [6] and charge pumps [7, 8]. Packaged multi-channel gate drivers [9, 10] that incorporate isolated floating DC supplies are commercially available. The disadvantages of those packaged drivers include high cost and the lack of flexibility for topology extension.

By extending the bootstrap capacitor technique, this paper presents a simple and low-cost alternative that requires only one DC voltage source to supply power to all the single channel driver circuits needed by the multilevel inverter. This design can significantly reduce the floating power supply requirements but still maintain adequate performance. The proposed design is based on a 3-cell 8-level floating source MOSFET multilevel inverter. In Chapter II, the 3-cell 8-level floating source multilevel inverter topology will be reviewed. In Chapter III, the detail design of the proposed gate driver will be explained. The lab validation of the proposed design will be shown in Chapter IV. Chapter V and VI will discuss and conclude the proposed gate driver design.

II. FLOATING SOURCE MULTILEVEL INVERTER

The per-phase topology of a 3-cell floating voltage source inverter is shown in Fig. 1. The traditional voltage ratio of the dc voltage sources is set to $V_1:V_2:V_3=1:2:3$, which yields 4 line-to-ground voltage levels as shown in Table I. As can be seen from Table I, there are several redundant switching states which allows capacitor banks to be used to form the two inner loop dc sources with balanced dc voltages [2]. It was discovered [5] that if one set the dc voltage ratio to $V_1:V_2:V_3=1:3:7$, 8 line-to-ground voltages can be generated as shown by Table II. As can be seen from Table II, each switching state refers to one line-

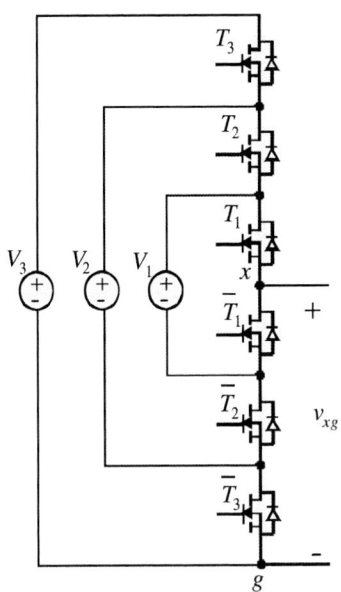

Figure 1. 3-Cell Floating Source Inverter Topology

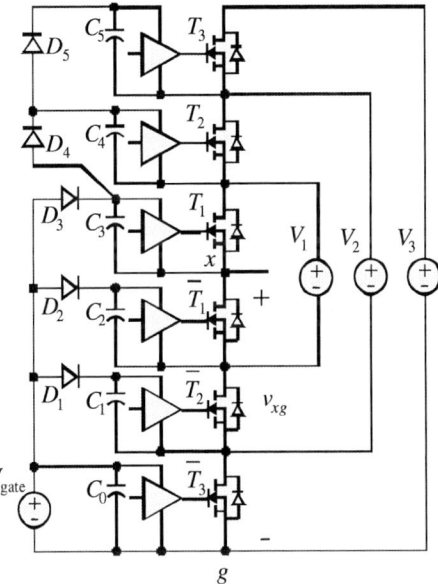

Table I. 3-Cell 4-Level Inverter

T_1	T_2	T_3	v_{xg}
0	0	0	0
0	0	1	$v_{dc}/3$
0	1	0	$v_{dc}/3$
0	1	1	$2v_{dc}/3$
1	0	0	$v_{dc}/3$
1	0	1	$2v_{dc}/3$
1	1	0	$2v_{dc}/3$
1	1	1	v_{dc}

Note: $V_3 = v_{dc}$, $V_1:V_2:V_3 = 1:2:3$

Table II. 3-Cell 8-Level Inverter

T_1	T_2	T_3	v_{xg}
0	0	0	0
0	0	1	$4v_{dc}/7$
0	1	0	$2v_{dc}/7$
0	1	1	$6v_{dc}/7$
1	0	0	$v_{dc}/7$
1	0	1	$5v_{dc}/7$
1	1	0	$3v_{dc}/7$
1	1	1	v_{dc}

Note: $V_3 = v_{dc}$, $V_1:V_2:V_3 = 1:3:7$.

III. PROPOSED LOW-COST GATE DRIVER DESIGN

As can be seen from Fig. 1, there are six power switches involved in each phase. The related gate drivers require six floating power supplies to properly switch the power MOSFETs. One may use six independent power sources to meet the requirements, but this is obviously bulky and expensive. Figure 2 shows an alternative low cost gate driver design for the 3-cell, 8-level floating source multilevel inverter. The proposed design only requires one independent dc voltage source for all the single channel gate drivers involved in the multilevel inverter. The other floating sources are formed by the bootstrap capacitors C_0-C_5. In this design, the diodes D_1-D_5 play important roles in creating the proper dc voltage reference. As a gate driver switches a MOSFET, the charge in the corresponding bootstrap capacitor is depleted. That charge needs to be periodically replenished for the capacitor to stay at a certain voltage level needed by the MOSFET gates. As can be seen, the diode configuration for the top two capacitors and bottom four capacitors are slightly different. This is caused by the two different charging approaches for the related capacitors, which is the key for the proper operation of the driver system. Figure 3 shows the charging methods for the ibottom four capacitors, the details of which are explained as follows:

- To drive the lowest MOSFET $/T_3$ is relatively simple. The capacitor C_0 is in parallel with the gate driver dc power supply V_{gate}. The capacitor voltage can remain approximately fully charged at all time.

to-ground voltage level and as a result, the power converting quality can be improved dramatically with the same amount of switches. Due to the lack of redundant switching states, fixed dc sources such as batteries or rectifiers have to be used to form the inner loop dc sources. In this paper, the 3-cell, 8-level floating voltage source inverter is used to demonstrate the proposed gate driver design. However, the proposed design can also be suitable to the flying capacitor or diode-clamped types of multilevel inverters.

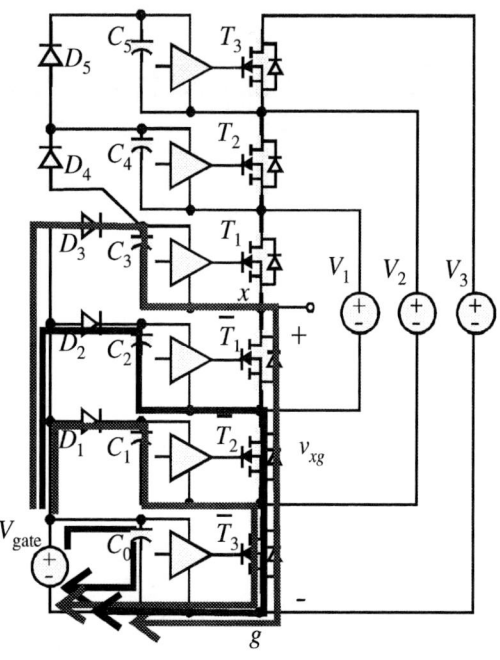

Figure 3. Capacitor charging path for $C_1 \sim C_3$

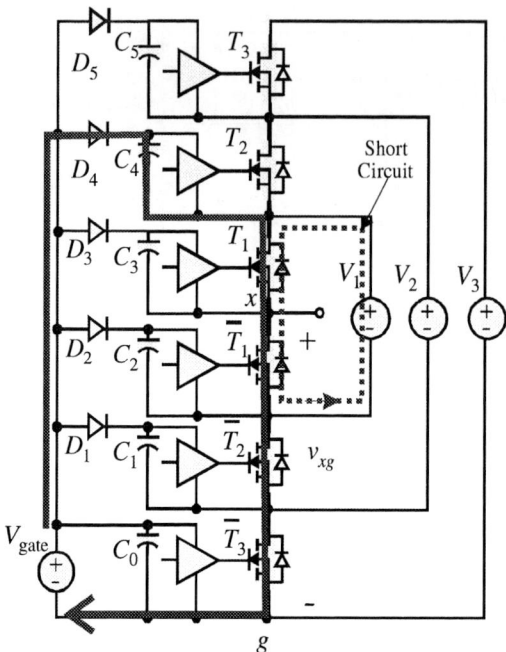

Figure 4. Wrong charging method for C_4 that will lead to short circuit on V_1.

- To drive the MOSFET $/T_2$, C_1 is used to form the floating supply. C_1 is charged from V_{gate} through D_1 if and only if the bottom of C_1 is grounded. That grounding action is easily accomplished when the bottom switch $/T_3$ is turned on. D_1 guarantees the single power flow direction to give the gate of $/T_2$ a proper voltage reference.
- To drive the MOSFET $/T_1$, C_2 is used to form the floating supply. C_2 is charged from V_{gate} through D_2 in a similar manner as C_1. The grounding of the bottom of C_2 is accomplished by simultaneously turning on the $/T_2$ and $/T_3$.
- To drive the MOSFET T_1, C_3 is used to form the floating supply. C_3 is charged from V_{gate} through D_3 in a similar manner as C_2. The grounding of the bottom of C_3 is accomplished when $/T_3$, $/T_2$ and $/T_1$ are all on.

For the capacitor voltage sources of the top two gate drivers, It should be noted that one can not use the same charging method as $C_1 \sim C_3$. In an attempt to charge C_4 by turning on $/T_3$, $/T_2$, $/T_1$ and T_1 results in shorting the power supply V_1 through switches T_1 and $/T_1$. This can be demonstrated by Fig. 4. To avoid the short circuit situation, one has to guarantee that T_1 and $/T_1$ are a complementary switching pair, which eliminates the possibility of the grounding reference of C_4 by turning on the bottom four MOSFETs. It is easy to see that charging C_5 has the similar problem. Therefore, a different charging strategy has to be developed. The proposed solution to this problem is to rearrange the diodes in the upper two driver stages into the configuration illustrated by D_4 and D_5 as shown in Fig. 2 and 3. Once that modification is made, C_3

becomes a supply not only for the gate driver of T_1, but also a supply for charging C_4 and C_5. The following explains the operation of the proposed gate driver circuit for the upper two MOSFETs:

- To drive the MOSFET T_2, C_4 is used to form the floating supply and C_3 become the source to charge C_4. The charging process of C_3 has been explained earlier. Once C_3 is charged, C_4 can then be charged by C_3 through D_4 when T_1 is on.
- To drive the MOSFET T_3, C_5 is used to form its floating supply. If T_2 and/or T_1 are turned on, then C_5 can be charged from C_3 and/or C_4, respectively.

The process of first charging $C_1 \sim C_3$ and then charging C_4 and C_5 from C_3 can be considered as a "charging cycle". For many applications, to repetitively command the charging cycles is necessary to guarantee the balanced voltage levels of $C_1 \sim C_5$. However, a potential problem with switching the power devices as a means to charge the bootstrap capacitors is the fact that the switching states are being subjected to the inverter output voltage levels. When enforced charging cycles are commanded, inverter output voltage waveform may show some unexpected jumps or dips. This potential problem is minimized by two factors. Firstly, the small size of the bootstrap capacitors gives a short charging time which allows them to be recharged faster. Secondly, for many non-critical applications, the small distortion of the inverter output voltage can be negligible as far as the load is concerned. Further investigation revealed the possibility of not having to include charging cycles at all in a certain application. It was found that if multilevel PWM was included in the control scheme of the inverter, then after the inverter has been started, the capacitor voltages can naturally be kept within a certain levels that are sufficient enough to switch the MOSFETs. This eliminates the necessity of the enforced charging cycles that tend to distort the inverter output voltages. However, the natural balancing of the gate driver capacitors is sensitive to the load condition and PWM modulation levels. One has to closely monitor the

variation of the capacitor voltages to make sure it won't fall down below the needed voltage levels. Otherwise, periodic insertion of charging cycles has to be included in the switching commands.

IV. GATE DRIVER VALIDATION

One phase leg of a 3-cell 8-level floating voltage source inverter was constructed to validate the proposed design and the special bootstrap capacitor charging method. A photo picture of the prototype can be found from Fig. 5. The functional schematic of the inverter is identical to that shown in Fig. 2. An 8-level step control was first adopted to command the switching action of the inverter. To guarantee the balancing of the bootstrap capacitors C_1~C_5, the extra charging command are injected on purposely. The inverter line-to-ground voltage waveform is shown in Fig. 6. As can been, some jumps and dips are included in the voltage waveform caused by the extra charging command. However for most of the applications, these distortions are insignificant compared to the overall multilevel inverter performance.

The 8-level PWM technique is also applied to the multilevel inverter prototype. The multilevel PWM signals were generated by comparing seven stacked in-phase triangle carriers to a sinusoidal reference wave [2, 5]. A graphical representation of the method with the modulation index equal to 1.0 is shown in Fig. 7. When using the 8-level inverter to drive an inductive load, the testing result for the line-to-ground voltage is shown in Fig. 8, where the fundamental reference frequency was set to 60 Hz and the triangle carrier frequency was set to 5 kHz. Explicit charging cycles (explained in the previous section) were omitted in order to evaluate the feasibility of relying on the PWM switching to naturally charge the bootstrap capacitors. The measured voltages for the bootstrap capacitors C_0~C_5 during approximately three cycles of the output voltage are shown in Fig. 9. As can be seen, the capacitor voltages were able to be maintained

Figure 5. Lab prototype of a 3-cell 8-level inverter applying the proposed gate driver design

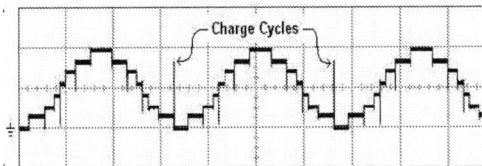

Figure 6. Inverter line-to-ground voltage waveform with 8-level step control and extra charging cycle command

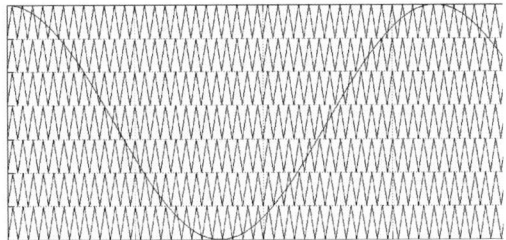

Figure 7. Graphic Representation of the 8-level PWM

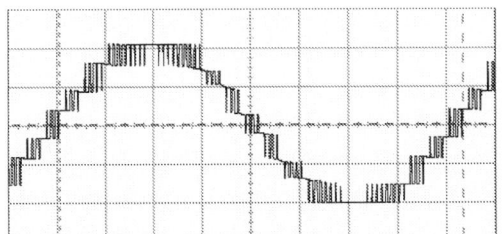

Figure 8. Testing results for the line-to-ground voltage applying PWM (Without extra charging command).

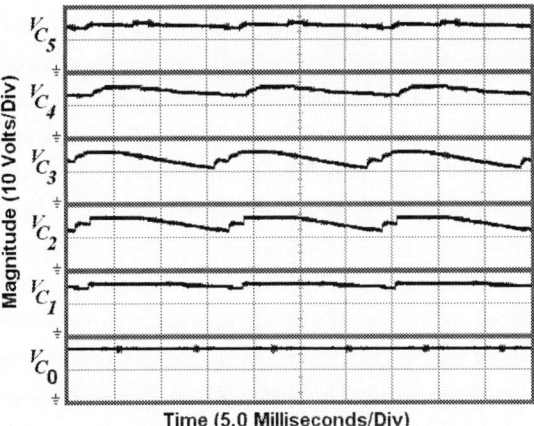

Figure 9. Voltages of the bootstrap capacitors applying PWM (Without extra charging command).

within a certain range naturally by the PWM switching commands. The most severe drop in voltage occurred across C_3, which make sense because C_3 is also used to charge the top two capacitors C_4 and C_5 as explained earlier. Despite the voltage variations, the bootstrap capacitors were adequate as floating voltage supplies when the load is not too heavy and the reference frequency is not too small. However, the drop on the gate driver floating power supplies will limit the MOSFETs' current conducting ability. Also, when the method applied to the IGBT types of applications, bigger voltage drop on the bootstrap capacitors may cause the IGBT to operate in the linear region and overheat quickly. If that happens, extra charging cycle commands have to be included in addition to the natural PWM switching to maintain the proper operation of the multilevel inverter.

V.   ANALYSIS

A number of factors need to be considered when sizing the bootstrap capacitors. First, as the inverter's fundamental output frequency decreases, the speed at which the inverter cycles through the sequence of switching states also decreases. Since the sequence is what helps maintain the charge on the bootstrap capacitors, there is a theoretical minimum frequency at which the inverter can reliably operate without the addition of charging cycles. A second consideration is the PWM frequency.   When the PWM frequency decreases, the switching resolution decreases. With a low enough resolution, the sequence may change such that the virtual charging cycles are no longer present and therefore the bootstrap capacitors may no longer have adequate charge. A third factor in bootstrap charge level is the PWM modulation index. As the PWM modulation index decreases, the virtual charge cycles may again be missing from the commanded sequence and, again, bootstrap capacitor charge levels may drop.

The size of the bootstrap capacitors is also important. They must be small enough to charge quickly, but large enough to discharge slowly. More information on bootstrap capacitor sizing is available [11].

Although the method presented in this paper is extendable to inverters with higher numbers of output levels, the extendibility is not limitless. If N is the number of power switches in a floating source multilevel inverter, the $N/2 + 1$ capacitor is the one supplying charge to all the capacitors above itself. That means the charge across the $N/2 + 1$ capacitor will be split $N/2$ ways. For this reason, as the method is extended, the size of the $N/2 + 1$ capacitor becomes more critical.

VI.   CONCLUSION

This paper has presented an interesting low-cost driver stage design for multilevel inverters, incorporating a novel method of charging bootstrap capacitors which require only a single power supply for all of the single channel gate drivers.   The method is applicable to floating voltage source or flying capacitor inverters with other voltage levels. A prototype 3-cell 8-level inverter has been built to verify the proposed method.

REFERENCES

[1] A. Nabe, I. Takahashi, and H. Akagi, "A New Neutral-Point Clamped PWM Inverter," *IEEE Transactions on Industry Application,* vol. 17, pp. 518–523, Sept./Oct. 1981.

[2] T.A. Meynard, H. Foch, "Multi-level Conversion: High Voltage Choppers and Voltage-source Inverters," *Proceedings of the IEEE Power Electronics Specialist Conference*, pp. 397-403, Toledo, Spain, 1992.

[3] J. S. Lai and F. Z. Peng, "Multilevel Converters—A New Breed of Power Converters," *IEEE Transaction on Industry Application*, vol. 32, pp. 509–517, May/June 1996.

[4] M. D. Manjrekar, P. Steimer, and T. A. Lipo, "Hybrid Multilevel Conversion System: A Competitive Solution for High Power Applications," *IEEE Transactions on Industry Application,* vol. 36, pp. 834–841, May/June 2000.

[5] X. Kou, K. A. Corzine, and Y. Familiant, "Full Binary Combination Schema for Floating Voltage Source Multilevel Inverters," *IEEE Transaction on Power Electronics*, vol. 17, pp. 891-897, Nov. 2002.

[6] G. Walker, G. Ledwich, "An isolated MOSFET Gate Driver," *Australasian Universities Power Engineering Conference*, AUPEC'96, vol. 1, pp. 175-180, Melbourne, Australia, 1996.

[7] B. A. Welchko, M. B. Correa, T. A. Lipo, "A Three-Level MOSFET Inverter for Low Power Drives", *IEEE Transactions on Industrial Electronics*, vol. 51, part 3, pp. 669-674.

[8] Jonathan Adams, "Bootstrap Component Selection for Control ICs," DT-98-2a, International Rectifier, El Segundo, CA.

[9] CT-Concept Technology, Ltd, "Datasheet 6SD10E, Six-Channel IGBT/MOSFET Driver".

[10] CT-Concept Technology, Ltd, "Datasheet 6SD10E, Dual IGBT/MOSFET Driver".

[11] G. Carrara, S. Gardella, M. Marchesoni, R. Salutari, and G. Sciutto, "A new multilevel PWM method: A theoretical analysis," *IEEE Transaction on Power Electronics*, vol. 7, no. 3, pp. 497-505, May 1992.

John. J. Graczkowski received B.S.E.E. degree from the University of Wisconsin – Platteville in 2004. He is currently employed in the hardware design group at Cray Inc. in Chippewa Falls, Wisconsin, USA.

Kevin Neff received B.S.E.E. degree from the University of Wisconsin – Platteville in 2004. He is currently a graduate student at Mayo College of Medicine, Minnesota, USA. His research interests include analog computation and molecular simulation.

Xiaomin Kou received his B.S.E.E. degree from Chong Qing University, Chong Qing, China, in 1995. He received his M.S. degree from the University of Wisconsin - Milwaukee in 2001 and the Ph.D. degree in 2003. In the fall of 2003, Xiaomin joined the faculty at the University of Wisconsin – Platteville as an assistant professor. His research interests include power electronics, electrical machinery and motor controls.

2006 5th International Power Electronics and Motion Control Conference

An Effective Method to Suppress Resonance in Input LC Filter of a PWM Current-Source Rectifier

Y.W. Li*, B. Wu*, N. Zargari**, J. Wiseman**, and D. Xu*

*Dept. Electrical & Computer Engineering, Ryerson University, Toronto, Canada
**MV Drive R&D, Rockwell Automation, Cambridge, Ontario, Canada
yunweili@ee.ryerson.ca, bwu@ee.ryerson.ca, nrzargari@ra.rockwell.com, jwiseman@ra.rockwell.com,
dxu@ee.ryerson.ca

Abstract— The PWM Current-Source Rectifier (CSR) is becoming a preferred choice to provide a DC current source for DC loads or current source fed drives. However the control of the resonance of the input LC filter is one of the main challenges. To dampen the LC resonance with minimum control and implementation efforts, a two-step Posicast controller for modulation signal shaping is first examined, which can be easily inserted into the existing CSR control loop. The parameters of the two-step Posicast controller are fixed by the lightly damped plant, which makes it unsuitable at high power levels due to the inaccurate compensation associated with the low switching frequency. A three-step compensator is then designed in this work. By selecting its parameters according to the existing control algorithm, the three-step compensator is very effective for the low switching frequency CSR with precise resonance compensation and easy implementation. Detailed design and analysis are presented. Simulation and experimental results are provided to verify the effectiveness of the proposed damping approach.

Keywords- current source rectifier; active damping; Posicast controller; three-step compensator

I. INTRODUCTION

The PWM Current-Source Rectifier (CSR) shown in Figure 1 is becoming a preferred choice to provide a DC current source for DC loads or current source fed drives, due to its input power factor regulation, line current harmonic mitigation and DC current control capabilities. At the high power medium voltage level, the most common switching devices for a PWM CSR are symmetric gate-turn-off thyristors (GTOs) or gate-controlled thyristors (GCTs), which are normally used with a switching frequency below 1 kHz.

As shown in Figure 1, an LC filter is normally required at the input of a PWM CSR, to assist in the switching devices commutation as well as mitigating the line current harmonics. At high power levels, passive damping of the resonant mode of this filter using resistors is impractical. Several different approaches have been developed to dampen the resonance actively, which can be categorized

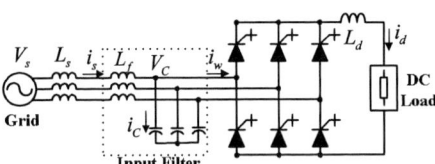

Figure 1. Medium voltage three-phase PWM CSR.

as the virtual damping resistor [1, 2], control signal compensation by feeding forward the LC filter model [3], shaping the control signal with different compensators [4-7]. Among them, the modulation signal shaping method is identified as being very effective and requires minimum design effort by being simply embedded into the existing control loop (compared to the virtual resistor approach which requires an additional feedforward loop) and with reduced sensitivity to high frequency noises (unlike the plant model inversion method).

In this paper, a two-step Posicast controller is first investigated. With parameters fixed by the lightly damped plant, the Posicast controller is effective with high control system sampling frequency [7-10]. But it is subject to inaccurate modulation signal shaping when implemented in the high power low switching CSR system, unless the CSR control algorithm is modified accordingly. The proposed three-step compensator, on the contrary, has significant implementation advantages in medium voltage CSR with more parameter selection freedom. The delay times for the second and the third step in the three-step compensator can be chosen first according to the existing CSR control algorithm, and the magnitudes for each step can be calculated subsequently. As a result, the easily implemented three-step compensator can compensate the resonance very precisely. Detailed design, analysis and comparison of the compensators are presented in this paper. The effectiveness of the proposed damping method has been verified by simulation using Matlab/Simulink and experimentally using a 10 kVA low voltage PWM CSR prototype.

II. CONTROL SCHEME OF THE CSR

Figure 2 illustrates a simple control scheme for the CSR, where a DC current i_d is measured and controlled in

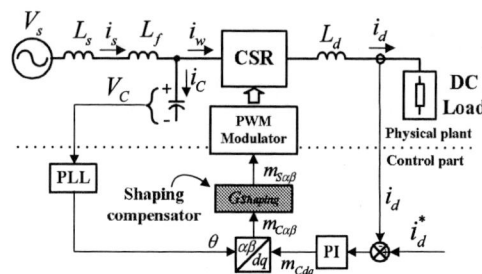

Figure 2. Proposed damping control scheme for the CSR.

a closed-loop manner. Input filter capacitor voltage V_C are measured to compute the reference angle (θ) for frame transformations. Note that voltage V_C can also be fed forward to compensate the input power factor displacement caused by the filter capacitor [11]. More complicated multi-loop control schemes for DC current control and power factor regulation with more measured variables are not studied here, since the focus of this work is for resonance damping with minimum control effort and feedback variables.

For the DC current control, the measured DC current i_d is compared with its reference i^*_d and the error is fed into a PI controller. Output of the DC current controller is subsequently transformed to the stationary α-β reference frame and fed to a modulation signal compensator (a two-step Posicast controller or a three-step compensator) which is inserted here to mitigate the resonance excited by a disturbance within the control loop by properly shaping the modulation signal. Finally the regulated modulation signal is fed to a PWM modulator for switching the CSR. In this work the space vector PWM modulation (SVPWM) is employed. In the following sections, the two modulation signal shaping compensators will be compared.

III. TWO-STEP POSICAST CONTROLLER

The two-step Posicast controller functions to split up a step input command into two smaller segments of steps. It is proposed to be used within a feedback control loop with reduced sensitivity in [7], where it is called "half-cycle" Posicast. Figure 3 illustrates the principle of the two-step Posicast control, where a unit step is separated into two steps with the first one leading the second by a time T_d. The magnitude M_1 of the first step and the transfer function of the two-step Posicast controller is expressed in (1) and (2) respectively.

$$M_1 = \frac{1}{1+\delta} \tag{1}$$

$$G_{tsp} = \frac{1}{1+\delta} + \frac{\delta}{1+\delta}e^{-sT_d} \tag{2}$$

where δ is the lightly damped system's step response overshoot and T_d is half of the resonance period. By separating a single unit step command into two using (1)

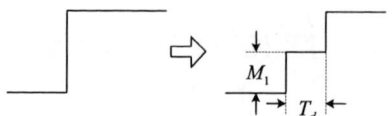

Figure 3. Principle diagram of two-step Posicast controller.

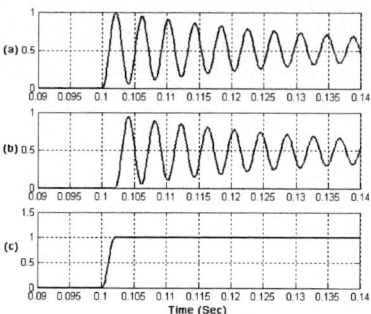

Figure 4. Step responses of the two-step Posicast controller: (a) first step, (b) second step, (c) summation of the two steps.

and (2), the lightly damped system's transient terms become zero after the second step is applied. A step response example is shown in Figure 4, where it can be seen that when the system's response of first step command reaches its peak value, the second step command is added. The system's response to second step has the same oscillatory effect but in the opposite direction and would therefore cancel the oscillations excited by the first step command, giving a fast and smooth response with specific settling time of T_d and without overshoot.

Frequency domain analysis of the two-step Posicast controller G_{stp} can be found in [7], where the Posicast controller exhibits multiple-notch filter characteristic with infinite number of zeros spaced at the odd multiples of the lightly damped system's resonant frequency. The first pair of zeros functions to cancel the dominant pair of poles introduced by the LC filter. However inaccurate knowledge of the plant dominant poles will result in residual oscillation for the Posicast control. To mitigate the imperfect compensation effects, the Posicast controller can be inserted within a feedback closed loop (as in this work) to obtain a more robust option. Sensitivity analysis of Posicast in feedback loop can be found in [7].

For digital implementation, the form of two-step Posicast controller in (2) should be discretized to (3):

$$G_{Ztsp} = \frac{1}{1+\delta} + \frac{\delta}{1+\delta}Z^{-\text{int}(T_d/T_s)} \tag{3}$$

where T_s is the control system sampling frequency and int( ) means the integer number of delayed samples. If T_d/T_s is not an integer, the truncation of it might result in serious delay time deviation for the second step command, unless a higher sampling frequency to resonant frequency ratio is available for higher resolution. But unfortunately, usually this is not the case in high power medium voltage

CSR. For example, with a system resonant frequency of 245 Hz and the DSP sampling frequency of 1080 Hz, the number of delayed samples is 2 with in a delay time deviation about $\Delta t = 0.2$ ms, which is large enough to affect the damping performance if the existing control algorithm (sampling frequency and switching frequency) is not modified (to achieve an integer T_d/T_s).

IV. THREE-STEP COMPENSATOR

As discussed, parameters for the two-step Posicast controller are fixed by the lightly damped plant, and only an integer T_d/T_s can provide accurate resonance compensation, which causes the implementation of Posicast controller to a medium voltage CSR troublesome since much more effort has to be made to modify the control algorithm accordingly to provide an integer T_d/T_s.

Inspired by the Posicast compensation principle, separating a step response into three steps would yield more freedom for delay time selection, since a first oscillatory response can be cancelled by two additional oscillations here with different magnitudes. Figure 5 illustrates the principle diagram of such a three-step compensator, where a unit step is separated into three segment steps with magnitudes of A_1 A_2 A_3 respectively and delay times of T_d and $2T_d$ for the second and the third step respectively. The expression of such a three-step compensator is shown in (4).

$$G_{Th}(s) = A_1 + A_2 e^{-T_d s} + A_3 e^{-2T_d s} \qquad (4)$$

To provide significant implementation advantage, the delay time can be first fixed according to the existing control algorithm, and the remaining task is to compute the magnitudes of the three segment steps. For a standard second order system (such as the resonant plant in (6)), the magnitudes of each step of the three-step compensator can be calculated using (5), where ω_n and ζ are the plant natural undamped frequency and damping ratio respectively.

$$
\begin{cases}
A_1 = \dfrac{1}{1 - 2\cos\delta \cdot e^{-\zeta \omega_n T_d} + e^{-2\zeta \omega_n T_d}} \\[2mm]
A_2 = \dfrac{-2\cos\delta \cdot e^{-\zeta \omega_n T_d}}{1 - 2\cos\delta \cdot e^{-\zeta \omega_n T_d} + e^{-2\zeta \omega_n T_d}} \\[2mm]
A_3 = \dfrac{e^{-2\zeta \omega_n T_d}}{1 - 2\cos\delta \cdot e^{-\zeta \omega_n T_d} + e^{-2\zeta \omega_n T_d}}
\end{cases}
\qquad (5)
$$

Observing the parameters in (5), it can be found that under the conditions of an undamped system with $\zeta = 0$ and a delay time $T_d = T_r/6$ (($T_r = 2\pi/\omega_r$ is the resonant period and $\omega_r = \omega_n(1 - \zeta^2)^{1/2}$ is the resonant frequency)), the magnitudes of three segment steps are $A_1 = A_3 = 1$ and $A_2 = -1$. This is actually the unity three-step method proposed in [6], where a reference unit step is followed by two additional unit steps (one negative and one positive) with delays of $T_r/6$ and $T_r/3$ respectively for oscillatory effect cancellation. The settling time ($T_r/3$) using this

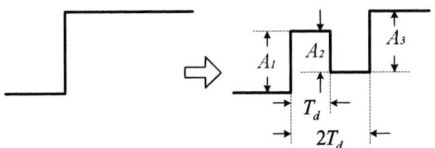

Figure 5. Principle diagram of proposed three-step compensator.

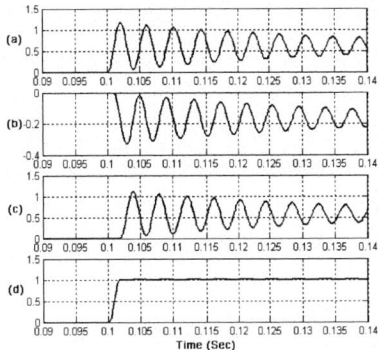

Figure 6. Step responses of the three-step compensator: (a) first step, (b) second step, (c) third step, and (d) summation of the three steps.

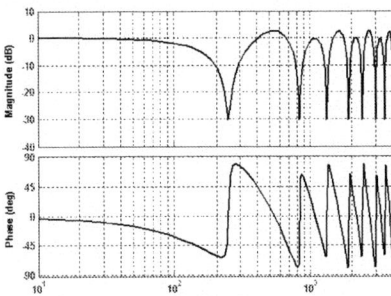

Figure 7. Bode plot of the three-step compensator.

unity three-step method is faster than the two-step Posicast ($T_r/2$). But the truncation problem mentioned must be more serious since both two parameters of the unity three-step method (T_d and $2T_d$) fixed by the lightly damped plant are related to the time delay (compared to δ and T_d of the two-step Posicast). Therefore, to use this unity three-step compensator, the control system sampling frequency and device switching frequency has to be changed according to the resonant frequency for an integer T_d/T_s as was done in [6]. Furthermore, this unity three-step method only considers the ideal case with an undamped system, where the oscillatory term does not decay with time.

It can also be revealed from (5) that for a three-step compensator with $T_d = T_r/4$, the second step becomes zero and this special form is actually the two-step Posicast controller in (2).

One can further find that a delay time of $T_d < T_r/4$ results in a negative second step with $A_2 < 0$. The second step becomes positive when $T_d > T_r/4$. Note that a larger delay time can still eliminate the resonance, but with longer setting time, as the response will only reach steady state once the last step is added. Furthermore, a large delay might affect the control loop dynamics. Therefore a desired delay time should be in the range of $T_d < T_r/4$, which can help to achieve shorter settling time and can be easily satisfied with the sampling frequency larger than four times the LC resonance frequency.

Step response of the three-step compensator is shown in Figure 6, where the top three step responses have the same oscillatory term (determined by the lightly damped plant) but with different magnitudes and delay times (defined by the three-step compensator). Once the third step is added, the response reaches the steady state resulting in a smooth summation response to a unit step with settling time of $2T_d$, as shown in the last trace in Figure 6.

Bode plot of the three-step compensator is drawn in Figure 7, where the three-step compensator also exhibits multiple-notch filter like characteristic, by having infinite number of zeros spaced at $nT_r/T_s\pm1$ (where n is an integer) times the plant resonant frequency. Again, the first pair of zeros functions to cancel the dominant pair of poles introduced by the LC filter. Similar to the two-step Posicast controller, the three-step compensator has limited high frequency gains and therefore is more robust to high frequency noises compared to model inversion pole cancellation method, which typically exhibits noise sensitivity due to the increasing gains at high frequency.

V. IMPLEMENTATION OF THE COMPENSATORS

A. CSR System Model

Assuming the ratio of switching frequency to fundamental frequency is high enough, the modulation signal of each leg of a CSR can be represented as a sinusoid m. From Figure 2, the resonant model transfer function from m to line current can be written in (6), and the DC current open loop transfer function can be expressed in (7), where R_f, R_s represent the filter and line resistance, i_s, i_d and V_s are the line current, DC current and grid voltage respectively, R_{load} is the equivalent DC load, V_{C_dq} and m_{dq} are the d or q components of the filter capacitor voltage and modulation signal respectively (V_{C_q} is zero with a frame synchronous to V_C, see Figure 2).

$$G_{res} = \frac{i_s}{m} = \frac{i_d}{C_f(L_f + L_s)s^2 + C_f(R_f + R_s) + 1} \quad (6)$$

$$\frac{i_d}{m_d} = \frac{V_{C\_d}}{L_d s + R_{load}} \quad (7)$$

B. Plant Nonlinearity Compensation

A complication anticipated when implementing the compensator to the medium voltage CSR is the nonlinear resonant plant model as expressed in (6). With a large enough input filter resonant frequency, the dynamics of the resonance compensation is much faster than that of the DC current control loop, therefore the DC current in the numerator of (6) could be approximated as a constant

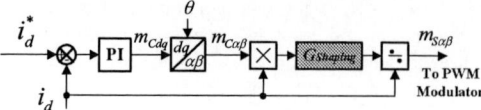

Figure 8. Compensation of plant model nonlinearity.

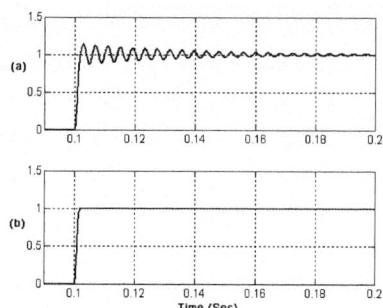

Figure 9. Step response of plant with (a) two-step Posicast controller and (b) three-step compensator.

when designing the compensator. But for a high power CSR system with a low switching frequency and large filter components, the LC resonant frequency is quite limited, and the dynamics of the DC current loop would be coupled with that of the control signal shaping compensator. The compensation of the plant nonlinear effects can be achieved by feeding forward the measured DC current as shown in Figure 8, where the shaping compensator acts on the product of modulation signal and the DC current (instead of shaping only the modulation signal as shown in Figure 2), and the output is normalized by the same DC current before fed to the PWM modulator. By this means, the DC current control loop dynamics can be compensated and the compensator can eliminate the resonance more effectively.

C. Performance Comparison

As has been discussed, the proposed three-step compensator can be implemented very easily without any modification of the existing control algorithm and can precisely eliminate the LC resonance excited by the reference command. On the contrary, the truncation of T_d/T_s into integer associated with two-step Posicast controller results in inaccurate resonance compensation. A comparison of these two compensators is illustrated in Figure 9. The plant has a resonant frequency of 245 Hz, and the high power CSR is controlled with a switching frequency of 540 Hz and a sampling frequency of 1080Hz. Therefore the Posicast controller has to use two sample delays for the second step, which generates 0.2 ms delay time deviation and causes residual resonance as shown in Figure 9(a). For the three-step compensator, with a defined one control sample delay, the magnitudes of three steps can be precisely calculated according to the resonant plant. The response of plant with such a three-step compensator is therefore very smooth without any residual oscillations as can be seen in Figure 9(b).

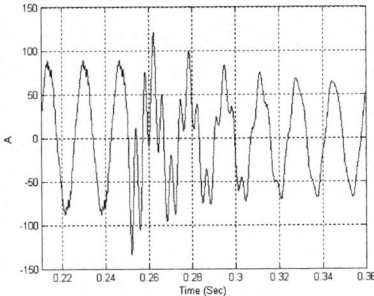

Figure 10. Line current (phase C) without damping (simulated).

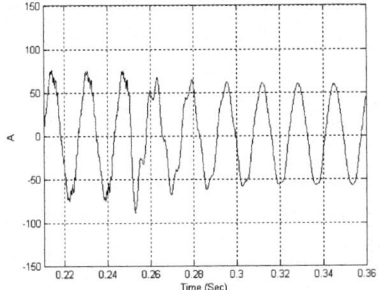

Figure 11. Line current (phase C) with two-step Posicast controller (simulated).

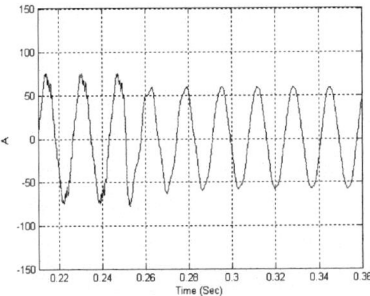

Figure 12. Line current (phase C) with proposed three-step compensator (simulated).

VI. SIMULATION RESULTS

Simulation results obtained using Matlab/Simulink are shown and discussed in this section. System parameters are listed in TABLE I. To excite the LC resonance, the DC current reference is suddenly changed from 180 A to 90 A at t=0.25 s. Without any damping method implemented, the line current has significant transient resonance as shown in Figure 10. This resonance, if not dampened properly, may degrade the CSR DC current control.

With the two-step Posicast controller implemented, the line current resonance shown in Figure 11 is highly suppressed. But as expected, the truncation of T_d/T_s into integer results in delay time deviation in the Posicast control and this inaccurate compensation causes residual resonance as can be noticed from Figure 11. Finally, the three-step compensator is tested with the line current shown in Figure 12. It can be seen that by properly

TABLE I SIMULATION PARAMETERS

Parameter	Value
Nominal grid line-line voltage (rms)	4160 V
Nominal power	1 MVA
Nominal Frequency	60 Hz
Switching frequency	540 Hz
Sampling frequency	1080 Hz
Equivalent input filter inductance $(L_s + L_s)$	0.2 p.u.
Equivalent input filter resistance $(R_s + R_s)$	0.03 p.u.
Input filter capacitance (C_i)	0.3 p.u.
DC link inductance (L_d)	2 p.u.
DC load (R_{load})	0.35 p.u.

selecting the delay time of the three-step compensator (one sample delay here) and calculate the respective magnitude of each step accordingly, the three-step compensator can effectively eliminate the resonance by precisely shaping the modulation signal.

In passing, it can be observed that the line current magnitude is reduced when the two-step Posicast controller or three-step compensator are implemented. This is because in the experiment, no compensation is made for the CSR system input power factor displacement caused by the filter capacitor. The delays in modulation signal shaping compensators cause the CSR to draw positive reactive power, which partly compensates the negative reactive power absorbed by the filter capacitor.

VII. EXPERIMENTAL VERIFICATION

A scaled 10 kVA hardware prototype of the PWM CSR is implemented for experimental verification. For the experimental system, the source voltage is provided by a three-phase programmable AC source at 208 V. The input LC filter is comprised of three filter inductors of 2.5 mH each and three delta-connected filter capacitors of 180 uF each. The experimental system parameters are summarized in TABLE II (filter capacitance in the table is equivalent star-connected value). The rectifier, using ABB GCTs switching at 540 Hz, is controlled by a DSP-FPGA control system, with a TMS F2812 fixed-point DSP for implementation of the control algorithms and the FPGA for the PWM signal generation and system protection.

In the experiment, the reference DC current is changed from 15A to 5A. The line current waveforms are shown in Figures 13-15. It can be seen that without damping control, the line current in Figure 13 has serious transient resonance under the DC reference change. This transient resonance is mitigated by the two-step Posicast controller. But as illustrated in Figure 14, a transient overshoot near 18A is still noticeable due to the inaccurate two-step compensation. Finally the best transient of line current is shown in Figure 15, where the precise three-step compensator is implemented.

Similarly, reduced current magnitude can be observed in the experiment when the modulation signal shaping compensator is implemented. Again, this is caused by the compensator delays, which compensate the reactive power drawn by the filter capacitor. Also note that compared to the simulation results, the LC resonance in the line current

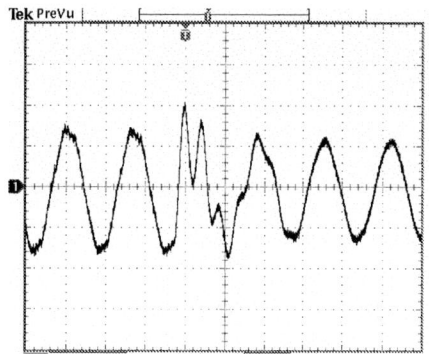

Figure 13. Line current (phase C) without damping (experimental, magnitude: 10A/div, time: 10ms/div).

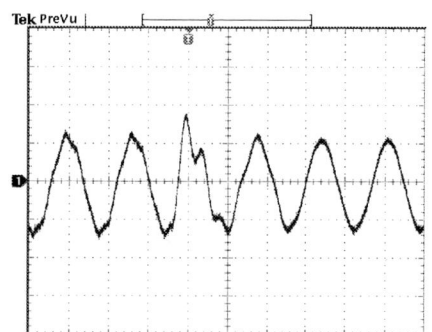

Figure 14. Line current (phase C) with two-step Posicast controller (experimental, magnitude: 10A/div, time: 10ms/div).

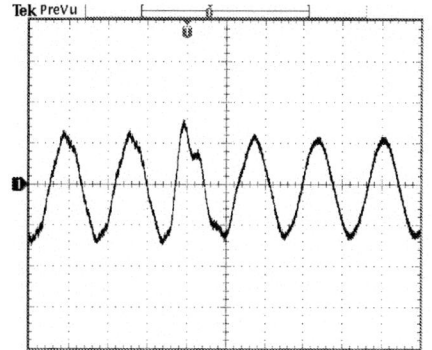

Figure 15. Line current (phase C) with proposed three-step compensator (experimental, magnitude: 10A/div, time: 10ms/div).

is less serious in experiment. This is due to the larger filter resistance at the lower power rating, which provides more damping in the LC circuit.

VIII. CONCLUSION

In this paper a three-step modulation signal shaping compensator is designed and proposed to improve the LC resonance damping in a PWM CSR system. Unlike the two-step Posicast controller, whose parameters are fixed by the resonant frequency, the delay units in the proposed three-step compensator can be designed in accordance with the existing CSR control sampling frequency.

TABLE II EXPERIMENT PARAMETERS

Parameter	Value
Nominal grid line-line voltage (rms)	208 V
Nominal power	10 kVA
Nominal Frequency	60 Hz
Switching frequency	540 Hz
Sampling frequency	1080 Hz
Equivalent input filter inductance ($L_f + L_s$)	0.22 p.u.
Equivalent input filter resistance ($R_f + R_s$)	0.06 p.u.
Input filter capacitance (C_f)	0.3 p.u.
DC link inductance (L_d)	2.6 p.u.
DC load (R_{load})	1.5 p.u.

Therefore the three-step compensator can be implemented conveniently with accurate modulation signal shaping capability. The proposed method has been tested in Matlab/Simulink simulation and experimentally using a 10 kVA GCT based CSR prototype. Close agreement between the simulation and experimental results verifies the effectiveness of the proposed damping method.

REFERENCES

[1] J. C. Wiseman and B. Wu, "Active damping control of a high-power PWM current source rectifier for line-current THD reduction," *IEEE Trans. Ind. Electron.*, vol. 52, pp. 758–764, Jun. 2005.

[2] Y. Sato and T. Kataoka, "A current-type PWM rectifier with active damping function," *IEEE Trans. Ind. Applicat.*, vol. 32, pp. 533–541, May/Jun. 1996.

[3] M. Salo and H. Tuusa, "A vector controlled current-source PWM rectifier with a novel current damping method," *IEEE Trans. Power Electron.*, vol. 15, pp. 464-470, May 2000.

[4] V. Blasko and V. Kaura, "A novel control to actively damp resonance in input LC filter of a three-phase voltage source converter," *IEEE Trans. Ind. Applicat.*, vol. 33, pp. 542–550, Mar/Apr. 1997.

[5] M. Liserre, A. Dell'Aquila, and F. Blaabjerg, "Stability improvements of an LCL-filter based three-phase active rectifier," in *Proc. IEEE-PESC'02*, 2002, pp. 1195-1201.

[6] Y. Neba, "A simple method for suppression of resonance oscillation in PWM current source converter," *IEEE Trans. Power Electron.*, vol. 20, pp. 132-139, Jan. 2005.

[7] J.Y. Hung, "Feedback control with Posicast," *IEEE Trans. Power Electron.*, vol. 50, pp. 94-99, Feb. 2003.

[8] Q. Feng, J.Y. Hung, and R. M. Nelms, "Digital control of a boost converter using Posicast," in *Proc. IEEE-APEC*, 2003, pp. 990-995.

[9] P. C. Loh, D. M. Vilathgamuwa, S. K. Tang, and H. L. Long, "Multilevel dynamic voltage restorer," *IEEE Letters Power Electron.*, vol. 2, pp. 125-130, Dec. 2004.

[10] Y.W. Li, P. C. Loh, F. Blaabjerg, and D. M. Vilathgamuwa, "Investigation and improvement of transient response of DVR at medium voltage level," in *Proc. IEEE-APEC*, 2006, pp. 1074-1080.

[11] N. R. Zargari, Y. Xiao, and B. Wu, "A near unity input displacement factor PWM rectifier for medium voltage CSI based AC drives," in *Proc. IEEE-APEC*, 1997, pp. 327-332.

2006 5th International Power Electronics and Motion Control Conference

Topological and Modulation Design of Three-Level Z-Source Inverters

P. C. Loh[*], F. Gao[*] and F. Blaabjerg[**]

[*] School of Electrical and Electronic Engineering, Nanyang Technological University, Singapore
[**] Institute of Energy Technology, Aalborg University, Denmark
epcloh@ntu.edu.sg; gaof0001@ntu.edu.sg; fbl@iet.aau.dk

Abstract—This paper presents the development of two three-level cascaded Z-source inverters, whose output voltage can be stepped down or up unlike a traditional buck three-level inverter. The proposed inverters are designed using two three-phase voltage-source inverter bridges, supplied by two uniquely designed Z-source impedance networks and cascaded at either their dc sides to form a dc-link-cascaded Z-source inverter or ac outputs using single-phase transformers to form a dual Z-source inverter. For controlling both inverters, various modulation schemes are designed with their performances verified experimentally using an implemented laboratory prototype.

Keywords-Three-level inverters, voltage-source inverters, buck-boost, pulse-width modulation.

I. INTRODUCTION

Two-level Z-source inverters of both voltage-source (VS) [1-3] and current-source (CS) [4, 5] types are recent single-stage buck-boost power converters proposed for use in fuel cell energy conversion systems [1] and motor drives [2]. Topologically, the only difference noted between traditional (buck only) and Z-source inverters is the presence of a Z-source impedance network connected between the input dc source and inverter circuitry, as shown in Fig. 1 for a VS-type Z-source inverter. This unique X-shaped impedance network is implemented using two inductors (L_1 and L_2) and capacitors (C_1 and C_2) with the inductors effectively appearing in series between the capacitors and inverter circuitry when any two switches from the same phase-leg (e.g. SA and SA' in Fig. 1) are turned ON to boost the ac output voltage in a defined shoot-through state. For voltage-boosting, the Z-source inverter should therefore be controlled using a modified pulse-width modulation (PWM) scheme with shoot-through states inserted in between active and null states [1-5].

Extending from the two-level Z-source topological concept, a three-level Z-source neutral-point-clamped (NPC) inverter has recently been proposed in [6] with an appropriate pulse-width modulation (PWM) scheme designed for its control. To provide more three-level topological alternatives for consideration, this paper now proposes two modified Z-source cascaded inverters, implemented using two **three-phase** inverter bridges

(different from the series connection of **single-phase** H-bridges in a traditional cascaded inverter), for comparison with the NPC topology.

For the designed inverters, the two inverter bridges are either connected at their dc links or at their ac outputs using three single-phase transformers to give the dc-link-cascaded and dual Z-source inverters respectively. Using two Z-source impedance networks, both inverters can again boost their output voltages by carefully shooting through the networks with equal time durations (termed as balanced voltage-boosting). To achieve that, a number of different PWM schemes are designed in the paper. Specifically, a modified carrier disposition scheme is formulated for controlling both inverters with alternate Z-source shoot-through, and a modified phase-shifted-carrier (PSC) scheme is proposed for controlling the dual Z-source inverter with simultaneous Z-source boosting and a better switch utilization. Interestingly, the designed PSC scheme also allows only a single Z-source network to be used for powering both the three-phase bridges in a dual Z-source inverter, hence halving its realization cost. These modulation schemes have been tested experimentally using an implemented laboratory prototype that can be configured as either the dc-link-cascaded or dual Z-source inverter.

II. DC-LINK-CASCADED Z-SOURCE INVERTER

The dc-link-cascaded Z-source inverter is shown in Fig. 2, where two isolated power sources are connected through two Z-source impedance networks to a uniquely configured inverter circuitry. Obviously, this proposed inverter uses only twelve switches with no clamping diodes required (unlike a Z-source NPC inverter). Its operational principles and modulation technique are now described in details before a thorough comparison with the Z-source NPC inverter can be pursued.

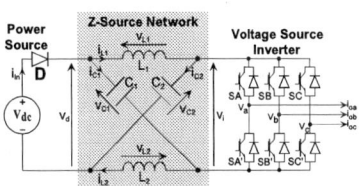

Fig. 1. Topology of two-level Z-source voltage-source inverter.

1-4244-0448-7/06/$25.00 ©2006 IEEE

TABLE I. SWITCHING STATES OF DC-LINK-CASCADED Z-SOURCE INVERTER (X = A, B OR C, AND $V_{i(U)} = V_{i(G)} = V_i$)

State Type	ON Switches	Voltage
Non-Shoot-Through	$SX_{(U)}$, $SX_{(G)}$, $SW_{(U)}$, $SW_{(G)}$	$+V_i$
Non-Shoot-Through	$SX'_{(U)}$, $SX_{(G)}$, $SW_{(U)}$, $SW_{(G)}$	0
Non-Shoot-Through	$SX'_{(U)}$, $SX'_{(G)}$, $SW_{(U)}$, $SW_{(G)}$	$-V_i$
*Non-Shoot-Through	$SX_{(U)}$, $SX'_{(G)}$, $SW_{(U)}$, $SW_{(G)}$	$-V_i$
Upper Shoot-Through (U)	$SX_{(U)}$, $SX'_{(U)}$	0
Lower Shoot-Through (G)	$SX'_{(U)}$, $SX_{(G)}$, $SX'_{(G)}$	0

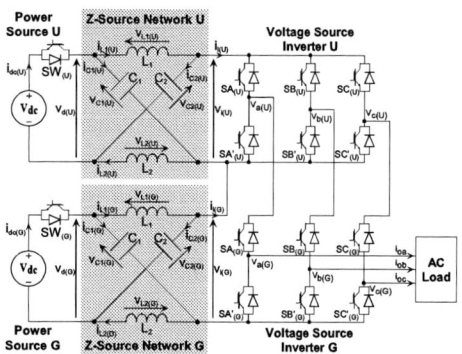

Fig. 2. Topology of dc-link-cascaded Z-source inverter.

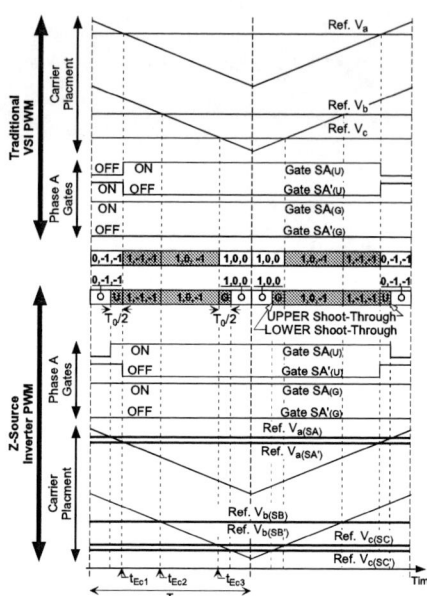

Fig. 3. Modulation of traditional and Z-source inverters when the reference phasor is in triangle 3 on the three-level vector diagram shown in Fig. 4.

A. Principles of Operation

As illustrated in Fig. 2, the dc-link-cascaded inverter is configured such that the output terminals of the upper three-phase bridge are used as independent positive rails for the lower three-phase bridge. In addition, the negative dc rail of the upper Z-source network is linked to the positive rail of the lower Z-source network to form a common neutral point with a potential of 0V. Using this topological connection, the proposed Z-source inverter can generate three distinct output voltage levels, as illustrated in Table I, where the different switch combinations are grouped under shoot-through and non-shoot-through states. During shoot-through, the inverter output voltage is boosted by either short-circuiting the upper or lower Z-source network (distinctly identified as "U" and "G" in Table I) with the only limitation being that the boosting durations of the two networks must be equal to ensure balanced voltage boosting. Alternatively, shooting through of both Z-source networks simultaneously by turning ON all switches is possible, but modulation analysis presented in the next section would reveal that this state introduces a volt-sec error and therefore should be avoided. Another point to note while shooting through the Z-source networks is that the corresponding input switches $SW_{(U)}$ and $SW_{(G)}$ must be turned OFF appropriately to disconnect the dc sources from the Z-source networks during voltage boosting, as noted in [1-3, 6] (for unidirectional operation, where only input diodes are used, the diodes will reverse-bias accordingly depending on which Z-source network is shorted).

The stored inductive energy during shoot-through can next be transferred to the ac load by transiting to any of the non-shoot-through states (see Table I), where the switches now are grouped into two complementary pairs ($\{SX_{(U)}, SX'_{(U)}\}$ and $\{SX_{(G)}, SX'_{(G)}\}$) per phase. A feature noted with the non-shoot-through states is that the voltage level of $-V_i$ (assuming $V_{i(U)} = V_{i(G)} = V_i$) can be generated by two distinct switch combinations, but closer observation would reveal that for single step transition between 0V and $-V_i$, transiting from $\{SX'_{(U)}, SX_{(G)}\}$ ON to $\{SX_{(U)}, SX'_{(G)}\}$ ON (indicated with an asterisk in Table I) would give rise to two switch commutations, as compared to only a single commutation for all other state transitions. Therefore, to minimize switching losses, only the state with $\{SX'_{(U)}, SX'_{(G)}\}$ ON should be used for generating $-V_i$. In other words, no phase-leg redundancy is available within the dc-link-cascaded topology for equalizing losses among the semiconductor switches.

B. Modulation Development

Fig. 3 shows appropriate PWM sequences for modulating a traditional three-level and a dc-link-cascaded inverter when the reference phasor is in triangle 3 on the vector diagram shown in Fig. 4. For both inverters, the Phase Disposition (PD) carrier arrangement is used, where the upper and lower carriers are in phase, and the gate pulses obtained from the reference and upper (lower) carrier comparison are used for driving switch $SX_{(U)}$ ($SX_{(G)}$). Generally, the PD approach is preferred for controlling three-level inverters with no redundant states (including the NPC and dc-link-cascaded inverters) since it always ensures that the nearest three vectors are used and therefore has a better performance, as compared to the alternative phase opposition disposition (APOD) technique [7, 8], where the two carriers are phase-shifted by 180°. For the PD method, it is also convenient to view inverter modulation as reproducing the effective reference phasor drawn in the bold hexagon on the vector diagram

1108

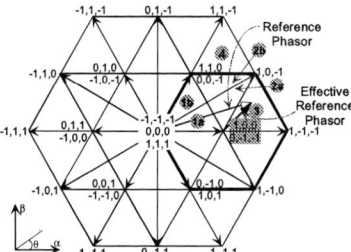

Fig. 4. Vector diagram of a three-level inverter.

TABLE II. SWITCHING STATES AND SWITCH COMBINATIONS OF DUAL Z-SOURCE INVERTER (X = A, B OR C, AND $V_{I(U)} = V_{I(G)} = V_i$)

State Type	ON Switches	Voltage
Non-Shoot-Through	SX$_{(U)}$, SX'$_{(G)}$, SW$_{(U)}$, SW$_{(G)}$	+V_i
Non-Shoot-Through	SX$_{(U)}$, SX$_{(G)}$, SW$_{(U)}$, SW$_{(G)}$	0
Non-Shoot-Through	SX'$_{(U)}$, SX'$_{(G)}$, SW$_{(U)}$, SW$_{(G)}$	0
Non-Shoot-Through	SX'$_{(U)}$, SX$_{(G)}$, SW$_{(U)}$, SW$_{(G)}$	−V_i
Upper Shoot-Through (U)	SX$_{(U)}$, SX'$_{(U)}$	0
Lower Shoot-Through (G)	SX$_{(G)}$, SX'$_{(G)}$	0

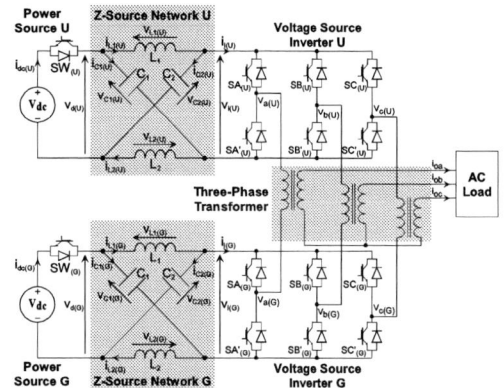

Fig. 5. Topology of a dual Z-source inverter with only a single Z-source impedance network.

in Fig. 4, with its origin and vertices representing "equivalent null" (E-null) and "equivalent active" (E-active) states respectively. Using this representation, three-level modulation is explicitly simplified to a two-level modulation problem that can readily be solved using the rich pool of existing two-level PWM knowledge.

Comparing the sequences shown in Fig. 3, it is noted that the only difference between them is the insertion of an upper and a lower Z-source shoot-through to the left of E-active state {1,-1,-1} and right of E-active state {1,0,-1} respectively within each half carrier period. Insertion of shoot-through states at these instants will not result in additional switching since (for example) the transition from {0,-1,-1} to {1,-1,-1} can be progressed by turning ON SA$_{(U)}$ at an advanced time of $t_{EC1} - T_0/2T$, while the turning OFF of SA'$_{(U)}$ occurs at the same time instant of t_{EC1} (see gate signals of phase A in Fig. 3). Therefore, from $t_{EC1} - T_0/2T$ to t_{EC1}, a shoot-through of the upper Z-source network is introduced with both SA$_{(U)}$ and SA'$_{(U)}$ intentionally turned ON. Note that the inserted upper shoot-through "U" state appears similar to the E-null {0,-1,-1} state since phase A is short-circuited while the other two phases remain unchanged at -1.

Similarly, a short interval for shooting through the lower Z-source network, with SC'$_{(U)}$, SC$_{(G)}$ and SC'$_{(G)}$ turned ON simultaneously, can be introduced by delaying the turning OFF of SC'$_{(G)}$ to $t_{EC3} + T_0/2T$ during the transition from {1,0,-1} (SC$_{(U)}$ and SC'$_{(G)}$ of phase C ON in this state) to {1,0,0} (SC'$_{(U)}$ and SC$_{(G)}$ ON). The resulting "G" state then appears similar to the E-null {1,0,0} state since phase C is shorted while the other two phases remain unchanged. Inserting of "U" and "G" states within the E-null intervals with the E-active intervals kept constant would therefore ensure that the same normalized volt-sec average, as for traditional PWM, is applied across the external ac load. In view of this volt-sec constraint, it is commented that the simultaneous shooting through of both Z-source networks cannot be performed since it gives rise to an intermediate state of {0,0,0} (all phases shorted), which would give rise to a volt-sec error when used.

Using carrier-based implementation, the time advance and delay described earlier can be commanded by using two additional modified references (for example, $V_{a(SA)}$ and $V_{c(SC')}$ in Fig. 3) for turning ON SA$_{(U)}$ (from the phase with the maximum amplitude) earlier and turning OFF SC'$_{(G)}$ (from the phase with the minimum amplitude) later

respectively. For the other switches, the original set of three-phase sinusoidal references is used for gating them without modification. Therefore, linearly mapping the horizontal time offset of $T_0/2T$ to vertical amplitude offset of T_0/T (note the factor of two introduced by this normalized mapping from a time range of 0↔1 to an amplitude range of -1↔1), the resulting set of modified references for controlling the dc-link-cascaded Z-source inverter can be summarized as:

$$\begin{cases} V_{max(SX)} = V_{max} + V_{off} + T_0/T \\ V_{max(SX')} = V_{max} + V_{off} \end{cases} \begin{cases} V_{mid(SX)} = V_{mid} + V_{off} \\ V_{mid(SX')} = V_{mid} + V_{off} \end{cases}$$

$$\begin{cases} V_{min(SX)} = V_{min} + V_{off} \\ V_{min(SX')} = V_{min} + V_{off} - T_0/T \end{cases} \quad (1)$$

where $X = A, B$ or C and V_{off} is a triplen offset added to give certain performance advantages such as the reduction of harmonic distortion achieved by maintaining equal E-null intervals at the start and end of every half carrier period [8, 9]. Equation (1) can also be used for the rising edge half carrier period, and is equally applicable when the reference phasor rotates to other triangles on the three-level vector diagram.

III. DUAL Z-SOURCE INVERTER

The second cascaded topology designed in this work is the dual Z-source inverter, whose circuitry is shown in Fig. 5. In that figure, two dc sources are again connected to the inverter circuitry through two Z-source impedance networks. Comparing with the dc-link-cascaded and NPC inverters, the dual inverter has some redundant states, and therefore additional modulation approaches can be used for controlling it with a better equalization of switching losses among the semiconductor devices achieved. The

1109

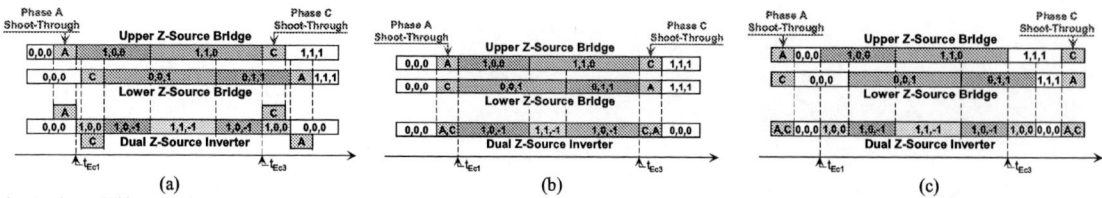

Fig. 6. Phase Shifted Carrier state sequences for (a) non-edge shoot-through insertion with $V_{off} = 0$, (b) non-edge insertion with $V_{off} = -0.5(V_{max} + V_{min})$ and (c) edge insertion with any value of V_{off} for dual Z-source inverter.

operational principles and modulation schemes for controlling the dual Z-source inverter are now described, as follows.

A. Principles of Operation

As its name implies, a dual Z-source inverter is implemented through the series connection of two three-phase Z-source bridges using three single-phase transformers, whose secondary terminals are wye-connected for supplying all generic three-phase loads with either three- or four-wire configuration. Unlike the traditional cascaded H-bridge topology, the use of single-phase transformers at the ac side of the inverter allows only two Z-source networks to be used instead of three needed by a traditional cascaded configuration. In addition, smaller dc ripple currents are expected to flow into the three-phase bridges, as compared to those flowing into the single-phase H-bridges of a traditional cascaded inverter.

Within the dual inverter, each Z-source bridge is controlled to switch between shoot-through and non-shoot-through states with the former corresponding to the short-circuiting of a phase-leg and the later corresponding to the eight switching states of a traditional VS inverter. Operating the bridges with an appropriate phase-shift then gives rise to three distinct voltage levels, as given in Table II, together with the upper and lower Z-source shoot-through conditions. A feature noted from Table II is that the output voltage of zero can effectively be generated by two redundant switch combinations, which in turn can be used for equalizing switching losses among the semiconductor devices, and to achieve other advantages highlighted next while developing the appropriate PWM schemes.

B. Modulation Development

The dual Z-source inverter proposed in this section can also be controlled using the modified PD modulation scheme presented in Section II(B) with the only modification needed being that the reference and upper carrier comparison is used for modulating $SX_{(U)}$, while the reference and lower carrier comparison is used for gating $SX'_{(G)}$ (not $SX_{(G)}$). Unfortunately, using the PD scheme results in only one switch combination ($SX'_{(U)} = SX'_{(G)} =$ OFF, third row in Table II) being used for producing the 0V output. For a better utilization of the redundant switching states, the alternative Phase-Shifted-Carrier (PSC) modulation approach should be used although it has

a poorer spectral performance, as compared to PD PWM [7, 8].

In principle, the PSC modulation scheme controls the upper Z-source bridge using the same reference set expressed in (1) together with a common triangular carrier, similar to that previously used for two-level Z-source inverter control [3]. Using the same carrier, but now the negation of (1) as the references, another set of gate pulses can be mapped out for controlling the lower bridge. With this modulation approach, example cases showing the relative placements of the generated state sequences are given in Fig. 6(a) and (b) for $V_{off} = 0$ (no triplen offset added) and $V_{off} = -0.5(V_{max} + V_{min})$ (standard offset used for two-level centered space-vector-modulation) respectively. A feature noted from the sequences in Fig. 6(b) is that their active intervals start and end at the same time instants of t_{EC1} and t_{Ec3} respectively. Inserting of shoot-through states immediately to the left and right of their total active intervals then results in simultaneous boosting of the two Z-source bridges although they are shorted by different phase-legs (e.g. phase A of the upper bridge and phase C of the lower bridge at the first shoot-through instant in Fig. 6(b)). Simultaneous boosting, together with transformer isolation in this case, allows both bridges of the dual Z-source inverter to be powered in parallel by a single Z-source impedance network (rather than two) with a significant saving in cost. Because of this topological merit, the PSC modulation scheme with $V_{off} = -0.5(V_{max} + V_{min})$ is recommended for dual inverter control even though its spectral performance is slightly inferior, as compared to PD modulation.

Besides the PSC sequences shown in Fig. 6(b), an alternative method for implementing PSC PWM with only a single dc source is illustrated in Fig. 6(c). Intuitively, the sequences in Fig. 6(c) can be implemented using three sinusoidal references (per inverter bridge) for dividing the switching period among the different inverter switching states, and two linear lines, expressed as $V_{lin(U)} = 1 - T_0/T$ and $V_{lin(G)} = -1 + T_0/T$, for inserting shoot-through states when the triangular carrier is above (below) the upper (lower) line $V_{lin(U)}$ ($V_{lin(G)}$). An advantage noted for this method is that the inserted shoot-through states of the two inverter bridges are always aligned, allowing simultaneous boosting of the Z-source networks regardless of the triplen offset added. Unfortunately, it is expected to have higher switching losses since it requires two additional device commutations per half carrier cycle. In view of this

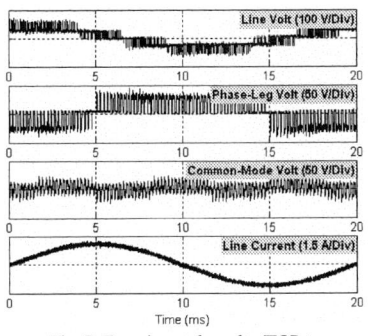 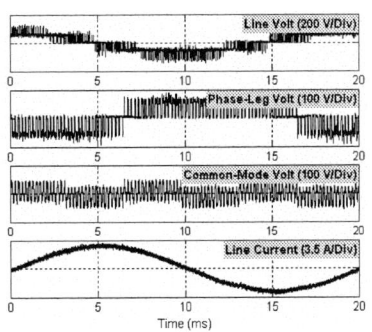 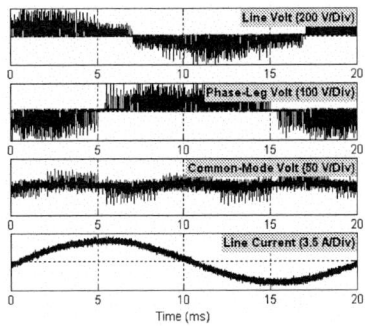

Fig. 7. Experimental results (TOP to BOTTOM): line voltage, phase-leg voltage, common-mode voltage and line current of dc-link-cascaded or dual Z-source inverter with PD modulation, $M = 0.7$ and $T_0/T = 0$.

Fig. 8. Experimental results (TOP to BOTTOM): line voltage, phase-leg voltage, common-mode voltage and line current of dc-link-cascaded Z-source inverter with PD modulation, $M = 0.7$ and $T_0/T = 0.3$.

Fig. 9. Experimental results (TOP to BOTTOM): line voltage, phase-leg voltage, common-mode voltage and line current of dual Z-source inverter with PSC modulation, $M = 0.7$ and $T_0/T = 0.3$.

disadvantage, the sequences presented in Fig. 6(b) are preferred, and are used for testing in the following section.

IV. EXPERIMENTAL RESULTS

For verifying the Z-source topologies designed and modulation concepts presented, simulations were firstly performed using PSIM with Matlab/Simulink coupler before a laboratory prototype that can be configured as either a dc-link-cascaded or dual Z-source inverter was constructed for physical testing.

With the constructed laboratory prototype configured as a PD-modulated dc-link-cascaded inverter, the captured experimental waveforms are shown in Fig. 7 and Fig. 8 for shoot-through durations of $T_0/T = 0$ and 0.3 respectively (modulation ratio M set to 0.7). Comparing the two figures, it is observed that the insertion of shoot-through states effectively boosts the dc voltage by 2.5 times ($= (1 - 2T_0/T)^{-1}$ as given in [1-5]) from $V_{dc} = 30$ V to $V_{i(U)} = V_{i(G)} = V_i = 75$ V (reflected by the vertical height of the phase-leg voltage pulses). This boosting action is also reflected by an increase of current from 1.3 A (peak) in Fig. 7 to 3.3 A (peak) in Fig. 8, which again gives a boost ratio of 2.5. The same observations are noted when the PD modulation scheme with the same shoot-through interval is used for controlling the dual Z-source inverter, but its results are not shown here due to space limitation.

With the DSP reprogrammed to perform single-source PSC PWM, Fig. 9 show the corresponding waveforms produced by a dual inverter using the same shoot-through interval of $T_0/T = 0.3$. Again, the same boosting effects are observed, but the inverter line voltage now does not switch between adjacent voltage levels, indicating a suboptimal spectral performance. In addition, with only a single Z-source network powering the dual inverter bridges, current ripples circulating around the system are larger during voltage boosting, and therefore, a slightly varying dc link voltage V_i is noted at the inverter input (reflected by the varying voltage waveforms in Fig. 9, as compared to those in Fig. 8). The above degraded performance is however more than compensated by the possibility of using fewer passive LC components and the

possibility of generating a lower common-mode voltage (the latter being a characteristic feature of PSC PWM [7]).

V. CONCLUSION

This paper presents two three-level Z-source inverters, whose ac output voltages can be stepped down or up, as desired. Both inverters use two three-phase bridges, which can either be connected at their dc sides for forming the dc-link-cascaded Z-source inverter, or ac sides using single-phase transformers for forming the dual Z-source inverter. Control wise, the dc-link-cascaded inverter can only be modulated using the modified PD PWM scheme because of its lack of redundant switching states, while the dual inverter can be controlled using both modified PD and PSC schemes. Using the PSC scheme, further analysis has revealed that with an appropriate triplen offset added, the resulting dual inverter can be powered using only a single Z-source impedance network to significantly minimize the system costs.

REFERENCES

[1] F. Z. Peng, "Z-source inverter", *IEEE Trans. Ind. Applicat.*, vol. 39, pp. 504-510, Mar/Apr. 2003.

[2] F. Z. Peng, A. Joseph, J. Wang, M. Shen, L. Chen, Z. Pan, E. Ortiz-Rivera and Y. Huang, "Z-source inverter for motor drives", *IEEE Power Electron. Trans.*, vol. 20, pp. 857-863, Jul. 2005.

[3] P. C. Loh, D. M. Vilathgamuwa, Y. S. Lai, G. T. Chua and Y. W. Li, "Pulse-width modulation of Z-source inverters", *IEEE Trans. Power Electron.*, vol. 20, pp. 1346-1355, Nov. 2005.

[4] P. C. Loh, D. M. Vilathgamuwa, L. T. Wong and C. P. Ang, "Z-source current-type inverters: Digital modulation and logic implementation", in *Proc. IEEE-IAS'05*, 2005, pp. 940-947.

[5] P. C. Loh, "Buck-boost thyristor-based PWM current source inverter", *IEE-EPA Proc.*, in press, 2006.

[6] P. C. Loh, F. Blaabjerg, S. Y. Feng and K. N. Soon, "Pulse-width modulated Z-source neutral-point-clamped inverter", in *Proc. IEEE-APEC'06*, 2006, pp. 431-437.

[7] P. C. Loh, D. G. Holmes, Y. Fukuta and T. A. Lipo, "Reduced common-mode modulation strategies for cascaded multilevel inverters", *IEEE Trans. Ind. Applicat.*, vol. 39, pp. 1386-1395, Sep/Oct. 2003.

[8] B. P. McGrath, D. G. Holmes and T. A. Lipo, "Optimized space vector switching sequences for multilevel inverters", *IEEE Trans. Power Electron.*, vol. 18, pp. 1293-1301, Nov. 2003.

[9] F. Wang, "Sine-triangle versus space-vector modulation for three-level PWM voltage-source inverters", *IEEE Trans. Ind Applicat.*, vol. 38, pp. 500-506, Apr. 2002.

2006 5th International Power Electronics and Motion Control Conference

Investigation of Power Supplies for a Piezoelectric Brake Actuator in Aircrafts

Rongyuan Li, Norbert Fröhleke, Hermann Wetzel, Joachim Böcker

Institute of Power Electronics and Electrical Drives

University of Paderborn 33098, Germany

Abstract - In order to facilitate the use of a piezoelectric brake actuator for direct drive applications in airplanes, a power supply system in the power range of several kW is developed, characterised by a high power density, high dynamic and low weight. It feeds the mechanical vibration and reactive power to a multi-mass ultrasonic motor. This contribution puts focus on the power supply and control architecture of the underlying PIBRAC project, whose objective is to study, design and test a novel type of piezoelectric brake actuator and its control electronics. The yield should be pronounced cuts in total weight and peak power demand, when compared to electromagnetic brake actuators.

Keywords - Ultrasonic motor; Ultrasonic driving scheme; Power supply, filtering, DC--AC; Modulation

I. INTRODUCTION

Current aircraft brakes are equipped with hydraulic actuators exhibiting environment and fire risks and high maintenance costs for required high dispatch ability, while mass, inertia and peak power demand is low. Hence the ultimate objective to eliminate all hydraulic components led to the development of electromechanical actuators (EMA). These are equipped with

Fig. 1 Prototype of piezoelectric actuator

electromagnetic motors and reduction gear, resulting in weight increase and even more important of very high peak electric power consumption. Most of this energy is wasted by the kinetic energy requirement due to the inertia of motor rotor and reduction gears during antiskid operation. However Boeing and Airbus decided to utilize electro magnetic brakes in currently developed air crafts.

This is regarded as a first phase in the conversion from hydraulic to EMA. Emerging high power piezoelectric vibration motor technology, thanks to their high torque/force – low speed characteristic and very low inertia, will hopefully allow overcoming latter mentioned drawbacks of EMA, facilitated with electromagnetic motors, if the power delivered by the MM-USM is advanced in the range of several kW and if it is designed to comply with aircraft specifications.

This contribution puts focus on the power supply and control architecture of the underlying PIBRAC project [1], whose objective is to study, design and test a novel type of piezoelectric brake actuator and its control electronics. The configurations of rotational piezoelectric motor structures shown in Fig. 1 are studied, based on a realization degree bound to previous work done by project partners [3].

Over the last decade research and development of power supplies for ultrasonic motors has been conducted, proposing several solutions mostly in applications restricted to small-sized low power drives. In applications of some ten to hundred Watts resonant operated power supplies were developed to drive ultrasonic motors. Drawbacks of these so called resonant inverters (pulse no./halfcycle =1) are the large volume and heavy weight of the resonant filter magnetic components such as transformer and inductor [9]. A model based control of a resonant inverter with LC-filter type is given in [4], outlined in more detail in [7]. Consequences of varying piezoelectric capacitances are investigated in [6], which led to the development of a resonant inverter with LLCC-filter type [2]. Originally it was introduced in [5]. A detailed comparison by Schulte revealed in [6] that the LLCC-filter type resonant inverter shows advanced characteristics and best suited properties in respect to efficiency, stationary and dynamic behaviour, control and commissioning efforts, except volume and weight aspects for ultrasonic resonating actuators with low electromechanical quality factor M.

Additionally in [6] a non-resonant PWM controlled inverter (pulse no./ halfcycle > 1) with LC-filter type was investigated in order to reduce the size and weight of the inductive filter component. The development is focused on driving schemes for piezoelectric transducers again with low electromechanical quality factor and a weakly

1-4244-0448-7/06/$25.00 ©2006 IEEE

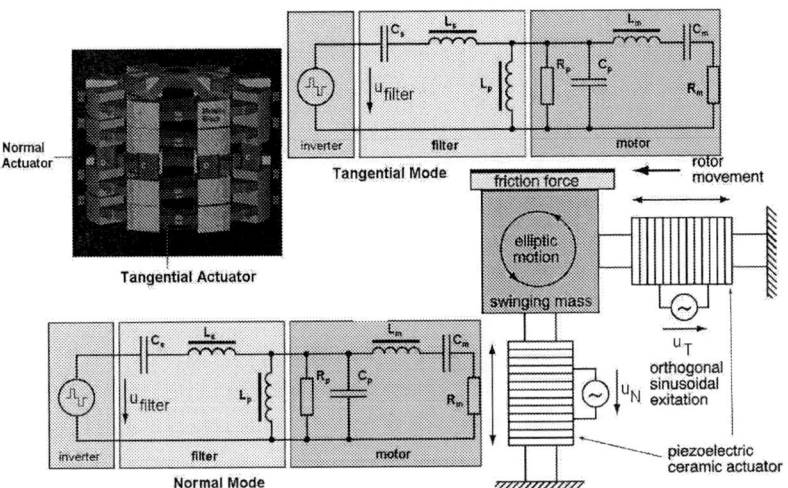

Fig. 2 Operating Principle of a Multi-Masse Ultrasonic Motor

opposite polarization at its mechanical eigenfrequency designed at 35kHz (called tangential mode yielding thrust), so that each metallic block oscillates in the plane of the ring.

Additionally this structure is able to oscillate also orthogonally to the surface of the stator rings and rotor disc (called normal mode yielding temporary clamping of the discs) at the same frequency as the tangential mode, but with an appropriate phase shift. The resulting elliptical movement by combining the two modes of operation drives the rotor shaft via friction. This 2nd energy conversion stage, relying on frictional contact mechanism between stator and rotor shows nonlinear behaviour as known from travelling-wave-type USM.

Here, the load is represented by the well known equivalent circuit of the transducer, depicted in Fig. 2, where the capacitance of the piezoelectric material is represented by Cp, dielectrical losses within the ceramics are represented by Rp and can be neglected. The mechanical vibration system is described by a series resonant circuit Lm-Cm-Rm. Voltage fed inverters, composed by a half- or full-bridge topology are employed to generate AC voltage feeding of the resonant filter. LLCC-filter type resonant circuits are proposed again as in [6] due to various advantages.

influenced (robust) output filter characteristic by the varying capacitance of the actors piezo stacks. These perturbations result from temperature, operational condition and sample deviation. From these results we conclude that PWM inverters are not recommended for driving piezoelectric resonating systems with high damping. Fewer publications [8] as compared to the supply and control of travelling wave-type USM address the class of resonant type (as compared to the quasi static mode of operation) lowly damped piezoelectric vibration systems such as bond sonotrodes. And even less publications exist in the field of superimposed sonotrode assisted ultrasonic drilling, cutting, chiselling, milling of tooling machines or atomizers for the production of fine granular powder, which can be grouped into the class of medium damped piezoelectric vibration systems. The MM-USM belongs into this latter class, too.

A novel PWM controlled power supply is presented, feeding a multi mass ultrasonic motor (MM-USM) via a resonant LLCC-type filter, which is designed in a way to reduce the total harmonic distortion (THD) of the motor voltage and to compensate locally the reactive power requirement of the motor. By combining a PWM controlled inverter and aforementioned resonant filter, the driving voltage of MM-USM can be varied in a suitable frequency range, though the output filter shows an optimized filter performance at minimised volume and weight, when compared to classical resonant or non resonant operated power inverters.

III. POWER SUPPLY DESIGN CONSIDERATIONS

The motor capacitance, originating from the high number of piezo stacks of the MM-USM, varies with temperature, resulting from differing operating conditions. This obviously complicates the power supply design. However, the voltage of MM-USM is not only governed by the inverter, but also influenced by the output filter, see Fig. 6. Under power supply aspects the aforementioned circuit should thus provide a robust and high dynamic operating behavior in order to stabilize the proper operation within a certain frequency and capacitance range. Thus, design of control and filter should take into account the above mentioned requirements.

A. Inverter Topology and Modulation Schemes

From the operating principle explained in chapter II, we know that the MM-USM is to be supplied out of a system with two phases for tangential and normal mode. A driving voltage amplitude of the piezoelectric actuators is required of 270VAC at a frequency of 35kHz. The input DC-link voltage of 270VDC is supplied from the

II. PRINCIPLE OF MOTOR OPERATION AND DRIVING SCHEME

The target motor principle s. Fig. 2 consists of two pairs of stator rings squeezing two rotor discs connected to the shaft. The stator ring houses eight metallic blocks alternating with multilayer piezoceramic blocks. This structure is excited by neighbouring piezo stacks of

aircraft power grid. The total required power by one motor is up to 1.5kW.

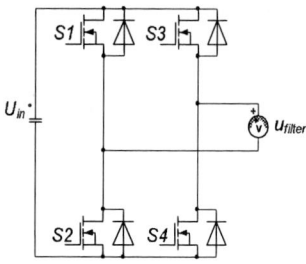

Fig. 3: Inverter topology

In order to limit the voltage and current stresses of power semiconductor switches, the full-bridge topology is selected for the inverter stage shown in Fig. 3. Piezoelectric actuators can advantageously be fed by resonant controlled modulation or by non-resonant pulse width modulation (PWM). In order to characterize these different modulation schemes, a comparison including weight and volume of magnetic components, total harmonic distortion (THD) of filter voltage and losses of power inverters is conducted.

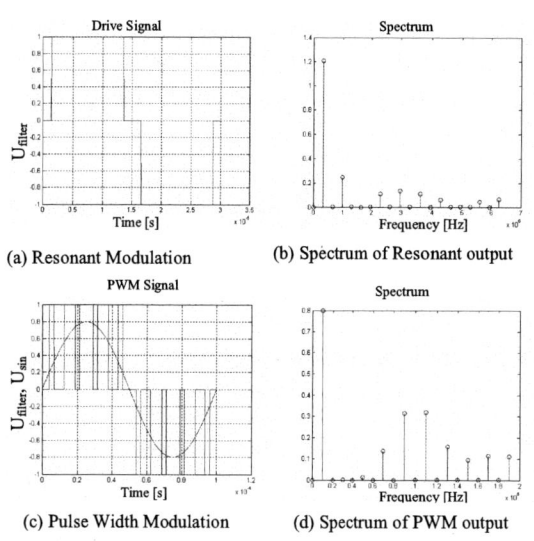

Fig. 4 Modulation schemes and spectrum of inverter voltage

The inverter with resonant modulation shown in Fig. 4(a - b) is operated at low switching frequency, approximately the mechanical resonant frequency of 35 kHz, when compared with the PWM controlled inverter (Fig. 4(c - d)). Hence, the switching losses of the power semiconductor components of resonant inverters are lower than for PWM controlled inverters at equal power level. But the essential drawbacks of resonant inverters are its relative large and heavy filter component, which limit its application in the case that volume and weight of the power supply are restricted, as for example the equipment in aircrafts or handheld devices. Note, that inverter voltage represents the input voltage of the filter.

B. Filter Topology

The inverter output filter plays an important role as coupling between power supply and actuator. Cable lengths of up to 10 meters, connecting the power supplies (mounted above the landing gear) with the brake actuators cannot be neglected during filter design. The model based control design has to consider the filter model when designing the internal control loops in order to regulate amplitude, phase and frequency, while the outer loops take care of the objectives for brake control.

Two types of filter topologies are shown in Fig. 5, with transformers being integrated into resonant circuits to provide galvanic isolation.

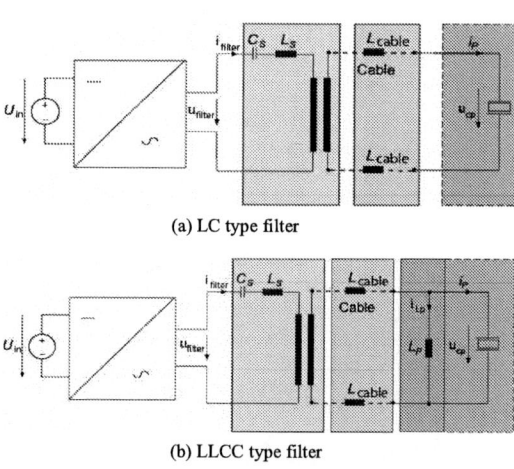

(a) LC type filter

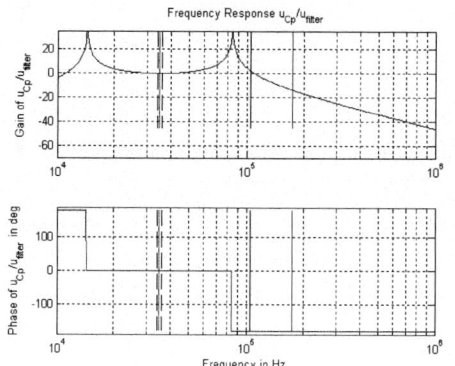

(b) LLCC type filter

Fig. 5 Filter Topologies

Fig. 6 Frequency response of LLCC-filter

Both filter circuits can be operated with resonant modulation and PWM modulation. For the LC-filter type with resonant modulation the largest size and weightiest inductive component is determined. In contrast the PWM controlled inverter with LC-filter type is improved by reducing the size and weight of the inductive filter component. But unluckily this inverter shows a poorer efficiency and larger transformer due to the missing reactive power compensation, when driving ultrasonic motors, cleaning baths or other applications.

The frequency response of LLCC-filter is shown in Fig. 6. By its inspection the advantages of LLCC- filters, such as robustness to parameter fluctuations of e.g. the piezo capacitance and simple controllability compared with LC-filters become obvious. The second electrical resonant point of the filter can be set at higher frequency than the inverter with resonant modulation due to low THD of the output voltage of the PWM inverter; thus smaller and lighter filter components can be used, and the dynamic response is improved.

The LLCC filter is designed in the following way:

First consider C_p as equivalent capacitance of the loading MM-USM. In order to compensate the reactive power of the capacitive load, the parallel inductor L_p is calculated by

$$L_p = \frac{1}{C_p \omega_m^2}, \qquad (1)$$

where ω_m is the mechanic resonant frequency.

For keeping the filter components small and increasing the dynamic, the second electrical resonant frequency ω_{02} is selected by $\omega_{02} \approx 3\omega_m$, see [6].

In order to attain robustness of the LLCC-filter, the geometric mid-frequency of the series and parallel resonant tank are designed to be of the same value as ω_m. Hence, the series resonant components are calculated by

$$L_s = \frac{L_p}{L_p C_p \omega_{02}^2 - 1},$$
$$C_s = C_p \frac{L_p}{L_s}, \qquad (2)$$

whereby a largely reduced L_s results (only 1/4 of the resonant controlled case).

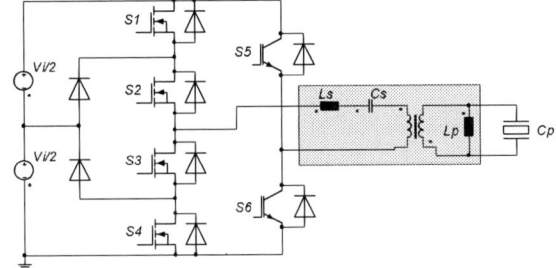

Fig. 7 Hybrid three levels PWM inverter plus LLCC-filter

C. Advanced PWM inverter topologies and modulations

The novel power supply scheme is supplied by a DC-voltage and consists of a 3-level inverter, which generates a PWM-controlled voltage applied across the LLCC-filter and transformer.

The power supply shown in Fig. 7 consists of a hybrid three level PWM inverter followed by a LLCC-filter with an isolating transformer being integrated as part of the filter. The left leg of the PWM inverter is composed of four MOSFETs (S1 – S4) used to generate high frequency PWM output voltage in order to achieve low THD (as shown in Fig. 8 - 10). S5 and S6 forming the right leg of the PWM inverter are employed to switch only the fundamental frequency of the mechanical vibration in order to maximize the efficiency.

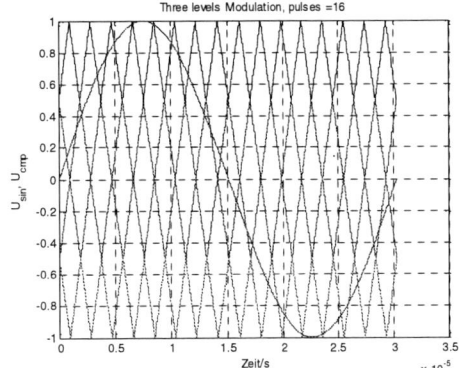

Fig. 8 Modulation scheme

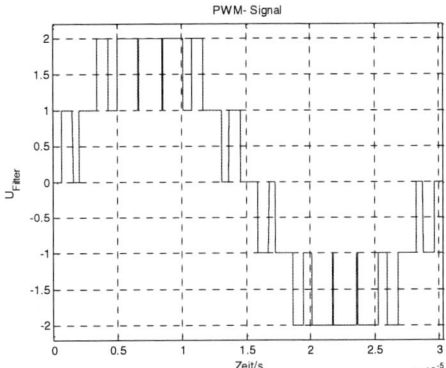

Fig. 9 Nominal output voltage of PWM inverter

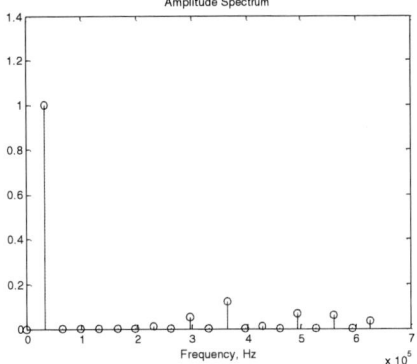

Fig. 10 Spectrum of PWM inverter output

(d) Control scheme design

A preliminary control scheme was developed shown in Fig. 11. Control algorithms will be designed and realised using FPGA. The single input braking command from the Braking Command Unit (BCU) is sampled and converted by ADC. PWM signals for driving power switches are generated directly by a controller, supplemented by a motor and actuator control on the control board.

By this means the efficiency and flexibility for design of control algorithm are improved, and hard real-time performance is attained. By review of the state-of-the-art technologies of embedded micro controller a FPGA was selected and tested for the control design.

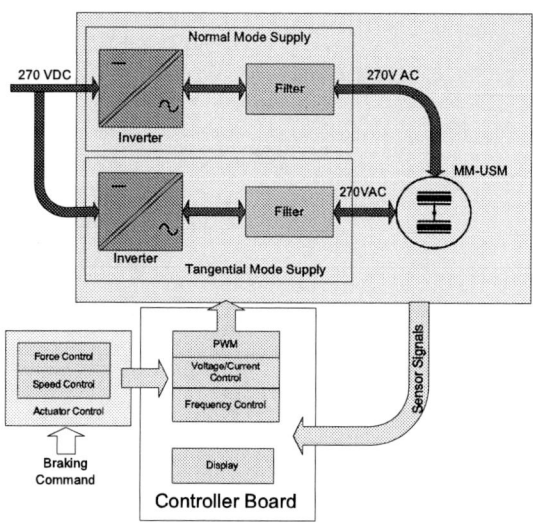

Fig. 11 Control scheme

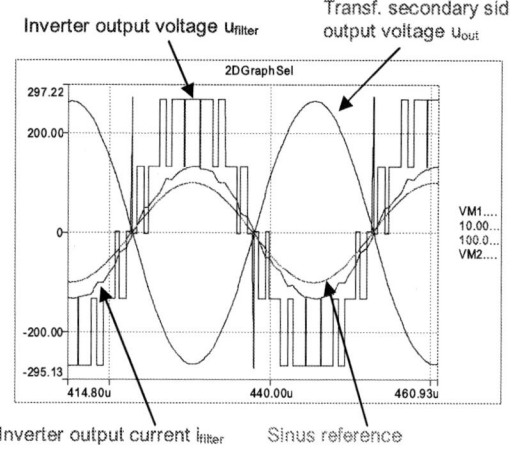

Fig. 12 Simulation results

IV. SIMULATION ANALYSIS

Simulation models of power supply composed of inverter, transformer, resonant filter, and models of the piezoelectric motor were built in order to facilitate the control design of the MM-USM.

Fig. 12 shows one simulation result of the designed LLCC-filter PWM inverter at steady state operation. It can be observed that a smooth voltage at the cable input is obtained, which is a prerequisite for low losses and p.d.

in cables caused by extra ohmic and "inductive" losses by dielectric currents. This in conjunction with local reactive power compensation by Lp (see Fig. 5) yields multistage filtering, making use of transformer, cable and motor parasitics.

V. CONCLUSION

A multi mass ultrasonic motor derived from known travelling wave-type USM is described, which is well qualified for airborne applications such as a brake actuator. Thus, it was selected for the European project PIBRAC. The power supply and control architecture feeding the MM-USM are presented in this contribution. The power supply is composed out of a simplified three-level inverter and a LLCC-filter due to its ability to compensate the reactive power of the load locally and robustness to parameter variations, easing power electronics and control design. Investigations on latter circuitry were conducted to minimise total harmonic distortion (THD) of motor voltage to ensure increased lifetime of the actor and total weight and losses of the power supply scheme, wherefrom few aspects could be outlined, supplemented by design guidelines. The operation of the power supply system is verified by simulation, yet.

ACKNOWLEDGEMENT

Thanks belong to the Europe Community for funding the PIBRAC project under AST4-CT-2005-516111 as well as our project partners

REFERENCES

[1]. Homepage of the project PIPRAC, www.pibrac.org

[2]. T. Schulte, N. Fröhleke, "Development of power converter for high power piezoelectric motors", Aupec 2001.

[3]. J. Audren, , D. Bezanere, "Vibration motors", United States Patent 6628044, Sep 2003

[4]. J. Maas, T. Schulte, N. Frohleke, "Model-based control for ultrasonic motors", IEEE/ASME Transactions on Mechatronics, Volume 5, Issue 2, June 2000 Page(s):165 - 180

[5]. F.-J. Lin, R.-Y. Duan, H.-H. Lin, "An Ultrasonic Motor Drive Using LLCC Resonant Technique." Proc. of IEEE Power Electronics Specialists Conference (PESC) '99, vol. 2, pp. 947-952

[6]. Thomas Schulte, "Stromrichter und Regelungskonzept für Ultraschall- Wanderwellenmotoren". Fortschritt Berichte VDI, Reihe 21 Nr.363; Paderborn, 2004.

[7]. J. Mass, "Modellierung und Regelung von stromrichtergespeisten Ultraschall-Wanderwellenmotoren". Fortschritt Berichte VDI, Reihe 21 Nr.278; Paderborn, 1998.

[8]. C. Kauczor, N. Fröhleke, "Inverter Topologies for Ultrasonic Piezoelectric Transducers with High Mechanical Q-Factor" Proc. of IEEE Power Electronics Specialists Conference (PESC) '2004.

[9]. H.-D. Njiende, N. Fröhleke, "Optimization of Inductors in Power Converter Feeding High Power Piezoelectric Motors", Aupec 2001.

2006 5th International Power Electronics and Motion Control Conference

A Line Power-Supply for LED Lighting using Piezoelectric Transformers in Class-E Topology

F.E. Bisogno[*], S. Nittayarumphong[*], M. Radecker[*], A. V. Carazo[**] and R. N. do Prado[***]

[*] Fraunhofer Institute für Autonome Intelligente Systeme - AIS, Sankt Augustin, Germany
[**] Face Electronics, LC – Norfolk, USA
[***] GEDRE – PPGEE – Santa Maria Federal University, Santa Maria, Brazil
fabio.bisogno@ais.fraunhofer.de

Abstract— An up to 5 Watts wide range power supply LED driver demonstrator with a Piezoelectric Transformer (PT) has been built to show the static and dynamic behavior as an off-line power supply [1],[2],[5],[6],[7]. The target of the power converter controller using a PT, with 2.3 mm thickness and 17 mm diameter at frequency of 155-175 kHz, designed in a resonant circuit of Class-E topology, is to maintain a constant output current of 290mA/DC with variable output voltage drop at wide input voltage range of 85-250 V/AC. Also, the ZVS (Zero-Voltage-Switching) condition is achieved over the full operation range. The used 1200 V Fieldstop IGBT (1 A-Type) shows losses around 200 mW over the full operating range for constant output current. The paper shows the regulation method for off-line power supplies using PT in Class-E topology in constant current mode regulation (CCMR). The regulation algorithm has been programmed in C language operated by a 32-bit DSP (TMS320F2812). Besides, in the Class-E application, the duty cycle of the switching frequency needs to be controlled in order to achieve the ZVS during the operation at different load and input condition [1]. This goal is achievable only by Phase Locked Loop (PLL) feedback in a satisfying way [3]. To solve this problem, the idea of PT with an auxiliary tap has been applied, and it is shown in Fig.3. The application is suited for low size LED line adapters, to be integrated in bulb sockets. The results are applicable for any controller design of resonant PT step-down topologies. The several suited optimum control concept is investigated in this paper, considering the circuit expense being minimized.

Keywords- LED; Class-E; Piezoelectric Transformer; Phase Locked Loop

I. INTRODUCTION

At this moment, emerging technologies such as light emitting diodes are a great topic, mainly in energy savings, and new applications on this large and diverse marketplace.

The applications where market adoption has started and is poised to grow include automobile safety and signal lights, aircraft passenger reading lights, airport taxiway edge lights, commercial advertising signs and holiday lights [15][16].

While greater energy efficiency is an important aspect of LED sources, there are several qualities that compel lighting users to adopt this technology over conventional light sources. These features enable users to enjoy better service, extended operating life, and enhanced safety. The following list describes some of these benefits: Reduced Energy Consumption, Long Operating Life, Durability, Reduced Heat Production, Smaller Package Size, Safety Improvements, Light Control [15].

As with other light source technologies, such as fluorescent and high intensity discharge, lighting systems using LEDs can be thought of as having a light source (the individual LED sources), ballast for LEDs, often called driver, and a luminaire. Unlike traditional lighting systems with few light sources, LED systems will likely contain arrays of many individual light sources in near future [17].

An LED driver performs a function similar to ballast for discharge lamps. It controls the current flowing through the LED.

Individual LEDs are low voltage devices. Single indicator LEDs requires 2 to 4 volts of direct voltage. A device containing multiple elements connected in series will require higher voltage corresponding to the larger number of individual elements in the device. The forward current is proportional to the light output of an LED over a large operating range, typically several hundred miliamperes, so dimming can be achieved with reductions in the forward current. As the LEDs can be rapidly turned on and off with no harmful effects, dimming can also be accomplished using a method called pulse width modulation. By adjusting the relative duration of the pulse and the time between pulses, the apparent light intensity of the LED can be dimmed. This must be done with high enough frequency that the LED appears to be continuously lighted, or else the rapid flickering will be distracting [17].

In some applications, like residential application, the difference between the line and the voltage for supplying an arrangement of LEDs is big. Some solutions include a transformer to step down the voltage or a series inductor where the difference of voltage is applied.

1-4244-0448-7/06/$25.00 ©2006 IEEE

The inductive transformers increase the size of the system and sometimes there is not enough space inside of the bulb of the system. The Piezoelectric transformers technology allows designing very small and light transformers comparing to inductive transformers. Power converters with Piezoelectric transformers allow for potential size and cost reduction in low power applications, such as CCFL backlighting inverters, DC-DC converters, and off-line power supplies, while suitable resonant topologies guarantee a high efficiency.

The PT works with electromechanical energy conversion, so the EMI are greatly reduced in comparison to inductive transformers, thus reducing the risks of impacting or damaging the other components of the system.

In this paper an off-line LED driver with Class E Converter using piezoelectric transformer is presented. As the output voltage of this converter has a sinusoidal waveform an anti-parallel arrangement of the LEDs is used. This converter has full control of the output current that is desirable for supplying the LEDs.

II. CLASS-E CONVERTER

The Class-E converter can be used to drive piezoelectric-transformers supplying LED. The piezoelectric transformer can be also driven by inductor-less half-bridge or by inductor half-bridge converters. The main three topologies are shown in Figure 1. These topologies are suitable to drive piezoelectric transformers because of the intrinsic equivalent resonant circuit of the piezoelectric transformer. The advantages of the class-E converter compared to the main alternatives are the following:

- comparing with the inductor-less half-bridge, the class-E converter has a wider range of control achieving zero-voltage-switching operation;
- Comparing with the inductor half-bridge, the class-E needs only one switch;
- Comparing with both half-bridge topologies, the class-E converter does not need a high side gate driver.

The design parameters of a class-E converter can be obtained solving the differential equations that describe the behaviour of each linear operating mode. The link between each mode is done using boundary conditions of the system, assuming the optimum switching behaviour defined by the turn off at zero voltage and current [6]. Some important normalized results to evaluate the class-E design can be seen the Figure 2. Figure 2 (a) shows the power-transfer-ratio ($P_0 R / V_{in}^2$) across duty-cycle Dc for the class-E converter considering different input resonance parameters ($A_3 = 1/\left(2\pi Dc\sqrt{L_f C_{d1}}\right)$) and for the inductor-less half-bridge, respectively. Figure 2 (b) shows the limitation of capacitor C_{d1} that represents the restriction in the design to achieve zero-voltage switching operation.

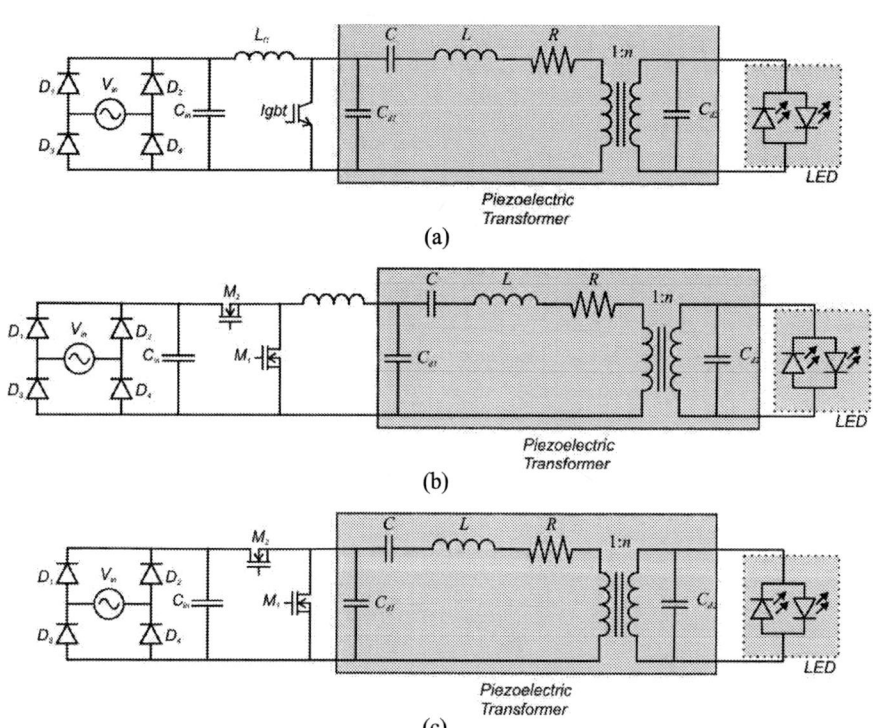

Figure 1 Main topologies suitable to drive piezoelectric transformers: (a) Class-E converter; (b) Inductor Half-Bridge converter; (c) Inductor-less Half-Bridge converter.

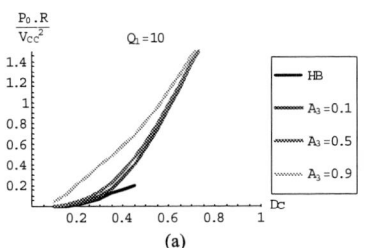 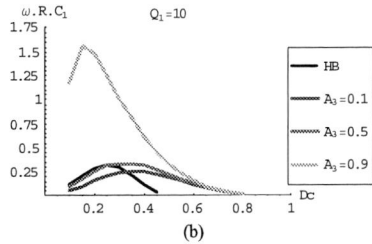

Figure 2 Class-E parameters (for A3=0.1; 0.5; 0.9), and inductor-less Half-bridge parameters (HB), as function of duty cycle (Dc)

III. CONTROL CONCEPTS

Resonant power converters, especially those with narrow frequency bandwidth, when controlled with a simple output voltage or current feedback to maintain constant output voltage, are hard to be optimized without auxiliary feed-forward loops or other additional measures when used over a wide range of load current or input voltage.

The solution can be, for instance, a consequent burst-mode concept like demonstrated in [12], or alternatively a pre-regulation by feed-forward open-loops, where duty-cycle tracking or frequency is pre-adjusted, e.g. by the input voltage [11]. In case that non-symmetric (e.g. single-switch) topologies as class-E are used, the tracking of the duty-cycle with frequency is required to achieve ZVS all the time [8]. In half-bridge topologies, the duty-cycle can be handled easier, e.g. to be constant or pre-regulated separately, independent of the frequency, but dependent from other parameters, like input voltage [10]. A second feed-back loop observing the load current can be used to lock the phase and the frequency to a certain point, but the degree of freedom in regulation and synchronization is smaller over the operation range if the input current of the PT is used as a reference, e.g. from an input resonant

inductor in a half-bridge topology [12].

The most efficient way to regulate the above mentioned systems is a PLL (phase-locked loop) feedback [8],[9]. With a simple and un-expensive tap for the observation of the motional current e.g. in a PT, applied to a resonant converter, the control problems can be solved satisfying as shown in this contribution. The tap of a PT provides the opportunity of a PLL synchronization signal for all topologies as class-E, or half-bridge, or others. The motional current (load current) amplitude is further suited to be controlled by a feed-back loop, when compared with a constant or variable reference. In this case, the equivalent control model scheme shows that the overall control loop response time can be accelerated against all other control methods, because the time delay of the plant (the resonant converter) is the only system to be stabilized in high frequency control feedback in its inner loop, regarding input voltage deviations. Thus, an outer control loop of the low frequency regulation, due to the larger output time constant of a possible output filter, becomes faster without lack of stability. In this case, the inner loop is sufficient to control the output current at a nearly constant level with a constant tap reference as shown in Figure 3, because the output filter is not there in this application. There is no other control loop needed. The

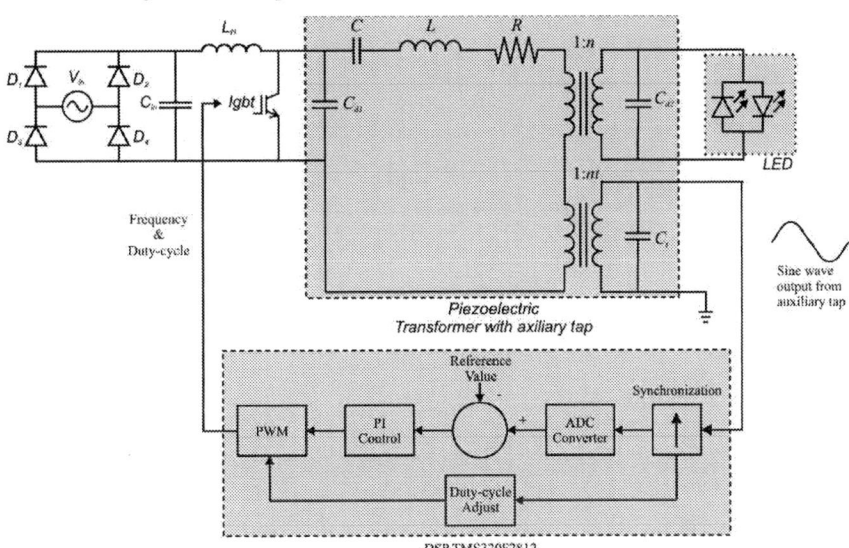

Figure 3 Class-E converter using tapped piezoelectric-transformer tapped with CCM regulation method principle and LED output load

feedback does not need to have auxiliary shunts in the LED branches to control the average current in the LED, and thus, the light intensity can be kept constant. In previous solutions [14], the decoupling of the LED branches is done by additional capacitors or resistances, which means additional expense and losses. Using the tap feedback control, no resistive losses will occur. The output voltage of the PT adapts automatically to the overall forward voltage drop of the LED's. This means that the power will increase if more LED's are added. Further, if galvanic isolation between the LED chains and the line input is required, this solution provides a simple system without feedback components like opto-couplers.

IV. DEMONSTRATOR IMPLEMENTATION AND MODELLING, RESULTS

A digital microcontroller, the 32-bit DSP TMS320F2812, was chosen as a controller for this application evaluation because of 7 ns time step resolution allows for high resolution of frequency and duty-cycle. Regulation above the resonant frequency of PT is done with the PI control concept in the DSP. In the chosen regulation method, the regulation is accomplished against the sine wave output voltage from the auxiliary tap of PT (Fig.3.). The DSP controller has been programmed to be synchronized with the sine wave output from the auxiliary tap. The amplitude of this sine wave is detected and compared with the reference value. The error from the comparison is fed to the PI controller to generate a suitable frequency to maintain a constant output current for a certain load with the varying input voltage. The information from the synchronization is also used to adjust the duty cycle to switch on at the correct moment to achieve ZVS over the operation range. The results show, that, with the synchronization of auxiliary tap, the turn on moment occurs near to the end of the reverse current period at the IGBT switch. Compared to the synchronization, the problem of ZVS condition was solved in the Class-E application concept without any further adjustment. The proportionality between output voltage and low loaded tap voltage would be given at lighter loads RL with

$$\frac{\hat{U}_2}{\hat{U}_3} = \frac{Nt}{No} * \frac{C3}{C2} * \frac{1}{\sqrt{1 + \frac{1}{(\omega C2 \operatorname{Re} q)^2}}} \ . \qquad (1)$$

This regulation regime can not be used here. The system is designed for full load instead, and we assume to achieve the maximum efficiency of the PT [2]. For full load we derive

$$I_{out} \approx \hat{U}_3 \frac{Nt}{No} * \frac{C_3}{C_2} * K_{FL} \ . \qquad (2)$$

The complete linearized, Laplace-transformed model is show in Fig. 4.

The function $K_{av} * F_{av}(s)$ has to be linearized by the differential equation for different cases of load and output voltage as well as the dynamical Class-E model, presented in [7].

The output of the tap voltage is compared with a constant reference value, adjusting the LED average current. The difference is applied to the PI controller to generate an input signal for the VCO driving the IGBT. The information from an auxiliary tap is also used to define a correct duty cycle regarding the correct turn-on moment of the switch (PLL). Fig. 4 shows the linearized model, also allowing for stability analysis and dynamical or transient simulation.

Measurement results are shown in Fig. 5. The average output current was 290mA. Table I demonstrates the measured losses of the components. Due to the high efficiency, the solution can be implemented in a small size without overheating.

TABLE I.
LOSSES MESUREMENTS

Input Voltage: 85V			
component	Δtemperature	Losses	Power
Switch R_th=25.5K/W	5° K	196mW	
Input Inductor R_th=70K/W	4° K	57mW	
Piezo-Transformer R_th=40K/W	9° K	225mW	
LEDs	6 x 0.95W		5.7W
Efficiency (without the input bridge)			92%
PT Efficiency			97%

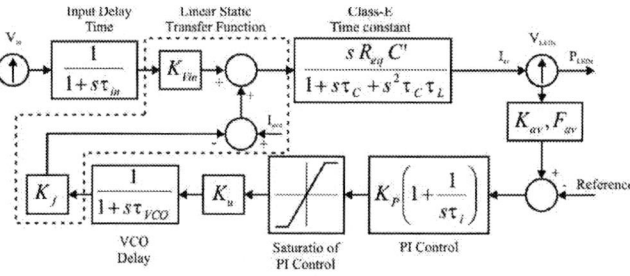

Figure 4 Linearized control model

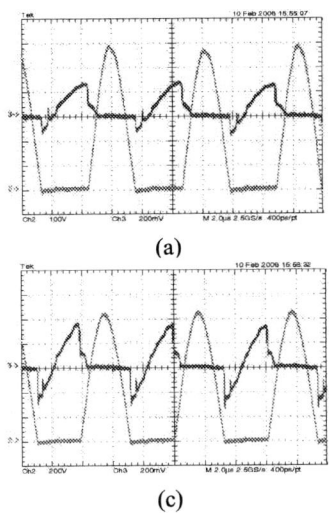

(a)

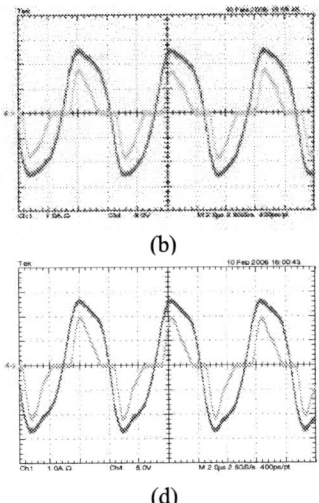

(b)

(c)

(d)

Figure 5 Experimental results (output power: 5.7W – 3 x 1W Luxeon LED LXHL-PW01 per branch of the bridge): Input line voltage: 120V (a) Switch voltage and current (100V/div, 200mA/div, 2µs/div); (b) LED current and voltage; Input line voltage (5V/div, 1A/div, 2µs/div): 230V (c) Switch voltage and current (200V/div, 200mA/div, 2µs/div); (d) LED current and voltage (5V/div, 1A/div, 2µs/div)

V. CONCLUSION

This paper presents a LED driver using a class-E converter with a piezoelectric transformer tap. The control concept using the piezoelectric transformer to keep the LED average current constant avoids the output feedback by shunts. The prototype was implemented using the 1200 V Fieldstop IGBT (1 A-Type), a piezoelectric transformer form FACE Electronics, LC, A digital microcontroller (32-bit DSP TMS320F2812), and the 1W Luxeon LED LXHL-PW01. The experimental results show the feasibility of the proposed driver using the tap feedback concept presenting a LED current error of up to 7% for the input voltage range from 120-230V AC. The solution can be extended to higher LED power applications.

REFERENCES

[1] Sadachai Nittayarumphong, et al. *Dynamic behaviour of PI controlled Class-E Resonant Converter for Step-Down Applications Using Piezoelectric Transformers*, EPE 2005, September 2005, Dresden, Germany.

[2] Fabio E. Bisogno, et al. *Comparison of Resonant Topologies for Step-Down Applications Using Piezoelectric Transformers*, IEEE PESC 04, June 2004, Aachen, Germany

[3] Matthias Radecker et al. *Design and Comparison of Hard-Switching and Soft-Switching Topologies for Off-Line Power Supplies*, IEEE PESC 05 Tutorial, June 2005, Recife, Brazil.

[4] Zaitsu Toshiyuki, et al. *AC/DC converter with a piezoelectric transformer*, European Patent Application EP 0 854 562 A2, 22/07/1998.

[5] Marian K. Kazimierczuk, Jacek Jozwik. *Resonant dc/dc Converter with Class-E Inverter and Class-E rectifier*, IEEE VOL.36, NO.4, November 1989.

[6] Marian K. Kazimierczuk, Krzysztof Puczko. *Exact Analysis of Class-E Tuned Power Amplifier at any Q and Switch Duty Cycle*, IEEE VOL. CAS-34, No.2, February 1987.

[7] Fabio.E Bisogno, et al. *Dynamical Modeling of Class-E Resonant Converter for Step-Down Applications Using Piezoelectric Transformer*, IEEE PESC 05, June 2005, Recife, Brazil.

[8] M. Harold., *Single-input phase locking piezoelectric transformer driving*, US patent S5866968, February, 02, 1999.

[9] Yan Yin; Zane, R.; *Digital controller design for electronic ballasts with phase control*, IEEE Power Electronics Specialists Conference, 2004. PESC 04. 2004 35th Vol 3, 20-25 June 2004 Page(s):1855 - 1860 Vol.3.

[10] Sanchez, A.M.; Sanz, M.; Alou, P.; Prieto, R.; Cobos, J.A.; *Experimental validation of an optimized piezoelectric transformer design with interleaving of electrodes*, IEEE Power Electronics Specialists Conference, 2004. PESC 04, Vol 2, 20-25 June 2004 Page(s):841 - 846 Vol.2.

[11] Redl, R.; Molnar, B.; *Design of A 1.5 MHz Regulated DC/DC Power*, PCI/Motorcon September 1983 Proceedings.

[12] Prieto, M.J.; Diaz, J.; Martin-Ramos, J.A.; Nufio, J.; *Closing a second feedback loop in DC/DC converters based on piezoelectric transformers*, Power Electronics Specialists Conference, 2004. PESC 04. 2004 IEEE 35th Annual, Volume 6, 20-25 June 2004 Page(s):4682 - 4688 Vol.6

[13] Patent application DE 102005023686.3.

[14] Clauberg et al., *Light Emitting Diode Driver*, United State Patent US 6,853,150 B2, February 2005;

[15] BOWERS, B. *Historical Review of Artificial Light Sources*. IEE Proceedings, v. 127, n. 2, p. 127-122, April 1980.

[16] COOK, B. *New Developments and Future Trends in High-Efficiency Lighting*. Engineering Science and Education Journal, v. 9, 5th ed., p. 207-217, Oct. 2000.

[17] MUTHU, S.; GAINES, J. Red, *Green and Blue LED-Based White Light Source: Implementation Challenges and Control Design*. Conference Record of the 38th IAS Annual Meeting, v. 1, p. 515 – 522, Oct. 2003.

Integrating Large Wind Farms into Weak Power Grids with Long Transmission Lines

Richard Piwko, *Fellow IEEE*, Nicholas Miller, *Fellow IEEE*,
Juan Sanchez-Gasca, *Fellow IEEE*, Xiaoming Yuan, Renchang Dai, James Lyons

Abstract -- In China, as in other parts of the world, many of the best resources for wind generation are located far away from load centers. Large generating facilities connected to distant load centers by long ac transmission lines face numerous technical challenges, regardless of the type of generating facility. This paper addresses some of the most significant challenges for wind generation facilities, including voltage control, reactive power management, dynamic power-swing stability, and behavior following disturbances in the power grid.

Wind generation technology has evolved significantly over the past several years, and proven solutions to these technical challenges now exist. Controls integrated into the power electronics and mechanical controls of individual wind-turbine-generators, combined with integrated wind-farm control systems, have the capability of controlling numerous wind turbines so that they act as one unified generating plant at the point of interconnection with the power grid. This advanced hierarchical control of both real and reactive power output can provide dynamic performance that is, in many cases, superior to that achievable with modern conventional synchronous generation. This paper describes:

a. Wind farm control functions, including performance for controlling grid voltage in quasi-steady-state and dynamic conditions.

b. Low-voltage ride-through characteristics, including performance following severe system disturbances

c. Dynamic power control functions within wind turbine-generators, including transient and dynamic performance for power swings.

Index Terms—Wind Generation, Wind Farm Integration, Low Voltage Ride Through, LVRT, Voltage Regulation, Power Swings.

I. INTRODUCTION

Integration of large wind farms into bulk power systems presents multiple challenges to system operation and security. One particular challenge to system security is vulnerability to common-mode tripping due to transmission system faults. Wind generators may have to be disconnected from the grid once the system has a disturbance, such as a short circuit fault, lightning strike on transmission lines, etc. Tripping generators normally has a negative impact on system stability, especially when wind farms have considerable penetration. This is a major concern throughout the world. Furthermore many wind farms, including proposed large offshore projects, are geographically remote and have relatively weak transmission systems. The presence of wind farms in such weak systems raises serious concerns about system stability, voltage regulation, and post-fault power swings.

R.J. Piwko is with GE Energy, One River Road, Schenectady, New York, USA (email: richard.piwko@ge.com)

Host utilities require that, during normal operation, wind farms should be capable of regulating voltage or reactive power to maintain a smooth voltage profile at the point of interconnection, protecting against voltage flicker caused by wind gusts. With the penetration of wind farms increasing, most host utilities also require that wind farms must tolerate system disturbances. For instance, wind farms must not trip during faults and other system disturbances, and they must remain stable during post fault electromechanical swings in the power grid.

These requirements are being addressed by the latest generation of wind turbine-generator (WTG) equipment. Wind Farm Management Systems (WFMS) and Low-Voltage Ride-Through (LVRT) technology for wind turbine generators can provide much improved system performance compared to more traditional wind generation equipment. LVRT technology is now able to eliminate most concerns about tripping during system voltage events and allows for the rapid and well-behaved recovery of the wind farm and the grid when system faults are removed. Wind farms controlled by WFMS can provide extremely fast initial response to system events and wind induced perturbations, and voltage and reactive power response similar to that of conventional synchronous generation.

This paper presents dynamic performance of GE wind turbine-generators with LVRT and WFMS technology. The information presented includes dynamic simulation results from existing power systems with large wind farm interconnections, and actual field measurements from operating systems. The paper also presents control design philosophy, innovative control designs and relevant control diagrams.

II. WIND TURBINE-GENERATOR CHARACTERISTICS

This section presents an overview of a dynamic model of the GE 1.5 MW WTG. A detailed description of the model, including pertinent parameter values, is provided in [1]. WTG models are continually being updated and improved, as wind generation technology evolves. GE regularly updates wind modeling documents.

A simple schematic of the WTG major components is shown in Figure 1. The GE WTG generator is unusual from a system simulation perspective. Physically, the machine is a relatively conventional technology wound rotor induction machine. However, the key distinction is that this machine is equipped with a solid-state AC excitation system. The AC excitation is supplied through an ac-dc-ac converter. The fundamental frequency electrical dynamic performance of the WTG is completely dominated by the field converter. In practice, the electrical behavior of the generator and converter is that of a current-regulated voltage source inverter. Like other voltage source inverters (e.g. a BESS or a STATCOM), the converter will make the WTG behave like a voltage behind a reactance that results in the desired active and reactive current being delivered to the device terminals. Conventional aspects of generator performance related to internal angle, excitation voltage, and synchronism are largely irrelevant. These characteristics have significant implications from the standpoint of power-swing performance and modal interactions.

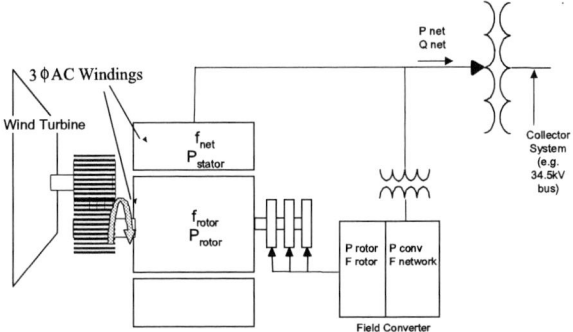

Figure 1. GE WTG Major Components.
Source: GE Energy ©2005. Used with Permission.

It should be noted that the wind turbine model used in this paper is based on presently available design information, test data and extensive engineering judgment. This model was developed specifically for the GE 1.5 and 3.6 MW WTGs. This model is not designed for, or intended to be used as, a general purpose WTG. There are substantial variations between models and manufacturers.

The overall WTG model consists of four major elements, as shown in Figure 2:

   i) Generator/Network Interface,
   ii) Electrical Control,
   iii) Wind Turbine, and
   iv) Wind Power Model.

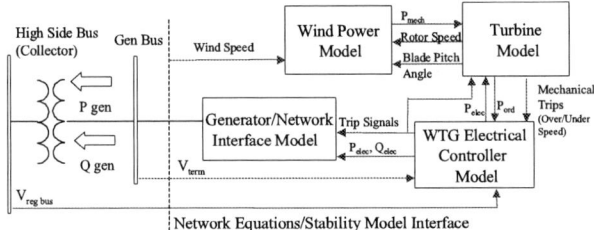

Figure 2. GE WTG Basic Dynamic Models and Data Connectivity
Source: GE Energy ©2005. Used with Permission.

A. Generator/Network Interface

This element is the physical equivalent of the generator and converter hardware. It provides the interface between the WTG electrical controller and the network, and contains no control functions or user settable functions.

B. Electrical Controller

The WTG Electrical Controller dictates the active and reactive power to be delivered to the system based on input from the turbine model and power system conditions. The model is greatly simplified, but maintains those aspects that are crucial to capturing the dynamic performance of interest to the system.

C. Wind Turbine

The wind turbine model provides a simplified representation of a very complex electro-mechanical system. The turbine control is designed to deliver power over a range of wind conditions, taking advantage of the variable speed capability of the machine. The controller enforces the power-speed relationship shown in Figure 3. Above about 75% rated power, the power levels of primary interest for stability studies, the controller works in two distinct regions. When the available wind power is above the equipment rating, the blades are pitched to reduce the mechanical power (Pmech) delivered to the shaft down to the equipment rating (1.0 p.u.), thereby returning the machine to the reference speed for full power operation, 120% of synchronous speed. When the available wind power is less than rated, the blades are fixed to maximize the mechanical power, and speed control is accomplished by adjusting the generator electrical power. The dynamics of the pitch control are moderately fast, and can have significant impact on dynamic simulation results. The block diagram for the model is shown in Figure 4.

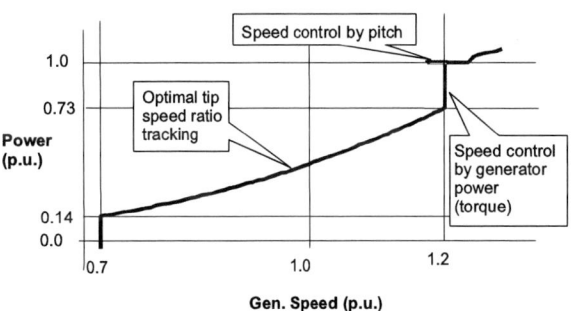

Figure 3. Power vs. Speed Steady State Curve
Source: GE Energy ©2005. Used with Permission.

The wind turbine model represents all of the relevant controls and mechanical dynamics of the wind turbine. The model accepts the machine terminal active power from the WTG Electrical Control Model and the mechanical power calculated by the Wind Power Model. The turbine control model sends a power order to the electrical control for the converter to deliver the requested power to the grid. The electric power actually delivered to the grid is returned to the turbine model, for use in the calculation of rotor speed.

The speed controller does not differentiate between shaft acceleration due to increase in wind speed or due to system faults. In either case, the response is appropriate and relatively slow compared to the electrical control.

The turbine control acts so as to smooth out electrical power fluctuations due variations in shaft power. By allowing the machine speed to vary around its rated value (120%), the inertia of the machine functions as a buffer to mechanical power variations.

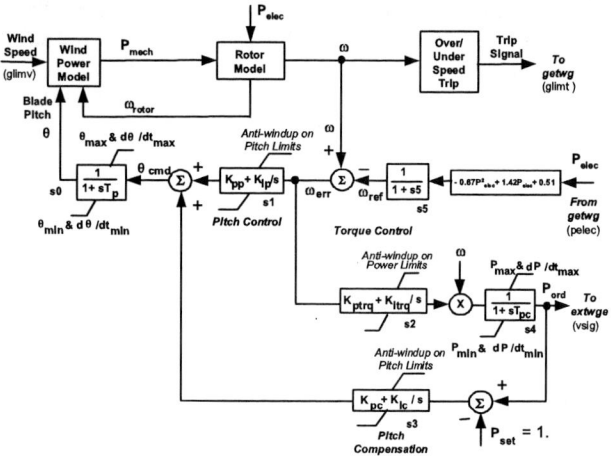

Figure 4. Wind Turbine Model Block Diagram
Source: GE Energy ©2005. Used with Permission.

D. Wind Power Model

The function of the wind power module is to compute the wind turbine mechanical power (shaft power) from the energy contained in the wind. The well-known relationship:

$$P = \frac{\rho}{2} A_r\, v^3\, C_p\,(\lambda, \theta)$$

is used for this purpose. P is the mechanical power extracted from the wind, ρ is the air density in kg/m^3, A_r is the area swept by the rotor blades in m^2, v is the wind speed in m/sec, and C_p is the is the power coefficient and is a function of λ and θ. λ is the ratio of the rotor blade tip speed and the wind speed (v_{tip}/v), θ is the blade pitch angle in degrees. C_p is a characteristic of the wind turbine and is usually provided as a set of curves (C_p curves) relating C_p to λ, with θ as a parameter. The C_p curves for the GE WTG model are fit with a fourth order polynomial on θ and λ.

III. Voltage Regulation with WFMS

For many wind farms, including large off-shore and remote and isolated projects, traditional approaches to managing reactive power are no longer viable. Voltage and reactive power control performed by WFMS minimizes voltage flicker, improves system stability, provides voltage regulation, reduces the risk of voltage collapse, and minimizes the impact of system disruptions. WFMS provides tight closed loop control of utility system voltages. This provides two major benefits: First, the impact of active power fluctuations from wind variation on the grid voltages are minimized; second, the fast and precise voltage control effectively strengthens the grid, improving the overall power system's resilience to large disruptions.

Figure 5 shows the simulated response of a wind farm of 108 GE 1.5 MW wind turbine generators (WTGs) to ten minutes

of highly variable wind near rated wind speed. The red traces show the system response with WFMS, and the black traces show the system response with conventional fixed power factor control. The fixed power factor control is local to each individual WTG.   At the utility bus (the point of interconnection), the system voltage with conventional power factor (pf) control exhibits unacceptably high variation. By comparison, the WFMS controlled system voltage exhibits very small variations. The voltage flicker index, Pst, is less than 0.02 for this high stress condition – well within industry requirements.

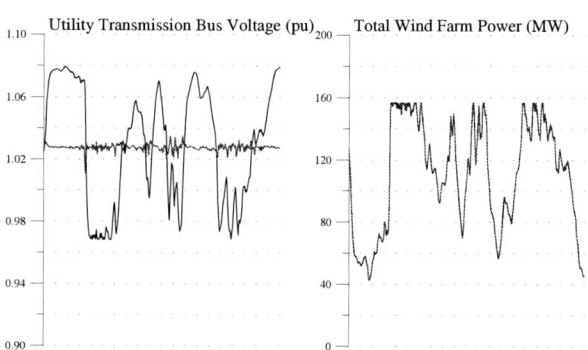

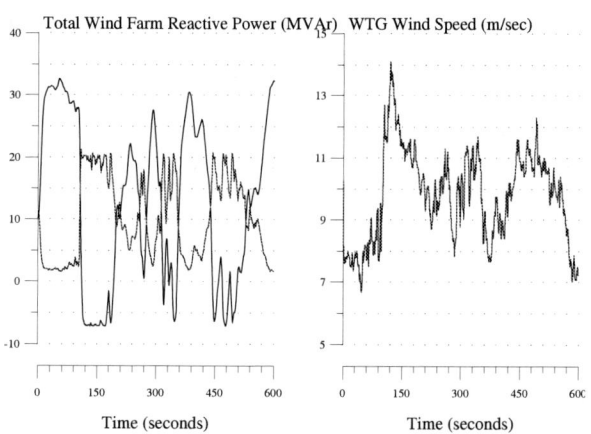

Figure 5. Simulated utility response of a large wind farm.
Black traces are without WFMS (conventional fixed PF control).
Red traces are with voltage control by WFMS.
Source: GE Energy ©2005. Used with Permission.

The variables plotted are as follows:

Utility voltage. This is the voltage at the point-of-common coupling (PCC) in p.u. For this system, the PCC is approximately 75 km from the wind farm.

Wind farm power. This is the total power delivered to the PCC, accounting for collector system and transmission losses.

Wind farm reactive power. This is the total reactive power delivered to the PCC, also accounting for the reactive losses of the collector and transmission system, which are significant.

Wind speed. This is wind profile for this simulation. All the individual machines are subjected to the same wind profile in this case, which is very conservative.

The behavior of one of the WTGs for this wind profile is

shown in Figure 6. The machine reactive power and terminal voltage are actively maneuvered by commands from the WFMS to produce the desired performance at the utility bus. Variables plotted include the following:

WTG terminal voltage, Vt. This is the p.u. voltage at the terminals of one of the individual wind turbine generators.

WTG speed, SPD. This is individual WTG speed variation due to wind speed variation and subject to the GE turbine control, which optimizes energy capture.

WTG reactive power, Qg. This is the individual reactive power produced by the machine.

WTG active power, Pg. This is the individual WTG active power output. Note that it is the same either with or without the WFMS.

Figure 7 shows actual operation of the WFMS voltage regulation function at a large wind farm in the western USA. The traces cover a one-hour period where wind generation

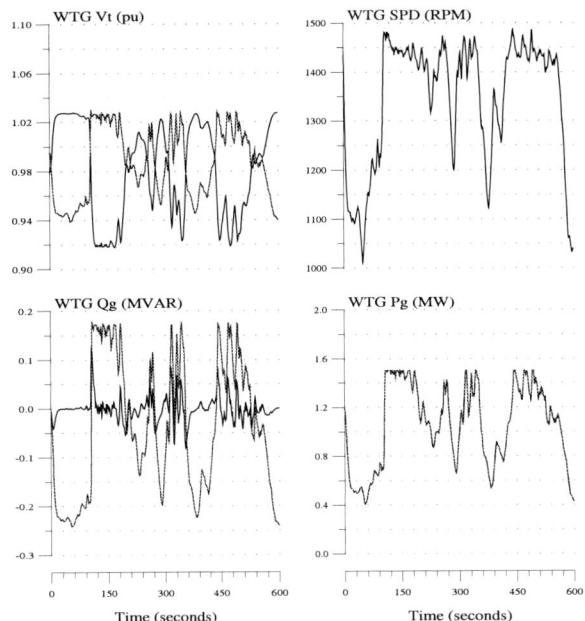

Figure 6. Simulated WTG response of a large wind farm.
Black traces are without WFMS (conventional fixed PF control).
Red traces are with voltage control by WFMS.
Source: GE Energy ©2005. Used with Permission.

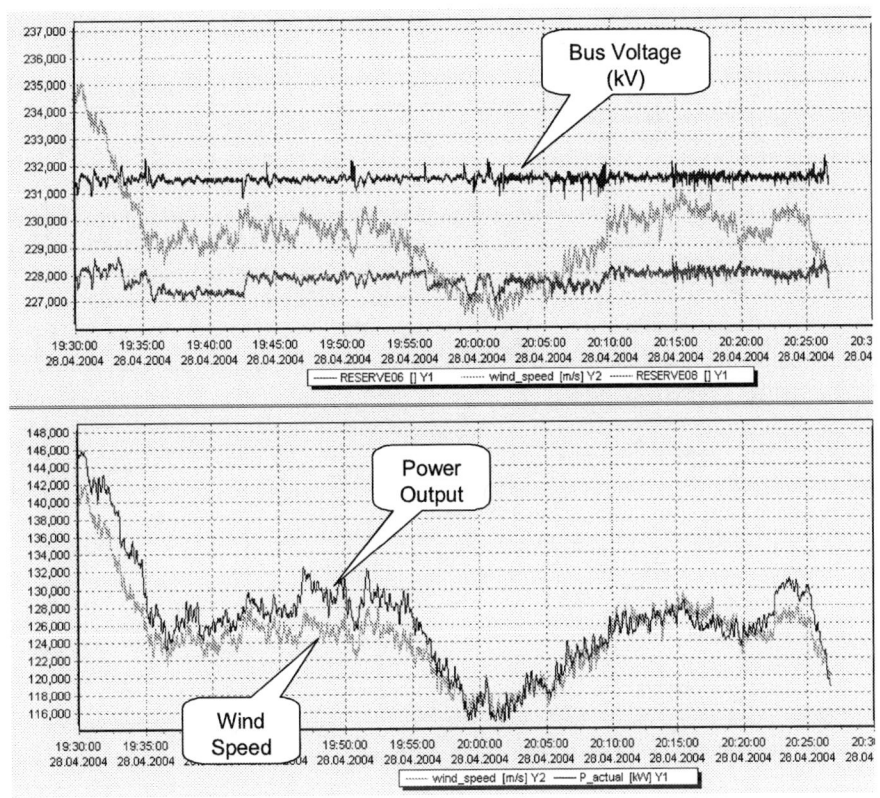

Figure 7. Measurements from Wind Farm in Western USA, April 28, 2004, 19:30 to 20:30
Source: GE Energy ©2005. Used with Permission.

1125

varied over a range of 114 to 146 MW. During this period, the bus voltage at the 230kV point of interconnection remained within 1 kV of its setpoint.

For transmission system events, the response of the WFMS promotes fast and stable recovery of system voltages following fault clearing.

IV. DISTURBANCE RESPONSE WITH LVRT

Provision of uninterrupted service has grown in importance as wind farms increase in size and comprise a larger portion of total generation on the grid. Transmission system events – lightning strikes, equipment failures, and downed power lines – are common on utility grids around the world. Transmission planners and system operators expect generators to tolerate and, ideally, recover from system events. Until recently, wind turbines have been designed to trip offline in response to instantaneous voltage drops in order to protect the wind turbine equipment until the grid recovers. In tightly interconnected grids with significant penetration of wind generation, however, this response can lead to cascading failure of the entire system. GE's LVRT feature renders this over-sensitive response obsolete by improving generator and control system design. Before LVRT technology, wind turbines would trip off-line on any voltage sag below 70%. Now they are designed to ride through severe grid disturbances, as illustrated by the test graphs in Figure 8.

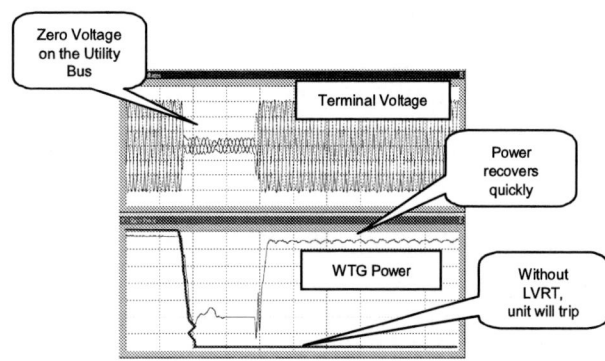

Figure 8. Equipment tests of LVRT performance.
Source: GE Energy ©2005. Used with Permission.

Unbalanced faults present a particularly difficult challenge, both in terms of equipment design and system simulation. Validated simulation models are essential tools for proper equipment design and performance analysis. The traces of Figure 9 compare a factory test measurement [red traces] of LVRT for a severe unbalanced fault against the simulation model [black traces]. The results demonstrate the capability of LVRT to handle unbalanced faults and a high level of fidelity in the digital simulations and models.

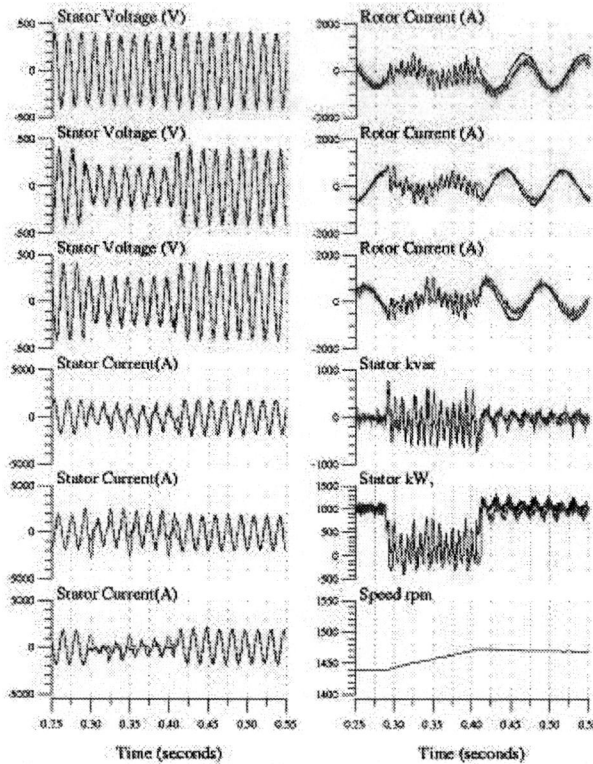

Figure 9. Unbalanced fault on GE 1.5 MW WTG – Tests and Simulation
Source: GE Energy ©2005. Used with Permission.

V. POWER-SWING STABILITY

Electromechanical oscillations, or power swings, often arise between areas in large interconnected power networks and are related to the dynamics of interarea power transfers. Figure 10 shows the two-area four-generator system described in [2] that has been widely used to evaluate power swing dynamics in transmission grids. This example shows the results obtained when a wind farm, represented by an equivalent WTG, replaces one of the four generators.

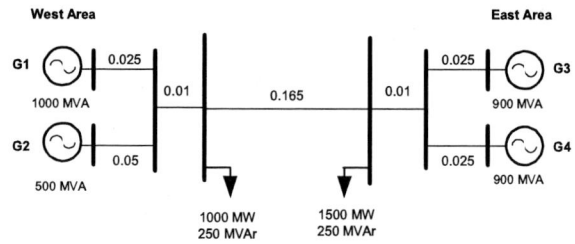

Figure 10 Four Generator Test System

In this study, G1, G3 and G4 are synchronous generators; G2 is modeled as either a synchronous generator or as a WTG. In the latter case, the single WTG represents a large wind farm. Representation of a wind farm consisting of many units by a single WTG and transformer with MVA ratings equal to the individual unit ratings, times the number of units in the farm, gives a reasonable equivalent for bulk system studies, since the impedance of a typical collector system is relatively small compared to the impedance of the unit transformer,

The rating of the wind farm was selected to represent a 33% penetration in the West Area. The model data are provided in [2].

A. Study Approach

The test system was analyzed at four different operating conditions. For each operating condition, the 500 MVA machine, G2, located in the West Area, was first modeled as conventional synchronous generator and then as a single WTG.

Each operating condition corresponds to a different power generation from machine G2. The power output from the WTG reflects the amount of power available for a given wind speed. The initial power output of G2 was set to 150 MW, subsequent generation levels considered were 250 MW, 350 MW and 450 MW. The dispatch from generators G1 and G4 was kept constant at 900 and 650 MW, respectively.

B. Analysis

The results of this investigation show that the interarea mode tends to become more stable as the real power dispatch of the WTG approaches its nominal value (450 MW), whereas the opposite is true for the system with the synchronous generator. These trends are shown as root locus plots in Figure 11. The inter-area mode locus for the case with the WTG is shown as a solid line. The locus for the case with the synchronous generator is shown as a dashed line. From a practical standpoint, the damping ratios for both cases exhibit a comparable range of variation (1%-2.5%) as depicted in Figure 12.

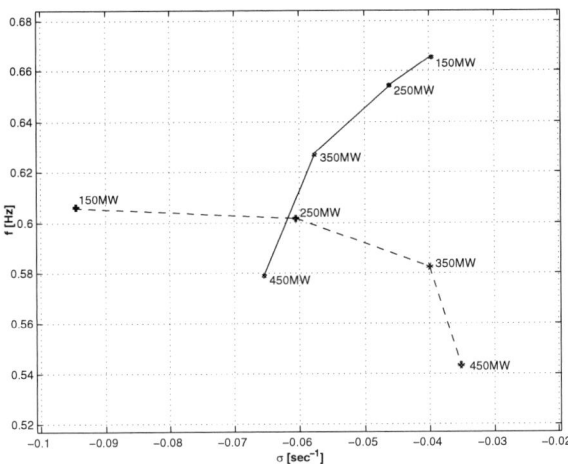

Figure 11. Root locus of interarea mode (--- WTG,   - - - Synch. Gen.)
Source: GE Energy ©2005. Used with Permission.

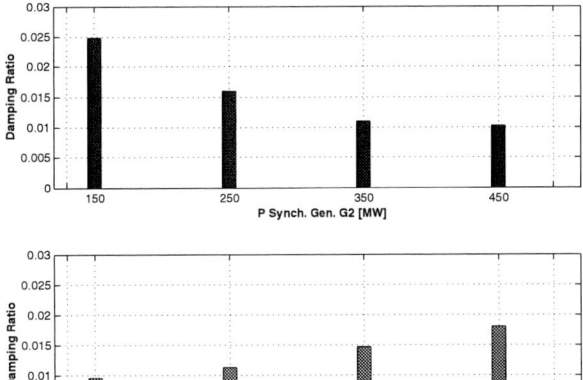

Figure 12. Damping ratios for interarea mode as G2 power output increases.
Source: GE Energy ©2005. Used with Permission.

The most significant differences in the modal characteristics between the systems with the WTG and the synchronous generator arises from the fact that the WTG appears to the rest of the system as a voltage source behind an impedance, and does not interface with the network through an internal angle as a synchronous generator does. This translates in the absence of the mechanical states, speed and angle, from the right eigenvector (mode shape) associated with the inter-area mode. Furthermore, the West Area local mode does not exist when G2 is a WTG. This is illustrated in Figures 13 and 14. Figure 13 shows the local modes of the West and East Areas for the four generation levels considered, when G2 is modeled as a synchronous generator. When G2 is modeled as a WTG, only the local mode associated with the East Area is present (Figure 14).

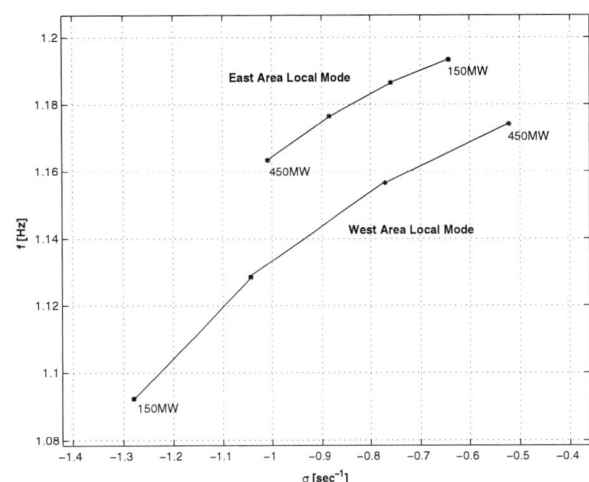

Figure 13. Local Modes when G2 is a Synchronous Generator
Source: GE Energy ©2005. Used with Permission.

For the system and cases considered, the local mode associated with the East Area tends to become less stable as the power output from G2 as a synchronous generator increases from 150 MW to 450 MW (Figure 13). With G2 as a WTG this condition does not arise since the East Area local mode is not present.

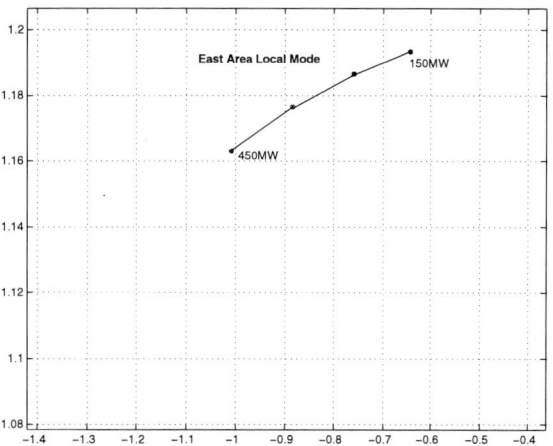

Figure 14. Local Mode when G2 is a WTG
Source: GE Energy ©2005. Used with Permission.

VI. ECONOMIC VALUE

Development of multiple large wind projects in the Hydro Québec system presents significant transmission interconnection challenges. Hydro-Québec's system, which has a winter peak of 35,000 MW, is geographically dispersed. Major hydro generation is up to 1000km remote from major load centers. The entire grid is asynchronous with neighboring systems in Canada and the US. The first region in Québec to be developed for wind generation is the Gaspé peninsula (Figure 15), which is several hundred kilometers from major load centers, has weak existing transmission infrastructure, and very limited local generation.

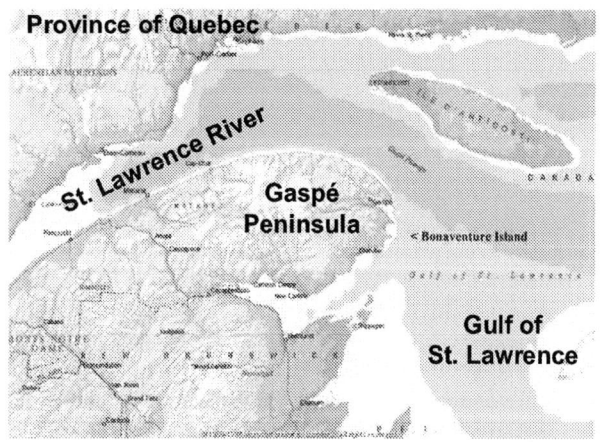

Figure 15. Gaspé Peninsula in eastern Quebec, Canada.

The initial wind tender solicited up to 1000 MW of new wind generation projects for the region. Initial transmission planning studies performed by Hydro Québec TransÉnergie showed requirement for $288M US in transmission upgrades to accommodate this new wind generation. This included $104M US for series, shunt and dynamic compensation. At the conclusion of the tendering process, GE wind generation was selected for all projects, totaling 990 MW. Transmission planning studies were then repeated, using updated models and wind generation controls consistent with the selected technology. The WTG equipment (including voltage regulation and LVRT features) significantly improved system performance, thereby reducing the need for other system reinforcements. The initial cost estimate for transmission reinforcement was reduced by about $95M US [3].

VII. CONCLUSIONS

Recent developments in wind generation technology have solved several of the serious problems posed by large wind farms connected to weak ac transmission grids. Coordinated voltage regulation of all individual WTGs in a large wind farm maintains constant voltage at the interconnecting bus, regardless of variations in wind power generation. LVRT technology enables wind farms to continue operation during and after severe faults or voltage depressions on the power grid. Power-electronic conversion and control technology incorporated into the generating system enables variable speed operation, while eliminating electromechanical power swing interaction with the grid. The combination of these features enable wind power plants to achieve stability performance that can exceed that of conventional synchronous generations of the same rating, installed at the same locations.

VIII. REFERENCES

[1] N. W. Miller, J. J. Sanchez-Gasca, W. W. Price, R. W. Delmerico, "Dynamic Modeling of GE 1.5 and 3.6 MW Wind Turbine-Generators for Stability Simulations", Proc. Power Engineering Society General Meeting. Toronto, Ontario. Canada, July 2003.

[2] M. Klein, G.J. Rogers, P. Kundur, "A Fundamental Study of Inter-Area Oscillations in Power Systems", IEEE Trans. on Power Systems, Vol. 6, No. 3, Aug. 1991, pp. 914-921.

[3] R. Champagne, M. Lamothe, S. Paquette, "Grid Connection of Large Wind Power Plants in Hydro-Quebec's System," WindPower '05, Denver Colorado, USA, May 16, 2005.

IX. BIOGRAPHIES

Richard Piwko (M'76, F'96) is a Principal Consultant with GE Energy in Schenectady, NY. His responsibilities include management of large-scale system studies, power plant performance testing, control system design, and analysis of interactions between turbine-generators and the power grid. He recently contributed to GE's development of the Variable Frequency Transformer (VFT), a new technology for transferring power between asynchronous power grids. Mr. Piwko is a Fellow of the IEEE and has served as chairman of the IEEE HVDC & FACTS subcommittee and as chairman of the IEEE Transmission and Distribution Committee.

2006 5th International Power Electronics and Motion Control Conference

Turn-on Condition and Characteristics of High-power Semiconductor Switch RSD

Y. M. Zhou, Y. H. Yu, H. G. Chen, L. Liang

Department of Electronic Science and Technology, Huazhong University of Science and Technology
Wuhan, Hubei, China
E-mail: ym_chow@163.com

Abstract—RSD (Reversely Switched Dynistor), a high-power semiconductor switch, is similar in design to thyristor with *pnpn*-regions. Instead of a conventional gate structure, RSD's anode consists of alternating n^+ and p^+ sections. In this paper, the RSD turn-on condition is systematically investigated. Analyses and simulations reveal that the trigger charge must be sufficient to assure RSD normal turn-on. To explore the potential application of RSD, several turn-on characteristics are experimentally evaluated. Results indicate that RSD has advantages in fast voltage-falling, great capability of handling high current and high *di/dt*.

Keywords-RSD; critical charge value; high current; di/dt

I. INTRODUCTION

Pulse power technology has been widely used in many industry areas such as waste water and gas cleaning, material modification and biotechnology[1-3], etc. As the key component in pulsed power applications, switch is the development bottleneck of pulsed power technology. Previously, mechanical switch and spark gaps have been the preferred candidates, but these devices usually have a short lifetime and insufficiently stable performance. Semiconductor switches have drawn special attention due to their superior properties in compactness, light weight, low cost and high efficiency[4-7]. RSD (Reversely Switched Dynistor), a high-power semiconductor switch proposed by I V Grekhov and his colleagues in 1980s, is initiated by a thin electron-hole plasma layer, which makes RSD turn-on simultaneously and uniformly over semiconductor wafer area[8].

In this paper, the RSD turn-on condition is investigated by simple analysis and device simulation. Experiments have been conducted to evaluate the RSD turn-on characteristics. Results from this work will help in design and development of RSD in pulsed power applications.

II. RSD TURN-ON CONDITION

A. Operating mechanism of RSD

An RSD is a four-layer device similar to a thyristor,

but the anode is made up of thousands of alternating transistor and thyristor components (Fig. 1). These components have a common collector junction J_2 which blocks the main discharge voltage. When the switch S is closed down, a trigger voltage is applied to the cathode, which makes the low-voltage emitter junction J_3 broken down. Then a trigger current pulse flows through transistor components, which is accompanied by an electron-hole plasma injection to the common *n*-base region for the transistor and neighboring thyristor components. After a delay of about 2μs, the magnetic switch (MS) saturates and the main voltage polarity returns to the initial state. This makes the electrons and holes of the near-collector layer go to the *n*- and *p*-base region respectively, which results in the injection of minority carriers from emitter layers, and subsequently RSD turns on uniformly over the device area. Such character of this process offers a possibility of switching high current pulse.

B. Turn-on condition of RSD

The trigger current makes the common *n*-base region full of plasma and a higher concentration plasma layer P_1 assembling at J_2 junction (Fig. 2). The charge of P_1 layer is expressed as

$$Q_1 = \frac{b}{b+1} Q_R. \qquad (1)$$

Where $b=\mu_e/\mu_h$ is the electron-to-hole mobility ratio in a weak field, Q_R is the total trigger charge duration of the trigger time t_R.

Under the drive of the main voltage, the electrons in P_1 are injected into the *n*-base, which forms the extracting current J_{extr}. At the same time, holes are

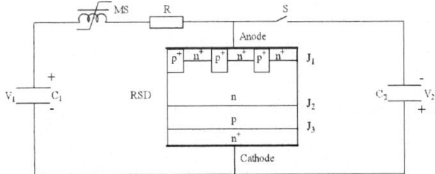

Figure 1. Structure and schematic circuit of RSD

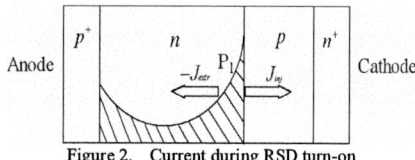

Figure 2.   Current during RSD turn-on

injected into the p-base, which results in the electrons injected from the n^+ region, thus an injection current J_{inj} forms. Nevertheless, Compared with J_{extr}, the generation of J_{inj} will take a little time because of the behavior of electrons passing through the p-base region. So the difference of Q_1 is

$$\frac{dQ_1}{dt} = J_{inj} - J_{extr}. \qquad (2)$$

A numerical result from above equation is shown in Fig. 3. It is noted that Q_1 decreases at first and then increases. Curve 1 is the case when the trigger charge Q_R is not enough, then J_{extr} exhausts the carriers in P_1, however, J_{inj} is too late to supplement carriers. In this case, a high voltage drop will occur due to the absence of carriers at J_2 junction to carry current, and device tends to fail. So a critical value of trigger charge Q^{cr} is required to assure RSD turn-on:

$$Q^{cr} = \frac{dJ_F}{dt}\frac{b+1}{b}\left(\frac{\tau_*}{\nu_1}+\frac{1}{b+1}-1\right)^{-1}\frac{\tau_*^4}{2\nu_1^2}. \qquad (3)$$

Where dJ_F/dt is the switched current rising rate, τ_* is the collector current rising time constant, ν_1 is the time of electron diffusion through the heavily doped p-base.

Curve 2 in Fig. 3 is the case when Q_1 is equal to Q^{cr}, where t_{min} is determined by $J_{inj(tmin)}=J_{extr(tmin)}$. Curve 3 is desirable for RSD application, where Q_1 is more than Q^{cr}, thus RSD will turn on with a low on-state voltage.

In the generally practical case, letting $\tau_*=0.9\nu_1$ and $\nu_1=10^{-7}$ leads to

$$Q^{cr} = 3.4 * 10^{-14}\frac{dJ_F}{dt}. \qquad (4)$$

An experimental test is performed on a 16mm RSD device to validate the analysis above. The test circuit is schematically shown in Fig. 1.

After the switch S is closed, the voltage of C_2 is

$$V_2(t) = -V_2\cos\frac{t}{\sqrt{L_2C_2}}. \qquad (5)$$

Because the trigger circuit is separated from the main circuit by the MS, the current of trigger circuit is expressed as

$$i_R(t) = V_2\sqrt{\frac{C_2}{L_2}}\sin\frac{t}{\sqrt{L_2C_2}}. \qquad (6)$$

The total trigger charge in a wafer area S after a delay time τ_s is

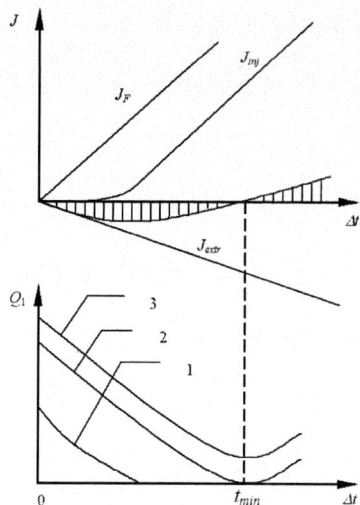

Figure 3.   Trend of Q_1 changing

$$Q_R = \frac{V_2C_2}{S}(1-\cos\frac{\tau_s}{\sqrt{L_2C_2}}). \qquad (7)$$

In the test, the delay time of magnetic switch is set to 2μs and the trigger voltages are 100V and 1000V. The corresponding on-state characteristics of 2000V main voltage are shown in Fig. 4. Under this condition, the required critical charge is calculated to about 9.83μC/cm^2 according to (4). From (7), the trigger charges offered by trigger voltages of 100V and 1000V are 5.22μC/cm^2 and 52.2μC/cm^2 respectively. We note that the trigger charge offered by 100V is lower than Q^{cr}, and it can be seen from Fig. 4 that a voltage-jumping occurs when the trigger voltage is 100V, whereas there is no jump for the 1000V trigger voltage.

To further reveal the effect of trigger charge on RSD turn-on, extensive simulations are carried out using the device simulator of Silvaco International on a $pnpn$ structure alternating 50μm n^+-section and 250μm p^+-section in the anode, the device thickness is 450 μm. A

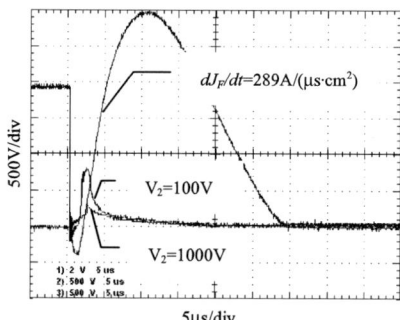

Figure 4.   On-state characteristic of RSD

triangle voltage waveform approximating the trigger pulse is applied to form the trigger source. Fig. 5 shows the plasma distribution curves obtained when the trigger final value is 40V and 100V respectively. Another triangle voltage with a final value of 300V is used to imitate the main voltage pulse. The corresponding plasma distributions after turn-on are shown in Fig. 6. From the diagram it can be seen that the plasma concentration is low after the trigger of 40V. In this case, the plasma level at J_2 junction after turn-on is below the doping concentration, which indicates an occurrence that carriers are exhausted. Nevertheless, the carriers are not exhausted when the trigger voltage is 100V.

III. RSD TURN-ON CHARACTERISTICS

A. Voltage-falling time

An RSD device of 420μm in thickness is used to test the voltage-falling time. The voltage waveform is measured by a 1000:1 high voltage probe. The voltage-falling process is shown in Fig. 7, and it can be noted that the voltage-falling time is about 50ns.

In our experiments, we notice that the voltage-falling time is related to silicon wafer parameters such as thickness and doping concentration. The voltage-falling time of lower doping level is shorter than that of higher

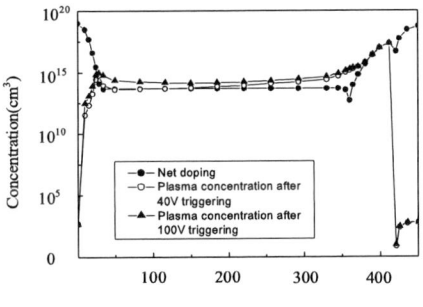

Figure 5. Plasma concentration after trigger

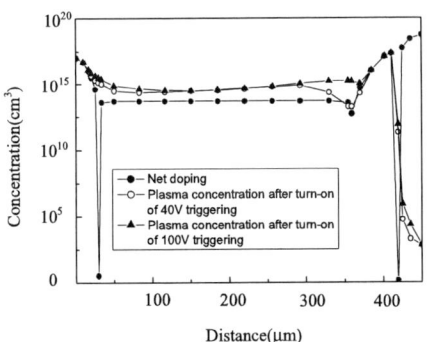

Figure 6. Plasma concentration after turn-on

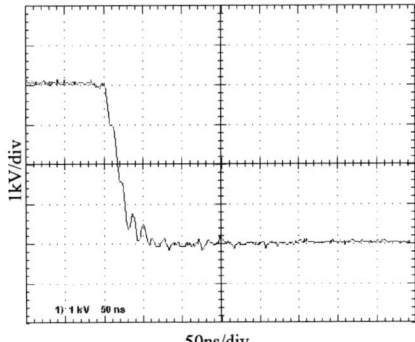

50ns/div

Figure 7. Process of voltage falling

doping level. This is because the probability of carriers scattered by crystal lattice is less in the case of low doping. Compared with to a thicker n-base, carriers take less time to penetrate through a thinner n-base.

B. Capability of handling high current

In this experiment a testbed is employed with characteristic parameters of a 0.4Ω resistor and a 21.4μF main capacitor. An RSD stack consisting of five 2.5kV 24mm RSD devices is used to switch the main current. The peak current is monitored by a current shunt of 246A/V, the waveform is recorded by a TDS210 scope. When the main voltage is 8kV, a pulsed current shown in Fig. 8 is achieved. It can be seen that the peak current is about 10kA with a basewidth of 34μs and di/dt of 2kA/μs.

RSD has no gate, so there is no loss of commutation area. The switching potentiality of RSD is two to three times higher than that of the best pulsed thyristors with the same dimensions of device structures.

C. Di/dt characteristic

A 16mm RSD device is used in this evaluation. Some measures are taken to achieve higher di/dt, including considerably reducing the inductance of the testbed, substituting 0.22μF capacitor for 21.4μF, shorting the

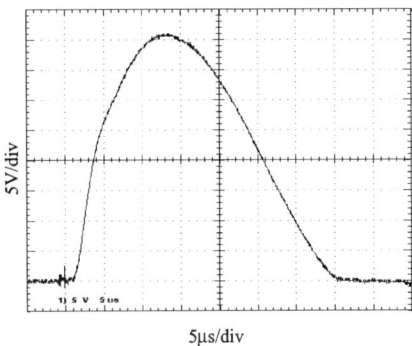

5μs/div

Figure 8. High current waveform of RSD

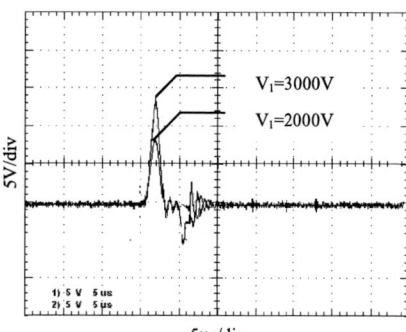

Figure 9. Di/dt characteristic of RSD

primary load, replacing the magnetic switch with another of higher saturation magnetic flux and higher residual magnetic flux. The trigger voltage is high enough to assure the device normal turn-on. The ratio of current shunt is 609A/V. Waveforms are shown in Fig. 9. When the main voltage is 2kV, a current pulse with a peak value of 5kA, di/dt of 4kA/µs and basewidth of 2.5µs is obtained. When the voltage rises to 3kV, peak current is 8.5kA and di/dt is 7.2kA/µs. After the test, the device keeps unbroken down.

It is well known that thyristors have typically required some type of saturable reactors to limit the current flow during the early stages of turn-on because the spreading rate of horizontal plasma lags behind the rising rate of current. So higher di/dt tends to make thyristor fail. However, for RSD, the whole chip area is switch on instantaneously as a result of fast spreading of the abundant plasma generated by the trigger current, therefore RSD has great ability to deal with high di/dt.

IV. CONCLUSIONS

In this paper, RSD turn-on condition is investigated and RSD turn-on characteristics are experimentally evaluated. It is demonstrated that RSD turn-on condition is that the charge in J_2 junction can not be exhausted, otherwise, RSD will block the main voltage in advance, and then commutation will fail. Through testing, RSD shows excellent characteristics in voltage-falling time, capability of handling high current and high di/dt. Such results reveal that RSD is a promising high-power switch in pulsed power applications.

ACKNOWLEDGMENT

This work is supported by a grant from the National Natural Science Foundation of China (No. 50277016 and 50577028) and Specialized Research Fund for the Doctoral Program of Higher Education (No. 20050487044).

REFERENCES

[1] A. Pokryvailo, Y. Yankelevich, M. Wolf, E. Abramzon, S. Wald, A. Welleman, "A high-power pulsed corona source for pollution control applications", IEEE Trans. Plasma Sci., vol. 32, pp. 2045-2054, Oct. 2004.

[2] E. J. M. van Heesch, K. Yan, A. J. M. Pemen, S. A. Nair, G. J. J. Winands, "Repetitive pulsed power to serve nanotechnology, sustainability and hydrogen production", in *Proc. 14th IEEE Int. Pulsed Power Conf.*, Dallas, TX, June 15-18, 2003, pp. 441-444.

[3] K. Yatsui, W. Jiang, H. Suematsu, "Pulsed-power applications to materials science", in *Proc. 14th IEEE Int. Pulsed Power Conf.*, Dallas, TX, June 15-18, 2003, pp. 29-34.

[4] E. Spahn, G. Buderer, E. Ramezani, "High voltage thyristor switch for pulse power applications", in *Proc. 21th Int. Power Modulator. Conf.*, Costa Mesa, CA, June 28-30, 1994, pp. 93-96.

[5] T. F. Podlesak, H Singh, S Schneider, S Behr, "High peak current burst repetitive operation of a 125 mm thyristor", in *Proc. 11th IEEE Int. Pulsed Power Conf.*, Baltimore, MD, June 29-July 2, 1997, pp. 396-404.

[6] A. Welleman, E. Ramezani, U. Schlapbach, "Semiconductor switches replace thyratrons and ignitrons", in *Proc. 13th IEEE Int. Conf. Pulsed Power*, Las Vegas, NV, June 17-22, 2001, pp. 325-328.

[7] R. Petr, J. Freshman, N. Orozco, "Solid-state pulsed power for driving a high-power dense plasma focus x-ray source", *Rev. Sci. Instrum.*, vol. 71, pp. 1360-1362, Mar. 2000.

[8] A. V. Gorbatyuk, I. V. Grekhov, A. V. Nalivkin, "Theory of quasi-diode operation of reversely switched dinistors", *Solid-state Electronics*, vol. 31, pp. 1483-1491, Oct. 1988.

2006 5th International Power Electronics and Motion Control Conference

The analysis and simulation of power circuits for high voltage converter

S. I. Volskiy*, Y. Y. Skorokhod**, V. V. Shergin**

* Moscow State Aviation Institute (Technical University), Moscow, Russia

** Ltd Transconverter, Moscow, Russia

e-mail: skorohod@transconverter.ru, volsky@ultranet.ru

Abstract — **The power converter with an unstable input high voltage (2000...4000 V DC) and stable output three-phase voltage 380 V, 50 Hz AC is presented. Various converter types with high frequency principle and 50 Hz principle of the electrical energy transformation are investigated. The main emphasis is given to the analysis procedure of power losses in semiconductors devices. Basis equations of power losses in semiconductors devices, characteristics and results of computer simulation in CASPOS are obtained and discussed.**

Keywords - high voltage power converter; power losses in semiconductor device; local train.

I.   INTRODUCTION

Specific technical properties and features of the power converter application for the local train are considered and analyzed. Basic requirements are the following:

- Wide range of the input high voltage (2000...4000 V, DC) with permissible enormous voltage peaks (up to 10000 V within 10 ms);

- Improved robustness and high electromagnetic compatibility;

- High efficiency, reliability, simple and compact electrical and mechanical design, using of the modular assembly;

- Effective mass-size indices;

- Wide range of the temperature (- 50°C...+ 40°C).

It is known that using of the high frequency principle of

the electrical energy transformation is the effective and attractive way for use in the power converters. Advantage of this principle consists in reduced weight and size.

This is why this principle is widely used in power converters of different types [1...3]. Power circuit of the one of such kind of converter is represented at Fig. 1. It consists of the following components:

- power module (A1) which increases input voltage (U_{in}) up to the value required (U_{out1}) and provides its stability;

- power module (A2) which transforms input (U_{out1}) into high-frequency alternating voltage (U_{out2});

- high frequency power transformer (TV1) which decreases U_{out2} down to the value required;

- output rectifier (AR1);

- three-phase module (A3) which transforms direct input voltage U_{inA3} into three-phase voltage 380 V, 50 Hz.

It should be mentioned that in the modules A1 and A2 power IGBT-transistors have been designed for use at 6500 V voltage level. Such high voltage IGBT-transistors has rather high switching and on-state power dissipation.

Other type of power converter circuit is represented at Fig. 2. It consists of the following components:

- four power modules (A11..A14) having functions and power scheme of power module A1 (Fig. 1);

- four power module (A21..A24) having functions and power scheme of power module A2 (Fig. 1);

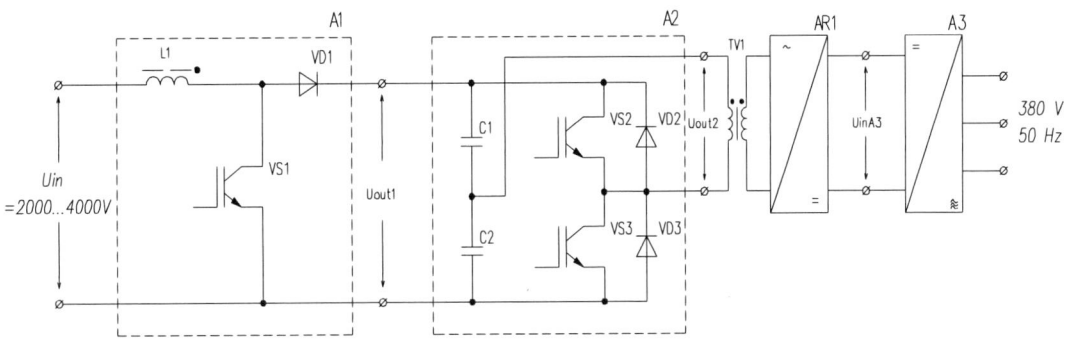

Figure 1.   Power circuit of the first converter

1-4244-0448-7/06/$25.00 ©2006 IEEE

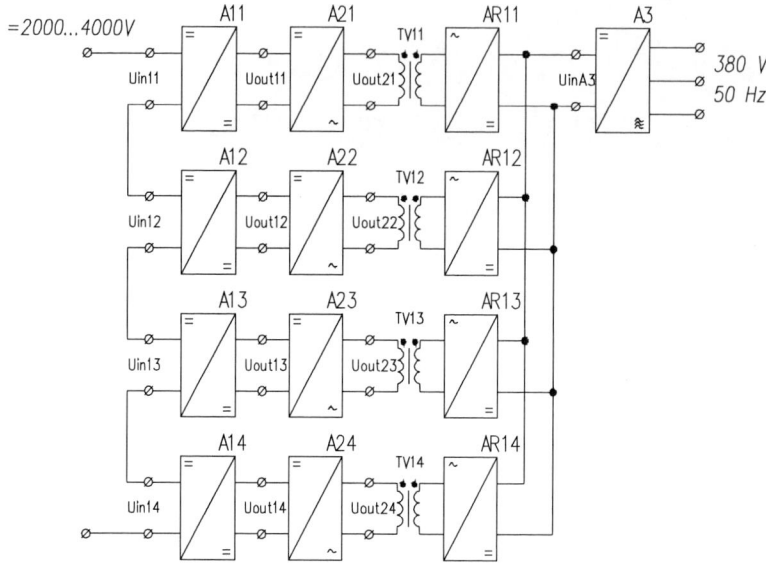

Figure 2.   The power circuit of the second converter

-     four high-frequency transformers (TV11...TV14);

-     three-phase module A3 which has output voltage 380 V at frequency 50 Hz.

It should be mentioned that in the power modules A11..A14 and A21..A24 IGBT-transistors designed for implementation of 1700 V voltage are used. These IGBT-transistors are distinguished by rather low switching and on-state power dissipation.

However, using of the high frequency principle of the electrical energy transformation leads to the number of problems. The most important problem is power dissipation in semiconductors devices, which is increasing as long as the frequency increase.

Besides, converters under consideration have five stages (A1, A2, TV1, AR1 and A3, Fig.1 or A11...A14, A21...A24, TV1...TV14, AR11...AR14 and A3, Fig. 2) of electric energy conversion which determine complexity of power scheme, rather low efficiency factor and

reliability in combination with increased summary power dissipation in semiconductor devices.

In the capacity of the converter alternative version using 50 Hz principale of the electrical energy transformation power circuit have been presented (Fig. 3). It consists of the following components:

-     three-phase module (A3) which transforms high input voltage (2000...4000V) into three-phase voltage at output frequency 50 Hz;

-     three-phase power transformer (TV2) which reduces three phase input voltage to required value 380 V at 50 Hz frequency.

It also should be mentioned that 6500 V voltage IGBT are used in the three-phase power module A3 after the analogy of the power modules A1 and A2 (Fig. 1).

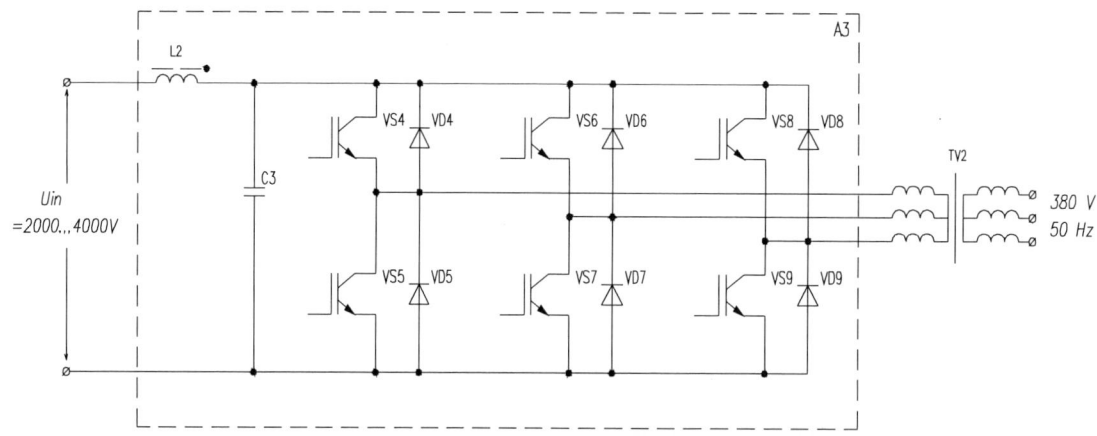

Figure 3.   The power circuit of the third converter

II. ENERGY DISSIPATION IN THE POWER SEMICONDUCTOR DEVICES OF HIGH VOLTAGE CONVERTER

Generally, on-state power dissipation in the power IGBT or power diode over one period at 50 Hz output frequency is determined according to equations:

$$S_{vs} = f_{out} \int_{o}^{t_{pvs}} i_{vs} \cdot u_{satvs} dt , \qquad (1)$$

$$S_{vd} = f_{out} \int_{o}^{t_{pvd}} i_{vd} \cdot u_{vd} dt , \qquad (2)$$

where:

f_{out} = 50 Hz - converter output voltage frequency;

t_{pvs} - duration of IGBT on-state within one period of switching;

i_{vs} - IGBT instant current value;

u_{satvs} - IGBT instant voltage saturation value;

$t_{pvd} = T_{vs} - t_{pvs}$ - power diode opened position duration within one IGBT period T_{vs} of switching;

i_{vd} - power diode current instant value;

u_{vd} - power diode instant forward voltage drop value at on-state position;

Switching losses in the power IGBT over one period 50 Hz output frequency are determined as:

$$P = \left(\frac{E_{on} \cdot I_{on} \cdot U_{on}}{I_{nvs} \cdot U_{nvs}} + \frac{E_{off} \cdot I_{off} \cdot U_{off}}{I_{nvs} \cdot U_{nvs}} \right) \cdot f , \quad (3)$$

where:

E_{on}, E_{off} – IGBT-transistor energy turn-on and turn-off switching dissipation taking in account I_{nvs} current value and U_{nvs} voltage value;

I_{on}, I_{off} - IGBT turn-on and turn-off instant current value;

U_{on}, U_{off} - IGBT-transistor turn-on and turn-off instant voltage value;

f - IGBT-transistor switching frequency.

In the power semiconductor devices switching and on-state analysis the following assumptions are taken into consideration:

1. Each power IGBT and power diode current at i-number interval of switching is of constant value;

2. Three-phase module A3 output current is of sine wave form due to use of external three-phase sine-wave filter;

3. Converter three-phase load is of resistive nature;

4. Conductive power dissipation at each of the power IGBT and the diode in shut position are equal to zero;

5. Forward direction voltage drop at each of the power IGBT and the diode in opened position is of constant value;

We have investigated and analyzed power semiconductor devices dissipation for all converters in consideration power modules taking in account assumptions of above mentioned. As a result we have obtained equations which can be used for calculation of switching power dissipation in IGBT VS1 and diode VD1 (power module A1, Fig. 1):

$$S_{vs1} = \frac{Q \cdot U_{vs1}}{U_{outA1} \cdot \eta} \cdot \frac{\gamma_{vs1}}{1 - \gamma_{vs1}} , \qquad (4)$$

$$S_{vd1} = \frac{Q \cdot U_{vd1}}{U_{outA1} \cdot \eta} , \qquad (5)$$

where:

Q – converter output power;

U_{vs1} - forward direction voltage drop at IGBT VS1 at current value I_{vs1} ;

η -converter efficiency factor;

$\gamma_{vs1} = \dfrac{t_{pvs1}}{T_{vs1}}$ - transistor VS1 duty ratio;

t_{pvs1} - transistor on-state duration VS1;

T_{vs1} - VS1 switching period;

U_{outA1} - power module A1 output voltage;

U_{vd1} - forward direction voltage drop at power diode VD1 at current value I_{vd1} .

Additionally power module A1 output voltage as well as on-state IGBT current I_{vs1} and diode current I_{vd1} can be determined using the following equations:

$$U_{outA1} = \frac{U_{in}}{1 - \gamma_{vs1}} , \qquad (6)$$

$$I_{vs1} = I_{vd1} = \frac{Q \cdot \eta}{U_{outA1} \cdot (1 - \gamma_{vs1})} , \qquad (7)$$

where:

U_{in} - converter input voltage.

We have also obtained equation which can be used for - transistor VS1 switching power dissipation calculation:

$$P_{vs1} = (E_{onvs1} + E_{offvs1}) \cdot \frac{I_{vs1} \cdot U_{outA1} \cdot f_{vs1}}{I_{nvs1} \cdot U_{nvs1}} , \quad (8)$$

where:

E_{onvs1}, E_{offvs1} –transistor VS1 turn-on and turn-off energy switching losses at the current I_{nvs1} and the voltage U_{nvs1} ;

f_{vs1} - transistor VS1 switching frequency.

Taking in account assumption (3) we come to conclusion that on-state dissipation of the power diodes

VD2 and VD3 (power module A2, Fig. 1) have zero value. Also, IGBT transistor VS2 and VS3 switching and on-state dissipation can be calculated using following equations:

$$S_{vs2} = S_{vs3} = \frac{2 \cdot Q \cdot U_{vs2} \cdot \eta}{U_{outA1}}, \qquad (9)$$

$$P_2 = P_3 = \left(E_{on2} + E_{off2}\right) \cdot \frac{I_{vs2} \cdot U_{outA1} \cdot f_{vs2}}{I_{nvs2} \cdot U_{nvs2}}, \qquad (10)$$

where:

$U_{vs2} = U_{vs3}$ - IGBT-transistor VS2 (VS3) forward voltage drop at the current $I_{vs2} = I_{vs3}$;

$I_{vs2} = \dfrac{2 \cdot Q \cdot \eta}{U_{outA1}}$ - VS2 (VS3) on-state current;

E_{on2} , E_{off2} – transistor VS2 (VS3) turn-on and turn-off switching energy dissipation at the current I_{nvs2} and the voltage U_{nvs2} ;

f_{vs2} - transistor VS2 (VS3) switching frequency.

In the second converter scheme (Fig. 2) power modules A11...A14 and A21...A24 linked in series relative to input voltage. Therefore every power module A11...A14 output voltage was determined according to equation:

$$U_{outA11} = \frac{U_{in}}{4 - \gamma_{vs11}}, \qquad (11)$$

where:

$\gamma_{vs11} = t_{pvs11} / T_{vs11}$ - power module A11...A14 transistor VS11 duty ratio;

t_{pvs11} - IGBT-transistor VS11 on-state duration;

T_{vs11} - IGBT-transistor VS11 switching period.

In order to calculate conduction losses in the IGBT-transistor VS11 and diode VD11 (power modules A11...A14, Fig. 2) S_{vs11} and S_{vd11} accordingly, (11) have been used to obtain necessary equations:

$$S_{vs11} = \frac{Q \cdot U_{vs11}}{U_{outA11} \cdot \eta} \cdot \frac{\gamma_{vs11}}{4 - \gamma_{vs11}}, \qquad (12)$$

$$S_{vd11} = \frac{Q \cdot U_{vd11}}{U_{outA11} \cdot \eta}, \qquad (13)$$

where:

U_{vs11} - transistor VS11 voltage drop at the current value I_{vs11} ;

U_{vd11} - power diode VD11 forward voltage drop at the current I_{vd11} ;

$$I_{vs11} = I_{vd11} = \frac{Q}{U_{outA11} \cdot \eta} \cdot \frac{1}{4 - \gamma_{vs11}}, \qquad (14)$$

We have also obtained equation for calculation IGBT-transistor VS11 switching power dissipation:

$$P_{11} = \left(E_{on11} + E_{off11}\right) \cdot \frac{I_{vs11} \cdot U_{outA11} \cdot f_{vs11}}{I_{nvs11} \cdot U_{nvs11}}, (15)$$

where:

E_{on11} , E_{off11} – IGBT-transistor VS11 turn-on and turn-off switching power dissipation at the current value at I_{nvs11} and the voltage value U_{nvs11} ;

f_{vs11} - IGBT-transistor VS11 frequency of switching.

Expressions for calculating IGBT-transistor VS21 and VS31 (power modules A21...A24, Fig. 2) switching and conducting losses are identical in general view to (9) and (10) equations.

As a result of the investigation undertaken equations have been obtained which are intended for calculation of losses in three-phase power module A3 power semiconductor devices. The same equations can be used in the course of analysis of any other converter under consideration.

Also, as a result of investigation and analysis procedure we have obtained equations for calculation IGBT-transistors VS4...VS9 power dissipation for the output voltage period:

$$S_{vs4} = \sum_{i=1}^{n} \frac{\pi \cdot Q \cdot U_{vs4}}{\sqrt{8} \cdot U_{outA3} \cdot \eta_{A3}} \cdot \gamma_{ivs4} \cdot B1, \qquad (16)$$

where:

$$B1 = \sin\left[\frac{\pi}{4}(i - 0.5)\right];$$

n - number of IGBT-transistor VS4...VS9 switching within half-period (10 ms) of converter output voltage;

i - power IGBT-transistor VS4...VS9 ordinal switching number;

U_{vs4} - power IGBT-transistor VS4...VS9 voltage drop at the current value I_{vs4} ;

$\gamma_{vs4} = t_{pvs4} / T_{vs4}$ - power IGBT-transistor VS4...VS9 duty ratio at i switching number period;

t_{ipvs4} - power IGBT-transistor VS4...VS9 on-state duration at i switching commutation period;

T_{vs4} - power IGBT-transistor VS4...VS9 commutation period;

U_{outA3} - power module A3 output line voltage.

Power IGBT transistor VS4…VS9 or power diode VD4…VD9 voltage drop should be determined from data sheet of this semiconductor devices at the current I_{vs4} value calculated according to equation:

$$I_{vs4} = \frac{Q}{1.11 \cdot U_{outA3}}, \qquad (17)$$

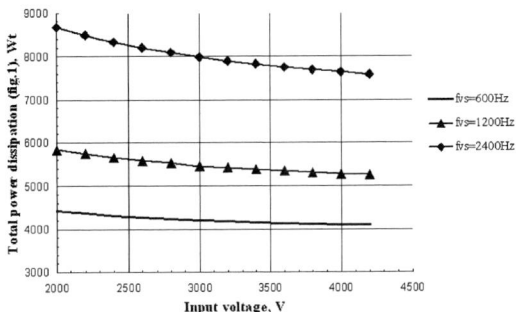

Figure 4. Total power dissipation in transistors of the first converter at different switching frequencies depending on input voltage

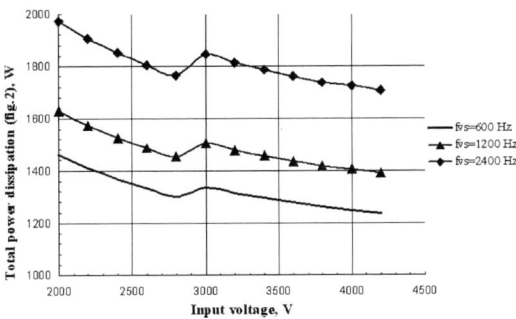

Figure 5. Total power dissipation in transistors of the second converter at different switching frequencies depending on input voltage

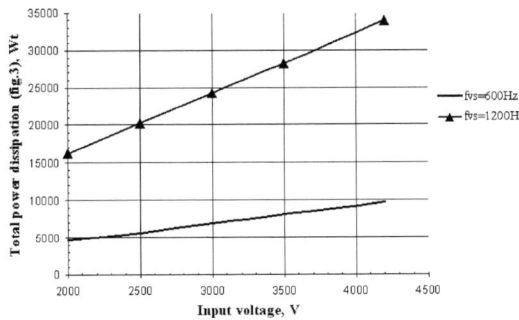

Figure 6. Total power dissipation in transistors of the third converter at different switching frequencies depending on input voltage

Power IGBT-transistor VS4…VS9 switching power dissipation within converter output voltage period should be determined according to the following equation:

$$P_{vs4} = \sum_{i=1}^{n} \frac{\pi}{\sqrt{24}} \cdot \frac{Q \cdot U_{inA3} \cdot f_{vs4}}{I_{nvs4} \cdot U_{nvs4} \cdot U_{outA3}} \cdot B2 , \quad (18)$$

where:

$$B2 = \left(E_{onvs4} + E_{offvs}\right) \cdot \sin\left[\frac{\pi}{n}(i - 0.5)\right];$$

$E_{onvs}4$, $E_{offvs}4$ – power IGBT-transistor VS4…VS9 VS4…VS9 turn-on and turn-off power dissipation at the current value I_{nvs4} and the voltage value U_{nvs4};

$f_{vs4} = 1/T_{vs4}$ - power IGBT-transistor VS4…VS9 switching frequency.

III. ANALYSIS OF POWER DISSIPATION IN HIGH VOLTAGE CONVERTER SEMICONDUCTOR DEVICES

The comprehensive analysis of presented power circuits is carried out for a wide range of different values of supply voltage (2000…4200 V) and power semiconductors devices switching frequency (600, 1200, 2400 Hz), see Fig. 4, 5, 6. These curves clearly show that the scheme of the converter given on the second figure is the most preferable.

IV. CONCLUSIONS

In the paper submitted principal equations and power dissipation analysis for semiconductor devices of different types used in the in the high input voltage converters (up to 2000…4000 V) have been presented. Power transistors and diodes power dissipation curves pertaining to different values of input voltage and switching frequency for semiconductor devices of different types have also been revealed.

Equations, relations and curves presented in this paper can be very useful for the designers of the high-voltage converter (up to 4000 V).

V. REFERENCES

[1] S. I. Volsky, V. I. Chuev, S. V. Aleshin, E. A. Lomonova, 1998, "Development and Test of the Power Converter for the Commuter Train", PCIM'98, Proceedings, Nuremberg, Germany, p.p. 527-537.

[2] A. G. Polykarpov, E. F. Sergienko, "Pulse regulators and converters of direct voltage", Moscow, Publishing house of MPEI, 1998.

[3] P. R. Martin, U. Nicolai "Application power modules", Semikron International. ISLE, 2000.

2006 5th International Power Electronics and Motion Control Conference

A novel IGCT-based Half-controlled Bridge Type Fault Current Limiter

Wanmin Fei and Yanli Zhang

School of Electrical and Automation Engineering, Nanjing Normal University, Nanjing, P. R. China
Email:feiwanmin@njnu.edu.cn

Abstract—A novel IGCT-based half-controlled bridge-type Fault Current Limiter is proposed in this paper. By substituting the half-controllable SCR in the rectifier bridge with self-turn-off device IGCT, the uncontrolled time of the Converter Bridge can be reduced from half a cycle to the delay time of the current measuring circuit. If the maximum current in the dc reactor is preset, the inductance, volume, weight and cost of the DC reactor can be reduced to one 32th of what it is in the SCR-based bridge type FCL in three phase power systems approximately. The magnetization time of the DC reactor is reduced. The control method is very simple. The dynamic performance of the proposal FCL can be improved. Topology and control strategies of the proposal FCL are described in detail. Simulations under each short circuit fault mode are carried out. An experimental model is constituted and tested. Simulation and experiment results proved the practicability and validity of the new FCL.

Keywords- Self-turnoff device, half-controlled rectifier bridge, Short circuit fault current Limiter, Control strategy

I. INTRODUCTION

As electric power systems expand and become more interconnected, at some points, the available fault currents levels may exceed the maximum short-circuit ratings of the switchgear. Conventional solutions to fault current over-duty such as major substation upgrades, splitting existing substation buses or multiple circuit breaker upgrades could be very expensive and require undesirable extended outages and result in lower system reliability. The best solution probably is short circuit Fault Current Limiting techniques [1]-[8]. Application of FCL in electric power systems can not only suppress the amplitudes of short circuit fault currents but also enhance the power system stability and voltage quality [1].

Several kinds of FCL have been proposed in recent years. Among these FCLs, the solid state bridge type FCL based on SCR has advantages over others in multi-operation capability, lower voltage drop, lower cost and higher reliability [9]-[11].

Fig.1 shows a single-phase SCR-based Bridge type FCL with bypass reactor [11], which is composed of a SCR rectifier bridge, a DC reactor L1, a bypass reactor L2, a ZnO arrestor, and a breaker S. Us, and Z is an AC source and load impedance respectively. In normal status,

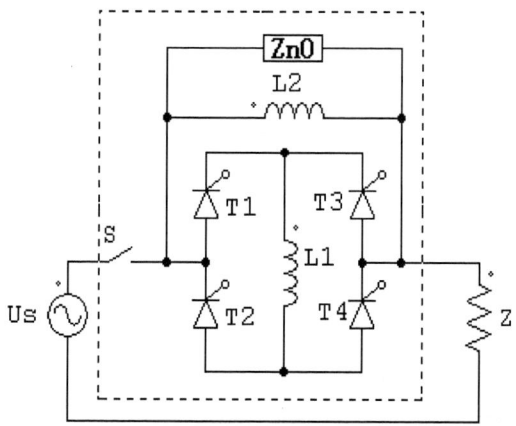

Fig. 1 SCR bridge type FCL with bypass reactor

the SCR set are gated and remain full conduction. When the power is just on, the line voltage will magnetize the DC limiting reactor L1. The current of L1 reaches the peak value of the sinusoidal load current and almost remains constant afterwards. When a short circuit fault occurs, the fault current reaches the reactor current instantaneously, and then rises at a proper speed designed so that in half a cycle the current will not exceed twice the peak value of the rated load current. When the control circuit detects the fault, the gate signals of T3 and T4 remain on, and the gate signals of T1 and T2 are removed. The current in the reactor L1 can freewheel through T3 and T4 until it falls to zero gradually. Then the entire fault current will flow through the bypass reactor L2 with a value meeting the requirement of relay protection. Because the inductive fault current flows through bypass reactor L2 and is finally cut off by the circuit breaker S1, a ZnO arrestor is employed to eliminate the operating over-voltage. The current rating of SCR and DC limiting reactor L1 is twice the peak value of load current. So the cost and the reliability of this kind FCL is superior to the classical SCR bridge type FCL described in reference [9] [10]. However, because the bridge is composed of SCR, when short circuit fault occurs and is detected by the measuring system, it will take half a cycle for the bridge to take action. So the inductance of the DC reactor L1 should be great enough [11].

1-4244-0448-7/06/$25.00 ©2006 IEEE   1138

$$L_{1op} = \frac{2U_s}{\omega I_r} \qquad (1)$$

Where, L_{1op} is the optimal value of L1. U_s, I_r and ω is the line voltage, the current rating and the line radian frequency respectively. If we substitute SCR in the converter bridge with self-turn-off device such as IGCT, the uncontrollable time of the bridge converter can be reduced to several hundred microseconds. If the maximum value of the current in L1 is preset, the inductance of L1 can be reduced greatly. So the Volume, weight and cost of L1 and FCL can be reduced, and the dynamic performance can also be improved.

II. TOPOLOGY OF THE NEW FCL

The topology of the new single-phase FCL is shown by Fig. 2, the power switches T1 and T2 are substituted with self-turn-off device IGCT. T3 and T4 are substituted with two diodes D1 and D2 to simplify the circuit and control method. Turn off T1 and T2 whenever a short circuit fault is detected.

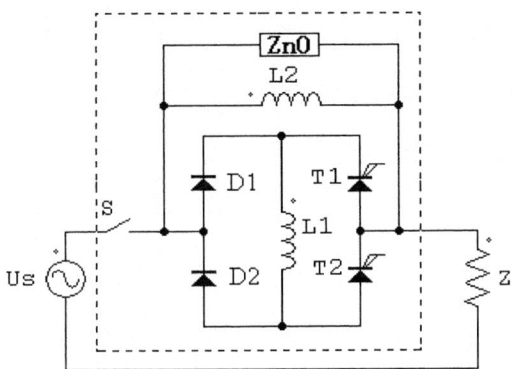

Fig 2 Topology of the proposal single-phase FCL

Install one such single phase FCL in every phase line, a three phase FCL scheme shown by Fig.3 can be obtained. The advantages of this scheme include simple control method, independence of operation, and convenience of location select and so on. The drawback is that for a three phase power system, 6 IGCT devices, three bypass reactor and three dc limiting reactors are needed.

In order to simplify the system and reduce the volume, weight and cost of FCL, a compact structure of three phase IGCT-based bridge type FCL which is shown by Fig.4 is proposed. The proposal three phase FCL is

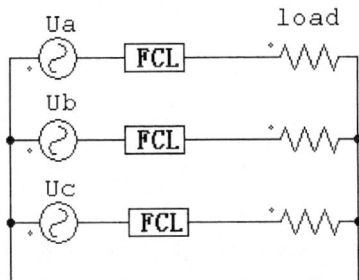

Fig.3 Three-phase FCL composed of single phase unit

consisted of 6 IGCT devices T1-T6, 2 diodes D1 and D2, 3 bypass reactors L2, L3, L4, 3 ZnO arrestors and only one DC limiting reactor L1. Sa, Sb and Sc are circuit breaker.

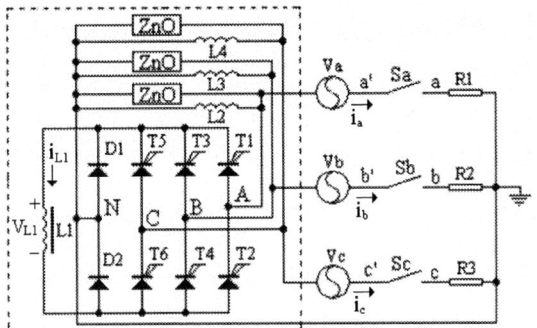

Fig 4 Topology of the proposal three-phase FCL

III. CONTROL STRATEGY

In normal condition, all the 6 IGCT devices are gated and remain full conduction. In steady state and if the on state voltage drops of IGCT devices and power diodes and the loss of the DC reactor L1 can be ignored, the current in L1 keeps constant and the voltage across L1 is zero, the three points a, b and c in Fig.4 are connected together equivalently, the existence of the proposal FCL has no harmful effect on the power system. In practice, voltage drop of IGCT devices and power diodes and the loss of the DC reactor result in waveform distortion near the top of line current, especially during starting or load increasing. The contents of harmonics decrease with the decreasing of the inductance of the DC reactor L1. So IGCT-based bridge type FCL has less harmonic and better dynamic performance than SCR-based bridge type FCL. Using a DC bias on the DC reactor to compensate the voltage drops of switch devices and the loss of the DC reactor L1, and pump the DC reactor current to a level above the peak value of rated load current, the problem of waveform distortion and the dynamic performance can be solved completely.

For three-phase four-wire systems shown by Fig 4, there are three short circuit fault modes: single phase, two phase and three phase short circuit. But the control method is very simple. Remove the gating signals of the IGCT devices in the fault phase or phases when a short circuit fault happens and is detected regardless the fault mode. The current of L1 freewheels through D1 and D2 afterwards. Then fault current will flow through the bypass reactor or reactors. Turn off the corresponding circuit breaker or breakers when the fault needs to be cut off. The normal phase or phases can not be affected by the fault phases or phase in three-phase four-wire power systems.

IV. PARAMETER DESIGN AND OPTIMIZATION

The bypass reactors L2, L3 and L4 are used for cooperation with relay protection and should withstand an over-current which is 5 to 10 times of the rated load current. So the current and inductance of them are as follow:

$$I_{sc} = k_1 I_r \tag{2}$$

$$L_2 = L_3 = L_4 = \frac{U_s}{\omega I_{sc}} \tag{3}$$

Where, k_1, I_r, U_s and ω is over-current times, rated line current, phase voltage and radian frequency respectively.

The most serious fault situation is three-phase short circuit fault. Take this serious fault situation for instance to design the parameter of the proposal FCL. Design aim of L1 is to minimize its size. When three-phase short circuit fault occurs, line voltage will be applied to DC reactor L1 and result in increasing of L1 current. Considering that the minimum size corresponding to a minimum energy stored in the reactor and the later is proportional to the inductance and the square of current of the reactor, the parameters of DC reactor L1 can be designed as follow:

$$I_{L1} = \sqrt{2} I_r + \int_{t_1}^{t_2} \frac{\sqrt{3} U_{sm} \sin(\omega t)}{\omega L_1} dt \tag{4}$$

Where t_1 is the time when the short circuit fault occurs, t_2 is the time when the rectifier bridge turns to invert mode. For SCR-based FCL, it takes half a cycle for the bridge to take action. Considering the symmetry characteristics, the current of L1 can be expressed as:

$$I_{L1} = \sqrt{2} I_r + \frac{3\sqrt{6} U_s}{\omega L_1} \tag{5}$$

The energy stored in L1 is:

$$E = \frac{L_1}{2} \left(\sqrt{2} I_r + \frac{3\sqrt{6} U_s}{\omega L_1} \right)^2 \tag{6}$$

When E is minimized, the size of L_1 is also reaches minimum. The optimal value of L_1 can be written as following:

$$L_{1op} = \frac{3\sqrt{3} U_s}{\omega I_r} \tag{7}$$

Substitute equation (7) into equation (5), the maximum value of the current in L_1 and T1-T6 can be obtained:

$$I_{L1m} = I_{IGCTr} = 2\sqrt{2} I_r \tag{8}$$

For IGCT-based FCL, the interval of (t_1, t_2) includes the time delay of the current sensor and the execute circuit and is about 300 microseconds with redundancy. The moment of t_1 and t_2 should be selected so that the integral of voltage of the AC power supply is maximum. So:

$$I_1 = \sqrt{2} I_r + \frac{0.0942\sqrt{6} U_s}{\omega L_1} \tag{9}$$

The energy stored in L1 is:

$$E = \frac{L_1}{2} \left(\sqrt{2} I_r + \frac{0.0942\sqrt{6} U_s}{\omega L_1} \right)^2 \tag{10}$$

When the E is minimized, the size of L_1 is also reaches minimum. The optimal value of L_1 can be written as following:

$$L_{1op} = \frac{0.0942\sqrt{3} U_s}{\omega I_r} \tag{11}$$

Substitute equation (11) into equation (9), the maximum values of the current in L_1 and T1-T8 can be obtained:

$$I_{L1m} = I_{IGCTr} = 2\sqrt{2} I_r \tag{12}$$

Compared equation (11) with equation (7), we can find that the inductance of L_1 in the proposal IGCT-based bridge type FCL can be reduced to about one 32th of what it is in the SCR-based bridge type FCL.

V. SIMULATION

In order to verify the validity of the proposed FCL, Simulations based on PSIM 6.0 for the circuit shown by Fig.4 are accomplished. The simulation waveforms are shown in Fig.5, where i_a, i_b, and i_c is three phase source current respectively. Parameters the simulations based on are listed: three phase line voltage 220V/50Hz; rated load current 220A; load resistance 1Ω; inductance of DC reactor L1=0.6mH; inductances of bypass reactor L2=L3=L4=0.5mH; preset short circuit fault current I_{sc} is 7 times of the rated load current. Power is turned on at t=0, single-phase short circuit fault occurs at t=0.04485, two-phase and three-phase short circuit fault occur at t=0.0428s.

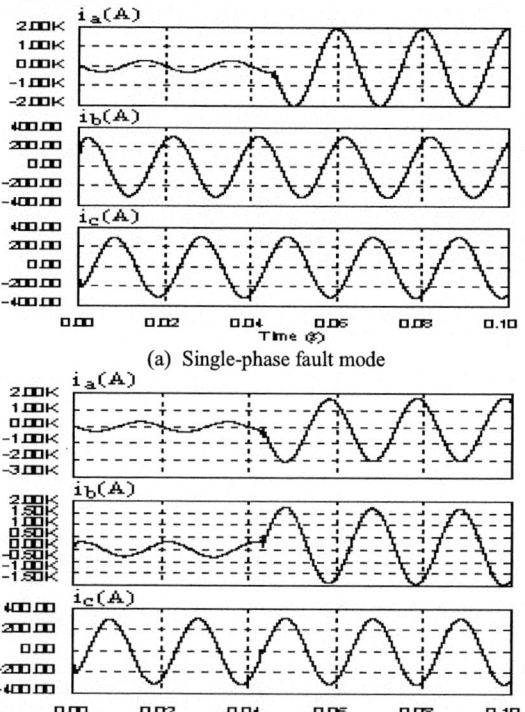

(a) Single-phase fault mode

(b) Three-phase fault mode

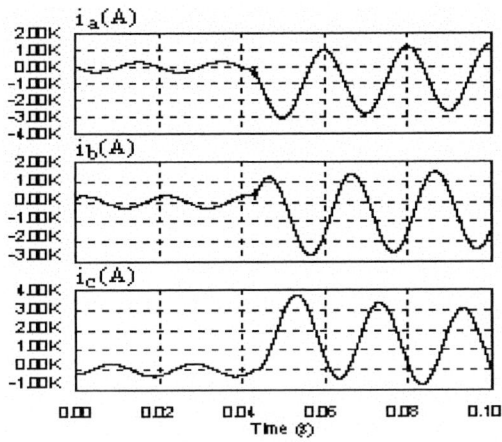

(c) Two-phase fault mode

Fig 5 Simulation result of the proposal FCL

VI. EXPERIMENT

To further verify the validity of the new FCL, an experimental model of the proposal single-phase FCL shown by Fig 2 is constituted. Fault current is preset to 5A. Other parameters are listed in Table I . The experiment waveform is shown by Fig 6. The source is a single-phase transformer, and its saturation causes the distortion of the waveform.

TABLE I

PARAMETERS OF THE EXPERIMENTAL MODEL

Item	Parameter
Source	50V/50Hz
Rated load current	1A
L1	11mH
L2	27mH
Z	50Ohm, 0.01mH

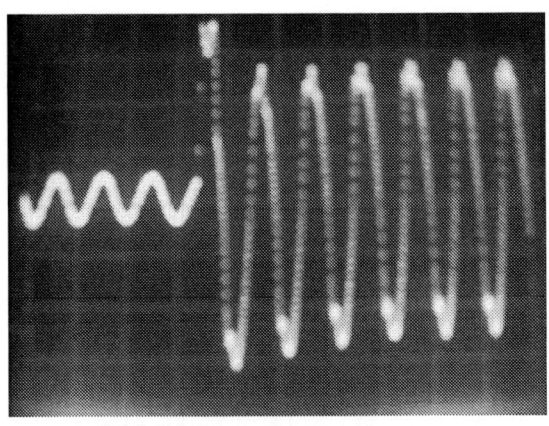

Horizontal: 20ms/div; Vertical: 2A/div

Fig 6 Experiment waveform of source current

VII. CONCLUSION

A novel IGCT-based half-controlled bridge-type Fault Current Limiter is proposed in this paper. By substituting the half-controllable SCR in the converter bridge with self-turn-off device IGCT, The uncontrolled time of the Converter Bridge can be reduced from half a cycle to the delay time of the current measuring circuit. If the maximum current in the DC reactor is preset, the inductance, volume, weight and cost of the DC reactor can be reduced to one 32th of what it is in the SCR-based FCL in three phase power systems approximately. The magnetization time of the DC reactor is reduced. The dynamic performance of the proposal FCL can be improved. Topology and control strategies of the proposal FCL are described. Simulations under each short circuit fault mode are carried out. An experimental model is constituted and tested. Simulation and experiment result proved the practicability and validity of the new FCL.

ACKNOWLEDGMENT

The authors would like to acknowledge the discussions with Prof. Zhengyu Lu and Prof. Zhaolin Wu of Zhejiang University, P. R.China.

REFERENCES

[1] Takahiro Nomura, Mitsugi Yamaguchi and Satoshi Fukui et al. Single DC reactor Type Fault Current Limiter for 6.6KV Power System. IEEE trans. On applied superconductivity, Vol.11, No.1, March 2001:2090~2093

[2] Lin Ye, LiangZhen Lin and Klaus-Peter Juengst. Application Studies of Superconducting Fault Current Limiters in Electric Power Systems. IEEE TRANSACTIONS ON APPLIED SUPERCONDUCTIVITY, VOL. 12, NO.1, MARCH 2002, Pages:900~903

[3] Leonard Kovalsky, Xing Yuan and Kasegn Tekletsadik et al. Applications of Superconducting Fault Current Limiters in Electric Power Transmission Systems. IEEE TRANSACTIONS ON APPLIED SUPERCONDUCTIVITY, VOL. 15, NO.2, JUNE 2005, Pages:2130~2133

[4] S. S. Choi, T. X. Wang, and Mahinda Vilathgamuwa. A Series Compensator with Fault Current Limiting Function. IEEE TRANSACTIONS ON POWER DELIVERY, VOL.20, NO.3, JULY 2005, Pages: 2248~2256

[5] E. Calixte, Y. Yokomizu and H. Shimizu et al. Reduction of rating required for circuit breakers by employing series-connected fault current limiters. IEE Proc. –Gener. Transm. Distrib., Vol. 151, No. 1, January 2004, Pages: 36~42

[6] Kenji Yasuda, Ataru Ichinose and Akio Kimura et al. Research & Development of Superconducting Fault Current Limiter in Japan. IEEE TRANSACTIONS ON APPLIED SUPERCONDUCTIVITY, VOL. 15, NO.2, JUNE 2005, Pages:1978~1981

[7] Harmonic Analysis and Improvement of a New Solid-state Fault Current Limiter. IEEE TRANSACTIONS ON INDUSTRY

APPLICATIONS, VOL. 40, NO. 4, JULY/AUGUST 2004, Pages: 1012~1019

[8] Seungje Lee, Chanjoo Lee and Tae Kuk Ko et al. Stability Analysis of a Power System with Superconducting Fault Current Limiter Installed. IEEE TRANSACTIONS ON APPLIED SUPERCONDUCTIVITY, VOL. 11, NO.1, MARCH 2001, Pages:2098~2101

[9] Wu Zhao-lin. A Type of Short Circuit Protection Circuit. Chinese Patent, ZL 96123001.0

[10] Wu Zhao-lin. Three Phase Short Circuit Current Limiting Transformer, Chinese Patent, ZL 00206596.7

[11] Zhengyu Lu, Daozhuo Jiang and Zhaolin Wu. A New Topology of Fault-Current Limiter and Its Parameters Optimization. IEEE Power Electronics Specialists Conference, PESC03.

Influence of Proton Irradiation dose on the Performance of Local

Lifetime Controlled Power Diode with

Proximity Gettering of Platinum

B.D. Han, D.Q. Hu, S.S. Xie, Y.P. Jia, B.W. Kang

College of Electronic & Control Engineering, Beijing University of Technology,

Beijing 100022, China

Abstract— Based on proximity gettering of platinum by vacancy defects which are induced by proton irradiation, local platinum doping is obtained. It is used as a local lifetime control technology in high-power diodes. The theoretical dependence of electrical active Pt concentration *Cpts* on irradiation induced defects concentration *Cv* is also studied. The diodes' reverse performance parameters are measured. They are functions of irradiation dose. For low proton irradiation dose, the gettered quantities of platinum by irradiation induced defects are enhanced when the irradiation dose increases. This can improve the performances of the device. But when the proton irradiation dose is high enough, the peak concentration of gettered platinum tends saturation. Further more, for deep junction device, the side effect brought by high dose irradiation will decrease the platinum gettering efficiency and the performances of the device degenerated under higher irradiation dose. On the base of theoretical study, we improved the device structure and manufacture process, a higher peak concentration is obtained. The recovery speed has been improved further.

Keywords: *localized platinum doping; platinum gettering; irradiation dose; gettering efficiency; power diode;*

1. INTRODUCTION

High power P-i-N diodes play an important role in most power circuits. The criteria for suitable diodes in various applications are low static and dynamic losses, low recovery charge and a soft recovery behavior. For these purpose, various structure and process methods are developed to optimize the devices' performances [1-2]. It is generally recognized that one of the most efficient concepts for an improved diode performance is the appropriate design of the on-state excess carrier distribution.

Theoretically, the most efficient method for plasma distribution control is local lifetime engineering [3-5].

The local lifetime control is realized mainly by light ion irradiation[6].Normally, the effective defect level induced by irradiation defects is at $E_T=E_C$-042eV. (E_T and E_C are conduct band level and any trap level, respectively).It's quite near the center of the band gap. Thus, the device with local lifetime controlled by such trap level has the disadvantage of high reverse leakage current and bad thermal stability[7, 8]. Comparatively, the recombination center generated by platinum is localize at $E_T=E_C$-0.23eV. It is far from the center of the band gap. The device which uses platinum as a lifetime killer has the advantage of low leakage current and good high temperature stability and reliability. It is the key process to obtain localized platinum doping for the manufacture of fast soft recovery performance diode with low leakage current.

On the base of proximity gettering of platinum[9, 10], we realize localized platinum doping in the local lifetime control power diode. The influence of proton irradiation dose on the performances of the device is examined. With this understanding, the device structure and platinum doping are improved.

2. EXPERIMENTS

The first group of sample diodes has P^+-N^--N^+ structure with 72 μm depth junction. On P^+ side, PtSi layer is produced by sputtering and sintering (450° C, 60 min). Then the devices were irradiated with proton at doses ranging from 5×10^{11}~2.7×10^{14}cm^{-2} through the anode. The energies 2.8 MeV was chosen. The sample devices were annealed at 700° C for 15min to promote the platinum diffusion from the PtSi contact to the position of the maximal damage.

To improve the switch speed, the emitter efficiency controlled structure is used in the second group samples. The diodes have P^+PNN^+ structure. The P^+ region has surface concentration of $1.8\times10^{19}cm^{-3}$ with 0.9μm deep. P region are $4.43\times10^{17}cm^{-3}$ and 4.5μm respectively. On P^+ side PtSi layer is produced by sputtering and sintering (300 °C, 60 min). Then the devices were irradiated with proton doses range of $1\times10^{13}\sim5\times10^{14}cm^{-2}$ through the anode. The implantation energies 550 KeV was chosen. The sample devices were annealed at 700° C for 15min.

3. RESULTS AND DISCUSS

(1) *For the First Sample Diodes*

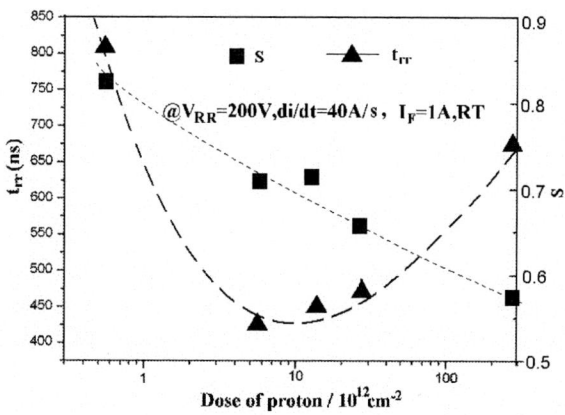

Fig. 1 Dependence of S and t_{rr} on irradiation dose for first group samples

The dependence of S and t_{rr} on irradiation dose for the first group of sample is depicted in Fig.1. Where V_{RR} is reverse recovery voltage, I_F is forward leakage current, t_{rr} is reverse recovery time and S is softness factor. From fig.1, one can see that the reverse recovery time decreases greatly as the irradiation dose increases at low proton irradiation dose. As the proton irradiation dose approaches $10^{13}cm^{-2}$, the reverse recovery time change gently with the irradiation dose. When the proton irradiation dose is more than $2\times10^{13}cm^{-2}$, the variation tendency reverses——the reverse recovery time increases as proton irradiation dose increases. Normally, t_{rr} is determined by the carrier lifetime and the lifetime is determined by localized active platinum concentration. Dependence of t_{rr} on irradiation dose embodies the

dependence of gettered active platinum on proton irradiation dose. That is the gettered active platinum concentration increases with proton irradiation dose at low irradiation dose and decreases with proton irradiation dose at high irradiation dose.

Generally, there are two configurations of platinum in silicon, the interstitial configuration (Pt_I) and the substitutional (Pt_S) one. The Pt_I guides the transport until it is transformed into Pt_S. In the later configuration, platinum is electrically active with an acceptor level at $E_T=E_C$-0.23 eV and a donor level at $E_T=E_V +0.32$ eV.(E_C , E_V , and E_T denote the energy positions of the conduction band, valence band, and any deep level, respectively.) They are good recombination centers as a lifetime killer. The transformation to the preferred substitutional configuration can be realized by the Frank–Turnbull mechanism:

$$Pt_I + V \Leftrightarrow Pt_S \qquad (1)$$

where the interstitial platinum recombines with a vacancy (V), or via the kick-out mechanism:

$$Pt_I \Leftrightarrow Pt_S + I \qquad (2)$$

where the interstitial platinum kicks out a silicon lattice atom thereby creating a silicon self-interstitial(I).

In irradiation silicon, there are various void defects (include void-void pair, void-O, void-impurity and so on). The Frank–Turnbull mechanism dominates the transformation [13-14]. For reaction （1）, there exists a equilibrium for the reactant concentration, that is:

$$C_{PtI} \times C_V = K(T)C_{PtS} \qquad (3)$$

Where C_{PtI}, C_V and C_{PtS} are concentration of interstitial Pt, voids and substitution Pt, $K(T)$ is a temperature related parameter.

Also:

$$C_{PtI} = C_{Pt} - C_{PtS} \qquad (4)$$

where C_{Pt} is total concentration of Pt, it relates to platinum solubility and voids degree in silicon under certain temperature. Using （3）and （4）, we obtain:

$$C_{PtS} = C_{Pt}\left(1 - \frac{K(T)}{K(T) + C_V}\right) \qquad (5)$$

Equation (5) shows that the concentration of substitutional Pt_S is mainly determined by the void concentration when C_{Pt} is determined. The more C_V is, the more C_{PtS} is. If C_V is high enough, C_{PtS} tends to C_{Pt}, but not decreases as C_V increases.

The experiment result of active Pt degree decreases as irradiation dose increases has been observed in alpha particles implantation[12]. It was attributed to the fact that in-diffusing platinum interstitials were transformed into the substitutional configuration on their way to the region of maximal damage. As the concentration of defects towards the surface increased, so did the number of interstitial platinum atoms that was transformed within this region. The degradation of softness factor also supports this point. Another possible reason is that the high dose irradiation implantation may cause voids-Pt complexes which have different activity from substitutional Pt. We discussed this in other place.

To validate equation (5) and above analysis, we design second group of diodes with shallower junction. Meanwhile, high temperature Pt pre-diffusion is made to increase C_{Pt}.

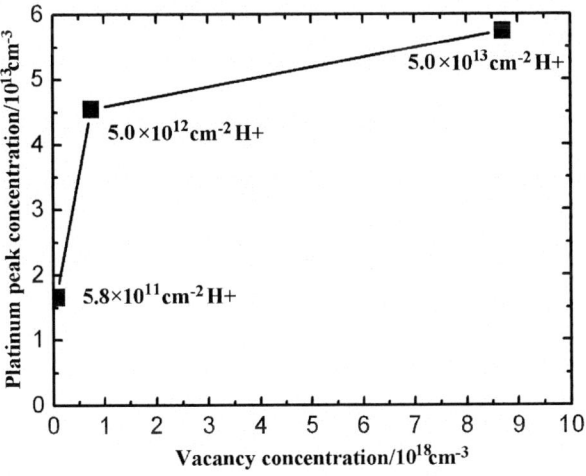

Fig. 2 Dose dependence of the peak concentration of the active Pt with H^+ irradiation [9]

Reference [9] gave the dose dependence of peak concentration of platinum with H^+ irradiation followed by

700° C annealing (see Fig.2). We can extract the value of parameter $K(T)$ at T=700° C. It is about 5×10^{17}cm^{-3}. Put it in equation (5), we can see that when voids concentration reaches 5×10^{18}cm^{-3} (the corresponding proton irradiation dose is 10^{13}cm^{-2}), the substitutional Pt is at 0.90 peak value. Under such consideration, the proton irradiation dose range $1\times10^{13}\sim5\times10^{14}$cm^{-2} is chosen for second experiment.

(2) *For the Second Sample Diode*

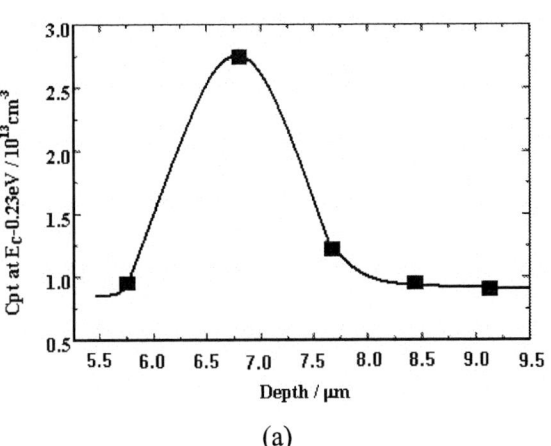

(a)

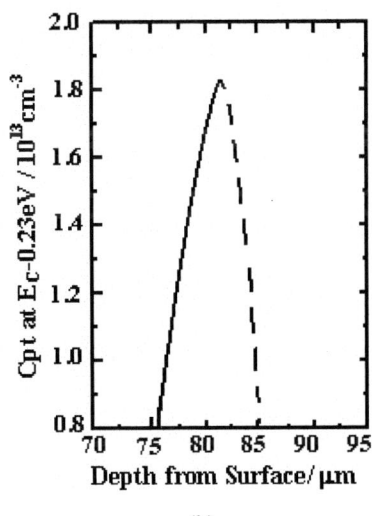

(b)

FIG. 3 Concentration vs. depth profiles of the platinum acceptor trap at $E_T=E_C-0.23$eV
(a) 550keV/ 1×10^{13}cm^{-2} proton irradiation followed by 700°C/15min annealing process.
(b) 2.8MeV/2.7×10^{13}cm^{-2} proton irradiation followed by 700° C/15min annealing process.

First, the influence of pre-diffusion of Pt on the

gettered peak concentration of active Pt is compared. It is depicted in Fig.3. It's very clear the pre-diffusion sample has higher gettered peak concentration. That is the existence of voids make C_{Pt} be higher than the platinum solubility in silicon. It can be improved by platinum pre-diffusion before proton irradiation.

The measured results of sample performance parameter are list in table-1. It's very clear that when the irradiation dose is more than 1×10^{13} cm^{-2}, its increasing influences little on the improvement of reverse recovery time. The gettering saturation tendency is very obvious. Comparing the data with Fig.1 one can see that performance of the second group sample is superior than the fist group's. (1) Under the same irradiation dose, the reverse recovery time of second group sample is almost half of the first group's. (2) The phenomenon what reverse recovery time increases with irradiation dose under high irradiation dose eliminates. Shallower junction and platinum pre-diffusion are effective. (3) Lower reverse recovery time t_{rr} with higher softness factor indicates that the reduction of t_{rr} is due to the improvement of gettered platinum peak concentration caused by platinum pre-diffusion, not pre-diffusion itself.

Tab.1 Datasheets of diode performance parameters for the second samples

parameters Proton Doses	I_R (μA)	t_{rr} (ns)	t_a (ns)	S
1×10^{13}cm^{-2}	0.564	252	126	1.00
8×10^{13}cm^{-2}	0.838	246	148	0.66
5×10^{14}cm^{-2}	1.04	242	140	0.73

Note: t_a is reverse falling time; I_R is reverse leakage current.
Test condition: I_R@V_R=100(V), T=125°C; t_{rr}@I_F=1A, V_R=30V, di / dt=-20A/μs, RT

5. CONCLUSION

On the proximity gettering of platinum by proton irradiation induced defects, we realize localized platinum doping in the local lifetime control power diode. The relationship of substitution platinum concentration, total platinum concentration and void concentration is given. The variation tendency of substitutional platinum with irradiation dose is studied. The experiments indicates that when irradiation dose approach 1×10^{13}cm^{-2} the gettered substitution platinum has saturation tendency. This limits the improvement of switch speed using such local lifetime control. Using shallow junction anode and platinum pre-diffusion, the faster and softer diodes are obtained.

REFERENCES

1 H. Schlangenotto, J. Serafin, F. Sawitzki, and H. Maeder, "Improved recovery of fast power diodes with self-adjusting p emitter efficiency," *IEEE Electron Device Letters*, vol. 10, pp. 322-324, July 1989.

2 S. Sawant and B. J. Baliga, "Comparative study of high voltage (4kV) power rectifiers PiN/MPS/SSD/SPEED", *IEEE International Symposium on Power Semiconductor Devices and ICs (ISPSD)*, Toronto, Canada, 1999: pp.153-156

3 O. Humbel, N. Galster, T. Dalibor, T. Wikstrom, F.D. Bauer, and W. Fichtner, "Why is Fast Recovery Diode Plasma-Engineering With Ion-Irradiation Superior to That With Emitter Efficiency Reduction," *IEEE Transactions on Power Electronics*, Vol. 18, pp. 23-29, Jan. 2003.

4 E. Napoli, A. G. M. Strollo, and P. Spirito, "Numerical Analysis of Local Lifetime Control for High-Speed Low-Loss P-i-N Diode Design," *IEEE Transactions on Power Electronics*, Vol. 14, pp. 615-620, April 1999.

5 J. Vobecky, P. Hazdra, and J. Homola, "Optimization of Power Diode Characteristics by means of Ion Irradiation," *IEEE Transactions On Electron Devices*, Vol. 43, pp. 2283-2289, Dec. 1996.

6 P. Hazddra, J. Vobecky, and K. Brand, "Optimum Lifetime Structuring in Silicon Power Diodes by means of Various Irradiation Techniques," *Nuclear Instruments and Methods in Physics Research*, Vol. B186, pp. 414-418, Jan. 2002.

7 R. Siemieniec and J. Lutz, "Possibilities and Limits of Axial Lifetime Control by Radiation Induced Centers

in Fast Recovery Diodes," *Microelectronics Journal*, Vol. 35, pp. 259-267, March 2004.

8 P. Hazdra, J. Vobecky, N. Galster, and O. Humbel, "New Degree of Freedom in Diode Optimization: Arbitrary Axial Lifetime Profiles by means of Ion Irradiation" *Proceeding of the 12th Intern. Symp. on Power Semiconductor Devices & ICs*, pp. 123-127, 2000.

9 D.C. Schmidt, B. G. Svensson, N. Keskitalo, S. Godey, E. Ntsoenzok, J. F, Barbot, and C. Blanchard, "Proximity Gettering of Platinum in Proton Irradiated Silicon," *Journal of Applied Physics*, vol. 84, pp. 4214-4218, Oct. 1998.

10 A. Cacciato, C. M. Camalleri, G. Franco, V. Raineri, and S. Coffa, "Efficiency and Thermal Stability of Pt Gettering in Crystalline Si," *Journal of Applied Physics*, Vol. 80, pp. 4322-4327, Oct. 1996.

11 D. C. Schmidt, B. G. Svensson, J. F. Barbot, and C. Blanchard, "Stability of Proximity Gettering of Platinum in Silicon Implanted with Alpha Particles at Low Doses, "*Appl. Phys. Lett.* Vol. 75, pp. 364-366, 1999.

12 D. C. Schmidt, B. G. Svensson, S. Godey, E. Ntsoenzok, J. F. Barbot, and C. Blanchard, "The Influence of Diffusion Temperature and Ion Dose on Proximity Gettering of Platinum in Silicon Implanted with Alpha Particles at Low Doses," *Appl. Phys. Lett.* Vol. 74, pp. 3329-3331, May 1999.

2006 5th International Power Electronics and Motion Control Conference

IMPLEMENTATION OF A HIGHER QUALITY DC POWER CONVERTER

Barsoum, N.N. & YII, M.L.

Curtin University of Technology, Miri, Sarawak, Malaysia
Tel: 6085 443821, Fax: 6085 443837, Email: nader.b@curtin.edu.my

Abstract: Many single and three-phase converters are well developed, and covered up in most of electric markets. It is used in many applications in power systems and machine drives. However, an exact definite output signal from the dc side still not recognized. The waveforms of output voltage and current demonstrate an imperfect dc signal and constitute losses, harmonic distortion, low power factor, and observed some ripples. An approximately perfect rectifier bridge is the aim of this research. Perhaps it gives the ability to identify the parameters of the converter to obtain, as much as possible, a perfect dc signal with less ripple, high power factor and high efficiency. Design is implemented by simulation on Power Simulator PSIM, and practically, a series regulator LM723 is applied to provide regulating output voltage. Comparisons of both simulation and hardware results are made to observe differences and similarities.

Keywords: Simulation, Microcontroller, Converter, Feedback, Power factor, Design

I. INTRODUCTION

In industrial application, Direct Current (dc) is used for controlling application such as Programmable Logic Controller (PLC), Microcontroller, dc motor and many commercial and domestic appliances [1,6,8,9]. Rectifier has become a popular power source [2,3,5] for these appliances because of its reduced cost and relatively low sensitivity to supply voltage variations under normal operating conditions. Stability is the key issue in these applications as it involves certain precision and decision make. These systems are usually employed in automotive and aerospace applications [3,7] where tiny little error in the supply (unstable dc) might cause a large disaster

The processed output voltage, current as well as frequency, will be as desired by the load. If the power processor's output can be regarded as a voltage source, the output current and the phase angle relationship between the output voltage and the current depend on the load characteristic. Also, normally, a feedback controller will be provided to perform comparison to the output of the power processor unit with a desired reference value, and the controller minimizes the error between the two.

Generally, design of converter that can be considered as a good design, should cover few essential aspects [10]. One of them will have to be efficiency, which can be considered as the ultimate goal of design in power electronic. Besides that, a few current issues can directly and indirectly affected the course of a design of a converter. One of these significant issue is the line quality, whereby is critical to ensure that the utility lines and transformer would supply undistorted wave voltage to customers. The source

and line inductances play an important role in the line quality issue. With the presence of the reactive power to the line, it increases the volt-ampere rating. Thus, the input ac line voltage becomes distorted from the higher peak currents. As a result, high reactive components are being used. This is a drawback because a poor power factor causes heavy expenses to the user. Besides that, the growing concern regarding harmonic pollution of the power distribution system has create awareness for clean ac line current and a power factor close to unity. The phase angle of the fundamental harmonic current [4] with respect to the line voltage is a very important parameter that determines the power factor. These issues are some of the critical aspects that should be taken into consideration when designing a good converter.

Power factor correction is achieved by the addition of capacitor in parallel with the connected motor circuits and can be applied at the starter, or applied at the switchboard or distribution panel. The resulting capacitive (leading) current is used to cancel the inductive (lagging) current flowing from the supply.

The microcontroller program performs this. The microcontroller is programmed in such a way that it will perform the checking the phase difference every 1/2MHz second (microcontroller running on 2MHz crystal) and calculate out the phase different between the current and voltage waves. The calculated value will be converted into signal to activate appropriate capacitor to correct the power factor [11,12].

Capacitors are installed parallel to the source, waiting for the signal from the microcontroller. When microcontroller sends signal, the signal will be amplified via MOSFET and activate a relay. The relay hence will turn on the capacitor. The feedback circuit is designed in such a way that the system is allowed to be a capacitive system but not an inductive system.

II. IMPLEMENTATION OF SINGLE PHASE CONVERTER
2.1. Step-up Converter with Feedback Control

To further improve the output signal generated, by means of current shaping, it can be archived through step-up converter with feedback control applied at dc side of the rectifier to replace to LC filtering component. With this kind of circuit arrangement, it is possible to shape the input current drawn by the rectifier bridge to be sinusoidal and in phase with the input voltage. For the purpose of better illustration, figure 1 shows the circuit configuration. At the input side, the input current i_s is desired to be sinusoidal and in phase with input voltage Vs, also, at the full bridge rectifier output, i_L and absolute value of Vs will have the same waveform as well. For the theoretical analysis below, the power loss of the rectifier bridge and the step-up converter will be neglected due to the fact that the losses are somewhat small

---

1-4244-0448-7/06/$25.00 ©2006 IEEE   1148

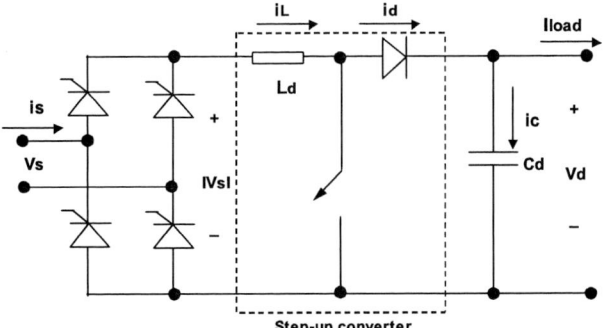

Fig. 1: Step-up converter for current shaping

Thus, we have, for $\hat{V}_S = \sqrt{2}V_S$ and $\hat{I}_S = \sqrt{2}I_S$, the input power can be expressed as:

$$P_{in}(t) = \hat{V}_S |\sin \omega t| \, \hat{I}_S |\sin \omega t| = V_S I_S - V_S I_S \cos 2\omega t$$

The average value of current I_d can be expressed as:

$$I_d = I_{load} = \frac{V_S I_S}{V_d}$$

Also, the current through the capacitor is:

$$i_C(t) = -\frac{V_S I_S}{V_d} \cos 2\omega t = -I_d \cos 2\omega t$$

From these expressions, the ripple in V_d can be determined by means of estimation, which is shown as below:

$$V_{d,ripple}(t) \approx \frac{1}{C_d} \int i_C \, dt = \frac{I_d}{2\omega C_d} \sin 2\omega t$$

The step-up converter in the figure 1 is operating in current-regulated mode, as our main purpose is to shape the input current of the step-up converter. The feedback control, represented in block diagram, is shown in figure 2. This feedback control serves the purpose of comparing the output generated with a reference value, in order to minimize the error between these two.

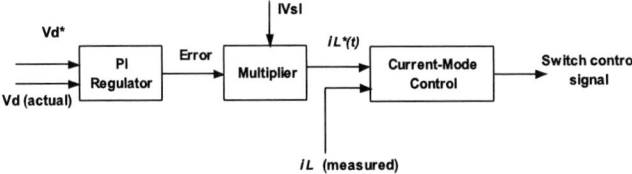

Fig. 2: Feedback Control block diagram

$i_L{}^*$ shown in figure 2 is the reference value of the current i_L in the step-up converter. The amplitude of $i_L{}^*$ should be such as to maintain the output voltage at a reference level of $V_d{}^*$, in spite of the variation of load and the fluctuation of the line voltage from its nominal value. The waveform of $i_L{}^*$ is obtained by means of measuring absolute value of Vs, by a resistive potential divider and multiplying it with the amplified error between the reference value $V_d{}^*$ and the actual measured value of V_d. On the other hand, the actual current i_L is sensed, usually by measuring the voltage across a small resistor inserted in the return path pf i_L. The status of the switch in the step-up converter is controlled by comparing the actual current i_L and $i_L{}^*$.

If constant frequency is applied for this feedback control, the ripple current can thus be expressed as:

$$I_{rip} = -\frac{(V_d - |V_S|)|V_S|}{f_S L_d V_d}$$

In terms of maximum ripple current, it can be expressed as:

$$I_{rip,max} = \frac{V_d}{4 f_S L_d}$$

The step-up converter topology is well suited for the input current shaping method because when the switch is off, the input current directly feeds the output stage.

2.2. Complete Design with Step-up Converter at dc side

The feedback controller shown above operates by comparing the output generated with a reference value set. In simulation, the actual output current is sensed by a current sensor connected at dc side of the rectifier, and this actual value of I_d is sent to negative probe of the summer. The reference value of I_d is transmitted through positive probe of summer. The comparison of signal will take place at the comparator, with signal generated from PI controller and triangular wave.

For better reference, figure 3 shows the complete design of single-phase controlled rectifier with step-up converter and feedback control implemented in PSIM

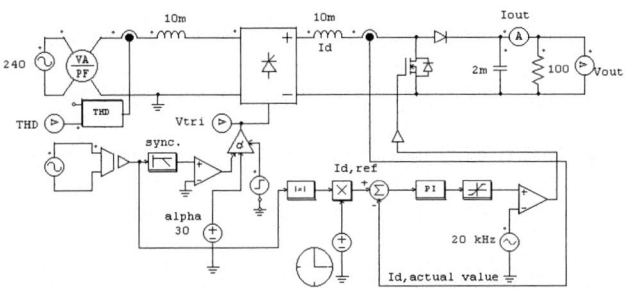

Fig. 3: Thyristor converter with step-up converter

2.3. Simulation Results

Simulation is carried out based on similar parameters applied on rectifier design without step-up converter in order to observe the difference between these two designs. Similarly, for better illustration, simulation results corresponding to firing angle of 30° are shown in figure 4.

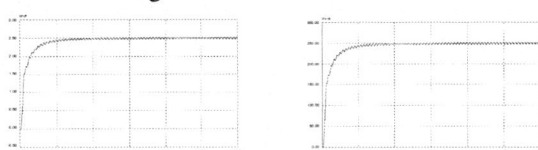

Fig. 4: Waveforms for output voltage and output current

2.4. Comparison of Results

Comparison between the two types of rectifier design configuration is given in table 1. It shows some of the essential parameters corresponding to firing angle of 0°

The table indicates that thyristor converter with step-up converter and feedback control shows an improvement in overall

1149

aspects. In terms of THD, it improved from 49% to 21%. While as for power factor produced, it increased from 0.81 to 0.98.

Parameters	Types of rectifier configuration	
	Without Step-up Converter	With Step-up converter and feedback control
Output voltage	200V	246.5V
Ripple voltage	4.3V (2.15%)	3V (1.21%)
THD	49%	21%
Power Factor	0.81	0.98

Table 1: Comparison of results

2.5. Design of Hardware

In hardware manner, the variable dc output will be controlled by means of applying a regulator chip LM723 to control the output voltage generated. This design method is rather different than the design approach implement in simulation. One of the reasons for this is that an exact solution for hardware implementation based on design in simulation has not yet been found; these reasons will be outlined in more details in problems encountered and suggestions for further development. For this practical design, in terms of rectifying component, Diode Bridge will be applied for this hardware implementation. For better illustration, the circuit diagram for the practical design is shown in figure 5.

For this particular design, the input voltage Vs is equal to 18V, which is stepped down by a transformer, not shown in figure. Diode D1, D2, D3 and D4 forms the diode bridge, KP206G. Also, another diode, IN539 is applied at dc side. One of the functions for this diode is to act as a feedback blocker, whereby it steers any current that might be coming from the device under power around the regulator to prevent the regulator from damages. These sorts of reverse current usually occur when the rectifier is been powered down.

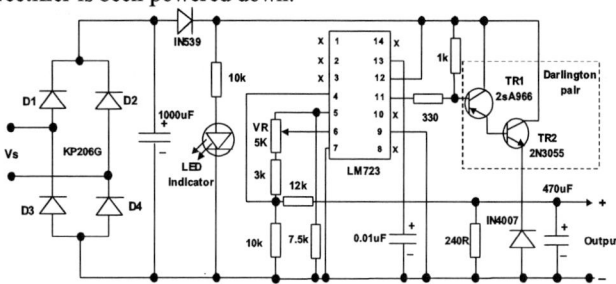

Fig. 5: Circuit diagram of practical converter design

2.6. Operation Analysis

Basically, this practical rectifier design applies a different approach as compare to implement in simulation. Referring to the figure, as the transformer, to an ac secondary voltage of 18V, steps down the input voltage. The voltage rectified by the diode bridge to produce unfiltered dc output voltage. This unfiltered dc output voltage will contain big ripple and is pulsating. This pulsating output voltage will then been filtered by the capacitive

filter of 1100 μF capacitor in order to manageable for the regulator.

As notice in figure 5, there's only capacitive filtering applied to the design, this is due to the fact that in low-power applications, the inductor required for rectification design could be a costly item, that's the main reason most low-power converters dispense with the inductor and apply an direct capacitive filtering method. With no load, the dc voltage across the terminals of the filter is going to be 18 to 30 volts.

The regulation is obtained using the Darlington pair (TR1 and TR2). They in turn are controlled by the 723 regulator. 723 has its own internal highly regulated voltage reference supply (pin 6). Internally the 723 compare this reference voltage to the output of the power supply and it is varied by means of variable resistor VR, shown in figure 5. This sets the output voltage. The regulation process evolves around pin 11 and pin 6 of LM723 regulator. Pin 11 of 723-regulator control voltage supply, and this V_C will trigger the base of TR1, which is 2SA966, PNP transistor, which will act as a simple amplifier to increase the current available to drive the base of the pass transistor, i.e. TR2 2N3055.This explains the function of Darlington transistor pair applied.

Capacitance of 0.01μF connected to frequency component (pin 13) of 723-regulator function as a transient response improver, which improves the response of the regulator when it is operating during high frequencies. Regulated dc output, will be filtered again by capacitance of 470μF to produce an output voltage that contains a minimum ripple and close to pure dc voltage. Necessary protective device are all been installed in this practical design such as fuses, metering component for voltage and current, not shown in the figure. Note here, these metering component are meant for as a guideline as the accuracy of these meters might be a ±1V difference for the case of voltmeter. Therefore, for better accuracy, a multimeter should be used.

Figure 6 shows an overall view of practical design. Note, TR2 and 2N3055 are mounted on a huge heat sink. This step is necessary, as the heat sink installed will helps to dissipate the massive flow of heat generated to the pass transistor.

2.7. Comparison of Results

As design implemented from hardware manner, which applied the diode bridge as main conversion component, for comparison purpose, we compare results from design in simulation corresponding to firing angle of 0°. Table 2 shows results from both of the design. It shows that the design implemented in simulation use 20V as input voltage, as main concern is on low power application. While in hardware design, the input voltage is in 18V, which is stepped down by a step down transformer. 18V secondary side voltage is one of the common rates of voltage used in terms of low voltage application.

Results shown clearly indicate that the simulation have gain advantages over the practical design. These are shown through power factor produced and total harmonic distortion created. However, the difference in terms of power factor and ripple voltage didn't show a big difference. The only major difference comes from the total harmonic distortion.

Note that when comes to comparison of results from simulation and practical manner, some slight discrepancies should be taken

1150

into account. This is due to the fact that from simulation point of view, generally ideal components are used and so they have theoretical constraints associated with them, which is possible to differ slightly from the actual physical component. As for practical converter, there might be losses in the process of stepping down voltage as well as where conversion from ac to dc takes place. Besides that, each of the components, particularly resistors and diode used, all contribute to losses generated.

Figure 6: Overall view of practical design implemented

Parameters	Design method	
	Design from simulation (firing angle = 0°)	Practical Design
Input Voltage	20V	18V
Output Voltage	37.7V	15.8V
Ripple voltage	1.24V (3.3%)	800mV (5.06%)
Output Current	0.37A	0.18A
Output power	13.9W	2.84W
THD	30.2%	47.7%
Power Factor	0.96	0.90

Table 2: Comparison of results

III. IMPLEMENTATION OF THREE PHASE CONVERTER
3.1. Feedback Control Loop

As far as power factor and total harmonic distortion is concern, inductors are added at the source to compensate the capacitive value produced by the capacitor at the filtering device. This solution is being taken one step further to install a feedback control device to compensate the inductive value produced by the load in case the converter is used for a dc motor. Thus, a feedback control loop is proposed to improve the power factor of the system.

Power factor correction is achieved by adding capacitors in parallel with the connected motor circuit and can be applied at the starter or the switchboard or distribution panel. The resulting capacitive current is leading current and it is used to compensate the lagging inductive current flowing from the supply.

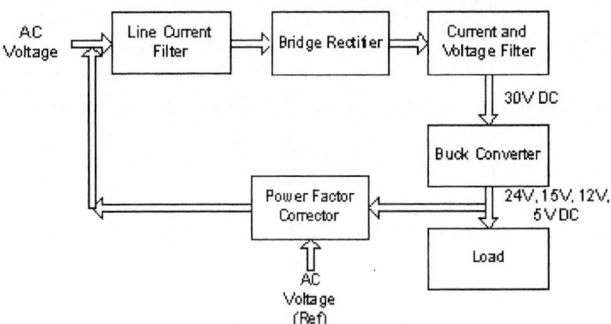

Fig. 7: Feedback controlled three phase converter

This correction task is carried out by micro-controller, which is specially designed to perform the checking and correcting the incoming voltage and current. External comparator MCP6024 is used to convert the analogue signal to 2-bit digital signal. The converted signal is then input into the microcontroller to perform phase difference analysis.

Signals that are about to be input into the microcontroller are phase-shifted voltage and the voltage that represents the current. Phase-shifted voltage is obtained near the load for the voltage at the load will represent the voltage's phase shift. Voltage's phase is taken as variable while the current's phase is taken as constant. If the load is capacitive, the voltage's phase is lagging the current's phase. If the load is inductive, the voltage phase is leading the current phase.

The microcontroller will observe the voltage phase. When the microcontroller detected a '1' for voltage, it will then observe the current phase (represented by voltage across the load). If the current is not detected, it will return a '1' and activate capacitor A from the capacitor bank. After that, it will check both the voltage and current phases again. If the current phase is again not detected when the voltage phase is detected, it will return another '1' and now, activate capacitor B. The whole process will repeat until the capacitor E in the capacitor bank is activated, or, the current is detected when the voltage phase is detected. The process algorithm is shown in Figure 8.

Every time when there is a change in the load (manually), then the power factor corrector has to be reset in order for it to function properly.

The power factor correction algorithm is being implemented in MATLAB simulation. The program enables the user to input the load inductance and the program will return the number of capacitor of the dedicated value needed. It can also display the uncompensated and the compensated waveforms. The output is shown in Figure 9.

In Figure 9, the load used is 10 kΩ and the load inductor is 1000 H. There are 99 capacitors rated 100 pF needed to compensate the inductance. It is understood that the compensation could not be 100% because the signal is being sampled with Nyquist sampling rate. The program is written in such a way that the microcontroller will ensure that the system is capacitive rather then inductive.

1151

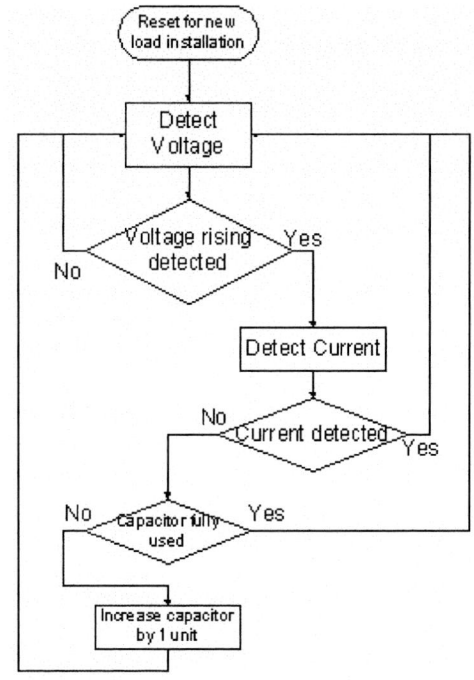

Fig. 8: Power factor corrector flowchart.

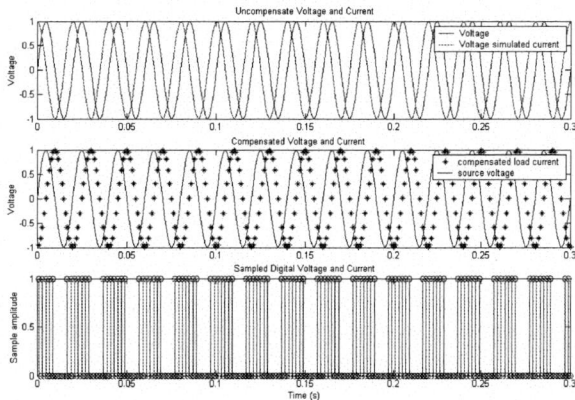

Fig. 9: Power factor correction made onto the inductive system

3.2. Integration of Designs

The designs are being integrated together to form a complete ac-dc converter with high power factor, low harmonic distortion and low voltage ripple. This is shown in figure 10

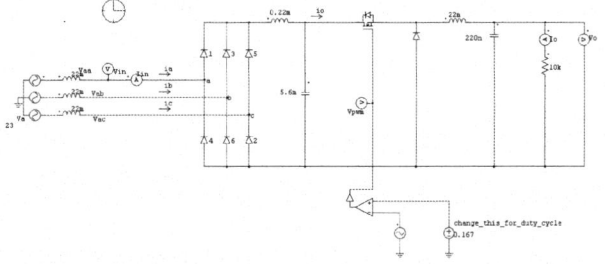

Fig. 10: Integration of the rectifier and Buck Converter.

It is noticed that the 1M Ω resistor at the output of the rectifier circuit is being removed. This is to allow the current to flow completely into the load circuit without any additional current drawn over unnecessary load.

The duty cycle D of the Buck converter is set to four mode of operation, namely, 0.167, 0.4, 0.5 and 0.8. These duty cycles will give 5V, 12V, 15V and 24V respectively.

It is also noticed that the source rms line voltage is being step up to 23V. This is because, the diodes used are no longer ideal diode, and a 0.7V voltage drop is introduced across the diode. Increasing the source voltage to 23V rms will ensure that the four duty cycle operation modes will achieve the desired output.

3.3. Simulation

At D = 0.8, the desired output is 24V and 10kΩ sample load is placed at the output of the ac-dc converter. The output current and voltage is shown in Figure 11:

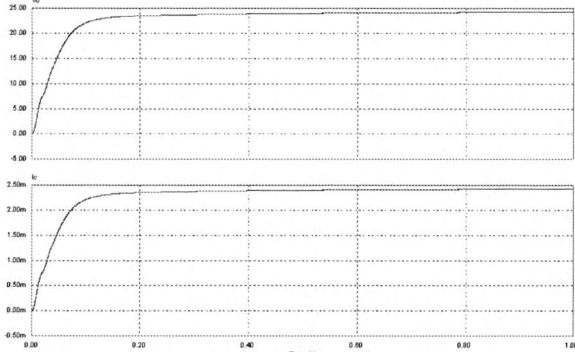

Fig. 11: Output current and voltage waveform with D = 0.8.

The output is being evaluated in frequency domain as is shown in figure 12:

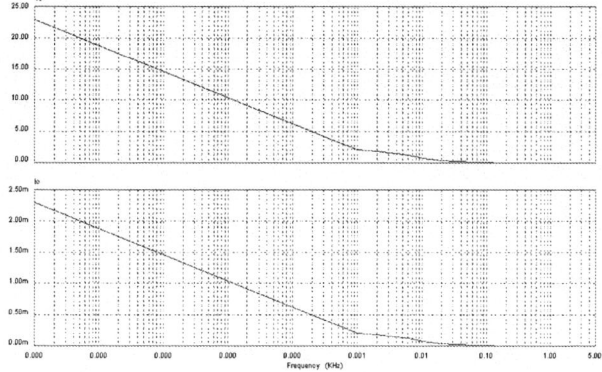

Fig. 12: Output current and voltage in frequency domain with D = 0.8.

It is noticed that the output current and voltage shows an extremely low ripple and the rise time is less than 0.4 seconds. In frequency domain, the frequency component is less than 200Hz, which means, the harmonic of the output is extremely low, even the fundamental frequency (300Hz) is almost completely reduced.

3.4. Implementation

Implementation is done up to the LC filter stage. The schematic that is being implemented is shown in Figure 13 with resistance value of 1.2M Ω. The constructed hardware is given in figure 14

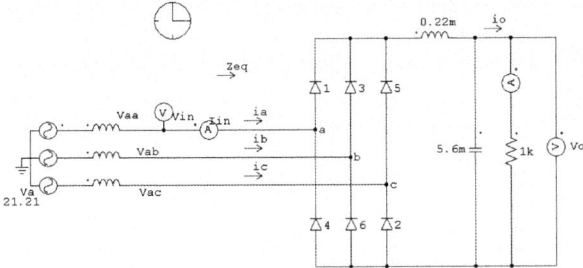

Fig. 13: Three-Phase rectifier with THD reduction

Fig. 14: Hardware implementation of the rectifier

Table 3 shows the input voltage, output voltage, ripple voltage, and the efficiency:

Vin (rms)	V_{out} (V)	V_r(mV)	V_r	Efficiency
2	1.3	12.96	0.978%	46.87%
4	4.2	41.18	0.992%	73.38%
6	7.0	67.81	0.972%	82.22%
8	9.8	93.61	0.955%	86.64%
10	12.6	123.88	0.981%	89.29%
12	15.5	152.06	0.984%	91.06%
15	19.7	197.31	1.002%	92.83%

Table 3: Tabulated simulation output.

It is noticed that the efficiency is increasing while the input rms voltage increase. The voltage drop is to overcome the cut-in voltage across the diode. There is some voltage drop observed across the filter inductor due to the change of the current. The voltage drop across the inductor is very small as the current stabilized over a short period of time.

As the input voltage increases, the efficiency increases because the voltage drop across the diode is a fix value.

The difficulty encountered during the simulation is that the inductor used could not sustain high power such that the inductor will blow off when the input rms voltages exceed 15V. This can be improved by using bigger inductor. The capacitor filter used could sustain 35V.

IV. CONCLUSION

The design development stages in simulation from studies of fundamental circuits to the development of final design, which utilizes the feedback controller at dc side of the converter to further improve the current waveform and thus producing an output signal with less distortion and better power factor. This design is then been use as a reference to implement a design in practical manner

This proves that the feedback controller applied with the step-up converter can help in producing a better output signal.

The goal to achieve low harmonic distortion, high power factor and low ripple voltage can be achieved. The implementation is of low cost and the components used are easy to obtain.

The development of the power factor corrector could be used in some other aspect such as three-phase air-conditional circuit, or three-phase induction motor. In order for the power factor corrector to function on the ac-dc converter circuit, more research needed to be done.

The converter's components are designed in such a way that it can sustain the load resistance from 10Ω to 100kΩ. The rise time is maintained to be less than 0.5 seconds and the output voltage and current are critically damped. This is to ensure that transient output would not damage the equipment connected to it.

V. REFERENCES

1. Mohan N., Undeland T. M., Robbins W. P., Power Electronics, Converters, Applications and Design, 2003, John Wiley & Sons, Inc.
2. J. Schaefer, Rectifier Circuits: Theory and Design. John Wiley & Sons, Inc.
3. Van Der Sluis, L. Transients in Power Systems, 2002, John Wiley & Sons, Ltd.
4. Arrillage, J., Watson, N. R., Power System Harmonics, 2nd ed., 2004, John Wiley & Sons, Ltd.
5. Nilsson, J. W., Riedel, S. A., Electrical Circuit, 6th ed., 2001, Prentice Hall International.
6. Spasov, P., Microcontroller Technology: The 68HC11, 3rd ed., 1999, Prentice Hall International.
7. Serway, R.A., Beichner, R. J., Physics for Scientists and Engineers with Modern Physics, 5th ed., 2000, United State.
8. Chapman, S. J., Electric Machinery Fundamentals, 3rd ed., 1999, McGraw-Hill International.
9. Ogata, K., Modern Control Engneering, 4th ed., 2002, Prentice Hall International.
10. Neamen, D. A., Semiconductor Physics and Devices: Basic Principles, 3rd ed., 2003, McGraw-Hill International.
11. http://www.seits.org/features/pwrsup.htm
12. http://www.uoguelph.ca/~antoon/circ/vps.htm

2006 5th International Power Electronics and Motion Control Conference

Design of a Digital Programmable Control IC for Single-Phase Controlled Rectifiers

Ming-Fa Tsai, *Member IEEE*, Fu-Jing Ke, Ying-De Lin, and Jui-Kum Wang

Department of Electrical Engineering, Minghsin University of Science and Technology, Hsinchu, Taiwan, China

mftsai@must.edu.tw

Abstract—This paper presents the design of a digital programmable control IC which is intended for controlling thyristors to produce a constant current source via a single-phase controlled rectifier. The architecture of the control IC consists of three major parts: a single-phase controlled triggering circuit, an inverse cosine look-up table, and a proportional-integral current controller. All the three parts have been designed using VHDL language. The simulation model of a constant current source based on a single-phase controlled rectifier has also been constructed by using Simulink and ModelSim cosimulation tool. The designed control IC circuit has been implemented on an Altera Flex 10K FPGA logic device. Further, the control parameters in this IC are programmable and the variables can be observed via an SPI interface to a monitoring processor. Simulation and experimental results are shown to verify the viability of the proposed control IC properly.

Keywords: digital programmable control IC, current control, phase-controlled rectifier, Simulink and ModelSim cosimulation.

I. INTRODUCTION

Phase-controlled rectifiers have been widely used in dc motor drives, lighting controls, battery chargers, dc current power supplies, and as preregulators for ac motors for several years [1]-[3]. They have also been used in high-power and large-current industrial processes, such as electrolytic and welding system [4]-[6]. In some of these applications, the main requirement for the rectifier is that it must have the capability to deliver a controlled dc current in a range of 0% - 100% of rated value. Recently, with successively improving reliability and performance of microprocessors and digital signal processors (DSP), digital control techniques have predominated over their analog counterparts in the past decade. Microprocessor-based digital control schemes which have the advantages of flexibility and low cost have been applied to the current and phase control of the rectifiers [1][5][7]. However, it suffers the disadvantages of computational time delay and computation load in the control algorithm including the high sampling rate current control and the thyristor firing signal generation to achieve a wide bandwidth performance. Therefore, engineers turn to the technology of digital control ICs, because the concurrent and high-speed hardwire logic can enhance the computation capability and thus relieve the computation load of microprocessors. [8]-[9]

Some of the phase control IC of the controlled rectifiers, such as TCA785, UAA145, U2008B, and U2010B, have been available in the commercial market [10]-[13]. However, these control ICs are designed in analog and digital mixed bipolar technology and are

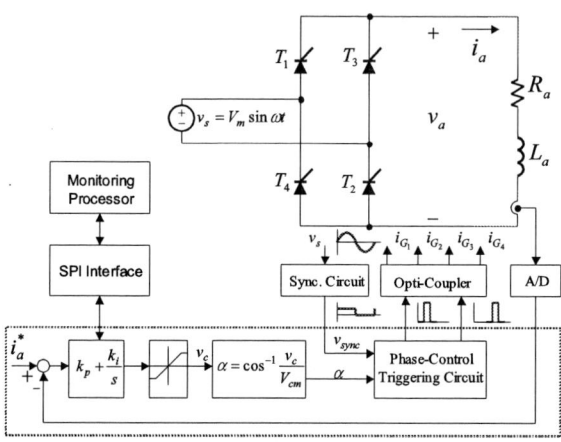

Fig. 1. Block diagram of the digitally single-phase controlled rectifier system

commonly used to generate the firing signal given the phase angle command. If the current control loop is required, the system needs additional circuits to implement the current controller. A fully digital current plus phase control IC for single-phase controlled rectifiers that have the advantage of reducing the series number of semiconductor devices at any time and can be expected to result in high operating efficiency has not yet been available.

Owing to the progress of microelectronics, the emergence of FPGA (field programmable gate array) or CPLD (complex programmable logic device), which utilizes digital CMOS technology, has drawn much attention due to its short design cycle and high density [14]. Thus, employing FPGA or CPLD to realize the control strategies to be a programmable digital control IC for phase-controlled rectifiers provides advantages such as rapid prototyping and simple hardware and software design, higher switching frequency, and relieving the computation load of the microprocessors. Realization of digital phase control scheme for phase-controlled rectifiers by using the digital CMOS technology thus provides advantages of easy use, low cost, and high performance. Therefore, it is the purpose of this paper to realize a digital programmable phase control IC by using an FPGA device for single-phase controlled rectifiers so as to achieve a constant current source under load variations.

II. ARCHITECTURE AND DESIGN OF THE DIGITAL CONTROL IC OF THE PHASE-CONTROLLED RECTIFIER

Fig. 1 shows the block diagram of the digitally single-phase controlled rectifier system, which is controlled by an

1-4244-0448-7/06/$25.00 ©2006 IEEE   1154

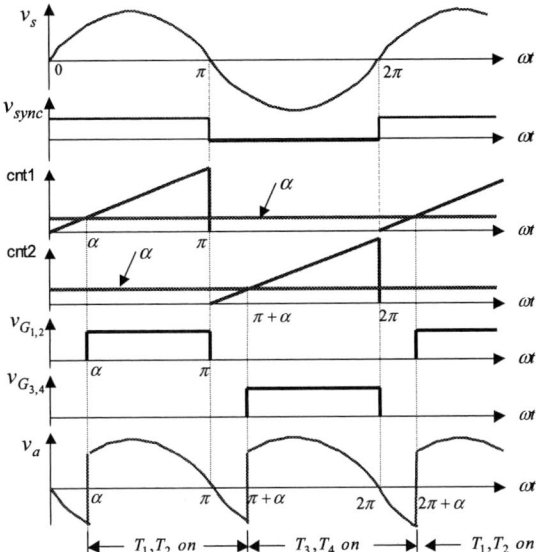

Fig. 2. The phase-controlled rectifier input, internal control, and output waveforms.

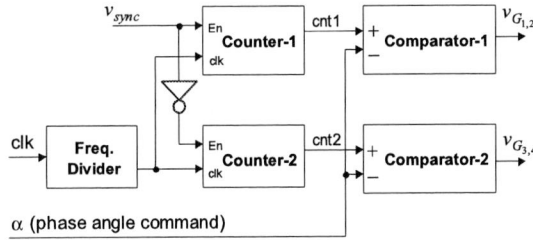

Fig. 3. The block diagram of phase-control triggering circuit.

$$V_{dc} = \frac{1}{\pi} \int_{\alpha}^{\pi+\alpha} V_m \sin(\omega t) d(\omega t) = \frac{2V_m}{\pi} \cos \alpha \qquad (1)$$

where V_m is the amplitude of the input ac source. The transfer characteristics from the phase angle command to the average output voltage is nonlinear. Hence, the use of this phase-controlled rectifier as a component in the current-feedback control system will cause an oscillatory response. A control technique to overcome this nonlinear characteristics is given by modifying the control input to be

$$\alpha = \cos^{-1}(\frac{v_c}{V_{cm}}) \qquad (2)$$

where v_c is the control input and V_{cm} is the maximum of the absolute value of the control voltage. Then it can be derived that the gain of the single-phase controlled rectifier is

$$K_r = \frac{V_{dc}}{v_c} = \frac{2}{\pi} \frac{V_m}{V_{cm}}. \qquad (3)$$

The inverse cosine function in (2) can also be programmed as a look-up table by using the *when-else* statement in VHDL language.

C. Design of Current Controller

A current source can be realized with the phase-controlled rectifier by incorporating a current feedback loop. For the controller design, the rectifier can be modeled as

$$G_r(s) = K_r e^{-T_r s} \approx \frac{K_r}{1+T_r s}, \qquad (4)$$

where T_r is the triggering delay. The delay may be treated as one half of the thyristor switching period [17]. For the 60-Hz frequency power system, the delay time in second can be computed as

$$T_r = \frac{180/2}{360}(\frac{1}{60}) = \frac{1}{240} \qquad (5)$$

With the resistive and inductive load combination, a proportional plus integral current controller can be designed. By using pole-zero cancellation technique, the controller parameters can be obtained as

$$\frac{K_p}{K_i} = T_r \qquad (6)$$

and the closed-loop transfer function of the current loop can be derived as

FPGA-based control IC.

The architecture of the control IC mainly consists of three parts: a phase-control triggering circuit, a proportional-integral current controller, and an inverse cosine look-up table, which is in series between the previous two parts to produce a linear constant gain for the phase-controlled rectifier. The controller parameters in this IC are programmable and the variables can be observed via an SPI interface to a monitoring processor. Design of the control IC is briefly described as follows.

A. Design of Phase-Control Triggering Circuit

Fig. 2 shows the input, internal control, and output waveforms of the single-phase controlled rectifier. The v_{sync} and α signals are the inputs to the phase-control triggering circuit. As shown in Fig. 3, the triggering circuit has been designed to produce the thyristors firing signals, $v_{G_{1,2}}$ and $v_{G_{3,4}}$. The circuit has been designed in VHDL language and simulated by using Simulink and ModelSim cosimulation tool [15][16]. Given the R-L load parameters of $R = 20\Omega$ and $L = 0.5$H, the simulation results for the phase angle command of $\alpha=\pi/4$ and $\alpha=\pi/2$ is shown in Fig. 4(a) and (b), respectively. As can be seen, the case of Fig. 4(a) works in the continuous conduction mode, but the case of Fig 4(b) works in the discontinuous conduction mode. The amplitude of the current ripple in each of the two cases is about 0.4 A.

B. Design of the Inverse Cosine Look-Up Table

Assuming the load current is continuous, the average output voltage is derived as

1155

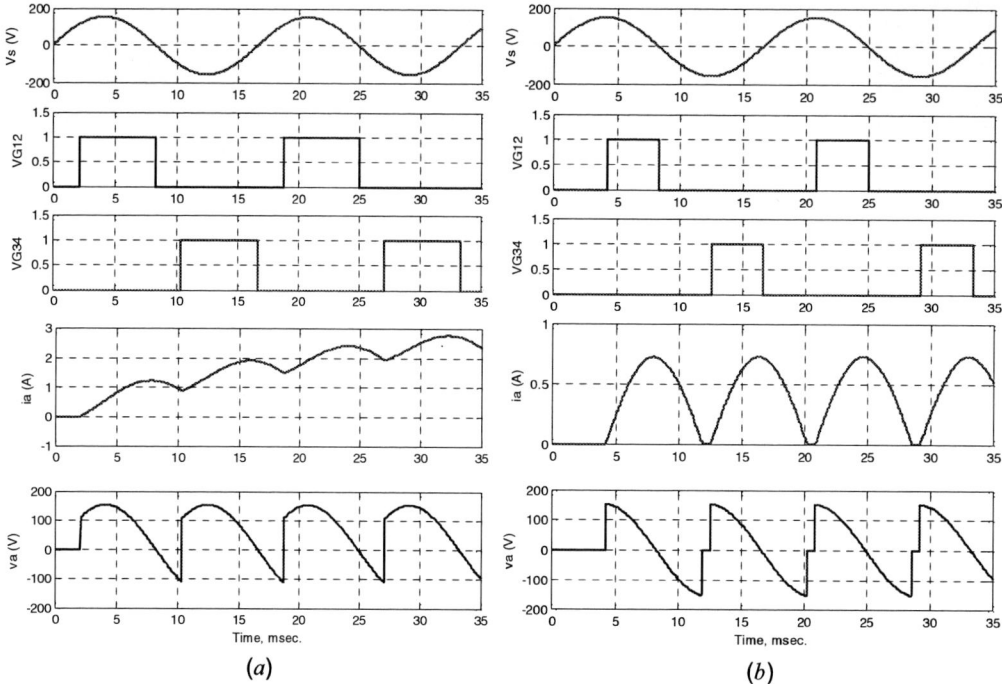

Fig. 4 The phase control simulation results: (a) phase angle command of $\alpha=\pi/4$, (b) $\alpha=\pi/2$.

$$G_{cl}(s) = \frac{K_i K_r}{L_a s^2 + R_a s + K_i K_r}. \qquad (7)$$

The K_i value then can be determined by the current response specifications. With phase margin of 52°, K_i is equal to 800,.and the K_p value can be obtained from (6).

For digital control, the PI control function in discrete form by using the backward rectangular rule can be expressed as follows:

$$y(k) = y(k-1) + K_p[e(k) - e(k-1)] + K_i T \cdot e(k), \qquad (8)$$

where T is the sampling time (1/120 second), and $e(k)$ is the error signal between the current reference and feedback signals. The authors have constructed a data path, which contains a 12-bit adder/subtractor, a 12-bit multiplier with Q-format selection, and a limiter, for the computation of the current control algorithm in (8), and a control unit using the finite-state machine to control the operation of the data path.

By combining the above three main parts of the design circuits into an integral VHDL control hardware IP (Intellectual Property), the closed-loop current control simulation model in the Simulink and ModelSim cosimulation environment is shown in Fig. 5. It can be seen that the control parameters K_p and K_i can be tuned for different R-L load. Fig. 6 shows the closed-loop current control simulation results with the current command of 3 A. As can be seen, the load current reaches the 3-A steady-state response, but has about ±0.4A current ripple, which is consistent with the open-loop phase control response in

Fig. 4.

III. HARDWARE REALIZATION AND EXPERIMENTAL RESULTS

The proposed control IC has been implemented by an Altera Flex 10K100 FPGA device to control a single-phase controlled rectifier, where four SGS-THOMSON TYN690 SCRs are employed for the power switches. Fig. 7(a) and (b) illustrate the experimental results for the phase angle command of $\alpha=\pi/4$ and $\alpha=\pi/2$, respectively. As can be seen, the rectifier works in the continuous conduction mode for the phase angle command of $\alpha=\pi/4$ and in the discontinuous mode for the phase angle command of $\alpha=\pi/2$, which are consistent with the simulation results in Fig. 4.

In the experimental system, a 12-bit high-speed A/D device, ADS7844 [18], has been employed for the load current sampling. The load current is sensed with a Hall-effect transducer so that one Ampere current is transferred to 230 mV voltage and the 12-bit digital value of 285. Because of using fixed-point operation, the multiplication in the PI controller is selected with Q8 format for tuning the value of K_p and K_i, respectively, as an integer number.

Fig. 8 shows the experimental result for the closed-loop current control response with the current command of 3 A. As can be seen, the step response has current ripples with amplitude about 0.4 A, which is consistent with the simulation result in Fig. 6(a).

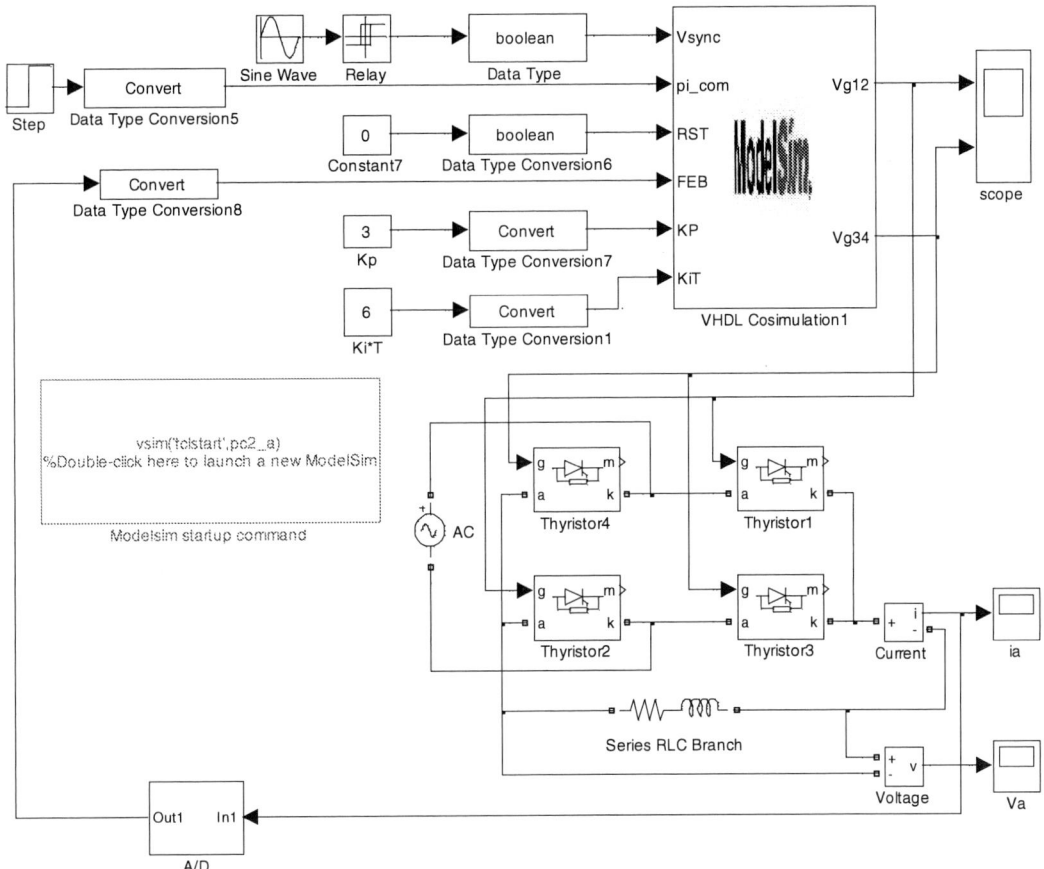

Fig. 5 The closed-loop current-control simulation model of the phase-controlled rectifier using Simulink and ModelSim cosimulation tool.

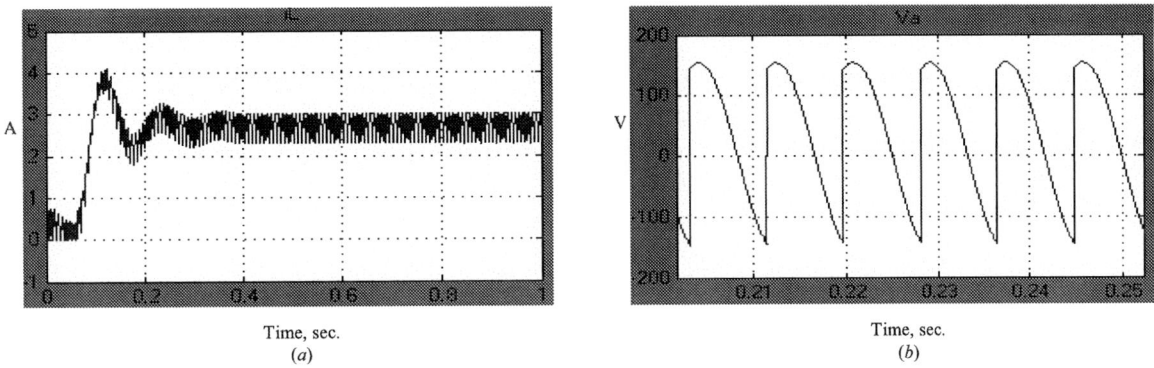

Time, sec.
(a)

Time, sec.
(b)

Fig. 6 The closed-loop current control simulation result with current command of 3A. (a) load current response, (b) output voltage response.

The controller parameters such as K_p and K_i in this IC are programmable from a monitoring processor via a serial bus interface to a bank of 12-bit registers, so that the control IC can be used in different power frequency and load system. The variables can also be observed via the interface. The serial bus interface includes three signals: slave select (*SS*), serial clock (*SCL*), and serial data (*SDA*), to meet the MOTOROLA serial peripheral interface (SPI) protocol for access the data in the IC [19].

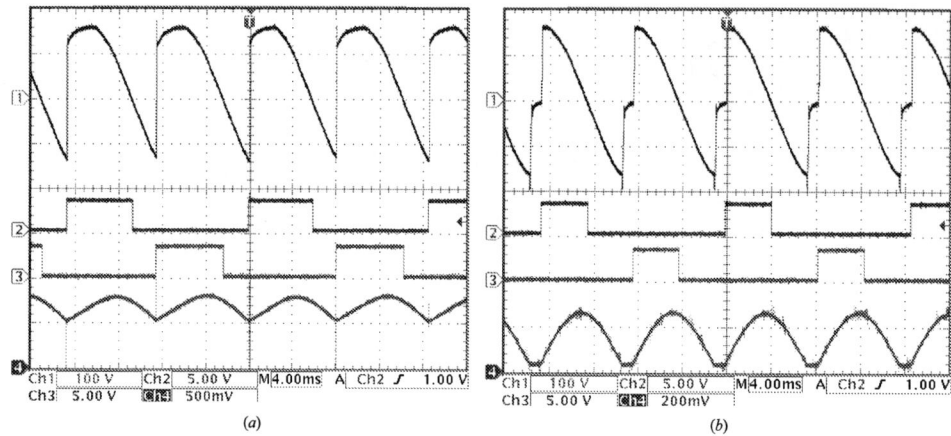

Fig. 7 Experimental results for open-loop phase control: Ch1 is v_s, Ch2 is $v_{G_{1,2}}$, Ch3 is $v_{G_{3,4}}$, Ch4 is i_a (4.35 A/V). (a) phase angle command $\alpha=\pi/4$, (b) $\alpha=\pi/2$..

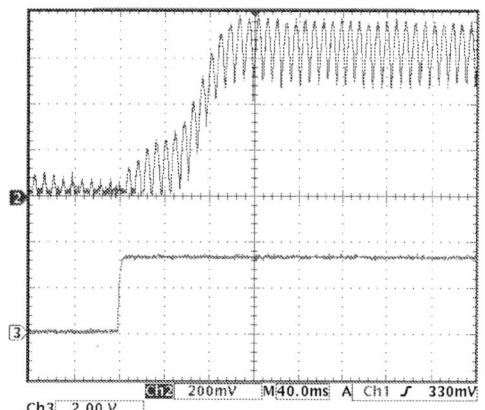

Fig. 8 Step response of the closed-loop current control. Ch2: current response (4.35A/V). Ch3: current command (0.91A/V).

IV. CONCLUSION

This paper presents the design of a digital programmable control IC for the current control of a single-phase controlled rectifier. The architecture of the control IC mainly consists of three parts: a phase-control triggering circuit, a proportional-integral current controller, and an inverse cosine look-up table, and has been integrated into a control IP in VHDL language. The closed-loop current control behavior has been simulated in the Simulink and ModelSim cosimulation environment. The experimental results have been shown to verify the control performance. To reduce the current ripple, three-phase controlled rectifier will be studied in the near future.

REFERENCES

[1] R. J. Hill and F. L. Luo, "Microprocessor-based control of steel rolling mill digital dc drives," *IEEE Trans. Power Electron.*, vol. 4, no. 2, April. 1989.

[2] R. M. Stephan, "A simple model for a thyristor driven dc motor considering continuous and discontinuous current modes," *IEEE Trans. Education.*, vol. 34, no. 4, Nov. 1991.

[3] J. A. Pomilio, D. Wisnivesky, and A. C. Lira, "A novel topology for the bending magnets power supply at LNLS," *IEEE Trans. Nuclear Science*, vol. 39, no. 5, Oct. 1992.

[4] J. R. Rodriguez, J. Pontt, C. Silva, E. P. Wiechmann, P. W. Hammond, F. W. Santucci, R. Alvarez, R. Musalem, S. Kouro, and P. Lezana, "Large current rectifiers: state of the art and future trends," *IEEE Trans. Ind. Electron.*, vol. 52, no. 3, June 2005.

[5] T. C. Manjunath, S. Janardhanan, and N. S. Kubal, "Simulation, design, implementation, and control of a welding process using Micro-controller,"
ascc2004.ee.mu.oz.au/proceedings/papers/P121.pdf.

[6] A. Borisavljevic, M. R. Iravani, and S. B. Dewan, "Digitally controlled high power switch-mode rectifier," *IEEE Trans. Power Electron.*, vol. 17, no. 6, Nov. 2002.

[7] Teccor Electronics Thyristor Product Catalog, Phase Control Using Thyristor, *AN1003*, 2002.

[8] Hewlett Packard, General purpose motion control ICs, Technical data, HCTL-1100 series.

[9] S. -L. Jung, M. -Y. Chang, J. -Y. Jyang, L. -C. Yeh, and Y. -Y. Tzou, "Design and implementation of an FPGA-based control IC for AC-voltage regulation," *IEEE Trans. Power Electron.*, vol. 14, no. 3, pp. 522-532, May. 1999.

[10] Siemens data sheet, TCA 785 Phase Control IC, Semiconductor Group, 1994.

[11] Temic Semiconductors, UAA145 phase control circuit for industrial applications, May 1996.

[12] Almel, U2008B low-cost phase-control IC with soft start, Oct. 2005.

[13] Almel, U2010B phase-control IC with current feedback and overload protection, Oct. 2005.

[14] S. Brown and J. Rose, "Architecture of FPGAs and CPLDs: a tutorial," *IEEE Design and Test of Computers*, vol. 13, no. 2, pp. 42-57, 1996.

[15] Mathworks Inc., Link for ModelSim User's Guide, 2003.

[16] U. Hatnik and S. Altmann, "Using ModelSim, Matlab/Simulink and NS for simulation of distributed systems," International Conference on Parallel Computing in Electrical Engineering (PARELEC'04), pp. 114-119, 2004.

[17] R. Krishnan, *Electric Motor Drives, Modeling, Analysis, and Control*, Prentice Hall, 2001.

[18] Burr-Brown Corporation, data sheet of ADS7844, 1998.

[19] Motorola, MC141547P2, 1998.

2006 5th International Power Electronics and Motion Control Conference

Feasibility Study of AlGaN/GaN HEMT for Multi-megahertz DC/DC Converter Applications

Yang Gao, Alex Q. Huang

Semiconductor Power Electronics Center (SPEC)
North Carolina State University, Raleigh, NC 27695, USA
E-mail: ygao3@ncsu.edu

Abstract—The DC and AC characteristics of a 30 V AlGaN/GaN HEMT was investigated by numerical simulations. By properly model the 2DEG in the AlGaN/GaN interface, we obtain a maximum transconductance of 221 mS/mm, a saturation current density of 1.28 A/mm, a specific Rdson of 2.5 mΩ -mm^2, a specific Qgd of 0.62 nC/ mm^2 and the value of FOM of 1.6 mΩnC. The comparison with of the state-of-the-art Si-LDMOS and Si-Trench MOSFET were carried out and the results indicate that HEMTs could be a good candidate for very high frequency DC/DC converter applications.

Keywords-AlGaN/GaN; HEMT; power; switch;

I. INTRUDUCTION

AlGaN/GaN high electron mobility transistors (HEMTs) have attracted attention in recent years due to their excellent high current drive capability, and high microwave power performance [1-3]. Due to the piezoelectric and spontaneous polarization, a two-dimensional electron gases (2DEG) with sheet carrier concentrations in the order of $1 \cdot 10^{13}$ cm^{-2} and an electron mobility in the order of 10^3 cm^2/Vs exists at the AlGaN/GaN heterojunction interface [1]. A low Rdson is therefore expected. Furthermore, being a lateral device, the HEMTs have a relatively small gate-drain capacitance Cgd which leads to the small gate-drain charge Qgd. The conduction and switching power loss depends on Rdson and Qgd in a DC/DC converter and the Figure-Of-Merit: FOM = Rdson·Qgd is generally used in order to describe the switching performance of a power device. It is therefore expected that AlGaN/GaN has low FOM. This, coupled with high critical electric field due to GaN's wide energy bandgap, imply that the AlGaN/GaN HEMTs could be a good candidate as a switch in high frequency DC/DC converter applications. However, to our knowledge no extensive experimental investigation of switching application of the low voltage HEMTs has been reported in the literature so far. In this paper, we report our numerical investigation results on a 30 V AlGaN/GaN HEMTs on sapphire, with a peak transconductance of 221 mS/mm, a saturation current density of 1.28 A/mm, Rdson_sp of 2.5 mΩ-mm^2, Qgd_sp of 0.62 nC/ mm^2 and a FOM of 1.6 mΩnC.

II. DEVICE STRUCTURE

The structure of studied AlGaN/GaN-HEMT is shown in Figure 1. The device consists of a sapphire substrate, an AlN buffer, 1.25 um GaN, a 30 nm undoped AlGaN cap layer. The two-dimensional electron gases (2DEG) with sheet carrier concentrations of $1.1 \cdot 10^{13}$ cm^{-2} and an electron mobility of 1100 cm^2/V-s exists at the AlGaN/GaN heterojunction interface [1-2]. The device has a gate-length of 0.2 um and a source-drain spacing of 2 um.

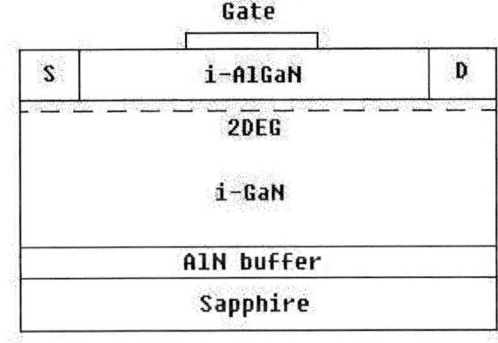

Fig. 1: Schematic drawing of investigated HEMT structure with the gate length of 0.2 μm and source-drain spacing of 2μm

III. SIMULATION AND RESULTS

The device is simulated using ISE-TCAD. The major parameters used in the simulation are summarized in Table I.

TABLE I.
KEY PARAMETERS USED IN SIMULATION

Material parameters	AlGaN	GaN
affinity (eV)	3.82	3.4
Eg300 (eV)	3.96	3.47
permittivity	9.5	9.5
mun (cm^2/V-s)	600	1500
mup (cm^2/V-s)	10	20
vsat (10^{7}cm/s)	2	2
Nc300 (10^{18}/cm^3)	2.07	2.65
Nv300 (10^{19}/cm^3)	1.16	2.5

1-4244-0448-7/06/$25.00 ©2006 IEEE

Hydro-dynamic models with field dependent mobility are used. The mobility for the 2DEG layer is adjusted to $1100 \text{cm}^2/\text{V-s}$ according to the reported results in literatures [3].

The calculated output characteristics of the HEMT are represented in Figure 2. The gate was biased from 1 V to –9 V in a step of –2 V. The maximum current I_{max} was about 1280 mA/mm at a gate bias of 1V and a drain bias of 30 V. Specific Rdson,sp was extracted from the curves as 2.5 mΩ-mm^2 at Vgs= 0 V, which is much smaller than that of a typical Si-Trench MOSFET, which is normally ~10 mΩ-mm^2. The transfer characteristics and the corresponding transconductance are shown in Figure 3. The peak extrinsic transconductance (g_m) was 221 mS/mm at V_{gs}= -6.2 V and V_{ds}= 8 V. By defining the threshold voltage (V_{th}) as the gate bias intercept of the

extrapolation of drain current at the point of peak gm, the Vth was determined to be –7.4V in our simulation. Simulated BV is higher than 30 V.

A small signal analysis for the HEMT was carried out and the specific gate-drain capacitance (Cgd,sp) was extracted out. The results are shown in Figure 4 with a sweep frequency of 1 MHz. For comparison, the Cgd,sp for a typical Si-Trench MOSFET is also shown with dashed line. Cgd,sp for HEMT is about 30% less than that of a typical Si-Trench MOSFET, which implies that the gate-drain charge Qgd in the HEMT is much smaller.. The simulation for Qgd,sp was carried out at a load current of 3 A and bus voltage of 11 V using the circuit shown in figure 5. The transient simulation results are shown in Figure 6. Qgd,sp was determined as 0.62 nC/mm^2 at the condition of Vds= 11 V, which is smaller than a typical value for Si-trench MOSFET. A faster switching speed and lower switching loss is thus expected for the HEMTs.

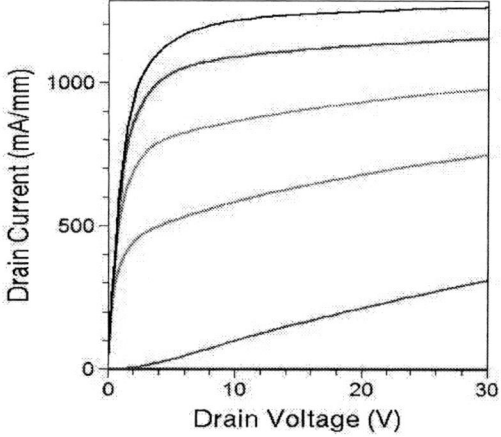

Fig 2. I-V characteristics of a 0.2 μ m AlGaN/GaN HEMT with a gate width of 100 μ m. The Vgs is swept from 1V to –9V

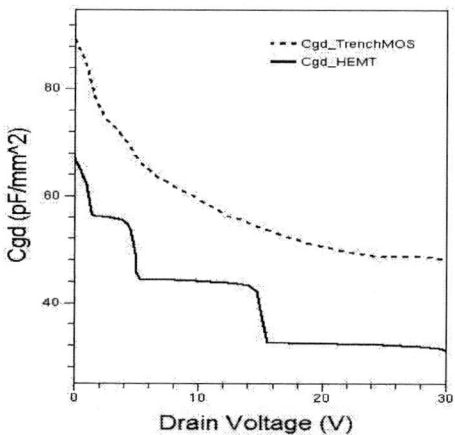

Fig 4: Specific gate-drain capacitance of a 0.2 μ m AlGaN/GaN HEMT (solid line) and of a typical 30V Si-Trench MOSFET (dashed line)

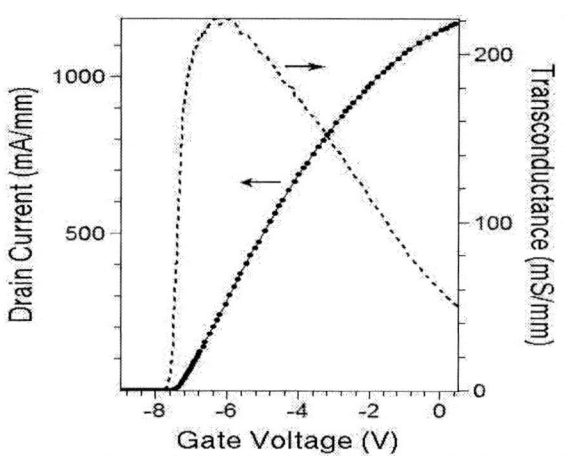

Fig 3. Transfer characteristics of a 0.2 μ m AlGaN/GaN HEMT at a drain bias of 8 V

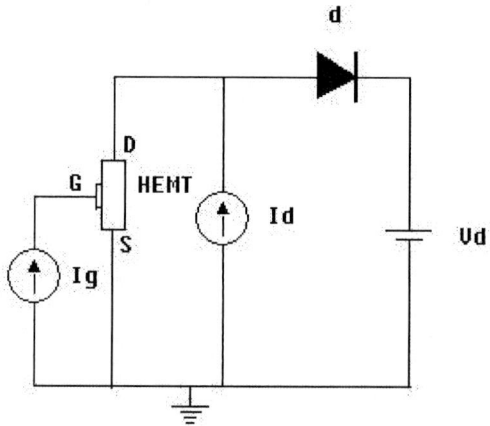

Fig 5. Circuit schematic for Qgd with I g=1 mA, Id=3 A and Vd=10.3 V

1160

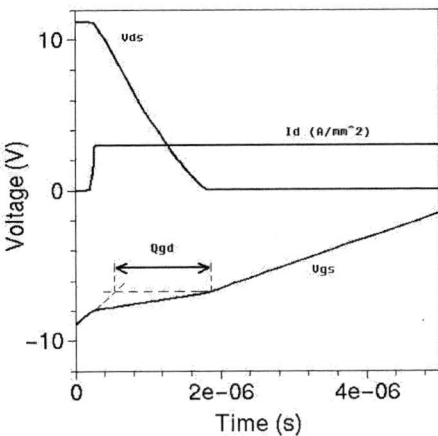

Fig 6. Simulated gate charge waveform for a 0.2 μm $\times$100 μm AlGaN/GaN HEMT

REFERENCES

[1] O.Ambacher etc. "Two-dimensinal electron gases induced by spontaneous and piezoelectric polarization charges in N- and Ga-face AlGaN/GaN heterostructures", J.Appl. Phys., 85, 3232 (1999)

[2] Shreepad Karmalkar and Umesh K.Mishra, "Enhancement of Breakdown Voltage in AlGaN/GaN High Electron Mobility Transistors Using a Field Plate," IEEE Trans. Electron. Devices, vol 48, pp. 1515-1521, Aug. 2001

[3] R. Li, S.J.Cai etc., "An Al0.3Ga0.7N/GaN Undoped Channel Heterostructure Field Effect Transistor with Fmax of 107GHz" IEEE Elec. Dev. Letters, 20 pp 323-325, Jul. 1999

[4] Norio Yasuhara etc. "Low Gate Charge 30 V N-channel LDMOS for DC-DC Converters," Proc. of the 14th-17th ISPSD 2003 pp 186-189

[5] M.A.A in 't Zandt, E.A. Hijzen, R.J.E Hueting, and G.E.J. Koops, "Record-low 4 mΩ- mm2 specific on-resistance for 20V Trench MOSFETs, "Proc. of the 14th-17th ISPSD 2003 pp 32-35

Table 2 summarizes the key characteristics of AlGaN/GaN HEMT, in comparison with best reported Si-LDMOS and Si-Trench MOSFET [4] [5]. It can be seen that HEMT has both better Rdson,sp and Qgd,sp performance comparing to the other two. The FOMs of HEMT is 1.6 mΩnC, which is only 1/6 of LDMOS and 1/11 of Trench MOSFET.

TABLE 2.

ELECTRIC CHARACTERISTICS OF THE INVESTGATED HEMTs, TOSHIBA SI-LDMOS and PHILIPS Si-TRENCH MOSFET

	HEMTs	Toshiba Si LDMOS	Philips Si Trench MOSFET
FET area	2 mm^2	5.9 mm^2	—
BV	>35V	35 V	21 V
Vth	-7.4 V	1.5 V	—
Vgs,max	1 V	12.5 V	—
Rdson,sp	2.5 mΩ- mm^2 @ Vgs=0 Id=7.5	12.2 mΩ- mm^2 @ Vgs=7 Id=7.5	5 mΩ- mm^2 @ Vgs=10 Id=7.5
Qgd,sp	0.62 nC/ mm^2 @ Vdd=11	0.82 nC/ mm^2 @ Vdd=11	3.6 nC/ mm^2
Rdson·Qgd	1.6 mΩnC @ Vgs=0 Vds=11	10 mΩnC @ Vgs=7 Vds=11	18 mΩnC @ Vgs=10

I.   IV. CONCLUSION

Numerical simulation of a 30 V AlGaN/GaN-HEMT has been carried out. The comparison with Si-trench MOSFET and Si-LDMOS shows AlGaN/GaN-HEMTs has a better characteristics for both Ron,sp and Qgd,sp. The value of FOMs for HEMT is1.6 mΩnC, which is 5 to 10 times better than those of "conventional" silicon LDMOS and trench MOSFET. The results indicate AlGaN/GaN-HEMTs could be a good candidate for high frequency DC/DC converter applications.

2006 5th International Power Electronics and Motion Control Conference

The Mechanism Analysis of IGBT Module Invalidation

Xu Aide [1], Fan Yinhai[1], Wang Xinxin[1] and Liu Yuanyuan[1]

[1]Dalian Maritime university, Dalian, China

aidexu@newmail.dlmu.edu.cn

Abstract—The main purpose of this paper is to make a brief introduction of the applicable scope and the testing condition of IGBT module, to analyze the various reason causing IGBT module to be invalid which mainly involves drive part trouble, excessive heat, mechanical hurts, over voltage, over current and inductance that exists in the leads of circuit, and to make a general summary of the cautions which should be paid much attention to preventing IGBT module from being invalidated.

Keywords—Insulated Gate Bipolar Transistor module; invalidation; gate drive voltage; excessive heat; mechanical hurts; over voltage; over current; inductance exists in the leads of circuit

I. INTRODUCTION

With the rapid development of electric power and electronic technology, the IGBT module has been applied to practice more and more extensively while the occurrence of its invalidation may be inevitable in the practical application. The intention of this paper is trying to make a thorough discussion about the mechanism invalidation of IGBT module.

The discussion is based on the certain conditions listed below:

a) The load connected to IGBT shall be inductive.

b) The internal lead inductance of IGBT module shall be ignored.

c) While measuring the voltage waveforms of IGBT module, the ports of IGBT must be measured directly. As is illustrated in Fig.1, A and B should be measured directly when we intend to get the voltage between the terminal of C and D, otherwise false result will occur by

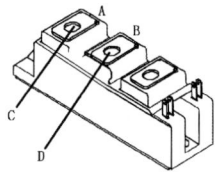

Figure 1. Measure the voltage of IGBT

measuring C and D. Similarly, direct measurements on IGBT ports are required while measuring the signal of

controller for the measurement of the signal on control panel usually can not reflect the reality.

II. THE ANALYSIS OF VARIOUS MECHANISM INVALIDATION OF IGBT

A. Invalidation caused by Drive Part Trouble

1) The decline of gate drive voltage

The decline of gate drive voltage leads the rise of saturated conducting voltage of IGBT, causing the increase of dissipation power imposed on IGBT, and then destroy the module by overheating.

2) The over high of gate drive voltage

If the gate wire is too long, oscillation and increment of stray induction will occur. The former can cause a sharp voltage on the gate drive, the latter will lead to an increase of L*di/dt. Both results will exceed the allowable rang of gate drive voltage and so as to damage the IGBT module [1][2].

Countermeasures:

- The drive wire shall be as short as possible,
- The connecting wires of both gate and emitter should be twisted tortuously,
- Zener diode can be inserted between the terminal of C and E to limit the drive voltage.

B. Invalidation caused by overheat

Phenomenon of overheat are chiefly caused by following reasons.

1) The increase of thermal resistance caused by
- the insufficient clamping force between IGBT and radiator
- the uneven surface of radiator
- the insufficiency, overuse, or uneven daubing of coolant

2) The decrease of radiating ability caused by
- the blocking up of the aperture of radiator
- the decrease of rotation rate or the cease of rotation of aged fan
- the overhigh surrounding temperature

3) The increase of static loss caused by
- the increment of saturating voltage drop caused by insufficient drive voltage

1-4244-0448-7/06/$25.00 ©2006 IEEE

● over current

4) The increase of dynamic loss caused by

● the increment of the switch frequency of IGBT

● the increment of turn-on loss caused by

    a) the increase of turn-on time due to the excessive gate drive resistance

    b) the excessive conducting time

    c) the increase of turn-off loss caused by the increment of turn-off time due to the excessive gate drive resistance

C. Invalidation caused by mechanical damage

1) The screw used for main terminal is too long, so that it will damage the sub-structure of terminals. (see figure 2 (a) and (b))

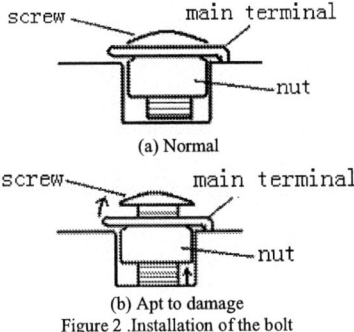

(a) Normal

(b) Apt to damage
Figure 2 . Installation of the bolt

2) The clamping moment between main terminal and external connection wire is too big.

3) The tightening torque between module and radiator is too big.

4) The unbalanced tightening torque between module and radiator leads to the deformation of insulating substrate of IBGT module.

5) The welding temperature is too high, so that it will melt the outer resin materials and the internal welding point of IGBT module.

D. Invalidation caused by over voltage

1) Excessive direct voltage

● The voltage of power source exceeds the maximum endurable voltage of module due to incorrect selection of type.

● Without the proper function of brake resistance during the process of regenerative action of brake by motor, the voltage of direct current bus wire climbs up abnormally.

2) Excessive spike voltage

● The module will be damaged because the surge voltage generated between the points of C and E during the turning-off of IGBT exceeds the limit of RBSOA (Reverse Bias Safe Operating Area).

The mechanism analysis of surge voltage is as follows.

As is shown in figure 3, the load of circuit is inductive, Ls is the lead inductance of the main circuit, the initial state of circuit is Q1 conducting, and current Ic flows through the load.

At this moment, turning-off Q1 and Q2 is conducting to supply continuous current loop, the changing process of current in the course of turning-off of Q1 is illustrated in figure 4.

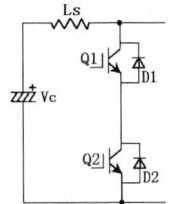

Figure 3 The mechanism analysis of surge voltage

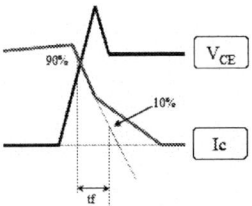

Figure 4. The current and voltage waveform while turning off IGBT

The decrease rate of current in the course of turning-off of IGBT is very fast, that is , the decrease rate of current flowing through the lead inductance is very fast, so that the electromotive force ($VLs=Ls*di/dt$) on Ls that overlaid on the points of IGBT is very large.

● As is shown in figure 5, the condition of D2 is turn-on, so that the voltage on Q1 and D1 is Vc+VLs. The module will be damaged by over voltage while the voltage exceeds the limit of RBSOA particularly in the case of over current or short cut of turning-off of IGBT. At this moment, soft way shall be adopted to turn off the IGBT to avoid the damage of module.

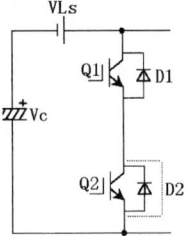

Figure 5. The equivalent circuit of VLs

Countermeasures

Reduce the lead inductance as practical as possible and enlarge the gate drive resistance appropriately. (To acquire a balance between the reducing of di/dt and the actual cooling condition, the proper plan of design is required because the increase of gate drive resistance will enlarge the switch loss.)

- During the reverse recovery of the diode parallelled with IGBT, surge voltage will be produced by inductance which may damage the module if it exceeds the limit of RBSOA.

The mechanism analysis of surge voltage is as follows.

As is shown in Figure 3, the load of circuit is inductive, Ls is the lead inductance of the main circuit, the initial state of circuit is that Q1 is closed, and D2 is conducted.

In this case, Q1 is conducting. D2 is supplied with reverse voltage. Because enough time is required during the reverse recovery of D2, it is impossible to be closed immediately. Therefore, a reverse current flows through Q1 to D2. The waveform illustrating reverse recovery of D2 is shown in Figure 6.

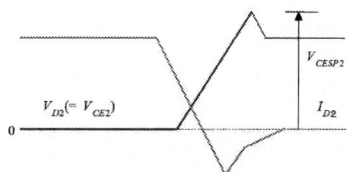

Figure 6. The waveform illustrating reverse recovery of D2

Electromotive force (VLs=Ls*di/dt) will be generated by high changing rate of current of Ls due to the large amount of di/dt in the process of reverse recovery of D2. The equivalent circuit is illustrated in Figure 7.

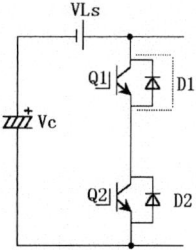

Figure 7. The VLs equivalent circuit for reverse recovery of D2

In this case, Q1 is conducting. Therefore, the voltage (Vc+VLs) is applied to the points of Q2 and D2. The module will be damaged when the voltage exceeds the limit of RBSOA.

- The slight disturbance gap of drive signal leads to the burning of module by over voltage.

As is shown in figure 8, very large surge voltage will be generated on Q2 when the slight disturbance gap (if the during of the gap is less than 1 us) appears on the drive pulse of Q1. Once it exceeds the limit of RBSOA, the module will be damaged(see figure 9).

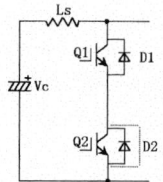

Figure 8. Disturb signal diagram

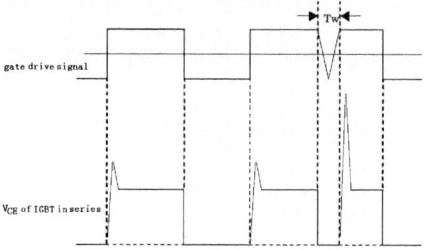

Figure 9.The waveform of disturb signal

The mechanism analysis of surge voltage is as follows.

Q1 is in the condition of normal conducting, while D2 is blocked. D2 is conducting during the interval of ephemeral interruption of Q1 drive signal, and then Q1 drive signal recovers immediately. As a result, extremely large di/dt will be generated due to the rapid enlarging of depletion layer by reason of that D2 turns into reverse recovery state after a very short conducting time without a sufficient quantity of carrier stored, so that the electromotive force (VLs=-Ls*di/dt) will be generated from the lead inductance Ls and applied to the terminal of Q2 and D2.

Therefore, the module will be destroyed by surge voltage which exceeds the range of RBSOA or the over voltage of Q1 and D1 caused by the slight gap of drive pulse of Q2 which is usually the consequence of transient interruption of drive signal that resulted from the phenomena such as electromagnetism disturbance.

Countermeasures

Reduce the lead inductance as practical as possible and shorten the length of gate drive signal wire, as well as twist the connecting wires between C and E tortuously.

E. Invalidation caused by overcurrent

The short circuit of tandem connection branch

- Insufficient dead-time

The dead zone time is usually longer than toff (max). As is shown in Figure 10, to turn off the IGBT safely, sufficient dead-time is required. If the drive is conducted before another tandem connected IGBT turned off, a short circuit between the upper and lower bridge arms will occur.

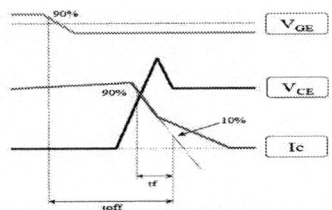

Figure 10. The waveform of dead time

- Incorrect performance of IGBT caused by dv/dt

As is shown in figure 5-2, the initial condition of Q1 is blocked, while D2 is flowing continuously. At this moment, the voltage of Q2 between C and E equals to the conducting voltage of D2 (Vd2 is approximately 0). In this case, trigger Q1, D2 will be in the condition of reverse recovery. If D2 is blocked, the voltage of Q2 between C and E will go up to saturated conducting voltage Vc-Q1（approximately equals to Vc）, that is, the voltage of Q2 between C and E will go up from 0 to Vc.

As can be seen from Figure 11, under the function of dv/dt, current Icg flows through reverse transmitting capacity Cress and gate resistance RG of Q2 and will general voltage V=Rg*Icg on the terminal RG. At this moment , the voltage of Q2 between C and E is V-Vcg. Once it reachs Vgeth of IGBT, Q2 will be conducted, and Q1 is also in the condition of conducting, so as to cause the phenomenon of short circuit between the upper and lower current branch.

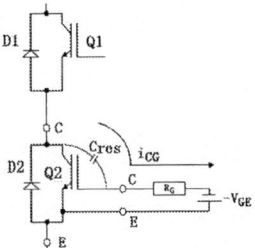

Figure 11.The diagram of short circuit for dv/dt

Countermeasures

Increase the reverse bias voltage of gate drive (-15V is recommended) and reduce the lead inductance as far as possible.

- The short circuit between Output and ground wire caused by the damage of load insulation or human error such as the mistake of wiring.

F. The analysis of circuit lead inductance

It is obvious that the reduced lead inductance plays a significant role on the safe performance of IGBT module. Detailed analysis of the composing of lead inductance is as follows.

The typical three-phase circuit is showed in Figure 6-1.

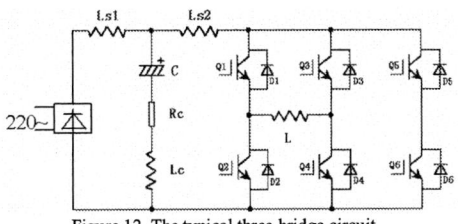

Figure 12. The typical three-bridge circuit

As can be seen from Figure 12, Ls1 is the lead inductance of connecting wires between rectifying bridge and filtering capacitor, Ls2 is the lead inductance between filtering capacitor and IGBT module, Lc is the stray inductance of filtering capacitor, Rc is the internal resistance of filtering capacitor, L is the inductance of the load.

1) Rectifying output voltage is larger than terminal voltage of capacitor.

In this case, the rectifying circuit charges the capacitor and at the same time supplies the current for IGBT module and both of the two parts of current flows through lead inductance Ls1. From the above analysis, we can draw a conclusion that inductive electromotive force which is in the same direction with voltage will be generated in Ls1 and Ls2 if the current of the main circuit is reducing and the di/dt is very large. The increase of electromotive force in Ls1 is confined by filtering capacitor, so that Ls1 will continue to flow through filtering capacitor.

Due to the voltage on resistance Rc that exists in the capacitor, the voltage on the terminal of capacitor goes up combined with the confinement of current of capacitor by inductance Lc which will hinder the continuous current flowing. Therefore, the filtering capacitor with less internal resistance and stray inductance is strongly recommended. Meanwhile, the length of the conducting wire between rectifying bridge and filtering capacitor should be shortened as far as possible to reduce the surge to the filtering capacitor so as to prevent the voltage on the terminal of capacitor from rising up abnormally.

2) The voltage of rectifying output is less than that of capacitor

In this case, the capacitor supplies current for IGBT. Electromotive force which is in the same direction with the voltage of capacitor will be generated in Ls2 when the current of main circuit decreases and the di/dt has a large amount. At the same time, the electromotive force Lc*di/dt will also be generated by inductance Lc of capacitor. Both of these two electromotive forces and the voltage of capacitor are applied to IGBT module. To avoid the over voltage of module, the connecting wire between filtering capacitor and IGBT module should be shortened as far as possible and the filtering capacitor with less inductance should be selected.

The inductance of connecting wires between modules should be reduced as far as possible when parallel connection of modules is applied to form the three-phase inverter circuit. As can be seen from Figure 13, inductances Ls3 and Ls4 have been added to the main circuit. When the current in circuit is decreasing rapidly,

1165

very large electromotive force will be generated in the inductance. As a result, the further the module is apart from the filtering capacitor, the easier it will be destroyed.

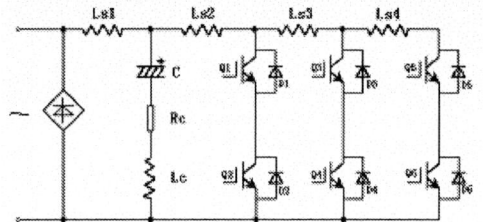

Figure 13.Three bridge inverter circuit in parallel connection

The surge voltage on the nearest IGBT from filtering capacitor is Vc+Ls2*di/dt (in Figure 14), while the farthest IGBT from filtering capacitor increases to Vc +Ls2*di/dt+Ls3*di/dt+Ls4*di/dt (in Figure 15).

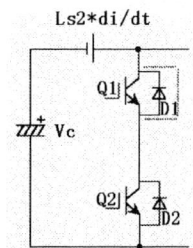

Figure 14.The surge voltage of the nearest module

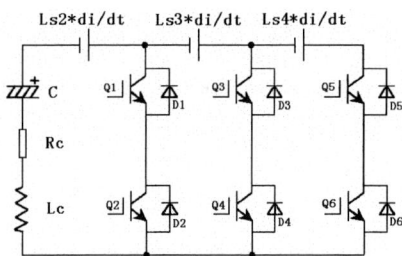

Figure 15.The surge voltage of the furthest module

III. SUMMARY

The analysis in this paper may also be applied to the transforming circuit composed of discrete IGBT. In this case, the connecting lines of IGBT and the internal lines of buffering circuit, as well as the connecting lines between IGBT and buffering circuit should be shortened as far as possible. Furthermore, the resistance and capacitor applied to the buffering circuit shall be free of inductance.

The items to be paid special attention in the practical use of IGBT are summarized as follows.

1) The inductance of gate drive line and main circuit line should be decreased as far as possible.

2) When FWD instead of IBGT is in use (for example in the chopping circuit), the reverse bias voltage with a value of $-5V$ or above ($-15V$ is recommended, but maximum $-20V$) should be applied to the unused IGBT between the terminal of G and E. Otherwise, the IGBT may be destroyed by erroneous triggering caused by dv/dt during the reverse recovery of FWD due to insufficient reverse bias voltage.

3) To prevent the IGBT of contraposition branch from erroneous triggering by overlarge conducting dv/dt, the most proper gate drive conditions ($+V_{GE}$,-V_{GE},R_G) is required.

4) To test the drive voltage(V_{GE}) and the pulse voltage of turning-on and turning-off on terminals of modules.

5) Be sure to install proper fuse with suitable capability and auto-breaker between the power source and IGBT module to avoid the spreading of malfunction.

6) To adopt cooling fan with auto-detecting function to ensure the reliability of system by reminding the processor of immediate action to be taken while the radiating ability of fan is falling down or totally invalid.

7) Neither in the conservation process nor in the actual practice can the gate circuit of IGBT be opened.

8) Particular attention shall be paid to the destructive effect on IGBT module caused by electrostatic in the practical use.

9) Try not to install the module in a vibrating circumstance because acute vibration will result in the loose of fastening bolt between module and radiator, which will cause the increasing of radiating resistance so as to burn the module by overheat in the extreme condition.

10) The filtering capacitor must be replaced if its capacity has been decreased too much after being used for a certain period of time.

REFERENCES

[1] Wei sanmin,Li fahai, "High power IGBT drive and protection circuit," J T singhua Univ (Sci & Tech), 2001, Vol. 41, No. 9.

[2] Hu junda, "The research of IGBT driving and protection circuit," Electrical engineering and electric appliances technology,2003,No.6.

[3] Liu xizhen,Zhou wenju,You jia,Zhang li, "State Identification of IGBT Junction Temperature on Line," Journal of Tian Jin University ,Vo l. 35 No. 2 Mar. 2002.

[4] Wuchen Wu, Guo Gao, Limin Dong. "Thermal Reliability of Power Insulated Gate Bipolar Transistor (IGBT) Modules," Twelfth IEEE SEMI-THERMTM Symposium 1996.

[5] Xu ping ,"The new structure and new development of manufacture on IGBT,"Electric Power and Electronics,2005,Vol 3,No 1.

[6] D. Braun D. Pixler P. LeMay. IGBT Module Rupture Categorization and Testing. *IEEE* Industry Application Society hnual Meeting New Orleans, Louisiana, October 5 - 9, 1997.

[7] Zhao zhengyi,Yang chao,Zhao liangbing, "Research on Two Snubbers for Three-level IGBT Converters," Proceedings of the CSEE, Vol. 20 No. 12 Dec. 2000.

[8] FUJI Electric Co. Ltd. The Fuji IGBT application manual, 2004,5, RCH

2006 5th International Power Electronics and Motion Control Conference

A New Injection Efficiency Controlled GTO

Wang Cailin   Gao Yong   Zhang Ruliang

Xi'an University of Technology, Department of Electronics Engineering, Xi'an, P.R.China

e-mail: wangcailin@xaut.edu.cn

Abstract—A new GTO structure called the injection efficiency controlled GTO (IEC-GTO) is proposed, and its anode injection efficiency can be controlled via an additional thin oxide layer located in short anode contact region of short anode GTO(SA-GTO). The operation mechanism is analyzed, and the conducting, blocking and switching characteristics are simulated by MEDICI simulator. The results show that injection efficiency of IEC-GTO varies with the anode current, thus it has lower gate trigger current and better switching characteristic, similar the conducting and blocking characteristics to the conventional SA-GTO, and hardly increase the complexity of process simultaneity. Lastly, the key structural parameter of IEC-GTO is optimized, and the results show the effective width of anode region of the IEC-GTO is slight smaller than that of SA-GTO.

Keywords-injection efficiency; power semiconductor devices; Gate Turn-off Thyristor (GTO); Gate Commutated Thyristor(GCT); short anode; ohmic contact; pn junction

I. INTRODUCTION

For power semiconductor devices, it is normally difficult to obtain high blocking voltage, low on-state voltage drop and fast switching speed with the same design parameters. In order to compromise the relations of conducting characteristic, switching characteristic and blocking characteristic, the anode injection efficiency should be enhanced during on state and weakened or ideally eliminated during turn-off. This can be achieved by the short anode structure in Gate Turn-Off Thyristor (GTO) [1], the transparent anode structure in Gate Commutated Thyristor (GCT) [2], and the double gate structure in DG-GCT [3]. However, the short anode used in GTO enables the gate trigger current to increase, the transparent anode used in GCT is a shallow junction, and needs an additional n buffer layer, it enables the fabrication process to be complicated; and the DG-GCT devices need control accurately the time phase of the switching signals applied to both gates during the turn-off process, it enables the drive circuit to be complicated. In this paper, based on short anode gate turn-off thyristor (SA-GTO), under the enlightenment of an injection efficiency controlled IGBT (IEC-IGBT) structure [4], and Injection Enhanced Gate Transistor (IEGT), an injection

efficiency controlled GTO (IEC-GTO) structure is proposed, its operation mechanism is analyzed and its realization method is given and the blocking, conducting and switching characteristic are analyzed and compared with that of SA-GTO.

II. DEVICE STRUCTURE AND OPERATION

A. Basic Structure and Equivalent Circuit

The basic structures and equivalent circuits of SA-GTO and IEC-GTO are shown in Fig.1. Seen from the figure 1(a), the partial p^+ anode region is replaced by n^+ short region in SA-GTO, and the ratio of the short region and the anode region is about 20% in SA-GTO [5]. The IEC-GTO in figure 1(b) is very similar to a SA-GTO. Comparatively, the anode contact area of the IEC-GTO is significantly reduced, and the n^+ region is floating and isolated from the anode contact via a thin silicon dioxide layer. If removing the additional thin oxide layer in the IEC-GTO, the structure of the IEC-GTO is similar to SAGTO shown in Fig. 1(a), and they have identical doping profiles and dimensions in all regions except the reduced anode contact area.

B. Operation

During the operation of the IEC-GTO, the partial

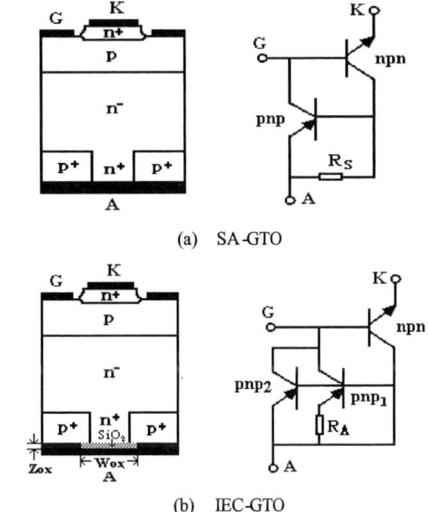

(a)   SA-GTO

(b)   IEC-GTO

Figure 1.   The comparison of the structures and equivalent circuits of IEC-GTO and SA-GTO

---

[a] Project supported by the Research Foundation of Xi'an University of Technology for Outstanding Doctor

electrons in n⁻ region drift toward the p⁺ anode and n⁺ floating regions under the external electric field, because of the existing the oxide layer, the electrons cannot be collected by the anode contact directly and accumulate at the oxide interface area, resulting in the increase of electron concentration in the n⁻ base and n⁺ floating regions. So it enables the electronic potential fall here. To maintain electrical neutrality in the n⁻ base and n⁺ floating regions, the holes injection from the p⁺ anode to n⁻ base region is enhanced. Because the p⁺ anode region of the IEC-GTO is narrow cross section, the resistor of this region, R_A, is significant. Also, the doping concentration in the n⁺ floating region is considerably higher than that of the n⁻ base region. Therefore the pnp transistor in the IEC-GTO can be taken as a two-emitter transistor as shown in Fig. 1(a), that is, the p⁺ anode, n⁻ base and p base form pnp₁ and the p⁺ anode, n⁺ floating, n⁻ base and p base form pnp₂. Seen from the equivalent circuit as shown in fig.1(b), the emitter-base voltage drop of pnp₂ transistor can be expressed as the following formula:

$$V_{be(pnp2)} = V_{be(pnp1)} + R_A I_{e(pnp1)}. \qquad (1)$$

As a positive voltage just applies at the anode and cathode of the IEC-GTO, the low anode current, I_A, flows mainly through pnp₁. With the increase of anode voltage V_A and thus I_A increases, the voltage drop across R_A, V_F, rises, the $V_{be(pnp1)}$ fall, causing the injection of holes from pnp₁ to fall and more current to flow through pnp₂. The pnp₁ has higher injection efficiency than pnp₂ because the n base region is adjacent to the p⁺ anode in pnp₁ and the pnp₂ is wide base transistor. So the anode injection efficiency of IEC-GTO varies with the anode current.

C. How to realize

Based on above analysis, the special IEC-GTO structure consists of the narrow p⁺ anode region with the reduced anode ohm contact area and an additional oxide layer based on the SA-GTO structure, the manufacture process of SA-GTO is suitable for the IEC-GTO, and the diffusion parameters can be unchanged, it only add an photolithography step forming the local oxide layer at the anode before fabricating anode ohm contact. Thus the manufacture process of the IEC-GTO structure is very simple. This shows that the IEC-GTO has not only adjustable anode injection efficiency as GCT, but also simpler process than the GCT with transparent anode.

III. ANALYSIS OF CHARACTERISTIC

In order to evaluate the performance of the IEC-GTO, a structural model of IEC-GTO as shown in the fig.1(b) is set up, the W_{ox} and Z_{ox} express respectively the width and thickness of the oxide layer. The key structural parameters are shown in table.1. Based on this model, the forward blocking, conducting and switching characteristics of IEC-GTO are analyzed by MEDICI simulator [6]. The doping distribution used in simulation is shown in fig.2

TABLE I.
THE MODEL PARAMETERS OF IEC-GTO STRUCTURE

regions	parameters	values
n⁻ base	Concentration N_D(cm⁻³)	2.3×10^{13}
	Thickness W_{n-} (μm)	400
p base	Concentration N_{ps}(cm⁻³)	3×10^{18}
	Thickness W_p (μm)	70
p⁺ anode	Concentration N_{as} (cm⁻³)	1×10^{19}
	Thickness t_a (μm)	30
	Width W_a (μm)	4000
n⁺ floating	Concentration N_{ns} (cm⁻³)	1×10^{19}
	Width W_n (μm)	1000
Oxide layer	Thickness z_{ox}(μm)	1
	Width W_{ox} (μm)	1200

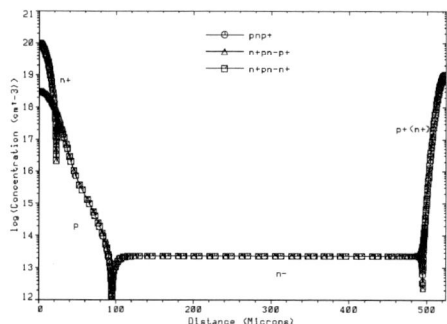

Figure 2. The doping distribution used in simulation

A. Blocking

The forward blocking characteristics of IEC-GTO and SA-GTO are simulated as shown in Fig.3. Seen from this figure, the forward breakdown voltage of IEC-GTO is close to that of the SA-GTO, but the blocking characteristic curve of IEC-GTO is little softer. This is because IEC-GTO structure has the silicon dioxide, and the movable charges in this oxide layer will affect the blocking characteristics of the IEC-GTO, and results in the softer blocking characteristic curve. This shows that the IEC-GTO has similar blocking characteristic under the thinner oxide layer to that of the conventional SA-GTO.

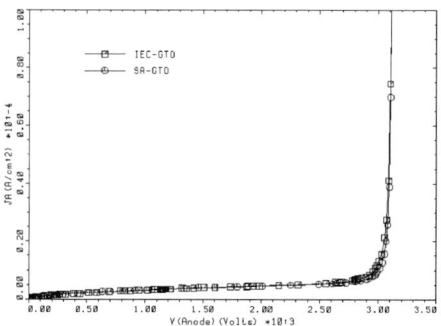

Figure3. The Comparison of the forward blocking characteristics of IEC-GTO and SAGTO

B. Conducting

The I-V characteristics during conducting of IEC-GTO and SA-GTO are simulated as shown in Fig.4. Seen from this figure, the forward voltage drops of the IEC-GTO is very close to that of the AS-GTO. When the current density is a certain value, the forward voltage drop of the IEC-GTO has a very tiny fall. This shows that the IEC-GTO has similar conducting characteristic to that of the conventional SA-GTO.

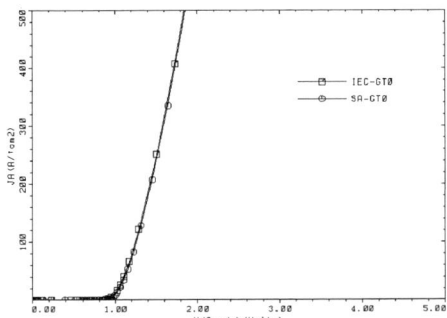

Figure 4. Comparison of the I-V characteristic of IEC-GTO and AS-GTO

C. Switching

The switching characteristic of the IEC-GTO devices is also analyzed. Firstly, the electron vector distribution of IEC-GTO during turn-off is simulated. A comparison of the electron vector distributions of IEC-GTO and SA-GTO during turn-off are shown in Fig.5. Seen from this figure, the plentiful non-equilibrium carriers stored in their n⁻ base region during conducting must disappear gradually when the device turns off at higher current density. For the SA-GTO, the electron can directly reach anode electrode by a low resistance path of n⁺ short region, as shown in fig. 5(a). But for IEC-GTO, due to a thin oxide layer located in anode, the plentiful electrons cannot be collected directly by the anode contact, and accumulated at near oxide in n⁺ floating and p⁺ anode regions. This enables the electrons interior n⁻ base region to be removed, and it is helpful to the device resuming quickly off state.

Secondly, the anode injection efficiency, γ, varied with anode current density, J_A, during turn on is also simulated. A comparison of the variations of γ with J_A of IEC-GTO and SAGTO is shown in fig.6, in which the electron injection efficiency, γ_e, and the hole injection efficiency, γ_h, can be expressed by J_n/J_A and J_p/J_A, respectively. Seen from this figure, the injection efficiency of the IEC-GTO has a variation with the J_A rising, (by symbol o); but the injection efficiency of the SAGTO (by symbol $\triangle$) has hardly any variation. In particular, the holes injection efficiency of the IEC-GTO at lower current is very high, and $\gamma_h \approx 1$; and then the γ_h decreases with the increase of the anode current, and $\gamma_h < 1$.

This is because the partial holes injected from the anode can combine with the electron in n⁺ floating region when a

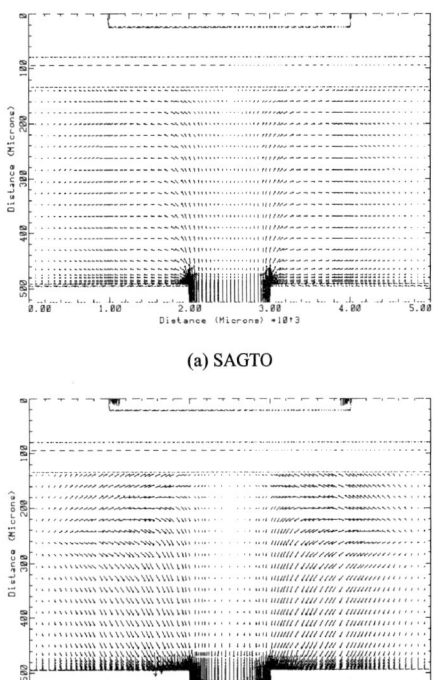

(a) SAGTO

(b) IEC-GTO

Figure 5. Comparison of the electron vector distribution during turn-off of three different anode GTO structures

positive voltage applies to across the anode and cathode of the IEC-GTO, it results the holes injection enhancement. And then, the voltage drop of the R_A increases with the rise of the J_A, the injection efficiency of the IEC-GTO is determined by pnp$_2$ transistor of which has an inherent low efficiency. But for the SAGTO, because of not existing oxide layer, the hole injection only increase quickly with the external positive voltage. And then the injection efficiency of the SAGTO is independent of the J_A but related to the anode effective area. So the hole injection efficiency of the SAGTO at lower J_A is increase quickly with the rise of J_A.

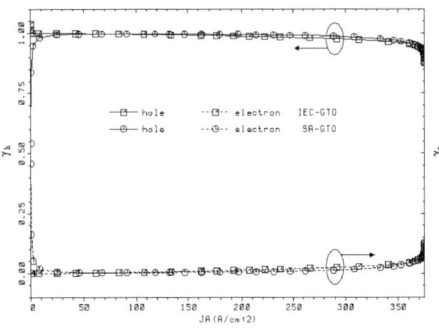

Figure 6. The comparison of the injection efficiency varied with anode current density for IEC-GTO and SAGTO

The comparison of the switching characteristics shows that the anode injection efficiency of IEC-GTO varies with J_A, that is, the γ_h is higher at lower J_A, and low at higher J_A. The higher γ_h can reduce the required gate trigger current while the device turns on; and the low γ_h is helpful to the device turns off at higher J_A. Although the short anode region of the SA-GTO can provide a path for electron during turn-off and this is helpful to the turn-off, however, the lower anode current gain will increase obviously the required gate trigger current. This shows that the IEC-GTO can improve further the switching characteristic and reduce the gate trigger current. This is similar to transparent anode of GCT [7], only the effect of electron injection at higher current for the IEC-GTO is weaker than that of the GCT. The results validate the correctness of the theoretical analysis for the IEC-GTO.

IV. OPTIMIZATION

The effective area of anode region determines anode the injection efficiency that affect on the conducting and switching characteristics. The larger anode contact area can enable on-state voltage drop to decrease and it is helpful to conducting characteristics, but at same time it can enable the IEC effect to weaken; the hole injection efficiency at low current to decrease, and it makes against the trigger characteristics. So the anode contact area of the IEC-GTO needs the optimum.

The fig.7 indicates the influence of the width of oxide layer, W_{ox}, of the IEC-GTO on conducting characteristics. Seen from the figure, the W_{ox} is smaller, i.e. the effective area of anode region is larger, the voltage drop is lower under the certain current density, the conducting characteristics is better, that is, the influence of the oxide layer is smaller.

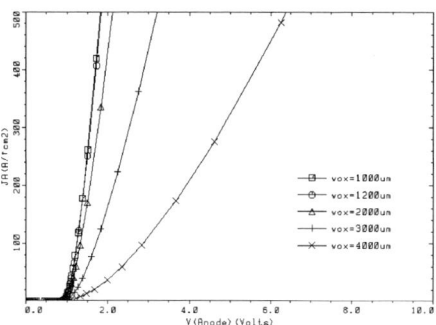

Figure 7. The influence of W_{ox} of the IEC-GTO on its conducting characteristics

The fig.8 indicates the influence of W_{ox} of the IEC-GTO on the anode injection efficiency. Seen from the figure, the injection efficiency has variation with J_A. When W_{ox} is more than the width of short region of SA-GTO, W_n, the $\gamma_h \approx 1$, and the γ_h is lower at higher current with the increase of the W_{ox}.

The above results show that the conducting characteristic and switching characteristic can obtain the trade-off when W_{ox} is about 1000~2000μm. This shows that the W_{ox} of the IEC-GTO is slight larger than the W_n of SA-GTO, but it can't too large.

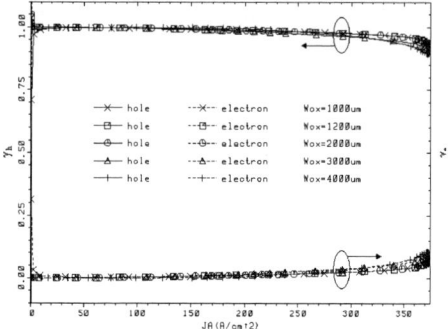

Figure 8. The influence of W_{ox} of the IEC-GTO on the anode injection efficiency

V. CONCLUSION

This paper presents a new IEC-GTO structure based on SA-GTO. The special structure consists of the narrow p^+ anode and the reduced anode contact area; its anode injection efficiency is high at low anode current density and low at high anode current density. So the IEC-GTO has low gate trigger current, better switching, simple manufacture process and similar conducting characteristic and blocking characteristic compared with the corresponding SA-GTO. This makes the IEC-GTO device very suitable for high-power applications.

REFERENCES

[1] H. E. Gruening, A. Zuckerberger, "Hard Drive of GTOs: Better Switching Capability through Improved Gate-Units", IEEE IAS, 1996, pp. 1474 – 1480

[2] Eric Carroll, Sven Klaka, Stefan Linder, "IGCTs: A New Approach to High Power Electronics" IEMDC, 1997

[3] Oscar Apeldoorn, Peter Steimer , Peter Streit, Eric Carroll, Andre Weber," High Voltage Dual-Gate Turn-off Thyristors" , The 36th IEEE Industry Applications Society Annual Meeting, IAS'2001, October 2001, Chicago, USA.

[4] S. Huang, G. A. J. Amaratunga, and F. Udrea. The Injection Efficiency Controlled IGBT. IEEE Electron Device Letters, Vol. 23, No. 2, 2002

[5] Kekura, M.; Akiyama, H.; Tani, M, .et, al. 8000-V 1000-A gate turn-off thyristor with low on-state voltage and low switching loss. IEEE Transaction on Power Electronic, 1990, pp.430-435

[6] Medici Two-Dimensional Device Simulation Program, Version4.0 User's Manual (AVANT), 2000

[7] Wang Cailin GaoYong Ma Li, et al, Analysis of Mechanism and characteristic for the Transparent Anode in Gate Commutated Thyristors, ACTA PHYS SIN-CH ED，2005, No.5. P2296-2301

2006 5th International Power Electronics and Motion Control Conference

Implementation and Analysis of 3-phase Voltage Sourced Regenerative Rectifier

Rui Chen，Qiongxuan Ge，Shijie Li

Institute of Electrical Engineering, Chinese Academy of Sciences, Beijing, 100080, P. R. China

Tel: 86-010-62637395-608 Email: raychen@mail.iee.ac.cn

Abstract—This paper describes an implementation of a 3-phase voltage sourced regenerative rectifier under direct current control strategy. General mathematical model of the converter is first established. Based on the model, the principle of the direct current control strategy is presented, simulation results are given, and circuit operations in steady state are analyzed. Based on SVPWM, the analysis explains the cause of dc voltage ripple, ac side current harmonics, and over modulation. Finally, parameters of a prototype are listed, and experimental results are presented.

Keywords—PWM rectifier; over modulation; fifth and seventh harmonics; unity power factor; direct current control;

I. INTRODUCTION

Nowadays, AC-DC converters are increasingly used in various industrial applications as an essential module. A large part of these converters are uncontrolled diode rectifiers or phase-controlled thyristor rectifiers. The advantages of these structures are high reliability, low complexity and good power-factor. However, such structures have inherent drawbacks that the power transfer can only occur from AC side to DC side. Furthermore, the non-sinusoidal input currents bring severe harmonics pollution to the power grid. By using the dual converters (12 thyristor) in an inverse parallel configuration, power regeneration is possible, but the input current distortion still exists.

The requirements on bidirectional power flow, fast response and low current harmonics have drawn much attention on force commutated pulse-width modulated (PWM) converters. Many papers have been published in this area, which focused on mathematical modeling, control strategies design, and performance optimizing. Paper [3] firstly establishes the mathematical model with switching function. Paper [1][4] present different ways in dealing with the nonlinearity of the model for controller design. Paper [5][6][7] propose several control strategies for voltage sourced converter. All these control schemes achieved the same steady-state goal of controlling power factor and fundamental current waveform, but their performances are not the same. Paper [1] discusses the methods to analysis the static and dynamic performances of the system.

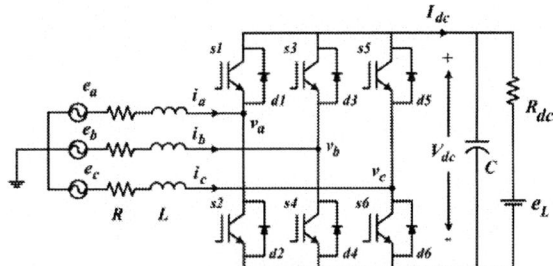

Figure 1.   Schematic diagram of a voltage source rectifier

In this paper, several aspects are discussed concerning the implementation of a voltage sourced converter under double closed-loop direct current control. Then the simulation and experimental results are reported, which confirm the validity of the analysis.

II. MATHMATICAL MODEL

The schematic diagram of a voltage sourced converter is shown in Figure 1.

The ac source voltages are e_a , e_b and e_c . The ac currents are i_a, i_b and i_c . The ac terminal voltages of the converter are v_a , v_b and v_c .The dc voltage and current are V_{dc} and I_{dc} . The ac-side impedance is modeled as an inductor L in series with a resistor R ,which is the equivalent sum of line resistor and switch resistor, thus is very small even could be neglected. The dc side capacitor is C_{dc} and the dc load is equivalent to a resistance R_{dc} in series with an electromotive force e_L .

The load model can represent two operation modes of the converter, rectifying mode $(e_L < V_{dc})$, and power regenerating mode $(e_L > V_{dc})$.

Analysis in this paper will base on the assumption that the utility is a three phase balanced, sinusoidal voltage source in the form: $e_a = E\cos(\omega t + \phi_0)$

By defining switching function s_a , s_b and s_c as follows:

$$\begin{cases} s_i = 1 & \text{when upper switch is on} \\ s_i = 0 & \text{when lower switch is on} \end{cases}, i = a, b, c$$

A general converter model could be established:

1-4244-0448-7/06/$25.00 ©2006 IEEE   1171

$$\begin{cases} L\dfrac{di_a}{dt} + Ri_a = e_a - V_{dc}(s_a - \dfrac{1}{3}\sum s_i) & (1) \\[2mm] L\dfrac{di_b}{dt} + Ri_b = e_b - V_{dc}(s_b - \dfrac{1}{3}\sum s_i) & (2) \\[2mm] L\dfrac{di_c}{dt} + Ri_c = e_c - V_{dc}(s_c - \dfrac{1}{3}\sum s_i) & (3) \\[2mm] C\dfrac{dV_{dc}}{dt} = i_a s_a + i_b s_b + i_c s_c - \dfrac{V_{dc}-e_L}{R_{dc}} & (4) \end{cases}$$

This model is very useful in computer simulation to get an accurate time response of the converter. However, because of its time variant and nonlinear nature, not only is it difficult to get analytical closed-form solutions, but it also hard to design the controller. To solve the time variant problem, transformation to a synchronized rotating frame of reference is effective. The transformation is defined as follows:

$$\begin{bmatrix} x_a \\ x_b \\ x_c \end{bmatrix} = \begin{bmatrix} cos(\theta) & -sin(\theta) \\ cos(\theta-2\pi/3) & -sin(\theta-2\pi/3) \\ cos(\theta-4\pi/3) & -sin(\theta-4\pi/3) \end{bmatrix} \begin{bmatrix} x_d \\ x_q \end{bmatrix} \quad (5)$$

The quantity x represents voltage, current or switching function. In (5) $\theta = \omega t + \phi_0$, thus the direct axis is aligned with the supply voltage phasor $\overline{E}$. The transformation results in a time invariant dq-axis model:

$$\begin{cases} L\dfrac{di_d}{dt} = e_d - (V_{dc}s_d + Ri_d - \omega Li_q) & (6) \\[2mm] L\dfrac{di_q}{dt} = e_q - (V_{dc}s_q + Ri_q + \omega Li_d) & (7) \\[2mm] C\dfrac{dV_{dc}}{dt} = \dfrac{3}{2}(s_d i_d + s_q i_q) - \dfrac{V_{dc}-e_L}{R_{dc}} & (8) \end{cases}$$

However, the nonlinearity still exists. To solve this problem, two methods are proposed in paper [1] and [2].

In paper [1], by using Fourier analysis, the switching function could be expressed as:

$$s_i = d_i + \sum_{n=1}^{\infty} (-1)^n \frac{2}{n\pi} sin(nd_i\pi)cos(n\omega_s t)$$

$$\sum_{i=a,b,c} s_i = \sum_i d_i + \sum_{n=1}^{\infty} \left[\sum_i (-1)^n \frac{2}{n\pi} sin(nd_i\pi) \right] cos(n\omega_s t) \quad (9)$$

where d_i is the average value (or duty ratio) of the switching function s_i within one switching period, and ω_s is the switching angular frequency. Then the model (1)~(4) could be divided into two models: a low-frequency model and a high frequency model, and the boundary separating them is the switching frequency. From the point of pulse width modulation, the duty ratio d_i could be expressed as:

$$d_i = \rho \cdot cos\left[\omega t + \phi_1 - (i-1)\frac{2\pi}{3} \right] + \frac{1}{2} \quad (10)$$

where ρ is the voltage utility ratio. By SPWM, $\rho = m/2$, and by SVPWM, $\rho = m/\sqrt{3}$, m is the modulation index.

So when the low frequency model is transformed in d-q coordinate, the nonlinear term $(s_d i_d + s_q i_q)$ in original model becomes

$$(d_d i_d + d_q i_q) = \rho \left[cos(\phi_1 - \phi_0)i_d + sin(\phi_1 - \phi_0)i_q \right] \quad (11)$$

Because in paper [1], the control strategy is Phase and Amplitude Control, ϕ_1 is a control signal. Thus, finally by using Small Signal Linearization, a linear low frequency model with analytical solutions around the steady state operating point is obtained, which also facilitate controller design.

Meanwhile, the high frequency model is very useful in analyzing the current harmonics and the DC voltage ripple.

By rewriting the state equation with different state variable, another method to acquire linearity is presented in paper [2][4].

In original model, the equation (4) is based on the Kirchhoff's current law. When considering the power flow relationship, the equation (4) could be modified.

Supposing $e_L = 0$, the power consumed on the DC side can be expressed as

$$P_{dc} = V_{dc}C\frac{d}{dt}V_{dc} + \frac{V_{dc}^2}{R_{dc}} \quad (12)$$

The power delivered from the AC can be written as

$$P_{ac} = \frac{3}{2}(e_d i_d + e_q i_q) \quad (13)$$

If the power loss in switching action and AC side resistance is neglected, then from (12) and (13)

$$V_{dc}C\frac{d}{dt}V_{dc} + \frac{V_{dc}^2}{R_{dc}} = \frac{3}{2}(e_d i_d + e_q i_q) \quad (14)$$

If V_{dc}^2 is taken as a state variable, instead of V_{dc}, (14) could be used as a dynamic equation instead of (8). Meanwhile, because in equation (6) and (7), the control inputs s_d, s_q are coupled with state variable V_{dc}, a nonlinear input transformation $v_d = s_d V_{dc}, v_q = s_q V_{dc}$ can be used to modify the original inputs s_d, s_q to new inputs v_d, v_q.

As a result, the new model could be written as

$$\begin{cases} L\dfrac{di_d}{dt} = e_d - Ri_d + \omega Li_q - v_d & (15) \\[2mm] L\dfrac{di_q}{dt} = e_q - Ri_q - \omega Li_d - v_q & (16) \\[2mm] C\dfrac{dV_{dc}^2}{dt} = 3(e_d i_d + e_q i_q) - \dfrac{2V_{dc}^2}{R_{dc}} & (17) \end{cases}$$

With this linear model, linear controller can be easily designed to improve system's response performance.

III. CONTROL STRATEGY AND SIMULATION

Based on the equations (1) ~ (4), a simulation model is constructed in Simulink/MATLAB. Then three control strategies are compared by simulation, which includes the Phase and Amplitude control, Current Hysteresis control, and the direct current control. The result shows

a) Current Hysteresis control can achieve best performance both in steady state and dynamic, however, the switching frequency is variable.

b) Phase and Amplitude control is depend on precise parameter of component, and its dynamic response is not satisfied.

c) Direct current control achieves both fast response and good steady state waveform.

Thus, direct current control is finally chosen to be implemented in prototype.

Figure 2 illustrates the control structure of the direct current control, in which double closed-loop PI controllers are used to regulate dc voltage and ac input currents. The d-axis current's reference signal is obtained from the output of dc voltage regulator. Feed forward decouple terms are added to the output of d-q current regulators to form the expecting modulation vectors v_d^*, v_q^*.

$$\begin{cases} v_d^* = -(k_p + k_i \int dt) \cdot (i_d^* - i_d) + \omega L i_q + E_d \\ v_q^* = -(k_p + k_i \int dt) \cdot (i_q^* - i_q) - \omega L i_d \end{cases} \quad (18)$$

Figure 3 is the steady state dc voltage, and phase voltage/current waveform, which shows low dc voltage ripple (± 0.3v at $V_{dc} = 600$), and both sinusoidal current and high displacement factor.

Figure 4 shows the dynamic response to a 750v step up

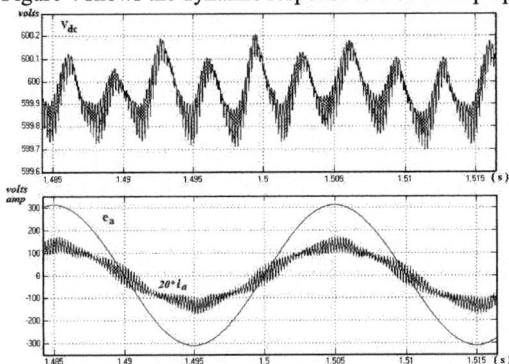

Figure 3.   DC voltage and Phase voltage/current Waveform

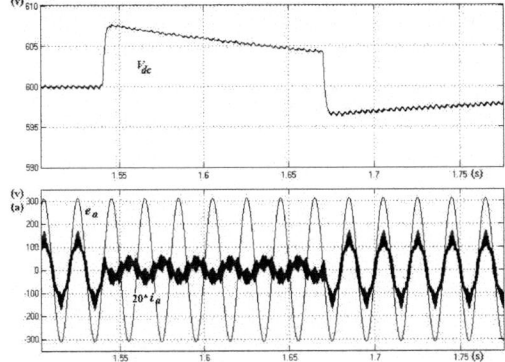

Figure 4.   Dynamic response between rectifying and regenerating

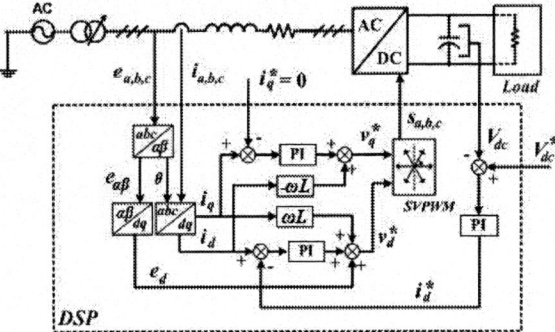

Figure 2.   Control Algorithm of Direct Current Control

dc voltage. The electromotive force e_L is set to be 0 at the start of simulation, and then step up to 750 volts at $t = 1.54s$ and step down to 0 at $t = 1.67s$. Switching between the two operation modes could be observed in Figure 4, and the dc voltage rise is less than 8 volts under this situation.

Simulation parameters are listed in Table 1.

TABLE I.   SIMULATION PARAMETER

Parameter	Symbol	Value	Units
AC phase voltage amplitude	E	311	V
AC side inductor	L	10	mH
AC side resistor	R	0	Ω
DC side capacitor	C	500	μF
Load resistor	R_{dc}	300	Ω
Electromotive force	e_L	0,750	V
DC voltage reference	V_{dc}^*	600	V
q-axis current reference	i_q^*	0	A
Sampling frequency	f_s	5	kHz
Switching frequency	f_{sw}	5	kHz
Triangular wave amplitude		1.0	
Modulation mode		SPWM	
Voltage regulator parameter			
Proportional factor	Kp_v	0.4	
Integral factor	Ki_v	2.0	
Integral Limit		7	
Output Limit		8	A
Current regulator parameter			
Proportional factor	Kp_idq	0.2	
Integral factor	Ki_idq	1.0	
d-axis Integral Limit		5	
d axis PI Output Limit		6	
q-axis Integral Limit		0.8	
q axis PI Output Limit		1	

IV.   CIRCUIT OPERATION AND PARAMETER ANALYSIS

To analyze the relationship between parameters and performance of the system, circuit operation in unity power factor rectifying mode with SVPWM is discussed in this part.

In Figure 5, the steady state operation is shown by a phasor diagram in complex plain.

In the diagram, $\vec{V_s^*}$ presents the fundamental vector of the three phase PWM signals, and $\vec{E}$ presents the voltage vector of the three phase power supply. Relationship between the two can be expressed as:

$$\vec{E} = \vec{V_s^*} + j\omega L \vec{I} \quad (19)$$

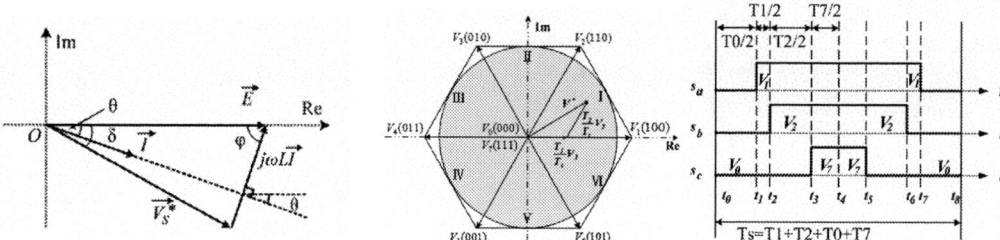

Figure 5. Steady state operation phasor diagram

Figure 6. General concepts of SVPWM

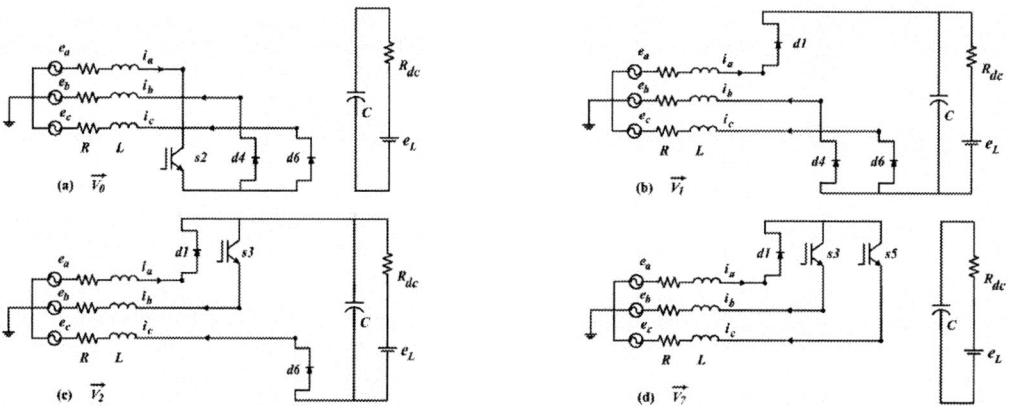

Figure 8. Circuit operation while the phase angle of $\vec{E}$ is 1.48°~30°

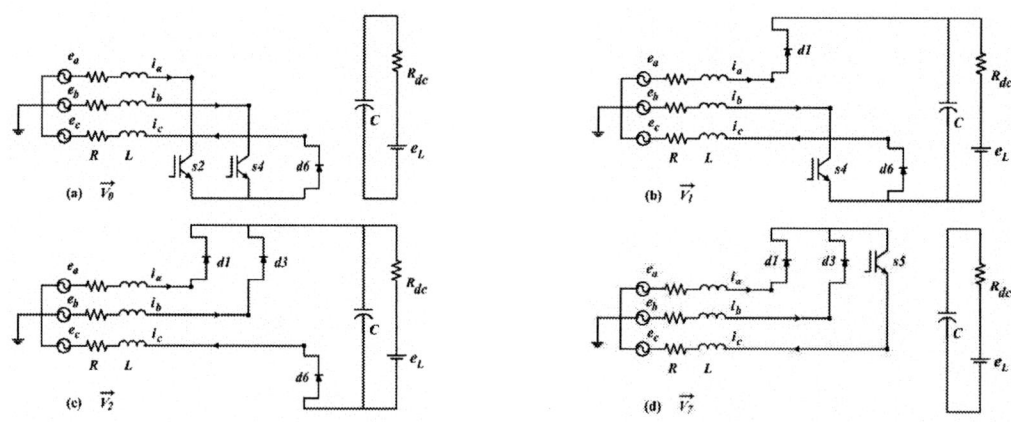

Figure 7. Circuit operation while the phase angle of $\vec{E}$ is 30°~61.48°

If $\vec{E}$ leads $\vec{V_s^*}$, the converter works in rectifying mode, else in regenerating mode. In simulation, $\vec{E}$ leads $\vec{V_s^*}$ $\delta = 1.48°$ at steady state.

While modulation mode is Space-Vector PWM, the reference vector $\vec{V_s^*}$ is synthesized by two adjacent space vectors in every switching cycle. Because vector duty time calculations in six sectors are identical, here only the circuit operations in sector 1 are analyzed. Considering in sector 1 the sequence of applied space-vector is V0-V1-V2-V7-V2-V1-V0, which is shown in Figure 6, V0-V1-V2-V7 operation is analyzed in Figure 7 and Figure 8

Suppose $e_a = Ecos(\omega t + \phi_0)$, and $\vec{E}$ leads $\vec{V_s^*}$ at $\delta = 1.48°$, thus when $(\omega t + \phi_0)$ is 1.48°~61.48°, $\vec{V_s^*}$ is

in sector 1.

First considering $(\omega t + \phi_0)$ is 1.48°~30° in Figure 7, $\vec{V_s^*}$ is in the first half part of sector 1. During this period, $i_a > 0, i_b < 0, i_c < 0$, $T1 \geq T2$.

Because capacitor is discharged through the resistor load in full cycle, the V_{dc} decrease caused by discharge through resistor load is omitted in following discussion.

Note: ↑ means increase, ↑↑ means increase with high speed, ↑↓ means in the whole range (1.48°~30°) the change at first is increase, but at a time point it becomes decrease, (×) means the change is in the counter direction of the unity power factor rectifying requirement, and V_{dc} ↑ (i_a) means the capacitor is charged by current i_a.

1174

Step 1: (a) $t_0 \sim t_1$, $i_a \uparrow\uparrow (\times), i_b \downarrow (\times), i_c \downarrow$

Step 2: (b) $t_1 \sim t_2$, $i_a \downarrow, i_b \uparrow, i_c \uparrow (\times) \downarrow, V_{dc} \uparrow\uparrow (i_a)$

Step 3: (c) $t_2 \sim t_3$, $i_a \uparrow (\times), i_b \downarrow (\times), i_c \uparrow (\times), V_{dc} \uparrow (-i_c)$

Step 4: (d) $t_3 \sim t_4$, $i_a \uparrow\uparrow (\times), i_b \downarrow (\times), i_c \downarrow$

Then considering $(\omega t + \phi_0)$ is 30°~61.48° in Figure 8, $\vec{V_s^\cdot}$ is in the second half part of sector 1. During this period, $i_a > 0, i_b > 0, i_c < 0$, $T1 \le T2$ (after $(\omega t + \phi_0) > 31.48°$).

Step 1: (a) $t_0 \sim t_1$, $i_a \uparrow (\times), i_b \uparrow, i_c \downarrow\downarrow$

Step 2: (b) $t_1 \sim t_2$, $i_a \downarrow, i_b \uparrow\uparrow, i_c \downarrow, V_{dc} \uparrow (i_a)$

Step 3: (c) $t_2 \sim t_3$, $i_a \uparrow (\times) \downarrow, i_b \downarrow (\times), i_c \uparrow (\times), V_{dc} \uparrow\uparrow (-i_c)$

Step 4: (d) $t_3 \sim t_4$, $i_a \uparrow (\times), i_b \uparrow, i_c \downarrow\downarrow$

These results explain the reason of periodic dc voltage ripple, and current harmonics in simulation waveform.

Note: in following discussion, the equations referred as increase or decrease rate are approximate value.

a) DC voltage ripple

First, considering the *low frequency periodic* dc voltage ripple, while $(\omega t + \phi_0)$ is 1.48°~30°, at first, because the duty ratio of V1 is high and the charge current (i_a) is around its maximum, dc voltage increases, and the maximum increase rate is achieved when V1 is applied:

$$\frac{dV_{dc}}{dt} = \frac{1}{C}(I - \frac{V_{dc}}{R_{dc}}) \qquad (20)$$

where I is the amplitude of ac side phase current.

Then the average increase rate during several switching cycles decreases because both the duty ratio of V1 and (i_a) decrease. V_{dc} achieves maximum around at $(\omega t + \phi_0)$ is 31.48°. Then it decrease, because the average charge rate becomes smaller than the discharge rate. The discharge rate is always the same:

$$\frac{dV_{dc}}{dt} = -\frac{1}{C}\frac{V_{dc}}{R_{dc}} \qquad (21)$$

While $(\omega t + \phi_0)$ is 30°~61.48°, as both the duty ratio of V2 and the charge current $(-i_c)$ increases, the speed of dc voltage decrease slows down. V_{dc} achieves the minimum around at $(\omega t + \phi_0)$ is 61.48°.

Thus the cycle of low frequency DC voltage ripple is around 60°, which results in an approximate six-order harmonics in dc voltage. Because the three phase input voltage of the converter are acquired by modulation of dc voltage, thus three-phase input voltages can be written as:

$$v_i = \rho \cdot cos\left[\omega t + \phi_1 - (i-1)\frac{2\pi}{3}\right] \cdot \left[V_{dc} + \widetilde{V_{dc}}cos(6\omega t + \phi_3)\right]$$

$$= v_i^* + \frac{\rho}{2} \cdot \widetilde{V_{dc}}cos(7\omega t + \phi_4) + \frac{\rho}{2} \cdot \widetilde{V_{dc}}cos(5\omega t + \phi_5) \qquad (22)$$

This equation shows that the fifth and seventh order current harmonics in ac side are generated by the sixth order dc voltage ripple.

Because $T1 \le Ts$ in each switching cycle, the maximum amplitude of the ripple can be estimated as:

$$\widetilde{V}_{max} = \frac{Ts}{C}(I - \frac{V_{dc}}{R}) \qquad (23)$$

b) AC side current harmonics

The characters $(\times)$ in the analysis results of circuit operation explain the cause of ac side *high frequency* current harmonics.

While $(\omega t + \phi_0)$ is in 1.48°~61.48°, to achieve unity power factor rectifying, i_a should decrease, i_b should increase, and i_c should decrease. However, the requirements of a boosted dc voltage conflict with these current changes, thus $(\times)$ appears.

Take i_a for example, in every switching cycle during this period, i_a decrease only in steps when V1 is applied, and its maximum decrease rate is:

$$\frac{di_a}{dt} = \frac{1}{L}(E - IR - \frac{2}{3}V_{dc}) \qquad (24)$$

where E is the phase amplitude of ac voltage source.

While in the rest steps, i_a keeps increasing, and the maximum increase rate (marked as $\uparrow\uparrow$) is achieved in steps when V0, V7 are applied in 1.48°~30° range:

$$\frac{di_a}{dt} = \frac{1}{L}(E - IR) \qquad (25)$$

The analysis explains the reason why the amplitude of *high frequency* current harmonics is very high when $(\omega t + \phi_0)$ is in 1.48°~30°.

Considering phase-b, there is another $\uparrow\uparrow$ which occurs in i_b when V1 is applied in the 30°~61.48° range, the maximum increase rate is:

$$\frac{di_b}{dt} = \frac{1}{L}(\frac{1}{2}E - \frac{1}{2}IR + \frac{1}{3}V_{dc}) \qquad (26)$$

While in other steps, the maximum decrease rate of i_b which occurs at steps when V2 is applied in the 1.48°~30° range could be estimated as:

$$\frac{di_b}{dt} = \frac{1}{L}(-\frac{1}{2}E + \frac{1}{2}IR - \frac{1}{3}V_{dc}) \qquad (27)$$

For i_c, when V0,V7 are applied in the 30°~61.48° range, the maximum decrease rate is achieved:

$$\frac{di_c}{dt} = \frac{1}{L}(-E + IR) \qquad (28)$$

While in other steps, the maximum increase rate of i_c appears at steps when V2 is applied in the 1.48°~30° range, which could be estimated as:

$$\frac{di_c}{dt} = \frac{1}{L}(-\frac{1}{2}E + \frac{1}{2}IR + \frac{2}{3}V_{dc}) \qquad (29)$$

To sum up, high frequency current harmonics are most serious at the time when one phase voltage achieves peak value or crosses zero point, and the maximum di/dt rates are estimated in equation (24)~(29), which would be helpful in inductor design.

c) Over modulation

Considering the steady state in simulation, the equation (19) could be written as:

$$\begin{cases} \rho V_{dc}sin(\delta) = \omega LIcos(\theta) \\ \rho V_{dc}cos(\delta) = E - \omega LIsin(\theta) \\ \dfrac{V_{dc}^2}{R_{dc}} = \dfrac{3}{2}EIcos(\theta) \end{cases} \text{ with } \begin{cases} V_{dc} = 600v \\ \theta = 0° \\ E = 311v \end{cases}$$

The solution is $\rho = 0.519, \delta = 1.484°, I = 2.57A$

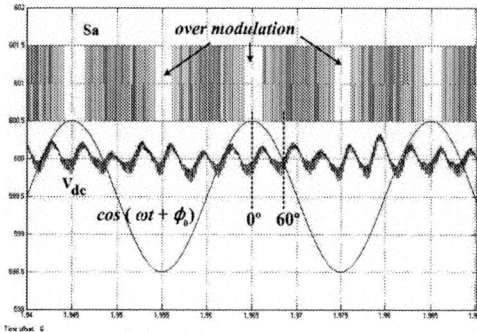

Figure 9.   Over modulation $V_{dc} = 600v$ (**0. 5V/div**)

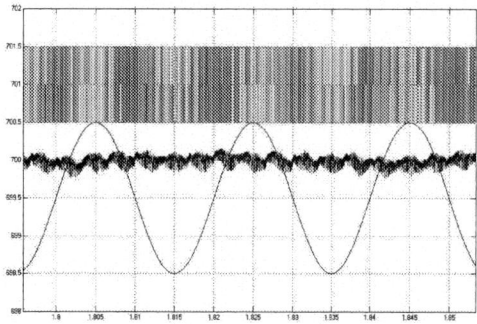

Figure 10. Normal modulation $V_{dc} = 700v$ (**0. 5V/div**)

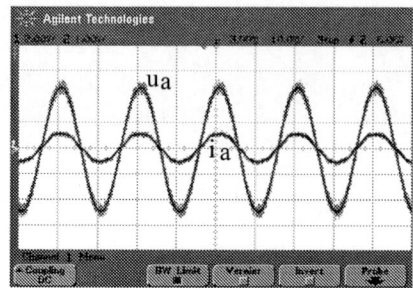

Figure 11. Phase voltage and current

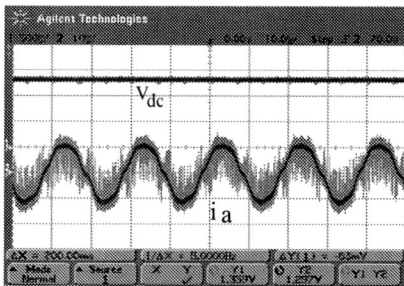

Figure 12. DC voltage and phase current

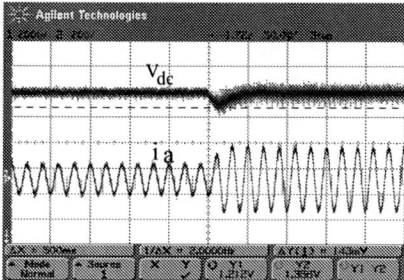

Figure 13. Load increase 100%

As the maximum of voltage utility ratio in SPWM is 0.5, while $\rho = 0.52 > 0.5$, thus with the parameter used in simulation, over modulation exists in steady state. Its effects could be illustrated by Figure 9, in which the switching command signal s_a and a function $cos(\omega t + \phi_0)$ have been put together with dc voltage waveform.

Considering when $(\omega t + \phi_0)$ is in 0°~60°, due to over modulation, the upper switch of phase A is forced to keep open in several full switching cycles around at 0°. Thus in dc side, the capacitor is continuously charged by i_a or $(-i_c)$, which results V_{dc} increases to a larger quantity compared with normal operation, then the amplitude of sixth order dc voltage ripple became larger. Finally, the fifth and seventh order current harmonics increase, and thus the THD value increase.

If set $V_{dc}^* = 700v$, with the same parameter, system's steady state solution is $\rho = 0.446, \delta = 2.025°, I = 3.5A$. Thus over modulation is avoided. The waveform is shown in Figure 10, the dc voltage ripple is much lower than $V_{dc}^* = 600v$, which proves the validity of the analysis.

V.   EXPERIMENT RESULTS

A prototype has been built to evaluate the performance of the unity power factor rectifying, and the control algorithm is implemented on a TI F2407 DSP. The parameters of the prototype are listed in Table 2, and experimental waveforms are reported in Figure (10) ~ (13).

System's steady state can be expressed as

$$E = 311v, \theta = 0°, V_{dc} = 650v, R_{dc} = 100\Omega$$
$$\rho = 0.48, \delta = 4.184°, I = 9.06A$$

TABLE II.   PROTOTYPPE PARAMETER FOR $V_{DC} = 650v$

Parameter	Symbol	Value	Units
AC phase voltage amplitude	E	311	V
AC side inductor	L	8	mH
AC side resistor	R	0	Ω
DC side capacitor	C	1700	μF
Load resistor	R_{dc}	50~100	Ω
Electromotive force	e_L	0	V
DC voltage reference	V_{dc}^*	650	V
q-axis current reference	i_q^*	0	A
Sampling frequency	f_s	10	kHz
Switching frequency	f_{sw}	10	kHz
Modulation mode	SVPWM		
SEMICRON IGBT	Three 150A/1700V/2U		
Voltage regulator parameter			
Proportional factor	Kp_v	0.75	
Integral factor	Ki_v	0.02	
Integral Limit	25A		3.125 p.u.
Output Limit	25A		3.125 p.u.
Current regulator parameter			
Proportional factor	Kp_idq	0.625	
Integral factor	Ki_idq	0.01	
d-axis Integral Limit	$650/\sqrt{3}\ v$		4.691 p.u.
d axis PI Output Limit	$650/\sqrt{3}\ v$		4.691 p.u.
q-axis Integral Limit	$650/\sqrt{3}\ v$		4.691 p.u.
q axis PI Output Limit	$650/\sqrt{3}\ v$		4.691 p.u.

Figure 11 shows phase voltage e_a and current i_a waveform which is obtained from a DA converter output channel.

The waveform of V_{dc} and i_a in Figure 12 is measured by high voltage differential probe and current clamp. The measure ratio of the voltage probe is $1/500$, and the measured width of voltage ripple is $\Delta y = 63mv$, thus the p-p amplitude of V_{dc} ripple is $63mv*500 = 31.5v$. The current clamp measure range is set to $1mv/1A$, and the measured amplitude of i_a is near 10mv, which is accordant with the predicted value $I = 9.06A$. The THD value measured simultaneously is 2%.

The waveform in Figure 13 is the result of sudden load increase experiment. When the load resistor R_{dc} was suddenly reduced from 100Ω to 50Ω, V_{dc} decreases nearly $143mv*500 = 71.5v$, and the regulation time is about $60ms$.

VI. CONCLUSIONS

In this paper, analysis for circuit operations in steady state explains several problems met in the course of implementation. By simulation and experiment, the validity of the analysis has been proved. The relationship between parameters and system performance is revealed through a set of estimate equation, which is helpful in system design.

REFERENCES

[1] R. Wu and S. B. Dewan, "Analysis of an AC-to-DC voltage source converter using PWM with phase and amplitude control," IEEE TRANSACTIONS ON INDUSTRY APPLICATIONS, VOL. 27, NO. 2, MARCH/APRIL 1991.

[2] Y. Ye and M. Kazerani, "Modeling, Control and Implementation of Three-Phase PWM Converters," IEEE TRANSACTIONS ON POWER ELECTRONICS, VOL. 18, NO. 3, MAY 2003.

[3] A. W. Green, J. T. Boys, and J. F. Gates, "3-phase voltage sourced reversible rectifier," IEE PROCEEDINGS, Vol. 135, Pt. B, No. 6, NOVEMBER 1988.

[4] D. C. Lee, G. M. Lee, and K. I. Lee, "DC-Bus Voltage Control of Three-Phase AC/DC PWM Converters Using Feedback Linearization," IEEE TRANSACTIONS ON INDUSTRY APPLICATIONS, VOL.36, NO.3, MAY/JUNE 2000.

[5] M. P. Kazmierkowski, and L. Malesani, "Current Control Techniques for Three-Phase Voltage-Source PWM Converters: A Survey," IEEE TRANSACTIONS ON INDUSTRY ELECTRONICS, VOL.45, NO.5, OCTORBER 1998.

[6] R. Wu, S. B. Dewan, and G. R. Slemon, "A PWM AC-to-DC Converter with Fixed Switching Frequency," IEEE TRANSACTIONS ON INDUSTRY APPLICATIONS, VOL.26, NO.5, SEPTEMBER/OCTORBER 1990.

[7] J. W. Dixon, and B. T. Ooi, "Indirect Current Control of a Unity Power Factor Sinusoidal Current Boost Type Three-Phase Rectifier," IEEE TRANSACTIONS ON INDUSTRY ELECTRONICS, VOL.35, NO.4, NOVEMBER 1988.

[8] V. Blasko, "Analysis of a Hybrid PWM based on Modified Space-Vector and Triangle-Comparison Methods," IEEE TRANSACTIONS ON INDUSTRY APPLICATIONS, VOL.33, NO.3, MAY/JUNE 1997.

2006 5th International Power Electronics and Motion Control Conference

Design and Implementation of Electronic Ballast for Fluorescent Lamps with Low Lighting Flicker

Yang-Sheng Lin, Chun-An Cheng, Jiann-Fuh Chen, Tsorng-Juu Liang and Wei-Shih Liu

Power Electronics Laboratory

Department of Electrical Engineering

National Cheng-Kung University, Taiwan, China.

e-mail: junan813@ms64.hinet.net

Abstract—This paper proposes an electronic ballast circuit for fluorescent lamps with low lighting flicker. The electronic ballast with constant-frequency control, in which the DC-bus voltage is fed by a power-factor corrector, has a low-frequency ripple, which can cause lighting flickers in fluorescent lamps. Precise image-detecting systems require lighting sources (such as fluorescent lamps) with low levels of lighting flicker. The presented ballast operates by detecting the DC-bus voltage ripple as a feed-forward signal in order to modulate the switching frequency of the inverter. Thus, the voltage gain of the resonant tank is adjustable. As a result, the lighting flicker caused by the DC-bus voltage ripple can be decreased. Finally, the experimental and simulation results are in agreement, so the functionality of the proposed circuit is verified.

Keywords- Electronic Ballast, Fluorescent lamps.

I. INTRODUCTION

The DC bus fed by a power-factor corrector has a low frequency ripple, which is twice the level of the utility-line frequency [1]-[4]. If the utility-line frequency is 60 Hz, it can be observed that the DC bus has a ripple of 120 Hz. In the electronic ballast for fluorescent lamps under constant-frequency control, as shown in Fig. 1, the low-frequency ripple causes the lamp voltage to have a 120Hz variation in amplitude, as shown in Fig. 2; therefore, there is a 120Hz variation in intensity of the illuminant. For precise image-detecting systems, the light reflected from the object can be detected in order to adjust the overall quality. Therefore, precise image detecting requires a light source with low lighting flicker [5]-[9].

The proposed ballast uses the DC-bus ripple as a feed-forward signal for modulating the frequency of inverters. The designed voltage gain of the resonant tank adjusts the switching frequency of the inverter in order to obtain a suitable gain for the lamp without any low-frequency variation in voltage and power; therefore, the illuminant becomes stable.

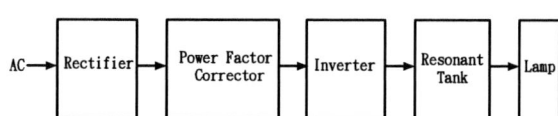

Fig. 1. Electronic ballast with constant-frequency control.

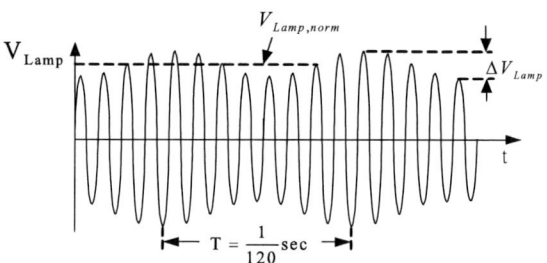

Fig. 2. Lamp voltage with 120Hz variation in amplitude.

II. OPERATIONAL PRINCIPAL OF PROPOSED BALLAST

Since the series capacitor of the resonant tank can resist the DC part of the input power, the input voltage of the resonant tank is equivalent to the waveform shown in Fig 3. This waveform is composed of a high-frequency square with a low-frequency variation; the high frequency part is created by the inverter and the DC-bus ripple effect of the low-frequency variation. The Fourier series of V_{HB}, where C_n is the coefficient value, is shown in Fig. 4.

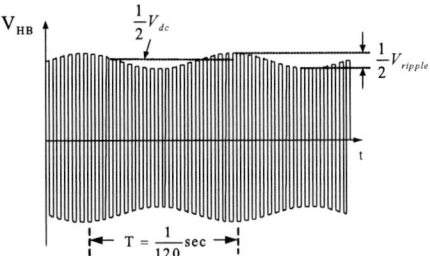

Fig. 3. The input voltage of the resonant tank.

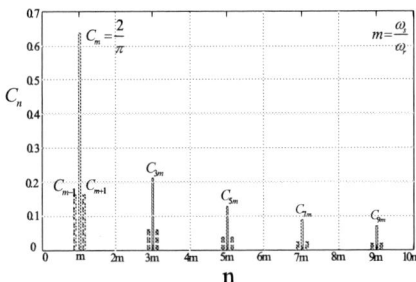

Fig. 4. The Fourier series of V_{HB}.

To allow for stable levels of lamp voltage and power, this paper proposes a new control scheme: The DC-bus ripple is detected in order to change the frequency of the inverter so that the gain of the resonant tank can be adjusted. The block diagrams of this paper are shown in Fig. 5.

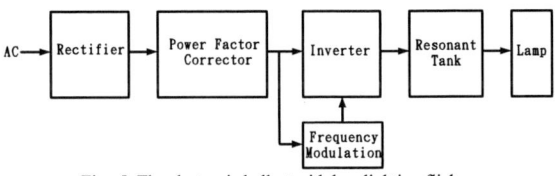

Fig. 5. The electronic ballast with low lighting flicker.

Ignoring the high-frequency part of the lamp voltage, and making the lamp voltage constant, the idea gain is shown as follows:

$$Gain(V_{dc} + \Delta V, f) = \frac{\pi}{2} \cdot \frac{V_{Lamp,norm}}{V_{dc} + \Delta V(t)}. \quad (1)$$

If the relationship between gain and ΔV is shown as (1), the lamp voltage and power will be stable. However Equation (1) is not a linear equation -- that is difficult to realize. Equation (1) can also be expressed in the following:

$$Gain(V_{dc} + \Delta V, f) = \frac{\pi}{2} \cdot V_{Lamp,norm} \cdot \frac{\frac{1}{V_{dc}^2}[V_{dc} - \Delta V(t)]}{1 + \left[\frac{\Delta V(t)}{V_{dc}}\right]^2}. \quad (2)$$

In the practical situation, the value of V_{dc} is several hundreds, and ΔV_{max} is much less than V_{dc}. Thus $\left(\frac{\Delta V_{max}}{V_{dc}}\right)^2 \ll 1$; therefore, $1 + \left(\frac{\Delta V_{max}}{V_{dc}}\right)^2 \approx 1$. Equation (2) is equivalent to the following equation:

$$Gain(V_{dc} + \Delta V, f) = \frac{\pi}{2} \cdot V_{Lamp,norm} \cdot \frac{V_{dc} - \Delta V(t)}{V_{dc}^2}. \quad (3)$$

The relationship between gain and ΔV is inverse in (3), as shown in Fig. 6.

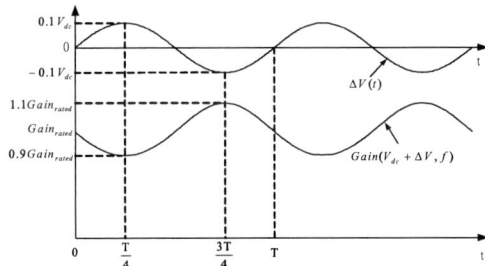

Fig. 6. The relationship between gain and DC-bus ripple.

III. DESIGN PROCEDURE

The relationship between the gain of the resonant tank and the frequency is a non-linear curve, but exists in a narrow region, such as a few kHz. This region can be seen as approximately linear, as shown in Fig. 7.

Thus, the relationship between gain and frequency is obtained by the following:

$$Gain(f) = a \cdot f + b, \quad (4)$$

where a and b are the coefficients of the linear equation.

1179

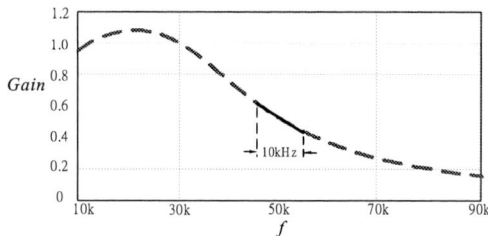

Fig. 7. The relationship between gain and frequency.

From Equations (3) and (4), the frequency of the inverter is given by:

$$f(V_{dc} + \Delta V) = \frac{1}{a}\left[\frac{\pi}{2} \cdot V_{Lamp,norm} \cdot \frac{V_{dc} - \Delta V(t)}{V_{dc}^2} - b\right]. \quad (5)$$

Assuming $f_{op,center} = \frac{1}{a}\left[\frac{\pi}{2} \cdot V_{Lamp,norm} \cdot \frac{1}{V_{dc}} - b\right]$ and

substituting this into (5) yields:

$$f(V_{dc} + \Delta V) = f_{op,center} - \left(f_{op,center} + \frac{b}{a}\right) \cdot \frac{\Delta V(t)}{V_{dc}}, \quad (6)$$

where $f_{op,cenetr}$ is the center frequency of the inverter.

Equation (6) shows the relationship between the frequency of the inverter and ΔV, which can be illustrated in Fig. 8.

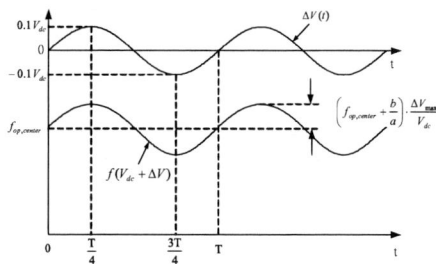

Fig. 8. The relationship between frequency and DC-bus ripple.

IV. DESIGN EXAMPLE AND EXPERIMENTAL RESULTS

An electronic ballast prototype has been built to meet the following specifications.

1) DC bus voltage: $V_{dc} = 310$ V

2) Steady-state fluorescent lamp voltage: $V_{lamp,rms} = 85$ V

3) Central frequency of the inverter: $f_{center} = 50$ kHz

4) Steady-state fluorescent lamp power: $P_{lamp} = 36$ W

5) Steady-state fluorescent lamp resistance: $R_{lamp} = 200\ \Omega$

6) DC-link capacitor: 6.8 μF

The waveform of lamp voltage/current at steady state operation is shown in Fig. 9. It can be obtained that lamp voltage and current are in phase. Relationships between the DC-bus voltage and the designed operating frequency are shown in Fig. 10 (a) and 10 (b). Referring to Fig. 10, when the DC-bus voltage is at its maximum value, that is, approximately 340 V, the switching frequency is designed to be 53 kHz. On the contrary, the switching frequency is set at 47 kHz when the DC-bus voltage is at its minimum value, that is, 280V.

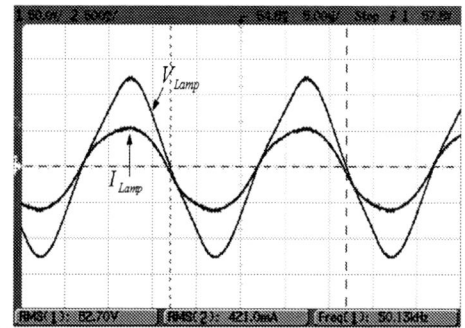

Fig. 9. The lamp voltage and current waveform. (50V/div, 0.5A/div, 5μs/div)

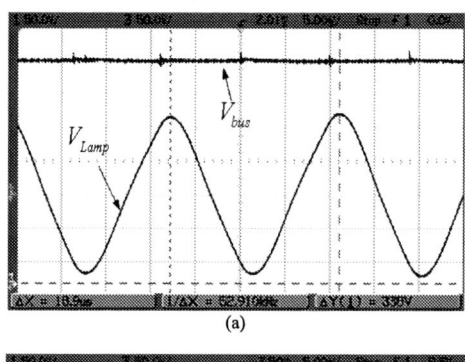

(a)

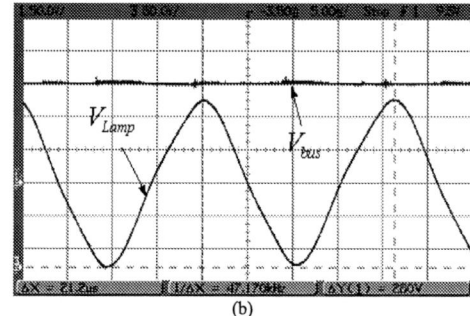

(b)

Fig .10. DC-bus voltage vs. operating frequency: (a) at maximum DC-bus voltage and (b) at minimum DC-bus voltage. (50V/div, 50V/div, 5μs/div)

The experimental waveforms of lamp voltage and current with constant / frequency modulation control are shown in Fig. 11. The crest factor (CF) of the lamp current with constant frequency is 1.61, which is greater than the 1.51 obtained for the one with frequency modulation. The measured results prove that the frequency-modulation control scheme, as proposed in this paper, reduces the CF, thus increasing the life of the lamp. The lamp voltage and DC-bus voltage with constant / frequency modulation are also shown in Fig. 12. The experimental results prove that frequency modulation is able to create a stable illuminant without influence from the DC-bus ripple.

The lamp power and DC-bus voltage with constant / frequency modulation are shown in Fig. 13. The experimental results prove that the frequency modulation controller creates a stable illuminant without influence from the DC-bus ripple.

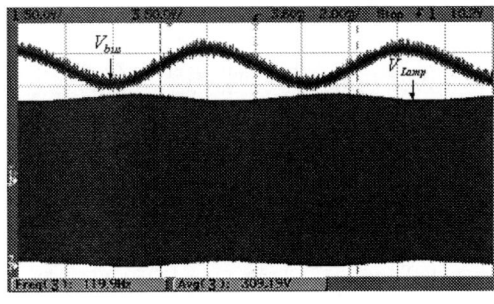

(a)

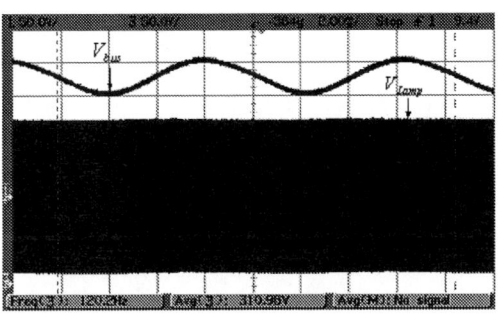

(b)

Fig .12. The lamp voltage and DC bus voltage (a) with constant / frequency, and (b) modulation control. (200mA/div, 2ms/div)

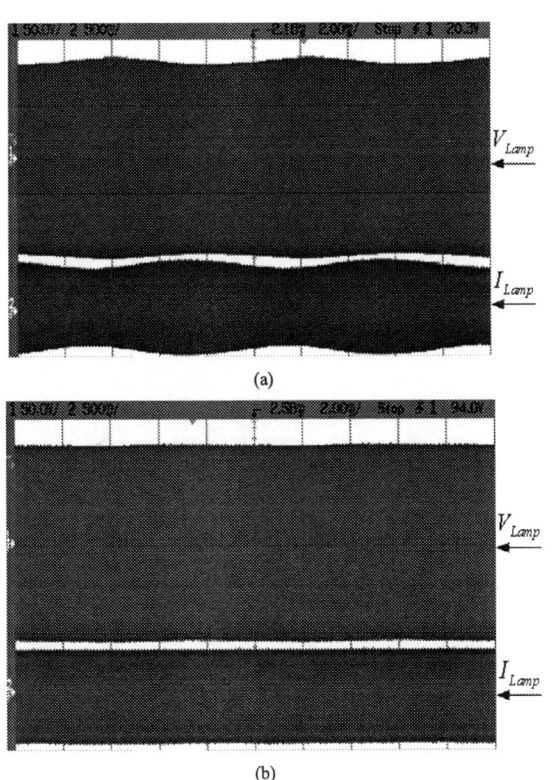

(a)

(b)

Fig .11. The lamp voltage and current (a) with constant-frequency control, and (b) frequency modulation control. (200mA/div, 2ms/div)

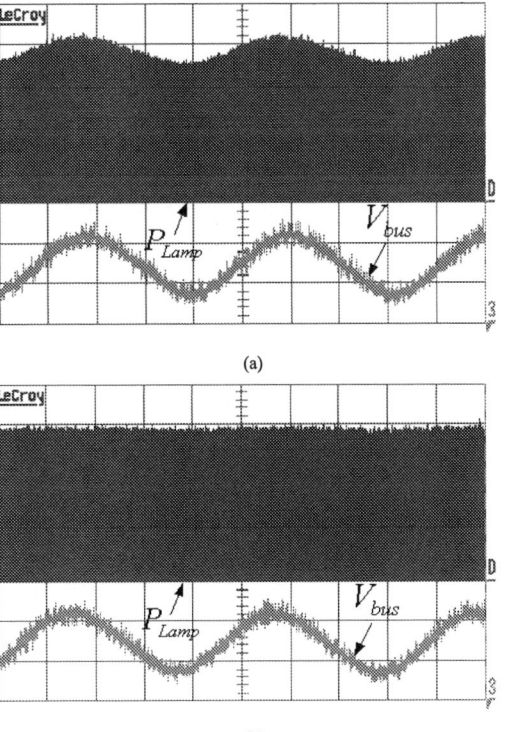

(a)

(b)

Fig .13. The lamp power and DC-bus voltage with constant-frequency, and (b) frequency modulation control. (200mA/div, 2ms/div)

V. CONCLUSIONS

Due to the utility-line frequency of 60 Hz, the full-bridge rectifier, the lamp voltage and power would have a 120Hz ripple whether or not the electronic ballast uses a power-factor corrector. This paper eliminates the low-frequency vibration of the lamp voltage and power by detecting the DC-bus voltage in order to modulate the switching frequency, and then changes the voltage gain of the resonant inverter to remove the low-frequency lighting flicker. The experimental results prove that frequency modulation control is better than constant-frequency control for decreasing the low-frequency vibration of lamp voltage and power.

REFERENCES

[1] F. D. Kieferndorf, M. Forstar and T. A. Lipo, "Reduction of DC-bus capacitor ripple current with PAM/PWM converter," *IEEE Trans. Ind. Applicat.*, vol. 40, March/April 2004, pp. 607–614.

[2] D. K. Jackson and S. B. Leeb, "Feedforward ripple cancellation for a full-bridge converter," *Proceedings of IEEE APEC'00,* 2000, pp..347–352.

[3] Y. Fukuta and G. Venkataramanan, "DC bus ripple minimization in cascaded h-bridge multilevel converters under staircase modulation," *Proceedings of IEEE IAS Annu. Meeting,* 2002, pp. 1988–1993.

[4] Hyun-Lark Do and Bong-Hwan Kwon, "Single-stage line-couple half-bridge ballast with unity power factor and ripple-free input current using a couple inductor," *IEEE Trans. Power Electron.*, vol.50, Dec. 2003, pp. 1259-1266.

[5] S. S. T. Lee, H. S. H. Chung and S. Y. R. Hul, "A comparative study of random switching for eliminating visible striations in fluorescent lamps," *Proceedings of IEEE PESC* 2003, pp.1006–1011.

[6] Y. Yin, R. Zane, R. Erickson and J. Glaser, "Dynamic analysis of frequency-controlled electronic ballasts," *Proceedings of IEEE-IAS Annu. Meeting,* 2002, pp. 685–691.

[7] L. Laskai, P. N. Enjeti and I. J. Pitel, "White-noise modulation of high-frequency high-intensity discharge lamp ballasts," *IEEE Trans. Ind. Applicat.*, vol 34. March/April 2004, pp. 607 –614.

[8] Y. Yin, R. Zane, J Glaser and R. W. Erickson "Small-signal analysis of frequency-controlled electronic ballasts," *IEEE Trans. Circuit and Systems I: Fundamental Theory and Applications*, vol 50, Aug. 2003, pp.1103 –1110.

[9] Chang-hua Lin and Kai-Jun. Pai, "Difference-integral dimming controller for the single-stage back-lighting electronic ballast," *Proceedings of IEEE PESC*2003, pp. 1000–1005.

2006 5th International Power Electronics and Motion Control Conference

A Floating-point Coprocessor Configured by a FPGA in a Digital Platform Based on Fixed-point DSP for Power Electronics

Haibing HU, Tianjun Jin, Xianmiao Zhang, Zhengyu LU, Zhaoming Qian

National Key Lab. of Power Electronics, Zhejiang University, Hangzhou, China

huhaibing@163.com

Abstract —A configurable floating-point coprocessor by a FPGA is designed to enhance the computational capability of the digital platform based on the fixed-point DSP, with which the platform will be competent to implement intensively computational tasks. Detailed design procedures of the coprocessor are presented. A new division algorithm is proposed by combining the lookup-table algorithm and multiplicative algorithm in order to reduce the number of LEs(Logic Element in FPGA) and latency. Error analysis of the proposed algorithm shows that the maximum absolute approximate error is less than 2ulp(Unit in Last Place). The coprocessor speed can reach up to 25 MFLOP(Million Floating-point Operations). FFT algorithm is adopted to test the computational efficiency of the floating-point units. Experimental results show the computation time by FPU is five times less than that of DSP algorithms.

Keywords-Floating-point;FPGA;Power Electronics;FFT

I. INTRODUCTION

The computational complexity of modern power electronics applications is steadily increasing. Meanwhile, the computation accuracy is in great demand for high performance applications. In such areas as Active Power Filter(APF), high performance motor servo drive and sensorless speed control techniques for motor drive, floating-point digital signal processors(DSPs) have their advantages over their counterparts, fixed-point processors, to accomplish these computational intensive tasks[1]. There are two typical digital platforms for these computational demanding applications.

1) Some researchers configured their digital platforms based on floating-point DSPs with extension circuits of A/D converters, D/A converters, PWM generators and specific peripherals for motor driver and power electronics[2,3,4]. This method not only increased the system cost greatly but also reduced the system reliability with so many "discrete" components added to the platforms.

2) Another typical digital platform uses both fixed-point DSP and floating-point DSP[5], combining the benefits of floating point DSPs for their high

computation capability and accuracy with the advantages of fixed-point DSPs dedicated to power electronics and motor drive applications. This type platform also has its drawbacks: (1)High cost; (2)Programming would be complicated; (3)Communication between two DSPs would be a headache problem.

However the fixed-point DSPs have continued to be the mainstay of the industry. The reason, of course, is the cost. With the advancement of FPGA technology, floating-point units configured by FPGA are affordable. In this paper, a new digital platform for power electronics and motor drive is proposed to overcome above problems. The platform used a fixed-point DSP(TMS320f2812) as a core processor and floating-point units(FPU) configured by a FPGA as a coprocessor to enhance the computation capability. The coprocessor can implement floating-point adder, multiplier and divider in 40 nanoseconds with the precision less than 2ulps.The coprocessor can be viewed as a peripheral of the fixed-point DSP and can be accessed via external interface of DSP.

II. PRINCIPLE OF FLOATING-POINT UNITS

The single precision numbers in the binary IEEE standard are formed as shown in Fig.1. The most significant bit is the sign bit, which indicates a negative number if it is set to 1. The following field denotes the exponent with a constant bias added to it. As shown in Fig.1, the remaining part of the number is normalized to have one non-zero bit to the left of the floating point[6].

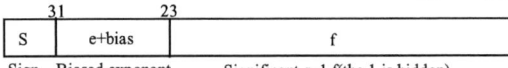

31		23		0
S	e+bias		f	

Sign   Biased exponent   Significant s=1.f(the 1 is hidden)

Figure 1. Format of IEEE single floating point number

Therefore, the value given by the standard format can be expressed using following expression.

$$m = (-1)^{\text{Sign}} \times 2^e \times 1.f \qquad (1)$$

The range of single precision floating-point number varies from -3.4028236 e+38 to -1.1754944 e-38 and from +1.1754944 e-38 to +3.4028236 e+38.

III. FLOATING-POINT ALGORITHMS

A. Addition algorithm[6,7]

The floating-point addition and subtraction algorithm studied here is similar to what is done in most traditional

---

Project supported by National Nature Science Foundation of China (50237030ZD)

1-4244-0448-7/06/$25.00 ©2006 IEEE   1183

processors. The layout of the floating-point addition implementation is demonstrated in Fig. 2. As can be seen

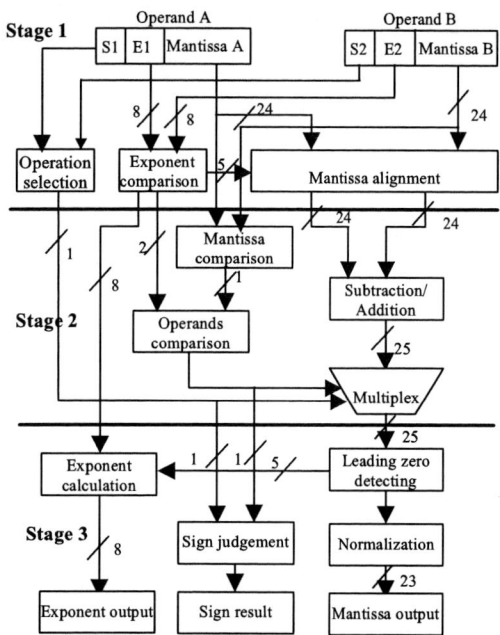

Figure 2. Flow chart of a 32-bit floating-point adder

from Fig.2, the floating-point adder can be divided into three stages. The numbers above lines in the flow chart stand for the data width. The computation procedures for each stage are as follows.

Stage 1:

● The larger value of exponents is found by comparing the exponents of operand A and operand B.

● If E1 is larger than E2, subtract E2 from E1 in order to calculate the number of positions to shift the mantissa B to right so as to make mantissa A and mantissa B aligned before addition and subtraction in next stage.

Stage 2:

● If S1 equals S2, add mantissa A to mantissa B, otherwise subtract mantissa A from mantissa B or conversely, which is decided by absolute values of operand A and operand B.

● According to signs and absolute values of operand A and operand B, route one of three results obtained in above step to stage 3.

Stage 3:

● The result obtained in above stage is shifted to the left until the highest order bit is a one, which is so called "leading zero detecting".

● Normalize above result to comply with the mantissa part of floating-point format.

● New exponent is calculated by subtracting larger exponent from the number of positions which is the result of leading zero detecting.

● New sign can be obtained according to signs and

absolute values of operand A and operand B.

B Multiplication algorithm

Floating point multiplication is much easier than floating point addition and is much like integer multiplication. The block diagram of 32-bit floating-point multiplier is shown in Fig.3. As can be seen from Fig.3, three elements(sign, exponent and mantissa) of a floating point expression can be easily obtained by following procedures.

Sign:

A new sign bit can be easily accomplished by a "xor" gate.

Mantissa:

A 24-bit multiplier is used for multiplying the mantissas of operand A and operand B. Only 25 bit production is produced by the multiplier, which will reduce chip area and latency significantly, since 25-bit production is accurate enough to obtain the new mantissa due to the fact that the mantissa is greater than 1.0 and less than 2.0. Therefore the leading zero detecting and normalization process are also much easier than those of floating-point addition.

Exponent:

New exponent is obtained by adding these two exponents of operand A and operand B and one bit carry from the leading zero detecting block.

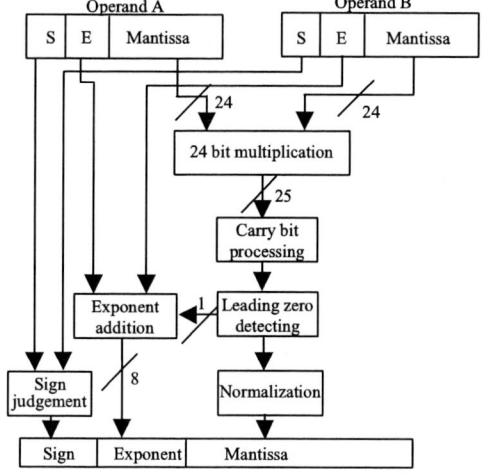

Figure 3. Flow chart of a 32-bit floating-point multiplier

C Division algorithm

In general, floating point division is accomplished in hardware by an algorithm belonging to one of three general classes: (1) Lookup table; (2)Subtractive;(3) Multiplicative[6]. On considering the precision requirements for power electronics applications and in order to reduce the number of LEs and latency, a new division algorithm is proposed by combining the lookup-table algorithm and multiplicative algorithm. The block diagram of 32-bit floating-point divider is shown in Fig.4. This algorithm can be expressed as follows.

1184

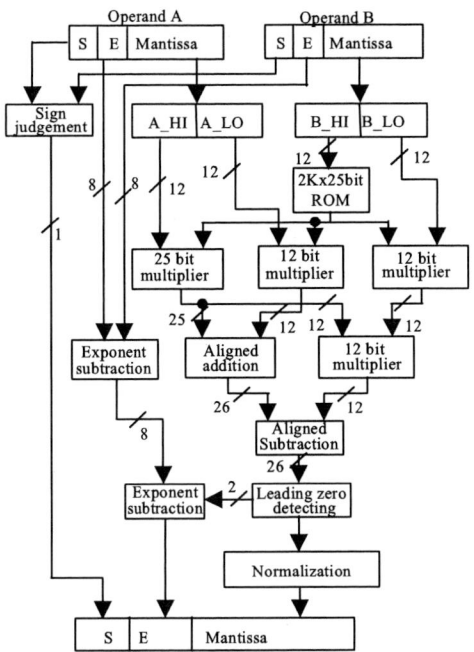

Figure.4 Flow chart of the 32 bit floating point divider

Suppose dividend and divider with single precision are in the following format.

$$m1 = (-1)^{s1} \times 2^{E1} \times 1.f_1 \qquad (2)$$

$$m2 = (-1)^{s2} \times 2^{E2} \times 1.f_2 \qquad (3)$$

so the quotient q can be expressed as (4).

$$q = \frac{m1}{m2} = (-1)^{S1-S2} \times (2)^{E1-E2} \times \frac{1.f_1}{1.f_2} \qquad (4)$$

Let $A = 1.f_1$, $B = 1.f_2$. Both A and B can be divided into high parts and low parts respectively as shown in Fig.5.

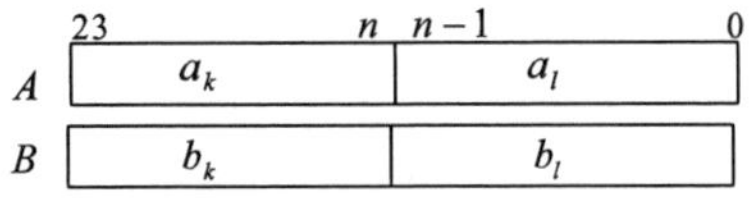

23	n	$n-1$	0
A	a_k		a_l
B	b_k		b_l

Fig.5　Split A,B into high parts and low parts

Therefore A and B can be expressed as (5) and (6).

$$A = a_h + a_l \qquad (5)$$

$$B = b_h + b_l \qquad (6)$$

Substituting above equations into (4), the last term of the expression can be rewritten as follows.

$$\frac{1.f_1}{1.f_2} = (\frac{a_h}{b_h} + \frac{a_l}{b_h})(\frac{1}{1 + b_l / b_h}) \qquad (7)$$

We can expand the term $\dfrac{1}{1 + b_l / b_h}$ into a Taylor series given by

$$\frac{1}{1 + b_l / b_h} = 1 - (\frac{b_l}{b_h}) + (\frac{b_l}{b_h})^2 - \cdots \qquad (8)$$

By properly selecting n, we can neglect the second-order and higher terms in above Taylor series with rational approximation. $\dfrac{1}{1 + b_l / b_h} \approx 1 - \dfrac{b_l}{b_h}$.

So expression (4) can be approximated by equation (9).

$$\frac{1.f_1}{1.f_2} = \frac{a_h}{b_h} + \frac{a_l}{b_h} - \frac{a_h}{b_h} \times \frac{b_l}{b_h} - \frac{a_l}{b_h} \times \frac{b_l}{b_h} \qquad (9)$$

For the same reason mentioned above, the last term in(9) can also be neglected.

The value of $\dfrac{1}{b_h}$ is stored in a lookup table. So by using this algorithm, the division algorithm only requires four multiplications and two additions, which will reduce the chip cost and latency greatly.

D Error analysis

It is obvious that only division algorithm will cause approximate error. Whether the approximate error can be acceptable or not? We can use the mathematical method to find out the maximum absolute approximate error. The following inequalities(10) and (11) are validated for this infinite Taylor series due to the facts:

$$0 < \frac{b_l}{b_h} < 1 \quad \text{and} \quad 1.0 < b_h < 2 \cdot$$

$$(\frac{b_l}{b_h})^2 - (\frac{b_l}{b_h})^3 + (\frac{b_l}{b_h})^4 - \cdots < (\frac{b_l}{b_h})^2 < (b_l)^2 \quad (10)$$

$$\frac{a_l}{b_h} \times \frac{b_l}{b_h} < a_l \times b_l \qquad (11)$$

So the maximum absolute approximate error can be estimated using the equation(12).

$$E_{\max} = (b_l)^2 + a_l \times b_l \qquad (12)$$

While the values of a_l and b_l can be expressed in the following format.

$$a_l = \underbrace{0.0 \cdots 0}_{24-n} \underbrace{xx \cdots xx}_{n} \qquad (13)$$

$$b_l = \underbrace{0.0 \cdots 0}_{24-n} \underbrace{xx \cdots xx}_{n} \qquad (14)$$

Using above expressions, the values of $(b_l)^2$ and $a_l \times b_l$ can be obtained as the following format.

$$(b_l)^2 = \underbrace{0.0 \cdots 0}_{2 \times (24-n)-1} \underbrace{xx \cdots xx}_{2n} \qquad (15)$$

$$a_l \times b_l = \underbrace{0.0 \cdots 0}_{2 \times (24-n)-1} \underbrace{xx \cdots xx}_{2n} \qquad (16)$$

Therefore the maximum absolute approximate error can be expressed as (17).

$$Error_{max} = \underbrace{0.0\cdots0}_{2\times(24-n)-2}\underbrace{xx\cdots xx}_{2n+1} \qquad (17)$$

Suppose $n=11$, the maximum absolute approximate error will be less than 1ulp(unit in last place). With the decrease in the number of n, the lookup table will increase exponentially. In this paper, we made a tradeoff between the size of look-up table and approximate error by selecting $n=12$. According to expression(17),the maximum absolute approximate error is less than 2ulp,which is accurate enough for power electronic applications.

IV. SIMULATION AND ITS IMPLEMENTATION

A Simulation

With the introduction of high level hardware description language such as VHDL, verilog HDL and etc, rapid design of floating point units has become possible. According to algorithms mentioned above, three arithmetic units(floating point adder, multiplier and divider) were developed on the powerful platform of QuartusII4.2 provided by Altera using VHDL. To verify the correctness of the above algorithms, only synthesis and simulation tools are used. Due to the size of VHDL source code it is not presented here. To illustrate the correctness of the algorithms, several simulation diagrams of floating point arithmetic units are shown in Fig. (6).

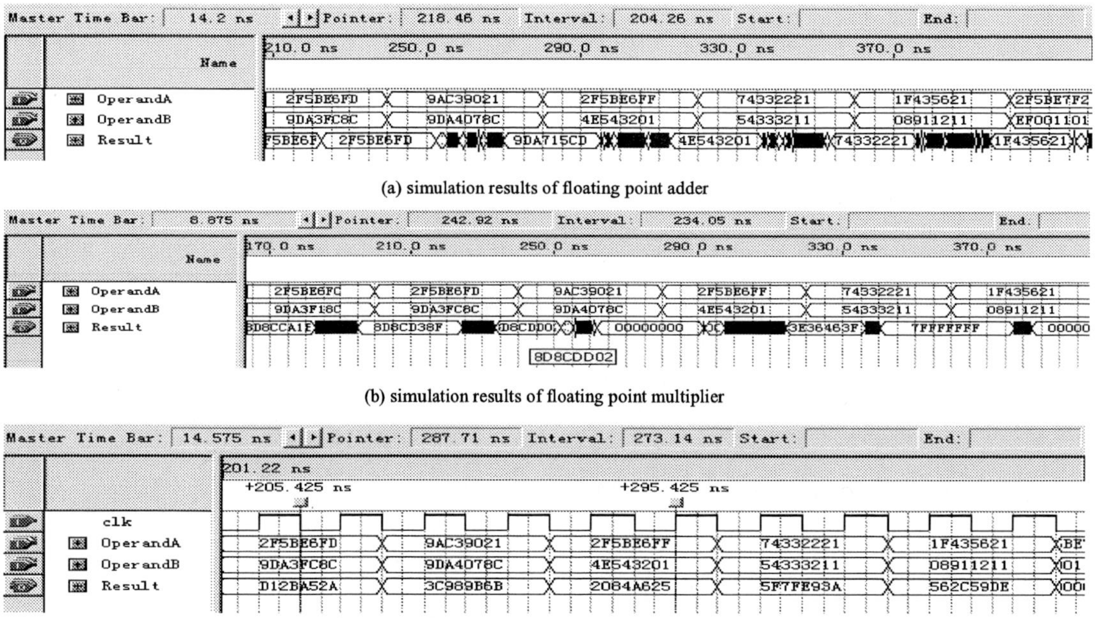

(a) simulation results of floating point adder

(b) simulation results of floating point multiplier

(c) simulation results of floating point divider

Figure 6. Simulation results of floating point units

B Implementation

For this project, we have chosen an Altera Cyclone FPGA EP1C6, featuring 5890 LEs and a total embedded RAM of 92160 bits, of which 98 pins are available to the user. After allocating the pins using assignment editor, a programming file can be successfully compiled by

Tab.2 Resources used in the arithmetic units

Arithmetic units	LE	Memory bits	PLL
Adder	883	0	0
Multiplier	995	0	0
Divider	1505	51200	1

the QuartusII software, which can directly download to FPGA via a JTAG using QuartusII programmer. In Tab(2) are summarized the resources dedicated to floating-point units respectively.

In order to overall verify the floating-point units, we have developed a program to compare the result calculated by DSP and that obtained by FPGA. Since

DSP (TMS320F2812) is fixed point by nature, the floating point operations are implemented by a run-time supporting library(rtl2800.lib). The flow chart of this program is shown in Fig (7). The DSP was operating under 100MHz with external interface access less than 40ns, that means the floating point units can operate under 25MHz.As you see from Fig (7), we randomly produced the operands A and B, then send them to FPGA and in next operation retrieve the result from FPGA, then calculated the results using DSP algorithms, after finishing computation, compared these two results. If the difference between two results were great than 2ulps (from above analysis, the error tolerance is less than 2ulps), DSP would halt to produce a fatal error, otherwise continue to do testing routine. In this test, we have run the program for several hours continuously. The test indicated the floating-point units were validated by hardware.

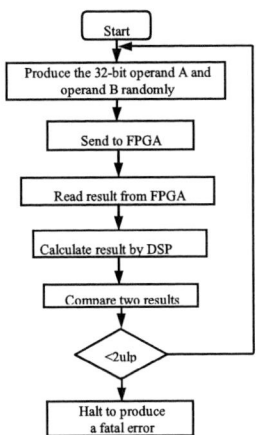

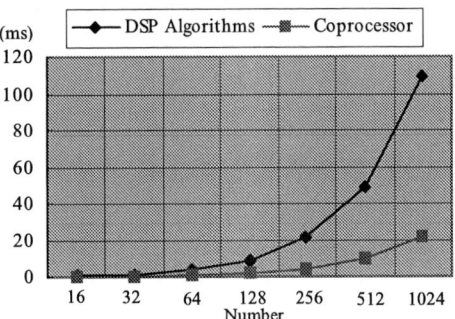

(ms)

Figure 9. Computation time of FFT between FPU and DSP algorithms

Figure 7. Flow chart for testing the floating -point units using DSP

V. VERIFICATION AND COMPARISON STUDY

FFT algorithm was adopted to verify the correctness of the results obtained by the FPU and make a comparison between the FPU and DSP algorithms(supported by run time library) in computation efficiency. All these

Figure 8. Proposed digital platform for testing FPUs

experiments were carried in the digital platform shown in Fig(8). The results of FFT calculated by FPU were exported by using "data saving" feature available in CCS and then analyzed in Matlab. The results of FFT obtained by FPU are totally in consistence with those calculated by Matlab.

To study the computation efficiency between DSP algorithms and FPU, FFT algorithm with different points was adopted. The time consumed by DSP algorithms and FPU was measured by the interval between the very beginning of FFT calculation and the end using DSP T2 timer. Notice that all the operation environments were set the same for all these tests. Computation time of two methods was shown in Fig(9). From this figure, it is very clear that the computation time by FPU is five time less than that of DSP algorithms. If the FPU were interfaced to the DSP with 32 bit data width or integrated into the DSP like other peripherals, the computation efficiency would further improved, since at least two-thirds time are wasted in organizing, feeding and retrieving data to and from FPU.

VI. CONCLUSION

In this paper, a floating-point coprocessor is successfully configured by FPGA to enhance the computation capability and flexibility of the digital platform for power electronics applications. The coprocessor can operate under 25MFLOP.The computation precision of the coprocessor, though lower than IEEE 574 standard in which 0.5ulp is required, is less than 2ulps, which is also sufficient accurate for common applications. The platform is being used for prototype development and implementation of computationally intensive algorithms for various power electronics applications.

REFERENCES

[1] Mongkol Konghirun, Longya Xu Jennifer Skinner Gray. "Quantization Errors in Digital Motor Control Systems".Power Electronics and Motion Control Conference,Aug. 2004,pp.1421-1426.

[2] D.D.Bester, J.A. du Toit J.H.R Enslin. "High Performance DSP/FPGA controller for Implementation of Computationally Intensive Algorithms".IEEE International Symposium on Industrial Electronics,Jul.1998,pp.240-244.

[3] Habib-ur Rehman, Richard J. Hampo. " A flexible high performance advanced controller for electric machines". Applied Power Electronics Conference and Exposition, Feb. 2000,pp.939-943.

[4] Joep Jacob,Dirk Detjen, etc. "Rapid Prototyping Tools for Power Electronic Systems: Demonstration with Shunt Active Power Filters". IEEE Trans. on Power Electronics,Vol.19(2),2004,pp:500-507.

[5] Wangjun Lei, Fang Zhuo,etc. "Development of 100KVA Active Filter with Digital Controlled Multiple Parallel Power Converters". IEEE Power Electronics Specialists Conference, Jun. 2004,pp.1121-1126.

[6] Albert Austin Liddicoat, "High-performance Arithmetic for Division and the Elementary Functions". Ph.D Dissertation, Stanford University. 2002.

[7] Nabeel Shirazi, Al Walters, Peter Athanas."Quantitative Analysis of Floating Point Arithetic on FPGA based Custom Computing Machines". IEEE Symposium on FPGAs for Custom Computing Machines. Apr. 1995,pp.155-162.

1187

2006 5th International Power Electronics and Motion Control Conference

An Analytical Model for 4H-SiC Super-Junction Devices

L.C. Yu and K. Sheng

Department of Electrical and Computer Engineering, Rutgers University, New Jersey, U.S.A.

Abstract—In this paper, for the first time, a new and easy-to-implement analytical model is developed for the breakdown voltage and specific on-resistance of 4H-SiC super-junction devices. The model features simple analytical equations while is still capable of predicting the device characteristics accurately for a large variety of device structural parameters. Accuracy of this model is verified by multi-dimensional numerical simulation.

Keywords- Model, 4H-SiC, Super-junction, Power device

I.   INTRODUCTION

One of the main objectives in the design of power semiconductor devices is to obtain high breakdown voltage (V_{BR}) at off-state and low specific on-resistance (R_{SP_ON}) at on-state at the same time. However, as stated in the well-known power law, $R_{SP\_ON} \propto V_{BR}^{2.5}$ [1], high V_{BR} and low R_{SP_ON} require conflicting device parameters. There exists a theoretical limit on the figure-of-merit (FOM), which is defined as V_{BR}^2/R_{SP_ON} to evaluate the superiority of a unipolar device. In the last a few years, 4H-SiC material has attracted extensive attention for power semiconductor applications. Experimentally demonstrated specific on-resistances of 4H-SiC power devices are quickly approaching the theoretical limit [2, 3]. The super-junction structure has been proposed to break this limit [1, 4]. Simulation results show that, with currently available technologies, device FOM exceeding the 4H-SiC theoretical limit by 200% is possible [4]. An analytical model for this 4H-SiC super-junction structure is therefore desired to predict the V_{BR} and R_{SP_ON}. A full two-dimensional analytical model was proposed by R. Ng et al for silicon super-junction structures [5]. However the model involves complicated series that make it difficult to implement for practical use. In this paper, a simple analytical model is developed, which is capable of accurately predicting the V_{BR} and R_{SP_ON} of 4H-SiC super-junction structures.

II.   4H-SiC SUPER-JUNCTION STRUCTURE

The cross-sectional view of a 4H-SiC super-junction voltage-sustaining region is shown in Figure 1. Compared with conventional structures, the homogeneous drift region is replaced by alternating N and P doped stripes. By taking advantage of the charge compensation concept [1], this structure is capable of blocking high voltage at off state with a drift region doping orders of magnitude higher than a traditional structure, which brings drastic

improvements on R_{SP_ON} when used in unipolar devices. When the structure is in voltage-blocking mode, both N and P stripes will be fully depleted and a two-dimensional electric field will be developed in these two regions.

Figure 2 shows the lateral electric field distributions at different applied voltages along the center of the drift region which corresponds to y=L/2 and x from –W to W in Figure 1. As the applied voltage increases, the N/P stripes are depleted more and the E_x distribution is similar as in a 1-D pn junction structure. It is worth noting that once the N/P stripes are fully depleted, the E_x distribution stops changing. Therefore in Figure 2, E_x distributions are identical for V_a=300V, 1000V and 1900V. The vertical electric field distributions along the middle of P stripe at different applied voltages are shown in Figure 3. Before full depletion, E_y distribution has a triangular shape which is also similar as in a 1-D pn junction structure. According to Poisson Equation,

$$\frac{\partial E_x}{\partial x} + \frac{\partial E_y}{\partial y} = \frac{qN}{\varepsilon} \qquad (1)$$

E_y distribution has a flat portion after full depletion, which is the reason why super-junction structure can give higher V_{BR} at the same doping concentration compared with conventional structures. Due to symmetry of the super-junction structure, E_y distributions along the middle of N stripe are the same as along the middle of P stripe, given that the charges are balanced.

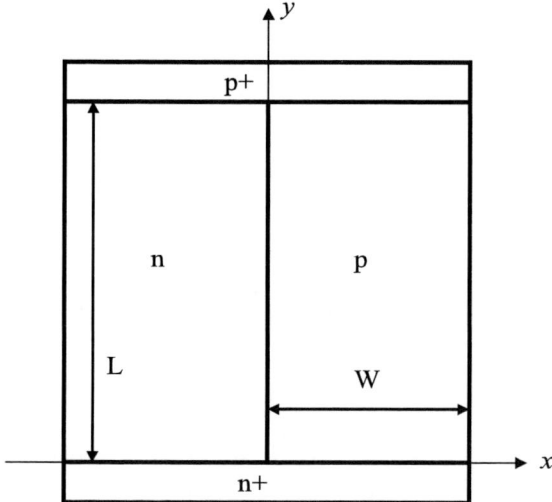

Figure 1. Cross-sectional view of a unit cell of the super-junction structure. Only half width of the N and P stripes are shown.

1-4244-0448-7/06/$25.00 ©2006 IEEE   1188

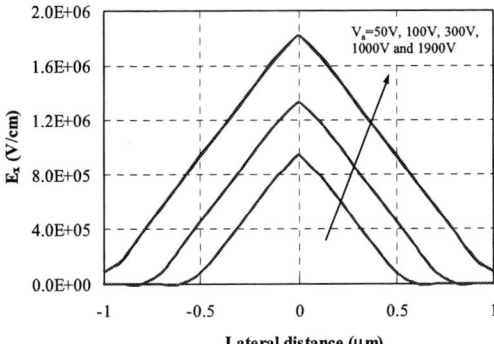

Figure 2. Lateral electric field distribution along the center of the drift region (y=L/2) of a super-junction structure (L=10μm, W=1μm, N=10¹⁷cm⁻³) at different applied voltages.

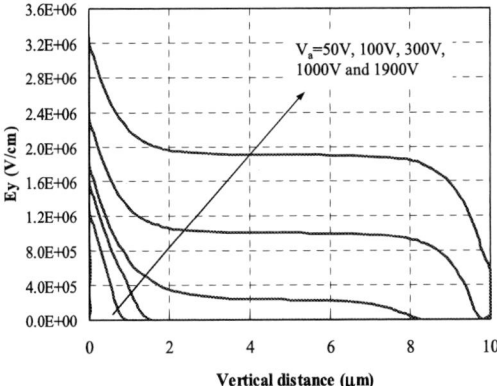

Figure 3. Vertical electric field distribution along the middle of P stripe (x=W) of a super-junction structure (L=10μm, W=1μm, N=10¹⁷cm⁻³) at different applied voltages.

In the following sections, analytical modeling of both the voltage-blocking and current conduction characteristics of this structure will be described.

III. ANALYTICAL MODEL FOR CURRENT CONDUCTION

As the super-junction structure is mostly used in its unipolar mode, only forward conduction in the unipolar mode will be modeled in this paper. In this mode, current only flows in the un-depleted part of N stripe. The specific on-resistance can be calculated as

$$
\begin{aligned}
R_{SP\_ON} &= \frac{1}{q \cdot \mu_n(N) \cdot N} \cdot \frac{L}{W_{n\_undep} \cdot z} \cdot 2W \cdot z \\
&= \frac{2L}{q \cdot \mu_n(N) \cdot N} \cdot \frac{1}{1 - \dfrac{W_{n\_dep}}{W}}
\end{aligned}
\tag{2}
$$

where q is the electron charge, N is the N stripe doping, W_{n_undep} is the undepleted N stripe width, Z is the third dimensional depth of the device, and $\mu_n(N)$ is the electron mobility in 4H-SiC that can be modeled as [6]

$$
\mu_n(N) = 40 + \frac{950 - 40}{1 + (\dfrac{N}{2e17})^{0.76}} \qquad (\text{cm}^2\text{V}^{-1}\text{s}^{-1}) \tag{3}
$$

.With N/P stripe doping equals to each other, the depletion width in N stripe is

$$
W_{n\_dep} = \frac{1}{2}\sqrt{\frac{2\varepsilon_r\varepsilon_0(V_{bi} + V_F)}{q \cdot \dfrac{N \cdot P}{N + P}}} = \sqrt{\frac{\varepsilon_r\varepsilon_0 V_{bi}}{q \cdot N}} \tag{4}
$$

where $\varepsilon_r\varepsilon_0$ is the permittivity of 4H-SiC, V_{bi} is the P/N junction built-in voltage, V_F is the forward biasing voltage (neglected in the equation above since $V_F \ll V_{bi}$ with reasonable forward current densities). Based on this model, the specific on-resistance versus doping curves for different L/W combinations are plotted in Figure 4 where they are compared with results obtained from 2-D numerical simulation. In the numerical simulation, super-junction MOSFET structures are used to simulate the R_{SP_ON} of the super-junction layer by minimizing the MOS channel resistance. It can be seen that this analytical model fits well with numerical simulation data, except for some minor discrepancies at very low doping levels, which are caused by JFET effect in the N stripe.

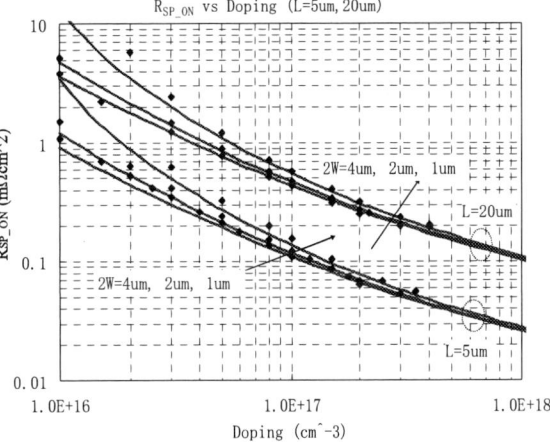

Figure 4. R_{SP_ON} versus base doping for 4H-SiC super-junction devices with different L/W combinations. Line: analytical model, symbol: numerical simulation.

IV. ANALYTICAL MODEL FOR VOLTAGE BLOCKING

By examining [5], it is clear that analytically solving the two-dimensional electric field distribution in a super-junction structure is a procedure involves excessive mathematics. In order to avoid a complicated electric field expression so that the device breakdown voltage can be modeled easily, simplifications have to be made.

A. Impact ionization coefficients for 4H-SiC

In the voltage-blocking mode, the condition for device avalanche breakdown is when the following is satisfied along any path between the two voltage blocking terminals (N⁺ and P⁺ regions in Figure 1).

1189

$$\int \alpha_{eff}\, dx = 1 \quad , \text{where} \quad \alpha_{eff} = CE^{\,7} \tag{5}$$

where α_{eff} is the effective impact ionization coefficient for 4H-SiC. As the reverse bias voltage increases, the impact ionization integral also increases. At a certain bias voltage level, the integral reaches unity along a specific path and avalanche breakdown takes place. This path is referred to as the 'critical path'. In this proposed model, Fulop's form [7] is used for its simplicity.

In order to build a good model for the voltage blocking capability of the 4H-SiC super-junction devices, it is critical that the impact ionization coefficients are modeled accurately. In the literature, two separate teams have systematically measured and reported the impact ionization coefficient for 4H-SiC [8, 9]. While small differences exist between the two sets of impact ionization data, the data and its expression in [8] are found to give more accurate prediction to experimental results [10].

The constant C in equation (5) needs to be determined so that the Fulop's form in equation (5) fits well with the impact ionization coefficient expressions provided in [8]. The value of 6×10^{-42} cm^{6}V^{-7} is found to provide the best accuracy for the effective impact ionization coefficient.

B. Electric field distribution along the critical path in the super-junction structure

It has been observed that in most cases, because of the electric field peaks at the top-left and bottom-right corners of the voltage blocking region, breakdown occurs at $x = \pm W$ in Figure 1, where the lateral component of the electric field (E_x) is zero. As the electric field along the $x=W$ and $x=-W$ lines are symmetrical, the ionization integration can be carried out along either of these two critical paths to calculate the device breakdown voltage. At those locations the total electric field equals to its vertical component (E_y). This further simplifies the model for breakdown voltage calculation.

In order to predict the device breakdown voltage, the electric field distribution along the critical path has to be modeled accurately. The two-dimensional Poison Equation (equation (1)) is what governs the electric field distribution in this structure.

Along the critical path ($x=W$, $E_x=0$), as y increases from 0 to y_f (in Figure 5), the slope of electric field ($\partial E_y/\partial y$) decreases because of the increasing lateral depletion effect that gives rise to an increasing $\partial E_x/\partial x$ term in equation (1). In other words, the two dimensional charge compensation effect helps decreasing the slope of E_y and making E_y more uniform. This can be clearly seen in both Figure 3 and Figure 5. At y=0, because of the lack of lateral depletion effect ($E_x=0$ and $\partial E_x/\partial x=0$), $\partial E_y/\partial y=qN/\varepsilon$ which is solely determined by the vertical N$^+$-P junction. As y increases beyond certain value (y_f, to be determined later), the lateral depletion term $\partial E_x/\partial x$ reaches the value of qN/ε and, according to equation (1), E_y remains constant.

In this model, based on the above observation, E_y at $x=W$ is derived as

$$E_y(y) = E_f + \int_y^{y_f} \frac{dE_y}{dy'}\,dy'$$

$$= E_f + \int_y^{y_f} k_{Ey}(y_f - y')^2\,dy'$$

$$= E_f + \frac{1}{3}k_{Ey}(y_f - y)^3, \quad y \in (0, y_f)$$

$$E_y(y) = E_f = V_a/L \qquad y \in (y_f, L) \tag{6}$$

where y_f is the point where the electric field becomes flat (E_f, see Figure 5) and k_{Ey} is given by the following equation.

$$k_{Ey} = qN/(\varepsilon \cdot y_f^2) \tag{7}$$

As the electric fields in the region $y \in (0, y_f)$ are the highest, they give major contribution to the integration in equation (5). In equation (6), a simple quadratic term is used to model the slope of E_y along y direction ($\partial E_y/\partial y$) for this region. The simple quadratic form is found to give sufficient accuracy to the electric distribution for a large variety of structures with different L, W and doping combinations.

Based on the quadratic equation, equation (7) is obtained simply by using $\partial E_y/\partial y=qN/\varepsilon$ at y=0.

Because of the nature of the lateral depletion effect on the electric field along the critical path, the distance at which E_y becomes flat (y_f) is mainly determined by the width of the P stripe (2W) for devices with a reasonably long drift region (longer than y_f). It has been observed through extensive numerical simulation that the value of y_f can be well approximated by $y_f = 2.3 \cdot W$.

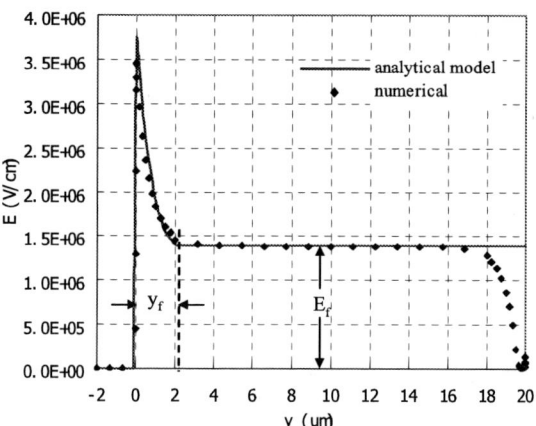

Figure 5. Electric field distribution along the drift region at x=W for a super-junction structure L=20μm, W=1μm and doping=1e17cm^{-3}

Based on equations (6) and (7), the electric field along the critical path (x=W) can be obtained analytically for any voltage bias that is large enough to fully deplete the N and P stripes. As an example, the modeled electric field distribution along the critical path for a super-junction structure with L=20μm, W=1μm and N=P=10^{17}cm^{-3} at breakdown is plotted in Figure 5 along with what has been obtained from two-dimensional numerical simulations. An excellent fitting has been achieved between the model and the numerical simulation. The small fall-off in the field at the right hand end for the numerical simulation was not accounted for in the model as it has insignificant effect on the device breakdown voltage.

C. Breakdown voltage calculation

Once the electric field distribution has been modeled, the breakdown voltage of a structure can be calculated by substituting equation (6) in to equation (5). A quick inspection of these two equations reveals that equation (5) will become a polynomial equation after the substitution. Solving the polynomial equation will result in the breakdown voltage at which the impact ionization integral in equation (5) reaches unity for any combinations of L, W, N and P.

Breakdown voltages for such sets of structures have been calculated based on the proposed analytical model. Results for drift region length L=5μm and 20μm, stripe width 2W=4μm, 2μm and 1μm, N and P doping varying from 10^{16}cm^{-3} to 5×10^{17}cm^{-3} have been shown in Figure 6 where they are compared to those obtained from numerical simulations. It can be seen that the analytical model is able to reproduce the numerical results with good accuracy.

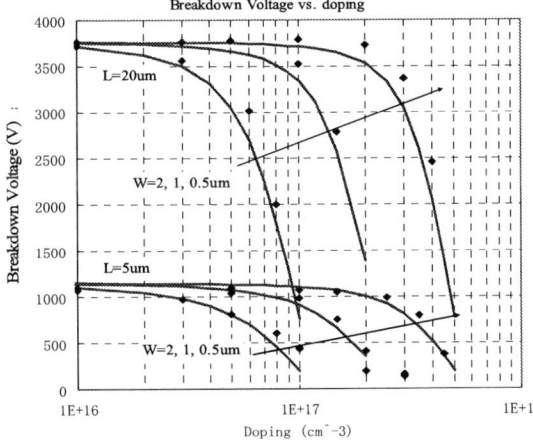

Figure 6. Breakdown voltage versus N doping for 4H-SiC super-junction devices with different L/W combinations. Line: analytical model, symbol: numerical simulation.

It is worth noting that the proposed analytical model is drastically simpler than the previous super-junction model reported in the literature [5] and therefore has a much faster simulation speed *without* sacrificing accuracy. Implementation of such a model in various software will be much easier without the complication of convergence problems. With the proposed analytical model, calculation of breakdown voltages for all the structures shown in Figure 6 takes less than one minute. This is very useful when compared to the numerical simulation which can take weeks of simulation time.

V.   SUMMARY

In this paper, an analytical model has been proposed for super-junction structures on 4H-SiC for the first time. The model includes both current conduction and voltage blocking characteristics. Dependence of carrier mobility on doping has been considered to give accurate device specific on-resistance (R_{SP_ON}) prediction. Electric field distribution on an avalanche critical path within the device structure has been analytically modeled. Ionization integral along this critical path was then used to calculate the device breakdown voltage. Results of the analytical model have been compared to 2-D numerical simulation for a variety of L, W, N and P combinations. It is found that the model is in good agreement with the numerical simulation on both the R_{SP_ON} and the V_{BR} for any L, W, N and P combination studied.

In addition, this model is drastically simpler and much easier for practical implementation than a full 2-D analytical model previously reported, without sacrificing accuracy. As the processing technology on 4H-SiC develops towards maturity, the model will provide useful guidance on super-junction device designs.

REFERENCES

[1] G. Deboy, et. al, "A new generation of high voltage MOSFETs breaks the limit line of silicon", Proc. of the IEDM, 1998, p. 683

[2] J. H. Zhao, K. Tone, X. Li, P. Alexandrov, L. Fursin and M. Weiner, "3.6 mm-cm2, 1726V 4H-SiC Normally-off Trenched-and-Implanted Vertical JFETs", ISPSD 2003, Cambridge, UK, pp. 50-53.

[3] J. Zhang, J.H. Zhao, P. Alexandrov, and T. Burke, "Demonstration of first 9.2 kV 4H-SiC bipolar junction transistor", IEE Electronics Letters, Vol. 40, Issue 21, pp. 1381-1382

[4] L.C. Yu and K. Sheng, "Breaking the Theoretical Limit of SiC Unipolar Power Device - a Simulation Study", Proc. of the ISDRS, 2005, FA 5-01.

[5] R. Ng, F. Udrea, and G. Amaratunga, "An analytical model for the 3D-RESURF effect", Solid-State Electronics 44, 2000, pp. 1753-1764

[6] M. Roschke, and F. Schwierz, "Electron Mobility Models for 4H, 6H and 3C SiC", IEEE Transactions on Electron Devices, Vol.48, No.7, pp. 1442-1447

[7] W. Fulop, "Calculation of avalanche breakdown of silicon p-n junctions", Solid-State Electron., Vol. 10, 1967, pp. 39–43

[8] A. O. Konstantinov, Q. Wahab, N. Nordell, and U. Lindefeltd, "Ionization rates and critical fields in 4H silicon carbide", Applied Physics Letters, Vol. 71, No. 1, pp. 90-92

[9] R. Raghunathan, B.J. Baliga, "Measurement of electron and hole impact ionization coefficients for SiC", IEEE International Symposium on Power Semiconductor Devices and IC's, 1997. pp. 173-176

[10] X. Li, "Design and simulation of high voltage 4H silicon carbide power devices", Rutgers University, Thesis collection, 2005.

2006 5th International Power Electronics and Motion Control Conference

Architecture Implementation of Class-D Amplifiers Using Digital-Controlled Multiphase-Interleaved PWM Technique

Yu-Tzung Lin, Chi-Yang Lee, and Ying-Yu Tzou, *Member, IEEE*

Power Electronics IC Design & Digital Power Lab., Dept. of Electrical and Control Engineering
National Chiao Tung Univ., Hsinchu, Taiwan, China

Abstract—The class-D amplifier has the advantages of having a high efficiency, a small-size and gains its wide applications in battery-powered audio systems. However, its output power is limited by the rating of its output transistors. This paper presents the development of a new architecture for the realization of a class-D audio amplifier using a digital-controlled multi-phase interleaved PWM converter to enhance its output power. This interleaved technique is widely used in DC-DC converters. A digital current control technique with interleaved PWM generation was developed to reduce the total harmonic distortion (THD) of the output current under lower PWM switching frequency. A comparative evaluation of a single-phase and a three-phase class-D amplifier structure is presented. Furthermore, the design and implementation of a digital control IC, including an ADC synchronous sampler, a digital compensator, and a phase-shift PWM generator, is introduced. Computer simulations and experimental verifications are given to illustrate the feasibility and performance of the proposed technique.

Keywords—*interleaved class-D amplifier, digital control, digital audio IC.*

I. INTRODUCTION

Operating principles of switching DC-DC converters in applications of audio amplifiers, the class-D amplifiers, have been known to us for a long time, but have only recently become realizable due to the economic manufacturing of high-quality switching power devices. Combined with digital signal processing techniques, the audio performance of modern class-D amplifiers can even exceeds that of conventional class-AB amplifiers. The high efficiency of class-D amplifiers provides advantages to reduce heatsink size, increasing output power, and/or lengthening battery life. Multi-channel surround audio amplifiers and portable audio-vide devices are the perfect applications for class-D amplifiers.

Fig. 1 shows a typical class-D audio amplifier. This switching amplifier consists of a PWM generator, a switching power converter, and a low-pass filter. An audio signal is compared with a high frequency triangle or saw-tooth waveform to generate the PWM signal. This generated signal is then used to drive a half or full bridge

This work was supported by National Science Council, Taipei, Taiwan, China. Project no. NSC94-2213-E-009-146.

The authors are the Department of Electrical and Control Engineering, National Chiao Tung University, 30050, Hsinchu, Taiwan, China.

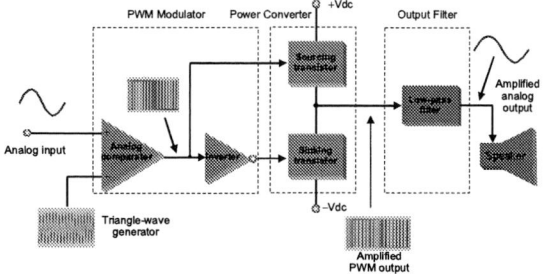

Fig. 1.   Typical class-D audio amplifier architecture.

power stage, creating the amplified digital signal. This signal is then finally applied to a low pass filter, which retrieves the sinusoidal audio signal. A well-designed class-D amplifier with an efficiency higher than 90% can be achieved [1]. Due to its high efficiency and simple implementation, the class-D amplifiers are preferred instead of the linear amplifiers in the audio market.

There are several methods for the generating of high-frequency PWM signals for a digital PWM (DPWM) generator, i.e, the fast-clock-counter method [2], the tapped-delay-line method [3], and a counter-delay-line method [4], etc. Usually, a high PWM resolution and fast switching frequency is necessary to reduce of output distortion. However, increasing the PWM frequency and resolution will require a very high clock frequency, and this imposes a design constraint for both the high-frequency digital circuit design and the cost. Higher switching frequency also accompanies with higher switching losses and electromagnetic interference (EMI). Therefore, it becomes a design trade-off between the required PWM switching frequency, control resolution, control scheme, and circuit topology. The interleaved PWM technique has been developed for high-density DC-DC converter to enhance its current output capability in VRM applications [5]. The phase-shift interleaved PWM scheme can effectively reduce the output ripple and distortion with its lower switching frequency [6].

Fig. 2 shows the block diagram of a three-phase interleaved class-D audio amplifier system. The core of the digital audio system is a digital multi-phase interleaved PWM control IC with a synchronously controlled multiplexed ADC converter. The input analog audio signal is first converted to an equivalent digital signal and is then

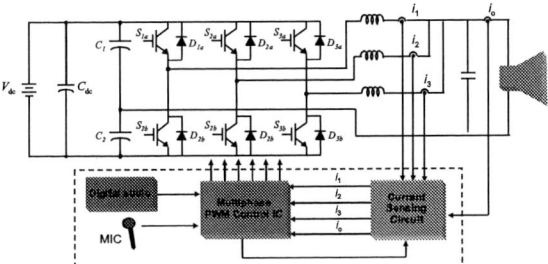

Fig. 2. Block diagram of a three-phase interleaved class-D amplifier audio system.

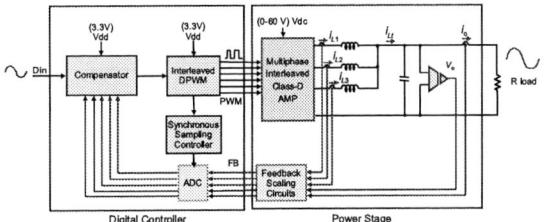

Fig. 3. Functional block diagram of a closed-loop regulated class-D amplifier.

transformed into a PWM signal via a digital PWM generator. An auxiliary phase-shift controller is used to generate a set of even phase-shifted PWM signals. A programmable dead-time controller is designed to prevent shorted-circuit operations of the phase leg. A digital compensator is designed for the loop compensation of the switching amplifier. Fig. 3 shows the functional block diagram for the realization of the proposed digital multi-phase interleaved PWM controller.

This paper proposes a digital realization technique for a multi-phase interleaved PWM inverter with digital closed-loop regulation for class-D audio amplifier to reduce its output THD through a lower PWM switching frequency. The paper is organized as follows. In Section II, the inductor current and output voltage ripple between single-phase and three-phase class-D amplifier are analyzed and compared at their open–loop steady-state. In Section III, we illustrate the digital control IC implement method, including an ADC synchronous sampling, a digital compensator, and a mainly phase-shift PWM generator. Some simulations and experimental results are given as illustrations in Section IV and finally, some conclusions are given in Section V.

II. OPEN-LOOP STEADY STATE ANALYSIS

Fig. 2 shows the power stage of a three-phase interleaved class-D audio amplifier. Its benefit is to reduce the total inductor current ripples, and output voltage ripples. A detailed analysis is as follows.

A. Inductor Current Ripple Analysis

First, assume the single-phase class-D amplifier inductor current ripple maximum value is as follows:

$$\Delta i_{L,1P,\max} = \frac{V_{dc}}{4L_{1P}f_s} \tag{1}$$

where V_{dc} is the input DC voltage by the full leg, f_s is the PWM switching frequency, L_{1P} is the single-phase class-D amplifier output inductor, and the n-phase interleaved PWM class-D amplifier total output inductor current ripple maximum value is as follows:

$$\Delta i_{Lt,nP,\max} = \frac{V_{dc}}{4L_{1P}(nf_s)} = \frac{\Delta i_{L,1P,\max}}{n} \tag{2}$$

From (2) can be find that the n-phase interleaved PWM class-D amplifier total output inductor current ripple maximum value is reduced by n, and the switching frequency can increased by n at the total output inductor current.

B. Capacitor Voltage Ripple Analysis

Assume the single-phase class-D amplifier output voltage ripple maximum value is as follows:

$$\Delta v_{o,1P,\max} = \frac{V_{dc}}{32L_{1P}C_{1P}f_s^2} \tag{3}$$

where C_{1P} is the single-phase class-D amplifier output capacitor, and the n-phase interleaved PWM class-D amplifier total output voltage ripple maximum value is as follows:

$$\Delta v_{o,nP,\max} = \frac{V_{dc}}{32L_{1P}C_{1P}(nf_s)^2} = \frac{\Delta v_{o,1P,\max}}{n^2} \tag{4}$$

From (4) can be found that the n-phase interleaved PWM class-D amplifier total output voltage ripple maximum value is reduced by n^2, and that the switching frequency can be increased by n^2 at the total output voltage.

C. Example

Now we can obtained results from section of A and B to determine the amounts of the n-phase interleaved PWM class-D amplifier inductor per phase and the output capacitor same inductor current ripples and voltage ripples with a single-phase class-D amplifier and an n-phase interleaved PWM class-D amplifier. Aside from this, we can reduce the n-phase interleaved PWM class-D amplifier switching frequency.

We first need to define two parameters:

$$f_{s,1P,eff} = f_{s,1P} \tag{5}$$

$$f_{s,nP,eff} = nf_{s,nP} \tag{6}$$

where $f_{s,1P,eff}$ is the effective switching frequency at the single-phase class-D amplifier inductor current where it can be equivalent to the PWM switching frequency $f_{s,1P}$, $f_{s,nP,eff}$ is the effective switching frequency at n-phase interleaved PWM class-D amplifier total output inductor current, and $f_{s,nP}$ is the PWM switching frequency per phase. For example, assume a single-phase class-D amplifier associated parameter as: $V_{dc} = 100\text{V}$, $f_{s,1P} = 150\text{kHz}$, $f_{s,3P} =$

1193

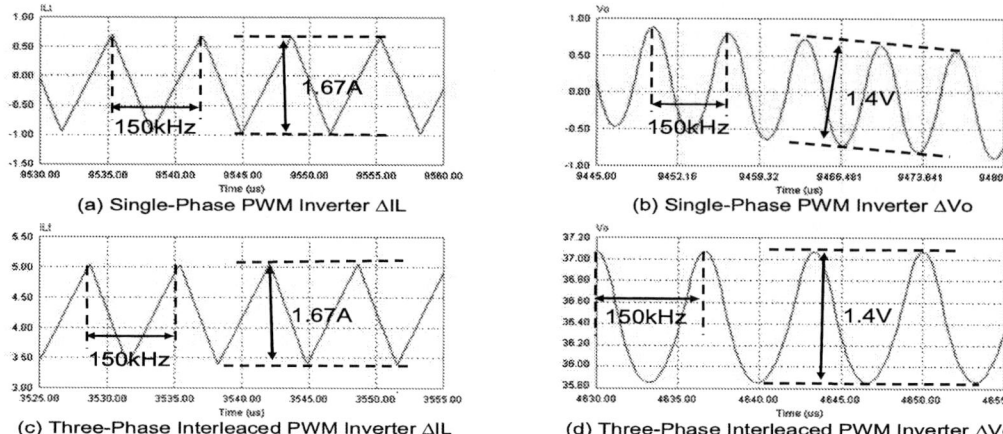

Fig. 4.   Simulation results of the single-phase and three-phase ΔI_L and ΔV_o.

50kHz, $L_{1P} = 100uH$, and $C_{1P} = 1uH$, then the single-phase class-D amplifier inductor current ripple maximum value is as follows:

$$\Delta i_{L,1P,\max} = \frac{V_{dc}}{4L_{1P}f_{s,1P,eff}} = 1.67A \qquad (7)$$

To determine the three-phase interleaved PWM class-D amplifier of the same output total inductor current ripple with a single-phase class-D amplifier, and to reduce the interleaved PWM switching frequency per phase, the three-phase class-D amplifier output total inductor current ripple is as follows:

$$\Delta i_{Lt,3P,\max} = \frac{V_{dc}}{4L_{3P}(3f_{s3P})}$$
$$= \frac{V_{dc}}{4L_{3P}f_{s3P,eff}} = 1.67A \qquad (8)$$

then the inductor can be determined as $L_{3P} = L_{1P} = 100uH$.

The single-phase class-D amplifier output voltage ripple maximum value is as follows:

$$\Delta v_{o,1P,\max} = \frac{V_{dc}}{32L_{1P}C_{1P}f_{s1P}^2} = 1.4V \qquad (9)$$

To determine the three-phase interleaved PWM class-D amplifier of the same output voltage ripple with a single-phase class-D amplifier, and to reduce the interleaved PWM switching frequency per phase, the three-phase class-D amplifier output voltage ripple is as follows:

$$\Delta v_{o,3P,\max} = \frac{V_{dc}}{32L_{3P}C_{3P}(3f_{s3P})^2}$$
$$= \frac{V_{dc}}{32L_{3P}C_{3P}(f_{s3P,eff})^2} = 1.4V \qquad (10)$$

Based on (10), the capacitor can be determined as $C_{3P} = C_{1P} = 1uH$.

So, with the same inductor current and voltage ripple with a single-phase class-D amplifier and n-phase interleaved PWM class-D amplifier condition, we can reduce the n-phase interleaved PWM class-D amplifier switching frequency, witch has the same inductance value per phase and capacitance value. The verification of simulation results are shown in Fig. 4.

III.   DIGITAL CONTROLLER HARDWARE IMPLEMENTATION

A.   Phase-Shift Interleaved DPWM Generator

There are some practical methods from the literature (i.e, counter-comparator [2], delay-line [3], and hybrid one [10]-[11], etc.) that can be used to generate digital PWM. For easy implementation, we chose the first method to realize this block.

Fig. 5 is the proposed DPWM block diagram. This phase-shift interleaved digital PWM (DPWM) generator block can generates up to six phases in real time. The DPWM was designed to apply flexibility to different power applications. The flexible features of the DPWM block include programmable dead-time setting, symmetric or asymmetric reference carrier alternating, PWM switching frequency setting, and the numbers of the phase are used. The DPWM can be clocked to the maximum speed of 200MHz.

There are two independent PWM reference carrier generators which can produce symmetric and asymmetric reference carrier waveforms. By setting the "SYM/ASYM" pin, we can decide which one to pass on to the comparator. This can, according to the phase amount, auto-calculate the delay angle per phase. This result is then sent to the reference carrier generator and, compared with the modulate signal, resulting in a PWM signal. The PWM signal finally passes through a dead-time generator, resulting in a dead time output. The pin descriptions are listed in Table I.

B.   Digital Compensator

Hardware description languages (HDL) are very convenient in implementing complex digital control schemes at the s-domain designed controller discretization

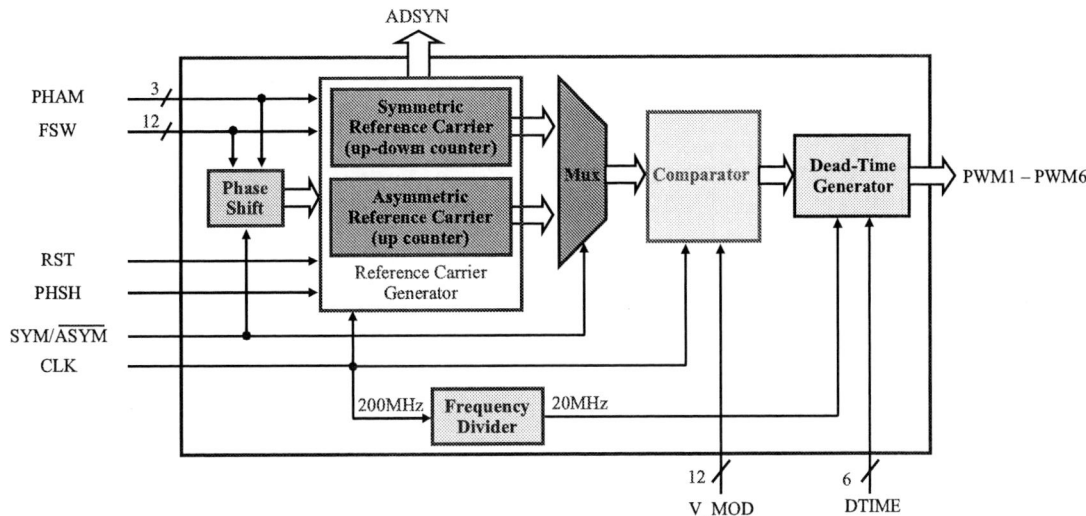

Fig. 5.   Functional diagram of programmable interleaved DPWM generator.

TABLE I   PIN DESCRIPTIONS OF THE PHASE-SHIFT
INTERLEAVED DPWM GENERATOR

Name	Status (In/Out)	Range	Description
PHAM[2:0]	In	1 ~ 6	The numbers of phase setting 001: 1 phase, 010: 2 phase 011: 3 phase, 100: 4 phase 101: 5 phase, 110: 6 phase
FSW[11:0]	In	0 ~ 4095	PWM switching frequency seting Symmetric: $f_{sw} = \text{CLK}/(2 \times \text{FSW})$ Asymmetric: $f_{sw} = \text{CLK}/\text{FSW}$
RST	In	0 or 1	Reset all PWM output 1: Disable PWM output 0: Enable PWM output
PHSH	In	0 or 1	Enable PWM output have phase shift 1: Enable PWM output phase shift 0: Disable PWM output phase shift
SYM / $\overline{\text{ASYM}}$	In	0 or 1	Reference carrier kind setting 1: Symmetric (up-down counter), 0: Asymmetric (up counter)
CLK	In	0 or 1	External reference clock input, maximum clock frequency is 200MHz
V_MOD[11:0]	In	0 ~ 4095	PWM modulation signal input
DT[5:0]	In	0 ~ 63	Dead-time setting $t_{dead} = \text{DT/CLK}$
PWM1-PWM6	Out	0 or 1	PWM output, including complementary pair output, can have Six phase PWM output
ADSYN[11:0]	Out	0 ~ 4095	Reference carrier signal output, output is 12-bit per phase

or at the z-domain designed controller through the direct form I or the direct form II [8]. In this paper, the first order digital compensator is implemented in the direct form II, providing a difference equation which is derived as follows:

$$D(z) = \frac{y(n)}{x(n)} = \frac{A_0 - A_1 z^{-1}}{1 + -B_1 z^{-1}} \quad (11)$$

$$\Rightarrow \begin{cases} y(n) = A_0 \cdot w(n) + A_1 \cdot w(n-1) \\ w(n) = x(n) + B_1 \cdot x(n-1) \end{cases}$$

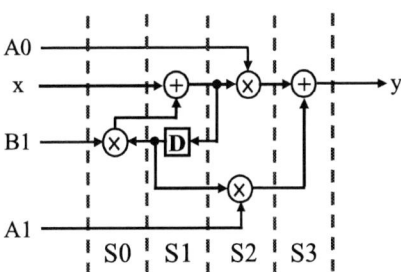

Fig. 6.   Execution sequence of first-order digital compensator.

In order to achieve minimum circuit realization and in consideration of reduction of numerical computation error by fixed-point arithmetic, the realization of the digital controller is based on the concept of state machine by using minimum common system resources.

There are different forms of a digital compensator in regulation of a control loop, such as PI controllers, PID controllers etc. Fig. 6 shows the state diagram in realization of a first-order digital filter using direct form II structure.

C.   Synchronous Sampling Scheme

The class-D amplifier output voltage and current usually follows the PWM switching frequency resulted ripple, which is a resulting spike in the on/off switch, in order to prevent the sampling of an incorrect signal. It can be used for a synchronous sampling scheme to overcome this important problem. On the one hand, it can prevent sampling spike signals and resulting controller error operations, but on the other hand it can also be used to sample average values of one switching period, which can then effectively eliminate ripples. This scheme is to sample the center of a symmetrical modulation waveform in order to derive an arithmetic, average noise-free current sampling. In this paper, we propose that the synchronous sampling

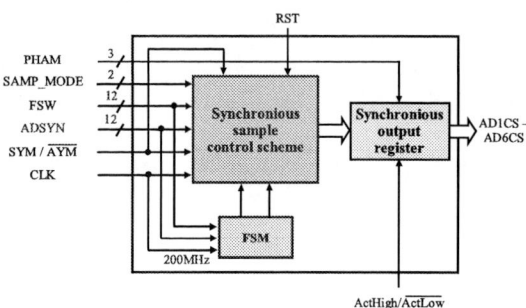

Fig. 7. Functional diagram of the synchronous sampler.

TABLE II PIN DESCRIPTIONS OF THE SYNCHRONIZE
SAMPLING SCHEME

Name	Status (In/Out)	Range	Description
RST	In	0 or 1	Reset all PWM output 1: Disable PWM output 0: Enable PWM output
PHAM[2:0]	In	1 ~ 6	The numbers of phase setting 001: 1 phase, 010: 2 phase 011: 3 phase, 100: 4 phase 101: 5 phase, 110: 6 phase
SAMP_MODE [1:0]	In	1 ~ 3	Select sampling mode 01:To sample at the waveform rising 10:To sample at the waveform falling 11:To sample at the waveform rising and falling
FSW[11:0]	In	0 ~ 4095	PWM switching frequency seting Symmetric: $f_{sw} = CLK/(2 \times FSW)$ Asymmetric: $f_{sw} = CLK/FSW$
ADSYN[11:0]	In	0 ~ 4095	Reference carrier signal output, output is 12-bit per phase
SYM / $\overline{ASYM}$	In	0 or 1	Reference carrier kind setting 1: Symmetric (up-down counter), 0: Asymmetric (up counter)
CLK	In	0 or 1	External reference clock input, maximum clock frequency is 200MHz
ActHigh/ActLow	In	0 or 1	Select controller outputs at active high mode or at active low mode 1: Active high, 0: Active low
AD1CS -AD6CS	Out	0 or 1	To 6 phases AD converter sampling clock.

scheme can be realized in a six-phase synchronous sampling for the interleaved class-D amplifier applications.

A switching converter can generate significant current ripples due to low value of inductance or switching frequency. The increasing of inductance will result larger magnetic core as well as a reduction of the current slew rate. On the other hand, an increase of the switching frequency will result higher switching losses and electromagnetic interference. One design strategy is to optimize the converter performance with a specified switching frequency.

The purpose of a current loop regulator is to control the average current injected into the load, in the example of a buck converter, this current is the average inductor current. The sampling of the current signals plays an important role in the realization of the digital controller for high bandwidth power converter. Fig. 7 shows the functional block diagram of the synchronous sampler and its pin

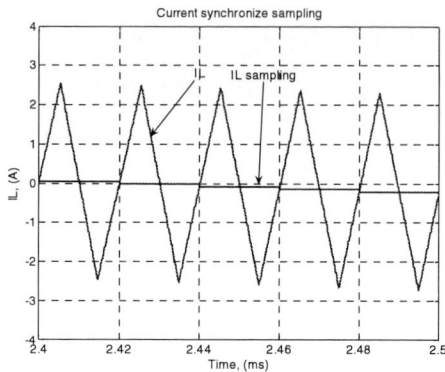

Fig. 8. Simulation results of the current synchronous sampling.

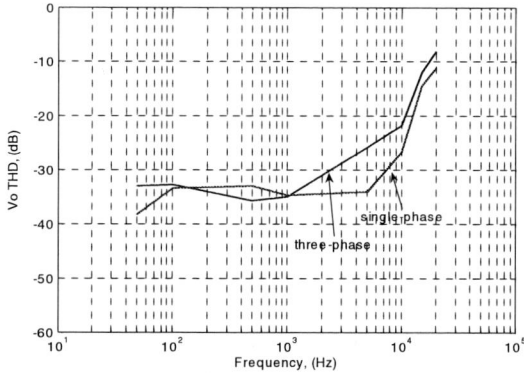

Fig. 9. Simulation results of a comparison between the single-phase and three-phase output voltage THD.

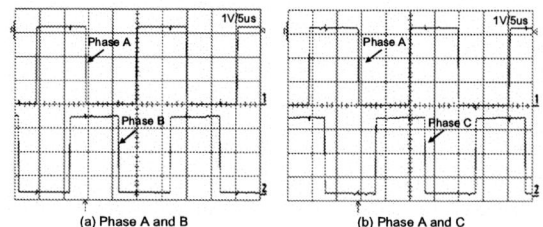

Fig. 10. Three-phase interleaved PWM signals differ by 120 degrees.

descriptions are listed in Table II. This synchronous sampling scheme can set-up three sampling modes through the pin "SAMP_MODE": to sample at the waveform rising slope, to sample at the waveform falling slope, and to sample at the waveform rising and falling slope. The pin of the "ADSYN" is a synchronized sampling signal input, causing the DPWM generator to send out to this pin through asynchronous sampling control scheme, which determines the sampling time point. This result is then saved to the synchronized output register. Further consider that the market condition's existing AD converters have active low and active high operations, so, the pin "ActHigh/ActLow" can be set to satisfied this requirement.

1196

The simulation result of the synchronous sampling controller is illustrated in Fig. 8.

IV. SIMULATION AND EXPERIMENTAL ANALYSIS

The simulation platform is constructed by using co-simulation software with Matlab/Simulink and Modelsim. The Simulink provides a system-level simulation of the multi-phase half-bridge inverter of the digital class-D amplifier and the Modelsim provides a device–level simulation for the VHDL realization of the digital controller. The experiment platform is constructed based on an Altera EP2C35 FPGA development system with a multi-phase half-bridge inverter with a rated output power of 100W.

The programmable feature of FPGA allows designers to construct complex digital circuits easily. Therefore, it is applicable to employ FPGA as a digital controller for power conversion. One special feature of the EP2C35 is that it provides a virtual embedded processor, the NIOS II. It processes many soft IPs such as RAM or ROM controllers, communication interfaces, etc. Designers can employ these IPs in their digital circuits to build a communication bridge to the host computer or other multiple applications. Fig. 9 shows simulation results of a comparison between the single-phase and three-phase output voltage THD. The exhibited the three-phase class-D amplifier output voltage THD is very well at frequencies from 100Hz to 1KHz. Fig. 10 shows experimental results of the three-phase interleaved PWM signals differing by 120 degrees, and verified the proposed interleaved DPWM generator can result from the precision phase-shift PWM signal.

V. CONCLUSION

This paper presents a multi-phase interleaved PWM inverter architecture for the realization of a class-D audio amplifier using a digital closed-loop regulation to reduce its output THD with lower PWM switching frequency. The digital control IC includes an ADC synchronous sampler, a digital compensator, and a phase-shift PWM generator. The proposed multi-phase control scheme can increase the output capability of a class-D amplifier by parallel operation and as well as in reduction of the output current ripples by phase-shifting current control. The synchronous sampling technique can minimize the current detection error under large current ripples and large common-mode noise spikes induced by switching of power devices. The digital current scheme provides a dead-beat control effect for the output current regulation and can achieve wide bandwidth as well as in reduction of the THD over the designed bandwidth. The implementation of the digital class-D amplifier controller was realized by using an Altera's FPGA device EP2C35. Simulation and experimental results are given to verify are given the feasibility and performance of the proposed technique.

REFERENCES

[1] J. S. Chang, M. T. Tan , Z. H. Cheng, and Y. C. Tong, "Analysis and design of power efficient class D amplifier output stages,"

IEEE Trans. Circuits and Systems I: Fundamental Theory and Applications. vol. 47, no. 6, pp. 897-902, 2000.

[2] Albert M. Wu, Jinwen Xiao, Dejan Markovic, and Seth R. Sanders, "Digital PWM control: application in voltage regulation modules," *IEEE Power Electronics Specialists Conf. Rec.*, vol. 1, pp.77-83, July, 1999.

[3] A. Dancy and A. P. Chandrakasan, "Ultra low power control circuits for PWM converters," *IEEE Power Electronics Specialist Conf. Rec.*, pp.21 - 27 , 1997.

[4] R. F. Foley, R. C. Kavanagh, W. P. Marnane, and M. G. Egan, "An area-efficient digital pulsewidth modulation architecture suitable for FPGA implementation," *IEEE Applied Power Electronics Conf. Rec.*, vol. 3, pp. 1412-1418, March, 2005.

[5] D. R. Garth, W. J. Muldoon, G. C. Benson, and E. N. Costague, "Multi-phase, 2 Kilowatt, high voltage, regulated power aupply," *IEEE Power Conditioning Specialists Conf. Rec.*, pp.110 - 116 , 1971.

[6] F. Estes, S. Lentijo, and A. Monti, "A FPGA-based approach to the digital control of a class-D amplifier for sound applications," *IEEE Power Electronics Specialists Conf. Rec.*, pp.122–126, June, 2005.

[7] Ned Mohan, Tore M. Undeland, and William P. Robbins, *Power Electronics*, 3rd., *Willy*, 2003.

[8] Alan V. Oppenheim, Ronald W. Schafer, and John R. Buck, *Discrete-Time Signal Processing* , 2ed., *Prentice-Hall*, 1999.

[9] P. Kollig, B. Al-Hashimi, K. M. Abbott, "Design and implementation of digital systems for automatic control based on behavioural descriptions," *IEE Colloquium*, pp. 2/1-2/4, Feb.1996.

[10] G. Y. Wei and M. Horowitz, "A fully digital, energy-efficient, adaptive power-supply regulator," *IEEE J. Solid-State Circuits.*, vol. 34, pp. 520-528, Apr. 1999.

[11] A. P. Dancy, R. Amirtharajah, and A. P. Chandrakasan, "High-efficiency multiple output dc-dc conversion for low-voltage systems," *IEEE Trans. VLSI Syst.*, vol. 8, no. 3, pp. 252-263, June 2000.

2006 5th International Power Electronics and Motion Control Conference

Integrated IC-like Thyristor–based Switching Structure for Pulse Current Generation to Electronic Ignition

C L Zhang, K S Jeon, C H Ahn, J D Park, E D Kim*, Na Zhi**and Yong Gao**

Semiwell Semiconductor Ltd., Bucheon, Korea

*Korea Eletrotechnology Research Institute, Changwon, Korea

**Xi'an University of Technology, Xi'an, China

E-mail:clzhang@semiwell.com

Abstract—A novel IC-like integration device structure and manufacture process for a thyristor-based capacitor discharging circuit applicable to pulse current generation are studied in this paper. Two structures are compared by means of different design of diode which used as for trigger main thyristor. Substitution of a Zener diode with a p+n avalanche diode in the equivalent circuit can give more stability of pulse-voltage temperature property. Simulation results by Silvaco commercial simulation software indicate that a shallow junction of the avalanche diode has a much improvement in the manufacture process as well as the stable operation at elevated temperature.

Keywords-thyristor; IC; pulse; electronic ignition

I. INTRODUCTION

Fig.1, shows the basic capacitor–discharge application circuit for of low–frequency pulse current generation for the electronic ignition of gas grills and boilers in the home application.

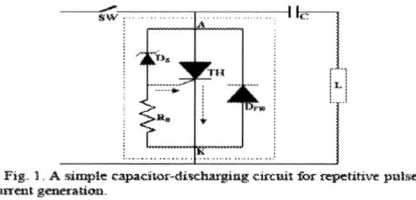

Fig. 1. A simple capacitor-discharging circuit for repetitive pulse current generation.

Fig.1, A simple capacitor-discharging circuit for repetitive pulse current generatoration

In this figure, the thyristor TH is turned on by voltage drop at the parallel resistor RB between gate and cathode of thyristor. This voltage drop is formed by current flowing through the Zener diode DZ as soon as the reverse breakdown happens at the point of capacitor-charging voltage reached to high enough across the DZ. When the thyristor turns –on the capacitor C will discharge to form the pulse current to the coil L so as to develop the high voltage. However, the high current flowing in the Zener diode at the reverse breakdown biased will, of course, assure the triggering thyristor. Again, the thyristor will be turned–off automatically if the anode current becomes smaller than its holding current IH. Therefore the capacitor-charging voltage develops again, so as to result in a repetitive pulse-wave from the circuit.

Although ST Microelectronics has reported that such commercialized in the series of such thyristor switch

1-4244-0448-7/06/$25.00 ©2006 IEEE

integrated monolithically [1], however detailed device structures and manufactured processes of such switch have not been announced. Also, the discrete Zener diode attached to the dedicated switch for commercial product was used.

In this paper, we suggest that IC-like full integrated planar structure design including inner avalanche diode forming by specific P+N junction to replace of Zener diode. The key design viewpoint in this work is based on the better stability in an elevated temperature 125˚C operation to consider the increase the margin of breakdown voltage about 100V in between avalanche diode and the thyritor via the shallow P+N diffusion of 7um for the diode. We believe that with respect to such design it can guarantee the normal operational functions at hot temperature over 125˚C for this monolithic integrated structure.

II. DESIGN TOPICS

The main design topic in this paper focuses on IC-like device of a vertical structure i. e. the monolithic vertical structure compatible with the modern planar process. The doping distribution profile is simulated by a commercial process simulation tool-Athena of Silvaco International Co. USA [2].

We will stress the design on the substitution of Zener diode with the avalanche diode in order to realize the full integration of IC-like device. In generally, the Zener diode is belong to quantum mechanical tunneling of electron in theoretically. In fact, the real Zener effectiveness becomes significant when its depletion width is small enough, i. e. $d < 100A˚$. Therefore, an ideal Silicon Zener diode with doping concentrations in both side higher than 1017 has a breakdown voltage V_{BR} lower than 4.5V[5]. Since the breakdown voltage of Zener diode in the above circuit (Fig.1) limits the capacitor-charging voltage, it is necessary to increase the V_{BR} of Zener diode. However, it is not so easy to make

Zener diode with V_{BR} higher than 7V in reality. Even one can adopt the series stacking of Zener diode to get the high charging voltages, it is so hard to integrate the series-connection of Zener into the monolithic vertical structure we proposal in this study. From other hand, the negative temperature coefficient of V_{BR}[5] will certainly decrease the pulse energy and increase the pulse repetition frequency in the circuit. Therefore, we will propose a novel monolithic structure with different P+N junction depth of avalanche-type diode to act as the triggering turn-on function of main thyristor. It is hopefully that such design can result in good stability during high temperature operation for electronic ignition application.

III. EXPERIMENT

We have proposed the structure 1 as shown in Fig. 2 which has the similar equivalent circuit as in Fig.1. As shown in Fig.2, the P^+N junction avalanche diode (line: a→b) monolithically integrated into the P-base of thyristor part. This avalanche diode replaces of Zener diode to set the turn-on voltage 0.7V at the resistance RB when the avalanche diode breakdown happens. Our first design (hereafter referred to Structure 1) has been conceptive illustration in Fig. 2 which has a deeper P+ well of the triggering diode in compared to the shallow P-base width of thyristor (line: c→d). Such P+ well design has been founded in the recently new power device such as MOSFETs or IGBTs [6].

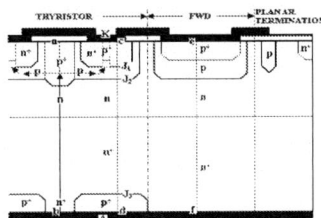

Fig.2, Thyristor-based switch (Structure 1)

When a forward biased voltage is applied to the

thyristor, a reverse bias voltage is formed at the P+n junction on the current paths (arrow) through anode shorting parts i. e. n+-substrate/n/p+/p-base flowing to the cathode shorting holes. The J_2 junction of thyristor part(n+pnn+p+) as well as the pn junction of p+pnn+ of free-wheeling diode (FWD) are reverse-biased simultaneously. In this case, the highest electric field at the p+n junction is, however, developed immediately because of a steeper doping profile with high concentration at p+-well side in the avalanche diode part. The higher the forward applied voltage has, breakdown of the p+n junction of avalanche diode will be happen faster. And therefore, the hole currents are injected into the thyristor p-base and flowing into the p+ cathode shorting holes. When the forward voltage drop at the n+p emitter junction via this hole current is over 0.7V, thyristor will immediately turn-on.

Since the triggering current flowing through the reverse -biased p+n junction, if this current so high it may put the junction under a dangerous condition that is permanent destruction. In this study it is controlled by adjustment of the lateral resistance Rb in the p-base. Suitable Rb can be achieved by proper p-base doping and width.

Fig.3 shows the simulation results on doping profile for Zener diode (avalanche diode), thyrsitor and FWD part, respectively.

It is clear that the p+ depth of avalanche diode (X_{jp+}=30um) is deeper than the p-base width of thyristor (X_{jp+}=25.24um). The deep p+ well can be formed by Boron implantation and drive-in on the n/n+ epi wafer. Since the thyristor and FWD should have the high breakdown voltage compared to avalanche diode, it is necessary to have a low doping p-base concentration formed by Al diffusion with very high diffusivity.

Finally, the n+ -cathode layer is formed by phosphorus diffusion. This structure 1 would need an additional process to form p+ deep well in comparison

with reverse-conducting GTO thyristor but same numbers of masks as those for anode-shorted IGBT. Table I lists the simulation results on breakdown voltage for the equivalent doping profiles in Fig.3.

TABLE I.

Breakdown voltage VBO for structure 1

Avalanche diode with P+ well X_{jp+}=30um	Thyristor part with P-base width X_{jp+}=25.24um	FWD with P diffusion depth X_{jp+}=25.41um
V_{BO}=77V	V_{BO}=133V	V_{BO}=125V

From this table, it is clear that the avalanche diode (V_{BO}=77V) does prematurely breakdown than the thyristor part (V_{BO}=133V). However, the margin of these two values of breakdown voltage is so small that it may result in abnormal operation at elevated temperature such as over 100°C for this integrated structure.

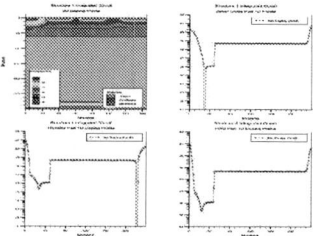

Fig.3, Doping profile for structure 1

The novel structure is proposed in Fig.4 so called structure 2 , are basically similar to the above structure 1 but its manufacture process is different from that of first one because of shallow p+n junction for the avalanche diode. First of all, the p-base depth of thyristor is fixed to constant at Xjp+=37.75um in low doping concentration formed by aluminum diffusion. It has a high value of p-base sheet resistance around Rb=2kΩ determining from circuit of Fig.1. In order to obtain low breakdown voltage of avalanche diode, the shallow p+ depth of formed by boron implantation is needed. For optimization p+ depth, different p+ depth (Xjp+=7um, 13um, 18um) effect to breakdown voltage VBO are simulated by silvaco simulation. Fig.5 indicates the simulated doping profile, electric field Emax and breakdown voltage VBO with p+ depth Xjp+=13um for avalanche diode. In this case, breakdown voltage VBO of avalanche diode has been down to 36V. In the Fig.6, it is founded that the breakdown voltage VBO of thyristor has been raised to 150V. Therefore, the big margin of breakdown voltage 114V between thyristor and avalanche diode has been realized via the structure 2 in our design. Such design structure will have the cost-effective manufacture process for a shallower P+ as well as the bigger margin of breakdown voltage.

The variations of breakdown voltage with junction depth are summarized in table II.

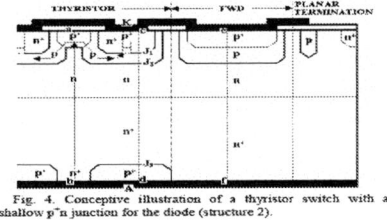

Fig. 4. Conceptive illustration of a thyristor switch with a shallow p⁻n junction for the diode (structure 2).

Fig.4, Conceptive illustration of a thyristor switch with a shallow P+n junction for the diode(structure 2)

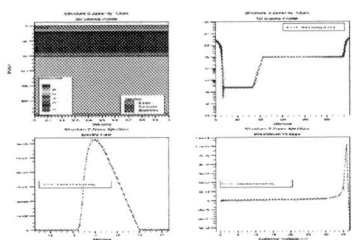

Fig.5, **p**+ depth Xjp+=13um for avalanche diode

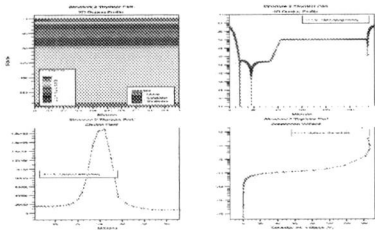

Fig.6, **p**-base depth Xjp=36.6um for thyristor

TABLE II.

Variation of VBO with junction depth of avalanche diode, thyristor and FWD

avalanche Diode Xjp+=13um	avalanche Diode Xjp+=18um	Thyristor Xjp=36.6um	FWD Xjp=36.2um
36V	47V	150V	180V

IV. CONCLUSIONS

Monolithically integration thyristor switch structure has been studied in this work. It is recommended an avalanche diode instead of the Zener diode to be used as for triggering main thyristor. Throughout the comparison of two modified structures, the structure 2 having a shallower p+ well for the avalanche diode is preferred owing to a big margin of breakdown voltage between the diode and thyristor. It is expected that bigger margin will guarantee the stable operation of this switch even though high transient temperature in the working environment.

REFERENCES

[1] Home page: http://www.us.st.com/products

[2] Athena(2D), Simulation Manual, Silvaco International Co., USA.

[3] Atlas(2D), Simulation Manual, Silvaco International Co., USA.

[4] H. Punter et al, "Method of diffusion Aluminum", U S Patent, 4381957, May 3, 1983

[5] C D. Todd, Zener and Avalanche Diodes, pp.1-13, John Wiley &Sons, Inc., New York, 1970.

[6] B J Baliga, Power Semiconductor Devices, Charp.8, PWS pub. Co. Boston, 1996.

[7] P. Streit, "high Power Reverse Conducting GTO", Power Semiconductor Devices and Circuits, Editor by A Jaecklin, pp.63, Plenum Press, New York, 1992.

[8] J. H. Park, "Insulated Gate Bipolar Transistors Having Built-in Freewheeling Diodes", U S Patent, 6051850, April 18, 2000.

2006 5th International Power Electronics and Motion Control Conference

A Wide Bandwidth Current Probe Based on Rogowski Coil and Hall Sensor

Dong Li [*], Guiyou Chen [**]

[*] School of Engineering and Technology, Shandong University of Technology, Zibo, China
[**] School of Control Science and Engineering, Shandong University , Jinan, China
lidong@sdut.edu.cn

Abstract— **A new wide bandwidth current probe for power applications based on a Hall sensor as well as on a Rogowski coil is presented in this paper. Employing these two sensors allows to eliminate the integrator circuit needed for the Rogowski signal. This paper shows that using a special way to merge the output of the Hall sensor with the Rogowski coil, a current probe can be realized which combines the advantages of both sensors. With this new probe DC currents as well as current transients with di/dt's of several kA/us can be measured.**

Keywords-Wide Bandwidth current probe, Sensors, Power semiconductor devices, New devices, Rogowski coil, Hall sensor

I. INTRODUCTION

As a result of the ongoing power semiconductor development power devices, like IGBT, allow to switch currents of some thousand amps in less than a microsecond. To keep the induced voltages as low as possible a low-inductance setup and short interconnections between the power elements are needed. A crucial problem is the current measurement in these low-inductance circuits needed for the detection of faults or for feedback control like the monitoring of paralleled IGBT [1]. Therefore a current probe must be small in size and should not influence the power current signal. The probe should be isolated and capable of measuring DC currents of several hundred amps as well as transients in the order of several kA/us. A current probe was developed which satisfies these requirements. It consists of two well known current measuring principles: a Hall sensor for the low frequencies and a Rogowski coil for the high frequencies.

II. THEORY

The main drawback when using a Hall sensor is its inability to measure high frequency current components. Drift and offset compensated Hall sensors such as spinning current sensors, have a low-pass characteristic with a cut-off frequency of some kHz. Hall sensors are used as stand-alone devices as well as in slotted cores [2] as part of a current compensated

probe. These probes show a good performance at low frequencies but their main disadvantage appears when measuring higher frequencies where the compensating circuitry has to be as fast as the current to be measured. Rogowski coil current probes are based on Faraday's induction law and can therefore not measure DC currents. According to Faraday's law, the output signal is proportional to the time derivative of the current to be measured. To obtain a signal which is proportional to the monitored current, the output signal of the coil must be integrated. This is a major disadvantage when constructing a Rogowski coil for current measurement [3,4,5,6].

Up to now the output signals of these two sensors have never been merged. This paper shows that using a special way to merge the output of the Hall sensor with the Rogowski coil, a current probe(Fig. 1) can be realized which combines the advantages of both sensors.

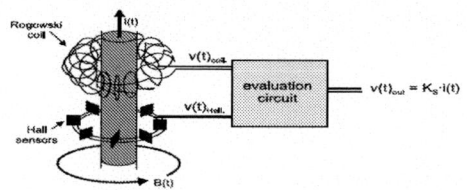

Figure 1.   Current probe overview

A.   Principle of operation

The output signal of a Hall sensor is proportional to the monitored current while the output signal of a Rogowski coil is proportional to the time derivative of the monitored current. Therefore we can model the Hall sensor as a low-pass and the coil as a differentiator. Fig. 2 shows a manner of merging both signals to get an output signal v(t) which is proportional to i(t) independent of the frequency components of i(t).

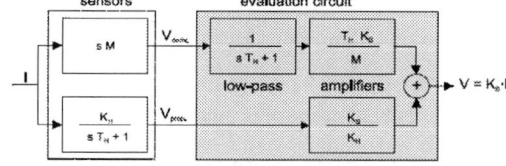

1-4244-0448-7/06/$25.00 ©2006 IEEE

Figure 2. Ideal current probe measuring principle

M is the mutual induction between the primary current and the Rogowski coil. K_H and T_H are the model parameters of the Hall sensor. K_H is the Hall sensor sensitivity which depends on the distance from the copper bar as well as on the supply current. The corner frequency of the evaluation circuit's low-pass is matched to the corner frequency of the Hall sensor $\omega_H = 1/T_H$. K_S/K_H is the sensitivity of the Hall signal path and $T_H * K_S/M$ is the sensitivity of the coil signal path. Both signals are added to form the output signal v(t) of the probe. K_S is the sensitivity factor of the probe and it is considered to be $K_S=1$ V/1000 A in this paper. The transfer function (TF) of the probe (1) is a constant: $H(jw)=K_S$.

$$\frac{V(s)}{I(s)} = K_S\left[\frac{sT_H}{sT_H+1}\left(\frac{M}{M}\right)+\frac{1}{sT_H+1}\left(\frac{K_H}{K_H}\right)\right]=K_S \quad (1)$$

We called this principle of low-passing, scaling and adding principle, which shows a proportional TF. It can always be applied if the same source is measurable with sensors having a low-pass and a high-pass or derivative behavior. However, in the case of current measurement, it is not possible to construct a coil which demonstrates only a derivative behavior for all frequencies. A real coil has some resonance frequencies, but as shown later, they may be neglected in some specific cases.

B. Coil

Several coils of different topology have been realized as multi layer PCB's. These low cost coils showed an ideal inductive behavior till frequencies in the range of tens of MHz. Every winding is normal to the flux generated by the central current bar and is connected to the next winding on the outer circumference. Since the flux decreases with 1/r the winding geometry should be highly regular on the inner circumference. The return of the winding is placed in the inner layer as it is usual for Rogowski coils as shown in Fig. 3.

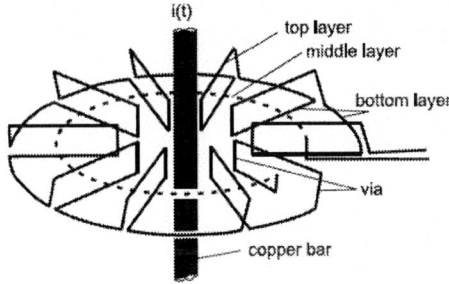

Figure 3. Coil topology

A model of the coil which approximates the coil's impedance till its first resonance frequency is given by (2).

$$Z(s) = \frac{sLc + Rc}{s^2 CcLc + sCcRc + 1} \quad (2)$$

Lc represents the self-inductance, Cc the interwinding capacitance and Rc the coil's wire resistance. For the simulation the following coil parameters have been used: Lc=1.55 uH, Cc=12.45 pF, Rc=3.4 Ohm.

The equation linking the primary current i(t) with the coil's output voltage v_{emf}(t) can be approximated by (3) where Rd represents the load resistance of the following circuit, which should be in the range of some tens of kOhm. A low value of Rd reduces the resonant peak of the coil but at the same time the output signal differs from that of an ideal derivative signal [9]. M is the mutual inductance and its value is about M=12.2 nH for the described coil.

$$\frac{V_{emf}(s)}{I(s)} = \frac{sM}{s^2 LcCc + s(\frac{Lc}{Rd} + RcCc) + \frac{Rc}{Rd} + 1} \quad (3)$$

The electrical model of the coil is given by the series connection of Lc, Rc and a voltage source. This branch is in parallel with Cc and Rd [9]. The voltage source is given by v_{emf}(t)=M*di(t)/dt.

C. Hall sensor

Hall sensors can be approximated by a low-pass with a sensitivity of K_H and a corner frequency of $1/T_H$ (4). The corner frequency of the sensor depends on the semiconductor material and on the device type: a Hall sensor can have corner frequencies of some hundred kHz or higher while spinning current Hall sensors and sensors with an amplification stage show better temperature and linearity behavior but have a corner frequency of some kHz. For our setup a KSY 44 Hall sensor was used with a cutoff frequency of 1 MHz as advised by the manufacturer.

$$\frac{V_{Hall}(s)}{I(s)} = \frac{K_H}{sT_H+1} \quad (4)$$

When using a Hall sensor in a current compensated circuit [2] the effect of stray magnetic fields on the measurement is low because of the magnetic core except when the stray field is strong enough to saturate the core locally. The monitored current can also saturate the core or alter its residual magnetism by a surge current. Since a probe is wanted which is capable to measure up to high frequencies and which will not be damaged by surge currents, the Hall sensor is placed close to the current bar (Fig. 1) and no core material is used. The drawback of not concentrating the magnetic field through a core is that the sensor is prone to measure stray magnetic fields, especially the ones from the return bar. If the geometry of the magnetic field distribution is known, one or few Hall sensors can be used for an accurate measurement. When the field distribution is unknown, more Hall sensors are needed. The exact number of sensors to be placed around the current bar can be computed by applying Biot-Savart's law.

D. Simulation

The TF of the primary current i(t) to the output voltage $v_{emf}(t)$ of the coil can be approximated by (3) with the specified values of the components. The TF of the Hall sensor is given by (4) with $T_H=1/(2*pi*1 \text{ MHz})$. The value of K_H and M are of no interest in the simulation since they are canceled out in (5). K_H, M and K_S are of interest when developing the evaluation circuit. They define the amplification value of the active low-pass. When several Hall sensors are used to increase the measurement accuracy and each sensor has a slightly different cutoff frequency, the block diagram of Fig. 2 must be changed to the one of Fig. 4

. Comparing Fig. 2 with Fig. 5 the following changes can be noted:

The low-pass of the evaluation circuit has been moved past the addition block. This is a significant improvement when designing the evaluation circuit but the Hall sensor signal is now also fed through this low-pass.

The corner frequency of the evaluation circuit's low-pass was changed to $\omega_V < \omega_H$.шω_V should be about 2 or more decades lower than the one of the Hall sensor so that its influence on the Hall signal can be neglected. The sensitivity of this low-pass is, with $T_V=1/(2*\pi*1 \text{ kHz})$ and the above given values, $T_V*K_S/M=13$ representing a realizable amplification.

The ideal coil has been replaced by a more realistic model where $T_2=\text{sqrt}(Lc*Cc)$, $T_1=Cc*Rc+Lc/Rd$ and $T_0 =Rc/Rd+1$.

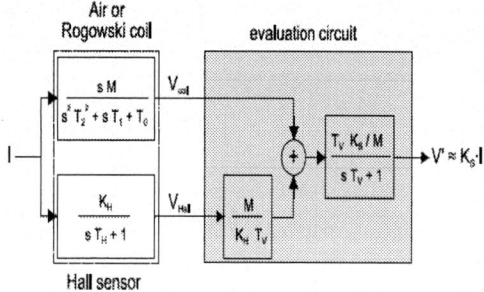

Figure 4.   Practical realization

In this case the principle is not ideal anymore but the TF of the probe will be K_S till at least one decade before the resonance frequency of the coil. As long as neither the coil's TF nor the Hall sensor's TF differ from a constant value (in this case below 20 MHz) the TF of the current probe is constant.

$$\frac{V'(s)}{I(s)} = K_s \left[\frac{sT_V}{sT_V+1} \left(\frac{1}{s^2T_2^2 + sT_1 + T_0} \right) \right] \quad (5)$$

Simulations of the expected TF (5) show this behavior in Fig. 5. The dashdotted line represents the low-pass and the dotted line with its peak at about 36 MHz is the TF of the coil. The dashed line shows the TF based on (5).

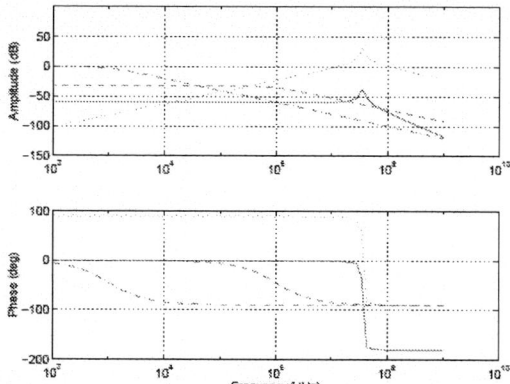

Figure 5.   Bode plot of the probe's transfer function

E. Measurement error

The real Rogowski coil shows a second order system behavior. The TF of the probe will not be constant over the whole frequency domain as it is in the ideal case (1). In the following the influence of the resonance frequency of the coil will be analyzed enabling us to predict the theoretical accuracy class of the probe based on a trapezoidal input signal. This signal was chosen since it is a very common current signal in power electronic applications. The accuracy class definition is given by (6) where $I_{FSprobe}$ is the full-scale (FS) value, i_{output} is the value delivered by the probe and $i_{truevalue}$ is the value obtained by a perfect instrument.

$$i_{output} = i_{truevalue} \pm I_{FSprobe} \left(\frac{Class\_Index}{100} \right) \quad (6)$$

The class index is a number representing the accuracy class of an instrument. The definition (6) always refers to the maximal measurable value, the full scale value. Based on (6) we can therefore define a time dependent class index as follows:

$$Class\_Index(t) = \frac{\left| i(t) - i_{output}(t) \right|}{I_{FSprobe}} 100 \quad (7)$$

The class index of the probe is the maximum of (7) over one period of the current signal Fig. 7. For frequencies above the cutoff frequency ω_V of the low-pass, the TF of the probe can be approximated by (8). This can be seen when analyzing the Bode plot for higher frequencies: the –20 dB/dec asymptote of the low-pass is compensated by the +20 dB/dec asymptote of the coils derivative behavior. The Hall sensor contribution to the final signal can be neglected for high frequencies.

$$H(s) = \frac{K_s}{s^2 T^2 + s T_1 + T_0} \qquad (8)$$

We can further assume that the coil's undamped natural frequency is higher than the cutoff frequency of the low-pass. This was always the case for all constructed PCB coils. Since we want to use the coil as a derivative element, and not as a current transformer, the load resistor Rd should be chosen to be of some kOhm. Since Rc represents the coils wire resistance, which is in the order of some Ohms, the term T_0 is in the following assumed to be $T_0=1$.

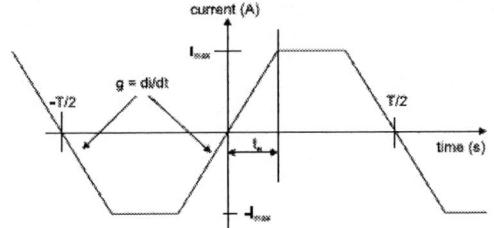

Figure 6.   Trapezoidal waveform

Some more terms are introduced for convenience like the damping ratio d, the undamped natural frequency ω_n and the damped natural frequency ω_d of the coil, the time constant T_a as well as the term α and t_e (9).

$$d = \frac{T_1}{2T_2} \qquad \omega_n = \frac{1}{T_2} \qquad \alpha = \sqrt{1-d^2}$$

$$\omega_d = \omega_n \alpha \qquad T_\alpha = \frac{1}{d\omega_n} \qquad t_e = \frac{I_{max}}{g} \qquad (9)$$

By using the Laplace transforms the response of the system (8) to the ramp r(t)=g*t, with g=di/dt being the slope of the ramp, was computed (10).

$$y(t) = gK_s \left[t - \frac{2d}{\omega_n} + \frac{1}{\omega_n \alpha} e^{\left(\frac{t}{T_\alpha}\right)} \sin(\omega_\alpha t + \varphi) \right]$$

$$\varphi = \arctan\left(\frac{2d\alpha}{2d^2 - 1}\right) \qquad (10)$$

The probe's output signal will follow the ramp after a delay of $2*d/\omega_n$. To keep this delay as small as possible the damping ratio should be small while the undamped natural frequency should be high.

With d=0.1 and ω_n =2*π*36 MHz the delay is below one nanosecond. To reduce the influence of the third term as much as possible, the amplitude of the sine should be made small. This can be accomplished by a high ω_n and a small d as well as by reducing the time constant T_a by a high value of ω_n and d. This short analysis demonstrates that a coil with a high ω_n is always a good choice while d should be kept small. In any case d has to be smaller than 1. Knowing the ramp response of the probe, the absolute measurement error is defined by (11).

$$err(t) = \left| gt - \frac{1}{K_s} y(t) \right| \qquad (11)$$

Note that the output signal of the probe must be back scaled by the factor 1/Ks. From (8) we can now compute the class index as a function of time (12).

$$Class\_Index(t) = \frac{err(t)}{I_{FSprobe}} 100 \qquad (12)$$

To keep the calculations simple, instead of computing the maximum over one trapezoidal waveform, only the error of a ramp signal has been considered, thus the transition of the signal from the steady state to the slope g. The worst case value for a ramp is given by (13).

$$Class\_Index = 100 \frac{g}{\omega_n I_{FSprobe}} (0.66d^2 + 1) \qquad (13)$$

III.   MEASUREMENT

In Fig. 7 on the top side the low frequency performance is demonstrated. The output of the HOKA probe is compared against three well known current probes.   On the bottom side the high frequency performance is demonstrated. The measured di/dt is about 2.5 kA/us. In both cases the relative error was computed with (14) choosing one of the standard current sensors as reference.

$$err(t) = \left(\frac{I_{reference} - I_{HOKA}(t)}{200A} \right) 100\% \qquad (14)$$

For the low frequency signal the ILA SMZ 200 and for the high frequency the LEM 25/10 coaxial shunt has been chosen as reference. The maximum error was +/- 5% in both cases referring to the sensors nominal current of 200

A. This means that the sensor belongs to the accuracy class 5.

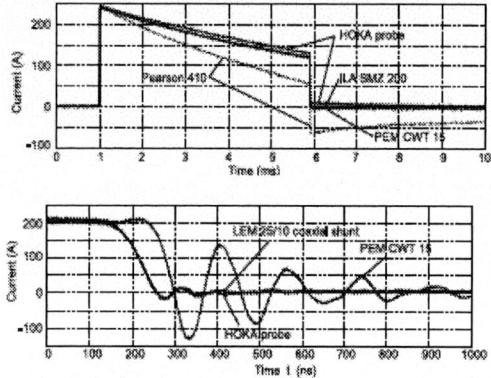

Figure 7.   Current measurement comparison

IV.   CONCLUSION

An isolated current probe based on the new measurement principle HOKA is presented. Merging the output signal of a Hall sensor with a Rogowski coil, enlarges the bandwidth of each sensor. This new probe can measure DC currents as well as di/dt of several kA/us. A relationship between accuracy class, di/dt and undamped natural coil frequency is given. The evaluation circuit, which merges the output signals of the two sensors consists of an active low-pass. A current probe prototype based on this principle was built for a nominal current of 200 A. The probe demonstrated an accuracy class 5 behavior for DC signals and transients up to 2.5 kA/us. Further research will be performed in order to increase the probe's performance.

REFERENCES

[1]   P.HOFER-NOSER and N.KARRER. "Monitoring of paralleled IGBT/diode modules", *IEEE Trans. on Power Electronics,* May 1999, in press.

[2]   *Magnetic Sensors – Data Book*, Siemens, München 81541, Germany, pp. 66-68, July 1996.

[3]   K.HEUMANN. "Messung und oszillographische Aufzeichnung von hohen Wechsel- und schnell veränderlichen Impulsströmen", *Techn. Mitt. AEG-TELEFUNKEN,* vol. 60, no. 7, pp 444-448, 1970.

[4]   A.RADUN. "An alternative low-cost current-sensing scheme for high-current power electronics circuits", *IEEE Trans. on Industrial Electronics,* vol. 42, no. 1, pp.78-84, February 1995.

[5]   W.F.RAY. "Rogowski transducers for high bandwidth high current measurement", *IEE Colloquium on Low Frequency Power Measurement and Analysis,* pp. 10/1-10/6, November 1994.

[6]   W.F. RAY. "Wide bandwidth Rogowski current transducer, part II: The integrator", *EPE Journal,* vol. 3, no.2,pp.116-122,June 1993.

[7]   A.P.CHATTOCK. "On a magnetic potentiometer", *Philosophical Magazin and Journal of Science,* vol. XXIV, 5th Series, pp. 94-96, Jul-Dec 1887.

[8]   W.ROGOWSKI and W.STEINHAUS . "Die Messung der magnetischen Spannung", *Archiv für Electrotechnik,* vol. 1, no. 4, pp. 140-150, 1972.

[9]   N.KARRER and P.HOFER-NOSER. "A new current measuring principle for power electronic applications", *Proc.ISPSD 99,* Toronto,1999, in press

2006 5th International Power Electronics and Motion Control Conference

Voltage Dip Detection Based on an Efficient Least Squares Algorithm for D-STATCOM Application

Thip Manmek, Chathura P. Mudannayake, and Colin Grantham
School of Electrical Engineering and Telecommunication,
The University of New South Wales, Sydney, AUSTRALIA.

Abstract—**This paper presents a fast voltage dip detection technique that is suitable for use in a Distribution Static Synchronous Compensator (D-STATCOM) in compensating balanced dip and unbalanced voltages in power systems. The proposed voltage dip detection method is based on an efficient least squares algorithm which offers structural simplicity and less computational complexity while maintaining dynamic performance and accuracy. It is also robust against distortions present in voltage waveforms. The proposed method extracts the active and reactive parts of the positive- and negative-sequence component for generating reference values of current that need to be injected into the point of connection D-STATCOM in order to compensate the voltage errors. The effectiveness of the voltage dip detection method in the D-STATCOM application has been verified by simulation results.**

Keywords- Distribution Static Synchronous Compensator; voltage dip detection; least squares algorithm.

I. INTRODUCTION

Voltage dips are a common problem in industrial power systems and can be characterized in terms of magnitude, mainly caused by short circuits and starting of large motors. These Voltage dips can cause severe disturbances to industrial equipments. For example, in lower case a clean room process, a power interruption of only a very short duration and magnitude can adversely affect a process. In fact, duration as low as 100 ms with a voltage drop of 25% may only be perceivable as a questionable blink of lights, yet such a voltage dip can initiate a chain reaction of industrial shutdowns and failures that can be catastrophic to a facility's daily profitability [1].

Recently, the distribution static synchronous compensator (D-STATCOM) has been introduced to distribution networks to manage the system reactive power to regulate the voltage at the distribution buses. A D-STATCOM usually consists of a voltage source converter (VSC) connected to the grid in shunt. This system can be used to inject a controllable current into the grid. By injecting a current into the point of common connection (PCC), a shunt-connected VSC can also boost the voltages at that point during a voltage dip.

Furthermore, an unbalance correction can be added to this function [2].

The extraction and tracking technique of voltage dip is the core of the D-STATCOM mitigating control strategy. In order to obtain the required information to control the D-STATCOM further processing on a low voltage signal is required and the processed magnitude is required to be updated as fast as possible. Moreover, the choice of techniques for voltage dip detection is highly dependent on the real-time implementations, on the available computational hardware, and on the amount of computational effort.

Typical standard information tracking or detection methods such as the Fourier transform or the practical digital implementation of it, the discrete Fourier transform (DFT) or Fast Fourier transforms (FFT) and phase-locked-loop (PLL) are normally. Nevertheless the main drawback of DFT and FFT methods is less efficiency in tracking the signal dynamics and that it relies on a uniform window, of the frequency components distribution. Moreover, the DFT is not a fast technique since it needs at least one cycle of the fundamental when dip has commenced before information regarding the magnitude and phase can be assumed accurate. The PLL emerged to overcome the DFT problems that can track changes in supply phase. However this method has a slow response; it takes more than two cycles of the fundamental period [3].

This paper presents an alternative voltage dip detection system for extracting the fundamental component and positive and negative-sequence component of the voltage to generate a current reference signal for the D-STATCOM to compensate voltage dips. The proposed method is based on an efficient least square algorithm and provides fast simple and straightforward techniques. The advantages of the proposed method include a fast transient time response, remarkably accuracy in measurement and its insensitivity to the distortions present in the measuring signal. It has less computational complexity and therefore is suitable for real-time implementation.

In this paper simulation results are provided to demonstrate that the proposed balanced and unbalanced voltage dip detection system is an excellent tool for

1-4244-0448-7/06/$25.00 ©2006 IEEE

extracting the required information for a D-STATCOM application.

II. VOLTAGE DIP DETECTION

A. Overview of the Proposed Efficient Least Squares Method

The instantaneous voltages waveforms in a three-phase power system can be generally expressed as:

$$v_{Sa}(t) = \sum_{i=1,5,7,...}^{k} V_{Sai} \cos(\omega_i t + \beta_{ai}) \quad (1)$$

$$v_{Sb}(t) = \sum_{i=1,5,7,...}^{k} V_{Sbi} \cos(\omega_i t + \beta_{bi}) \quad (2)$$

$$v_{Sc}(t) = \sum_{i=1,5,7,...}^{k} V_{Sci} \cos(\omega_i t + \beta_{ci}) \quad (3)$$

In which V_{Sai}, V_{Sbi} and V_{Sci} are the unknown magnitude values of the three-phase voltages, and β_{ai}, β_{bi}, and β_{ci} are the unknown phase angles. Subscript $i = 1,5,7,...k$ refers to the fundamental and harmonic components respectively. The symbol ω_i is known angular frequency of the ith harmonic.

The proposed method is used to extract the fundamental component of the three-phase voltages. The discrete-time version of three-phase voltage can be written in matrix notation as

$$y = \mathbf{A}x \quad (4)$$

$\mathbf{A}$ is a complex rotation matrix, x is a complex vector consisting of magnitude and phase angle of the input signal. The vector y is the input signal. The proposed method is based on the conventional least squares algorithm. The aim is to solve the system equations without inverting any matrix with real number elements. The proposed least squares algorithm can calculate harmonic components by simply multiplying each set of input signals by a constant matrix. Consequently, the proposed method is immune to transient distortions and unbalanced conditions.

The vector x which consists of a complex signal of fundamental and harmonic component can be found using the proposed method as follows:

$$x = \left((\mathbf{R})^\top \mathbf{R}\right)^{-1}(\mathbf{R})^\top y = \mathbf{C}y \quad (5)$$

where

$$\mathbf{R} = \begin{bmatrix} H_1^0 & H_{-1}^0 \\ H_1^1 & H_{-1}^1 \\ H_1^2 & H_{-1}^2 \\ \vdots & \vdots \\ H_1^l & H_{-1}^l \end{bmatrix}$$

H_1^l is the rotation matrix , ie. $\begin{bmatrix} \cos \omega_1 nT & -\sin \omega_1 nT \\ \sin \omega_1 nT & \cos \omega_1 nT \end{bmatrix}$

l is the number of measured samples (ie. $l>2k$)

$\mathbf{C}$ is a constant matrix, ie. $\left(\mathbf{R}^\top \mathbf{R}\right)^{-1}\mathbf{R}^\top$

It may be noted that the conventional linear least square algorithms requires $(2k)^3 + 8lk^2 + l$ number of multiplications and additions. The proposed method is computationally efficient, since it only performs one matrix multiplication per sample time. The size of the matrix is $2k \times l$ and hence only $2k \times l$ multiplication and addition operations are required. The response time to detect changes in magnitude and phase angle of the fundamental component depends on the sampling period (T) and the number of measured samples (l), can be found as $T \times l$.

B. Detection for Three-Phase System

The fundamental component in term of the complex signals can thus be obtained as follows:

$$\begin{bmatrix} V_{Sa1}\cos(\omega_1 t + \beta_{a1}) \\ V_{Sa1}\sin(\omega_1 t + \beta_{a1}) \\ \vdots \\ V_{Sb1}\cos(\omega_1 t + \beta_{b1}) \\ V_{Sb1}\sin(\omega_1 t + \beta_{b1}) \\ \vdots \\ V_{Sc1}\cos(\omega_1 t + \beta_{c1}) \\ V_{Sc1}\sin(\omega_1 t + \beta_{c1}) \end{bmatrix} = \mathbf{C} \times \begin{bmatrix} v_{Sa}(n) \\ v_{Sa}(n-l+1) \\ \vdots \\ v_{Sb}(n) \\ v_{Sb}(n-l+1) \\ \vdots \\ v_{Sc}(n) \\ v_{Sc}(n-l+1) \end{bmatrix} \quad (6)$$

This method is capable of identifying the fundamental component accurately even though the point of common coupling voltage is strongly corrupted by voltage harmonics.

The identified Sin and Cos terms in the vector shown in left hand side can be used to determine the instantaneous positive-and negative-sequence components as shown in Fig.1.

The fundamental positive- and negative- sequence components are extracted by using equations (7) and (8) respectively.

$$\begin{bmatrix} V_a^+\cos(\omega_1 t + \beta_{a+}) \\ V_a^+\sin(\omega_1 t + \beta_{a+}) \\ V_b^+\cos(\omega_1 t + \beta_{b+}) \\ V_b^+\sin(\omega_1 t + \beta_{b+}) \\ V_c^+\cos(\omega_1 t + \beta_{c+}) \\ V_c^+\sin(\omega_1 t + \beta_{c+}) \end{bmatrix} = \frac{1}{3}\begin{bmatrix} 1 & a & a^2 \\ a^2 & 1 & a \\ a & a^2 & 1 \end{bmatrix}\begin{bmatrix} v_a^f(t) \\ v_b^f(t) \\ v_c^f(t) \end{bmatrix} \quad (7)$$

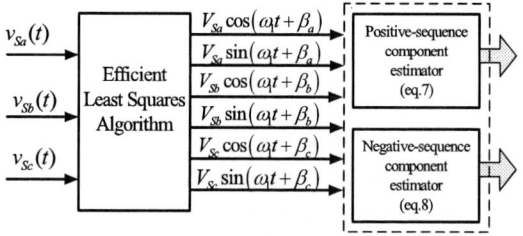

Figure 1. Block diagram of the proposed balanced and unbalanced voltage sequence component detection system.

1208

$$\begin{bmatrix} V_a^- \cos\left(\omega_1 t + \beta_{a-}\right) \\ V_a^- \sin\left(\omega_1 t + \beta_{a-}\right) \\ V_b^- \cos\left(\omega_1 t + \beta_{b-}\right) \\ V_b^- \sin\left(\omega_1 t + \beta_{b-}\right) \\ V_c^- \cos\left(\omega_1 t + \beta_{c-}\right) \\ V_c^- \sin\left(\omega_1 t + \beta_{c-}\right) \end{bmatrix} = \frac{1}{3}\begin{bmatrix} 1 & a^2 & a \\ a & 1 & a^2 \\ a^2 & a & 1 \end{bmatrix}\begin{bmatrix} v_a^f(t) \\ v_b^f(t) \\ v_c^f(t) \end{bmatrix} \quad (8)$$

where $a = e^{j120^\circ}$ is a 120° phase-shift in the time-domain. In (7) and (8), the superscript f identifies the fundamental component which is determined by the proposed least square method from (6), and $+$ denotes positive-sequence component and $-$ denotes negative-sequence component.

III. APPICATION OF PROPOSED VOLTAGE DIP DETECTION METHOD IN D-STATCOM

A. System Configuration

The basic idea of the voltage dip mitigation using a shunt-connected voltage source inverter is to dynamically inject a current of desired amplitude, frequency and phase into the grid line. The configuration of the proposed D-STATCOM is illustrated in Fig. 2. The D-STATCOM consists of a three-phase voltage source converter (VSC), a dc-side capacitor C_{dc} with its leakage resistance R_{dc}, and filter inductance L_F on the ac-side of the inverter. A shunt capacitor with capacitance C_F and L_{tr} are added to the ac-side of the voltage source converter to form a LCL filter that helps to filter out the switching ripple in the voltage effectively. The three-phase supply is represented as an ideal source and an inductance L_S and a resistance R_S, which characterizes the transformer and power cable inductance and resistance respectively. The analysis and design of the D-STATCOM controller are conducted in the rotating reference frame which is synchronized with the voltage vector. This reference frame is usually represented in the $d-q$ reference frame and all the electrical quantities are converter in to $d-q$ axes components.

As may be seen from Fig. 2, the proposed D-STATCOM controller consists of two cascade loops. The V_{dc}^2 outer loop controller regulates the dc-link voltage to a required level. Two $d-q$ current control loops force the converter currents i_{Fd} and i_{Fq} to follow command currents i_{Fd}^* and i_{Fq}^* respectively. The command i_{Fd}^* to the d-axis current loop is obtained by summing the DC-link controller output and d-axis component of the reactive power controllers output i_{Cd}. The command i_{Fq}^* to the $q-$ axis current loop is obtained from the $q-$ axis component of reactive power controllers output i_{Cq}. The modulation signals m_d and m_q to the PWM generator are produced from the decoupling block.

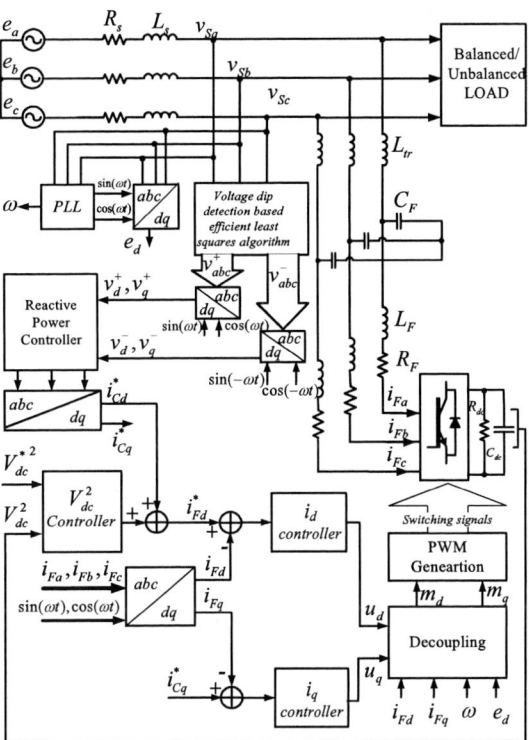

Figure 2. Proposed balanced and unbalanced voltage dip compensation for D-STATCOM.

B. Modelling of Three-Phase PWM Converter

The currents of the three-phase PWM converter in the synchronous reference frame can be obtained by applying Kirchoff's Voltage Law (KVL) and Kirchoff's Current Law (KCL) to the ac-side of the converter. These are represented in (9) and (10).

$$\frac{d}{dt}i_{Fd} = -\frac{R_F}{L_F}i_{Fd} + \omega i_{Fq} - \frac{V_{dc}}{2L_F}m_d + \frac{1}{L_F}e_d \quad (9)$$

$$\frac{d}{dt}i_{Fq} = -\frac{R_F}{L_F}i_{Fq} - \omega i_d - \frac{V_{dc}}{2L_F}m_q \quad (10)$$

where

i_{Fd}, i_{Fq} are $d-q$ axis components of the converter current,

m_d, m_q are $d-q$ axis components of modulation signals,

e_d is $d-$ axis component of PCC voltage,

V_{dc} is dc-link voltage, and

ω is angular frequency in rad/s.

The power delivered from the ac-side of the converter must be balanced with the power received at the dc-side. By neglecting losses in the converter, the following equation for dc-link voltage can be written.

$$\frac{dV_{dc}}{dt} = -\frac{V_{dc}}{R_{dc}C_{dc}} + \frac{3e_d i_{Fd}}{2C_{dc}V_{dc}} \quad (11)$$

C. Design of d-q Current Controllers

Equations (9) and (10), which described the dynamics of the d–q axes currents are coupled to each other (i.e. i_q depends on i_d and visa versa). It is necessary to decouple them for proper control design. Decoupling can be achieved by introducing new terms u_d and u_q to (9) and (10) respectively. The terms u_d and u_q are shown in (12) and (13).

$$u_d = -\omega i_{Fd} - \frac{V_{dc}}{2L_F} m_q \qquad (12)$$

$$u_q = \omega i_{Fq} - \frac{V_{dc}}{2L_F} m_d + \frac{1}{L_F} V_{mp} \qquad (13)$$

Fig. 3 shows the transfer function (G_T) for the decoupled converter in the synchronous reference frame together with the PI (proportional plus integral) controller (G_c) for both d–q axes. The closed loop transfer function of the d–q current feedback loops (inner loops) can be described by:

$$G_{cloop}(s) = \frac{G_c G_T}{1 + G_c G_T} = \frac{K_P s + K_I}{s^2 + \left(\frac{R_F}{L_F} + K_P\right)s + K_I} \qquad (14)$$

By comparing the denominator of (15) with the optimum coefficients of ITAE (integral of time multiplied by absolute magnitude of the error) criterion for a ramp input for 2nd-order transfer function in [4], the parameters of the PI controller (G_c) can be selected as follows:

$$K_P = 3.2\omega_n - \frac{R_F}{L_F} \qquad K_I = \omega_n^2 \qquad (15)$$

where ω_n is the natural frequency of the closed-loop response.

K_P, K_I are proportional gain and integral gain respectively. The dynamic response of the current controller depends on the natural frequency (ω_n), and hence the value of ω_n is chosen for the desired dynamic response.

D. Positive-Negative-Synchronous Referece Frame Controller(Reactive Power Controllers)

In the case of balanced three-phase voltage, the direct transformation of abc voltages into the d–q reference frame will result in dc-quantities. Hence, the D-STATCOM can use a conventional PI-controller to control the injected reactive currents.

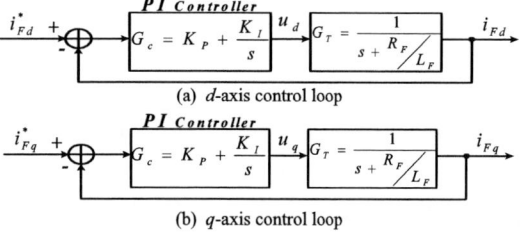

(a) d-axis control loop

(b) q-axis control loop

Figure 3. Equivalent block diagram of the two d–q control loops.

However, if the grid voltage or the load voltages are unbalanced, a ripple of double the grid frequency will occur in the d–q reference frame. In the case of unbalance three-phase voltages, breaking the voltage signals into positive- and negative- sequence components and then transforming into d–q synchronous reference frame results in dc-quantities and these d–q feedback signals allow in control design reactive power control.

Transformation of three-phase balance voltages with unity magnitude into the positive- and negative- d–q synchronous reference frame results in dc-quantities with following values.

$$v_d^+ = 1\ \text{pu}, \quad v_q^+ = 0\ \text{pu},$$

$$v_d^- = 0\ \text{pu}, \quad v_q^- = 0\ \text{pu}$$

where

v_d^+, v_q^+ - dq components of positive-sequence voltage, and

v_d^-, v_q^- - dq components of negative-sequence voltage.

The voltage dips (both balanced and unbalanced) in power systems can be corrected by regulating positive- and negative-synchronous dq components to the values corresponding to the balanced case given above. Three controllers are utilized to regulate these dq components of the positive- and negative- sequence voltages as shown in Fig. 4. As shown in Fig. 4 v_d^+ is regulated to 1 pu via i_q^{+*} in the positive-sequence synchronous reference frame. v_d^- and v_q^- are regulated to zero via the i_q^{-*} and i_d^{-*} in the negative-sequence synchronous reference frame. These dq - current components in the positive- and negative-synchronous reference frames are converted into three axes components and added those together to obtain three-phase currents that need to be injected to the line in order to compensate for the voltage dips.

These abc current commands are then converted to the positive- sequence synchronous reference frame dq components to generate the current commands (i.e i_{Cd}^* and i_{Cq}^*) for the current controllers. All these are indicated in Fig. 4.

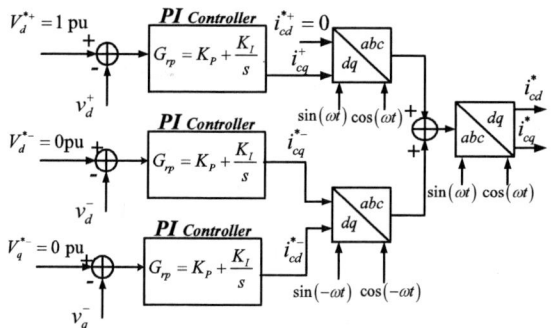

Figure 4. Block scheme of reactive power controller for D-STATCOM.

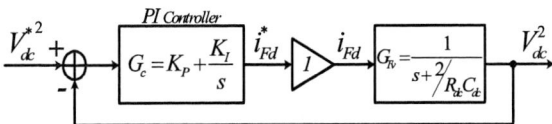

Figure 5. Equivalent block diagram of the voltage control loop.

E. Design of Outer Loop DC Voltage Controller

The voltage control loop regulates the dc-link voltage at a required level. Equation (11) which describes the relationship between V_{dc} and i_{Fd}, is a non-linear equation. This equation is rearranged so that it can be treated with linear control theory. Rearranging(11),

$$\frac{d(V_{dc})^2}{dt} = -\frac{2(V_{dc})^2}{R_{dc}C_{dc}} + \frac{3e_d i_{Fd}}{C_{dc}} \qquad (16)$$

Now, V_{dc}^2 instead of V_{dc} can be used for control design. This does not cause any technical problems since V_{dc} is unidirectional [5]. The block diagram of the voltage control loop is illustrated in Fig. 5. The d-axis inner loop is assumed to be very fast compared to the outer voltage loop so that the inner loop can be replaced with unity gain.

IV. SIMULATION RESULTS

The configuration of the proposed D-STATCOM system shown in Fig. 2 is modeled in MATLAB/SIMULINK as shown in Fig. 6. The D-STATCOM system has been tested for performance under balanced and unbalanced voltage dips with the system parameters displayed in Table I.

Simulations are carried out on two different cases: three-phase 75% balanced voltage dip and three-phase 75% voltage dip with 20% amplitude unbalance. In both cases, the unbalance occurs from t=0.05 to 0.35 sec. The extracted dq component voltages in the positive- and negative- sequence synchronous reference frame for 75% balance voltage dip using the proposed efficient least square method are shown in Fig. 7.

TABLE I.

DESIGN SPECIFICATIONS AND CIRCUIT PARAMETERS

Three-phase supply, phase voltage: (V_s) (rms)	120V=1 pu, 50Hz
Sampling frequency (f_s)	5000 Hz
Converter Switching frequency (f_{sw})	5000 Hz
DC-link voltage (V_{dc})	700 V
Source inductance (L_S)	6 mH
Filter inductance (L_F)	2 mH
Capacitor (C_F)	60 μF
Leakage inductance (L_{tr})	4 mH
Natural frequency d-q control loop (ω_n)	942.47 rad/s
Natural frequency V_{dc}^2 control loop (ω_{n_dc})	15.7 rad/s
Sampling frequency of voltage dip detector (T)	4000 Hz
Number of measured sample (l)	40

As can be seen from Fig. 7 the amplitude of the d-component of positive-sequence reduced by 0.25 p.u and the q-component of positive- sequence is zero. Also, the amplitude of the dq component of negative-sequence remains zero.

Fig. 8 shows the extraction of instantaneous voltage dq components of positive- and negative- sequence synchronous reference frame during 75% voltage dip with 20% unbalance. The amplitude of the d-positive-sequence component decreases to 0.75 p.u. and the q-component remain zero. In negative-sequence, the d-component has amplitude of 0.1 p.u. and has zero q-component during unbalanced voltage dip.

It can be noted from Figs 7 and 8 that the proposed efficient least square method is capable of tracking symmetrical components accurately within less than half a cycle of a fundamental period.

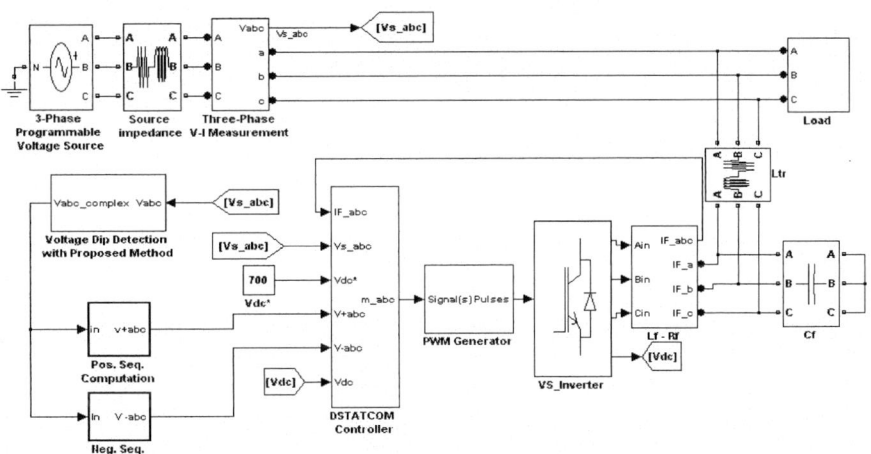

Figure 6. MATLAB/SIMULINK model for the proposed D-STATCOM.

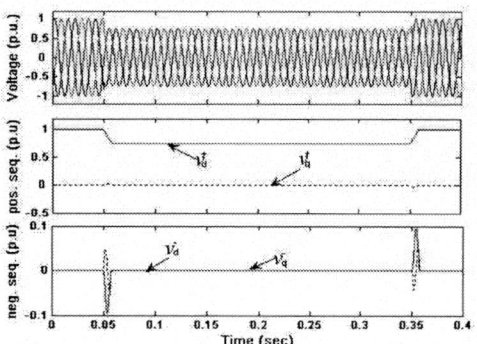

Figure 7. Identification of dq component in v_d^+ and v_d^- sequence synchronous reference frame for 75% balanced voltage dip.

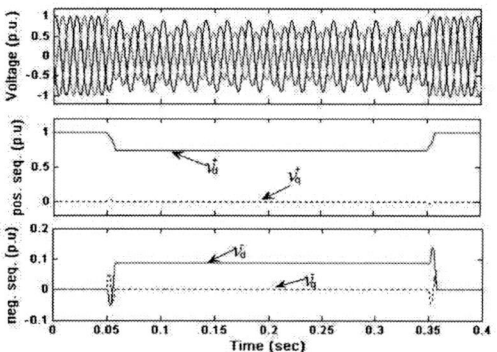

Figure 8. Identification of dq component in v_d^+ and v_d^- sequence synchronous reference frame for 75% voltage dip with 20% unbalance.

Figs. 9 and 10 show the voltage waveforms after being compensated by the proposed D-STATCOM in the cases of balanced and unbalanced voltage dips that have been depicted in Fig. 7 and Fig. 8 respectively. The load voltage is constant at 1p.u before, during and after dip, apart from the transient at the start and the end of the dip. The error in the voltage magnitude is also shown in Fig. 9 and Fig. 10. As can be seen the error is very small in the steady state, while during the transient at the start and the end of the dip, error less than 0.2 p.u. appears for a short duration.

It should be noted that the LCL filter does introduce a delay in the controller response time. However, as the controller response is limited to 1 cycle, this delay has little effect on the overall controller performance.

V. CONCLUSIONS

This paper described an alternative method for determining the supply voltage parameters for the D-STATCOM. The proposed efficient least squares algorithm is used to determine the positive- and negative-sequence component such that the control scheme can be realized in real-time. In order to handle balanced and unbalanced voltage dips, the control strategy is based on separating the positive- and negative-sequence and controlling them separately.

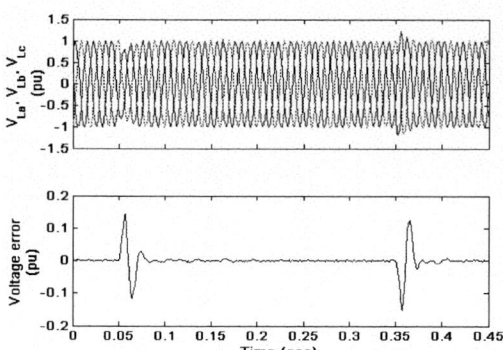

Figure 9. Compensated voltage waveforms and error in the voltage magnitude in the case of 75% balanced voltage dip.

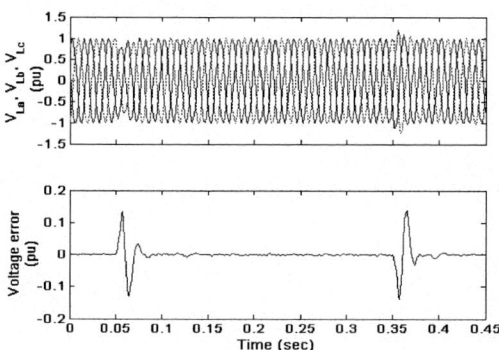

Figure 10. Compensated voltage waveforms and error in the voltage magnitude in the case of 75% voltage dip with 20% unbalance.

Through analysis and simulation results, it is shown that the proposed algorithm requires less than half a cycle period to regain the accurate tracking supply voltage parameter after a sudden change in actual supply waveform. The simulation results also demonstrate that the D-STATCOM maintains a balanced voltage at 1 p.u. before and after the system disturbance. Therefore, the proposed voltage dip detection system is a suitable tool for extracting the required control information for a D-STATCOM application.

REFERENCES

[1] T. Davis, G.E. Beam, C.J. Melhorn, "voltage sags: their impact on the utility and industrial customers," *IEEE Trans. Industry Applications,* vol. 34, no.3, pp. 549-558, May-June 1998.

[2] Clark Hochgraf and Robert H. Lasseter, "Statcom controls for operation with unbalanced voltages," *IEEE Trans Power Delivery,* vol. 13, no. 2, pp.538-544, April 1998.

[3] Vikram Kaura, Vladimir Blasko, "Operation of a phase locked loop system under distorted utility conditions," *IEEE Trans. Industry Applications,* vol. 33, no. 1, pp. 58-63, January 1997.

[4] William S. Levire, Control Handbook, Boca Raton, Fl:[New York,NY] CRC Press Handbook, pp. 169-173, 1996.

[5] Y. Ye, M. Kazerani and V. H. Quintana, "Modeling control and implementation of three-phase PWM converter," *IEEE Trans. Power Electron.,* vol. 18, no. 3, pp. 857-864, May 2003.

2006 5th International Power Electronics and Motion Control Conference

Optimal Design and Analysis on Bearingless Permanent Magnet-type Synchronous Motors Using Finite Element Method

Chang Jiang , Huangqiu Zhu and Zhenyue Huang

Jiangsu University / School of Electrical and Information Engineering, Zhenjiang 212013, China

nuaajiangchang@sohu.com

Abstract—There are complicated relationships among the radial suspension forces, configuration of windings, permanent magnet thickness and currents in bearingless permanent magnet-type synchronous motors (BPMSM), so researching these relationships has important reference value for designing and optimizing BPMSM. Based on the principle of producing radial suspension forces in BPMSM, the mathematics models of radial forces are deduced. The gap magnetic circuits of BPMSM are studied using Finite Element Method when the currents in radial force windings are changed. The demagnetization of permanent magnets is considered. The most critical area in the permanent magnets is made clear for both torque and radial force generations. The relationship between the radial suspension forces and permanent magnet thickness is calculated and analyzed when the permanent magnet thickness is changed under the fixed motor gap. The radial suspension force and the Maxwell force of the additional 2-pole radial windings and additional 6-pole radial windings are compared under the 4-pole motor windings. The relationship between radial suspension force and current is tested on prototype machine with p_M=2 and p_B=3 under the state of static suspension; the experiment conclusions have proved that the account results are accurate by using ANSYS software.

Keywords- bearingless motor; permanent-type magnet motor; Finite Element Method; radial suspension force

I. INTRODUCTION

Magnetic bearings have been applied to high-speed motors such as machine tools and turbomolecular pumps [1]. The advantages of magnetic bearings are no mechanical contacts, no lubrication and no wear, etc. However, long shaft is required in these high-speed motors with magnetic bearings because magnetic bearings are arranged in the both ends of the motor. Therefore, the critical speeds are decreased, and increasing speed of the motor is limited some extent. Bearingless motors which combine the functions of magnetic bearings and electric motors can be used to realize high-speed and high power

Project supported by National Natural Science Foundation of China (50275067), High technology research of Jiangsu Province (BG2005027), and by SRF for ROCS, SEM.

drives. Radial force windings are wound together with conventional motor windings to produce radial magnetic forces. Bearingless motors have some advantages over conventional high-speed motors with magnetic bearings as follows: a) the motor shaft length can be short, so the critical speed is high if the output power is equal. b) High power can be achieved if shaft lengths are the same. c) DC exciting current of conventional magnetic bearings is not needed [1]-[7].

Recently, permanent magnet synchronous motors have been developed due to developments of magnet material with strong magnetization and mechanical configuration to hold magnets even though rotors are rotated at high speed [3]. Permanent magnet-type synchronous motors have considerable possibilities in super high-speed and high power applications. In this paper, an analysis is done for a prototype bearingless permanent magnet-type motor.

In the bearingless permanent magnet-type motor, the thickness of the permanent magnets on rotor surface is one of important keys in designing bearingless motors. Thick permanent magnets have high equivalent MMF of exciting currents. Radial magnetic forces are proportional to the equivalent motor exciting current of a permanent magnet. Therefore, relationships of radial suspension forces, radial force wingdings currents and airgap flux density should be derived to determine thickness of permanent magnets. The design and selection of permanent magnet thickness and airgap length between the rotor and stator iron cores are some of the most important aspects [3].

In this paper, the expressions between radial suspension forces and currents in the radial force windings are deduced. The relationships between the permanent magnet thickness and radial suspension forces are also educed. A limitation of demagnetization is analytically derived. The flux densities on the surfaces of permanent magnets are obtained as a function of the motor and radial force winding currents. The optimal designed point in producing radial suspension forces is found. One prototype permanent magnet motor with p_M=2 and p_B=3 was designed and constructed for optimal radial force production. The motor is tested to confirm the theoretical results.

II. Principles of Radial Force Production

In order to produce controllable radial forces, the pole pairs relationship between torque windings p_M and radial force windings p_B should be $p_B = p_M \pm 1$ [2]. Fig. 1 shows the principles of radial force production. Additional 2-pole radial force windings N_x and N_y are wound in the stator slots together with conventional 4-pole motor windings N_a and N_b. The radial forces can be produced by the unbalanced flux density in the airgap caused by the interactions between 4-pole excitation fluxes Φ_p of permanent magnets and the fluxes generated by 2-pole radial force winding currents i_x and i_y. For example, if the positive radial force winding current i_x exits in N_x winding as shown in Fig. 1, the 2-pole fluxes Φ_x are generated. Therefore, the flux density is increased in the airgap 1 while decreased in the airgap 3. Radial force F is generated in the positive direction of x-axis. If the radial force winding current is negative, a radial force can be produced in the negative direction of x-axis. With current in the N_y winding, y-axis direction force can be produced.

It can be seen that the paths of 2-pole fluxes Φ_x pass through the permanent magnets A and B. It is well known that the permeability of permanent magnets is approximately equal to permeability in the air. Therefore, the thick permanent magnets result in large 2-pole MMF requirements to produce radial force. Consequently, thin permanent magnets are preferred to generate radial force efficiently. However, thick permanent magnets have advantages to achieve reasonable flux density for motor performance as well as to avoid demagnetization. So there exists an optimum thickness of permanent magnet to produces radial forces most efficiently. The relationships between the radial forces and permanent magnet thickness are described in the next section.

III. Mathematics Model of Radial Forces

In this paper, a two-phase machine model is used for simplicity, though a three-phase is practical. The current and flux linkage of BPMSM are shown in a synchronous rotating reference frame. The relationship between the radial suspension forces of the BPMSM and the currents of radial force windings can be expressed as

$$\begin{cases} F_{ix} = (K_M \pm K_L) \cdot (i_{2d} \cdot \psi_{1d} + i_{2q} \cdot \psi_{1q}) \\ F_{iy} = (K_L \pm K_M) \cdot (i_{2q} \cdot \psi_{1d} - i_{2d} \cdot \psi_{1q}) \end{cases} \quad (1)$$

F_{ix}, F_{iy} are the radial suspension forces which are composed of Maxwell forces and Lorentz forces. K_M is Maxwell forces constant. K_L is Lorentz forces constant. i_{2d}, i_{2q} are current components of radial force windings. ψ_{1d}, ψ_{1q} are the airgap flux linkages components of motor windings.

In addition, according to the Theory of Electromagnetic Field, when the rotor is out of the center, another radial force will exist. This effect is known as the magnetic tensile force in the electromagnetic field of electrical

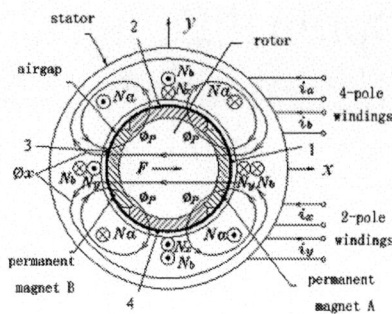

Figure 1. Principles of radial force production

motor. The generated Maxwell forces F_{sx}, F_{sy} are in proportion to the displacement. They are inherent forces and can be written as

$$\begin{cases} F_{sx} = k_s x \\ F_{sy} = k_s y \end{cases} \quad (2)$$

where $k_s = k \cdot \dfrac{\pi r l B^2}{\mu_0 \delta}$; k_s is the force-displacement coefficient. μ_0 is the vacuum permeability. δ is the airgap length. k is the attenuation factor, $k \approx 0.3$.

So the radial suspension force F_x and F_y of x- and y-direction can be expressed as

$$\begin{cases} F_x = F_{ix} + F_{sx} \\ F_y = F_{iy} + F_{sy} \end{cases} \quad (3)$$

Substituting (1), (2) into (3), while (3) can be written as

$$\begin{cases} F_x = (K_M \pm K_L) \cdot (i_{2d} \cdot \psi_{1d} + i_{2q} \cdot \psi_{1q}) + k_s \cdot x \\ F_y = (K_L \pm K_M) \cdot (i_{2q} \cdot \psi_{1d} - i_{2d} \cdot \psi_{1q}) + k_s \cdot y \end{cases} \quad (4)$$

When $p_B = p_M + 1$, (4) can be written as

$$\begin{cases} F_x = (K_M + K_L) \cdot (i_{2d} \cdot \psi_{1d} + i_{2q} \cdot \psi_{1q}) + k_s \cdot x \\ F_y = (K_L + K_M) \cdot (i_{2q} \cdot \psi_{1d} - i_{2d} \cdot \psi_{1q}) + k_s \cdot y \end{cases} \quad (5)$$

When $p_B = p_M - 1$, (4) can be written as

$$\begin{cases} F_x = (K_M - K_L) \cdot (i_{2d} \cdot \psi_{1d} + i_{2q} \cdot \psi_{1q}) + k_s \cdot x \\ F_y = (K_L - K_M) \cdot (i_{2q} \cdot \psi_{1d} - i_{2d} \cdot \psi_{1q}) + k_s \cdot y \end{cases} \quad (6)$$

The stator flux linkage equation is as follows

$$\begin{cases} \psi_{1d} = L_d i_{1d} + \psi_r \\ \psi_{1q} = L_q i_{1q} \end{cases} \quad (7)$$

where ψ_{1d} and ψ_{1q} are airgap flux linkages. ψ_f is rotor flux linkages. L_d and L_q are the self-inductance of motor windings in the 2-phase coordinate, respectively.

IV. FEM ANALYSIS FOR PROTOTYPE MOTOR

In this paper, a bearingless motor having additional sets of 6-pole windings in 4-pole motor stator to produce radial force is proposed. A Finite Element Method analysis is used for a prototype permanent magnet bearingless motor.

Fig. 2 (a) and Fig. 2 (b) show the model and mesh of surface-mounted permanent magnet-type bearingless motor. Both 4-pole motor windings and 6-pole radial force winding are wound in the stator slots for torque and radial suspension forces generations, respectively. The diameter of each wire is 0.63 mm. The rated line current is set to 2.86 A. The outer and inner diameter of stator iron core is 155 mm and 98 mm, respectively. The stack length of the iron core is 105 mm. The iron core is made of laminated silicon steels. Permanent magnets with thickness of 3 mm are pasted on the surface of rotor iron. The outer diameter of rotor iron is 88 mm. Thus, the airgap length between the stator inner surface and outer surface of permanent magnets is 2 mm. NbFeB permanent magnets are employed in the prototype machine. The permanent magnets are pasted to produce four-pole flux distribution.

Fig. 3 shows the flux distribution at no load. The rotor is positioned at the center of the stator. No current are supplied to the stator windings. Symmetrical four-pole fluxes are generated by permanent magnets.

Fig. 4 shows the flux distribution with radial force windings current of 15 A. The 4-pole symmetrical flux distribution, generated by permanent magnets as shown in Fig. 3, is unbalanced by 6-pole fluxes. As shown in Fig. 4, it is obvious that the fluxes in the negative direction of x-axis are fewer than that in the positive direction of x-axis. Therefore, the radial force is produced in the positive direction of x-axis due to the unbalance of the airgap flux densities. The principles of radial force production are confirmed intuitively by the ANSYS software.

In Fig. 5, the radial forces on the surface of rotor are shown. Fig. 5 (a) shows the radial forces in x-direction. Fig. 5 (b) shows the radial forces in y-direction. It can be seen that the radial force is produced in the positive direction of x-axis. Radial forces in y-direction are equal to zero.

In the surface-mounted permanent magnet-type machines, thin permanent magnets and small airgap lengths are preferred to generate the radial forces effectively. However, thin permanent magnets are easy to be demagnetized. Thus, it is very important to consider demagnetization of permanent magnets. It is undesirably possible that both the radial force fluxes and torque ones pass through a permanent magnet against the pre-magnetization direction. This condition is the most critical for demagnetization. The flux densities in the critical area

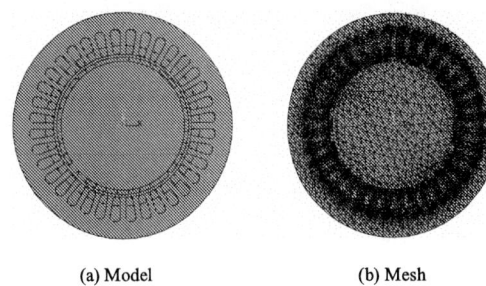

    (a) Model   (b) Mesh

Figure 2.   Cross-section of the test machine

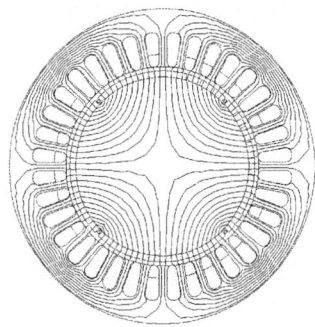

Figure 3.   Flux distribution at no current

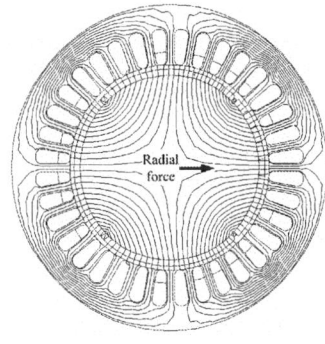

Figure 4.   Flux distribution with radial force

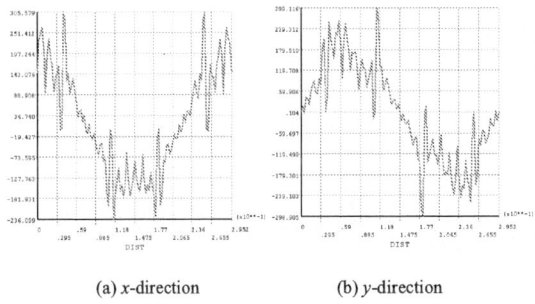

    (a) x-direction   (b) y-direction

Figure 5.   Schematic diagram of radial forces on the surface of rotor

for demagnetization are decreased as the current is increased [5].

Fig. 6 (a) shows torque and radial force fluxes. In this case, the torque current flows in the positive direction in the motor winding N_b, then the torque fluxes Φ_4 are generated. The rotor is rotated in clockwise direction. In the areas A, B, C and D, the torque fluxes pass through the permanent magnets opposite to the pre-magnetization direction. Thus, these four areas are critical for demagnetization. The resultant radial force fluxes Φ_6 are generated as shown in Fig. 6 (a). Both the torque fluxes Φ_4 and radial force fluxes Φ_6 pass through the area A opposite to pre-magnetization direction of the permanent magnet. Thus, the area A is the most critical area for demagnetization. In this flux distribution, a radial force is generated in the direction of F as shown in Fig. 6 (a) due to the unbalance of the airgap flux densities.

Fig. 6 (b) shows flux densities on the surfaces of permanent magnets. The surface of the permanent magnets is the critical part of demagnetization. It can be seen that the flux densities are more than 0 T as the function of the rated torque current of 2.86 A and the rated radial force current of 2.86 A. Thus, there is no demagnetization.

As mentioned in chapter III, given a motor's airgap, permanent magnet has an optimal thickness to produce the most effective radial force. Fig. 7 shows the relationship of radial force produced by unite current in radial force windings and permanent magnet thickness l_m. It can be seen that radial force changes according to the thickness of permanent magnets when the airgap is given. But there exists an optimal thickness to produce the maximum radial force. In this paper, the motor's airgap length is 2 mm. The radial force for unit current arrives at the maximum value of 45 N/A when permanent magnet thickness is around 1.8 mm.

From the principle of bearingless motor [6], radial forces consist of Maxwell force and Lorentz force. When the pole pairs relationship between torque windings p_M and suspension windings p_B is $p_B = p_M + 1$, Maxwell force and Lorentz force have the same direction. Therefore, radial forces are the sum of them. When $p_B = p_M - 1$, Maxwell force and Lorentz force have the reverse direction. Radial forces are the difference of them. In Fig. 8, the pole pairs of motor windings are 4. Curve 1 shows the relationship between radial forces and 6-pole radial force windings current. Curve 2 shows the relationship between radial Maxwell force and 6-pole radial force windings current. Curve 3 shows the relationship between Maxwell force and 2-pole radial force windings current and curve 4 shows the relationship between radial forces and 2-pole radial force windings current. As shown in Fig. 8, it can be seen that radial forces generated by 6-pole radial force windings are larger than that of 2-pole radial force windings under the same mechanical parameters. The results indicate that radial forces consist of Maxwell

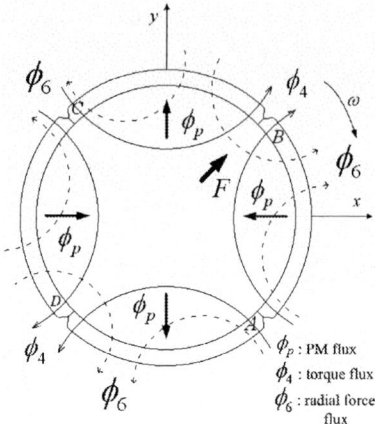

(a) Torque and radial force fluxes

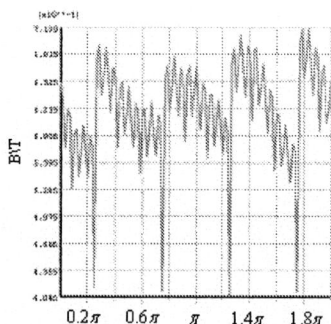

(b) Flux densities on the surfaces of permanent magnet

Figure 6.   Flux orientation and demagnetization

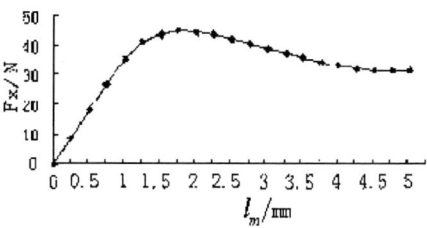

Figure 7.   Relationships of radial force and thickness of permanent magnet

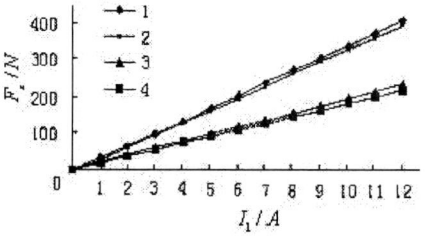

Figure 8.   Relationships of the radial force, Maxwell force and current

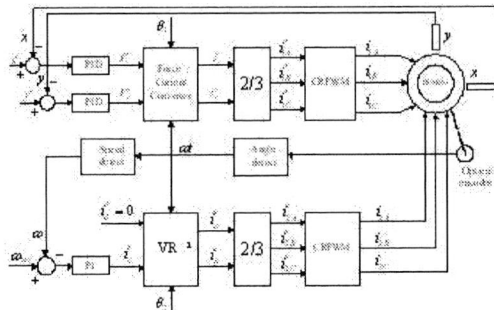

Figure 9. Control system of BPMSM

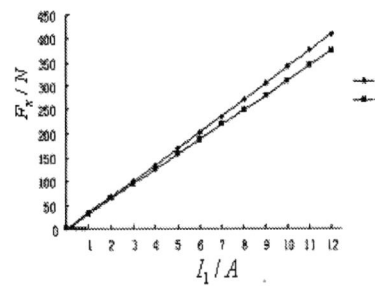

Figure 10. Relationships between theory and actual measurement radial forces and currents

force and Lorentz force. Maxwell force is much larger than Lorentz force.

V. EXPERIMENT RESULTS OF SYSTEM CONFIGURATION

Specifications of the prototype machine have been mentioned in chapter IV. Fig. 9 shows the control system of BPMSM. The relationship between radial suspension forces and current is tested on prototype machine with $p_M=2$ and $p_B=3$ under the state of static suspension. The loads are applied to the BPMSM in the x-direction. Radial force is measured when the radial force windings current I_1 is applied from 1 A to 12 A. In Fig. 10, Curve1 shows the values of radial force analyzed by using ANSYS software. Curve 2 shows the experimental values. It is found that theoretical values are 9% larger than the experimental values. The experiment conclusions have proved that the account results are accurate by using ANSYS software.

VI. CONCLUSION

In this paper, a prototype of bearingless permanent magnet-type motor is set up. The relationships between radial force and radial force windings current are analytically derived. The demagnetization of permanent magnets is considered. The most critical area in the permanent magnets is made clear for both torque and radial force generations. There is no demagnetization. The permanent magnet thickness to produce the maximal radial force with 2 mm airgap length is calculated. The studied results indicate that radial force is different with the different pole pairs of the redial force windings. Under the condition of the same radial force windings current and two pole pairs of motor windings, radial force of motor with three pole pairs is larger than that of motor with one pole pairs. It is confirmed by one prototype motor. The results provide the basis for optimizing and designing bearingless permanent magnet-type motors.

REFERENCES

[1] M. Ooshima, A. Chiba, T. Fukao, "Characteristics of a permanent magnet type bearingless motor," *IEEE Trans. Indus. Application*, vol. 32, pp: 363-370, April 1996.

[2] J. Bichsel, "The bearingless electrical machine,"in *Proc. Int. Symp. Magn. Suspension. Technol. NASA Langley Res. Center*, Hampton, 1991, pp. 561-573.

[3] M. Ooshima, A. Chiba, T. Fukao, et al, "Design and analysis of permanent magnet-type bearingless motors," *IEEE Trans. Indus. Electr*, vol. 43, pp: 292-299, April 1996.

[4] Huangqiu Zhu, Zhiquan Deng, Yangguang Yan, et al, "Principles of bearingless motors and research status," *Micromotors*, vol. 33, pp. 29-31, December 2000.

[5] A. Chiba, M. Ooshima, S. Miyazawa, et al, "An analysis of a Prototype Permanent Magnet-Type Bearingless Motor using Finite Element Method," in *Proc. 5th International Symposium on Magnetic Bearings*, Kanazawa, 1996, pp. 351-356.

[6] Xianxing Liu, Yuxin Sun, Huangqiu Zhu, et al, "Development, application and Prospect of Bearingless Permanent Magnet-type Motors," *China Mechanical Engineering*, vol. 15, pp: 1594-1597, July 2004.

[7] M. Ooshima, S. Miyazawa, A. Chiba, et al. "Performance evaluation and test results of a 11 000 rpm, 4kW surface-mounted permanent magnet-type bearingless motor," In Proc. *7th International Symposium on Magnetic Bearings, ETH Zürich*, August 2000, pp. 23-25.

2006 5th International Power Electronics and Motion Control Conference

The Restrain of Harmonic Circulating Currents between Parallel Inverters

Yu Zhang[*], Shanxu Duan[*], Yong Kang[*], and Jian Chen[*], *Senior Member, IEEE*

[*] Huazhong University of Science and Technology/Electrical and Electronic Engineering College, Wuhan, China

zyu1126@public.wh.hb.cn

Abstract—In inverters, distortions of output sine voltages caused by factors like dead-times of power switches can be well compensated by instantaneous voltages and/or currents feedback regulators such as PID or repetitive controllers. However, in our experiments, we found that the dead-time differences between inverters would result in significant harmonic circulating currents, and instantaneous regulators had limited effects on restraining them, esp. in large parallel power supply systems where the filter reactors and output impedances in each inverters must be small. In this paper, for analyses of harmonic circulating currents, an inverter model that considered the bias in SPWM voltage as a disturbance was established. The research on restraining of harmonic circulating currents by instantaneous feedback regulators indicated that there had been relationships in inverters between output voltage distortions and harmonic circulating currents. The experiments and simulations based on the duel loop voltage and current feedback controls demonstrated that instantaneous feedback regulators that achieve good waveforms could also lead to good restraining on the harmonic circulating currents.

Keywords-inverter; parallel; instantaneous voltage feedback regulation; harmonic circulating current

I. INTRODUCTION

In inverters, distortions in output sine voltages caused by dead-times of power switches are severe, but they can be perfectly compensated by the instantaneous voltages and/or currents feedback regulations [1-2]. By far, the instantaneous feedback regulations with good effects on restraining the distortions include: PID control, dead-beat control, dual-loop controls with instantaneous current and voltage feedbacks, repetitive control, and so forth [3-4]. However, in our experiments of parallel inverters, the dead-time differences in power switches between inverters resulted in large harmonic circulating currents, and the restraining of them by instantaneous feedback regulations are limited because the filter inductances and parallel reactors in high power inverters must be small [5]. The traditional dynamic models of inverters tend to consider dead-time effects of power switches as equivalent resistances, which are included in r as shown in Fig. 1.

National Natural Science Key Foundation of China supports the project in this paper. Contact number is 50237020

However, these models could not explain the circulating currents caused by dead-times, because it will be concluded that larger dead times can restrain the circulating currents better. Since the dead-time effects are the bias voltages in SPWM voltage that output from inverter-bridges [2], so in this paper, we consider these bias voltages as disturbances in inverter models and call them harmonic disturbances. The dynamic inverter model based on the harmonic disturbance is established. Based on this model, the harmonic circulating currents and the restraining of them by instantaneous feedback regulations are researched. The analyses based on duel loop instantaneous voltages and currents feedback controls demonstrated that the instantaneous regulators achieve better waveforms in output sine voltages would also lead to better restraining of the harmonic circulating currents.

II. THE INVERTER MODEL BASED ON HARMONIC DISTURBANCES

In full bridge inverter shown in Fig. 1, the dead-times of power switches must be set large for safety operations when outputting large load currents. This will lead to large dispersive of dead-times in each inverter due to the

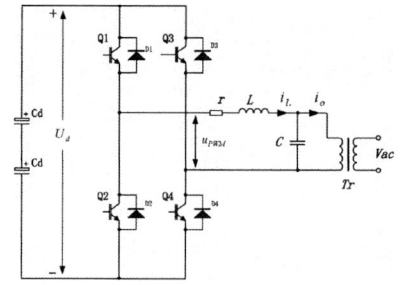

Figure 1. Full bridge single-phase inverter

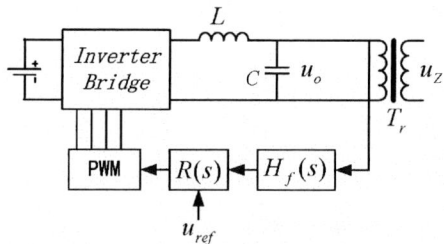

Figure 2. The instantaneous voltage feedback regulation

1-4244-0448-7/06/$25.00 ©2006 IEEE

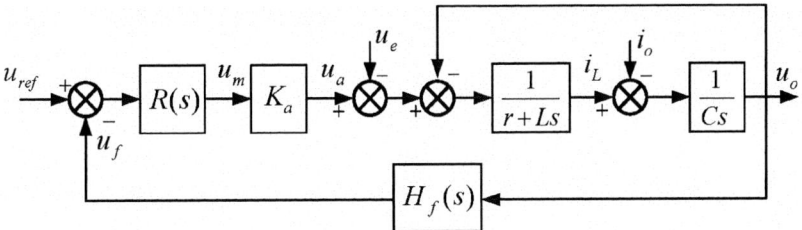

Figure 3. The transfer function block of single-phase inverter

dispersive of hardware devices.

The instantaneous feedback regulation can realize the stability and good waveform of output voltage in inverters. The instantaneous voltage feedback regulation is shown in Fig. 2. The filter capacitor voltage $U_o(s)$ as the feedback variable will be compared with the reference voltage $U_{ref}(s)$ and then the modular signal $U_m(s)$ is got through the regulator $R(s)$. It is compared with tri-angle wave to get the SPWM drive signal.

From Fig. 2, we deduce the transfer function block that is shown in Fig. 3, where the r only represent dissipations in filter inductance and cables. If switch frequencies are high enough, the ideal inverter bridges can be treated as linear amplifiers, which will amply the modular signal $U_m(s)$ by K_a time. However, dead-times set in IGBT drivers will lead to biases in SPWM voltages output from inverter bridges [1][6]. Resistances were normally equivalent to dead-time effects in traditional dynamic inverter models [2].

The dead-time effects can be equivalent to a resistance r that is series to the filter inductance L shown in Fig. 1. In fact, the bias voltages $U_e(s)$ caused by dead-times are rectangle waves, which will be determined by dc bus voltages and switch frequencies, and they are anti-phase with currents in filter inductances [7]. The bias voltages will cause harmonics distortion in output voltage, so it can be considered as disturbances, and in this paper, for analysis of the circulating currents caused by dead times, we consider $U_e(s)$ as harmonic disturbances, as shown in Fig. 3. The fundamental components in $U_e(s)$ will cause voltage drops in output voltage $U_o(s)$, and their harmonic components will cause the distortions in $U_o(s)$.

Several other facts like over-modulation and the delay in drive circuits of IGBT will cause bias voltages $U_e(s)$ too, but the dead-times are the main reason. It will bring large low order harmonic components to $U_e(s)$ [8].

From Fig. 3, the close loop transfer function can be derived in (1).

$$U_o(s) = \frac{K_a R(s)}{LCs^2 + rCs + K_a H_f(s)R(s) + 1} U_{ref}(s)$$
$$- \frac{1}{LCs^2 + rCs + K_a H_f(s)R(s) + 1} U_e(s) \qquad (1)$$
$$- \frac{Ls + r}{LCs^2 + rCs + K_a H_f(s)R(s) + 1} I_o(s)$$

If we make:

$$B_e(s) = LCs^2 + rCs + K_a H_f(s)R(s) + 1 \qquad (2)$$

Then, Equation (1) can be expressed as (3).

$$U_o(s) = \frac{K_a R(s)}{B_e(s)} U_{ref}(s) - \frac{1}{B_e(s)} U_e(s) - \frac{Ls + r}{B_e(s)} I_o(s) \qquad (3)$$

In (3), the distortions in output voltages $U_o(s)$ caused by $U_e(s)$ can be expressed as:

$$X_e(s) = U_e(s)/B_e(s) \qquad (4)$$

The $X_e(s)$ represent the error voltages caused by the $U_e(s)$. Equation (4) indicates that the fundamental component in $U_e(s)$ will lead to steady drop in output voltage $U_o(s)$, and harmonic components in $U_e(s)$ will cause harmonic distortions in output voltage $U_o(s)$.

Equation (3) shows that instantaneous voltage feedback regulations can lower the steady drops in output voltage $U_o(s)$. More over, adding an average voltage feedback regulation loop can completely get rid of them. However, the instantaneous voltage feedback regulation can restrain the harmonic distortions in output voltage $U_o(s)$, but adding an average voltage feedback regulation loop can't diminish it. We can get rid of the distortion only by diminishing the $U_e(s)$. But for high power inverters, reductions of dead-times are limited for the safe operation of power switches. Then, we can increase the gain of $B_e(s)$ which will reduce the distortions caused by the $U_e(s)$. To increase the $B_e(s)$, we can increase gains of regulator $R(s)$ and $H_f(s)$, or increase the filter reactor L and filter capacitor C. But in high power inverters, the L must be low for consideration of the cost and voltage drop, and in the mean time, when increase the gain of regulator the stable of close loop system should be considered.

From (1), Nonlinear loads will cause large harmonic currents in output currents $I_o(s)$, these are very important facts to bring distortion in output voltage, but in this paper, only the distortions caused by $U_e(s)$ are discussed, and the loads of parallel inverter system in our experiments are linear.

III. THE HARMONIC CIRCULATING CURRUNTS BETWEEN PARALLEL INVERTERS

The circuit of two parallel single-phase inverters is shown in Fig. 4, where the r_{o1} and r_{o2} are the equivalent resistance of dissipation in filter inductance and cables of each inverter. The Z_1 and Z_2 are parallel impedances in

1219

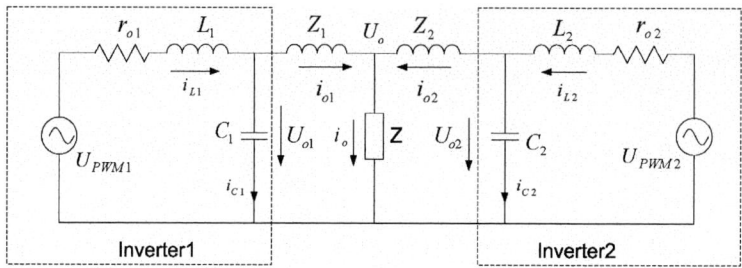

Figure 4. The parallel single phase inverters

each inverter, they can be realized by leakage impedances of output transformers T_r. We assume that two inverters have the same parameters, so $r_{o1} = r_{o2} = r$.

The circulating current is defined as (5) [10]:

$$i_H(t) = [i_{o1}(t) - i_{o2}(t)]/2 \qquad (5)$$

If $Z_1 = Z_2 = sL_K$, then, from Fig. 4, we can get:

$$I_H(s) = \frac{U_{o1}(s) - U_{o2}(s)}{2 \cdot s \cdot L_K} \qquad (6)$$

The s in (6) is the Laplace Operator. Equation (6) shows us that load currents will not affect the circulating currents. Then based on Fig. 4, take (3) into (5), we get:

$$U_{o1}(s) - U_{o2}(s) = \frac{K_a R(s) \cdot [U_{ref1}(s) - U_{ref2}(s)]}{B_e(s)}$$
$$- \frac{U_{e1}(s) - U_{e2}(s)}{B_e(s)} - \frac{(Ls+r) \cdot [I_{o1}(s) - I_{o2}(s)]}{B_e(s)}$$
$$\qquad (7)$$

Combine (7) with (5) and (6), and we let:

$$\Delta U_e(s) = [U_{e1}(s) - U_{e2}(s)]/2$$
$$\Delta U_{ref}(s) = [U_{ref1}(s) - U_{ref2}(s)]/2$$

We will get the expression of circulating currents between two parallel inverters:

$$I_H(s) = \frac{K_a \cdot R(s)}{Ls + r + sL_K \cdot B_e(s)} \Delta U_{ref}(s)$$
$$- \frac{1}{Ls + r + sL_K \cdot B_e(s)} \Delta U_e(s) \qquad (8)$$

If $U_{ref1}(s) = U_{ref2}(s)$, we further get:

$$I_H^e(s) = \frac{1}{(Ls+r) + sL_K \cdot B_e(s)} \Delta U_e(s) \qquad (9)$$

So, from (8) and (9), we will find that even with equal reference sinusoidal voltages in each inverter, i.e. $U_{ref1}(s) = U_{ref2}(s)$, the circulating currents $I_H^e(s)$ will

also exist, and they are proportional to $\Delta U_e(s)$. This means that dead-times differences of power switches between parallel inverters would lead to harmonic circulating currents. In this paper, we call the $I_H^e(s)$ as bias circulating currents.

The bias circulating currents contain fundamental and harmonic components, esp. the large low order harmonic components. The fundamental component of $I_H^e(s)$ can be eliminated by regulating the $U_{ref}(s)$ in each inverter, however, to eliminate harmonic components is difficult because the $U_e(s)$ are intrinsic in inverters and it is hard to superpose suitable harmonic components in $U_{ref}(s)$.

Compare (4) with (9), we will find that lifting the gain of $B_e(s)$ can restrain both the harmonic distortion $X_e(s)$ and the bias circulating currents $I_H^e(s)$. So, if an instantaneous voltage feedback regulation has better effect in restraining harmonic distortion in output voltages of inverters caused by the $U(s)$, it also can better restrain harmonic circulating currents between parallel inverters caused by $\Delta U(s)$. More over, if raising a parameter of inverter can better restrain distortions caused by $U(s)$, it can also restrain the circulating currents between parallel inverters caused by $\Delta U(s)$ better.

However, in traditional model of inverters, the dead-time effects were always been equivalent to resistances, which is series with filter inductances. This will lead to a conclusion that larger dead-times have better effects on restraining harmonic circulating currents. However, these equivalences can't interpret the experiment results that the differences of dead-times between inverters can lead to large harmonic circulating currents in parallel inverter systems. In this paper, we consider the $U(s)$ as a disturbance in inverter, and the harmonic circulating currents between parallel inverters can be interpreted and researched properly.

From (8), we find an important fact that the $B_e(s)$ is

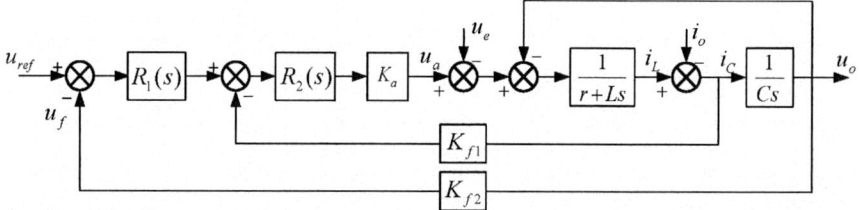

Figure 5. The dual loop control with inner capacitor current feedback loop and outer capacitor voltage feedback loop

multiplied by L_k, so, the lower L_k will make the instantaneous voltage feedback regulations have less effects on restraining the harmonic circulating currents. When the parallel impedances L_k are cancelled, the instantaneous voltage feedback regulations have no effects on harmonic circulating currents. However, the parallel impedance in large power inverters must be small for consideration of the cost and voltage drops, and this will lead to large harmonic circulating currents.

The $\Delta U_e(s)$ between two inverters can be eliminated by using high precision devices in hardware circuits, but it will increase the cost. The digital control technologies based on microprocessors are good solutions for lower the $\Delta U_e(s)$ between parallel inverters.

IV. THE RESTRAIN OF HARMONIC CIRCULATING CURRENTS BASED ON DUAL LOOP CONTROL

Many analyses and researches had shown us that an instantaneous feedback control system with an inner capacitor or inductance current control loop and an outer capacitor voltage feedback loop will result in a successful operation of the inverters [10]. These control schemes possess very fast dynamic responses and lend themselves to adapt both linear and nonlinear load applications.

The dual loop control with an inner capacitor current control loop and an outer capacitor voltage feedback loop is shown as Fig. 5.

From Fig. 5, we can get close loop transfer function:

$$U_o(s) = \frac{K_a R_1(s)R_2(s)U_{ref}(s)}{K_a R_2(s)[K_{f1}Cs + K_{f2}R_1(s)] + LCs^2 + rCs + 1}$$
$$- \frac{U_e(s)}{K_a R_2(s)[K_{f1}Cs + K_{f2}R_1(s)] + LCs^2 + rCs + 1}$$
$$- \frac{(Ls+r)I_o(s)}{K_a R_2(s)[K_{f1}Cs + K_{f2}R_1(s)] + LCs^2 + rCs + 1} \tag{10}$$

If we make:

$$B_{e2}(s) = K_a \cdot R_2(s) \cdot [K_{f1}Cs + K_{f2}R_1(s)] + LCs^2 + rCs + 1 \tag{11}$$

Then, (10) can be expressed as:

$$U_o(s) = \frac{K_a R_1(s)R_2(s)}{R_{e2}(s)}U_{ref}(s) - \frac{U_e(s)}{B_{e2}(s)} - \frac{Ls+r}{B_{e2}(s)}I_o(s) \tag{12}$$

Furthermore, from (12) and (6), we can get expression of circulating currents shown in (13).

$$I_H(s) = \frac{K_a R_1(s)R_2(s)}{Ls+r+sL_K \cdot B_{e2}(s)}\Delta U_{ref}(s)$$
$$- \frac{1}{Ls+r+sL_K \cdot B_{e2}(s)}\Delta U_e(s) \tag{13}$$

TABLE I.
THE PARAMETERS OF INVERTER

Parameter	Quantity
L	0.43 mH
C	140 μ F
L_K	0.1 mH
Dead Time	3 μ s

Compare (3) with (11), we can find that the gain of $B_{e2}(s)$ can be larger than $B_e(s)$ due to the gain of inner current regulator $R_2(s)$, and result in a better restraining to $U_e(s)$ and $\Delta U_e(s)$.

V. SIMULATION AND EXPERIMENT

A parallel inverter system contains two 20 kVA inverter prototypes were built for experiments, which had the structure as Fig. 1. The dual loop control scheme with an inner capacitor current control loop and an outer capacitor voltage feedback loop were used in each inverter, which is shown in Fig. 5. The dead times of each power switch are 3 μ s. The parameters of inverter are shown in Table I.

Firstly, We make $U_{ref1}(s) = U_{ref2}(s)$ and $R_1(s) = 1$, and no dead time differences between inverters. The circulating current in experiment is shown in Fig. 7.

Then, we changed dead time of Q_2 in inverter1 to 4 μ s. The circulating current caused by 1 μ s difference in Q_2 is

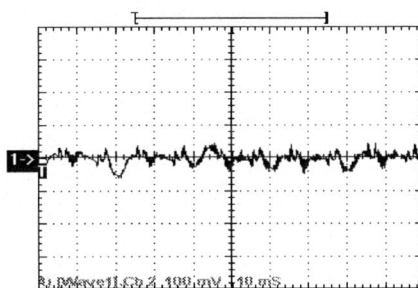

Figure 6. The circulating current without dead-time differences at 70A resistant load (Vertical: 20 A/div, Horizontal: 10 ms/div)

shown as Fig. 7. The circulating currents are saw tooth waves. In spite of the instantaneous feedback regulation, the magnitude of circulating current is 15A at 40A

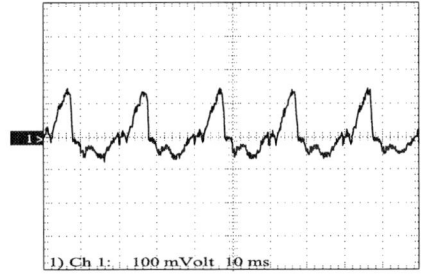

Figure 7. The circulating current with dead-time differences at 40A resistant load (Vertical: 20 A/div, Horizontal: 10 ms/div)

1221

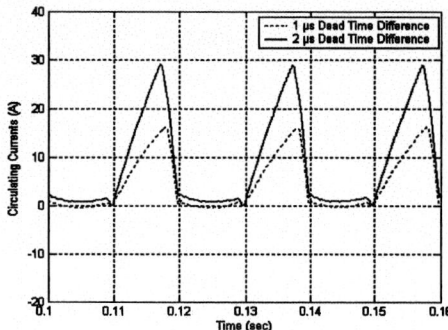

Figure 8. The circulating currents with dead-time differences at 30A resistant load (Vertical: 20 A/div, Horizontal: 10 ms/div)

resistant load. They contain large low order harmonic components, which will bring noise and make inverters easy to arise the overload protection.

The simulations can ensure that the only difference between two inverters is the dead time. The simulations of circulating currents are based on Matlab/Simulink. The circulating current with 1μ s dead time difference in Q_2 is shown as the imaginary line in Fig. 8. If we increase the difference to 2μ s, the circulating current is shown as the real line in Fig.8. The magnitude of circulating currents will get 30A at 30A resistant load.

We keep the 1μ s dead time difference between Q_2 of two inverters and make the $R_1(s) = K_r$. If we make the K_r to be 1, 5 and 10, the corresponding simulation results of circulating currents are shown in Fig. 8.

The simulation shows that the magnitude of circulating current caused by 1μ s dead time difference between two inverters will decrease from 5A to 15A when the K_r increased from 1 to 10. And when inverter run alone, the THD of the output voltage from inverter will decrease from 0.84 % to 0.55 % with K_r from 1 to 10.

VI. CONCLUSION

For large power inverter in parallel system, the filter inductances and parallel impedances must be small. The dead time differences between inverters will cause large harmonic circulating currents, which can't be interpreted by the traditional inverter dynamic models where the dead time effects are equivalent to resistance. Moreover, the restraining of harmonic circulating currents by the instantaneous feedback regulation is limited.

In this paper, the bias voltages in SPWM voltage output from inverter bridges are treated as intrinsic harmonic disturbances in inverters. And the dynamic inverter model based on harmonic disturbance is established to analyze the harmonic circulating currents. This model indicates

that instantaneous feedback regulations in each inverter can restrain the circulating currents better by raising gains in regulators while stability is guaranteed. The instantaneous feedback regulators that can achieve better output waveforms will lead to better effects in restraining harmonic circulating currents. However, the reduction of the impedances of parallel reactors will lower these effects. The experiments and simulations on instantaneous dual loop feedback regulations validated above conclusions.

ACKNOWLEDGMENT

The National Natural Science Key Foundation of China supports the project in this paper. The contact number is 50237020.

REFERENCES

[1] Jeong Seung-Gi, Lee Bang-Sup, Kim Kyung,Park Min-Ho, "The analysis and compensation of dead time effects in PWM inverter," *IEEE Trans. On Industry Electronics,* 1991, vol. 38, no. 2, pp. 108-114.

[2] M. J. Ryan, W. E. Brumsickle, R. D. Lorenz, "Control topology options for single phase UPS inverters," *IEEE Trans. on Industry Applications,* vol. 33, no. 2, 1997, pp. 493-501.

[3] Y. Y. Tzou, R. S. Ou, S. L. Jung, et al, "High performance programmable AC power source with low harmonic distortion using DSP based repetitive control technique," *IEEE Trans. on Power Electronics,* 1997, vol. 12, no. 7, pp. 715-725.

[4] O. Kukrer, "Deadbeat control of a three-phase inverter with an output LC filter," *IEEE Trans. on Power Electronics,* 1996, vol. 11, no. 1, pp. 16-23

[5] Yu Zhang, Xikun Chen, Yong Kang, Jian Chen, "The restrain of the dead time effects in parallel inverters," *IEEE International Electric Machines and Drives Conference* (IEMDC'05), May, 2005, Texas, USA, pp. 797-802.

[6] N. Mohan, T. M. Undeland, W. P. Robbins, *Power Electronics – Converters, Applications, And Design.* 3rd Edition, John Wiley & Sons Inc, Hoboken, 2003.

[7] B. K. Bose, *Modern Power Electronics and AC Drives.* Prentice Hall, Englewood Cliffs, NJ, 2002

[8] M. Naser Abdel-Rahim, John E. Quaicoe, "Analysis and design of a multiple feedback loop control strategy for single-phase voltage-source UPS inverters," *IEEE Transactions on Power Electronics,* 1996, vol. 11, no. 4, pp. 532-541

[9] M. Prodanovic, T. C. Green, H. A. Mansir, "A survey of control methods for three-phase inverters in parallel connection," *8th International Conference on Power Electronics and Variable Speed Drives* (IEE Conf. Publ. No. 475), 2000, pp. 472-477.

[10] D. M. Brod, D. W. Novotny, "Current control of VSI-PWM Inverters," *IEEE Trans. on Industry Applications,* 1985, vol. IA-21, pp. 562-570.

2006 5th International Power Electronics and Motion Control Conference

Simulation of Permanent Magnet Synchronous Motor with Dual Closed Loop by Time-Stepping Finite Element Model

Xinhua Liu, Jianzhong Jiang, Yu Gong and Ye Ding

Shanghai University, Shanghai, 200072, China

Abstract--**This paper presents the process of circuit and field coupled finite element (FE) modeling of permanent magnet synchronous motor (PMSM). Compared with traditional centralized parameters model the presented model can consider many effects to the dynamic performance of PMSM, such as the effects of saturation of iron materials, the eddy current, the relative movement of teeth and slots and high-order harmonics in both time and space domains. Dual closed control loop can also be coupled into the FE model to form the global system simulation, if we make the step time very small. An application of the proposed method to a PMSM with dual closed loop is given, the results validate work of this paper.**

Keywords--finite element method, permanent magnet synchronous motor, time-stepping method and circuit and field coupled model.

I. INTRODUCTION

Recently with the development of permanent magnet material, power electronic and control method, permanent magnet synchronous motor (PMSM) has been widely used in servo system due to its small inertia, high efficiency and high torque density. This leads researchers to pay much attention to the dynamic characteristics of PMSM. Traditional dynamic simulation of PMSM is based on centralized parameters mathematical model. This model has some disadvantages, because parameters of this model vary with many factors during the operation of the motor and it can't consider the effects of saturation of iron material, the eddy current, the relative movement of teeth and slots and high-order harmonics both in time and space domains to the dynamic performance of the motor.

Recent years circuit and field coupled finite element method (FEM) analysis has been rapidly developed [1]-[3]. In FEM model magnetic field equations are directly coupled with the circuit equations, and we can get current and field results, which vary with time, simultaneously by using the time-stepping method. And if we make the step time very small, it is very convenient to add control loop to the simulation system to simulate the performance of the global system. It does provide researchers a very useful tool to simulate the dynamic characteristics of motors.

In this paper the authors present the FEM modeling process of PMSM with speed and current closed loop system. Then give an example of application of this method, results show the validity of the presented method.

II. CIRCUIT AND FIELD COUPLED FEM MODEL OF PMSM

In 2D filed, after using the moving boundary method to process the boundary conditions between stator and rotor, the basic electromagnetic filed equation of motor can be descript as following [4]:

$$\frac{\partial}{\partial x}\left(v\frac{\partial A}{\partial x}\right)+\frac{\partial}{\partial y}\left(v\frac{\partial A}{\partial y}\right)=-j \qquad (1)$$

Where A is the axial component of magnetic vector potential; V is the relativity of material; j is the total current density.

A. In the region of air and iron core

$$j = 0 \qquad (2)$$

After applying a standard Galerkin procedure we can get the field equations in the region of air and iron core as following:

$$[C_{11}]\cdot[A]=[0] \qquad (3)$$

Where C_{11} is $N_1\times N_1$ coefficient matrix; here A is a $1\times N_1$ column vector and represents the magnetic vector potential of N_1 nodes in the region; 0 represents $1\times N_1$ zero column vector; N_1 is the number of mesh nodes in the region of air and core.

B. In the stator conductor region

One phase stator circuit is shown as Fig.1:

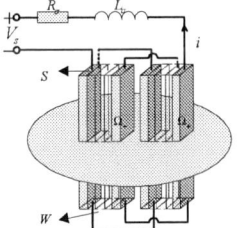

Figure 1. One phase stator circuit

1-4244-0448-7/06/$25.00 ©2006 IEEE   1223

One phase stator circuit equation is:

$$V_s = e - Ri - R_\sigma i - L_\sigma \frac{di}{dt} \qquad (4)$$

Where V_s is the impressed voltage ; i is the phase current; R is the resistance of straight part of one phase winding; L_σ is the inductance of the end winding; R_σ is the resistance of end winding; e is the electromotive force (*EMF*) of one phase, and according to reference [1]:

$$e = -\frac{l}{S}(\iint_{\Omega-} \frac{\partial A}{\partial t} d\Omega - \iint_{\Omega+} \frac{\partial A}{\partial t} d\Omega) \qquad (5)$$

Where l is the axial length of the iron core ; S is the area of one stator conductor; Ω^+ and Ω^- are the cross-sectional areas of 'go' and 'return' side of the phase conductors of the coils respectively.

Substituting (5) into (4), at the same time noticing $R = 2Wl / \sigma S$, we can get the current equation:

$$i = -\frac{\sigma}{2wl}\left[S(V_s + R_\sigma i + L_\sigma \frac{di}{dt}) + l(\iint_{\Omega+} \frac{\partial A}{\partial t} d\Omega - \iint_{\Omega} \frac{\partial A}{\partial t} d\Omega) \right] \quad (6)$$

Where σ is the conductivity of stator conductor; W is the turns of one phase stator winding;

So in stator conductor region the field equation is:

$$\frac{\partial A}{\partial x}\left(v \frac{\partial A}{\partial x} \right) + \frac{\partial A}{\partial y}\left(v \frac{\partial A}{\partial y} \right) = \mp j = \mp \frac{i}{S} \qquad (7)$$

Applying Galerkin procedure to (7), we can get field equations as following:

$$[C_{11} \quad C_{12}] \cdot \begin{bmatrix} A \\ i \end{bmatrix} = [0] \qquad (8)$$

Applying Galerkin procedure to (6), we can get circuit equations as following:

$$[0 \quad C_{22}] \begin{bmatrix} A \\ i \end{bmatrix} + [D_{11} \quad D_{22}] \begin{bmatrix} \frac{\partial A}{\partial t} \\ \frac{di}{dt} \end{bmatrix} = [P_2] \qquad (9)$$

Where P_2 is the column vector related to impressed voltages.

Applying the same method to other phase regions we can get other phase current equations, which are similar to equations (9).

Adding (8) and (9) together, we get the basic equations of circuit and field coupled in stator conductor region:

$$\begin{bmatrix} C_{11} & C_{12} \\ 0 & C_{22} \end{bmatrix} \begin{bmatrix} A \\ i_s \end{bmatrix} + \begin{bmatrix} 0 & 0 \\ D_{21} & D_{22} \end{bmatrix} \begin{bmatrix} \frac{\partial A}{\partial t} \\ \frac{di_s}{dt} \end{bmatrix} = \begin{bmatrix} 0 \\ P_2 \end{bmatrix} \quad (10)$$

Where i_s represents the $1 \times m$ column vector of phase current; m is the phase number of the motor.

C. In the permanent magnet region

Usually there are two models which are commonly used to represent permanent magnets [5]: one is magnetization vector and the other is equivalent current sheet. In this paper we adopt the first method. According to Fig.2, the relationship between B and H of any point P on the demagnetization curve of permanent magnet is:

$$H = v(B - B_r) \qquad (11)$$

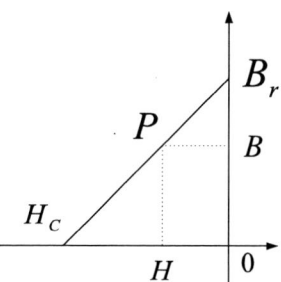

Figure 2. Demagnetization curve of permanent magnet

Using $rotH = J$ and $rotA = B$, and at same time considering the eddy current effect in the permanent magnet $J = -\sigma_{pm}\frac{\partial A}{\partial t}$, we can write:

$$rot(vrotA) = -\sigma_{pm}\frac{\partial A}{\partial t} + rot(v_{pm}B_r) \qquad (12)$$

So the basic field equation in the permanent magnet is:

$$\frac{\partial}{\partial x}(v \cdot \frac{\partial A}{\partial x}) + \frac{\partial}{\partial y}(v \cdot \frac{\partial A}{\partial y}) = \sigma_{pm}\frac{\partial A}{\partial t} + v_{pm}(\frac{\partial B_{rx}}{\partial y} - \frac{\partial B_{ry}}{\partial x}) \quad (13)$$

Applying Galerkin procedure to (13), we can get the matrix as following:

$$[C_{11}] \cdot [A] + [D_{11}]\left[\frac{\partial A}{\partial t} \right] = [P_1] \qquad (14)$$

Where P_1 is the column vector related to permanent magnet.

D. Whole equations

Adding (3), (10) and (14) together, we can get the whole equation of circuit and field coupled FEM model of PMSM as following:

$$\begin{bmatrix} C_{11} & C_{12} \\ 0 & C_{22} \end{bmatrix} \begin{bmatrix} A \\ i_s \end{bmatrix} + \begin{bmatrix} D_{11} & 0 \\ D_{21} & D_{22} \end{bmatrix} \begin{bmatrix} \frac{\partial A}{\partial t} \\ \frac{di_s}{dt} \end{bmatrix} = \begin{bmatrix} P_1 \\ P_2 \end{bmatrix} \quad (15)$$

III. TIME-STEPPING METHOD

Using the Backward Euler's method to discretize the time variable of (15), we can get:

$$
\begin{bmatrix}
C_{11}+\dfrac{D_{11}}{\Delta t} & C_{12} \\[2mm]
\dfrac{D_{21}}{\Delta t} & C_{22}+\dfrac{D_{22}}{\Delta t}
\end{bmatrix}
\begin{bmatrix}
A^{+\Delta} \\[2mm]
i_s^{t+\Delta t}
\end{bmatrix}
=
\begin{bmatrix}
P_1 \\[2mm]
P_2
\end{bmatrix}
-
\begin{bmatrix}
\dfrac{D_{11}}{\Delta t} & 0 \\[2mm]
\dfrac{D_{21}}{\Delta t} & \dfrac{D_{22}}{\Delta t}
\end{bmatrix}
\begin{bmatrix}
A \\[2mm]
i_s^{t}
\end{bmatrix}
\quad (16)
$$

Where Δt is the step time; $t+\Delta t$ is the time of current step and t is the time of last step.

Using Newton-Raphson iteration method to solve the nonlinear problem of equation (16) and through ICCG method, we can get the field and current results of the current step time simultaneously as we know the values of last step time.

IV. MOVEMENT EQUATION

As we know, the movement equation of motor is:

$$
J\frac{d\omega}{dt} = T_e - T_L \quad (17)
$$

Where T_L is the load torque; J is the inertia of motor; ω is the mechanical angular velocity; T_e is the electromagnetic torque, in 2D field the electromagnetic torque is usually calculated by the Maxwell Stress Tensor method [5], that is:

$$
T_e = \frac{L_{ef}}{\mu_0} \oint r^2 B_r B_\theta \, d\theta \quad (18)
$$

With the results of (14) we can easily work out the electromagnetic torque. The integral route is an arbitrary closed curve located in the air-gap.

The discretion iteration format of equation (17) is:

$$
\omega^{t+\Delta t} = \frac{T_{em}-T_L}{J}\Delta t + \omega^{t} \quad (19)
$$

According to (19) we can work out the current time speed $\omega^{t+\Delta t}$ and the movement increment $\Delta\theta$ of the rotor. Then we rotate the FEM rotor mesh according to the movement increment $\Delta\theta$ to simulate the actual movement of motor.

V. STRUCTURE OF PMSM WITH DUAL CLOSED LOOP SYSTEM

In motor FEM model voltages are the input data at each step, so if each step time is very small (for example in this paper the step time is 0.01ms), it is very convenient to simulate voltage-driven motor in spite of the voltage form. That is to say control method can be attached to the motor FEM model together. Fig.3 shows the block diagram of a PMSM control system with dual closed loop. In Fig.3 PMSM is simulated by a FEM model as shown. Speed control can be obtained through the outside speed closed loop. The difference between the reference speed and the feedback speed is modulated through a PI controller, and the result is of the given reference of current closed loop. Current loop control is implemented by current hysteresis loop method. The output of the current hysteresis loop and the rotor position signals are used together to form the control signal of the inverter.

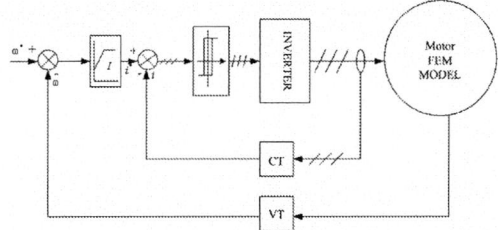

Figure.3 PMSM with dual closed loop control system

VI. APPLICATION

In this section a 4-pole surface-mounted PMSM with dual closed control loop is analyzed by the presented method. The motor has 6 slots, and the stator out diameter and the effective length of the motor are 54.5mm and 14mm respectively. Reference speed is $1000\,\text{r/min}$. Load torque is $0.06\,\text{N}\cdot\text{m}$. Parameters of the PI controller are K_P=2, T_S=0.01 respectively, and the maximum output is 3. Hyrestesis band is 0.01, and the step time is fixed as 0.01ms. Fig.4 is the mesh of the computed region of the motor, which is half of the motor. Fig.5 and Fig.6 show the flux distribution at two different time during the movement.

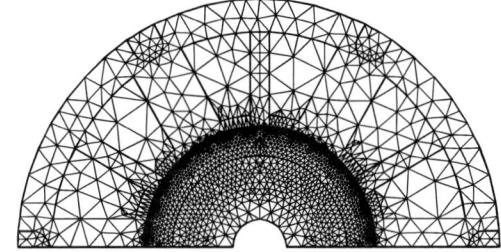

Figure 4. Mesh of calculated region of the motor

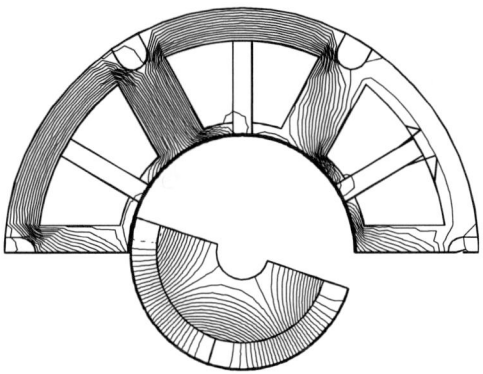

Figure.5 Computed flux distribution at one step

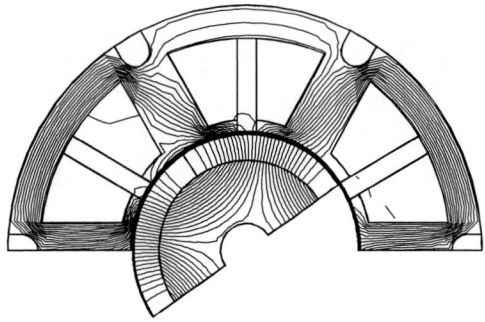

Figure.6 Computed flux distribution at another step

Fig.7, 8 and 9 are the computed results of speed vs. time, electromagnetic torque vs. time and three phase stator current vs. time respectively. Fig.10 is the contrast of one phase current at steady state between computed and test results. Fig.11 is the contrast of one phase back-*EMF* between computed and test results. It can be seen that the computed results are quite closed to test results, which validates this paper's work.

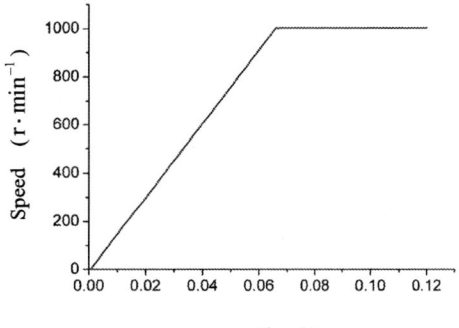

Figure.7 Computed speed vs. time

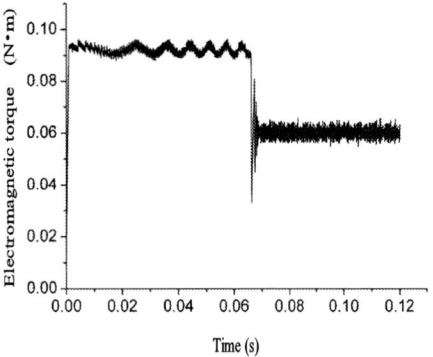

Figure.8 Computed electromagnetic torque vs. time

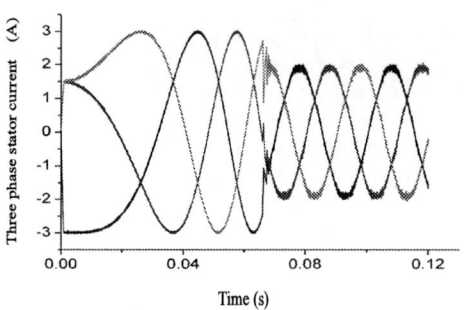

Figure.9 Computed three phase current vs. time

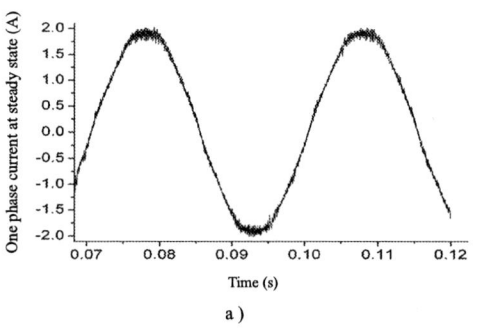

a)

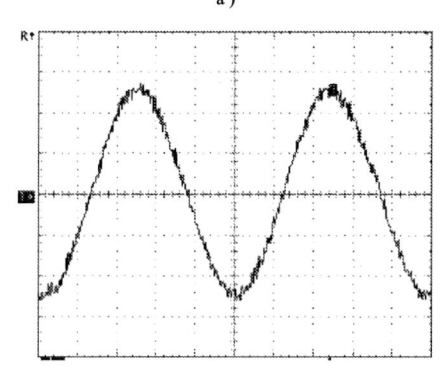

b) (0.8A/div 5ms/div)
Figure.10 One phase current vs. time at steady state
a) computed results   b) test results

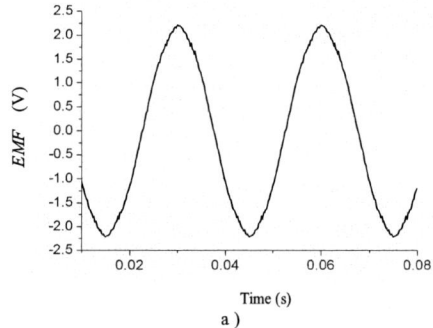

a)

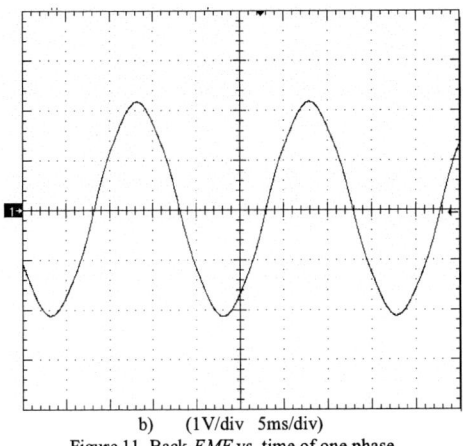

b)   (1V/div 5ms/div)

Figure.11 Back-*EMF* vs. time of one phase
a) computed results   b) test results

VII. CONCLUSION

In this paper the process of circuit and field coupled FE modeling of PMSM is presented. Because voltage is input data in this model, if we take the time stepping very small, it is very convenient to add control method to this model. In the paper a PMSM with dual closed loop system is studied by the presented method. The computed results show good accordance with the test results, which validates this paper's work.

REFERENCES

[1] S. L. Ho and W. N. Fu, "A Comprehensive Approach to the Solution of Direct Coupled Multislice Model of Skewed Rotor Induction Motors Using Time-stepping Eddy-current Finite Element Method," *IEEE Trans. Magnetics*, vol. 33, no. 3, May 1997, pp.2265-2273.

[2] G. H. Jang, J. H. Chang, D. P. Hong and K. S. Kim, "Finite-element analysis of an electromechanical field of a BLDC motor considering speed control and mechanical flexibility", *IEEE Trans. on Magnetics*, Vol. 38, no. 2, March, 2002, pp.945-948.

[3] S. L. Ho, W. N. Fu, H. L. Li, H. C. Wong and H. Tan, "Performance Analysis of Brushless DC Motors Including Features of the Control Loop in the Finite Element Modeling," *IEEE Trans. Magnetics*, vol. 37, no. 5, September. 2001, pp.3370-3374.

[4] Sun Yutian, Yang Ming and Li Beifang, "The moving problem in the dynamic FEM of electric machines," *Large Electric Machine and Hydraulic Turbine*, no. 6, 1997, pp.35-39.

[5] Sheppard J. Salon, Finite element analysis of electrical machines. Boston: Kluwer Academic Publishers, 1995

Online Dynamic Parameter Estimation of Transformer Equivalent Circuit

M. Reza Feyzi and Mehran Sabahi

Faculty of Electrical and Computer Engineering
University of Tabriz, Tabriz, Iran
feyzi@tabrizu.ac.ir   sabahi@tabrizu.ac.ir

Abstract. **The dynamic parameters of a transformer, including its inductances and resistances, are estimated by the application of recursive least squares routine on the measured terminal voltages and currents. The harmonic content of the measured input parameters can be included in the proposed method. The estimation is carried out while the transformer is under load. Therefore, the estimated parameters are obtained in real time conditions taking the saturation effects into account. The procedure needs no additional data and/or initial assumption. As a stable RLS routine, the U.D. Bierman's method is used to ensure long period convergence of the parameter estimation procedure. Forgetting factors are proposed and used to accelerate the convergence of the estimation and improve the accuracy of the parameters.**

Keywords- Parameter Estimation; Transformer

I. INTRODUCTION

Transformers are one of the main components of any power system. An accurate estimation of system behavior, including load flow studies, protection, and safe control of the system calls for an accurate equivalent circuit parameters of all system components such as generators, transformers, etc. Different models have been proposed for dynamic and transient structure of transformers [1-2].

Different long time off-line estimation methods have been proposed using frequency response analyses [3], steady state analyses [4] and recently genetic algorithms [5].

In this paper, an on line time domain estimation routine is proposed to estimate the dynamic equivalent circuit parameters of the transformer at its actual operating point. The proposed method allows including the effect of possible unbalancing, saturation, or nonlinear loads and harmonic content of the voltages and currents. Therefore, the direct sampling of input and output voltages and currents of a transformer are needed to provide the required information for the estimation mechanism. The parameters of each phase can be considered separately and therefore the unbalancing of the system will have no effect on the accuracy of the estimated parameters.

The method is applied on a typical transformer as the model. The transformer equivalent circuit of this transformer is represented as two time discrete equations in the terms of measurable values in the form of "Auto Regressive Moving Average" (ARMA) equations which are suitable for online estimations methods. Then required information is obtained from the simulation of the modeled transformer using PSCAD software [6]. The acquired data from the simulation is launched to the estimation mechanism in order to adjust the estimated parameters by minimizing the error between the actual model parameters and the corresponding estimated parameters.

II. ESTIMATION ALGORITHM

Based on output of the observations of a system and their comparison with estimated values from the model, a so called "cost function" is defined. It is essential that the selected model to be observable so that the input signals could be exited in every modes of the model. Input signals may be regular system variations or random disturbances.

Least squares algorithm is basically an offline method and counted as an explicit routine for fixed parameters, not a suitable algorithm for online estimations with variable parameters. By some modification, the Recursive Least Square (RLS) algorithm is obtained. The new algorithm is capable of online estimation and allows gradual variations of system parameters. In order to apply RLS algorithm on an adaptive or self tuning systems, the estimation of parameters should be in a recursive relation with data from the previous steps so that the estimated parameters are updated every time step. Fig. 1 shows the procedure as a simple block diagram.

If k is assumed to be the time step number, the RLS algorithm performs the following stages in $(k+1)^{th}$ step:

i. The new input vector, $\mathbf{x}(k+1)$ is formed by using the newly sampled data,

ii. New estimation error is calculated from (1).

$$\varepsilon(k+1) = y(k+1) - \hat{y}(k+1)$$
$$where: \quad \hat{y}(k+1) = \mathbf{x}^T(k+1)\hat{\mathbf{\Theta}}(k) \tag{1}$$

iii. New value of covariance matrix $\mathbf{P}(k+1)$ is obtained from (2) where I_{np} represents the unit matrix,

$$\mathbf{P}(k+1) = \mathbf{P}(k)\left[\mathbf{I}_{np} - \frac{\mathbf{x}(k+1)\mathbf{x}^T(k+1)\mathbf{P}(k)}{1+\mathbf{x}^T(k+1)\mathbf{P}(k)\mathbf{x}(k+1)} \right] \tag{2}$$

iv. The estimated parameters are updated using (3).

$$\hat{\Theta}(k+1) = \hat{\Theta}(k+1) + \mathbf{P}(k+1)\mathbf{x}(k+1)\varepsilon(k+1) \quad (3)$$

v. Finally, wait to receive new observations and return to the first step.

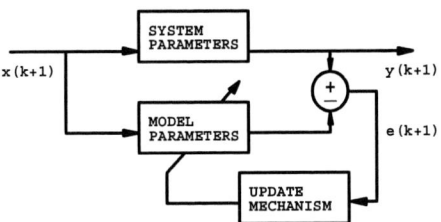

Figure 1. Recursive estimation algorithm

The parameters in recent equations are as follows:

$\varepsilon(k+1)$:   Estimation error

$y(k+1)$:   Real system output

$\mathbf{x}^T(k+1)$:   Data vector

$\hat{y}(k+1)$:   Estimation model output

$\hat{\Theta}(k)$:   Estimation parameter vector

Selection of the time step period, T_S, depends on the maximum frequency of output data. Theoretically the minimum frequency of sampling should be at least two times as large as the highest harmonic frequency of the input data. There is no explicit rule to select the sampling period. A self tuning system, where the sampling period is selected adaptively, is an appropriate method to study frequency response of such models.

III. APPLICATION OF THE PROPOSED METHOD

In practice, an actual transformer can be used to gather exact values of currents and voltages by online sampling. In this work, the equivalent circuit of a 2400/240 V, 15 kVA transformer is used for this purpose. The main parameters of the transformer, all refereed to the low voltage side, are given in table 1. Fig. 2 shows the most popular dynamic equivalent circuit of a transformer. This model is simulated by using PSCAD software to obtain the required data for estimation mechanism. The input voltage of the transformer is assumed to contain some 3rd and 5th harmonics in order to generalize the procedure. Fig. 3 shows the distorted secondary voltage of the transformer.

TABLE I.   SELECTED MODEL VALUES

$R_1 [\Omega]$	$R_2 [\Omega]$	$R_m [\Omega]$	$L_1 [\mu H]$	$L_2 [\mu H]$	$L_m [H]$
0.0245	0.02	1050	100	71	0.29

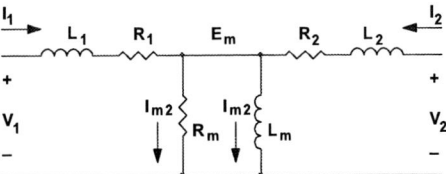

Figure 2. Transformer dynamic equivalent circuit

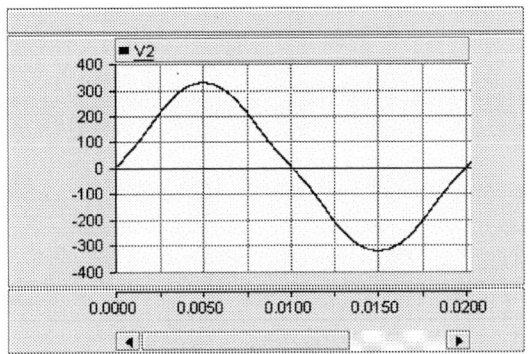

Figure 3. Secondary voltage, V_2

IV. TRANSFORMER ESTIMATION MODEL

Only four variables, including the primary and secondary voltages and currents are readily available for sampling in an actual transformer. The same parameters are extracted and sampled from the equivalent circuit of the modeled transformer during the estimation process. Application of KVL on the secondary loop of the equivalent circuit results in equation (4) as follows:

$$v_1 = L_1 \frac{di_1}{dt} + R_1 i_1 - R_2 i_2 - L_2 \frac{di_2}{dt} + v_2 \quad (4)$$

An approximation as equation (5) is used to convert equation (4) to a discrete time domain:

$$\frac{df(t)}{dt} \approx \frac{1}{2T_s}[F(k+1) - F(k-1)] \quad (5)$$

T_S is the sampling time period. Using (4) and (5), the first estimation equation in the ARMA form is obtained as (6):

$$I_1(k+1) - I_1(k-1) = \theta_{11}(V_1(k) - V_2(k)) + \theta_{12}I_1(k) + \theta_{13}(I_2(k+1) - I_2(k-1)) + \theta_{14}I_2(k) \quad (6)$$

Therefore, the standard form of estimation can be defined as follows:

$$y_1(k+1) = I_1(k+1) - I_1(k-1) \quad (7)$$

1229

$$\mathbf{x}_1(k+1) = \begin{bmatrix} V_1(k) - V_2(k) \\ I_1(k) \\ I_2(k+1) - I_2(k-1) \\ I_2(k) \end{bmatrix} \tag{8}$$

The first estimated output is driven as (9):

$$\hat{y}_1(k+1) = \mathbf{x}_1^{T}(k+1)\hat{\boldsymbol{\theta}}_1(k) \tag{9}$$

where:

$$\hat{\boldsymbol{\theta}}_1(k) = \begin{bmatrix} \hat{\theta}_{11}(k) \\ \hat{\theta}_{12}(k) \\ \hat{\theta}_{13}(k) \\ \hat{\theta}_{14}(k) \end{bmatrix} \tag{10}$$

By reverse solving of estimation vector, the estimated values of the transformer parameters except for R_m and L_m are obtained as follows:

$$\hat{L}_1(k) = \frac{2T_s}{\hat{\theta}_{11}(k)} \quad ; \quad \hat{R}_1(k) = -\frac{\hat{\theta}_{12}(k)}{\hat{\theta}_{11}(k)} \tag{11}$$

$$\hat{L}_2(k) = \frac{2T_s}{\hat{\theta}_{11}(k)}\hat{\theta}_{13}(k) \quad ; \quad \hat{R}_2(k) = \frac{\hat{\theta}_{14}(k)}{\hat{\theta}_{11}(k)} \tag{12}$$

In order to extract the values of R_m and L_m, the value of E_m in the equivalent circuit of Fig. 2 is required where it cannot be determined via direct measurement. Therefore, a state observer is used for this purpose. The observer can be structured from the estimated values, obtained from the previous time step by (11) and (12).

The value of E_m may be estimated from the primary loop of the equivalent circuit as follows:

$$\hat{E}_m{}'(k) = V_1(k) - \hat{R}_1(k)I_1(k) \\ -\frac{1}{2T_s}\hat{L}_1(k)\big(I_1(k+1) - I_1(k-1)\big) \tag{13}$$

A similar equation may be extracted from the secondary loop of the equivalent circuit:

$$\hat{E}_m{}''(k) = V_2(k) - \hat{R}_2(k)I_2(k) \\ -\frac{1}{2T_s}\hat{L}_2(k)\big(I_2(k+1) - I_2(k-1)\big) \tag{14}$$

Using (13) and (14), E_m at k^{th} time step is estimated as:

$$\hat{E}_m(k) = \frac{\dfrac{\hat{E}_m'(k)}{\sigma_1^2} + \dfrac{\hat{E}_m''(k)}{\sigma_2^2}}{\dfrac{1}{\sigma_1^2} + \dfrac{1}{\sigma_2^2}} \tag{15}$$

σ_1 and σ_2 are standard deviations of the measurement errors at primary and secondary side of the transformer respectively.

The discretised form of KVL for the middle branch of the equivalent circuit at k^{th} time step can be written as:

$$I_1(k) + I_2(k) - I_1(k-2) - I_2(k-2) = \\ \theta_{21}\big(\hat{E}_m(k) - \hat{E}_m(k-2)\big) + \theta_{22}\hat{E}_m(k-1) \tag{16}$$

$$y_2(k) = I_1(k) + I_2(k) - I_1(k-2) - I_2(k-2) \tag{17}$$

$$\mathbf{x}_2(k) = \begin{bmatrix} \hat{V}_m(k) - \hat{V}_m(k-2) \\ \hat{V}_m(k) \end{bmatrix} \tag{18}$$

The next estimated output is given by (19):

$$\hat{y}_2(k+1) = \mathbf{x}_2^{T}(k+1)\hat{\boldsymbol{\theta}}_2(k) \tag{19}$$

$$\hat{\boldsymbol{\theta}}_2(k) = \begin{bmatrix} \hat{\theta}_{21}(k) \\ \hat{\theta}_{22}(k) \end{bmatrix} \tag{20}$$

Again by reverse solving of the estimation vector; the estimated values of R_m and L_m can be obtained as follows:

$$\hat{L}_m(k) = \frac{2T_s}{\hat{\theta}_{22}(k)} \quad ; \quad \hat{R}_m(k) = \frac{1}{\hat{\theta}_{21}(k)} \tag{21}$$

Finally, an appropriate forgetting function, as described in the following section, is defined to fade out the effect of the previous values.

V. EXTENDED RLS AND FORGETTING FACTORS

Input noise and disturbances may penetrate the estimation routing during data acquisition. On the other hand, estimation routines show high sensitivity to the noise level. Therefore, the noise level can be considered as major problems for online estimation. A high *signal to noise* ratio is required for a successful convergence of an estimation mechanism. In practice, high efficiency band pass filters are required in order to block the noise penetration into estimation routine. In the meantime, software digital filtering may be used for this purpose .Then the white input noise is converted into colored noise, where it may introduce a dc bias on the estimated parameters. An extended RLS routine can be used to solve this problem. For this purpose, some additional parameters are added to the parameter vector Θ. At the same time, the estimations errors from the previous steps are inserted to the data vector $\mathbf{X}$ as given by (22).

1230

$$\mathbf{x}_{ERLS}(k) = \begin{bmatrix} [\mathbf{x}_{RLS}(k)] \\ \varepsilon(k) \\ \varepsilon(k-1) \\ \ldots \end{bmatrix} \qquad (22)$$

In order to fade out the effect of the previous values, a function so called *forgetting factor* $0 < \lambda(k) \le 1$ is defined as follows [7]:

$$\lambda(k) = (1-\sigma)\left(1 - \exp\left(-\frac{kT_s}{\tau_f}\right)\right) + \sigma \qquad (23)$$

Where τ_f is forgetting time constant and σ is a value between zero and unit. Therefore, the new form of (2) is written as follows:

$$\mathbf{P}(k+1) = \frac{\mathbf{P}(k)}{\lambda(k)}\left[\mathbf{I}_{np} - \frac{\mathbf{x}(k+1)\mathbf{x}^T(k+1)\mathbf{P}(k)}{\lambda(k) + \mathbf{x}^T(k+1)\mathbf{P}(k)\mathbf{x}(k+1)}\right] \qquad (24)$$

Another forgetting factor could be used to trace the gradual variations of the system parameters. In this case forgetting factor would be simply a number close to one like 0.99, or a function as:

$$\lambda(k) = \sigma + (1-\sigma)\, e^{-k_f \eta_f(k)} \qquad (25)$$

where:

$$\eta_f(k) = \sqrt{(\hat{L}_m(k) - \hat{L}_m(k-1))^2} \qquad (26)$$

This factor is not used in this work. Initial value for covariance matrix $\mathbf{P}$ is taken as:

$$\mathbf{P}(0) = r\mathbf{I}_{np} \qquad ; \qquad r \gg 1 \qquad (27)$$

The main method which is normally used to stabilize the estimation routine is to determine the covariance matrix as a positive definite matrix. This can be defined as the multiplication of two matrices as (28):

$$\mathbf{P}(k) = \mathbf{S}(k)\mathbf{S}^T(k) \qquad (28)$$

Compared to other existing estimation routines, such as Givens routine [7] the U.D. Bierman algorithm needs less computation time. Therefore, this is selected as an appropriate estimation routine in this work.

VI. Estimation Results

In practice, an actual transformer can be used to gather exact values of currents and voltages by online sampling. In this work, the equivalent circuit of a 2400/240 V, 15kVA transformer is used for this purpose.

As described earlier, the simulation results from the equivalent circuit of a 2400/240 V, 15kVA transformer is used as the representative of direct measurement result of an actual transformer. The simulation is carried out in PSCAD-4 environment and fed into the estimation routine which is programmed on MATLAB- 6.5. The corresponding results are summarized in Table 2.

TABLE II. ESTIMATION RESULTS

Parameter	L1 [µH]	L2 [µH]	Lm [H]
Real value	100	71	0.29
Estimated value	107.43	63.387	0.2902
Error percentage	7.434	-10.7228	0.0576

Parameter	R1 [Ω]	R2 [Ω]	Rm [Ω]
Real value	0.0245	0.02	1050
Estimated value	0.0245	0.0194	1046.4
Error percentage	2.9457e-4	-3.2074	-0.3425

Sampling intervals is taken as short as 10 [µSec]. Today, such fast sampling rate is quite practical in analog to digital converters and DSP processors. In the meantime, longer time intervals were examined successfully. Forgetting factor time constant is assumed to be about 0.001 second. Noise effect is incorporated by adding a random number to the original information. It is also assumed that the sampling devices is equipped with an appropriate band pass filters to eliminate noise and unwanted information in the working frequency regions. Figures 2 to 9 illustrate the convergence procedure of the estimated parameters.

VII. Conclusion

It was shown that RELS method can be used to estimate the dynamic equivalent circuit parameters of a transformer at its actual operating point. The proposed routine needs only four terminal variables of the transformer which can be easily sampled in any actual transformer; no external excitation signals and/or no initial information are required. Since the actual sampled data is directly fed to the routine no variable conversion such as reference frame conversion is needed. In this work, the simulation results from PSCAD were substituted as the representative of sampled data from an actual transformer. Using the current data along with the data from the previous steps, a so called "cost function" was defined. In the mean time, a forgetting factor was used to fade out the effect of previously acquired data. The proposed estimation algorithm can be used to trace any gradual variation of the parameters It is also capable

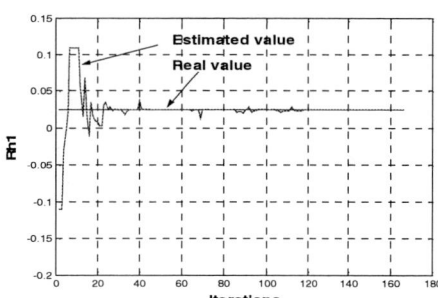

Figure 4. Convergence procedure of the $\hat{R}_1$

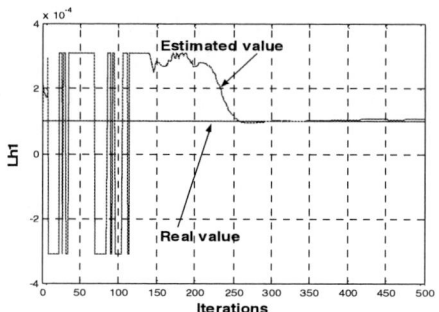

Figure 7. Convergence procedure of the $\hat{L}_1$

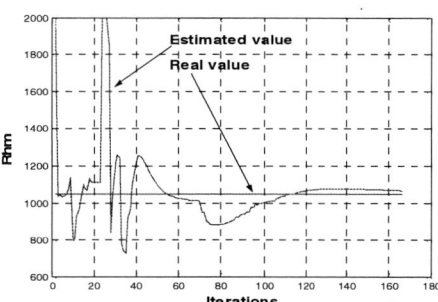

Figure 5. Convergence procedure of the $\hat{R}_2$

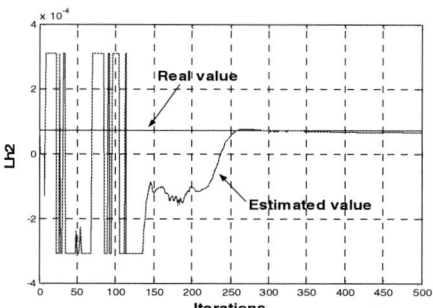

Figure 8. Convergence procedure of the $\hat{L}_2$

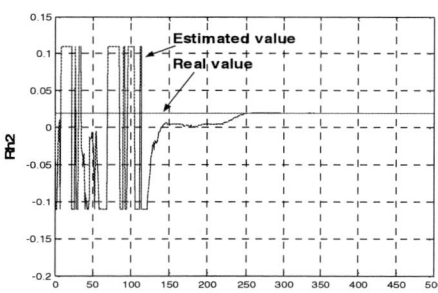

Figure 6. Convergence procedure of the $\hat{R}_m$

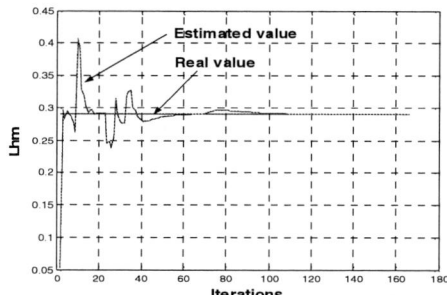

Figure 9. Convergence procedure of the $\hat{L}_m$

to include the effects of unbalancing, nonlinearity, saturation, and harmonic effects of the transformer. The procedure showed a fast and successful convergence in the estimation of required parameters.

VIII. REFRENCES

[1] M. Sabahi, P. Karimifard, G. Gharehpetian, ''RLS Method Estimation of the Transformer Transient Parameters by Using of Power System Transient Disturbances'' ,13th Iranian Conference on Electrical Engineering, Univesity of Zanjan, May 2005

[2] M.H. Nazemi, G.B. Gharehpetian, M. Shafiee, A. R. Allami, "Transformation of Detailed Model Transformer Winding to it's Terminal Model" , ISH 2003, Delft, Netherlands, August 2003.

[3] R.C. Degeneff, "A General Method for Determining Resonance in Transformer Winding'' IEEE Trans. on Power App. and Sys., Vol. PAS-96, No.2, March/Apr,1977, pp.423-430

[4] J. Christian. K.Feser, Usundermann and T. Leibfried, "Diagnostics of Power Transformers by Using the Transfer Function Method", 11,th International Symposium on High Voltage Engineering, London, England, 22-27 August 1999 ,Vol.I, No.467, pp.37-40

[5] S. H. Thilagar, G. S. Rao, "Parameter Estimation of Three-Winding Transformers Using Genetic Algorithm", Elsevier Science Ltd. , Engineering Applications of Artificial Intelligence, 2002, pp. 429-437

[6] Manitoba HVDC research center Inc., "PSCAD/EMTDC Version 4," 2003

[7] P. E. Wellstead, M.B. Zarrop, "Self tuning systems control and signal processing" , Wiley 1991

2006 5th International Power Electronics and Motion Control Conference

Worst-Case Tolerance Analysis for a Power Electronic System by Modified Genetic Algorithms

Toshiji Kato, Kaoru Inoue, Kazuya Nishimae
Department of Electrical Engineering, Doshisha University
Kyotanabe, Kyoto, 610-0321, JAPAN
E-mail: tkato@mail.doshisha.ac.jp

Abstract— It is necessary to consider variations of circuit element values for power electronic converter design to satisfy a given specification for a frequency characteristic or a dynamic transient response because the variations of element values generate those of the designed characteristic. Therefore element values and their variation ranges should be designed to give a satisfactory characteristic within an allowable specification range or a tolerance. This analysis is so-called tolerance or worst-case analysis. The analysis is a sort of nonlinear minimum and maximum (min/max) problem with respect to circuit parameters. However its computation cost is large and practically the problem is converted to a simpler and easier one with various techniques. This paper proposes an approximate method which finds a quasi-min/max solution by applying the Genetic Algorithm (GA). According to the GA, the higher the fitness value is, the more likely the genetic code remains. The solution is improved by generation and generation efficiently and the solution at the final generation is considered to be optimum. This paper proposes modified GA methods, a sharing-GA and a relay-search-GA, which are suitable for more precise tolerance analysis by utilizing diversity of solutions. Output voltage analysis examples of three types of converters for element parameter variations are investigated.

Keywords— tolerance analysis, worst-case analysis, genetic algorithm, Relay-Search GA, CAD, power converter

I. INTRODUCTION

It is necessary for power electronic circuit design to consider variations of element values to satisfy a given specification for transient or frequency characteristics. Usually a tolerance characteristic band or region is allowed as a design specification and the characteristic for any possible parameters should be set to lie inside the region by a proper design method. In this process it is necessary to find relation between the parameter and characteristic regions. This computation is called the worst-case tolerance analysis or simply tolerance analysis in this paper[1] and the process is repeated until a given specification is satisfied.

Generally this analysis is an optimization problem which finds the maximum or the minimum values by considering dependence and nonlinearities among parameters The problem often can be solved precisely by the interval analysis[2] − [4]. However its computation cost increases exponentially with increase of number of parameters and there is no good method for a transient problem at present. It is practically approximated to a simpler and easier problem with various techniques[5], [6]. For example, sampling methods like the Monte Carlo method are general and practical. However these are statistical analysis and are not the worst case one. Therefore they are not discussed in this paper.

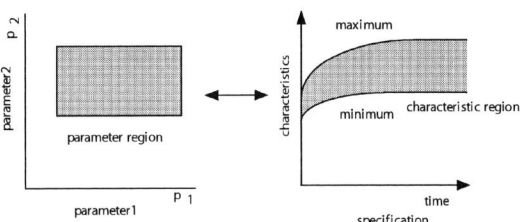

Figure 1. Tolerance analysis.

This paper proposes a new worst case tolerance analysis method by the genetic algorithm (GA) which is based on an engineering model of biological evolution processes. Some of the analysis methods have been already proposed[5], [7]. This paper develops them by introducing a metric or distance concept and new algorithms like the relay-search type for more precise analysis. As application examples, two types of converters are analyzed and investigated for the results.

II. TOLERANCE ANALYSIS OF A POWER ELECTRONIC CIRCUIT

In a power converter design, a specified characteristic should be considered with variations of circuit parameters. An allowable or tolerance characteristic band is given as a design specification for this purpose. Therefore it is necessary to analyze a correspondence from a possible element parameter region to its characteristic region in Fig.1. This is called the tolerance analysis which have two types; one is the statistical analysis which considers a statistical distribution of parameters and the other is the worst case analysis which is discussed in this paper and the process is repeated until a given specification is satisfied.

For example, a Ćuk converter circuit in Fig.7 have five passive elements which may have parameter variations which should be considered in a manufacturing process. It is to compute, for example, an output characteristic region which corresponds to a given circuit parameter variation region.

Generally an objective circuit characteristic f is determined for a given i-th set P_i with n parameters first in the tolerance analysis.

$$P_i = [p_{i1}, \ p_{i2}, \cdots, \ p_{in}]^t \tag{1}$$

$$f_i = f(x, p_{i1}, p_{i2}, \cdots, p_{in}) \tag{2}$$

1-4244-0448-7/06/$25.00 ©2006 IEEE

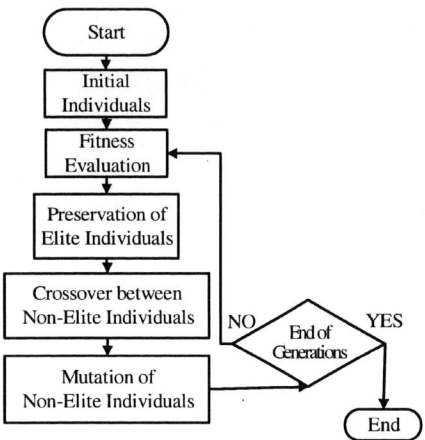

Figure 2. Flow-chart of the basic genetic algorithms(S-GA, SH-GA).

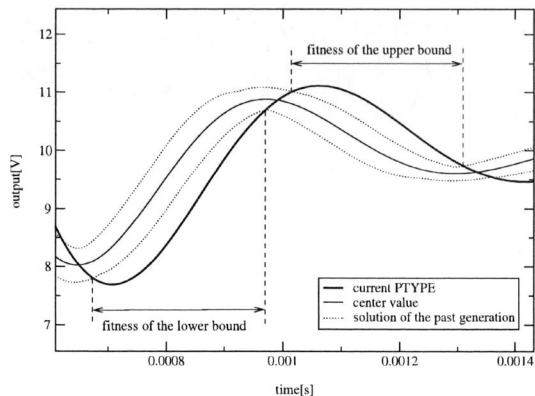

Figure 3. Fitness evaluation.

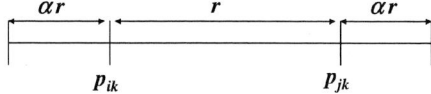

Figure 4. Blend crossover(BLX-α).

where x is, for example, time or frequency. The worst case tolerance analysis is to find the smallest characteristic regions, the minimum and maximum characteristic boundaries, which cover any f_i for P_i. A parameter set which gives the minimum or maximum characteristic value is complicatedly dependent on time or frequency. The tolerance analysis is a sort of optimization problem with respect to element parameters.

III. GENETIC ALGORITHMS

A. Simple Genetic Algorithm

A simple genetic algorithm (S-GA) is a sort of optimization technique which utilizes an engineering model of life evolution processes. It searches the optimum solution efficiently by iteratively generating high fitness individuals more likely for next generations or by improving current solutions by generation and generation. Its basic flow in shown in Fig.2.

Initially genetic parameters of individuals are generated at random. Fitness values of individuals are evaluated. According to them, the following genetic operations are executed. A rate of elite individuals are preserved. Another small rate of individuals are mutated. The other rate of them are used for crossover where parents are selected according to a probabilistic distribution and children are generated. The above steps are iterated until a generation termination condition is satisfied.

A main problem of the GA, or of general optimization methods, is possibility of only local solutions. It is possible to improve this possibility by keeping diversities of individuals with the mutation process. However additional techniques are also needed.

B. Fitness Function in Tolerance Analysis

It is important how to select a fitness function in GAs. It is selected to be suitable for the worst-case tolerance analysis as in Fig.3. A time (or frequency) characteristic is computed with a discrete algorithm and consists of num-

ber of discrete steps. The fitness value is a number of time steps which are improved by a result of an objective individual over the currently accumulated tolerance band just updated at the previous step. The fitness value is a number of time steps improved by the individual.

The worst values, or the minimum and maximum values, at the i-th step $x_j(j = 1, \cdots, m)$ are as $f_{min}(x_j)$ and $f_{max}(x_j)$ where a number of the total steps is denoted as m. Then the following count q_i is computed by comparing the band values with a characteristic value $f(x_j, P_i)$ of the individual P_i and it is accumulated to its fitness value $F_S(P_i)$ over all the steps.

$$q_j = \begin{cases} 1 & (f(x_j, P_i) < f_{min}(x_j)) \\ 0 & (f_{min}(x_j) \le f(x_j, P_i) \le f_{max}(x_j)) \\ 1 & (f(x_j, P_i) > f_{max}(x_j)) \end{cases} \quad (3)$$

$$j = 1, \cdots, m$$

$$F_S(P_i) = \sum_{j=1}^{m} q_j \quad (4)$$

The tolerance band values, $f_{min}(x_j)$ and $f_{max}(x_j)$, are updated to consider improvement of the solution by generation and generation.

C. Sharing-GA(SH-GA)

The largest disadvantage point of GA is that its solution is likely to be only locally optimum and it is inevitable from its principle. One modification technique in the S-GA is mutation. Another technique is the sharing-GA (SH-GA) which can find multiple solutions at the same time and has possibility of wider solution searches. The principle is that its fitness is modified to consider diversity of solutions

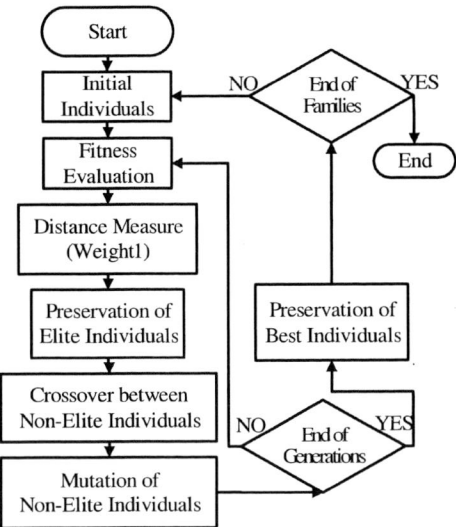

Figure 5. Flow-chart of the genetic algorithm(RS-GA).

by estimating similarity between solutions with a distance measure.

A sort of modification or correction coefficient $S(P_i, P_j)$ is defined with the following equation where d is a distance between individuals P_i and P_j and N is the total number of individuals.

$$d = d(P_i, P_j) = \sqrt{\sum_{k=1}^{n} (\frac{p_{ik} - p_{jk}}{p_k})^2} \qquad (5)$$

$$S(P_i, P_j) = \begin{cases} 1 - (\frac{d(P_i, P_j)}{\sigma})^{\beta} & (d < \sigma) \\ 0 & (d \geq \sigma) \end{cases} \qquad (6)$$

where β, σ are user-defined constants and p_k is the k-th nominal parameter. The fitness function $F_{SH}(P_i)$ in the SH-GA is as follows.

$$F_{SH}(P_i) = \frac{F_S(P_i)}{\sum_{j=1}^{N} S(P_i, P_j)} \qquad (7)$$

This SH-GA is a simple extension of the previous S-GA by considering the metric correction factor for diversity preservation. Its basic flow is the same as that of the S-GA in Fig.2.

D. Relay-Search-GA (RS-GA)

The above SH-GA is a suitable technique to preserve diversity. However individuals tend to have distances from others due to the metric fitness factor and sometimes they do not converge to desirable solutions. It is not good to to keep these mutual interferences. When a local quasi-optimum solution is found, it is effective to find another candidate by resetting the situation. The relay-search GA

(RS-GA) is based on this iterative search for different solutions and its flow is shown in Fig.5.

After the generation process of initial individuals, a small part of elite individuals are preserved without any genetic operation, and the others are generated by crossover or mutation operations. This process is repeated for a designated number of generations which is called one generation family or simply a family, and the best solution for this family is memoried. This process is repeated and a new solution different from memoried solutions is searched for a number of families. A proposed fitness function F_{RS} for the RS-GA is derived as follows by considering all distances d' from memoried best solutions.

$$F_d(P_i, P_j) = \begin{cases} \beta' d'^{\gamma} & (d' < \sigma') \\ 0 & (d' \geq \sigma') \end{cases} \qquad (8)$$

$$d' = d'(P_i, P_j) = \sum_{k=1}^{n} \frac{\sqrt{(p_{ik} - p_{jk})^2}}{e_k \ p_k} \qquad (9)$$

$$F_{RS}(P_i) = F_S(P_i) \prod_{j=1}^{N} F_d(P_i, P_j) \qquad (10)$$

where β', γ, σ' are user-defined constants, e_k is a variation ratio of the k-th parameter, and $F_d(P_i, P_j)$ is a fitness correction factor by a distance d' between P_i and P_j.

The above derivation is for general cases. As for the worst case tolerance analysis, a tolerance band at a generation in a family can be considered to include all past information of accumulated best solutions. It is not necessary to compute a distance correction factor F_d and the restart process is enough for the RS-GA which is very suitable for the analysis.

IV. BASIC BENCHMARK PROBLEMS

A. Benchmark Problem for Multiple Solutions

A benchmark problem for multiple solutions is selected to make clear differences among three types of GAs (S-GA, SH-GA, RS-GA). It is the following fitness function which has five multiple solutions for $0 \leq x \leq 1$.

$$F_1(x) = \sin^6(5\pi x) \qquad (11)$$

The analytical solutions are $x = 0.1, 0.3, 0.5, 0.7, 0.9$. Results by the three GAs are shown in Fig.6(a)-(c). The computation condition for the S-GA and the SH-GA is 50 individuals for 30 generations. That for the RS-GA is 10 individuals for 30 generations, and for 5 families to make the total number of individuals equal to that of the two GAs. The result by the S-GA is shown in the figure (a) where solutions converge to $x = 0.5$. In the S-GA, the diversity is monotonously reduced as the evolution proceeds and all tend to converge to one solution. From another set of initial individuals, they may converge to another solution. Anyway they converge to only to one solution for one S-GA run.

The result by the SH-GA is shown in the figure (b). Individuals seem to be distribute over the interval and the five

1235

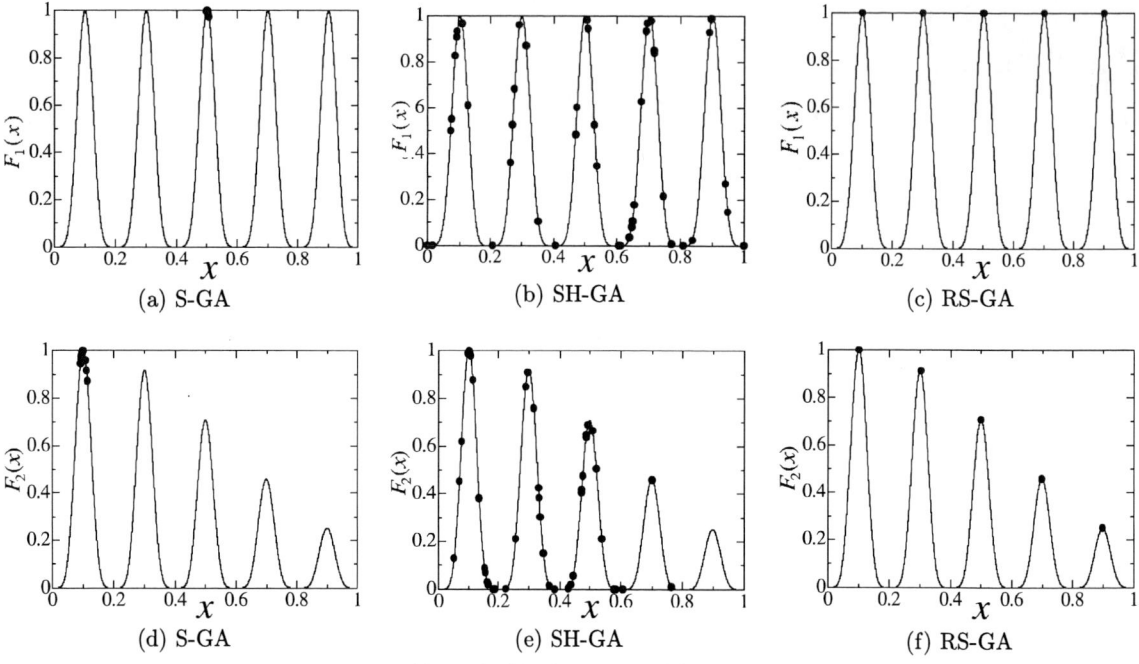

Figure 6. Two basic test problems.

solutions are found. This is because individuals are estimated to have differences from others to keep the diversity and they seem to distribute uniformly as the result. In the SH-GA, a large number of individuals should be evoluted to find multiple solutions with better precision.

The result by the RS-GA is shown in Fig.6(c) and all of the 5 solutions are efficiently computed. One solution is found for the first family, and another solution, which is different from the known solutions, is sequentially found for every family restart. This GA is suitable for this type of problem.

B. Benchmark Problem for Local Solutions

Another benchmark problem with the following fitness for local solutions is investigated also for the three GAs(S-GA, SH-GA, RS-GA) under the same condition.

$$F_2(x) = \exp[-2\log(2)\{(\frac{x - 0.1}{0.8})^2] \sin^6(5\pi x) \qquad (12)$$

The solution is unique at $x = 0.1$. However there are four local or quasi-optimum solutions at $x = 0.3, 0.5, 0.7, 0.9$. Solutions are converged only to the optimum in Fig.6(d) for the S-GA which are not rich in diversity. They are distributed almost uniform in Fig.6(e) for the SH-GA. However no solution is distributed around the last peak. They are converged to all possible local solutions in Fig.6(f) for the RS-GA whose convergence process is similar to the previous problem.

V. APPLICATION EXAMPLES TO POWER ELECTRONIC CONVERTERS

A. Ćuk converter

First the proposed worst-case tolerance analysis is applied to a Ćuk converter in Fig.7. It is analyzed by the three methods for ±10% variations of all parameters. As GA operation parameters, the crossover and mutation rates are 90% and 10% respectively. The computation condition for the S-GA and the SH-GA is 400 individuals for 10 generations. That for the RS-GA is 80 individuals for 10 generations, and for 5 families. These conditions are the same in the next converter case.

All results are shown in Fig.8(a). There are three groups of curves; the rigid curve in the middle is the center value for the nominal circuit parameters, the upper and lower curve groups are the maximum and minimum worst-case values by the three methods. The figure is partly enlarged in (b) which shows small differences of them. The results by the S-GA and the SH-GA are the same and the GA parameters should be selected more carefully. That by the RS-GA is a little wider and the result by the RS-GA is more precise.

B. Series Resonant Converter

Finally a series resonant converter in Fig.9 is analyzed for its output voltage as an example with more parameters ($n = 12$) by the three methods. The results are shown in Fig.10. The results by the S-GA and the SH-GA are the same and that by the RS-GA is slightly wider as in the above example. This shows an advantage of the RS-GA over the others.

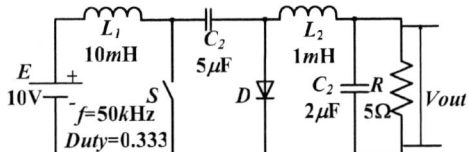

Figure 7. Ćuk Converter.

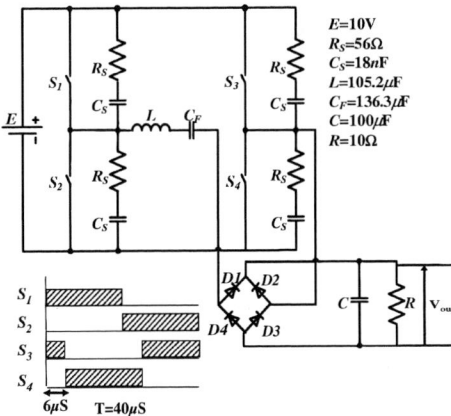

$E=10V$
$R_S=56\Omega$
$C_S=18nF$
$L=105.2\mu F$
$C_F=136.3\mu F$
$C=100\mu F$
$R=10\Omega$

Figure 9. Series resonant converter.

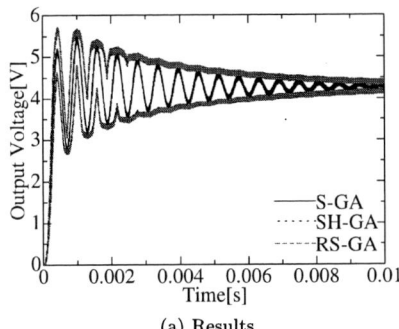

(a) Results

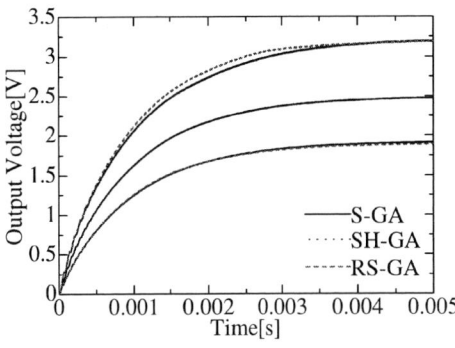

Figure 10. Analyzed result of output voltage of the series resonant converter.

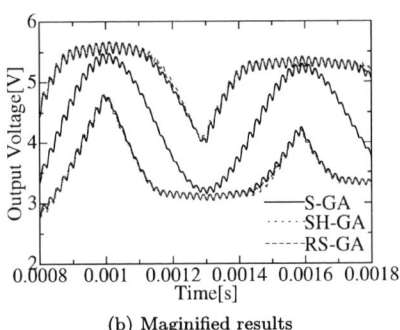

(b) Maginified results

Figure 8. Analyzed result of output voltage of the Ćuk converter.

VI. CONCLUSIONS

A worst-case tolerance analysis method for a power electronic converter by optimization techniques based on genetic algorithms was proposed in this paper. Especially the SH-GA and RS-GA are investigated to modify a main disadvantage of the S-GA of local solutions. The SH-GA improved the disadvantage and showed good performances in the benchmark problems. However it did not show advantages in application examples. It needs a large number of individuals and has difficulties for proper GA parameter selections.

The RS-GA showed best performances in both benchmark and application examples. In the worst case tolerance analysis, a tolerance band at a generation in a family can be considered to include all past information of accumulated best solutions. It is not necessary to compute a distance correction factor and the restart process is enough for the RS-GA which is very suitable for the analysis. One remark for the RS-GA is that it always tries to find different solutions and it cannot find a satisfactory solution when evolution effects are mature for a small number of generations.

Results by GAs are heavily dependent on GA parameters which should be selected properly and are the future subjects.

REFERENCES

[1] R. Spence and R.S. Sion : *Tolerance design of electronic circuits*, Imperial College Press, (1997)

[2] R.E. Moore : *Methods and applications of interval analysis*, SIAM Studies in Applied Mathematics, (1979)

[3] L.V. Kolev : *Interval methods for circuit analysis*, World Scientific, (1993)

[4] A. Cirillo, N. Femia and G. Spabnuolo : "An interval mathematics approach to tolerance analysis of switching converters ," *IEEE PESC Record*, vol.2, pp.1349-1355, (1996)

[5] N. Femia and G. Spagnuola : "Genetic optimization of interval arithmetic-based worst case circuit tolerance analysis," *IEEE Trans. on Circuit and System - I : Foundamental Theory and Applications*, VOL.46, No.12, pp.1441-1456, (1999)

[6] T.Fukuyama and T.Kato : "Tolerance computation of a power electronic circuit by higher order sensitivity analysis method," *Trans. IEE of Japan*, Vol.121-D, No.8, pp.835-840, (2001)

[7] N. Femia and G. Spabnuolo : "True worst-case tolerance analysis using genetic algorithms and affine arithmetic," *IEEE Trans. on Circuits and Systems - I : Fundamental Theory and Applications*, Vol.47, No.9, pp.1285-1296, (2000)

2006 5th International Power Electronics and Motion Control Conference

The Reduction of Force Ripples of PMLSM Using Field Oriented Control Method

Yu-wu Zhu, Kun-seok Jung, Yun-hyun Cho

Dong-A University/Electrical Engineering Servo Machine Lab., Pusan, Korea

Ywzhu1980@gmail.com

Abstract—The significant drawback of the permanent magnet linear synchronous motor (PMLSM) is force ripples, which are generated by the distortion of the stator flux linkage distributions, cogging forces caused by the interaction of the permanent magnet and the iron core and the end effects. This will deteriorate the performance of the drive system in high precision applications. The PMLSM and its parasitic effects are analyzed and modeled using the complex state-variable approach. To minimize the force ripple and realize the high precision control, the components of force ripples are extracted first and then compensated by injecting the instantaneous current to counteract the force ripples. And this method of the PMLSM system is realized by field oriented control method. In order to verify the validity of this proposed method, the system simulations are carried out and the results are analyzed. It can be seen the effective of the proposed force ripples reduction method according to the comparison between the compensation and non-compensation cases.

Keywords-force ripple; PMLSM; field oriented control; current injection

I. INTRODUCTION

Linear motor drives are more and more used in factory automation and numerical control system, because it can be operated without indirect coupling mechanisms, such as gear boxes, chains and screw coupling [9], then the high precision control can be achieved. Permanent magnet linear synchronous motor (PMLSM) drives are probably the most suitable to applications involving high speed and high precision motion control among the linear motors. It has certain unique features such as large air-gap, open-wide slots, pole and interpole configurations [8]. The main benefits of a PMLSM include the high force density in the air-gap, a rapid dynamic response, low thermal losses and simple structure.

In spite of such advantages, PMLSM drives may exhibit some drawbacks. A significant and well-known one is the phenomenon of force ripple. The effects of force ripple are particularly undesirable in some demanding motion control and machine tool applications. They lead to speed oscillations which cause the deterioration in the performance [2]. The force ripples change periodically as the mover advances during its motion. The resulting force ripples contain the flux linkage harmonics, cogging harmonics and time harmonics.

Flux linkage harmonics result from the non-sinusoidal flux linkage and from current waveform distortions. And the reluctance torque is very low for permanent magnet machines, especially it can be ignored for PMLSM using field oriented control method [2]. Cogging force arises from the interaction of the permanent magnets and the ferromagnetic core. This force exists even in the absence of any winding current and it exhibits a periodic relationship with respect to the position of the mover relative to the magnets [6], [8]. Time harmonics are caused by current waveform distortions in the feeding power converter. To minimize their effect, the switching frequency must be high, and the pulse-width modulation scheme must not generate subharmonics currents [11]. This can be satisfied by program of field control method in 20 [kHz]. The field oriented control algorithm will enable real-time control of force by controlling q-axis current component. This control structure, by achieving a very accurate steady state and transient control, leads to high dynamic performance in terms of response times and power conversion [5].

In this paper we will extract the force ripples and give explicit function to them. And then only control the q-axis current to cancel out the force ripples. First, the structure and operating principle of the PMLSM are described in detail. Second, the space vector modulation is introduced. Thus, appropriate models are then defined for the ripple and a concept for the compensation of force ripple by an adaptive control system is presented, and its effectiveness is demonstrated by some simulation results.

II. MODELING OF PMLSM

PMLSMs may be classified into the short primary type (long PM poles) and short secondary type (short PM poles) according to their structural features. Fig. 1 shows the basic structure of a short primary type PMLSM. There are alternant N-pole and S-pole permanent magnets

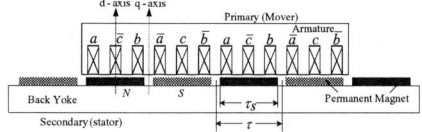

Figure 1.   Structure of PMLSM

1-4244-0448-7/06/$25.00 ©2006 IEEE   1238

fixed on the stator (secondary) of PMLSM, and the mover (primary) comprises the armature and windings, the mover will move with the cable which supplies power by a current-controlled pulse width modulated (PWM) voltage source inverter (VSI) to PMLSM.

In this paper the space harmonics and force ripple induced by the rotor slots are neglected in a first approach. The following derivation is based on these assumptions: Linear magnetic conditions (no saturation); no temperature or frequency dependence of the resistances and inductances.

Since the flux density field in the air gap is created with permanent magnets attached to the stator, this field will never be perfectly sinusoidal shown in Fig. 2.

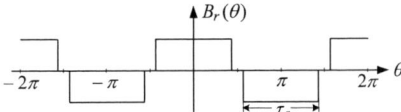

Figure 2. Flux Density in air-gap

So the induced flux linkage from the stator magnets in the mover winding can be expressed as a sum of odd cosines where the coefficients decrease rapidly [4]. Thus

$$\psi_{m,a}(\theta) = \psi_1\cos\theta + \psi_3\cos(3\theta) + \psi_5\cos(5\theta) + \cdots \quad (1)$$

where the amplitude of each flux linkage harmonic depends on the specific design of the stator winding.

We have already known that the PMLSM flux velocity υ_r [1],

$$\upsilon_r = 2\tau_p f_s \quad (2)$$

$$\omega_r = \pi\upsilon_r / \tau_p \quad (3)$$

where υ_r is the electrical velocity, ω_r is the rotating speed, τ_p is pole pitch and f_s is the source frequency.

Based on the upper equations and the machine is wye-connected with an isolated neutral point, we can derivate the following equations in synchronous rotating reference frame considering the non-sinusoidal components as follows [4]:

$$v_d = R_s i_d + L_d \frac{di_d}{dt} - \frac{\pi}{\tau}\upsilon_r L_q i_q + \frac{\pi}{\tau}\upsilon_r\psi_{d6}\sin(6\theta_r) + \frac{\pi}{\tau}\upsilon_r\psi_{d12}\sin(12\theta_r) + \cdots \quad (4\text{-a})$$

$$v_q = R_s i_q + L_q \frac{di_q}{dt} + \frac{\pi}{\tau}\upsilon_r L_d i_d + \frac{\pi}{\tau}\upsilon_r\psi_m + \frac{\pi}{\tau}\upsilon_r\psi_{q6}\cos(6\theta_r) + \frac{\pi}{\tau}\upsilon_r\psi_{q12}\cos(12\theta_r) + \cdots \quad (4\text{-b})$$

where the equations have been truncated so that they only contain harmonics of up to order 12. Also, $\varphi_m = \varphi_1$ in (4-b). The general expressions for the flux linkage harmonics in (4) are

$$\psi_{dk} = -(k-1)\psi_{k-1} - (k+1)\psi_{k+1} \quad k = 6,12,\ldots \quad (5\text{-a})$$

$$\psi_{qk} = -(k-1)\psi_{k-1} + (k+1)\psi_{k+1} \quad k = 6,12,\ldots \quad (5\text{-b})$$

The electromagnetic power can be calculated using the well-known relation

$$P_e = F_x \upsilon_e \quad (6)$$

where υ_e is the mechanical speed of PMLSM. Since P_e is the electromagnetic power, the force expression becomes [4]

$$F_x(\theta_r) = \frac{3}{2}\cdot n_p \cdot \frac{\pi}{\tau}[\psi_m i_q + (L_d - L_q)i_d i_q + (\psi_{d6}\sin(6\theta_r) + \psi_{d12}\sin(12\theta_r))\cdot i_d + (\psi_{q6}\cos(6\theta_r) + \psi_{q12}\cos(12\theta_r))\cdot i_q] \quad (7)$$

In this force equation, it only considers the harmonics till the 12th, because the higher component of harmonics has very small effects.

And for the surface-mounted PMLSM, the d-axis inductance and q-axis inductance are almost the same, in this paper the field oriented control method is used, which means the d-axis current component i_d is zero, so the thrust can be finally expressed as

$$F_x(\theta_r) = \frac{3}{2}\cdot n_p \cdot \frac{\pi}{\tau}[\psi_m i_q + (\psi_{q6}\cos(6\theta_r) + \psi_{q12}\cos(12\theta_r))\cdot i_q] \quad (8)$$

The difference between the standard model of the PMLSM and the proposed one is that the electrical force of the proposed motor is now a function of the mover position. This leads to force oscillations which have to be minimized in order to control the machine exactly.

In this part we analyzed the model of PMLSM and gave its mathematics representation only considering the non-sinusoidal harmonics, for more detail analysis of harmonics and its compensation it will be analyzed in next part.

III. REDUCTION OF FORCE RIPPLES

In the vector control, a special control is making the stator sinusoidal MMF wave quadrature to the permanent magnet fundamental exciting flux, and this is usually called 'field oriented control'. The field oriented control algorithm will enable real-time control of torque and speed. As this control is accurate in every mode of operation (steady state and transient), no oversize of the power transistors is necessary. The transient currents are constantly controlled in amplitude. Moreover, almost no force ripple appears when driving this sinusoidal BEMF motor with sinusoidal current (ideal model).

The basic goal of the control algorithm in this paper is to control the current so that the force ripple is cancelled out. Force variations of very low frequency are normally eliminated by the speed control system. Force harmonics of higher frequency can be compensated in principle by generating an inverse force component through appropriate modulation of the mover current. Because field oriented control method is used, the modulation of the mover current is in a matter of fact to control the amplitude of q-axis current component i_q.

The design of a permanent magnet motor drive for high performance and minimum force ripple is based on a machine model representing the harmonic effects. So in this paper we first give an appropriate function of force ripples which contains non-sinusoidal flux linkage harmonics and cogging harmonics. Then using the control algorithm compensates these harmonics.

If only considering the ideal model of PMLSM, the back EMF and force will not contain the harmonics items and (4-a) , (4-b) and (8) will be written as:

$$v_d = R_s i_d + L_d \frac{di_d}{dt} - \frac{\pi}{\tau} \upsilon_r L_q i_q \tag{9-a}$$

$$v_q = R_s i_q + L_q \frac{di_q}{dt} + \frac{\pi}{\tau} \upsilon_r L_d i_d + \frac{\pi}{\tau} \upsilon_r \psi_m \tag{9-b}$$

$$F_x = \frac{3}{2} n_p \frac{\pi}{\tau} \psi_m i_q = k_f i_q \tag{10}$$

where $k_f = \frac{3}{2} n_p \frac{\pi}{\tau} \psi_m$.

No considering the force ripple, the mechanical operating equation can be represented as

$$F_x = M \frac{d\upsilon_r}{dt} + D\upsilon_r + F_l \tag{11}$$

where F_x is the electrical force, M is the total mass of moving element system, D is the viscous friction and iron-loss coefficient, F_l is the external load force.

Based on (9) ~ (11), the signal flow graph of PMLSM is visualized in Fig. 3

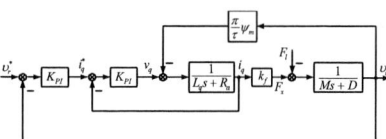

Figure 3.   Control system block diagram

Fig. 3 shows the adaptive speed control graph of PMLSM. In the close control loop, the feedback speed υ_r compares with the reference speed value υ_r^*, and then using PI controller adjusts the error between υ_r and υ_r^* to be zero. In this case, it does not consider the harmonics, for high precision and accurate control, we have to analyze and give an appropriate function of harmonics, and then compensate these harmonics.

A.   Identification of Flux Linkage Harmonics

The effect of flux linkage harmonics is already analyzed and from (8) using field oriented control method this force harmonics can be seen as

$$F_{x-\psi h}(\theta) = \frac{3}{2} n_p \frac{\pi}{\tau} [\psi_{q6} \cos(6\theta_r) + \psi_{q12} \cos(12\theta_r)] \cdot i_q \tag{12}$$

So the compensated flux linkage force harmonics component force $F_{q-\psi h}$ which is generated by q-axis current component is

$$F_{q-\psi h} = -F_{x-\psi h} \tag{13}$$

This means

$$\frac{3}{2} n_p \frac{\pi}{\tau} \psi_m \cdot i_{q-\psi h} = -\frac{3}{2} n_p \frac{\pi}{\tau} [\psi_{q6} \cos(6\theta_r) + \psi_{q12} \cos(12\theta_r)] \cdot i_q$$

The compensated current $i_{q-\psi h}$ for flux linkage force harmonics is

$$i_{q-\psi h}(\theta_r) = -\frac{\psi_{q6} \cos(6\theta_r) + \psi_{q12} \cos(12\theta_r)}{\psi_m} \cdot i_q \tag{14}$$

Because the flux linkage harmonics current $i_{q-\psi h}$ is far smaller than the reference current i_q , so this approximation is accurate enough.

B.   Identification of Cogging Harmonics

Cogging force comprises two components due to slotting and the finite length of the armature core. It depends on the position of mover and varies periodically. This cogging harmonic force function $F_{x-ch}(\theta_r)$ is indirectly acquired by operating the machine at very low speed. Because the cogging force is independent of the input current, it is only dependent on the structure of motor, so the measuring of cogging force is taken in a no-load and low speed case. In this case, from equation (14) it can be seen that the flux linkage harmonics are very small, the cogging force is the predominant. Considering the harmonics, the dynamic equation of PMLSM can be represented as:

$$F_x = M \frac{d\upsilon_r}{dt} + D\upsilon_r + F_l + F_h \tag{15}$$

where F_h is the total harmonics of PMLSM. It is

$$F_h = F_{x-\psi h} + F_{x-ch} \tag{16}$$

To obtain the cogging harmonics, it has to calculate the total harmonics first.

In the measuring process, because of setting the motor in a no-load operation, then F_l is zero. It keeps the amplitude of input stator current i_q constant, so according (10) the thrust force F_x is constant. Then using linear encoder to measure the mover position and speed, finally the total harmonics can be calculate by (15)

$$F_h = F_x - M \frac{d\upsilon_r}{dt} - D\upsilon_r \tag{17}$$

In the test, its ac content is the force ripple, and the dc offset is the load force including friction of the motor. After getting the total force ripple, just minus the flux linkage force ripple, the cogging harmonics is obtained, and then the corresponding compensating q-axis current component is

$$i_{q-ch}(\theta_r) = \frac{F_h - F_{x-\psi h}}{k_f} \tag{18}$$

For the upper analysis, both flux linkage and cogging force harmonics vary with the mover position θ_r . Using field oriented control method the control diagram can be drawn as Fig 4. The shaded part is the compensation part.

In the control diagram of a PMLSM, first it samples the phase currents i_a and i_b , then transfers them into

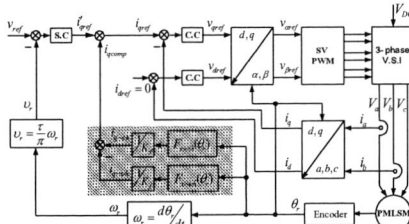

Figure 4. Control diagram of PMLSM

rotor-fixed frame components i_d and i_q by Clarke and Park transformation. Secondly, according the reference speed and the sampled mover speed of PMLSM to calculate the q-axis current component i'_{Sqref} by PI (Proportional-Integral) speed controller, considering the harmonics so the reference current i_{qref} is the sum of i'_{qref} and i_{qcomp} , . Because the field oriented control algorithm is used, the d-axis reference current is $i_{dref} = 0$. Thirdly, using PI current controllers calculates d-axis reference voltage v_{dref} and q-axis reference voltage v_{qref} in terms of the errors between i_{qref} and i_q , i_{dref} and i_d respectively. Finally using the Park^{-1} transformation gets $v_{S\alpha ref}$, $v_{S\beta ref}$ and modulates the space vector PWM to offer the motor suitable voltage to maintain the constant speed of PMLSM.

IV. SIMULATED RESULTS

The specification of PMLSM are shown in table I.

TABLE I.
SPECIFICATION OF PMLSM

Parameter	Value	Parameter	Value
Rotor magnet material	NeFeB	LSM width	52.8 mm
Rated voltage (DC)	50 V	Pole pitch	30 mm
Rated current	5 A	Magnetic pole pitch	25 mm
Rated force	160N	Air-gap length	2 mm
Rated velocity	1.2 m/s	Slot numbers	12
Winding turns per pole per phase	40 turns	Slot numbers per pole per phase	1
Number of pole pairs	2	Armature resistance	0.75 Ω
Mover mass	2.0 Kg	d-axis inductance	0.85mH
Fundamental flux linkage	103.9 mWb	q-axis inductance	0.85 mH

Since we have known the parameters, we can first simulate the performance of PMLSM and get some results in theory.

In this paper, the PI current controllers and the PI speed controller parameters are:

$$K_{pd} = K_{pq} = 4$$
$$K_{id} = K_{iq} = 400$$

$$K_{pv} = 0.15$$
$$K_{iv} = 6.0$$

The simulation is realized by C++ language program and the results are shown in Matlab. This simulation contains the rated speed 1.2 [m/s] command responses, speed switch responses and the load switch responses. In the simulation, the comparisons between the ripple non-compensated and compensated are taken, it can be seen that the results of the compensation are much effective.

A. Rated Speed 1.2 [m/s] Command Responses

In this case, the motor was taken into rated speed 1.2 [m/s] command from standstill without load. We observed the phase current, speed, and force, respectively.

It can be seen that the current in Fig. 5(b) has much higher component than that in Fig. 5(a), which is the injected current to counteract the force ripple. The

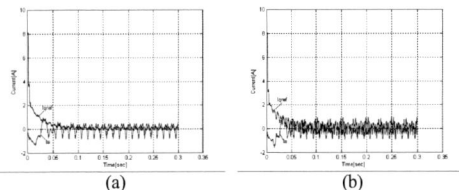

(a)   (b)

Figure 5. Current Waveform of PMLSM. (a) No compensation;
(b) Compensation

corresponding speed and force responses can be seen in Fig. 6. It takes the compensation command at 0.6 [sec], and Fig. 6 (b) and (d) are the blow-up of Fig. 6 (a) and (c)

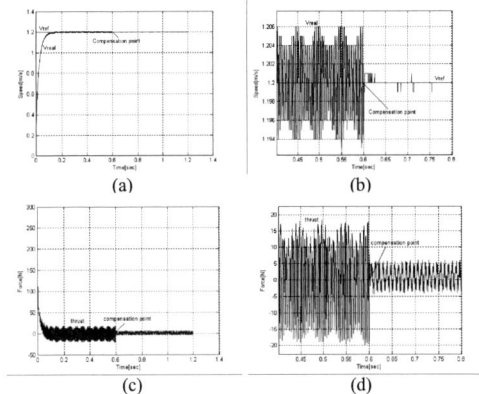

(a)   (b)

(c)   (d)

Figure 6. Waveforms under Rated Speed Command. (a) Speed;
(b) Speed blow-up; (c) Force; (d) Force blow-up

around compensation point. It can be seen that the speed and force ripples are significantly suppressed.

B. Reference Speed Switch Command Responses

In this section the speed command is given at 0.8 [m/s] and kept 0.4 [s] at the beginning, then changes into the rated speed command 1.2 [m/s] and also keeps 0.4 [s], finally changes into 0.8 [m/s] again. We can see the responses waveform in Fig. 7. There is large force to

1241

make the mover achieve the command speed when the speed command changes, and the force ripples are reduced under any speed command.

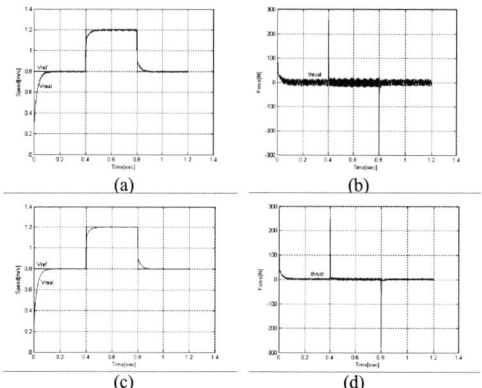

Figure 7. Waveforms under different speed commands. (a) Speed no-compensation; (b) Force no-compensation; (c) Speed compensation; (d) Force compensation.

C. Load Switch Command Responses

Fig. 8 shows the responses under the condition that the PMLSM operates at rated speed 1.2 [m/s] and adds the 1/2 rated load 80 [N] during the period from 0.4 [s] to 0.8 [s]. In the comparison it can be seen that the ripples are minimized. Especially in the no load case, the force ripple is significant reduced and the compensation is very effective to smooth this ripple. For adding the load case, although the force ripple is obvious after compensation, but it is ignorable to the load because there is almost no effect to speed.

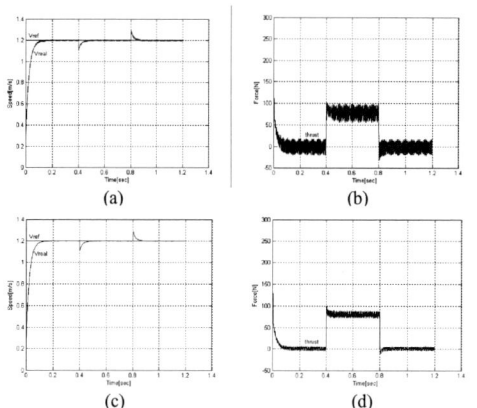

Figure 8. Waveforms under No-Load and Load Commands. (a) Speed no-compensation; (b) Force no-compensation; (c) Speed compensation; (d) Force compensation.

According the upper simulation results, it proves this force ripple compensation method is effective under any cases.

V. CONCLUSION

In this paper we first give an appropriate function of force ripples which contain non-sinusoidal flux linkage harmonics and cogging harmonics. It is feasible to implement the compensation for the explicit function easily using field oriented control method. The simulation results, under different speed and load conditions, verified that the proposed algorithm is very effective.

ACKNOWLEDGMENT

This work has been supported by KESRI (R-2003-B-271) which is founded by MOCIE.

REFERENCES

[1] FAA-JENG LIN, KUO-KAI SHYU, and CHIH-HONG LIN, "Incremental Motion Control of Linear Synchronous Motor," *IEEE. Trans. on Aerospace and Electronic System*, vol. 38. No. 3 July 2002.

[2] Joachim Holtz, "Identification and Compensation of Torque Ripple in High-Precision Permanent Magnet Motor Drives," *IEEE. Trans. on Industrial Electronics,* vol. 43. No.2, April 1996.

[3] B.-J. Brunsbach, G. Henneberger, and Th. Klepsch, "Compesation of Torque Ripple," *University of Technology Aachen, Germany,* pp. 588-593.

[4] Oskar Wallmark, "Modelling of Permanent-Magnet Synchronous Motors Machines with Non-Sinusoidal Flux Linkage," Chalmers University of Technology, Sweden.

[5] "Field Oriented Control of 3-Phase AC-Motor," Texas Instruments Europe, February 1998.

[6] K.K. Tan, S.N. Huang, and T.H. Lee "Robust Adaptive Numerical Compensation for Friction and Force Ripple in Permanent-Magnet Linear Motor," *IEEE. Trans. on Magnetics*, vol. 38. No. 1 January 2002.

[7] Nicola Bianchi, Silverio Bolognani, and Alessandro Dalla Francesca Cappello, "Back E.M.F Improvement and Force Ripple Reduction in PM Linear Motor Drives," *35th Annual IEEE Power Electronics Specialiists Conference. Achen, Germany,* 2004.

[8] P.J. Hor, Z.Q. Zhu, D. Howe, J. Rees-Jones, "Minimization of Cogging Force in a Linear Permanent Magnet Motor," *IEEE Trans. on Magnetics*, vol. 34, No. 5, September 1998.

[9] K.K. Tan, S. Zhao, "Adaptive Force Ripple Suppression in Iron-core Permanent Magnet Linear Motors," *Proceeding of the 2002 IEEE International Symposium on Intelligent Control, Vancouver, Canada,* October, 2002.

[10] N. Bodika, R.J. Cruise, and C.F. Landy, "Design of a PI Controller to Counteract the Effect of Cogging Forces in a Permanent Magnet Synchronous Linear Motor," *IEEE, University of the Wiwatersrand, Johannesburg, South Africa,* 1999.

[11] J. Holtz, "pulsewidth modulation—A Survey,' IEEE Trans. Ind. Electron, vol. 39, No.5, pp. 410-420, October 1992.

2006 5th International Power Electronics and Motion Control Conference

Analysis and Design of Signal Stage AC/DC Converter with Resonant Model PFC

Weiping Zhang, Liangrui Lin, Dongyan Zhang and Xusen Zhao
North China University of Technology/Lab of Green Power & Energy, Beijing, China
Email: zwp@ncut.edu.cn

Abstract—A single stage AC/DC converter with resonant model PFC has been deeply studied in this paper by mathematic analysis. The following results have been revealed: (1).The working conditions for the circuit shown in Fig.1 have been derived and can be expressed by the formulas, which can be used to select the parameters of the circuit; (2). It has been proved that the bus voltage decreases if output power increases and the turn-on time is kept constant; (3). The optimal controlling strategy for regulating the output voltage has been put forward. In order to verify the theoretic analysis conclusions, the simulation results and experimental results have been given out. The theoretic analysis conclusions and math formulas have provide a new tool for designing the circuit.

Keywords- PFC; AC/DC; power; model

I. INTRODUCTION

Power factor correction (PFC) techniques have become attractive since several regulations have been effected recently. Many PFC converters have been given out [1~4]. They usually can be divided into two categories: the two-stage approach and the single-stage approach. The two-stage approach actually includes two power converter processes. The first stage is a PFC converter that has two proposes: one is to make the input current harmonics to meet the requirements of IEC 6100-3-2 Class D harmonic regulation; the other is to regulate its output voltage. And the second stage is a DC/DC converter or DC/AC converter to regulate the output voltage. This approach has good performances for power factor and output voltage regulation. The main disadvantage is high cost due to increase a PFC stage. The single –stage approach integrates the PFC stage and DC/DC converter or DC/AC converter into one stage, which is defined as single stage PFC+AC/DC converter [5]. In order to tightly regulate the output voltage, an internal energy storage capacitor is required so that the output voltage is free of line ripple. The switches in the single stage PFC+AC/DC converters suffer from high DC bus voltage stress at light load with high line input [1-3]. The reason of producing high DC bus voltage is the followings: because both the PFC stage and DC/DC converter share a common converter, only

Project supported by Beijing Natural Science Foundation (No.4052011)
Project supported by National Natural Science Foundation of China (N0.50477054)

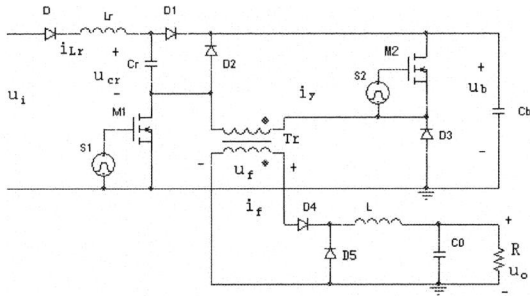

Figure 1. The signal stage AC/DC converter with resonant model PFC

one controller can be applied to control one variable. Usually, the output voltage is required to be tightly regulated by the voltage loop, the DC bus voltage of the bulk energy-storage capacitor is not regulated and it is required that this kind of converter must has an inherit PFC function.

According to the IEC 6100-3-2 Class D harmonic requirements, the designer should pay more attention to meet the Class D harmonic requirements and less attention to improve the power factor. The designers also have another one of the greatest concerns problem that is the DC bus voltage tress at light load with high line input. To reduce the DC bus voltage, DC bus voltage feedback technique is introduced to the single stage PFC+AC/DC converter [4].

In this paper, the signal stage AC/DC converter with resonant model PFC targets at IEC6100-3-2Class D applicants up to 300W. It have good performances, such as high power, lower DC bus voltage, high power factor and lower cost.

II. THE OPERATIONAL PRINCIPLE AND WORKING CONDITIONS OF THE CIRCUIT

Fig.1 shows the topology of single stage AC/DC converter with resonant model PFC, in which L_r, C_r forms the resonant tank. The output voltage is regulated by controlling the duty circle of switches M1 and M2. While M1 and M2 are turned on, the voltage across the bulk capacitor C_b is applied to the primary of the transformer, and the transformer delivers the power into the load. This implements the function of DC/DC conversion. At the same time, the resonant tank begins to be charged to draw power from the line. While M1 and M2 are turned off, the

1-4244-0448-7/06/$25.00 ©2006 IEEE   1243

DC/DC converter stops delivering power into its load, and the output voltage will be maintained by the storage energy in the output filter inductor and capacitor. In this interval, the energy stored in the resonant components will move to the bulk capacitor C_b. So with the charge and discharge of resonant tank, it makes that energy transmits in high speed, and the peak of input current follows the movement of input voltage wave. As a result, the power factor is improved.

Fig.2 shows the waveforms in single switching period. The resonant inductor L_r operates in DCM, that is, the current through it is discontinuous between switching periods. The input voltage source is half-sine wave with period T and peak value U_m, produced by rectifying. To analyze the steady-state circuit behaviors, the following assumptions are made:

(a). The operational frequency is sufficiently so higher than the line's frequency that the input voltage can be considered as keeping constant during a switching period T_e. So the amplitude of input voltage is   in the nth switching period.

(b). The bus voltage u_b is considered constantly as its average value U_b, and the output voltage u_o is considered constantly as its average value U_o.

(c). Only the load costs power in the whole circuit; the iron-core transformer is a perfect coupling transformer.

A switch cycle can be divided into three stages. S1 and S2 is the driving signals of M1 and M2. To represents the switch-on interval while T_{off} represents the switch-off interval. The ratio $p=T_e/T_o$, the integer $h=[T/T_e]$. In the transformer, N is the transformation ratio and L_y is primary inductance.

The first stage: $t \in [t_o, t_1]$

At t_o, M1 and M2 are turned on, D are turned on. Resonant angle frequency is that $\omega_s(= 1/\sqrt{L_r C_r})$. During this interval, the current through the resonant inductor and the voltage across the resonant capacitor are increased from zero as the following:

$$i_{Lr} = \sqrt{C_r/L_r} U_m \sin \omega t_n \sin \omega_s t . \qquad (1a)$$

$$u_{cr} = U_m \sin \omega t_n (1 - \cos \omega_s t). \qquad (1b)$$

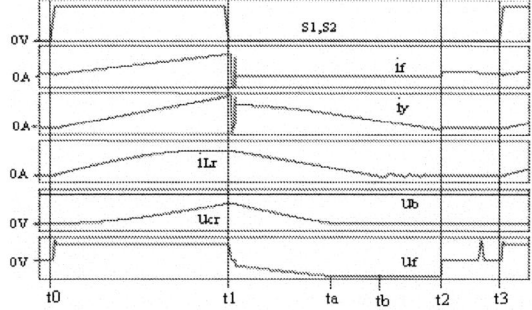

Figure 2. The simulated waveforms in switching period

The voltage (u_{cr}) across the resonant capacitor is keeping lower than U_b to prevent D1 to turn on in this interval. The voltage u_{cr} and the current i_{Lr} are supposed to increase continuously to ensure that the storage energy in the resonant components is increased in this interval. To do this, the switch-on interval must not be longer than a quarter of the resonant period:

$$\pi/2\omega_s = \pi\sqrt{L_r C_r}/2 \geq T_o. \qquad (2)$$

In this interval, D4 is turned on and the voltage U_f across the secondary side of the transformer is equal to NU_b, the current through the filter inductor increases linearly from zero if the inductor works in DCM:

$$i_f = t(U_f - U_o)/L .$$

By the above formula, the maximum value I_f of the inductor is $T_o(U_f - U_o)/L$.

According to the characteristic of perfect coupling transformer, the primary current (i_y) of the transformer is as the following,

$$i_y = \frac{U_b t}{L_y} + \frac{N(U_f - U_o)t}{L} = (\frac{U_b}{L_y} + \frac{N^2 U_b - NU_o}{L})t .$$

Because the output power is equal to the average power produced by secondary side of the transformer,

$$P_o = U_o^2/R = U_f I_f/2p^2 = T_o U_f(U_f - U_o)/2p^2 L .$$

So the relationship between output voltage and bus voltage is built up,

$$U_o^2 + (U_o - NU_b)NU_b T_o R/2p^2 L = 0,$$

$$U_o = \frac{2NU_b}{1 + \sqrt{1 + 8Lp^2/T_o R}}. \qquad (3)$$

It shows that the bus voltage is direct proportional to its output voltage. While the filter inductor become smaller, the output voltage will be bigger, nearly equalling to the peak value of secondary voltage if $L \ll T_o R/8d$. Based on (3), the average output power is available as

$$P_o = \frac{2T_o N^2 U_b^2}{T_o R + 4Lp^2 + \sqrt{T_o^2 R^2 + 8T_o RLp^2}}. \qquad (4)$$

The second stage: $t \in [t_1, t_2]$

At t_1, M1 and M2 are turned off. D1, D3 and D5 are turned on as well as D4 is turned off. The two rectifying diode, which are connected to secondary side, is freewheeling period.

During the interval from t_1 to t_b, the energy stored in the inductor L_r has been transformed to C_b and the current i_{Lr} linearly decrease to zero, that is

$$i_{Lr}(t) = \sqrt{C_r/L_r} U_m \sin \omega t_n \sin \omega_s T_o - \frac{U_b - U_m \sin \omega t_n}{L_r} t . \qquad (5)$$

At t_b, D and D1 are turned off.

The length of the interval can be determined by

$$T_b = t_b - t_1 = \frac{\sqrt{C_r L_r} U_m \sin \omega_s T_o \sin \omega t_n}{U_b - U_m \sin \omega t_n}.$$

Through the resonant inductor, the bulk capacitor C_b can draw the energy (E_n) from the line in a switching period.

$$E_n = U_b \sqrt{C_r / L_r} U_m \sin \omega t_n \sin \omega_s T_o \bullet T_b / 2. \quad (6)$$

To make L_r operate on DCM, the maximum length of the interval should be

$$T_{bm} = \frac{\sqrt{C_r L_r} \sin \omega_s T_o}{U_b / U_m - 1} = \sqrt{C_r L_r} \frac{\sin T_o / \sqrt{C_r L_r}}{U_b / U_m - 1} < T_{off}. \quad (7)$$

At t_1, the magnetizing inductor of the transformer starts to resonate with the resonant capacitor C_r and the energy stored in C_r can be transformed to bulk capacitor. At t_a, C_r has no any energy and the resonant stop. After that, D2 is turned on; the current i_y linearly is reduced. D2 is turned off at t_2.

During the interval from t_1 to t_a, the formulas of the voltage across C_r and current through the magnetizing inductor are

$$u_{cr} = -\sqrt{(u_{cr}(t_1) - U_b)^2 + (U_b T_o \omega_y)^2} \sin(\omega_y t + \theta) + U_b.$$

$$i_y = \sqrt{C_r / L_y} \sqrt{(u_{cr}(t_1) - U_b)^2 + (U_b T_o \omega_y)^2} \cos(\omega_y t + \theta).$$

Where, $\theta = \arctan \dfrac{U_b - u_{cr}(t_1)}{U_b T_o \omega_y}$, $\omega_y = 1 / \sqrt{L_y C_r}$.

Thus, to make sure that the voltage across Cr becomes zero, it should be

$$1 \le \sqrt{\left(\frac{(1 - \cos \omega_s T_o) U_m}{U_b} - 1\right)^2 + T_0^2 \omega_y^2} \le \sqrt{\left(\frac{U_{cr}(t_1)}{U_b} - 1\right)^2 + T_0^2 \omega_y^2}.$$

$$C_r \le \frac{T_o^2}{1 - ((1 - \cos \omega_s T_o) U_m / U_b - 1)^2} \frac{1}{L_y}. \quad (8)$$

Based on the formula of u_{cr}, T_a can be calculated by

$$T_a = t_a - t_1 = \frac{\arcsin\left(1 / \sqrt{(U_{cr}(t_1) / U_b - 1)^2 + T_0^2 \omega_y^2}\right) - \theta}{\omega_y}.$$

The maximum time length, it takes up for the magnetizing current to reduce to zero, is as the following,

$$T_m = t_2 - t_1 = \frac{1}{\omega_y} \arctan \frac{U_b T_o \omega_y}{U_b - U_m (1 - \cos \omega_s T_o)} < T_{off}. \quad (9)$$

Base on the above analysis, the following conclusions can be make up: (a) if (7) is satisfied, the resonant inductor Lr works on DCM; (b) if (8) is satisfied, the initial voltage of the resonant capacitor Cr is zero when the switches M1 and M2 begin to be turned on; (c) if (9) is satisfied, the transformer can work well. These formulas have given up the working conditions of the proposed circuit shown in Fig.1.

The third stage: $t \in [t_2, t_3]$

Every diode maintains the status quo, until the next switching period begin.

III. THE RELATIONSHIP OF POWER, DC BUS VOLTAGE AND LOAD AND THE CONTROLLING STRATEGY

Generally, the DC bus voltage is a function of input voltage and the load as well as the output power. When the system reaches to steady state, the bus voltage is regarded as the medium of energy transmittal. Its total energy of drawing from the line in all the first stage should be equal to its total that of delivering to its load in all the second stage. Based on this, it can be proved that the DC bus voltage will increase as the load becomes lighter. In this section, we will discuss the optimal controlling strategy for regulating output voltage.

If T_o is keeping constant and (6) is employed, C_b's average power in a line period, which can obtained from the line through the resonant inductor, can be calculated by the following equation,

$$P_{Lr} = \sum_{n=1}^{h} E_n / T$$
$$= \frac{C_r U_b U_m^2 \sin^2 \omega_s T_o}{2T} \sum_{n=1}^{h} \frac{\sin^2 \omega t_n}{U_b - U_m \sin \omega t_n}.$$

Because the turn-off time (T_o) is far less than the line period, we can apply integration to express the above equation, then

$$P_{Lr} = \frac{C_r U_b U_m \sin^2 \omega_s T_o}{2T \omega T_e} \int_0^\pi \frac{\sin^2 \omega t}{U_b / U_m - \sin \omega t} d\omega t = \quad (10)$$
$$\frac{C_r U_m^2 \sin^2 \omega_s T_o}{2\pi T_e} \left(\frac{U_b^3}{U_m^3} \int_0^\pi \frac{d\omega t}{U_b / U_m - \sin \omega t} - \frac{\pi U_b^2}{U_m^2} - \frac{2 U_b}{U_m} \right)$$

The key waveforms in one line period are shown in Fig.3. In Fig.3, the low-frequency amplitude envelop of input current has same frequency and same phase with the input voltage, so the power factor is nearly unity.

The average input power on the half line period is as that

$$P_i = \overline{u_i \bullet i_{Lr}} = \overline{U_m \sin \omega t \bullet I_1 \sin \omega t} \Big|_{\omega t = 0}^\pi = U_m I_1 / 2.$$

$$I_1 = \frac{2}{T} \int_0^T i_{Lr} \sin \omega t dt = \frac{2}{T} \sum_{n=1}^{h} \int_{n0}^{nb} i_{Lr} \sin \omega t dt$$
$$= \frac{2}{T} \sum_{n=1}^{h} \sin \omega t_n \left(\int_{n0}^{n1} i_{Lr} dt + \int_{n1}^{nb} i_{Lr} dt \right)$$

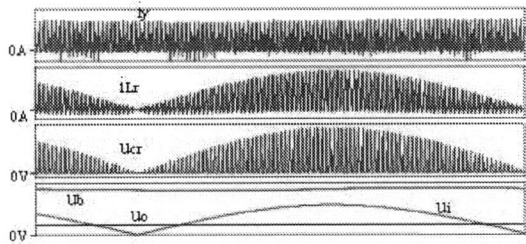

Figure 3. The simulated waveforms in switching period

Where, t_{no}, t_{n1} and t_{nb} represent respectively t_o, t_1 and t_b in the n switch period.

If (1) and (5) are applied, the following equations can obtained,

$$\int_{t_{n0}}^{t_{n1}} i_{Lr}dt = C_r U_m \sin\omega t_n (1-\cos\omega_s T_o) \cdot$$

$$\int_{t_{n1}}^{t_{nb}} i_{Lr}dt = \sqrt{C_r/L_r}\,U_m \sin\omega t_n \sin\omega_s T_o \bullet T_b/2$$

$$= \frac{C_r\left(U_m \sin\omega_s T_o \sin\omega t_n\right)^2}{2\left(U_b - U_m \sin\omega t_n\right)}$$

$$I_1 = \frac{2C_r U_m}{T}\sum_{n=1}^{h}\left((1-\cos\omega_s T_o)\sin^2\omega t_n + \frac{U_m \sin^2\omega_s T_o \sin^3\omega t_n}{2\left(U_b - U_m \sin\omega t_n\right)}\right).$$

So, the input power is as that,

$$P_i = \frac{C_r U_m^{\,2}\sin^2\omega_s T_o}{2\pi T_e}\left(\frac{\pi}{1+\cos\omega_s T_o} - \int_0^\pi \frac{\sin^3\omega t}{\sin\omega t - U_b/U_m}d\omega t\right)$$

$$= \frac{C_r U_m^{\,2}\sin^2\omega_s T_o}{2\pi T_e}\left(\frac{U_b^{\,3}}{U_m^{\,3}}\int_0^\pi \frac{d\omega t}{U_b/U_m - \sin\omega t} + \frac{\pi}{1+\cos\omega_s T_o}\right.$$

$$\left. -\frac{\pi U_b^{\,2}}{U_m^{\,2}} - \frac{2U_b}{U_m} - \frac{\pi}{2}\right) \tag{11}$$

Comparing (10) with (11), the difference is the other part of average power C_b, which draws from resonant capacitor.

$$P_{Cr} = \frac{C_r U_m^{\,2}\sin^2\omega_s T_o}{2\pi T_e}\left(\frac{\pi}{1+\cos\omega_s T_o} - \frac{\pi}{2}\right) \cdot$$

$$= \frac{C_r U_m^{\,2}(1-\cos\omega_s T_o)^2}{4T_e}$$

According to assumption (c), the efficiency of this circuit is 1, $P_i = P_o$. The first derivative of the input power and output power is greater zerao,

$$\frac{dP_o}{dU_b} = \frac{dP_i}{dU_b} = \frac{U_m}{2}\frac{dI_1}{dU_b} < 0 \cdot$$

Where, $\left(\int_0^\pi \dfrac{\sin^3\omega t}{\sin\omega t - U_b/U_m}d\omega t\right)'\Bigg|_{U_b}$

$$= \int_0^\pi \frac{\sin^3\omega t}{U_m\left(\sin\omega t - U_b/U_m\right)^2}d\omega t > 0$$

It is obvious that bus voltage decreases if input power and out power increases and the turn-on time is kept constant. That is to say, the DC bus voltage will reach maximum at light load with high voltage.

Because $P_i = P_o$, according to (4) into (11), the expression of U_b is obtained

$$\frac{U_m^{\,2}}{U_b^{\,2}}\int_0^\pi \frac{\sin^3\omega t}{U_b/U_m - \sin\omega t} + \frac{1}{1+\cos\omega_s T_o}d\omega t$$

$$= \frac{4\pi T_e T_o N^2}{C_r \sin^2\omega_s T_o\left(T_o R + 4Ld + \sqrt{T_o^{\,2}R^2 + 8T_e RL}\right)} \tag{12}$$

Based on above formulas and (3), the optimal controlling strategy for regulating the output voltage is to

change the switching period and maintain the turn-on time. The bus voltage and output voltage drops as switch period increases. When the load becomes lighter, it needs to increase switch period to keep the output voltage and bus voltage steady.

IV. THE SIMULATION AND EXPERIMENTAL RESULTS

Table 1 shows the simulation results. P_{ia} is average input power calculated by (11) with the help of MATLAB. P_o is the average output power in circuit simulation by Pspice. In the simulation, U_m=200V, T_o=2µs, N=1/1.73, L_y=150µH, L=100µH.

If the current-program control is as inter controlling loop and the voltage control is as out controlling loop, the DC bus voltage will be reduce to be less than 360volts. The experimental result shows in Fig.4.

V. CONCLUSION

The single stage AC/DC converter with resonant model PFC is analyzed by math approach in this paper. The working conditions for the circuit shown in Fig.1 have been derived and can be expressed by (7), (8) and (9). If the parameters of this circuit satisfy (7), (8) and (9), it will operate well.

Through analyzing, the following results have been made up: (a). It has been proved that the bus voltage decreases if input power and out power increases and the turn-on time is kept constant; (b). The optimal controlling strategy for regulating the output voltage has been put forward, it is to change the switching period and maintain the turn-on time.

In order to verify the theoretic analysis conclusions, the simulation results and experimental results have been given out. The theoretic analysis conclusions and math formulas have provide a new tool for designing the circuit.

TABLE I.
THE SIMULATION RESULTS

R (Ω)	Lr (µH)	Cr (µF)	Te (µs)	Ub (V)	Pia (W)	Po (W)
40	80	0.05	8	360	98	92
60	50	0.08	5	409	230	209
60	100	0.04	5	337	134	128
60	200	0.02	5	289	81	80
200	200	0.02	5	384	60	56

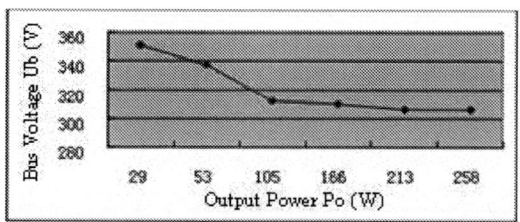

Figure 4. The experimental relation of bus voltage and output

REFERENCES

[1] M. Madigan, R. Erickson and E. Ismail, "Integrated high quality rectifier regulators", *IEEE Power Electronics Specialists Conference*, 1992, pp.1043~1051.

[2] R. Redl, and L. Balogh, "Design consideration for single stage isolated power factor corrected power supplies with fast regulation of the output voltage", *IEEE Applied Power electronics Conference*, 1995 , pp.454-458 .

[3] M. M. D.M Tsang and F. C. Lee, "Reduction of voltage stress in integrated high-quality rectifier-regulators by variable frequency control", *IEEE Power Electronics Conference*, 1994,pp.569-575.

[4] Weiping Zhang , Three bandwidth analysis for PFC circuit, *ACTA Electronica Sinica* , VOL.25.N0.11，1997.

[5] Weiping Zhang et al，A signal stage AC/DC converter with resonant model PFC,*IEEE PESC*, pp. 2004 .

[6] Robert W. Erickson and Dragan Maksimovic, *Fundamentals of Power Electronics （Second Edition）*, 2001, Kluwer Academic Publisher.

[7] Cai Xuansan, Gong Shaowen, *High Frequency Power Electronics*, China Science Publisher,1993.

[8] R. Erickson, et al, Design of a simple high-power-factor rectifier based on the fly-back converter, *Proceedings of APEC'90*, pp.792~801.

2006 5th International Power Electronics and Motion Control Conference

Low Frequency Model for the Metal Halide Lamp

Weiping Zhang , Yuanchao Liu , Xiaoqiang Zhang, Hongtao Li and Wenji Liu
North China University of Technology /Lab of Green Power & Energy System, Beijing, China
Email: zwp@ncut.edu.cn

Abstract—A new Modeling approach for the metal halide (abbrev. MH) lamp has been put forward in this paper. Our works focus on the followings: (1) Based on the low frequency behaviors of MH lamp, a new model for the MH lamp has been proposed. This model is simple but very useful; (2) The voltage waveform of the lamp is not sinusoid but looks like a square waveform, a method for estimating the voltage value has been developed; (3) In order to verity the model for the MH lamp, we calculate the inductor's current using the new model. The calculation results are agreed with theoretic results. (4) The five prototypes, which power range is 575W, 1200W, 2500W, 4000W and 6000W, have been made. The entire five prototypes work well. (5) Five experiments have been provided in this paper. The experimental results are good agreed with theoretic results, the maximum error is less than 2.5%. So, the new model provides useful tools for analyzing and designing electromagnetic ballasts.

Keywords- MH lamps; ballast; model

I. INTRODUCTION

High-Intensity-Discharge (HID) lamp has been paid more attention in residential as well as commercial lighting applications. Among various kinds of HID lamps, metal halide (abbrev. MH) lamp is the most popular light source with outstanding features, such as good color rendering, high lighting efficiency, stable luminance as well as long lifetime. Because the increment impendence of MH lamp is negative, the current through the lamp will tend to run away unless ballast is connected in series with the lamp. Conventionally, an inductor, called as electromagnetic type ballast, is employed to drive the lamps [1]. The electromagnetic type ballast has a lot of advantages, such as high stability and reliability. For the inductor is used for restricting the lamp current, the lamp will work well when the inductance matches the lamp. If inductance is too small, it will cause the lamp over power and reduce its life, and if inductance is too large, it is difficult to start the lamp. So, a proper designing value of the inductor is the most important issue. In order to design the inductor, the modeling of MH lamp is necessary to describe its behaviors. In this paper, we will provide a

Project supported by Beijing Natural Science foundation (No. 4052011)

low-frequency model of the lamps and based on this model, we also put forward a simple and useful approach for designing the inductor.

II. MODELING FOR THE MH LAMP

Fig.1 shows the current through the MH lamp and the voltage across the MH lamp. In this figure, the current through it is a sinusoid wave; however, the voltage across the lamps is not a sinusoid wave. It is difficult to calculate the output power. Usually, the lamp is considered a pure resistance. This pure resistance model cannot represent the behaviors of the MH lamp.

In this paper, a new model has been studied. In this model, the MH lamp is a capacitive load. Table.1 shows the experimental data of lamps provided by OSRAM. In the table, U_{in} represent the input voltage of the ballast, I is the current of the lamp, U_{MH} is the voltage of the lamp, and PF means the power factor of the lamp.

To model the behaviors of the steady- state MH lamps, the following assumptions are made:

(1). The voltage across the lamp is a period wave, employing the first-order Fourier series approximation; this voltage can be replaced by the fundamental wave;

(2). The inductor and AC voltage source are ideal;

(3). Based on the table 1, the average value of power factor $\cos\varphi$ is employed, $\cos\varphi$=0.86, very close to $\cos(\pi/6)$=0.866. So we choose φ=($\pi/6$) as the approximate value of the lamp's phase to calculate the circuit.

The new model of the MH lamp is consisted of a capacitor, a resistor and a back-to-back Zener. The breakdown voltage is equal to the normal value of lamp's voltage. Fig.2 shows the theoretic model.

Employing the theoretic model of the MH lamp and its vector plot of the analytical model shown in Fig.3, we can calculate the inductor's designing value to verify the model of the MH lamp. Where, φ=($\pi/6$), $\angle 1$=($\pi/2$), then $\angle 2$=($2\pi/3$), the voltage of the inductor U_L can be calculate by the followings equation, which derive from the cosine formula:

$$U_R^2 + U_L^2 + U_R \times U_L = U_{in}^2. \qquad (1)$$

Where, U_L is the voltage across the inductor (rms), U_R is the fundamental component of lamp's voltage, U_{in} is the line voltage (rms).

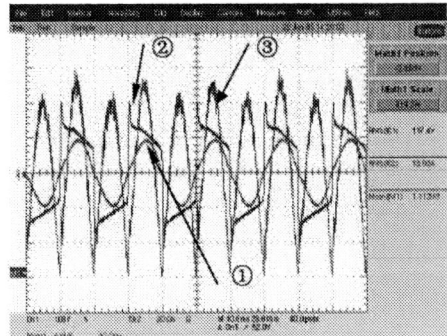

Figure 1. The current through the lamp (①), the voltage of the lamp(②) and the power of the lamp(③)

Table.1.Data of lamps distinguished by power

Power of lamp	U_{in}(V)	I(A)	U_{MH}(V)	$PF(\cos\varphi)$
575W	220	7	95	0.865
1.2kW	220	13.8	100	0.8696
2.5kW	220	25.6	115	0.8492
6kW	220	55	123	0.887
4kW	380	24	200	0.83

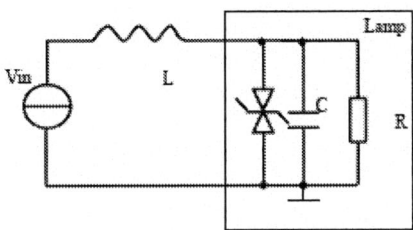

Figure 2. The theoretic model

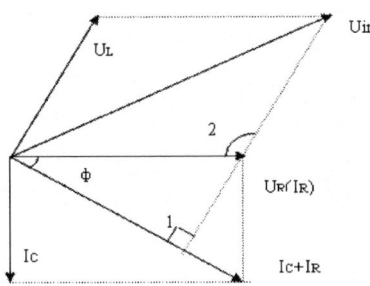

Figure 3. The vector plot of the analytical mode

Then the value of the inductor,

$$L = U_L / (2\pi f \cdot I). \qquad (2)$$

Where f is the line frequency, I is the normal value of the lamp, U_L is the voltage across the inductor, which can compute through (1). In Fig.1, the voltage across the lamp looks like a square waveform, so, we use a square waveform to replace the voltage of lamp, and the amplitude of the square waveform is average value Amax, shown in Fig.4.

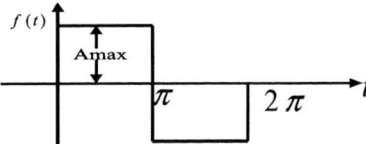

Figure 4. The voltage waveform of the inductor

Employing the Fourier series represent the square waveform,

$$U_R(t) = \frac{4A_{max}}{\pi}(\sin\omega t + \frac{1}{3}\sin 3\omega t + \frac{1}{5}\sin 5\omega t + \cdots).$$
$$(3)$$

Where $\omega = 2\pi f$, f is the line frequency.

In the Fig.1, the current waveform of the lamp is a sinusoid, and then the only fundamental component of the lamp's voltage is useful to produce the power of lamps. So,

$$U_R(t) \approx 4A_{max}\sin\omega t/\pi. \qquad (4)$$

The above formula can be applied to calculate the value of the voltage across the lamp. The voltage value across the inductor can be computed by (1).

III. CALCULATION THE INDUCTOR'S CURRENT USING THE NEW MODEL

Experiment 1. The output power Po of the lamp is 1200W, the AC input U_{in} is 220Vrms, the static state inductance of the experimental inductor is about 44.5mH. If a 1200W lamp is used, U_{in} =220Vrms, the normal inductance is 35mH, based on the new model and (1), the voltage across the inductor U_L is equal to 152V, the current through the lamp is I=13.8A, the absorb power of the lamp is 1200W. During this experiment, an inductor with 44.5mH is employed, because the voltage across the inductor keeps constant, the voltage will be 152V. That is to say, for a given lamp, the voltage across the inductor is independent with the inductance. The follow formula is employed to estimate the current.

$$I_L = U_L / 2\pi f L = 152/(44.5 \cdot 2\pi \cdot 50) = 10.87(A).$$

The waveform of the experiments is shown in Fig.5. The figure shows that the measure current is 10.6A. We use the follow formula to calculate the error,

$$(10.87/10.6 - 1) \times 100\% = 2.5\%.$$

The result show that the calculate value is good agreed with the experimental value.

Experiment 2. The normal absorb power Po of the lamp is 575W, the AC input U_{in} is 220Vrms, the static state inductance of the experimental inductor is about 85mH, the normal inductance is 71.2mH, the normal current of lamp is 7A, the voltage across the inductor U_L is equal to 156.6V.

According to above data,

$$I_L = U_L / 2\pi fL = 156.6/(85 \cdot 2\pi \cdot 50) = 5.87(A).$$

The waveform of the experiments is shown in Fig. 6. The figure shows that the measure current is 5.97A and the error is -1.7%.

Experiment 3. The normal absorb power Po of the lamp is 2500W, the AC input U_{in} is 220Vrms, the static state inductance of the experimental inductor is about 21mH, the normal inductance is 17.25mH, the normal current of lamp is 25.6A, and the voltage across the inductor U_L is equal to 138.67V.

According to above data, the current through the inductor is equal to 21.02A. The waveform of the experiments is shown in Fig.7.The figure shows that the measure current is 21.5A and the error is -2.2%.

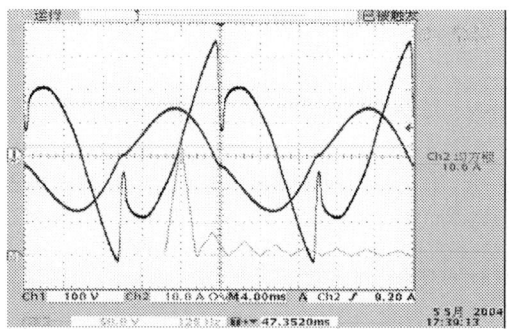

Figure 5. The experiments waveform (The output power is 1200W)

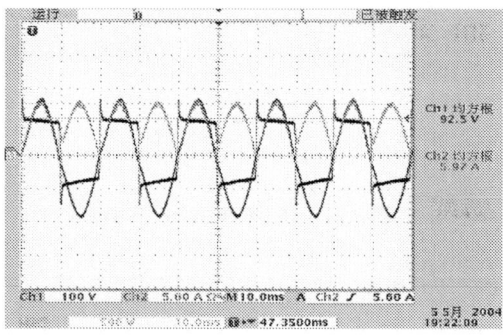

Figure 6. The experiments waveform (The output power is 575W)

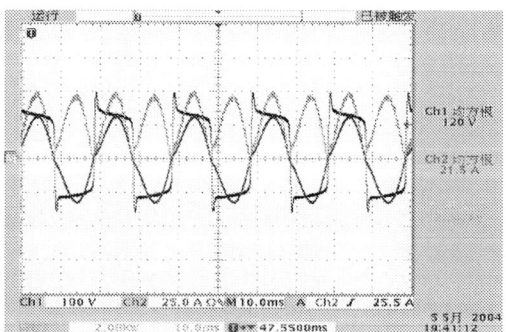

Figure 7. The experiments waveform (The output power is 2500W)

IV. DESIGN INDUCTOR USING THE NEW MODEL

In the table2, U_{MH} represents the normal value of the lamp, U_{in}, I are the normal line voltage and the normal lamp current respectively; U_L is the voltage of the inductor. The equation (1) can be employed for computing U_L and the inductance can be developed by (2). As for each kind lamp from 575W to 6000W, we can calculate every value under different working condition, shown in Table2.

V. TESTING AN ACTUAL INDUCTOR IN THE MAGNETIC BALLAST

By using the approach in this paper, a series of ac inductors, whose power range is from 575W to 6KW, has been designed, shown in table2 and made up practically. We use these actual inductors to test and verify the theorist value.

Our testing procedure is as the followings: applying an oscilloscope TEK5052 with software Power-3 measures the dynamic inductance of new magnetic ballast, the voltage and the current of the lamp. Because the voltage waveform across the lamp is not a sinusoid, as discussion in section II, we use a square waveform to replace the voltage of the lamp, the measuring value by the oscilloscope is considered as the amplitude (Amax)of the equivalent square waveform. The equation (4) is used to estimate the voltage value (rms) across the under test inductor. The voltage value of the inductor can be computed by (1) and the current through the inductor can be calculated by (2). The lamp's current mainly determines the lamp's power. So, we use the current through lamp or the inductor as standard, comparing the calculating value and measuring value can verify the new model and new approach.

Table.2. Calculate value of lamps

Power of lamp	U_{in}(V)	I(A)	U_L(V)	U_{MH} (V)	L(mH) f=50Hz	L(mH) f=60Hz
575W	220	7	156.6	105.5	71.2	59.33
	240	7	178	105.5	81	67.6
1.2kW	220	13.8	152.2	111	35	29.17
	240	13.8	179.4	111	41.4	34.5
2.5kW	220	25.6	138.67	127.7	17.25	14.4
	240	25.6	160.9	127.1	19.9	16.6
4kW	380	24	238.23	222.1	31.6	26.33
6kW	220	55	131	136.6	7.58	6.32
	240	55	153.6	136.6	8.9	7.4

Experiment 4.1200W/240V magnetic ballast is under test.

The inductor value in the static state is 38mH, measured by LCR meter. Fig.8 shows the current through the lamp (①), the voltage of the lamp (②) and the power of the lamp(③). Fig.9 shows the input power of the ballast (①), the input voltage of the ballast (②) and the input current of the ballast (③). The measured data are as follows: I=13A, U_{MH}=117.4V, U_{in}=231V, P_{MH}=1312W.

The current through the inductor is that

$$I = U_L / 2\pi f L = 13.3(A).$$

Where U_L is computed by (1), here U_L=159.2V. The error is 2.3%.

Experiment 5.2500W/240V magnetic ballast is under test.

Fig.10 shows the current through the lamp (①), the voltage of the lamp (②) and the power of the lamp (③). Fig.11 shows input power of the ballast (①), the input voltage of the ballast(②), and the input current of the ballast(③). The measure data are as follows: I=25.4A, U_{MH}=126V, U_{in}=227V, P_{MH}=2716W, L=19mH.

The current through the inductor is that $I = 24.78(A)$.

Where U_L is computed by (1), here U_L=147.9V. The error is 2.44%.

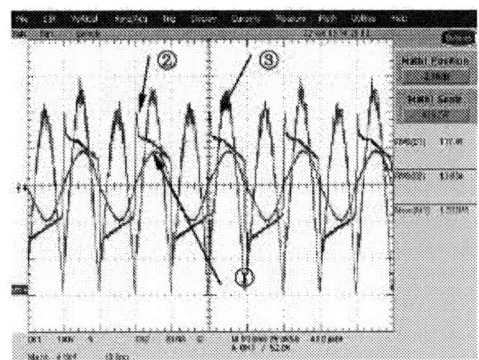

Figure 8. The current through the lamp (①), the voltage of the lamp(②) ,and the power of the lamp(③)

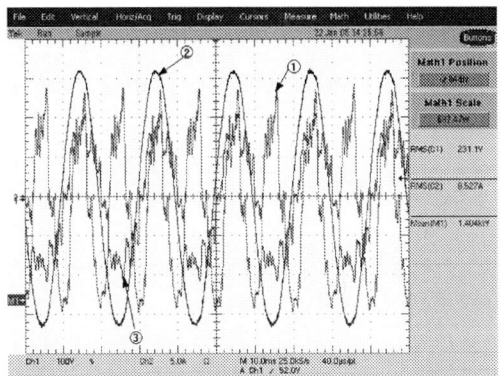

Figure 9. The input power of the ballast (①) , the input voltage of the ballast(②), and the input current of the ballast (③).

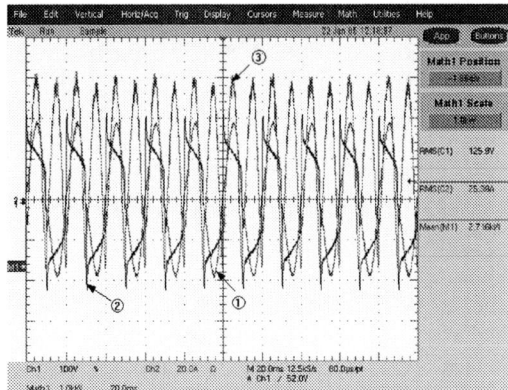

Figure 10. The current through the lamp (①), the voltage of the lamp (②) and the power of the lamp (③)

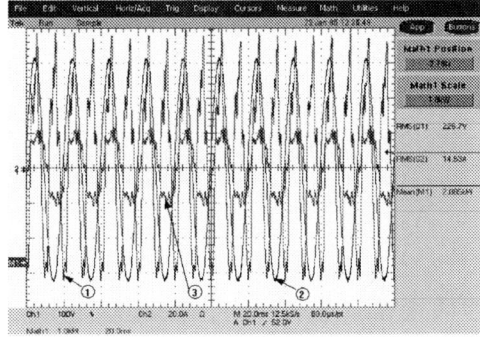

Figure 11. The input power of the ballast (①), the input voltage of the ballast (②), and the input current of the ballast (③)

VI. CONCLUSION

In this paper, a new model of MH lamp and designing approach for the inductor has been proposed. The experimental results are agreed with the theoretic results. Some conclusions can be made as following:

(1)The new model for MH lamp is very simple but useful to describe the low-frequency behaviors. This model is fundamental to analyzing and designing electromagnetic ballast.

(2) In order to verity the model for the MH lamp, the experiment 1~3 shows that applying new model, the current through the inductor can be estimated and the theoretic error is less than 2.5%.

(3)Based on the new model, a new designing approach for ballast has been developed. This approach provides a simple method for designing the inductor. The experiment 4~5 shows that theoretic error is less than 2.5%.

REFERENCES

[1] E.Rasch and E. Statnic, "Behavior of Metal Halide lamps with conventional and Electronic Ballast," J.of the IES, pp.88-96,Summer,1991

[2] R.L. Steigerwald, "A Comparison of Half –bridge Resonant Converter topologies," IEEE Trans. On PE.,Vol. 3, No. 2, Apr. 1988, pp 174-182

[3] A. K. S. B hat. " Fixed-frequency PWM Series-Parallel Resonant Converter," IEEE IAS Annual Meeting Conf. Proc., 1989,pp1115-1121.

2006 5th International Power Electronics and Motion Control Conference

H_∞ Robust Controller Based on Local Feedback Recurrent Neural Network for Permanent Magnet Linear Synchronous Motor

Junyou Yang, Naiguang Fa, Ruijuan Chen

School of Electrical Engineering, Shenyang University of Technology, Shenyang, 110023, P. R. China.
Tel&Fax: 86 24 25692800 E-mail: junyouyang@sut.edu.cn

Abstract—An H_∞ robust controller based on local feedback recurrent neural network (RNN) is proposed for the position tracking control of the permanent magnet linear synchronous motor (PMLSM). The proposed local feedback RNN with a simple-architecture is employed to estimate dynamic mapping of lumped element uncertainty. Based on the result of RNN, the quadric form function of the H_∞ robust controller is chosen and the performance index stabilizing the closed loop is obtained. The simulated results show that the PMLSM control system with large parameter changes and external disturbances has good robustness and control performances.

Keywords-H_∞ control; permanent magnet linear synchronous motor; local feedback recurrent neural network

Ⅰ.INTRODUCTION

Permanent magnet linear synchronous motors (PMLSMs) have high performances and have been widely applied to the industrial and servo drive fields [1]. However, the servo performances of the PMLSMs are greatly affected by the uncertainties, which include thrust ripple, parameter variations, friction force, and external load disturbances [2, 3]. Therefore, to compensate the uncertainties, an appropriate control strategy plays an increasingly important role.

As the recurrent neural network (RNN) internal feedback network can treat the time-varying input and output through delays, RNN has a better dynamic performance than feed-forward neural network (FNN). The RNN comprises both feed-forward and feedback connections [4], so it has dynamic property and demonstrates good control performances in the uncertainty control system. On the other hand, strong robustness is always an important property that a good controller should achieve and H_∞-norm has been widely used to measure the robustness for a given feedback control system. The H_∞-norm bound can be achieved if an associated Hamilton–Jacobi inequality admits a positive-definite solution, and the smaller H_∞-norm means the better robustness [5].

For the purpose of real-time control, a local feedback RNN with simple network structure is proposed in this paper, the function of the RNN is to learn the dynamic mapping of the uncertainties. For the significant effect of the PMLSM parameters changes and external disturbances on the control system, good control performances can be obtained by applying dynamic conversion characteristic of this RNN. By dynamic back propagation algorithm, its on-line learning law can be obtained. The proposed H_∞ robust controller can choose the quadratic form function from the solution of the RNN, then the Hamilton–Jacobi inequality is reduced to the Riccati inequality under the choice [6], the performance index stabilizing the closed loop can be obtained based on the energy cost theory, back propagation algorithm, and RNN on-line parameter training rule. The proposed controller is easy to construct and suitable for practical applications, and this controller applied to the PMLSM control system with large parameter changes and external disturbances has good robustness and control performances.

Ⅱ.MATHMATICS MODEL

A. PMLSM Modeling

Fig. 1 shows the reference space vector of PMLSM.

The mathematic model of a PMLSM can be described in synchronous rotating reference frame as follows:

$$u_d = R_s i_d + p\lambda_d - \omega_r \lambda_q , \qquad (1)$$

$$u_q = R_s i_q + p\lambda_q + \omega_r \lambda_d , \qquad (2)$$

where $p = d/dt$, and

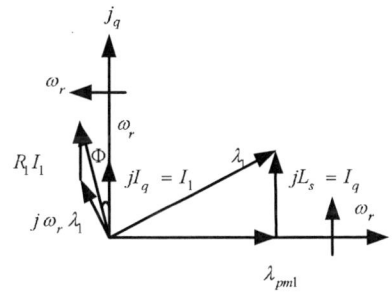

Fig.1 Reference space vector of PMLSM

Supported by National Natural Science Foundation of China (50375102) and Program for Liaoning Excellent Talents in University (RC-04-14)

1-4244-0448-7/06/$25.00 ©2006 IEEE   1253

$$\lambda_d = L_d i_d + \lambda_{PM}, \tag{3}$$

$$\lambda_q = L_q i_q, \tag{4}$$

$$\omega_r = \pi v / \tau, \tag{5}$$

where u_d, u_q are d and q axis voltages, i_d, i_q are d and q axis currents, R_s is phase-winding resistance, L_d, L_q are d and q axis inductances, ω_r is the angular velocity of the mover, λ_{PM} is permanent-magnet flux linkage, P_n is the number of primary poles and p denotes differential operator, v is linear velocity, τ is pole pitch.

Keeping the total power fixed, the electromagnetic force is

$$F_e = \frac{\pi}{\tau} n_P (\lambda_d i_q - \lambda_q i_d) = \frac{\pi}{\tau} n_P [\lambda_{PM} i_q + (L_d - L_q) i_d i_q]. \tag{6}$$

The mover dynamic equation is

$$F_e = M\dot{v} + Dv + F_L, \tag{7}$$

where M is the total mass of the moving element system, D is the viscous friction and iron-loss coefficient and F_L is external disturbance.

With the implementation of vector control, if $i_d = 0$, the mathematics model of PMLSM can be simplified as follows:

$$F_e = K_F i_q^*, \tag{8}$$

$$K_F = \pi P_n \lambda_{PM} / \tau, \tag{9}$$

$$H_p(s) = 1/(Ms + D) = b/(s + a), \tag{10}$$

where K_F is thrust coefficient, i_q^* is the command of the thrust current and s is the Laplace transform.

B. Controller Design

In this section, there are two mathematics models, that is, the model of local feedback RNN and H_∞ robust controller.

At first, the field-oriented PMLSM can be formulated by rewriting (6) and (7) as follows:

$$\ddot{d}(t) = -\frac{\overline{D}}{M}\dot{d}(t) + \frac{K_f}{M}i_q^*(t) - \frac{1}{M}F_L, \tag{11}$$
$$\approx A_p \dot{d}(t) + B_p U_A(t) + D_p F_L$$

where $A_p = -\overline{D}/\overline{M}$; $B_p = K_f/\overline{M}$; $D_p = -1/\overline{M}$, $U_A(t) \approx i_q^*$ is the control effort.

Define the tracking error:

$$E = \begin{bmatrix} d_m & \dot{d}_m \end{bmatrix}^T - \begin{bmatrix} d & \dot{d} \end{bmatrix}^T = \begin{bmatrix} e & \dot{e} \end{bmatrix}^T, \tag{12}$$

where d_m represents a desired trajectory specified by a reference model. An ideal control law can be obtained:

$$U_A^*(t) = B_p^{-1} \left[-A_p \dot{d}(t) - D_p F_L + \ddot{d}_m(t) + K^T E \right], \tag{13}$$

where $K = \begin{bmatrix} k_2 & k_1 \end{bmatrix}^T$, k_1 and k_2 are constant gains. Substituting (13) into (11) gives:

$$\ddot{e} + k_1 \dot{e} + k_2 e = 0. \tag{14}$$

However, since the parameter variations and external load disturbance of the system are difficult to measure in practical applications, (11) is rewritten as follows:

$$\ddot{d}(t) = A_p \dot{d}(t) + B_p U_A(t) - \omega_L(t), \tag{15}$$

where $\omega_L(t)$ is called the lumped uncertainty and defined as:

$$\omega_L(t) = \left(\Delta \overline{M} \ddot{d} + \Delta \overline{D} \dot{d} \right) / \overline{M} - D_p F_L. \tag{16}$$

To counteract its effect on the system performance, a local feedback RNN is introduced to learn the dynamics of the lumped uncertainty.

C. Recurrent Neural Network

The structure of the RNN is shown in Fig. 2.

There are three layers in the local feedback RNN, an input (the i layer), a hidden (the j layer) (local feedback layer), and an output layer (the k layer).

In this study, the input of the RNN is the tracking error vector E^*, which is composed of $\eta, e, \dot{e}$, where $\dot{\eta} = e$, used to achieve a zero steady-state error. $i = 3$ and the output $O_k^3(N)$ of the RNN is an estimated value $\hat{\omega}_L(t)$ of the lumped uncertainty.

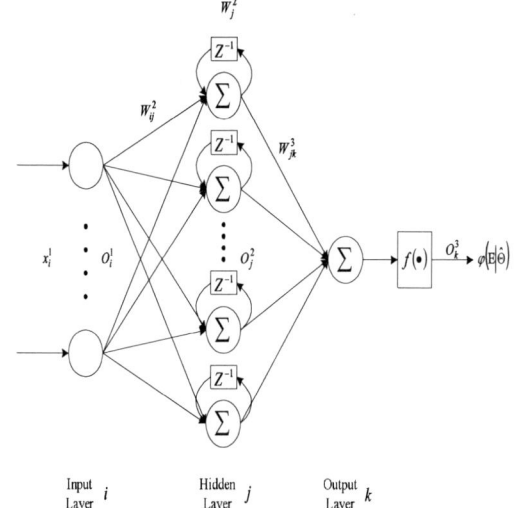

Fig.2 Structure of the RNN

$$O_k^3(N) = f_k^3 \left(\sum_j \omega_{kj}^3 * f_j^2 \left(\omega_j^2 O_j^2(N-1) + \sum_i \omega_{ij}^2 O_i^1(N) \right) \right)$$
$$= f_k^3 \left(\sum_j \omega_{kj}^3 * f_j^2 \left(x_i^1(N-1), x_i^1(N), \omega_{ij}^2, \omega_j^2 \right) \right) \quad , (17)$$

where, N denotes the number of iterations, ω_j^2 are the recurrent weight for the units in the hidden layer, ω_{ij}^2 are the connective weights between the input layer and the hidden layer, ω_{jk}^3 are the connective weights between the hidden layer and the output layer, f_j^2 is the activation function of the hidden layer, which is a sigmoidal function, defined as:

$$f_i^2(z) = 1/(1 + \exp(-z)). \quad (18)$$

In the input layer, the activation function is also a sigmoidal function, f_k^3 is the activation function, which is set to be unit,

$$f_k^3(z) = z. \quad (19)$$

$x_i^1(N)$ represents the i th input to the node of output layer.

Moreover, $\hat{\omega}_L(t)$ can be rewritten as follows:

$$\hat{\omega}_L(t) = \varphi(E^*/\hat{\Theta}) = W_A^T \Xi, \quad (20)$$

where $\quad W_A^T = \begin{bmatrix} \omega_{11}^3 & \omega_{21}^3 & \cdots & \omega_{l1}^3 \end{bmatrix}$,
$\Xi = \begin{bmatrix} x_1^3 & x_2^3 & \cdots & x_l^3 \end{bmatrix}^T$. $\hat{\Theta}$ is the collections of the adjustable parameters $\begin{pmatrix} \omega_{jk}^3 & \omega_j^2 & \omega_{ij}^2 \end{pmatrix}$ of local feedback the RNN. Therefore, the lumped uncertainty $\omega_L(t)$ can be rewritten as:

$$\omega_L(t) = \varphi(E^*/\hat{\Theta}) - \left[\varphi(E^*/\hat{\Theta}) - \varphi(E^*/\Theta^*) \right] - \varepsilon, \quad (21)$$

where ε denotes the estimation error due to the use of finite dimension RNN.

D. Online Learning Laws of the Parameters in RNN

Selection of parameters for the connecting and recurrent weights of the local feedback RNN has a significant effect on the network performance. In order to train the RNN effectively, the online learning laws of the parameters in the RNN $\dot{\hat{\Theta}} \begin{pmatrix} \omega_{jk}^3 & \omega_j^2 & \omega_{ij}^2 \end{pmatrix}$ is proposed, which can be computed using the backpropagation algorithm as follows:

$$\dot{\omega}_{jk}^3 = \Gamma X^T P B_2 \left[\frac{\partial \varphi(E/\hat{\Theta}) \partial net_k^3}{\partial net_k^3 \partial \omega_{jk}^3} \right] = \Gamma X^T P B_2 x_j^3, (22)$$

$$\dot{\omega}_j^2 = \Gamma X^T P B_2 \left[\frac{\partial \varphi(E/\hat{\Theta})}{\partial net_k^3} \frac{\partial net_k^3}{\partial O_j^2} \frac{\partial O_j^2}{\partial net_j^2} \frac{\partial net_j^2}{\partial \omega_j^2} \right], (23)$$
$$= \Gamma X^T P B_2 \omega_{jk}^3 P_j^2$$

$$\dot{\omega}_{ij}^2 = \Gamma X^T P B_2 \left[\frac{\partial \varphi(E/\hat{\Theta})}{\partial net_k^3} \frac{\partial net_k^3}{\partial O_j^2} \frac{\partial O_j^2}{\partial net_j^2} \frac{\partial net_j^2}{\partial \omega_{ij}^2} \right], (24)$$
$$= \Gamma X^T P B_2 \omega_{jk}^3 Q_{ij}^2$$

where $P_j^2 = \frac{\partial O_j^2}{\partial \omega_j^2}$ $\quad Q_{ij}^2 = \frac{\partial O_j^2}{\partial \omega_{ij}^2}$, and Γ is the adaptive learning rate in the RNN, $\Gamma > 0$. Not only are the connecting weights between the layers adjusted on line but also the recurrent weights.

E. H_∞ Controller

The feedback control law is in the following form:

$$U_A = B_p^{-1} \left[-A_p \dot{d}(t) + \ddot{d}_m(t) + K^T E - \upsilon + \varphi(E^*/\hat{\Theta}) + U_C \right], (25)$$

where υ is the outer loop control to be further specified later, and $\varphi(E^*/\hat{\Theta})$ is the output of RNN structure and U_C compensating controller. With (25), (21), (15), the tracking error dynamics are expressed as:

$$\ddot{e} = \upsilon - K^T E + \left[\varphi(E^*/\Theta^*) - \varphi(E^*/\hat{\Theta}) \right] - \varepsilon - U_C. \quad (26)$$

Then, taking the Taylor series expansion of (26), it can be obtained that:

$$\varphi(E^*/\Theta^*) - \varphi(E^*/\hat{\Theta}) = (\Theta^* - \hat{\Theta}) \left[\frac{\partial \varphi(E^*/\hat{\Theta})}{\partial \hat{\Theta}} \right] + H(\Theta^* \quad \hat{\Theta}), (27)$$

$$\ddot{e} = \upsilon - k_1 \dot{e}(t) - k_2 e(t) + \Phi(\Theta) - \varepsilon, \quad (28)$$

where $\Phi(\Theta) = (\hat{\Theta}^* - \hat{\Theta})^T \left[\partial \varphi(E/\hat{\Theta})/\partial \hat{\Theta} \right] + H(\Theta^*, \hat{\Theta}) - U_C$.

The error dynamics can be described by the state-space representation:

$$\begin{bmatrix} \dot{e}_1 \\ \dot{e}_2 \end{bmatrix} = \begin{bmatrix} 0 & 1 \\ -k_2 & -k_1 \end{bmatrix} \begin{bmatrix} e_1 \\ e_2 \end{bmatrix} + \begin{bmatrix} 0 \\ 1 \end{bmatrix} \Phi(\hat{\Theta}) + \begin{bmatrix} 0 \\ -1 \end{bmatrix} \omega + \begin{bmatrix} 0 \\ 1 \end{bmatrix} \upsilon, (29)$$

where $e_1 = e$ $e_2 = \dot{e}$ $\omega = \varepsilon$.

With the integral action $\dot{\eta} = e$, the objective of the augmented error dynamics is to find the H_∞ controller of the form, $\dot{\hat{\Theta}} = \alpha(X, \hat{\Theta}), \upsilon = \beta(X, \hat{\Theta})$.

So that the resulting closed-loop system is stable.

$$\dot{X} = A_1 X + B_2 \Phi(\hat{\Theta}) + B_1 \omega + B_2 \beta(X, \hat{\Theta}), \quad (30)$$

$$\dot{\hat{\Theta}} = \alpha(X, \hat{\Theta}), \quad (31)$$

$$z = C_A X + D_A \beta(X, \hat{\Theta}). \quad (32)$$

1255

And the H_∞-norm constraint is satisfied for the value of γ as small as possible.

$$\int_0^T \|z(t)\|^2 dt \le \gamma^2 \int_0^T \|\omega(t)\|^2 dt \quad \forall T \ge 0, \quad (33)$$

where z is the penalty variable. And

$$X = \begin{bmatrix} e_1 \\ e_2 \\ \eta \end{bmatrix} \qquad A_1 = \begin{bmatrix} 0 & 1 & 0 \\ -k_2 & -k_1 & 0 \\ 0 & 0 & 0 \end{bmatrix}$$

$$B_1 = \begin{bmatrix} 0 \\ -1 \\ 0 \end{bmatrix} \qquad B_2 = \begin{bmatrix} 0 \\ 1 \\ 0 \end{bmatrix}$$

$$C_A = \begin{bmatrix} c_1 & 0 & 0 \\ 0 & 0 & 0 \end{bmatrix} \qquad D_A = \begin{bmatrix} 0 \\ d_1 \end{bmatrix}.$$

The overall control system with the RNN-based H_∞ controller is shown in Fig. 3.

III. SIMULATION

In simulation experiment, the idea is to simulate pure H_∞ controller and H_∞ robust controller based on local feedback RNN applied to PMLSM system, then compare the two results.

The proposed parameters of the PMLSM system are

$$m = 0.121 kg, a = 1.65, b = 6.2. \quad (34)$$

In addition, the gains of the H_∞ controller are given in the following:

$$k_1 = 100, k_2 = 20, \gamma = 0.9645, \Gamma = 12. \quad (35)$$

All the gains in the H_∞ controller are chosen to achieve the best transient control performance in the simulation considering the requirement of asymptotical stability. To show the effectiveness of this control system with small number of neurons, the RNN has 3, 18, and 1 neurons at the input, hidden, and output layers respectively.

To investigate the effectiveness of the proposed controllers, two kinds of disturbances, the variation of the mover mass and external force disturbance, are considered here.

The simulated tracking responses of two control systems due to periodic sinusoidal command are shown in Fig. 4 and Fig. 5. It can be seen that, in the case of periodic sinusoidal command, perfect tracking responses and robust control characteristics can be improved using the H_∞ robust control system based on local feedback RNN. And the thrust and speed are shown in Fig. 6 and Fig. 7, where the external load is added in the system at 0.2 s.

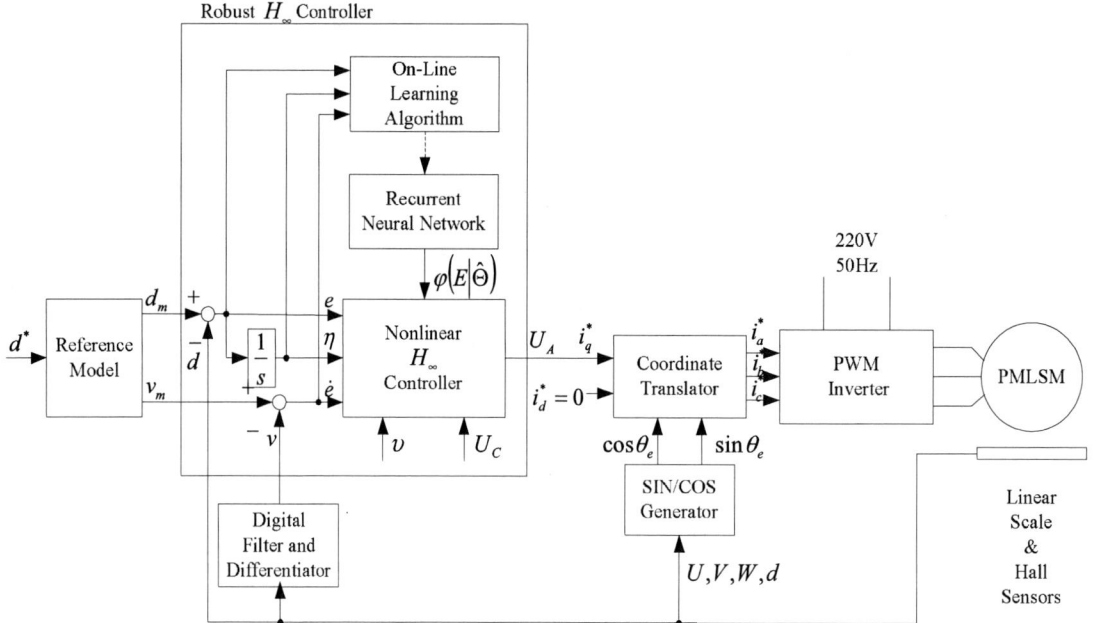

Fig.3 Block diagram of control system

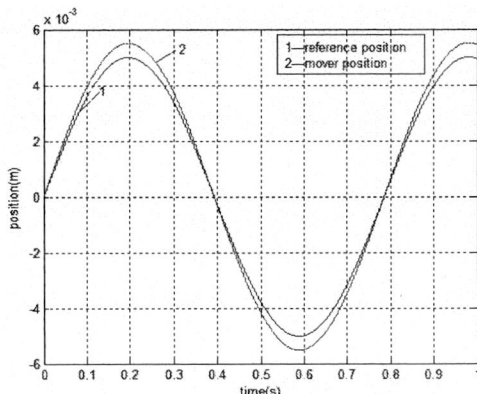

Fig.4 The tracking response of pure H_∞ control system

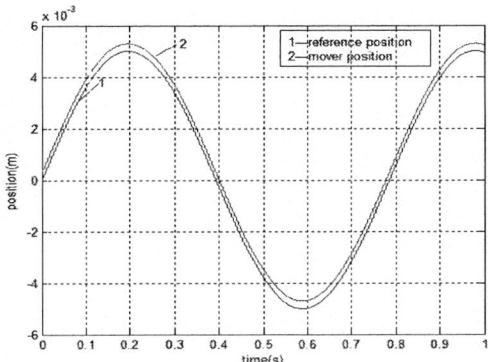

Fig.5 The tracking response of H_∞ robust control system based on local feedback RNN

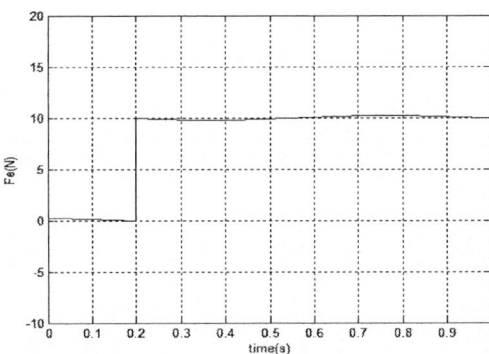

Fig.6 The electromagnetic force of H_∞ robust control system based on local feedback RNN

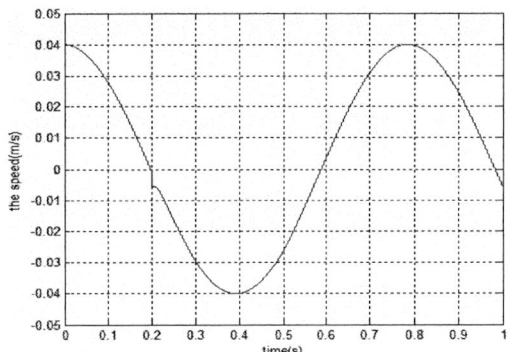

Fig.7 The speed of H_∞ robust control system based on local feedback RNN

IV.CONCLUSIONS

From the model of the proposed controller and the simulated results, the conclusion as follow can be predicted, compared to other H_∞ controllers, the H_∞ robust controller based on local feedback RNN, which applied to PMLSM existed large parameter varieties and large external disturbances, has improved dynamic tracking performances, and the proposed control system exhibits good dynamic response and strong robustness. Meanwhile, the controller proposed in this paper is capable of tracking both step and sinusoidal command inputs.

REFERENCES

[1] Qingding Guo, Chengyuan Wang, Meiwen Zhou and Tingyu Sun. *Linear AC Servo System Precise Control Technology.* Beijing, *China Machine Press,* 2000.

[2] I Boldea, S A Nasar. *Linear Electric Actuators and Generators.* Cambridge*, Cambridge University Press,* 1997.

[3] F.J.Lin, T.S.Lee and C.H.Lin. "Robust H_∞ Controller Design With Recurrent Neural Network for Linear Synchronous Motor Drive", *IEEE Trans Ind Electron,* vol.50, no.3, June 2003.

[4] Shuang Cong, Yi Dai. "Structure of recurrent neural networks." *Computer Application,* vol.24, no.8, pp.18-20, Aug.2004.

[5] Tielong Shen. H_∞ *Control Theory and Application.* Beijing, *Tsinghua University Press,* 1996.

[6] J.Doyle, K.Glover, P.P.Khargonekar and B.A.Francis. "State-space solutions to standard H_2 and H_∞ control problem." *IEEE Trans. Automat. Contr.,* vol. 34, pp. 831–847, Aug. 1989.

Parameter Estimate Modeling of Electronic Transformer

Jiaju WU* , Hidehiko SUGIMOTO** and Changkun Wang*

* Nan Chang Institute of Aeronautical Technology/Nondestructive Test Key Laboratory of Ministry Education, China
** Fukui University/Department of Electrical and Electronics Engineering, Fukui ,Japan

Abstract—In the plane-topology description of multi-dimensional dynamic coupling relations between the distributed parameters of electronic transformer, the distributed inductance, capacitance and resistance are often lumped. An indirect method based on the step response of full-loaded electronic transformer, is developed for the lumped parameters estimate of leakage inductance, equivalent capacitance, iron loss equivalent resistance and coil resistance of the transformer. Results of experiments show that both the estimate accuracy and the simulation confidence are satisfactory.

Keywords-Electronic transformer;Mathemati model; Parameter estimate

I. INTRODUCTION

The properties of electronic transformer have great influence on increasing power density, improving output waveform, and raising efficiency and the price ratio of performances. As a part of control system, the adaptation of a transformer model and accuracy of the parameters obtained by the model play an important role in improving system dynamic and static performance and shortening research and construction cycle. When the structure of a coil and that of the iron-core are consistent with what is required in related document, some parameters of a few types of transformer can be calculated according to the related documents [1] and [2]. However, the ferromagnetic material and the structure of transformer vary with the type of transformer, and the slight change of the property of material, circuit and magnetic structure may have notable influence on the transformer parameters; Under high-frequency switching condition, the transformer parameters present multivalued, nonlinear, and distributed characteristic, skin effect and proximity effect rising high, whose values are not only related to the magnitude of current and the material of iron-core coil, but also related to the impulse voltage width, the change rate of current and the initial point of magnetic flow. In this situation, the equivalent circuit of electronic transformer and its test method are different from those of power frequency transformer. Maxwell equations theoretically describe the operation of electronic transformer in 3-dimensional time-varying field[3]. Its corresponding mathematic model is set of partial differential equations with varying coefficients.The 2dimentional planetopology

model, which is based on the lumped resistance, capacitance and inductance, is introduced into engineering analysis of the multi dynamic coupling relations between the distributed parameters of the transformer[4-8]. The corresponding simulation model is a (set of) differential equation(s) with constant coefficients, i.e., $dx/dt=f(x,t,A)$. The uniqueness of equation solution x_i , is rather sensitive to the uncertainty of parameter A. The simulation accuracy and confidence strongly rely on the model parameters, or the simulation is not so robust. Topology likelihood approximation is available for the former question, meanwhile, for the latter, more studies on parameter estimate are necessary[9][10]. Though some parameters for a few transformers with standard structures are already available by computations, it is still meaningful, according to following reasons, to conduct further studies on parameter estimate of electronic transformers[2-8].

1) It is very difficult to meet the strict conditions asked by the computation, because various ferromagnetic materials and coil structures are used in transformers.

2) Tiny fluctuations, either in electric circuit, or in magnetic circuit of the transformer, may strongly affect its characteristics.

3) Due to the multi-valued non-linearity, the distribution effect, skin effect and proximity effect in the transformer, the estimated results rely not only on the magnetizing current intensity, but also on the pulse width, the current fluctuation rate and the start point of magnetizing.

Researchers have long been puzzled by the mathematic models of electronic transformers and by the parameter estimate[11]. To overcome the bottleneck, more and more attentions have been paid on the studies[2-8][11]. The model presented in Reference [12] consists of 80-150 elements. Compared with magnetic resistance model, as well as with the SPICE software, better results are available. But, due to the huge number of elements, much stronger skills are asked in the computation. Besides, the equivalent capacitance still remains no consideration in the model. New ways of parameter measurement were introduced in Reference [4], at the expense of the secondary being open-circuited, and of no consideration of leakage inductance. The estimated iron loss equivalent resistance at power frequency for the motor operates in stationary state was shown in References [13], but the estimate of iron loss equivalent resistance in switching operation, as well as in electronic transformer operation,

Project Supported by National Natural Science Foundation of China（50467003）

were not seen. Meaningful studies in power electronics on the Buck-Cascaded Push-pull Converter modeling and simulation were conducted in Reference [14], but parameter estimate was not included.

The electronic transformer model is a multi parameter system. The parameters vary along with the changes of V, dV/dt, i, di/dt, Φ and $d\Phi/dt$. But conventional methods are in essence single-parameter direct measurement. Various conditions, shown as follow, are introduced into the measurement[2-8][11-14].

1) The measurement is carried out at a specified frequency, with small input signal and the coil being open-circuited or short-circuited.

2) Effect of iron loss is not taken into consideration.

3) With particular coil configuration, to obtain other leakage inductances, the leakage inductance of the coil most close to iron core is ignored.

4) Without considering skin effect and proximity effect.

5) In inductance estimate, the effect of voltage fluctuation on distributed capacitance is ignored, or further, the distributed capacitance is totally ignored.

6) Ignoring leakage inductance in capacitance estimate.

7) The actual μ value is substituted by the μ value at the neighborhood of zero point in B-H chart. As a result, the fluctuation of μ value on leakage inductance is not taken into consideration.

Therefore, the results obtained by conventional methods can not be directly introduced into the quantitative analysis on actual operation of electronic transformer.

Based on the residual magnetizing information, an indirect multi-parameter estimate method in the modeling of electronic transformer is presented. Prototype response is at first obtained under conditions as close as possible to the actual operation, then homomorphically mapped on the simulation model for further parameter estimate.

Main features of the estimate method are as follow.

1) Compared with the open/short-circuit method mentioned above, test conditions in the method, i.e. the values of V, dV/dt, di/dt, Φ, $d\Phi/dt$, together with the magnetizing start point and the pulse width, are as close as possible to the actual transformer operation.

2) The method is emphasized to show the combinational behaviors of the distributed parameters of the transformer. Instead of obtaining only a single parameter in conventional methods,in our method, more equivalent lumped parameters are available by a single test.

3) Both leakage inductances and leakage capacitances are obtained at the same time, instead of ignoring one to obtain another.

II. MAIN CIRCUIT AND EQUIVALENT CIRCUIT OF ELECTRONIC TRANSFORMER

In "An Instantaneous Voltage Drop Compensator" [15], an electronic transformer was used in DC-AC conversion to obtain 220V_{AC} output. The primary parameters are: N_1=12 turns, 24V_{DC} input with 10 kHz PWM switching

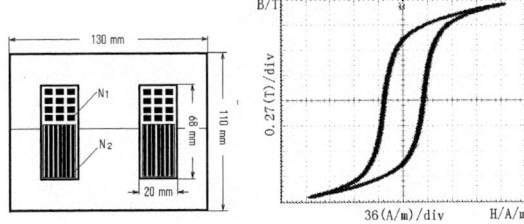

Fig.1 Configuration of Transforme   Fig.2 B -H Loop of transformer

frequency. The secondary parameters are: N_2 =206 turns , 220V 50 Hz AC output. Fig.1 shows the transformer configuration, where the iron core thickness is 70 mm. Fig.2 is the B-H characteristics at 50 Hz power frequency.

For an electronic transformer with typical 2-coil configuration, there will be 8-9 energy-storage elements in the equivalent circuit as the distributed parameters are lumped [2][11]. The pulse operation of the transformer can be, in principle, expressed by a 9th-order differential equation. It is very difficult to find its analytical solution, as well as to obtain applicable engineering solutions.

Numbers of energy-storage elements can be reduced significantly by dividing the pulse operation into several stages. The simplified equivalent circuit for pulse rising edge in the up-step transformer is shown in Fig.3(a), while Fig.3(b) shows the equivalent circuit of unloaded transformer in pulse duration response. (1) indicates the secondary voltage in Fig.3.

$$V_o(s) = \frac{V_i}{s} \cdot \frac{a_0}{b_2 s^2 + b_1 s + b_0} \qquad (1)$$

$$a_0 = R_2^{''}; \quad b_2 = C_S^{'} L_S R_2^{''}; \quad b_1 = C_S^{'} R_1 R_2^{''} + L_S;$$

$$b_0 = R_1 + R_2^{''} ; \qquad R_2^{''} = R_2^{'} // R_m = \frac{R_2^{'} R_m}{R_2^{'} + R_m}$$

where V_i is the input voltage, V_o the output voltage, R_m the iron loss equivalent resistance, k_{12} the voltage transformation ratio, $V_o^{'}$ the voltage normalized to primary, R_1 the coil resistance normalized to primary, $R_2^{'}$ the load resistance normalized to primary, $C_S^{'}$ the lumped capacitance normalized to primary, L_S the lumped inductance normalized to primary, respectively.

The beginning of the step response is dominated by L_S , because magnetizing inductance L_m is much greater than leakage inductance L_S. As a result, the simplification can be carried out by ignoring L_m, or let $L_m \to \infty$.

In some cases, iron loss equivalent resistance R_m was also ignored while ignoring magnetizing inductance L_m[4]. But in our point of view, as the transformer is in a lower-loaded state, ignoring iron loss equivalent resistance R_m

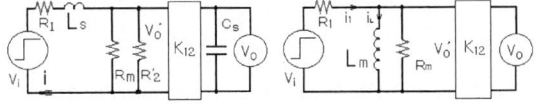

a. (Pulse rising edge)   b. (Pulse duration)
Fig. 3: Step response equivalent cicuit

may lead to significant deviation. At the rising edge of the step pulse, though the wattless current just begins its increase from zero, but owing to the very fast change of the magnetic flux, as well as the very rich high frequency components, neither magnetic hysteresis loss nor eddy current loss can be ignored. The MATLAB simulated results and corresponding experimental waveforms will show this late.

According to Fig.3(a), we have (2), where L_m is the magnetizing inductance.

$$v_o(t) = \frac{R_m}{R_1 + R_m} V_i e^{-\frac{t}{T_1}}$$
$$\approx \frac{R_m}{R_1 + R_m} V_i \left(1 - \frac{t}{T_1}\right) \qquad (2)$$
$$T_1 = \frac{L_m(R_1 + R_m)}{R_1 R_m}$$

III. ESTIMATE OF PARAMETER

Fig.4 shows the step response test circuit for electronic transformer operates with rated voltage and rated load. The initial condition $B|_{H=0}$ shown in Fig.2 is regulated by E, E' and R_1. A MOSFET(IXTH13N80) was used as the power switch in the Fig.4, Fig.5 and fig.6.

A.. Test condition and quantitative analysis on related parameters

Test conditions and quantitative analysis on related parameters are as follow.

1) The longest oscillation cycle (about 10^{-3} Sec.) belongs to the loop in relation to primary magnetizing inductance L_m.

2) The oscillation cycle of L_S related loop is about 10^{-5} S

3) The loop consisting of equivalent inductance due to the coil leads and the oscilloscope probe capacitance, takes an oscillation cycle of 10^{-7} S.

4) At different coil segments, dV/dt takes various values in different stages of the step response, accordingly, the dynamically-distributed capacitance is no longer a constant. But to highlight its physical meaning, the equivalent capacitance C_S, which is a constant lumped from the distributed capacitance, was used in the test.

5) The di/dt value keeps changing due to proximity effect and skin effect. It means that R_1 value is not a constant. But to highlight the physical meaning, the lumped equivalent value(a constant) was employed in the test.

6) Both thermal effect and mechanical effect caused by electro-magnetic activities in the transformer operation,

were not taken into consideration.

7) Electro-magnetic energy radiation between transformer and surrounding was ignored.

8) Time hysteresis between B and H, together with corresponding effect, were ignored.

9) The DC power supply was taken as an ideal voltage source.

10) All initial conditions but the magnetizing start point are zero .

B. Estimate of iron loss equivalent resistance R_m

Primary magnetizing inductance L_m strongly relies on the magnetizing start point and pulse width, as well as on the magnetic flux density increment ΔB and its fluctuation dB/dt. With same magnetizing current, the higher the power frequency is, the larger the area of B-H loop is, the more similar to an ellipse the loop is, too. Besides, the dB/dt value is in detail not a constant. Values of L_m and R_m are also varying during the step response. To emphasis the physical meaning and to simplify the computation, when primary voltage, secondary load, pulse width and magnetizing start point have been determined, the lumped equivalent values L_m and R_m areused. Besides, the iron core saturation is seen in an abnormal state, and not taken into the estimate. There are two test circuits for the estimate of R_m, according to the demagnetizing mode of main circuit, as shown in Fig.5 and Fig.6. The lumped equivalent parameters started from any point in Fig.2 can be estimated based on Fig.4, Fig.5 and Fig.6. Among them, Z stands for the load, R_1 and R_2 the adjustment of magnetizing start point and magnetizing current, respectively, while K the co-operative switch. Further notes are, (a) A synchronous signal to the storage oscilloscope V_0 is needed; (b) Power electronic devices are suggested to eliminate arc interferences

1) The $T_{off} > T_{on}$ demagnetizing mode   A new cycle is beginning as the magnetic intensity H returned to zero. Both start point $(0, B_{ri})$ and finish point (H_i, B_{rii}) of magnetic flux density ΔB are in quadrant I. Fig.5 shows the test circuit. The primary voltage is shown in Fig.8. According to Fig.4 and (2), we obtain (3).

$$R_m = \frac{V_o(0)}{V_i - V_o(0)} R_1 \qquad (3)$$

Test conditions of Fig.5 and Fig.6 are as follow. (a) V_i is DC voltage source, the serial resistance has already been known, based on the output voltage at $t = 0$ and the oscillation cycle 10^{-6} Sec, we obtain R_m=167Ω. (b) The magnetizing current shows the effect of pulse width on magnetizing inductance. In Fig.4, with 10kHz PWM

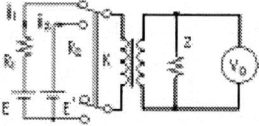

Fig.4 The test circuit with rated voltage and rateed

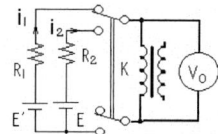

Fig.5 Iron loss estimate,circuit 1

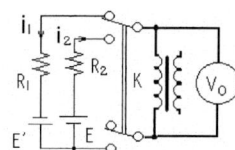

Fig.6 Iron loss estimate,circuit 2

switching frequency in $T_{off}>T_{on}$ demagnetizing mode, the pulse duration T_{on} will be $T_{on}<50\mu S$, therefore, $i_L= i_L(V_i ,t)= i_L(24v, 50\mu S) \approx 1A$. To show the tested voltage drop more clearly, as the magnetizing current keeps unchanged or $i_1=1A$, the DC voltage and the serial resistance are regulated to $V_i= 42.5V$ and $R_1= 33\Omega$, respectively. (c) To obtain the primary magnetizing inductance L_m, there is (4) based on (2) and (3).

$$L_m = \frac{R_1 R_m^2 V_i t_k}{R_m V_i (R_1 + R_m) - (R_1 + R_m)^2 V_o} \tag{4}$$

2) The specially-designed demagnetizing circuit In switching operation, the sustaining time of inverse H makes contribution to the demagnetizing results. When the demagnetizing has completed, the finish point of ΔB will be at quadrant I, with its start point at quadrant III. Fig.6 shows the test circuit.

Following are the two cases may happen.

(1) The forward magnetizing begins when the ith switching inverse demagnetizing current has returned to zero. In this case, the magnetic field intensity is zero, the residual magnetic flux density is $-B_{ri}$, while the magnetizing start point is $(0,-B_{ri})$.

(2) The forward magnetizing begins when the ith switching inverse demagnetizing current has not returned to zero. In this case, the demagnetizing field intensity is $-H_i$, the magnetic flux density is $-B_{ri}$, while the magnetizing start point is $(-H_i,-B_{ri})$. Suppose that the abnormal magnetic saturation region is taken no consideration, the multi-valued nonlinear B-H loop can be seen as a rectangle, or there is $-B_{ri} \approx -B_{rii}$. It means that with a specially-designed demagnetizing circuit, the magnetizing start point can be approximately seen as $(0, -B_{ri})$. There will be different tested waveforms for different magnetizing start points (Fig.7, Fig.8). According to (3), corresponding R_m values are also different. It indicates that there are different iron losses for different magnetizing start points. Detailed computations are omitted .

3) R_m and R'_m It should be pointed out that the iron loss equivalent resistance R_m which is obtained based on Fig.8 and (3), only corresponds to the iron loss between the magnetizing start point $(0,B_{ri})$ and a finish point (H_i,B_{rii}) in B-H chart.

There are following notes in practical applications.

(1) When there is no oscillation at the rising edge of the secondary step response, the loaded secondary step response will finish during the monotone increasing of H_i starting from zero. In this case, the R_m values can be used directly.

(2) When there are oscillations at the rising edge of the secondary step response, for the tested electronic transformer, relations between the magnetic flux density and magnetic field intensity have formed a loop in B-H chart. In those cases, both the iron loss occurred in the increase of magnetic field intensity and the iron loss occurred in the decrease of magnetic field intensity should be taken into consideration. It means that it is necessary to find the revised value R'_m based on R_m. According to reference [2] and [11], the iron loss occurred in the magnetic field intensity increase is approximately half of the overall iron loss, or $R'_m \approx 0.5 R_m=83\Omega$.

(3) For fully-loaded transformers, even there is no oscillation at the rising edge of secondary step response , $R'_m \approx 83\Omega$ will also be used in the analysis, without taking the transformer load state into consideration, since the iron loss is a small quantity compared with the transformer load, and can be ignored.

(4) R'_m is taken as the iron loss equivalent resistance in the step response of the transformer. The actual B-H loop can then be shown as follows, 10000/50 = 200 tiny ellipses are attached on the curve in Fig.2, with point $(0,B_{ri})$ as the reference.

C. Estimate of L_S , R_l and C'_S

1) The secondary step response in loaded transformers
As the electronic transformer operates with rated voltage and rated load, the tested secondary step response is shown in Fig.9. R_m , L_S and C'_S values keep no longer constants because the values of V, dV/dt, i, di/dt and Φ, $d\Phi/dt$ are varying. But in our discussion, based on the overall effect analysis of those parameters on the step response, R_m , L_S and C'_S values are conditionally seen as equivalent constants.

2) Mathematic model of secondary step response
The characteristic (1) takes the form $b_3 S^3+b_2 S^2+b_1 S+b_0 = 0$. There will be three different real roots according to the step response of the full loaded transformer ($t\in[0, 18\mu S]$, See Fig.9). As a result, the step response ofthe transformer secondary can be obtained by (5) and (6).

$$V_o(s) = \frac{x_1}{s-x_2} + \frac{x_3}{s-x_4} + \frac{x_5}{s} \tag{5}$$

$$V_o(t) = x_1 e^{x_2 t} + x_3 e^{x_4 t} + x_5 \qquad t \geq 0 \tag{6}$$

In the modeling of Fig.9, fitting result and the expression are to be considered. Constraints are introduced into the estimate method to deal with the expression.

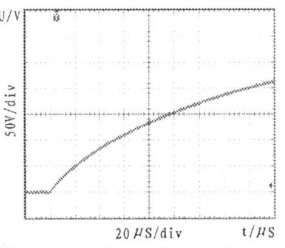

Fig. 9 Experimental waveform

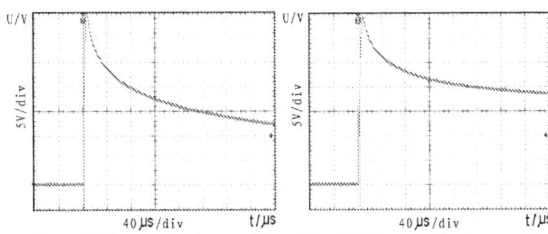

Fig. 7 Voltage of primary (1)   Fig. 8 Voltage of primary (2)

Based on (1) and (5), by means of the method of equating coefficients, the constraint Equations are obtained, and shown by (7) and (8).

$$x_1 + x_3 + x_5 = 0 \qquad (7)$$

$$x_5(x_2 + x_4) + x_2 x_3 + x_1 x_4 = 0 \qquad (8)$$

Also, based on residue method and the initial conditions of energy storage elements, following constraint Inequalities are obtained.

$$x_2 < 0, \ x_4 < 0, \ x_1 x_3 < 0, \ x_5 > 0$$

3). Curve fitting   The curve fitting is briefly introduced as follows.

(1) Sample space   Under conditions of 10 kHz PWM frequency and resetting the magnetizing start point by $T_{off} > T_{on}$, 10000 groups data are sampled by the oscilloscope for the curve in Fig.10, and stored in array files. More attentions should be paid to the data of $t \in (0^-, 20\mu S)$.

(2) Target function The modeling of the step response of full-loaded transformer is in fact to find its optimum solution as shown by (9), where V_{oi} stands for the ith sample of the secondary step response at time t_i

$$d = \sum_{i=1}^{n} \{U_o(t_i) - H(t_i)\}^2 = \sum_{i=1}^{n} U_o^{\,2}(t_i) -$$

$$2\sum_{i=1}^{n} U_o(t_i)H(t_i) + \sum_{i=1}^{n} H^2(t_i) = \frac{1}{\Delta t}\sum_{i=1}^{n} U_o^{\,2}(t_i)\Delta t -$$

$$\frac{2}{\Delta t}\sum_{i=1}^{n} U_o(t_i)H(t_i)\Delta t + + \sum_{i=1}^{n} H^2(t_i) \approx$$

$$\frac{1}{\Delta t}\int_{T_0}^{T_K} U_o^{\,2}(t)\mathrm{d}t - \frac{2}{\Delta t}\int_{T_0}^{T_K} U_o(t)H(t)\mathrm{d}t + \sum_{i=1}^{n} H^2(t_i)$$

$$----------- \quad (9)$$

(3) Solutions of $x_1, x_2, \dots x_5$ In order to find the optimum solutions of $x_1, x_2, \dots x_5$, a set of nonlinear equations based on (9) are to be solved. Numerical solutions of $x_1, x_2, \dots x_5$ are available either by means of MATLAB, or by the program developed. Following are the numerical solutions of (6) and (10), the numerical form of (9).

$$V_o'(t) = 346 - 346.02497e^{-8860t}$$

$$+ 0.02497e^{-120000000t} \quad ------- \quad (10)$$

$$x_5 = 346 \ , \quad x_3 = -346.02497 \ , \quad x_1 = 0.0247$$

$$x_4 = -8860 \quad , \quad x_2 = -120000000$$

4) Solution of the set of equations   Based on (1) and (5), by means of the method of equating coefficients, (11), (12) and (13) are obtained.

$$\frac{V_i}{C_S' L_S} = x_2 x_4 x_5 \qquad (11)$$

$$\frac{R_1}{L_S} + \frac{1}{C_S' R_2''} = -x_2 - x_3 \qquad (12)$$

$$\frac{R_2'' + R_1}{C_e' L_e R_2''} = x_2 x_4 \qquad (13)$$

It has been known that V_i is DC voltage source, $R_2'' =$

$= R_2' \, // \, R_m' = R_2 (k_{12})^2 // R_m'$, $k_{12} = 0.0581 = N_1/N_2$, $R_2 = 48.4\Omega$, and $R_m' \approx 83\Omega$ when the residual magnetizing is of the same direction as magnetizing. According to those known quantities given above, combined with the solutions of x_1, $x_2, \dots x_5$, the unknown $R_1 = 0.032\,\Omega$, $R_m = 167\,\Omega$, $R_m' = 83\,\Omega$, $L_S = 22.0\,\mu$H and $C_S' = 51$nF are obtained.

VI. EXAMINATION OF THE ESTIMATE METHOD AND ESTIMATED RESULTS

A. Comparison between the estimate method and conventional methods

Based on the output response of the transformer corresponding to the PWM regulated step input, the estimate method takes more parameters than conventional methods into consideration at the same time. Compared with experimental waveforms, better simulations are obtained by the estimate method.   Table 1 shows the comparison between the estimate method and conventional methods

Table1 Comparison of the estimate method with conventional methods

Items	The estimate method	Conventional Methods
Operating principle	Indirect, soft measurement, A number of parameters Obtained by a single measurement	Usually direct, hard measurement, Several measurements for a single parameter
Number of tests	2	>8
Magnetizing start point	Considered	No consideration
Coil voltage and voltage gradient du/dt	With rated step pulse input, Close to actual operation state	With low voltage sine input, Far away from actual operation
Current i and di/dt, Fluctuation of μ	Rated current intensity used, Fluctuation of μ considered	Open circuit, or short circuit with lower voltage, fluctuation of μ ignored
Parameters ignored	Leakage capacitance and leakage inductance obtained the same time	Leakage capacitance ignored in obtaining leakage inductance
Frequency spectrum	Rich in step response	Single frequency
Computation	Complicated at the beginning, but easier as programmed	A great number of simple calculations, easier lead to errors
Iron loss	Considered	No consideration
Skin effect and proximity effect	Both considered	Considering proximity effect while skin effect considered, not seen
Simulation results	Better approximations to experimental waveforms, with unloaded, half-loaded, full-loaded transformers	Simulation results for unloaded, as well as for full-loaded trans- formers in a single report, not yet seen
Equipments need	Storage oscilloscope, switches	Bridge, signal genera- tor, amplifier, conven- tional electric meters

B. Comparison examination of the estimated results

1) Results comparison analysis   YD2816 digital bridge, EM1463 signal generator and amplifier are used in the conventional measurement of primary and secondary

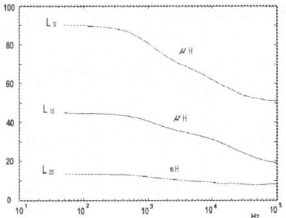

Fig.10. Leakage inductance by conventional method

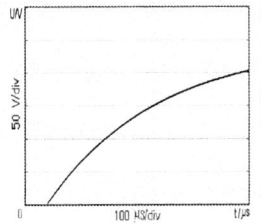

Fig.11. Simulated waveform (Full-load)

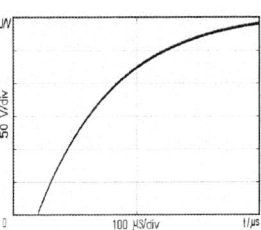

Fig.12. Simulated waveform (Half-load)

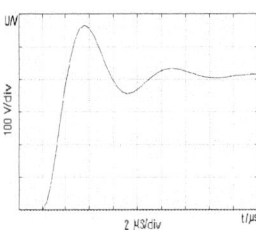

Fig.13. Simulated waveform with R_m (Unload)

parameters of the transformer. The measured values are normalized to corresponding elements shown in Fig.3

(1) Coil resistance R_l The primary DC resistance $r_1 = 0.0154\Omega$, while $r_2 = 1.928\Omega$ for the secondary. In Fig.3 and Fig.4, the normalized coil DC resistance $R_l = r_1 + r_2 k_{12}^2 = 0.0219\Omega$. Though the AC impedance for given frequency is available based on DC resistances r_1 and r_2, but the approach effect should not be undervalued. It is so far seldom seen in coil resistance measurement that both skin effect and approach effect are taken into consideration at the same time.

(2) Leakage inductance L_S In conventional methods[7], as the distributed capacitance is ignored, 25 sine inputs from 50Hz through 100kHz are used in the measurement. The primary leakage inductance L_{1S} is obtained as the secondary of the transformer has been short-circuited. The secondary leakage inductance L_{2S} is obtained as the primary has been short-circuited. The equivalent leakage inductance is then $L_S = L_{1S} + (k_{12})^2 L_{2S}$. Fig.10 shows the results. It is obvious that, except few cases, the leakage inductance decreases as the voltage frequency increases. But how to find the leakage inductance under the conditions of PWM input, which is of very rich frequency components, still remains to be solved. Besides, as the iron core is made of materials of very high magnetic conductivities ($\mu \gg 1$), the leakage inductance will be increased further[4]. Furthermore, magnetic start point, current intensity, as well as sustaining time of the current, also make contributions to μ. But they are usually not taken into consideration in conventional methods.

(3) Iron loss equivalent resistance R_m The iron loss equivalent resistance in transient state of transformer is often ignored [4 - 8]. For comparison, a normalized iron loss equivalent resistance $R_m = 3.235\ \Omega$ is obtained by measuring the iron loss power consume of unloaded transformer with conventional Wattmeter, with rated voltage of 50Hz power frequency[1]. It should be pointed out that the iron loss equivalent resistance is obtained in the stationary sine state. As a result, it can not be directly used to describe energy consume occurs in the transient state of high-frequency switching circuit.

(4) Equivalent capacitance C_S In conventional methods[2][11], signals with continuously-adjustable

frequency are used to find the resonance frequency of the tested circuit. Three capacitances are tested at first, the capacitance across the high-voltage lead and the low-voltage lead of the primary, the capacitance across the high-voltage lead and the low-voltage lead of the secondary, and the capacitance across the two high-voltage leads. Based on the 3 capacitances, together with the transformation ratio, the equivalent capacitance is $C_S' = 85.1$ nF (detailed computation omitted). It is worth knowing that the equivalent capacitance is obtained with sine input on unloaded transformer. But in switching state, due to the different voltage distribution, different current load, and different di/dt distribution in the coil, the voltage fluctuation rate will directly affect the dynamic capacitance of the transformer[2][11].

Table 2 shows the results comparison of the estimate method with conventional methods.

Table 2 Results comparison

Parameter	Unit	Our method	Conventional methods
R_l	Ω	0.032	0.0219(DC resistance)
R_m	Ω	167	3.235(with 50Hz power frequency)
R_m'	Ω	83	
L_S	μH	22	90 (with 50Hz power frequency and rated voltage)
C_S'	nF	51	85.1 (by resonance)

2). Simulation analysis on the estimate method
Based on the estimated parameters and the circuit in Fig.3, MATLAB is introduced to obtain simulated waveforms. The simulation conditions are: $T_{off} > T_{on}$ demagnetizing mode, both start point and finish point of ΔB are in Quadrant I (positive Y axis included), with rated DC voltage applied on the primary. Fig.11 and Fig.12 are the MATLAB simulated waveforms for full load and half load cases, respectively. Fig.13 and Fig.14 are the simulated waveforms for unload cases with or without considering the iron loss resistance R_m', respectively. Experiments are also conducted under same conditions as for the MATLAB simulation. Fig.9, Fig.15 and Fig.16 are the experimental waveforms for full load, unload and half load cases, respectively.

The detailed analysis is as follows.

(1) The likelihood computation is extremely difficult,

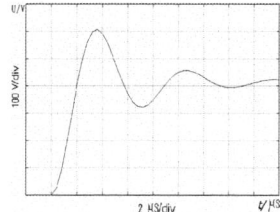 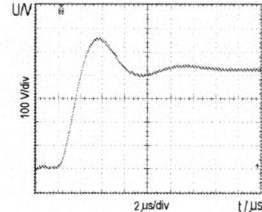 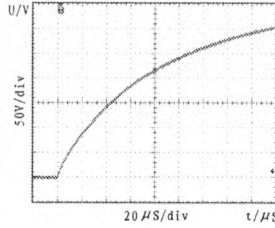 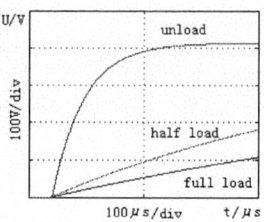

Fig.14. Simulated by
waveform without R_m (Unload)

Fig.15. Experimental waveform
(Full-load)

Fig.16. Experimental waveform
(Half-load)

Fig.17. Simulated waveform
for conventional methods

since are too many things to be considered in the computation between the estimation model and the actual electronic transformer. To be more simply and practically, the method of equating coefficients is introduced into the likelihood analysis.

(2) Above figures show that the MATLAB simulated waveforms are fairly accordant with the experimental waveforms, regardless the transformer is full loaded, half loaded, or unloaded. It indicates that the equivalent parameters estimated from the secondary step response under conditions of rated load and rated voltage, are more applicable than those by conventional methods.

(3) By Fig.13 and Fig.14, the unload simulated waveforms considering iron loss resistance R'_m fits the experimental waveforms well. By Fig.13 and Fig.15, or for the unload simulated waveforms without considering iron loss resistance R'_m, it can be clearly seen that there is a significant deviation between the simulation and the experiment. Quantitatively, the experimental overshooting is $550-425=125V$, and $600-425=175V$ for the simulated waveform.

(4) Electronic transformer is a nonlinear, time-varying system with 3-dimentional distributed parameters. The simulation model is based on 2-dimentional lumped parameters. Therefore, it is impossible to completely approximate the dynamic coupling characteristics of the transformer by energy transformation within lumped parameters. That is why there will always be differences between the simulated waveforms and the tested waveforms.

(5) Another reason leads to the deviation between MATLAB simulation and experiment is that in the simulation model, the serial dissipation caused by the "flows" such as copper lost and eddy current lost, are taken into consideration, but the parallel dissipation caused by the "leakages" such as magnetic hysteresis and the energy radiations, etc. are ignored.

3). Simulation analysis on conventional methods

Results obtained by conventional methods(See Table 2) are introduced into the circuit as shown in Fig.3, with unload, half load and full load conditions, corresponding MATLAB simulated waveforms are obtained and shown in Fig.17. Compare these waveforms with those in Fig.10, Fig.15 and Fig.16 respectively, significant deviations between corresponding curves can be seen clearly.

C. Metrological review on the accuracy of the elements

Fig.3 and Fig.4 show the 2-dimentional topology structures consist of resistive element R, reactive elements L and C. With PWM step inputs, the characteristic study of electronic transformer can be conducted by means of comparing the output responses of the model with those of the actual transformer. The differential equation reflects not only the feedback mechanism in a system, but also the output response of the system. The MATLAB simulated outputs reflect the functional relations of the system in a visual and continuous way. Comparison of the simulation model with actual transformer is shown in Table 3.

Table 3 Comparison of simulation model with
actual electronic transformer

Items	Actual Transformer	Simulation Model
Nature of elements	Distributed parameters	Lumped equivalent parameters
Quantity of elements	Numerous	Limited
Topology feature	3-Dimentional	2-Dimentional
Mathematically	A set of partial differential equations with varying coefficients	Linear differential equations with constant coefficients
Transformation within elements	Energy exchange	Information exchange
Output of dynamic processing	Tested waveforms of voltage, current, etc	Simulated waveforms of voltage, current, etc
Dissipation	$\nabla P=-Df$	I^2R
Potential energy storage	$\nabla f=-E_P\partial P/\partial t$	$0.5CU^2$
Flow energy storage	$\nabla P=-E_f\partial f/\partial t$	$0.5LI^2$

In Table 4, ∇P is the potential gradient, D the coefficient of energy transferred to heat, f the flow, E_P the potential accumulation across unit length, E_f the flow accumulation across unit length, respectively.

The simulation model is a homomorphic mapping of a transformer. There are considerable differences in the quantity and characteristics of the elements, as well as in the combinational order structure. The likelihood analysis between the model and the actual transformer is based on the outputs corresponding to the PWM step inputs. Neither comparison with standard devices, nor direct measurement by conventional electric meters is available in the accuracy analysis of the resistance, inductance and capacitance used in the simulation model. The metrological accuracy analysis of the elements should be based on the likelihood properties under specified conditions. More practically, the accuracy analysis of the lumped equivalent parameters should be based on the results obtained under conditions that V, dV/dt, i, di/dt and Φ, $d\Phi/dt$, as well as the magnetizing start point and the pulse width, are as close as possible to the actual operation state.

Based on the estimated parameters, satisfactory results are obtained in the absorption circuit design of push-pull inverters, as well as in the design of low pass filers.

V. CONCLUSIONS

Comparison analysis based on experimental waveforms indicates that the estimate method presents a better approximation to electronic transformer in pulse operation. To obtain better results in the application of the estimate method, it is helpful to put more attentions on following notes.

1). To find the simulation model by means of homomorphic mapping, the meterage of the elements should be carried out under conditions as close as possible to the actual operation. The prototype response is at first mapped on the model, corresponding elements can then be obtained quantitatively.

2). According to the analysis on the secondary step response of electronic transformer with full load and rated voltage, equivalent parameters such as leakage inductance, equivalent coil resistance and equivalent capacitance, can be obtained quantitatively. In the estimate method based on PWM step inputs, the transformer parameters such as voltage and its fluctuation rate, current and its fluctuation rate, magnetic flux and its fluctuation rate, as well as magnetizing start point and pulse width, are more closed to the actual operation conditions. As a result, the equivalent parameters obtained will be more closed to the actual values.

3). In the curve fitting of the secondary step response, to meet the constraints mentioned before, also to obtain higher curve fitting accuracy, the sampling interval should be at least 4 times larger than the conduction time of the actual transformer within one cycle.

4). In the estimate of leakage inductance, coil resistance and equivalent capacitance based on the equations from the fitting curve, ignoring magnetizing inductance brings benefits to analytical solution. Meanwhile, taking iron loss equivalent resistance into consideration will lead to lower simulation deviation for underloaded transformers.

ACKNOWLEDGMENT

The authors would like to thank Prof. CHEN XUAN, for his work and support.

REFERRENCES

[1] *Electrical Engineering Handbook.* The corporate judicial person Institute of Electrical Engineers of Japan, Tokyo,JAPAN, 2001

[2] *Electronic transformer Handbook.* Electronic transformer Specialized committee,Shengyang, CHINA,1998,

[3] P.Lorrain, D.R.Corson, *Electromagnetic fields and waves,* The people educate the publishing house, Beijing, CHINA, 1981

[4] L.M. Reduction, E.Margato and J.F.Silva, Rise Time Reduction in High-Voltage Pulse Transformers Using Auxiliary Windings, *IEEE* TRANSATION ON POWER ELECTRONICS. Vol.17, No.2, pp196-205, MAR, 2002

[5] J.M. Lopera, M.J. Prieto, A.M. Pern i a and F. Nu ň o, *A Multiwinding Modeling Method for High Frequncy Transformers and Inductors,* IEEE TRANSATION ON POWER ELECTRONICS. Vol.18, No.3, pp896-906, MAY ,2003.

[6] Johan Tjeed Strydom, Jacobus Daniel van Wyk, *Eletromagnetic Design Optimization Tool for Resonant Integrated Spiral Planar Power Passivers (ISP³),* IEEE TRANSACTION ON POWER ELECTRONICS, Vol.20,No.4, pp743-752, JULY, 2005

[7] Hirokazu YOSHIDA, Kenji NAKAMURA, Osamu ICHINOKURA, *Analysis of Operating Characteristics of Ferrite Orthogonal-Core Type Variable Inductor on Three-Dimensional Nonlinear Magnetic Circuit Considering,* T.IEE Japan,Vol.123-D,No.4, pp.386-391, APR, 2003

[8] Hiroshi HASEGAWA, Sunt SRIANTHUMRONG, *Hirofumi AKAGI, DC Magnetic Deviation, and its Suppression, of a Matching Transformer in a Series Active Filter,* T.IEE Japan,Vol.. 122-D ,No.7, pp.744 -751,JULY,2002

[9] WEN Chuan-yuan, *Exploration on Similarity Theory for System Simulation* , JOURNAL OF SYSTEM SIMULATION, Vol.17, No.1 , pp1- 6, JAN, 2005

[10] ZHOU Mei-li, Similarity and Complexity of Modeling for Simulation Systems, JOURNAL OF SYSTEM SIMULATION, Vol.16, No.12, pp2664-2672, DEC, 2004

[11] ZHANG Zhan-song, CAI Xuan-san, *Switching power supply principle and design,* Publishing House of Electronics Industry, Beijing,CHINA,2004

[12] J.M. Lopera, M.J. Prieto, A.M. Pern i a and F. Nu ň o, *A Multiwinding Modeling Method for High Frequncy Transformers and Inductors,* IEEE TRANSATION ON POWER ELECTRONICS. Vol.18,No.3,pp896-906, MAY ,2003.

[13] Hai Yan Lu, S.Y.Ron-hui, *Experimental Determination of Stray Capacitances in High Frequency Transformers,* IEEE TRANSATION ON POWER ELECTRONICS. Vol.18, No.3, pp1105-1112, MAY ,2003.

[14] Xiaolin Gao, Raja Ayyanaar, A High-Performance, *Integrated Magnetics Scheme for Buck-Cascaded Push-Pull Converter, IEEE* POWER ELECTRONICS LETTERS. Vol.2, No.1, pp29-33, MAR , 2004.

[15] WU Jia-ju, WANG Niu-bao, *An Instantaneous Voltage Drop Compensator,* POWER ELECTRONICS, Vol.33, No.6, pp7-10, DEC, 1999.

2006 5th International Power Electronics and Motion Control Conference

Analysis and Design of Boost DC-DC Converters for Intrinsic Safety

SHU-LIN LIU *, JIAN LIU *, Senior Member, HONG MAO **

* School of Electrical and Control Engineering, Xi'an University of Science & Technology, Xi'an, 710054, CHINA
** Astec Power, 35 New England Business Center, Andover, MA 01810, USA
e-mail: slliu100@xust.edu.cn, edliu@bylink.com.cn

Abstract—An improved Boost DC-DC converter meeting the requirements of intrinsic safety is presented. The maximum values of Output Short-circuit Discharged Energy (OSDE) in CCM and DCM are analyzed, respectively. It is pointed out that the OSDE of a Boost converter is the summation of the energy from the source and the inductor and the energy stored in the capacitor. It is proved that the Maximal value of OSDE (MOSDE) in CCM is the largest within the total operating range. Once MOSDE is less than the minimum ignition discharged energy, the converter meets the requirement of output intrinsic safety. The design method of inductor and capacitor is proposed according to the requirement of output intrinsic safety and the desired output voltage ripple level within the total ranges of input voltage and the load. Experiment results are in positive with the analysis showing the feasibility of the described methods.

Keywords-Intrinsic safety; boost converters; the minimum ignition discharged energy; output ripple voltage

I. INTRODUCTION

Because of its small size and high efficiency, switching power supply is widely applied in many fields, such as posts and telecommunication, military apparatus and industrial equipment. However, the power supplies applied in flammable and explosive conditions, such as coal mine and petro-chemical industries, must meet anti-explosive requirements [1]. The published anti-explosive power supplies are usually linear ones with hulking anti-explosive shells and low efficiency. Intrinsic safety is the optimal means for anti-explosive, the intrinsically safe electronic equipments without hulking anti-explosive shell have four remarkable merits: the highest safe level, the smallest size, the lightest weight and the lowest cost. Consequently, Intrinsic safety switching power supply will be competitive in the future [2-4].

With the development of the electronic technology, a low voltage power source is more and more applied in the electronic equipment. But the voltage grades of the system

Supported by Industrialization Foster Fund of Shannxi, China (05JC19); Key Research Project of Science and Technology of Xi'an, China (GG05047).

are various. Therefore, there is no doubt that Boost converters will be widely used in these systems [5-6]. In order to meet the requirement of intrinsic safety, we must make sure that both the electric spark and calorific domino effect produced in normal or abnormal conditions cannot ignite prescriptive explosive admixture. In fact, as for any kinds of switching converters, the output-short-circuit spark energy is mainly determined by two factors: one is the filter inductor and capacitor of the converter; the other is the input power supply of the converter.

As for the energy stored in the inductor and capacitor of the converter, reference [7] and [8] suggest to add a current-limiting resistance or a voltage-stabilizing and current-limiting measure to the output of the converters to limit short-circuit current of the capacitor. But the efficiency of the converter is decreased.

As for the energy from the input power supply, owing to its inherent structure of the Boost converter, we can't cut off the power supply thoroughly by turning off the main switch of a Boost converter while other converters, such as Buck, Buck-Boost and isolating converters, can. Therefore, we must improve the structure of Boost converter.

We present an improved structure of Boost DC-DC converter, which can realize intrinsic safety. By analyzing the output short-circuit discharged energy of the Boost DC-DC converter and using the ignition curves for capacitors and inductors as the criterion of the intrinsic safety, a design method of inductor and capacitor is obtained according to the requirement of the output intrinsic safety and the desired output voltage ripple level in this paper.

II. IMPROVED BOOST CONVERTER FOR INTRINSIC SAFTY

A. A Basic Boost Converter

A basic Boost converter is shown in figure 1. If the converter works in Continuous Conduction Mode (CCM), the relationship between the input voltage (V_i) and the output voltage (V_O) is

$$V_O = V_i /(1-d) \qquad (1)$$

Where d is the conducting ratio of the power switch.

1-4244-0448-7/06/$25.00 ©2006 IEEE

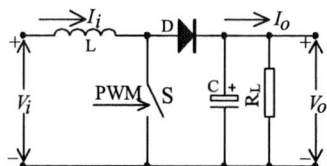

Figure 1. A basic Boost DC-DC converter

The boundary condition of CCM and DCM is [9]

$$L_C = \frac{R_L d(1-d)^2}{2f} \qquad (2)$$

Where L_C is the critical inductance of CCM and DCM, f is the switching frequency, R_L is load resistance. When $L > L_C$, the converter will be in CCM. Otherwise, it is in DCM.

If the Boost converter is in CCM and the inductance is large enough, the output voltage ripple is described by[10]

$$V_{PP} = \frac{dTI_O}{C} = \frac{dV_O}{R_L Cf} = \frac{V_O - V_i}{R_L Cf} \qquad (3)$$

From figure 1, we can find out that we can't isolate the energy from the input in case of a short-circuit occurring in the output. In another word, even if S is turned off by the protection circuit, much energy will still be transferred from the input power supply to the output. Obviously, the basic converter can't meet the requirement of intrinsic safety. Thus we must improve it.

B. Improved Intrinsically Safe Boost Converter

An improved Boost converter meeting the requirement of intrinsic safety is shown in figure 2. Compared to figure 1, an isolating switch SK and a diode D1 are added to the input terminal. When the output short-circuit occurs, the protection circuit of the converter will be triggered immediately. As a result, SK will be shut off and S will keep on conducting. Thus the input is isolated from the output thoroughly while the current is through the inductor L and D1, and the energy stored in the inductor can be isolated effectively. Therefore, the short-circuit spark energy is only the energy stored in the capacitor.

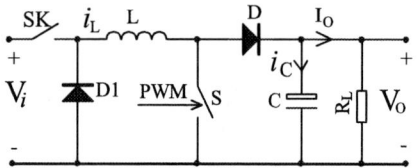

Figure 2. An improved Boost DC-DC converter

C. Operating Region and Operating Mode

Supposing that the input voltage range of the converter is from $V_{i,min}$ to $V_{i,max}$, and the load range is from $R_{L,min}$ to $R_{L,max}$. As shown in figure 3, operating region of the converter is a rectangle in R_L-V_i plane. In figure 3, curve (a) shows that the converter is in DCM within the whole operating range. Curve (b) shows that the converter is in CCM in case of larger output power. Curve (c) indicates

that the converter is in CCM within its whole operating range.

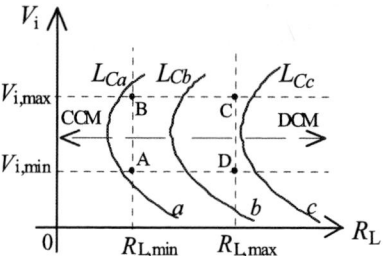

Figure 3. CCM and DCM region on R_L-V_i plane

From figure 3, we can deduce that each curve of critical inductor has a corresponding minimum critical resistance. Therefore, we can get the operating region according to such resistance. According to (1) and (2), the critical resistance R_{LC} is determined by

$$R_{LC} = \frac{2V_O^3 Lf}{V_i^2(V_O - V_i)}. \qquad (4)$$

When $V_i = 2V_O/3$ is within the input voltage range, the maximum value of $V_i^2(V_O - V_i)$ is $4V_O^3/27$ in case of $V_i = 2V_O/3$. Thus, the minimum critical resistance $R_{LC,min}$ is

$$R_{LC,min} = 13.5Lf \qquad (5)$$

III. OUTPUT SHORT-CIRCUIT DISCHARGED ENERGY

The total output-short-circuit energy W of the converter includes two parts: W_C and W_s, that is

$$W = W_C + W_S \qquad (6)$$

Where W_C is the energy stored in the filter capacitor; W_s is the energy coming from the input power source and the inductor. Because the inductor peak current in CCM is different from that in DCM, it is necessary to study the output short-circuit discharged energy of boost converter in CCM and DCM, respectively.

A. Output Short-circuit Discharged Energy in CCM

The peak current through the inductor in CCM is[9]

$$I_{LP} = \frac{V_o^2}{R_L V_i} + \frac{V_i(V_O - V_i)}{2V_O Lf} \qquad (7)$$

Since $V_i(V_O - V_i) = \frac{V_O^2}{4} - (V_i - \frac{V_O}{2})^2 \le \frac{V_O^2}{4}$, therefore, the maximal peak current through the inductor within the whole operating region is

$$I_{LP,max} = \frac{V_O^2}{R_{L,min}V_{i,min}} + \frac{V_O}{8Lf} \qquad (8)$$

Supposing that the short-circuit protection circuit could make the switch S turn on and the switch SK shut off within a very short time (t) after the output short-circuit occurs. Because the transient process of output short-circuit is rather complicated, we have to make some simplifications. We consider the short circuit-spark

1268

voltage V_H (in general, $V_H < V_i$) as a constant value within the time of t.

When the output short-circuit occurs, the current through the inductor depends on the state of S. If S is on, the inductor current will increase continuously along with its original rising trend. If S is off, the inductor current will change from the falling trend to the rising trend. Clearly, the maximum current (I_{LH}) through the inductor occurs in the former case. Moreover, I_{LH} gets the largest if the peak current through inductor just reaches its maximum value $I_{LP,\max}$ when the output short-circuit occurs. Therefore, we have

$$I_{LH} = I_{LP,\max} + \Delta I_L \qquad (9)$$

Where $\Delta I_L = \dfrac{V_i - V_H}{L} t$. To simplify the analysis, we just consider the most serious case. We use $V_H = 0$ in calculation of I_{LH}, thus

$$I_{LH,\max} = I_{LP,\max} + \Delta I_L = I_{LP,\max} + \frac{V_i}{L} t \qquad (10)$$

We can approximately regard the product of $I_{L,\max}$, V_H and t as the energy transferring from the input power source and the inductor to the output after the output short-circuit occurs. To simplify the analysis, we use $V_H = V_i$ in calculation of the output short-circuit energy. Therefore, the maximum energy $W_{s,\max}$ from the input power source and the inductor is

$$W_{s,\max} = V_H I_{LH,\max} t = V_i I_{LH,\max} t = \frac{V_i^2}{L} t^2 + V_i I_{LP,\max} t \qquad (11)$$

Considering that the output ripple voltage is rather small compared with the V_O, the maximum energy stored in the capacitor is $W_C \approx CV_O^2/2$. Therefore, the maximum value of the total output short-circuit discharged energy in CCM is

$$W_{CCM} = W_{s,\max} + W_C = \frac{V_i^2}{L} t^2 + V_i I_{LP,\max} t + \frac{CV_O^2}{2} \qquad (12)$$

B. Output Short-circuit Discharged Energy in DCM

When a Boost converter is in DCM, the relationship among V_i, I_O and d is [9]

$$d^2 = 2I_O Lf \left(\frac{V_O}{V_i} - 1 \right) \frac{1}{V_i} \qquad (13)$$

Therefore, the peak current through inductor I_{LP}' is

$$I_{LP}' = \frac{dV_i}{Lf} = \sqrt{\frac{2V_O(V_O - V_i)}{R_L Lf}} \qquad (14)$$

In DCM, the minimum critical resistance $R_{LC,\min}$ is determined by (5), considering $V_i(V_O - V_i) \le V_O^2/4$, the maximum value of peak current through inductor is

$$I_{LP,\max}' = \frac{V_O}{\sqrt{27}Lf} \qquad (15)$$

According to the similar way mentioned above, the maximum output short-circuit energy in DCM is

$$W_{DCM} = W_{s,\max}' + W_C = \frac{V_i^2}{L} t^2 + V_i I_{LP,\max}' t + \frac{CV_O^2}{2} \qquad (16)$$

C. Maximum Output Short-circuit Energy

Since the output voltage of the converter is constant, the energy stored in the filter capacitor in CCM is the same as that in DCM. But the energy from the input power source and the inductor is different in CCM and in DCM.

According to (8) and (15), we have

$$\frac{I_{LP,\max}}{I_{LP,\max}'} = \frac{\sqrt{27}LfV_O}{R_{L,\min}V_{i,\min}} + \frac{\sqrt{27}}{8} > \frac{5Lf}{R_{L,\max}} + \frac{5}{8}$$

Usually, $Lf > 3R_{L,\min}/40$, i.e., $I_{LP,\max} > I_{LP,\max}'$. According to (12) and (16), we have

$$W_{CCM} > W_{DCM} \qquad (17)$$

Equation (17) indicates that the maximum output short-circuit energy of the Boost converter in CCM is the largest within the whole operating region. According to (8) and (12) and using $V_{i,\max}$ as V_i, we obtain the maximum output short-circuit discharged energy is

$$W_{\max} = \frac{V_{i,\max}^2}{L} t^2 + \left(\frac{V_O}{R_{L,\min}V_{i,\min}} + \frac{1}{8Lf} \right) V_{i,\max} V_O t + \frac{CV_O^2}{2} \qquad (18)$$

IV. IDENTIFICATION OF OUTPUT INTRINSIC SAFETY

As for the identification of output intrinsic safety, the output of the Boost converter is a capacitive circuit. According to the minimum ignition voltage curve of capacitive circuit [11], we can obtain a definite minimum ignition voltage U for a given capacitance C_B. In order to guarantee intrinsic safety of the converter, we must consider an adequate safety factor K. That means the definite minimum ignition voltage U corresponding to the output voltage V_O is

$$U = KV_o \qquad (19)$$

Therefore the minimum ignition discharged energy W_B corresponding to the output voltage V_O and the capacitance C_B is

$$W_B = 0.5 C_B V_o^2 \qquad (20)$$

Therefore, as for a Boost converter, if the maximum short-circuit discharged energy within its whole operating region is less than W_B, we call such converter as an output intrinsic safe boost converter. The corresponding identification condition is

$$W_{\max} < W_B \qquad (21)$$

V. Design Of Output Intrinsically Safe Boost Converters

A. Design of capacitor

According to (3), the minimal value of the capacitor is determined by the desired output ripple voltage, i.e.,

$$C_{min} = \frac{V_O - V_{i,min}}{R_{L,max} m V_O f} \qquad (22)$$

Where $m = V_{PP}/V_O$.

The maximum value of the capacitor is determined by the ignition curve of capacitive circuit [11]. Take safety factor K into consideration, the capacitance corresponding to the voltage of $U = KV_O$ can be obtained from the curve, which is used as the maximum capacitance C_{max}.

B. Design of inductor

The operating mode of a boost DC-DC converter is always designed as figure 3(b) or (c). Because the peak current in DCM is larger than that in CCM with the same averaged current. So we can determine the value of inductor according to this rule.

On the one hand, the minimum value of inductor is determined by the converter's operating mode. Generally, we may design the converter to be in CCM in case of the current through inductor larger than a certain value I_A. Thus

$$L > V_i^2 (V_O - V_i) R_A / 2V_O^3 f \qquad (23)$$

where $R_A = V_O/I_A$. Since the maximum value of $V_i^2(V_O - V_i)$ is $4V_O^3/27$ in case of $V_i = 2V_O/3$, we have $L > 2R_A/27f$ according to (23). Therefore,

$$L_{min1} = 2R_A / 27f = 2V_O / 27 f I_A \qquad (24)$$

On the other hand, to meet the output intrinsically safe requirement, the value of inductor must meet (21). Taking safety factor K into consideration, V_O is replaced by KV_O. According to (18) and (21), we have

$$\frac{V_{i,max}^2}{L} t^2 + K V_{i,max} V_O \left(\frac{KV_O}{R_{L,min} V_{i,min}} + \frac{1}{8Lf} \right) t + \frac{CK^2 V_O^2}{2} < W_B \qquad (25)$$

So we have

$$L_{min2} = \frac{8 f V_{i,max}^2 t^2 + K V_{i,max} V_O t}{8 f W_B - 4 f K^2 V_O^2 \left(\dfrac{2V_{i,max}}{R_{L,min} V_{i,min}} t + C \right)} \qquad (26)$$

The larger one between L_{min1} and L_{min2} is the minimum value of inductor, which is written as L_{min}.

VI. An Example

As for a Boost converter applied in the class I explosive circumstance with the parameters as: $V_i = 10\sim14$V, $V_O = 18$V, output ripple voltage $V_{PP} = 2\% V_O$, $R_L = 36\sim180\Omega$, $f = 80$kHz, $C = 100\mu$F, $K = 1.5$ and the responding time of short-circuit protection circuit is $t = 5\mu$s.

A. Check of output intrinsic safety

Based on above parameters, $V_i = 2V_O/3 = 12$V is involved in the input voltage range. According to (5), the corresponding point of the minimum critical resistance is (162Ω, 12V), so we can conclude that the converter operates in the region of figure 3(b).

The minimum ignition energy corresponding to $V_O = 18$V is $W_B = 1.62$mJ according to the minimum ignition voltage curve of capacitive circuit [11]. Take above parameters into (18), we get $W_{max} = 16.3$mJ$> W_B$. Obviously, this converter doesn't meet the requirement of the output intrinsic safety.

B. Desiogn of capacitor and inductor

We increase the switching frequency to $f = 100$kHz, according to (22), the minimum capacitance is $C_{min} = 6.2\mu$F. According to the minimum ignition voltage curve of the capacitive circuit [11], the maximum capacitance is $C_{max} = 10.0\mu$F. Therefore, we design $C = 8.2\mu$F. Supposing that the converter is desired to be in CCM when the output current is larger than 250mA. According to (24) and (26), we get $L_{min1} = 53.3\mu$H and $L_{min2} = 110\mu$H, so $L_{min} = 110\mu$H. Therefore, we design $L = 120\mu$H.

C. Experiment results

Based on above calculation, the parameters of the converter are $f = 100$kHz, $L = 120\mu$H, $C = 8.2\mu$F.

The experimental results are shown in figure 4, where i_L and V_{i1} are the current through inductor and the voltage behind SK after output short-circuit occurs. It can be seen from figure 4(a) that SK turns off immediately and V_i decreases to zero rapidly due to the fast triggering of protection circuit. However the current through the inductor continues to increase in the switch-off period of SK because of the output short-circuit. Nevertheless the turn-off period of SK is transitory, so the increase of the current through the inductor is very small. Therefore, the energy from the input power source and the inductor is isolated efficiently, except for a little energy transferred to the output.

The waveforms of output current i_O, output voltage V_O and the output short-circuit discharged energy W_O are shown in figure 4(b). We can see that the dominating discharge period of the capacitor is about 20μs with the discharged energy of 1.02mJ which is much less than the maximum output short-circuit energy (i.e., 1.48 mJ) obtained by theoretical calculation, according to (18). That means the output short-circuit energy from the power source and the inductor is very small.

In addition, 1.02mJ is also much less than the corresponding minimum ignition energy of 1.62mJ, indicating that the converter is intrinsically safe.

The waveforms of inductor current i_L and output ripple voltage V_{PP} are shown in figure 4(c) in case of the minimum input voltage of V_i=10V and the minimum load of R_L =36Ω. The output voltage ripple is V_{PP} =340mV, which meets the requirement of the desired ripple voltage level.

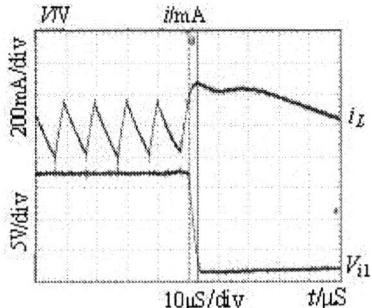

(a) Experiment results of i_L and V_{i1}

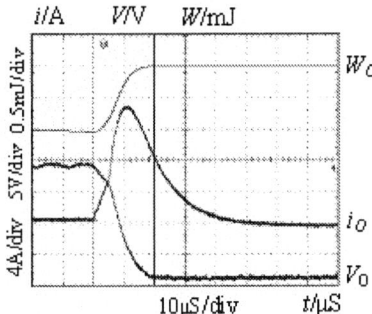

(b) Experiment results of i_O, V_O and W_O

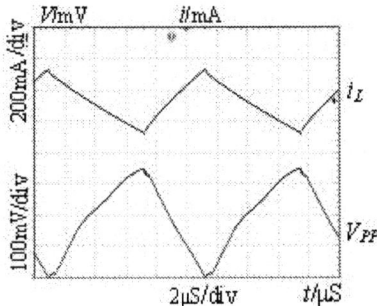

(c) Experiment results of i_L and V_{PP}

Figure 4. Waveforms in case of the output short-circuit

VII. CONCLUSIONS

The improved intrinsic safety Boost converter can isolate the energy from the power source and the inductor efficiently.

The output short-circuit discharged energy of a boost converter is the summation of the energy from the source and the inductor and the energy stored in the capacitor. The maximal output short-circuit discharged energy in CCM is the largest within the total ranges of input voltage and the load. If it is less than the corresponding minimum ignition discharged energy, the converter meets the requirement of output intrinsic safety.

The design method of inductor and capacitor is proposed according to the requirement of output intrinsic safety and the desired output voltage ripple level within the total ranges of input voltage and the load.

The feasibility of the proposed methods are proved by experiment results.

REFERENCES

[1] Da Li, Xinyuan Fan, The Summarize of Intrinsically Safe System, Automation in petro-chemical industry, No. 6, pp..8~10,Jun 2000, in Chinese

[2] Liu Shulin, Liu Jian, Yang Yinlin, Zhong Jiuming Design of Intrinsically Safe Buck DC/DC Converters, Proceedings of the Eighth International Conference on Electrical Machines & Systems, 2005.9, pp. 1327-1331

[3] Liu Shulin, Liu Jian, Zhong Jiuming. Analysis of Output Short-circuit Discharged Energy and Optimal Design of Output Intrinsically Safe Buck Converters, Proceedings of Asia Pacific Symposium on Safety, 2005.11, pp. 1978-1984

[4] Liu Jian, Liu Shulin, Yang Yinling, Zhang Yanmei. Output Intrinsically Safe Behavior of buck Converters and Its Optimal Design [J]. Proceedings of the CSEE, 2005, Vol.25(19), pp. 52-57. in Chinese.

[5] Sujuan Fang, Yubiao He, The Technology Development of the Low Voltage and Low Power Loss CMOS ASIC. Semiconductor technology, No.3, pp. 16-19, Jun 1997, in Chinese

[6] Zhenghai Zhou, Xiancan Deng, Xiangxiong Lou, The design of the low voltage and high frequency PWM DC-DC Converter[J], Research &Progress of SSE, Vol.24(4), pp. 462-465, Aug. 2004.

[7] Xiaoqiang Liu, On Intrinsically safe DC switching power supply and its reserve power supply[D],Doctoral Dissertation of china University of Ming &Technology,2001, in Chinese.

[8] Jianfei Li, Zhixin Xu, Heqing Zhong, Constant voltage restricted current and constant current restricted voltage DC-DC converter[J], Power Electronic,No.1, pp. 42-44, Feb.1999, in Chinese

[9] Zhansong Zhang, Xuansan Cai, Theory and Design of Switching power supply, Publishing house of electronics industry, Beijing, 1998, in Chinese

[10] Xinbo Ruan, Yangguang Yan, Soft switching technology of DC switching power supply, Science press, Beijing, 2000, in Chinese

[11] Compilation of China National Standard[S]——GB 3846.4—2000[S],China standard press,Beijing,2000, in Chinese.

Liu Shu-lin: received his master degree in 1988. He is a professor and Ph.D student in School of Electrical and Control Engineering at Xi'an University of Science and Technology (XUST). His research interests focus on intrinsic safety circuit and power electronics.

Liu Jian (SM'01): received his Ph.D degree in 1997. He is a professor of Xi'an University of Science and Technology (XUST). His research interests focus on switching converters and control.

Modeling and Fuzzy Logic with Integrator Control for the ZVZCS PWM DC/DC Converter

Shen Hong[1,2], Wan Jianru[1], Yang Xiaobo[2], Wu Weiyang[2] and Wang Xiaohuan[2]

1. Tianjin University, Tianjin, China
2. Yanshan University, Qinhuangdao, China
shenhong@ysu.edu.cn

Abstract—In this paper, a small signal model of a novel ZVZCS FB-PWM DC/DC converter is built and analyzed. A double closed loop controller is employed: the inner loop is a PI controller and the outer loop is a fuzzy logic with integrator controller. The novel control strategy improves the performance of the converter greatly, such as stability, control precision and dynamic response. The dramatic feature of topology and the superiority of control strategy are proved by experiment results on a 3kW prototype.

Keywords-ZVZCS; Fuzzy control; Integrator

I. INTRODUCTION

In recent years, many ZVZCS-FB topologies have been proposed [1-4], operating with zero-voltage switching in leading legs and zero-current switching in lagging legs. In this paper, a ZVZCS-FB PWM DC/DC converter with secondary auxiliary circuit which consists of one small capacitor and two small diodes to provide ZVZCS conditions to primary switches as well as to clamp secondary rectifier voltage without any additional passive and active clamp circuits is adopted [5], as shown in Fig. 1. This topology has many remarkable advantages such as the small duty-cycle loss, the low rectifier voltage, low current stress and low cost. Therefore, it suits to high power, high efficiency and high power density requirements. To study the dynamic character of converter and to design the controller effectively, mathematic model should be built. There are two ways to built the mathematic model: emulation method and analytic modeling method. The advantage of the former is high accuracy. The response characteristic, the waveform of the small signal perturbation and the large signal motivation can be obtained. the disadvantage of this method is indistinct sense. The latter is expressed by analytic formulae. Its' sense is very distinct, and its' parameters are convenient to be adjust. And it is not easy to put into effect. Analytic modeling include discontinue method and average sequence method. State-Space Averaging Method that is one of average sequence method was first proposed and perfected by R.D.Middlebrook in 1976. It preserves the advantage of analytic modeling method and can be corrected by Bode diagram. The method has been used widely.

Fuzzy control is based on fuzzy set theory, fuzzy

[a] This paper is supported by the Key Program National Nature Science Foundation of China (No. 50237020) and the Nature Science Foundation of Tianjin (05YFJMJC11500)

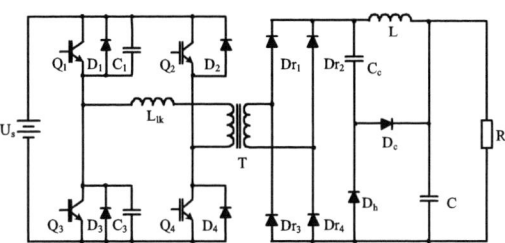

Figure 1. ZVZCS FB converter

langrage variable and fuzzy logic inference. Since fuzzy control does not need an exact mathematical model of the system, it is well suited to non-linear, time-variant, ill-defined systems in which models are difficult to be obtained or even not exist [6]. Two dimension fuzzy controller is similar to a PD controller. It has good dynamic characteristic, but bad static characteristic. In this paper, two-dimension fuzzy controller with integrator is proposed to solve this problem.

In this paper, two controlled sources are added to the small signal equivalent circuit model of Buck converter. Small equivalent circuit model of ZVZCS FB-PWM circuit is deduced and analyzed. The scheme of double loop control which is composed of fuzzy controller with integrator and PI controller is adopted to control the whole system. A prototype (input 220 V, output 600 V, power 3kW) is built. Theoretical analysis and experimental results indicate that the proposed topology and the modulation scheme have the promising features.

II. SMALL SIGNAL MODEL

The small signal model of Buck converter is shown in Fig. 2. In the ZVZCS PWM DC/DC converter proposed in this paper, There are 8 operating mode in the half of circuit operating cycle. The duty cycles belonged to every operating mode are recorded as d1, d2, …d8 in turn. The specific operating procedure is shown in paper [5]. To calculate seven differential equations is complicated. The circuit topology can be looked as a derivation of basic Buck circuit. By defining effective duty cycle found introducing feedback addition items on the basis of small signal model of Buck converter, small signal model of ZVZCS PWM DC/DC converter is built [7]. According to paper [5], the voltage of output inductance is obtained as:

$$u_L(t) = u_{rec}(t) - u_o. \qquad (1)$$

In this paper, the output voltage, the current of output

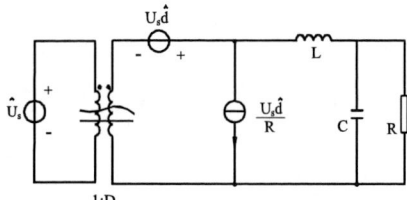

Figure 2.   The small signal model of Buck converter

inductance and the input voltage are expressed by u_o, i_L, u_s. The steady state values of variables are expressed by U_o, I_L, U_s. In the steady state, according to the voltage-second balance equation, to calculate the integral of u_L in the half of cycle [5], we can get:

$$\int_0^{T_s'} \left(u_{rec}(t) - u_o\right) dt. \qquad (2)$$

or,

$$\int_0^{d_1 T_s'} 0 dt + \int_{d_1 T_s'}^{d_2 T_s'} (\frac{u_o}{n} - u_o)[1 - \cos \omega_n(t) + u_o] dt +$$

$$\int_{d_2 T_s'}^{d_3 T_s'} \frac{u_s}{n} dt + \int_{d_3 T_s'}^{d_4 T_s'} (\frac{u_s}{n} - \frac{i_L}{n^2 C_{eq}} t) dt + \int_{d_4 T_s'}^{d_6 T_s'} 2(\frac{u_s}{n} - u_o) dt + \qquad (3)$$

$$\int_{d_6 T_s'}^{d_8 T_s'} [2(\frac{u_s}{n} - u_o) - \frac{i_L}{C_c} t] dt + \int_{d_7 T_s'}^{T_s'} 0 dt = \int_0^{T_s'} u_o dt.$$

Define

$$d_{eff} = n \frac{u_o}{u_s} \qquad (4)$$

Combined (3) and (4), we can get:

$$\frac{2 d_{eff}^2 (C_c + n^2 C_{eq})}{n^2 i_L} - \frac{d_{eff} u_s [n T_s' i_L + 4(C_c + n^2 C_{eq}) u_s]}{n^2 i_L}$$

$$+ \frac{u_s[2 n d T_s' i_L + (4 C_c + 5 n^2 C_{eq}) u_s]}{2 n^2 i_L} = 0. \qquad (5)$$

Neglecting the impact of $C_{eq} = C_1 + C_3, d_{eff}$ is obtained as:

$$d_{eff} = \frac{n T_s' i_L + 4 C_c u_s}{4 C_c u_s} -$$

$$\frac{\sqrt{n T_s' i_L (n T_s' i_L + 8 C_c u_s - 8 C_c D u_s)}}{4 C_c u_s}. \qquad (6)$$

Notice that: not only the d but also the u_s and the change of i_L can impact the effective duty ratio of the ZVZCS FB-PWM circuit.

Define: $d = D + \hat{d}$, $u_s = U_s + \hat{u}_s$, $i_L = I_L + \hat{i}_L$, and the perturbations of d_{eff} to d is $\hat{d}_d$, to u_s is $\hat{d}_v$, to i_L is $\hat{d}_i$. From (6) we can get:

$$d_{eff} = \hat{d}_d + \hat{d}_v + \hat{d}_i \cong \hat{d} + G_v \hat{u}_s + G_i \hat{i}_L. \qquad (7)$$

Where

$$G_v = \frac{\sqrt{n T_s' i_L} [n I_L T_s' - 4 C_c (D - 1)]}{4 C_c U_s^2 \sqrt{n I_L T_s' - 8 C_c (D - 1)}} - \frac{n T_s' I_L}{4 C_c U_s^2}. \qquad (8)$$

and

$$G_i = -\frac{\sqrt{n T_s'} \left[\sqrt{n I_L T_s' - 8 C_c (D - 1)^2} - \sqrt{n I_L T_s'} \right]^2}{8 C_c U_s \sqrt{I_L} \sqrt{n I_L T_s' - 8 C_c (D - 1) U_s}}. \qquad (9)$$

Introducing the analysis to small signal equivalent circuit model of Buck converter, we can get the small equivalent circuit model of ZVZCS FB-PWM circuit.

The effects of $\hat{d}_v$ and $\hat{d}_i$ can be represented by two controlled source, shown in Fig. 3.

From Fig. 3 the control-to-output voltage transfer function is obtained by:

$$G_{ud}(s) = \frac{u_s}{n R L C s^2 + n(L + R R_d C) s + n(R + R_d)}. \qquad (10)$$

Where, $R_d = -G_i U_s / n$.

The control-to-inductor current transfer function is given by:

$$G_{id}(s) = \frac{\dfrac{u_s}{n} s + \dfrac{u_s}{n R L C}}{L C s^2 + (\dfrac{L}{R} + R_d C) s + (1 + \dfrac{R_d}{R})}. \qquad (11)$$

The "built-in" current feedback can be seen, and G_i has a negative value. This character can be expressed by the equation $d_{eff} = \hat{d} + G_i \hat{i}_L$. For instance, when $\hat{i}_L$ has a positive perturbation, because G_i is negative, the effective duty cycle will decrease so the perturbation is restrained.

Compared with Buck converter, with same parameters, coefficient of damping of ZVZCS FB-PWM circuit is higher than the former. The damp is not only decrease the resonance peak value of $G_{ud}(s)$, but also increase the phase angle margin of resonance angular frequency.

III.   CONTROL SCHEME

A strategy of double closed loop that PI controller as inner loop and Fuzzy logic with integrator controller as external loop is adopted in this paper. Principle block diagram of double closed loop is shown in Fig. 4.

A.   Design of Inner Loop

The block diagram of current closed loop system is shown in Fig. 5. Where, $W_1(s)$ is a current regulator, $K_{PWM}/(0.5 T_s + 1)$ presents delay caused by converter and $1/(L_s + R_L)$ is inductor transfer function.

The transfer function of current loop PI controller is

$$W_1(s) = K_p \frac{\tau_i + 1}{\tau_i s}. \qquad (12)$$

Where, K_p is proportionality coefficient of current

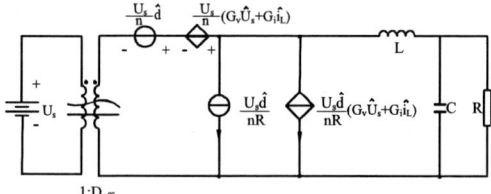

Figure 3.   Small signal model of ZVZCS FB-PWM circuit

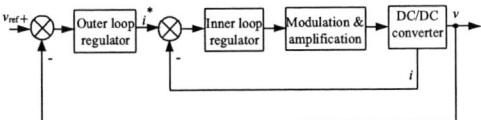

Figure 4. Principle block diagram of double closed loop

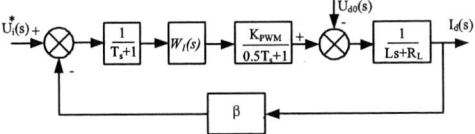

Figure 5. The block diagram of current closed loop system

regulator and τ_i is leading characteristic time of current regulator. To eliminate the zero of regulator and the large characteristic time of controlled device, we can choice that

$$\tau_i = \frac{L}{R_L}. \tag{13}$$

By simplified, dynamic block diagram of current closed loop system is shown in Fig. 6.

B. Design of Outer Loop

1) Construction of fuzzy controller: Two-dimension fuzzy controller is adopted. The fuzzy input variables are chosen to be the voltage error, $E = V_{ref} - V_o$, and the derivative of the voltage error, ΔE. The output variable of the fuzzy controller is selected to be U. Two-dimension fuzzy controller is similar to PD controller. It has good dynamic characteristic, however the static characteristic is poor[8]. So in this paper, two-dimension fuzzy controller with integrator is proposed to solve this problem. The block diagram of fuzzy control system is shown in Fig. 7.

2) Fuzzy language: {NB, NM, NS, ZE, PS, PM, PB}.

3) Rule of fuzzy control: Rule is the core of fuzzy controller. 49 pieces of rule are chosen in this paper and is shown in TABLE I.

4) Region of variable and scale coefficient: Assumption

TABLE I.
Rule Of Fuzzy Control

U\E E	NB	NM	NS	ZE	PB	PM	PS
NB	NB	NB	NB	NB	NM	NS	ZE
NM	NB	NB	NB	NM	NS	ZE	PS
NS	NB	NB	NM	NS	ZE	PS	PM
ZE	NB	NM	NS	ZE	PS	PM	PB
PS	NM	NS	ZE	PS	PM	PB	PB
PM	NS	ZE	PS	PM	PB	PB	PB
PB	ZE	PS	PM	PB	PB	PB	PB

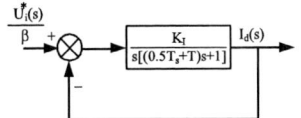

Figure 6. The simplified block diagram of current closed loop system

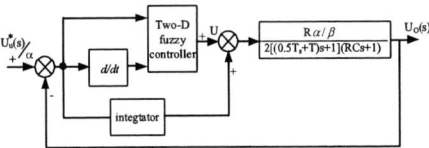

Figure 7. The block diagram of fuzzy control system

that the region of E is $\{-n_1, -(n_1-1), ..., 0, 1, ..., n_1-1, n_1\}$, the region of ΔE is $\{-n_2, -(n_2-1), ..., 0, 1, ..., n_2-1, n_2\}$, the region of U is $\{-m, -(m-1), ..., 0, 1, ..., m-1, m\}$, we can get that

$$a_e = \frac{n_1}{|e_{max}|} \tag{14}$$

$$a_c = \frac{n_2}{|e_{max}|} \tag{15}$$

$$a_u = \frac{m}{|e_{max}|} \tag{16}$$

Where, a_e is proportionality coefficient of E, a_c is proportionality coefficient of ΔE, a_u is proportionality coefficient of U. The region of E and ΔE are divided into 13 levels, {-6, -5, -4, -3, -2, -1, 0, 1, 2, 3, 4, 5, 6}. The region of U is {0~6}.

5) Degree of membership: Degree of membership for E and ΔE is shown in Fig. 8. Degree of membership for U is shown in Fig. 9.

6) Inference relationship: The inference relationship of fuzzy control represents the relationship of input and

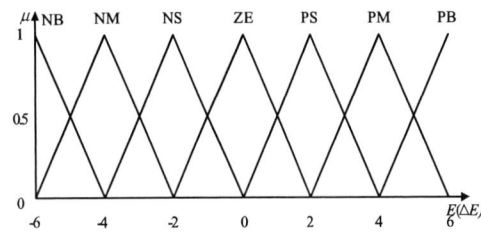

Figure 8. Degree of membership for E and ΔE

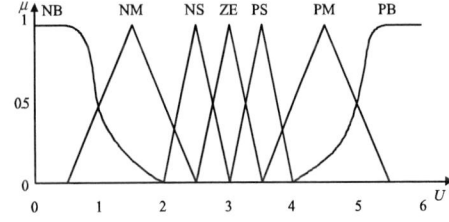

Figure 9. The block diagram of fuzzy control system

output. It can be described by

$$R = \bigcup_{1,2} (E_i \times E_j \times u_{ij}) \qquad (17)$$

Where, E_i is i^{th} fuzzy state of E, E_j is j^{th} fuzzy state of ΔE and u_{ij} is state of output corresponding to i^{th} fuzzy state of E and j^{th} fuzzy state of ΔE .

According to the fusion calculation of fuzzy assemble, the fusion calculation of two-dimension fuzzy controller output is

$$u = (E \times \Delta E) \circ R \qquad (18)$$

7) *Defuzzification*: The maximum degree of membership method is adopted. Fuzzy assemble which is get by inference is changed into the actual value.

IV. EXPERIMENTAL RESULTS

According to measurement, $L = 1.35$ mH, $R_L = 3.5$, $\tau_i = L/R_L = 3.86 \times 10^{-4}$. K_P is depended on ω_{c} which is barrier frequency of current loop and dynamic performance index. In general, overshoot $\sigma\% < 5\%$, $\zeta = 0.707$, $K_I (0.5T_s + T) = 0.5$, T_s=50 μ s are adopted. So $K_I = \omega_{c} = 1/(T_s + 2T)$, $K_P = L/[(Ts + 2T)K_{PWM}\beta]$, $K_{PWM}=nU_s$=770, $\beta = 0.1$,

$$W_I(s) = 0.03 \frac{3.86 \times 10^{-4} + 1}{3.86}$$

A TMS320LF2407A DSP and a CPLD from ALTERA are employed to realize the proposed arithmetic. The specifications and designed components are summarized in TABLE II.

TABLE II.
SPECIFICATIONS AND DESIGNED COMPONENTS

INput Voltage	220V
Output power	2kW
Switches Q_1-Q_4	HGTG30N60A
Transformer	PC40(EE85)N1:N2=1:3.1
L_{lk}	1.6uH
Output Voltage	500V
Switch frequency	20kHz
C_1, C_3	2.2n
C_o	2000uF
C_c	0.2uF
L_f	PC40 EE55 1.3mH
Controller	TMS320LF2407A

Waveforms of the primary current and voltage are shown in Fig. 10. Waveforms of the primary current and the secondary rectifier voltage are shown in Fig. 11. Response of output voltage and output current waveforms to step load without integrator is shown in Fig. 12. Response of output voltage and output current waveforms to step load with integrator is shown in Fig. 13. The voltage is 250 V/div and the current is 1 A/div in Fig. 12 and Fig. 13.

From Fig. 12 and Fig. 13, it is indicated that there is steady state error in output voltage without integrator, but there is steady-state error in output voltage with integrator. It proves that it is very necessary to put integrator into fuzzy controller.

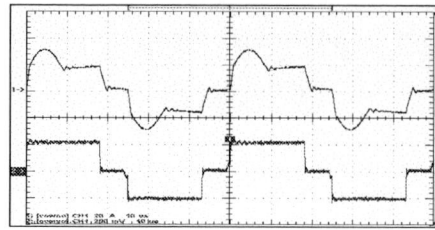

Figure 10. Waveforms of the primary current and voltage

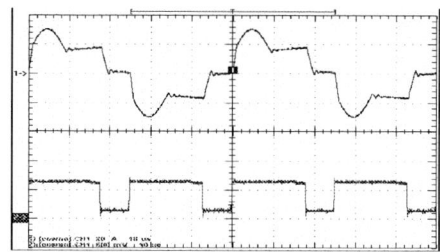

Figure 11. Waveforms of the primary current and the secondary rectifier voltage

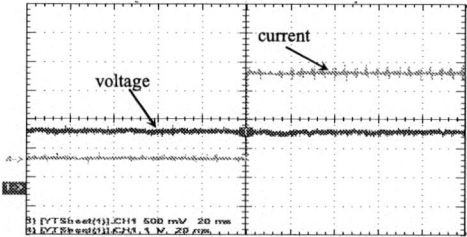

Figure 12. Response of output voltage and output current waveforms to step load without integrator

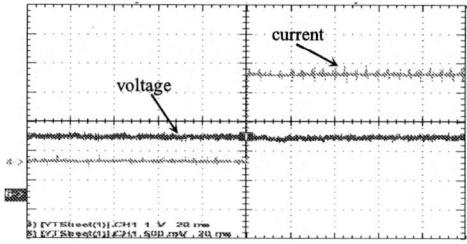

Figure 13. Response of output voltage and output current waveforms to step load with integrator

V. CONCLUSION

In this paper a small signal model of a novel ZVZCS FB PWM DC/DC converter is built and analyzed. A strategy of double closed loop that PI controller as inner loop and Fuzzy logic with integrator controller as external loop is adopted. System is realized by DSP and CPLD. The dramatic feature of topology and the superiority of control strategy are proved by experimental results.

REFERENCES

[1] Keming Chen, Tomas A. Stuart, "A Study of IGBT Turn-off Behavior and Switching Losses for Zero-Voltage and Zero-Current Switching," IEEE *Applied Power Electronics Conference and Exposition*, 1992, pp. 411-418.

[2] C. Cuadros, C. Y. Lin, D. Boroyevich, ect., "Design procedure and modeling of high power, .High Performance, Zero-Voltage Zero-Current Switched, Full Bridge PWM Converter," *IEEE Applied Power Electronics Conference and Exposition*, 1997, pp. 790-798.

[3] Xinbo Ruan, Yangguan Yan, "A novel zero-voltage and zero-current switching PWM full-bridge converter using two diodes in series with the lagging leg," IEEE Trans on Power Electronics, 2001, vol. 48, No. 4, pp. 777-785.

[4] Yang Xiaobo, Zhou Huisheng, and Shen Hong, "Analysis and design of an improved zero-voltage and zero-current switching (ZVZCS) full bridge DC-DC PWM converter," IPEMC. Xi'an, 2004, No. 8, pp. 90-95.

[5] R. J. Marks II, "Fuzzy logic technology and applications," *IEEE Press*, 1994, pp. 128-136.

[6] Yang xiaobo, Wu Weiyang, and Shen Hong, "Analysis and modeling of the ZVZCS full bridge PWM DC-DC converter using a secondary auxiliary circuit," *IEEE PESC*, 2005, pp.1018-1023.

[7] Rukonuzzaman M, Nakaoka M, "Fuzzy logic current controller for threephase voltage source PWM-inverters," *Conference Record of IEEE*, 2000, No. 2, pp. 1163-1169.

2006 5th International Power Electronics and Motion Control Conference

ZVS DC-DC Converter with Parallel-Connected Current Doubler Rectifier

Bor-Ren Lin, *Senior Member, IEEE,* Shuh-Chuan Tsay, Chun-Sheng Yang and Chien-Lan Huang
National Yunlin University of Science and Technology, Yunlin 640, Taiwan, China

Abstract—A soft switching converter with parallel-connected full-wave rectifiers is presented. In the proposed converter, the primary windings of two transformers are connected in series. Two full-wave rectifiers with ripple current cancellation are connected in parallel at the output side to reduce the current stress of the secondary winding of transformer. The clamp circuit based on an auxiliary switch and a clamp capacitor is connected in parallel with the primary side of the transformer to recycle the energy stored in the leakage inductance. The leakage inductance of transformers, the magnetizing inductance and the clamp capacitance is resonant to achieve zero-voltage switching (ZVS) of auxiliary switch. The resonance between the leakage inductance of transformer and output capacitance of switch will achieve ZVS operation for the main switch in the proposed converter. The pulse-width modulation technique is adopted to regulate the output voltage. Experimental results for a 200W (5V/40A) prototype are given to demonstrate the effectiveness of the proposed converter.

Keywords- zero voltage switching, dc-dc converter, current doubler rectifier.

I. INTRODUCTION

Soft switching techniques for dc-dc converter have been proposed in the switching mode power supplies to increase circuit efficiency and power density and to reduce voltage/current stress of switching devices. Zero voltage switching (ZVS) and zero current switching (ZCS) techniques [1] have been proposed to reduce the switching losses of power switches and to increase circuit efficiency. The active clamp techniques [2-3] based on auxiliary switch and clamp capacitor were proposed to reduce the voltage stress of main switch and to increase efficiency of the converter. The phase-shift pulse-width modulation (PWM) techniques [4-5] have been presented to regulate the output dc voltage and to achieve ZVS operation. The leakage inductance of transformer and output capacitance of switching devices are used to resonant such that the ZVS conditions of switching devices are satisfied. The asymmetrical PWM techniques [6-7] were proposed to achieve ZVS turn-on and to increase circuit efficiency.

The new active clamp converter with parallel-connected full-wave rectifiers based on current doubler topology is proposed to achieve ZVS operation for switching devices. In the proposed converter, the primary windings of two transformers are connected in series. Two current doubler rectifiers are connected in parallel at the converter output side to share the load current. Compared with the center-

tapped rectifier topology, the current doubler rectifier has the advantages of one diode conduction drop, ripple current cancellation on the output inductors and the low current rating of secondary winding of transformer. The leakage inductance of transformer and output capacitance of switching devices are used to implement the resonant behavior in order to release the energy stored in the output capacitance of switches and achieve ZVS operation for switching devices. The circuit configuration, principle of operation and design considerations of the proposed converter are presented. Finally, experimental results based on a 200W (5V/40A) prototype circuit are presented to verify the circuit performance.

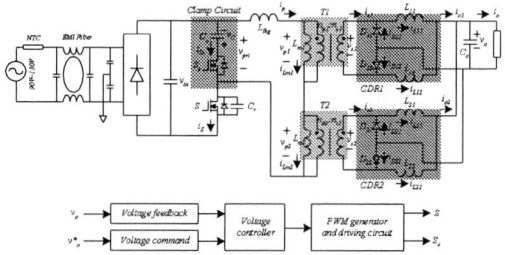

Fig. 1 Circuit configuration of the proposed converter.

II. CIRCUIT CONFIGURATION

The circuit configuration of proposed converter is shown in Fig. 1. The input voltage v_{in}, isolation transformers $T1$ and $T2$, and switch S are basic components in the primary side of the proposed converter. A clamp circuit based on a clamp switch S_c and a clamp capacitor C_c is connected in parallel with the primary winding of series-connected transformers. The clamp circuit is used to recycle the surge energy stored in the leakage inductance of transformer and to clamp the voltage stress of main switch S to input voltage v_{in} and clamp voltage v_C. L_{m1} and L_{m2} are the magnetizing inductances of isolation transformers $T1$ and $T2$ respectively. The inductance L_r is the resonant inductance which includes the leakage inductance of two transformers and the external inductance. The capacitance C_r is equal to the parallel combination of the junction capacitance of switches S and S_c and the parasitic capacitance across the primary winding of two transformers. C_c is the clamp capacitor. The current doubler rectifier, CDR1, included D_{11}, D_{12}, L_{11} and L_{12} is used to achieve full-wave rectification and ripple current cancellation. Two current doubler rectifiers, CDR1 and CDR2, are connected in

1-4244-0448-7/06/$25.00 ©2006 IEEE   1278

parallel to share the load current. In the proposed converter switch S and clamp switch S_c are all turned on at ZVS based on the resonance during the commutation interval.

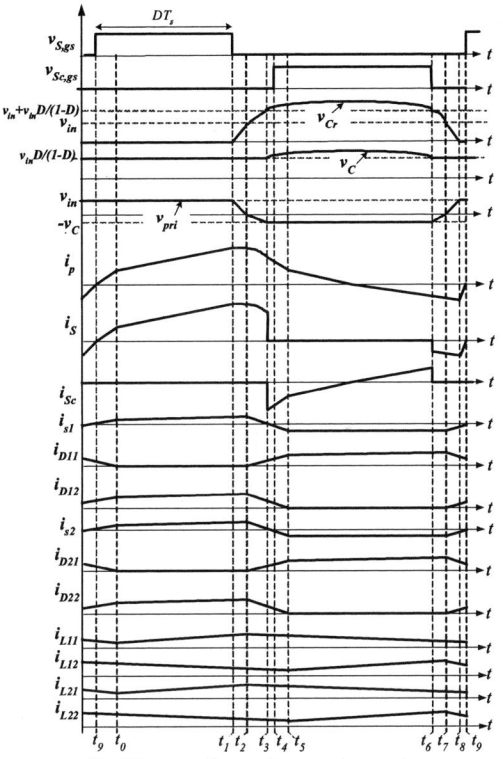

Fig. 2 Key waveforms of proposed converter.

III. OPERATION PRINCIPLE

Before the system analysis, some assumptions of the proposed converter are made as follows: (1) The clamp capacitance C_c is large enough; (2) The transformer leakage inductance at the secondary side is small enough to neglect in the system analysis; (3) The output filter inductance $L_{11}\sim L_{22}$ is large enough to be ripple free; (4) The resonant inductance L_r is smaller than the magnetizing inductances L_{m1} and L_{m2} ($L_{m1}=L_{m2}$); (5) The turn ratio between the primary winding turn of transformer and the secondary winding turn is $n=n_{p1}/n_{s1}=n_{p2}/n_{s2}$. The resonant inductance and output capacitance of power switches are used to achieve ZVS operation for switching devices. Fig. 2 shows the key waveforms of power switches, primary side voltage, primary current, and output inductor currents. There are ten operating modes as shown in Fig. 3 during one switching period in the proposed converter.

Mode 1 ($t_0<t<t_1$; Fig. 3(a)): In this mode, switch S is turned on. The magnetizing inductors L_{m1} and L_{m2} and leakage inductor L_r are charged by the input voltage v_{in}. The capacitor voltage $v_{Cr}=0$. The primary side voltage $v_{pri}=v_{in}$. The primary winding voltages of transformers $v_{T1}=v_{T2}=mv_{in}$ ($m=L_m/(2L_m+L_r)$, $L_m=L_{m1}=L_{m2}$). The magnetizing currents increase linearly. The secondary

winding voltages of transformers $v_{s1}=v_{s2}=mv_{in}/n$. The rectifier diodes D_{12} and D_{22} are tuned on, and diodes D_{11} and D_{21} are reverse-biased. The inductor voltages v_{L12} and v_{L22} equals $-v_o$ and the inductor currents i_{L12} and v_{L22} decreases linearly. For the steady state condition, each current doubler rectifier supplies half of load current. The diode currents at the secondary side $i_{D11}=0$, $i_{D21}=0$, $i_{D12}=i_{o1}$ and $i_{D22}=i_{o2}$. The switch current $i_S=i_{pri}$. This mode is ended at time $t=t_1$ when the switch S is turned off.

Mode 2 ($t_1<t<t_2$; Fig. 3(b)): At time $t=t_1$, switch S is turned off. The primary current is positive and charges capacitor C_r from 0 to v_{in}. The resonant inductor L_r, magnetizing inductors L_{m1} and L_{m2} and resonant capacitor C_r are resonant with the resonant frequency $f_{r1}=1/2\pi\sqrt{(L_r+2L_m)C_r}$. The primary current is larger enough to charge capacitor C_r linearly. The primary side voltage $v_{pri}=v_{in}-v_{Cr}$. The secondary side voltages $v_{s1}=v_{s2}\approx(v_{in}-v_{Cr})/2n$. The diode currents at the secondary side $i_{D11}=0$, $i_{D21}=0$, $i_{D12}=i_{o1}$ and $i_{D22}=i_{o2}$. At time $t=t_2$, the capacitor voltage $v_{Cr}=v_{in}$, and the transformer magnetizing inductor voltage $v_{T1}=v_{T2}=0$. The secondary side diodes $D_{11}\sim D_{22}$ are all turned on. The inductor currents $i_{L11}(t_2)$ and $i_{L21}(t_2)$ are reaching the maximum values $I_{L11,max}$ and $I_{L21,max}$ respectively. The magnetizing currents $i_{Lm1}(t_2)=I_{Lm1,max}$ and $i_{Lm2}(t_2)=I_{Lm2,max}$.

Mode 3 ($t_2<t<t_3$; Fig. 3(c)): At time $t=t_2$, the transformer primary voltage is zero. The secondary side voltage $v_{s1}=v_{s2}=0$ and the rectifier diodes $D_{11}\sim D_{22}$ are all turned on. The slopes of the magnetizing currents are zero. The inductor voltages $v_{L11}=v_{L12}=v_{L21}=v_{L22}=-v_o$ so that the inductor currents decrease linearly. To ensure ZVS turn-on of clamp switch S_c, the capacitor voltage v_{Cr} should greater than $v_{in}+v_C$ before the end of this mode. At time t_3 the resonant capacitor voltage v_{Cr} equals $v_{in}+v_C$ and the intrinsic diode of S_c is turned on. This operating mode is ended at $t=t_3$ when $v_{Cr}=v_{in}+v_C$.

Mode 4 ($t_3<t<t_4$; Fig. 3(d)): At $t=t_3$, the anti-parallel diode of switch S_c is turned on and the rectifier diodes at the secondary side are all turned on. The secondary side voltages $v_{s1}=v_{s2}=0$. Since the clamp capacitance C_c is much larger than resonant capacitance C_r, most of the current i_p flows through intrinsic diode of clamp switch S_c and charges clamp capacitor. In this mode the current i_{Sc} is negative and the clamp switch S_c can be turned on at the ZVS condition. The switch current i_S is zero in this mode. The inductor currents $i_{L11}\sim i_{L22}$ decrease linearly. The voltages across the magnetizing inductors $v_{T1}=v_{T2}=0$. The resonant tank in this mode includes L_r and C_c. In this mode $i_{D12}<i_{L12}$, $i_{D22}<i_{L22}$, $i_{D11}>i_{L11}$ and $i_{D21}>i_{L21}$, and i_{s1} and i_{s2} are negative. Before the primary current i_p becomes negative value, the switch S_c should be turned on to achieve ZVS operation. This mode is ended at $t=t_4$ when switch S_c is turned on.

Mode 5 ($t_4<t<t_5$; Fig. 3(e)): In this mode, the clamp switch S_c is turned on at ZVS. Since the diode currents $i_{D11}<i_{o1}$ and $i_{D21}<i_{o2}$ such that the rectifier diodes at the secondary side are all turned on. The operation principle in this mode is similar to the operation in mode 4. Since

1279

the primary current i_p is positive, the energy stored in the resonant inductor L_r is released to charge capacitor C_c. This mode is ended at $t=t_5$ when rectifier diodes D_{12} and D_{22} are turned off. At time t_5, $i_p(t_5)=I_{Lm1,max}-I_{L12,min}/n$ and $I_{L12,min}=i_{L12}(t_5)$.

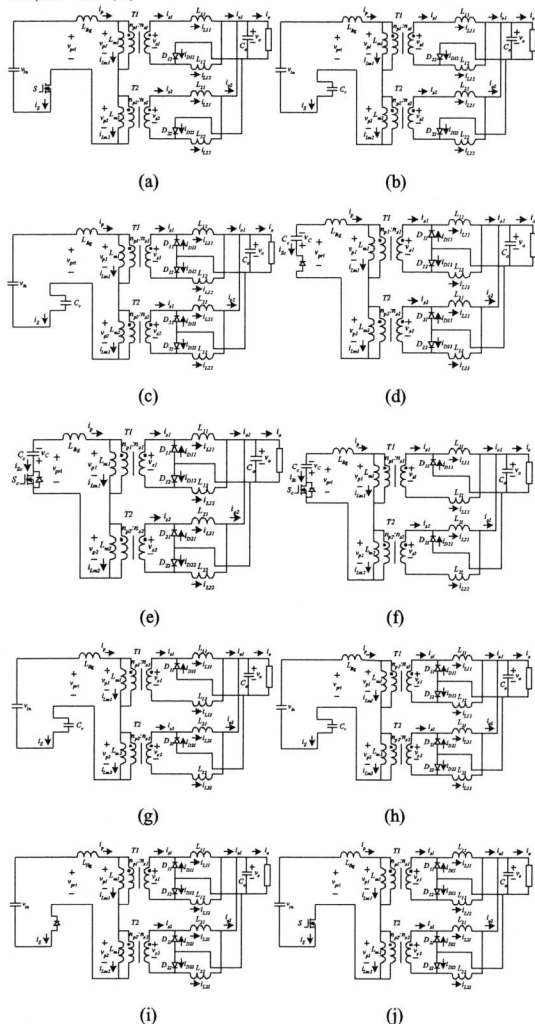

(a)   (b)

(c)   (d)

(e)   (f)

(g)   (h)

(i)   (j)

Fig. 3 Ten operation states of proposed converter (a) mode 1 [t_0-t_1] (b) mode 2 [t_1-t_2] (c) mode 3 [t_2-t_3] (d) mode 4 [t_3-t_4] (e) mode 5 [t_4-t_5] (f) mode 6 [t_5-t_6] (g) mode 7 [t_6-t_7] (h) mode 8 [t_7-t_8] (i) mode 9 [t_8-t_9] (j) mode 10 [t_9-t_0].

Mode 6 ($t_5 < t < t_6$; Fig. 3(f)): In this mode, switch S_c is still turned on and secondary diodes D_{11} and D_{21} are turned on. The rectifier diode currents $i_{D12}=i_{D22}=0$, $i_{D11}=i_{o1}$ and $i_{D21}=i_{o2}$. The clamp capacitor C_c is discharged. The primary voltage $v_{pri}=-v_C$. The resonant circuit in this mode includes L_r, L_{m1}, L_{m2} and C_c. The inductor currents i_{L11} and i_{L21} decreases linearly, and the inductor currents i_{L12} and i_{L22} increase linearly. This operating mode is ended when the clamp switch S_c is turned off.

Mode 7 ($t_6 < t < t_7$; Fig. 3(g)): In this mode switches S and S_c are all turned off. The operation of this mode is similar

to the operation in mode 2. The primary side voltage v_{pri} equals $v_{in}-v_{Cr}$. The primary current i_p discharges the capacitor C_r from $v_{in}+v_C$ to v_{in} in this mode. The inductor currents i_{L11} and i_{L21} decreases linearly, and the inductor currents i_{L12} and i_{L22} increase linearly in this mode. At time $t=t_7$, the capacitor voltage $v_{Cr}=v_{in}-v_C$, and the primary side voltage $v_{T1}=v_{T2}=0$. The rectifier diodes at the secondary side are in free wheeling mode. The inductor currents $i_{L12}(t_7)$ and $i_{L22}(t_7)$ are equal to the maximum values. The magnetizing currents equal the minimum values.

Mode 8 ($t_7 < t < t_8$; Fig. 3(h)): At time $t=t_7$, the transformer primary voltage $v_{T1}=v_{T2}=0$. The secondary side voltage $v_{s1}=v_{s2}=0$ and the secondary side diodes are all turned on. The diode currents i_{D11} and i_{D21} decrease linearly and diode currents i_{D12} and i_{D22} increase linearly. The output inductor currents i_{L11}~i_{L22} decrease linearly. The inductor L_r and capacitor C_r are resonant in this mode. To ensure ZVS operation to turn on switch S, the capacitor voltage v_{Cr} should reach zero before the end of this mode. Therefore the energy stored in the resonant inductor L_r must be greater than the energy stored in the resonant capacitor C_r. This mode is ended at $t=t_8$ when $v_{Cr}=0$.

Mode 9 ($t_8 < t < t_9$; Fig. 3(i)): At time $t=t_8$ the resonant capacitor voltage $v_{Cr}=0$ and the intrinsic body diode of switch S is turned on. The secondary side rectifier diodes are still in the free wheeling mode. The inductor currents i_{L11}~i_{L22} decrease linearly in this mode. The leakage inductor voltage v_{Lr} equals input voltage v_{in} and the primary side current increases. Before the primary current i_p becomes positive, the switch S should be turned on to achieve ZVS operation. This mode is ended at $t=t_9$ when switch is S turned on.

Mode 10 ($t_9 < t < t_0$; Fig. 3(j)): This operating mode starts at time $t=t_9$ when switch S is turned on. The secondary side diode currents i_{D12} and i_{D22} are still increasing until $i_{D12}=i_{o1}$ and $i_{D22}=i_{o2}$, and diode currents i_{D11} and i_{D21} are decreasing until to zero. At this moment the mode 10 is ended and the circuit goes to the operating mode 1. This mode is ended at $t=t_0$ when diode currents i_{D11} and i_{D21} are decreasing to zero.

IV. STEADY STATE ANALYSIS AND DESIGN EQUATION

We neglect the duty cycle loss during the transition interval between two switches S and S_c at modes 2-5 and 7-10. When power switch S is turned on and switch S_c is turned off in mode 1, the transformer primary voltages $v_{T1}=v_{T2}\approx v_{in}/2$. The output inductor voltages $v_{L11}=v_{L21}\approx v_{in}/2n-v_o$ and $v_{L12}=v_{L22}=-v_o$. When switch S is turned off and switch S_c is turned on in mode 5, the primary voltages of transformers $v_{T1}=v_{T2}\approx -v_C/2$. The output inductor voltages $v_{L11}=v_{L21}=-v_o$ and $v_{L12}=v_{L22}\approx v_C/2n-v_o$. Based on the voltage-second balance across the magnetizing inductances L_{m1} and L_{m2}, one can obtain the clamp voltage v_C.

$$v_C=Dv_{in}/(1-D) \qquad (1)$$

where D is the duty cycle of switch S. When duty cycle is less than 0.5, the clamp voltage $v_C < v_{in}$. When duty cycle is greater than 0.5, the clamp voltage $v_C > v_{in}$. Due to the voltage-second balance across the output inductors $L_{11} \sim L_{22}$, the output voltage can be expressed as:

$$v_o = D v_{in} / (2n) \qquad (2)$$

Based on (1) and (2), the clamp voltage can be rewritten as:

$$v_C = 2 n v_o / (1-D) \qquad (3)$$

When switch S is turned on, the current ripple on the magnetizing inductance is given as follow.

$$\Delta i_{Lm1} = \Delta i_{Lm2} \approx DT v_{in} / (2L_m) \qquad (4)$$

where T is the switching period. From (2) and (4), the current ripple on the magnetizing inductance can be further expressed as:

$$\Delta i_{Lm1} = \Delta i_{Lm2} \approx \frac{n v_o T}{L_m} \qquad (5)$$

The ripple currents on output inductors $L_{11} \sim L_{22}$ are given as:

$$\Delta i_{L11} = DT(v_{in}/2n - v_o)/L_{11} = (1-D)T v_o / L_{11}, \Delta i_{L12} = \frac{D v_o T}{L_{12}},$$

$$\Delta i_{L21} = DT(v_{in}/2n - v_o)/L_{21} = (1-D)T v_o / L_{21}, \Delta i_{L22} = \frac{D v_o T}{L_{22}} . (6)$$

The ripple currents on the output inductors L_{11} and L_{12} are equal if $L_{11}/L_{12} = (1-D)/D$. From (6) one can obtain the output inductances if the ripple currents on the output inductors are given.

$$L_{11} = \frac{(1-D)v_o T}{\Delta i_{L11}}, \ L_{12} = \frac{D v_o T}{\Delta i_{L12}}, \ L_{21} = \frac{(1-D)v_o T}{\Delta i_{L21}}, L_{22} = \frac{D v_o T}{\Delta i_{L22}} . (7)$$

The average currents on the output inductors are $I_{L11} = I_{L21} = I_{L12} = I_{L22} \approx I_o/4$. The root mean square currents on the output filter inductors are

$$I_{L11,rms} \approx \frac{I_o}{4} \sqrt{1 + \frac{(\frac{\Delta i_{L11}}{I_{L11}})^2}{12}}, \quad I_{L12,rms} \approx \frac{I_o}{4} \sqrt{1 + \frac{(\frac{\Delta i_{L12}}{I_{L12}})^2}{12}},$$

$$I_{L21,rms} \approx \frac{I_o}{4} \sqrt{1 + \frac{(\frac{\Delta i_{L21}}{I_{L21}})^2}{12}}, \quad I_{L22,rms} \approx \frac{I_o}{4} \sqrt{1 + \frac{(\frac{\Delta i_{L22}}{I_{L22}})^2}{12}} . (8)$$

The maximum currents on the output inductors are given as:

$$I_{L11,max} \approx \frac{I_o}{4}(1 + \frac{\Delta i_{L11}}{2I_{L11}}), \ I_{L12,max} \approx \frac{I_o}{4}(1 + \frac{\Delta i_{L12}}{2I_{L12}}),$$

$$I_{L21,max} \approx \frac{I_o}{4}(1 + \frac{\Delta i_{L21}}{2I_{L21}}), \ I_{L22,max} \approx \frac{I_o}{4}(1 + \frac{\Delta i_{L22}}{2I_{L22}}) \quad (9)$$

The average currents on the rectifier diodes $D_{11} \sim D_{22}$ are expressed as:

$$I_{D11,av} = I_{D21,av} = (1-D)I_o / 2,$$

$$I_{D12,av} = I_{D22,av} = DI_o / 2 \qquad (10)$$

The root mean square currents of the rectifier diodes are given as:

$$I_{D11,rms} = I_{D21,rms} = \frac{I_o}{2} \sqrt{1-D},$$

$$I_{D12,rms} = I_{D22,rms} = \frac{I_o}{2} \sqrt{D} \qquad (11)$$

The voltage stresses of the rectifier diodes are given as:

$$V_{D11,stress} = V_{D21,stress} = v_o/D, \qquad V_{D12,stress} = V_{D22,stress} = v_o/(1-D) \ (12)$$

The turn ratio between the transformer secondary side and primary side is equal to

$$n = \frac{n_{p1}}{n_{s1}} = \frac{D_{max} v_{in,min}}{2 v_o} \qquad (13)$$

where D_{max} is the maximum duty cycle when input voltage v_{in} is minimum. The voltage stresses of switches S and S_c are approximately equal to $v_{in}/(1-D_{max})$. The maximum currents of switches S and S_c are approximately expressed as:

$$i_{S,max} \approx I_{Lm1,max} + I_{L11,max}/n = \frac{I_o}{4n} + \frac{\Delta i_{Lm1} + \Delta i_{L11}/n}{2},$$

$$i_{Sc,max} \approx -I_{Lm1,min} - I_{L12,max} = \frac{I_o}{4n} + \frac{\Delta i_{Lm1} + \Delta i_{L12}/n}{2} (14)$$

The root mean square currents of switches S and S_c are approximately expressed as:

$$i_{S,rms} \approx \sqrt{D[(I_o/4n)^2 + \frac{(\Delta i_{Lm1} + \Delta i_{L11}/n)^2}{12}]},$$

$$i_{Sc,rms} \approx \sqrt{(1-D)[(I_o/4n)^2 + \frac{(\Delta i_{Lm1} + \Delta i_{L12}/n)^2}{12}]} \ (15)$$

In stage 8 the energy stored in the resonant inductance must be greater than the energy stored in the resonant capacitance to ensure the ZVS operation for switch S.

$$L_r \geq \frac{C_r (v_{in,max})^2}{i_p(t_7)^2} \qquad (16)$$

The resonant angular frequency in states 3 and 8 is $\omega_r = 1/\sqrt{L_r C_r}$. To achieve ZVS operation, the time period t_d at $t_1 \sim t_3$ and $t_6 \sim t_8$ should be equal to $1/4f_r$. If the resonant capacitance C_r is given, the resonant inductance L_r can be expressed as $L_r = 4t_d^2/(C_r \pi^2)$. Based on the time delay t_d, the voltage ratio between the input and output voltages should be modified as:

$$v_o/v_{in} = (D - D_d)/(2n) \qquad (17)$$

V. EXPERIMENTAL RESULTS

For experimental evaluation of the proposed converter, the circuit was built. The converter was designed with the following parameters: v_{in}: 130V~180V, v_o=5V, $I_{o,max}$=40A, f_s=150kHz (switching frequency), D_{max}=0.6, and P_o=200W. The parameters used in this experiment are as follows: $L_{m1} = L_{m2}$=80µH; the turns ratio n_p:n_s=13:5; the filter inductances $L_{11} = L_{12} = L_{21} = L_{22}$=12µH; the output filter capacitance C_o=5400µF; and the clamp capacitance C=0.132µF. The 2SK2645 and DSSK60-0045B are used as switching devices and rectifier diodes, respectively. The equivalent resonant capacitance C_r=400pF. The selected resonant inductance L_r=30µH. The experimental voltage and current waveforms of the adopted converter operating at 200W output power is given in Fig. 4. From Fig. 40(a), it can be seen that the drain voltages $v_{S,ds}$ and $v_{Sc,ds}$ reach zero before the gate voltages $v_{S,ds}$ and $v_{Sc,ds}$ are changed from low to high voltage level. Therefore the ZVS conditions for switches S and S_c are achieved. The measured waveforms of gate voltage $v_{S,gs}$, primary side voltage v_{pri}, primary side current i_p and clamp capacitor voltage v_c are given in Fig. 4(b). When switch S is turned on, the primary side voltage $v_{pri}=v_{in}$. When switch S turns off and switch S_c turns on, the primary side voltage $v_{pri}=-v_C$. At this condition, the clamp capacitance C is resonant with magnetizing inductance. The experimental results of gate voltage $v_{S,gs}$, primary side current i_p and secondary side currents i_{s1} and i_{s2} are shown in Fig. 4(c). The secondary winding currents i_{s1} and i_{s2} are balanced and equaled. Fig. 4(d) gives the measured results of secondary side winding current i_{s1} and rectifier diode currents i_{D11} and i_{D12}. During the free wheeling states, the rectifier diodes are both turned on. When the gate voltage $v_{S,gs}$ is positive at the steady state, the secondary side winding current i_{s1} is positive. The diode current i_{D12} equals half of load current and diode current i_{D11} equals zero. On the other hand, the diode current i_{D11} equals half of load current and diode current i_{D12} equals zero when the secondary side winding current i_{s1} is negative at the steady state. Fig. 4(e) and 4(f) give the experimental waveforms of secondary side current i_{s1}, diode current i_{D11} and i_{D12}, and output inductor currents i_{L11} and i_{L12}. The duty cycle of switch S is larger than 0.5 so that the current ripple on the inductor L_{12} is greater than the current ripple on the inductor L_{11}. Fig. 4(g) and 4(h) show the measured results of secondary side current i_{s2}, diode current i_{D21} and i_{D22}, and output inductor currents i_{L21} and i_{L22}. The system efficiency of the proposed converter is about 88% at the rated output power (5V/40A).

VI. CONCLUSION

The system analysis, circuit design consideration and the implementation of a soft-switching converter with parallel-connected full-wave rectifiers are presented in this paper. In the adopted circuit, the leakage inductance of the transformer is used to as one part of resonant inductance to achieve ZVS operation of switching devices. The active clamp topology is used in the circuit to reduce the voltage stress of main switch and achieve ZVS operation. The current sharing topology with full-wave rectifiers parallel-connected at the secondary side of transformer and two transformer series-connected at the primary side is used to balance the output currents of two current doubler rectifiers. Finally the experimental results based on a prototype circuit with 5V/40A output are provided in this paper. From the experimental results, the ZVS operations of switching devices are achieved, the inductor currents on the output side are almost balanced, and the current rating of two current doubler rectifiers are equal.

ACKNOWLEDGMENT

The authors would like to thank the National Science Council, Taiwan,China, for supporting this project under grant NSC 95-ET-7-224 -004 –ET.

REFERENCES

[1] Hua, G. and Lee, F. C., 'Soft-switching techniques in PWM converters', *IEEE Transactions on Industrial Electronics*, vol. 42, no. 6, pp. 595-603, 1995.

[2] Watson, R., Lee, F. C. and Hua, G. C., 'Utilization of an active-clamp circuit to achieve soft switching in flyback converters', *IEEE Transactions on Power Electronics*, vol. 11, no. 1, pp. 162-169, 1996.

[3] Lee, Y. S. and Lin, B. T., 'Adding active clamping and soft switching to boost-flyback single-stage isolated power-factor-corrected power supplies', *IEEE Transactions on Power Electronics*, vol. 12, no. 6, pp. 1017-1027, 1997.

[4] Yungtack, J., Jovanovic, M. M. and Chang, Y. M., 'A new ZVS-PWM full-bridge converter', *IEEE Transactions on Power Electronics*, vol. 18, no. 5, pp. 1122-1129, 2003.

[5] Chan, H. L., Cheng, K.W.E. and Sutanto, D., 'ZCS-ZVS bi-directional phase-shifted DC-DC converter with extended load range', *IEE Proceeding - Electric Power Applications*, vol. 150, no. 3, pp. 269-277, 2003.

[6] Choi, B. and Lim, W., 'Current-mode control to enhance closed-loop performance of asymmetrical half-bridge DC-DC converters', *IEE Proceedings - Electric Power Applications*, vol. 152, no. 2, pp. 416-422, 2005.

[7] Chen, T. M. and Chen, C. L., 'Analysis and design of asymmetrical half bridge flyback converter', *IEE Proceedings - Electric Power Applications*, vol. 149, no. 6, pp.433-440, 2002.

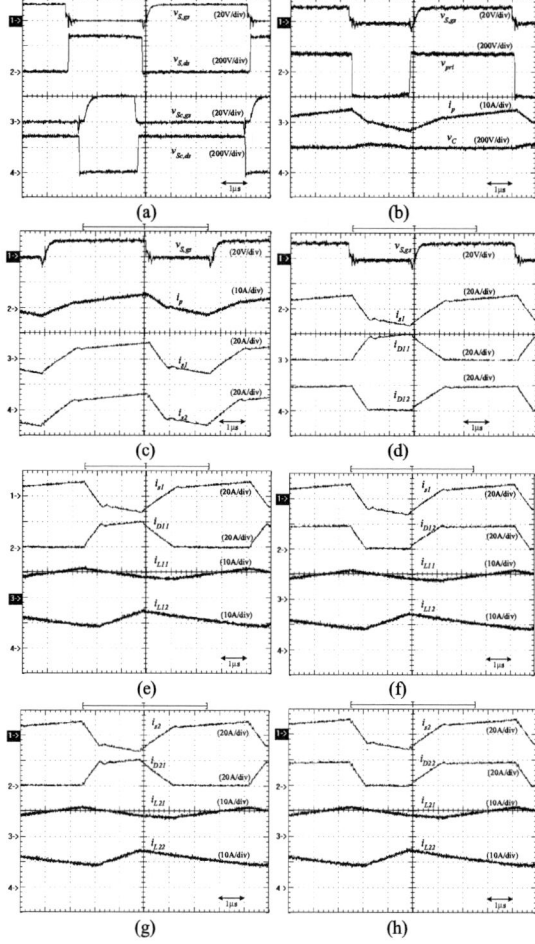

Fig. 4 Experimental voltage and current waveforms of proposed converter for 200W load (v_o=5V and i_o=40A).

Study on the Dynamical Model and Analytical Method for DC-DC Switching Converter

Li-Li Wang[*,**], Yu-Fei Zhou[*,#], Jun-Ning Chen[*]

[*] Department of Microelectronics, Anhui University, Hefei, China, 230039;
[**] Department of Computer, Anhui institute of architecture & industry, Hefei ,China, 230032
[#]zhouyf@ahu.edu.cn

Abstract—DC-DC converters are the kernel parts in the switching power supplies. There are abundant dynamical behaviors in them. In order to discover all these linear and nonlinear phenomena, it is demanded to put forward some appropriate modeling methods. This article discusses the modeling methods, which aimed at some familiar DC-DC converters under the control of voltage and current mode. They include exact simulation model and discrete map. All results are brought with experiments. The study methods and process in this paper are suitable to the other converters operating under the control of voltage or current mode.

Keywords- DC-DC converters; exact simulation mode; discrete map; characteristic multiplier

I. INTRODUCTION

DC-DC converters are the kernel parts in the switching power supplies, which are widely used in the communications, military affairs, computers, atomization and etc [1-2]. There are abundant complex behaviors in them, such as bifurcation and chaos [3-6]. In order to explore all these linear and nonlinear phenomena, it is demanded to put forward some appropriate modeling methods. At the present time, there are always two methods used. One is to build state equations and exact simulation model corresponding to the topology of the system, the other is to construct the discrete map .The former can lead to all the operating behavior in the converters, but it is only a kind of numerical method aim to exact simulation model, and need tremendous computation workload, while the latter can be used to analyze the stability of the system, but it is only suitable to the analysis of first bifurcation. In a word, they supplement each other.

Buck converter under the control of voltage mode and Boost converter under the control of current mode are the important family in the studying of complex behavior in power electronics [7]; they are widely used in reality. This article discusses the methods of modeling and analysis, and all results are brought with experiments. The study methods and process in this paper are suitable to the other converters operating in the control of voltage mode or current mode.

This work was supported by NSFC (60402001) and postgraduate creative project of Anhui university.

II. BUCK CONVERTERS UNDER THE CONTROL OF VOLTAGE MODE

A. Exact State Equations

Denoting the Buck converter under the control of voltage mode as shown in Fig.1(a), supposing it works in the CCM, the operation of the system can be described as follows. The converter is controlled via a simple pulse-with modulation(PWM) scheme, in which the error between the out voltage v_o and reference voltage V_{ref} get across the amplifier, and get a control voltage V_{con}:

$$V_{con} = A(V_o - V_{ref}) \qquad (1)$$

Then V_{con} is compared with a sawtooth signal to generate a pulse-width modulated signal that drives the switch, as shown in Fig.1(b). The sawtooth signal of the PWM generator is defined as

$$V_{ramp}(t) = V_L + (V_U - V_L)(\frac{t}{T} \bmod 1) \qquad (2)$$

where V_L and V_U are the lower and upper voltage limits of V_{ramp}, and T is the switching period. The PWM output is "1" when the control voltage is greater than V_{ramp}, and "0" otherwise.

The presence of the switch G allows two possible switch states. The state equations corresponding to these two switch states can be written as

$$\begin{aligned} \dot{x} = A_{on}x + B_{on}E \qquad \text{for G on} \\ \dot{x} = A_{off}x + B_{off}E \qquad \text{for G off} \end{aligned} \qquad (3)$$

where E is the input voltage, x is the state vector defined as, $x = [v_C, i_L]^T$, and $A's$ and $B's$ are the system matrices, given by

$$A_{on} = A_{off} = \begin{bmatrix} 1/RC & 1/C \\ 1/L & 0 \end{bmatrix}$$

$$B_{on} = \begin{bmatrix} 0 \\ 1/L \end{bmatrix}, B_{off} = \begin{bmatrix} 0 \\ 0 \end{bmatrix}$$

thus we have:

$$\dot{x} = \begin{bmatrix} \dot{v}_C \\ \dot{i}_L \end{bmatrix} = \begin{bmatrix} 1/RC & 1/C \\ 1/L & 0 \end{bmatrix} \begin{bmatrix} v_C \\ i_L \end{bmatrix} + \begin{bmatrix} 0 \\ E/L \end{bmatrix} u \qquad (4)$$

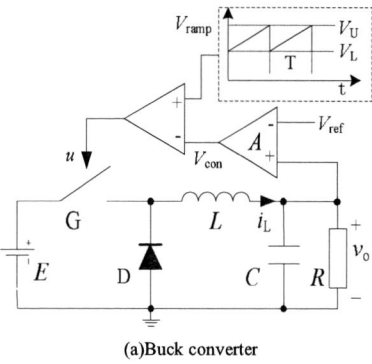

(a)Buck converter

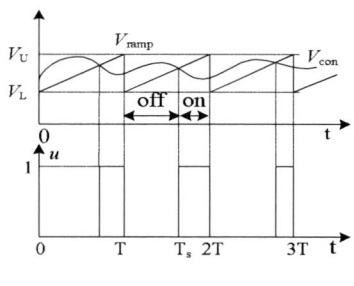

(b)The key waveforms

Figure 1.   Buck converter under the control of voltage

So, based on the equations shown above, the exact simulation model can be constructed.

Let $E = 12 \sim 33$V, $L = 20$mH, $C = 47$ μF, $R = 22$ Ω, $T = 0.4$ ms, $V_{ref} = 11.3$ V, $A = 8.4$, $V_L = 3.8$V, $V_U = 8.2$V, we sample the state variables stroboscopically at the beginning of each switching cycle, so as to build up the Poincaré section, we can obtain the bifurcation diagram with E as parameter as shown in Fig.2. The diagram shows that the converter experiences a typical period-doubling bifurcation and eventually enter chaos when $E \approx 32.3$V. There are 3 pieces of coexisting attracters observed and 2 among them can be seen with clear bifurcation structure [4]. We also capture the phase

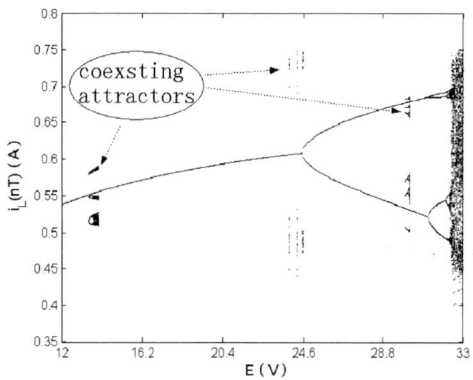

Figure 2.   Bifurcation diagram with E as bifurcation parameter

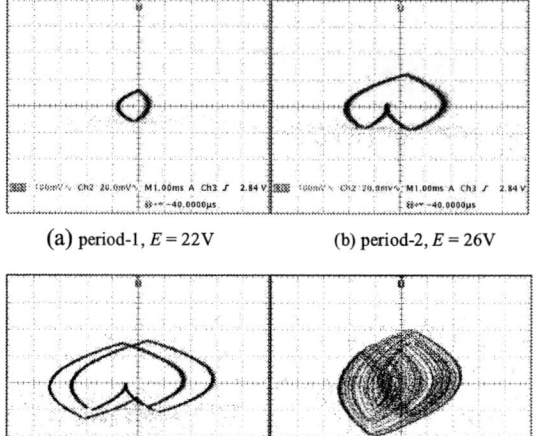

(a) period-1, $E = 22$V   (b) period-2, $E = 26$V

(c) period-4, $E = 30$V   (d) chaos, $E = 33$V

Figure 3.   Phase trajectories for Buck converter

trajectories for some periodic and chaotic orbits as shown in Fig.3.

B.   Discrete Map Model

The exact model afore can be used to achieve all the operating behavior in the converters, but it is only a kind of numerical method, need tremendous computation, and cannot be used to analysis the stability of the converter.

The iterative discrete map model is widely used to analysis the stability of the system [5]. It can be obtained by sampling the system every nT.

Define here

$$x_n = x(nT) = \left[v_C(nT), i_L(nT) \right]^T, n = 1,2,3 \ldots \quad (5)$$

where subscript n denotes the value at the beginning of the nth switching cycle .As shown in Fig.1 (b), under the control of voltage mode, the converter will commute between two different linear time-invariant configurations, i.e. phase 1 and phase 2:

(1) Phase 1: for T ~ Ts, G is switched off, define x_m as the state vector when this phase finishes:

$$x_m = f_{off}(x_n, \bar{d}_n) = N_{off}(\bar{d}_n)x_n + M_{off}(\bar{d}_n)E \quad (6)$$

(2) Phase 2: for Ts ~ 2T, G is switched on, define x_{n+1} as the state vector when this phase finishes:

$$x_{n+1} = f(x_m, \bar{d}_n) = N_{on}(1 - \bar{d}_n)x_m + M_{on}(1 - \bar{d}_n)E \quad (7)$$

Thus we can get the iterative discrete-map:

$$x_{n+1} = f(x_n, \bar{d}_n) = N_{on}(1 - \bar{d}_n)N_{off}(\bar{d}_n)x_n$$

$$+[N_{on}(1\text{-}\bar{d}_n)M_{off}(\bar{d}_n)+M_{on}(1\text{-}\bar{d}_n)]E \quad (8)$$

where

$\bar{d}_n = 1\text{-}d_n$; d_n is the duty cycle; T_s is the switching instant;

$N_{on}(1\text{-}\bar{d}) = e^{A_{on}dT}$; $M_{on}(1\text{-}\bar{d}) = A_{on}^{1}[N_{on}(1\text{-}\bar{d})\text{-}I]B_{on}$;

$N_{off}(\bar{d}) = e^{A_{off}\bar{d}T}$; $M_{off}(\bar{d}) = A_{off}^{1}[N_{off}(\bar{d})\text{-}I]B_{off}$.

To find the defining equation of the duty cycle, we need to find the relationship between the instant of switch and state vector. Now define s as:

$$s(\bar{d}_n) = V_{con}(\bar{d}_n T) - V_{ramp}(\bar{d}_n T) \quad (9)$$

Thus, when $s(\bar{d}_n)<0$, switch is on, otherwise off, i.e., equation $s(\bar{d}_n) = 0$ defines the switching instant.

Following we will study the stability of the converter by the iterative discrete map. The Jacobian plays an important role in the capture of the dynamics in the small neighborhood of the equilibrium point.

The Jacobian of the discrete map (8) evaluated at the equilibrium can be written as follow:

$$\Gamma(x_n,\bar{d}_n) = \frac{\partial x_{n+1}}{\partial x_n} = \frac{\partial f}{\partial x_n} - \frac{\partial f}{\partial \bar{d}_n}\left(\frac{\partial s}{\partial \bar{d}_n}\right)^{-1}\frac{\partial s}{\partial x_n} \quad (10)$$

where

$\frac{\partial f}{\partial x_n} = N_{on}(1-\bar{d}_n)N_{off}(\bar{d}_n)$

$\frac{\partial f}{\partial \bar{d}_n} = [\frac{\partial N_{on}(1-\bar{d}_n)}{\partial \bar{d}_n}N_{off}(\bar{d}_n)$

$+ N_{on}(1-\bar{d}_n)\frac{\partial N_{off}(\bar{d}_n)}{\partial \bar{d}_n}]x_n$

$+ [\frac{\partial N_{on}(1-\bar{d}_n)}{\partial \bar{d}_n}N_{off}(\bar{d}_n)$

$+ N_{on}(1-\bar{d}_n)\frac{\partial M_{off}(\bar{d}_n)}{\partial \bar{d}_n}$

$+ \frac{\partial M_{on}(1-\bar{d}_n)}{\partial \bar{d}_n}]E$

$= [-A_{on}TN_{on}(1-\bar{d}_n)M_{off}(\bar{d}_n)$

$+ N_{on}(1-\bar{d}_n)A_{off}TN_{off}(\bar{d}_n)]x_n$

$= [-A_{on}TN_{on}(1-\bar{d}_n)M_{off}(\bar{d}_n)$

$+ N_{on}(1-\bar{d}_n)N_{off}(\bar{d}_n)B_{off}T$

$- N_{on}(1-\bar{d}_n)B_{on}T]E$

$\frac{\partial s}{\partial \bar{d}_n} = [A \quad 0][A_{off}TN_{off}(\bar{d}_n)x_n + TN_{off}(\bar{d}_n)B_{off}E]$

$- (V_U - V_L)T$

$= [A \quad 0]N_{off}(\bar{d}_n)(A_{off}x_n + B_{on}E)T - (V_U - V_L)T$

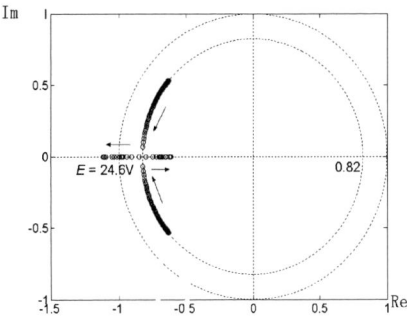

Figure 4.   Loci of characteristic multipliers as E varies

$$\frac{\partial s}{\partial x_n} = [A \quad 0]N_{off}(\bar{d}_n)$$

To find the characteristic multiplier, we solve the following equation:

$$\det[\lambda I - \Gamma(x_n,\bar{d}_n)]_{x_n=x_q,\bar{d}_n=\bar{d}_Q} = 0 \quad (11)$$

where x_Q and $\bar{d}_Q$ are the equilibrium values. The characteristic multipliers can now be calculated, as shown in Fig.4. With input voltage E increasing two characteristic multipliers get close along a circle of radius less than 1(0.82) which indicate stable operation of the converter. Further increase E, two characteristic multipliers touch the real axis, and then move apart. The converter will maintain stable operation till $E = 24.6V$, when one characteristic multiplier leave the unit circle and the converter loses stability with experiencing period-doubling at this point. This process of losing stability agrees with Fig.2 exactly.

III.   BOOST CONVERTER UNDER THE CONTROL OF CURRENT MODE

A.   Exact State Equations

Denoting the Boost converter under the control of current mode [6] as shown in Fig.5(a), when the converters is operating in continuous condition mode(CCM), the presence of the switch G allows a total of two possible switch states. The state equations corresponding to these two switch states can be written as:

$$\begin{aligned}\dot{x} &= A_{on}x + B_{on}E \quad \text{for G on} \\ \dot{x} &= A_{off}x + B_{off}E \quad \text{for G off}\end{aligned} \quad (12)$$

where E is the input voltage, x is the state vector defined as, $x = [v_C, i_L]^T$, and $A's$ and $B's$ are the system matrices given by

1285

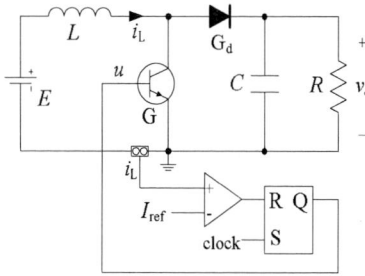

(a)Boost converter

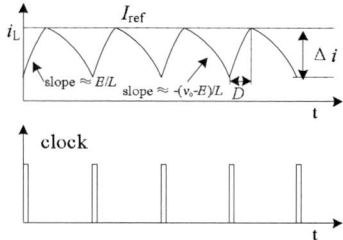

(b)The key waveforms

Figure 5.   Boost converter under the control of current

$$A_{on} = \begin{bmatrix} 0 & 0 \\ 0 & 1/RC \end{bmatrix}, A_{off} = \begin{bmatrix} 0 & -1/L \\ 1/C & -1/RC \end{bmatrix},$$

$$B_{on} = B_{off} = \begin{bmatrix} 1/L \\ 0 \end{bmatrix} \qquad (13)$$

thus we have:

$$\dot{x} = \begin{bmatrix} \dot{i}_L \\ \dot{v}_o \end{bmatrix} = \begin{bmatrix} 0 & -1/L \\ 1/C & -1/RC \end{bmatrix}\begin{bmatrix} i_L \\ v_o \end{bmatrix} + \begin{bmatrix} \dfrac{v_o}{L} \\ -\dfrac{i_L}{C} \end{bmatrix}u + \begin{bmatrix} 1/L \\ 0 \end{bmatrix}E \quad (14)$$

So we can construct the exact simulation model based on the equation (14).

Let $E = 10$V, $L = 1$mH, $C = 12$ μF, $R = 20$ Ω, $T = 0.1$ms, $I_{ref} = 0.6\sim5.5$A, $f_s = 10\,kHZ$. We sample the circuit state variables on every clock pulse to build up the

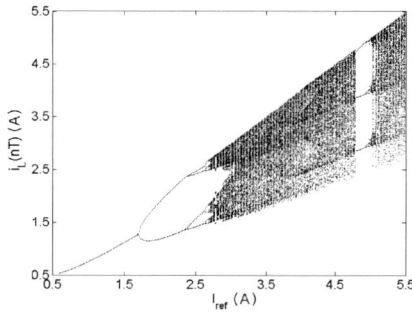

Figure 6.   Bifurcation diagram with i_L as bifurcation parameter

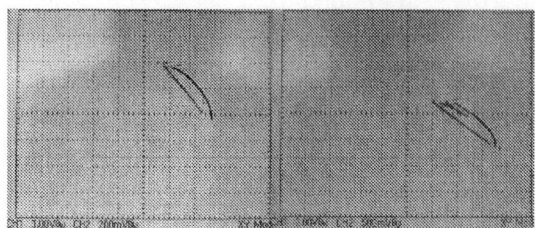

(a) period-1, $I_{ref} = 1.2$A   (b) period-2, $I_{ref} = 2.2$A

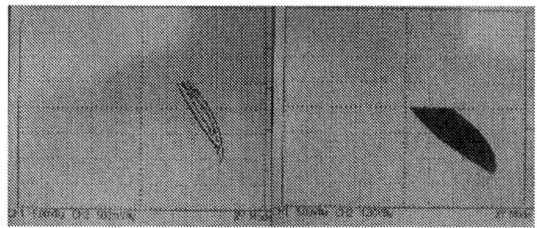

(c) period-4, $I_{ref} = 2.5$A   (d) chaos, $I_{ref} = 4$A

Figure 7.   Phase trajectories for Boost converter

Poincaré section, thus obtain the bifurcation diagram with I_{ref} as bifurcation parameter as shown in Fig.6. The bifurcation diagram shows the converters experiences the typical period-doubling bifurcations with I_{ref} increasing, and eventually enter chaos when $I_{ref} = 2.7$A.We also capture the phase trajectories for some periodic and chaotic orbits as shown in Fig.7.

B.   *Discrete Map Model*

In order to study the dynamical behavior such as stability of Boost converter under the control of the current mode, we are demanded to construct the interactive discrete map. From Fig.5 (b), we get

$$\frac{I_{ref} - i_n}{DT} = \frac{E}{L} , \quad \frac{I_{ref} - i_{n+1}}{(1-D)T} = \frac{v_o - E}{L} \qquad (15)$$

where D is duty cycle, thus

$$i_{n+1} = \left(1 - \frac{v_o}{E}\right)i_n + \frac{v_o I_{ref}}{E} - \frac{(v_o - E)T}{L} \qquad (16)$$

We only focus on the small neighborhood of equilibrium point. The equation above can be expressed with a small perturbation as follows:

$$\delta i_{n+1} = \left(\frac{-D}{1-D}\right)\delta i_n + O(\delta i_n^2) \qquad (17)$$

henceforces, the characteristic multiplier is

$$J = \frac{-D}{1-D} \qquad (18)$$

When $J \in (-1,1)$, the converter operates in regular period-1; and when J exceeds -1, period-doubling bifurcation can be observed and the converter is settled in period-2. So the stable principle of the Boost converter under the control of current mode will be $J < -1$, or $D < 0.5$ correspondingly.

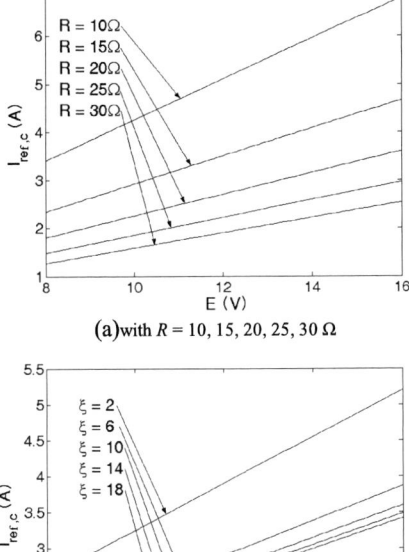

(a) with R = 10, 15, 20, 25, 30 Ω

(b) with ξ = 2, 6, 10, 14, 18

Figure 8. $I_{\text{ref,c}}$ calculated from (22) verses E

We can also drive the follow from the power-balance equation:

$$\left(I_{ref} - \frac{\Delta i}{2}\right)E = \frac{v^2}{L} \qquad (19)$$

Combining with $\dfrac{v_o}{E} = \dfrac{1}{1-D}$, we get

$$\left(I_{ref} - \frac{\Delta i}{2}\right)E = \frac{E^2}{(1-D)^2 R} \qquad (20)$$

where Δi is the scope of the fluctuate in i_L, $\Delta i = \dfrac{DTE}{L}$ .Thus ,As follows we get the stable principle for I_{ref} of   Boost converter under the control of current.

$$I_{\text{ref,c}} = \frac{E}{R}\left[\frac{DR}{2\xi} + \frac{1}{(1-D)^2}\right]_{D=0.5} = \frac{E}{R}\left[\frac{R}{4\xi}+4\right] \qquad (21)$$

where $\xi = L/T$, i.e., when $I_{ref} < I_{\text{ref,c}}$ the converter is stablized in period-1.

Circuit parameters are critical to the value of $I_{\text{ref,c}}$, Fig.8 shows these influence which are concluded as follows:

(1) The critical value of I_{ref} assuming stability, i.e. $I_{\text{ref,c}}$, increases with the increasing of E.

(2) The critical value $I_{\text{ref,c}}$ decreases with the increasing of R.

(3) The critical value $I_{\text{ref,c}}$ decreases with the increasing of ξ .

All these conclusions exactly agree with the influences of circuit parameters which are made by a large amount of simulations. Thus it is very useful for the practical design of power converters.

IV.   CONCLUSION

There are two main methods to study the complex behaviors in power converters, i.e., exact simulation model and iterative discrete map. They are supplementing each other. The former is based on the state equation and can lead to all the operating behaviors in the converters, but it is only a kind of numerical method aim to exact circuit model, and need tremendous computation. While the latter can be used to analyze the stability of the system, but it is only suitable to the first bifurcation of losing stability. This article gives an example of the Buck converter under the control of voltage mode and the Boost converter under the current mode, and discusses the modeling methods, including exact simulation model and iterative discrete map. All results are brought with experiments. The study methods and process in this paper are suitable to the other converters simulations and operating under the control of voltage mode or current mode, based on which some useful results for practical design are concluded.

REFERENCES

[1] A El Aroudi, R. Leyva. Quasi-periodic route to chaos in a PWM voltage-controlled DC-DC boost coverter. *IEEE Trans. Circuits and Systems-I*, vol.48, no.8, pp. 967-978, 2001.

[2] C. K. Tse, M Di Bernardo. Complex behavior in switching power converter. *Proceedings of IEEE*, vol.90, no.5, pp. 768-781, 2002.

[3] J. H. B. Deane, D. C. Hamill, Instability, subharmonics, and chaos in power electronic systems, *IEEE Trans. Power Electron.*, vol.5, no.3, pp.260-268, 1990.

[4] Yu-fei Zhou, Jun-ning Chen. Simulation and Experimental Studies on Coexisting Attractors in DC-DC Switching Converter. *Proceedings of the CSEE*, vol.21, no.21, pp.96-101, 2005.

[5] M Di Bernardo, F. Vasca. Discrete-time Maps for the Analysis of Bifurcations and Chaos in Dc/Dc Converters. *IEEE Trans. Circuits and Systems-I*, vol.47, no.2, pp.130-143, 2000.

[6] C. K. Tse, et al. Control of bifurcation in DC/DC converters: an alternative viewpoint of ramp compensation. *Int. Conference on Industrial Electronics, Control and Instrumentation (IECON'2000)*, Nagoya, Japan, 2000. 2413-2418.

[7] Yufei Zhou, C. K. Tse, et al.. An Improved Resonant Paramentric Perturbation for Controlling and Anti-Controlling Chaos in DC-DC Converters. *Int. Symp. on Nonlinear Theory and Its Applications(NOLTA '02)*, Xian, China, Oct. 2002: 151-154.

2006 5th International Power Electronics and Motion Control Conference

A Novel Topology Family of Single-stage Parallel Mode
Uninterruptible AC/DC Converter with PFC

Xuejun Ma, Hongxia Wu, Congsheng Huang, Xuwen Huang

Department of Electrical Engineering, Huangshi Institute of Technology Huangshi,435003,P.R.China

Email:hsmxj@tom.com

Abstract—**A circuit topologies family of single-stage parallel uninterruptible AC/DC converter with high power factor are presented. The topology includes four sections, i.e. full bridge rectifier, parallel converter with power factor correction(PFC), uninterruptible power supply and fast output voltage regulation. The converter has lots of advantages such as: single stage power conversion, uninterruptible power supply, high power factor, electrical isolation between battery and AC input, fast regulation speed of the output voltage. The operational modes, control principle, and design method of key parameters are proposed. A prototype of 1kW is designed. The performance of the proposed scheme is verified through the experimntal result.**

Keywords-AC/DC converter; UPS; PFC; Single-stage parallel mode;

I. INTRODUCTION

There are two obvious defects in traditional DC switching power supply .One of them is that DC power supply is interrupted when AC input is interrupted. Usually, an UPS is demanded to ensure the reliability. Obviously ,the cost will increase .The other is that EMI is serious without PFC. In order to decrease EMI a PFC circuit is usually applied.

The uninterruptible DC switch power supply circuit referred in [2]~[4] has some disadvantages such as complicated circuit topology, serious harmonic pollution, low output power(<50W), no electric isolation between battery and AC electrical grid.

In this paper, a single –stage parallel mode uninterruptible AC/DC converter with high power factor is presented. There are high power factor(PF), electrical isolation between battery and AC electrical grid. And fast output voltage regulation can be realized.

II. THE STRUCTURE AND TOPOLOGY FAMILY

A. Circuit topology

Fig.1 shows the structure of single –stage parallel mode uninterruptible AC/DC converter with high power factor .It consists of three modules: main power unit DC converter (used for power factor adjusting),

auxiliary unit DC converter and charger. Power DC converter works on DCM mode used for adjusting power factor. C_1 paralleled with load transferred its whole energy to load. Auxiliary DC unit converter works on CCM or DCM mode, This unit can realize fast regulation of output voltage. Battery charging module is Buck converter, working on CCM mode, Although there are three units in the converter circuit, 50% output power is directly transferred from main power unit to load, and the other power is transferred from auxiliary unit and charge unit, so it is still similar to single stage circuit structure.

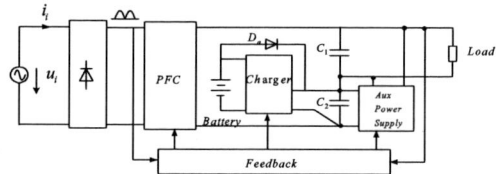

Fig.1 The structure of single –stage parallel mode uninterruptible AC/DC converter with high power factor

B. Circuit topology family

In Fig.2,main power unit is flyback topology, auxiliary unit type can be as following topologies: Buck-boost, Boost-Buck, forward ,push-pull, forward half-bridge, Full-bridge, Boost, flyback etc.

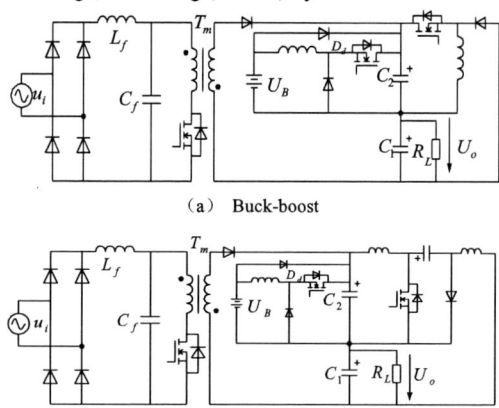

(a) Buck-boost

(b) Boost-buck

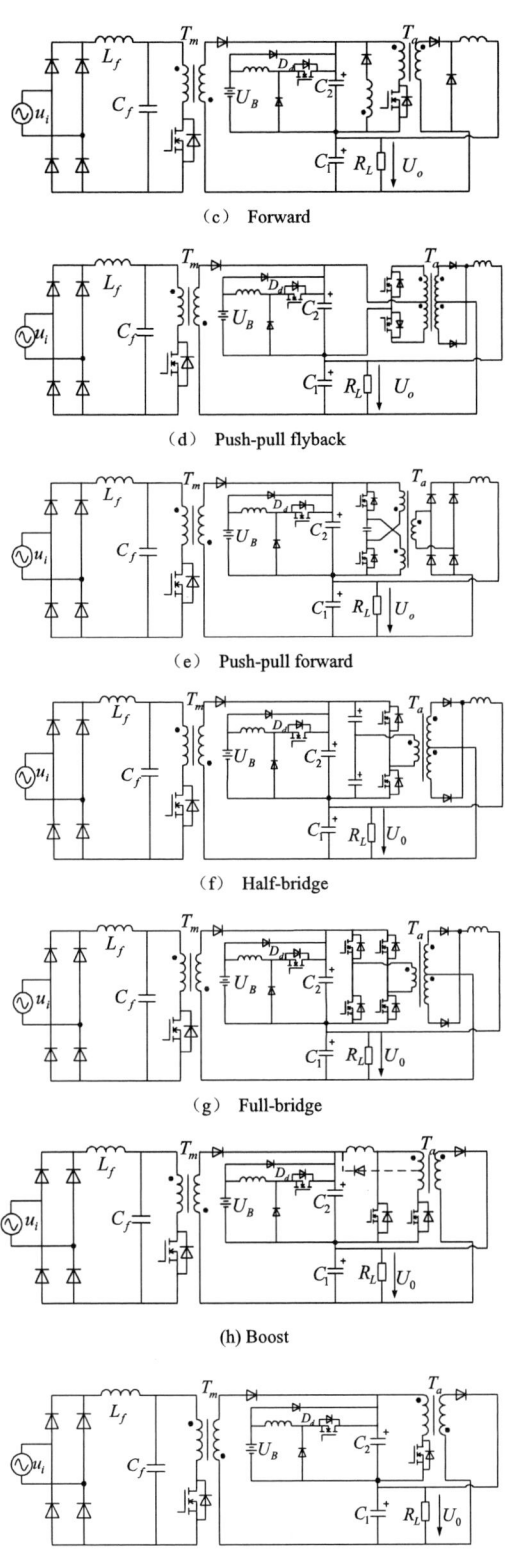

（c）Forward

（d）Push-pull flyback

（e）Push-pull forward

（f）Half-bridge

（g）Full-bridge

（h）Boost

（i）Flyback

Fig.2 Circuit topologies family of single –stage parallel mode uninterruptible AC/DC converter with high power factor

III. OPERATION PRINCIPLE AND CONTROL METHOD

In order to analyze the operation principle,Fig.2(a) is considered as a sample.

A .Three operation mode

There are three operation modes in the presented topology.

Normal mode: In this mode AC electrical grid transfers power to load. The equivalent circuit when the converter working on normal mode is shown on Fig.3(a).The control signal of S_1,S_2 is same, when S_1,S_2 is turned on , D_1,D_2 are turned off, L_p,L_2 store energy. When S_1,S_2 is turned off,D_1,D_2 is turned off, the energy in L_p is transferred to C_1,C_2. C_1,C_2 is charged, the energy in L_1 is transferred to C_1,R_L.

Backup mode: Battery is charged from AC electrical grid and battery transfer power to load. The equivalent circuit when the converter working on backup mode is shown as Fig.3(b). When the AC input is power-off or excessively low , the voltage of C2 is lower than the voltage of battery 0.7V, then Dd is turned on,battery supplies power to load as auxiliary power unit.

Charge mode: Its' equivalent circuit is shown as Fig.3(c). Under this mode, as the power supply of Buck converter C2 supplied power to battery and the battery is charged.

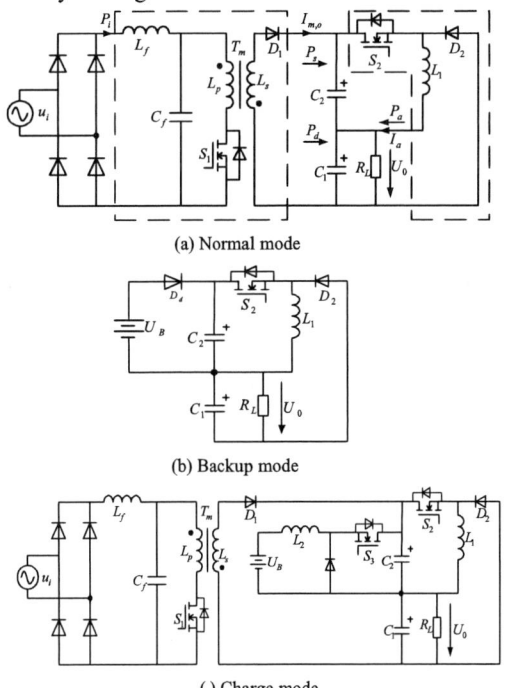

(a) Normal mode

(b) Backup mode

( ) Charge mode

Fig.3 Three operation modes

Under stable state, the waveforms of input voltage and current of main power unit is shown as Fig.4 . $P_{m,o}= P_i$,and its frequency is twice as that of AC input voltage.

Output voltage of the main power unit is the sum

of the voltage of C_1 and C_2. If the power of the main power is $P_{m,o}$, P_d is the power directly transferred to load, P_s is the power stored in C_2, then $P_{m,o}=P_d+P_s$. If the power of auxiliary power unit is P_a, then $P_a=P_o-P_d$. Fig.4 shows the power waveform of each unit of the converter.

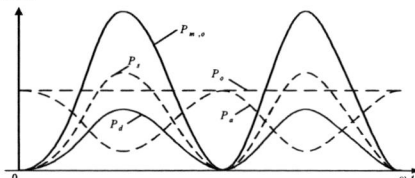

Fig.4 power waveforms of converter

B. Control method

In order to control the output voltage, a PI compensator is applied. The control block diagram is shown as Fig.5. Under this method, PFC is realized in normal mode and charging mode, and real online UPS is realized in backup mode.

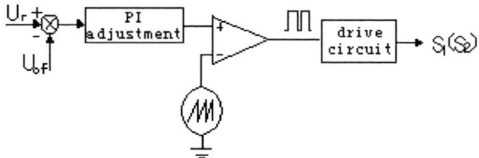

Fig.5 The control block diagram

IV. KEY PARAMETERS DESIGN

A. C_1, C_2 extremity voltage ratio

The power of main power unit $P_{m,o\ I}$ can be expressed as:

$$P_d(\omega t) = \frac{U_{C1}}{U_{C1}+U_{C2}} \cdot P_{m,o}(\omega t) \tag{1}$$

$$P_s(\omega t) = \frac{U_{C2}}{U_{C1}+U_{C2}} \cdot P_{m,o}(\omega t) \tag{2}$$

In order to increase the efficiency, P_d should be as high as possible. Supposed the transfer efficiency is 100%, for the out power is equal is equal to the average output power of the main power unit, then the maximum allowed value of P_d is 50% of P_i, then:

$$U_{C2} \gg U_{C1} \tag{3}$$

When $U_{c2}=U_{c1}$, $P_d=P_s=P_{m,o}/2$, converter may get a high efficiency, and the volume of auxiliary unit can be decreased.

Supposed $\alpha = U_{c1}/U_{c2}$, the efficiency of main power unit is η_m, the efficiency of auxiliary is η_a then the whole transfer efficiency:

$$\eta = P_o/P_i = (P_d+P_a)\eta_m/(P_d+P_s) = \eta_m\frac{\alpha+\eta_a}{\alpha+1} \geq \eta_m\eta_a \tag{4}$$

When $\alpha = 0$, $\eta = \eta_m\eta_a$, the converter is equal to traditional PFC; when $\alpha = 1$, $\eta = \eta_m（1+\eta_a）/2$, the converter has the best transfer efficiency. Usually α

is designed to be equal to 1.

B. Critical continuous inductance

In order to make main power unit work on DCM mode,

$$L_p \leq \frac{R_L T_s (N_1/N_2)^2}{4\left(1+\dfrac{U_o N_1/N_2}{\sqrt{2}U_{i,\min}}\right)^2} \tag{5}$$

C. Power switch voltage and currency stress

Normal mode: power device voltage and current stress.

Currency stress of S_1

$$I_{S1pk} = \sqrt{2}U_{i,\min}D_{\max}T_S/L \tag{6}$$

Voltage stress of S_1, S_2

$$U_{DS1}=(U_{C1}+U_{C2})N_1/N_2+\sqrt{2}\ U_i \tag{7}$$

$$U_{DS2}=U_{C1}+U_{C2} \tag{8}$$

Voltage stress of D_1, D_2

$$U_{D1}=(U_{C1}+U_{C2})+\sqrt{2}\ U_i N_2/N_1 \tag{9}$$

$$U_{D2}= U_{C1}+U_{C2} \tag{10}$$

Backup mode: power device voltage and current stress
Voltage stress of S_2

$$U_{DS2} = U_B + U_o \tag{11}$$

Currency stress of S_2

$$I_{S2pk} = I_{Lo,pk} = I_1' + \frac{1}{2}\Delta I = \frac{P_o}{DU_B\eta_a} + \frac{U_B DT_s}{2L_o} \tag{12}$$

Voltage stress of D_2

$$U_{D2} = U_B + U_o \tag{13}$$

Current stress of D_2

$$I_{D2} = I_{LO,pk} \tag{14}$$

Choose normal mode voltage and currenct stress for S_1, D_1; choose normal mode voltage stress and backup mode current stress for S_2, D_2.

V. EXPERIMENTAL RESULTS

In order to verify the principle of the converter, a prototype is made with following parameters: AC input voltage $U_i=220 \pm 20\%$V 50Hz, DC Output voltage $U_0=U_{C1}=48$V, $U_B=12$V $\times$ 4, $P_o=120$W, swiching frequency $f_s=50$kHz, $L_p/L_s=405\mu$H/405μH, $L_1=680\mu$H, $C_1=250\mu$F, $C_2=100\mu$F. The key experimental waveforms are shown as Fig.6 on normal mode, with rated voltage and load.

(1)No obvious phase shifting between input voltage and input current waveforms, current waveform is good, PF reaches 0.98;

1290

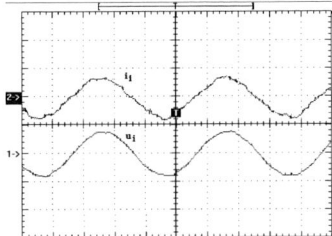

(a) Waveforms of input voltage and current
（1A/div , 400V/div , 5ms/div）

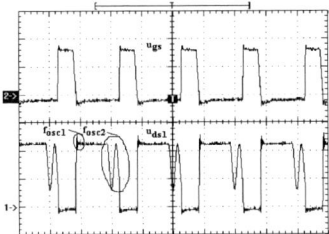

(b) Waveforms of ugs and   uds1
（10V/div , 200V/div , 10 μ s/div）

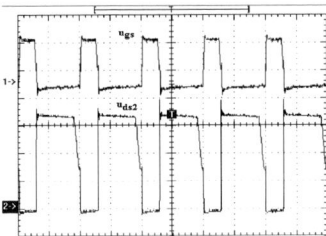

(c) Waveforms of ugs and uds2
（10V/div , 50V/div , 10 μ s/div）

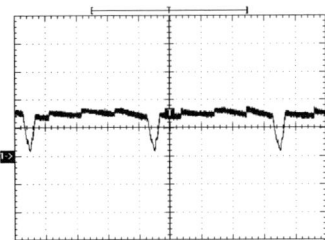

(d) Waveform of uC of clamping capacity
（100V/div ,   2.5ms/div）

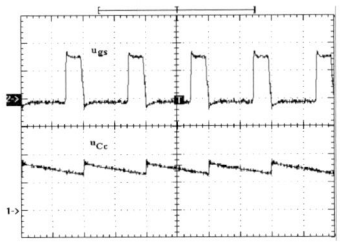

(e) Waveforms of uC and ugs
(100V/div , 10V/div , 10 μ s/div)

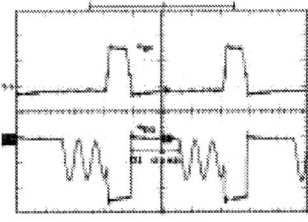

(f) Waveformof ugs and uD1 with high input voltage
（10V/div , 200V/div , 5 μ s/div）

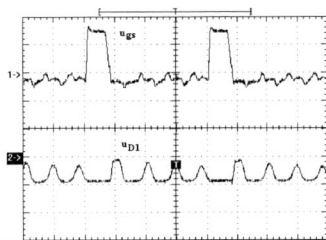

(g) Waveformof ugs and uD1 with low input voltage
（10V/div , 100V/div , 5 μ s/div）

Fig.6   Experimental waveforms under normal mode

(2)RCD clamping circuit can control peak leakage power voltage, f_{osc1} is the resonant frequency of L_{lk} and MOSFET output capacity. After S_1 is turned off , f_{osc2} is the resonant frequency of both storage transfer power unit induactance L_p and   MOSFET output capacity after D_1 close.

(3) The frequency of Voltage of the energy storage transformer primary winding is twice as the AC input voltage frequency, as well as the leakage inductance and the voltage of clamping capacity C_2.

(4)When voltage is high, D_1 is turned on and work on DCM mode after S_1 is turned off,.When voltage is low, L_p and leakage inductance of storge transfer L_{lk} can not be charging Uc enough, and D_1 can not   be turned on in a one period.

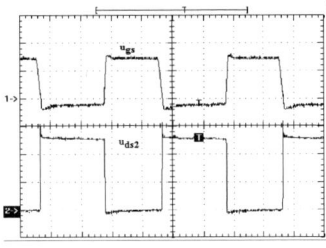

Waveforms of ugs and uds2
（10V/div , 40V/div , 5 μ s/div）

Fig.7 Backup mode experimental waveform

Fig.7 shows the experimental waveform on backup mode. Meanwhile, auxiliary converter work on CCM mode.

1291

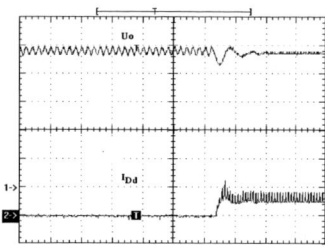

(a) From normal mode to backup mode.

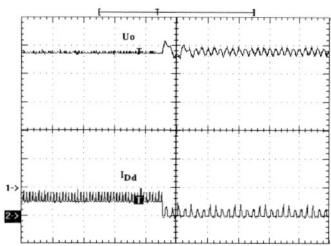

(b) From backup mode to normal mode.

put voltage u_i & currency of D_d

Fig.8 Waveforms of Uo and I_{Dd} under two mode

Fig.8 shows Waveforms of Uo and I_{Dd} under two mode the waveform of. Obviously , the converter can automatically changing its mode in very short time.

Fig.9 shows the efficiency curve. The efficiency of the converter is high. And the efficiency is decreasing while input voltage is increasing on normal mode,and the efficiency on normal mode is higher than on backup mode.

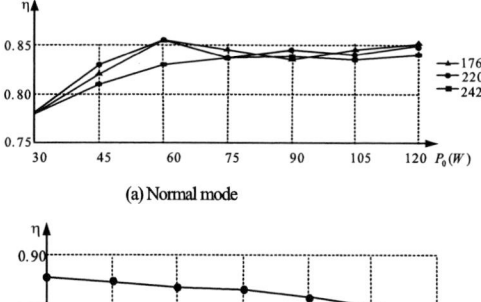

(a) Normal mode

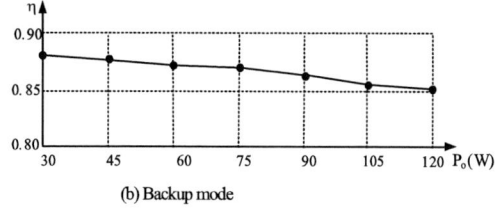

(b) Backup mode

Fig.9 The efficiency curve of the converter

VI. CONCLUSION

Experimental results show that presented converter have three operation modes: normal mode, backup mode, charge mode.The shift between the modes and voltage regulation is very fast. Besides above ,there are high power density and PF with this converter and good dynamic characteristic. It will have a good prospect in the field of computer, communication, etc.

REFERENCES

[1] Chen Daolian , "DC-AC inverting technology and its application" ,Beijing，China Machine Press，2003.11.

[2] Kwok-wai Ma, Yim-shu Lee, "A novel uninterruptible dc-dc converter for UPS Applications",.IEEE Trans.on IA , vol. 28(4),1992，pp. 808-815.

[3] Kwok-wai Ma, Yim-shu Lee, "An integrated flyback converter for DC uninterruptible power supply",IEEE Trans.on PE，vol. 11(2), 1996, pp.318-327.

[4] E. Rodriguez, O. Garcia, J.A. Cobos, et al. , "A single-stage rectifier with PFC and fast regulation of the output voltage", PESC,1998, pp.1642～1648

2006 5th International Power Electronics and Motion Control Conference

Analysis and Design of an Automatic-Current-Sharing Control Based on Average-Current Mode for Parallel Boost Converters

Wenxun Xiao, Bo Zhang, Dongyuan Qiu

Electrical power college, South China University of Technology (SCUT), Guangzhou 510640, Guangdong province, China.

E-mail: epxwx@yahoo.com.cn

Abstract—**This paper presents an automatic-current-sharing control method based on average-current mode for parallel boost converters. With the proposed method, the second-order system of the external loop of boost converter is reduced to a first-order in low frequency, which makes the design of the control circuit easy and improves the dynamic response of parallel boost converters. This paper presents detail analysis and improved design guidelines for the proposed control system. The multi-module parallel number is limitless in theory. Two 40V/120V/4.2A Boost converters have been designed for parallel experiment, verifying the correctness of the theory analysis.**

Keywords-Average-current mode; Automatic-current-sharing technique; Three-loop control

I. INTRODUCTION

The current-sharing technology is a research focus in DC-DC converter field, for the demand of large power application and reliability. Reference [1] presented an automatic-current-sharing (ACS) control method based on voltage-mode control, in which the voltage-loop gain bandwidth of boost converter is limited by the right-half-plane (RHP) zero, which causes poor dynamic response [2]. The control technologies reported in [3-5] were based on current-mode control. The voltage-loop gain bandwidth can be set wider to improve the dynamic response of the cell module. But all modules use a common voltage rectifier, which reduces the reliability of the system [3-4]. And the current-sharing-loop gain bandwidth is limited by the voltage-loop gain bandwidth, so that the parallel power system can't quickly respond to the load change [5]. Reference [6-7] improved above problems, with the current-sharing loop set inner the voltage loop to improve the dynamic current-sharing response of the parallel system.

With the ACS control method based on average-current mode (ACM) proposed in [7], this paper further realizes that the current loop compensates the external loop, and

the second-order system of the external loop is reduced to a first-order system in low frequency, so that the control circuit is easy to be designed for improving the static and dynamic properties of parallel system.

The theory of the current loop compensating the external loop is analyzed, the improved design guidelines of the ACS control based on ACM are presented, and parallel boost converters based on this control method is researched for practical application. Two 40V/120V/4.2A boost converters have been designed for parallel experiment, verifying the control properties of good disturbance rejection, redundancy, static and dynamic current share.

II. AUTOMATIC-CURRENT-SHARING CONTROL BASED ON AVERAGE-CURRENT MODE

A. Analysis of the presented control

The ACS control based on ACM is shown in Fig. 1 [7], it is actually a three-loop control system: the current loop is inner loop which includes the current sensor $H_e(s)$ and the current error amplifier A_1; the current-sharing loop is middle loop which includes the share resistor R_S, the share bus and the current-sharing amplifier A_2; the voltage loop is outer loop which includes the voltage sensor β and the voltage error amplifier A_3.

The small signal model of control system for one of the parallel modules is shown in Fig. 2. Where, $F_V(s)$ is duty cycle disturbance to output voltage transfer function; $F_I(s)$ is duty cycle disturbance to inductor current transfer function; β is voltage feedback coefficient; $H_e(s)$ is current to voltage signal transfer function; $G_v(s)$ is transfer function of voltage amplifier; $G_s(s)$ is transfer function of current-sharing amplifier; $G_c(s)$, $G_{cs}(s)$ and $G_{cv}(s)$ are current amplifier transfer functions for current loop, current-sharing loop and voltage loop respectively; F_m is PWM modulator transfer function.

According to Fig. 1 and Fig. 2, for n modules in parallel, if the share resistors R_{S1} --- R_{Sn} in each module are equal, the current-sharing error between current signal in sub-module (e.g. module n) and sharing-bus signal is derived by [7]

The Project Sponsored by Natural Science Foundation of China (60474066) and Natural Science Foundation of Guangdong (05103540).

1-4244-0448-7/06/$25.00 ©2006 IEEE

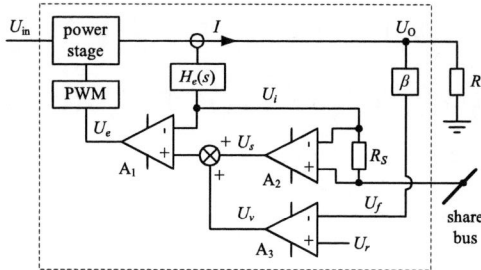

Fig. 1. ACS control based on ACM

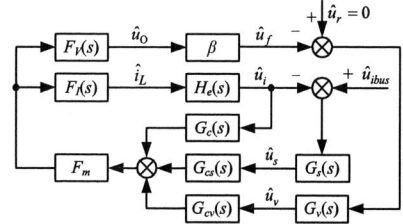

Fig. 2. The small signal model of control system

$$\hat{u}_{sn} = G_s(\hat{u}_{ibus} - H_e(s)\hat{i}_{Ln}) = -\frac{n-1}{n}G_s H_e(s)\hat{i}_{Ln} +$$

$$G_s \frac{H_e(s)}{n}(\hat{i}_{L1} + \hat{i}_{L2} + \cdots + \hat{i}_{Ln-1}), \qquad (1)$$

where $\hat{i}_{L1} \sim \hat{i}_{Ln}$ are inductors current. $\hat{u}_{ibus}$ is the voltage signal of the share bus. Signal in the share bus reflects the average of inductor current in each module.

The second term on the right-hand side of (1) represents the interactions among the modules. For the module n, it can be looked as an external disturbance, if the external disturbance could be neglected, the current-sharing error can be rewritten as

$$\hat{u}_{sn} \cong -\frac{n-1}{n}G_s H_e(s)\hat{i}_{Ln}. \qquad (2)$$

The current-sharing error is sent to the positive port of the amplifier A_1. The sum of the current-sharing error signal and the voltage error signal makes the current reference of the current loop. If one module's output current increases, the current-sharing error will decrease, then the current reference of the current loop will decrease, which leads this module's output current to decrease, thereby the parallel power system will get to current share. If all the output currents of each module are equal, the current sense signal is equal to the signal in the share bus, and no current flows through the current-sharing resistor R_S, the current-sharing error is zero. In this case, the control circuit of each module can be looked as two-loop average-current-mode control.

According to Fig. 2, we obtain the open-loop gains of the current-sharing loop T_S, the current loop T_C, and the voltage loop T_V:

$$T_S = \frac{n-1}{n}F_m F_I(s)H_e(s)G_s(s)G_{cs}(s). \qquad (3)$$

$$T_C = F_m F_I(s)H_e(s)G_c(s). \qquad (4)$$

$$T_V = F_m F_V \beta G_v(s)G_{cv}(s). \qquad (5)$$

Then the full-loop gain T_1 and external-loop gain T_2 can be derived by

$$T_1 = T_C + T_V + T_S. \qquad (6)$$

$$T_2 = \frac{T_V}{1 + T_C + T_S}. \qquad (7)$$

B. Improvement of the design guidelines

The proposed control method is based on ACM. According to ACM control theory, the current loop is in the voltage loop. In order to decrease the influence to the voltage loop accused by the current-loop phase shift, the crossover frequency of the current loop is usually designed much higher than that of the voltage loop. In low frequency the voltage loop makes primary contribution. In high frequency the current loop makes primary contribution. To avoid the dip in the overall loop gains that can cause the system to be unstable, don't make the phases of the two loops in the opposite directions when the two loops cross over [8]. Under the system stable condition, the current-loop crossover frequency should be set as high as possible to improve dynamic response [9].

The main object of the current-sharing control is keeping output dc currents of every module even, even if there is a little error in it is acceptable. So, for simplifying the control system design, the current-sharing amplifier is designed as a proportional amplifier. Under the system stable, the amplifying multiple of the proportion loop is designed as large as possible.

According to multi-loop theory, in order to avoid the unstable caused by the interaction of loops, the crossover frequency of the three loops should be different. For the system stable, the phase and gain margins of the full-loop gain T_1 and external-loop gain T_2 should be larger than 45° and 6dB respectively. In order to improve the anti-input and anti-output disturbance capacity of parallel system, the closed-loop audio-susceptibility and output impedance gains should be as low as possible.

III. ANALYSIS AND DESIGN OF PARALLEL BOOST CONVERTERS BASED ON THE PROPOSED CONTROL

For demonstrating the properties of the ACS control based on ACM, this paper presents parallel boost converters based on this control method for example.

The close-loop small signal model of parallel boost converters operating in continuous conduction mode (CCM) is developed based on Fig. 2 and PWM three-terminal switch model [10], and shown in Fig. 3. The voltage transfer function $F_V(s)$, current transfer function $F_I(s)$, open-loop output impedance $Z_{po}(s)$ and open-loop audio-susceptibility $A_p(s)$ are derived by

$$F_I(s) = \frac{\hat{i}_L}{\hat{d}} = \frac{U_O(s + \frac{2}{RC})}{LS_1}. \tag{8}$$

$$F_V(s) = \frac{\hat{u}_O}{\hat{d}} = \frac{U_O R_C[\frac{R(1-D)^2}{L} - s](\frac{1}{CR_C} + s)}{R(1-D)S_1}. \tag{9}$$

$$Z_{po}(s) = \frac{\hat{u}_O}{\hat{i}_O}\bigg|_{\hat{u}_{in}=\hat{u}_v=0} = -\frac{R_C(s + \frac{1}{R_C C})[s + \frac{U_O(T_C+T_S)}{LF_I(s)}]}{S_1 + s\frac{U_O(T_C+T_S)}{LF_I(s)} + 2\frac{U_O(T_C+T_S)}{LCRF_I(s)}}, \tag{10}$$

where

$$S_1 = s^2 + s[\frac{1}{CR} + \frac{R_C(1-D)^2}{L}] + \frac{(1-D)^2}{LC}.$$

$$A_p(s) = \frac{\hat{u}_O}{\hat{u}_{in}}\bigg|_{\hat{i}_O=\hat{u}_v=0} = \frac{U_O(T_C+T_S)}{F_I(s)RC(1-D)(s+\frac{2}{RC})[s+\frac{U_O(T_C+T_S)}{LF_I(s)}]} \tag{11}$$

The external-loop gain can be rewritten as

$$T_2 = \frac{F_m\beta G_v G_{cv}(s)U_O R_C(\frac{1}{CR_C} + s)[\frac{R(1-D)^2}{L} - s]}{R(1-D)(s+\frac{2}{RC})(s+\omega_P)}, \tag{12}$$

where $\omega_P = \frac{U_O F_m H_e(s)(G_c(s) + G_{cs}(s)G_s(s))}{L}$ is high frequency pole. T_2 is at least first-order system up to RHP zero, which is caused by the compensation of the current loop.

And then, the close-loop output impedance $Z_{poc}(s)$ and close-loop audio-susceptibility $A_{pc}(s)$ are derived by

$$Z_{poc}(s) = \frac{\hat{u}_O}{\hat{i}_O}\bigg|_{\hat{u}_{in}=0,\hat{u}_v\neq0} = \frac{Z_{po}(s)}{1+T_2}. \tag{13}$$

$$A_{pc}(s) = \frac{\hat{u}_O}{\hat{u}_{in}}\bigg|_{\hat{i}_O=0,\hat{u}_v\neq0} = \frac{A_p(s)}{1+T_2}. \tag{14}$$

The topologies and parameters of the compensated nets of the current loop, voltage loop and current-sharing loop have been designed, shown in Fig. 4.

The transfer function of the compensated net of the current loop is

$$G_c(s) = \frac{K_C}{s+\omega_{CP}},$$

where $K_C = \frac{1}{R_{C1}C_C}$. $\omega_{CP} = \frac{1}{R_{C2}C_C}$ is high-frequency pole which is set to eliminate high-frequency disturbance.

The compensated net of the voltage loop is designed as pole-zero topology. Its transfer function is

$$G_v(s) = \frac{K_V(s+\omega_{VZ})}{s(s+\omega_{VP})},$$

where $K_V = \frac{1}{R_{V1}C_{V2}}$. $\omega_{VZ} = \frac{1}{R_{V2}C_{V1}}$ is zero which is set higher than 100π to keep power line ripple attenuated down to a very low level at the output [9]. $\omega_{VP} = \frac{C_{V1}+C_{V2}}{R_{V2}C_{V1}C_{V2}}$ is pole which is set to counteract the zero caused by the equivalent series resistance in the output capacitor and then drastically attenuate the open-loop gain of the voltage loop within high frequency.

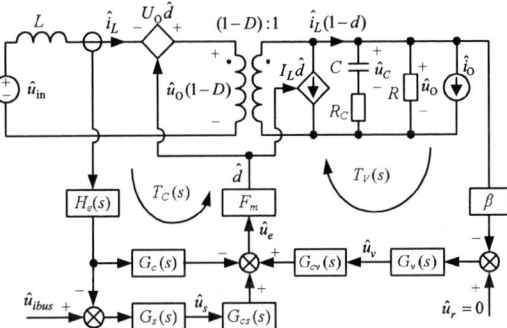

Fig. 3. The close-loop small signal model of parallel boost converters

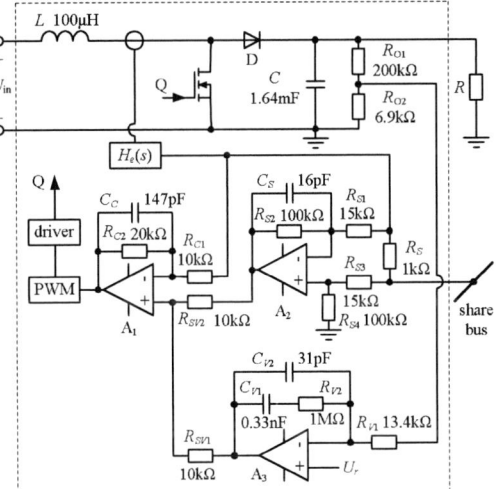

Fig. 4. Circuit diagram of each module

The compensated net of the current-sharing loop is designed as proportional adjustor. Under the system stable, its gain is designed as high as possible. In order to reject the disturbance of the share bus, a small capacitance C_S is set in parallel with the feedback resistance R_{S2}.

IV. SIMULATION AND EXPERIMENT

Two 40V/120V/4.2A Boost converters have been designed, and the circuit diagram of each module is shown in Fig. 4. The maximum power of two-module parallel system is 1 kW. The operation frequency is 50 kHz, the current-sharing error is below 5%, and the output voltage ripple is below 0.6V.

A. Frequency characteristic simulation

The frequency characteristics of T_1, T_2, $A_{PC}(s)$, and $Z_{POC}(s)$ of the parallel boost converters are shown in Fig.5 (a)~(d) respectively.

We can see from Fig.5 that when two boost converters is in parallel ($n=2$), the crossover frequency, gain margin and phase margin of T_1 are about 28 kHz, 27 dB and 55° respectively. The crossover frequency, gain margin and phase margin of T_2 are about 773 kHz, 13 dB and 54° respectively. The maximum gains of $A_{pc}(s)$ and $Z_{poc}(s)$ are about -64 dB and -12.6 dB respectively within the whole frequency. The close-loop system meets the stability guidelines given above. And it has fast dynamic response, good input and output disturbance rejection characteristics.

It is obvious from Fig. 5 that the infinite-module parallel system ($n=\infty$) maintains a good performance which is close to that of the two module parallel system. So in the proposed control technology, the parallel-module number has little effect to the performance of the close-loop system, and it is unlimited in theory.

B. Experimental results

The experimental static-current data of the two parallel boost converters are shown in table I, when the load changes from 25% to 100%. It is obvious that the current-sharing error is less than 5%.

The dynamic waveforms of the two modules output currents and the system output voltage when the load steppes up from 50% to 100% load are shown in Fig. 6. It is obvious that the output current is equally distributed and the overshoots are inconspicuous during the dynamic state, and the ripple of output voltage is less than 0.6V. So the system has a good dynamic current sharing property.

Fig. 7 presents the transient waveforms of the output currents during turning off one module. During the dynamic state, the system can still work normally.

V. CONCLUSION

An ACS control method based on ACM for parallel boost converters has been presented and analyzed. And the small-signal model of parallel boost converter based on this control method has been developed. The analytical result demonstrates that the second-order system of the external loop of boost converter is reduced to a first-order in low frequency, which makes the design of the control circuit easy and improves the dynamic response of parallel boost converters. This paper presents analysis method of the proposed control technology and its design guidelines, offering a theory basis and design method for the application of the ACS control method based on ACM.

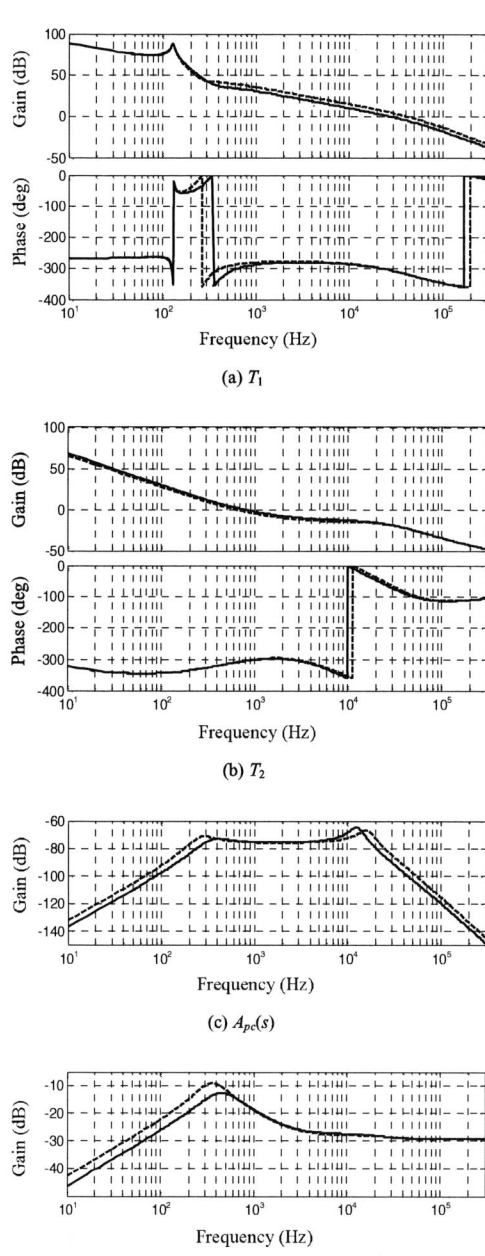

(a) T_1

(b) T_2

(c) $A_{pc}(s)$

(d) $Z_{poc}(s)$

Fig. 5. Frequency characteristics of close-loop system ($n=2$ —, $n=\infty$ --)

TABLE I.
CURRENT-SHARING PERFORMANCE OF TWO PARALLEL MODULES

Total current (A)	Module 1 current (A)	Module 2 current (A)	Error rate (%)
2.09	1.07	1.02	4.8
4.17	2.05	2.12	3.4
6.05	2.98	3.07	3.0
7.94	3.92	4.02	2.5

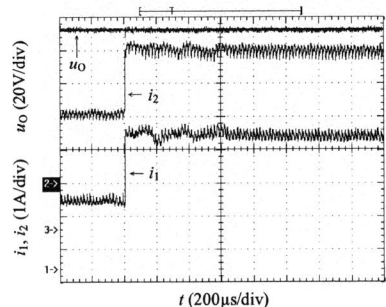

Fig. 6. Transient responses of output voltage and currents for load changing (4-8A)

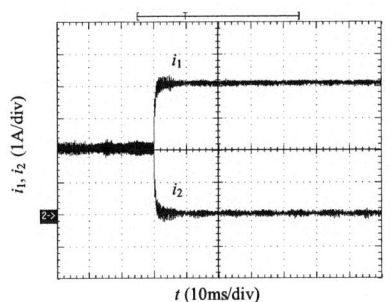

Fig. 7. Transient waveforms of two modules output current during turning off one module

REFERENCES

[1] M. Jordan, *UC3907 Load Share IC Simplifies Parallel Power Supply Design*, Unitrode Corp., Merrimack, NH, Unitrode Application Note U-129, 1993-1994.

[2] Zhang Fanghua, Zhu Chenghua, Yan Yangguang, "The controlled model of bi-directional DC-DC converter," *Proceedings of the CSEE*, Vol. 25, No. 11, pp. 46-49, 2005.

[3] Gabriel Garcerá, Marcos Pascual, Emilio Figueres, "Robust average current-mode control of multimodule parallel DC-DC PWM converter systems with improved dynamic response," *IEEE Trans. on Industrial Electronics*, Vol. 48, No. 5, pp. 995-1005, 2001.

[4] G. Garcerá, M. Pascual, E. Figueres, J.M. Benavent, "Analysis and design of a robust average current mode control loop for parallel Buck DC-DC converters to reduce line and load disturbance," *IEE Proc.-Electr. Power Appl.*, Vol. 151, No. 4, pp. 414-424, 2004.

[5] K. T. Small, "Single-wire current-share paralleling of power supplies," U.S. Patent 4734844, Mar. 1988.

[6] Qiu Dongyuan, Zhang Bo, Wei Congying, "Study of paralleled Buck converters with improved automatic current-sharing technique," *Transactions of China electrotechnical society*, Vol. 20, No. 10, pp. 41-47, 2005.

[7] Chang-Shiarn Lin, Chern-Lin Chen, "Single-wire current-share paralleling of current-mode-controlled DC power supplies," *IEEE Trans. on Industrial Electronics*, Vol. 47, No. 4, pp. 780-786, 2000.

[8] B. Choi, B.H. Cho, and F.C. Lee, "Three-loop control for multi-module converter systems," *IEEE Trans. Power Electron.*, Vol. 8, pp. 466-474, 1993.

[9] Abraham I. Pressman, *Switching Power Supply Design*, 2nd ed., New York: McGraw-Hill Companies, Inc. 1998, pp. 438-439.

[10] Vatché Vorpérian, "Simplified analysis of PWM converters using model of PWM switch, part I: continuous conduction mode," *IEEE Trans. on Aerospace and Electronic Systems*, Vol. 26, No. 3, pp. 490-496, 1990.

2006 5th International Power Electronics and Motion Control Conference

A Novel Digital Charge Control
for DC-DC Converters

Shi Wenqing[*],[**], Xu Haiping[*], Wen Xuhui[*] and Wen Wei[*],[**]
[*] Institute of Electrical Engineering, Chinese Academy of Sciences, Beijing, China
[**] Graduate School of the Chinese Academy of Sciences, Beijing, China

Abstract—**This paper presents a novel digital charge control method for DC-DC converters. The analog controller is replaced by the digital controller, which contains a programmed DSP. In the DSP control program, the switching cycle is divided into some equal time slices. Products of the electrical current and the time of each slices is added up one by one. The result is the charge number, which is the controlled parameter. The estimating technique is applied in the program and the defect of duty-cycle jumping, which may occur in the digital control, is eliminated. Experimental results confirm that the novel charge control method can be used for DC-DC converters.**

Keywords-charge control; DC-DC converters

I. INTRODUCTION

In recent years, PWM controlled DC-DC converters have been widely used, and control methods of DC-DC converters have been developed too. The voltage mode control, the current mode control and the average current mode control are used on various converters. Charge control is a new control method of DC-DC converters. This method has been presented in a paper in the early of 1990's [1]. A paper of our team published before has analyzed the advantage of this control method on the constant power load [2].

Above papers are all based on analog circuits. As we all know, DSP controllers are more flexible than analog controllers. In this paper, a novel digital charge control principle is presented and turned into a real DSP controller. The estimating technique is applied in the control program to eliminate the defect of duty-cycle jumping, which may occur in the digital control method. A TI DSP TMS320LF2407 is used to compose the digital controller for a Boost converter. DSP control programs are presented in detail by flow charts.

In section II, the charge control principle based on analog circuit is reviewed and the basic digital charge control principle is presented. In section III, the novel digital charge controller is described by flow charts. The experimental results are compared in section IV. At last, conclusions are given in section V.

II. DIGITAL CHARGE CONTROL PRINCIPLE

Before introducing the digital charge control, let's review the charge control principle based on analog circuit. Fig. 1 shows a charge controlled Boost converter. The constant switching period T_s is decided by the clock. At the beginning of each switching cycle, the active switch Q in the power stage is turned on. When the switch is on, the DC source v_s, the inductance L and the active switch Q compose a closed loop. The switch current i_s equals to the switch current i_Q. The capacitor C_T is charged by i_Q through a current sensor as the switch Q is on. The voltage $v_T(t)$ rises from 0V. Supposing that the beginning time of a switching cycle is kT_s and the switch is keeping on to the time $kT_s + t_0$, then

$$\int_{kT_s}^{kT_s+t_0} i_Q(t)dt = v_T(kT_s + t_0)C_T . \quad (1)$$

When the voltage $v_T(t)$ reaches the reference voltage v_r, the switch Q is turned off, and the switch across capacitor C_T is turned on to discharge C_T. C_T is totally discharged before the next switching cycle starts. In a switching cycle,

$$\int_{kT_s}^{kT_s+dT_s} i_Q(t)dt = v_r C_T = I_Q T_s . \quad (2)$$

In (2), d is the duty-cycle and I_Q is the average current of the switch current for the whole switching cycle. The reference voltage v_r determines the charge passing

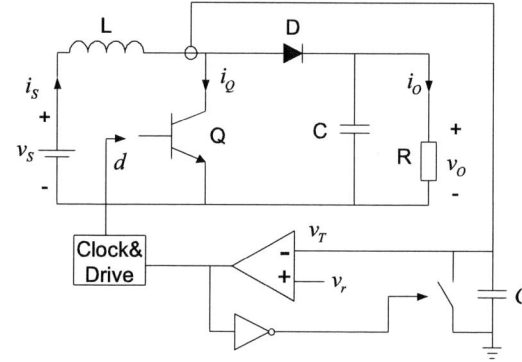

Figure 1. Charge control based on analog circuit

through the switch Q in one cycle. The charge is in proportion to I_Q, as the switching period T_s is constant. Then v_r determines I_Q.

The charge control principle based on analog circuit is reviewed above. To obtain the digital control principle, a switching cycle is divided into some equal time slices. Supposing that the time of one slice is T_o, we have the switch current graph in Fig. 2. The switch current is sampled every T_o by the sensor from the beginning time of a switching cycle kT_s.

Supposing that at time kT_s, $kT_s + T_0$, $kT_s + 2T_0$ $\cdots$ $kT_s + m\tau \cdots$, the corresponding switch currents are i_{k0}, i_{k1}, $i_{k2} \cdots i_{km} \cdots$. Sampling switch currents can be used to get the current integral. At time $kT_s + T_0$, $kT_s + 2T_0$, $kT_s + 3T_0$, $kT_s + m\tau$, the integrals of currents are

$$Q_1 = \frac{1}{2}(i_{k0} + i_{k1})T_0, \tag{3}$$

$$
\begin{aligned}
Q_2 &= \frac{1}{2}(i_{k0} + i_{k1})T_0 + \frac{1}{2}(i_{k1} + i_{k2})T_0 \\
&= Q_1 + \frac{1}{2}(i_{k1} + i_{k2})T_0,
\end{aligned}
\tag{4}
$$

$$
\begin{aligned}
Q_3 &= \frac{1}{2}(i_{k0} + i_{k1})\tau + \frac{1}{2}(i_{k1} + i_{k2})T_0 + \frac{1}{2}(i_{k2} + i_{k3})T_0 \\
&= Q_2 + \frac{1}{2}(i_{k2} + i_{k3})T_0,
\end{aligned}
\tag{5}
$$

$$
\begin{aligned}
Q_m &= \frac{1}{2}(i_{k0} + i_{k1})T_0 + \frac{1}{2}(i_{k1} + i_{k2})T_0 + \cdots \\
&+ \frac{1}{2}(i_{k(m-1)} + i_{km})T_0 = Q_{m-1} + \frac{1}{2}(i_{k(m-1)} + i_{km})T_0.
\end{aligned}
\tag{6}
$$

The integral is done after every sampling, and the integral is compared to the reference $v_r C_T$. If the integral is less than the reference, the active switch continues being on. If the integral is larger than or equals to the reference, the active switch is turned off immediately. Supposing that the switch is turned off after nT_0 from the beginning of a switching cycle, nT_0 equals to dT_s approximately.

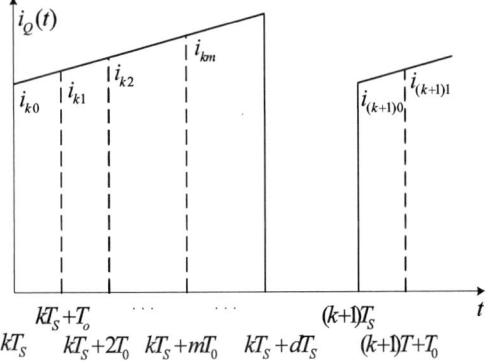

Figure2.   Switch current of a Boost converter

III.   THE NOVAL DSP CHARGE CONTROL PROGRAME

From the digital charge control principle, a DSP control program can be obtained directly. The DSP used here is TMS320LF2407. The control program contains three parts. The first part is the main program, the second part is the General-Purpose Timer 1(GP T1) period interrupt service program, and the last part is the General-Purpose Timer 2(GP T2) period interrupt service program.

In the main program, the DSP system is initialized. Then the program circulates and waits for the GP T1 period interrupt. The GP T1 period interrupt service program and the GP T2 period interrupt service program are shown in Fig. 3.

In this program, the variable QSET represents the

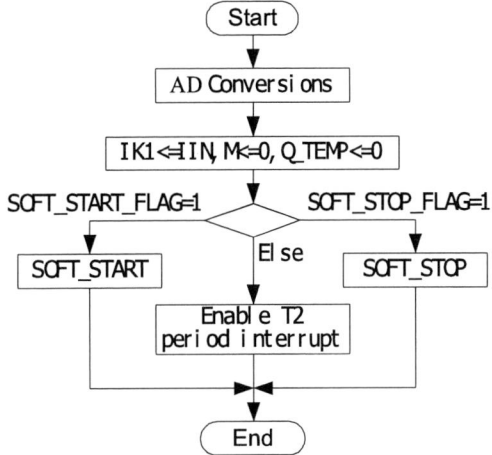

(a)   GP T1 period interrupt service program flow chart

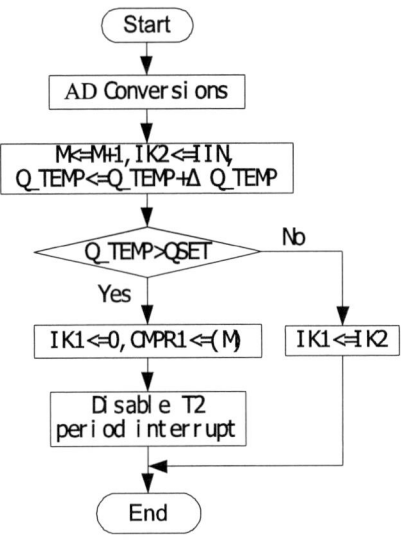

(b)   GP T2 period interrupt service program flow chart

Figure3.   Interrupt service program flow charts

reference $v_r C_T$, Q_TEMP is the current integral, ΔQ_TEMP is the current integral of one time slice, M is the number of time slices passing by, IIN is the sampling current, IK2 is the sampling current this time slice, and IK1 is the sampling current last time slice.

The period of the T1 interrupt is the switching period T_s , and the period of the T2 interrupt is the time slice T_o . The T1 period interrupt is enabled in the main program and is not disabled in the whole program. While the T2 period interrupt is enabled in the T1 period interrupt service program and disabled in the T2 period interrupt service program when Q_TEMP arrived at QSET.

This control program corresponds to the digital control principle directly, but adjacent duty-cycles may vary obviously when the output voltage is regulated, because it is the integral multiple of the time slice. The obvious duty-cycle variety is called duty-cycle jumping here. Duty-cycle jumping has disadvantageous effect on the stability of the system.

An improved program with estimating technique has been designed to eliminate the duty-cycle jumping. The improved program contains three parts too. The main program and the GP T1 period interrupt service program are not changed, while the GP T2 period interrupt service program is different, which is shown in Fig. 4.

In this program, ΔQ_TEMP1 is the estimated current integral of the coming time slice, M1 is a fraction of the coming time slice.

When the current integral Q_TEMP is less than the reference QSET, we estimate the current integral Q_NEXT at the end of the coming time slice, then compare Q_NEXT and QSET. If Q_NEXT is less than QSET, the program does the same as the program of Fig.3 (b). If Q_NEXT is larger than QSET, the active

switch should be turn off in the coming time slice. We estimate the turnoff time and it is M adding M1. The program for estimating M1 is showed in Fig.5. In this program, QSET,Q_TEMP and M1 is the same as above. M1_MIN, M1_MAX, Q_LIM, Q_EST, Q_MIN, Q_MAX and Q_MID are all transitional parameters.

The main part of the duty-cycle comes from current integral, so the program maintains the advantage of fast dynamic response in charge control. The duty-cycle does not jump, because it is not the integral multiple of the time slice any more.

IV. EXPERIMENT RESULTS

The experimental system is shown in Fig.6. The input voltage of the Boost converter is 40V, and the output voltage is set on 60V. The switching period is 204.8 μs , and a switching cycle is divided into 16 time slices. The resistor load is 20 Ω . The DSP controller is composed on a TI DSP TMS320LF2407 and peripherals.

The digital charge controllers without estimating program and with estimating program are applied in the system respectively. Switching signals of two controllers can be seen in Fig. 7. The figure confirms that the duty-

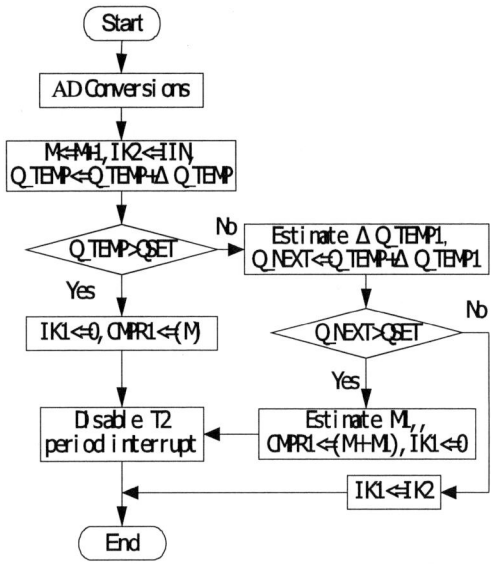

Figure4. GP T2 period interrupt service program flow chart of the improved DSP control program

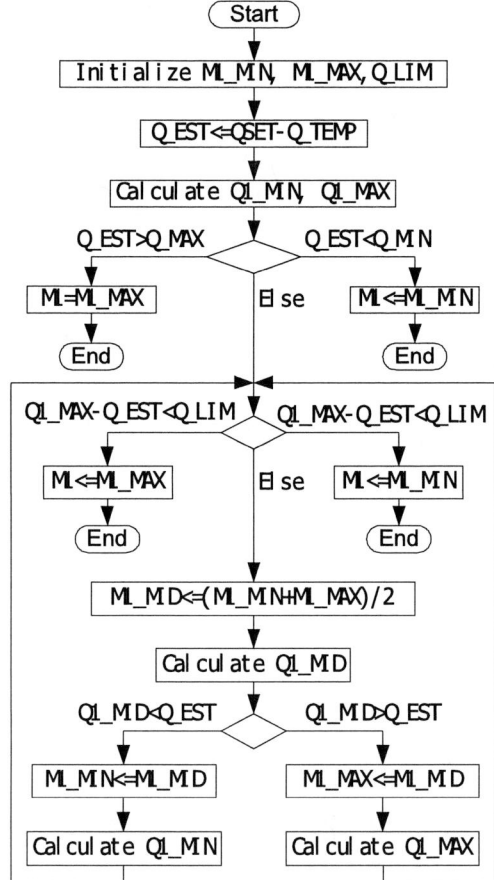

Figure5. The program for estimating M1

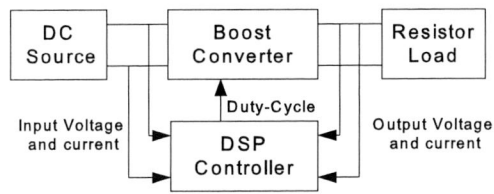

Figure6.   The experiment system

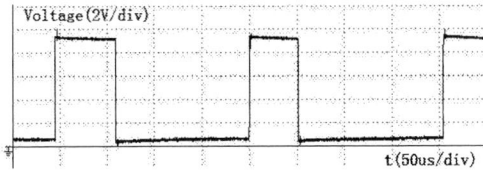

(a)   Switching signal in the controller without estimating program

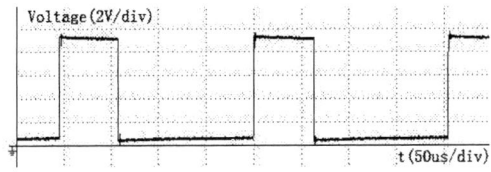

(b).   Switching signal in the controller with estimating program

Figure7.   Comparing of switching signals

cycle jumping will arise using the digital charge control program, and the digital charge controller with the estimating program can eliminate this defect.

V.   CONCLUSIONS

The charge control method of DC-DC converters based on an analog circuit is reviewed in this paper, then the digital charge control principle is presented based on the time slice division. From this principle, the DSP control program is obtained directly. This program has the defect of the duty-cycle jumping, because the duty-cycle is always the integral multiple of the time slice. To eliminate this defect, the improved DSP control program with estimating technique is designed. The experiment results confirm that the novel charge control method of DC-DC converters can be realized by digital circuits. Appling the novel DSP control program, the advantage of fast dynamic response of charge control can be maintained and the defect of duty-cycle jumping can be eliminated. The power stage of the paper is a Boost converter, while the novel digital charge control method presented here can also be applied for other converters.

REFERENCES

[1] Wei Tang, F. C. Lee, R. B. Ridley, and Isaac Cohen, "Charge control: modeling, analysis, and design," *IEEE Trans. Power Electron.*, vol. 8, no. 4, Oct. 1993.

[2] Shi wenqing, Xu Haiping, Wen Xuhui, Wen Wei, "One-Cycle Controlled DC-DC Converters Operating with Nonlinear Power Load," *ICEMS.*, vol. II , pp.1361~1365, Sept.2005

2006 5th International Power Electronics and Motion Control Conference

An Asymmetrical Switched Capacitor and Lossless Inductor Quasi-Resonant Snubber-Assisted ZCS-PWM DC-DC Converter with High frequency Link

Khairy Fathy[1], Keiki Morimoto[2], Toshimitsu Doi[2], Hyun Woo Lee[1] and Mutsuo Nakaoka[1],[3]

[1]Kyungnam University, Masan, Republic of Korea
[2] Daihen Corporation,Osaka, Japan
[3] Industerial College of Technology University, Hyogo, Japan
Email: khairy@ieee.org

Abstract – **In this paper, a novel type of auxiliary switched capacitor assisted edge resonant soft switching PWM resonant DC-DC Converter with two auxiliary edge resonant lossless inductor snubbers. The operation principle of this converter is described using the switching mode equivalent circuits. This newly developed multi resonant DC-DC Converter can regulate its dc output AC power under a principle of constant frequency edge-resonant soft switching commutation by asymmetrical PWM control scheme. The high frequency power regulation and actual power conversion efficiency characteristics of the proposed soft switching PWM series load resonant DC-DC Converter are evaluated. The operating performances of the newly proposed soft switching inverter are discussed based on simulation a results from an application point of view.**

Keywords- DC-DC converter, series capacitor compensated transformer leakage inductance, auxiliary lossless snubbing inductor, auxiliary switched capacitor, zero current soft switching, auxiliary edge resonant snubber

III. INTRODUCTION

The advanced developments of a variety of soft switching PWM DC-DC power converter circuits with or without a high frequency transformer link for the purpose of improving their actual efficiency, minimized power density in a volumetric size and achieving higher performances as high quality waveforms and quicker responses have actively been introduced. In addition, the current ringing caused by parasitic parameters and high di/dt and dv/dt dynamic stresses in the power semiconductor switches are more indispensable in accordance with high frequency switching pulse modulation. Some topologies of soft switching pulse modulation controlled voltage source full bridge inverter type DC-DC converters with a high frequency transformer have been developed so far and evaluated for the telecommunication DC feeding power plants, automotive power supplies and new energy-related power conditioners

In this paper, a novel circuit topology of voltage source multi resonant ZCS DC-DC converter with constant frequency PWM control strategy using active auxiliary quasi-resonant lossless inductor sunbbers and switched

capacitor snubber is newly proposed, which additionally includes practical outstanding features. The operating principle of the proposed DC-DC converter topology incorporating ZCS-PWM control scheme is illustrated and evaluated on the basis of simulation results and the effectiveness of this proposed DC-DC high frequency ZCS inverter using IGBTs is substantially proved. The voltage source type ZCS DC-DC converter and its modifications match the practical operating requirements mentioned previously. The proposed ZCS with PWM control scheme Can be able to regulate its output power under constant frequency PWM control strategy.

IV. PROPOSED ZCS PWM DC-DC CONVERTER WITH ACTIVE EDGE RESONANT SNUBBER

A. Circuit Configuration

Fig. 1 shows the newly developed multi-resonant ZCS-SEPP PWM DC-DC converter circuit using the latest trench gate IGBTs and operating with constant frequency PWM control strategy. This voltage-fed ZCS PWM DC-DC converter circuit consists of two main switches of reverse conducting IGBTs $Q_1(SW_1/D_1)$ and $Q_2(SW_2/D_2)$, a single auxiliary switch $Q_3(SW_3/D_3)$ in series with auxiliary edge-resonant switched capacitor C_r as an active snubber in parallel with Q1 and L_S, One ZCS-assisted lossless inductor snubbers L_S connected in series with the main switches Q_1 and Q_2, power factor compensated series load resonant capacitor C_s, high frequency centre-tapped transformer and dc filter L_o- C_o output filter is connected in parallel with the DC load. that used to smooth out the DC output voltage. The proposed voltage source ZCS-PWM DC-DC converter is configured by a few circuit components and power semiconductor devices as three controlled active switches are used.

B. Gate Pulse Timing Sequences

The high frequency AC output power of the proposed DC-DC converter circuit, can be continuously regulated by a constant frequency asymmetrical PWM (duty cycle) control scheme under a condition of zero current soft switching commutation mode. The proposed gate voltage pulse timing PWM sequences for the active

1-4244-0448-7/06/$25.00 ©2006 IEEE

power switches Q_1, Q_2, and the auxiliary power switch Q_3 are shown schematically in Fig. 2. The main active power switch Q_1 is firstly switched on during period of time T_{on1} and before the main switch Q_1 is turned off by a time of T_o the auxiliary switch Q_3 is turned on for a period T_{on3} inserting a an overlapping time of T_o between the switches Q_1 and Q_3. Then, the main switch Q_2 is turned on after turning off the auxiliary switch Q_3 by a dead time of T_{d1}. The main switch Q_1 is again switched on after a dead time T_{d2} as another period starts as depicted in Fig. 2.

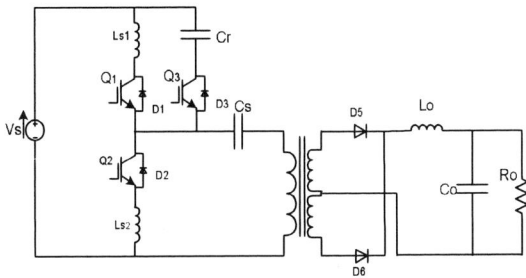

Fig. 1 Proposed single ended push-pull ZCS PWM DC-DC power converter

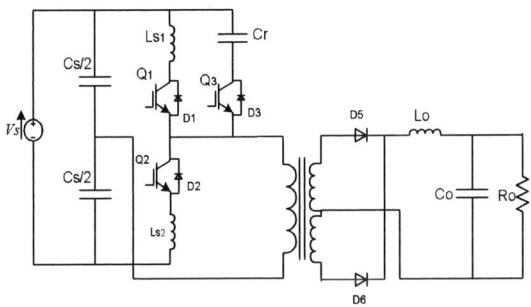

(b) Half-bridge topology with divided series capacitor

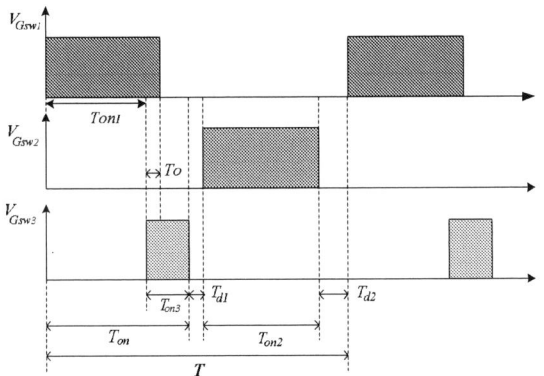

Fig. 2 The Proposed PWM gate pulse timing sequences.

C- Control Variables

By adjusting the constant frequency asymmetrical PWM control duty cycle, which is defined as the sum of the conduction time T_{on1} of the main active switch Q_1 and conduction time T_{on3} of the auxiliary power switch

Q_3 to the total switching period operating time T of the high switching, the proposed high frequency ZCS-PWM inverter can regulate its high frequency output power continuously under a condition of soft switching. The conduction time T_{on1} of the main active switch Q_1 can be controlled while keeping the conduction time T_{on3} of the auxiliary active switch Q_3, the overlapping time T_o and the dead time T_{d1} constants. As a control variable, the duty cycle D is defined as

$$D = (T_{on} + T_{d1})/T \qquad (1)$$

The proposed voltage source ZCS PWM DC-DC converter with two lossless inductor snubbers and a single switched capacitor can not only be controlled by the constant frequency asymmetrical PWM technique for high power settings, but also it can be controlled by a constant high frequency pulse density modulation (PDM) technique at low power settings. In addition, by using a dual mode hybrid control of asymmetrical PWM and PDM at a constant high frequency, soft switching operating range can be effectively expanded from high power to low power settings.

III. PRINCIPLE OF SOFT SWITCHING COMMUTATION

At the beginning of each switching cycle, the high side main power switch SW_1 of Q_1 is now conducting, During this time, the primary side energy is supplied to the load R in the secondary circuit through the high frequency transformer HF-T. After the switch current i_{SW1} through SW_1 of Q_1 naturally commutates to the switch anti-parallel diode D_1 of Q_1 by quasi-resonance due to ZCS-assisted high side inductor snubber L_{S1}, in series with the switch Q_1, together with the auxiliary series inductive primary side of HF-T tuned capacitor C_s, the auxiliary active power switch SW_3 of Q_3 is turned on and the main power switch SW_1 of Q_1 is turned off. As a result, a ZCS commutation at a turn-off switching mode transition can be achieved by the arbitrarily timing processing when turning off the main power switch SW_1 of Q_1. At this mode, since an auxiliary resonant current i_{SW3} flows through the switch SW_3 of Q_3 and increases softly, a ZCS commutation at a turn-on switching mode transition can be achieved for SW_3 of Q_3. Then, after i_{SW3} is commutated to the anti-parallel diode D_3 of Q_3 by the resonance formed by C_r, L_t leakage inductance of transformer circuitry and power factor series load compensated capacitor C_s, a ZCS soft switching commutation at a turn-off switching mode transition can be performed by turning off SW_3 of Q_3. While the auxiliary power switch SW_3 of Q_3 is conducting, the voltage v_{Q2} across the low side main switch SW_2 of Q_2 decreases toward zero. Before the low side main switch SW_2 of Q_2 is turned on as soon as the diode D_2 of Q_2 becomes reverse biasing state and begins to conduct naturally. While the diode D_2 continues conducting, the current flowing through D_2 of Q_2 is naturally commutated to SW_2 of Q_2. Therefore, a complete ZVS and ZCS (ZVZCS) hybrid commutation transition can be actually achieved for SW_2 of Q_2.

On the other hand, after the current i_{SW2} through the low side main switch SW_2 of Q_2 is naturally commutated

1303

to D_2 of Q_2 with the aid of low side ZCS-assisted inductor snubber L_{S2}, and load power factor compensation series load resonant tuned capacitor C_s, ZCS commutation at a turn-off switching mode transition can be performed by turning off the switch SW_2 of Q_2. While the diode D_2 of Q_2 is conducting, the current i_{D2} flowing through D_2 is commutated to the switch SW_1 of Q_1 by turning on the switch SW_1 of Q_1 when a second switching cycle starts. At this mode, a ZCS turn-on switching commutation can be realized with the aid of ZCS-assisted inductor snubber L_{S1}. The proposed edge resonant ZCS PWM DC-DC converter offers a complete ZCS for all the main and auxiliary switches and achieves ZVZCS hybrid commutation at turn-on switching mode transition for the switch SW_2 of Q_2.

IV. OPERATION MODES

The switching operation mode equivalent circuits of the proposed zero current soft switching DC-DC converter in steady state during one switching cycle are shown in Fig. 3. The current and voltage operating waveforms of each element and the relevant operating modes in steady state during one switching period are illustrated in Fig. 4. for a duty cycle $D = 0.36$. This multi resonant high frequency soft switching DC-DC converter circuit includes eleven operating switching modes as shown in Fig. 3. The operation principle of the proposed zero current soft switching DC-DC converter circuit will be explained in the following by using the corresponding switching mode equivalent circuits,

Mode 1: Gating signal is applied to main switch Q_1, the snubber capacitor start to discharge through snubber inductor and diode D_3.

Mode 2: The power now supplied to the load through Q_1, C_s, high frequency transformer HFT, diode D_5, and capacitor C_r still discharging

Mode 3: Diode D3 commutates and power supplied to the load through the main switch Q1

Mode 4: As soon as the gating signal is applied to Q3 the current flows in D1 and Q1 turned off at ZVZCS

Mode 5: Current will supplied to the load through the auxiliary switch Q3 and diode D6 in secondary circuit conducts

Mode 6: The diode D2 of the main switch Q2 start to conducts

Mode 7: The gating signal is removed from Q3 and D3 conducts so SW3 turns off at ZVZCS

Mode 8: Although the gating signal is applied to Q2 the diode D2 continue conducting and the stored energy in the high frequency transformer and compensating capacitor is supplied to the load

Mode 9: The auxiliary switch SW3 of Q3 turns on at ZVZCS

Mode 10: SW2 of Q2 still conducting and D5 in the secondary circuit commutates

Mode 11: D5 conducts and SW2 is turned off.

Mode 12: No conduction in the primary circuit and current circulates in the secondary circuit.

Thereafter, the aforementioned operating processes are repeated in sequence during each switching cycle.

V. SIMULATION ANALYSIS AND OPERATING WAVEFORMS

The prospective experimental setup assembly implementation will be by using trench gate reverse conducting IGBTs with low saturation voltage to validate the steady state performance evaluations of the proposed zero current soft switching high frequency dc-dc converter circuit. The design specifications and circuit parameters used in the experimental breadboard setup are respectively indicated in Table1. The circuit parameters of this high frequency ZCS-PWM DC-DC converter are determined by considering the operating condition of zero current soft switching commutation condition and the required output power ranges.

Although the proposed high frequency dc-dc converter operating under a principle of asymmetrical PWM control or duty cycle control can achieves a complete soft switching commutation operation in the high output power settings, it becomes a partially hard switching commutation operation in a certain low power settings and its actual efficiency for this reason might be substantially reduced. However, this high frequency DC-DC converter can still operate under a considerable soft switching operation. For example, in the case of specific duty cycle $D = 0.17$, under a condition of the minimum low output power setting requirement of duty cycle PWM control scheme it is still soft switching operation.

In addition, the zero current soft switching operating range of the proposed high frequency DC-DC converter can be actually extended by the use of high frequency pulse density modulation (PDM) control scheme under the low power settings or by the use of dual mode implementation of PWM control. Therefore, zero current the soft switching operation can be completely realized over all the output power regulation ranges including in low power setting conditions and PWM in high power settings. The output voltage regulation characteristics versus the duty cycle is shown in Fig. 5

TABLE 1. DESIGN SPECIFICATIONS AND CIRCUIT CONSTANTS

Item	Symbol	Value
DC Source Voltage	V_s	220 V
Switching Frequency	f_{sw}	50 kHz
ZCS-assisted Inductor	L_{s1}, L_{s2}	0.2 μH
Auxiliary Quasi-resonant Capacitor	C_r	0.2 μF
Compensation series tuned capacitor	C_s	5 μF
Output filter inductance	Lo	50 μH
Output filter capacitance	Co	5 μF
High frequency transformer turns Ratio	N1:N2:N3	1:2:2
HFT magnetizing inductance	L_m	5mH
HFT primary inductance	L_1	1 μH
HFT secondary and tertiary inductance	L_2, L_3	μH

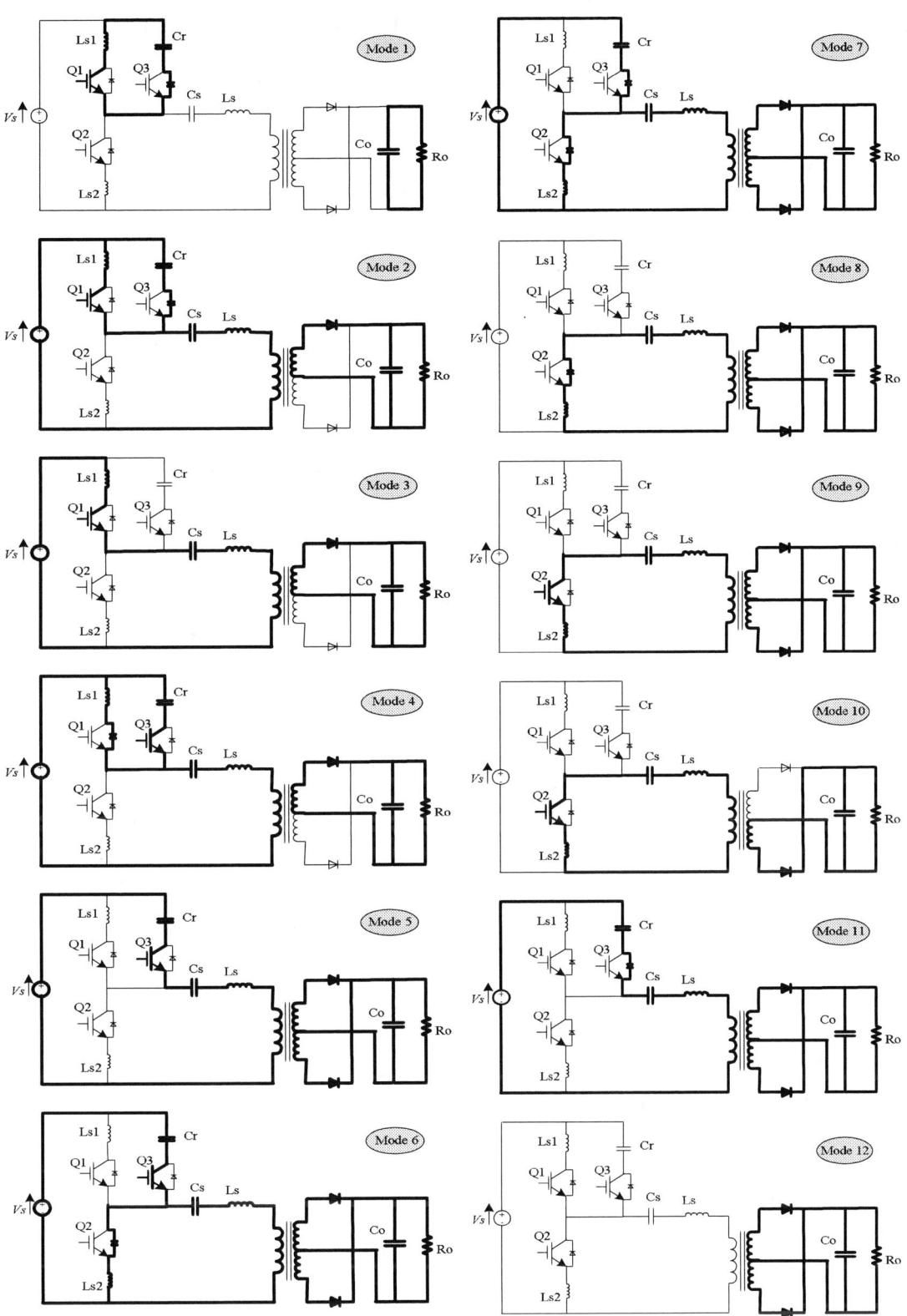

Fig. 3. Operating mode transitions and equivalent circuits
in steady state during one switching cycle.

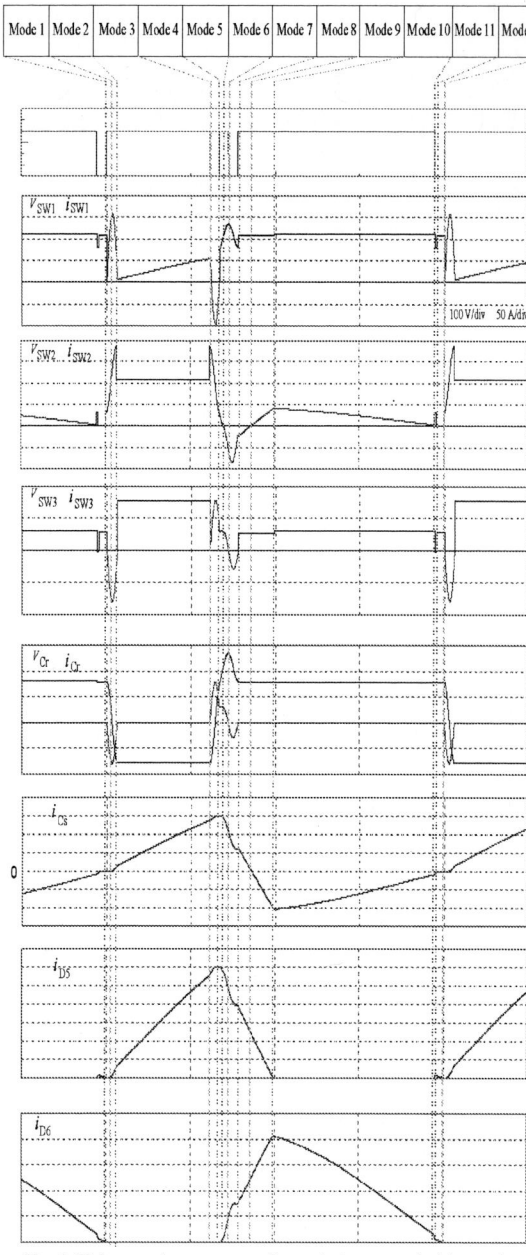

Fig. 4. Voltage and current waveforms during one switching cycle operating modes for a duty cycle D= 0.36.

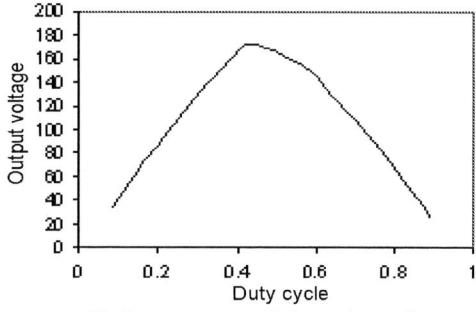

Fig. 5 output power regulation vs. duty cycle.

VI. CONCLUSIONS

In this paper, a new topology of active auxiliary quasi-resonant snubber-assisted voltage source type ZCS-PWM DC-DC power converter using IGBTs, with is composed of an active auxiliary switched snubber capacitor and two lossless snubber inductors has been proposed and developed The high frequency operation principle, switching mode transitions and the operating characteristics of the PWM controlled soft switching DC-DC converter have been illustrated and evaluated on the basis of simulation. The practical effectiveness of the newly-proposed voltage source type ZCS-PWM high frequency multi resonant DC-DC converter have been proved on the basis of the simulation. A wide soft switching commutation operation range has been obtained as compared with the previously developed voltage source type ZVS-PWM DC-DC converter.

The high frequency power regulation strategy of this converter could be efficiently supplied from full power to relatively small power settings. Applying a dual mode pulse modulation control strategy of the asymmetrical PWM, the output power of this soft switching pulse modulated multi resonant DC-DC converter could be regulated under a condition of expanded stable soft switching operation ranges as compared with previously developed ZVS-PWM high frequency DC-DC converter. Therefore, the newly proposed dual mode ZCS PWM DC-DC converter could actually achieve higher efficiency, high performance and wider soft switching operating ranges.

In the future the new prototype topology of ZCS-PWM DC-DC converter with high frequency link proposed here should be actually evaluated for PEM fuel cell stack power conditioner. This ZCS-PWM DC-DC power converter using 100 kHz IGBTs or 150kHz IGBTs developed by IXYS corporation should be introduced and discussed for some high frequency transformer structures.

REFERENCES

[1] K. Morimoto, T. Doi, H. Manabe, M. Nakaoka, N.A. Ahmed, H.W. Lee, E. Hiraki, T. Ahmed, "Next Generation High Efficiency High Power DC-DC Converter incorporating Active Switch and Snubbing Capacitor Assisted Full-Bridge Soft-Switching PWM inverter with High Frequency Transformer for Large Current Output", Proceedings of IEEE-APEC, pp1549-1555, March, 2005

[2] Hang-Seok Choi; Bo Hyung Cho; "Novel zero-current-switching (ZCS) PWM switch cell minimizing additional conduction loss", Industrial Electronics, IEEE Transactions on, Volume 49, Issue 1, Feb. 2002 Page(s):165 – 172

[3] H. Tanimatsu, H. Sadakata, T. Iwai, H.Omori, Y.Miura, E.Hiraki, H.W.Lee, and M. Nakaoka, " Quasi-Resonant Inductive Snubbers-Assisted Series Load Resonant Tank Soft Switching PWM SEPP High-Frequency Multi Resonant Inverter with Auxiliary Switched Capacitor" The 6th International Conference on Power Electronics (ICPE'04), pp II-299, Busan, Korea, October, 2004 acknowledgment

[4] Wakabayashi, F.T.; Bonato, M.J.; Canesin, C.A.; "Novel high-power-factor ZCS-PWM preregulators", IEEE Transactions on Industrial Electronics, Volume 48, Issue 2, April 2001 Page(s):322 – 333.

2006 5th International Power Electronics and Motion Control Conference

A Divided Voltage Half-Bridge High Frequency Soft-Switching PWM DC-DC Converter with High and Low Side DC Rail Active Edge Resonant Snubbers

Khairy Fathy[1], Keiki Morimoto[2], Toshimitsu Doi[2], Hiroyuki Ogiwara[3], Hyun Woo Lee[1] and Mutsuo Nakaoka[1],[4]

[1]Kyungnam University, Masan, Korea
[2]Daihen Corporation, Osaka, Japan
[3]Ashikaga Institute of Technology, Tochigi, Japan
[4]Industrial College of Technology University, Hyogo, Japan
khairy@ieee.org

Abstract— This paper presents a new circuit topology of dc bus line switch-assisted half-bridge soft switching PWM inverter type dc-dc converter for arc welder. The proposed power converter is composed of typical voltage source half-bridge high frequency PWM inverter with a high frequency transformer link in addition to dc bus line side power semiconductor switching devices for PWM control scheme and capacitive lossless snubbers. All the active power switches in the half-bridge arm and dc bus lines can achieve ZCS turn-on and ZVS turn-off commutation operation and consequently the total turn-off switching losses can be significantly reduced. As a result, a high switching frequency of using IGBTs can be actually selected more than about 20 kHz.

Keywords: Arc welding machine, DC-DC power converter, DC rail active quasi-resonant snubbers, High frequency transformer link, Soft switching PWM.

I. INTRODUCTION

A. Research Background

Recently, saturable inductor assisted ZVS-PWM full-bridge high-frequency inverter link dc-dc power converter [1] and lossless capacitors and transformer parasitic inductive components assisted soft switching dc-dc power converter with phase-shifted modulation control scheme in secondary-side of high frequency transformer [2]-[5] have been developed and evaluated. These power converter circuit topologies are suitable for handling high output power more than about several kW, especially for high voltage and low current applications as new energy related power supplies. However, secondary magnetic switches or transformer secondary side semiconductor switching devices in these converter circuit topologies may cause large conduction loss when these power circuit topologies are adopted for low voltage and large current application as arc welding power supplies. Therefore, for the low voltage and large current application required for arc welding power supplies, a soft switching dc-dc power converter with active switches in the primary side of high frequency transformer is

considered to be more suitable and acceptable. As a circuit topology to meet this requirement for arc welder, authors developed the novel circuit topology of voltage source full-bridge type soft switching PWM inverter in which all the active switches can actively achieve ZCS turn-on and ZVS turn-off commutation operation.

B. Research Objectives

This paper presents a novel circuit topology of voltage source half-bridge type soft switching PWM inverter. Under the newly-proposed high frequency inverter link dc-dc power circuit, all the active switches in the half-bridge arm and dc bus lines can actively achieve ZCS turn-on and ZVS turn-off commutation operation.

The steady state operating principle of the proposed soft switching PWM dc-dc power converter is described with its remarkable features.

II. NEW SOFT SWITCHING DC-DC CONVERTER TOPOLOGY

Fig. 1 shows the proposed soft switching PWM dc-dc converter circuit using a novel type half-bridge soft switching PWM inverter with high frequency transformer link. Proposed converter is composed of typical voltage source half-bridge inverter with a active PWM switch $Q_3(S_3/D_3)$ in positive dc bus line and a active PWM switch $Q_4(S_3/D_3)$ in negative dc bus line and two lossless snubbing capacitors C_1, C_2 and two diodes D_5, D_6. Two centre points of two capacitors C_1, C_2 and two diodes D_5, D_6 are connected to a mid point of two voltage sources E_1, E_2 and one of terminals of primary winding of high frequency transformer. The voltage of two voltage sources E_1, E_2 and capacitance of capacitors C_1, C_2 are designed so as to be equal ($E_1=E_2=E$, $C_1=C_2=C$). The main active switches $Q_1(S_1/D_1)$ or $Q_2(S_2/D_2)$ can be turned on and turned off in accordance with modified PWM control circuit similar to conventional half-bridge type hard switching PWM inverter. Under the proposed soft switching dc-dc converter, the switches in the half-bridge type inverter can perform ZVS turn-off transition due to the presence of the active PWM switches Q_3 or Q_4 which

1-4244-0448-7/06/$25.00 ©2006 IEEE

are turned off and the snubbing capacitors C_1 or C_2 are completely discharged before the active switches Q_1 or Q_2 in half-bridge type inverter arms are turned off. In addition, the inverter switches can also perform ZCS at a turn-on transition with the aid of inductance L_S, as a parasitic leakage inductance of high frequency transformer.

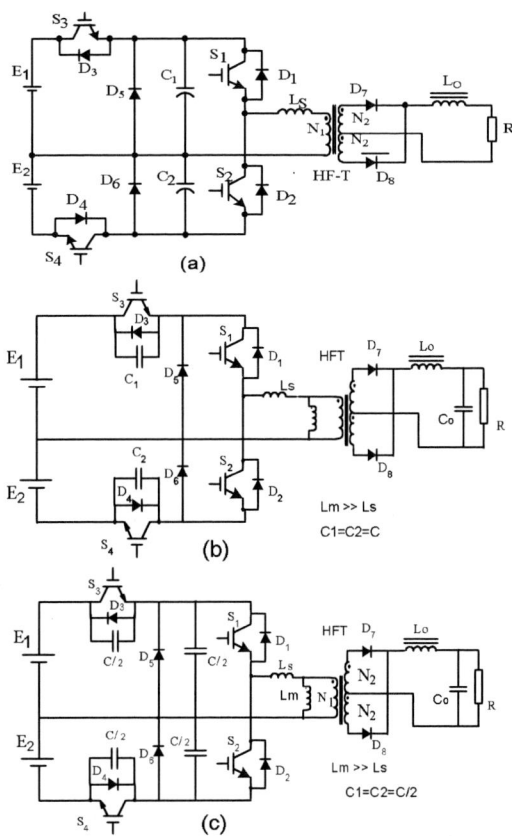

Fig. 1 A novel half-bridge Soft-switching PWM dc-dc power converter.

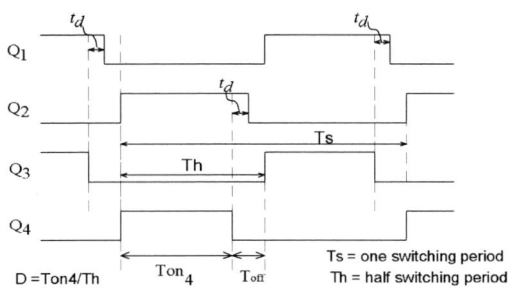

Fig. 2 Pattern sequences of switching gate driving pulses.

As for the active PWM switches Q_3 or Q_4 in series with the dc bus line side, the PWM controlled switches can achieve ZVS at a turn-off mode transition due to the lossless snubbing capacitors C_1 or C_2. These active PWM switches Q_3 or Q_4 can also achieve ZVS/ZCS at a turn-on mode transition due to the lossless snubbing capacitors C_1 or C_2, which have been charged up to the same voltage as

the half voltage E_1 or E_2 of dc power bus line voltage source by the energy storage in the leakage inductance L_S after the half-bridge type inverter switches are turned off completely. Although the conduction power loss of the additional switches may increases the total power loss a little, the total turn-off switching loss of half-bridge type PWM inverter can be significantly decreased with the optimum aid of dc bus line series switches Q_3 or Q_4 and the lossless snubbing capacitors C_1 or C_2.

III. PRINCIPLE OF OPERATION

A. Gate Voltage Pulse Timing Sequences

Fig. 2 shows timing pattern sequences of switching gate driving pulses. The gate voltage pulse signals for the inverter switches Q_1 or Q_2 in voltage source half-bridge inverter arms are the same as PWM signal sequences of conventional half-bridge inverter. Regarding the turn-on gate voltage pulse signals to the dc bus line side series switches Q_3 or Q_4, the signals are applied to Q_3 or Q_4 at the same timing as the turn-on signals to Q_1 or Q_2, respectively. As for the turn-off signals to Q_3 or Q_4, the signals are delivered to Q_3 or Q_4 before the predetermined length of time t_d on the basis of the time when the turn-off signals are applied to Q_1 or Q_2. In other words, the turn-off pulse signals are applied to Q_1 or Q_2 after the turn-off gate signals are supplied to Q_3 or Q_4 by a time t_d.

B. Duty Cycle Control Scheme

The studied inverter arc welding machine uses the current control algorithm. . The Output current control signal is obtained from secondary current of planar transformer, thereafter it is compared with current setting reference signal. Hence this process, output current controller generates the PWM switching function of the converter.

The control circuit for newly-developed circuits can be implemented easily by modifying the conventional PWM signal processing circuit using common PWM control IC(μPC494) with dual alternating output switches and dead time control.

C. Operation Modes

Fig. 3 illustrates the relevant operating waveforms in a complete switching period for the pulse pattern of gate drive timing sequences shown in Fig. 2. The operation modes are divided into twelve operation mode from mode 1 to mode 12 in accordance with operational timing and each operation principle is described hereafter. The equivalent circuits to each mode are shown in Fig. 4.

1) Mode 1 : Before time t_0, the switches Q_1 and Q_3 are turned on. At this time, i_{t1} flows through the primary winding of transformer HF-T. Also, i_{s1} flows through Q_1 and i_{s3} flows through Q_3. In this period, all currents i_{t1}, i_{s1} and i_{s3} are equal and the voltage v_{C1} across the capacitor C_1 is the same as the dc bus line voltage E_1.

2) Mode 2: At time t_0, the turn-off signal is applied to Q_3. At this time, the series switch Q_3 in dc bus line can be turned off with ZVS because the current i_{s3} through Q_3 is immediately cut off due to the lossless snubbing capacitor

1308

C_1. After time t_0, the voltage v_{C1} across the capacitor C_1 discharges constantly toward zero voltage from E1=E voltage. At this time, the voltage v_{C1} across the lossless snubber capacitor C_1 is estimated as,

$$v_{C1}(t) = E - (i_{t1}/C)t \qquad (1)$$

where, i_{t1} is a primary current of high frequency transformer. The more the current i_{t1} though primary winding of transformer HF-T is large, the more the discharging time for capacitor C_1 is short. On the other hand, the more the current i_{t1} is small, the more the discharging time is long. Under the condition of the maximum i_{t1} and the maximum output current. In this case, switches Q_1 or Q_2 can achieve ZVS turn-off transition completely. If we need to enlarge the complete ZVS operation range at the turn-off commutation for the switches Q_1 or Q_2, the delay time t_d should be varied according to the value of current i_{t1}.

3) Mode 3: At time t_1, the voltage v_{C1} becomes zero. In the interval from t_1 to t_2, the diodes D_5 is turned on and the current i_{t1} through transformer primary winding flows through the circulation loop; $L_S \rightarrow D_5 \rightarrow S_1 \rightarrow L_S$.

4) Mode 4: At time t_2, the turn-off gate pulse signal (see Fig.2) is applied to Q_1. At this time, the switch Q_1 can be turned off with ZVS because the voltage v_{C1} was already zero during last half operation cycle and the diodes D_2 of Q_2 are immediately turned on. After that, the capacitor C_2 is charged up to the same voltage as dc bus line voltage E_2. At this mode, the condition that the capacitor C_2 is just charged up to the same value as dc busline voltage E_2, it is estimated by eq. (2).

$$(1/2)CE^2 = (1/2)L_S(i_{t1})^2 \qquad (2)$$

However, as described after in mode 6, circuit parameters should be designed to meet the condition of $(1/2)CE^2 \leq (1/2)L_S(i_{t1})^2$ in order to achieve ZVS commutation at turn-on transition of Q_4.

5) Mode 5: Under a condition of $(1/2)CE^2 < (1/2)L_1(i_{t1})^2$, after the voltage v_{C2} reaches the dc bus line voltage E_2, the voltage v_{C2} across the snubber capacitor C_2 is clamped to dc bus line voltage E_2 because the diode D_4 of Q_4 is turned on and the energy stored into leakage inductance L_S is back to dc bus line voltage source E_2.

6) Mode 6: No current flow in the primary circuit.

7) Mode 7: At time t_5, the turn-on gate pulse signals are applied to the switches Q_2 and Q_4. At this time, the switches Q_2 can be turned on with ZCS due to the parasitic inductance L_S of transformer HF-T. And more the series switch Q_4 in the bus line achieves a complete soft-switching ZVS/ZCS at turn-on because the voltage v_{C2} is the same voltage as dc power bus line voltage E_2.

8) Mode 8: the turn-off signal is applied to Q_4. At this time, the series switch Q_4 in dc bus line can be turned off with ZVS because the current i_{s4} through Q_4 is immediately cut off due to the lossless snubbing capacitor C_2. The voltage v_{C2} across C_2 discharges constantly toward zero.

9) Mode 9: the voltage v_{C2} becomes zero, the diodes D_6 is turned on and the current i_{t1} through transformer primary winding flows through the loop; $L_S \rightarrow S_2 \rightarrow D_6 \rightarrow L_S$.

10) Mode 10: the turn-off gate pulse signal is applied to Q_2. At this time, the switch Q_2 can be turned off with ZVS because the voltage v_{C2} was already zero during last half operation cycle and the diodes D_1 of Q_1 are immediately turned on. After that, the capacitor C_1 is charged up to the same voltage as dc bus line voltage E_1.

11) Mode 11: after the voltage v_{C1} across the snubber capacitor C_1 reaches the dc bus line voltage E_1, it is clamped to dc bus line voltage E_1 because the diode D_3 of Q_3 is turned on and the energy stored into leakage inductance L_S is back to dc bus line voltage source E_1.

12) Mode 12: at this mode, all operations are stopped in the primary circuit.

IV. EXPERIMENTAL RESULTS AND DISCUSSIONS

A. Total System Implementations

The experimental setup circuit is shown in Fig. 5. In the setup implementation, the maximum output voltage and current are 36V, 400A, respectively. The maximum output of this experimental setup is 36V, 400A (14.4 kW). The IGBT modules are mounted on the heat sink and connected by the printed circuit board which the capacitors C_1 and C_2 are mounted on and the capacitors C_3 and C_4 are directly connected by the printed circuit board enables to minimize the stray inductance at connections.

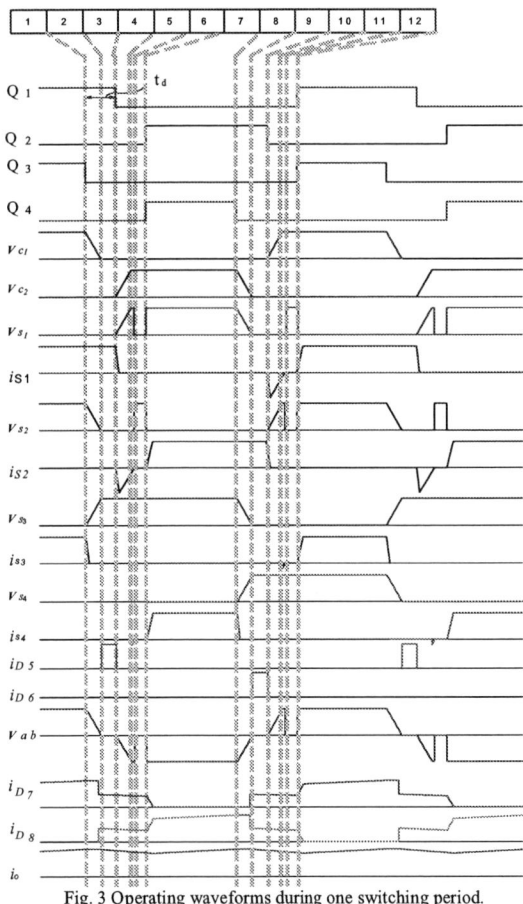

Fig. 3 Operating waveforms during one switching period.

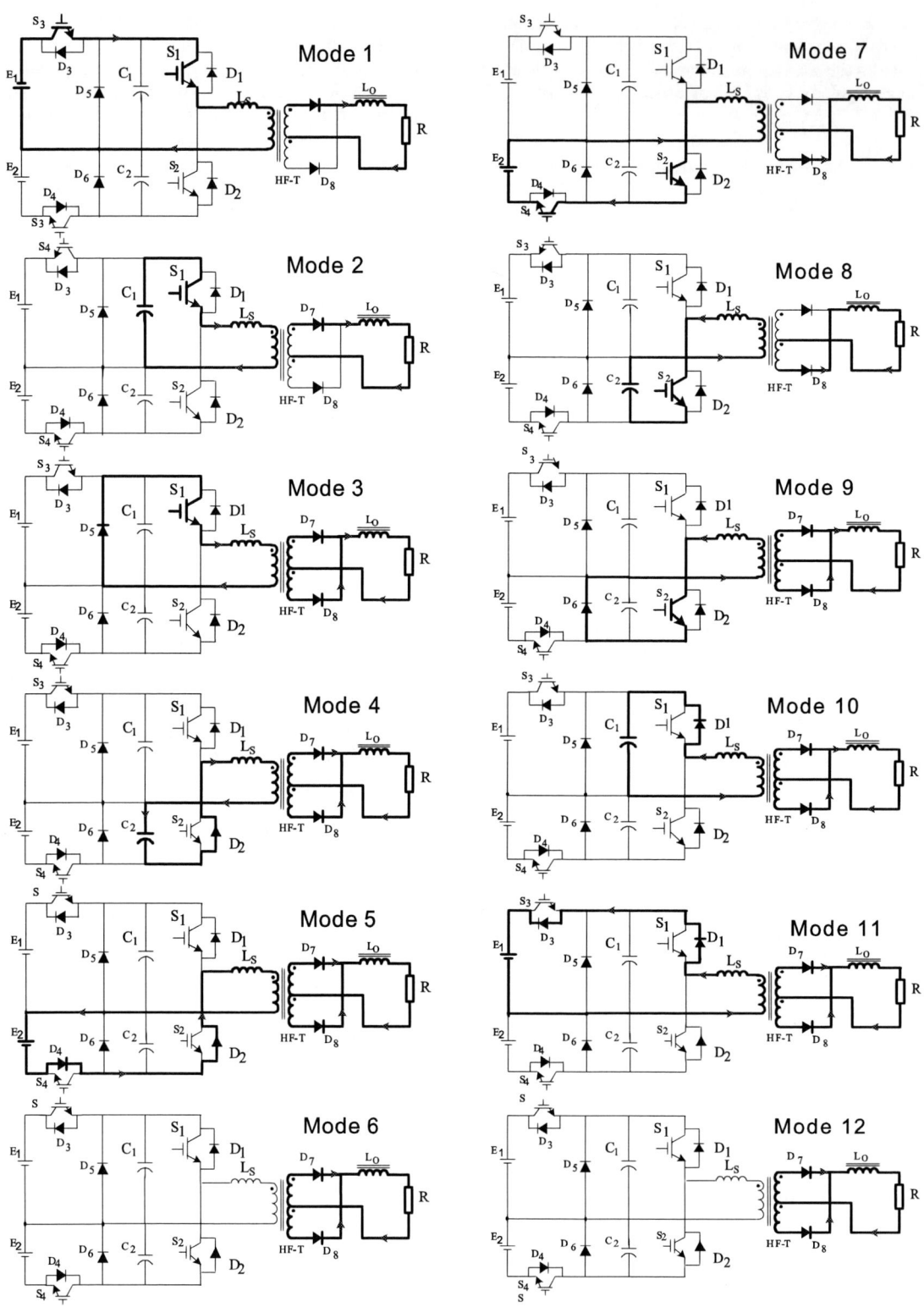

Fig. 4 Equivalent circuits and operation modes for one switching Cycle.

B. Measured Voltage and Current Switching Waveforms

The switching operating waveforms for voltage and current when the switch Q_1 is turned on and turned off are shown in Fig. 6. Observing these waveforms, the switch Q_1 is turned on with ZCS and is turned off with ZVS. The switching waveforms for voltage and current when the switch Q_3 is turned on and turned off are shown respectively in Fig 6. Observing the operating waveforms, the switch Q_3 is turned on with ZVS/ZCS and is turned off with ZVS. However, at the turn-off transition for Q_1 and Q_3, some power loss still exists due to tail current characteristic of IGBTs.

C. Comparative Results of Power Los Analysis

In power loss analysis and evaluations shown in Fig.7, the total power loss of all the switches including Q_3 and Q_4 in the proposed dc-dc power converter circuit is compared with that of all the switches in conventional hard switching one. When the switching frequency is 10 kHz, the total power losses for both converter circuits are almost equal each other. The more the switching frequency of inverter increases, the more this proposed converter circuit has remarkable advantages as for the power conversion efficiency and power density as compared with the conventional hard switching dc-dc power converter.

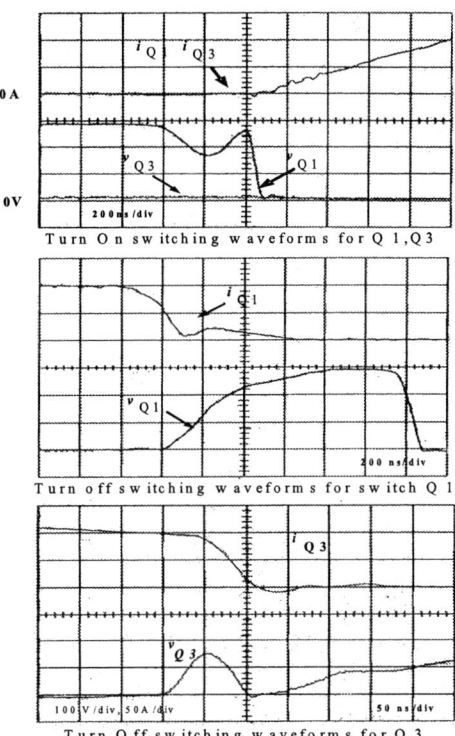

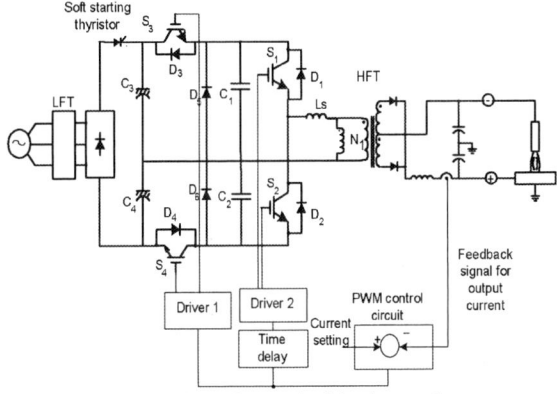

Fig. 5 Experimental setup circuit implementation.

V. CONCLUSIONS

In this paper, the novel circuit topology of half-bridge soft switching PWM dc-dc power converter with a high frequency link was presented. The operating principle and switching pattern of the half-bridge soft switching PWM dc-dc power converter described for 40 kHz, 36V, 400A output are illustrated and discussed for low voltage and large current output applications. The power loss analysis of the proposed soft switching power converter was discussed and evaluated as compared with hard switching PWM dc-dc power converter with a high frequency link. Under the simple circuit which has two additional semiconductor switching devices and two passive components to the typical half-bridge inverter, all the active switching devices achieve ZVS turn-off or ZCS turn-on commutation.

Fig. 6 Measured waveforms for active power switches Q_1 and Q_3

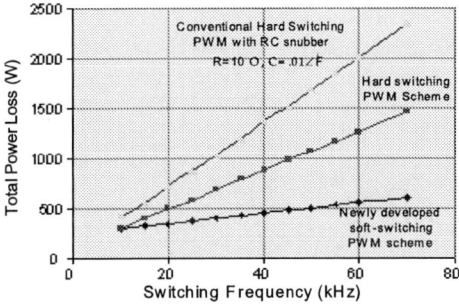

Fig. 7 Comparative power loss analysis between newly developed and conventional hard switching converters.

REFERENCES

[1] S. Hamada, M. Nakaoka, "Saturable Inductor-Assisted ZVS-PWM Full-Bridge High-Frequency Link DC-DC Power Converter Operating and Conduction Losses", Proceedings of IEE International Conference on Power Electronics and Variable-Speed Drives 1994, pp.483-488.

[2] O. D .Patterson and D. M. Divan, "Pseudo-Resonant Full Bridge DC/DC Converter", Records of IEEE-PESC, pp.424-430, June,1987

[3] M. Michihira, M. Nakaoka, "A Novel Quasi-Resonant DC-DC Converter using Phase-Shifted Modulation in Secondary-Side of High-Frequency Transformer", Records of IEEE-PELS Power Electronics Specialists Conference, Vol.1, pp100-105, June, 1996

[4] S. Moisseev, S. Hamada, M. Nakaoka, "Novel Soft-Switching Phase-Shift PWM DC-DC Converter", Proceedings of Japan Society Power Electronics , Vol.28, pp107-116, 2003.

[5] K.Morimoto, T. Doi, H. Manabe, N.A. Ahmed, H.W Lee, M. Nakaoka, "Advanced High Power DC-DC Converter using Novel Type Half-Bridge Soft Switching PWM Inverter with High Frequency Transformer for Arc Welder", Proceeding of IEEE-PEDS, pp113-118, November, 2005.

2006 5th International Power Electronics and Motion Control Conference

Dynamic Analysis of a Current Source Inductively Coupled Power Transfer System

Wenqi Zhou, Hao Ma

College of Electrical Engineering, Zhejiang University, Hangzhou, China

mahao@cee.zju.edu.cn

Abstract—In this paper, dynamic analysis of a current source ICPT system is performed based on the circuit equations of each stage in an operation cycle. From the solution, frequencies of different stages are discussed. Besides a common frequency of the circuit, the other two are resonant frequencies of each side of the transformer. A simplified form of frequency expression is obtained. It is found that in certain circumstance, the resonant frequencies of primary and secondary sides are different and depend on elements of their respective sides. Some suggestions about the parameter selection are provided. The analytical analysis and considerations are verified by a practical ICPT system.

Keywords- ICPT; dynamic; resonant frequency; current source

I. INTRODUCTION

As a relatively new research area, inductively coupled power transfer (ICPT) systems have received ever-growing attention. In ICPT system, power is transferred from one device to the other through electromagnetic field with a separated transformer. If one or both of the devices is mobile, or in certain environment that requires electric isolation, ICPT technology is advantageous over the conventional conductive method using wires and connectors, as it is unaffected by dirt, dust, water, or other chemicals and is maintenance free. Because of all these potential enhanced safety, reliability and convenience, it can be used in non-contact battery charging for electrical vehicles [1][2], material handling systems, mining or under water working environments [3][4], and some medical applications where energy can be transferred to the implanted circulatory assist devices through the intact skin of a patient [5].

In all these applications, the transformer has to be separate, so that there is a large air gap between the primary and secondary windings of the transformer. Consequently, the leakage inductance is much larger, the magnetizing inductance is smaller and the coupling coefficient is very small compared with common ones. To improve power transfer capability, resonant converters that involve the leakage inductance as a circuit component is a good choice.

Fig. 1 shows a current source ICPT system, where both primary and secondary are parallel compensated. I_d is a current source; a completely coupled phase splitter L_{a1},

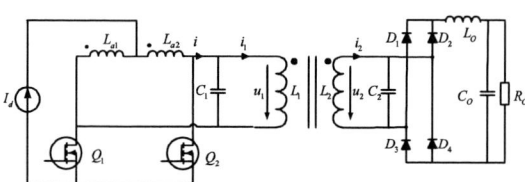

Fig. 1 Current source ICPT system

L_{a2} (much larger than the primary inductance L_1), and the complementarily operated switches constitute a push-pull inverter, which produces a low distortion square current i at the input of the resonant tank. So power could be efficiently transferred to the load through primary and secondary resonance.

The analyses of resonant circuits have begun more than a decade ago. They are almost based on closely coupled transformer. Recently attention has been paid to ICPT systems. By far there are a number of articles that have discussed the steady state and dynamic characteristics of the circuit. For the voltage source ICPT, [6] presents a circuit model, which is composed of 3 stages, and the voltage gain is obtained as a function of frequency and related elements; [7] gives a detailed stage analysis of an entire operation cycle; [8] and [9] identify all the steady-state operational modes of a series load ICPT in both below- and above-resonance operation. Detailed dynamic and steady-state solutions are obtained, and boundaries between CCM and different DCM modes are determined. But for the current source topology, [10] presents a DC analysis technique for the secondary pick-ups, where DC equivalent circuit model and the analytic transfer functions are developed, by the assumption that pick-up tuning are perfect and all the diodes have continuous current; [11] has discussed the steady state operation using a power flow balance analysis between the inverter and the resonant tank; [12] presents a bifurcation phenomena of the resonant frequency using load model, where the load is assumed to be linear; however, about the dynamic stage analysis without linear assumption and the discussion of the resonant frequency based on it for the current source ICPT, few publications can be found.

In this paper, some design procedure is proposed based on dynamic equations of the circuit. In Section Ⅱ, stage analysis is performed and the dynamic circuit equations are obtained. Based on the equations, operation frequency is analyzed in detail in Section Ⅲ. A simplified form of resonant frequencies in both sides of the transformer is

1-4244-0448-7/06/$25.00 ©2006 IEEE

presented, and some limit conditions are suggested. Finally in Section IV, experimental results are shown to validate what have been discussed.

II. CIRCUIT MODEL AND DYNAMIC ANALYSIS

Based on the configuration of Fig. 1, a simplified circuit model is built as Fig. 2 shown, where the transformer is replaced by its "T" equivalent model, and all the secondary component values must be referred to the primary side. To simplify the analysis, following assumptions are made:

1) i_s is taken as an ideal square current source;

2) The capacitors, diodes and inductors are all ideal components;

3) The load current i_0 remains constant, i.e. output filter are considered ideal;

4) The losses in transformer are negligible.

In this T model, turns ratio n is defined as n_1/n_2, where n_1 and n_2 are primary and secondary turns, respectively. The magnetizing inductance and both sides leakage inductances referred to the primary side are[13][14]:

$$L_m = M \cdot n \qquad (1)$$
$$L_{s1} = L_1 - L_m = L_1 - M \cdot n \qquad (2)$$
$$L_{s2}' = n^2(L_2 - M/n) \qquad (3)$$

where M is the mutrual inductance of the two winding inductances L_1, L_2. And M is given as

$M = k\sqrt{L_1 L_2}$, where k is the coupling coefficient.

Fig. 3 gives the key waveforms of the circuit at resonant frequency. When primary voltage and current are near sinusoidal at this frequency, secondary waveforms have some distortion because of the loosely coupled transformer and the nonlinear rectifier of the secondary. Different diodes of the rectifier bridge will conduct depending on the polarity of secondary capacitor voltage. Since the secondary capacitor voltage is discontinuous, there is an additional interval during which secondary circuit remains shorted with all the diodes conducting. Here turns ratio n is supposed to be larger than 1 and the value of compensation capacitors are close to each other. It is found that these factors will affect the waveforms so that when they change, the waveform will be distorted to other shapes. The reason why they are so decided will be discussed later in Section □.

Therefore, a steady-state period T can be decomposed into six stages according to the state of the input current and the conduction of different diodes. State variables are

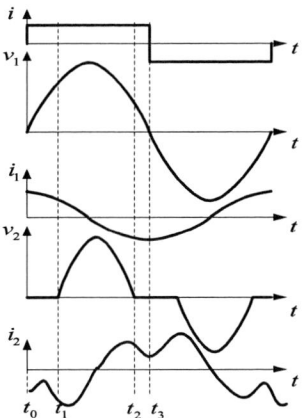

Fig. 3 key waveforms of ICPT at resonant frequency

selected as the inductance current i_1, i_2 and capacitor voltage u_1, u_2. Only the first half of a period T is considered because analysis of the other half of the period is similar to this for the symmetry of the circuit.

Stage 1(t_0-t_1): At this time interval, input square current is positive, but secondary is shorted by all the diodes of the rectifier bridge. Power stored in L_0 will circulate in load and diodes. The equivalent circuit is shown as Fig. 4.

Circuit equations are:

$$i_s = C_1 \frac{du_1}{dt} + i_1 \qquad (4)$$

$$u_1 = L_k \frac{di_1}{dt} \; ; \; u_2' = 0 \qquad (5)$$

where
$$L_k = L_{s1} + L_m /\!/ L_{s2}' \qquad (6)$$

Written in matrix:

$$
\begin{bmatrix} \dfrac{di_1}{dt} \\ \dfrac{di_2'}{dt} \\ \dfrac{du_1}{dt} \\ \dfrac{du_2'}{dt} \end{bmatrix} =
\begin{bmatrix} 0 & 0 & \dfrac{1}{L_k} & 0 \\ 0 & 0 & \dfrac{L_k - L_{s1}}{L_{s2}'L_k} & 0 \\ -\dfrac{1}{C_1} & 0 & 0 & 0 \\ 0 & 0 & 0 & 0 \end{bmatrix}
\begin{bmatrix} i_1 \\ i_2' \\ u_1 \\ u_2' \end{bmatrix} +
\begin{bmatrix} 0 & 0 \\ 0 & 0 \\ \dfrac{1}{C_1} & 0 \\ 0 & 0 \end{bmatrix}
\begin{bmatrix} i_s \\ i_0' \end{bmatrix} \quad (7)
$$

or $\dot{x} = A_1 x + B_1 u$, with the initial value $x(t_0)$ (8)

Stage 2 (t_1-t_2): At this time interval, D_1 and D_3 conduct. Equivalent circuit is shown in Fig. 5. Circuit equations are:

$$i_s = C_1 \frac{du_1}{dt} + i_1 \qquad (9)$$

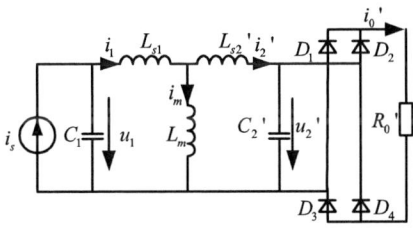

Fig. 2 ICPT circuit model

Fig. 4 equivalent circuit of Stage 1

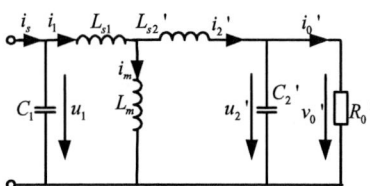

Fig. 5 equivalent circuit of Stage 2

$$u_1 = L_{s1}\frac{di_1}{dt} + L_m\frac{di_m}{dt} \qquad (10)$$

$$i_m = i_1 - i_2' \qquad (11)$$

$$L_m\frac{di_m}{dt} = u_2' + L_{s2}'\frac{di_2'}{dt} \qquad (12)$$

$$C_2'\frac{du_2'}{dt} = i_2' - i_0' \qquad (13)$$

(9)- (13) can be simplified and written in matrix:

$$\begin{bmatrix} \frac{di_1}{dt} \\ \frac{di_2'}{dt} \\ \frac{du_1}{dt} \\ \frac{du_2'}{dt} \end{bmatrix} = \begin{bmatrix} 0 & 0 & \frac{1}{L_a} & -\frac{1}{L_b} \\ 0 & 0 & \frac{1}{L_c} & -\frac{1}{L_d} \\ -\frac{1}{C_1} & 0 & 0 & 0 \\ 0 & \frac{1}{C_2'} & 0 & 0 \end{bmatrix} \begin{bmatrix} i_1 \\ i_2' \\ u_1 \\ u_2' \end{bmatrix} + \begin{bmatrix} 0 & 0 \\ 0 & 0 \\ \frac{1}{C_1} & 0 \\ 0 & -\frac{1}{C_2'} \end{bmatrix} \begin{bmatrix} i_s \\ i_0' \end{bmatrix} \qquad (14)$$

or $\dot{x} = A_2 x + B_2 u$ with the initial value $x(t_1)$. (15)

where

$$L_a = \frac{L_{s1}L_{s2}' + (L_{s1} + L_{s2}')L_m}{L_{s2}' + L_m} \qquad (16)$$

$$L_b = L_c = \frac{L_{s1}L_{s2}' + (L_{s1} + L_{s2}')L_m}{L_m} \qquad (17)$$

$$L_d = \frac{L_{s1}L_{s2}' + (L_{s1} + L_{s2}')L_m}{L_{s1} + L_m} \qquad (18)$$

Stage 3 (t_2-t_3): in this stage, the equivalent circuit is the same as stage 1, as shown in Fig. 4, and the circuit equation is the same as (10). Here only the subscript is changed to keep consistency:

$$\dot{x} = A_3 x + B_3 u, \text{ but the initial value is } x(t_2). \qquad (19)$$

The general solution of a state equation of stage k can be expressed as:

$$x(t) = e^{A_k(t-t_{k-1})}x(t_{k-1}) + \int_{t_{k-1}}^{t} e^{A_k(t-\tau)}B_k u(\tau)d\tau \qquad (20)$$

where $x(t_{k-1})$ is the initial value of this stage, namely the final value of last stage. And the final value of stage k is given as:

$$x(t_k) = e^{A_k(t_k-t_{k-1})}x(t_{k-1}) + \int_{t_{k-1}}^{t_k} e^{A_k(t-\tau)}B_k u(\tau)d\tau \qquad (21)$$

Thus, with circuit equations of stage 1 to stage 3 and their initial values, dynamic equation of each state variable can be obtained. But as solutions of a group of 4-order equations, the results are a little too complicated to give explicit formulas. To understand the dynamic circuit characteristics, numerical solutions may be helpful. However, the resonant frequency can be easily obtained from the equations, in next section the operation frequency will be discussed in detail.

III. RESONANT FREQUENCY ANALYSIS

From (8) and (19) it is known that, in stage 1 and 3, there are only one frequency component:

$$\omega_1 = \frac{1}{\sqrt{C_1 L_k}} \qquad (22)$$

While in stage 2, there are two frequency components in the solution (obtained from the characteristic roots of (15)):

$$\omega_{2,3} = [\frac{1}{2}(\frac{1}{C_1 L_a} + \frac{1}{C_2' L_d} \pm r)]^{1/2} \qquad (23)$$

where

$$r = (\frac{1}{C_1^2 L_a^2} + \frac{1}{C_2'^2 L_d^2} + \frac{4}{C_1 C_2' L_b L_c} - \frac{2}{C_1 C_2' L_a L_d})^{1/2} \quad (24)$$

This is why secondary waveforms have distortion of a higher frequency component apart from the primary dominant resonant frequency.

Since the result is too complicated, a question is raised whether the expression can be simplified in some condition.

Fig. 6 shows the plots that how ω varies when different parameter changes. It is interesting to find from the figure that when the value of inductances is fixed, the frequencies are mainly dependent on one capacitor. The lower frequency ω_2 is mainly relevant to C_1, and the higher frequency ω_3 is mainly relevant to C_2, if the values of capacitors are not too small. Similarly, when capacitor values are fixed, the frequencies seem to only relevant to one inductance: ω_2 mainly depends on L_1 and ω_3 mainly depends on L_2, when the inductances are not chosen too small, i.e. larger than 10μH. The variety ranges of all the parameters are chosen to enable a likely operation frequency of the circuit.

Let's look at the frequency expression carefully. Actually, in our leakage transformer model, the secondary parameters have to be referred to the primary side. When n is larger than 1, especially 2 or larger, C_2' becomes much smaller, hence if C_1 and C_2 are in the same order, the terms in the expression including C_1 are relatively smaller and can be neglected, which results in a simpler equation:

$$\omega_{2,3} = [\frac{1}{2}(\frac{1}{C_1 L_a} + \frac{1}{C_2' L_d} \pm \frac{1}{C_2' L_d})]^{1/2} \qquad (25)$$

It is clear that the frequencies are approximated to:

$$\omega_2 = \frac{1}{\sqrt{2C_1 L_a}}, \quad \omega_3 = \frac{1}{\sqrt{C_2' L_d}} \qquad (26)$$

1314

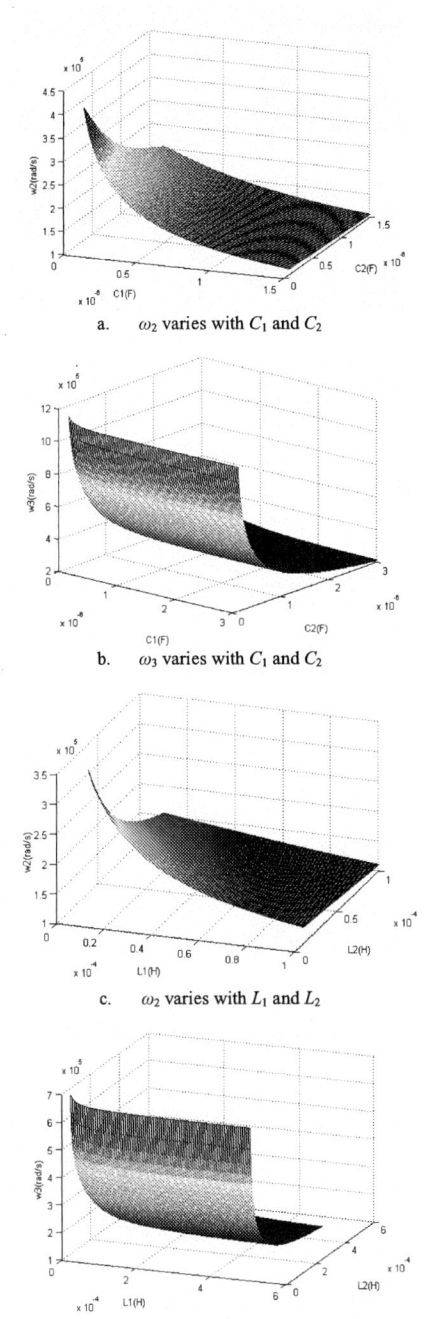

a. ω_2 varies with C_1 and C_2

b. ω_3 varies with C_1 and C_2

c. ω_2 varies with L_1 and L_2

d. ω_3 varies with L_1 and L_2

Fig. 6 ω varies when different parameters change

To give a specific explanation of L_a and L_d, let's examine the definitions of them in (16) and (18). It is not difficult to find the meanings of L_a and L_d from the definitions by some equivalent transformation. So (16) and (18) can be written as:

$$L_a = L_{s1} + L_m \mathbin{/\!/} L_{s2}{}'\tag{27}$$

$$L_d = L_{s2}{}' + L_m \mathbin{/\!/} L_{s1}\tag{28}$$

Expression (27) and (28) show that L_a is the equivalent inductor seeing from the primary when secondary shorted.

Accordingly, L_d is the equivalent inductor seeing from the secondary when primary shorted. According to transformer "T" model in (1)-(3), (27)-(28) are equivalent to:

$$L_a = (1-k^2)L_1\tag{29}$$

$$L_d = n^2(1-k^2)L_2\tag{30}$$

Therefore, L_a and L_d are related to the inductance of their respective sides, when k and n are decided.

After all the analysis before, the result is exciting, which means that the frequency of each side only depends on parameters of their own resonant tank in certain circumstances. The condition can be generalized as: C_1 and C_2 are in the same order, n is larger than 2, and L_1, L_2 can not be too small (larger than 10µH).

It is interesting to compare the frequency of stage 1, 3 and stage 2. In fact, from (6) and (27), it can be found that L_k is equal to L_a, so that if the condition is satisfied, ω_1 and ω_2 will have the relationship as following:

$$\omega_1 = \sqrt{2}\,\omega_2\tag{31}$$

As a result, the practical operation frequency is often between the two frequencies when primary is resonant, usually in the middle of them. In next section it will be illustrated by experimental results.

Now let us discuss the turns ratio n briefly. For all the analysis before, n is chosen larger than 1 (better larger than 2). In this circuit, resonant frequency of primary side is often a dominant frequency component to enable good power transfer ability. If n is smaller than 1, secondary side resonant frequency will be lower than the primary side, and the peak resonant voltage and power transferred from primary will be greatly reduced at such a higher operation frequency. So turns ratio discussed here is supposed to be larger than 1 in consideration of power transfer ability.

IV. EXPERIMENTAL RESULT

A prototype with Fig. 1 configuration has been built in the lab, whose parameters are chosen according to the analysis in Section II and III. Parameters about the transformer are shown in TABLE I, and in TABLE II some parameters of the circuit are listed.

TABLE I.
TRANSFORMER PARAMETERS FOR DIFFERENT k

Test No.	No.1	No.2	No.3
k	0.603	0.631	0.694
L_1(µH)	55.2	52.80	61.9
L_{1s}(µH)	10.5	10.5	8.83
L_2(µH)	16.6	14.23	17.2
L_{2s}(µH)	9.1	7.2	7.6
L_m(µH)	44.7	42.3	57.1
M(µH)	18.3	17.29	23.4
Turns ratio n_1:n_2	22: 9	22: 9	22: 9

TABLE II.
PARAMETERS OF THE CIRCUIT

L_{a1}, L_{a2} (mH)	2
$C_1(\mu F)$	0.4
$C_2(\mu F)$	0.6
$Ro(\Omega)$	5.3
Core type	EE55
Q1,Q2	SPW17N80C3

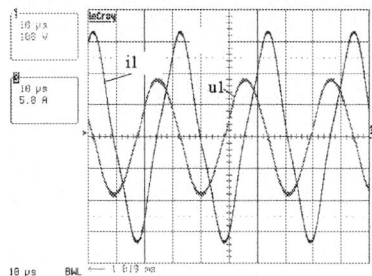

Fig. 7 primary voltage and current

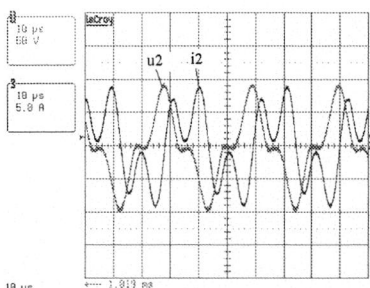

Fig. 8 secondary voltage and current

TABLE III.
THEORETICAL AND EXPERIMENTAL RESONANT FREQUENCY

	No.1	No.2	No.3
$\omega_1(10^3 rad/s)$	217.9	229.0	220.8
$\omega_2(10^3 rad/s)$	167.0	171.0	156.1
ω_2(approximated)$(10^3 rad/s)$	154.1	161.9	152.2
$\omega_3(10^3 rad/s)$	506.2	561.2	553.1
ω_3(approximated)$(10^3 rad/s)$	486.5	540.2	529.4
Experimental main angular frequency $(10^3 rad/s)$	191.5	195.3	188.4
Experimental frequency(kHz)	30.5	31.1	30.0

Fig. 7 and Fig. 8 show typical waveforms of the circuit at main resonant frequency, when primary voltage and current are near sinusoidal.

The theoretical and experimental resonant frequencies are compared in TABLE III. Both (23) and (26) are calculated to verify the approximation. As expected, the approximated frequencies are close to their analytical values. And it can be seen that the practical operation frequency is between ω_1 and ω_2, usually in the middle of them, as discussed before.

V. CONCLUSION

Dynamic analysis of a current source ICPT system is carried out in this paper. Based on the dynamic circuit equations of each stage in an operation period, the general solution is discussed. From the solution, frequencies of different stages are discussed, and totally three frequency components in the circuit are calculated. Besides a common frequency of the circuit, the other two are resonant frequencies of each side. A simplified form of frequency expression is obtained. It is found that in certain circumstance, the resonant frequencies of primary and secondary are different and depend on elements of their respective sides. The analytical analysis and considerations are confirmed by a practical ICPT system.

REFERENCES

[1] A. Esser, "Contactless charging and communication for electric vehicles," *IEEE Industry Applications Magazine,* vol. 1, no. 6, pp.4-11, Nov.-Dec. 1995

[2] K. W. klontz, A. Esser and P. J. wolfs, "Converter selction for electric vehicle charger systems with a high-frequency high-power link," Power Electronics Specialists Conference 1993, pp. 855-861, Jun. 1993

[3] T. Kojiya, F. Sato , H. Matsuki and T. Sato, "Automatic power supply system to underwater vehicles utilizing non-contacting technology," Oceans '04. MTS/IEEE TECHNO-OCEAN '04, vol. 4, pp.2341 – 2345, 2004

[4] Jia Junlin, Liu Weigang and Wang Haiqun, "Contactless power delivery system for the underground flat transit of mining," Electrical Machines and Systems 2003, ICEMS 2003. vol. 1, pp.282 – 284, Nov. 2003

[5] J. Diaz, J. M. Lopera and J. V. Comas, "Pulmonary blood flood flow regulation with contactless energy transmission system," proceeding of the second EMBS/BMES conference, pp. 1538-1539, Oct. 2002

[6] Wonseok Lim, Jaehyun Nho, Byungcho Choi and Taeyoung Ahn, "Design and implementation of low-profile contactless battery charger using planar printed circuit board windings as energy transfer device," *IEEE Trans. Indus. Electro.,* vol. 51, no. 1, pp. 140 – 147, Feb. 2004

[7] Yungtaek Jang and M. M. Jovanovic´, "A contactless electrical energy transmission system for portable telephone battery chargers," *IEEE Trans. Indus. Electro.,* vol. 50, no. 3, pp. 520-527, Jun. 2003

[8] Hai-Jiang Jiang and G. Maggetto, "Identification of steady-state operational modes of the series resonant DC–DC converter based on loosely coupled transformers in below-resonance operation," IEEE *Trans. Power Electron.,* vol. 14, no. 2, pp. 359-371, Mar. 1999

[9] Hai-Jiang Jiang, G. Maggetto, and P. Lataire, "steady-state analysis of the series resonant DC–DC converter in conjunction with loosely coupled transformer-above resonance operation," *IEEE Trans. Power Electron.,* vol. 14, no. 3, pp. 469-480, May. 1999

[10] J. T. Boys, G. A. Covic and Yongxiang Xu, "DC analysis technique for inductive power transfer pick-ups," *IEEE Power Electron. Letters,* vol. 1, no. 2, pp. 51-53, Jun. 2003

[11] Chwei-Sen Wang, Grant A. Covic, Oskar H. and Stielau, "Investigating an LCL load resonant inverter for inductive power transfer applications," *IEEE Trans. Power Electron.,* vol. 19, no. 4, pp. 995-1002, Jul. 2004

[12] C. S. Wang, G. A. Covic and O. H. Stielau, "Power transfer capability and bifurcation phenomena of loosely coupled inductive power transfer systems," *IEEE Trans. Indus. Electron.,* vol. 51, no. 1, pp. 148-157, Feb. 2004

[13] J. G. Hayes, N. O'Donovan, M. G. Egan and T. O'Donnell, "Inductance characterization of high-leakage transformers," Applied Power Electronics Conference and Exposition 2003, APEC '03, vol. 2, pp. 1150 – 1156, Feb. 2003

[14] A. Ghahary and B. H. Cho, "Design of a transcutaneous energy transmission system using a series resonant converter," Power Electronics Specialists Conference 1990, PESC '90, pp.1-8, Jun. 1990

2006 5th International Power Electronics and Motion Control Conference

A New Topology of Capacitor-Clamp Cascade Multilevel Converters

Anees Abu Sneineh, Ming-yanWang and Kai Tian

Harbin Institute of Technology/School of Electrical Engineering and Automation, Habin, China

enganees2002@yahoo.com, mingyan@hit.edu.cn, tk2002_0@163.com

Abstract— **Use of multilevel converter has become popular in recent years. This paper will present a new topology of capacitor-clamp cascade multilevel converter that is derived from two popular topologies. The new concept of the novel capacitor-clamp cascade converter is based on the connection of multiple three-level capacitor-clamp converter modules. Nine level waveform of the proposed multilevel converter is synthesized by adding of each converter output voltage. Sub-harmonic PWM method is employed in the new topology. The proposed converter is also verified by computer simulation using MATLAB-Simulink. Simulation results are also presented in this paper.**

Keywords- Cascade; Capacitor-clamp; Multilevel converter; Topology; Sub-Harmonic PWM

I. INTRODUCTION

Multilevel power conversion has become increasingly popular in recent years due to its advantages [1-6]. The general concept of the researches in power conversions involves producing ac waveform from small voltage steps by utilizing isolated dc sources or a bank of series capacitors. The small voltage steps yield waveforms with low harmonic distortion as well as low dv/dt. The advantages of multilevel converters if they are compared with conventional two-level converters are the capability of increasing the output voltage magnitude and reducing the output voltage and current harmonic content, the switching frequency and the voltage supported by each power semiconductors.

The most famous three types of multilevel converters are diode-clamp, flying-capacitor, and cascade converter with separated DC sources have been developed. In fact, the cascade multilevel converter with separated DC sources has been shown many advantages over the other two. Specially, the modularized circuit layout and packaging are possible. This makes the cascade multilevel converter feasible for manufacturing.

Fig.1 shows a schematic of a single-phase cascade converter in which two cell of traditional two-level power converters with separated DC sources are series-connected [1, 2]. The output waveform is synthesized by adding of each converter output voltage. Assuming the DC bus voltage of each converter is E. Based on switch combinations, five output voltage levels can be synthesized (0, ±E, ±2E).

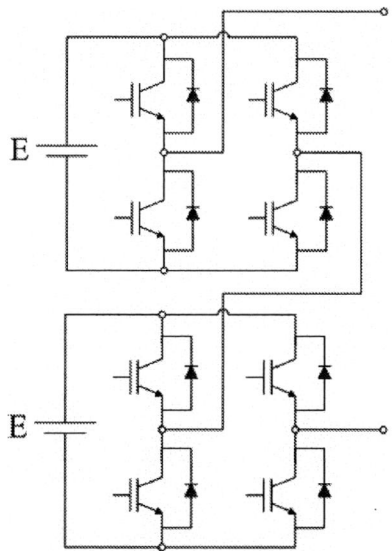

Fig.1. Single phase Cascade multilevel convener

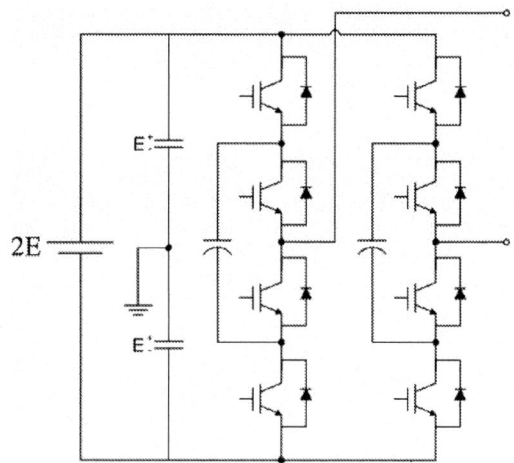

Fig.2. Capacitor-clamp multilevel convener

In traditional cascade converter, the dc bus voltage of each module has the same value, and switching frequency and voltage blocking capability of all switches are the same.

Obviously by using proper control method applied to this topology, the stepped output waveform can be approximate to a sinusoidal waveform. Here, many algorithms can be employed, such as optimized stepped

1-4244-0448-7/06/$25.00 ©2006 IEEE   1318

waveform method, Sub-harmonic elimination PWM method, etc. [3, 4].

In general, the output voltage of a given multilevel converter can be calculated [5] from:

$$V_o = (S - \frac{n-1}{2})E \qquad (1)$$

Where V_o is the output of the multilevel converter, n is the number of the output levels; S is the switching state that ranges from 0 to (n-1). E is the minimum voltage level the multilevel converter can produce. For example, when S= 0, 1, 2, 3, 4, then from (1) five output levels can be synthesized respectively.

Fig.2 shows a schematic of a single phase Capacitor-clamp multilevel converter. Assuming the DC bus voltage of the converter is 2E, it can be easily found that there are five-levels in the output waveform, according to (1), obtains:

$$V_o = (S - 2)E \qquad (2)$$

Where E is the minimum voltage level, S=0, 1, 2, 3, 4.

For S select 0, 1, 2, 3, 4, five different values of the output voltage can be achieved, i.e. (0, ±E, ±2E) accordingly. Approximate sinusoidal output waveform can be achieved by applying proper control methods.

II. A NEW CAPACITOR-CLAMP CASCADE MULTILEVE CONVERTER

The basic concept of the cascade converter shown in Fig.1 is based on the connection of multiple two-level traditional power converter modules; the output waveform is synthesized by adding of each converter output voltage. The proposed new capacitor-clamp cascade topology is based on the traditional cascade and capacitor-clamp converter that mentioned in part I. The novel concept of the proposed topology is based on the connection of several 3-level capacitor-clamp converter modules, and the multilevel waveform is synthesized by adding of each converter output voltage. Unlike converters for different topologies proposed in ref. [6].

To clearly explain the proposed converter, an example of the novel converter is shown in Fig.3. It consists of two capacitor-clamp multilevel converters that are series connected together. Suppose the DC bus voltage of each capacitor-clamp converter is 2E, five output voltage levels of each cell in the proposed converter can be synthesized, i.e. (0, ±E, ±2E). The synthesized output waveform of the whole converter is nine levels in total (0, ±E, ±2E, ±3E, ±4E). According to (1) we get:

$$V_{ac} = (S - \frac{9-1}{2})E = (S - 4)E \qquad (3)$$

Where S=0, 1, 2, 3, 4, 5, 6, 7, 8, E is minimum voltage level the converter produces. From (3), the output level of the proposed converter can be defined by the switching states S, such as, when S=7, from (3), V_{ac}= +3E.

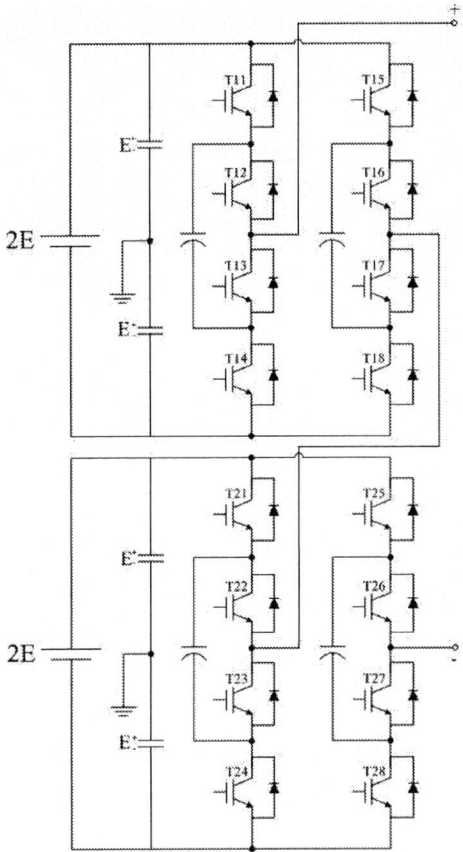

Fig.3. New capacitor-Clamp Cascade Multilevel convener

Table I shows a group of switch combinations of the proposed converter .Because of more than one group of switch combination can be used to produce some certain voltage levels. For example, there are two switch combinations when yielding +E of cell one. So, Table I only shows a group of switching states. Obviously, more combination choices lead to complexity of control.

The advantage of this topology is that it provides flexibility for expansion of the number of levels easily without introducing undue complexity in the power circuit [7]. Moreover, it requires the same number of switches as in diode-clamped cascade topology to achieve a given number of voltage levels [8]. One of the disadvantages in using a diode-clamped inverter is that the required voltage blocking capability of the clamping diodes varies with the levels. This may result in the requirement of multiple series diodes at the higher voltage levels.

Another advantage of this topology is that there is no unbalanced capacitor voltage problem because of its independent voltage source structure. However, a problem associated with this approach is the requirement of a complicated control strategy to regulate the floating capacitor voltages [9, 10].

Table I
Switching combinations states of the proposed converter

S	Output	Cell One	Cell Two	T₁₁	T₁₂	T₁₃	T₁₄	T₁₅	T₁₆	T₁₇	T₁₈	T₂₁	T₂₂	T₂₃	T₂₄	T₂₅	T₂₆	T₂₇	T₂₈
8	+4E	+2E	+2E	1	1	0	0	0	0	1	1	1	1	0	0	0	0	1	1
7	+3E	+2E	+E	1	1	0	0	0	0	1	1	1	1	0	0	0	1	0	1
		+E	+2E	1	1	0	0	0	1	0	1	1	1	0	0	0	0	1	1
6	+2E	+2E	0	1	1	0	0	0	0	1	1	0	1	1	0	0	1	1	0
		+E	+E	1	1	0	0	0	1	0	1	1	1	0	0	0	1	0	1
		0	+2E	0	1	1	0	0	1	1	0	1	1	0	0	0	0	1	1
5	+E	+2E	-E	1	1	0	0	0	0	1	1	0	0	1	1	1	0	1	0
		+E	0	1	1	0	0	0	1	0	1	0	1	1	0	0	1	1	0
		0	+E	0	1	1	0	0	1	1	0	1	1	0	0	0	1	0	1
		-E	+2E	0	0	1	1	1	0	1	0	1	1	0	0	0	0	1	1
4	0	+2E	-2E	1	1	0	0	0	0	1	1	0	0	1	1	1	1	0	0
		+E	-E	1	1	0	0	0	1	0	1	0	0	1	1	1	0	1	0
		0	0	0	1	1	0	0	1	1	0	0	1	1	0	0	1	1	0
		-E	+E	0	0	1	1	1	0	1	0	1	1	0	0	0	1	0	1
		-2E	+2E	0	0	1	1	1	1	0	1	1	1	0	0	0	0	1	1
3	-E	-2E	+E	0	0	1	1	1	1	0	1	0	1	1	0	0	0	1	0
		-E	0	0	0	1	1	1	0	1	0	0	1	1	0	0	1	1	0
		0	-E	0	1	1	0	0	1	1	0	0	0	1	1	1	0	1	0
		+E	-2E	1	1	0	0	0	1	0	1	0	0	1	1	1	1	0	0
2	-2E	-2E	0	0	0	1	1	1	1	0	1	0	0	1	1	1	0	1	0
		-E	-E	0	0	1	1	1	0	1	0	0	0	1	1	1	1	0	0
		0	-2E	0	1	1	0	0	1	1	0	0	0	1	1	1	1	0	0
1	-3E	-2E	-E	0	0	1	1	1	1	0	0	0	0	1	1	1	0	1	0
		-E	-2E	0	0	1	1	1	0	1	0	0	0	1	1	1	1	0	0
0	-4E	-2E	-2E	0	0	1	1	1	1	0	0	0	0	1	1	1	1	0	0

III. SUB-HARMONIC PWM

Sub-harmonic PWM (SHPWM) is a conventional control method suit for multilevel converter [3, 4]. The control principle of the Sub-harmonic PWM method is to use several triangular carrier signals with only one modulation wave per phase. For an n-level inverter, (n-1) triangular carrier of the same frequency f_c, and the same peak-to peak amplitude A_c, are disposed so that the bands they occupy are contiguous. The zero reference is placed in the middle of the carrier set. The modulation wave is a sinusoid waveform of frequency f_m and amplitude A_m. At every instant each carrier is compared with the modulation waveform.

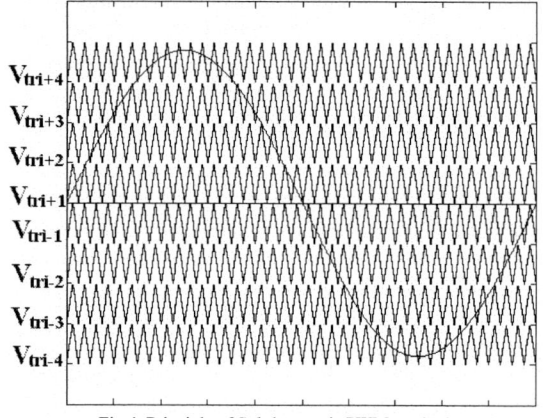

Fig.4. Principle of Sub-harmonic PWM method

Each comparison switches the device on if the reference signal is greater than the triangular carrier assigned to that device level; otherwise, the device switches off. For example, a nine-level inverter shown in Fig.3 is considered. Eight triangular carriers are compared with a sinusoid modulation waveform shown in Fig.4, when $V_{sin} > V_{tri+4}$ the switching state S is 8, the output voltage of the proposed converter V_{ac} is +4E, when $V_{tri+4} > V_{sin} > V_{tri+3}$, the switching state S is 7, V_{ac} is +3E, when $V_{Sin} < V_{tri-4}$, the switching state S is 0, the output voltage V_{ac} is -4E.

Different switch combination can be selected according to the switching state S. For example, approximate sinusoidal can be achieved by applying Sub-harmonic PWM method as well as the switching states in Table I.

IV. VALIDATION

In order to validate the proposed converter, computer simulation using MATLAB-Simulink has been created. The nine-level converter shown in Fig.3 is considered in the simulation. Sub-harmonic method is applied to trigger the power switches for controlling the voltage levels generated on the ac side. The DC bus of each cell is 0.5kV and 110Ω resistor and 100mH inductor were used as the load. The frequency of the triangle carrier waveform f_c is 2 kHz.

The synthesized output simulation waveform of the two cells is shown in Fig.5. Fig.6 shows Fourier analysis of the waveform shown in Fig.5. It can be seen from Fig.6 that the significant harmonics are near 2 kHz, and the components of these harmonics are low. Fig.7 shows the current simulation waveform of the load. Fig.8 shows the Fourier analysis of Fig.7. It can be seen that the outstanding load current waveform is achieved. Through calculation the THD of load current is 1.555%. The rated voltage of the switches in the topology is only 0.25kV, while the DC bus voltage is 0.5kV. Also the phase output peak-to-peak voltage can reach near 2kV.

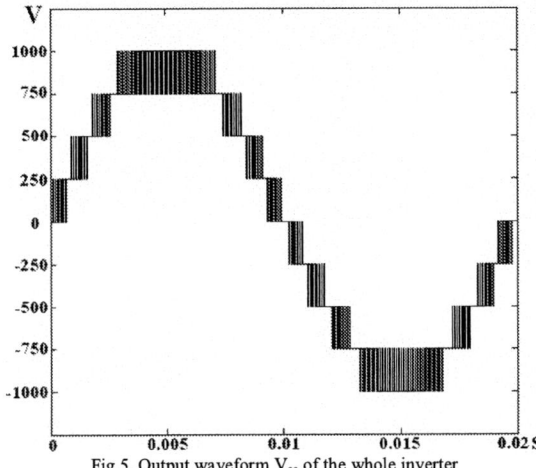

Fig.5. Output waveform V_{ac} of the whole inverter

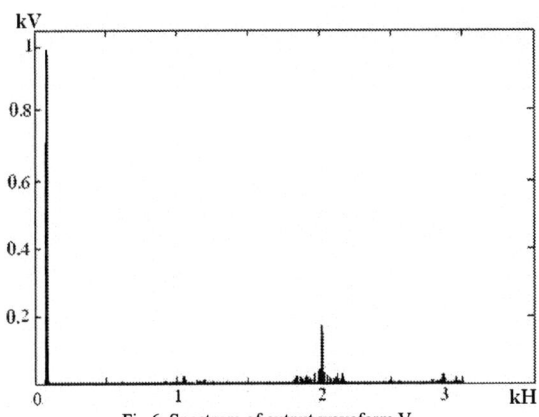

Fig.6. Spectrum of output waveform V_{ac}

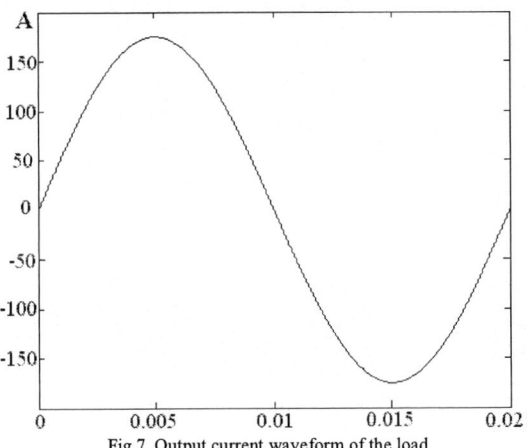

Fig.7. Output current waveform of the load

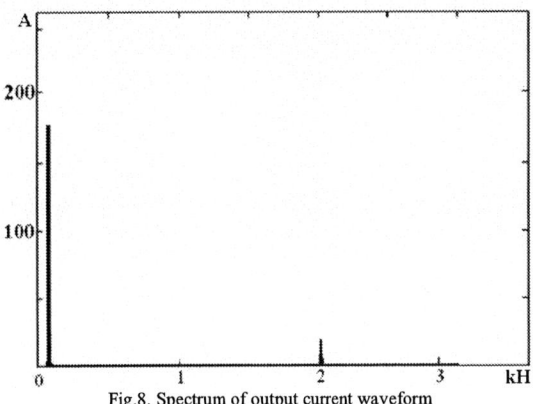

Fig.8. Spectrum of output current waveform

As can be seen from Fig.5, nine-voltage levels are achieved, yielding an outstanding power quality. Furthermore, the output voltage peak-to-peak voltage can reach much higher when higher rated voltage devices and higher DC Bus voltage are involved; with the development of the micro-electronics the control method applied on the novel topology is more readily implemented on a DSP or double-DSP system.

V. CONCLUSION

This paper has presented a new Capacitor-Clamp Cascade multilevel converter. The advantage of this topology is that it provides flexibility for expansion of the number of levels easily without introducing undue complexity in the power circuit. Another advantage of this topology is that there is no unbalanced capacitor voltage problem because of its independent voltage source structure.

The novel concept of the proposed converter is based on the connection of several three level Capacitor-clamp converter cells. The output waveform is synthesized by adding of each capacitor-clamp converter output voltage.

The new converter was verified by Simulation using two cells of a capacitor-clamp converter. To clearly explain, a nine-level converter is studied. Sub-harmonic PWM method is applied to the new topology to trigger the power switches for controlling the voltage levels generated on the output. Simulation results are also presented.

To synthesize nine-level voltage in the conventional capacitor-clamp converter topologies, reported in literatures, needs 16 switches and 28 capacitors. It may be observed that in a simple comparison between the conventional capacitor-clamp converter and the capacitor-clamp cascade multilevel converter that the proposed converter offers the same number of levels at the output with the same number of switches, and the number of capacitors is reduced to four.

REFERENCES

[1] Peter W. Hammond, "Medium Voltage PWM Drive and Method," U. S. Patent 5 625 545, Apr.1997.

[2] M.D. Manjrekar and T.A. Lipo, "A generalized structure of multilevel power converter," in *Proc. IEEE PEDES'98*, 1998, pp. 62–67.

[3] Sung-Jun Park, Feel-Soon Kang, et al. "A New Single-Phase Five-Level PWM Inverter Employing a Deadbeat Control Scheme," *IEEE Trans. on Power Electronics*, Vol. 18, No. 3, MAY 2003, Page 831-843.

[4] Richard lund, Madhav D.Manjrekar, et al. "Control Strategies for a Hybrid Seven-level Inverter," *EPE'99* Conference proceedings, 1999.

[5] Keith Corzine, Yakov Familiant, "A New Cascaded Multilevel H-Bridge Drive," *IEEE Trans. on Power Electronics*, Vol. 17, No. 1, 2002, page 125-131.

[6] Madhav D. Manjrekar, Thomas A. Lipo, "A hybrid multilevel inverter topology for drive applications," *IEEE APEC'98*, 1998, page 523-529.

[7] L. Zhang, S. J. Watkins, W. Shepherd, "Analysis and Control of A Multi-level Flying Capacitor Inverter," IEEE CIEP, Oct. 2002 Page 66-71.

[8] Kai Ding, Yun-ping Zou, et al. "A Novel Hybrid Diode-Clamp Cascade Multilevel Converter for High Power application," *IEEE Industry Applications Conference, 39th IAS Annual Meeting*, Vol. 2, Oct. 2004, Page 820-827.

[9] J.L. Duarte, P.J.M. Jullicher, L.J.J. Offringa, and W.D.H. Groningen, "Stability Analysis of Multilevel Converters with Imbricated Cells," *EPE'97 Conference Proceedings*, page 4.168-4.174, 1997.

[10] Roberto Rojas and Tokuo Ohnishi, "PWM Control Method with Reduction of Total Capacitance Required in a Three-level Inverter," *COBEP'97 Conference Proceedings*, page 103-108, 1997.

Anees Abu Sneineh was born in 1976 in Hebron, Palestine. He received the B.E. degree in Industrial Automation Eng. from Palestine Polytechnic University (PPU), Hebron, Palestine, in 2000, and the M.S. degree in Mechatronics Eng. from Harbin Institute of technology, China, in 2004. He is currently working toward the Ph.D. degree at Harbin Institute of technology, China.

His research interests are modeling, design, and control of power conversion systems of power electronics.

Ming-yan Wang was born in 1957 in Heilongjiang province, China. He received B.E. and M.S. and Ph.D. degree in Electrical engineering and Automation from Harbin Institute of technology, China, in 1982, 1988 and 2002, respectively. In 1999 he becomes the director of department of electrical engineering in Harbin Institute of technology. In 1999 he becomes professor in Electrical engineering and Automation.

His research interests include medium and high voltage rectifiers and inverters, multilevel converters.

Kai Tian was born in Hebei province, China, in 1980. He received the B.E. degree in Electrical engineering and Automation from Harbin Institute of technology, China, in 2004. He is currently working toward the master degree in power electronics and electrical drives at Harbin Institute of Technology, China.

His research interests include medium and high voltage rectifiers and inverters, multilevel converters and modern digital devices.

2006 5th International Power Electronics and Motion Control Conference

Evaluation of Semiconductor Losses in Cryogenic DC-DC Converters

C. Jia* and A. J. Forsyth**
*University of Birmingham, Birmingham, UK
cxj239@bham.ac.uk
**University of Manchester, Manchester, UK
Andrew.Forsyth@manchester.ac.uk

Abstract—A detailed experimental evaluation is reported of DC-DC power electronic converters operating at cryogenic temperatures; the motivation for the study being the eventual integration of a converter with a superconducting electrical machine or energy storage device, SMES, with the benefit of increasing power density. The performance is examined of 120 V and 500 V, 500 W prototypes operating at temperatures down to 20 K; in particular the semiconductor losses are analyzed. Several designs are considered for each voltage level involving different power MOSFET devices, different power diodes, including a silicon carbide Schottky diode, a diode-less synchronous rectifier topology, and a zero voltage resonant transition DC-DC converter. The overall reduction of semiconductor losses at cryogenic temperatures compared with room temperature was up to 85%, depending on the particular circuit.

Keywords-cryogenic power electronics

I. INTRODUCTION

As the development of high-temperature superconductor technology approaches commercial applications, for example superconducting machines for marine propulsion, superconducting fault current limiters and superconducting magnet energy storage [1] [2] [3], it is timely to consider the operation of power electronic converters at cryogenic temperatures, with a view to the possible integration of the power converter in a cryostat with a superconducting device. A number of authors [4] [5] [6] [7] [8] have reported studies into the characteristics of power devices and circuits at very low temperatures, achieving significant performance gains for many device types in terms of both conduction and switching losses. Furthermore, passive component losses are also reduced at very low temperature [9] [10].

Therefore, by integrating a power converter in a cryostat with a superconducting device, the potential exists to achieve substantial gains in converter performance and to increase power density, especially important in transport applications. However, these benefits must obviously be traded against the increased size and power requirements of the refrigeration system.

This paper makes a contribution to the understanding of power converter operation at cryogenic temperatures by studying in detail the performance of several DC-DC converters.

In particular the semiconductor losses are examined in three 500 W buck converter topologies: hard-switched, synchronous rectifier and soft-switched, over a temperature range extending down to 20 K. In the following sections the circuit topologies themselves are first introduced followed by a description of the experimental method and then the results are summarized.

II. CIRCUIT TOPOLOGIES AND DESIGN

The three buck converters are shown in figures 1, 2 and 3 along with their principal waveforms. The basic hard-switched converter, figure 1 operates in the continuous inductor current mode with a comparatively small inductor ripple current I_{Lpp}. In the synchronous rectifier converter of figure 2, the freewheeling diode D has been replaced by a second MOSFET, S2, which is driven in anti-phase with the main MOSFET S1. Therefore, when the main MOSFET, S1, turns off, the inductor current will pass in the reverse direction through the channel of S2, providing that the channel resistance of the device is sufficiently low. However if the product of inductor current I_L and on-state resistance R_{DSon} for device S2 is greater than the forward conduction voltage of the MOSFET body diode, then the body diode will conduct instead, and the synchronous rectifier operation will be lost. Furthermore, to prevent the conduction of the body diode of S2 immediately before the turn on of S1, and therefore to avoid a reverse recovery transient, the device signals for S1 and S2 included virtually no dead time [11].

The soft-switching converter, figure 3, is a constant frequency zero-voltage transition topology in which the freewheel diode is again replaced by a second MOSFET S2. The filter inductor value is significantly reduced in size, resulting in a large ripple current I_{Lpp}, and the inductor current is seen to reverse for a short period of time each cycle. The MOSFETs are driven in anti-phase, but with significant dead times, therefore at the turn off instant, each MOSFET commutates a positive drain-source current, which then divides between the snubber

1-4244-0448-7/06/$25.00 ©2006 IEEE

capacitors, C1 and C2, giving rise to an orderly and well controlled transition of the MOSFET drain-source voltages. When the voltage transition is complete, the current transfers to the body diode of the incoming MOSFET, allowing the device to be turned on under lossless zero-voltage conditions. Furthermore, the controlled rise of each MOSFET voltage at turn off allows the turn off losses to be virtually eliminated. In practice, the snubber capacitors will be partly formed by the MOSFET output capacitances.

The three converters were designed to operate from a 120 V DC source and to deliver 480 W to a 60 V resistive load. The average inductor current was 8 A in all converters, and the peak-to-peak ripple was set at 19 A in the soft-switched topology and 1.37 A for the other circuits. All three prototypes used an IRFB31N20D MOSFET as the main device operating at 50 kHz. This is a 200 V device with a continuous current rating of 31 A and an on-state resistance of 0.082 Ohms at 25°C. The forward breakdown voltage of the devices will fall by around 20% at cryogenic temperatures [5], and an appropriate allowance was made in the component selection. Three different diodes were used in the basic hard-switched converter of figure 1,

- A 200 V, 20 A silicon Schottky diode, MBR20200CT
- A 600 V, 15 A ultra-fast diode, MUR1560
- A 600 V, 10 A silicon carbide Schottky diode, CSD10060

In the synchronous rectifier topology, figure 2, and the soft-switched circuit, figure 3, the lower device, S2, was formed using a second IRFB31N20D, the same type as used for S1. Snubber capacitors of 2.2 nF were added in parallel with the MOSFETs in the soft-switched converter.

To allow the power devices to be operated at cryogenic temperatures, devices S1, S2 and D were mounted on a thermally conducting substrate, which was fixed to the cold head of a cryocooler, a Leybold 120T. Standard plastic packaged devices were used. The remaining power circuit components, the input capacitor and the output filter inductor and capacitor were mounted outside the cold chamber. The gate drive circuit, an IR2110, was mounted close to the devices inside the chamber, although not in contact with the cold head. Figure 4 shows a photograph of the synchronous rectifier converter devices mounted on the cold head. The electrical feedthrough for the power connections is shown on the left and the feedthrough for the device signals and measurements is shown on the right.

The cold head is surrounded by a stainless steel vacuum jacket to prevent convection heating of the devices and condensation problems. The cold head was supplied through a sealed liquid helium cooling circuit from a separate compressor.

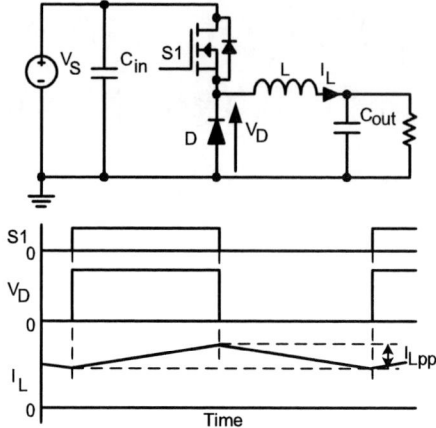

Figure 1. Hard switching buck converter

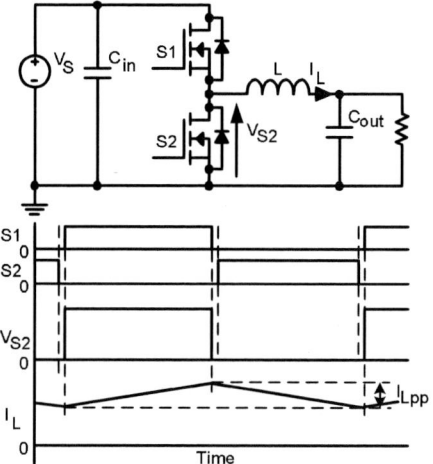

Figure 2. Synchronous rectifier buck converter

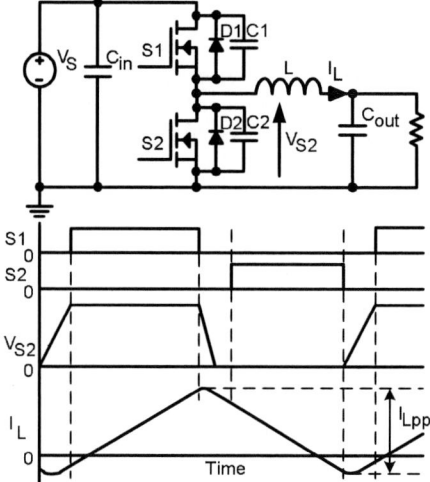

Figure 3. Zero-voltage switching buck converter

III. EXPERIMENTAL MEASUREMENTS

Before examining the operation of the converters, the static characteristics of the individual devices were measured over the temperature range 20-300 K. The on-state resistance of the MOSFET with a gate source voltage of 12 V and a drain current of 8 A is plotted in figure 5, showing that the resistance falls to approximately 15% of its room temperature value at 50 K. The increase in the resistance at temperatures below 50 K was attributed to carrier freeze out effects. Figure 6 shows the on-state voltages of the three diodes that were used in the prototypes, again measured with a forward current of 8 A. The overall trend seen in each of the three sets of measured data is an increase in voltage at low temperature; the proportionate increase is greatest for the silicon Schottky and smallest for the silicon carbide Schottky. The detailed shape of these characteristics depends on the relative sizes of the two components that contribute to the total on-state voltage, the voltage across the blocking drift region, which has a positive temperature coefficient, and the metal-semiconductor voltage drop, which has a negative temperature coefficient.

By operating the converters with the input capacitor outside the cold chamber, the device waveforms could be monitored, allowing the device on-state and switching losses to be calculated directly. The switching losses were obtained by multiplying and integrating the device voltages and currents, whilst the on-state losses were calculated from a knowledge of the currents and the static characteristics in figures 5 and 6. The measured device losses were confirmed by measurement of the converter input and output powers using high resolution multimeters and measurement of the inductor losses.

The measurements were taken at a set of temperature points over the range 20-300 K, taking care to ensure that thermal equilibrium had been reached inside the chamber at each temperature point. The temperature of the devices was measured using a temperature calibrated sensing diode.

Figure 7 shows the total semiconductor losses in the five converters, whilst figures 8 and 9 show the breakdown of the semiconductor losses into conduction and switching. Figure 7 shows that the soft-switched and the synchronous rectifier prototypes offer the lowest semiconductor losses. At room temperature the synchronous rectifier has a slight advantage due to its significantly lower conduction losses, figure 8; the peak current levels in the soft-switched converter are much higher due to the very large inductor ripple current. However, at very low temperature the conduction losses in both these prototypes reduce substantially due to the reduction in MOSFET R_{DSon}, figure 5, but the switching losses remain unchanged at around 2 W for the synchronous rectifier, figure 9, and virtually zero for the soft-switched circuit. The switching losses were too low to measure in the soft-switched circuit and so are not plotted in figure 9. Therefore at cryogenic temperatures the soft-

Figure 4.   Prototype converters inside cold chamber

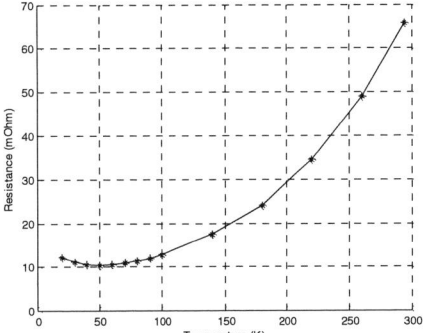

Figure 5.   Measured R_{DSon} of the IRFB31N20D MOSFET

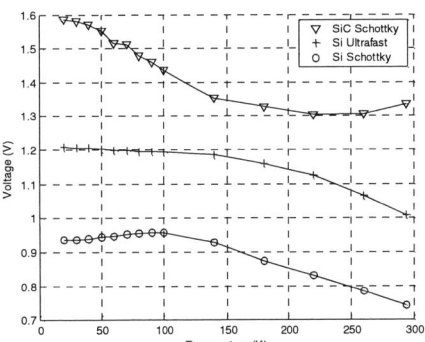

Figure 6.   Measured diode on-state voltage drops at 8 A

switched prototype has superior performance, the semiconductor losses being approximately 15% of the room temperature value, furthermore, the absolute value of the losses could easily be reduced by adding additional MOSFETs in parallel.

The hard-switched prototypes with the silicon diodes have much higher semiconductor losses, and exhibit a smaller reduction in loss at low temperatures. Whilst there is a significant reduction in switching loss, principally due to lower diode reverse recovery, this is off-set by the increase in the diode on-state losses, figure 6. The semiconductor loss against temperature characteristics of the hard-switched converters will clearly vary with transistor duty-ratio due to the changing durations of MOSFET and diode conduction; the reduction in losses at low temperature will become smaller with lower values of duty-ratio.

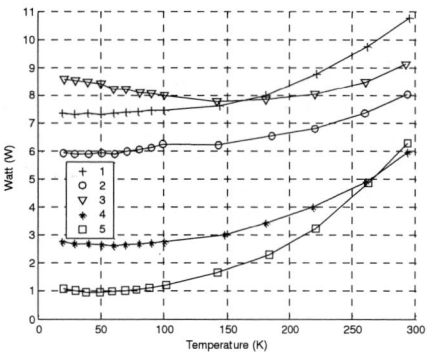

Figure 7.   Total semiconductor losses for the 120 V prototypes

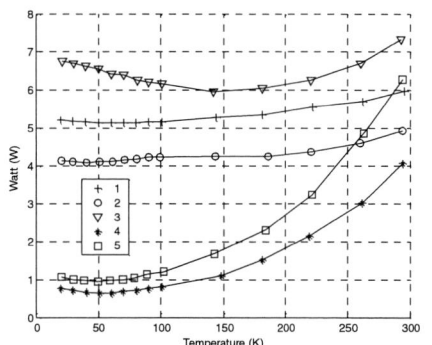

Figure 8.   Semiconductor conduction losses for the 120 V prototypes

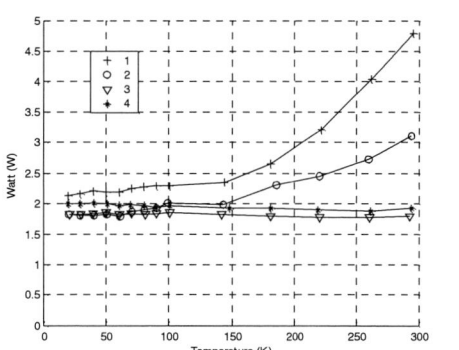

Figure 9.   Semiconductor switching losses for the 120 V prototypes

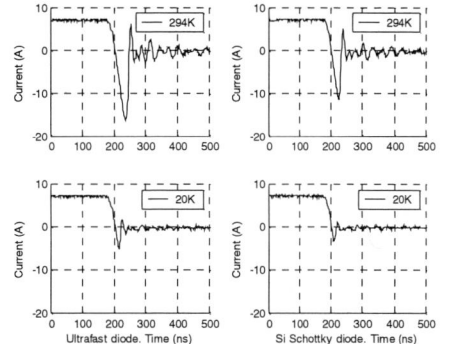

Figure 10.   Turn off current waveforms for the silicon diodes

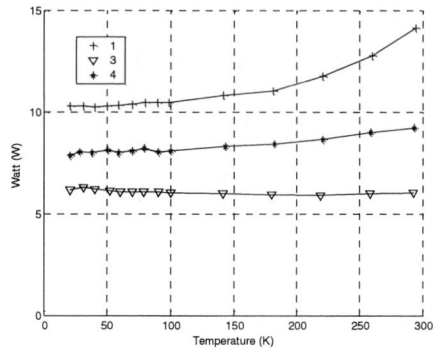

Figure 11.   Total semiconductor losses for the 500V prototypes

TABLE I. PROTOTYPES LIST

1	Hard-switched with ultrafast diode
2	Hard-switched with Si Schottky diode
3	Hard-switched with SiC Schottky diode
4	Synchronous rectifier
5	Zero-voltage transition

The reduction in the reverse recovery transients of the silicon diodes is illustrated in figure 10, which shows the turn off currents for both diodes at room temperature and 20 K. The reverse recovery charge for the ultrafast diode is reduced by almost an order of magnitude at 20 K compared with room temperature and the reverse recovery time is reduced by over 50%. A similar improvement is seen in the Schottky diode waveforms and this is the dominant factor in the reduction of switching losses at low temperatures in the hard-switched prototypes figure 9.

The semiconductor losses in the hard-switched converter with the SiC Schottky diode are amongst the highest of the 120 V prototypes due to the high on-state voltage of the diode, however repeating the work at 500 V reveals a different pattern to the results. Figure 11 shows the total semiconductor losses in three prototype buck converters operating from a 500 V source and delivering 500 W to a 250 V load. The switching frequency was again 50 kHz. A SPP20N60C3 600 V, 20 A MOSFET with an on-state resistance of 0.19 Ohms at 25°C was used in the three converters. The first prototype used the same ultrafast silicon diode as used previously (MUR1560, 600 V, 15 A), the second used the CSD10060 silicon carbide Schottky diode and the third was a synchronous rectifier topology, using two of the SPP20N60C3 MOSFETs.

The results in figure 11 show that the SiC-diode-based prototype yielded the lowest semiconductor losses, and furthermore the losses varied very little with temperature, the reduction in MOSFET conduction losses at low temperature being counteracted by the increased diode conduction losses. There was virtually no noticeable change in the switching losses with temperature. The performance of the synchronous rectifier topology at low temperatures was again limited by semiconductor switching losses.

IV. CONCLUSION

Whilst MOSFET conduction losses reduce by almost an order of magnitude as the temperature is reduced from 300 K to cryogenic levels, the switching losses in hard-switched converters do not reduced in proportion and are a limiting factor on the performance of power electronic converters operating at very low temperatures. Zero-voltage soft-switching techniques virtually eliminate switching losses and allow very low semiconductor losses to be obtained in MOSFET-based converters at cryogenic temperatures. In the 120 V soft-switching prototype developed in this work, the semiconductor losses at cryogenic temperatures were reduced to 15% of the room temperature level.

REFERENCES

[1] D.U. Gubser, "Superconductivity: an emerging power-dense energy-efficient technology," Applied Superconductivity, IEEE Transactions on Volume 14, Issue 4, Dec. 2004 Page(s):2037 – 2046

[2] I.J. Iglesias, J. Acero, A. Bautista, "Comparative study and simulation of optimal converter topologies for SMES systems," Applied Superconductivity, IEEE Transactions on Volume 5, Issue 2, Part 1, Jun 1995 Page(s):254 – 257

[3] S. Woodruff, H. Boenig, F. Bogdan, T. Fikse, L. Petersen, M. Sloderbeck, G. Snitchler, M. Steurer, "Testing a 5 MW high-temperature superconducting propulsion motor," Electric Ship Technologies Symposium, 2005 IEEE 25-27 July 2005 Page(s):206 - 213

[4] B. Ray, S.S. Gerber, R.L. Patterson, I.T. Myers, "Low-temperature operation of a buck DC/DC converter," Applied Power Electronics Conference and Exposition, 1995. APEC '95. Conference Proceedings 1995., Tenth Annual Issue 0, Part 2, 5-9 March 1995 Page(s):941 - 946 vol.2

[5] R. Karunanithi, A. K. Raychaudhuri, Z. Szücs and G. V. Shivashankar, "Behavior of power MOSFETs at cryogenic temperatures," Cryogenics, Volume 31, Issue 12, December 1991, Pages 1065-1069

[6] A.I. Gardiner, S.A. Johnson, E. Schempp, "Operation of Power Electronic Converters at Cryogenic Temperatures for Utility Energy Conditioning Applications," Energy Conversion Engineering Conference, 1996. IECEC 96. Proceedings of the 31st Intersociety Volume 4, 11-16 Aug. 1996 Page(s):2209 - 2214 vol.4

[7] M.E. Elbuluk, S. Gerber, A. Hammoud, R.L. Patterson, "Characterization of low power DC/DC converter modules at cryogenic temperatures," Industry Applications Conference, 2000. Conference Record of the 2000 IEEE Volume 5, 8-12 Oct. 2000 Page(s):3028 - 3035 vol.5

[8] O.M. Mueller, K.G. Herd, "Ultra-high efficiency power conversion using cryogenic MOSFETs and HT-superconductors," Power Electronics Specialists Conference, 1993. PESC '93 Record., 24th Annual IEEE 20-24 June 1993 Page(s):772 – 778

[9] R.L. Patterson, A. Hammond, S.S. Gerber, "Evaluation of capacitors at cryogenic temperatures for space applications," Electrical Insulation, 1998. Conference Record of the 1998 IEEE International Symposium on Volume 2, 7-10 June 1998 Page(s):468 - 471 vol.2

[10] A. Hammoud, S. Gerber, R.L. Patterson, T.L. MacDonald, "Performance of surface-mount ceramic and solid tantalum capacitors for cryogenic applications," Electrical Insulation and Dielectric Phenomena, 1998. Annual Report. Conference on 25-28 Oct. 1998 Page(s):572 - 576 vol. 2

[11] Hongrae Kim; Jahns, T.M.; Venkataramanan, G., "Minimization of reverse recovery effects in hard-switched inverters using CoolMOS power switches," Industry Applications Conference, 2001. Thirty-Sixth IAS Annual Meeting. Conference Record of the 2001 IEEE, Volume 1, 30 Sept.-4 Oct. 2001 Page(s):641 - 647 vol.1

[12] Henze, C.P.; Martin, H.C.; Parsley, D.W., "Zero-voltage switching in high frequency power converters using pulse width modulation," Applied Power Electronics Conference and Exposition, 1988. APEC '88. Conference Proceedings 1988, Third Annual IEEE 1-5 Feb. 1988 Page(s):33 – 40

2006 5th International Power Electronics and Motion Control Conference

Design and Performance Evaluation of a 10-kW Interleaved Boost Converter for a Fuel Cell Electric Vehicle

G. Calderon-Lopez, A. J. Forsyth and D. R. Nuttall

The University of Manchester/School of Electrical and Electronic Engineering, Manchester, UK

gerardo.calderon-lopez@postgrad.manchester.ac.uk

Abstract— A step-up converter is described to interface the low, poorly regulated fuel cell output voltage (70-120 V) with the higher voltage supercapacitor bank and traction drive system on a small fuel cell powered electric vehicle. The paper focuses on the design of an interleaved boost converter, in particular comparing a two inductor circuit with an alternative topology comprising a single inductor plus interphase transformer (IPT). Whilst the two circuits have identical inductive stored energies, the lower AC excitation of the inductor in the IPT topology offers performance benefits.

Keywords- electric vehicle; interleaved boost converter

I. INTRODUCTION

The integration of a fuel cell energy source into the drive train of an electric vehicle presents a number of technical challenges: handling the low voltage high current fuel cell output efficiently and stepping up to the traction system DC-link voltage, accommodating the wide variation in fuel cell voltage, and providing an energy storage buffer between the fuel cell and the traction system to source the high instantaneous power required during acceleration and to absorb the regenerated braking energy [1].

The application under consideration in this paper is a small factory or airport vehicle, and at this low power level (10 kW), a low cost, low maintenance system is required, therefore, a battery-less architecture has been chosen. The system diagram is shown in Fig. 1, where it is seen that the supercapacitor energy storage buffer is connected directly to the DC-link of the traction motor drive, and a uni-directional DC-DC converter is used to control the power flow from the fuel cell and to step up the fuel cell output to the supercapacitor voltage. The output voltage of the DC-DC converter must therefore vary with the state of charge of the supercapacitors.

The fuel cell input voltage ranges between 120 V on no load to 70 V on full load, whilst the supercapacitor voltage may vary between 140V and 210 V; the lower limit is set by the traction motor drive and the voltage rating of the supercapacitors fixes the upper limit.

The supercapacitor bank is sized to have a usable energy storage capacity that is just greater than the

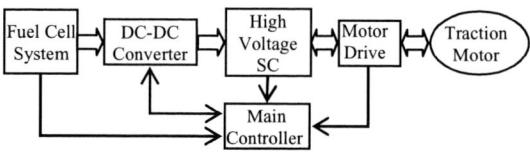

Fig. 1. Block diagram of the system

maximum kinetic energy of the fully laden vehicle. The main controller of the system regulates the state of charge of the supercapacitor bank according to the vehicle speed; the capacitors are fully charged when the vehicle is stationary and at maximum vehicle speed the capacitor voltage tends to the minimum operational level.

The design and performance of the DC-DC converter employed to interface the fuel cell with the supercapacitor bank and the power train of the vehicle is described in this paper.

II. INTERLEAVED BOOST CONVERTER TOPOLOGY

The dual interleaved boost converter was chosen for this system due to its simple robust topology and the inherent cancellation of ripple currents at input and output. The basic topology is shown in Fig. 2 along with idealised current waveforms for the two inductors, and the overall input current, $I_{in(2)}$. The transistors Q_1 and Q_2 operate with equal duty-ratios, but the operation of one device is delayed behind the other by half a switching cycle. The resultant input current ripple is reduced in amplitude compared with the ripple current in the individual inductors and is at twice the switching frequency. For the special case of a duty-ratio of 0.5, $I_{in(2)}$ will be perfectly smooth. A similar effect is seen in the output capacitor current waveform.

Several authors have considered the possibility of integrating the separate input inductors onto a single core to reduce the overall magnetic component size and weight, however the optimal solution appears to require non-standard cores, which may include two different types of core material [2, 3]. In order to reduce the size and weight of the magnetic components without using special cores, the use of an interphase transformer and single input inductor, Fig. 3 is considered in this paper. The windings of the interphase transformer in Fig. 3 have equal numbers of turns.

1-4244-0448-7/06/$25.00 ©2006 IEEE

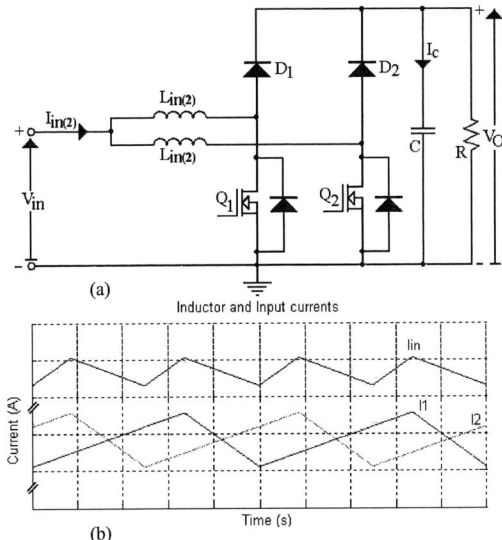

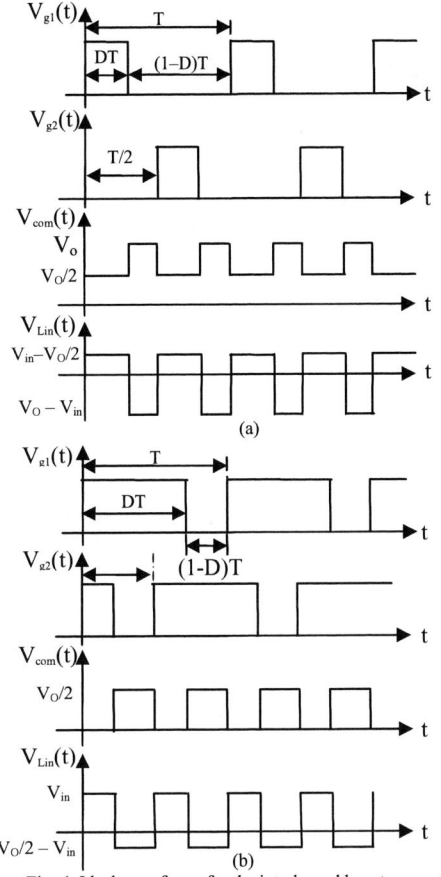

Fig. 2. (a) Dual interleaved boost converter; (b) Ideal input current waveforms

Fig. 4 shows the idealised waveforms for the interleaved boost converter with the interphase transformer for two values of duty ratio, below 0.5 and greater than 0.5. The converter is assumed to be in the continuous conduction mode with the inductor current $I_{in(1)}$ dividing equally between the two windings of the interphase transformer. The figure shows the gate-source voltages, V_{g1} and V_{g2} for the two MOSFETs, the voltage V_{com} at the centre-tap point of the interphase transformer and finally the voltage $V_{Lin} = V_{in} - V_{com}$ across the input filter inductor.

When both MOSFETs are off their drain-source voltages and also the voltage V_{com} will be equal to V_O. Conversely when both transistors are switched on V_{com} will be zero. However, when one transistor is on and the other is off the voltage V_{com} will be equal to $V_O/2$. The resultant inductor voltage waveforms are seen to be at twice the switching frequency, and the expressions for the peak-to-peak inductor ripple current $\Delta I_{in(1)}$ can be shown to be given by:

$$\Delta I_{in(1)} = \begin{cases} \dfrac{V_{in}DT}{2L_{in(1)}} * \left[\dfrac{1-2D}{1-D}\right] & \text{for } 0 < D \leq 0.5 \\[2ex] \dfrac{V_{in}DT}{2L_{in(1)}} * \left[\dfrac{1-2D}{D}\right] & \text{for } 0.5 < D < 1 \end{cases} \quad (1)$$

Fig. 4. Ideal waveforms for the interleaved boost converter with IPT; (a) D < 0.5; (b) D > 0.5

The voltage conversion ratio expression for the converter may be obtained by equating the average of the V_{com} waveform to V_{in}, and it may be shown that the resultant expression is $V_O/V_{in} = 1/(1-D)$ for both small and large values of duty-ratio, identical to the expression for a simple boost converter.

In addition to the common mode component of current in the interphase transformer windings, equal to $I_{in(1)}/2$, a second component of current, I_{diff}, will flow due to the voltage V_{diff} across the total transformer winding. The differential current I_{diff}, will be maximum when the transistor duty-ratio is 0.5 and the input voltage is one half the output voltage, since under these conditions the V_{diff} waveform will be a square wave of $\pm V_O$. At other values of duty-ratio the peak current I_{diff} will become smaller as there will be zero-voltage periods in the V_{diff} waveform. Fig. 5 shows for both small and large duty-ratios the differential voltage and current waveforms V_{diff} and I_{diff}, the input inductor current $I_{in(1)}$ and the current in one of the transistors, I_{Q1}, illustrating how the total transistor current (in solid lines) comprises two components, $I_{in(1)}/2$ (in dotted lines) and I_{diff}. The differential current will produce a small increase in the rms values of the transistor currents.

Fig. 3. Interleaved boost converter with interphase transformer

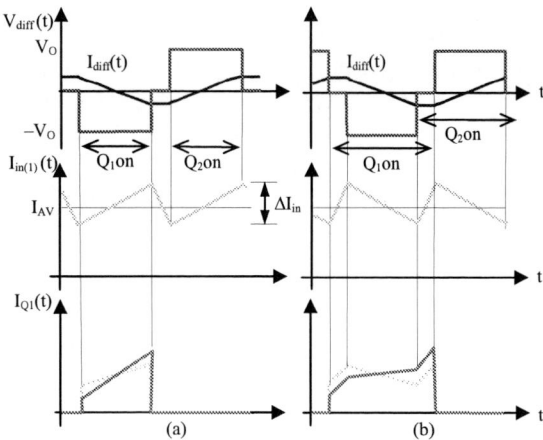

Fig. 5. Differential voltage, differential current, input current and transistor current waveforms for the interleaved boost converter with IPT (a) D < 0.5 and (b) D > 0.5

To compare the input ripple current with that of a single-transistor boost converter, the expressions in (1) are divided by the peak-to-peak ripple current expression for a simple boost converter, namely $\Delta I = V_{in}DT/L_{in(1)}$, and the resulting expressions for the normalised ripple current of the interleaved converter are plotted against D in Fig. 6. The ripple current is substantially reduced for the interleaved converter, especially for mid-range duty-ratios.

By analysing the input inductor currents for the two inductor converter in Fig. 2, the peak-to-peak input ripple current $\Delta I_{in(2)}$ may be expressed as:

$$\Delta I_{in(2)} = \begin{cases} \Delta I_{(2)} * \left[\dfrac{1-2D}{1-D}\right] \text{ for } 0 < D \le 0.5 \\[2mm] \Delta I_{(2)} * \left[\dfrac{1-2D}{D}\right] \text{ for } 0.5 < D < 1 \end{cases} \quad (2)$$

where $\Delta I_{(2)}$ is the peak-to-peak ripple current in the individual inductors. It is interesting to note that the overall ripple currents drawn by the two converters vary with duty-ratio in identical ways.

By equating the input ripple current expressions for the two converters, (1) and (2), and substituting for $\Delta I_{in(2)}$ in terms of the input inductor value, $L_{in(2)}$, reveals that for

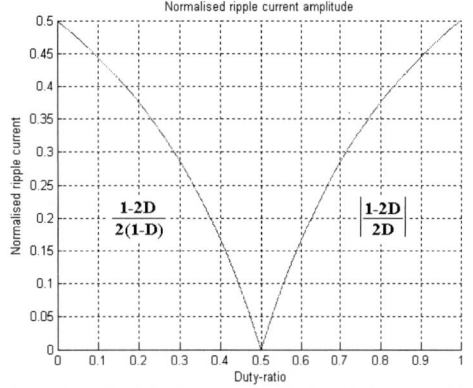

Fig. 6. Normalised ripple current amplitude of the interleaved boost converter with IPT

identical input currents:

$$L_{in(1)} = \frac{L_{in(2)}}{2} \quad (3)$$

The input inductor in the IPT circuit of Fig. 3 must be half the value of those in the circuit of Fig. 2 for the same input ripple. Since the inductor in IPT circuit carries twice the DC current of those in two inductor circuit, the DC inductive stored energy in the two circuits is identical for the same input ripple currents.

However, the design of the inductors for the two circuits will differ since the inductor in the IPT converter is subjected to a much lower AC excitation. Comparing the ratio of LΔI for the inductors in the two circuits, and assuming the total input ripple currents are the same ($\Delta I_{in1} = \Delta I_{in2}$) results in

$$\frac{L_{in(1)}\Delta I_{(1)}}{L_{in(2)}\Delta I_{(2)}} = \frac{L_{in(1)}\Delta I_{in(1)}}{2L_{in(1)}\Delta I_{in(2)}\left(\dfrac{1-D}{1-2D}\right)} = \frac{1-2D}{2(1-D)} \quad (4)$$

for $0 < D \le 0.5$, which is the function plotted in Fig. 6. The corresponding function is obtained by considering $0.5 < D < 1$.

The LΔI is much lower in the IPT converter, especially for mid-range values of duty-ratio, however the inductor excitation in this converter is at twice the switching frequency. The LΔI is important since it is related to the AC flux, the core area and the number of turns, which will affect the size of the inductor and the core losses.

The lower LΔI in the IPT converter inductor must be traded against the requirement for the IPT, which will have a high AC excitation, however, since the component has zero DC flux, a high frequency ferrite core may be used to minimise the AC losses.

III. CONVERTER DESIGN AND COMPARISON

To compare the performance of the dual interleaved boost converter with and without the interphase transformer, prototypes of each topology were designed and constructed for the system specification summarised in the introduction. The maximum power throughput is 10 kW with an input voltage of 70V and the output voltage must vary over the range 140V to 210V. A switching frequency of 40 kHz was used for the two prototypes. Metglas amorphous metal tape-wound C-cores were used for the inductors in the prototypes, having a high saturation flux density of 1.56T [4], and the windings were made from copper strip. The inductors were designed using the Metglas DCC core design program [5]. An ETD59 ferrite core was used for the IPT.

The inductors were designed to give an overall input ripple current of 19A peak-to-peak under worst-case conditions (V_{in}=70V, V_O=210V), and the IPT was designed to have a maximum peak-to-peak differential current of 6.5 A under worst case conditions (D = 0.5). 300V power MOSFETs type number APT30M17JLL

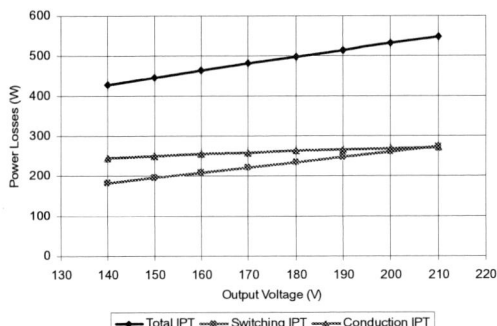

Fig. 7. Semiconductor device losses vs. output voltage.
Vin=70V, output power = 10 kW.

were used in the prototypes, having an on-state resistance of 17 mΩ at 25 °C.

Fig. 7 shows a plot of the calculated total semiconductor losses in the IPT converter plotted against the converter output voltage for a power throughput of 10 kW. The conduction and switching losses are approximately equal to each other and the total semiconductor losses are 550W under worst case conditions of V_O = 210V. The plot for the two inductor converter is virtually identical since the IPT differential current was chosen to be very small.

Table 1 summarises the performance of the magnetic components in the two prototypes. The total magnetic component weight in the two-inductor converter is lower by 0.3 kg; however the losses in the inductors are high at 160W, almost twice the total magnetic component losses in the IPT-based converter.

The predicted results in Fig. 7 and Table 1 were confirmed by experimental measurements on the prototypes. The efficiency of the IPT-based converter was measured to be 94% when delivering 9.6 kW to a 200V load, implying a total loss of 610W. The distribution of the losses was confirmed by temperature measurements of the heatsinks.

Table 2 shows the effect of re-designing the inductors in the two prototypes to obtain a larger peak-to-peak input ripple current of 30A. The inductor for the IPT converter is significantly lighter, but has increased losses, principally due to the higher AC excitation. The inductor losses in the two inductor converter are massively increased, again due to core losses and the component weight is also increased.

Finally, Table 3 shows the effect of designing the inductors for a lower peak-to-peak input ripple current of 10A. The total magnetic component weight is the same for the two circuits, but the losses are still considerably higher in the two inductor circuit.

TABLE 1. MAGNETICS WEIGHT AND LOSSES WITH V_O= 210 V, POWER = 10 kW, ΔI_{IN} = 19A

	Two-inductor	IPT plus one inductor
Inductor design	Core type = AMCC26S Turns = 18	Core type = AMCC26S Turns = 15
Weight	Total = 1.14 kg	Inductor = 0.88 kg IPT = 0.55 kg Total = 1.43 kg
Losses	Total = 160 W	Inductor = 40 W IPT = 35 W Total = 75 W

TABLE 2. MAGNETICS WEIGHT AND LOSSES WITH V_O= 210 V, POWER = 10 kW, ΔI_{IN} = 30A

	Two-Inductor	IPT plus one inductor
Inductor design	Core type = AMCC367S Turns = 4	Core type = AMCC25 Turns = 6
Weight	Total = 3.58 kg	Inductor = 0.63 kg IPT = 0.55 kg Total = 1.18 kg
Losses	Total = 642W	Inductor = 53 W IPT = 35 W Total = 88 W

TABLE 3. MAGNETICS WEIGHT AND LOSSES WITH V_O= 210 V, POWER = 10 kW, ΔI_{IN} = 10A

	Two-Inductor	IPT plus one inductor
Inductor design	Core type = AMCC50 Turns = 13	Core type = AMCC63 Turns = 10
Weight	Total = 1.8 kg	Inductor = 1.28 kg IPT = 0.55 kg Total = 1.83 kg
Losses	Total = 108 W	Inductor = 31 W IPT = 35 W Total = 66 W

IV. CONCLUSIONS

The performance has been compared of two alternative dual interleaved boost converters for an electric vehicle application. The first circuit requires two inductors whilst in the second, one of the inductors is replaced by an IPT. The total stored energy in the two circuits was shown to be identical for the same total input ripple currents, however the much larger AC excitation of the inductors in the two-inductor circuit tends to increase the total magnetic component losses and or weight over the values obtainable in the IPT-based circuit. The prototype IPT-based circuit had an efficiency of over 94% at full load conditions.

ACKNOWLEDGEMENTS

G. Calderon-Lopez thanks to CONACYT for his PhD sponsorship and the IPN (COTEPABE), Mexico. All the authors are grateful to the European Union for sponsorship of the INTELLICON project and to the staff at HILTech Developments and Vrije Universiteit Brussel for their technical advice.

REFERENCES

[1] N. Schofield, H. T. Yap, and C. M. Bingham, "A hydrogen fuel cell-high energy dense battery hybrid energy/power source for an urban electric vehicle," *EPE- PEMC 2004, Riga, Latvia*, 2004.

[2] A. Fratta, G. Griffero, S. Nieddu, G. M. Pellegrino, and F. Villata, "New hybrid iron-ferrite-core coupling reactors upgrade effectiveness of h-bridge-based power conversion structures," *IEEE International Symposium on Industrial Electronics, ISIE 2002*, vol. 3, pp. 884-889, 2002.

[3] S. Chandrasekaran and L. U. Gokdere, "Integrated magnetics for interleaved dc-dc boost converter for fuel cell powered vehicles," *35th Annual IEEE Power Electronics Specialists Conference, Aachen, Germany*, pp. 356-361, 2004.

[4] Metglas, "Powerlite high frequency distributed gap inductor cores," 2006: Metglas Inc., 2005.

[5] Metglas, "DC choke core design tool," v. 3.0.6 Metglas, 2004. downloaded from http://www.metglas.com/ design/

2006 5th International Power Electronics and Motion Control Conference

Analysis of Abnormal Phenomenon in Common-Source-type Forward Converter with Self-driven Synchronous Rectifier

Kentaro Fukushima[*], Takayoshi Hashimoto[*], Tamotsu Ninomiya[*] and Takeshi Segawa[**]

[*] Kyushu University, Fukuoka, Japan
[**] Impulse Inc., Tokyo, Japan

E-mail fukushima@ckt.ees.kyushu-u.ac.jp
Tel +81-92-642-3938
Fax +81-92-642-3957

Abstract— Many DC-DC converters use soft-switching techniques with active-clamp snubber in order to reduce switching surge and common-mode noise. In particular, common-source type active-clamped DC-DC converter is one of the most useful circuit topology for reduction of common-mode noise. Furthermore, in low-voltage and large-current DC-DC converters, synchronous rectifier circuits are used to improve the power efficiency. Self-driven synchronous rectifier circuits are generally used because of isolation and easy control.

Recently, it has been noticed that an abnormal phenomenon occurs under some conditions in the common-source type forward converter with self-driven synchronous rectifier circuit.

This paper describes the analysis of this abnormal phenomenon and clarifies the conditions of the abnormal phenomenon with experimental results.

Keywords-abnormal phenomenon; common-source-type active-clamped; synchronous rectifier

I. INTRODUCTION

A lot of DC-DC converters use soft-switching techniques with active-clamp snubber in order to reduce switching surge and common-mode noise. In particular, common-source type active-clamped DC-DC converter is one of the most useful circuit topology for reduction of common-mode noise. Furthermore, in low-voltage and large-current DC-DC converters, synchronous rectifier circuits are used to improve the power efficiency. Self-driven synchronous rectifier circuits are generally used because of isolation and easy control.

However, it has been noticed, recently, that an abnormal phenomenon occurs under some conditions in

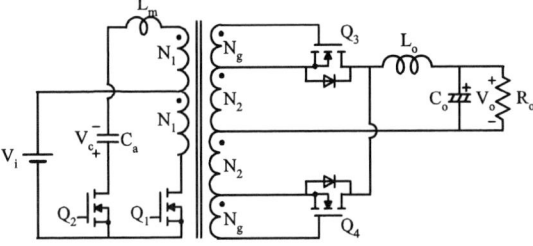

Fig. 1. Common-source type active-clamped forward converter with self-driven synchronous rectifier.

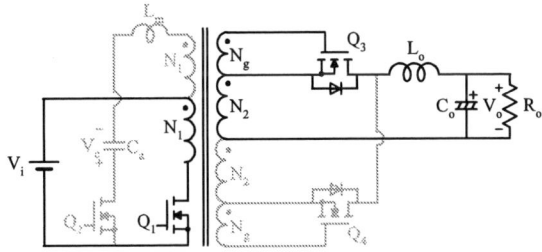

(1) State 1 (Q_1, Q_3: ON; Q_2, Q_4: OFF)

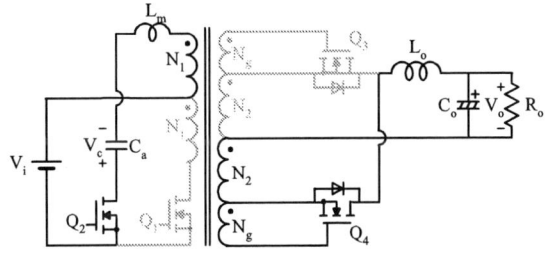

(2) State 2 (Q_1, Q_3: OFF; Q_2, Q_4: ON)
Fig. 2. Equivalent circuits representing the current paths.

1-4244-0448-7/06/$25.00 ©2006 IEEE

the common-source type forward converter with self-driven synchronous rectifier circuit.

This paper describes the analysis of this abnormal phenomenon and clarifies the conditions of the abnormal phenomenon with experimental results.

II. CIRCUIT TOPOLOGY AND OPERATION

Figure 1 shows a common-mode source type forward converter. Gate signal to synchronous rectifier MOSFET (Q_3 and Q_4) is provided from secondary side auxiliary transformer.

A. State 1 (Q_1, Q_3: ON; Q_2, Q_4:OFF)

This state is when main MOSFET Q_1 and secondary side MOSFET Q_3 are turned on. Figure 3(1) shows the equivalent circuit for analysis of state 1.

State equation on state 1 is given by:

$$V'_i = v_m \cdots (1)$$

$$v_{L_o} = V'_i - \hat{v}_o - r_s \hat{i}_L \cdots (2)$$

$$i_C = \hat{i}_L - \frac{1}{R}\hat{v}_o \cdots (3)$$

$$V'_i = \frac{N_2}{N_1}V_i \cdots (4)$$

B. State 2 (Q_1, Q_3: OFF; Q_2, Q_4:ON)

This state is when auxiliary MOSFET Q_2 and secondary side MOSFET Q_4 are turned on. Figure 3(2) shows the equivalent circuit for analysis of state 2.

State equation on state 2 is given by:

$$V_i + \hat{v}'_{C_a} = -v_m \cdots (5)$$

$$v_{L_o} = V_i + \hat{v}'_{C_a} - \hat{v}_o - r_s \hat{i}_L \cdots (6)$$

$$i_C = \hat{i}_L - \frac{1}{R}\hat{v}_o \cdots (7)$$

$$\hat{v}'_{C_a} = \frac{N_2}{N_1}\hat{v}_{C_a} \cdots (8)$$

From above sates, the steady state characteristics written as follow by using state space averaging method:

$$V_o = \frac{N_2}{N_1} \cdot \frac{2D}{1 + \frac{r_s}{R}}V_i \cdots (9)$$

$$V_{C_a} = \frac{2D-1}{1-D}V_i \cdots (10)$$

Here, duty ratio of Q_1 defines D.

Furthermore, voltage of gate-driven for Q_3 and Q_4 ($V_{gs}(Q_3)$ and $V_{gs}(Q_4)$) are given by:

$$V_{gs}(Q_3) = \frac{N_g}{N_1}V_i \cdots (11)$$

$$V_{gs}(Q_4) = \frac{N_g}{N_1}\left(V_i + V_{C_a}\right) = \frac{N_g}{N_1} \cdot \frac{D}{1-D}V_i \cdots (12)$$

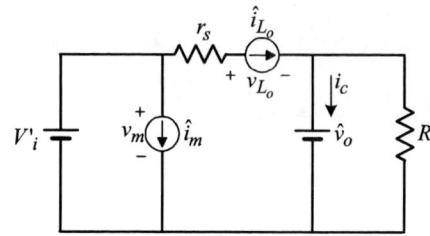

(1) State 1.

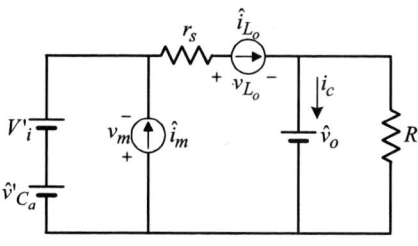

(2) State 2.

Fig. 3. Equivalent circuits for analysis.

III. EXPERIMENTAL RESULTS

A. Settings of circuit parameter values

In order to evaluate the performance of the circuits, the experimental circuits are implemented using the specifications and parameters in Table1.

B. Consideration of the abnormal phenomenon

An abnormal phenomenon of the common-source type active-clamped forward converter with self-driven synchronous rectifier circuit is examined. Input voltage and load resistance are set to constant which are 30[V] and 50[Ω]. And the waveforms of v_{Ca}, v_{ds}, v_{gs}, i_L are observed by changing of duty ratio $Q_1 - v_{Ca}$ is voltage of clamp capacitance, v_{ds} is drain-source voltage of Q_4, v_{gs} is gate-source voltage of Q_4, and i_L is current of output inductance.

Table 1. Circuit parameter values.

Symbol	Description	Value
Lm	Primary Inductance	93uH
Ca	Cramp Capacitor	1uF
Lo	Output Inductance	10.7uH
C	Smoothing Capacitor	470uF
N1	Turn times of primary side	15T
N2	Turn times of secondary side	15T
Ng	Turn times of gate-driven side	3T
fs	Switching Frequency	50kHz

1334

1) When D=0.30

Figure 4 show the experimental waveforms at D=0.30. An abnormal phenomenon is not occurred at this time. From the waveform of i_L, the period of 0 [A] is existed. This waveform means that the state of Q_4 is OFF. Gate-source voltage is not reached threshold voltage, so Q_4 is not turned ON. From the waveforms of v_{ds}, the appearance is observed. And v_{Ca} has little rippled observed from Fig. 4(b).

2) When D=0.34

Figure 5 show the experimental waveforms at D=0.34. An abnormal phenomenon is occurred at this time. From the waveform of i_L, the period of 0 [A] is existed irregularly. Further, from Fig. 5(b), v_{Ca}, i_L are oscillated low frequency.

v_{gs} is depended on v_{Ca} from Eq. (12). As the result, v_{gs} is oscillated nearby threshold voltage by v_{Ca}, so i_L is oscillated.

Furthermore, when an abnormal phenomenon is occurred, an abnormal noise is heard from the circuit transfer.

3) When D=0.38

Figure 6 show the experimental waveforms at D=0.38. An abnormal phenomenon is not occurred at this time. From the waveform of v_{ds}, it is observed that switching ON/OFF is going on. And v_{Ca} has little rippled observed from Fig. 6(b).

4) When D=0.45

Figure 7 show the experimental waveforms at D=0.45. An abnormal phenomenon is not occurred at this time and up. From the waveform of i_L, the period of the negative current is not existed at this time and up. Needless to say, it is observed that switching ON/OFF is going on, and v_{Ca} has little rippled observed from Fig. 7(b). Furthermore, v_{Ca} is nearby 0[V] observed from Fig. 7(a).

C. Area of the abnormal phenomenon

Used by the manufactured common-source type active-clamped forward converter with self-driven synchronous rectifier circuit, the area of the abnormal phenomenon is examined. Load resistance is 50 [Ω] and the other parameters are as same as Table 1.

From Fig. 8, it is shown the abnormal phenomenon has regular area. The phenomenon is mentioned D=0.34.

The threshold voltage of Q_4 used experiments is shown the line in Fig. 8 by changing input voltage from Eq. (12). As the results, the area of an abnormal phenomenon is nearby the threshold line. Therefore, the phenomenon is affected on threshold voltage of Q_4.

The area of abnormal phenomenon is shutdown at D=0.43 by irrespective of input voltage values. When duty ratio is 0.45, the waveform of i_L has little the period of negative current. From the result, it is gathered that

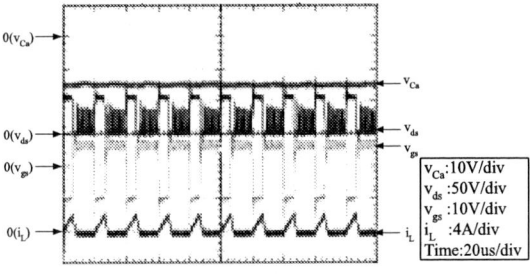

(a) Time division is 20us/div.

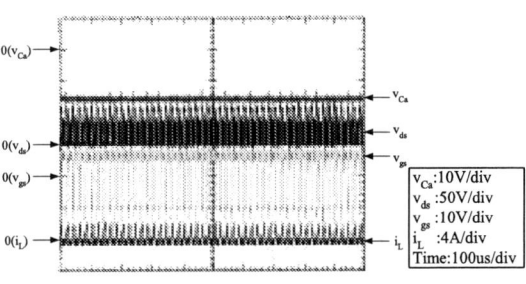

(b) Time division is 100us/div.
Fig. 4. Experimental waveforms at D=0.30.

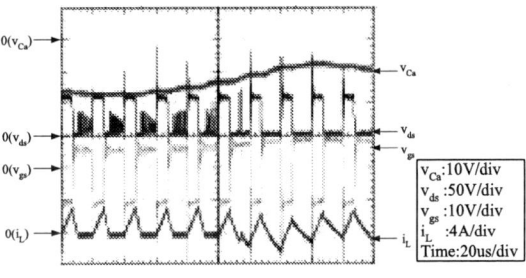

(a) Time division is 20us/div.

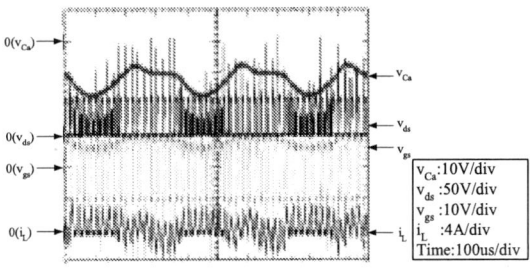

(b) Time division is 100us/div.
Fig. 5. Experimental waveforms at D=0.34.

the area of the abnormal phenomenon is limited the boundary at going in the negative current.

The analysis of the boundary at going in the negative current is given by:

$$4\frac{1-D}{R_o} = \frac{(2D-1)}{2L}DT_s + \frac{1}{R_o}$$

$$+ \sqrt{\frac{(2D-1)}{L}T_s\left\{\frac{(2D-1)D^2}{4L}T_s + \frac{3D-2}{Ro}\right\}} \cdots (13)$$

From the above equation, this boundary is not depended on input voltage. As the result constant duty ratio is found by assigned experimental values. From Eq. (13), duty ratio is D=0.46. This value is nearby the limit of the area.

Thus, the area of the abnormal phenomenon occurs the follow conditions:

1) gate-source voltage is nearby threshold voltage.

2) the period of the negative current is existed at i_L

IV. IMPROVEMENT METHOD

For the reduction of the abnormal phenomenon, the improvement method is suggested as follow.

A. The turn ratio of N_g vs. N_1 is increasing

The factor of the abnormal phenomenon is the shortage of the $V_{gs}(Q_4)$. Gate-source voltage of Q_4 is depended on duty ratio, input voltage and turn ratio of N_g vs. N_1. Therefore, what the turn ratio is increasing reduce the abnormal phenomenon.

However, as the turn ratio is increasing, gate-source voltage is going near the rated voltage of Q_4. In the case of the time when the variation of the input voltage is large, the setting of the turn ratio is paid attention, particularly.

Further, as the turn ratio is increasing, current of gate-driven is going larger. That occurs decreasing of power efficiency for the converter. Then, the large resistance goes into serial to the gate-driven transformer. But the large resistance is equal to the gate-driven resistance. As the results, the resistance may occur the strain of the gate-pulse.

B. Output inductor values is increasing

Second method is that output inductor is increasing so that the period for the negative current of i_L can reduce.

However, as output inductor is increasing, the size of converter is going larger.

Thus, for reduction of the abnormal phenomenon, settings of the parameters are paid attention carefully.

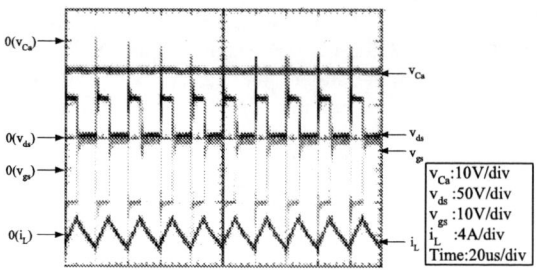

(a) Time division is 20us/div.

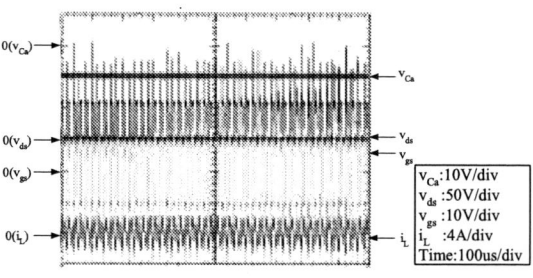

(b) Time division is 100us/div.
Fig. 6. Experimental waveforms at D=0.38.

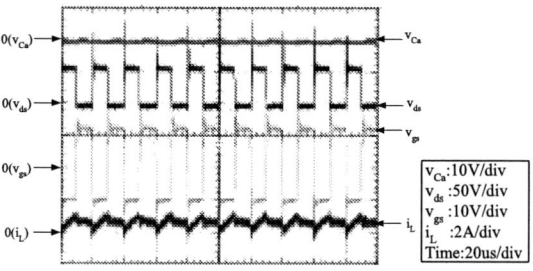

(a) Time division is 20us/div.

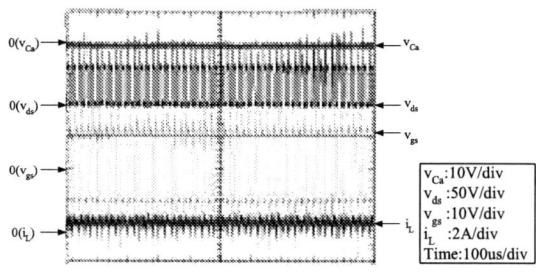

(b) Time division is 100us/div.
Fig. 7. Experimental waveforms at D=0.45.

V. CONCLUSION

The abnormal phenomenon in the common-source type active-clamped forward converter with self-driven synchronous rectifier circuit has been discussed on the basis of voltage and current waveforms. When the abnormal phenomenon occurs, the output inductor current oscillates, and the acoustic noise arises from the transformer. The abnormal phenomenon occurs dependently on the threshold voltage of the synchronous switch Q_4, and this phenomenon has been analyzed.

The method of the improvement for the abnormal phenomenon is suggested. However, settings of the parameters for converter are paid attention carefully.

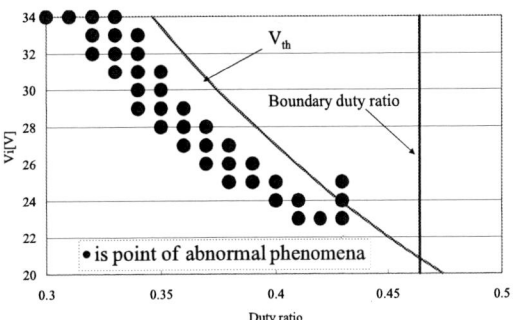

Fig. 8. Area of the abnormal phenomenon for common-source type forward converter.

REFERENCES

[1] B. Carsten : "Design techniques for transformers active reset circuits at high frequencies and power levels", HFPC'90 Proc. Pp.235-246 May.1990.

[2] E. Takegami : "Forward converter with active clamp circuit", US Patent#6,061,254 (Filed 1999-4,Issued; 200-5).

[3] Li, Ninomiya, Shoyama : "Noise reduction of Common-source type active-clamped DC-DC converters", IEEJ transactions on Industry Applications Vol. 123, No.9 pp.1037-1042 2003.

Power Quality Conditioning in Distributed Generation Systems

R.K. Járdán[*], I. Nagy[*,**]

[*]Budapest University of Technology and Economics, Department of Automation and Applied Informatics
Goldmann Gy. t. 3, H-1111, Budapest, HUNGARY, jrk@get.bme.hu
[**]Computer and Automation Institute, Hungarian Academy of Science, Kende 13-17, H-1111 Budapest, Hungary
nagy@get.bme.hu

Abstract – A novel solution for Power Quality Conditioning applied in Distributed Generation System is presented. The solution is exploiting the capabilities of the DC/AC converter of a system that has been developed to generate electric power by utilizing alternative, renewable and waste energies. The method described can also be applied in other fields of power electronics energy conversion. The solution of Power Quality Conditioning is based on the application of space vector theory, using an algorithm easy to implement. The theoretical analysis is confirmed by computer simulation results.

I. INTRODUCTION

The increasing amount of non-linear loads causes considerable problems by deteriorating the power quality of utility networks. The effect of non-linear loads can be significant in industrial environment; however, due to the large number of PCs, with their diode rectifier input stage, the harmonic distortion of the mains voltage waveform can be quite considerable even in a university campus. Generally, the adverse effects of the loads can be listed as *low displacement factor* (DF), caused by reactive power, *low power factor*, resulting, apart from low DF, by higher harmonic current components, *negative and zero sequence currents* in the three phase systems, caused by unbalanced and non-linear loads and finally *flicker*, resulting from periodically varied load currents. To tackle the problem, the standards worked out in the last decade, like IEC 61000 series, strictly limit the harmonic current emission of equipment as well as the flicker caused by fluctuating loads. Thanks to the intensive research activity in the area, both the theory and the practical application of equipment reached a high level and considerable selection of technical solutions and systems are available for the design engineer. The main types of equipment are Static VAR Compensators (SVC) that have been developed to compensate the reactive power [1].

Active Power Filters (APF) and Power Factor Correctors (PFC) that are designed and built to generate higher harmonic current components in order to minimize or eliminate harmonics drawn from the utility mains. Controlled power electronics circuits have been developed to improve the symmetry in the three phase systems by reducing the negative and zero sequence current components. The solutions to the problems of improving current waveforms can be based on the application of parallel connected converters or correcting the voltage waveforms by series connected converters [2]. These circuits, developed to improve power quality, can be integrated into the equipment of non-linear character or larger units can separately be applied to carry out compensation for a group of loads. The built-in PFC circuits can ensure that the equipment basically behaves like a resistive load, i.e. it draws fundamental harmonic current with unity PF. A considerable part of power electronics equipment, connected to the mains, at high power levels is equipped with PFC circuits ensuring that the unit will not deteriorate the power quality; however, with slight modification and negligible increase in cost, it could be achieved that these equipment would serve also as PFCs or APFs [3]. Examples are UPS systems and inverter-fed induction motor drives, generally systems, utilizing DC/AC converters.

In the paper a method and algorithm of power quality conditioning will be shown in connection with a special power generating system, developed for the utilization of renewable energy sources and waste energies [4]. The basic principle and operation will be shown briefly in the next chapter.

II. DESCRIPTION OF THE SYSTEM STUDIED

The operation of the system is described in more detail in [4], here only a short overview is given.

A simplified block diagram of the system, including the feature of PQC, is shown in Figure 1 for the case of parallel operation, assuming an application in a steam network. In this case the system is used to replace a throttling valve thus utilizing waste energy. The working medium (steam) is fed to a special high speed turbine through a control valve and the torque of the turbine drives the generator G. The electric power obtained at varying voltage level and frequency is fed to the utility mains via a DC link converter.

Research supported by the Control Engineering Research Group of the Hungarian Academy of Science and the National Research Fund (OTKA, T046240 and T049640).

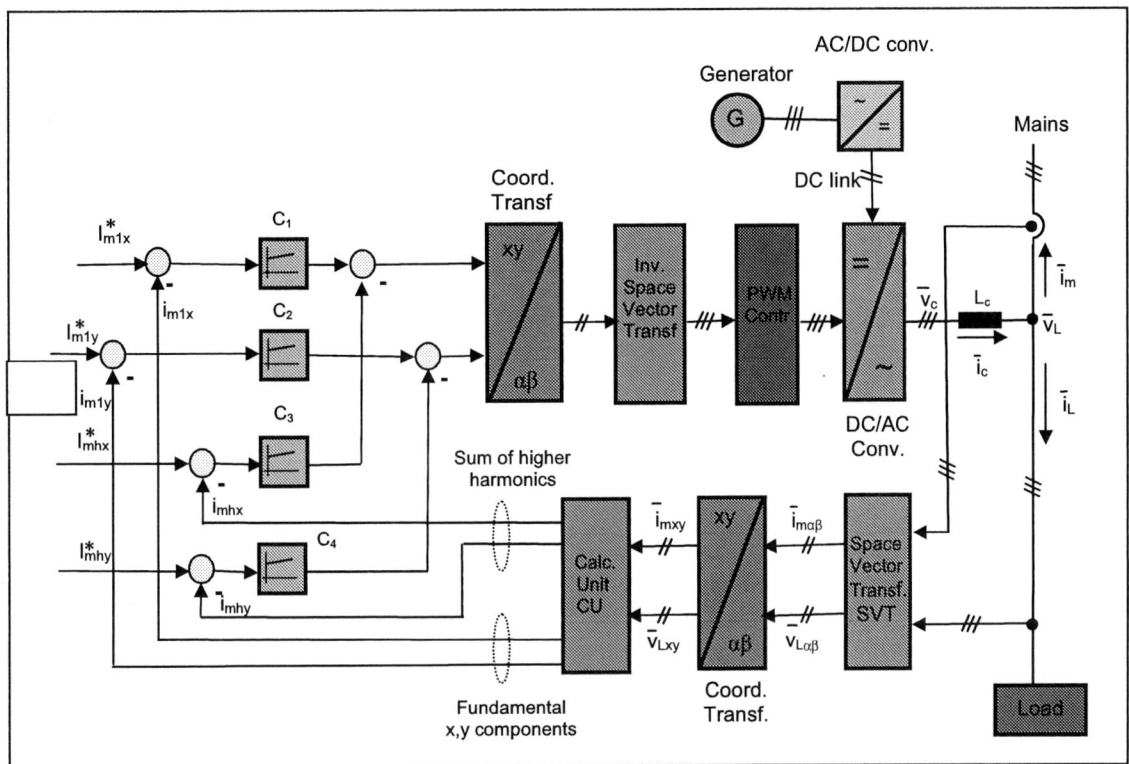

Figure 1. Block diagram of the PQC system including control loops

The three phase, full bridge PWM DC/AC converter, together with the AC/DC converter serve as a power interface between the high speed generator and the utility mains or the separate load in SAM operation.

The output power of the converter is controlled by a speed controller. The output voltage of the converter is synchronized to the mains and the control unit varies the amplitude and the phase of the fundamental converter internal voltage. As an active circuit, the DC link converter system is capable of producing reactive power for power factor correction, by changing the amplitude and phase of the converter internal voltage as compared to the mains voltage, the level of reactive power can independently be controlled to follow a reference signal.

The system can also be operated in stand-alone mode (SAM) to supply a group of loads. The two basic modes of operation require different overall control strategies to affect the pressure and the output power.

In SAM the output voltage of the AC/AC converter is kept constant and the load determines the output power of the converter. In this mode of operation the control valve is operated to maintain the power equilibrium, thus the outlet pressure cannot be controlled unless a parallel throttle valve is applied. As the primary energy source is independent of the mains, a special field of application of the system is possible, applying it as a UPS.

III. POWER QUALITY CONDITIONING

The three phase PWM DC/AC converter with high switching frequency, connected to the utility mains, makes it possible to utilize the inherent features of the system for Power Quality Conditioning:

By appropriate control of the converter it can be assured that, beside active power, the DC/AC converter also supplies reactive power that can be utilized for *DF correction*. Increasing the rating of the converter by a small degree, reactive power can be drawn from it in addition to the rated active power with a slight extra investment cost.

Exploiting the high frequency PWM operation of the converter, the output current can be controlled to produce required higher harmonic current components up to orders, limited only by the switching frequency, thus *PFC* can be realized.

Negative sequence current components can also be generated by the converter that makes it possible to compensate the effects of unbalanced loads.

In the paper the method of sensing the positive and negative sequence fundamental as well as higher harmonic current components that can be used in the controller of PQC is presented. The method of analysis is based on the use Space (Park's) vectors. An equivalent circuit of the system studied is shown in Figure 2.

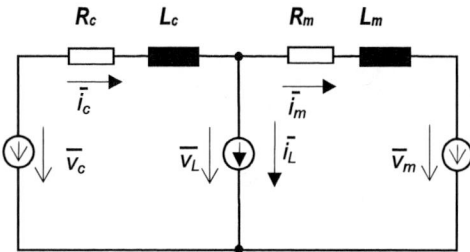

Figure 2. Equivalent circuit of the system studied

The meaning of the variables denoting currents and voltages can be seen in Figures 1 and 2. When the system is designed to serve also as a UPS, a selected group of loads can be separated by a static switch in case of mains failure or disturbance. The load is represented by a current source with a current $\bar{i}_L$, the load voltage is $\bar{V}_L$. The AC Link Inductor (Figure 1.) is characterized by L_c and R_c, while the mains represented by $\bar{V}_m$ behind R_m and L_m. $\bar{V}_c$ is the back emf of the converter that will have to be controlled by PWM to produce the required current components. In general case it can be assumed that the load is unbalanced i.e. it contains both positive p and negative n sequence as well as higher harmonic current components:

$$\bar{i}_L = \bar{i}_L^p + \bar{i}_L^n = \sum_{h=1}^{\infty} I_h^p e^{j(h\omega_1 t - \varphi_h^p)} +$$

$$+ \sum_{h=1}^{\infty} I_{Lh}^n e^{-j(h\omega_1 t - \varphi_h^n)} = I_{L1}^p e^{j(\omega_1 t - \varphi_1^p)} +$$

$$+ \sum_{h=2}^{\infty} I_{Lh}^p e^{j(h\omega_1 t - \varphi_h^p)} + I_{L1}^n e^{-j(\omega_1 t - \varphi_1^n)} +$$

$$+ \sum_{h=2}^{\infty} I_{Lh}^n e^{-j(h\omega_1 t - \varphi_h^n)} \tag{1}$$

Where:

h = the order of higher harmonics.

$\bar{i}_L^p$ = positive sequence load current component

$\bar{i}_L^n$ = negative sequence load current component

I_{Lh}^p, i_{Lh}^n = amplitude of hth positive and negative sequence load current components respectively

ω_1 = fundamental angular frequency

φ_h^p = phase of the hth positive sequence load current component and

φ_h^n = phase of the hth negative sequence load current component.

The mains voltage $\bar{V}_m$ behind the mains impedance can contain positive, negative and higher harmonic components, but in practice it is acceptable if only positive sequence, fundamental and higher harmonic components are taken into account. Using the above notions:

$$\bar{V}_m = \sum_{h=1}^{\infty} V_{mh} e^{j(h\omega_1 t - \varphi_{mh})} =$$

$$= V_{m1} e^{j(\omega_1 t - \varphi_{m1})} + \sum_{h=2}^{\infty} V_{mh} e^{j(h\omega_1 t - \varphi_{mh})} \tag{2}$$

The DC/AC converter makes it possible to generate three phase output voltages containing positive and negative sequence fundamental as well as higher harmonic components. In the system studied no zero sequence voltage component can be produced, thus the converter output voltage can generally be expressed as:

$$\bar{V}_c = V_{c1}^p e^{j(\omega_1 t - \varphi_{c1}^p)} + \sum_{h=2}^{\infty} V_{ch}^p e^{j(h\omega_1 t - \varphi_{ch}^p)} +$$

$$+ V_{c1}^n e^{-j(\omega_1 t - \varphi_{c1}^n)} + \sum_{h=2}^{\infty} V_{ch}^n e^{-j(h\omega_1 t - \varphi_{ch}^n)} \tag{3}$$

In the arrangement shown in Figure 1. the aim is power quality conditioning by modifying the current components of $\bar{i}_m$ possibly to contain only symmetrical three phase fundamental components. This goal can be achieved by generating proper current components in the converter output to cancel the unwanted components originating from the load current $\bar{i}_L$. For this purpose extraction of the various components in $\bar{i}_m$ is necessary in order to obtain feedback signals for the controllers designed to fulfill the task of PQC. Here the method of extraction of the positive sequence fundamental current and the higher harmonic current components will be presented.

The reference values for the fundamental active and reactive components are I_{m1x}^*, I_{m1y}^* and for the higher harmonics: $I_{h1x}^* = I_{h1y}^* = 0$.

For the practical application of the method described, the phase quantities of the mains currents and voltages have to be measured, while for the simulation program of the system, these time functions are to be calculated. Setting up Kirchhoff's equations for the circuit of Figure 2. and expressing the mains current and the load voltage:

$$\bar{i}_m = \Delta\bar{V} \frac{1}{R} \frac{1}{1+sT} - \bar{i}_L \frac{R_c}{R} \frac{1+sT_c}{1+sT} \tag{4}$$

$$\bar{V}_L = \bar{V}_m + \bar{i}_m R_m (1 + sT_m) \tag{5}$$

Where: $T = (L_m + L_c)/R$, $T_c = L_c/R_c$, $T_m = L_m/R_m$
$R = R_m + R_c$ and $\Delta\bar{V} = \bar{V}_c - \bar{V}_m$

In the simulation program the calculation of the variables $\bar{i}_m$ and $\bar{V}_L$ is carried out by a calculator block implementing (4) and (5).

The space vector time functions are generated by a Space Vector Transformation (SVT) unit. Following a coordinate transformation into a synchronously rotating reference frame (by Coord. Transf.) CU separates the fundamental and the sum of harmonic components. In steady state condition the trajectories of these variables are closed forms as illustrated in Figure 3. in stationary reference frame. Here a time function is composed of a fundamental, 5th, 7th, 11th and 13th harmonic current components, with $I_1 = 35$ A, $I_5 = 20$ A and $I_7 = 14$ A, $I_{11} = 4$ A, $I_{13} = 2.7$ A, and $\varphi_h = 0^o$.

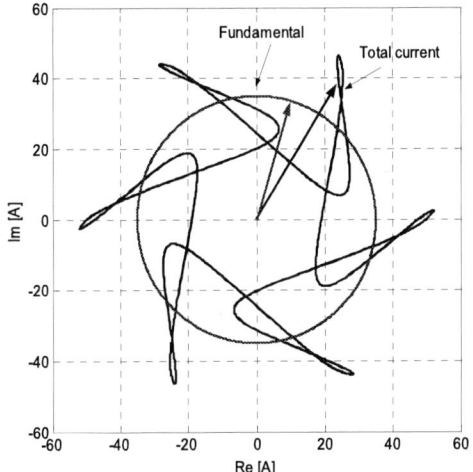

Figure 3. Space vector of the current in stationary reference frame

The algorithm of selecting any of the positive and negative sequence fundamental and or harmonic components of a periodic time function is as follows:

- Calculate the time function of space vector from the three phase quantities.

- Transform the time function of space vectors into a reference frame revolving synchronously by the angular velocity of the selected harmonic component. The transformation can be carried out by a Coordinate Transformation Unit. As a result of the coordinate transformation, the selected harmonic component will be a constant vector, while the sum of the rest of components will describe a closed trajectory around the origin of the coordinate system.

- In the third step calculation of the average value of the harmonic components in the revolving coordinate system will yield the selected component while the average value of all the other components will identically be zero. If separation of the fundamental component and the sum of higher harmonic components is needed, the coordinate transformation should apply a reference frame revolving with the angular velocity of the fundamental component.

- In the fourth step, if the value of the selected harmonic component and the sum of the rest of harmonic components are needed, the calculated value of the selected harmonic component has to be subtracted

from the original space vector time function to obtain the sum of the rest of harmonic components.

In the case of the previous example, where only positive sequence fundamental and higher harmonic components were assumed, a coordinate transformation with fundamental angular velocity results in a time function that will be characterized by a trajectory obtained by superposing a stationary fundamental component and the sum of higher harmonic components, revolving in a

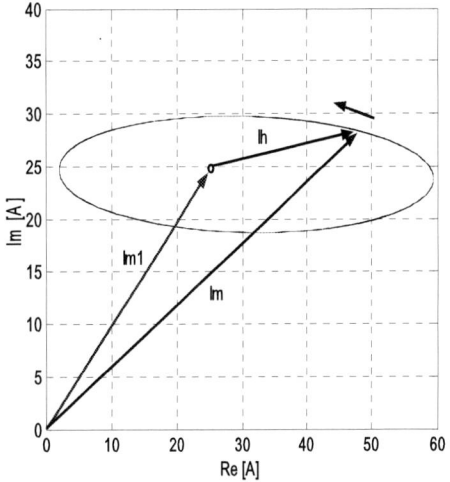

Figure 4. Trajectory of the current space vector in a synchronously rotating reference frame

closed path around the end point of the fundamental as shown in Figure 4. The fundamental component should point to the center of gravity of the closed path of the higher harmonics [5,6]. By calculating the average value of the time function $\bar{i}_m$ will yield the fundamental component as the average of the higher harmonics for a full period or multiple periods is zero, i.e.:

$$\bar{I}_{av} = \bar{I}_1 + \sum_{h=2}^{\infty} I_{hav} = \bar{I}_1 \qquad (6)$$

where:

$$\bar{I}_{hav} = \frac{1}{T_h} \int_0^{T_h} \bar{i}_h dt = 0$$

the average value of the hth harmonic components,

$\bar{I}_1$ = fundamental component,

T_h = cycle time of the hth harmonics.

Thus the fundamental component of the mains current can simply be obtained by calculating the average value of the current, $\bar{i}_m$ for a longer time, t $\gg$ T$_h$, i.e:

$$\bar{I}_{m1} = \frac{1}{t} \int_0^t \bar{i}_m dt \qquad (7)$$

and the sum of higher harmonics $\bar{i}_{mh}(t)$ can be extracted by realizing the equation:

$$\bar{i}_{mh}(t) = \sum_{h=2}^{\infty} \bar{i}_{mh}^{p}(t) + \sum_{h=2}^{\infty} \bar{i}_{mh}^{n}(t) = \bar{i}_{m}(t) - \bar{I}_{m1} \quad (8)$$

where $\bar{i}_{mh}^{p}(t)$ and $\bar{i}_{mh}^{n}(t)$ are positive and negative sequence current component of hth order.

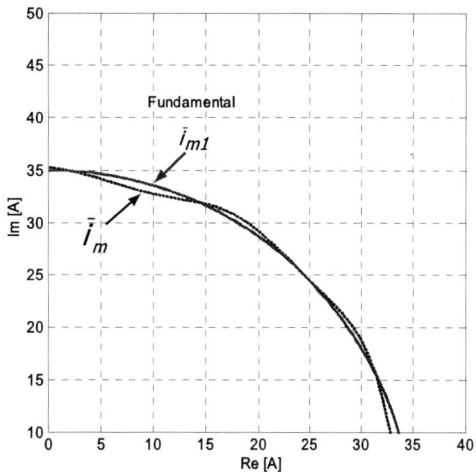

Figure 5. Higher harmonics components shown in Figure 3. are reduced by the PQC system

IV. SIMULATION OF THE SYSTEM

The system is simulated by Matlab/Simulink program, using the principles described above. The subsystems contain the DC/AC converter, mains, load, Coordinate Transformation Units, calculation of the fundamental and higher harmonics, controllers, inverse space vector transformation and the PWM controller of the DC/AC converter. Results of the simulation, carried out for a 30 kW system, loaded by a symmetrical load with a frequency spectrum given in the previous chapter are shown next. Control loops shown in Figure 1. are designed to keep the real and imaginary fundamental components of the mains current $\bar{i}_{m}$ at the level defined by the reference signals and at the same time reduce the higher harmonics. The absolute value of the fundamental mains current reference signal is 35 A, its components are: Real(Im) = 25 A, Imag(Im) = 25 A. The values of reference signals of the control loops for the higher harmonics are set to zero.

A simulation result representing the effectiveness of the PQC function of the system is shown in Figure 5, where a zoomed-in segment of the mains current $\bar{i}_{m}$ is plotted in a stationary reference frame. The result has to be compared with the trajectory of the load current shown in Figure 3. The Total Harmonic Distortion (THD) of the load current is 71% and this has been reduced in the mains current to 1.4%.

V. CONCLUSION

The paper presents a novel method for the extraction of higher harmonic components that can be used as feedback signals for the controller, carrying out power quality conditioning. The method described can also be applied in other fields of power electronic energy conversion. The solution of PQC is based on the application of space vector theory, using an algorithm easy to implement. The transient response of the solution is not fast, but the accuracy of the control is high and in a PQC system the speed is normally not relevant. Should it be necessary, the calculation of the average values could be realized by a faster numerical method at the cost of more sophisticated hardware and software. It is expected that the approach of the PQC suggested, can improve the economy i.e. reduce the payback time of the Turbine-Generator system by providing additional features at the price of a slight increase in the cost of material and the modification in the software of the control and supervisory part within the microprocessor.

VI. REFERENCES

[1] Bhattacharya, S. Po-Tai Cheng Divan, D.M. : Hybrid solutions for improving passive filter performance in high power applications, IEEE Transactions on Industry Applications, Volume: 33 Issue: 3, May-June 1997, pp.732-747

[2] Járdán, R.K.-Raaijen, E.: An Efficient and Economical Active AC Line Conditioner, Proceedings of Intelec'95 Conference October 29-November, 1. The Hague, The Netherlands, pp.664-670

[3] Leuchter, J., Bauer, P, Kurka, O., Polinder, H.: Generator-Converter Set for mobile Power Sources, 11th International Power Electronics and Motion Control Conference, Riga, Latvia, 2004.

[4] Járdán, R.K.-Nagy, I, Korondi, P, Nitta, T, Ohsaki, H.: Power Factor Correction in a Turbine-Generator-Converter System. IEEE-IAS Conference, October 8-12, 2000, Rome, Italy

[5] Rácz, I.: Recording and Harmonic Analysis of Three-Phase Vectors, Periodica Politechnica, Budapest, Vol.8. No.4. 1964. (In German).

[6] Járdán, R.K.- Dewan, S.B: - Slemon, G.R.: General Analysis of Three-Phase Inverters, IEEE Transaction on Industry and General Applications, Vol. IGA-5, No.6, Nov./Dec., 1969.pp. 672-679

[7] Mohammad N. Marwali, Jin-Woo Jung, Ali Keyhani "Control of Distributed Generation Systems- Part II: Load Sharing Control" IEEE Transaction on Power Electronics, USA, November 2004, Vol.19,No 6.,pp. 1551-1651.

[8] Sven Wendt, Frank Benecke, Henry Güldner, Jens Hampel, Joachim Zschernig, Ralf Briest, Norbert Ueffing "A Novel 500 kW High-Speed Turbine PM Synchronous Generator Set for a Power Generation System" The 2005 International Power Electronics Conference, IPEC-Niigata 2005, 4-8 April, 2005, Niigata, Japan. CD Rom ISBN: 4-88686-065-6

[9] W. Koczara: Methods of DC Voltage Control in Adjustable Speed Power Generation Systems, International Conference on Electrical Drives and Power Electronics, EDPE 2003, 24-26 September, High Tatras, Slovakia. pp.248-253.

2006 5th International Power Electronics and Motion Control Conference

Active Clamp Forward Converter Combined with Dither Voltage Generator for Poultry Stunning Applications

S. -Y. Tseng, H.-T. Wen, H.-H. Chang and J. -S. Kuo

Department of Electrical Engineering

Chien Kuo Technology University

Changhua, Changhua City, Taiwan, China

E-mail: sytseng@cc.ctu.edu.tw

Tel: 886-4-7111111 ext. 3234

Fax: 886-4-7111111 ext. 3200

Abstract

This paper presents an intelligent poultry stunner for improving carcass quality and animal welfare. In the stunner system, an active clamp circuit is used in a forward converter to reset the magnetizing energy and to recover leakage energy trapped in the transformer, increasing its conversion efficiency. In addition, a dither generator realized with a half-bridge inverter is adopted to fast breakdown the skin impedance of the poultry with which can reduce poultry stress and improve carcass quality during stunning interval. To simplify the circuit of the proposed stunner system, switches in the forward converter and the half-bridge inverter are integrated with the synchronous switch technique. With this approach, the proposed stunner system can achieve a higher conversion efficiency, less component counts, lighter weight and smaller size. Compared with the one without switch integration, the proposed one can raise efficiency of 5% under full load condition. Under a fixed frequency of 400 Hz, its output voltage can vary from 60 V to 140 V and its duty ratio is from 0.3 to 0.7. Performance measurements from a prototype have verified the feasibility of the overall system design. The designed system has contributed a lot to improve animal welfare and has attracted much attention.

I. Introduction

In recent years, the issue related to animal welfare attracted a lot of attention. In particular, livestock or poultry must be rendered unconscious and insensible to pain before they are exsanguinated with humane slaughter methods. In many countries, carbon dioxide (CO_2) and electrical stunning are the two typical methods used to stun livestock or poultry before slaughtering [1]-[10]. Since CO_2 method is subjected to many limitations and requires a higher cost, the electrical stunning method has been used more popular.

In general, there are three main parts to form neurons: soma, axon, and dendrite. Its illustration of impulse propagation between neurons is shown in Fig. 1. In Fig. 1, the nerve impulse propagation direction is, in turn, through one neuron to the other neurons, until the nerve impulse propagation is transmitted to a receiver of brain. To attain an effectiveness coma of livestock or poultry, livestock or

poultry must be to cause unconsciousness by inducing an epileptiform seizure, which includes two phases: a tonic phase and a clonic phase [1]. Degrees of an epileptiform seizure are dependent upon a mount of current passing the brain. To explain the relationship between electrical stunning for poultry and a suppression of nerve impulse propagation, an equivalent circuit for describing the impulse signal propagation between neurons is shown in Fig. 2. In Fig. 2, since synapses play a role of transducer, it can be considered as a switch Q_I. When current I_I passing the brain is large enough, it will induce a high potential V_I to turn on switch Q_I, resulting in the nerve impulse single bypassed. Thus, nervous system loses capability to propagate nerve impulse. Its detail description is explained in [11].

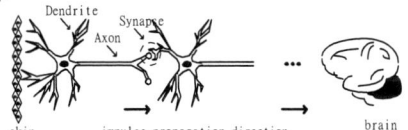

Fig. 1. Block diagram of procedure for impulse propagation between neurons.

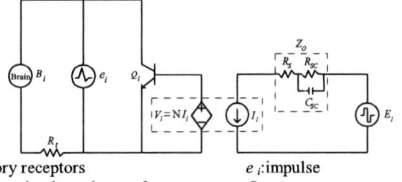

B_i: sensory receptors e_i: impulse
R_i: propagation impedance of neuron Q_I: synapse
V_I: potential of postsynaptic membrane I_I: current passing chicken
Z_0: equivalent impedance of chicken E_I: stunning voltage

Fig. 2. An equivalent circuit for describing the impulse signal propagation between neurons.

In poultry stunning, the minimum current and voltage required for stunning a chicken is about 40 mA and 60 V, respectively, and they must sustain at least 3 s. Conventionally, to generate the specified electrical waveforms, rectified line voltage or battery voltage is chopped into square waveforms with power switches and they are boosted through a low-frequency transformer. In addition, the conventional stunner only used two electrodes,

1-4244-0448-7/06/$25.00 ©2006 IEEE   1343

as shown in Fig. 3(a). As a consequence, it is in large volume, size and heavy weight, and there exist bone fractures and ecchymosis in chicken carcass, resulting in low meat quality. To solve above problems, a dc/dc converter with PWM control and a dither voltage generator for driving three electrodes are adopted for electrical stunning applications, which can properly limit current level, regulate output voltage and reduce chicken stress during stunning interval. The three-electrodes structure is shown in Fig. 3(b).

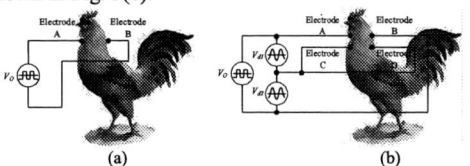

(a)   (b)

Fig. 3. Illustration of stunning chicken (a) with the conventional two-electrode configuration, and (b) with the proposed three-electrode configuration.

Since a poultry stunner system belongs to low power level applications, its power stage is usually realized with a forward converter, simplifying circuit structure and reducing capacity of output filter capacitor [12]. To recover the energy trapped in leakage and magnetizing inductances of the transformer in the forward converter, an active clamp circuit is introduced into the forward converter. Additionally, to generate dither voltage, a half-bridge inverter is used in the stunner system, as shown in Fig. 4. To combine the active clamp forward converter and the half-bridge inverter, their switches are integrated with the synchronous switch technique [13] to reduce their component counts, weight and size, and increase its conversion efficiency, as shown in Fig. 5.

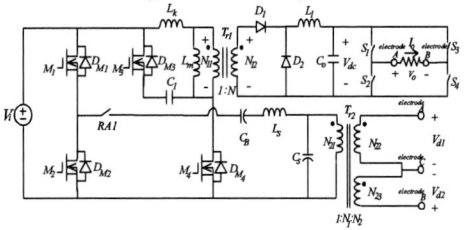

Fig. 4. Schematic diagram of an active clamp forward converter associated with a dither voltage generator.

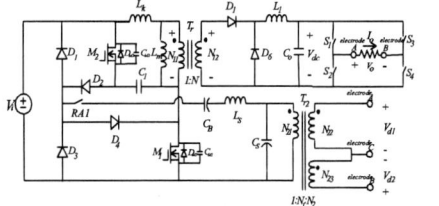

Fig. 5. Schematic diagram of the proposed stunner system with a three-electrode configuration.

II. Operational Principle of the Proposed Converter

The proposed stunner system consists of an active clamp flyback converter, a full-bridge inverter and a half-bridge inverter for generating the breakdown voltage of skin impedance, as shown in Fig. 5. To integrate the switches of

synchronous switch technique [13] is adopted. Fig. 6(a) shows an independently active clamp forward converter associated with a half-bridge inverter. According to theory of the synchronous switch technique, when switch M_3 and M_5, or M_4 and M_6 are operated in synchronous condition and there exist a common node between them, the derived circuit shown in Fig. 6(a) can be degenerated, as shown in Fig. 6(b). Since voltage stresses of switches M_3 and M_4 are respectively greater than switches M_5 and M_6, diodes D_1' and D_2' can be shorted, and diodes D_3' and D_4' can be removed, as shown in Fig. 6(c). From Fig. 6(c), it can be observed that the proposed converter can use the synchronous switch technique to decrease components of active switch, reducing its cost.

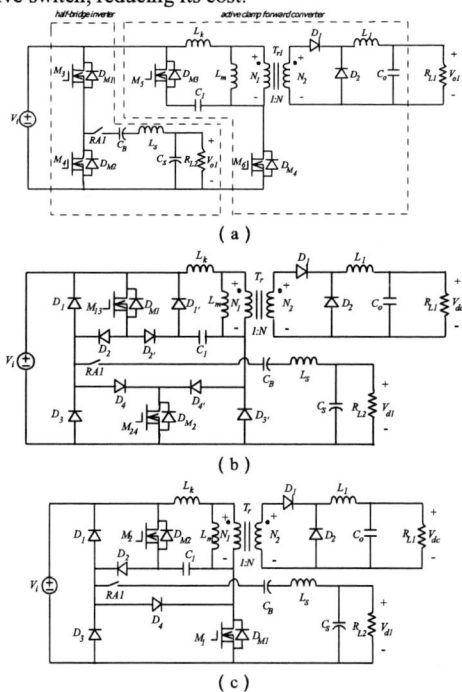

(a)

(b)

(c)

Fig. 6. Derivation of the proposed active clamp forward converter associated with a high-bridge inverter with the synchronous switch technique.

The proposed converter, as shown in Fig. 6(c), is formed by incorporating an active-clamp circuit into the basic forward topology to recover the energy trapped in magnetizing and leakage inductors of transformer T_r, and to achieve zero-voltage switching (ZVS) at turn-on transition. Then, the proposed one is also associated with a half-bridge inverter to generate dither voltage for causing breakdown of skin impedance. Its operation can be divided into 11 modes and their equivalent circuit and key waveforms are illustrated in Figs. 7. In the following, each operational mode is described briefly.

Mode 1 [Fig. 7; $t_0 \leq t < t_1$]: Before t_0, switches M_1 and M_2 are in the off state, and the body diode D_{M1}, diodes D_3 and D_6 are in forwardly bias. Voltage across winding N_1 is equal to input voltage V_i. At the same time, voltage across switch M_1 is equal to 0. When $t = t_0$, switch M_1 is turned on.

1344

switch current I_{DS1} varies from a negative value to 0, and inductor current I_{L1} is decreased linearly.

Mode 2 [Fig. 7; $t_1 \leq t < t_2$]: When $t = t_1$, switch M_1 is still kept in the on state. At this same time, switch current I_{DS1} is equal to 0 and leakage inductor current I_{LK} ($= I_{LM}$) also equals 0. The energies stored in magnetizing and leakage inductors are released to input source. Thus, diodes D_3 and D_6 still stay in forwardly bias. Within this time interval, inductor current I_{LK} ($= I_{Lm}$) is increased linearly. While, inductor current I_{LS} changes from a positive value to 0.

Mode 3 [Fig. 7; $t_2 \leq t < t_3$]: At time t_2, switch M_1 is still kept in the on state. At the same time, inductor current I_{N2} is equal to inductor current I_{L1}. Thus, diode D_6 is in reversely bias, while D_5 is in forward bias. Within this time interval, the input power is transferred to output load through transformer T_r. In addition, since inductor current I_{LS} still sustains in a positive value, diode D_3 is in freewheeling through inductor L_S.

Mode 4 [Fig. 7; $t_3 \leq t < t_4$]: When $t = t_3$, inductor current I_{LS} is equal to 0. Thus, diode D_3 is in reversely bias. While, diode D_4 is forwardly biased. Within this time interval, input voltage still supplies power to output load.

Mode 5 [Fig. 7; $t_4 \leq t < t_5$]: when $t = t_4$, switch M_1 is turned off. At this same time, the energies stored in leakage and magnetizing inductors of transformer T_r, and inductor L_S is released to junction capacitor C_{M1}. In addition, inductor current I_{LK} is equal to I_{LM}. Thus, inductor current I_{N1} is equal to 0. As a result, inductor current I_{N2} equals 0 and then, diode D_5 is in reversely bias. Within this time interval, diode D_6 is in freewheeling through inductor L_1.

Mode 6 [Fig. 7; $t_5 \leq t < t_6$]: At time t_5, the voltage across switch M_1 reaches input voltage V_i. Thus, diode D_4 is in reversely bias, while diode D_1 starts conducting. During this time interval, inductor current I_{L1} is decreased linearly, and I_{LS} is increased from a negative value to 0 by the resonant network of inductor L_S and capacitor C_S.

Mode 7 [Fig. 7; $t_6 \leq t < t_7$]: At time t_6, the voltage across switch M_1 reaches $V_i + V_{C1}$. The junction capacitor D_{M2} starts conducting. As a consequence, inductors L_K and L_m, and capacitor D_{M2} form a resonant network, and they begin to resonate. Within this time interval, since inductor current I_{LS} is a negative value, diode D_1 is in freewheeling through inductor L_S. In addition, diode D_6 is also in freewheeling through inductor L_1, and inductor current I_{L1} is decreased linearly.

Mode 8 [Fig. 7; $t_7 \leq t < t_8$]: When $t = t_7$, switch M_2 is turned on. Since the junction diode D_{M2} is forwardly biased before switch M_2 is turned on, switch M_2 can be operated with ZVS at turn-on transition. During this time interval, diodes D_1 and D_6 are in freewheeling through inductors L_S and L_1, respectively.

Mode 9 [Fig. 7; $t_8 \leq t < t_9$]: At time t_8, inductor current I_{LS} is equal to 0, As a result, diode D_1 is in reversely bias and diode D_2 is in forwardly bias. With this time interval, inductors L_K and L_m, and capacitor C_1 still sustain in the resonant condition. Inductor current I_{L1} is still decreased linearly.

Mode 10 [Fig. 7; $t_9 \leq t < t_{10}$]: At time t_9, switch M_2 is turned off. Inductor current I_{LK} ($= I_{Lm}$) is negative value and the

energy stored in junction capacitor C_{M1} can be released by inductor current I_{LK}. At this same time, since the voltage V_{C1} across capacitor C_1 is greater than that across junction capacitor C_{M1}, diode D_2 is reversely biased. While, diode D_3 is in freewheeling through inductor L_S. During this time interval, inductor current I_{L1} is decreased linearly.

Mode 11 [Fig. 7; $t_{10} \leq t < t_{11}$]: when $t = t_{10}$, the energy stored in junction capacitor C_{M1} is completely released. Thus, the body diode D_{M1} is forward biased. During this time interval, diodes D_3 and D_6 are in freewheeling through inductors L_S and L_1, respectively. When switch M_1 is turned on again at the end of time t_{11}, a new switching cycle will start.

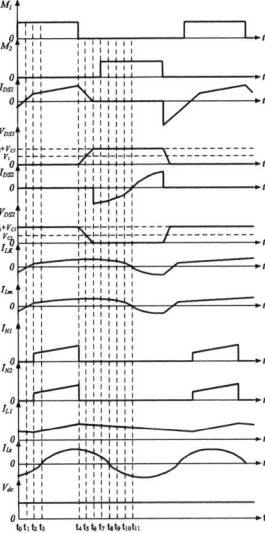

Fig. 7. Key waveforms of the proposed active clamp forward converter.

III. Design of the Proposed Converter

The proposed converter is composed of an active clamp forward converter and a half-bridge inverter. Since switches in the active clamp flyback converter and a half-bridge inverter are integrated with the synchronous switch technique, design of the proposed converter must satisfy requirements of each circuit. The half-bridge inverter is formed by a series-resonant parallel-loaded network. The quality factor Q of the resonant network is designed to be high enough to attain high input to output transfer gain before skin impedance breakdown, and it is usually designed to operate above resonance to achieve ZVS at turn-on transition. Therefore, design of the active clamp forward converter is only analyzed briefly in this section. In design of the active clamp flyback converter, determination of duty ratio D, transformer T_r, active clamp capacitor C_1 and output filter are important. In the following, their designs are analyzed briefly.

A. Duty Ratio D

To determine duty ratio, we must first attain input to output voltage transfer ratio M. Since the active clamp circuit only helps switch M_1 to achieve soft-switching feature, it does not affect transfer ratio M of the proposed as the conventional one. According to volt-second balance principle, the following equation can be obtained:

$$(nV_i - V_{DC})DT_s + (-V_{DC})(1-D)T_s = 0, \qquad (1)$$

where n $(= N_2/N_1)$ is the turns ratio of transformer T_r, V_{DC} is output dc-link voltage, and T_s is the period of the proposed converter. From (1), it can be found that transfer ratio M can be expressed as

$$M = \frac{V_{DC}}{V_i} = nD, \qquad (2)$$

Based on the operational condition of the half-bridge inverter, the optimized duty ratio of switch M_1 is 0.5, and switches M_1 and M_2 are operated in complementary. According to (2), a large duty ratio D corresponds to a smaller transformer turns ratio n, which results in a lower current stress imposed on switches M_1 and M_2, as well as voltage stress on freewheeling diode D_5. However, in order to accommodate variations of load, line voltage, component value, and the optimized operation of the half-bridge inverter, it is better to select an operating range as $D = 0.45 \sim 0.55$.

B. Transformer T_r

Once the duty is selected, the turns ratio of transformer T_r can be determined using (2), which yields

$$n = \frac{V_{DC}}{DV_i}, \qquad (3)$$

By applying the Faraday's law, N_1 of the transformer T_r can be given as

$$N_1 = \frac{DV_iT_s}{A_C\Delta B}, \qquad (4)$$

where A_C is the effective cross-section area of the transformer core and ΔB is the working flux density. According to (3) and (4), N_2 can be therefore determined.

To achieve a ZVS feature, the energy stored in inductor L_m must satisfy the following inequality:

$$\frac{1}{2}L_m(I_{Lm(tv9)} - I_{Lm(tv10)})^2 \geq \frac{1}{2}C_{M1}V^2_{DS1(max)}, \qquad (5)$$

Where $I_{Lm(tv9)}$ is the magnetizing inductor current at time t_9, $I_{LK(tv10)}$ is that at time t_{10}, C_{M1} is the junction capacitor of switch M_1 and $V_{DS1(max)}$ is the voltage across switch M_1 and its value is equal to $(V_i + V_{C1})$. According to volt-second balance principle, the voltage V_{C1} can be expressed by

$$V_{C1} = \frac{DV_i}{1-D}, \qquad (6)$$

Once C_{M1}, and $I_{Lm(tv9)}$ and $I_{Lm(tv10)}$ are specified, the inequality of the magnetizing inductor L_m can be expressed as

$$L_m \geq \frac{C_{M1}V_i^2}{(1-D)^2(I_{Lm(tv9)} - I_{Lm(tv10)})^2}, \qquad (7)$$

From (4) and (7), it can be seen that the magnetizing inductor L_m must be satisfy (4) and (7) simultaneously.

C. Output Filter L_1 and C_O

Since the proposed forward converter is operated in continuous conduction mode (CCM), output filter inductor L_1 must be large enough to maintain CCM operation. The inductance of L_1 must satisfy the following inequality:

$$L_1 \geq \frac{V_{DC}(1-D)T_s}{\Delta I_{...}}, \qquad (8)$$

where $\Delta I_{L1(max)}$ is the maximum ripple of output inductor current I_{L1}. When the maximum current ripple is specified, the minimum magnetizing inductance can be determined.

The output capacitor C_O is primarily designed for reducing ripple voltage. The ripple voltage across output capacitor C_O is determined as follows:

$$\Delta V_{rco} = \frac{\Delta Q_{CO}}{C_O} + \Delta I_{L1(max)} * ESR$$

$$= \frac{1}{C_O} \times \frac{1}{2} \times \frac{\Delta I_{L1(max)}}{2} \times \frac{T_s}{2} + \Delta I_{L1(max)} * ESR$$

$$= \frac{\Delta I_{L(max)}}{C_O}(\frac{1}{8f_s} + C_O * ESR), \qquad (9)$$

For alumimum electrolytic capacitors, the product of C_O*ESR (capacitance and equivalent series resistance) is much less than $1/8f_s$ and it can be neglected. Thus, capacitor C_O is selected as

$$C_O = \frac{\Delta I_{L1(max)}}{8f_s\Delta V_{rco}}, \qquad (10)$$

IV. Measured Results

To verify the performance of the proposed intelligent stunner, as shown in Fig. 5, a prototype with the following specifications was implemented.

(A) active clamp forward converter
  □input voltage V_i: 150 Vdc,
  □output voltage VDC: 100 Vdc,
  □switching frequency f_{s1}: 50 kHz, and
  □maximum output current I_{dc}: 2 A.

(B) dither voltage generator (half-bridge inverter)
  □input voltage V_i: 150 Vdc,
  □output voltage V_O: 20 ~ 200 Vac,
  □switching frequency f_{s2}: 50 kHz, and
  □maximum output power: 20 W.

(C) full-bridge inverter
  □input voltage V_{DC}: 100 Vdc,
  □output voltage V_O: ±100 V,
  □switching frequency f_{s3}: 400 Hz, □maximum
output current I_O: ±2 A, and
  □maximum output power: 200 W.

According to the specifications, components of the active clamp forward converter associated with a half-bridge inverter are determined as follow:
  □turns ratio of transformer T_r: 1,
  □magnetizing inductor L_m: 12 mH,
  □leakage inductor L_K: 20 µH,
  □transformer core: EI-42,
  □diode $D_1 \sim D_4$: UF104,
  □diode $D_5 \sim D_6$: UF304, and
  □switches $M_1 \sim M_2$: IRF840.

To generate ac voltage waveforms, switches of the full-bridge inverter are also determined as $S_1 \sim S_4$: IRF840.

Measured voltage V_{DS} and current I_{DS} of switch M_1 is shown in Fig. 8 under 50% of the full load, from which it can be seen that the proposed converter can achieve ZVS features. Fig. 9 depicts the efficiency comparison curve between the proposed converter with integration switches, and the conventional hard-switching one, illustrating that

conventional one, and its efficiency is 91% under full load condition. In addition, the efficiency of the overall stunner system is around 85% under full load condition. Measured waveforms of output voltage V_O and output current I_O without or with dither voltage during chicken stunning as shown in Figs. 10(a) and (b), respectively. From Fig. 10(a), it can be found that the poultry stunner system without dither voltage cannot generate enough stunning current. That is, the proposed stunner system with dither voltage not only can reduce stunning voltage but can generate enough stunning current. As a consequence, it can reduce chicken stress during stunning and increase carcass quality. Fig. 11 shows measured waveforms of dither V_{dI} and output current I_O, illustrating that when skin breakdowns, the dither voltage V_{dI} will drop. To reduce power loss, relay R_{AI} which is connected in series with dither-voltage path is controlled to open. From practical experimental results for stunning a chicken, it can observed that the coma time of chicken with voltage of ±100 V, current of ±100 mA, and stunning time of 6 s can sustain about 30 s, and it is long enough for bleeding.

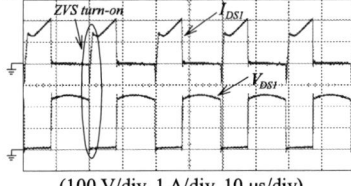

(100 V/div, 1 A/div, 10 µs/div)

Fig. 8. Measured voltage and current waveforms of switch M_I in the proposed converter with active clamp.

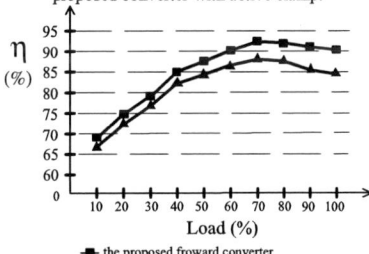

Fig. 9. Comparison conversion efficiency between the conventional hard-switching forward converter and the proposed one from light load to heavy load.

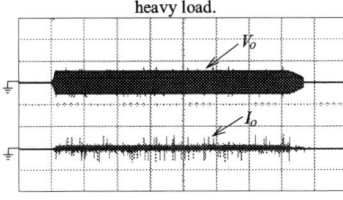

(200 V/div, 100 mA/div, 1 s/div)
(a)

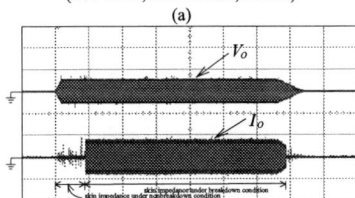

(b)

Fig. 10. Measured waveforms of output voltage V_O and output current I_O: (a) without dither voltage, and (b) with dither voltage.

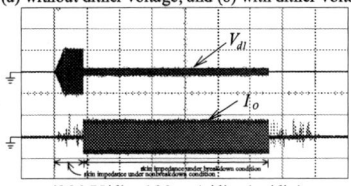

(200 V/div, 100 mA/div, 1 s/div)

Fig. 11. Measured waveforms of dither voltage V_{dI} and output current I_O during chicken stunning.

V. Conclusion

In this paper coma mechanism of poultry with electrical stunning has been briefly reviewed. Operational principle, steady-state analysis and design of the proposed active lamp converter associated with a half-bridge inverter, and a full-bridge inverter has been implemented to generate stunning electrical parameters, in which its current amplitudes is ±100 mA and its voltage amplitude is ±100 V. The proposed active clamp forward converter with switch integration can achieve the efficiency around 91% under full load condition, and the efficiency of overall stunner system is about 85%. From experimental results, it can also found that the proposed one can attain a good meat quality, while meet the regulation of animal welfare.

REFERENCES

[1] H. A. Channon, A. M. Payne and R. D. Warner, "Comparison of CO_2 Stunninng with Manual Electrical Stunning (50Hz) of Pig on Carcass and Meat Quality," *Trans. on Meat Science*, 2002, pp.63—68.

[2] S. B. Wotton and M. O. Callaghan, "Electrical Stunning of Pigs: the Effect of Applied Voltage on Impedance to Current Flow and the Operation of a Fail-Safe Device," *Trans. on Meat Science*, 2002, pp. 203—208.

[3] A. Velarde, *et al.*, "Effect of Electrical Stunning on Meat and Carcass Quality in Lambs," *Trans. on Meat Science*, 2003, pp. 35—38.

[4] H. A. Channon, A. M. Payne and R. D. Warner, "Effect of Stun Duration and Current Level Applied During Head to Back and Head only Electrical Stunning of Pigs on Pork Quality Compared with Pigs Stunned with CO2, " *Trans. on Meat Science*, 2001, pp. 1325—1333.

[5] E. Lambooij, *et al.*, "Some neural and behavioural aspects of electrical and mechanical stunning in ostriches," *Trans. on Meat Science*, 1999, pp. 339—345.

[6] E. Lambooij, *et al* l., "The Effects of Captive Bolt and Electrical Stunnin, and Restraining Methods on Broiler Meat Quality," *Trans. on Poultry Scien ce*, 1999, pp. 600—607.

[7] B. Savenije, *et al.*, "Electrical Stunning and Exsanguination Decrease the Extracellular Volume in the Broiler Brain as Studied with Brain Impedance Recordings," *Trans. on Poultry Science*, 2000, pp. 1062—1066.

[8] V. Sante, *et al.*, "Effect of Stunning Current Frequency on Carcass Downgrading and Meat Quality of Turkey," *Trans. on Poultry Science*, 2000, pp. 1208—1214.

[9] S.F. Bilgili, "Recent Advances in Electrical Stunning," *Trans. on Poultry Science*, 1999, pp. 282—286.

[10] W. D. McNeal, *et al.*, "Effects of Stunning and Decapitation on Broiler Activity During Bleeding, Blood Loss, Carcass, and Breast Meat Quality," *Trans. on Poultry Science*, 2003, pp. 163—168.

[11] S.-Y. Tseng, S.-H. Tseng and J. G. Huang, "High Step-up Converter with Partial Energy Processing for Livestock Stunning Applications," *Proceedings of the Applied Power Electronics Conference*, 2006, pp. 1537-1543.

[12] T.-F. Wu, *et al.*, "Unified Approach to Developing Single Stage Power Converters, " *IEEE Transactions on Aerospace and Electronic*

2006 5th International Power Electronics and Motion Control Conference

A Novel Zero-Voltage Switching Resonant Pole Inverter

Sanbo Pan and Junmin Pan

Department of Electrical Engineering, Shanghai Jiao Tong University, Shanghai, P. R. China
pansansjtu@hotmail.com

Abstract—**In order to realize a simple topology, high efficiency, high frequency, low voltage stress, easy to control soft switching three phase inverter, this paper introduces a novel resonant pole three phase inverter, which realized the zero-voltage switching of the main switches, zero-current switching of auxiliary switches and possesses the small power auxiliary circuit and full PWM capability. It avoids the bulk capacitor of the auxiliary resonant commutated pole inverter (ARCPI) and no center tap potential variation problem of ARCPI, unlike the delta or wye(Y) configured resonant snubber inverter, the inverter possesses the advantages of the decoupled resonation of each phase, it is easy to implement the various control schemes. The operation principle of one phase circuit has been analyzed and the equivalent circuits at different operation modes for both ZCS and ZVS operations are present. Simulation and experimental results are proposed to verify the theoretical analysis.**

Keywords- Resonant pole inverter, Soft switching, Zero current switching, Zero voltage switching.

I. INTRODUCTION

The hard switching inverters have many problems such as low switching frequency, high switch loss, severe turn on current spike and turn off voltage spikes, high electromagnetic interference (EMI) and acoustic noise, etc., to solve these problems, many zero-voltage switching (ZVS) or zero-current switching (ZCS) circuits present. The first resonant inverter was the resonant DC link inverter (RDCLI) proposed by Dr. Divan in 1980s [1]. Which has the advantage of simple topology, fewer components, high frequency etc., but it has disadvantages such as high voltage stress of the switches, high dc link voltage ripple, and large resonant inductor power losses. It is with discrete pulse modulation (DPM) control that it is hard to achieve real PWM control and will result in sub-harmonics. Then comes parallel resonant DC link inverter (PRDCLI) [2]-[3], Although they can solve the problems of RDCLI, DC link inverters make the DC voltage for inverters periodically resonant to zero, the resonant is coupled to three phases, thus inverters have DC link voltage losses and have low output voltages than traditionally hard switching inverters in the same control strategies, and the switches or resonant components in the DC link cause high losses [2], the AC side resonant inverter has not the short comes of DC link resonant inverter, there are many zero-voltage switching inverters such as the auxiliary resonant commutated pole inverter

(ARCPI) [4], the delta or star connected resonant snubber inverters[5] and the coupled-inductor inverters[6] or the transformer-aid inverters [7], which either need bulk capacitors or additional transformer or 3 phase are coupled together thus lead to complex circuit and complex control of PWM method. One kind of ZCS-ZVS inverter [8] has resonant inductor in DC link thus result in moderately inductor conduction loss. Among the various resonant inverter circuits, the ARCPI [4] and transformer-assisted bridge configuration inverters [9] have been mostly favored due to their small rating auxiliary circuit and flexible PWM application capability. But the ARCPI has following main problem: to avoid the variation of the DC link capacitor center tap voltage, the two capacitors should be in huge values. Thus the physic volume is huge. And the variation may reduce the system reliability [9]. The transformer-assisted bridge configuration inverters need additional transformer and the magnetic flux reset path [9], the circuit is complexity. So although there are many resonant inverter topologies exist now, the industry still hunger for a better resonant inverter topology.

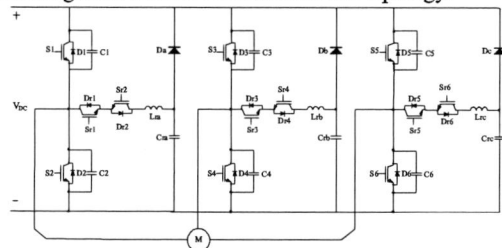

Fig.1. Proposed ZVS-ZCS resonant pole inverter

In this paper, a novel ZVS resonant pole inverter is proposed. As shown in Fig.1. The proposed inverter's resonant circuit operation is independent for each phase and suitable to apply the matured PWM modulation scheme, such as space vector pulse width modulation (SVPWM), each single phase resonant circuit is composed by one bidirectional switch unit, one resonant inductor, one small resonant capacitor and one clamp diode, avoid using the huge bulk capacitor and additional transformer, the novel ZVS inverter has the advantages of low switching power losses, low resonant voltage stress.

II. TOPOLOGY AND OPERATION PRINCIPLE

For simple analysis purpose, one single independent phase leg circuit of the novel ZVS inverter is discussed and the equivalent circuit is shown in Fig.2, where V_{DC} is the DC bus voltage, Io is the load current, I_{Lr} is the resonant inductor current, Vcr is the resonant capacitor voltage.

1-4244-0448-7/06/$25.00 ©2006 IEEE 1348

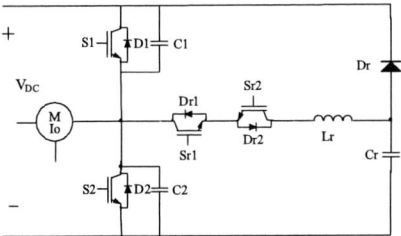

Fig.2. one phase equivalent circuit of proposed inverter

Assume the load current is positive, that is, the current is flow from inverter to load, and load current is considered constant during one switching cycle in this topology, the main switches turned off softly under ZVS by the effect of the parallel capacitor, to achieve good soft turn off effect and reduce the turn off losses greatly, the capacitor value should be moderate [7], thus if the load current is not big enough, the natural zero voltage turn on of switches by the load current effect, as denoted in [8], may not occur, but in this novel topology, the resonant circuit can overlap the load current to conduct the freewheeling diode, turn on switches under ZVS and need no detection the load current direction. The characteristic waveforms of gate driver signals, the voltage across switch, the resonant inductor current and the capacitor voltage under different operation modes in one period are illustrated in Fig.3. Corresponding equivalent circuits and current flow in each operation modes are shown in Fig.4.

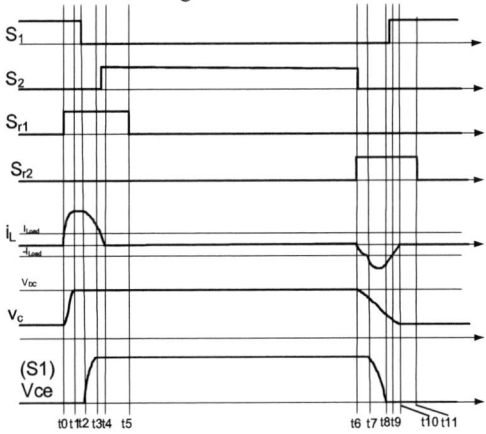

Fig.3. Operation modes and waveforms of circuit

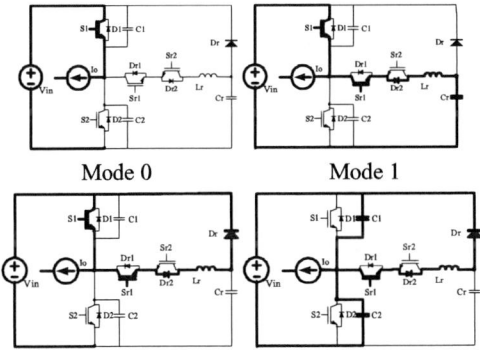

Mode 0   Mode 1

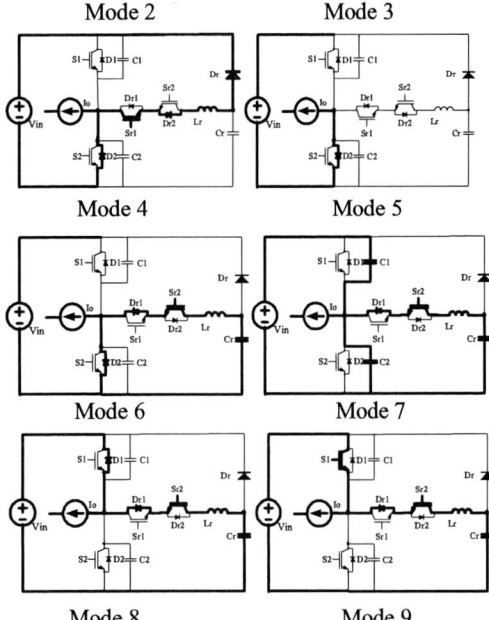

Fig.4. Equivalent circuits under different operation modes in full resonant mode

Operation modes analysis:

(a) Mode 0 [t -t_0]: Initial state. When upper switch S_1 is on and load current flows through S_1. The auxiliary switches are keeps off.

(b) Mode 1 [t_0-t_1]: At t_0, switch Sr_1 is on, current flows through S_1, Sr_1, Dr_2, L_r and C_r. L_r resonant with Cr_1, the resonant current charge Cr_1 and V_{cr} increasing, till time t_1, i_{Lr} reaches the threshold value I_{T1}, at that time V_{cr} equals to V_{T1}, this mode ended. Assume the initial voltage of V_{cr} is V_{T2}. If V_{T1} equals to V_{DC}, then this resonant mode is full resonant mode, and next mode is mode 2, otherwise, this resonant mode is half resonant mode, and next mode is mode 3.

(c) Mode 2 [t_1-t_2]: At time t_1, V_{cr} reaches V_{DC}, diode Dr turns on, current flow through S_1, Sr_1, Dr_2, Lr, Dr in cycle. i_{Lr} keeps I_{T1} unchanged, Vcr keeps V_{DC} unchanged.

(d) Mode 3 [t_2-t_3]: At time t_2, S_1 turns off, the load current Io and inductor current i_{Lr1} both charge the parallel capacitor C_1 and discharge C_2, resonant inductor Lr resonant with C_1,C_2. Assume C_1=C_2=C, if the circuit is in full resonant mode, where Vcr keeps the value of V_{DC} unchanged during that time. If the circuit is in half resonant mode, where Vcr is increasing from V_{T1} during that time.

(e) Mode 4 [t_3-t_4]: At time t_3, voltage across C_2 decreased to zero and freewheeling diode D_2 turned on. If in the full resonant mode, Vcr keeps Vin unchanged during t_3 to t_4, the voltage across Lr equals to V_{DC}, so i_{Lr} decrease lineally to zero at the time t_4. During that time turn on S_2 under ZVS. If in the half resonant mode, Lr continuous resonant with Cr, till the tome of t_4, i_{Lr} decreased to zero and Vcr raised to V_{T3}.

(f) Mode 5 [t_4-t_6]: At the time of t_4, the resonant stop for the reverse block of the diode Dr2, the resonant current keeps zero, turn off Sr_1 under ZCS at the time t_5. S_1 and auxiliary switches are keep off, S_2 is on, the load current

1349

flow through freewheeling diode D_2. It is the main power transfer period, the time of this period is determined by PWM control strategy.

(g) Mode 6 [t_6-t_7]: At the time of t_6, turn on Sr_2 and turn off S_2, S_1, D_1, Sr_1, Dr are all keeps off. Sr_2 is on. Cr resonant with Lr, i_{Lr} increasing in negative value. Until at time t_7, $i_{Lr}(t_7)$=-I_0, then D_2 block off, mode 6 is end..

(h) Mode 7 [t_7-t_8]: At t_7, only $Sr2$ and $Dr1$ are on. Lr resonant with C_1,C_2 and Cr. Similar to mode 3.At the time of t_8, Vc_2 increase to V_{DC}, and diode D_1 turns on.

(i) Mode 8 [t_8-t_9]: At the time of t_8, $D1$ turns on and Lr resonant with Cr, the state equations is the same as mode 1. Before i_{Lr} decreases to I_0 in absolute values, diode D_1 keeps on. At the time of t9, i_{Lr} =-I_0, then diode D_1 turn off. During this period turn on S1 under ZVS condition.

. (j) Mode 9 [t_9-t_{11}]: At the time t9, diode D_1 turns off, i_{Lr} is less than I_0 in absolute values, current flow through S_1 and Sr_2 to load. The state equations are the same as Mode 8. When at the time of t10, i_{Lr} reaches zero and resonant stopped, at the time of t_{11}, $Sr2$ turn off under ZCS condition. Mode 9 is ended.

Then followed by Mode 0, an entire PWM period under positive load current mode is analyzed. The operation modes of the circuit under negative load current condition are similar to the modes under positive load current condition. When upper switch is turned off the resonant inductor current discharge the capacitor parallel with the low switch to create ZVS condition for the low switch, So even in very light load current conditions, the resonant current can discharge low capacitor to create ZVS on condition for low switch. The same, when low switch turns off then the resonant inductor current discharge the upper capacitor and charge the low capacitor to create ZVS on condition for upper switch no matter the load current direction and value.

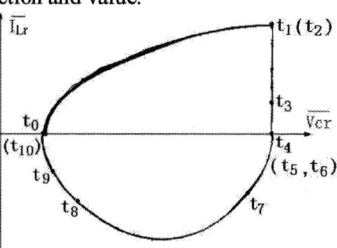

Fig.5 Phase plane representation of the resonance during a switching cycle

Assume the unified current $\overline{i_{Lr}} = i \cdot Z_r / V_{DC}$, unified voltage $\overline{v} = v_{Cr} / V_{DC}$, and unified time $\overline{t} = t \cdot \omega_0$. Based on the above analysis, the phase plane representation of the L_r current and Cr voltage loci during the whole switching cycle is shown in Fig.5. Curve t_0 to t_1 represents the resonance of mode 1, Line $t_1(t_2)$, t_3 - t_4 depicts the switch S_1 to diode D_2 commutation procedure, Curve t_4-t_9 depicts the diode D_2 to switch S_1 commutation procedure, the point $t_4(t_5,t_6)$ represents the main switch S_2 on procedure of mode 5, the point $t_0(t_{10})$ represents the main switch S_1 on procedure of mode 0, From the locus curve, the voltage stress is restricting to V_{DC}, which is benefit to choose the low voltage switch and diode.

III. DESIGN CONSIDERATION OF RESONANT PARAMETERS

The task of this section is to design the resonant parameters to form an experiment inverter, which including choose the parallel capacitor of main switches $C=C_1=C_2$, the resonant inductors Lr, the resonant capacitor Cr.

The parallel snubber capacitor C_1 and C_2 is used to reduce the turn off losses of S_1 and S_2, the value of C is inversely proportional to the rise rate of the switch voltage drop when turning off the lower main inverter switches. In order to maximize the benefits of soft switching, Cr be chosen to operate in the over snubbed mode. But if C is too large, the charge or discharger current and charge time should be bigger, thus result more effective values of resonant current and more losses. So there should be an trade off of choose the value of C [7]-[9]. The selection of the snubber capacitor can be determined as [7] denoted

$$C \geq \frac{I_{So\max}}{V_{DC}} \cdot (3-5)t_{off} \quad (1)$$

Where I_{somax} means the maximum output current of switch, t_{off} refers to the turn off time of the switch.

One limitation of the inductor Lr is to ensure the ZCS turn on of the auxiliary switch, so the inductance of Lr should be big enough to limit the rising rate of the auxiliary switch current. It should be at least the following values to limit the turn on losses of auxiliary switch [7].

$$Lr \geq \frac{V_{DC}}{I_{So\max}} \cdot (3-5)t_{on} \quad (2)$$

Where t_{on} is the turn on time of the switch.

One limitation of the inductor Lr is its cooperation with resonant capacitor Cr and C. Considerate the full resonant mode which represent the maxim ability of the circuit, in mode 1 the maximum resonant current should below the allowed maxim current of switches, Ism, and should be lower possible to reduce the losses.

$$i_{Lr}(\max) = \frac{V_{DC}}{Z_r} \leq Ism \quad (3)$$

Ensure to realize the ZVS of S_2, the energy stored in inductor should greater than the energy stored in capacitors, E_f is the energy feed back to bus, so

$$\frac{1}{2}Lr\left(\frac{V_{DC}-V_{T2}}{Z_r}\right)^2 - \frac{1}{2}LrI_o^2 \geq CV_{DC}^2 + E_f > CV_{DC}^2 \quad (4)$$

Also ensure to realize the ZVS of S_1 in mode 7, the energy stored in capacitor Cr should greater than the energy stored in snubber capacitors. So

$$\frac{1}{2}CrV_{DC}^2 \geq CV_{DC}^2 \quad (5)$$

From the mode analysis above, assume the capacitor voltage at the end of mode 3 and mode 8 is V_{T3} and V_{T8}, the resonant time is

Ta=T_1+T_2+T_3+T_4

$$\approx \pi\sqrt{Lr \cdot Cr}\big/2 + \pi\sqrt{Lr \cdot 2C}\big/2 + \sqrt{Lr \cdot Cr} \quad (6)$$

$$T_b = T_6 + T_7 + T_{8-10}$$

$$\approx \sqrt{L_r \cdot C_r} \cdot \arcsin \frac{I_o \cdot Z_r}{V_{T3}} + \pi \sqrt{L_r \cdot 2C}/2 +$$

$$\sqrt{L_r \cdot C_r} \cdot arctg \frac{V_{T3} - V_{T8}}{Z_r \cdot I_o} \qquad (7)$$

To reduce the losses, the circle current time T_2 should be zero, and the resonant time should be within 1/10 of the switching period time. The resonant parameters following above rules can realize the soft switching of switches.

IV. SIMULATION AND EXPERIMENT RESULTS

Simulation for the resonant ZVS inverter has been carried out by pspice software, the parameters follows:

Resonant inductor Lr: 5uH

Resonant capacitor Cr: 0.68uF

Snubber capacitor C: 0.1uF

DC link voltage V_{DC}: 300V

Load current (constant) Io: 10A

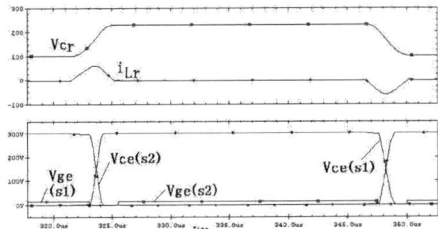

Fig.6 The simulation of S1 and S2 ZVS on at positive half load current under half resonant Mode

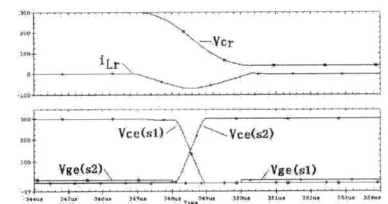

7(a) S1 ZVS on at negative full load current

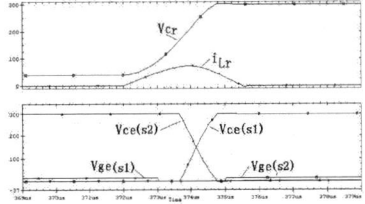

7(b) S2 ZVS on at negative full load current
Fig.7 The simulation under full resonant Mode

Simulation waveforms in Fig.6 and Fig.7 shows ZVS turn on features and corresponding inductor current and resonant capacitor voltage under different operation conditions. The one switching period waveforms of S2 and S1 ZVS on at positive half load current is shown in Fig. 6. Fig.7 shows the waveforms under full resonant

mode, the waveforms of S2 and S1 ZVS on at negative full load current are shown in Fig. 7(a) and Fig. 7(b).

The simulation results agree with the theoretical analyses. To S1, S2, from the waveforms of Vge and Vce, it clearly shows that when Vce decrease to zero, then the gate driver signals send out, that, the freewheeling diodes is on, then the switches turn on, the switches are turn on under ZVS conditions, no matter what the load current direction and resonant mode are.

In order to verify the theoretical analysis and simulation results, a proof-of concept one phase inverter as show in Fig.2 has been built. Two SEMIKON IGBT modules (SKM50GB123D, 1200V/50A) are employed as the main and auxiliary switches, one ultrafast recovery diode DSEP29-06B(600V/30A) is used as the clamp diode Dr, the snubber capacitor C using low-loss polypropylene o.1uF, the resonant inductor is 5uH in value(15 turns of wires with size AWG 15 Litz copper wire). Resonant capacitor choosing the high frequency high current ripple polypropylene film capacitor 943C6P33K (parallel four capacitors and get 1uF). For simple control, the driver signal of auxiliary switches Sr1 and Sr2 using the fixed time control, the turn on gate signals of them lead the turn off gate signals of main switches 1us, and lasted for 5us to cover one resonation time. Dead time of 2us is inserted interlocking the two main switches. Because the snubber capacitor is 0.1uF, the main switches driver is interfaced by an external zero-voltage detecting circuit as denoted in [9], which releases the gating signals when the Vce becomes zero, thus ensure the zero voltage switching on of main switches and avoid the switching on losses of snubber capacitor when the switches turn on not under ZVS conditions.

The test conditions are:

DC link voltage: 300V;

Rated load current: 10A;

Switching frequency: 10 KHz

The system is tested in half load and full load in positive load current, the voltage waveforms across the main inverter switches S1 and their gate signals in full and half load current are shown in Fig. 8(a). Fig. 8(b) shows the voltage waveforms across the auxiliary inverter switches Sr1and the flow currents in full load current.

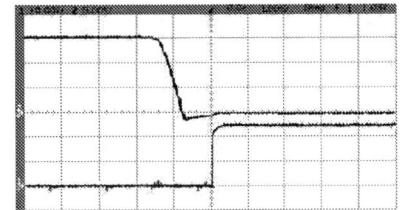

Fig. 8 (a) Vce (100V/Div) and Vge (5V/Div) waveforms of switch S1 under full load current

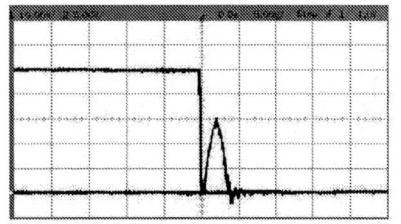

Fig. 8 (b)The voltage (100V/Div) and current (25A/Div) waveforms of auxiliary switch Sr1

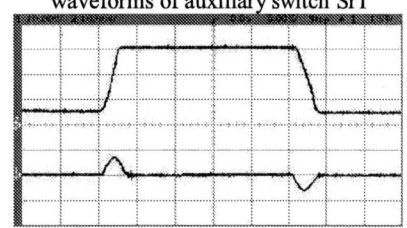

Fig. 8 (c) The inductor current (100A/Div) and resonant capacitor voltage (100V/Div) waveforms at full load

The inductor current and resonant capacitor voltage at full load current are shown in Fig. 8(c). It can be seen that the resonant pole inverter works well under various load currents, and the zero-voltage switching on occurs. The zero-current switching on and off of auxiliary switches occurs too.

V. CONCLUSION

A novel ZVS resonant pole inverter is proposed. Its operation principle is analyzed and resonant parameter select rules is given, the simulation and the experiment results verified the theory analysis. The circuit has following features: The circuit structure is simple, and the main switches and the auxiliary switch all are operations under ZVS or ZCS;The circuit has full PWM capability; No need to sense the load current direction, and even in light load, the circuit can realized ZVS of the main switches; The voltage stresses of Lr, Cr are limited within Vin, avoid the high voltage stress; The resonant make the diode blocked at the zero current time and lead to no reverse current and corresponding losses.

REFERENCES

[1]   M.D.Bellar, T.S. Wu, A.Tchamdjou, J.Mahdavi, M.Ehsani, A review of soft-switched DC-AC converters, *Industry Applications, IEEE Transactions on*, Vol.34, Issue 4, 1998, pp.847-860.

[2]   Zhi Yang Pan, Fang Lin Luo, Novel soft-switching inverter for brushless DC motor variable peed drive system, *IEEE Transactions on Power Electronics,* Vol.19, Issue 2, 2004, pp. 280-288.

[3]   Chen Guocheng, Xu Chunyu, Sun Chengbo, Qu Keqing; Katsunori, T, Characteristic analysis of PWM pattern for three-phase ZVS inverter, *The 4th International Power Electronics and Motion Control Conference*, Vol.2, 2004, pp. 936-941.

[4]   R.W. De Doncker, J. P. Lyons, The auxiliary resonant commutated pole converter, *IEEE Industry Applications Society Annul Meeting*, Vol.2 1990, pp. 1228–1235.

[5]   J.-S. Lai, R.W. Young Sr., G.W. Ott Jr., J.W. McKeever, and F. Z. Peng, A delta-configured auxiliary resonant snubber inverter, *IEEE Trans.Ind. Appl.*, Vol.32, No.3, 1996, pp. 518-525.

[6]   Jae-Young Choi, Dushan Boroyevich, Jerry Francis, and Fred C. Lee, A novel ZVT inverter with simplified auxiliary circuit, *Proc. of Applied Power Electronics Conf.*, Vol.2, 2001, pp. 1151-1157.

[7]   Zhi Yang Pan, Fang Lin Luo, Novel resonant pole inverter for brushless DC motor drive system, *IEEE Transactions on Power Electronics*, Vol.20, Issue 1, 2005, pp.173- 181.

[8]   Jianming Yao and Thomas A.Lipo, A novel soft-switching inverter with ZCS-ZVS features, in *Proc. IEEE Power Electronics Specialists Conf.*, 2001, pp. 1141-1146.

[9]   X.Yuan and I.Barbi, Analysis, design and experimentation of a transformer-assisted PWM zero-voltage switching pole inverter, *IEEE Trans. on Power Electronics*, Vol.15, No.1, 2000, pp.72-82.

2006 5th International Power Electronics and Motion Control Conference

Analysis of Three-Level ZVS PWM Inverter for Induction Heating Applications

A. Jangwanitlert[*], J. Songboonkaew[**], W. Thammasiriroj[***] and J.C. Balda[****]

[*] Department of Electrical Engineering, Faculty of Engineering,
King Mongkut's Institute of Technology Ladkrabang, Bangkok, Thailand
[**] Department of Electrical Engineering, Faculty of Engineering,
Thonburi College of Technology, Bangkok, Thailand
[***] Department of Power Engineering Technology, College of Industrial Technology,
King Mongkut's Institute of Technology North Bangkok, Bangkok, Thailand
[****] Department of Electrical Engineering, College of Engineering,
University of Arkansas, Fayetteville, AR, USA
Email: kjanuwat@kmitl.ac.th

Abstract—**This paper presents an operation analysis of a high frequency three-level (TL) PWM inverter applied for an induction heating applications. The feature of TL inverter is to achieve zero-voltage switching (ZVS) at above the resonant frequency. The circuit has been modified from the full-bridge inverter to reach high-voltage with low-harmonic output. The device voltage stresses are controlled in a half of the dc input voltage. The prototype operated between 70 and 78 kHz at the dc voltage rating of 580 V can supply the output power rating up to 3000 W. The iron has been heated and hardened at the temperature up to 800°C. In addition, the experiments have been successfully tested and compared with the simulations.**

Keywords-Three-level PWM inverter; zero-voltage switching; induction heating

I. INTRODUCTION

The full-bridge (FB) PWM converter [1-2] is one of the high power converter topologies that posses the basic desirable characteristics of both the hard switching PWM converters and the soft-switching ones, avoiding their drawbacks such as commutations losses and high conduction losses. However, the power switches of the FB-PWM converter are subjected to the maximum input voltage level. In the case of high input voltage, it may not be possible to get adequate switches [3-4]. In order to reduce the voltage stress of the switches, this paper proposes the TL-PWM inverter applied for induction heating applications. In this TL-PWM converter, the switching frequencies are considered. As switching frequencies increase, the switching losses associated with the turn on and turn off of the devices also increase. In switch-mode power supplies, these losses can be significant enough that the operation of the power supplies at very high frequencies can be prohibitive because of low conversion efficiencies, even when applying soft-switching techniques. In resonant-mode inverter, however, the switching losses are low, when the resonant converter operates at very high frequencies.

Therefore, the use of resonant converter remains an interesting option for some applications requiring the previous specifications, even though their conduction losses, e.g. circulating energy, and component stresses could be comparatively high. In resonant inverters, MOSFET devices are appropriate for high switching frequencies, and a ZVS operation is recommended. Series and parallel resonant inverters operate under ZVS for the active devices when the switching frequency is above the resonant frequency [1]. Therefore, this paper presents the TL-PWM inverter using a series resonant mode as shown in Fig.1.

II. SYSTEM CONFIGURATION

Fig.1 shows a system configuration of an induction heating system developed for the induction hardening. The power circuit of high frequency inverter consists of a voltage-source inverter using four MOSFETs. Each MOSFET power ratings are 500 V and 16 A. The lossless snubbing circuit consisting of only a turn-on parasitic capacitor of 440 pF is connected between drain and source of each MOSFET. The output terminal of the inverter is connected to a series resonant circuit with a matching transformer with the turn ratio of 3:1.

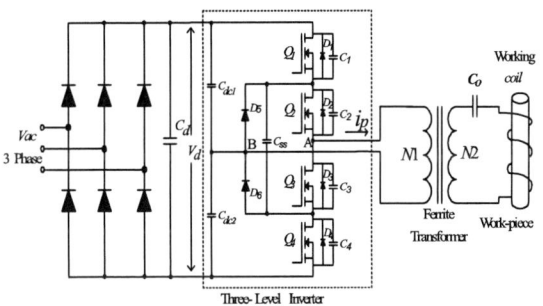

Figure 1. TL-ZVS-PWM inverter circuit for induction heating applications

1-4244-0448-7/06/$25.00 ©2006 IEEE   1353

III. PRINCIPLE OF OPERATION

An analysis of TL-ZVS-PWM inverter for induction heating applications is derived from the equivalent circuit of the induction coil. The induction coil and the work piece (iron bar) are designed using a transformer principle. The induction coil is similar to the primary winding and the work piece is the secondary winding that wound as the single turn. As shown in Fig. 1.

In Fig.2, the inductance L_s and the resistance R_s are the work piece, and the inductance L_p and the resistance R_P are the induction coil and high-frequency transformer. The parameters are transferred from secondary side to primary side. The equivalent inductance (L_{eq}) and resistance (R_{eq}) are expressed in (1) and (2). These parameters are added with C_O transferred to the primary side as expressed in (3), where N is the turn of the induction coil and n is the turn of the transformer.

$$L_{eq} = L_p + N^2 L_s. \tag{1}$$

$$R_{eq} = R_p + N^2 R_s. \tag{2}$$

$$C_{eq} = C_o n^2. \tag{3}$$

From the TL-ZVS-PWM inverter as shown in Fig.1, the circuit is the series resonant mode [5-6]. The capacitor C_o resonates with L_{eq}. The operation of TL-ZVS-PWM inverter for induction heating applications can be divided into ten modes in each cycle as shown in Fig.3.

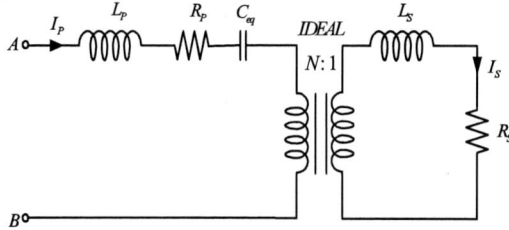

Figure 2. Equivalent circuit of the induction coil and the work piece

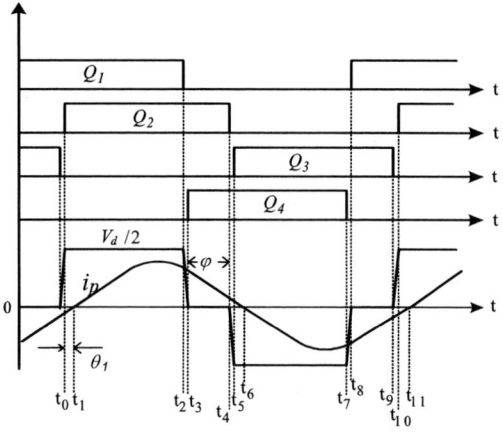

Figure 3. Operational waveforms, where φ is phase-shifted angle

Mode 1 ($t_1 - t_2$)
During this stage, Q_1 and Q_2 conduct and the input power transfers to the load that is the series equivalent circuit (C_{eq}, R_{eq}, and L_{eq}). The magnitude of the primary voltage or output voltage v_{ab} is equal to $V_d/2$ that is a half of the dc input voltage (V_d).

Mode 2 ($t_2 - t_3$)
At time t_2, Q_1 is turned off. The primary current charges C_1 and discharges C_4 through the flying capacitor C_{ss}. At the end of this stage, when the voltage across the capacitor C_1 reaches $V_d/2$, anti-parallel diode D_1 will start to conduct. Simultaneously, the voltage across the capacitor C_4 reaches zero and the anti-parallel diode D_4 starts to conduct. The circuit is shown in Fig.4 (b).

Mode 3 ($t_3 - t_4$)
C_1 and C_4 are completely charged and discharged, respectively. The primary current freewheels through diode D_5 and Q_2. At this time, Q_4 begins to conduct. The circuit is shown in Fig.4 (c).

Mode 4 ($t_4 - t_5$)
At t_4, Q_2 is turned off. The voltage across the capacitor C_2 rises up to the half of the input voltage V_d, and voltage across C_3 decreases to zero. Diode D_4 also conducts. The circuit is shown in Fig. 4(d).

Mode 5 ($t_5 - t_6$)
At t_5, the primary current freewheels through diode D_3 and D_4 as shown in Fig. 4(e). Therefore, Q_3 and Q_4 can be turned on under ZVS conditions.

Mode 6 ($t_6 - t_7$)
At t_6, the primary current reverses its polarity and begins to flow through Q_3 and Q_4 respectively. During this stage, the primary current reaches the output load current as shown in Fig. 4(f).

Mode 7 ($t_7 - t_8$)
At t_7, Q_4 is turned off, but Q_3 still conducts. The primary current charges C_4 and discharges C_1 through flying capacitor C_{ss} as shown in Fig. 4(g).

Mode 8 ($t_8 - t_9$)
C_4 and C_1 are completely charged and discharged, respectively. The primary current freewheels through diode D_6 and Q_3. At this time, Q_1 begins to conduct. The circuit is shown in Fig. 4(h).

Mode 9 ($t_9 - t_{10}$)
At t_9, Q_3 is turned off. The voltage across the capacitor C_4 rises up to the half of the input voltage V_d, and voltage across C_2 decreases to zero. Diode D_1 also conducts. The circuit is shown in Fig. 4(i).

Mode 10 ($t_{10} - t_{11}$)
At t_{10}, the primary current freewheels through diode D_1 and D_2 as shown in Fig. 2(j). Therefore, Q_1 and Q_2 can be turned on under ZVS conditions.

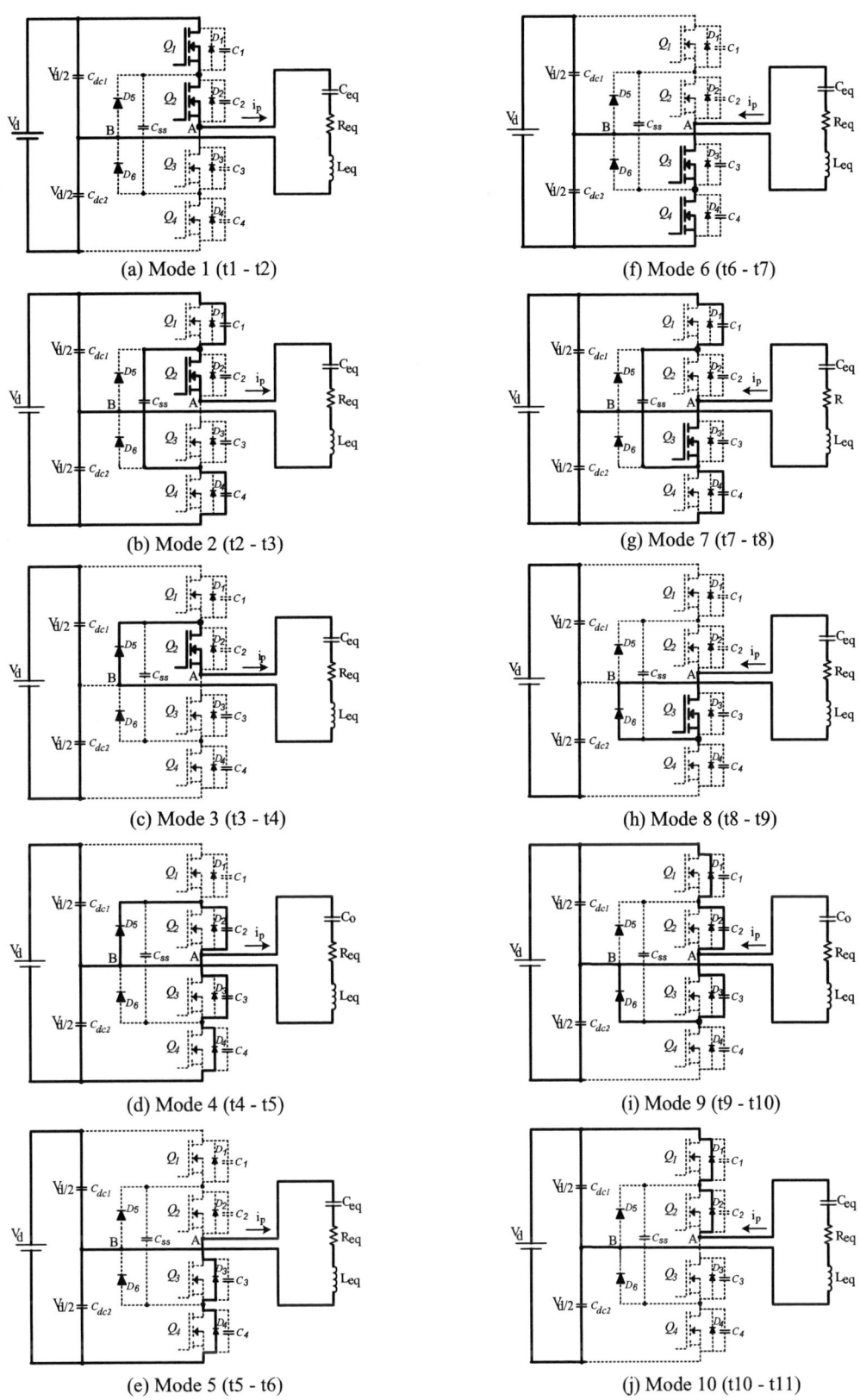

Figure 4. Modes of the operation of the TL-ZVS-PWM inverter

The primary-current (i_p) equation of all modes can be described as (4), where θ_1 is displacement power factor.

$$i_p = \frac{4(V_d/2)\cos(\varphi/2)\cos(\theta_1)}{\pi R_{eq}}\sin(\omega t - \theta_1). \quad (4)$$

Zero–phase–shifted angle condition

When the phase-shifted angle (φ) is adjusted to zero. Some operational modes differ from phase-shifted PWM strategy. The operational waveforms are shown in Fig. 5.

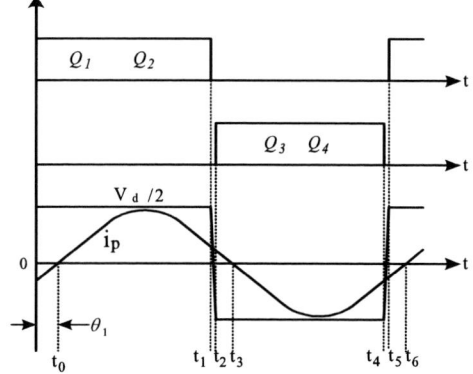

Figure 5. Operational waveforms: zero–phase–shifted angle condition

Mode 1 ($t_0 - t_1$)
Similar to mode 1 of phase-shifted PWM condition, Q_1 and Q_2 turn on simultaneously.

Mode 2 ($t_1 - t_2$)
At t_1, Q_1 and Q_2 are turned off. The primary current charge C_1, C_2 and discharge C_3, C_4. Some discharge-current of C_4 flow through the flying capacitor C_{ss}. The current path is shown in Fig. 6.

Mode 3 ($t_2 - t_3$)
Similar to mode 5 of phase-shifted PWM condition, the primary current flows through D_3 and D_4 and then Q_3 and Q_4 turn on simultaneously under ZVS conditions during the freewheeling.

Mode 4 ($t_3 - t_4$)
As same as mode 6 of phase-shifted condition.

Mode 5 ($t_4 - t_5$)
At t_4, Q_3 and Q_4 are turned off. The primary current charges C_3, C_4 and discharges C_1, C_2, respectively. Some discharge current of C_2 flows through the flying capacitor C_{ss}. The current path is shown in Fig. 7.

Mode 6 ($t_5 - t_6$)
Similar to mode 10 of phase-shifted PWM condition, the primary current flows through D_1 and D_2 and then Q_1 and Q_2 are turned on simultaneously under ZVS conditions during the freewheeling.

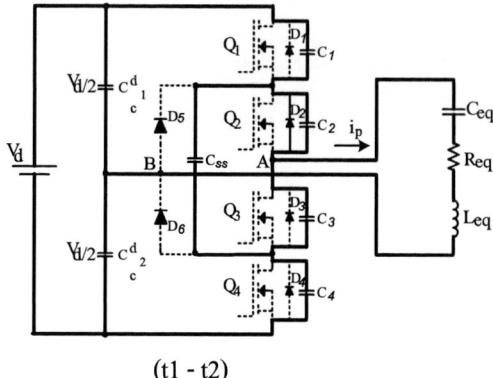

(t1 - t2)

Figure 6. Mode 2 of the zero–phase–shifted angle condition

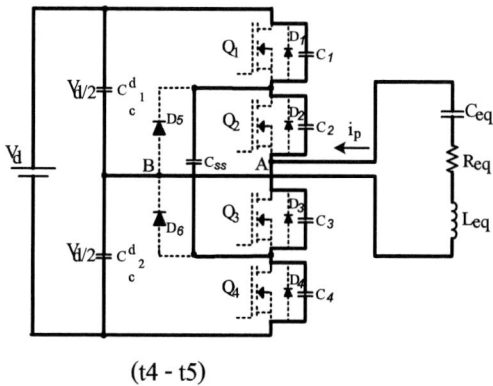

(t4 - t5)

Figure 7. Mode 5 of the zero–phase–shifted angle condition

IV. DESIGN AND EXPERIMENT

An experimental prototype of TL-ZVS-PWM inverter using a phase-shifted PWM control scheme was designed for the following specifications.

DC input voltage : 580 V
Maximum output power : 3 kW
Switching frequency (f): > 70 kHz
Switching device (Q_1-Q_4) : MOSFET (IRFP460)
L_{eq} = 0.22 mH
R_{eq} = 23 Ω
D_5, D_6 = 16CTU04
C_{ss} = 3 μF
C_o = 2.7 nF
$C_1 - C_4$: Parasitic capacitors of MOSFETs = 440 pF
n : Transformer turn ratio ($n = N_p/N_s = 3$)
Transformer cores: EE-80 × 2 cores.

To verify the proposed circuitry, a TL-ZVS-PWM inverter operating at above resonance frequency and using MOSFET were implemented in the laboratory. From the experiment, it shows that there exists an 800 ns dead time and a 0.48 duty cycle maximum in the power devices. As mentioned early, the waveforms of delay circuit (i_d lag v_{ds}) for achieving ZVS are shown in Fig. 8. It can be seen from Fig.8, which the zero-voltage

1356

switching features do exit in the sense that all MOSFETs do turn-on when they are operating at v_{ds} equaling to zero. The experimental results show the validity of the analysis and they are similar to the simulation results as shown in Figs.9 and 10. Figs.9 and 10 show the output voltage and primary current at phase-shifted angle φ = 45° and 0°, respectively. The primary current lags output voltage all the time that means the switching frequency is above resonant frequency. The switching frequencies for Figs.9 and 10 are 74 kHz and 70.5 kHz, respectively. Consequently, all MOSFETs can be achieved ZVS conditions all the time. The waveforms under a different angle φ also meet output power requirement for induction heating are shown in Fig.11. In addition, the waveform of an efficiency of system is not reduced so much when compared with temperature of a C-4 iron bar while tested under $\varphi = 0°$ as shown in Fig.12. The highest efficiency and output power at rated are 91 % and 3 kW, respectively.

The TL-ZVS-PWM inverter and phase-shifted PWM control scheme is helpful and can be applied to much more power devices and different power devices such as IGBT for induction heating applications.

V. CONCLUSION

The application of induction heating using the TL-ZVS-PWM inverter can increase output power by using high dc input voltage. The voltage across switches is a half of dc input voltage. The operational circuit is a series resonance, whose current is a sinusoidal waveform that provides a low harmonic current. The output power is 3 kW and efficiency is 91% at rated.

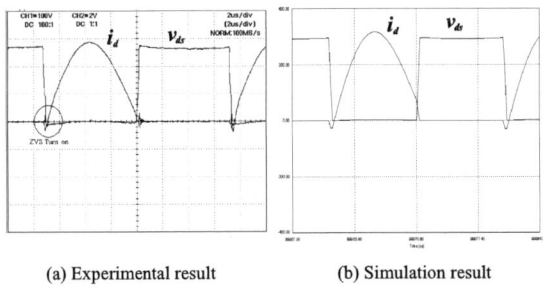

(a) Experimental result   (b) Simulation result

Figure 8.　Switch voltage v_{dS} and current i_d at $\varphi = 0°$

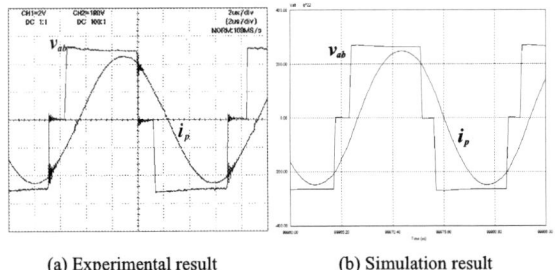

(a) Experimental result   (b) Simulation result

Figure 9.　Output voltage and primary current at $\varphi = 45°$

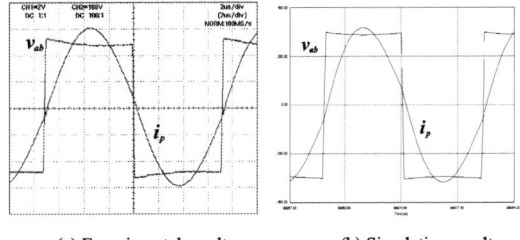

(a) Experimental result   (b) Simulation result

Figure 10.　Output voltage and primary current at $\varphi = 0°$

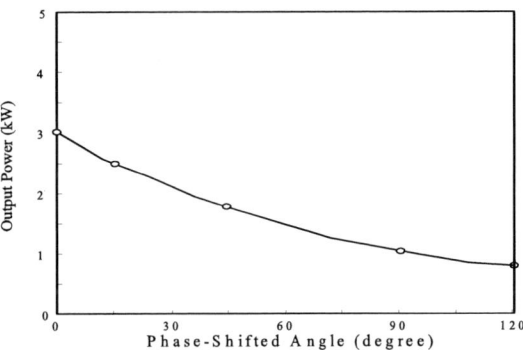

Figure 11.　Output power versus Phase-shifted angle φ

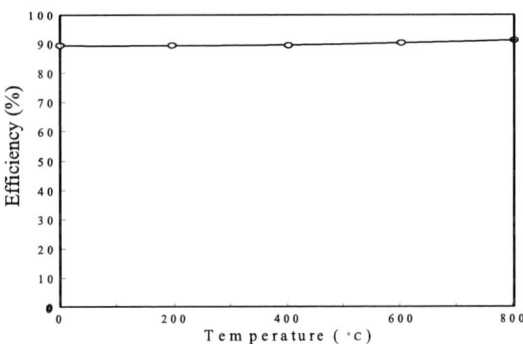

Figure 12.　Measured system efficiency versus temperature of iron bar

REFERENCES

[1] J. R. Pinheiro and I. Barbi, "Wide load range three-level zvs-pwm dc-to-dc converter," *Proc. IEEE PESC*, 1993, pp. 171-177.

[2] T. Song, N. Huang, A. Ioinovici, "A zero-voltage and zero-current switching three-level dc-dc converter with reduced rectifier voltage stress," *Proc. IEEE APEC*, Vol.2, 2004, pp. 1071 – 1077.

[3] F. Canales, P.Barbosa, F.C. Lee, "A zero-voltage and zero-current switching three-level dc/dc converter," *IEEE Trans. Power electron*, Vol. 17, No. 6, 2002, pp. 898-904.

[4] B.Song, R.McDowell, and J. Ennis, "A three-level dc-dc converter with wide input voltage operations for ship-electric-power-distribution systems," *IEEE Trans Plasma science*, Vol. 32, No. 5, 2004, pp.1856-1863.

[5] H.W. Koertzen, J.D Van, and J.A. Ferreira, "Design of the half-bridge, series resonant converter for induction cooking," *Proc. IEEE PESC* , Vol. 2, 1995, pp. 729- 735.

[6] D. Hart, *Introduction to Power Electronics*, Prentice Hall, 1997.

2006 5th International Power Electronics and Motion Control Conference

Dual Duty Cycle Controlled Voltage Source Soft-Switching High Frequency Inverter with AC Load Side Reverse Blocking Switched Resonant Capacitor

Khairy Fathy[1], Ju-Sung Kang[1], Hiroyuki Ogiwara[3], Bin Eiuo[3], Hideki Omori[3],

Hyun Woo Lee[1] and Mutsuo Nakaoka[1],[4]

[1]Kyungnam University, Masan, Korea
[2]Ashikaga Institute of Technology, Tochigi, Japan
[3]Matsushita Electric Industrial Co. Ltd., Osaka, Japan
[4]Industrial College of Technology, Hyogo, Japan
khairy@ieee.org

Abstract—**This paper presents a new ZVT-PWM high frequency inverter. The ZVT operation is achieved in the hole load range by using a simple auxiliary Reverse blocking switch in parallel with series resonant capacitor**
Dual Duty cycle is used to provide a wide range of output power regulation which is important in many high frequency inverter applications. It is more suitable for induction heating applications The operation and control principle of the proposed high frequency inverter are described and verified through simulated results.

Keywords-High frequency inverter; Zero Voltage soft switching, Reverse blocking switched capacitor; Dual duty cycle control; Induction heating

I INTRODUCTION

With great advances of high frequency power electronics, an efficient electromagnetic eddy current based induction heating (IH) technology is more acceptable for consumer food cooking and processing appliances such as cooking heater, rice cooker and warmer, hot water producer and steamer, along with super heated steamer. A variety of IH equipments not only fill the key demands of safety and cleanliness, but also has excellent advantages of very high thermal conversion efficiency, rapid heating, local spot heating, direct heating, high power density, high reliability, low running cost and non-acoustic noise. Aforementioned IH appliances using high frequency inverters make use of eddy current based Joule's heat and hysteretic loss heat due to Faraday's electromagnetic induction law and can supply high-frequency power to IH load, which consists of working coil and eddy current based heating materials. Some high-frequency inverters operating over power frequency ranges from 20kHz to several MHz need to be cost effective high efficiency and high power density. There are various high-frequency inverter circuit topologies, such as full bridge, half bridge, single-ended push-pull, center tap push-pull and boost half bridge. Of these, the voltage source type ZCS (Zero Current Soft switching)

SEPP (Single-Ended Push Pull) resonant and quasi-resonant hybrid high-frequency inverter, has unique features as simple configuration, high efficiency and wide soft commutation range.

As we know the concept of induction heating is employed of an IH cooker, this concept can be simplified as follows. First convert the utility AC voltage to DC using rectifier then connected to high frequency switching circuit to supply high frequency current to the working coil . there are several types and topologies of high frequency inverter used in consumer induction heating application , the proposed high frequency inverter consists of half bridge series resonant converter and bypass switch. The advantages of half bridge series resonant converter is stable switching. low cost and streamlined design.

II PROPOSED ZVT PWM HIGH FREQUENCY INVERTER

A. Circuit Configuration

Fig. 1 shows the newly developed duty cycle ZVT PWM high-frequency inverter circuit topologies using the latest trench gate IGBTs and operating with constant frequency PWM control strategy. This voltage-fed ZCS PWM high frequency inverter circuit consists of two main switches of reverse conducting IGBTs $Q_1(SW_1/D_1)$ and $Q_2(SW_2/D_2)$, a single reverse blocking auxiliary switch $Q_3(SW_3/D_3)$. The modified versions of the proposed ZVT inverter are illustrated in Fig. 1 (a)-(d).
The resonant circuit comprises of resonant inductance (Lr) and reso-nant capacitance (Cr). The capacitors, C1 and C2, are the lossless turn-off snubbers for these switches, S1 and S2.

B. High Frequency AC Power Control Scheme

The high frequency AC output power of the proposed inverter circuit, which is delivered to the IH load as IH cooking heater and rice cooker, can be continuously regulated by a constant frequency asymmetrical PWM

1-4244-0448-7/06/$25.00 ©2006 IEEE

(duty cycle) control scheme under a condition of zero current soft switching commutation mode. The proposed gate voltage pulse timing PWM sequences for the active power switches Q_1, Q_2, , and the auxiliary power switch Q_3 are shown in Fig. 3

C. Power Regulation Scheme

The output high-frequency AC effective power of the proposed inverter circuit in Fig.1 can be continuously regulated by a constant frequency Dual duty cycle PWM control scheme under a condition of zero voltage transition soft switching principle. The PWM gate pulse timing sequences are illustrated in Fig.2. By the constant frequency asymmetrical PWM control scheme which is based on varying the time ratio of total conduction times Ton of Q_1 , Q_2 and Q3 to the operating switching period T of high-frequency, the proposed high-frequency inverter circuit can control the high-frequency output power continuously. The conduction time Ton1 of Q_1 (SW_1/D_1) and the conduction time Ton3 of Q_3 can be varied. As a control variable in the proposed Dual duty cycle PWM schemes, duty Factor D1 is defined as D1 = Ton1/T, D2=Ton3/T
By varying the two duty factor D1, D2 the high-frequency output power of this inverter can be regulated continuously

D. Remarkable Features

The outstanding features of the newly developed soft switching high frequency inverter are summarized below;

(1) The DC component of the high frequency current through the working coil is zero because of series capacitor compensated resonant tank.

(2) The ZVT soft switching commutation range becomes much wider.

(3) The active efficiency of this high frequency inverter is much higher over wider power regulation range from high power settings to low power settings because of selective dual duty cycle mode PWM control scheme.

(4) Constant frequency operation can be implemented.

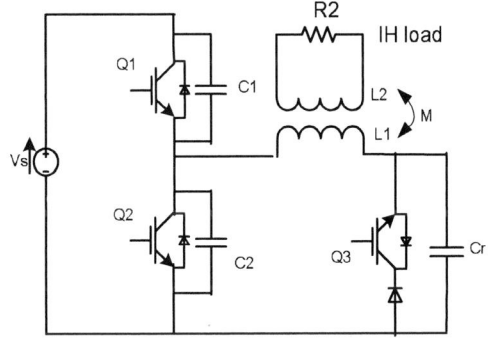

Fig. 1 Single ended ZVS inverter with induction heated load cascade switched capacitor

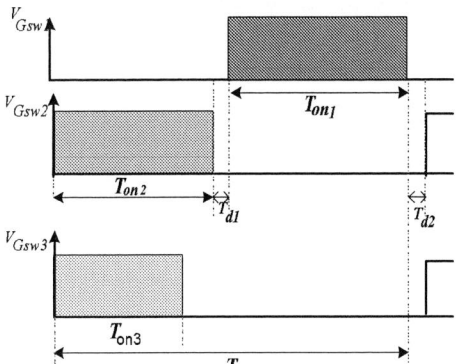

Fig 3 Gate pulse timing sequence pattern

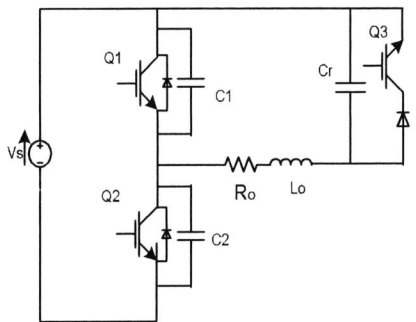

a) Single ended ZVT inverter with high side switched capacitor

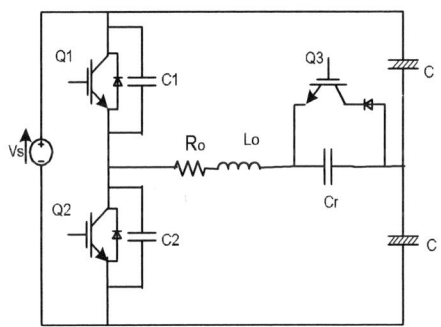

b) Divided electrolytic capacitor type half bridge ZVT inverter

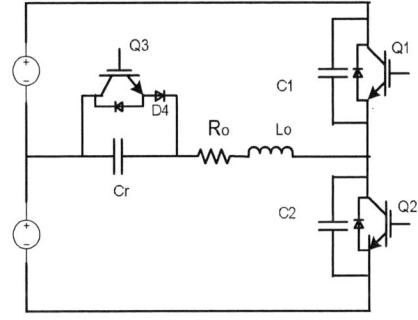

c) Divided voltage source type half bridge ZVT inverter

Fig. 3 Modified topologies of the circuit shown in Fig. 1

III PRINCIPLE OF SOFT SWITCHING OPERATION

A. Operating voltage and current waveforms

The current operating waveforms and the relevant operating modes of this inverter in steady state are illustrated in Fig. 4. The switching operation mode commutation transitions and their corresponding equivalent circuits of the proposed zero current soft switching high frequency inverter in steady state during one switching cycle are shown in Fig. 5-a,b. The switches of ZVT resonant converter turn on and off at zero voltage

The capacitor C is connected in parallel with switch S1,S2,S3 to achieve ZVT , the internal switch capacitance is added with the capacitor C and it effect the resonant frequency only , thereby contributing no power dissipation in the switch, the switch is implemented with transistor Q1,Q2 with anti parallel diode D1,D2, So the voltage across C is clamped by D1.D2 and these switches operate in half wave configuration . Also the voltage across Cr is clamped by D3. If the Diode D4 is connected in series with Q3 the voltage across Cr is oscillates freely and the switch is operated in a full wave configuration as shown in Fig 2 (a), (b), (c)

B. Operation principle in each mode

Mode 1 : this mode starts when SW2 of Q2 begin conducting , the current will flow through Ro, Lo , Cr until D4 is forward biased .

Mode 2: starts when D4 begin to conduct, the current passing through Ro , Lo, D4 ,Q3 , Q2 until Q3 is turned off at this moment Mode 3 starts

Mode 3: Current circulates in Cr, Lo, Ro , Q2 until Q2 is turned off.

Mode 4 : after Q2 turned off C1 will be in series with the parallel combination of (C2//Cr+Lo+Ro) . After dead time Td1 switch Q1 is turned on(mode 5).

Mode 5: switch Q1 is turned on, the supply current passing through SW1 and divided between (Cr+Ro+Lo) and C2 .

Mode 6: after the voltage across C2 is exceeded its maximum value diode D1 now is conducting.

Mode 7 : diode D2 conducts due to energy stored in Lo until Q2 start conducting (mode 1).

IV. POWER REGULATION CHARACTERISTICS

The input power (or output power) versus duty factor characteristics for the proposed voltage source type ZVT-dual duty cycle high-frequency inverter with PWM control scheme using the trench gate IGBTs is depicted Fig.6. this figure shows the output power regulation with the main duty cycle and the influence of the auxiliary duty cycle on the output power for constant main duty cycle. As we can see the auxiliary duty cycle can increase the output power especially at main duty =0.55 approximately. Also we find from the simulation results that the maximum output power drawn by the load take place at the main duty D1=0.55 and D2=0.5.

In the high-frequency inverter circuit proposed here, the input power of this inverter can regulate approximately from 0.2 kW to 3.4 kW at switching frequency 50 kHz under a principle of ZVT soft switching commutation. The soft switching operating range is relatively large in the proposed ZVT-PWM dual duty cycle high-frequency inverter., Fig. 7 illustrates the output power regulation due to an auxiliary control at certain value of main control scheme . Fig 8 depicts the output power regulation characteristics of the proposed inverter.

The soft switching high frequency inverter shown in Fig. 1 should be evaluated as compared with a new circuit topology depicted in Fig. 9,as a dual duty cycle voltage source high frequency inverter with reverse conducting switched resonant capacitor

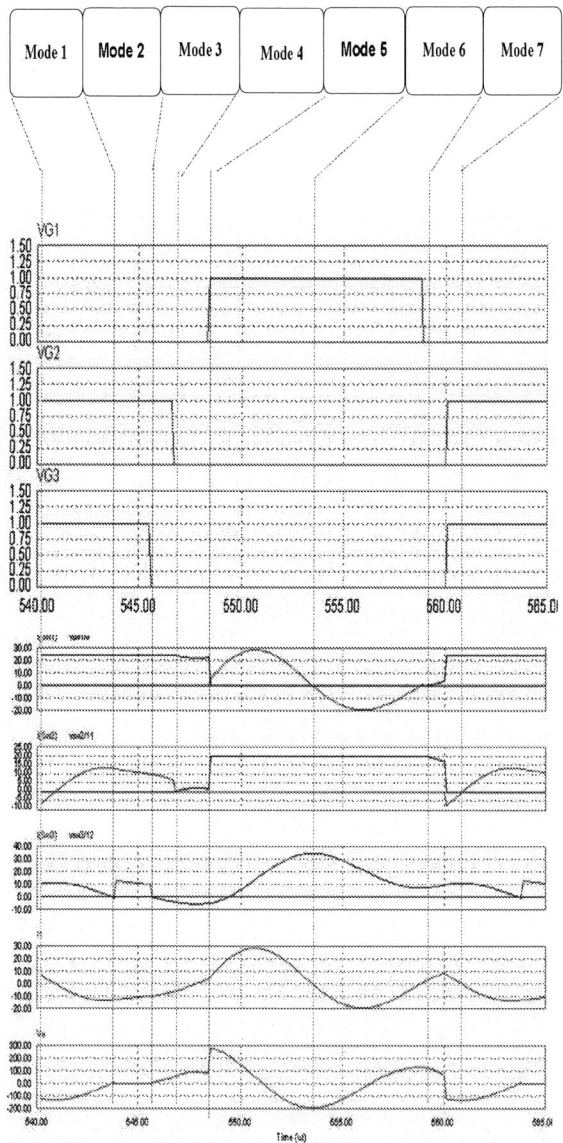

Fig. 4 Relevant voltage and current waveforms

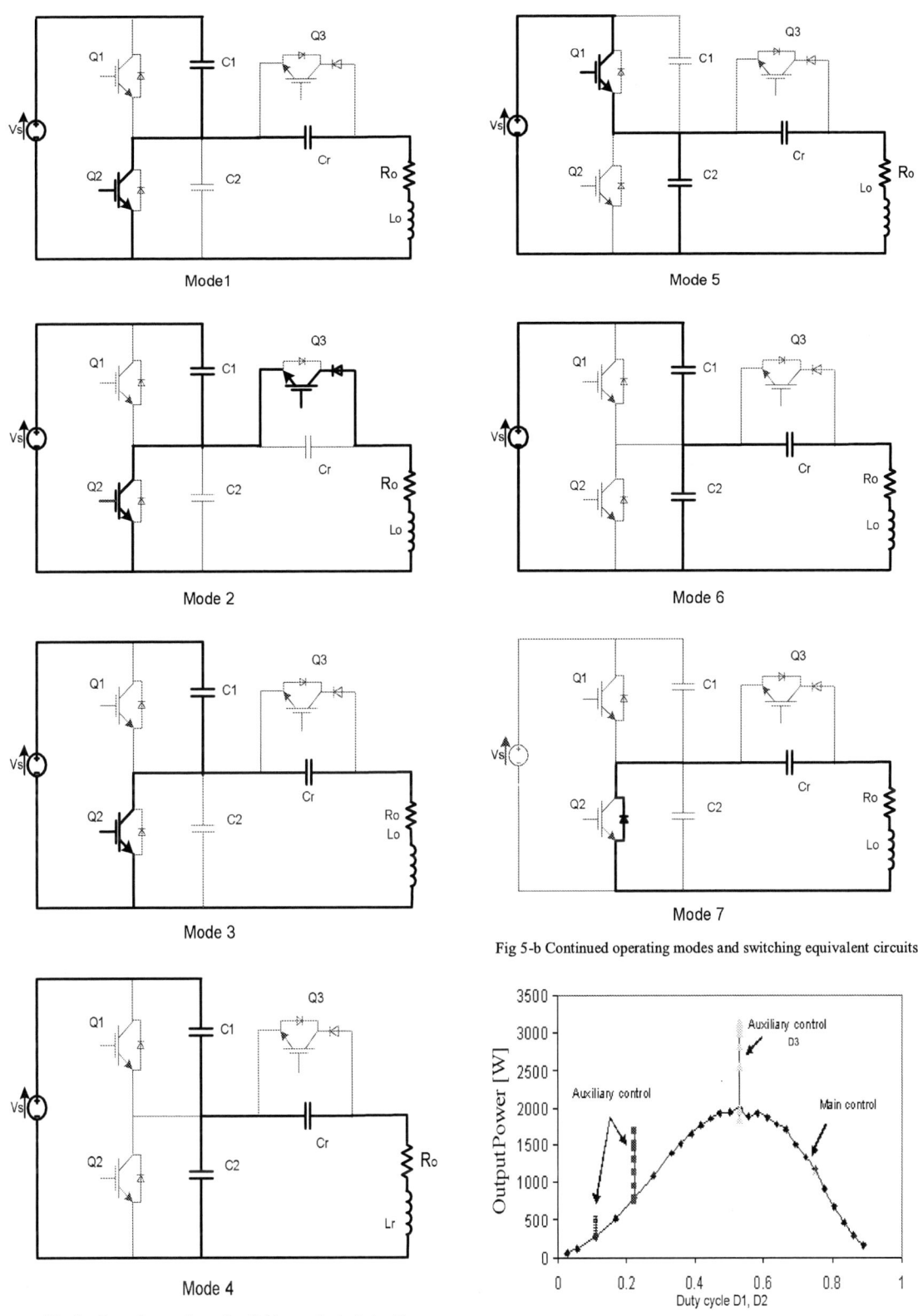

Fig. 5-a Operating modes and switching equivalent circuits

Fig 5-b Continued operating modes and switching equivalent circuits

Fig 6 Output power regulation characteristics

1361

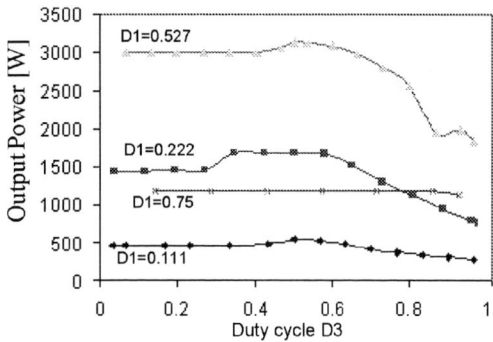

Fig 7 Output power vs Auxiliary control duty cycle

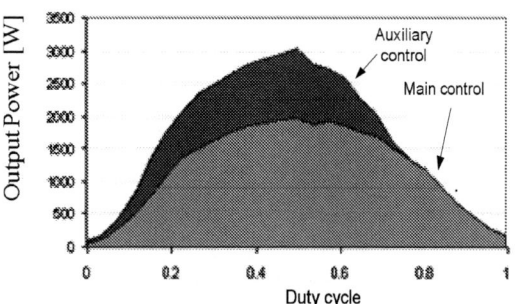

Fig. 8 Power output regulation characteristics

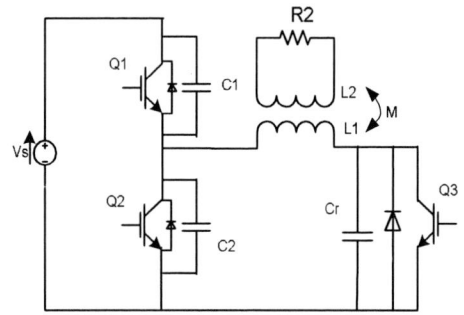

Fig. 9 dual duty cycle voltage source high frequency inverter with reverse conducting switched resonant capacitor

V. CONCLUSIONS

In this paper, a new topology of a Dual Duty Cycle Controlled ZVT High Frequency Inverter with Single Auxiliary Reverse Blocking Switch in Parallel with Series Resonant Capacitor The operating principle and the operating characteristics of the new high-frequency circuit treated here were illustrated and evaluated on the basis of simulation results. it was examined that the complete soft switching operation can be achieved even for low power setting by introducing the high-frequency dual duty cycle control scheme. In the proposed high-frequency inverter treated here, the dual mode pulse modulation control strategy of the asymmetrical PWM in the higher power

setting and the lower power setting, the output power of this high-frequency inverter could introduce in order to extend soft switching operation ranges. Therefore, it was proved from that the proposed ZVS soft switching PWM high-frequency inverter can actually achieve high efficiency, high performance and obtain wider soft switching range in selective Dual duty cycle PWM control scheme.

The operating principle and the operating characteristics of the high-frequency circuit treated here were illustrated and evaluated on the basis of simulation. For a consumer IH cooking heater and IH steamer, the practical effectiveness of the proposed voltage source ZVT-PWM dual duty cycle high-frequency inverter was proved on the basis of the simulation results. Then, the soft switching commutation range was much wider in spite of the pulse modulation control. The high-frequency power of this inverter could be efficiently supplied to the induction heated load as IH cooking heater from full power setting to small power setting. Furthermore, it was examined that the complete soft switching operation can be achieved even for low power setting by introducing the high-frequency dual duty cycle control scheme.. Therefore, it was proved from an experimental point of view that the proposed zero current soft switching PWM high-frequency inverter for IH cooking heater and IH steamer can actually achieve high efficiency, high performance and obtain wider soft switching range in selective dual duty cycle power control schemes.

REFERENCES

1] H. Rashid, Power Electronics Circuits, Devices and Applications, Third Edition

2] Tetsuya Etoh, Tarek Ahmed ,Eiji Hiraki, Keiki Morimoto, Khairy Fathy, Nabil A. Ahmed Hyun Woo Lee, Mutsuo Nakaoka, " 32V-300A/60kHz Edge Resonant Soft-Switching PWM DC/DC Converter with DC Rail Series Switch-Parallel Capacitor Snubber Assisted by High-Frequency Transformer Parasitic Components " page 1159-1165 Proceedings of The 31st Annual Conference of the IEEE Industrial Electronics Society IECON 2005

3] H. Tanimatsu, H. Sadakata, T. Iwai, H.Omori, Y.Miura, E.Hiraki, H.W.Lee, and M. Nakaoka, " Quasi-Resonant Inductive Snubbers-Assisted Series Load Resonant Tank Soft Switching PWM SEPP High-Frequency Multi Resonant Inverter with Auxiliary Switched Capacitor" The 6th International Conference on Power Electronics (ICPE'04), pp II-299, Busan, Korea, October, 2004 acknowledgment

4] Laknath Gamage, Tarek Ahmed, Hisayuki Sugimura, Srawouth Chandhaket and Mutsuo Nakoka," Series Load Resonant Phase Shifted ZVS-PWM High frequency Inverter with a Single Auxiliary Edge Resonant AC Load Side Snubber for Induction Heating Super Heated Steamer", Proceedings of 2003 International Conference on Power Electronics and Drive Systems (PEDS), Vol. 1, pp. 30-37, Singapore, November,2003.

5] Won-Suk Choi, Nam-Ju Park. Dong-Yun Lee and Dong-Seok Hyun. " A New Control Scheme for Class D inverter with induction Heating Jar Application by Constant Switching Frequency", Journal of Power Electroics, Vol. 5, No. 4, October 2005, pp272-281.

2006 5th International Power Electronics and Motion Control Conference

A Switched-Capacitor Lossless Inductor ZCS Snubber-Assisted Series Load Resonant High Frequency Inverter with Dual Mode Pulse Modulation Scheme

Khairy Fathy[1] , Takaaki Okude[2], Hideki Omori[2], Hyun Woo Lee[1] and Mutsuo Nakaoka[1],[3]

[1]Kyungnam University, Masan, Korea
[2]Matsushita Electric Industrial Co. Ltd., Osaka, Japan
[3]Industrial College of Technology, Hyogo, Japan
khairy@ieee.org

Abstract--In this paper, a novel type auxiliary active edge resonant snubber assisted zero current soft switching pulse modulation series load resonant inverter using IGBT power modules is proposed for cost effective consumer high-frequency induction heating (IH) appliances. Its operating principle in steady state is described by using each switching mode equivalent operating circuits. The new multi resonant high-frequency inverter with series load resonance and edge resonance can regulate its high frequency output power under a condition of a constant frequency zero current soft switching (ZCS) commutation principle on the basis of asymmetrical pulse width modulation (PWM) control scheme. The consumer brand-new IH products using proposed ZCS-PWM high-frequency inverter is evaluated and discussed as compared with conventional type inverter on the basis of experimental results. In order to extend ZCS operation ranges under a low power setting PWM as well as to improve efficiency, the high frequency pulse density modulation (PDM) strategy is demonstrated for high frequency multi resonant inverter. Its practical effectiveness is substantially proved from an application point of view.

keywordss—Auxiliary switched capacitor snubber, Consumer IH appliances, Dual mode PWM and PDM control, Lossless inductor snubbers, Zero current soft switching.

I. INTRODUCTION

A- Technical Backgrounds

With great advances of high frequency power electronics technologies, an efficient electromagnetic eddy current based induction heating (IH) approach is more acceptable for consumer food cooking and processing appliances such as cooking heater, rice cooker and warmer, hot water producer and steamer, along with super heated vapor steamer. A variety of small scale IH appliances not only fulfill the key demands of safety and cleanliness, but also has excellent advantages as high thermal conversion efficiency, rapid heating, local spot heating, direct heating, high power density, high reliability, low running cost and non-acoustic noise. Aforementioned IH home appliances using high frequency inverters make use of eddy current based Joule's heat and hysteretic loss heat due to Faraday's electromagnetic induction law and can supply high-frequency AC power to the IH load, which consists of

working coil and eddy current based heating materials. Some types of high-frequency multi resonant inverters operating over power frequency ranges from 20kHz to several MHz have some advantages as cost effective high efficiency and high power density. There are various high-frequency inverter circuit topologies, such as full bridge, half bridge, single-ended push-pull, center tap push-pull and boost half bridge. Of these, the voltage source type ZCS (Zero Current Soft switching) SEPP (Single-Ended Push Pull) resonant and quasi-resonant high-frequency inverter topologies developed by the authors have remarkable features as simple configuration, high efficiency and wide soft commutation range.

B- Research Objective

In this paper, a voltage source type ZCS-SEPP high-frequency multi resonant inverter with a constant frequency duty cycle PWM control function using active auxiliary quasi-resonant lossless sunbbers, which is composed of switched capacitor and two lossless inductors in series with the main switches, is newly proposed for consumer IH food cooking and processing applications. The operating principle of the proposed high frequency inverter circuit with PWM control scheme for IH heated power regulation is described herein by using switching mode equivalent circuits. The feasible operating characteristics of this high frequency inverter using IGBTs are illustrated and evaluated on the basis of experimental results and simulation ones. Finally, in order to extend soft switching operating ranges under low power setting condition, a pulse density modulation-based power regulation scheme is discussed for this ZCS-PWM controlled IH high frequency inverter.

II. ZCS-PWM CONTROLLED HIGH FREQUENCY INVERTER CIRCUIT TOPOLOGY

A. Circuit description

The newly developed voltage source type ZCS-PWM high-frequency inverter circuit for consumer induction heater is shown in Fig.1 and Fig.2, which includes of two lossless inductors-assisted series load resonant inverter with a single auxiliary switched capacitor.

1-4244-0448-7/06/$25.00 ©2006 IEEE

This multi-resonant ZCS-PWM high-frequency inverter circuit consists of the main active switches $Q_1(SW_1/D_1)$ and $Q_2(SW_2/D_2)$, a single auxiliary active switch $Q_3(SW_3/D_3)$ in series with auxiliary quasi-resonant capacitor Cr, ZCS-assisted two inductor snubbers L_{S1} and L_{S2} in series with Q_1 and Q_2, IH load power factor compensation series resonant capacitor C_S, and IH load with Ro and Lo series circuitry represented by equivalent inductive circuit model, Ro; Lo; are the effective equivalent resistance and inductance of IH Load. Because of constant frequency operation, the load parameters (Ro & Lo) are kept constant and seem to be no skin effect. It is noted that the proposed soft switching PWM high-frequency inverter circuit consists of a few circuit components and low cost circuit configurations.

Another Multi-resonant ZCS-PWM high frequency inverter topologies are shown in Fig. 2a Single-ended half bridge topology and Fig.2b divided capacitor single-ended half bridge high frequency inverter and Fig. 2c Divided Capacitor double-ended half-bridge topology the position of Q3(SW3/D3) can be interchanged to control the output to satisfy the load requirements as Fig.2d,e.

B. High Frequency AC Power Regulation Scheme

The high-frequency AC effective output power of the proposed inverter circuit in Fig.1 can be continuously regulated by a constant frequency asymmetrical PWM control scheme under a condition of zero current soft commutation principle. The PWM gate pulse timing sequences of this high-frequency inverter are illustrated in Fig.3. By the constant frequency asymmetrical PWM control scheme which is based on varying the time ratio of total conduction times T_{on} of Q1 and Q3 to the operating switching period T, the proposed multi-resonant inverter circuit can control the high-frequency AC output power continuously. The external gate signal driver only changes the conduction time Ton1 of Q1 (SW1/D1). The overlapping time T_O is determined by conduction time T_{on1} of Q1 and conduction time T_{on3} of Q3 is designed for a constant value. As a control variable in the proposed asymmetrical duty cycle PWM scheme, the Duty Factor D is defined as follows,

$$D = \frac{T_{on} + T_{d1}}{T} \qquad (1)$$

By varying the duty factor D as a constant variable, the high-frequency AC output power of this multi-resonant inverter can be regulated continuously. The voltage-fed ZCS-PWM SEPP high-frequency IH load resonant inverter with two lossless inductor snubbers and switched capacitor snubber can be controlled by not only constant frequency asymmetrical PWM scheme but also constant frequency dual mode PWM/PDM scheme. By using PDM control scheme illustrated in Fig.4, zero current soft switching operating range can be achieved even in low power setting ranges. As a control variable of PDM scheme, another Duty Factor Dp can be defined as:

$$Dp = \frac{T_{Pon}}{Tp} \qquad (2)$$

The high frequency AC effective power regulation can be achieved by means of changing the PDM ratio as T_{pon} to one cycle period T_p specified here.

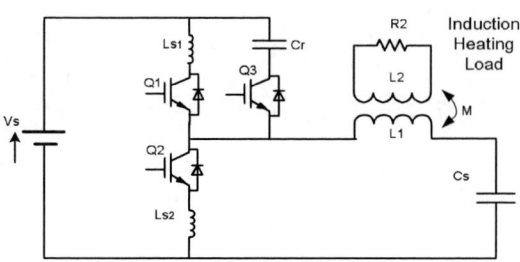

Fig. 1 Multi Resonant ZCS-PWM high frequency inverter topology (1) with low side load resonant circuit

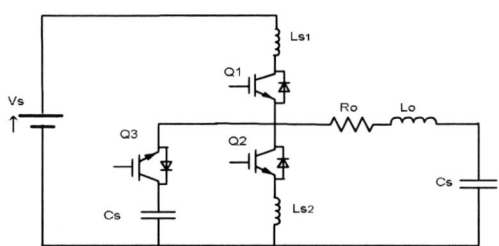

a) Single ended half bridge topology (2)

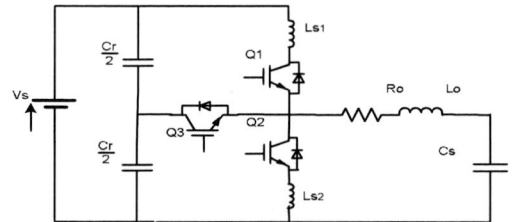

b) Divided Capacitor Single Ended half bridge Topology (3)

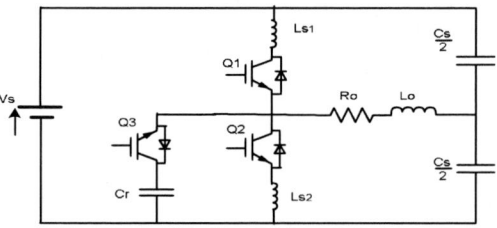

c) Divided Capacitor Double Ended half bridge Topology (4)

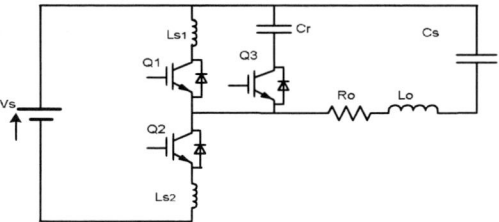

d) Single ended high side load resonant topology (5)

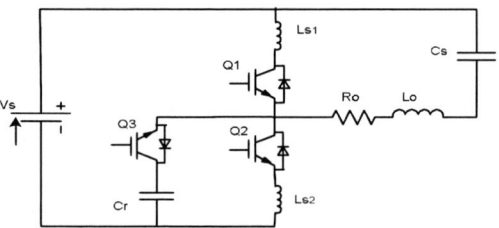

e) Single ended high side load resonant topology (6)

Fig. 2 Multi-resonant ZCS-PWM high frequency inverter topologies.

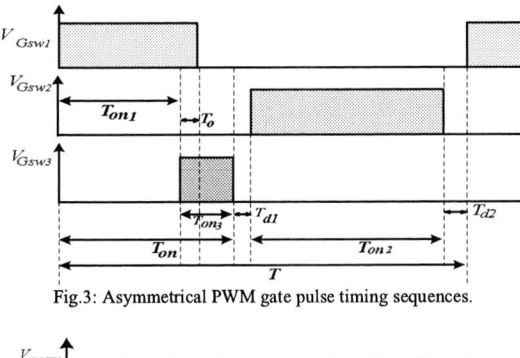

Fig.3: Asymmetrical PWM gate pulse timing sequences.

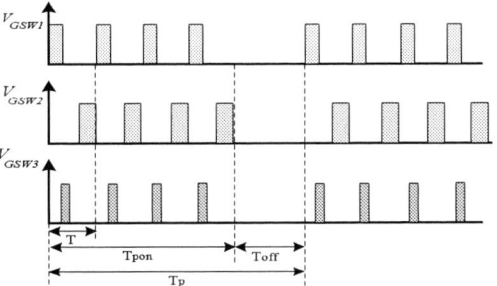

Fig.4: Dual mode PWM/PDM gate pulse timing sequences.

III. STEADY STATE OPERATING PRINCIPLE

A. Operating Waveforms

The operating mode transitions and equivalent circuits of the proposed high frequency inverter circuit in periodic steady state are represented in Fig.5. This multi-resonant high-frequency inverter circuit has eleven operating mode transitions repeatedly. The operation principle of the proposed inverter circuit is explained below.

B. Operating Principle

The zero current soft switching operation for all the active switches Q_1, Q_2, Q_3 in the proposed multi-resonant inverter circuit can be achieved under a gate pulse sequence pattern as depicted in Fig.3. Mode1, in the first place, the high side main switch SW_1 of Q_1 is now conducting and high-frequency AC power is supplied to the IH load. After the current i_{SW1} through SW_1 of Q_1 commutates to anti-parallel diode D_1 of Q_1 by quasi-resonance due to ZCS-assisted inductor snubber L_{S1} in series with Q_1 together with the auxiliary resonant capacitor Cr, the auxiliary active switch SW_3 of Q_3 is turned on and the main switch SW_1 of Q_1 is turned off. As a result, ZCS commutation at a turn-off mode transition can be implemented by the arbitrarily timing processing in turning off the switch SW_1 of Q_1. At this time, since an auxiliary resonant current i_{SW3} flows through the switch SW_3 of Q_3 and increases softly, ZCS commutation at a turn-on mode transition can be achieved for SW_3 of Q_3. Then, after i_{SW3} flowing through SW_3 is commutated to anti-parallel diode D_3 of Q_3 by the quasi-resonance together with Cr, R_O-L_O inductive load circuitry with a series power factor compensation tuned capacitor C_S, ZCS commutation at a turn-off mode transition can be performed by turning off SW_3 of Q_3.

While the auxiliary active switch SW_3 of Q_3 turns on, the voltage across the low side main switch SW_2 of Q_2 decreases toward zero. Before the low side main switch SW_2 of Q_2 turns on, D_2 of Q_2 begins to conduct.

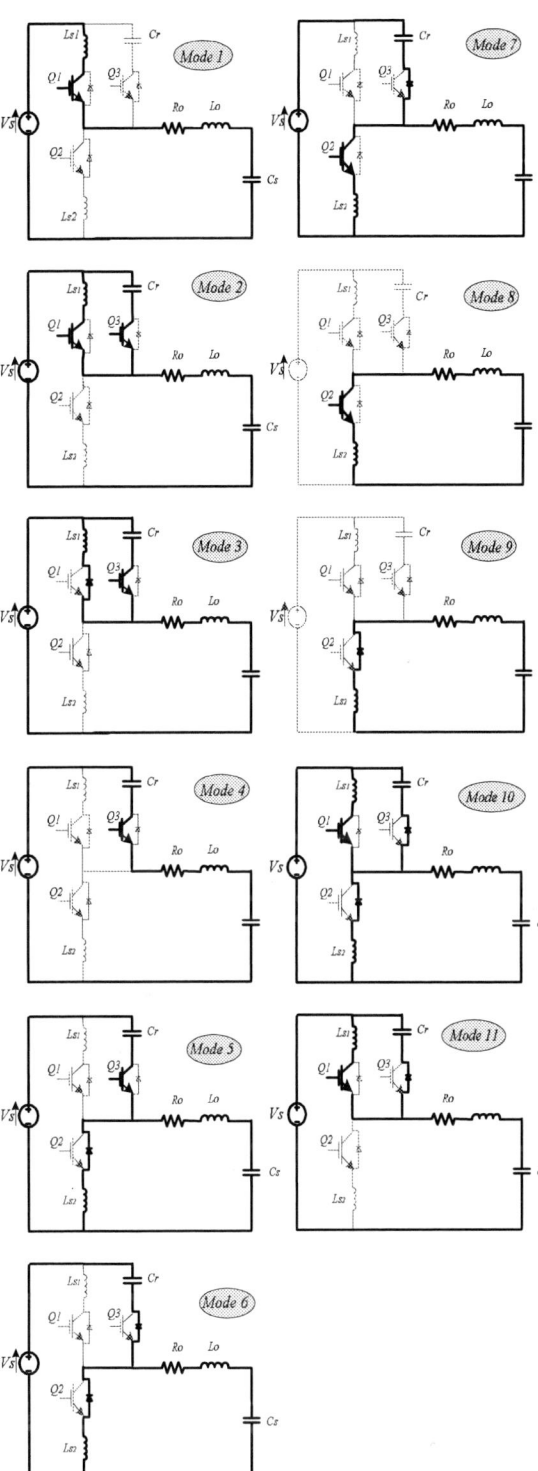

Fig. 5 Operating Mode Transition and equivalent circuits at steady state during one switching cycle

While D_2 of Q_2 continues conducting, the current flowing through the diode D_2 is naturally commutated to SW_2 of Q_2. Therefore, a complete ZVS and ZCS hybrid commutation can be actually achieved for SW_2 of Q_2. On the other hand, after the current i_{SW2} through low side main switch SW_2 of Q_2 is naturally commutated to D_2 of Q_2 with the aid of low side ZCS-assisted inductor snubber L_{S2}, induction heated load represented by R_O-L_O and power factor compensation capacitor C_s, the ZCS commutation at a turn-off transition can be performed by turning off the switch SW_2 of Q_2. While D_2 of Q_2 is now conducting, the current i_{D2} flowing through D_2 of Q_2 is commutated to the switch SW_1 of Q_1 by turning on the switch SW_1 of Q_1, a ZCS turn-on commutation can be realized with the aid the first inductor snubber L_{S1}.

IV. EXPERIMENTAL RESULTS AND EVALUATIONS

A. Design Specifications and Circuit Parameters

The evaluations and discussion in experiment prove the validity of the proposed multi-resonant high-frequency inverter circuit treated here. The design specifications and circuit parameters used in experiment are respectively listed in Table1. The proposed multi-resonant inverter circuit topology is designed for consumer IH food cooking heater applications in home and business applications. Therefore, the stainless steel pan with its bottom diameter 18cm is used for IH load as a heated object. This IH load consists of stainless steel pan, ceramic or plastic spacer, and litz wire working coil. The circuit parameters of this multi-resonant inverter are determined by considering the soft switching condition and output power ranges.

B. Measured Operating Waveforms

The steady state measured switching operating waveforms for duty factor D=0.34 under a condition of the input DC power 2.4 kW for built-in burner are represented in Fig.6. As can be seen in Fig.6, it is noted that all the active main and auxiliary power switches can operate under a principle of zero current soft switching PWM strategy. In particular, it can be recognized that complete ZVS and ZCS hybrid soft commutation at turn-on transition is performed for the main switch SW_2 of Q_2, because it is turns on during a conduction period of D_2 of Q_2. Because of zero current soft switching operation in all the power switches, in spite of additional auxiliary switch SW_3 of Q_3, a high efficiency power conversion can be achieved in proposed inverter circuit in Fig.1.

C. Power Regulation Characteristics

The input DC power (or output AC power) vs. duty factor characteristics for the proposed voltage source type ZCS-SEPP high-frequency multi-resonant inverter with PWM control scheme using the trench gate IGBTs is depicted Fig.7. In the high-frequency multi-resonant inverter circuit proposed here, the input DC power of this high-frequency inverter can regulate approximately from 0.5kW to 2.4kW under a principle of zero current soft switching commutation. The soft switching operating range is relatively large in the proposed ZCS-PWM-SEPP high-frequency inverter using IGBTs.

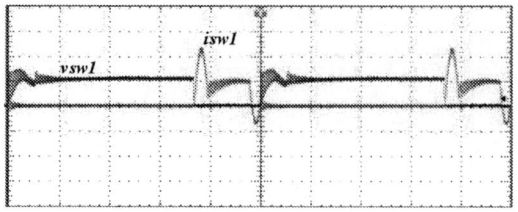

(a) Voltage and current waveforms of Q1 (250V/div, 40A/div, 10 μs/ div)

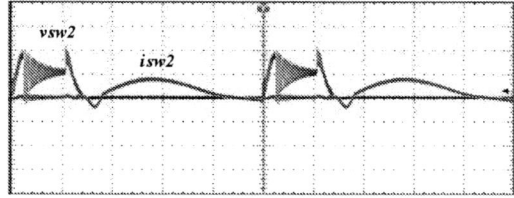

(b) Voltage and current waveforms of Q2 (250V/div, 40A/div, 10 μs/div)

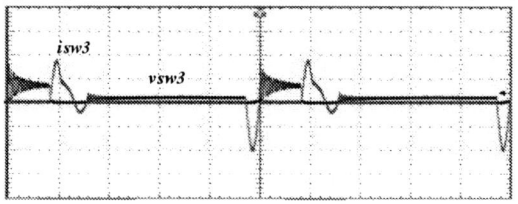

(c)Voltage and current waveforms of Q3 (250V/div, 40A/div,10 μs/div)

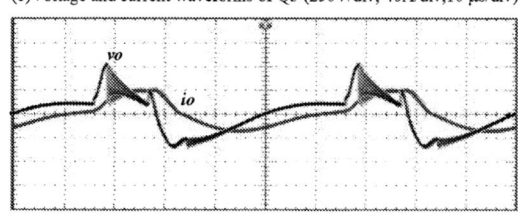

(d)Output voltage and current waveforms (250V/div, 40A/div,10 μs/div)

Fig. 6 Measured voltage and current waveforms in case of D=0.34

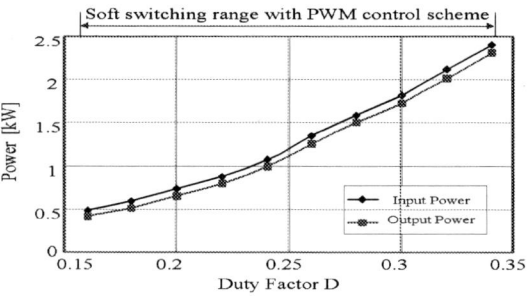

Fig. 7 Output power vs. duty factor characteristics

TABLE I
DESIGN SPECIFICATIONS AND CIRCUIT CONSTANTS

Item	Symbol	Value
DC Source Voltage	V_s	270 V
Switching Frequency	f_{sw}	20 kHz
ZCS-assisted Inductor	L_{s1}	2.01 μH
ZCS-assisted Inductor	L_{s2}	2.01 μH
Auxiliary Quasi-resonant Capacitor	C_r	330 nF
Compensation series tuned capacitor	C_s	0.802 μF
Load Resistance	R_o	2.1
Load Inductance	L_o	45 μH

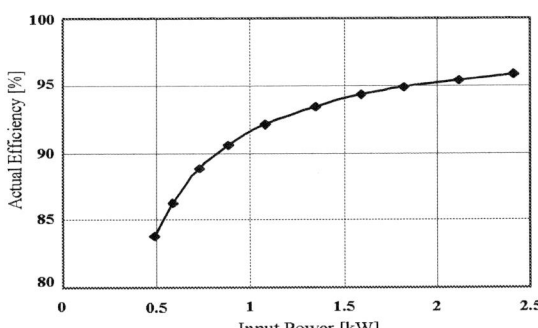

Fig. 8 Actual efficiency vs. input characteristic in case of PWM control scheme

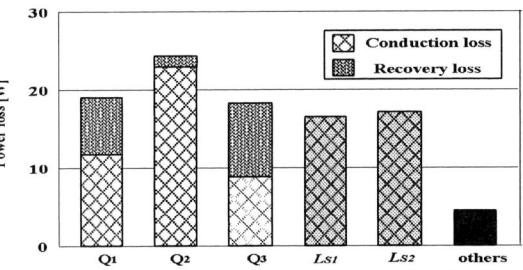

Fig. 9: Power conversion loss analysis in case of D = 0.34

D. Actual Efficiency Performances

The power conversion actual efficiency characteristics of the proposed ZCS-PWM inverter for the consumer IH cooking heater is shown in Fig.8. Under the rated output condition, the actual efficiency of the proposed inverter using IGBTs power module is experimentally estimated as about 96%. Since a zero current soft switching operation in this multi-resonant inverter can be completely achieved even in case of duty factor D=0.16 under the condition of the minimum output power for PWM control, the power conversion efficiency can be sufficiently kept to be 84%.

E. Power Loss Analysis and Circuit Evaluations

Power conversion loss analysis of the proposed high-frequency multi-resonant inverter in case of D=0.34 (Input power 2.41 kW) is shown in Fig.9. Conduction loss is composed of which due to saturation voltage of switches (Q_1, Q_2, Q_3) and resistivity of ZCS assisted inductors (L_{S1}, L_{S2}). The loss of Q_3 is conjected to be constant despite of Duty Factor, because it is generated by soft switching PWM control. As can been seen in Fig.10, peak voltage of switch Q_3 is lower than switch Q_1 and switch Q_2.

V. CONCLUSIONS

In this paper, a new topology of active edge resonant snubbers assisted voltage source type ZCS-PWM-SEPP high-frequency multi-resonant inverter with a series capacitor compensated IH load resonant tank was proposed for IH appliances. The operating principle and the operating characteristics of the new high-frequency circuit treated here were illustrated and evaluated on the basis of simulation and experimental results. It was examined that the complete soft switching operation can be achieved even for low power setting by introducing the high-frequency PDM control scheme. In order to extend soft switching operation ranges in the proposed high-frequency inverter, we used the dual mode pulse modulation control strategy of the asymmetrical PWM in the higher power setting and the PDM in the lower power setting. It was proved from an experimental point of view that the proposed ZCS PWM high-frequency inverter for IH appliances can achieve high efficiency, high performance and obtain wider soft switching range in selective PWM and PDM control scheme.

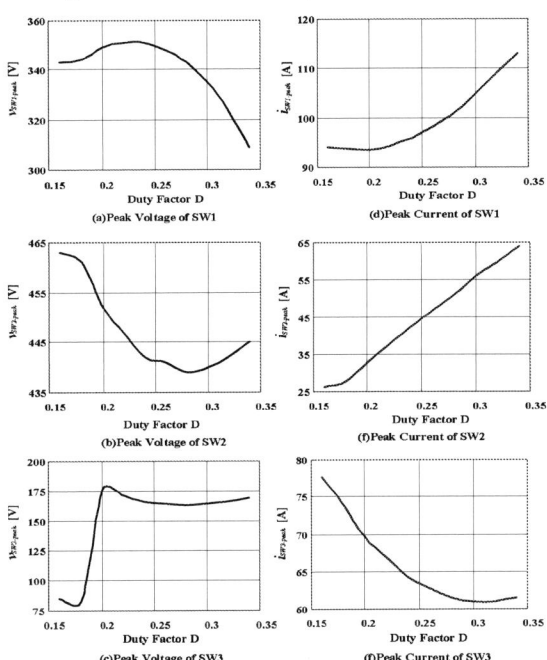

Fig. 10 Peak voltage and current characteristics

REFERENCES

[1] H. Tanimatsu, H. Sadakata, T. Iwai, H.Omori, Y.Miura, E.Hiraki, H.W.Lee, and M. Nakaoka, " Quasi-Resonant Inductive Snubbers-Assisted Series Load Resonant Tank Soft Switching PWM SEPP High-Frequency Multi Resonant Inverter with Auxiliary Switched Capacitor" The 6th International Conference on Power Electronics (ICPE'04), pp II-299, Busan, Korea, October, 2004

[2] H. Terai, H. Sadakata, H. Omori, H. Yamashita, and M. Nakaoka,"High Frequency Soft Switching Inverter for Fluid-Heating Appliance Using Induction Eddy Current-based Involuted Type Heat," Proceedings of IEEE Power Electronics Specialists Conference, Vol 4, pp. 1874-1878, Cairns, Australia, June, 2002.

[3] H. Terai, T. Miyauchi, I. Hirota, H. Omori, Mamun A. Al, and M. Nakaoka, "A Novel Time Ratio Controlled High Frequency Soft Switching Inverter using 4th Generation IGBTs," Proceedings of IEEE Power Electronics Specialists Conference, (PESC), Vol. 4, pp. 1868-1873, Vancouver, Canada, June, 2001.

[4] Laknath Gamage, Tarek Ahmed, Hisayuki Sugimura, Srawouth Chandhaket and Mutsuo Nakoka," Series Load Resonant Phase Shifted ZVS-PWM High frequency Inverter with a Single Auxiliary Edge Resonant AC Load Side Snubber for Induction Heating Super Heated Steamer", Proceedings of International Conference on Power Electronics and Drive Systems (PEDS), Vol. 1, pp. 30-37, Singapore, November,2003.

2006 5th International Power Electronics and Motion Control Conference

Topologies of Switch-Linear Hybrid Power Conversion & Special Operation States

Lu-sheng Ge, Qian-zhi Zhou and Wu bin

Anhui University of Technology school of electrical engineering and information

Ma'anshan China, 243002

e-mail: lsge@ahut.edu.cn

Abstract—**Typical topologies of Switch-Linear Hybrid (SLH) power converter are summed up by former research practices. The main frame, hybrid criteria, category classification and extension regulation for the development of SLH-topology are presented by practical examples. Different operation states by different devices are also discussed for applying the topologies reasonably. All the above will set up the foundation of the SLH to develop into a complete self-family in power conversion branches and obtain wide applications.**

Keywords-conversion; topology; SLH; robustness; operation state

I. INTRODUCTION

Relying on the features of the high input resistance and high output resistance attached to Emitter Follower (Or Source Follower), Switch-Linear Hybrid (SLH) power converters are not hard to satisfy simultaneously the requirements in high efficiency with good waveforms, meeting the needs of multi-kinds of loads, including non-linear one, undergoing load shock or parameters' changes with strong robustness and fast response. Hence they can be used where high steady and dynamic performances are emphasized, for instance in high stable supplies, ultra-low frequency motor control and improving network quality, etc.. These contents including SLH scheme, different configurations, efficiency estimation, theoretical and experiment analyses between SLH and other conversion modes have been described in a series of papers [1-4].

The recent developing tendency of SLH is 1) utilizing the present topologies proved by experiments for developing new products series; 2) applying them in suitable conditions for high performance requirements; 3) developing new branches of SLH-topologies according scientific regulations. Hence, based on previous work, it is necessary to classify the topologies into deferent categories scientifically, also compare their advantages and disadvantages, describe some key problems in detail for clearing some concepts, proposing a reasonable hybrid criterion for directing future's work on SLH. Except the specified illustration, the descriptions below will take sinusoidal ac conversion as examples, because dc conversion can be seen as a ac system with zero frequency, and arbitrary waveform conversion may be

treated as the conversion combined with multi-single frequency, where the function and concept are easy to be extended.

II. TOPOLOGIES OF SLH

2.1 Essential criteria for Hybrid

Fig.1 express the fundamental principle of SLH, where the main role to deliver the energy is a special linear unit, which presents the excellent steady and dynamic performances over the pure switch mode conversion, showing the essential feature of SLH.

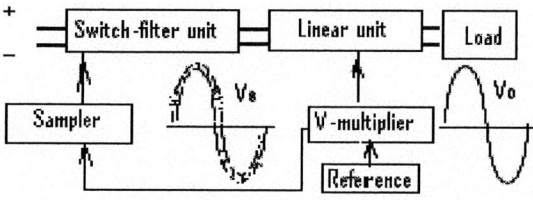

Fig.1 Fundamental principle of SLH

2.2 Typical Categories of SLH

2.2.1 PSLH

If the linear unit connects the switch-filter unit and the load parallel to form SLH, which is named after PSLH, in this situation, the power devices in the linear unit are fed by the positive and negative uni-polar waveforms respectively.

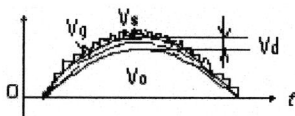

Fig.2 Voltage on power device inside the linear unit

1 OCL-PSLH

Fig.3 shows a topology that an OCL (Output Capacitorless) amplifier fed by two BUCK-SPWM filter units, so named after OCL-PSLH. There exists a common path to the Neutral point, so it need not require the LC parameters seriously symmetrical in positive and negative half cycle, or in three phases. The parameters' discreteness only affect the tooth numbers of the ripple voltage same or

1-4244-0448-7/06/$25.00 ©2006 IEEE   1368

not, do not affect the output performances of the converter. Also relying on the common neutral path, some interfere signals are easy by-passed, EMC measures will be not complex for designing a PSLH converter. So it acquires excellent THD values. A V/F open loop controlled OCL-PSLH inverter operated with a 4.5kW PM motor from 5Hz to 60Hz in no load experiment, pure sinusoidal voltages feeding the motor, running stable almost without noise, torque and speed ripples [5-6].

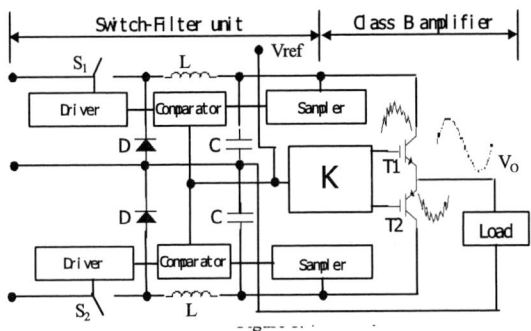

Fig. 3 OCL-PSLH

2 OT-PSLH

In a conventional Class B amplifier with Output Transformer (OT), just change the configuration from collector connection with the transformer winding into the emitter connection with the winding, producing a modified OT unit fed by a BUCK circuit, which creates an OT-PSLH shown in Fig.4(the modified form of OCL-PSLH).

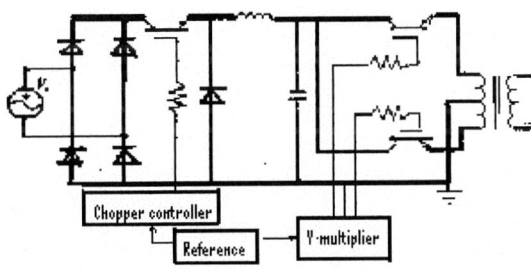

Fig. 4 OT-PSLH

2.2.2 SSLH

Considering the disadvantages of PSLH, particularly the safe operation, the topology of SSLH is constructed by connecting a pair of Emitter (Source) Followers, the switch-filter unit and the load in series, so named after SSLH, where two typical structures are shown in Fig. 5 a) and b) respectively[3]. Bi-polar ripple supply feeds the linear unit in this situation. The devices in the linear unit only bare low voltage drops about the saturation voltage of the device, regardless of the devices at on-state or at off- state (during off-state, the reverse current flows via the anti-paralleled diodes inside the device bodies). If the linear unit operates in breakdown, the switch unit in former stage will be still safe relying on the serial connections.

2.2.3 HSLH (BTL-SLH) [7-8]

A kind of SLH connects in serial-paralleled hybrid configuration (HSLH). For an example, If a conventional BTL (Balanced Transformerless) amplifier is fed by two half-cycle positive uni-polar waveforms, which has reversed the negative half cycle into positive one, named after BTL- SLH, where the power devices at on-state connects the load in series, then together connect the switch-filter unit in parallel. Fig.6 shows the operation principle (the anti-paralleled diodes are omitted in the diagram). BTL-SLH is also an extension of OCL-PSLH. the switch-filter unit can be realized by a single Buck circuit with a simple control strategy., sinusoidal voltage V_{gg} connected the upper devices while another rectangular signal pulse-pair in step of V_{gg} controlling the bottom devices, so that the Emitter Follower configuration symmetrically to the load can be still implemented.

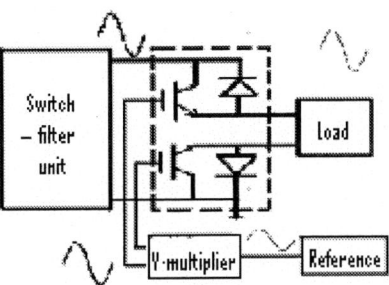

a) SSLH with the same channel devices

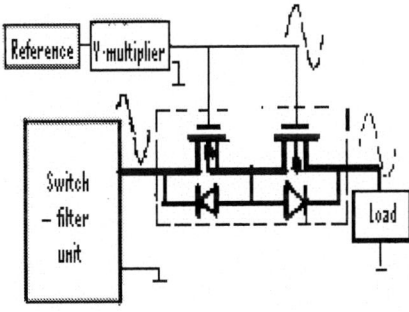

b) SSLH with the different channel devices

Fig. 5 Topology of SSLH

2.3 Extension

The research practices from the SLH idea have not yet ended the development of SLH-topologies, merely explored a branch direction to construct high performance converters. According to the previous criteria for hybrid, mentioned in 2.1, a series of SLH-topologies could be created continuously for special high performance applications. One way is based on conventional linear amplifiers with Emitter (Source) Follower matching to proper switch-filter circuits, just like previous OCL-PSLH. While another way is based on present ac switchers, changing the operation state of the devices from switching mode into linear-saturated critical state, then matching to specified switch-filter circuits, just like previous SSLH.

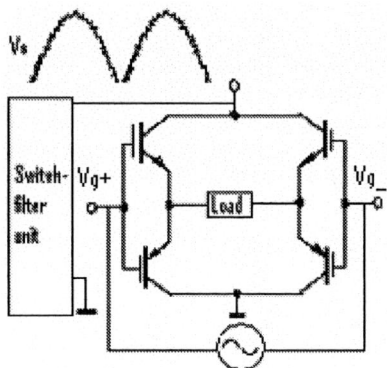

Fig. 6 Topology of HSLH (BTL-SLH)

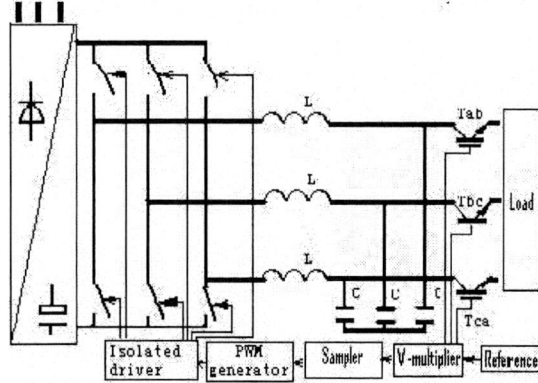

Fig.7 Topology of 3 phases of SSLH saved devices

2.4 Other Topologies for Hybridizing Switch mode & Linear Mode

Similar to SLH's idea, another kind of hybrid power conversion also attempts to combine the advantages both of switch converters and linear amplifiers for trade-off [9-11]. Fig.8 shows the topology different from the present SLH.

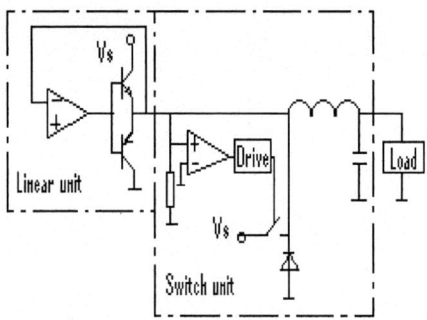

Fig.8 Another hybrid mode different from the present SLH

III. SPECIAL OPERATION STATES

In previous descriptions, typical topologies are introduced. However more importance is how to

implement them balanced in efficiency, steady and dynamic performance, particularly in robustness, etc., and get to an optimal operation. Due to the limitation in paper's length, some topics cannot be discussed one by one. The discussion below will emphasize the device state relative to robustness.

Fig.9and Fig.10 show the typical characteristics of IGBT and MOSFET respectively. If devices operate in the flat horizontal line area (II-region), almost near the critical curves, the situation will be just as above mentioned. In fact, the special supply feeds the linear unit with tooth-like waveforms (Fig.2), which would probably make the devices operate from II-region to I-region back and forth if the peak value difference between V_s and V_o is adjusted too small. Comparing the robustness between IGBT-SLH and MOSFET-SLH in this situation, what conclusion should be gotten? Evident difference will occur in I-region where IGBTs enters saturated state, exhibiting BJT's characteristics much more, while MOSFEFs operate in changeable resistance state with variable Trans-conductor g_m. In other words, Robustness of IGBT-SLH disappears in I-region while that of MOSFET-SLH still exists in I-region only not strong as that in II-region. Thus, it is obviously from Fig. 2 that during a load shock occurs just at the transient corresponding to the peak voltage of V_s and V_o, including their neighbor time period, the robustness of IGBT-SLH will be stronger than that of MOSFET-SLH.

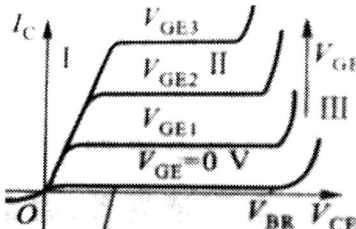

Fig. 9 IGBT- characteristics

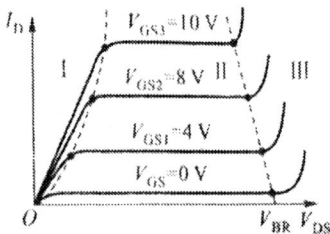

Fig. 10 MOSFET- Characteristics

An experiment proved that more special situation will occur in the system of Fig.7, where the load waveforms of line voltage and phase voltage are shown in Fig.12. Only take the waveform of phase V_{ab} for analyzing, suppose the system operating in resistive load condition, corresponding to the period $V_a > V_b$ and $V_{ab} > V_{bc}$, the current flows from phase A to phase B via Tab, load and the anti-paralleled diode inside Tbc back to the switch-filter unit, so the load voltage of V_{bc} in this period cannot follower the Gate voltage of Tbc without turning on, and the Emitter voltage of Tbc is higher than that of its Gate

1370

voltage due to the clamp of the diode, merely lower about 0.7V than that of its Collector, getting the waveform similar to the ripple voltage from the switch-filter unit. It is not hard to analyze the waveforms in the negative half cycle of V_{bc} and the whole cycle of V_{ca} and V_{ab}. For clear expression, other ripple voltage curves are omitted in the diagram. Consequently, the ripple voltage waveforms will occur in 0°~30° and 180°~210° period of every line voltage of the load. During the load shock appears in these periods, there is no robustness like in other SLH topologies. So the topology in Fig. 7 saves the device numbers at the price of partly losing the robustness regardless of what kind of device adopted. While in R-L load situation like in Fig.11, the ΔV_o corresponding to the zero current point must be larger than that in other SLH systems.

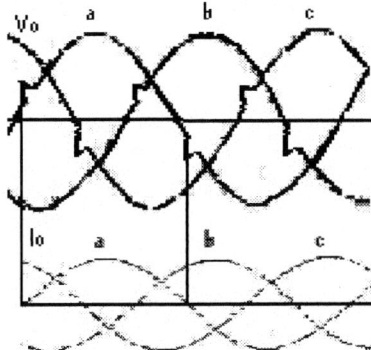

Fig. 11 ΔV_o corresponding to zero current points

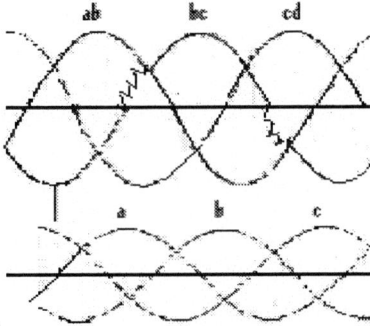

Fig. 12 Special waveforms of the topology in Fig.7

IV. CONCLUSION

First, typical topologies are introduced, compared for developing SLH conversion with synthetic excellent performances in high efficiency, good waveforms, fast response, particularly in robustness.

Secondly, scientific methods for extending new topologies of SLH are summed up as two ways: One is based on original conventional amplifiers fed by special switch-filter. Another is based on changing the operation state of the devices in present dc or ac switchers from switching state into special linear state, fed by special switch-filter.

Thirdly, the difference of robustness between IGBT-SLH and MOSFET-SLH is analyzed in detail. 3-phase SLH of saving devices possesses non-complete robustness

regardless what kind of devices are used because of the diode clamp. So the load parameters, the practical requirements and the efficiency should be balanced to consider for applying SLH-topology reasonably.

ACKNOWLEDGMENT

This work is supported by the natural science foundation of Anhui Province Education Bureau under Project 2001kj040zd.

REFERENCES

[1] Qianzhi Zhou, et al. Switch- linearity (CTA) conversion technique based on optimal waveform criterion. Proc. of 1th international power electronics and motion control conference (IPEMC'1994)[C], 2: 980~985

[2] Qianzhi Zhou, Luseng Ge. Switch-linearity hybrid power conversion (SLH) with low output resistance[C]. Proc. of IPEMC'2004, 1:96~98

[3] Zhou Qianzhi, Xu Haibin, Wu Dongsheng. Tracking type Active Power filter [P]. China Patent, No. ZL 00112420. X, 2003

[4] Qianzhi Zhou, Ding li, Hangdong Zhang. The analysis on SLH power conversion scheme and its validity [J]. Trans. of China Electrotechnical Society, 2002, 17(4): 75~79

[5] Qianzhi Zhou, Zhiguo Chen, Peng Jiang, et al. A novel high performance frequency changer with sine waves based on voltage comparing- tracking- amplifying (CTA) scheme. Proc. of 4th European conference on power electronics and application (EPE'91), Florence, 1991, 4: 324~329

[6] Peng Jiang, Zhong Yang, Weiqun Chen. C T A frequency changer controlled by a16-bit microprocessor[C]. Proc. of IPEMC' 94, Beijing, 1994, 2:607~612

[7] Urs Boegli, Remo Ulmi. Realization of a new inverter circuit for direct photovoltaic energy feedback into the public grid[J]. IEEE Trans. on Industry Application, 1986, 22(2):255~258.

[8] Zhou Qianzhi, Zhang Handong, Zhuang Kai. A single-three phase inverter adopting SLH scheme [J]. Journal of University of Electronics Science & Technology of China. 2003, 32(6):655~660

[9] Donato Vincenti, HuaJin and Phoivos Ziogas. Design and Implementation of a 25-kVA three-phase PWM AC line conditioner[J]. IEEE Trans. on Power Electronics, 1994, 9(4):384~389

[10] Andrés Barrado, Ramón Vázquez, etc Lázaro and Jorge Pleite. Theoretical study and implementation of a fast transient response hybrid power supply[J]. IEEE Trans. Power Electronics, 2004, 19(4), 1003~1009

[11] P. Midya. Linear switcher combination with novel feedback [C]. Proc. of Conf. PESC'00, 2000: 1425~1429

2006 5th International Power Electronics and Motion Control Conference

Single Reverse Blocking Switch Type Pulse Density Modulation Controlled ZVS Inverter with Boost Transformer for Dielectric Barrier Discharge Lamp Dimmer

Hisayuki Sugimura[1], Bishwajit Saha[1], Hideki Omori[2], Hyun-Woo Lee[1] and Mutsuo Nakaoka[1][3]

[1]Electric Energy Saving Research Center, Graduate School of Electrical and Electronics Engineering,
Kyungnam University, Masan, Korea
[2] Home Appliance Company, Matsushita Electric Industrial Co., Ltd., Osaka, Japan
[3]Department of Electrical & Electronics Engineering, Industrial College of Technology-University, Hyogo, Japan

Abstract- **This paper presents soft switching zero voltage switching high frequency inverter with reverse blocking single switch for dielectric barrier discharge lamp. The simple high-frequency inverter can completely achieve stable zero voltage soft switching (ZVS) commutation for wide its output power regulation ranges and load variations under its constant high frequency pulse density modulation (PDM) scheme. Its transient and steady state operating principle is originally described and discussed for a constant high-frequency PDM control strategy under a stable ZVS operation commutation, together with its output effective power regulation characteristics based on the high frequency PDM strategy. Its light dimming characteristics due to power regulation scheme are evaluated and discussed on the basis of simulation and experimental results. The feasible effectiveness of this high frequency inverter appliance implemented here is proven from the practical point of view.**

Keywords- Dielectric barrier discharge (DBD), Rare gas fluorescent lamp circuit modeling, Pulse density modulation (PDM); Zero voltage switching (ZVS); High frequency soft switching inverter; High frequency AC voltage boost transformer; Reverse blocking type IGBT switch; Consumer power electronics

I. INTRODUCTION

At present, the cold cathode fluorescent lamp (CCFL) using mercury lamp has been generally used for liquid crystal backlight source of personal computer and car

navigation and so on. This kind of lamp is more excellent on luminance performance and cost. However, the requirements of liquid crystal backlight due to a light source without mercury have been strongly increased from a viewpoint of the actual influence on environmental preservation and environmental recycling. As fluorescent lamp without mercury, Dielectric Barrier Discharge based rare gas fluorescent lamp (DBD-FL) using xenon (Xe) gas has been studied so far. This DBD lamp has no influence on the human body and environmental recycle. Its operating life is long because electrode is out.

The current source Royer type center tap push pull high frequency circuit is widely used for compact liquid crystal backlight. Its actual efficiency is relatively low. On the other hand, soft switching high frequency inverter with reverse blocking single switch can achieve high efficiency and high quality. This high frequency inverter circuit topology is composed of a few parts and its power regulation can operate under zero voltage soft switching using the PDM control implementation. This PDM control method enables to maintain the discharge sustaining voltage and achieve zero soft switching commutation over wide dimming control ranges.

In this paper, the simulation and experimental results of soft switching high frequency inverter with reverse blocking single switch as a high frequency power supply circuit for DBD-FL using Xe gas are comparatively evaluated and discussed from a practical point of view.

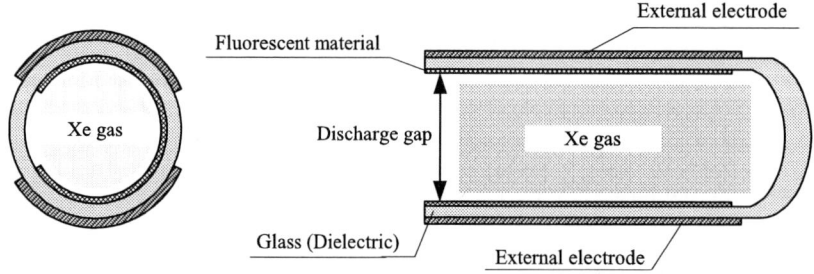

Figure 1. Schematic structure of rare gas fluorescent lamp

II. DIELECTRIC BARRIER DISCHARGE-BASED RARE GAS FLUORESCENT LAMP

A rare gas fluorescent lamp based on DBD principle is a kind of fluorescent lamp, which uses gas, and since it is not influenced upon temperature and constant actinography can be able to obtain. It is considered for light source in copy machine and scanner. Figure.1 shows a schematic structure of rare gas fluorescent lamp based on the dielectric barrier discharge principle. Two metal electrodes is set up around glass tube axis outside of the glass tube, phosphors are applied to inner surface of glass tube.

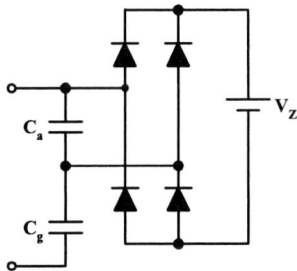

Figure 2. Equivalent circuit of rare gas fluorescent lamp as dielectric barrier discharge lamp

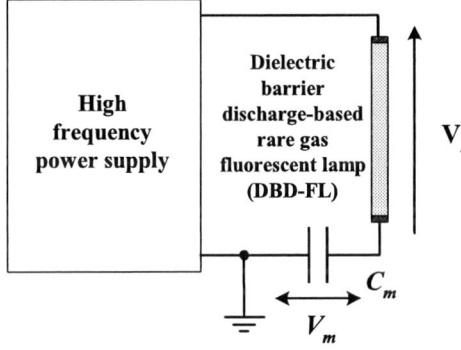

Figure 3. Circuit for measuring circuit parameters of DBD-FL

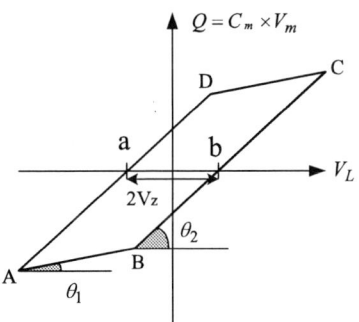

Figure 4. Q vs. V_L lissajous figure estimating circuit parameters of DBD-FL

This electric discharge phenomenon can be described on the basis of dielectric barrier discharge. At first, a high frequency AC high voltage is applied between two metal electrodes. And the dielectric polarization occurs and high voltage is applied for driving lamp. The silent discharge starts to generate when the voltage across substrate glass reaches up to breakdown voltage without sound. This silent discharge is based on the dielectric barrier discharge or silent electric discharge.

Moreover, the electric equivalent circuit represented by a nonlinear capacitive load and diode full bridge including the voltage source corresponding to the discharge sustaining voltage is shown in Figure 2. In this circuit model, the dielectric barrier discharge fluorescent lamp using rare gas can carry out the analysis of high frequency AC power supply circuit. C_a denotes capacitance between dielectric gaps with Xe gas, C_g denotes capacitance of substrate dielectric part of glass and V_z denotes dielectric sustaining voltage. C_a is connected in series with C_g during non-discharge, and the voltage across C_a is clamped to V_z during stable discharge.

Figure 3 shows the basic circuit for measuring circuit parameters of the equivalent circuit represented by a non-linear capacitive circuit. The auxiliary capacitor C_m is additionally connected in series with the DBD-FL and high frequency AC voltage is applied to a series circuit of DBD-FL and C_m. Then, the voltage V_L across the fluorescent lamp and V_m across C_m is respectively displayed as lissajous figure depicted in Figure 4 on oscilloscope. In this time, V_m multiplied by C_m given as electric charge Q. In Figure 4, the transition A-B denotes non-discharge period. Resonant capacitor C_{ag} series connected in series with C_a and C_g is obtained from eq.(1).

$$C_{ag} = \tan\theta_1 = \frac{\Delta Q_1}{\Delta V_{L1}} = \frac{C_m \Delta V_{C1}}{\Delta V_{L1}} \qquad (1)$$

The transition B-C denotes discharge period. C_g is obtained from eq.(2).

$$C_g = \tan\theta_2 = \frac{\Delta Q_2}{\Delta V_{L2}} = \frac{C_m \Delta V_{C2}}{\Delta V_{L2}} \qquad (2)$$

In addition, C_{ag} is represented by eq.(3), so that C_a is estimated by eq.(4).

$$C_{ag} = \frac{C_a C_g}{C_a + C_g} \qquad (3)$$

$$C_a = \frac{C_{ag} C_g}{C_g - C_{ag}} \qquad (4)$$

Moreover, V_z is equal to half of distance between point a and b on V_L axis.

III. CURRENT-FED ROYER TYPE RESONANT HIGH FREQUENCY INVERTER

The circuit topology the Royer type resonant high frequency inverter is shown Figure 5. This circuit is composed of 12V DC power supply, choke coil L_1, resonant capacitor C_r, and high frequency transformer TR. Bipolar transistor S1 and S2 are used for the switching devices. A rare gas fluorescent lamp of the load is shown by the equivalent circuit model which is a capacitive load as described in chapter II. As for the feature of Royer type inverter, the base signal of the transistors to switch is taken directly from the circuit. In general, the switching pulse signal generating circuit is necessary for the power conversion circuit in addition a main circuit. However, the Royer type resonant high frequency inverter is a method supplied the switching signal directly by the main circuit that is called the self-excited type. Therefore, because the number of circuit components decreases, low-cost can be achieved. In addition, it has the advantage of miniaturizing the circuit easily because the switching pulse signal generating circuit is unnecessary. The power regulation of this high frequency inverter, that is, the dimming control of the lamp is pulse amplitude modulation (PAM) control method. In general, the output is regulated with the DC-DC converter connected with the foreside of Royer type high frequency inverter shown Fig.6.

IV. Soft Switching PDM High Frequency Inverter Using Single Power MOSFET

A. Circuit Constructions

Soft switching PDM high frequency inverter using single power MOSFET is illustrated Figure 7. This circuit is composed of the 12V input DC power supply, high frequency transformer (consisted of exciting inductance L_m, leakage inductance L_l and ideal transformer), semiconductor power switching device Q(SW/D) (MOSFET: manufactured by IR, IRFP264), resonant capacitor C_r, and DBD-FL load.

B. Pulse Density Modulation Control

It proposes the PDM control for the power regulation of reverse conducting type high frequency inverter for driving DBD-FL. The principle of PDM Control is shown

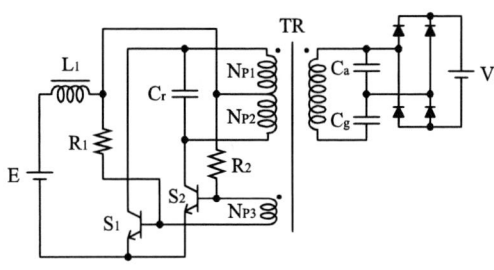

Figure 5. Royer type resonant high frequency inverter

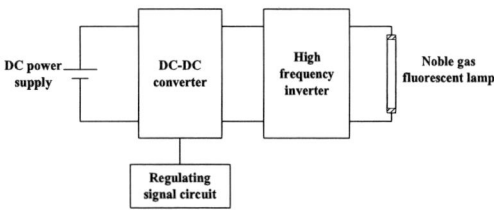

Figure 6. Power regulating system of Royer type inverter

in Figure 8. The PDM control is a power regulating method changing the ratio at power injection period and power non-injection period with constant frequency of the

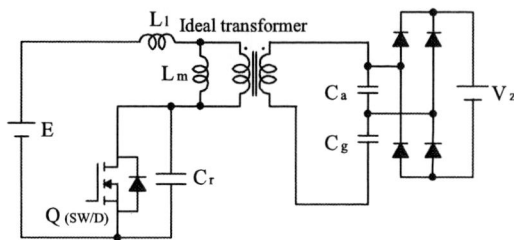

Figure 7. ZVS-PDM high frequency inverter using single switch

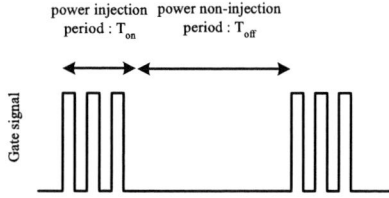

Figure 8. Gate signal pulse sequences of PDM Control

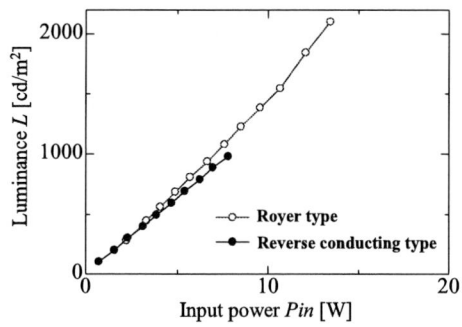

Figure 9. Luminance vs. input power characteristics

TABLE I.
DESIGN SPECIFICATIONS AND CIRCUIT PARAMETERS

Item	Symbol	Value
Input DC Voltage	E	12.0V
Leakage inductance	L_l	2.45μH
Magnetizing inductance	L_m	32.38μH
Turn ratio	-	1:25
Resonant capacitor	C_r	120nH
Gap capacitance	C_a	517pF
Dielectric capacitance	C_g	671pF
Discharge sustaining voltage	V_z	237.9V
Switching frequency	f_{sw}	25.0kHz
PDM frequency	f_{PDM}	100Hz

high frequency inverter. The PDM control variable D is defined in the eq.(5).

$$D = \frac{T_{on}}{T_{on} + T_{off}} = \frac{T_{on}}{T} \quad \ldots\ldots\ldots\ldots\ldots(5)$$

The variable is a ratio to one PDM signal cycle T ($T_{on} + T_{off}$) for the power injection period T_{on}. By controlling this variable D, output power of the high frequency inverter can be regulated. Because the discharge sustaining voltage V_z doesn't decrease when the high frequency inverter is controlled with PDM, the lamp can be discharged stably in a wide output power regulating range.

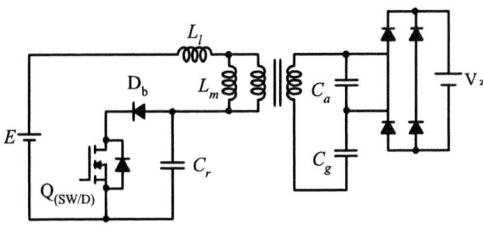

Figure 10. Soft switching high frequency inverter with reverse blocking single switch

C. Simulation and Experimental Results

Design specifications and circuit parameters are shown in Table 1. The luminance vs. input power characteristics

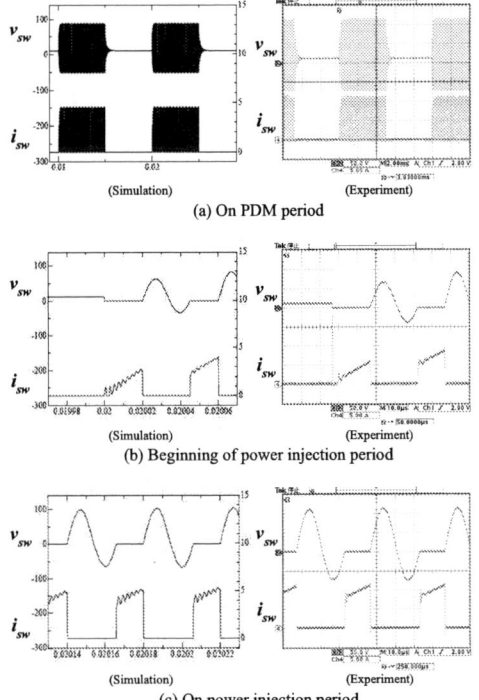

(a) On PDM period

(b) Beginning of power injection period

(c) On power injection period

Figure 11. Simulation and experimental waveforms of switch Q in case of D=0.5

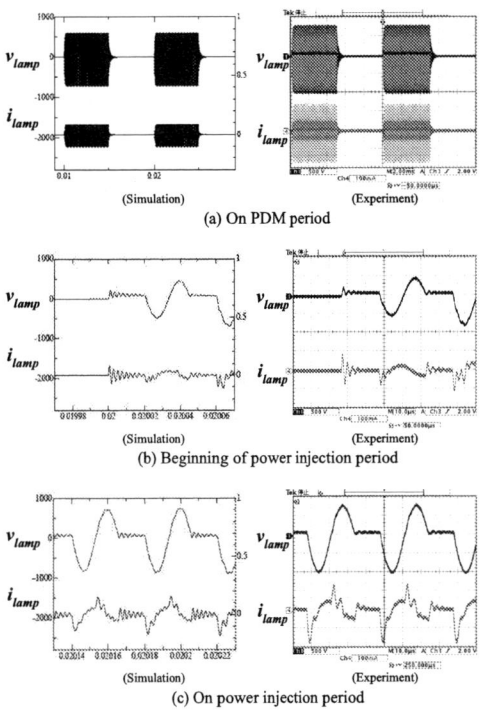

(a) On PDM period

(b) Beginning of power injection period

(c) On power injection period

Figure 12. Simulation and experimental waveforms of DBD-FL in case of D=0.5

TABLE II.
DESIGN SPECIFICATIONS AND CIRCUIT PARAMETERS

Item	Symbol	Value
Input DC Voltage	E	12.0V
Leakage inductance	L_l	4.94μH
Magnetizing inductance	L_m	72.26μH
Turn ratio	-	1:10
Resonant capacitor	C_r	180nH
Gap capacitance	C_a	517pF
Dielectric capacitance	C_g	671pF
Discharge sustaining voltage	V_z	237.9V
Switching frequency	f_{sw}	25.0kHz
PDM frequency	f_{PDM}	100Hz

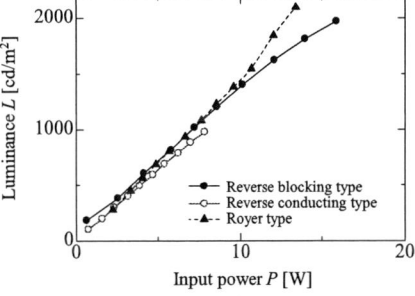

Figure 13. Luminance vs. input power characteristics

1375

is shown in Figure 9. As can be seen, soft switching PDM high frequency inverter using single power MOSFET can not be output compared with the Royer type inverter. In this inverter, the resonance period achieved high power output ends commutating from capacitor C_r to a reverse conducting diode of the switch Q. It is complicated to shorten the resonance period by this commutation operation.

In the above, it is considered soft switching PDM high frequency inverter using single power MOSFET has high controllability and can be lighted without flicker in the low luminance control range. However the high frequency inverter can not be achieved high luminance.

V. SOFT SWITCHING HIGH FREQUENCY INVERTER WITH REVERSE BLOCKING SINGLE SWITCH

In chapter III, soft switching PDM high frequency inverter using single power MOSFET is shown that a high luminance output is more difficult than the conventional Royer type inverter. Therefore, this chapter describes soft switching high frequency inverter with reverse blocking single switch that can output higher luminance.

A. Circuit Constructions

Soft switching high frequency inverter with reverse blocking single switch is illustrated Fig.10. The circuit topology of the high frequency inverter is the same as the inverter shown in Figure 7 excluding reverse blocking diode. The power regulation of the proposed inverter can be achieved by PDM control.

B. Simulation and Experimental Results

Design specifications and circuit parameters are shown in Table 2. Simulation and experimental waveforms of switch Q and DBD-FL in case of D=0.5 are depicted in Figure 11 and Figure 12. As can be seen in this figure, the switch is achieved a ZVS & ZCS turn on and ZVS turnoff commutation. The typical voltage and current operation waveforms of simulation and experimental ones have a good agreement within the slight error. However, switching serge current is flowed through the switch in case of beginning of power injection period. Because, the serge current is flowed by the discharge of the charge accumulated in the capacitor C_r on non-power injection period when the switch turns on.

The luminance vs. input power characteristics is shown in Figure 13. As can be seen, it is shown that the proposed circuit has higher luminance equal with the Royer type inverter and linear controllability of the reverse conducting switch type inverter.

VI. CONCLUSIONS

This paper presented the Royer type resonant high frequency inverter, soft switching PDM high frequency inverter using single power MOSFET and soft switching high frequency inverter with reverse blocking single switch. Moreover, these operation principles, circuit topologies and circuit characteristics were described here. In addition, characteristics of these circuits were compared the simulation with the experimental results. Soft switching high frequency inverter with reverse blocking single switch can be achieved higher luminance and linear controllability.

ACKNOWLEDGMENT

This work was financially supported by MOCIE through IERC program.

REFERENCES

[1] Y.L. Wang, "Research of PDM high frequency inverter for driving excimer lamp", The master thesis of Graduate School of Science and Engineering, Yamaguchi University, Yamaguchi Japan

[2] Masaaki Tanaka, Shigenori Yagi, Norikazu Tabata, "Measurement of electrical characteristics for silent discharge and lissajous graphic", IEE-Japan "discharge association document" ED-84-50 1984

[3] Norikazu Yamamoto, "Development of high illuminance electrodeless discharge lamp", NEC Technological Journal Vol.52 No.2, 1999

[4] Taichiro Minda, Akihiko Iwata, Masaaki Tanaka, "Measurement of ac-PDP discharge parameter with V-Q lissajous graphic", IEE-Japan "discharge association document" ED-96-274 1996

[5] "Light edge", Ushio Technological Information, No.15 November 1998

[6] Kudryavtsev Oleg, "Studies on Electrical Characteristic Evaluations of Dielectric Barrier Discharge Load and Development of Its Optimal Driving Power Supply", The doctoral thesis of Graduate School of Science and Engineering, Yamaguchi University, Yamaguchi Japan

[7] Y.Konishi, S.P.Wang, S.Shirakawa and M.Nakaoka, "Pulse density modulated high-frequency load resonant inverter for ozonizer and its feasible performance evaluations", IEEE-IAS Annual Meeting, pp.1313-1319, 1998

[8] S.P.Wang, M.Ishibashi, Y.Konishi, M.Nakaoka, "Series-compensated inductor type resonant inverter using pulse density modulation scheme for efficient ozonizer", International Conference on Power Electronics and Drive Systems, Vol.1, pp.19-23, 1997

[9] Kentaro Fujita, Hidekazu Muraoka, Tarek Ahmed, Eiji Hiraki, Mutsuo Nakaoka, Hyun Woo Lee, Kazunori Nishimura, "A Voltage Source Single-Ended High Frequency Edge Resonant Inverter with Tube Type Dielectric Barrier Discharge Lamp for Liquid Crystal Backlight", Proceedings of 2004 International Conference on Power Electronics (ICPE), II-41,2004.

2006 5th International Power Electronics and Motion Control Conference

PDM Controlled Series Load Resonant Soft Switching High Frequency Inverter for Induction Heated Toner Fixing Outer Roller with Inner Cylindrical Working Coil Stator

Hisayuki Sugimura[1], Hideki Omori[2], Hyun Woo Lee[1], Mutsuo Nakaoka[1][3]

[1]Electric Energy Saving Research Center, Graduate School of Electrical and Electronics Engineering,
Kyungnam University, Masan, Korea

[2]Home Appliance Company, Matsushita Electric Industrial Co., Ltd., Osaka, Japan

[3]Department of Electrical and Electronics Engineering, Industrial College of Technology-University, Hyogo, Japan

Abstract— This paper presents the two lossless auxiliary inductors-assisted voltage source type single ended push pull (SEPP) series resonant high frequency inverter for induction heated fixing roller in copy and printing machines. The simple high-frequency inverter treated here can completely achieve stable zero current soft switching (ZCS) commutation for wide output power regulation ranges and load variations under its constant high frequency pulse density modulation (PDM) control scheme. Its transient and steady state operating principle is originally described and discussed for a constant high-frequency PDM control strategy under a stable ZCS operation commutation, together with its output effective power regulation characteristics based on the high frequency PDM strategy. The experimental operating performances of this voltage source SEPP ZCS-PDM series resonant high frequency inverter using IGBTs are illustrated as compared with computer simulation results and experimental ones. Its power losses analysis and actual efficiency are evaluated and discussed on the basis of simulation and experimental results. The feasible effectiveness of this high frequency inverter appliance implemented here is proved from the practical point of view.

Keywords-Voltage source type series load resonant inverter, Lossless inductive snubbers, Zero current soft switching, Pulse density modulation, Induction heated outer roller, Consumer power electronics

I. INTRODUCTION

For industrial and consumer IH power applications in the next generation, the voltage-fed high frequency inverter with series capacitor resonant tank circuitry has been widely applied so far. The general method of output power regulation in this kind of high frequency inverter is based on pulse frequency modulation (PFM) scheme of its inverter frequency. The PFM strategy implies changing the working frequency of the inverter that has essentially some drawbacks for IH applications. That is to say, the

high frequency AC effective output power in case of PFM control strategy depends linearly to square root of the series load resonant inverter working frequency and inverter system efficiency decreases significantly for light load in copy machine, facsimile, scanner, data recorder and printer in stand-by mode. In addition to this, when two or more inverters are assembled in a set of equipment, the actual problem of acoustic noise due to the difference operating frequency of the inverters may occurs. Furthermore, the skin effect resistance of the IH fixing roller as well as the depth of the induced eddy current penetration depend on inverter working frequency have much worse influence on the temperature distribution characteristics of the IH fixing drum roller.

In this paper, the voltage source type half-bridge series resonant voltage-fed series load resonant inverter with two lossless inductor snubbers in series with each active switch is introduced, which can operate under a high frequency ZCS-PDM operation conditions. The high frequency power regulation characteristics of the developed high frequency series load resonant inverter which is based on a constant high frequency ZCS-PDM are presented in this paper, together with the performance evaluations of the power losses analysis or efficiency characteristics on the basis of the simulation.

II. INDUCTION-HEATED FIXING ROLLER EQUIPMENT

A. Schematic structure of induction-heated fixing roller

The cross sectional and physical structure of the experimental induction heated fixing roller is schematically shown in Fig.1. Presently, the main electric heating method for the fixing roller as light radiant heated roller in the copy and printing machines is introduced which can be heated directly by light emission from the halogen lamp. This scheme has some disadvantages such as relatively low efficiency, required maintenance, easy to temperature, non-recycle, quicker temperature response and short life. On the other hand, the fixing heat roller with an induction-heated working coil inserted inside the rolling drum made of stainless steel plate (SUS410) is depicted schematically in Fig.2. The titanium alloy and the carbon ceramic are effectively applied for the induction heated fixing heat roller in the copy machine and so forth.

B. Transformer circuit modelling of induction heated load

This IH fixing heat roller used in this paper is modelled by using the transformer-based circuit model represented in Fig.3. This transformer circuit model is also used for the IH load circuit analysis including the high frequency inverter. The resistor R_2 is the resistance related to the high frequency dependent skin effect and current

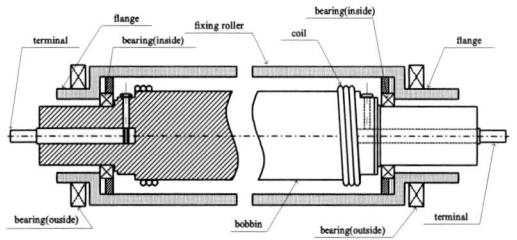

Fig.1 Sectional view of toner fixing roller in copy/printer equipment

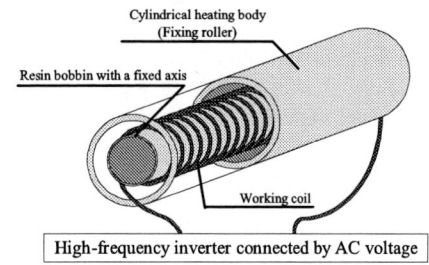

Fig.2 Induction heating fixing roller

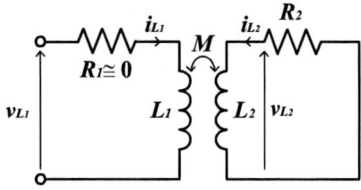

Fig.3 Transformer model of induction heating load

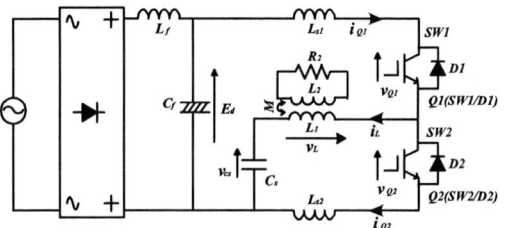

Fig.4 Voltage source SEPP ZCS-PDM high frequency inverter system for induction heated fixing roller

penetration depth that are based on the operating inverter frequency. In the circuit analysis of the induction heated fixing heat roller load delivering the paper, three parameters of self-inductance L_1 of cylindrically shaped working coil itself with an internal zero resistance or a slight resistance, load time constant $\tau = L_2/R_2$ and electromagnetic coupling coefficient $k = M/\sqrt{L_1 L_2}$ of the transformer model are introduced on the basis of measurable variables.

III. PDM CONTROLLED SERIES LOAD RESONANT HIGH FREQUENCY ZCS INVERTER

A. Total system description and some features

The overall high frequency power conversion system composed of the voltage-fed SEPP series load resonant ZCS-PDM controlled high frequency inverter using IGBTs is depicted in Fig.4. The voltage E_d is a DC voltage applied to the voltage source-fed high frequency inverter via single phase capacitor input type diode full bridge rectification of 200V/60Hz utility AC power source grid. The single phase PFC converter with boost chopper can be conveniently used in place of diode rectifier. For cost effective appliance design, a diode rectifier with non-smoothing filter is connected in the input side of high frequency ZCS-PDM inverter. This high frequency inverter consists of the active power switches Q_1 and Q_2 the reverse conducting switches due to the power semiconductor switches (IGBTs); SW_1 and SW_2 with antiparallel diodes; D_1 and D_2, Cr as a tuned resonant capacitor in series with IH load. Two inductors Ls_1 and Ls_2 as auxiliary ZCS-assisted inductive lossless snubbers connected in series with Q_1 and Q_2 and the IH fixing heat drum roller represented by the transformer circuit modelling. In this high frequency series load resonant inverter circuit, the active power switches Q_1 and Q_2 can operate completely under ZCS principle and its high frequency AC power regulation based on a variable pulse frequency modulation for both turn-on and turn-off mode

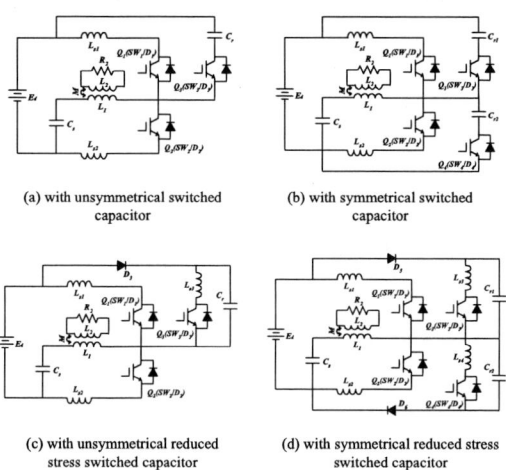

(a) with unsymmetrical switched capacitor

(b) with symmetrical switched capacitor

(c) with unsymmetrical reduced stress switched capacitor

(d) with symmetrical reduced stress switched capacitor

Fig.5 Variations of voltage soured SEPP ZCS-PWM and PDM high frequency inverter with switched capacitor

1378

zero current soft switching commutations. The effective AC output power of the high frequency inverter in Fig.4 can be newly regulated by a constant high frequency PDM control strategy on the basis of the pulse group modulation principle in Fig.6. The IH load surrounded by the dotted line is the transformer model parameters represented by the circuit parameters; four unmeasurable values (M, L_2, k, R_2) or three measurable values (L_1, $k = M/\sqrt{L_1 L_2}$, $\tau = L_2/R_2$) of the IH load comprised of the cylindrical working coil and induction heated fixing roller load displayed in Fig.2.

In addition, Fig.5 depicts the variations of voltage source SEPP ZCS-PWM and PDM high frequency inverter with asymmetrical and symmetrical switched capacitor.

B. Pulse density modulation controlled high frequency AC power regulation

As shown in Fig.6, the high frequency AC power regulation can be achieved by varying the pulse density modulation under time ratio during T_{on} period that the AC output power is injected into the induction heated load and a period T_{off} that the AC output power is non-injected into the induction heated load. With the changing the PDM time ratio, the applied pulse density ratio is taking place while the working frequency of the high frequency inverter is kept constant under a condition of zero current soft switching transition commutation.

The auxiliary inductive snubbers Ls_1 and Ls_2 ($Ls_1 = Ls_2 = Ls$) in series with the active switches provide ZCS commutation operation for Q_1 and Q_2 in the continuous load current mode which is based on the overlapping

TABLE II DESIGN SPECIFICATIONS AND CIRCUIT PARAMETERS

Quantity	Symbol	Value
Input DC voltage	E_d	280V
Series resonant capacitance	C_r	0.49µF
ZCS inductive snubber value	L_s	12.0µH
Self inductance of work coil	L_1	90.0µH
Time constant of the load	τ	9.23µsec
Electro Magnetic coupling co-efficient	k	0.48
IGBT(TO-3P)	V_{CE}	600V
	I_C	75A
Antiparallel diode (TO-3P)	V_{RM}	600V
	I_0	30A

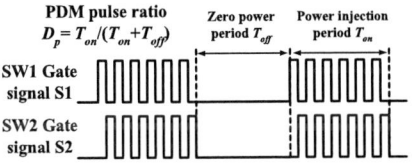

Fig.6 Principle of high frequency PDM control for series resonant inverter

current in (SW_1, D_2) and (SW_2, D_1).

Since a complete ZCS operation for Q_1 and Q_2 is provided over whole power regulating ranges, the high frequency leak current related electromagnetic noise and switching power losses for Q_1 and Q_2 are kept to be low. Furthermore, as compared with the series load resonant inverters driven by the other control methods of PFM, PWM and PAM for the conditions of the light induction heated load, almost no power losses in the ZCS-PDM scheme is consumed during the power non-injected period in this high frequency inverter. The switching power losses in ZCS-PDM scheme can be reduced even in power injection mode. Therefore, the high inverter efficiency is observed for heavy and light induction heated loads.

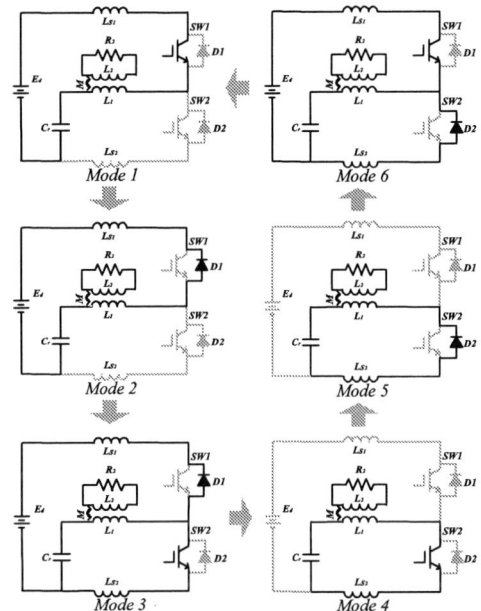

Fig.7 Operating mode transitions and switching mode equivalent circuits

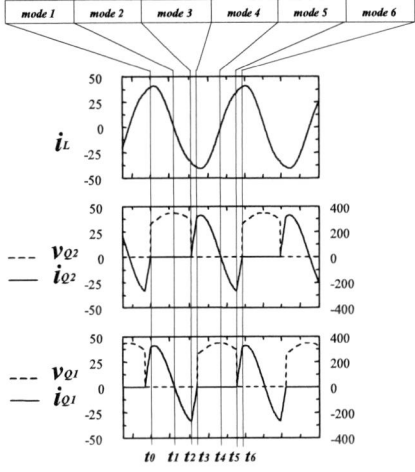

Fig.8 Simulation voltage and current waveforms in steady state of power injection

C. Circuit operation waveforms in steady state

There are the circuit mode transitions of proposed circuit as shown in Fig.7.

The voltage and current operation waveforms of the voltage source series load resonant ZCS-PDM inverter circuit shown in Fig.6 during the power injection period are illustrated in Fig.8.

The circuit parameters of the voltage source type PDM controlled high-frequency ZCS inverter using IGBTs are indicated in Table I. Two auxiliary inductive snubbers L_s are adjusted so as to be 12μH to provide the switch peak voltage 350V that includes ascertain tolerance to the limited standard rating parameters of the selected IGBTs. In this case, the dynamic switch current stress di/dt_{max} becomes 12.5A/μs and current overlapping time is set to 3.8μs.

IV. EXPERIMENTAL RESULTS AND EVALUATIONS

A. Pulse density modulated output voltage and current

The developed voltage source high-frequency series resonant ZCS-PDM inverter uses IGBTs (Mitsubishi Electric Co., Ltd., CT75AM-12) with soft recovery fast switching diodes (Origin Electric Co. Ltd, US30P) as the antiparallel fast recovery diodes. For pulse density modulation ratio D_p=0.2 and D_p=0.8 in a PDM control scheme, the measured operating waveforms of load current i_L and load voltage v_L are depicted in Fig.9. Observed voltage and current waveforms v_{Q1} & i_{Q1}, v_{Q1} & i_{Q1} for the active power switches Q_1 and Q_2 in switching

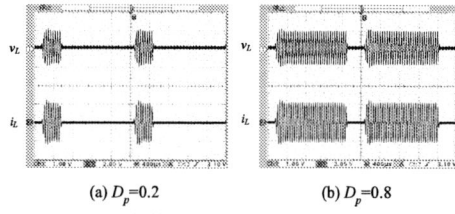

(a) D_p=0.2   (b) D_p=0.8

v_L:500[V/div], i_L:40[A/div], t: 400[μsec/div]
Fig.9 Experimental waveforms of v_L and i_L for PDM duty cycle

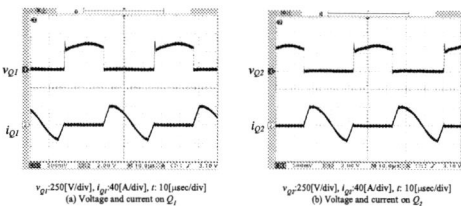

v_{Q1}:250[V/div], i_{Q1}:40[A/div], t: 10[μsec/div] v_{Q2}:250[V/div], i_{Q2}:40[A/div], t: 10[μsec/div]
(a) Voltage and current on Q_1 (b) Voltage and current on Q_2

Fig.10 Experimental waveforms of switch voltage and current

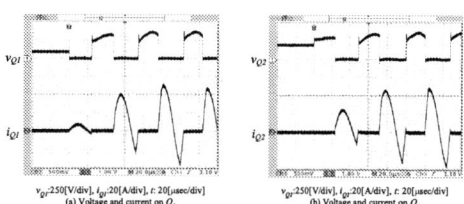

v_{Q1}:250[V/div], i_{Q1}:20[A/div], t: 20[μsec/div] v_{Q2}:250[V/div], i_{Q2}:20[A/div], t: 20[μsec/div]
(a) Voltage and current on Q_1 (b) Voltage and current on Q_2

Fig.11 Experimental waveforms of switch voltage and current at the beginning of the power injection period

arms of a voltage source type series resonant ZCS-PDM inverter are shown in Fig.10.

The validity of the transformer type circuit models parameters of the induction heated type fixing heat roller load in Fig.1 and Fig.2 is proven on the basis of these experimental results.

The voltage and current operating waveforms of the active power switches Q_1 and Q_2 during the beginning interval of the power injection period of PDM scheme are shown in Fig.11. It is clear that Q_1 and Q_2 can operate under a condition of ZCS principle for a PDM control implementation.

B. Output high frequency AC power regulation and efficiency characteristics

Figure 11 illustrates the pulse density modulation ratio D_p vs. output power characteristics and pulse density modulation ratio D_p vs. power conversion efficiency characteristics for voltage source type series resonant high frequency ZCS-PDM inverter in Fig.4. The high frequency AC output effective power of the high frequency inverter treated here can be regulated and linearly by changing the pulse density modulation ratio

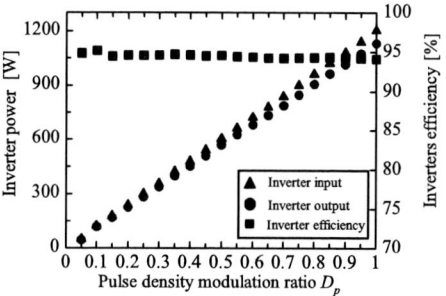

Fig.12 Power regulation characteristics in experiment

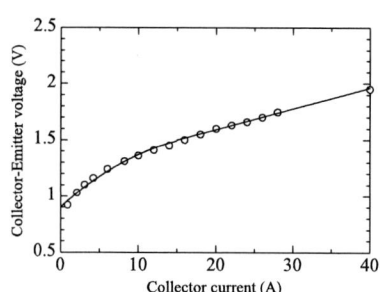

Fig.13 v-i characteristics of IGBT

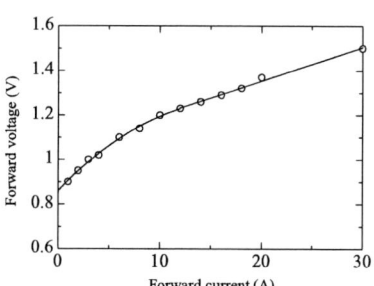

Fig.14 v-i characteristics of diode

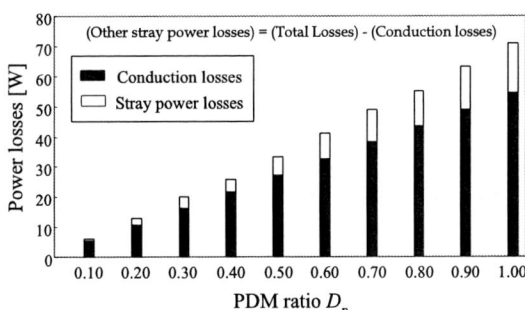

Fig.15 Results of power loss analysis

D_p. For output power regulation ranges from 5% to 100% of the maximum output power, the actual AC power conversion efficiency more than 94% can be obtained by the breadboard setup implementation. Especially, it is more important that actual efficiency more than 94% is able to be achieved even for both D_p=1.0 in copy machine printing mode and D_p=0.05 in its stand-by mode, which make the proposed voltage source series resonant ZCS-PDM controlled high-frequency inverter more effective for the induction heating type fixing heat roller applications in copy and printing machines.

C. Power loss analysis

The voltage-current characteristics of IGBTs & antiparallel diodes (constant temperature condition) are shown in Fig.13 and Fig.14. These switching characteristic curves in experiment can be approximated by quadratic polynomial on low power regulation area and linear function on high power regulation area. These experimental and numerical results are introduced into simulation algorithm. Namely, equations (1) and (2) are represented by using switching characteristic functions of Fig.13 and Fig.14.

$$\begin{cases} v_{I\_on} = -0.0015 i_{IGBT}^2 + 0.616 i_{IGBT} + 0.904 \, ; i_{IGBT} > 15A \, \cdots\cdots (1) \\ v_{I\_on} = -0.0185 i_{IGBT} + 1.223 \, ; i_{IGBT} \leq 15A \end{cases}$$

$$\begin{cases} v_{D\_on} = -0.00141 i_{Diode}^2 + 0.0477 i_{Diode} + 0.859 \, ; i_{Diode} > 12A \, \cdots (2) \\ v_{D\_on} = -0.0152 i_{Diode} + 1.05 \, ; i_{Diode} \leq 12A \end{cases}$$

The conduction losses of IGBTs Q_1 & Q_2 can be calculated approximately by using simulation results on the basis of these equations. In addition to this, the switching losses of the IGBTs Q_1 & Q_2 can be estimated from using total power losses of this high frequency inverter on the basis of experimental results.

The results of power loss analysis are shown in Fig.15. In this figure, power losses increase with PDM ratio D_p proportionately. In case of copy printing mode (rated power), conduction losses account for 80% of total power losses because of ZCS turn on transition by using lossless snubbing inductors L_{s1} & L_{s2}.

V. CONCLUSIONS

In this paper, the voltage-fed high-frequency single ended push-pull (SEPP) type series load resonant zero current soft switching inverter topology with ZCS-assisted two auxiliary inductive snubbers has been introduced for the induction-heated fixing roller in the copy and printing machines. Its steady state inverter operation under PDM control scheme has been evaluated and discussed on the basis of simulation and experimental data.

The high frequency AC power regulating characteristics and operating performances of this simple voltage source SEPP series resonant high frequency ZCS-PDM inverter using IGBT modules in steady state operation has been qualitatively evaluated in simulation and experiment. For the power loss estimations of this high frequency inverter, the transformer type circuit model of the induction-heated fixing roller in copy and printing machines has been used from a practical point of view.

The actual high efficiency more than 94% of the series load resonant ZCS-PDM high frequency inverter for IH roller in copy and printing machines has been observed for all the output AC power regulation ranges from 50W to 1200W with stable zero current soft switching operation processing and linear output power control characteristics under a condition of ZCS commutation. The voltage source SEPP type high frequency series resonant ZCS inverter with lossless inductive snubbers, which is based on a constant frequency PDM control scheme has provided its practical effectiveness for the stand-by mode and printing mode.

In the future, SiC-based new material power semiconductor switching devices such as SiC-SBD, SiC-MOSFET as well as the Si-based new structure power semiconductor switching devices such as ESBT, Super-Junction MOSFET, and reverse blocking IGBT in should be introduced and evaluated for the high frequency inverters treated here.

ACKNOWLEDGMENT

This work was financially supported by MOCIE through IERC program.

REFERENCES

[1] P.R.Palmer, A.N.Githiari, "An MCT Based Industrial Induction Cooker Circuit Using Zero Current Switching," Proceedings of EPE (European Power Electronics), pp.2.677-2.682, September, 1995.

[2] S.Kubota, Y.Hatanaka, "A Novel High Frequency Power Supply for Induction Heating," in Proc. of IEEE-PESC (Power Electronics Specialists Conference), Vol.1, pp.165-171, June, 1998.

[3] S.P.Wang, Y.Konishi, O.Koudriavtsev, M.Nakaoka, "Voltage-Fed Pulse Density and Pulse Width Modulation Resonant Inverter for Silent Discharge Type Ozonizer," Trans. on IEE of Japan, Vol.120-D (J-IAS), No.4, pp.587-592, April, 2000.

[4] H.Sugimura, A.M.Eid, H.W.Lee, M.Nakaoka, "Series Load Resonant Tank High Frequency Inverter with ZCS-PDM Control Scheme for Induction-Heated Fixing Roller", Proceedings of IEEE-ICIT (International Conference on Industrial Technology), pp.756-761, December, 2005.

2006 5th International Power Electronics and Motion Control Conference

Zero-Voltage and Zero-Current Switching Two-Transformer Full-Bridge Converter Using the Output-Voltage-Doubler

H.K. Yoon[*], E.S. Choi[*], S.K. Han[**], G.W. Moon[*] and M.J. Youn[*]

[*] KAIST/Dept. of Electrical Engineering, Daejeon, Republic of Korea
[**] Kookmin Univ./ Dept. of Electrical Engineering, Seoul, Republic of Korea
hkyoon@powerlab.kaist.ac.kr

Abstract— A zero-voltage switching (ZVS) and zero-current switching (ZCS) two-transformer full-bridge PWM converter using the output-voltage-doubler is proposed in this paper. It is based on the phase-shifted full-bridge with series-connected two transformers (TTFB) which act as an output inductor as well as a main transformer. The TTFB converter achieves ZVS for all the switches inherently even at a light load with the use of the energy stored in the magnetizing inductance of each transformer. However, it has several drawbacks such as circulating energy in freewheeling mode, loss of duty cycle, and high voltage stress across the output rectifier diodes. To overcome these drawbacks, the proposed converter employs an output-voltage-doubler which reduces the circulating current, the freewheeling interval, and the voltage stress across output rectifier diode. The operational principle, theoretical analysis, and design considerations are presented. Also, to confirm the operation, validity, and features of the proposed converter, experimental results from a 420W, 385V$_{dc}$/210V$_{dc}$ prototype are presented.

Keywords-component; ZVS, ZCS, Two-transformer, Full-bridge, Voltage-doubler

I. INTRODUCTION

Recently, the flat panel display (FPD) is widely spread on large area display applications. However, as the panel size increases, the consumed power is also increased with a subduplicate ratio. Nevertheless, the large area FPD tends to require the thinness, lighter weight, and fan-less system for the lower acoustic noise and vibration. Among the various FPD, the plasma display panel (PDP) has significant application, because it has advantages such as wide view angle, long life time, and high contrast over the conventional display. To carry out the successful driving for PDP, the operation of one sub-field can be divided into three periods such as resetting, addressing, and sustaining periods. Among them, during the sustaining period, high voltage sustaining pulses make PDP emit light by inducing gas discharge. Since most of the power for driving the PDP is consumed during this period, the power

module for sustaining is especially responsible for the overall system efficiency. Therefore, the PDP sustain power module (PSPM) requires the high power density, high performance and also high efficiency over the wide load range.

Among the various developed PWM DC/DC converters for middle/high power applications such as the PSPM over 400 W, the conventional Phase-Shift Full-Bridge (PSFB) converter is widely used, which provides all switching devices with the ZVS operation at the high frequency using the parasitic components. However, the PSFB converter has several drawbacks such as the narrow ZVS range of the lagging leg, circulating current and subsequent conduction losses during the transformer zero-voltage period, effective duty cycle loss, and serious voltage ringing across the output rectifiers [1].

To solve these problems, various soft switching schemes such as ZVS and ZCS have been applied on the conventional PSFB. To extend ZVS range, a PSFB converter with series-connected two transformers has been proposed [2]. It is easy to achieve the ZVS of the lagging leg under the light load since the magnetizing currents of the transformers also contribute to the ZVS condition. However, it still has a circulating current in freewheeling state, which causes not only the high conduction losses but also turn-off switching losses of the lagging leg switches. Consequently, to reduce a circulating current and a turn-off switching losses at the full-load condition, ZCS operation of the lagging leg is more attractive than ZVS operation. Therefore, many researches have been proposed to achieve ZCS operation of lagging leg. Adding an active clamp circuit or a snubber circuit in the

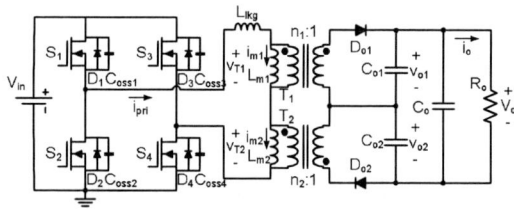

Figure 1. Circuit diagram of the proposed converter

1-4244-0448-7/06/$25.00 ©2006 IEEE

secondary side was proposed in [3][4]. Also, adding a blocking capacitor and saturable inductor with lagging leg switch was suggested in [5].

In this paper, a ZVS and ZCS two-transformer full-bridge PWM converter using the output-voltage-doubler is proposed as shown in Fig. 1. In the proposed converter, ZCS operation of lagging leg is achieved at the full-load to reduce the circulating current, while at the light-load ZVS operation is naturally achieved. Therefore, soft-switching of all switches in the proposed converter is achieved along the wide load range. Also, the rated voltage across the output rectifier-diode is reduced to be a half than that of the conventional TTFB due to the output-voltage-doubler. The proposed converter provides the simple structure, low rated voltage across the output rectifier diodes, high efficiency, and high reliability.

The operations, analysis and design considerations, and experimental results are presented to confirm the validity of the proposed converter.

II. OPERATIONAL PRINCEPLES

Fig.2 shows the operational key waveforms of the proposed converter. The basic operation of the proposed converter is identical to that of the conventional TTFB [2]. One cycle period of the proposed converter is divided into two half cycles, $t_0 \sim t_3$ and $t_3 \sim t'_0$. Because the operational principle of two half cycles are symmetric, only the first half cycle is explained. A half cycle can be divided into 3 modes and each equivalent circuit is shown in Fig.3. The switches of each leg (i.e. S_1/S_2 and S_3/S_4 are leading and lagging leg, respectively) turn on and off alternately with the constant duty ratio and the phase difference between both legs determines the operational duty cycle of the converter where $D_{eff}T_s$ is the operational conduction time.

The primary side of the proposed converter is same as

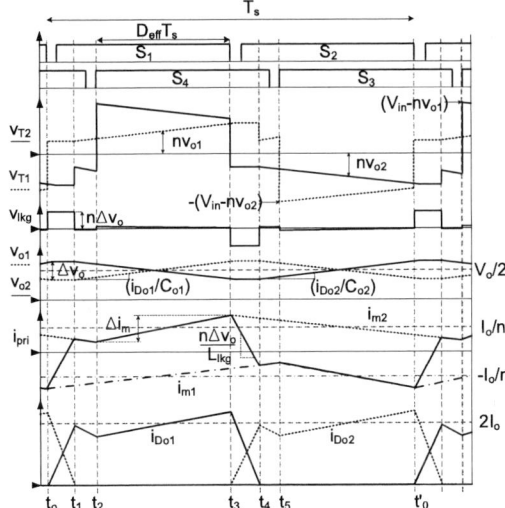

Figure 2. Operational Key waveforms of the proposed converter

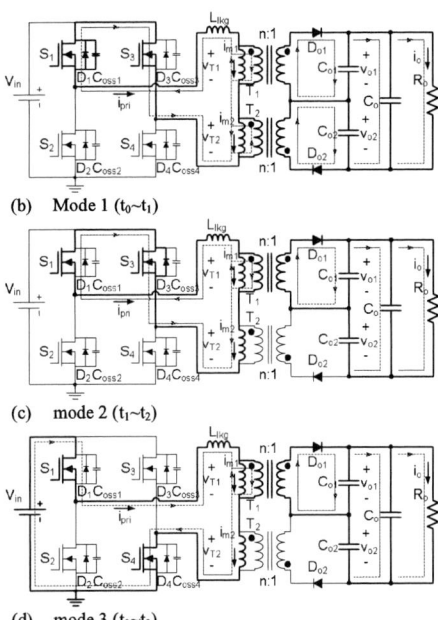

(b)   Mode 1 ($t_0 \sim t_1$)

(c)   mode 2 ($t_1 \sim t_2$)

(d)   mode 3 ($t_2 \sim t_3$)

Figure 3.   Equivalent circuits of the proposed converter

that of the conventional TTFB. It has series-connected two transformers acting as not only a main transformer but also an output inductor, performing alternately as a forward and flyback transformer. Namely, while one transformer transfers the input power to the secondary side of the converter, the other stores the power as a flux form like an inductor. Therefore, the output inductor of the conventional PSFB is not needed in the secondary side. On the other hand, the secondary side of the proposed converter consists of two diodes and three capacitors, which is operated for the output voltage-doubler. Therefore, the proposed converter has doubled output/input voltage conversion ratio than that of the conventional TTFB, which also makes doubled turn ratio and a half voltage rating of the output rectifier diodes.

For the convenience of the mode analysis in the steady state, several assumptions are made as follows:

▪ Power switches ($S_1 \sim S_4$) are ideal except for their internal diodes ($D_1 \sim D_4$) and output capacitors ($C_{oss1} \sim C_{oss4} = C_{oss}$).

▪ Both transformers have identical turn ratio, magnetizing and leakage inductances. ($n_1 = n_2 = n$ and $L_{m1} = L_{m2} = L_m$).

▪ L_{lkg} represents the total leakage inductance of the proposed converter, which is the sum of the leakage inductor of each transformer

Also, to understand the operation of the proposed converter easily, mode analysis is simplified with the linear modeling.

Mode 1 ($t_0 \sim t_1$): After S_2 is turned off at t_0, mode 1 begins. The primary current, i_{pri}, used to flow through S_2 flows in the direction of discharging C_{oss1}, so that the ZVS of S_1 is achieved. At the same time, since D_{o1} and D_{o2} begin to be commutated, the each voltage of the output-voltage-doubler, v_{o1} and v_{o2}, is reflected to the primary side of T_1 and T_2 in opposite polarity, respectively. Therefore, the voltage difference between nv_{o1} and nv_{o2}, $n\Delta v_o$, is applied to L_{lkg} so that the i_{pri} is steeply increased with the slope of $n\Delta v_o/L_{lkg}$. The primary current is expressed as follows:

$$i_{pri}(t) = i_{pri}(t_0) + \left(n\Delta v_o(t)/L_{lkg}\right)(t-t_0). \quad (1)$$

In the conventional TTFB, since the remained primary current on the leakage inductor circulates through S_1 and S_3, it causes high conduction losses. However, the remained primary current in the proposed converter is reduced rapidly, so that the circulating energy during the freewheeling state can also be reduced and the commutation between rectifier diodes is finished softly in this mode.

Mode 2 (t_1~t_2): When i_{pri} is equal to the magnetizing current of T_2, i_{m2}, at t_1, mode 2 begins. Since secondary current transferring through T_2, i_{Do2}, becomes 0, the commutation between D_{o1} and D_{o2} is completed. That is, T_2 acting as a forward type transformer before t_1 is changed to play a role as an inductor in the primary side. Therefore, it has an effect to determine the primary current. Then again, since D_{o2} is conducting, T_1 transfers the differential current between the i_{pri} and the magnetizing current of T_1, i_{m1}, to the secondary side through D_{o2}. As a result, i_{Do2} charges v_{o2} and discharges v_{o1}, respectively, which makes the voltage ripple of output voltage-doubler. The i_{pri}, i_{m1}, i_{Do1}, v_{o1}, and v_{o2} are expressed as follows:

$$i_{pri}(t) = i_{m2}(t) = i_{pri}(t_1) + \left(nv_{o1}(t)/L_{m2}\right)(t-t_1) \quad (2)$$

$$i_{m1}(t) = i_{m1}(t_1) + \left(nv_{o1}(t)/L_{m1}\right)(t-t_1) \quad (3)$$

$$i_{Do1}(t) = n\left(i_{m2}(t) - i_{m1}(t)\right) \quad (4)$$

$$v_{o1}(t) = v_{o1}(t_1) + \left(i_{Do1}/2C_{o1}\right)(t-t_1) \quad (5)$$

$$v_{o2}(t) = v_{o2}(t_1) - \left(i_{Do1}/2C_{o2}\right)(t-t_1) \quad (6)$$

In addition, since the current flowing through S_3 is already changed to the negative direction, the lagging leg switch S_3 is achieved ZCS operation before S_3 is turned off.

Mode 3 (t_2~t_3): When S_4 is turned on at t_0, mode 3 begins. The input power is transferred to the secondary side through S_1 and S_4. T_1 and T_2 still act as a forward-type transformer and an output inductor, respectively, like the operation at mode 2. Also, T_1 transfers the differential current between the primary current and the magnetizing currents of T_1 to the secondary side, that is, flows through D_{o1}. Moreover, i_{Do1} is split to charging C_{o1} and discharging C_{o2}, respectively. Therefore, v_{o1} and v_{o2} are increased and decreased, respectively, with the slope of $i_{Do1}/2C_{o1}$ like the operation at mode 2. Since D_{o1} is conducting before t_2, v_{o1} is reflected on the primary side of T_1, and V_{in}-nv_{o1} is impressed on $L_{m2}+L_{lkg}$. That is, the primary current is increased with the slope of (V_{in}-nv_{o1})/($L_{m2}+L_{lkg}$). The energy stored on L_{m2} should be discharged through D_{o2} in all modes except for this mode. Therefore, i_{pri}, i_{m1}, and i_{Do1} are expressed as follows:

$$i_{pri}(t) = i_{m2}(t) = i_{pri}(t_2) + \left\{\left(V_{in} - nv_{o1}(t)\right)/L_{m2}\right\}(t-t_2) \quad (7)$$

$$i_{m1}(t) = i_{m1}(t_2) + \left(nv_{o1}(t)/L_{m1}\right)(t-t_2) \quad (8)$$

$$i_{Do1}(t) = n\left(i_{m2}(t) - i_{m1}(t)\right) \quad (9)$$

Mode 3 ends at t_3 when S_1 is turned off.

Mode 4~6 (t_3~t_0'): The operations from mode 4 to mode 6 are the same as previous modes except for the direction of powering through switches S_2 and S_3.

III. ANALYSIS OF THE PROPOSED CONVERTER

A. Voltage Conversion Ratio

For the convenience of the analysis of the steady-state operation, several assumptions are made as follows:

- The dead time between the same lag switches is discarded.
- The primary current, i_{pri}, and the voltage across C_{o1} and C_{o2}, v_{o1} and v_{o2}, are increased and decreased linearly.
- The output capacitor C_o is large enough to be considered as a constant voltage source V_o.

By imposing the volt-second balance rule on the each magnetizing inductor L_m, the steady state equation can be obtained as

$$V_o = \left(2D_{eff}/n\right)V_{in} \quad (10)$$

As employing the output voltage-doubler, the proposed converter has a doubled voltage conversion ratio and makes a half voltage rating of the output rectifier diodes. The rated voltage across the output rectifier diodes is expressed as follows:

$$V_{Do1} = V_{in}/n \quad (11)$$

B. Analysis of the Soft-Switching by the Load Current and magnetizing inductance

The magnetizing inductor of the series-connected two transformers acts alternately as a forward transformer and a flyback transformer. As a result, each magnetizing inductor has magnetizing current offset in proportion to the output current, I_o, like a flyback transformer. The $i_{m2}(t)$ swings with a current ripple, Δi_m, biased a positive DC offset (I_o/n), while $i_{m1}(t)$ swings with Δi_m biased a negative DC offset (-I_o/n). Also, the primary current, i_{pri}, should swing only between $i_{m2}(t)$ and $i_{m1}(t)$, and follow them alternately. The ripple current of magnetizing inductor, Δi_m, is expressed as follows:

$$\Delta i_m = (V_{in} - nV_o/2)D_{eff}T/L_m = nV_o(1 - D_{eff})T_s/L_m \quad (12)$$

Moreover, the ZVS of the leading leg can achieved easily because the large magnetizing inductor with its current can discharge output capacitor of them easily. However, the soft-switching of lagging leg can be determined by the magnetizing current offset and ripple. Fig. 4 shows i_{pri}, $i_{m1}(t)$, and $i_{m2}(t)$ during the lagging leg transition, t_3~t_5. From this figure, the peak current, $i_{m1}(t_5)$, has an effect on the soft-switching of lagging leg, which can be represented as a function of the load current and magnetizing current ripple as follows:

$$i_{m1}(t_5) = \Delta i_m/2 - i_o/n \quad (13)$$

1384

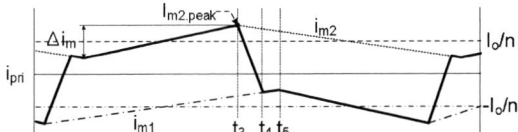

(a) ZCS operation of the lagging leg by the heavy load and large magnetizing inductance

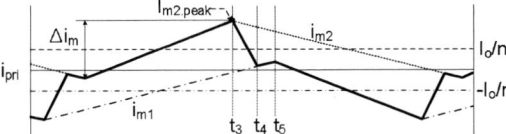

(b) ZVS operation of the lagging leg by the light load and small magnetizing inductance

Figure 4. Soft-switching analysis by load current and magnetizing inductance

C. ZCS conditon for the lagginge leg switches

As shown in Fig. 4(a), the heavy load condition or large magnetizing inductor can contribute to achieve the ZCS operation of lagging leg. Since $n\Delta v_o$ is impressed on the L_{lkg} during $t_3 \sim t_4$, the primary current decreases rapidly and the current flowing through the lagging leg is also changed to the opposite direction so that ZCS of lagging leg can be achieved. The ZCS operation means that the primary current flows through the lagging leg in opposite direction of powering when the lagging leg switches are turned off. That is, to achieve the ZCS of lagging leg, $i_{m1}(t_4)$ is a negative value of the powering current and $i_{pri}(t_3)$ should decrease to $i_{m1}(t_4)$ within $(0.5-D_{eff})T_s$. These mathematical conditions are expressed as follows:

$$i_{pri}(t_5) = i_{m1}(t_5) < 0 \qquad (14)$$

$$\Delta t_{34} = t_4 - t_3 < (0.5 - D_{eff})T_s \qquad (15)$$

where $\Delta t_{34} = \dfrac{L_{lkg}C_{o1}(4L_m + n^2 R_o (0.5-D_{eff})T_s)}{n^2(L_m T_s + L_{lkg}C_{o1}R_o)}$

$i_{pri}(t_3) = i_{m2.peak} = (2I_o + n\Delta i_m)/(2n)$

$i_{pri}(t_4) = -i_{m2}(t_3) + (0.5nV_o(0.5T_s + \Delta t_{34}))/L_m$

Consequently, to achieve the ZCS operation of lagging leg, the capacitances of output voltage-doubler, C_{o1} and C_{o2}, are restricted as follows:

$$C_{o1} \leq \left(n^2(0.5-D_{eff})T_s^2\right)/\left(4L_{lkg}\right) \qquad (16)$$

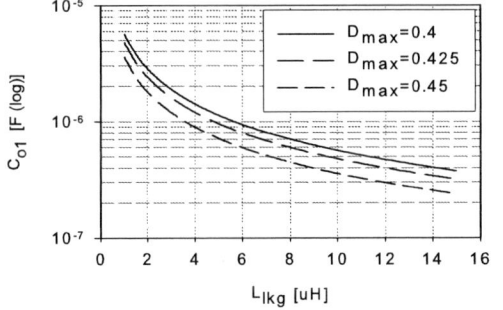

Figure 5. Maximum capacitance of the output voltage-doubler along the duty and leakage inductance for the ZCS operation lagging

Fig. 5 shows the maximum capacitance of the output voltage-doubler as a function of the duty and leakage inductance.

D. ZVS condition for the lagging leg switches

Based on the specifications of PSPM, it is very difficult for the conventional PSFB converter to achieve the ZVS of lagging leg only by a leakage inductor because of the low current and high voltage in the primary side of the transformer even at a full load. However, as shown in Fig. 4(b), the light load condition or small magnetizing inductor of the proposed converter can achieve the ZVS operation of lagging leg. Namely, when Δi_m, is larger than I_o/n, it is possible to achieve the ZVS of the lagging leg by the large magnetizing inductor. Furthermore, although the magnetizing inductor of the transformer is too large, ZVS of the lagging leg is achieved naturally when the load current decreases. These mathematical conditions for ZVS operation are expressed as follows:

$$i_{m1}(t_5) = i_{pri}(t_5) = 0.5\Delta i_m - I_o / n > 0 \qquad (17)$$

$$\left(4C_{oss}V_{in}^2\right)/3 \leq 0.5(L_{lkg} + L_m)i_{pri}^2(t_5) \qquad (18)$$

Therefore, the proposed converter is designed so as to achieve the ZVS of lagging leg at less than 50% of the load condition.

E. Maximum Voltage Ripple of the Output-Voltage-Doubler

For the convenience of the estimation of the maximum voltage ripple of the output voltage-doubler, it is assumed that constant I_o charges and discharges the output voltage-doubler. The maximum voltage ripple of the output voltage-doubler is obtained as follows:

$$\Delta v_o = v_{o2}(t) - v_{o1}(t) = \left(I_o T_s\right)/\left(2C_{o1}\right) \qquad (19)$$

As obtained differential voltage from eq.(19), Δv_o is transferred to the primary side with the turn ratio, n, of the transformer and impressed on L_{lkg}. As a result, the primary current is decreased rapidly and commutation of output rectifier diodes is completed softly. Meanwhile, these operations are rather significant and fast compared with the conventional schemes, because the turn ratio makes an applied voltage on L_{lkg} far larger.

Furthermore, although the output voltage-doubler has a voltage ripple, constant output voltage, V_o, is preserved. This is because the each output-voltage-doubler has contrary voltage ripple and the output capacitor, C_o, is enough large to remove the voltage ripple.

IV. EXPERIMENTAL RESULTS

The prototype of the proposed converter shown in Fig.3 is implemented with the specifications of V_{in}=385V_{dc}, V_o=210V_{dc}, rated power P_o=425W, L_{lkg}=8μH, L_{m1}=L_{m2}=404μH, C_{o1}=C_{o2}=330nF, C_o=470μF/250V, transformer turn ratio n=1.5, $S_1 \sim S_2$= IRF840, $S_3 \sim S_4$=IRG4PC30FD, D_{o1}=D_{o2}=10ETF06, and switching frequency =100kHz. The experimental key waveforms of the proposed converter at the full load are shown in Fig.6 and 7. Fig. 6(a) shows the voltages and currents of the switches S_2 and S_4. It can be seen in this figure that the ZVS turn-on of leading leg switch S_2 and ZCS turn-off of lagging leg switch S_4 are achieved at full load. The primary voltage of each transformer and primary current are shown in Fig. 6(b). Fig.7(a) shows the output voltage, the voltage of the output-voltage-doubler, and the voltage and current across the output rectifier. As described previously, voltage ripple has no place in the output voltage. Fig.7(b) shows that the ZCS of S_4 at 100% load and ZVS of S_4 at 50% load. From this figure, the ZCS of S_4 is achieved at larger than the 50% load, while the ZVS of S_4 is achieved at less than 50% load. Therefore, the soft-switching of S_4 is obtained along the whole load range. From these figures, it can be seen that all waveforms well coincide with the theoretical analysis in section II.

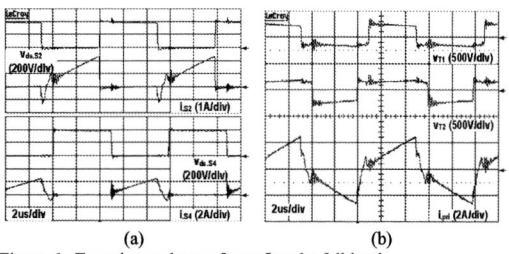

Figure 6. Experimental waveforms I at the full load
(a) $v_{ds.S2}$, i_{S2}, $v_{ds.s4}$, and i_{S4} at the full load
(b) i_{pri}, v_{T1}, and v_{T2} at the full load

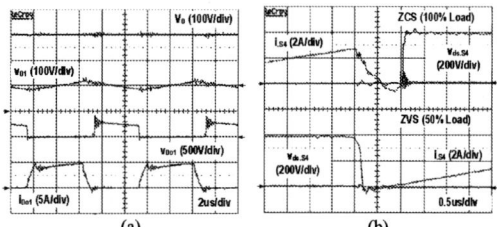

Figure 7. Experimental waveforms II
(a) V_o, v_{o1}, i_{Do1}, and v_{Do1} at the full load
(b) ZCS of S_4 at 100 % load and ZVS of S_4 at 50 % load

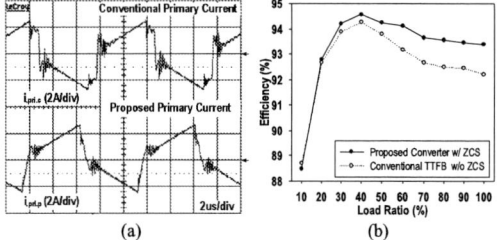

Figure 8. Comparison of primary current and measured efficiencies
(a) Primary current of the proposed converter and conventional TTFB at the full load
(b) Measured efficiency under the load variation

From Fig.8(a), the primary current of the proposed converter is compared with that of the conventional TTFB with the same magnetizing inductance. As shown in this figure, the primary current of the proposed converter is less than that of conventional converter during the freewheeling state due to the ZCS operation. Fig.8(b) shows the measured efficiency of the proposed converter is compared with that of conventional TTFB according to the load condition. The maximum efficiency of the proposed converter is as high as 94.5%, which improves about 2% over the conventional converters at the full-load condition.

V. CONCLUSIONS

In this paper, the zero-voltage switching and zero-current switching two-transformer full-bridge converter using the output-voltage-doubler has been proposed. Using the voltage ripple between the output-voltage-doubler, ZCS operation of lagging lag is achieved and circulating current in freewheeling state is reduced at the full load. Moreover, when the load current decreases, the soft-switching of lagging leg is changed to the ZVS operation. In addition, the commutation between output rectifier diodes is finished fast and the output-voltage-doubler structure also reduces the rated voltage across the output rectifier diodes.

A prototype has been designed to prove the validity of the proposed converter. The experimental results of the prototype converter have been presented for the specifications of 385 V_{dc} input, 210 V_{dc} output, and 420 W power. The measured efficiency is as high as above 93.5% over the wide load range and the maximum efficiency comes up to 94.5% at a rated load condition, which shows higher efficiency than a conventional converter. Therefore, the proposed converter is expected to be suitable for the high output voltage/power applications, e.g, PSPM because of the high reliability, low noise, simple structure, and high efficiency.

REFERENCES

[1] L.H.Loveday, H.Mweene, C.A.Wright, and M.F.Schlecht, "A 1kW 500 kHz Front-End Converter for a Distributed Power Supply System", *IEEE Trans. On Power Electronics*, Vol.6, No.3, 1991, pp.398-407.

[2] Gwan-Bon Koo, Gun-Woo Moon and Myung-Joong Youn, "Analysis and Design of Phase Shift Full Bridge Converter With Series-Connected Two Transformers", *IEEE Trans. on Power Electronics*, Vol.19, No.2, 2004, pp.411-419.

[3] J. G. Cho, C. Y. Jeong, and C. Y. Lee, "Zero-voltage and zero-current-switching full-bridge pwm converter using secondary active clamp", *IEEE Trans. on Power Electronics*, Vol.13, No.4, 1998, pp.601-607.

[4] E. S. Kim, K. Y. Joe, M. H. Kye, and Y. H. Kim, "An improved ZVZCS pwm FB DC/DC converter using the modified energy recovery snubber", in *Proc. IEEE PESC'97*, 1997, pp.227-232.

[5] X. Ruan and Y. Yan, "An improved phase-shifted zero-voltage and zero-current switching pwm converter", *in Proc. IEEE APEC'98*, 1008, pp. 811-815.

2006 5th International Power Electronics and Motion Control Conference

A Single-stage Boost-Flyback PFC Converter

Zhao Qinglin[1], Wen Yi[1], Wu Weiyang[1] and Chen Zhe[2]

[1] Institute of Electrical Engineering, Yanshan University, Qinhuangdao, Hebei, P. R. China
[2] Institute of Energy Technology, Aalborg University, Aalborg, Denmark
E-mail: wenyi19811219@hotmail.com

Abstract—This paper presents a novel single-stage single-switch power factor correction (PFC) converter. The proposed topology is derived by combining a boost cell and a flyback cell into one power stage. In this converter, the transformer in the flyback cell has two-coupling primary windings, which have the same turns. Two bulk storage capacitors are used to store the power energy from the input source, and then feed one primary winding of the transformer respectively. In addition, the two capacitors can absorb the leakage inductances energy and clamp the voltage of power switch. Compared with using one bulk capacitor, the voltage across each capacitor here is lower, only half of the line voltage. This converter will potentially have low switch voltage stress, good regulation capability and high efficiency. To verify the performance of the proposed converter, a design example is given with its experimental results.

Keywords-power factor correction (PFC); single-switch; single-stage; power factor

I. INTRODUCTION

With the development in advanced power semiconductor devices, more and more switching power supplies are used in modern power system. Due to their nonlinear behavior, distorted currents will be introduced from the line, resulting in high total harmonic distortion and low power factor [1], [2].

To deal with this problem, active power factor correction (APFC) techniques are used to achieve low THD and high power factor in most applications. Generally, the APFC techniques can be categorized into single-stage PFC scheme and two-stage PFC scheme according to their system configurations [3]. The two-stage PFC scheme is used most commonly [4], [5], in which a power factor correction circuit is placed in front of a dc/dc converter. The two stages can be controlled independently by using two controllers, and thus both stages can be optimized. Due to its characteristic, the two-stage scheme has such superior performance as high power factor, low current harmonics, and fast output voltage regulation. Nevertheless, it has obvious drawbacks. The complicated power stage topology and control circuits result in high cost and large size, particularly in low power applications. To overcome these shortcomings of two-stage scheme, many feasible single-stage circuits have been developed [6]-[10]. For the single-stage scheme, it

combines the PFC circuit and the dc/dc converter into one power stage, and typically uses only one controller and share power switches. Due to its simplified circuit configuration, the one-stage scheme makes it attractive in low cost and low power application. Unfortunately, this scheme suffers from such drawbacks as limited regulation capability in respect that the power switch performs both PFC and regulation purposes, as a result, the voltage on storage capacitor varies with the load and line variation.

A converter, which consisting of a boost cell and a flyback cell, is proposed in the paper. This converter utilizes two bulk storage capacitors to store power energy from input source, and then feed the two primary windings of the flyback transformer. In addition, the two capacitors can absorb the leakage inductances energy and clamp the voltage of power switch at the sum of two capacitors voltages. The experimental results for a 40W converter at a constant switching frequency of 50kHz have been obtained to verify the performance of the proposed converter.

The topology of this converter and the principle of operation are introduced in the next section. In section III, the design procedure of this converter will be given. Section IV presents the experimental results which are provided to verify the converter operation. Finally, a conclusion is given in section V.

II. CIRCUIT TOPOLOGY AND CIRCUIT OPERATION

The circuit topology of the proposed converter is shown in Fig.1. The boost cell is formed by diode D, inductor L_{boost} and power switch S. C_1 and C_2 are the same bulk storage capacitors, which store energy from input source, and then serve as a source of the flyback cell. The flyback cell is formed by transformer T and diodes (D_1, D_2, D_4), L_1 and L_2 represent the two-coupling primary windings inductances of flyback transformer and these two primary windings have the same turns, the secondary winding inductances is represented by L_3. Diodes D_1, D_2 are placed

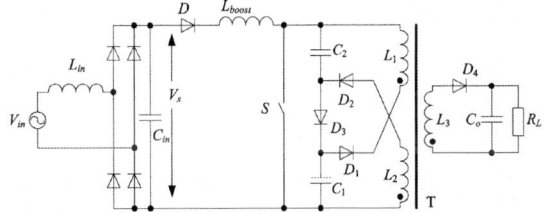

Figure 1. The circuit topology of the proposed converter

1-4244-0448-7/06/$25.00 ©2006 IEEE

here to avoid transformer on forward state in switching-off times.

The ideal waveforms and equivalent circuits of the proposed converter are depicted in Fig.2 and Fig.3.

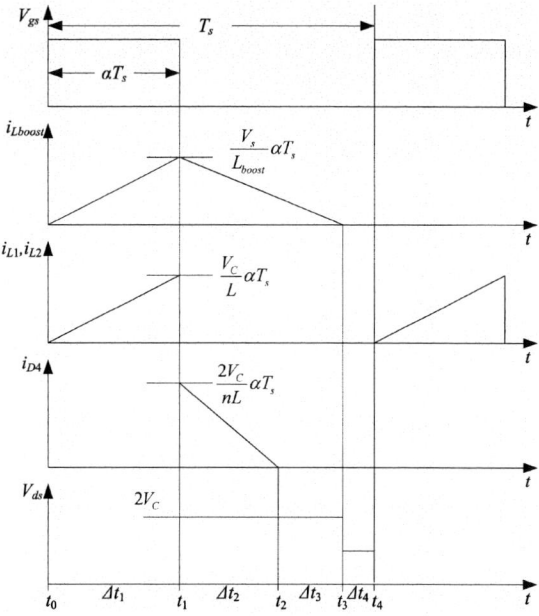

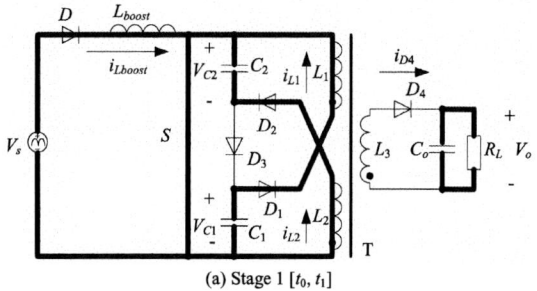

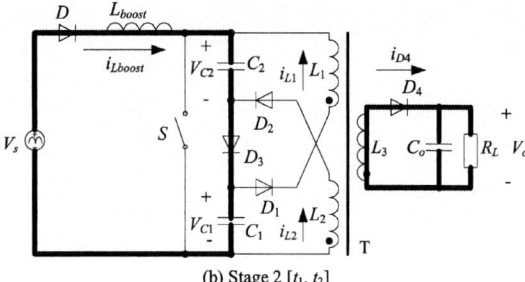

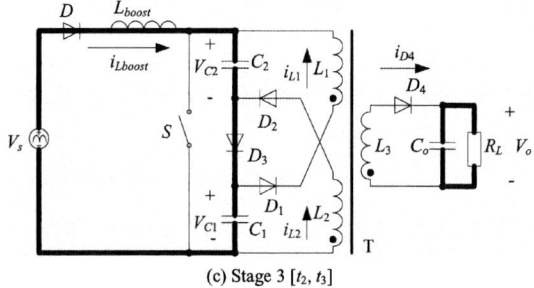

Figure 2. The ideal waveforms of the proposed converter

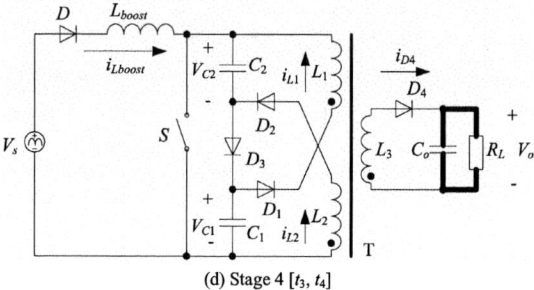

(d) Stage 4 [t_3, t_4]

Figure 3. Equivalent circuits of the four stages

According to Fig.2, in steady state, the converter has four operation stages during one switching cycle. Here is assumed that all semiconductors are ideal and the leakage inductances are neglected.

Because the switching frequency is much greater than the line frequency, the line voltage is assumed to be constant during the switching period. In the equivalent circuits, the line voltage is represented by V_s ($V_s = |V_{in}| = |V_{IN} \sin \omega t|$), which is a rectified sinusoidal voltage. Inductances of the transformer windings are represented by L_1, L_2, and L_3 ($L_1 = L_2 = L$), the turn ratio of primary windings referring to secondary winding is considered as 1: 1: n, so L_3 is $n^2 L$. Capacitors C_1 and C_2 are designed to be large enough and equal. In the steady state analysis, voltage cross each capacitor serves as a dc source of flyback cell ($V_{C1} = V_{C2} = V_C$).

The four stages of operation are discussed as follow:

(a) Stage 1 [t_0, t_1]

The stage as shown in Fig.3 (a) begins with the switch S turning on at $t = t_0$. As diode D conducting, the source voltage V_s is applied to inductor L_{boost}, because the voltage V_s is assumed to be constant, therefore the current i_{Lboost} increases linearly. During this stage, the inductor L_{boost} stores energy from input source. On the other hand, as the sources of flyback cell, the voltages on capacitors C_1 and C_2 are applied to windings inductors L_1 and L_2 respectively, so the currents through both inductors increase linearly, accordingly energy is stored in both inductors from capacitors. The induction voltage of secondary winding makes diode D_4 suffer from reverse bias voltage, so it is blocked, which results in no energy being transferred from the source to the load. This stage ends when the power switch turns off at $t = t_1$. In the stage the following relations hold:

$$i_{L_{boost}}(t) = \frac{V_s}{L_{boost}} \cdot (t - t_0) \qquad t \in [t_0, t_1] \quad (1)$$

$$i_{L_1}(t) = i_{L_2}(t) = \frac{V_{C_1}}{L_1} \cdot (t - t_0) = \frac{V_{C_2}}{L_2} \cdot (t - t_0) = \frac{V_C}{L} \cdot (t - t_0)$$

$$t \in [t_0, t_1] \quad (2)$$

The duration of this stage is:

$$\Delta t_1 = t_1 - t_0 = \alpha \cdot T_s \quad (3)$$

where α is the duty cycle.

(b) Stage 2 [t_1, t_2]

This stage is shown in Fig.3 (b), at $t=t_1$ the power switch is turned off, and diode D_3 is turned on, the voltage of the power switch is clamped by the voltages of capacitors C_1 and C_2 at $2V_C$. The source energy and the energy stored in inductor L_{boost} will be transferred to storage capacitors C_1 and C_2, so the energy loss of capacitors during the stage 1 will be recovered. As the flyback cell, when the power switch turns off, the currents i_{L1} and i_{L2} decrease to zero immediately and energy stored in primary windings of transformer will be all transferred to the secondary winding, here the induction voltage of secondary winding makes diode D_4 conduct, thus energy will be transferred to load. As shown in Fig.2, the current i_{D4} which through diode D_4 decreases linearly in this stage. At $t=t_2$, when i_{D4} decreases to zero, in other words, the energy stored in primary windings during stage 1 has been all transferred to load, this stage ends. The function of diodes D_1 and D_2 here is to avoid transformer on forward state. In this duration, some relations can be described as follow:

$$i_{L_{boost}}(t) = \frac{V_s}{L_{boost}} \cdot \alpha T_s - \frac{2V_C - V_s}{L_{boost}} \cdot (t - t_1) \quad t \in [t_1, t_3] \quad (4)$$

$$i_{D_4}(t) = \frac{1}{n}\left[i_{L1}(t_1) + i_{L2}(t_1)\right] - \frac{V_o}{L_3} \cdot (t - t_1)$$

$$= \frac{2V_C}{nL} \cdot \alpha T_s - \frac{V_o}{n^2 L} \cdot (t - t_1) \quad t \in [t_1, t_2] \quad (5)$$

The duration of this stage is:

$$\Delta t_2 = t_2 - t_1 = \frac{2n}{V_o} \cdot V_C \cdot \alpha T_s \quad (6)$$

(c) Stage 3 $[t_2, t_3]$

This stage is shown in Fig.3 (c). In this stage, the current through inductor L_{boost} continues to decrease linearly, capacitors C_1 and C_2 are still charged. Energy is prohibited from transfering to primary windings by diodes D_1, D_2 here. This stage ends at $t=t_3$ when inductor current i_{Lboost} reaches zero. The relation of i_{Lboost} is the same as (4).

The duration of this stage is:

$$\Delta t_3 = t_3 - t_2 = \left(\frac{V_s}{2V_C - V_s} - \frac{2n}{V_o}V_C\right) \cdot \alpha T_s \quad (7)$$

(d) Stage 4 $[t_3, t_4]$

This stage is shown in Fig.3 (d). This stage is known as a free wheeling stage, which is used for regulation purpose, it ends at $t=t_4$ when power switch is turned on.

The duration of this stage is:

$$\Delta t_4 = t_4 - t_3 = T_s - \left(\frac{2V_C}{2V_C - V_s}\right) \cdot \alpha T_s \quad (8)$$

The leakage inductances of transformer windings here are neglected in analysis of operation principle described above. In fact, the leakage inductances must exist. The current waveforms of the two primary windings are shown in Fig.4. During the period of power switch turning on as stage 1 described above, energy is transferred to primary windings of transformer from capacitors C_1 and C_2, and

some energy must be stored in leakage inductances. When power switch is turned off, energy stored in leakage inductance of L_1 will be delivered to capacitor C_2 through diodes D_1, D_3 quickly, and energy stored in leakage inductance of L_2 will be delivered to capacitor C_1 through diodes D_2, D_3 too. So the currents through primary windings decrease dramatically as shown in Fig.4.

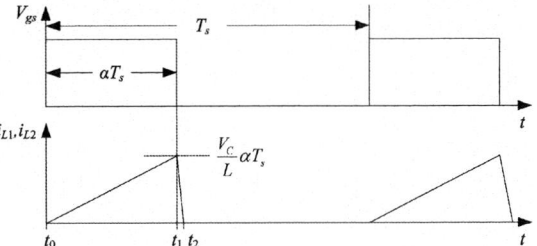

Figure 4. The current waveforms of the two primary windings when leakage inductances are considered

III. THE DESIGN PROCEDURE OF THE PROPOSED CONVERTER

A. Determine the maximal duty ratio α_{max} and design the inductor L_{boost}

In order to assure a high power factor, the inductor L_{boost} must operate in discontinuous conduction mode (DCM). In all range of rectified sinusoidal input voltage, the elementary condition of the inductor L_{boost} operating in DCM is that inductor L_{boost} operates in the critical state of continuous conduction mode (CCM) and DCM at $V_{in}=V_{IN}$.

Because the switching frequency is much greater than the line frequency, the line voltage is assumed constant during the switching period, so at $V_{in}=V_{IN}$, α_{max} can be obtained from (4) as follow:

$$\alpha_{max} = 1 - \frac{V_{IN}}{2V_C} \quad (9)$$

From (1), $I_{Lboost,max}$ is obtained:

$$I_{Lboost,max} = \frac{V_{IN}}{L_{boost}} \cdot \alpha_{max} T_s \quad (10)$$

According to the power balance, input power must be equal to output power. The equation is given by (11),

$$\frac{I_{Lboost,max} V_{IN}}{4} \approx \frac{P_o}{\eta} \quad (11)$$

where η is the efficiency of the converter.

L_{boost} is given by (12), which is derived from (10), (11),

$$L_{boost} \approx \frac{V_{IN}^2 \eta}{4P_o} \alpha_{max} T_s \quad (12)$$

B. Design the transformer

As described above, V_{C1} and V_{C2} are the sources of primary winding L_1 and L_2 respectively, so the inductances of each primary winding L is determined by:

$$L = \frac{V_{C,min} V_C \alpha_{max}^2 T_s}{P_o} \quad (13)$$

1389

The turn number of each primary winding can be determined by:

$$N_{pri} = \frac{V_{C,\min}\alpha_{\max}T_s}{B_{\max}A_e} \tag{14}$$

Because the flyback transformer is operating in DCM, the turn number of secondary winding is determined by:

$$N_{sec} = \frac{N_{pri}V_o\left(1-\alpha_{\max}\right)}{\sqrt{2}V_{C,\min}\alpha_{\max}} \tag{15}$$

The air gap is determined by:

$$l_{gap} \approx \frac{0.4\pi L_{pri}I_{L,\max}^2}{A_e B_{\max}^2}\times 10^8 \tag{16}$$

where $I_{L,\max}$ can be obtained from (2).

Design here, some specifications are given as follow:

$V_{in,rms} = 220\text{V-}50\text{Hz}$, $P_o = 40\text{W}$, $V_o = 15\text{V}$

$f_s = 50\times10^3\,\text{Hz}$, $\eta = 78\%$, $B_{\max} = 0.15\text{T}$

And the desired range of voltage V_C is 180V-220V.

According to the design procedure, the final selection of devices and parameters are:

$L_{boost} = 1.97\times10^{-3}\text{H}$

D, D_1, D_2, D_3: MUR8100

D_4: MBR3020. C_1, C_2: 220u/250V

C_o: 1000u/35V. S: IXTH25N60

Transformer core is: EI33, N_{pri}=46, N_{sec}=10, l_{gap}=0.06cm.

IV. EXPERIMENTAL RESULTS

In order to verify the performance of the proposed converter, a prototype of 40W, 220V AC RMS to 15V DC converter operating at 50kHz was built. The parameters and devices are described above.

Experimental waveforms are depicted in Fig.5. From Fig. 5 (a), it can be seen that the waveform of input current is also an approximate sine wave, which indicates this converter has a high power factor. The line voltage and inductor L_{boost} current, voltage of the power switch and driver signal, windings currents and driver signal , are shown in Fig.5 (b), (c), (d) respectively.

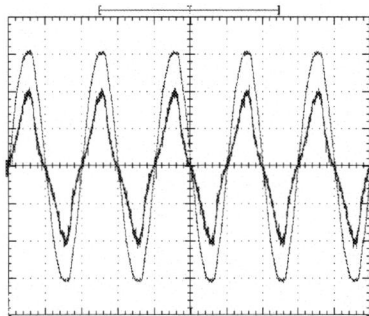

(a) Input voltage (blue) and input current (red)

(Y-axis:100V/div, 0.2A/div, X-axis:10ms/div)

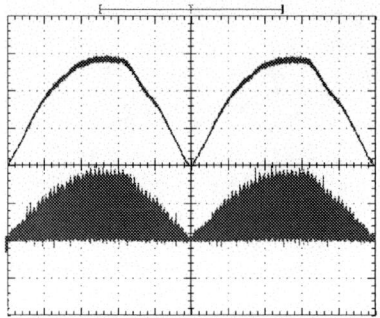

(b) Line voltage (red) and the inductor L_{boost} current (blue)

(Y-axis: 100V/div, 0.5A/div, X-axis:2ms/div)

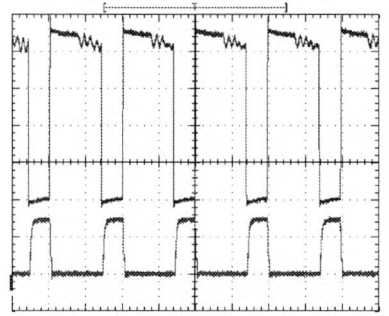

(c) Voltage of the power switch (red) and driver signal (blue)

(Y-axis: 100V/div, 10V/div, X-axis:10us/div)

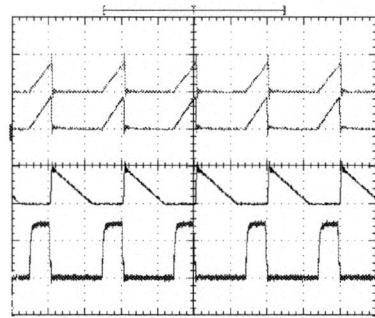

(d) L_1 (green), L_2 (blue) L_3 (black) currents and driver signal (red)

(Y-axis: 1A/div, 1A/div, 10A/div, 10V/div, X-axis:10us/div)

Figure 5. Experimental waveforms of the proposed converter

Fig.6 (a) shows the measured power factor, THD and efficiency against output power at V_{in}=220V RMS. Fig.6 (b) shows the measured power factor, THD and efficiency against input voltage at full load. It can be seen from Fig.6 that this converter obtains power factor of about 0.98, THD of 18.5% and efficiency of 79.01% under full load. Fig.7 (a) shows the measured voltage on capacitors C_1 and C_2 against output power, Fig.7 (b) shows the measured voltage on capacitors C_1 and C_2 against input voltage. From Fig.7 it can be known that the voltage on storage capacitor varies little with either the output power or the input voltage variation, so this converter has better regulation capabilities.

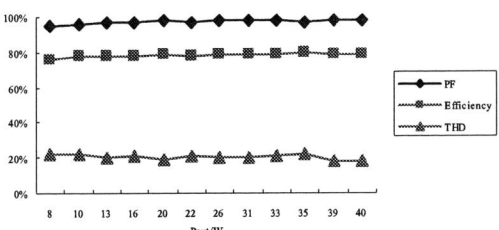

(a). Measured PF, efficiency and THD against output power

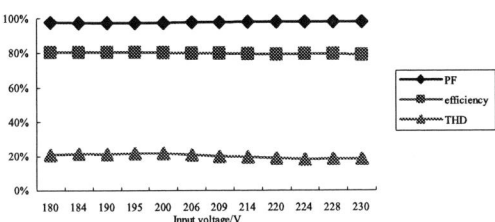

(b). Measured PF, efficiency and THD against input voltage

Figure 6. Measured PF, efficiency and THD

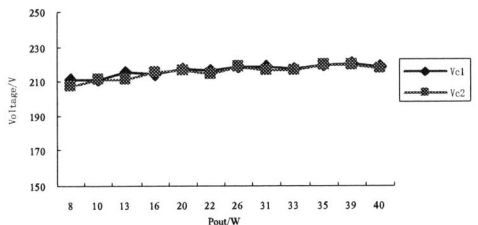

(a). Measured voltages on capacitors against output power

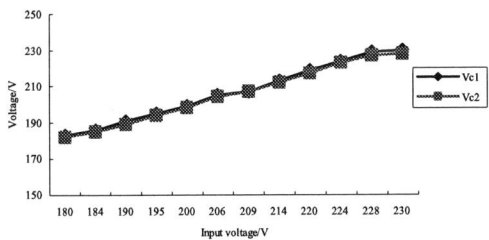

(b). Measured voltages on the capacitors against input voltage

Figure 7. Measured voltages on the capacitors

V. CONCLUSION

A single-stage single-switch power correction (PFC) converter has been presented in this paper, which combines a boost cell and a flyback cell into one power stage. The converter has only one power switch to achieve both power factor correction and regulation purposes.

Two storage capacitors are used to improve the performance, so the converter can obtain high power factor, low switch voltage stress, good regulation capabilities and high efficiency. The experimental results of a 40W converter at a constant switching frequency of 50kHz show that it achieves power factor of about 0.98, THD of 18.5% and efficiency of 79.01% under full load.

ACKNOWLEDGMENT

This work was supported by the National Natural Science Foundation of China, NO. 50237020.

REFERENCES

[1] J. S. Lai, D. Hurst, and T. Key, "Switch-mode supply power factor improvement via harmonic elimination methods," *Applied Power Electronics Conference and Exposition, 1991. APEC '91*, pp. 415-422.

[2] R. Redl, P. Tenti, and J. D. Van WYK, "Power electronics' polluting effects," *IEEE Spectrum*, pp. 32-39, May 1997.

[3] M. Orabi, and T. Ninomiya, "A unified design of single-stage and two-stage PFC converter," *Power Electronics Specialist Conference, 2003. PESC '03*, vol. 4, pp. 1720-1725.

[4] M. Orabi, and T. Ninomiya, "Novel nonlinear representation for two-stage power-factor-correction converter instability," *Industrial Electronics, 2003. ISIE '03*, vol. 1, pp. 270-274.

[5] M. Orabi, and T. Ninomiya, "Study of alternative regimes to analyze two-stage PFC converter," *Applied Power Electronics Conference and Exposition, 2004. APEC '04*, col. 3, pp. 1488-1494.

[6] Chongming Qiao, and K. M. Smedley, "A topology survey of single-stage power factor corrector with a boost type input-current-shaper," *Applied Power Electronics Conference and Exposition, 2000. APEC 2000*, vol. 1, pp. 460-467.

[7] A. Uan-Zo-li, F. C. Lee, and R. Burgos, "Modeling, analysis and control design of single-stage voltage source PFC converter," *Industry Applications Conference, 2005*, vol. 3, pp. 1684-1691.

[8] A. Lazaro, A. Barrado, J. Pleite, and E. Olias, "New power factor correction AC/DC converter with reduced storage capacitor voltage," *IECON 02*, 2002, vol. 1, pp. 353-358.

[9] M.H.L Chow, Yim-Shu Lee, and C.K. Tse, "Single-stage single-switch isolated PFC regulator with unity power factor, fast transient response, and low-voltage stress," *IEEE Trans. Power Electron*, vol. 25, pp. 156-163, Jan. 2000.

[10] W.Y. Choi, J.M. Kwon, H.-L. Do, and B.-H. Kwon, "Single-stage half-bridge converter with high power factor," *Electric Power Applications*, vol. 152, pp. 634-642, May 2005

2006 5th International Power Electronics and Motion Control Conference

Control Bifurcation in PFC Boost Converter under Peak Current-Mode Control

Yi-Jing Ke[#], Yu-Fei Zhou and Jun-Ning Chen
Department of Microelectronics, Anhui University, Hefei, Anhui, China
[#]kehehe@163.com

Abstract—**Modulating reference current with a compensating ramp is a traditional method to stabilize the operation of the peak current-mode controlled dc/dc Boost converter. In this paper, this method is applied to a PFC Boost converter for gaining the stable operation target. Time-domain waveforms obtained by computer simulations are provided to illustrate the control results. Analytical investigations confirm the results achieved by simulations, and the control parameter can also be calculated or predicted assuring the stable operation of PFC Boost converter.**

Keywords- power-factor-correction; peak current-mode control; bifurcation; instability

I. INTRODUCTION

Due to avoid unpredictable or sometimes undesirable consequences in systems, the control of bifurcation has now become an important topic. In particular, much research in this topic has been directed to the suppression or prevention subharmonic operations since conventional engineering designs always put "stability" and "reliability" as the top priorities.

Power-factor-correction (PFC) converters are widely used in power supplies for pre-regulating of power factor [1]. Generally speaking, any type of switching converters can be the candidate for PFC purpose. But in practical the Boost converter has been the favorable and popular choice when taking into account the factor of current stress and efficiency. As a typical nonlinear circuit system, PFC Boost converters are recently revealed to exhibit fast-scale instability, such as bifurcation and chaos operation, over the time of line cycle. These complex behaviors implying instability should be avoided from the viewpoint of traditional design principles, which can be realized by the changing of circuit parameters, or enclosing the accessional control method when the circuit parameters are fixed.

In this paper, we will consider a non-feedback control method, which adds a compensating ramp to the reference current, for controlling bifurcation in a peak current-mode controlled PFC Boost converter. This method of ramp compensation is original aimed at the stable design of

current-mode controlled Boost dc-dc converter, and which are also reexamined in the light of "avoiding bifurcation" [2,3]. In our studies follow, it is proved that the bifurcation control method of ramp compensation is also valid in controlling fast-scale instability in PFC Boost converter, and with analysis, the effective compensating ramp assuring stable operation can be concluded, which is expressed as a compact formula, based on which the influence on the effective compensating ramp inducing by some circuit parameters are discussed.

II. SYSTEM DESCRIPTION

The circuit schematic of PFC Boost converter under study is presented with the peak current-mode control as shown in Fig.1. The closed-loop system has an outer voltage loop and an inner current loop. The voltage loop provides the reference for the inner current loop. In this control configuration, the inductor current i_L is chosen as the programming variable and is compared to the reference current i_{ref} in order to generate the switching signal for switch S. The switch S is turned on periodically by the clock, and off according to the output of a comparator that compares the inductor current i with a current reference i_{ref}. Specifically, while the switch is on, the inductor current climbs up, and as it reaches i_{ref}, the switch is turned off, thereby causing the inductor current to ramp down until the next clock comes. Thus the action of the current-mode control prescribed before forces the peak of input current (inductor current) to follow cycle-by-cycle a predefined waveform proportional to the input voltage. Let v_{in} be the input voltage. If $v_{in} = \hat{v}_{in} \sin(2\pi f_m t) = \hat{v}_{in} \sin(\theta)$, where $\hat{v}_{in}$ is the peak input voltage, f_m is the line frequency, and θ is the phase angel varying from 0 to π corresponding to the time varying from 0 to half line cycle, which is defined as:

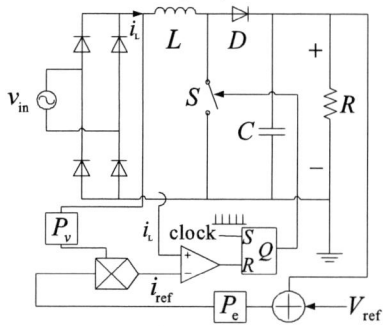

Figure 1. Circuit Schematic of PFC Boost converter.

This work was supported by the National Natural Science Foundation of China (60402001) and the Natural Science Research Foundation of the Education Committee of Anhui Province(Grant No.2005KJ054)

1-4244-0448-7/06/$25.00 ©2006 IEEE

$$\theta = 2\pi f_{\mathrm{m}} t = \omega_{\mathrm{m}} t \qquad (1)$$

Then the predefined waveform for the input current (reference current), i_{ref}, will be follow if the power factor approaches one,

$$i_{\mathrm{ref}} \approx \hat{i}_{\mathrm{L}} \sin(\theta) \qquad (2)$$

where $\hat{i}_{\mathrm{L}}$ is the peak inductor reference current. The peak of the input current (inductor current) will always follow the pre-set envelope, i_{ref}. Therefore, a near unity power factor is maintained. Usually, the switching frequency is much higher than the line frequency. This condition is assumed throughout the paper.

The PFC Boost converter described afore is revealed to exhibit instable operation of bifurcation and chaos [4]. We rebuilt the system by computer simulations performed with MATLAB and SIMULINK environment. The circuit parameters are selected as shown in Table 1 [4]. From the simulation waveforms in Fig.2 we find the fast-scale instability over the time of half line cycle, where the upper is the waveform of inductor current, while the lower is the same waveform sampled at the switching instant. Actually, this fast-scale instability show to be a period-doubling bifurcation with time increasing and decreasing.

By investigating Fig.2, two critical points corresponding to the first bifurcation from period-1 to

period-2 subharmonic can be clearly identified. Between these two points the inductor current follows accurately the sinusoidal shape of the reference current i_{ref}. We also note from Fig.2 that the converter operate chaotically near the two ends of the half line cycle. In the follow, we will consider the method of compensation ramp, which can decrease the two critical bifurcation points, and the bifurcation-free operation can be achieved when these two critical bifurcation points are decreased below 0, thus also achieve the purpose of bifurcation control.

III. BIFURCATION CONTROL IN PFC BOOST CONVERTER UNDER A CONVENTIONAL PEAK CURRENT-MODE CONTROL

It is well know that the operation of peak current-mode controlled dc/dc Boost converter becomes unstable when the duty cycle (designed steady-state value) exceeds 0.5. The usual remedy in practical is to modulate the reference current with a compensating ramp, as show in Fig.3. The compensating ramp used here is reexamined and considered to be a measure for controlling bifurcation. Here we will prove that this compensating slope originally coping with dc-dc converters still have its effect in the bifurcation-control of PFC Boost converter, and the effective compensating slope assuring stability of the converter can also be predicted by the analysis we introduced.

The critical duty cycle D_{c}, at which the first period-doubling bifurcation occurs, can be obtained by [2, 3]

$$D_{\mathrm{c}} = \frac{M_{\mathrm{c}} + 0.5}{M_{\mathrm{c}} + 1} \qquad (3)$$

where,

$$M_{\mathrm{c}} = \frac{m_{\mathrm{c}} L}{V_{\mathrm{in}}} \qquad (4)$$

and m_{c} is the negative slope of i_{ref} when i_{L} reaches i_{ref}. It is given by

$$m_{\mathrm{c}} = -\frac{di_{\mathrm{ref}}}{dt} \qquad (5)$$

Then from (3) and (4), we have

$$D_c = 1 - \frac{0.5 V_{\mathrm{in}}}{m_{\mathrm{c}} L + V_{\mathrm{in}}} \qquad (6)$$

A conclusion can be gained from (6) that the larger m_{c} we have, the bigger D_{c} we get. So in order to make the Boost

TABLE I.
CIRCUIT PARAMETERS USED IN SIMULATIONS

Component/Parameters	Values
Input voltage v_{in}	$110\sin(100\pi t)$ V
Reference output voltage V_{ref}	220 V
Inductance L	2 mH
Capacitance C	470 µF
Load resistance R	135 Ω
Switching period T_s	20 µs
Feedback gain P_e	1/60
Gain P_v	0.08

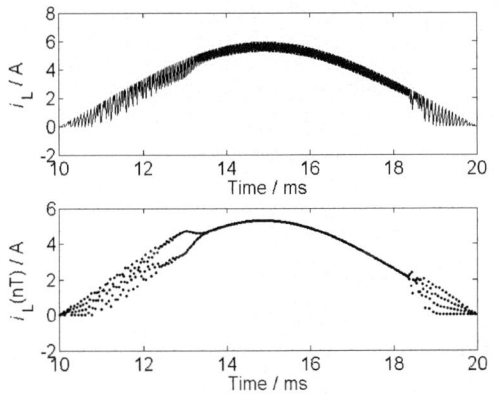

Figure 2. Stimulated inductor current time-domain waveform (upper) and same waveform sampled at the switching frequency (lower) with fast-scale instability.

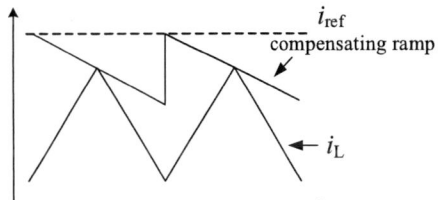

Figure 3. Inductor current with ramp compensating.

Converter operate stably with a bigger D_c, we should increase m_c.

The former conclusion achieved aiming at dc-dc Boost converter is the same as that achieved aiming at PFC Boost converter. Here the original current reference is given in (2), and the compensated current reference will be:

$$i_{ref}^* = \hat{i}_L \sin(\theta) - V_{ramp} \quad (7)$$

where V_{ramp} is the compensating ramp signal with slope being k, and thus the corresponding m_c of i_{ref}^* will be recalculated as:

$$m_c = -\frac{di_{ref}^*}{dt} = -\omega_m \hat{i}_L \cos\theta + k \quad (8)$$

When $k > 0$, current reference is compensated by a ramp with slope being positive, as the occasion shown in Fig.3. The value of m_c is increased in (8) compared with that in (6), so the critical value of duty cycle, D_c, is enlarged with the presence of this compensating ramp. That is to say, the compensating ramp with a positive slope can provide more margins for the converter keeping away from bifurcation of the operation state.

We show the former occasion of modulating the compensating ramp to current reference in Fig.4. Specifically, the frequency of the compensating ramp is the same as switching frequency, i.e. the frequency of clock.

Fig.5 is the control results with k selected as 1.1. From the waveforms of i_L and the sampled i_L, it is found that the converter operates in the bifurcation-free mode in the whole line cycle. The magnified waveforms of inductor current i_L and the compensated current reference i_{ref}^* are shown in Fig.6, from which the similar scenery is found as that in Fig.3 for the occasion of dc-dc Boost converter. The inductor current i_L touch its reference i_{ref}^* at a point with negative slope, which makes the converter operate more stable, accordingly avoid the emergence of bifurcation.

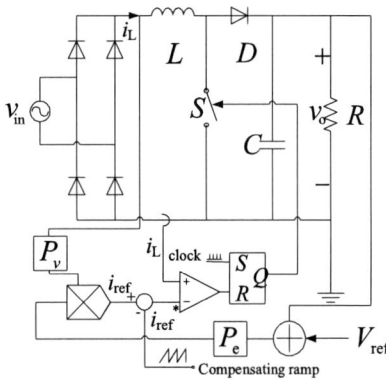

Figure 4. Circuit schematic of PFC converter with compensating ramp.

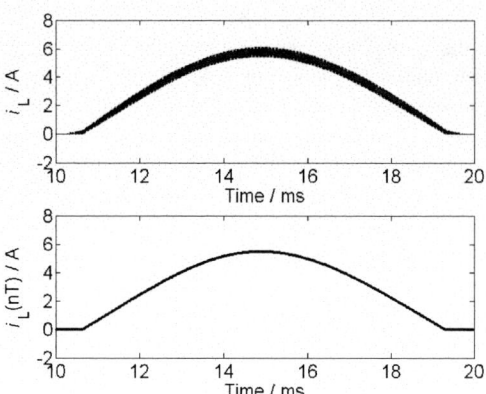

Figure 5. Simulated time-domain waveform of inductor current and the corresponding waveform sampled at the switching instant with bifurcation-free operation.

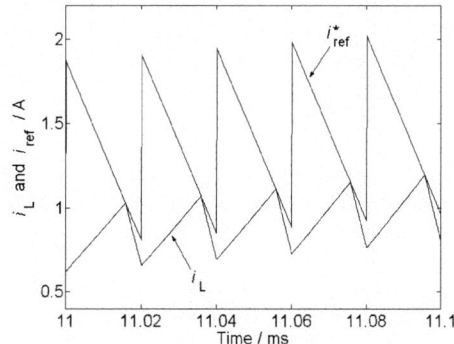

Figure 6. Closer look of the inductor current and reference current.

IV. THEORETICAL ANALYSIS

The aim of this section is to investigate how much the slope of the compensating ramp we should have to guarantee the stable operation of the PFC Boost converter. For analytical brevity, we utilize the definition of phase angle in (1) to indicate the critical bifurcation point. From (6) and (8) we obtain

$$D_c = 1 - \frac{0.5\hat{v}_{in}\sin\theta}{kL - \hat{i}_L \omega_m L \cos\theta + \hat{v}_{in}\sin\theta} \quad (9)$$

Note that the input-output voltage conversion ratio of Boost converter in continuous-conduction mode is [5]

$$\frac{V_{ref}}{\hat{v}_{in}\sin\theta} = \frac{1}{1-D} \quad (10)$$

So we can get

$$V_{ref} = 2kL - 2\hat{i}_L \omega_m L \cos\theta + 2\hat{v}_{in}\sin\theta \quad (11)$$

It can be concluded from Fig.2 that the PFC converter has asymmetric regions of stability at the left-hand and right-hand sides of one half line cycle. Specifically, the left-hand side shows to bifurcate with a rather large value of input voltage, while in the right-hand side bifurcation occurs with a rather small value of input voltage, i.e., the left-hand side is more unstable corresponding to the right-hand side, which is the subsequence induced by the slope of the input voltage v_{in}. Therefore our main task is to control the bifurcation existing in the left-hand side with a appropriate compensating ramp signal, and this compensating ramp of course is satisfied to the control of bifurcation in right-hand side. Consequently we conclude that if the bifurcation point of the left-hand side is decreased to $\theta = 0$, the PFC converter will operate in a bifurcation-free mode, i.e., the converter is stable in the overall line cycle. Based on this guidance, (11) would be follow with θ equaling 0,

$$V_{ref} = 2\left(kL - \hat{i}_L \omega_m L\right) \tag{12}$$

And the slope k of the compensating ramp demanded for the bifurcation control purpose can be found as:

$$k = \frac{V_{ref}}{2L} + \hat{i}_L \omega_m \tag{13}$$

Moreover, incorporating the power equality, i.e. $\hat{V}_{in}\hat{i}_L = 2V_{ref}^2/R$ (assuming 100% efficiency), k can be written as

$$k = \frac{V_{ref}}{2L} + \frac{2\omega_m V_{ref}^2}{R\hat{V}_{in}} \tag{14}$$

Thus, the corresponding amplitude A of the compensating ramp demanded can be calculated by

$$A = T_s k = T_s \left(\frac{V_{ref}}{2L} + \frac{2\omega_m V_{ref}^2}{R\hat{V}_{in}} \right) \tag{15}$$

To confirm the validity of (15) in control bifurcation of PFC Boost converter, we compare the values of A calculated from (15) with those found by tremendous simulations, as shown in Fig. 7, from which we can find they are consistent basically in values and all demand larger A with V_{ref} increasing. Actually, many parameters of the circuit will affect the effective values of A demanded for the control of bifurcation. We have summarized in Figs. 8 the dependence of the choice of the effective A of compensating ramp upon peak input voltage $\hat{V}_{in}$, load R, and inductor L.

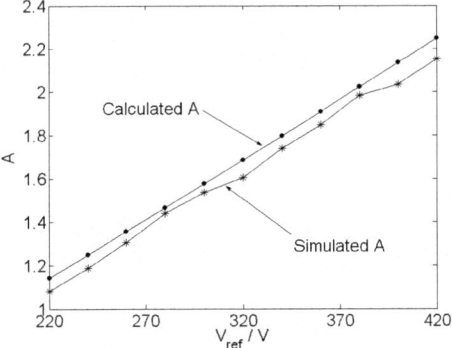

Figure 7. The value of A found by simulations and those obtained by analysis.

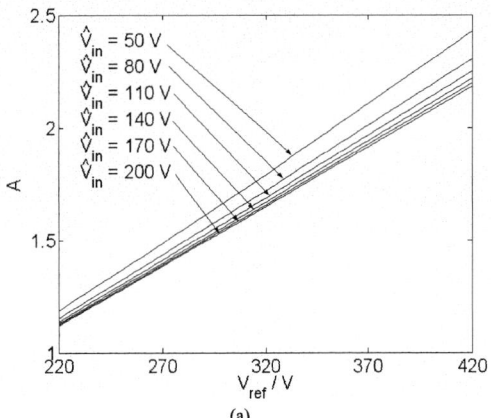

(a)

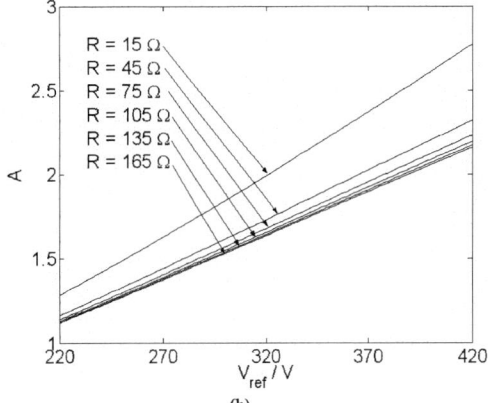

(b)

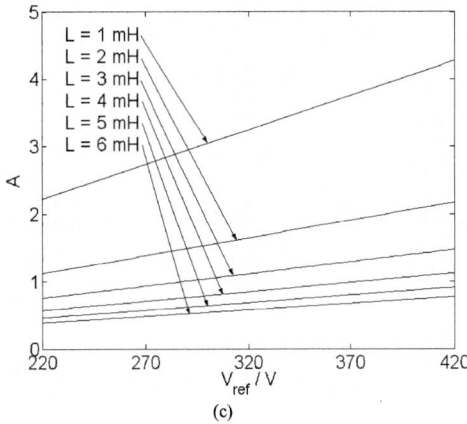

(c)

Figure 8. The amplitude A of the compensating ramp needed for bifurcation control versus V_{ref}. (a) with $\hat{V}_{in}$ = 50, 80, 110, 140, 170, 200 V; (b) with $R =$ 12, 16, 20, 24, 28, 32 Ω; (c) with $L = 1, 2, 3, 4, 5, 6$ mH.

V. CONCLUSION

PFC Boost converter under a conventional peak current-mode control is a kind of nonlinear system, which is reveal to exhibit complex fast-scale instability. Hence, to avoid this instability, the bifurcation control measure should be enclosed. In this paper a simple non-feedback bifurcation control method, i.e. modulating reference current with a compensating ramp, which is widely used in peak current-mode controlled dc/dc Boost converter, is applied to stabilize the operation of PFC Boost converter. Compared to the feedback control methods, the non-feedback chaos control methods are usually easier to apply, as demonstrated in the Boost PFC converter studied in this paper before. Furthermore, with the analysis we introduced here, the slope, and also the amplitude of the compensating ramp, can be calculated or predicted for guaranteeing the stable operation of the PFC Boost converter, which can therefore provide guidance for the practical design.

REFERENCES

[1] R. Redl, "Power-factor-correction in single-phase switching-mode power supplies — an overview," *Int. J. Electron.*, vol. 77, no. 5, pp. 555–582, 1994.

[2] C. K. Tse and Y. M. Lai, "Control of bifurcation in current-programmed DC/DC converters: a reexamination of slope compensation," *IEEE ISCAS,* Geneva Switzerland, pp. I-671–674, June 2000.

[3] C. K. Tse and Y. M. Lai, "Control of bifurcation," in *Nonlinear Phenomena in Power Electronics: Attractors, Bifurcations, Chaos, and Nonlinear Control*, S. Banerjee S. and G. Verghese, Eds. New York: IEEE Press, 2001, pp. 418–427.

[4] O. Dranga, C. K. Tse, H. H. C. Iu and I. NAGY, "Bifurcation behaviour of a power-factor-correction Boost converter", *Int. J. of Bifurcation and Chaos*, vol. 13, no. 10, pp. 3107–3114, 2003.

[5] A. S. Kislovski, R. Redl and N.O. Sokal, "Dynamical analysis of switching mode DC/DC converter", New York: Van Nostrand Reinhold, 1996.

2006 5th International Power Electronics and Motion Control Conference

Analysis and Design of One-Cycle-Controlled Dual-Boost Power Factor Corrector

Yue-feng Yao   Yuan-rui Chen

Electric Power College, South China University of Technology, Guangzhou, China

Email: yrchen@scut.edu.cn   yuefengyao@163.com

Abstract-In this paper, analysis and design for a dual-boost single-phase active power factor corrector with one cycle control (OCC) is presented. The dual boost topology combines rectification and power factor correction (PFC) together. Only one of the two power switches is operated respectively during each positive or negative half line cycles. The controller employs the one-cycle-control strategy, which features great simplicity and excellent stability. Stability analysis and some guidelines for the selection of the circuit parameters in practical application are provided in the paper. The proposed topology can achieve unit power factor theoretically and be very suitable for medium to high power applications. The theoretical analysis is verified by simulation or experiments based on a 400W OCC-PFC prototype.

Keywords-Power Factor Correction (PFC); One Cycle Control (OCC); Dual-Boost Converter; Power Quality Control; Rectifier

I INTRODUCTION

The traditional diode or thyristor type rectifiers with bulky capacitor filter usually draw pulse current from power lines, which results in poor power factor due to both the phase displacement and harmonic distortion factors. The harmonics not only lower the energy efficiency, but also lead to the harmful disturbance to the equipments connected to the line. In order to limit the total harmonic distortion (THD) of electronic equipments and improve the power quality, the power factor correction techniques must be considered.

The conventional single switch boost converter has been used widely in active power factor correction [1-3], which gets good effect. In article [4], a dual boost topology is presented, which also can be used in power factor correction.

In article [5], a one cycle control (OCC) technique is presented, the OCC-based PFC rectifier can be used in various power systems [6-8]. The OCC controller usually uses an integrator with reset to force the controlled variables to meet the control goal in each pulse width modulation (PWM) cycle. In this paper, a one cycle

This research work was supported in part by the key project of National Natural Science Foundation of China under Grant No.60534040, and in part by the Natural Science Foundation of Guangdong under Grant No.04020011.

controlled dual-boost PFC converter is presented, which can be implemented by two kinds of modulation way, one is trailing edge modulation, the other is leading edge modulation. Steady state design and stability analysis are detailed in the paper. All analyses are verified by simulation or experiments based on a 400W prototype.

II OPERATION PRINCIPLE ANALYSIS

A. Dual-Boost Topology Analysis

The dual-boost topology is shown in Fig.1. The two switches are driven by the same driver block, however, only one switch operates during the positive half cycle and the other, negative half cycle. When the input voltage source of the converter is in the positive half cycle, i.e., $V_g>0$, the parasitic body diode of switch S_2 is turned on for the entire half line cycle, so switch S_1 operates. Fig.2 shows the equivalent circuit during the positive half cycle. Similarly, when the input voltage source of the converter is in the negative half cycle, i.e., $V_g<0$, the equivalent circuit is illustrated in Fig.3.

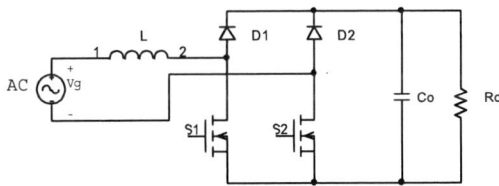

Figure 1.   Topology of a dual boost converter

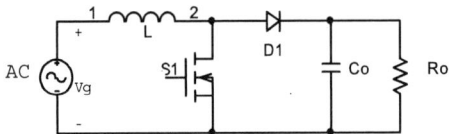

Figure 2.   The equivalent circuit when $V_g>0$

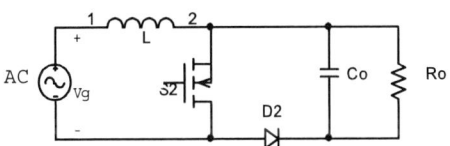

Figure 3.   The equivalent circuit when $V_g<0$

1-4244-0448-7/06/$25.00 ©2006 IEEE   1397

B. Control Circuit

During the positive half cycle ($V_g>0$), the inductor voltage V_L in each PWM cycle is given as follows:

$$\begin{cases} V_L=V_g, & \text{while } 0<t\leq dT_s, \text{ S}_2 \text{ is switched on;} \\ V_L=V_g-V_o, & \text{while } dT_s<t\leq T_s, \text{ S}_2 \text{ is switched off} \end{cases}$$

where d is the PWM duty-ratio, T_s is the switching cycle, V_o is the output voltage.

For constant PWM switching frequency operation and quasi-steady-state analysis, the average voltage-second product of inductor L is approximately balanced during each switching cycle, we have

$$V_g \cdot d + (V_g - V_o) \cdot (1-d) = 0 \quad \text{when } V_g>0$$

The similar equation during the negative half cycle can be obtained as

$$V_g \cdot d + (V_g + V_o) \cdot (1-d) = 0 \quad \text{when } V_g<0$$

For the converter under continuous conduction mode (CCM), combination of above two equations results in a relationship between duty ratio of switches and AC input voltage as well as DC output voltage, i.e.,

$$V_{ge} = V_o \cdot (1-d) \cdot \tag{1}$$

where $V_{ge} = \begin{cases} V_g, & V_g > 0 \\ -V_g, & V_g < 0 \end{cases}$

In order to achieve unity power factor, the control goal of PFC is to force the AC input current be in phase with the AC input voltage, i.e.,

$$V_g = R_e \cdot i_g \tag{2}$$

where R_e is the equivalent input resistor of the converter, i_g is the input current.

Define

$$\bar{i}_{ge} = \begin{cases} i_g, & V_g > 0 \\ -i_g, & V_g < 0 \end{cases}$$

then the control goal in (2) can be rewritten as

$$V_{ge} = R_e \cdot \bar{i}_{ge} \tag{3}$$

By substituting (3) to (1) and introducing a new parameter V_m, we obtain a key equation described in (4).

$$R_s \bar{i}_{ge} = V_m \cdot (1-d) \tag{4}$$

where R_s is the equivalent current sensing resistor and V_m is defined by

$$V_m = \frac{R_s V_o}{R_e}$$

The unity power factor can be achieved by controlling the total AC input current to satisfy (4). There are two modulation ways to realize (4): one is trailing edge modulation shown in Fig.4, the other is leading edge modulation shown in Fig.5.

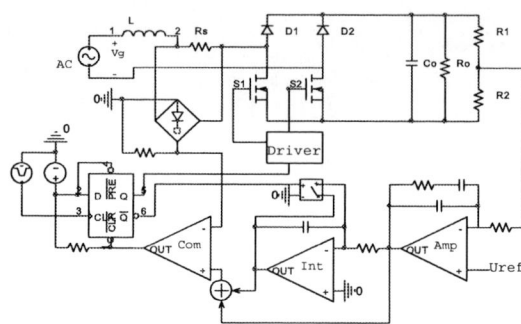

Figure 4. The trailing edge modulation PFC

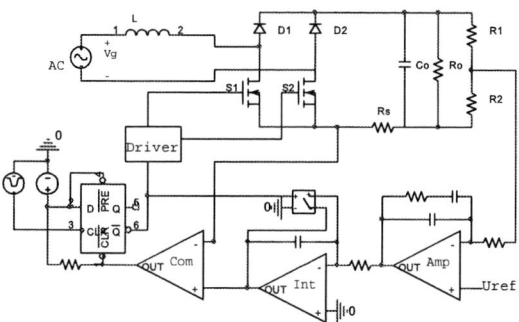

Figure 5. The leading edge modulation PFC

If trailing edge modulation is performed, the signal Q is used to drive the power switch, the duty ratio of the power switch is identical to the duty ratio of Q, the ON time of the switch is controlled. In this case, the input current must be sensed correctly. Because the direction of the input current is uncertain, so the bridge rectifier is required.

If leading edge modulation is performed, the signal $\overline{Q}$ is used to drive the power switch, the power switch duty ratio is identical to the duty ratio of $\overline{Q}$, the OFF time of the power switch is controlled. In this case, the integrator operates while the power switch is OFF and the input current passes through the DC bus. So, the input current can be sensed from the DC bus using a resistor as shown in Fig.5.

From the control schematic we can see that the leading edge modulation is much simpler in structure than the trailing edge modulation. So, the simulation and experiments in this paper will be performed with leading edge modulation.

III DESIGN CONSIDERATIONS

A. Capacitor Design

The output DC-link capacitor of the voltage source converter is determined by the output voltage ripple. The expression is given by

$$C \ge \frac{P_o}{2 \cdot f \cdot (V_{o\max}^2 - V_{o\min}^2)}$$

where P_o is the output power, f is the input voltage frequency, V_{omax} and V_{omin} are the maximum and minimum peak value of the output DC-link voltage ripple, respectively.

For example, suppose the power is 400W; the output voltage is 400V with 2% ripple. The line frequency is 50 Hz. The capacitance is calculated by

$$C \ge \frac{400}{2 \cdot 50 \cdot 400^2 \cdot [(1+0.02)^2 - (1-0.02)^2]} = 312.5uF$$

B. Selection of Inductance

The control key equation to realize the PFC is

$$R_s \bar{i}_{ge} = V_m \cdot (1-d) \quad \text{viz.} \quad \bar{i}_{ge} = \frac{V_m}{R_s} - \frac{V_m}{R_s T_s} dT_s.$$

For leading edge modulation, the operation waveform is shown in Fig.6.

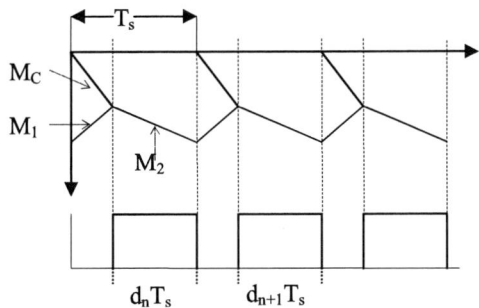

Figure 6.   Operation waveforms for leading edge modulation

Considering that the load current is with low frequency and the influence of load current can be neglected, we only concern the inductor current in the stability analysis.

In each switching cycle, we define

$$M_1 = \frac{V_o - V_g}{L} \; ; \; M_2 = \frac{V_g}{L}; \quad M_C = \frac{V_m}{R_s T_s}$$

where M_1 is the OFF slope of the inductor current and M_2 is the ON slope of the inductor current; M_C is the equivalent slope of the carrier signal $V_m(1-t/\tau)$, which is implemented by an integrator with reset.

In the two continuous cycles, n-th and $(n+1)$-th, the duty ratios are d_n and d_{n+1}, respectively. From Fig.6, the equation below can be derived:

$$M_C(1-d_n)T_s + M_2 d_n T_s = (M_C + M_1)(1 - d_{n+1})T_s$$

So

$$d_{n+1} = \frac{M_1}{M_C + M_1} + \frac{M_C - M_2}{M_C + M_1} d_n$$

Obviously, the duty ratio in each switching cycle can be calculated with the value in last cycle.

Define:

$$\mu = \frac{M_C - M_2}{M_C + M_1}$$

The equilibrium point d^* of the duty ratio can be found by equalizing d^* to $f(d^*)$, where

$$f(d) = \frac{M_1}{M_C + M_1} + \frac{M_C - M_2}{M_C + M_1} d_n.$$

Then :

$$d^* = \frac{M_1}{M_C + M_1},$$

$$d_n = d^*(1 - \mu^n) + \mu^n d_0$$

where d_0 is the duty ratio of cycle 0. The duty ratio converges to the steady-state point only if $|\eta| < 1$, i.e.,

$$M_C > \frac{1}{2}(M_2 - M_1)$$

According to the convergency condition we can get

$$V_m > \frac{R_s T_s}{2L}(2|V_g| - V_o)$$

By definition, we have

$$V_m = \frac{R_s V_o}{R_e}, \quad \frac{|V_g|}{V_o} = 1 - d.$$

So the stability condition for the leading edge modulation for the converter is:

$$d > \frac{1}{2} - \frac{L}{T_s R_e}$$

Suppose the energy efficiency is η. The input power should equal the output power based on energy balance, that is

$$P_{in} \cdot \eta = \frac{V_{grms}^2}{R_e} \cdot \eta = P_o$$

According to the stability condition, the value of the inductor can be calculated as follow:

$$L > (\frac{1}{2} - d) \cdot T_s \cdot R_e > \frac{1}{2} \cdot T_s \cdot \eta \cdot \frac{V_{grms}^2}{P_o}$$

For example, in the case with maximum output power $P_o = 400W$, PWM cycle $T_s = 20\mu s$, max(V_{grms})=265V, and efficiency $\eta = 90\%$, the minimum inductance is calculated as 1.58mH.

IV   SIMULATION RESULTS

The simulation in this paper is carried out with PSPICE. Simulation conditions are as follows: output power is 400W, the root-mean-square value of input voltage is 110V, output voltage is DC 400V, output capacitance is 470uF, inductance is 1.8mH, the current sensing resistance R_s is 0.33 Ohm. The waveforms of input voltage and input current are shown in Fig.7. The voltage of current sensing resistor R_s, carrier signal U_c and the driver signal $\overline{Q}$ are shown in Fig.8. After Fourier analysis, the spectrum of the input current is shown in Fig.9

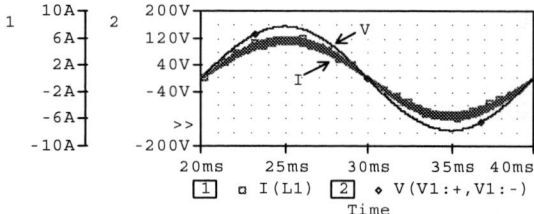

Figure 7. Waveforms of input voltage and input current

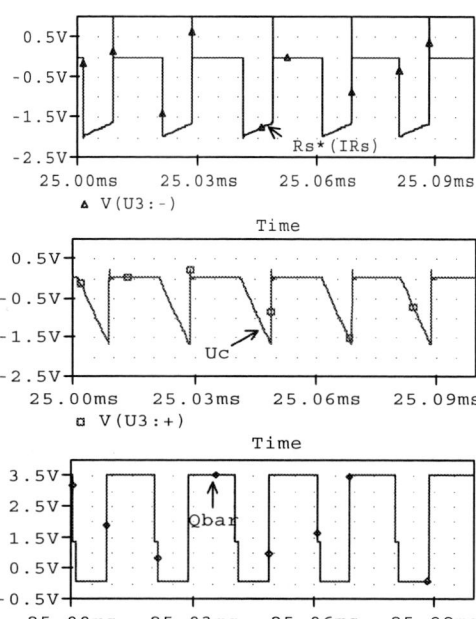

Figure 8. Waveforms of control block

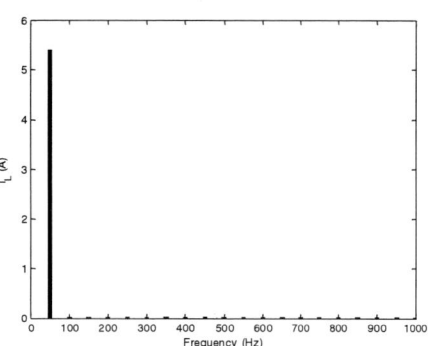

Figure 9. The spectrum of input current

According to the results of the Fourier analysis shown in Fig.9, we can know that the THD of input current is 0.94%, so, the proposed converter can achieve unit power factor approximately. The simulation results verify the analysis in the paper.

V CONCLUSIONS

In this paper, a dual boost PFC converter is presented. The operation principle and stability are analyzed in detail. Some guidelines for the selection of circuit parameters are provided. The theoretical analysis is verified by the simulation and experiments. The proposed converter in this paper can achieve a low total harmonic distortion and a high power factor at low cost.

REFERENCES

[1] Zhansong Zhang and Xuansan Cai, *Switchmode Power Supply Principle and Design*. Beijing: Publishing House of Electronics Industry, pp273-292, 1998.

[2] Zhimin Zhou, Jihai Zhou, and Aihua Ji, *Switchmode Power Supply PFC Design and Applications*. Beijing: People's Posts & Telecommunications Publishing House, Nov. 2004.

[3] Chunjiang Zhang, Xiuhong Zhang, Qinglin Zhao, and Junzhi Feng, "Analysis and Design of a Single Phase PFC with One-Cycle-Control," *Power Electronics*, 2002.

[4] A.Pietkiewicz and D.Tollik, "New high power single-phase power factor corrector with soft-switching," *Telec. Energy Conf.*, Oct. 1996, pp.114-119.

[5] K.M.Smedley and S.Cuk, "One-cycle control of switching converters," *Proc. Power Electronics Specialists Conf.*, 1991, pp. 888-896.

[6] Yaoping Liu and Keyue Smedley, "Control of a Dual Boost Power Factor Corrector for High Power Applications," *IEEE Industrial Electron. Soc.*, 2003, 29th *Annual Conf.*, vol. 3, pp. 2929-2932, Nov. 2003.

[7] Guozhu Chen and Keyue M.Smedley,"Steady State and Dynamic Study of One Cycle Controlled Three Phase Power Factor Correction," *IEEE Trans. Industrial Electron.*, vol. 52, no. 2, pp. 355-362, Apr. 2005.

[8] Chongming Qiao and Keyue M.Smedley, "A Single-Phase Active Power Filter With One-Cycle-Control Under Unipolar Operation," *IEEE Trans. Circuits and Systems-I: Regular Papers*, vol. 51, pp. 1623-1630, Aug. 2004.

2006 5th International Power Electronics and Motion Control Conference

A Novel Single-phase Buck PFC Converter Based on One-cycle Control

Chen Bing , Xie Yun-Xiang , Huang Feng and Chen Jiang-Hui
Electric Power College, South China University of Technology, Guangzhou, China

Abstract—In this paper, a novel single-phase Buck PFC converter based on one-cycle control is presented. In contrast to the conventional Boost PFC converter, some disadvantages of the Boost PFC converter are overcome, and the switching loss and the stress of the main switch are reduced. In addition, with the one-cycle control, the multipliers and the voltage sensors used in the traditional direct current control are eliminated, the control circuit is simple and high efficiency. The power circuit topology and the operation principle of the proposed converter are described in detail in this paper. An experimental example is presented, and the parameters of the power circuit and the control loop are designed and optimized, and the system stability is analyzed. The results of the simulation and experimentation are provided that the power factor is close to unity and the total harmonic distortion(THD) is low, the experimental results verify the feasibility and validity of the power circuit topology and control strategy of the proposed converter.

Keywords-Buck PFC converter; one-cycle control; compensation

I. INTRODUCTION

With significant development of power electronics technology, the plentiful applications of all kinds of power converters have resulted in severe harmonic contamination in power systems. At present, there are mainly two methods including Power Factor Corrector (PFC) and Active Power Filter (APF) which are used to compensate harmonic. The PFC technique is used to improve power quality through updating the structure and control strategy of DC-DC converter itself, which is fit for the fields of the middle/small power. Through the proper control to DC-DC converter, the input current of the uncontrollable rectifier keeps in-phase with the input voltage, to improve the power factor to close to unity. The traditional PFC technique is often based on Boost converter, and the direct current control strategies such as peak current mode control, average current mode control and hysteresis current mode control are commonly adopted. According to these control strategies, the input voltage, the input current and the output voltage of the converter must be detected, whereafter the control purpose can be achieved by the operation of the multiplier. However, in practical application the Boost PFC

converter based on the direct current control presents the following disadvantages [1]:

1) The output voltage exceeds the peak of the input voltage, which results in high switching stress of sequent converter.

2) There is the poor performance during the startup, the overloading and the non-load.

3) The high switching losses and EMI. They are resulted from the great ripple current flowing from the power switch and diode.

4) The nonlinear distortion of the multiplier may increase the input current harmonic.

5) The control circuit is complex and high cost.

In order to overcome these problems, in this paper, a novel single-phase Buck PFC converter based on one-cycle control is presented. The power circuit topology and control block diagram is shown in Fig.1. A power switch (T_2), an inductor (L) and a capacitance (C) are added parallelly between the uncontrollable rectifier and the Buck converter, the one-cycle control strategy is adopted, the additive power switch (T_2) and the power switch (T_1) of Buck converter are turned on complementarily, to meet the demand that the input

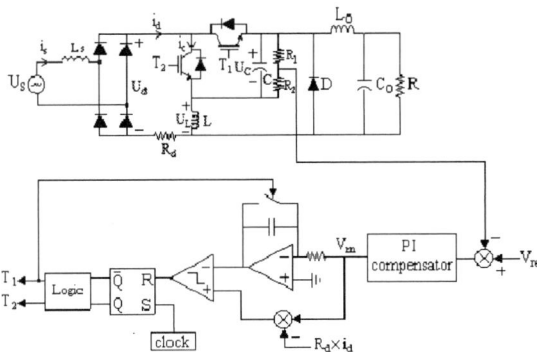

Figure 1.   The power circuit topology and control block diagram of the proposed Buck PFC converter

current of the rectifier tracks the input voltage with a high degree of accuracy, to reduce the input current harmonic distortion, and to achieve the power factor to close to unity. With one-cycle control, the control loop consists of one PI compensator, one integrator with reset, one RS trigger and some logic circuits, compared with the direct current control, three is no need to detect the input

1-4244-0448-7/06/$25.00 ©2006 IEEE

voltage and the output voltage of the Buck converter, the voltage sensors and the multiplier in the control loop are deleted, the control circuit is simple, robust and low cost, and the power circuit topology is new and overcomes the proposed disadvantages of the Boost converter. With proper design, the proposed Buck PFC converter can operate in a wider load range and higher efficiency.

This paper is organized as follows: The operation principles of the power circuit and control scheme are discussed in Section II, The parameters optimization design is analyzed in Section III. A design example is presented, and the results of the computer simulation and the experimental tests are provided in Section IV. Section V gives a conclusion.

II. OPERATION PRINCIPLE

The power circuit topology and control block diagram of the proposed Buck PFC converter is shown in Fig.1, the switch T_1 and T_2 are controlled complementarily, suppose the switching frequency is constant, and it is much more than the system frequency. The output voltage of the rectifier is given by

$$U_d = U_L \qquad 0 \langle\ t\ \langle\ DT_s \qquad T_1\ \text{OFF}, T_2\ \text{ON} \qquad (1)$$

and

$$U_d = U_C + U_L \qquad DT_s \langle\ t\ \langle\ T_s \qquad T_1\ \text{ON}, T_2\ \text{OFF} \qquad (2)$$

For constant-frequency operation and steady-state analysis, the average inductor voltage second is approximately balanced during each switching cycle, that is

$$U_L\ (\text{ON}) \times DT_s + U_L\ (\text{OFF}) \times (1-D)\,T_s = 0 \qquad (3)$$

Substitution of (1) and (2) into (3) yields

$$U_d = (1-D)\,U_C \qquad (4)$$

In order to achieve unity power factor, the AC input current of the rectifier is to obtain sine wave and to track the AC input voltage in-phase. i.e., the waveform and phase of the DC output current i_d of the rectifier are the same with that of the DC output voltage U_d. Therefore, for steady-state analysis, an equivalent resistor R_S is used to emulate the nonlinear load including the compensation network in back of the rectifier, that is

$$U_d = i_d \times R_s \qquad (5)$$

Combination of (4) and (5) yields

$$(1-D)\,U_C = i_d \times R_s \qquad (6)$$

To multiply the two side of equation (6) by a current sensing resistor R_d yields

$$i_d \times R_d = \frac{R_d \times U_C}{R_s} \times (1-D) = V_m \times (1-D) \qquad (7)$$

Define

$$V_m = \frac{R_d \times U_C}{R_s}$$

During each switching cycle T_S, to integrate the two side of equation (7)

$$\frac{1}{T_s} \int_0^{T_s} i_d \times R_d\, dt = \frac{1}{T_s} \int_0^{T_s} V_m \times (1-D)\, dt \qquad (8)$$

V_m and i_d are constant, thus equation (8) can be rewritten as

$$V_m - i_d \times R_d = \frac{1}{T_s} \int_0^{T_s} D \times V_m\, dt \qquad (9)$$

Therefore, in each switching cycle, if the duty ratio D is controlled to meet the equation (9), the equation (5) is realized, the control goal of achieving unity power factor is realized, too. This control strategy is called one-cycle control [2]-[4].

III. DESIGN CONSIDERATIONS

A. Output Filter Inductance Design

It is very important for switch operation safety and efficiency to select reasonably the output filter inductance(L_o). The larger L_o is, the bigger the volume of converter is, and the less the power density is. However, the less L_o is, the larger the current ripple quantity and peak value are, and the larger the output voltage ripple quantity is. The Buck converter can operate in continuous conduction mode(CCM) and discontinuous conduction mode(DCM), which mode is decided according to the output filter inductance(L_o). The boundary inductance who decides the operation mode of the Buck PFC converter can be computed by equation (10) [5], from equation(10), it can be seen that the boundary inductance is relative to the switching period, the DC voltage conversion ration and the load.

$$L = \frac{RT_s}{2} \frac{|\sin \omega t| - M}{2|\sin \omega t|^3} \frac{\pi - 2\sin^{-1} M + 2M\sqrt{1 - M^2}}{\pi}$$

$$(10)$$

where T_S is the switching period, M is the DC voltage conversion ration, and R is the load.

B. Output Filter Capacitor Design

Suppose the output filter capacitor(C_o) is infinite, the output voltage is a constant without ripple. In practice, the alternating current part of the output inductance current flows through the capacitor(C_o), thus the ripple voltage is produced, and it is relative to the capacitor(C_o). In order to achieve that the output ripple voltage is less than normal($\sigma \leq 2\%$), the output filter capacitor can be computed by equation (11).

$$C = \frac{U_d \times T_s^2}{8L \times \Delta U_o} \quad (11)$$

where U_d is the DC input voltage of the converter, ΔU_o is the ripple voltage, L is the output filter inductance, and T_S is the switching period.

C. Compensation Capacitor and Inductance Design

For the proposed Buck PFC converter, it is crucial to determine the compensation capacitor and inductance, which not only decides the correction effect, but also decides the practicability of the proposed converter. The compensation capacitor is determined by the range of the compensation capacitor voltage ripple and the compensation capacity. The compensation capacitor can be computed by equation (12) [6].

$$C > \frac{P}{2f\left(U_{C\max}^2 - U_{C\min}^2\right)} \quad (12)$$

where $U_{C\max}$ and $U_{C\min}$ are the maximum and the minimum of the compensation capacitor voltage, f is the system frequency, and P is the compensation capacity.

In theory, the larger the compensation current change ratio(di_c/dt) is, the better the compensatory effect is. However, the less the compensation inductance is , the larger di_c/dt is, and the larger the compensation current ripple also is, which influences the effect of the compensation. In order to guarantee the effect of the compensation, the minimum of the compensation current change ratio is more than the maximum of the load current change ratio [6], that is

$$\min\left[\max(K_1, K_2)_{T_s}\right]_T \geq |di_l / dt|_{\max} \quad (13)$$

where

$$K_1 = (U_d - U_c)/L \qquad T_1 \text{ ON}, T_2 \text{ OFF} \quad (14)$$

$$K_2 = U_d / L \qquad T_1 \text{ OFF}, T_2 \text{ ON} \quad (15)$$

Substitution of (14) and (15) into (13) yields the formula of computing the compensation inductance.

$$L \leq \min\left[\max(U_d - U_C, U_d)_{T_s}\right]_T / |di_l / dt|_{\max} \quad (16)$$

D. PI Compensator Design

PI compensator is designed as non-error compensator, which is shown in Fig.2, to keep the compensation

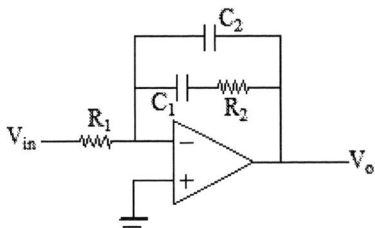

Figure 2. PI compensator

capacitor voltage constant. For designing PI compensator, the crossover frequency(f_c) is taken at one-fifth the switching frequency, a zero frequency(f_z) and a pole frequency(f_p) must be chosen, they will be chosen so that $f_c/f_z=f_p/f_c$, the farther apart f_z and f_p are, the greater the phase margin at f_c is. The phase margin at f_c is $45°$ so that the system is stable. f_z is less than f_c in order to eliminate steady-state error, and f_p is more than half of f_c so as to minimize high-frequency noise spikes. Therefore, from Fig.2, the capacitor C_1, C_2 and the resistor R_2 can be chosen by equation (17) and (18).

$$f_z = 1/2\pi R_2 C_1 \quad (17)$$

$$f_p = 1/2\pi R_2 C_2 \quad (18)$$

PI compensator transfer function is shown by equation (19).

$$H(s) = k\frac{1 + s/w_z}{s\left(1 + s/w_p\right)} \quad (19)$$

IV. SIMULATION AND EXPERIMENTAL RESULTS

According to the design considerations of Section III, the proposed single-phase Buck PFC converter is simulated, and the operation conditions are as follows: the AC input voltage V=110V; the switching frequency f_s=20kHz; the load R=10Ω; the output filter inductance L_o=1mH; the output filter capacitor C_o=470uf; the compensation inductance L=22uH; the compensation capacitor C=150uf. An inductance L_s is linked with the input of the rectifier so as to decrease the input current ripple, and L_s=150 uf; PI compensator design parameters: the crossover frequency f_c=4 kHz; the zero frequency f_z= 53Hz; the pole frequency f_p= 300kHz; R_1=1K Ω ;

R_2=10KΩ; C_1= 300nf; C_2= 53pf. Fig.3 shows the input voltage and current simulation waveforms of the rectifier without compensation. Fig.4 shows the input voltage and current simulation waveforms with compensation. From Fig.3~4, it can be seen that the input current is severe distortion without compensation, however, with compensation, the input current waveform is improved evidently to close to sine wave, and to keep in-phase with the input voltage waveform, so that the power factor is close to unity, and THD is low.

An experimental circuit has been built to verify the analysis and simulation described above. The circuit parameters are all the same as those of simulation. The experimental results are shown in Fig.5~6. They are identical to the simulation results. The theoretical analysis is confirmed by the results of the simulation and the experiment.

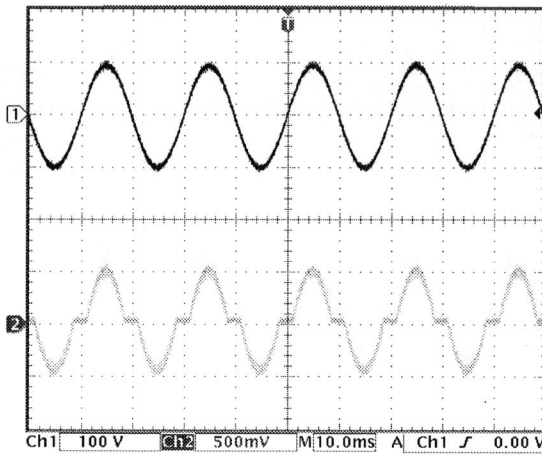

Figure 5. The input voltage and current experimental waveforms of the rectifier without compensation

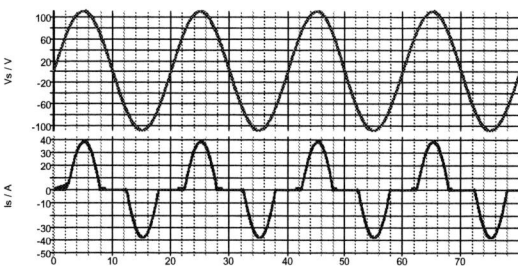

Figure 3. The input voltage and current simulation waveforms of the rectifier without compensation

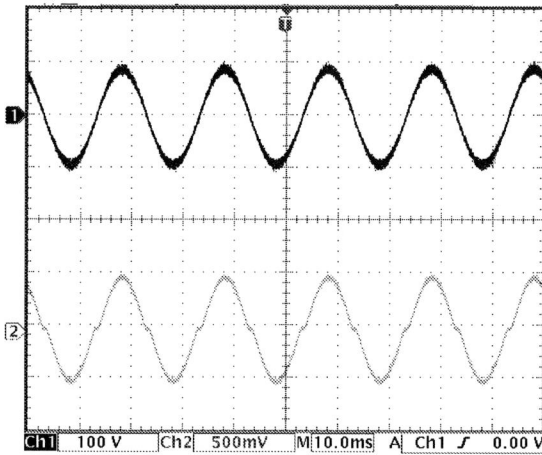

Figure 6. The input voltage and current experimental waveforms with compensation

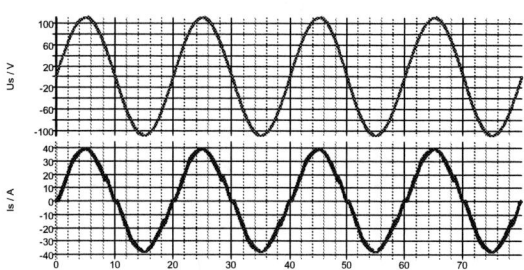

Figure 4. The input voltage and current simulation waveforms with compensation

V. CONCLUSION

In this paper, a new Buck PFC converter topology is presented, the one-cycle control method is adopted, no multipliers and voltage sensors are required in the control loop, an integrator with reset is employed as core component to control the duty ratio, so that the AC input current is sine wave and tracks closely the input voltage, the power factor is close to unity, the total harmonic distortion(THD) is low, this control method is simply and high efficiency. An experimental example is presented in this paper, the operation principle and

parameters design considerations are analyzed in detail. The results of the simulation and the experiment are presented, and the results verify the feasibility and validity of the proposed converter topology and control strategy.

REFERENCES

[1] Deng Chao-ping, Ling Zhi-bin, Yang Xi-jun. Power factor correction of single phase Buck converter in Discontinuous-capacitor-Voltage mode operation. Journal of ShangHai JiaoTong University. vol.38,no.8,pp:1296-1299;Aug. 2004.

[2] Xie PinFang, Du Xiong, Zhou Luowei. One cycle controlled DC side single phase active power filter. Transaction of China Electrotechnical Society. vol.18,no.4,pp:51-55,Aug.2003.

[3] Q.Chongming, K.M.Smedley and F.Maddaleno. A single-phase

active power filter with one-cycle control under unipolar operation. IEEE Trans. Circuit and Systems.vol.51,no.8, pp:1623-1629, Aug.2004.

[4] K.M.Smedley, Z.Luowei and Q.Chongming. Unified constant-frequency integration control of active power filters—steady-state and dynamics. IEEE Trans. Power Electronics, vol.16, no.3,pp:428-436. May.2001.

[5] J.Sebastian, J.A.Cobos, P.Gil and J.Uceda. The determination of the boundaries between continuous and discontinuous conduction modes in PWM DC-to-DC converters used as power factor preregulators. In IEEE Power Electronics Specialists Conf. 1992.pp:1061-1070.

[6] Du Xiong, Zhou Luowei, Xie PinFang. The relationship between compensation performance and main circuit parameter of DC side APF. Proceedings of the CSEE. vol.24,no.11,pp:39-42,Nov. 2004.

Modeling and Simulation of Three Phase High Power Factor PWM Rectifier

Yu Fang[*], Yong Xie[*] and Yan Xing[**]

[*] College of Information Engineering of Yangzhou University, Yangzhou, P.R.China

[**] Aero-Power Sic-tech Center, Nanjing University of Aeronautics and astronautics Nanjing 210016, P.R.China

yufang@nuaa.edu.cn

Abstract—Numerical simulation methodology based on high frequency mathematical model of high power factor PWM rectifier is proposed in this paper. And input to output transform function is deduced. With the help of input to output transform function, the design method of voltage regulator has been presented. The simulation for three-phase PWM rectifier is implemented with Runge-kutta, thus comprehensive simulation can be achieved with M file in MATLAB. Finally, the simulation results and experimental results are given. The complete agreement with the results of simulation and experiment manifests that the proposed simulation method in this paper is effective. Of course, the design period of three-phase high power factor PWM rectifier can be shortened by virtue of proposed simulation methodology.

Key words— power factor correction; SVPWM; Numerical simulation

I. INTRODUCTION

PWM rectifier is main technique applied in large unit power factor. Three phase PWM rectifier can be cataloged in voltage and current types, in which current type rectifier may employ many kinds of conventional PWM techniques, however, input line current is discontinuous and output voltage ripple is much larger. While voltage type topology not only adopt many appropriate control strategies but also implements bidirectional energy flow, and its main merit is that far dynamic response and simple configure. And smaller input filter can achieve lower EMI. Consequently, the voltage type PWM rectifier is usually served as power factor correction.

SVPWM is the best way to suppress harmonics [1-4], and used to adjust motor velocity. Especially it can attain low harmonics even in lower operating frequency. SVM can shape the input voltages of three phase bridge converter into rotating circular voltage vector. With the development of DSP, SVPWM have been easily implemented so far. And another merit employing SVPWM in PWM inverter can achieve 15% more fundamental component among output voltage. Hence, if employed SVPWM technique in PWM rectifier, the use efficiency of DC voltage will be higher close to unit. As the result that the voltage stresses across switches is decreased.

The best way to implement SVPWM is digital control. Due to that DSP have been commercialized, high performance digitalized SMPS come to true. However, it is hard to program to implement controller in DSP, which result in longer development period. This case is remarkable in the design of three phase power factor correction; because in this case many control variables need to control and double closed loops must be accomplished. In face of this point, the numerical simulation methodology based on high frequency mathematical model of high power factor PWM rectifier is proposed in this paper. The simulation program contains the voltage regulator, current controller and the SVPWM, so we can call it comprehensive system simulation for PFC. The total simulation is implemented with M file in MATLAB, hence we only translate the M file into the DSP language to achieve digitalized three phase power factor correction. Consequently the compensation parameters can be optimized in simulation and the design period will be greatly shortened design cost cut. It is clear that the proposed simulation methodology is valuable for fast design of three-phase PFC and may be used for reference to other digitalized SMPS design.

In this paper, the high frequency mathematical model is developed firstly, and then the control to output transformation function is deduced based on instantaneous power theory. It is nature that the voltage regulator is given and the PI regulator is testified to work here. In this way, we write these mathematic models into program in M file. And finally the simulation results and experimental results are given. The complete agreement with the results of simulation and experiment verify that the proposed simulation method in this paper is effective.

II. MODELING FOR PWM RECTIFIER

The model is important way to analyze the operation basic, dynamic and stable state of PWM rectifier. The studied topology is shown in fig.1.

Assuming that：
   (1) idea switches
   (2) ballanced input voltage source

(3) without switching dead time

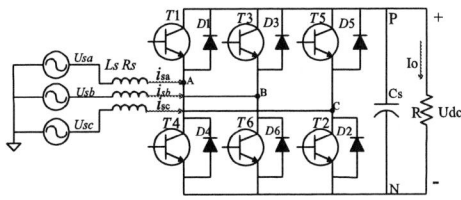

Fig.1 Three-phase PWM VRC topology

$$
\begin{cases}
U_{sa} = E_m \cos(\omega t) \\
U_{sb} = E_m \cos(\omega t - 2\pi/3) \\
U_{sc} = E_m \cos(\omega t + 2\pi/3)
\end{cases}
$$

$$
\begin{cases}
i_{sa} = I_m \cos(\omega t) \\
i_{sb} = I_m \cos(\omega t - 2\pi/3) \\
i_{sc} = I_m \cos(\omega t + 2\pi/3)
\end{cases}
\tag{1}
$$

where E_m (I_m)and ω are amplitude of the phase voltage(current) and angular frequency respectively.

A. High frequency mathematical model in ABC fixed coordinate

As shown in fig.1, U_{sa}、U_{sb}、U_{sc} are input phase line voltages, i_{sa}、i_{sb}、i_{sc} input line currents, R_s is resistence of input rail, L_s boost inductor，C_s fiter capacitor; U_{dc} output voltage，R effective load resistence and I_o load current.

Due to that the complmentary characteristic of switches in the same bridge leg, the following switching function can be defined.

$$
S_i = \begin{cases} 1 & \text{upper switch on in phase i} \\ 0 & \text{bottom switch on in phase i} \end{cases} \quad i=a,b,c \tag{2}
$$

Hence we can get the model：

$$\dot{X} = AX + BU \tag{3}$$

where,

$$
A = \begin{bmatrix}
-R_s/L_s & 0 & 0 & S^*-S_a \\
0 & -R_s/L_s & 0 & S^*-S_b \\
0 & 0 & -R_s/L_s & S^*-S_c \\
S_a/C_s & S_b/C_s & S_c/C_s & 0
\end{bmatrix}
$$

$$B = diag[1/L_s \quad 1/L_s \quad 1/L_s \quad 1/C_s]^T$$

$$X = [i_{sa} \quad i_{sb} \quad i_{sc} \quad U_{dc}]^T$$

$$U = [U_{sa} \quad U_{sb} \quad U_{sc} \quad -I_{do}]^T$$

$$S^* = (S_a + S_b + S_c)/3$$

Eq.（3）shows that each line current is composed of three phase switch function, so PWM rectifier is coupled

nonlinear time variation system. It can be concluded that the neutral point voltage level of output capacitor differs from line grid neutral one in three-phase PWM rectifier without neutral connection. As far as the high frequency is concerned, PWM rectifier is coupled with each phase. Eqs. (2)-(4) are the mathematic mode for three-phase PWM rectifier in high frequency state. Fig.1 can be employed in variable frequency converter, three-phase UPS, APF , SVG and so on. Fig.2 shows the equivalent electrical circuit.

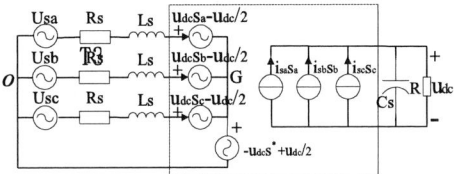

Fig.2 PWM rectifier high frequency equivalent circuit

B. High frequency mathematic model in α-β fixed coordinate

The above gives the high frequency PWM rectifier mathematical model in ABC fixed coordinate, while this section presents corresponding model in α-β fixed coordinate.Expression (5) is translation matrix from three dimension coordinate to two dimensions.

$$
T_{abc/\alpha\beta} = \sqrt{\frac{2}{3}} \begin{bmatrix} 1 & -1/2 & -1/2 \\ 0 & \sqrt{3}/2 & -\sqrt{3}/2 \end{bmatrix} \tag{5}
$$

Hence, the mathematical models in α-β fixed coordinate are as follows:

$$\dot{X}_{\alpha\beta} = A_{\alpha\beta}X_{\alpha\beta} + B_{\alpha\beta}U_{\alpha\beta} \tag{6}$$

where,

$$
A_{\alpha\beta} = \begin{bmatrix}
-R_s/L_s & 0 & -S_\alpha \\
0 & -R_s/L_s & -S_\beta \\
S_\alpha & S_\beta & 0
\end{bmatrix}
$$

$$X_{\alpha\beta} = [i_\alpha \quad i_\beta \quad V_{dc}]^T$$

$$B_{\alpha\beta} = [1/L_s \quad 1/L_s \quad 1/C]^T \tag{7}$$

$$U_{\alpha\beta} = [u_\alpha \quad u_\beta]^T$$

$$
\begin{bmatrix} i_\alpha \\ i_\beta \end{bmatrix} = T_{abc/\alpha\beta} \cdot \begin{bmatrix} i_{sa} \\ i_{sb} \\ i_{sc} \end{bmatrix}
$$

As seen from expression (7), the related variables have been decoupled. i_α、i_β are only effected by corresponding switching function S_α、S_β respectively in α-β fixed coordinate. However, voltage and current are still sinusoidal.

When switching frequency far more than that of line

grid, (4)and (7) the switching function can be replaced with duty cycle of upper switch in one switching period d_k(k=a,b,c), hence, the average model in one switching period can be gotten, i.e. low frequency mathematical model for PWM rectifier. The average model neglects switching process and simplifies PWM rectifier model. (8) presents the relations between control voltages and average switching function.

$$\begin{cases} u_{r\alpha} = S\alpha U_o \\ u_{r\beta} = S\beta U_o \end{cases} \qquad (8)$$

From (6), the output current is shown in (9).

$$i_o = S\alpha i_\alpha + S\beta i_\beta \qquad (9)$$

From (6) and (9), high frequency model for PWM rectifier in α-β fixed coordinate is drawn in fig.3. And output current contains ripple current in α-β fixed coordinate.

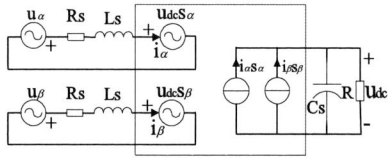

Fig.3 high frequency equivalent circuit in α-β fixed coordinate

III. CONTROL-TO-OUTPUT SIMULATION MODEL

In order to facilitate analysis, the control schematic block diagram is shown in fig.4, and the closed loop system block diagram in fig.5. Control variable is the amplitude of input current i_m^*, controlled variable is output voltage u_{dc}. When in steady state, i_m^* and u_{dc} are constant values I_m^* and U_{dc} respectively. $\hat{i}_m^*$、$\hat{u}_{dc}$ are corresponding small signal disturbance of I_m^* and U_{dc} respectively. $\hat{u}_{dcref}^*$ is output disturbance.

In fig.5 G(s) is control-to-output transfer function, and G_c(s) is voltage regulator model.

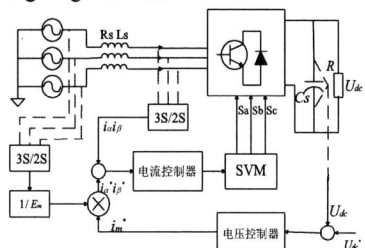

Fig.4 Three-phase HPF PWM rectifier control block diagram

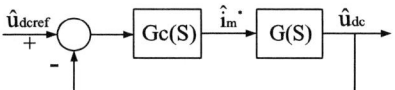

Fig.5 Closed loop system block diagram

A. small signal model for Control to output

If neglecting harmonic components in phase line currents, and assuming phase line currents in phase with phase line voltage, the following expression can be

gotten.

$$\begin{cases} i_a = I_m \cos(\omega t) \\ i_b = I_m \cos(\omega t - 2\pi/3) \\ i_c = I_m \cos(\omega t + 2\pi/3) \end{cases} \qquad (10)$$

With assumption that boost inductive is Ls, Rs is input rail resistance, output capacitor is Cs, and output load is R.

Assuming the output of voltage regulator $\hat{i}_m^* \ll I_m^*$, hence, the control current is as follows:

$$i_{con}^* = I_m^* + i_m^* \approx I_m^* \qquad (11)$$

From（10）we can get:

$$i_a^2 + i_b^2 + i_c^2 = 1.5 I_m^2 \qquad (12)$$

Due to input transient power equals output one, (13) is gotten:

$$P_{in} = P_L + P_R + P_C + P_o \qquad (13)$$

Where P_o is output power, P_L, P_R and P_C are powers of Ls , Rs and Cs respectively.

So the following expression can be achieved：

$$P_{in} = 0.5 L_s \frac{d \sum_{k=a,b,c} i_k^2(t)}{dt} + R_s \sum_{k=a,b,c} i_k^2(t) + u_{dc}(t) i_o(t) \qquad (14)$$

Where, i_o is output current including currents through Cs and load R. (15) is seen from input.

$$P_{in} = \sum_{k=a,b,c} u_{sk}(t) i_{conk}(t) = 1.5 E_m I_m \qquad (15)$$

In expression we have known that $i_{conk} \approx I_m$ .Combined with expression(12) we get (16). Where, P_L is power on boost inductor.

$$P_L = 0.5 L_s \frac{d \sum_{k=a,b,c} i_k^2(t)}{dt} = 1.5 L_s I_m \frac{d I_m}{dt} \qquad (16)$$

The power on Rs is:

$$P_R = R_s \sum_{k=a,b,c} i_k^2(t) = 1.5 R_s I_m^2 \qquad (17)$$

Substituting (17) into expression（14）and （15），we get (18).

$$i_o(t) = \frac{1.5}{u_{dc}(t)} \left(E_m i_{con} - L_s i_{con} \frac{di_{con}}{dt} - R_s i_{con}^2 \right) \qquad (18)$$

Because:

$$i_o(t) = C_s \frac{du_{dc}(t)}{dt} + \frac{u_{dc}(t)}{R} \qquad (19)$$

If considering small disturbances at steady state, the expression (20) can be gotten:

$$\begin{cases} i_o(t) = I_o + \hat{i}_o(t) \\ u_{dc}(t) = U_{dc} + \hat{u}_{dc}(t) \\ i_{con}(t) = I_m + \hat{i}_m(t) \end{cases} \qquad (20)$$

1408

Where I_o, U_{dc} and I_m are stable quantities(averaged), and $\hat{i}_o(t)$、$\hat{u}_{dc}(t)$ and $\hat{i}_m(t)$ are low frequency small disturbances. If substituting (20) into (18) and (19), we apart small signal quantities from expression, get (21).

$$\hat{i}_o(t) = \frac{1.5}{U_{dc}}\left(E_m\hat{i}_m - L_sI_m\frac{d\hat{i}_m}{dt} - 2R_sI_m\hat{i}_m\right) - \frac{\hat{u}_{dc}}{R} \quad (21)$$

$$\hat{i}_o(t) = C_s\frac{d\hat{u}_{dc}(t)}{dt} + \frac{\hat{u}_{dc}(t)}{R} \quad (22)$$

Substituting expression（22）into （21）, get:

$$\frac{d\hat{u}_{dc}(t)}{dt} + \frac{\hat{u}_{dc}(t)}{0.5RC_s}$$
$$= \frac{1.5}{C_sU_{dc}}\left((E_m - 2R_sI_m)\hat{i}_m - L_sI_m\frac{d\hat{i}_m}{dt}\right) \quad (23)$$

Expression (23) is small signal linear time domain model of three-phase HPF Boost PWM rectifier (control variable is $\hat{i}_m(t)$, and output variable $\hat{u}_{dc}$). Expression (24) is the Laplace transformation of (23).

$$G(S) = \frac{\hat{u}_{dc}(s)}{\hat{i}_m} = K\frac{1-T_zS}{1+T_pS} \quad (24)$$

Expression (24) is just the open loop control to output transfer function of three-phase HPF Boost PWM rectifier. Where:

$$\begin{cases} T_p = 0.5RC \\ T_z = \dfrac{L_s I_m}{E_m - 2R_s I_m} \\ K = \dfrac{3R(Em - 2Rs\,I_m)}{4U_{dc}} \end{cases} \quad (25)$$

As usual, $E_m \gg 2R_s I_m$, so the item $2R_s I_m$ can be omitted and espression(26) gotten:

$$\begin{cases} T_p = 0.5RC_s \\ T_z = \dfrac{L_s I_m}{E_m} = \dfrac{L_s}{R_i} \\ K = \dfrac{3RE_m}{4U_{dc}} \end{cases} \quad (26)$$

Where R_i is equivalent input resistance of three-phase rectifier.

As seen in the expression (26) there exists one zero point ($1/T_z$) on RHP of G(S), and this zero pole can not be neglected usually. Hence HPF Boost PWM rectifier belongs to non-minimum-phase system.

B. The design of voltage regulator $G_c(s)$

Simulation parameters are as follows: Rs=0.002Ω, Ls=7.8mH, Cs=2200 μF, input phase line voltage amplitude is $E_m = 50\sqrt{2}$ V, output DC voltage U$_{dc}$=150V, output power Po=1KW, and switching frequency

fs=10KHz.

Based on characteristic of G(s) the desired open loop model is specified as follows:

$$Q(s) = G_c(s)G(s) = \frac{m(1-T_zS)}{T_zS(1+T_zS)} \quad (27)$$

Hence Gc(s) is:

$$G_c(s) = Q(s)/G(s) = \frac{m(1+T_pS)}{KT_zS(1+T_zS)} \quad (28)$$

To guarantee input line currents can track down the line voltages completely, the bandwidth of voltage loop must be far lower than that of current loop. Hence 1/5 of switching frequency is specified as the bandwidth of voltage here by 160 rad/s, and equals to m/Tz as seen in expression (28), so we get m=0.16. due to selected output capacitor Cs much larger in this design and power smaller, $T_z \ll T_p$, T_z can be neglected, as a result, voltage regulator design is simplified as shown in expression (29).

$$G_c(s) = \frac{0.16(1+T_pS)}{KT_zS} \quad (29)$$

This is typical PI regulator. From (29) we can get P and I:

$$\begin{cases} K_p = \dfrac{0.16T_p}{KT_z} \\ K_i = \dfrac{0.16}{KT_z} \end{cases} \quad (30)$$

Substituting the simulation parameters given into (30), get: $K_p = 0.479$, $K_i = 19.34$. As shown in fig.6 phase margin is 80.79 degree and gain margin 6.25dB, hence the system is steady.

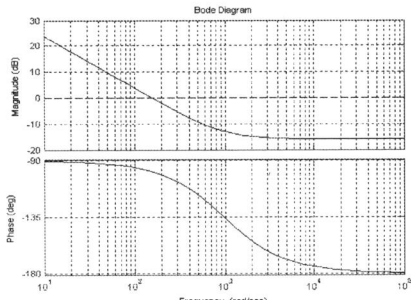

Fig.6　Amplitude and phase versus frequency

IV. SIMULATION AND EXPERIMENT

Voltage regulator, predictive current controller and SVPWM fast algorithm are all implemented in DSP in this experimental prototype. To simulate physical profile authentically, M file in MATLAB6.5 is employed to implement the control logic in which all control algorithms and power converter model are included. In M file three-phase PWM rectifier model is accomplished with aid of Runge-kutta method and the sample frequency is just switching frequency.

Simulation flow block diagram is shown in fig.7.

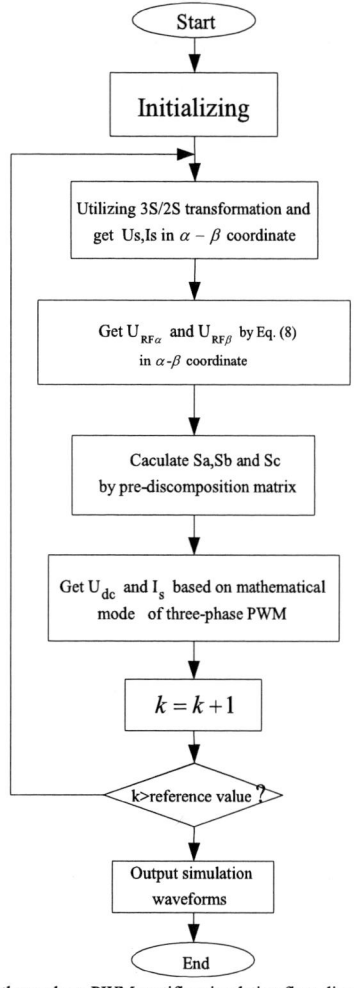

Fig.7 three-phase PWM rectifier simulation flow diagram

A. SVPWM waveforms

Fig.8 shows the simulation waveforms of the PWM outputs which is achieved after the carrier has been taken out with a low-pass filter. The upper and the bottom waveforms in fig.8 are two of the three PWM outputs. The waveform in the middle is the deference between the two, representing the line-to-line input voltage applied to three-phase-bridge rectifier. It is clear that the line to line voltage is still sinusoid and SVPWM waveforms can be equivalent to the sinusoidal waveforms plus the 3^{thd} harmonic. Hence over-modulation can not occur until modulation rate is high by 1.15. SVPWM is indeed the best modulation method to suppress harmonics and efficient use of DC supply is high, as close to 1 can attained.

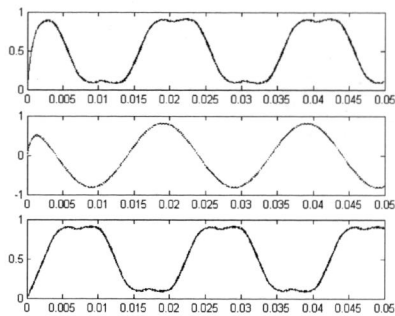

Fig.8   SVPWM simulation outputs with carrier filtered out

B. Waveforms in steady state

Simulation waveforms in steady state are seen in fig.9. If we let three-phase PWM rectifier start up from DC output voltage at zero, the small negative output voltage can be observed and then the output voltage turn positive up to nominal output voltage. This case proves the control loop of three-phase HPF PWM rectifier really exists zero in RHP. Seen from fig.9(a) and fig.10(a), input line currents achieve sinusoid in phase with phase line voltage. It can be verified that current controller is effective. Fig.10 (b) gives the experimental input line currents waveforms of phase A and B.

Fig.9(b) and fig.10(c) are the simulation waveforms and the experimental waveforms respectively, they are stable, hence, the voltage regulator design is effective.

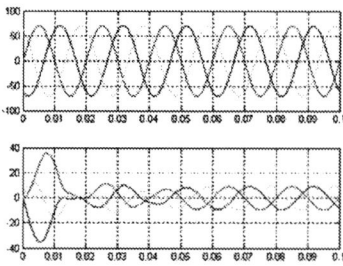

(a) Input voltage and current waveforms

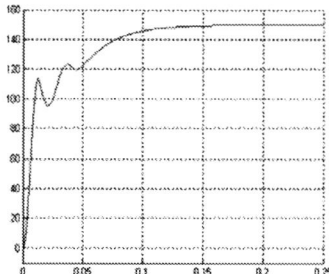

(b) Output direct voltage waveform

Fig.9   Main waveforms in steady state

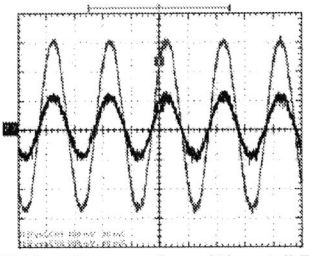

(a) CH1 The voltage waveform of Phase A (25V/div)
      CH2 The current waveform of Phase A (10A/div)

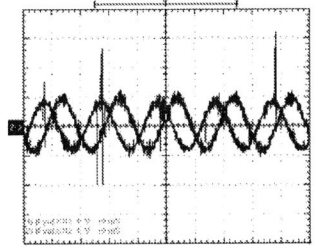

(b) CH1 The current waveform of Phase A (10A/div)
      CH2 The current waveform of Phase B (10A/div)

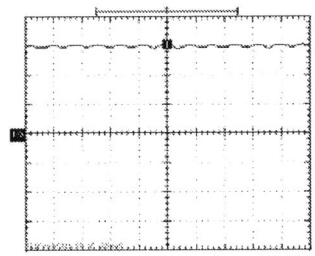

(c) The waveform of output voltage (50V/div)
      Fig.10   The waveforms in steady state

V.   CONCLUSION

Simulation and experimental results tell us hat the unit power factor can be accomplished based on SVPWM and PI voltage regulator. Simulation waveforms verify low frequency small signal model of VSR PWM rectifier. The complete agreements between simulation and experimental waveforms prove that the proposed simulation algorithm in MATLAB is effective in this paper and this simulation file combined power circuit model with control scheme can greatly work to instruct the design of three-phase HPF rectifier. Especially the control algorithm in M file can easily be transplant to the DSP program so that we can promptly implement control strategy in DSP. Furthermore, these simulation points can work in others digitalized SMPS design.

REFERENCES

[1] Thomas G. Habetler, "A Space Vector -Based Rectifier Regulator for AC/DC/AC Converters", IEEE Transactions on Power Electronics, Vol.8, No.1, pp.30-36,1993.

[2] Luigi Malesani, Paolo Tomasin, and Vanni Toigo, "Space Vector Control and Current Harmonics in Quasi-Resonant Soft-Switching PWM Conversion", IEEE Transactions on Industry Applications,Vol.32,No.2, pp.269-277,1996.

[3] K. Yamamoto, K. Shinohara, "Comparison between space vector modulation and sub-harmonic methods for current harmonics of DSP-based permanent-magnet AC servo motor drive system", IEEE Proc. Electr. Power Appl.,Vol.143,No.2, pp.151~156,1996.

[4] Yu Fang and Yan Xing , "A Fast Algorithm for SVPWM in Three Phase Power Factor Correction   Application", IEEE PESC'2004.

[5] Yang Degang, Liu Runsheng and Zhao Liangbing, "Current controller design of a three-phase high-power-factor rectifier", Transactions of China electrotechnical society, Vol.15, No. 2, pp.83~87,2000.

2006 5th International Power Electronics and Motion Control Conference

Effect of the Ripple Current on Power Factor of CRM Boost APFC

A. Abramovitz
Sami Shamoon College of Engineering,
Dept. of Electrical and Electronics Engineering,
Beer-Sheva, Israel
e-mail: alexabr@sce.ac.il

Abstract—**Boost Active Power Factor Corrector operating in a Borderline or Critical Conduction Mode is reexamined. Approximate analytical expressions of the inductor RMS current as well as estimation of Power Factor of the APFC are obtained. Based on the proposed theory, the paper suggests simple design criteria for the line filter.**

Keywords- Borderline; CRM; APFC; PFC

I. INTRODUCTION

Active Power Factor Correction (APFC) systems are generally designed around high frequency converters. The Boost power stage is perhaps the most popular converter used in single phase APFCs due to its simplicity, low part count and inherent ability to sustain input current through the deeps of the line voltage [1]. Whereas the Average Current Mode operation is preferred for high power levels, the simple control schemes operating on the Continuous-Discontinuous Inductor Current Boundary [2] are considered as a cost effective solution for the low power applications. This operating mode is also referred to as Borderline, Transition or Critical Conduction Mode (CRM).

Schematic diagram of the CRM APFC is shown at Fig. 1 (a). This scheme may be implemented with MC33262, UCC38050 or similar control ICs and requires only few external components. PWM signal is generated by window comparators. The Critical Mode (CRM) is attained as Zero Current Detection (ZCD) comparator initiates the power switch conduction at the zero inductor current, whereas the Threshold comparator terminates the switch conduction as the inductor peak current hits the threshold level established by the multiplier. Inductor current ramps from zero up to the programmed peak current, I_p, and then falls back to zero resulting in triangular input current as shown at Fig. 1 (b). The average output voltage of CRM APFC is regulated by a single voltage feedback loop. To comply with power requirements of the load the current reference, I_p, is derived modulating the line voltage by the voltage error amplifier signal. It is clear that the outer loop of the APFC must have a limited bandwidth so that the output voltage ripple will be heavily attenuated. Otherwise, while stabilizing the output voltage, the output voltage ripple will distort the input current by penetrating into the reference of the current loop. The limited loop bandwidth results in slow transient response.

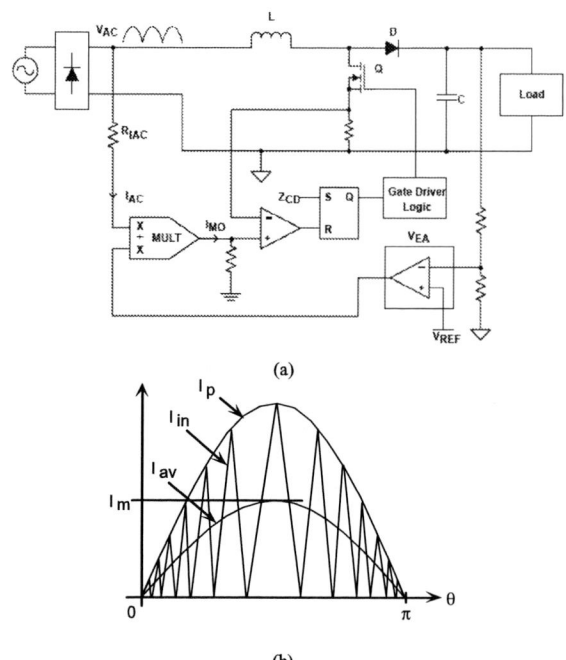

(a)

(b)

Figure 1. A Borderline APFC block diagram (a) and its input current (b).

The simplicity and unconditional stability of the current loop, rather high power factor and the favorable zero current switching (ZCS) of the power devices are of the obvious advantages of operation in CRM. However, CRM control strategy dictates high inductor ripple which causes higher magnetic as well as conduction losses in the core and inductor winding. Conduction losses of the semiconductors are also elevated. In addition, the switching frequency of CRM APFC varies in rather broad range. Variations of the switching frequency causes increased EMI problems. The high current ripple and the sluggish transient response to load and line variations are main drawbacks of CRM APFC.

The following paper presents an estimation of the inductor RMS current including the ripple and its impact on the main performance indexes of CRM APFC- the Power Factor (PF) and the Total Harmonic Distortion (THD).

1-4244-0448-7/06/$25.00 ©2006 IEEE

II. REVIEW OF CRM APFC OPERATION

A. Input Current Shaping

It is assumed that the APFC's input voltage is of a 'pure rectified sine' shape. To simplify the notation the analysis confined to one half of the power line period for which the rectified input can be expressed as:

$$v_i = V_m \sin \theta \qquad (1)$$

where $\theta = \omega_L t$ is the power line angle ($0 \leq \theta \leq \pi$) and ω_L is the power line angular frequency.

The Boost converter is operated in CRM generating instantaneous input current of a triangular wave-shape as shown at Fig. 1 (b). The current feedback network R_s senses the instantaneous input current, $i_i(\theta)$, with a varying peak value, $i_p(\theta)$. The PWM comparator terminates the switch conduction as the peak sensing signal $v_{sp}(\theta)$ at the output of the current sensing network:

$$v_{sp}(\theta) = i_p(\theta) R_s \qquad (2)$$

equals the threshold level $v_{cp}(\theta)$ established by the multiplier. The threshold level, $v_{cp}(\theta)$, is derived modulating the rectified line voltage, $v_i(\theta)$, by the slow varying voltage error amplifier signal, v_e, which will be considered as a constant throughout the line half cycle:

$$v_{cp}(\theta) = k_m v_e v_i(\theta) \qquad (3)$$

Here, k_m is the multiplier gain. Equating the eq. (2) and (3) above it could be seen that the peak input current $i_p(\theta)$ varies proportionally to the rectified line voltage:

$$i_p(\theta) = \frac{k_m}{R_s} v_e v_{in}(\theta) = \frac{k_m}{R_s} v_e V_m \sin \theta \qquad (4)$$

Since the input current is of a triangular shape, the average input current is:

$$i_{av}(\theta) = \frac{1}{2} i_p(\theta) = I_m \sin \theta \qquad (5)$$

Using Eq. (4) and (5) the amplitude of the average input current is:

$$I_m = \frac{k_m}{2 R_s} V_m v_e \qquad (6)$$

The average current (5) is in phase with the line voltage and follows its sinusoidal shape making high power factor possible. The average current amplitude and APFC's power level are linear functions of the Error Amplifier voltage, v_e:

$$P_{av} = \frac{1}{2} I_m V_m = \frac{1}{4} V_m^2 \frac{k_m}{R_s} v_e \qquad (7)$$

B. The Switching Frequency

The switching frequency of the APFC is much higher than that of the power line so the line voltage may be considered as constant throughout the switching cycle. As discussed above, see Fig. 1 (b), the peak inductor current, $i_p(\theta)$, follows the instantaneous line voltage:

$$i_p(\theta) \approx \frac{v_i(\theta) T_{on}}{L} = \frac{V_m T_{on}}{L} \sin \theta \qquad (8)$$

The conduction interval T_{on} of the main switch is proportional to the voltage error amplifier signal and to the power level:

$$T_{on} = \frac{L i_p(\theta)}{v_i(\theta)} = L \frac{k_m}{R_s} v_e = \frac{4 L P_{av}}{V_m^2} \qquad (9)$$

This result is obtained combining equations (5)-(8). Thus, if a ripples voltage error amplifier signal, v_e, is applied to the multiplier input, it generates a constant on-time, T_{on}, which results in a sinusoidal average line current (8) throughout the power line period. As the input inductor is discharged to the load, its current falls to zero at the end of the succeeding T_{off} interval so that:

$$i_p(\theta) = \frac{V_o - v_i(\theta)}{L} T_{off}(\theta) \qquad (10)$$

From the relationship (8) and (10) above, the off-time is given as:

$$T_{off}(\theta) = \frac{v_i(\theta)}{V_o - v_i(\theta)} T_{on} \qquad (11)$$

The switching cycle is given by:

$$T_s(\theta) = T_{on} + T_{off}(\theta) = \left(1 + \frac{v_i(\theta)}{V_o - v_i(\theta)}\right) T_{on} = \frac{V_o}{V_o - v_i(\theta)} T_{on} \quad (12)$$

thus, the switching frequency varies with the line voltage:

$$f_s(\theta) = \frac{1}{T_s(\theta)} = \left(1 - \frac{v_i(\theta)}{V_o}\right) \frac{1}{T_{on}} = \left(1 - \frac{V_m \sin \theta}{V_o}\right) f_{sh} \quad (13)$$

reaching its maximum value at the vicinity of zero crossings of the input voltage:

$$f_{sh} = \frac{1}{T_{on}} = \frac{V_m^2}{4 L P_{av}} \qquad (14)$$

The fact that the highest switching frequency f_{sh} is limited by the controller may be considered as an advantage over other hysteretic control methods. The lowest switching frequency f_{sl} occurs at high power level at the peak of the line voltage:

$$f_{sl} = \left(1 - \frac{V_m}{V_o}\right) f_{sh} = \left(1 - \frac{V_m}{V_o}\right) \frac{V_m^2}{4 L P_{av}} \qquad (15)$$

III. THE SWITCHING CURRENT RIPPLE OF THE APFC

The input current of the VF-APFC $i_i(\theta)$ is illustrated at Fig. 1 (b). The input current may be represented by its average component, $i_{av}(\theta)$, given by Eq. (5), with a superimposed high frequency ripple current, $i_r(\theta)$:

$$i_i(\theta) = i_{av}(\theta) + i_r(\theta) \qquad (16)$$

The ripple current is of a triangular shape, however, in the following discussion it will be approximated to a sinusoid with variable amplitude $I_{rm}(\theta)$ and variable phase:

$$i_r(\theta) = I_{rm}(\theta)\sin(\phi_s(\theta)) \qquad (17)$$

By inspection of Fig. 1 (b) it could be noticed that the amplitude of the ripple current equals to the average line current:

$$I_{rm}(\theta) = I_m \sin(\theta) \qquad (18)$$

whereas the phase may be calculated using the expression for $f_s(\theta)$ as given by Eq. (13):

$$\phi_s(\theta) = 2\pi \int_0^\theta f_s(\theta)\frac{d\theta}{\omega_L} = \frac{2\pi f_{sh}}{\omega_L}\int_0^\theta \left(1 - \frac{V_m \sin\theta}{V_o}\right)d\theta =$$

$$= \frac{\omega_{sh}}{\omega_L}\left(\theta - \frac{V_m}{V_o}(1-\cos\theta)\right) \qquad (19)$$

here $\omega_{sh} = 2\pi f_{sh}$. Substituting this result into Eq. (17) yields the analytical approximation for the ripple current:

$$i_r(\theta) = I_m \sin(\theta)\sin\frac{\omega_{sh}}{\omega_L}\left(\theta - \frac{V_m}{V_o}(1-\cos\theta)\right) \quad (20)$$

It is convenient to define the normalized ripple, $i_{rn}(\theta)$ as:

$$i_{rn}(\theta) = \frac{i_r(\theta)}{I_m} = \sin(\theta)\sin\frac{\omega_{sh}}{\omega_L}\left(\theta - \frac{V_m}{V_o}(1-\cos\theta)\right) \quad (21)$$

IV. ESTIMATION OF POWER FACTOR OF CRM APFC

The normalized ripple current, $i_{rn}(\theta)$, RMS value is designated as I_{RNrms}, and calculated by definition:

$$I_{RNrms} = \left[\frac{1}{\pi}\int_0^\pi i_{rn}^2(\theta)d\theta\right]^{\frac{1}{2}} =$$

$$= \left[\frac{1}{\pi}\int_0^\pi \sin^2(\theta)\sin^2\left(\frac{\omega_{sh}}{\omega_L}\left(\theta - \frac{V_m}{V_o}(1-\cos\theta)\right)\right)d\theta\right]^{\frac{1}{2}}$$

$$= \left[\frac{1}{4\pi}\int_0^\pi (1-\cos 2\vartheta)\left(1-\cos 2\left(\frac{\omega_{sh}}{\omega_L}\left(\theta - \frac{V_m}{V_o}(1-\cos\theta)\right)\right)\right)d\theta\right]^{\frac{1}{2}} =$$

$$= \frac{1}{2\sqrt{\pi}}\left[\int_0^\pi 1 - \cos 2\vartheta - \cos 2\left(\frac{\omega_{sh}}{\omega_L}\left(\theta - \frac{V_m}{V_o}(1-\cos\theta)\right)\right) + \cos 2\theta \cos 2\left(\frac{\omega_{sh}}{\omega_L}\left(\theta - \frac{V_m}{V_o}(1-\cos\theta)\right)\right)d\theta\right]^{\frac{1}{2}} \quad (22)$$

Though the integral (22) may look intimidating, applying the considerations described below yields a surprisingly simple result. First, notice that the integral of the second harmonic term $\cos(2\theta)$ over power line half cycle vanishes. Integration of the high frequency term $\cos 2\left(\frac{\omega_{sh}}{\omega_L}\left(\theta - \frac{V_m}{V_o}(1-\cos\theta)\right)\right)$ yields a factor $\frac{\omega_L}{2\omega_{sh}}$, which is negligibly small since the switching frequencies ω_{sh} is much higher than the power line frequency ω_L, thus this term may be neglected. The term $\cos 2\theta \cos 2\left(\frac{\omega_{sh}}{\omega_L}\left(\theta - \frac{V_m}{V_o}(1-\cos\theta)\right)\right)$ could be manipulated using trigonometry formula for cosine product which yields sum and difference of the high frequency term and the second harmonic of the line frequency. The frequency shifting caused by the low line frequency could be neglected. Once again, since the integrand is of high frequency, the integration result is proportional to the factor $\frac{\omega_L}{2\omega_{sh}}$ and could be neglected.

Thus, the resulting normalized rms ripple current is :

$$I_{RNrms} = \frac{1}{2\sqrt{\pi}}\left[\int_0^\pi d\theta\right]^{\frac{1}{2}} = \frac{1}{2} \qquad (23)$$

The Total Harmonic Distortion (THD) *due to the high frequency ripple* of the CRM APFC could be found as:

$$THD = \frac{I_{Rrms}}{I_{1rms}} = \frac{I_{Rms}}{\frac{1}{\sqrt{2}}I_m} = \sqrt{2}I_{RNrms} = \frac{1}{\sqrt{2}} = 0.707 \qquad (24)$$

Since the average current is in phase with the input voltage (the phase of the first harmonic of the average input current $\varphi_1 = 0$), the Power Factor (PF) of the CRM APFC *without filtering the current ripple component* is:

$$PF = \frac{I_{1rms}}{I_{rms}} = \frac{1}{\sqrt{1+THD^2}} = \sqrt{\frac{2}{3}} = 0.817 \qquad (25)$$

V. INPUT FILTER DESIGN CONSIDERATIONS

It is assumed here that the line filter of second order is sufficient. Approximated line filter's transfer function, see Fig. (2), is of a unity gain in within its pass-band up to the corner frequency f_0, -40db/dec roll-off in within the transition band and has a constant attenuation, H_{sb}, in within the stop band above the stop-band frequency f_{sb}:

1414

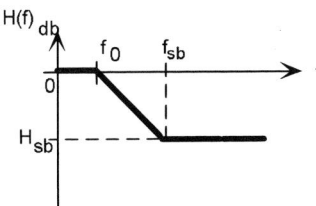

Figure 2. Line filter transfer function.

$$H(f) = \begin{cases} 1 & f < f_0 \\ H_{sb} & f > f_{sb} \end{cases} \qquad (26)$$

The filter should be designed so that all the current ripple components fall in within the stop-band. Consequently, the filtered ripple current, $i_{fr}(\theta)$, is proportional to the input inductor ripple current, $i_r(\theta)$:

$$i_{fr}(\theta) = H_{sb} i_r(\theta) \qquad (27)$$

Normalization of the filtered current ripple, Eq. (27), to the average current amplitude I_m gives:

$$I_{frN}(\theta) = \frac{i_{fr}(\theta)}{I_m} = H_{sb} \frac{i_r(\theta)}{I_m} = H_{sb} I_{RN}(\theta) \quad (28)$$

Following the same reasoning as above, see Eq. (24), yields the THD of the *filtered line current* as:

$$THD = \sqrt{2} H_{sb} I_{RNrms} = 0.707 H_{sb} \qquad (29)$$

According to Eq. (29), a negligible THD level of 0.7% could be achieved for $H_{sb} = 0.01$ or 40db filter attenuation.

To provide the required attenuation, the stop band frequency, f_{sb}, should be chosen just below the lowest harmonics of the ripple component $f_{sb} \le f_{sl}$. Lowest switching frequency, f_{sl}, occurs at full rated power, as given by Eq. (15). Filter corner frequency, f_0, should be placed about one decade below the stop-band frequency $f_0 \approx 0.1 f_{sb}$. It should be verified that the filter corner frequency is at least a decade above the line frequency to avoid displacement of the average component of the line current with respect to the line voltage. Otherwise the input filter could lower the PF by $\cos(\varphi_1)$. Several filter topologies and design procedure could be found in [3].

VI. EXPERIMENTAL RESULTS

Experimental CRM APFC was build and tested. The measurements were performed by LeCroy digitizing oscilloscope. Measured PF of *unfiltered current* was 0.856, see experimental results of Fig. 3 (a), in a good agreement with theoretical estimation obtained in Eq. (25). With line filter installed, the PF raised to 0.968, see Fig 3 (b). Crossover as well as third harmonic distortion could be observed.

VII. CONCLUSIONS

The paper estimated the impact of inductor current ripple on the main performance indexes of CRM APFC. As expected, the ripple current degrades the performance of the CRM APFC. The analysis above suggested a theoretical limit of THD as 70.7% and PF of 0.816 which may be considered as rather poor performance for a system which supposed to exhibit a near zero THD and near unity PF. Acceptable systems performance is achieved by filtering out the ripple current. Input filter design parameters, the required attenuation and corner frequencies were suggested. The theoretical prediction stands in good agreement with experimental results.

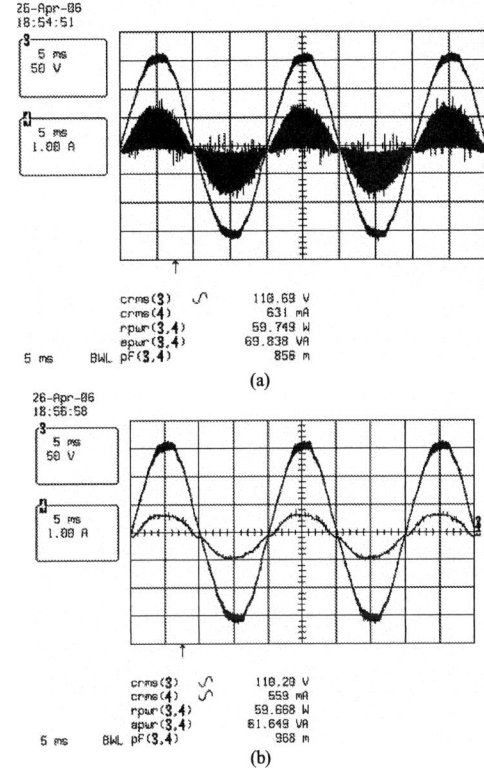

Figure 3. Measured line voltage and unfiltered line current (a), Line Voltage and filtered Line Current (b), notice the printout of measured

ACKNOWLEDGMENT

The credit for construction and experimental work goes to Mr. Mor M. Peretz, who excelled at his undergraduate project.

REFERENCES

[1] Qiao, K. Smedley, "A Topology Survey of Single-Stage Power Factor Corrector with a Boost Type Input-Current-Shaper", APEC 2000.

[2] D.S. Chen, J.S. Lai, "Design Consideration for Power Factor Correction Boost Operating at the Boundary of Continuous Conduction Mode and Discontinuous Conduction Mode", IEEE APEC 1993.

[3] V. Vlatkovic', D. Borojevic', and F.C. Lee, "Input Filter Design for Power Factor Correction Circuits", International Conference on Industrial Electronics, Control and Instrumentation, Nov. 1993.

2006 5th International Power Electronics and Motion Control Conference

Simulated Study of Three-Phase Single-Switch PFC Converter with Harmonic Injected PWM by MATLAB

Zhanlong Li and Yupeng Tang

School of Electrical Engineering; Beijing Jiaotong University; China

Abstract—**This paper describes the operation and simulation of a three-phase single-switch PFC converter with a sixth-order harmonic-injection PWM. The PFC converter operates in a discontinuous current mode (DCM). In proposed method, the sixth-order harmonic is injected in PWM so that the harmonic content of the input current meet IEC555-2(A) requirement. The theoretical analysis of the harmonic-injection is discussed. Furthermore, the model and simulation of PFC converter are made via MATLAB/SIMULINK software. Finally the simulated waveforms of the input current are also showed. And the performance of the PFC converter with harmonic injected PWM is verified on a simulation model, and a good consistency is achieved.**

Keyword—*power factor correction (PFC); harmonic injection; discontinuous conduction mode (DCM); continuous conduction mode (CCM); total harmonic distortion (THD)*

I. INTRODUCTION

Recently, three phase PFC converter have gained considerable attention due to the increasing demand to power quality. Based on the number of switches controlled, all 3-phase PFC converters can be divided into two groups: single switch PFC converters and multiple switches PFC converters. Though the multiswitch PFC converters are high-power high-performance application, the increased number of switches and the complexity of their control make them too expensive in medium power levels.

So the single-switch three-phase DCM boost converter is an attractive topology because of its simplicity, low cost and high efficiency. However, there are some disadvantages and advantages in this kind of topology. For example, the DCM operation is associated with a higher voltage or current stress. Several solutions to this problem have been presented in Ref. [7] and Ref. [8].The main switch has a reduced loss due to its zero current switching .Also the boost diode has no the reverse recovery problem for the same reason.

In DCM operation with constant frequency PWM, the single-switch three-phase PFC converters have a larger current distortion, so it's not suitable to high power level. According to IEC555-2(A) standard, the power level of converter can't be increased in order to meet the maximum permissible harmonic current. However, if in order to meet the IEC555-2(A) specifications, the output voltage of three phase rectifier must be boosted to 900V. Such designs increase the voltage stress of switch.

To reduce the harmonic current distortion, some methods have been proposed. One of the methods is to operate the converter on boundary between CCM and DCM. Namely, when the current in the boost inductors falls to zero, the main switch is turned on immediately. With variable switching frequency, this method has disadvantages like high switching losses resulted from the increased switching frequency, and have difficulty in designing the converter and the EMI filter.

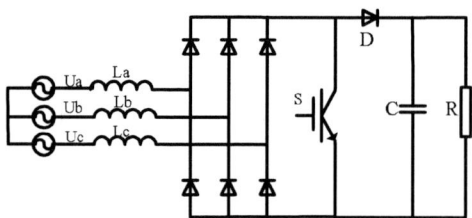

Fig.1 The main circuit

This paper analyzes the performance of a single-switch 3-phase PFC converter (Fig.1) that uses harmonic injected PWM by MATLAB/SIMULINK software. In section Ⅱ, the operation of converter is simply introduced. In section Ⅲ, the method for reducing harmonic content is analyzed. Finally, in section Ⅳ the waveforms of the input current by simulation are showed.

II. OPERATION OF CONVERTER

The operation of the three-phase single-switch PFC converter with a constant switching frequency is analyzed in this section. The waveform of input current in a period is showed in Fig.2. To simplify the analysis

1-4244-0448-7/06/$25.00 ©2006 IEEE   1416

of the converter system, the following assumptions are made.

- The 3 inductors operate in DCM.
- The system components are ideal, such as: resistors, inductors, capacitors.
- For the switching frequency is much higher than the line frequency, the input voltages are considered to constant within a switching period.

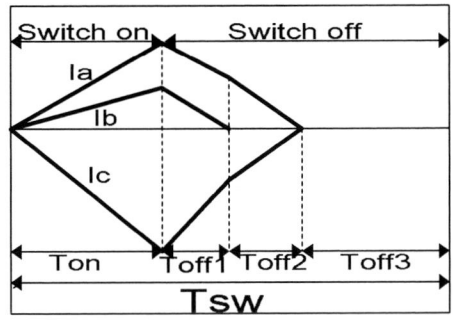

Fig.2 The waveform of input current in a period

According to Fig.2, the shape of the input current is triangular pulse, and a switching period consists of four time internals. When the switch is 'on', the input inductors are fed by the input voltage and the energy is accumulated in them. The peak of current is proportional to the input voltage. When the switch is 'off', the current start discharging until all of the phase currents falls zero. In this internal (Toff), there are three different states depending on the corresponding phase voltage. The first internal (Toff1), all the phases conduct until one of the phase currents falls to zero. The second internal (Toff2) is characterized by the two diodes. In the third internal (Toff3), there is no current in the inductors. Actually, as the input current is proportional to the input voltage, there exist no low frequency harmonics in the input currents. However, during the switch 'off interval, the phase currents fall to zero at different moments. Therefore, the phase currents are nonlinear functions of their phase voltages, then it contains several low frequency harmonics (Ref. [1] and Ref. [5]).

III. ANALYSIS OF HARMONIC REDUCTION WITH HARMONIC INJECTED PWM

Suppose 3-phase input voltage is equal in magnitude, purely sinusoidal, and with $120°$ mutual phase.

$$
\begin{aligned}
U_a &= U_m \sin(\omega t) \\
U_b &= U_m \sin(\omega t - \frac{2\pi}{3}) \\
U_c &= U_m \sin(\omega t - \frac{4\pi}{3})
\end{aligned} \quad (1)
$$

Where U_m is the peak voltage and $\omega = 2\pi f_L$, f_L is the line frequency. According to Ref. [2], the average input currents within a switching period are given by the following equations.

$$
\bar{i}_a = \frac{U_o D^2}{2 L f_s} \frac{\sin(\omega t)}{\sqrt{3}M - 3\sin(\omega t)} \quad (0 \le \omega t \le \frac{\pi}{6})
$$

$$
\bar{i}_a = \frac{U_o D^2}{2 L f_s} \frac{M\sin(\omega t) + \frac{1}{2}\sin(2\omega t - \frac{2\pi}{3})}{[\sqrt{3}M - 3\sin(\omega t + \frac{2\pi}{3})][M - \sin(\omega t + \frac{\pi}{6})]}
$$

$$
(\frac{\pi}{6} \le \omega t \le \frac{\pi}{3}) \quad (2)
$$

$$
\bar{i}_a = \frac{U_o D^2}{2 L f_s} \frac{M\sin(\omega t) + \sin(2\omega t + \frac{\pi}{3})}{[\sqrt{3}M + 3\sin(\omega t + \frac{2\pi}{3})][M - \sin(\omega t + \frac{\pi}{6})]}
$$

$$
(\frac{\pi}{3} \le \omega t \le \frac{\pi}{2})
$$

Where U_o is the output voltage, D is the duty ratio, L is the value of the input inductor, f_s is the switching frequency, and M is the ratio of U_o and $\sqrt{3}U_m$.

In Ref. [1] and Ref. [2], the reason for three-phase single-switch PFC converter with the sixth-order harmonic injected PWM has been discussed in detail. There is no more analysis here. Only the theory of harmonic injected PWM is showed

According to the frequency spectrum of input current, the fifth-order harmonic and the seventh-order harmonic are the dominant, if neglecting the higher order harmonics, the three-phase input currents can be approximated as:

$$
\begin{aligned}
\bar{i}_a &= I_1 \sin(\omega t) - I_5 \sin(5\omega t) \\
\bar{i}_b &= I_1 \sin(\omega t - \frac{2\pi}{3}) + I_5 \sin(5\omega t - \frac{\pi}{3}) \\
\bar{i}_c &= I_1 \sin(\omega t - \frac{4\pi}{3}) + I_5 \sin(5\omega t + \frac{\pi}{3})
\end{aligned} \quad (3)
$$

Now the THD of the input current is the following equation.

$$
THD = \frac{I_5}{I_1} \quad (4)
$$

The idea of harmonic-injection is that the suitable sixth-order harmonic is injected in PWM to modulate the duty cycle, so that decrease the fifth order harmonic content of the input current. The following (5) is the duty cycle after modulated.

$$
D_m(t) = D(t)[1 - m\cos(6\omega t)] \quad (5)
$$

If $D(t)$ of (2) is substituted by $D_m(t)$, it is not difficult to understand the following equation.

1417

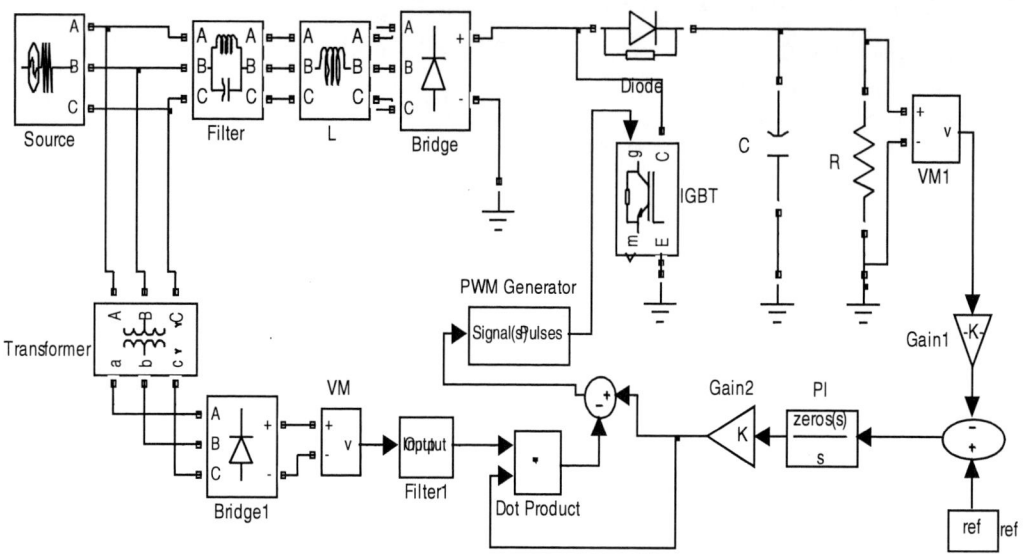

Fig.3   The simulated circuit model of PFC converter with harmonic injected PWM

$$\bar{i}_{a,\mathrm{mod}} = \bar{i}_a[1-m\cos(6\omega t)]^2$$

$$\bar{i}_{b,\mathrm{mod}} = \bar{i}_b[1-m\cos(6\omega t)]^2 \qquad (6)$$

$$\bar{i}_{c,\mathrm{mod}} = \bar{i}_c[1-m\cos(6\omega t)]^2$$

After simplifying the above equation, and ignoring the presence of m^2 ($m^2 \ll 1$) and high-order harmonics (n>7), then (6) is simplified as (7).

$$\bar{i}_{a,\mathrm{mod}} = I_1\sin(\omega t) - (I_5 - mI_1)\sin(5\omega t) -$$
$$mI_1\sin(7\omega t)$$

$$\bar{i}_{b,\mathrm{mod}} = I_1\sin(\omega t - \frac{2\pi}{3}) + (I_5 - mI_1)\sin(5\omega t - \frac{\pi}{3})$$
$$- mI_1\sin(7\omega t - \frac{2\pi}{3}) \qquad (7)$$

$$\bar{i}_{c,\mathrm{mod}} = I_1\sin(\omega t - \frac{4\pi}{3}) + (I_5 - mI_1)\sin(5\omega t + \frac{\pi}{3})$$
$$- mI_1\sin(7\omega t - \frac{4\pi}{3})$$

From (7), it can be seen that the fifth-order harmonic content is decreased, even though the amplitude of seventh-order harmonic is increased. However, the THD is improved, when the value of m is selected correctly.

$$THD_{\mathrm{mod}} = \frac{\sqrt{(I_5 - mI_1)^2 + (mI_1)^2}}{I_1} < \frac{I_5}{I_1} = THD \qquad (8)$$

IV. THE MODEL AND RESULTS OF SIMULATED CIRCUIT

The simulated circuit model of 3-phase single-switch PFC converter with harmonic injected PWM is showed in Fig.3. In proposed method, the generator of the sixth order harmonic consists of four important devices: a 3-phase transformer, a 3-phase diode rectifier bridge, a band-pass filter and a multiplier. The 3-phase transformer is used to reduce voltage and isolate with the main circuit. The function of filter is to eliminate the DC component and high order harmonics (n>6) of the voltage signal rectified by the 3-phase rectifier bridge. Then, the sixth order harmonic is gained, but is not the injected signal. Finally, the amplitude and the polarity are modulated by a multiplier and an adder itself respectively. Therefore, the desired injection signal is obtained.

Fig.3 with the following parameters has been simulated via MATLAB/SIMULINK software. The simulated results are also showed.

The main parameters of the simulated circuit:

The input phase voltage: U_{inP}=220V_{rms}/ f=50Hz

The switch frequency: f_s =10k Hz

The output: U_{out} =800V/P_{out}=6.4kW

The output capacitor: C_o=220uF

The input inductor: L=0.21mH

The EMI filter: L_f=6mH, C_f=4uF

Fig.4 shows the amplified input current waveform of one boost inductor. It can be seen that the input current has fallen to zero at the end of period, so the converter operates in DCM.

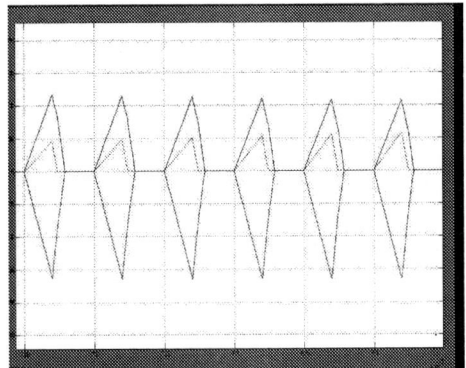

Fig.4 The amplified input current waveform without the EMI filter

The following figures show the input current in the case of with harmonic-injection and without harmonic-injection. According to the figures, with the sixth-order harmonic injected PWM, the fifth-order harmonic content is attenuated by adjusting the amplitude of the sixth-order harmonic, but the seventh-order harmonic content increases. However, the total current distortion is improved evidently. Furthermore, with the optimized m, the value of THD meets the IEC555-2(A) requirement(THD<10%).

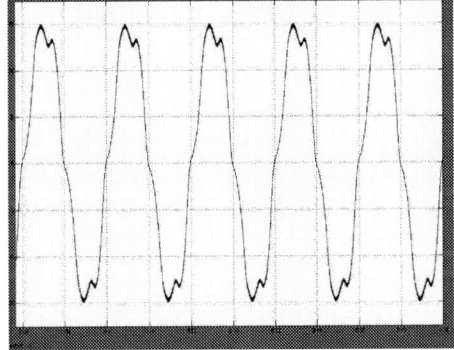

Fig.5.1   The waveform of input current without harmonic-injection

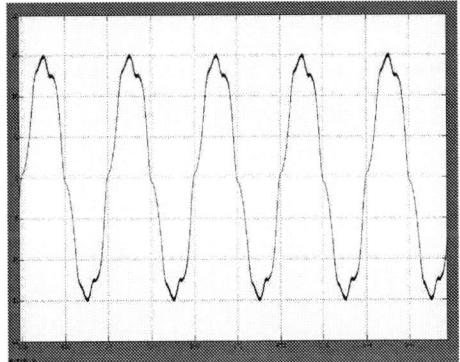

Fig.5.2   The waveform of input current with harmonic-injection

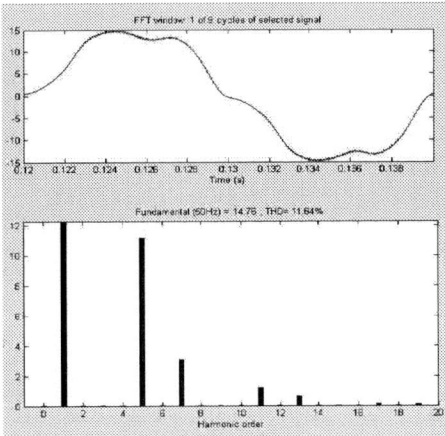

Fig.6.1 The frequency spectrum of the input current without harmonic-injection

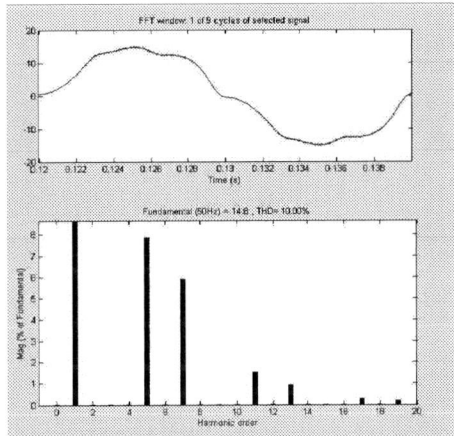

Fig.6.2 The frequency spectrum of the input current with harmonic-injection

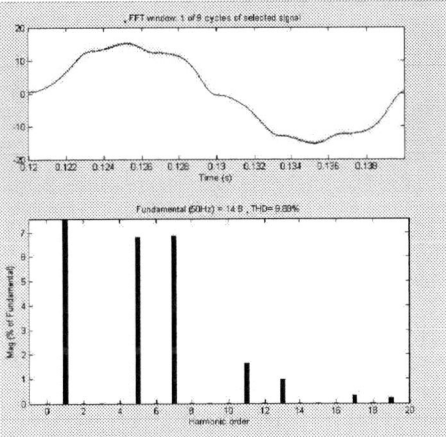

Fig.6.3 The frequency spectrum of the input current with the optimized harmonic-injection

V. CONCLUSION

In this paper, the sixth-order harmonic-injection method for a 3-phase single-switch PFC converter was presented. Using this method, undesirable high voltage transfer ratio can be avoided, without sacrificing the excellent power factor. The whole simulated model of the main circuit and the control circuit was made and analyzed. Finally, the analysis with the determined parameters is conformed by simulation using MATLAB/SIMULINK software. The simulated results are showed.

REFRENCES

[1]. Q. Huang and F. C. Lee, "Harmonic Reduction In A Single-Switch, Three-phase Boost Rectifier With High Order Harmonic Injected PWM" in *IEEE Power Electronics Specialists Conf. (PESC) Rec.*, 1996, pp 1266-1271.

[2]. Yungtaek. Jang, and Milan.M Jovanovic, "A Novel Robust Harmonic-injection with Method for Single Switch, Three-phase, DCM Boost Rectifiers", *IEEE Transactions on Power Electronics*, vol.28, No.1 March 2000,pp. 268-27.

[3]. Yungtaek. Jang, and Milan.M Jovanovic, "A comparative Study of Single-Switch Three-Phase High-Power-Factor Rectifier", *IEEE Transactions on Industry Application*, vol.34, No.6 November/December 1998, pp. 1327-1334

[4]. J. Sun, N. Frohleke, and H. Grotstollen, "Harmonic reduction techniques for single-switch three-phase boost rectifiers." in *conf. Rec IEEE-IAS Annu. Meeting*, 1996, pp. 1225-1232.

[5]. DaFeng Weng and S. Yuvarajan, "AC-DC Converter using second-harmonic-injected PWM" in *PESC'95 Rec.*, pp. 1001-1006,1995

[6]. Akiteru Ueda, Satoko Ando, Akihiro Torii, "Study on LC Filter Design Method for Boost Type Rectifier with Single Switching Device," *IPEC-Tokyo 2000*, pp.820-825

[7]. E. Ismail, C.Oliveira, and R. Erickson, "A low-distortion three-phase multi-resonant Boost rectifier with zero current switching ," in *Proc. of IEEE APEC'95*, pp. 849-855,1995

[8]. K. Chen, A. Elasser, D. A. Torrey, "A soft switching active snubber optimized for IGBTs in single switch unity power factor three-phase diode rectifiers," in *Proc. of APEC'94*, pp. 280-286, 1994

2006 5th International Power Electronics and Motion Control Conference

A Simple Digital Controller for Constant Instantaneous Input Power type Three-Phase Boost Rectifier under Unbalanced System

JIN Ai-juan*, LI Hang-tian** and LI Shao-long*

* College of Electrical Engineering, University of Shanghai for Science and Technology, Shanghai 200093, China.
** Guangdong Telecom, Guangzhou Branch, Guangzhou 510620, China
Email: ajjin@126.com

Abstract—Power Factor Correction (PFC) rectifiers are essential for load side harmonic and reactive power correction. In recent years, research and applications of PFC rectifiers have attracted more attention due to the increased energy awareness in the global. In this paper, a simple digital controlled three-phase six-switch H-bridge boost rectifier is proposed. The control objective is to operate the rectifier in the high power factor mode under normal operating condition, but to give an overriding priority to the constant instantaneous input power in the case of unbalanced input voltages. This results in that the low order harmonics of output DC voltage are eliminated whenever in balanced or unbalanced input voltages, and with this control method, the output capacitor can be significantly minimized. The controller has three loops: the inner loop implements resistor emulator type input average current strategy; the input voltage forward loop performs constant input power regulator; the output voltage loop performs magnitude scaling operations on input current control. The system is implemented in full digital, and one Texas Instrument's DSP TMS320F240 is used as the digital controller.

Keywords-digital control; power factor correction; three-phase; unbalanced system; constant instantaneous input power.

I. INTRODUCTION

With the development of power electronic technology, convertors are widely used. But the nonlinear rectifier causes large harmonic currents and thus highly pollutes the power grid, decreases its power factor and hence reduces the utility's effective capacity. The international standards presented in IEC1000-3-2 and EN61000-3-2 imposed harmonic restrictions to modern rectifiers, which have resulted in a focused research effort on the topic of unity-power-factor rectifiers. Three-phase PFC rectifiers are preferred for high power applications due to their symmetric current-drawing characteristics. Many topologies have been proposed recently, and the six-

This work is sponsored by Shanghai Municipal Education Commission development foundation program (No: 05EZ06).

switch H-bridge boost rectifier is a commonly used topology. And the control methods normally used for this type of rectifier are based on either hysteretic control or d-q transformation control. In those methods, it is assumed that the input voltages and the three phase circuit parameters are all balanced. If the input voltages are unbalanced, then that would cause abnormal even harmonics in the output dc voltage and odd harmonics in the input phase currents [1].

A few methods have been proposed [2-6] to solve the problem of harmonics under unbalanced input voltage condition. In [2] the input voltages are decomposed into symmetrical components so that the detected negative sequence components can be added to the positive sequence control voltages for balancing the currents. This method performs well, however it requires input voltage sensing and the current balanced function is executed in the open loop. Another method proposed in [3] computes the second-order harmonics in the dc output voltage and generates three independent current references for cancellation of the even harmonics. The input currents need not be balanced in this method. It also needs input voltage sensing and the implementation is based on variable switching frequency operation. The method proposed in [4] is complex and harmonics were only reduced instead of elimination. The method proposed in [5] is to operate the rectifier in the high power factor mode under normal operating condition, but to give an overriding priority to the current balanced function in the case of unbalanced input voltages, but the output dc voltage contents low order harmonics and the peaks of α and β axis current are difficult to achieve accurately [6]. The unified constant-frequency integration rectifier based on one cycle control which is proposed in [7] is a low cost analog controller and has good performances, literature [8] shown that under unbalanced input sources, the input currents keep low harmonic and linearly proportional to the non-zero sequence of the three-phase input voltages, it needs large output capacitor to reduce the second order harmonic in the output dc voltage.

In this paper a simple digital controlled three-phase three-wire six-switch H-bridge boost rectifier is proposed.

1-4244-0448-7/06/$25.00 ©2006 IEEE

The control objective is to operate the rectifier in the high power factor and constant instantaneous input power mode under balanced or unbalanced input sources. The proposed controller has the following advantages:

(1) This control algorithm provides good performance with much simpler control structure than the methods mentioned above ;

(2) High quality sinusoidal input currents and high quality DC output voltage;

(3) Whenever the input voltage is balanced or unbalanced, the inputs currents are controlled such that the input power is kept in constant instantaneous input power mode, the low order harmonics of the output voltage are eliminated. With this control method, the output capacitor can be minimized.

Experimental results are presented to verify the advantages of the proposed control strategy.

II. GENERAL SCHEME

A. Proposed Topology

Fig. 1 shows the topology of a three-phase six-switch H-bridge boost PFC controller. For analysis convenience, we assume that the input three-phase voltages are unbalanced, for example, phase b amplitude is 20% decreased, phase c is $\pi/6$ delay. The voltage waveforms are shown in Fig.2, and divided into 6 regions. It should be noticed that the starting points of regions may be different with the different unbalanced conditions of input voltages, and it will be described in the following "Region Determination block" section. According to literature [7], the three-phase PFC rectifier may be implemented by controlling the two switches in a region of linear cycle.

Assume that the input voltages are sine waves, the three phase circuit parameters are balanced, the forward impedances and all parasitic parameters of power devices are neglected. In region I, switch S_{bn} is kept always on, only switches S_{an} and S_{cn} need to be controlled, the three phase rectifier may be decoupled to parallel connected dual boost topology just like Fig.3, where U_p, U_n, L_p, L_n, L_t, T_p, T_n, D_p, D_n and D_t are symbolized voltages, inductors and diodes, whose corresponding components are listed in Table I[7].

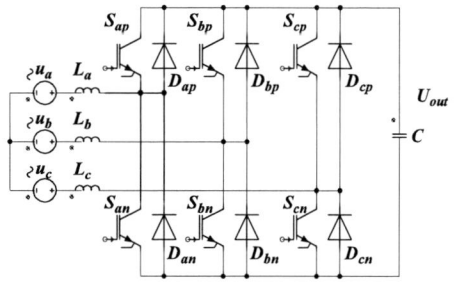

Figure.1 Topology of three-phase six-switch H-bridge boost PFC rectifier

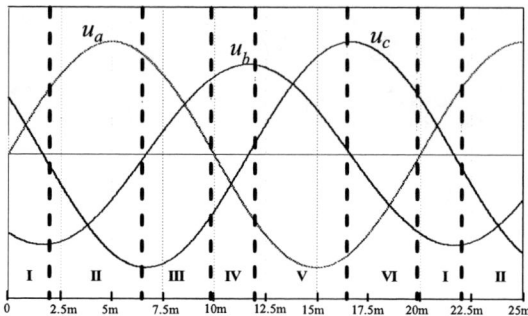

Figure.2 Waveforms under phase b 20% decreased, phase c $\pi/6$ delay.

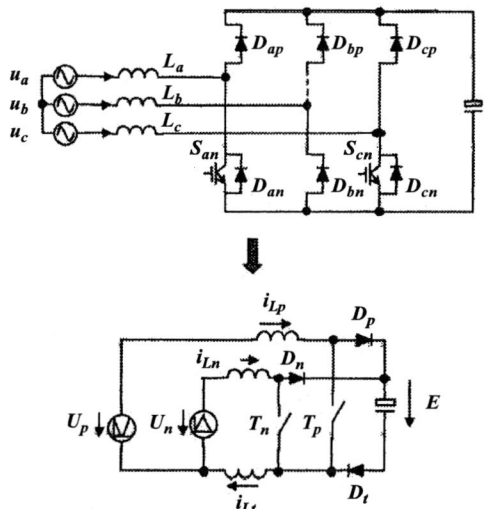

Figure.3 Change from three phase rectifier in Fig.1 to a parallel connected dual boost topology during region I.

TABLE I
CROSS-REFERENCE BETWEEN SIX-SWITCH BRIDGE THREE-PHASE RECTIFIER AND THE PARALLEL-CONNECTED DUAL-BOOST TOPOLOGY

	U_p	U_n	L_p	L_n	L_t	T_p	T_n	D_p	D_n	D_t
I	u_{ab}	u_{cb}	L_a	L_c	L_b	S_{an}	S_{bn}	D_{ap}	D_{cp}	D_{bn}
II	u_{ab}	u_{ac}	L_b	L_c	L_a	S_{bp}	S_{cp}	D_{bn}	D_{cn}	D_{ap}
III	u_{bc}	u_{ca}	L_b	L_a	L_c	S_{bn}	S_{an}	D_{bp}	D_{ap}	D_{cn}
IV	u_{bc}	u_{ba}	L_c	L_a	L_b	S_{cp}	S_{ap}	D_{cn}	D_{an}	D_{bp}
V	u_{ca}	u_{ba}	L_c	L_b	L_a	S_{cn}	S_{bn}	D_{cp}	D_{bp}	D_{an}
VI	u_{ca}	u_{cb}	L_a	L_b	L_c	S_{ap}	S_{bp}	D_{an}	D_{bn}	D_{cp}

B. Proposed Control Block

Fig. 4 shows the control block of the proposed system. There are three loops: the forward constant input power regulator decomposes the input voltage component, and thus forms u_{p_ref} and u_{n_ref}, the wave-shapes of the input currents, the output voltage regulator samples the output voltage, and then forms the input conductance G_{in}, which

1422

is multiplied with u_{p_ref} and u_{n_ref}, thus the expectation input currents i_{p_ref} and i_{n_ref} are achieved.

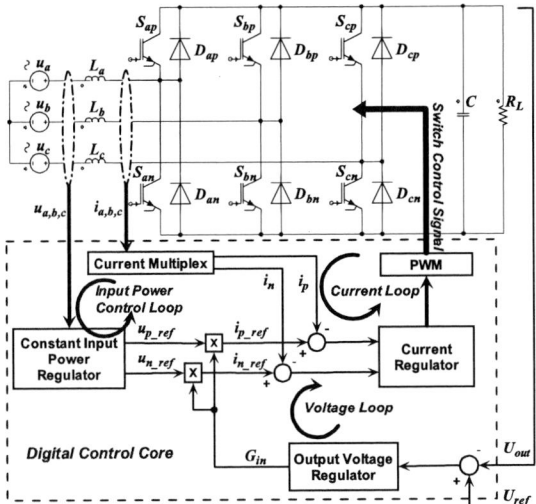

Figure.4 Control block of the proposed PFC rectifier.

III. CURRENT CONTROL STRATEGY

The three-phase currents are controlled such that the three-phase input power is constant. This control strategy is particularly beneficial to the load of the rectifier since the output filter can be minimized with a minimal output DC voltage ripple. To simplify the analysis, the instantaneous space vector representation is defined as [9]:

$$\vec{x} = x_\alpha + x_\beta = \sqrt{\frac{2}{3}}(x_a + x_b e^{j\frac{2\pi}{3}} + x_c e^{j\frac{4\pi}{3}}) \tag{1}$$

where $\vec{x}$ can be a voltage or current vector.

Supposing the input voltage vector is unbalanced, and the input current is controlled to be sinusoidal, they can be decomposed into positive and negative sequence components

$$\vec{U}_{in} = \vec{U}^+ + \vec{U}^- = U^+ e^{j(\omega t + \theta_u^+)} + U^- e^{j(-\omega t + \theta_u^-)} \tag{2a}$$

$$\vec{I}_{in} = \vec{I}^+ + \vec{I}^- = I^+ e^{j(\omega t + \theta_i^+)} + I^- e^{j(-\omega t + \theta_i^-)} \tag{2b}$$

where $\vec{U}^+$, $\vec{U}^-$ and $\vec{I}^+$, $\vec{I}^-$ are the input positive and negative sequence voltage and current vectors. Thus the input vector apparent power is

$$\begin{aligned}\vec{S}_{in} &= \vec{U}_{in}\vec{I}_{in}^* = P_{in} + jQ_{in}\\ &= U^+ I^+ e^{j(\theta_u^+ - \theta_i^+)} + U^- I^- e^{j(\theta_u^- - \theta_i^-)}\\ &\quad + U^+ I^- e^{j(2\omega t + \theta_u^+ - \theta_i^-)} + U^- I^+ e^{j(-2\omega t + \theta_u^- - \theta_i^+)}\end{aligned} \tag{3}$$

To make P_{in} instantaneously constant, the following equations must be satisfied

$$U^+ / U^- = I^+ / I^- \tag{4a}$$

$$\theta_u^+ - \theta_i^+ = \theta_i^- - \theta_u^- - \pi \tag{4b}$$

From (4a) we get

$$I^+ = G_{in} U^+ \tag{5a}$$

$$I^- = G_{in} U^- \tag{5b}$$

where G_{in} is the emulated conductance of the rectifier. In order to get better performance, we let positive sequence current in-phase with positive sequence voltage, that is

$$\theta_u^+ = \theta_i^+ \tag{6}$$

Substituting (6) into (4b), we get

$$\theta_i^- = \theta_u^- + \pi \tag{7}$$

Substituting (6) and (7) into (3), we get

$$P_{in} = U^+ I^+ - U^- I^- \tag{8a}$$

$$Q_{in} = (U^+ I^- + U^- I^+)\sin(2\omega t + \theta_u^+ - \theta_i^-) \tag{8b}$$

From (5)-(7), we can get

$$i_a = G_{in} u_a', \quad i_b = G_{in} u_b', \quad i_c = G_{in} u_c' \tag{9}$$

where u_a', u_b' and u_c' are delta positive and negative sequence components of input phase voltages (DPNSC), that is

$$u_a' = u_{ap} - u_{an}, \quad u_b' = u_{bp} - u_{bn}, \quad u_c' = u_{cp} - u_{cn} \tag{10}$$

and u_{ap}, u_{bp}, u_{cp}, u_{an}, u_{bn}, u_{cn} are the positive and negative sequence components of the input voltages and they can be calculated by the following equations[9]

$$u_{ap}(\omega t) = \frac{1}{3}(u_a(\omega t) + u_b(\omega t - \frac{4\pi}{3}) + u_c(\omega t - \frac{2\pi}{3})) \tag{11a}$$

$$u_{bp}(\omega t) = \frac{1}{3}(u_b(\omega t) + u_c(\omega t - \frac{4\pi}{3}) + u_a(\omega t - \frac{2\pi}{3})) \tag{11b}$$

$$u_{cp}(\omega t) = \frac{1}{3}(u_c(\omega t) + u_a(\omega t - \frac{4\pi}{3}) + u_b(\omega t - \frac{2\pi}{3})) \tag{11c}$$

$$u_{an}(\omega t) = \frac{1}{3}(u_a(\omega t) + u_b(\omega t - \frac{2\pi}{3}) + u_c(\omega t - \frac{4\pi}{3})) \tag{12a}$$

$$u_{bn}(\omega t) = \frac{1}{3}(u_b(\omega t) + u_c(\omega t - \frac{2\pi}{3}) + u_a(\omega t - \frac{4\pi}{3})) \tag{12b}$$

$$u_{cn}(\omega t) = \frac{1}{3}(u_c(\omega t) + u_a(\omega t - \frac{2\pi}{3}) + u_b(\omega t - \frac{4\pi}{3})) \tag{12c}$$

It is noted from (9) that only one parameter G_{in} is uncertain, it can be obtained from the output voltage regulator, and it will be described in the following section. Now when we keep input phase currents always satisfying with (9), the control motive of constant input instantaneous power can be achieved.

In ideal condition, the loss of power devices can be ignored, that is

$$p_{in} = p_{out} \tag{13}$$

Assume the load is linear type, that is

$$p_{out} = U_{out}^2 / R_L \tag{14}$$

From (13) and (14), we can realize that if we keep the input instantaneous power constant, the output voltage will always keeps constant and low order harmonics are eliminated too.

From (8)-(11), if the input voltages are balanced, then $\vec{U}^-$, $\vec{I}^-$ are all zero, this lead to unity-power-factor; if the input voltages are unbalanced, $\vec{U}^-, \vec{I}^-$ are all nonzero, the three types of power factor defined by IEEE std 1459-2000[12] are $P_{FV} = 1$, $P_{Fe} \leq P_{FA} < 1$, and it can be noted that the input power factor is non-unity.

IV. CONTROLLER IMPLEMENT

The implementation of digital controller is shown in Fig.5, it includes five functional blocks.

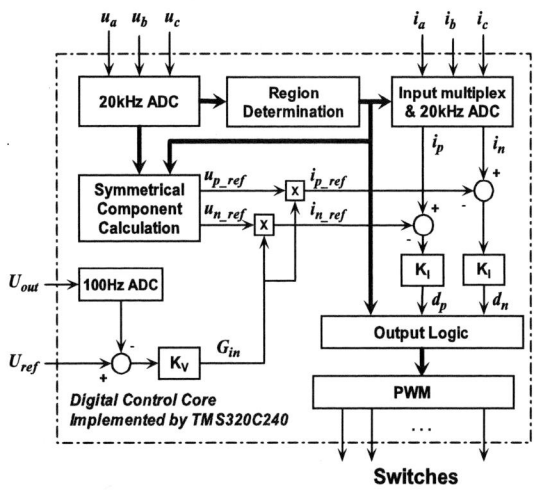

Figure.5 Digital Control block of the proposed rectifier.

TABLE II
THE REGION DETERMINATION ALGORITHM

conditions	region	u_{p_ref}	u_{n_ref}	i_p	i_n
$-u_{bn0} > u_{an0}$ $-u_{bn0} > u_{cn0}$	I	u'_a	u'_c	i_a	i_c
$u_{an0} > -u_{bn0}$ $u_{an0} > -u_{cn0}$	II	$-u'_b$	$-u'_c$	$-i_b$	$-i_c$
$-u_{cn0} > u_{bn0}$ $-u_{cn0} > u_{an0}$	III	u'_b	u'_a	i_b	i_a
$u_{bn0} > -u_{cn0}$ $u_{bn0} > -u_{an0}$	IV	$-u'_c$	$-u'_a$	$-i_c$	$-i_a$
$-u_{an0} > u_{cn0}$ $-u_{an0} > u_{bn0}$	V	u'_c	u'_b	i_c	i_b
$u_{cn0} > -u_{an0}$ $u_{cn0} > -u_{bn0}$	VI	$-u'_a$	$-u'_b$	$-i_a$	$-i_b$

A. Region Determination block

This block is to determine the operation region as well as the corresponding equivalent circuit at a given time. From Section II we realize that the line voltages are controlled to rectify the output dc voltage, it can be implemented by comparing the non-zero components of the input voltages. Now we assume that

$$
\begin{aligned}
u_0 &= (u_a + u_b + u_c)/3 \\
u_{an0} &= u_a - u_0 \\
u_{bn0} &= u_b - u_0 \\
u_{cn0} &= u_c - u_0
\end{aligned}
\tag{15}
$$

The region determination algorithm is listed in table II.

B. Symmetrical Component Calculation block

This block uses (12) to calculate instantaneous positive and negative sequence components of input phase voltages, uses (10) to calculate the delta positive and negative sequence components of input phase voltages (DPNSC): u'_a, u'_b and u'_c, then from table II to determine u_{p_ref} and u_{n_ref}.

C. Output Voltage-Loop Control Block

This block is to sample the output voltage, compare it with the reference voltage, and then achieve the parameter G_{in} of (9). For the design of the digital output voltage-loop controller, a PI regulator and a low frequency Analog-to-Digital convertor which is based on sampling the output voltage at twice the line frequency can be used. Sampling at this frequency provides filtering of the second order harmonic components. However, this low sampling frequency also limits the bandwidth of the voltage loop to be less than the ac line frequency.

D. Current-Loop Control Block

The current loop regulator design is based on digital redesign, using the pole-zero matched transformation technique [10]. A high switching (and current sampling) frequency in this case offers an additional advantage of simple implementation of the high-bandwidth current loop using a low-resolution computational unit [10], [11]. The limitations caused by sampling and processing delay are insignificant because of the high sampling frequency. The bandwidth of the current loop is approximately 20 kHz.

TABLE III
SWITCH CONTROL STATE IN DIFFERENT REGION

Region number	Switch control state					
	S_{ap}	S_{an}	S_{bp}	S_{bn}	S_{cp}	S_{cn}
I	off	d_p	on	off	off	d_n
II	off	on	d_p	off	d_n	off
III	off	d_n	off	d_p	on	off
IV	d_n	off	off	on	d_p	off
V	on	off	off	d_n	off	d_p
VI	d_p	off	d_n	off	off	on

E. Output logic and PWM

This block uses both the outputs of current-loop (the switch duty ratios: d_p and d_n) and what the region number lay to control the PWM block of the TMS320F240 outputting the PWM control signal. The relationship between switch control state and region is listed in Table III.

V. EXPERIMENTAL VERIFICATION

To verify the theoretical analysis, a 2kW three-phase PFC experimental system is built according to the main circuit topology shown in Fig.4, and the control block is shown in Fig.5. The system uses TMS 320F240 as main controller; input inductances are 560 μH ; output capacitor is 470 μF ; the switch components are MTY25N60E; rectifier diodes are MUR3080; The input phase voltages are: phase a is 110V, phase b is 88V(20% amplitude decreased), phase c is 110V, with $\pi/6$ delay; The output DC voltage is 475V; the switching frequency is 20kHz; load resistance is 150 Ω ; output power is 1.5 kW. The measured three-phase current waveforms and phase a voltage are shown in Fig.6. All waveforms are measured by Tektronix oscilloscope TDS520. The measured THD at full load is 5.4%, while the phase voltages have a THD of 1.5%.

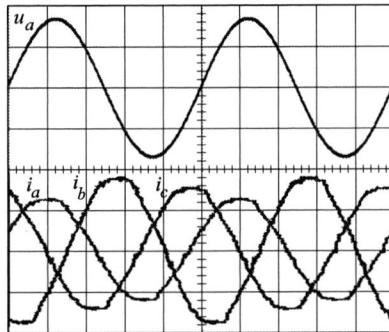

Figure.6 Waveforms of phase a voltage and input currents. Horizontal scale 4ms/div. Top: input phase voltage 100V/div. Bottom: input phase currents, 5A/div.

VI. CONCLUSION

In this paper, a new simple digital controller of three-phase power-factor-correction Rectifier has been described. The controller is derived from a parallel-connected dual-boost topology and uses the most common six-switch H-bridge topology.

With the proposed controller, low current THD and high power factor can be achieved whenever the input utility is balanced or unbalanced. Especially, the controller introduces a forward control loop that can keep the input instantaneous power constant, and thus make output DC voltage low harmonic eliminated. This results in that the output capacitor can be minimized with a minimal output DC voltage ripple. Furthermore, because only two switches are operated in high frequency during each region, switching losses are reduced significantly. The digital implementation of this control strategy is so simple that it is a low cost solution for realizing three-phase PFC.

REFERENCES

[1] L. Mom, P. D. Ziogas and G. Joos, "Design aspects of synchronous PWM rectifier-inverter system under unbalanced input voltage conditions," *IEEE Transactions on Industry Applications*, vol. 28, pp. 1286-1293, Nov. 1992.

[2] H. S. Kim, H. S. Mok, G. H. Choe, et al. "Design of current controller for 3-phase PWM converter with unbalanced input voltage". *IEEE PESC Conf. 1998*, vol. 1, pp. 503-509. May 1998.

[3] A. V. Stankovic and T. A. Lipo, "A novel control method for input output harmonic elimination of the PWM Boost type rectifier under unbalanced operating conditions". *IEEE Transactions on Power Electronics*, vol. 6, pp. 603-611, Sept 2001.

[4] P. Rioual, H. Pouliquen and J. P. Louis, "Regulation of a PWM rectifier in the unbalanced network state", *IEEE-PESC Conference 1993*, pp. 641-647, Jun. 1993.

[5] S. Chattopadhyay; V. Ramanarayanan, "A voltage sensorless control method to balance the input current of the boost rectifier under unbalanced input voltage condition," *IEEE 33rd Annual on Power Electronics Specialists Conf.*, Vol.4, pp. 1941-1946, June 2002.

[6] M. Malinowski, G. Marques, M. Cichowlas, et al, "New direct power control of three-phase PWM boost rectifiers under distorted and imbalanced line voltage conditions." *IEEE International Symposium on Industrial Electronics*, Vol.1, pp. 438-443, June 2003.

[7] C. Qiao and K. M. Smedley, "A General Three Phase PFC Controller for Rectifiers With a Parallel Connected Dual Boost Topology", *IEEE Trans. on Power Electronics*, vol. 17, pp. 925-934, Oct. 2002.

[8] T. Jin, C. Qiao, K. M. Smedley, "Operation of unified constant-frequency integration controlled three-phase active power filter in unbalanced system", *IEEE the 27th Annual Conf. of the Industrial Electronics Society 2001*, Vol. 2, pp. 1539-1545, Nov. 2001.

[9] L. Wei, Y. Matsushita and T. A. Lipo, "Investigation of dual-bridge matrix converter operating under unbalanced source voltages", *IEEE 34th Power Electronics Specialist Conf.*, vol.3, pp. 1293-1298, June 2003.

[10] G. F. Franklin, J. D. Powell, and M. L. Workman, Digital Control of Dynamic Systems, 3rd ed. New York: Addison-Wesley, 1998.

[11] D. Williamson, Digital Control and Implementation. Englewood Cliffs, NJ: Prentic-Hall, 1991, pp. 201-274.

[12] IEEE-SA Standards, IEEE Trial-Use Standard Definitions for the Measurement of Electric Power Quantities Under Sinusoidal, Nonsinusoidal, Balanced, or Unbalanced Conditions. pp. 18-25, Jan. 2000.

2006 5th International Power Electronics and Motion Control Conference

An Improved and Digital Current Control Strategy for One Cycle Control Based Three-Phase Boost Rectifier under Unbalanced System

LI Shao-long*, JIN Ai-juan* and LI Hang-tian**

* College of Electrical Engineering, University of Shanghai for Science and Technology, Shanghai 200093, China.
** Guangdong Telecom, Guangzhou Branch, Guangzhou 510620, China
Email: ajjin@126.com

Abstract—**A new tendency in development of power supplies is Power Factor Corrected (PFC) rectifiers which can work without harmonic pollution to source. Literature [1] proposes a new one-cycle controlled three-phase boost PFC rectifier which has many advantages[2]. But the rectifier is analog based and difficult to control input currents flexibly under unbalanced source. This paper improves the control strategy and introduces a new general current control equation which can control the input currents flexibly under unbalanced source. By selecting different current control parameters (CCP), we can achieve four control goals: (1) the same method as that in literature [1]; (2) input currents fully in-phase with its own input voltages; (3) constant input power; (4)balanced input currents. Under unbalanced input voltages, in the case of constant input power of controller, a new control strategy is presented to achieve minimum input line loss by adjusting the input currents. Experimental results are presented to verify the theoretical analyses. The system is implemented in full digital, and one Texas Instrument's DSP TMS320LF2811 is used as the digital controller.**

Keywords-digital control; power factor correction; three-phase; unbalanced system; flexible current control.

I. INTRODUCTION

In recent years, the harmonic pollution caused by convertors has been our growing concern. In order to overcome this problem, several PFC convertors have been developed and applied in practice[3,4]. And the control methods normally used for the six-switch bridge boost rectifier which is a common topology are based on either hysteretic control or d-q transformation control.

C. Qiao and K.M. Smedley have introduced a new one cycle controlled unified constant-frequency integration (UCI) six-switch bridges boost PFC rectifier[1], it is simple and reliable, fast response and high precision; and the input currents may keep low harmonic and linearly proportional to the non-zero sequence of the three-phase input voltages[2]. But it has two shortcomings. First, the

This work is sponsored by Shanghai Municipal Education Commission development foundation program (No: 05EZ06).

controller is analog based; Second, it fails to control input currents flexibly under unbalanced input source[4].

This paper proposes a full digital implemented controller, and improves the control strategy which is proposed by C. Qiao, introduces a new general current control equation that can control the input currents flexibly. By selecting CCP, we can get four control methods: (1) Constant CCP method. The system is the same as that in [1]; (2) In-phase currents control method. Each phase current strictly traces its own phase voltage; (3) Constant input power method. With this, the output filter can be minimized; (4) Balanced input currents method under balanced or unbalanced source.

In the case of constant input power of controller and under unbalanced source voltages, a new control strategy is presented to achieve minimum input line loss by ajusting the input currents. So the efficiency of source power is maximum.

II. MODELING AND CURRENT CONTROL EQUATION

Fig. 1 shows the topology of a three-phase six-switch H-bridge boost PFC controller. For analytical convenience, we assume that the input three-phase voltages are unbalanced, for example, phase b amplitude is 20% decreased, phase c is $\pi/6$ delay. The voltage waveforms are shown in Fig.2, and divided into 6 regions. It should be noticed that the starting point of regions may be different with the different unbalanced condition of input voltages, and it will be described in the following "Region Determination block" section. According to literature [1], the three-phase PFC rectifier may be implemented by controlling the two switches in a region of linear cycle.

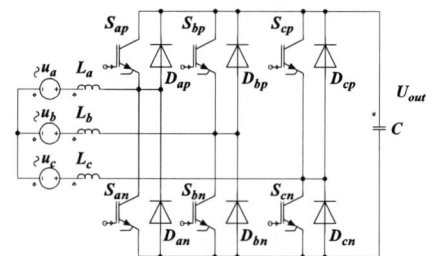

Figure.1 Topology of 3-phase six-switch H-bridge boost PFC rectifier.

1-4244-0448-7/06/$25.00 ©2006 IEEE   1426

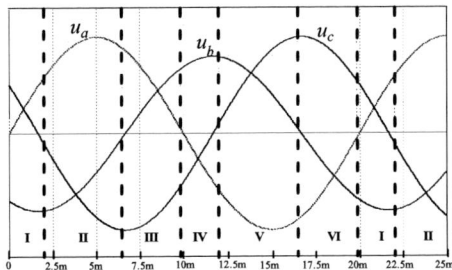

Figure.2 Waveforms under phase b 20% decreased, phase c $\pi/6$ delay.

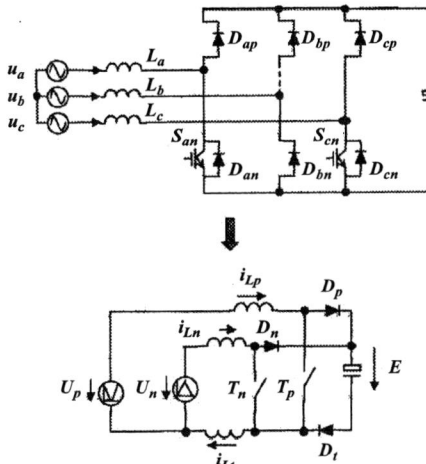

Figure.3 Change from three phase rectifier in Fig.1 to a parallel connected dual boost topology during region I.

TABLE I.
CROSS-REFERENCE BETWEEN SIX-SWITCH BRIDGE THREE-PHASE RECTIFIER AND THE PARALLEL-CONNECTED DUAL-BOOST TOPOLOGY

	U_p	U_n	L_p	L_n	L_t	T_p	T_n	D_p	D_n	D_t
I	u_{ab}	u_{cb}	L_a	L_c	L_b	S_{an}	S_{bn}	D_{ap}	D_{cp}	D_{bn}
II	u_{ab}	u_{ac}	L_b	L_c	L_a	S_{bp}	S_{cp}	D_{bn}	D_{cn}	D_{ap}
III	u_{bc}	u_{ca}	L_b	L_a	L_c	S_{bn}	S_{an}	D_{bp}	D_{ap}	D_{cn}
IV	u_{bc}	u_{ba}	L_c	L_a	L_b	S_{cp}	S_{ap}	D_{cn}	D_{an}	D_{bp}
V	u_{ca}	u_{ba}	L_c	L_b	L_a	S_{cn}	S_{bn}	D_{cp}	D_{bp}	D_{an}
VI	u_{ca}	u_{cb}	L_a	L_b	L_c	S_{ap}	S_{bp}	D_{an}	D_{bn}	D_{cp}

Assume that the input voltages are sine waves, the three phase circuit parameters are balanced, the forward impedances and all parasitic parameters of power devices are neglected. In region I, switch S_{bn} is kept always on, only switches S_{an} and S_{cn} need to be controlled, the three phase rectifier may be decouplled to parallel connected dual boost topology just like Fig.3, where U_p, U_n, L_p, L_n, L_t, T_p, T_n, D_p, D_n and D_t are symbolized voltages, inductors and diodes, whose corresponding components are listed in Table I[1]. Since the switching frequency is much higher than the line frequency, the inductor voltage-second balance is used

$$\begin{cases} U_p^{\bullet} d_n + (U_p^{\bullet} + \frac{1}{3}E)(d_p - d_n) + (U_p^{\bullet} - \frac{1}{3}E)(1 - d_p) = 0 \\ U_n^{\bullet} d_n + (U_n^{\bullet} - \frac{2}{3}E)(d_p - d_n) + (U_n^{\bullet} - \frac{1}{3}E)(1 - d_p) = 0 \\ U_t^{\bullet} d_n + (U_t^{\bullet} - \frac{1}{3}E)(d_p - d_n) + (U_t^{\bullet} - \frac{2}{3}E)(1 - d_p) = 0 \end{cases} \tag{1}$$

$$U_p^{\bullet} + U_n^{\bullet} - U_t^{\bullet} = 0 \tag{2}$$

This results in

$$\begin{bmatrix} 1 - d_p \\ 1 - d_n \end{bmatrix} = \frac{1}{E}\begin{bmatrix} 2 & 1 \\ 1 & 2 \end{bmatrix}\begin{bmatrix} U_p^{\bullet} \\ U_n^{\bullet} \end{bmatrix} \tag{3}$$

Where $\begin{bmatrix} U_p^{\bullet} & U_n^{\bullet} & U_t^{\bullet} \end{bmatrix} = \begin{bmatrix} U_p & U_n \end{bmatrix}\begin{bmatrix} 2/3 & -1/3 & 1/3 \\ -1/3 & 2/3 & 1/3 \end{bmatrix}$ (4)

As shown in Fig.1, the PFC rectifier is a three-phase three-line system, so at any time,

$$i_a + i_b + i_c = 0 \tag{5}$$

Now we assume that

$$u_{an0} = \lambda_a R_e i_a, u_{bn0} = \lambda_b R_e i_b, u_{cn0} = \lambda_c R_e i_c \\ u_a = \lambda_a' R_e i_a, u_b = \lambda_b' R_e i_b, u_c = \lambda_c' R_e i_c \tag{6}$$

where $\begin{cases} u_{an0} = u_a - (u_a + u_b + u_c)/3 \\ u_{bn0} = u_b - (u_a + u_b + u_c)/3 \\ u_{cn0} = u_c - (u_a + u_b + u_c)/3 \end{cases}$ (7)

and λ_a, λ_b, λ_c, λ_a', λ_b', λ_c' are CCP, R_e is the standard equivalent resistance. From (4), in region I, we get

$$U_p^{\bullet} = \frac{2}{3}U_p - \frac{1}{3}U_n = \frac{2}{3}u_a - \frac{1}{3}u_b - \frac{1}{3}u_c = \frac{2}{3}u_{an0} - \frac{1}{3}u_{bn0} - \frac{1}{3}u_{cn0} \tag{8-1}$$

$$U_n^{\bullet} = \frac{2}{3}U_n - \frac{1}{3}U_p = \frac{2}{3}u_c - \frac{1}{3}u_a - \frac{1}{3}u_b = \frac{2}{3}u_{cn0} - \frac{1}{3}u_{an0} - \frac{1}{3}u_{bn0} \tag{8-2}$$

$$\langle i_{Lp} \rangle = \langle i_{La} \rangle = i_a, \quad \langle i_{Ln} \rangle = \langle i_{Lc} \rangle = i_c \tag{8-3}$$

Substituting (5), (6) into (8), results in

$$\begin{bmatrix} U_p^{\bullet} \\ U_n^{\bullet} \end{bmatrix} = \frac{R_e}{3}\begin{bmatrix} 2\lambda_a' + \lambda_b' & \lambda_b' - \lambda_c' \\ \lambda_b' - \lambda_a' & 2\lambda_c' + \lambda_b' \end{bmatrix}\begin{bmatrix} \langle i_{Lp} \rangle \\ \langle i_{Ln} \rangle \end{bmatrix} \tag{9}$$

Substituting (9) into (3), results in

$$\begin{bmatrix} 1 - d_p \\ 1 - d_n \end{bmatrix} = \frac{R_e}{ER_s}R_s\begin{bmatrix} \lambda_a + \lambda_b & \lambda_b \\ \lambda_b & \lambda_c + \lambda_b \end{bmatrix}\begin{bmatrix} \langle i_{Lp} \rangle \\ \langle i_{Ln} \rangle \end{bmatrix} \tag{10-1}$$

Suppose $\quad U_m = ER_s / R_e \tag{10-2}$

Where R_s is the current sensor equivalent resistance, V_m is the output voltage of error compensator in feedback voltage loop. Now (10) may be expressed as

$$U_m\begin{bmatrix} 1 - d_p \\ 1 - d_n \end{bmatrix} = R_s\begin{bmatrix} \lambda_a + \lambda_b & \lambda_b \\ \lambda_b & \lambda_c + \lambda_b \end{bmatrix}\begin{bmatrix} \langle i_{Lp} \rangle \\ \langle i_{Ln} \rangle \end{bmatrix} \tag{11}$$

Using the same analytical method, the duty ratios in all regions may be expressed as

$$U_m\begin{bmatrix} 1 - d_p \\ 1 - d_n \end{bmatrix} = R_s\mathbf{T}\begin{bmatrix} \langle i_{Lp} \rangle \\ \langle i_{Ln} \rangle \end{bmatrix} = R_s\mathbf{T}'\begin{bmatrix} \langle i_{Lp} \rangle \\ \langle i_{Ln} \rangle \end{bmatrix} \tag{12}$$

Where the current control matrix (CCM) **T** in different regions is listed in Table II, and substituting λ_a, λ_b, λ_c to λ_a', λ_b', λ_c' we get the next current control matrix (NCCM) **T'**.

TABLE II.
MATRIX **T** IN DIFFERENT REGIONS

	T		**T**
I	$\begin{bmatrix} \lambda_a + \lambda_b & \lambda_b \\ \lambda_b & \lambda_c + \lambda_b \end{bmatrix}$	IV	$\begin{bmatrix} \lambda_c + \lambda_b & \lambda_b \\ \lambda_b & \lambda_a + \lambda_b \end{bmatrix}$
II	$\begin{bmatrix} \lambda_b + \lambda_a & \lambda_a \\ \lambda_a & \lambda_c + \lambda_a \end{bmatrix}$	V	$\begin{bmatrix} \lambda_c + \lambda_a & \lambda_a \\ \lambda_a & \lambda_b + \lambda_a \end{bmatrix}$
III	$\begin{bmatrix} \lambda_b + \lambda_c & \lambda_c \\ \lambda_c & \lambda_a + \lambda_c \end{bmatrix}$	VI	$\begin{bmatrix} \lambda_a + \lambda_c & \lambda_c \\ \lambda_c & \lambda_b + \lambda_c \end{bmatrix}$

Now (12) is the current control equation, and it shows that if the CCP λ_a, λ_b and λ_c or λ_a', λ_b' and λ_c' are ascertained, the three-phase PFC can be realized by controlling the duty ratio of switches and such that the linear combination of inductor currents $\langle i_{Lp} \rangle$ and $\langle i_{Ln} \rangle$ satisfy the current control equation.

III. FOUR METHODS OF INPUT CURRENTS CONTROL UNDER UNBALANCED SOURCE

It can be noted from (12) that this rectifier offers a great flexibility to support various current control options, this is an especial useful characteristic to achieve different control motive under unbalanced source.

A. Constant Current Control Parameters Method

The three input phase currents are controlled using constant CCP, and from table II, the CCM is constant also. This is the simplest control method and the same as which is proposed by C. Qiao[1].

Let $\lambda_a = \lambda_b = \lambda_c = 1$, since $u_{an0} + u_{bn0} + u_{cn0} \equiv 0$, from (6), we know that the phase current restriction equation $i_a + i_b + i_c \equiv 0$ is satisfied, now we get

$$U_m \begin{bmatrix} 1 - d_p \\ 1 - d_n \end{bmatrix} = R_s \begin{bmatrix} 2 & 1 \\ 1 & 2 \end{bmatrix} \begin{bmatrix} \langle i_{Lp} \rangle \\ \langle i_{Ln} \rangle \end{bmatrix} \tag{13}$$

$$[i_a \ i_b \ i_c] = \frac{1}{R_e}[u_{an0} \ u_{bn0} \ u_{cn0}] = \frac{U_m}{R_s E}[u_{an0} \ u_{bn0} \ u_{cn0}] \tag{14}$$

Eqn.(14) shows that under unbalanced source and by selecting constant CCP, the phase current will be linearly proportional to the non-zero sequence of the three-phase system u_{an0}, u_{bn0} and u_{cn0}, and the three phase currents are expected to be sinusoidal.

B. In-phase Currents Control Method

In some cases we require each phase current strictly traces its own phase voltage, that is each phase has a pure resistance load equivalently: R_a, R_b and R_c; thus $u_a = R_a i_a$, $u_b = R_b i_b$ and $u_c = R_c i_c$. To achieve this goal, we may let one group of CCP: λ_a', λ_b' and λ_c' reflect to input unbalanced voltage correspondingly.

For a three-phase three-line system, at any time t_0, t_1, t_2, ..., t_n, the phase currents must satisfy with (5), that is

$$u_a(t_0)/R_a + u_b(t_0)/R_b + u_c(t_0)/R_c = 0$$
$$u_a(t_1)/R_a + u_b(t_1)/R_b + u_c(t_1)/R_c = 0$$
$$\vdots \tag{15}$$
$$u_a(t_n)/R_a + u_b(t_n)/R_b + u_c(t_n)/R_c = 0$$

Due to the potential interference in the power grid, in order to have a more accurate result, the sampling voltages may be divided into several groups, for example, we select two groups that have M and N sampled points and the CCP may be given by calculating the two groups. According to (15), we have

$$\begin{cases} \dfrac{1}{R_b}\sum_{i=0}^{M}u_b(t_i) + \dfrac{1}{R_c}\sum_{i=0}^{M}u_c(t_i) = -\dfrac{1}{R_a}\sum_{i=0}^{M}u_a(t_i) \\ \dfrac{1}{R_b}\sum_{j=0}^{N}u_b(t_j) + \dfrac{1}{R_c}\sum_{j=0}^{N}u_c(t_j) = -\dfrac{1}{R_a}\sum_{j=0}^{N}u_a(t_j) \end{cases} \tag{16}$$

resulting in

$$\lambda_a' = 1; \lambda_b' = -\frac{\begin{vmatrix} \sum_{i=0}^{M}u_b(t_i) & \sum_{i=0}^{M}u_c(t_i) \\ \sum_{j=0}^{N}u_b(t_j) & \sum_{j=0}^{N}u_c(t_j) \end{vmatrix}}{\begin{vmatrix} \sum_{i=0}^{M}u_a(t_i) & \sum_{i=0}^{M}u_c(t_i) \\ \sum_{j=0}^{N}u_a(t_j) & \sum_{j=0}^{N}u_c(t_j) \end{vmatrix}}; \lambda_c' = -\frac{\begin{vmatrix} \sum_{i=0}^{M}u_c(t_i) & \sum_{i=0}^{M}u_b(t_i) \\ \sum_{j=0}^{N}u_c(t_j) & \sum_{j=0}^{N}u_b(t_j) \end{vmatrix}}{\begin{vmatrix} \sum_{i=0}^{M}u_a(t_i) & \sum_{i=0}^{M}u_b(t_i) \\ \sum_{j=0}^{N}u_a(t_j) & \sum_{j=0}^{N}u_b(t_j) \end{vmatrix}}$$
$$R_a = \lambda_a' R_e = R_e; R_b = \lambda_b' R_e; R_c = \lambda_c' R_e \tag{17}$$

It can be simply proved that when the input voltages are sinusoidal and steady, the CCP keeps stable at that moment, and thus the input phase currents are sinusoidal and will fully trace each own input voltages.

C. Constant Input Power Method

The three-phase currents are controlled so that the three-phase input power is a constant. This control strategy is particularly beneficial to the load of the rectifier since the output filter can be minimized with a minimal output DC voltage ripple.

As has been described in literature [5, 6], in order to archive constant input power, the input currents and voltages must satisfy with

$$|U^+|/|U^-| = |I^+|/|I^-| \tag{18}$$

$$\theta_u^+ - \theta_i^+ = \theta_i^- - \theta_u^- - \pi \tag{19}$$

From (18) we get

$$|I^+| = G_{in}|U^+| \tag{20}$$

$$|I^-| = G_{in}|U^-| \tag{21}$$

In order to get better performance, we let positive sequence current in-phase with positive sequence voltage, that is

$$\theta_u^+ = \theta_i^+ \tag{22}$$

From (19), we get

$$\theta_i^- = \theta_u^- + \pi \qquad (23)$$

Form (20)-(23), we can get

$$i_a = G_{in}(u_{ap} - u_{an}), i_b = G_{in}(u_{bp} - u_{bn}), i_c = G_{in}(u_{cp} - u_{cn}) \quad (24)$$

Where u_{ap}, u_{an}, u_{bp}, u_{bn}, u_{cp}, u_{cn} are the positive and negative sequence voltage which can be calculated by the method which is described in literature [6]. Now from (6), (24), the CCP λ_a, λ_b and λ_c can be calculated.

D. Balanced Input Currents Method

With balanced input currents control strategy, three phase currents are balanced. In order to get better performance, we assume that phase a current in-phase with phase a voltage, now we get

$$i_a(t) = u_{an0}(t) / \lambda_a / R_e$$
$$i_b(t) = u_{an0}(t - 2T/3) / \lambda_a / R_e \qquad (25)$$
$$i_c(t) = u_{an0}(t - 4T/3) / \lambda_a / R_e$$

Where T is the fundamental period of input voltage. Let $\lambda_a = 1$, from (6) and (25), we get

$$\lambda_a = 1$$
$$\lambda_b = (t - 2T/3)u_{bn0}(t) / u_{an0} \qquad (26)$$
$$\lambda_c = (t - 4T/3)u_{cn0}(t) / u_{an0}$$

From the above analyses, this CCP is time-variant, and using current control equation (12), we can control the input currents in balanced method with low harmonic.

IV. INPUT PHASE-VOLTAGE AND CURRENT CONTROL PARAMETER ESTIMATE

In practical application, the input power grid is always slight unbalanced, the phase angle delayed do not exceed $\pm 10°$ and the differences among the line-to-neutral voltages remain within the range of $\pm 10\%$[7]. Since the PWM frequency is far higher than the line frequency, and the unbalanced condition is changing slowly, we can use the phase currents which have been detected to calculate the phase voltages at the same moment, once one or two previous cycle voltages have been calculated and stored, we can substituting those voltage values into (17), (24) or (26) to estimate the CCP of the next PWM cycle. Now we describe how to use detected phase currents to calculate the phase voltages at the same moment.

From (4), we can get

$$\begin{bmatrix} U_p \\ U_n \end{bmatrix} = \begin{bmatrix} 2 & 1 \\ 1 & 2 \end{bmatrix} \begin{bmatrix} U_p^* \\ U_n^* \end{bmatrix} \qquad (27)$$

Substituting (27), (10-2) into (9), we can get

$$\begin{bmatrix} U_p \\ U_n \end{bmatrix} = \frac{ER_s}{U_m} \begin{bmatrix} \lambda_a + \lambda_b & \lambda_b \\ \lambda_b & \lambda_c + \lambda_b \end{bmatrix} \begin{bmatrix} \langle i_{Lp} \rangle \\ \langle i_{Ln} \rangle \end{bmatrix} \qquad (28)$$

Using the same analytical method, the duty ratios in all regions may be expressed as

$$\begin{bmatrix} U_p \\ U_n \end{bmatrix} = \frac{ER_s}{U_m} \mathbf{T} \begin{bmatrix} \langle i_{Lp} \rangle \\ \langle i_{Ln} \rangle \end{bmatrix} = \frac{ER_s}{U_m} \mathbf{T'} \begin{bmatrix} \langle i_{Lp} \rangle \\ \langle i_{Ln} \rangle \end{bmatrix} \qquad (29)$$

Once U_p, U_n have been calculated, from table I, we can calculate phase voltages.

V Minimum line loss control strategy

In a three-phase three-line system, under unbalanced source, the three phase input currents of PFC rectifiers may be adjusted to achieve minimum input line loss by detecting the phase voltages in the case of constant input power of controller.

The input power of controller is known as

$$\begin{aligned} P_{in} &= U_A I_A \cos(\theta_{uA} - \theta_{iA}) + U_B I_B \cos(\theta_{uB} - \theta_{iB}) \\ &\quad + U_C I_C \cos(\theta_{uC} - \theta_{iC}) \\ &= |U^+||I^+|\cos(\theta_u^+ - \theta_i^+) + |U^-||I^-|\cos(\theta_u^- - \theta_i^-) \end{aligned} \qquad (30)$$

The input line loss is

$$P_l = r(I_A^2 + I_B^2 + I_C^2) = r(|I^+|^2 + |I^-|^2) \qquad (31)$$

Substituting (30) into (31), we can get

$$P_l = r\left(|I^-|^2 + \frac{\sec^2(\theta_u^+ - \theta_i^+)(\cos^2(\theta_u^- - \theta_i^-)|U^-||I^-| - P_{in})^2}{|U^+|^2}\right) \quad (32)$$

Solve the partial derivative equations

$$\frac{\partial P_l}{\partial I^-} = 0, \quad \frac{\partial P_l}{\partial \theta_i^+} = 0, \quad \frac{\partial P_l}{\partial \theta_i^-} = 0 \qquad (33)$$

We may get

$$\theta_i^+ = \theta_u^+, \quad \theta_i^- = \theta_u^-, \quad |I^-| = \frac{|U^-|P_{in}}{(|U^-|^2 + |U^+|^2)} \qquad (34)$$

Substituting (34) into (31), we can get $|I^+|$. Because I^0 is zero in a three-phase three-line system, the currents I_A, I_B and I_C may be gotten by coordinate transformation. Here, the line loss is Minimum. In other words, the efficiency of source power is maximum.

VI EXPERIMENTAL VERIFICATION

To verify the theoretical analyses, a 2kW three-phase PFC experimental system is built according to the main circuit topology shown in Fig.1, and the control block is shown in Fig.5. The system uses TMS 320F2811 as main controller; input inductances are $560\,\mu H$; output capacitor is $470\,\mu F$; the switch components are MTY25N60E; rectifier diodes are MUR3080; The input phase voltages are: phase a is 110V, phase b is 88V(20% amplitude decreased), phase c is 110V, with $\pi/6$ delayed; The output DC voltage is 475V; the switching frequency is 10kHz; load resistance is 150 Ω; output power is 1.5 kW. The input current waveforms are shown in fig.5, where the voltage waveforms are 50v/div and the current waveforms are 5A/div; fig.5 (a) using constant CCP control method; fig.5 (b) using in-phase current control method; fig.5 (c) using constant input power control method; fig.5 (d) using balanced input currents control method. Fourier analysis is used in the current waveforms in fig.5, all the THD are lower than 3%. These verify the validity of the improved

1429

control strategy. All waveforms are measured by Tektronix oscilloscope TDS520. The measured THD at full load is 5.4%, while the phase voltages have a THD of 1.5%.

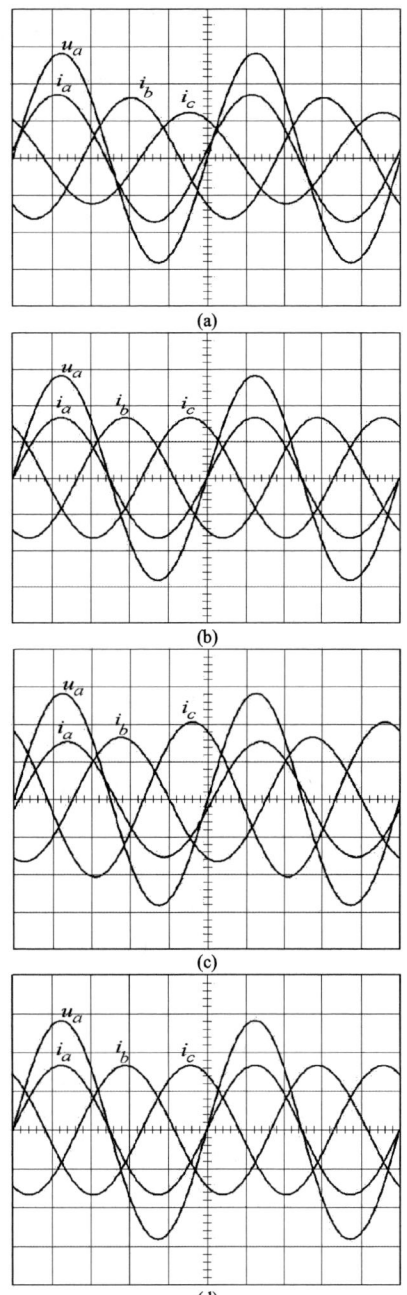

(a)

(b)

(c)

(d)

Fig. 5 Experimental waveforms under unbalanced source (phase a is 110V, phase b is 88V, phase c is 110V and with $\pi/6$ delayed) with four control methods: (a) Constant CCP control method; (b) In-phase currents control method; (c) Constant input power control method; (d) Balanced input currents control method.

V. CONCLUSION

This paper improves the new three-phase PFC boost rectifier control strategy based on one cycle control which is proposed by C. Qiao, and introduces a new general current control equation which can control the input currents flexibly to satisfy different situation. In the case of constant input power of controller and under unbalanced source voltages, a new control strategy is presented to achieve minimum input line loss by ajusting the input currents. By selecting different CCP, we get four methods to control the input currents in order to achieve our goals. Experimental results show that all the four methods keep the input currents in low harmonic, and verify that the modified control strategy is valid and has good performance. Since the CCP may be time-variant, the proposed controller needs to do some fast multiplication and division. As the rapid development of DSP technology, the cost of high performance DSP device is lower and lower, to implement this controller will also cost low now, and thus it can be widely used in many applications.

REFERENCES

[1] C. Qiao and K.M. Smedley, "A General Three Phase PFC Controller for Rectifiers With a Parallel Connected Dual Boost Topology", *IEEE Trans. on Power Electron.*, vol. 17, pp. 925-934, Nov. 2002.

[2] T. Jin; C. Qiao; K.M. Smedley, "Operation of unified constant-frequency integration controlled three-phase active power filter in unbalanced system", *Industrial Electronics Society, 2001. IECON '01.* Vol. 2, pp. 1539 – 1545, Nov 2001.

[3] T. Jin, L. Li, and K.M. Smedley; "A Universal Vector Controller for Three-phase PFC, APF, STATCOM, and Grid-Connected Inverter", *Applied Power Electronics Conf. and Exposition, 2004*, Vol. 1, pp. 594 – 600, Feb. 2004

[4] R. Zhang, F.C. Lee and D. Boroyevich, "Four-Legged Three-phase PFC Rectifier with Fault Tolerant Capability", *IEEE Power Electron. Specialists Conf. 2000.* Vol. 1, pp. 359-364. June 2000.

[5] A. V. Stankovic, T. A. Lipo, "A novel control method for input output harmonic ellmination of the PWM boost type rectifier under unbalanced operating conditions", *IEEE Trans. on Power Electronics*, Vol. 16, pp. 603-611, Sept. 2001.

[6] L. Wei, Y. Matsushita, and T. A. Lipo, "Investigation of dual-bridge matrix converter operating under unbalanced source voltages", *Power Electron. Specialist Conf., 2003. PESC '03. 2003 IEEE 34th Annual.* Vol. 3, pp. 1293-1298, June 2003.

[7] IEEE-SA Standards, IEEE Trial-Use Standard Definitions for the Measurement of Electric Power Quantities Under Sinusoidal, Nonsinusoidal, Balanced, or Unbalanced Conditions, pp.18-25, Jan. 2000.

2006 5th International Power Electronics and Motion Control Conference

Control Method for Power Quality Compensation Based on Levenberg-Marquardt Optimized BP Neural Networks

ZHOU MING[*], WAN JIAN-RU[*], WEI ZHI-QIANG[*] AND CUI JIAN[*]
[*]Tianjin University/Electrical Engineering & Automation, Tianjin, China
e-mail: wan_jr@eyou.com

Abstract—**Unified Power Quality Conditioner (UPQC) has the function of improving voltage supply, compensating load reactive power, suppressing harmonic current and increasing power factor; however, tradition control method has a certain extent limitation for such multiple input, multiple output, close coupling nonlinear issue. Artificial Neural Networks (ANN) can deal with data for multiple objective learning in parallel continuous way, the control of complex object is achieved through interactions between nerve cells. Levenberg-Marquardt algorithm optimized back propagation neural network has the characteristic of efficient learning and faster convergence; ANN outputs control signals for voltage and current compensation to UPQC through weights training. Simulation model is built in Matlab, load which is three phase unbalanced and has badly distorted current is simulated under the case of voltage sag. Simulation experiment indicates its compensation effectiveness is much more satisfying than traditional control method.**

Keywords-UPQC; neural network; harmonics compensation; voltage sag

I. INTRODUCTION

Recent years power quality issue draws our attentions higher and higher, Japanese scholar Akagi brought forward the conception of Unified Power Quality Conditioner (UPQC) which has been viewed as combination series and shunt Active Power Filters in [1]. The UPQC that installs at customer leading-in terminal be able to improve voltage supply, compensate load reactive power, suppress harmonic current and increase power factor. Since power distribution network is a huge miscellaneous system, also single load fluctuates continually and could hardly be forecasted; traditional control method would be influenced greatly by load fluctuation and could hardly compensate well when it is adopted to control UPQC. Artificial Neural Networks (ANN) technique that simulates biologic nervous system of human brain, can deal with a mass of data for multiple objective coordinated control in parallel way through continuous learning and training; so it is very suitable for such multiple input, multiple output, close coupling nonlinear issue in power system. Therefore, in this paper,

TianJin Fund for Natural Sciences (05YFJMJC11500)

ANN is applied to the design of UPQC controller; trap filter is used to extract the harmonic component of source current and load voltage and transfers them to ANN controller; ANN outputs control signals for voltage and current compensation to UPQC through weights training. At the same time, in order to reduce the influence of load fluctuation, against traditional UPQC topology some amelioration is made where parallel side is connected to source supply and series side to load. Parallel side provides path for harmonic current through filter compensation and improves sinusoidal level of source current waveform; series side enhances the ability resisting load fluctuation and provides customer with near sinusoidal voltage through filter compensation [2-4]. Simulation model is built in Matlab, load which is three phase unbalanced and has badly distorted current is simulated under the case of voltage sag.

II. ANN CONTROL CIRCUIT

A. Operating principles analyse

Fig.1 shows the equivalent circuit of ANN controlled UPQC. We denote the load voltage by v_L, source voltage by v_s and the voltage at the point of common coupling (PCC) by v_t. The shunt and series voltage source inverters are denoted as VSI.1 and VSI.2, respectively; these inverters are connected to DC energy storage absorption capacitor Cdc in their DC ends, and are connected to two transformers T1 and T2 which are joined in power network in their AC ends. The output of transformer T1 is connected in shunt with the feeder, while filter capacitor C1 absorbs the high frequency component of source current. The output of transformer T2 is connected in parallel with the capacitor C2 that is connected in series with the feeder and the load.

The idealized UPQC combines the current source i_f and the voltage source v_d. The purpose of the series voltage source of UPQC is to insert voltage v_d such that the load voltage v_L is a balanced sinusoid irrespective of unbalance or distortion in PCC voltage v_t. On the other hand, the purpose of the shunt current source is to inject current i_f such that the source current i_s is balanced and distortion free irrespective of the shape of the load current i_L[5].

According to Kirchhoff's current law (KCL), we get $i_s = i_l + i_f$; current transformer is used to measure source current i_s and transfer its value along with i_f to trap filter. The extracted harmonic component i_{sh} is carried to ANN control circuit, which outputs PWM control signals for VSI.1 inverter through weights training and learning. While $v_t = v_t + v_d$, potential transformer is used to measure load voltage v_L, and the extracted harmonic component v_{Lh} along with v_d are carried to ANN control circuit, which outputs PWM control signals for VSI.2 inverter.

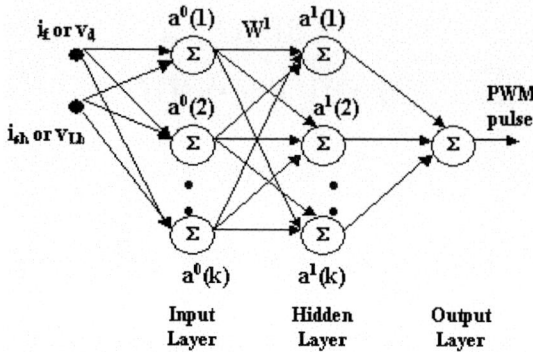

Figure 2. Feedforward ANN architecture

The gradient of V can be represented by

$$\frac{\partial V}{\partial \overline{x}} = J^T(\overline{x})e(\overline{x}). \qquad (3)$$

Where $\overline{x}$ is the vector of all the weights and biases, and $J(\overline{x})$ is the T*C Jacobian Matrix

$$J(\overline{x}) = \begin{bmatrix} \dfrac{\partial e_1(\overline{x})}{\partial x_1} & \dfrac{\partial e_1(\overline{x})}{\partial x_2} & \cdots & \dfrac{\partial e_1(\overline{x})}{\partial x_C} \\ \dfrac{\partial e_2(\overline{x})}{\partial x_1} & \dfrac{\partial e_2(\overline{x})}{\partial x_2} & \cdots & \dfrac{\partial e_2(\overline{x})}{\partial x_C} \\ \vdots & \vdots & \ddots & \\ \dfrac{\partial e_T(\overline{x})}{\partial x_1} & \dfrac{\partial e_T(\overline{x})}{\partial x_2} & \cdots & \dfrac{\partial e_T(\overline{x})}{\partial x_C} \end{bmatrix}. \qquad (4)$$

Well then, the weights and biases are updated using

$$\Delta \overline{x} = \left[J^T(\overline{x})J(\overline{x}) + \mu I \right]^{-1} J^T(\overline{x})e(\overline{x}). \qquad (5)$$

where μ is the step multiplier.

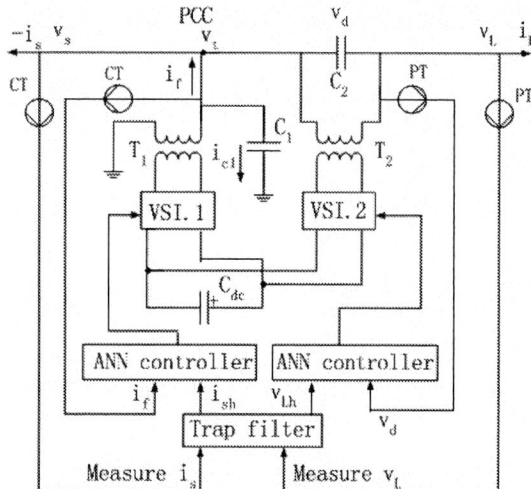

Figure 1. Equivalent circuit of ANN controlled UPQC

B. Trap filter

We adopt a Chebysbev 5th-order trap filter which has a lower stopband edge frequency set at 45Hz and an upper stopband edge frequency at 55Hz. This is to ensure that the fundamental component (50Hz) falls within the trap region. The working algorithm for this Chebysbev trap filter is based on a low-pass Chebysbev filter concept having a transfer function shown as:

$$H(s) = \frac{B(s)}{A(s)} = \frac{b(1)s^n + b(2)s^{n-1} + \cdots + b(n+1)}{s^n + a(2)s^{n-1} + \cdots + a(n+1)}. \qquad (1)$$

where n is the order of the low-pass filter and the transfer function will have (n+1) zeros b and (n+1) poles a.

C. Design of ANN controller

Levenberg-Marquardt optimized BP neural networks is well suited for ANN training where the performance function is mean square error. LMBP algorithm uses resulting derivatives for weigh updating, which possesses the characteristics of efficient learning, faster convergence and escaping local minimum [6]. The ANN controller designed here has two inputs and one output with a single hidden layer, shown as Fig.2.

The kernel steps in the LMBP algorithm are:

For the performance function in mean square error format

$$V = \frac{1}{2}\sum_{k=1}^{K} \overline{e}^T(k)\overline{e}(k) = \frac{1}{2}\sum_{k=1}^{K}\sum_{j=1}^{S_M} e_{j,k}^2. \qquad (2)$$

III. SIMULATION EXPERIMENTATION ANALYSES

A. Simulation Model and Parameter setting

In order to verify the above scheme of ANN controlled UPQC, we build the corresponding simulation model with Matlab neural network toolbox shown as Fig.3. We use triphase induction motor and biphase reactance through uniphase rectifier bridge to approximate load which is three phase unbalanced and has badly distorted current. Assume the system power supply is three phase balanced 380V and the simulation is started from 0 second, voltage sag occurs at 0.25 second which makes the line voltage drop to 200V and lasts for 0.2 second. The impedance value of the feeder is R=0.05, X= ωL=0.3 (per unit).

Other parameters are $V_{dc} = 1.5$,
$X_1 = \dfrac{1}{\omega C_1} = 7.02$, $X_2 = \dfrac{1}{\omega C_2} = 4.0$.

Some parameters setup in ANN training function trainlm are:
a) Maximal training times net.trainParam.epochs=300
b) Required training precision net.trainParam.goal=1e^{-5}
c) Training iteration process net.trainParam.show=5

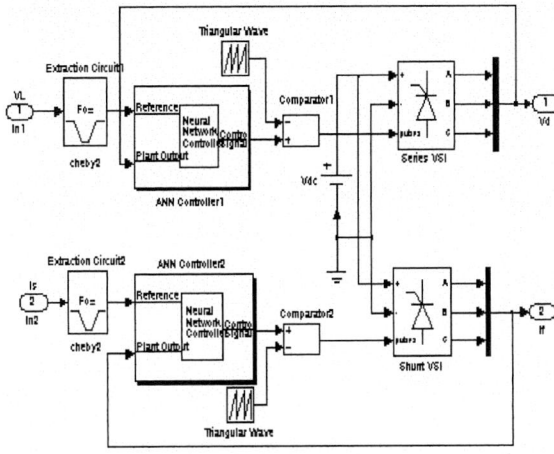

Figure 3. Model of ANN controlled UPQC

B. Simulation results

Configure the above parameters to the simulation model, and compare with the SPWM method which represents traditional control method in compensation performances of source current and load voltage.

(1) The waveform of load current of phase A before compensation is shown as Fig.4.

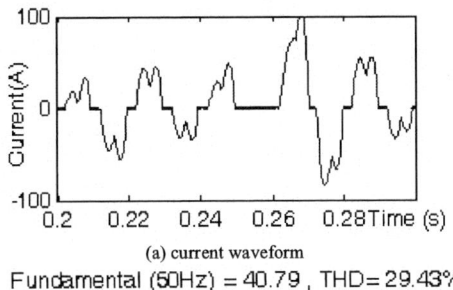

(a) current waveform

Fundamental (50Hz) = 40.79 , THD= 29.43%

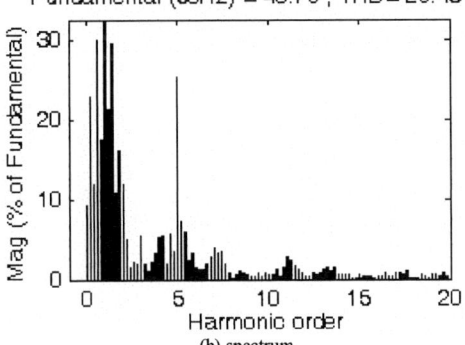

(b) spectrum

Figure 4. Load current waveform and spectrum of phase A

(2) The waveforms of source current and load voltage of phase A after SPWM compensation is shown as Fig.5 and Fig.6.

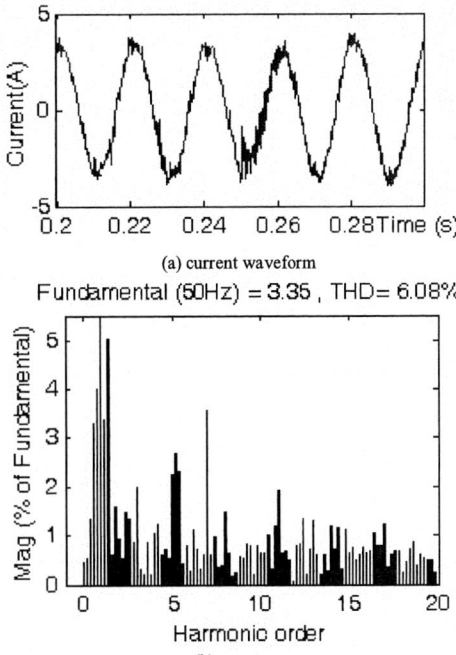

(a) current waveform

Fundamental (50Hz) = 3.35 , THD= 6.08%

(b) spectrum

Figure 5. Source current waveform and spectrum of phase A after SPWM compensation

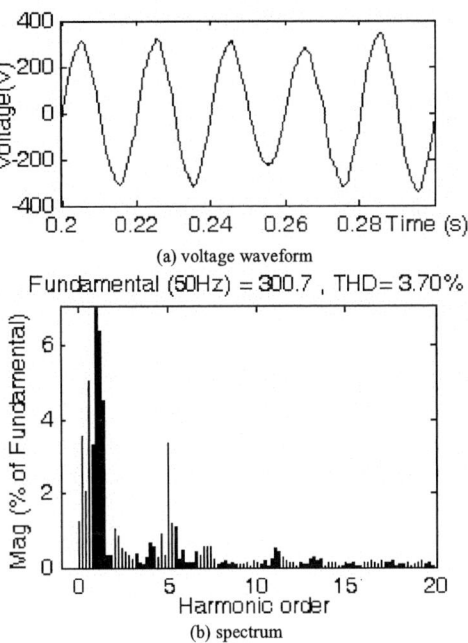

(a) voltage waveform

Fundamental (50Hz) = 300.7 , THD= 3.70%

(b) spectrum

Figure 6. Load voltage waveform and spectrum of phase A after SPWM compensation

(3) The waveforms of source current and load voltage of phase A after ANN compensation is shown as Fig.7 and Fig.8.

1433

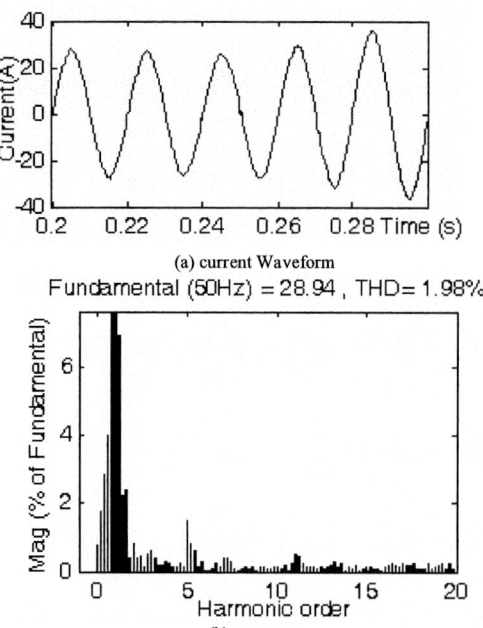

(a) current Waveform

Fundamental (50Hz) = 28.94 , THD= 1.98%

(b) spectrum

Figure 7.   Source current waveform and spectrum of phase A after ANN compensation

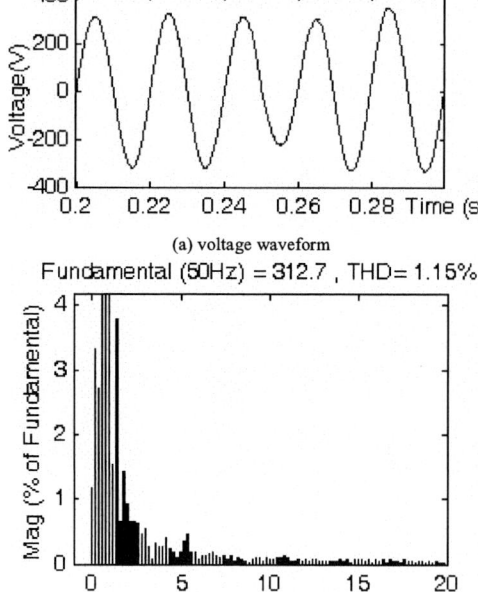

(a) voltage waveform

Fundamental (50Hz) = 312.7 , THD= 1.15%

(b) spectrum

Figure 8.   load voltage waveform and spectrum of phase A after ANN compensation

As shown in Fig.4 without compensation, the load current is badly distorted and three phase unbalance, and the Total Harmonic Distortion (THD) is 29.43%. In Fig.5 and Fig.6, the waveform of the source current after SPWM which represents traditional control method controlled UPQC compensation has an improved THD of 6.08%; under the case of voltage sag occurs at 0.25 second, the load voltage can recover in a primitive period

with an improved THD of 3.70%; the compensation performance suffers greater influence by load fluctuation. In Fig.7 and Fig.8, the waveform of the source current and load voltage after ANN controlled UPQC compensation has an improved THD of 1.98% and 1.15%, respectively. Therefore, simulation results indicate the compensation effectiveness of ANN controlled UPQC is more obvious than traditional control method.

IV. CONCLUSION

In this paper, the parallel side of UPQC that provides path for harmonic current through filter compensation, improves sinusoidal level of source current waveform; the series side of UPQC enhances the ability resisting load fluctuation and provides customer with near sinusoidal voltage through filter compensation [7-9]. ANN controlled UPQC, which adopts Levenberg-Marquardt optimized BP neural networks algorithm, improves learning efficiency and accelerates rate of convergence, make the system have faster response speed and lower harmonic content. Simulation results indicate the compensation effectiveness of ANN controlled UPQC is more obvious than traditional control method.

REFERENCES

[1] Hideaki Fujita, Hirofumi Akagi, "The unified power quality conditioner: the integration of series- and shunt- active fliters," *J.IEEE transactions* on power electronics, 1998, 13(2): 315-322.

[2] Wan Jian-ru, Pei wei, Zhang Guo-xiang. "Research on synchronization deadbeat control algorithm for unified power quality conditioner,"*J. Proceedings of the CSEE*,2005,25(13): 63-67.

[3] Zhu Pengcheng, Li Xun, Kang Yong, et al. "Study of control strategyfor a unified power quality conditioner," *J. Proceedings of the CSEE*, 2004, 24(8): 67-73.

[4] Toshiyuki noda, Guobin wang, Hiroyuki kita, et al. "Interior structure of UPQC-QCC at the low-voltage side in the FRIENDS"*J. Electrical Engineering in Japan*,vol 146,NO.3,2004 : 1157-1166.

[5] Arindam Ghosh, Gerard Ledwich. "A unified power quality conditioner for simultaneous voltage and current compensation"*J. Electric power systems research* 59(2001) : 55-63.

[6] L.H.Tey, P.L.So, Y.C.Chu,"Neural network-controlled unified power quality conditioner for system harmonics compensation"*C. Transmission and Distribution Conference and Exhibition 2002: Asia Pacific. IEEE/PES*, 2002: vol.2 1038-1043.

[7] Hee-Jung Kim, Byung-Yeul Bae, Byung-Moon Han, et al, "Performance analysis of UPQC with compensation capability for voltage interruption"*J. Transactions of the Korean institute of electrical engineers*，2003：279-286.

[8] Chen Guozhu, Chen yang, Sanchez Luis Felipe, et al, "Unified power quality conditioner for distribution system without reference calculations"*C. 4th international power electronics and motion control conference*，2004：1201-1206 vol.3

[9] Basu.M, Farrell.M, Conlon.M.F, et al, "Optimal control strategy of UPQC for minimum operational losses"*C. 39th international universities power engineering conference(UPEC)* 2004：246-250 vol.1

A Nonlinear Method for Hybrid Electromagnetic Suspension

Junwei Cui, Jianhui Wang

Department of Electric Engineering, Shanghai Jiao Tong University, Shanghai 200240, China

Abstract— This paper mainly presents a nonlinear control method for hybrid electromagnetic suspension. Targeting to keep a constant distance between the stator and load, model is researched and controlled in the perpendicular direction. Windings are added into the model to control the disturbance force; at last the load will reach a stable state and be kept in the appointed position in the zero-power control mode which is only permanent magnets' attractive force suspend the levitated body and the steady currents of the electromagnets converge to zero by adjusting the levitation gap length corresponding to its loaded mass. The usefulness of the method is that using springs to analogize the electromagnet system on the base of nonlinear character in common, and calculate the stable position with the Lyapunov functions. So it is a novel method to solve the magnet levitation problem effectively with approximately zero- power in one dimension. The approach is demonstrated to be effective to control the nonlinear system by simulation.

Keywords-electromagnetic suspension; Lyapunov Function; nonlinear control; zero power

I. INTRODUCTION

In magnetic levitation (maglev) technology, the electromagnetic force is used to support objects without mechanical contact. This technology can solve the problems, such as the noise emission, maintenance and limitation of rotating speed, which are caused by friction, abrasion, vibration and lubricating oil in the mechanical support system. It is expected as a key technology for constructing a novel conveyance system in industrial application.

There are many methods to realize the magnetic levitation. Electromagnetic suspension (EMS) technology, in which attractive forces between electromagnets and ferromagnetic materials are utilized as suspension forces, is widely used in the field of magnetic bearings, conveyance system, tool machines, because of the various advantages, such as little leakage flux, compactness, no need of superconductor. The EMS system, however, is substantially unstable; hence, an active control is necessary.

To construct a hybrid electromagnetic suspension system, a novel electromagnet, which can construct a levitation system solely by itself, is proposed. The suspension control method is investigated in this paper. It makes sure that at the steady state the load can reach a balanced position in which the load only has gravity and attraction force on it. The current of windings adjust the levitation gap length corresponding to its loaded mass and the steady currents of the electromagnets will converge to zero. The method controls the system with zero power[1][2]. And the other disturbance from outside has less effect on the system. Through some parameters including the load's gravity, the initial auxiliary current and the disturbance force, we study the steady states of system and find the method proposed keep system to be steady.

The paper begins by introducing the structure of the model and basic knowledge of electromagnet theory. The paper then introduces the nonlinear control method which makes an analogy between spring and the model. According to the spring's character, we solve the equation and calculate the balanced points. The discussion extends the solution to consider the general case of disturbance effects and ends by presenting some simulation results of the nonlinear system with MATLAB.

II. The structure of EMS

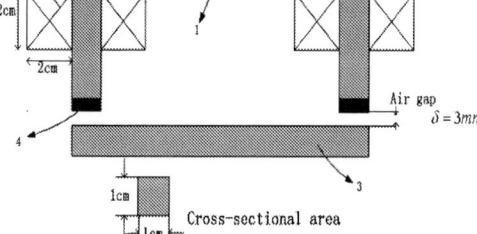

Fig .1 The structure of electromagnetic suspension
1. Core 2.Winding 3. Load 4.Permanent magnet

As the Fig.1 shown, the novelty of our design is that two pieces of permanent magnet (PM) are added in the core. The windings only play an auxiliary role to adjust the length of air gap. When the EMS reaches a stable state, the current flowing in the winding almost is zero and the load only has the attraction force which produced by PM to balance the gravity in the perpendicular direction.

III. The mathematic model

It is easily to get the magnetic flux picture as shown in the fig.2. For the sake of convenience to analyze the EMS, the mathematic model only consider the right side of physical model because it is symmetrical. The movement of model in the horizontal direction is ignored. An attraction force (f_1) will be generated when the right side winding has electrical current (i_1).The magnitude of f_1 is changed to overcome the disturbance until the length of air gap is proper and then the model will get a balanced state .In the situation the gravity will be equal to attraction force which is produced by the permanent magnet (PM). The attraction force, produced by electrified winding, only takes auxiliary effect in the course of control.

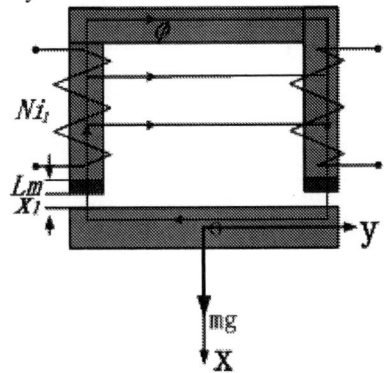

Fig.2.Electromagnetic flux of model

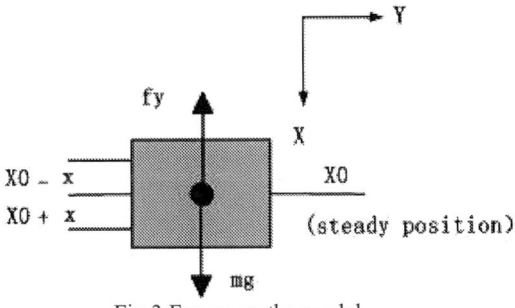

Fig.3.Forces on the model

A. The equation of system

For the sake of simplification, magnetic permeability of permanent magnet is supposed to equal to permeability of the air. ($\mu \approx \mu_0$). The magnetic motive potential of PM can be got as following[3]:

$$F_m = H_c L_m \qquad (1)$$

Where: H_c is coercive force of PM, L_m is the thickness of PM. The sum of magneto motive force (mmf) in the model is

$$F = Ni_1 + F_m = Ni_1 + H_c L_m \qquad (2)$$

Where: N is turns of winding, i_1 is current flowing in the right winding. The length of air gap are given by,

$$x_0 = (x_1 + L_m) \qquad (3)$$

where x_1 is the real air gap as shown in the Fig.2.

Magnetic flux density is

$$B = \mu_0 H \qquad (4)$$

So from equation (2) (3) (4), we easily get:

$$B = \mu_0 \frac{F}{x_0} = \mu_0 \frac{Ni_1 + H_C L_m}{x_1 + L_m} \qquad (5)$$

$$f_{mag} = \frac{B^2}{2\mu_0} \cdot 2S \qquad (6)$$

The attraction force can be written as:

$$f_{mag} = f(x,i) = \frac{\mu_0 S (F_m + Ni_1)^2}{(x_1 + l_m)^2} \qquad (7)$$

Then we make an equivalent conversion: the PM is equivalent to the winding whose turns are N and current is i_2 in the base of equal mmf. So the mmf of PM can be expressed as below:

$$F_m = H_c L_m = N i_2 \qquad (8)$$

The sum of mmf can be given by:

$$f_{mag} = \frac{\mu_0 S (Ni_1 + Ni_2)^2}{(x_1 + l_m)^2}$$

$$= \frac{\mu_0 SN (i_1 + i_2)^2}{(x_1 + l_m)^2} \qquad (9)$$

In the case of the steady state , the current flowing in the real winding and equivalent winding is i_0,

$$i_0 = i_1 + i_2 \qquad (10)$$

And the length of air gap is x_0,

$$x_0 = x_1 + l_m \qquad (11)$$

In the course of adjustment the length of air gap would have a small change, so the mmf can be expressed by,

$$f_{mag} = \frac{\mu_0 S N^2 (i_0 + i)^2}{(x_0 + x)^2} = k \frac{(i_0 + i)^2}{(x_0 + x)^2} \quad (12)$$

Where $k = \mu_0 S N^2 \qquad (13)$

The equation of load's movement can be written by,

$$m\ddot{x} = mg - f_{mag} - F_d \qquad (14)$$

Where the F_d is disturbance from outside, mg is gravity of load.

$$m\ddot{x} + k(\frac{i_0 + i}{x_0 + x})^2 = mg - F_d \qquad (15)$$

According to equation (15), the mathematical model of the system is shown as followings:

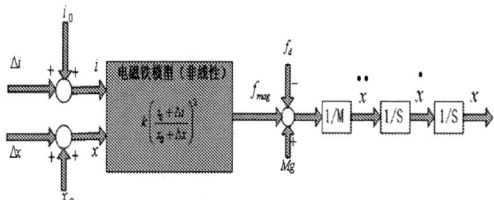

Fig.4. the mathematical model

If the disturbance F_d is ignored, then we get:

$$mg - f = m\ddot{x} \qquad (16)$$

When $i = 0, x = 0$, the sum of forces on the load is zero. The balanced point can be reached in the moment when the attraction force, produced by the permanent magnet, is equal to the gravity of load. And just then the load has no other forces in the other directions. So the current flowing in the winding (i_1) is approximately equal to zero. Using the equations (12) and (16), x_0 can be got. So the balanced point can be found and fixed.

IV.   nonlinear control method

So far, many researches are accomplished to analyze nonlinear system and mainly two effective methods are presented.

Linearization: using the Taylor Series Function to calculate the nonlinear function and take the main part of the series as new linear function. The way could satisfy the need of accuracy on the condition that the error is tolerant.

Lyapunov function: it is a method in the view of energy which mainly discusses the stability of model and make designs relative to system's stability[5].

The paper will apply the Lyapunov function method to investigate the stability.

$$V(x, \dot{x}) = \frac{1}{2} x^2 + \frac{1}{2} \beta \dot{x}^2,$$
$$\beta > 0 \qquad (17)$$

In order to make sure system to be steady, the lyapunov function must satisfy the condition as shown below:

$$\dot{V} = x\dot{x} + \beta \dot{x}\ddot{x} = \dot{x}(x + \beta \ddot{x}) \le 0 \qquad (18)$$

When the air gap x is not equal zero, the $\gamma \dot{x}$ will be not equal to zero. Where γ is a constant ($\ne 0$). According to the Lyapunov Function method, the system is steady under the condition that shows as below:

$$\dot{V} = \dot{x}(x + \beta \ddot{x}) \le -\gamma^2 \dot{x}^2 \qquad (19)$$

$$\dot{x}(x + \gamma^2 \dot{x} + \beta \ddot{x}) \le 0 \qquad (20)$$

From equation (18) and (20) we can get:

$$\dot{x}\left[x + \gamma^2 \dot{x} + \frac{\beta}{m}(mg - f) \right] \le 0 \qquad (21)$$

There are two kinds of situations of the equation (21) which needs to be considered and discussed.

(A) $\dot{x} \ge 0$,

$$\frac{\beta}{m}(mg - f) \le -(x + \gamma^2 \dot{x}) \qquad (22)$$

$$\frac{\beta}{m} f \ge (x + \gamma^2 \dot{x}) + \beta g \qquad (23)$$

$$\frac{\beta}{m}\left[k\left(\frac{i + i_0}{x + x_0} \right)^2 \right] \ge (x + \gamma^2 \dot{x}) + \beta g \qquad (24)$$

$$i \ge (x + x_0)\sqrt{\frac{m}{\beta k}\left[\beta g + (x + \gamma^2 \dot{x}) \right]} - i_0 \quad (25)$$

It is supposed that

$$p = (x + x_0)\sqrt{\frac{m}{\beta k}\left[\beta g + (x + \gamma^2 \dot{x}) \right]} - i_0 \quad (26)$$

(B) $\dot{x} < 0$,

$$\frac{\beta}{m}\left(mg - f\right) \geq -\left(x + \gamma^2 \dot{x}\right) \qquad (27)$$

$$\frac{\beta}{m} f \leq \left(x + \gamma^2 \dot{x}\right) + \beta g \qquad (28)$$

$$\frac{\beta}{m}\left[k\left(\frac{i+i_0}{x+x_0}\right)^2\right] \leq \left(x + \gamma^2 \dot{x}\right) + \beta g \qquad (29)$$

$$i \leq \left(x + x_0\right)\sqrt{\frac{m}{\beta k}\left[\beta g + \left(x + \gamma^2 \dot{x}\right)\right]} - i_0 \quad (30)$$

Conclusion can be made from previous analysis:

(1) when $\dot{x} \geq 0$, $i \geq P$;

(2) when $\dot{x} < 0$, $i \leq P$;

so the method is given by,

(1) when $\dot{x} \geq 0$, let $i = aP$;

$$(a \geq 1)$$

(2) when $\dot{x} < 0$, let $i = bP$;

$$(0 \leq b \leq 1)$$

Let $a = b = 1$, then we can get

$$i = \left(x + x_0\right)\sqrt{\frac{m}{\beta k}\left[\beta g + \left(x + \gamma \dot{x}\right)\right]} - i_0 = P$$

From the equation (18) and (26), we can get:

$$f' = mg - f_{mag} = -\frac{m}{\beta}x - \frac{m\gamma^2}{\beta}\dot{x} \qquad (31)$$

$$f' = -k_s \cdot x - k_v \cdot \dot{x} \qquad (32)$$

where $k_s = \dfrac{m}{\beta}$, $k_v = \dfrac{m\gamma}{\beta}$

It is easily to get conclusion that there is an analogy between the spring and the EMS. So the EMS can be equivalent to the spring whose elastic modulus is k_s and the damper coefficient is k_v [4].This is shown in Fig.5

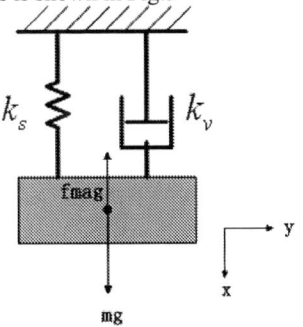

Fig.5. The spring and damper system

$$\beta = m / ks \qquad (33)$$

$$\gamma^2 = k_v / k_s \qquad (34)$$

When a force f_d exerts on the spring and damper system, the static displacement of spring can be given by,

$$x_d = f_d / k_s \qquad (35)$$

$$k_s = f_d / x_d \qquad (36)$$

Therefore, k_s is easy for us to get if the system has maximum displacement under the function of a definite force. The movement equation of the system can be written by:

$$f + f_d = -k_s x - k_v \dot{x} + f_d = m\ddot{x}$$

With Laplace transform function, transfer function of system can be given by,

$$X(s) = \frac{f_d}{k_s}\frac{\dfrac{k_s}{m}}{s^2 + \dfrac{k_v}{m}s + \dfrac{k_s}{m}} \qquad (37)$$

The standard transfer function is known as[6]:

$$X(s) = 1\frac{\omega_n^2}{s^2 + 2\xi\omega_n s + \omega_n^2} \qquad (38)$$

Through a comparison between equation 38 and 37, expressions of w_n and k_v is obtained as follows:

$$\omega_n = \sqrt{k_s / m} \qquad (39)$$

$$k_v = 2\xi\sqrt{mk_s} \qquad (40)$$

Where $\xi = 0.707$ or 10

According to Equation (33), (38), (39) and (40), some parameters can be obtained as follow:

$$\beta = \frac{m}{k_s} = \frac{mx_d}{f_d} \qquad (41)$$

$$\omega_n = \frac{1}{\sqrt{\beta}} \qquad (42)$$

$$\gamma = 2\xi\sqrt{\beta} \qquad (43)$$

V.   simulation results

In the section some results are used to explore the proposed method. The entire system is modeled and simulated using the SIMULINK toolbox. Fig.6 shows the block diagram of the EMS in SIMULINK environment.

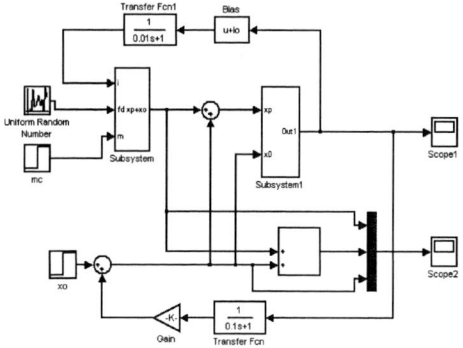

Fig.6.Block diagram of the EMS in SIMULINK.

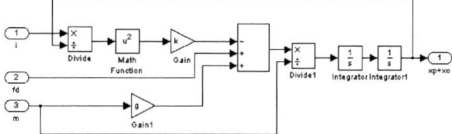

Fig.7.Block diagram of force

According to the equation (12) and movement equation of the system, the input parameters are the mass of load (m) and the disturbance force (F_d). When the system reaches a stable state, the displacement x and auxiliary current will become zero. Load has only the gravity and attraction force which produced by the PM.

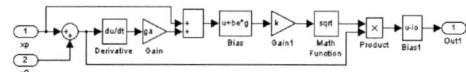

Fig.8.Block diagram of auxiliary current

Fig.8 shows the details of relationship between displacement and current flowing the winding.

$$ i = \left(x + x_0 \right) \sqrt{\frac{m}{\beta k} \left[\beta g + \left(x + \gamma \dot{x} \right) \right]} - i_0 $$

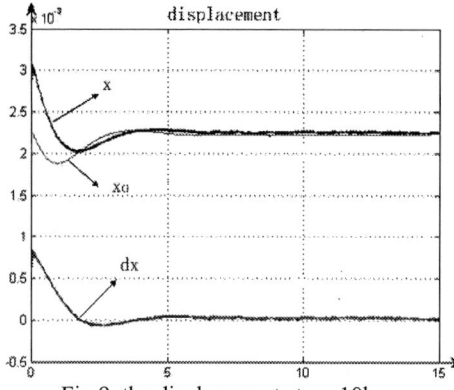

Fig.9. the displacement at m=10kg

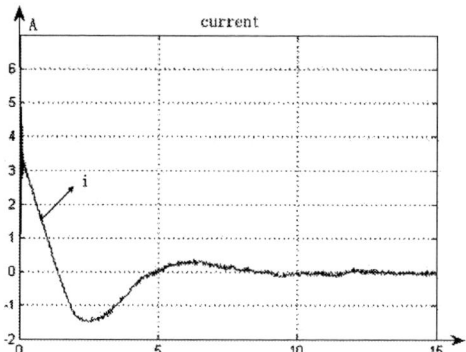

Fig.10. the steady current converged to zero

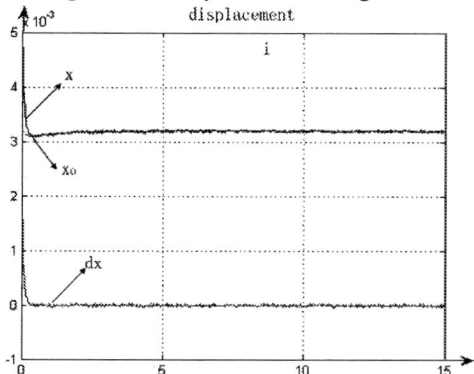

Fig.11 the displacement at m=5kg

In Fig.9, $x = x_0 + dx$.

When the model reaches a stable state, dx will converge to zero.

In Fig.10, the current flowing the winding only play an auxiliary role to help the system reach steady state. When the system becomes stable, the current will becomes zero. Then the power consumption is nearly zero.

In Fig.11, the mass of load is reduced and the system spent much less time to reach steady state.

VI. CONCLUSION

In this paper, a method based on the characters of spring and Lyapunov function is proposed and investigated. It is effective to judge and adjust the stability of nonlinear system. Performance of the proposed method is analyzed and tested with MATLAB Simulink.

REFERENCE

[1] Jiangheng Liu, Koji Yakushi, Takafumi Koseki and Satoru Sone, "**3 Degrees of Freedom Control of Zero- Power Magnetic Levitation for Flexible Transport System**", *The 16th International Conference on Magnetically Levitated Systems and Linear Drives, pp.382-386. June 7-10, 2000, Rio de Janeiro, Rrazil.*

[2] M. SHIDA, E. MASADA: "**A Hybrid Control Scheme for Electro-Magnetic Suspension System of HSST**", *The 16th International Conference on Magnetically Levitation System and Linear Drives. pp.185-188, June 7-10, 2000,*

Rio de Janeiro, Brazil.

[3] Fen Cixiang **"Electromagnetic field and wave,"** Shanxi, XiDian University Press

[4] Yu Lie **"Controllable rotor in magnetic levitation system，"** *Beijing, Science Press*

[5] Liu Bao **The theory of modern control，"** *Beijing, Machine Press*

[6] Wu Qi "The principle of automatic control," *Beijing, Tsinghua University Press*

2006 5th International Power Electronics and Motion Control Conference

new topology of multi - level - converter for harmonic reduction

Frank Grundmann and Jian Xie, University of Ulm
Frank.Grundmann@uni-ulm.de, Jian.Xie@uni-ulm.de

Abstract—**This paper deals with a new topology for a multi - level - converter without an input transformer. Multi - level - converter with an increased number of voltage steps can be realised. The number of valves is equal or smaller than the one of today present topologies.**

Index Terms—**Multi - level - converter, topology**

I. INTRODUCTION

TODAYS converter have to provide more than a suitable fundamental. The harmonics of the power consumption and production have to be taken more and more into account. Additional, the switching frequency has to be as low as possible to reduce the switching losses and to increase the live time.

The necessary quality is actually realised through high switching frequencies (by using standard two - point - converter) or very complex multi - level - converter. With the here presented topology a multi - level - converter can be build without additional switching elements. Thereby, neither a high switching frequency nor a complex multi - level - converter are necessary to achieve the intended quality.

Multi - Level - Inverter [1], [2], [3] can today be realised with for example NPC - topologies [4], flying capacitors [5] or cascaded inverters [6], [7], [8]. The results (spectrum) of the different multi - level - converter (with the same number of voltage level) are limited by the used calculation algorithm for the switching points [1], [9], [10]. The same algorithm and therefor the same switching points with the same spectrum can be used in most of the topologies. The difference lays within the complexity and thereby the costs of the circuit. All these topologies need special input transformers with additional secondary windings or complex circuit structures. The cascaded inverters work with separate DC voltages, which were separated by transformer windings with following inverter. The NPC - topologie needs either for each capacitor a separate secondary winding or a complex circiut structure.

The here presented topology only needs an transformer if the voltage level has to be changed (low voltage to high voltage transformer). In this case a standard transformer can be used (one winding per phase, not per capacitor). Otherwise only an inductor is necessary for decoupling and current control. Thereby the costs of the inverter can be slashed.

The number of valves is another cost aspect. Most of the today practicable topologies combine standard two or three - point - converter to reach a multi - level - converter. Thereby the number of valves will be multiplied. The here presented topology is an all in one design, which uses the valves in

different configuration. Thereby the number of valves can be reduced.

II. TOPOLOGY

The origin of the analysis is the known three - point - converter (figure 1)[11]. A symmetric intermediate circuit (voltage ratio 1:1) is used there, wherefor the potential $\frac{U_d}{2}$ can be actuated with two different switching states. In the presented new topology a asymmetric intermediate circuit is used. The advantage is, that with the same topology like the three - point - converter, a four - point - converter can be designed (figure 2).

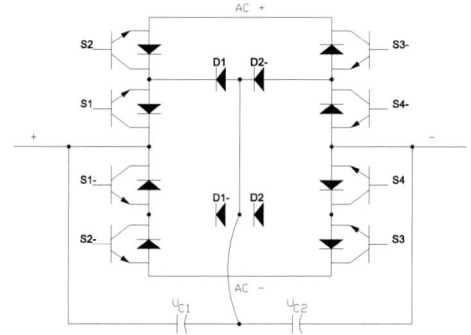

Fig. 1.   three - point - converter

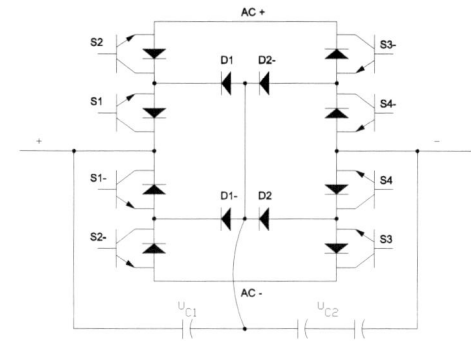

Fig. 2.   four - point - converter

With this modification an improvement in the harmonic behaviour can be achieved without additional switching elements or

---

1-4244-0448-7/06/$25.00 ©2006 IEEE

an increase in the switching frequency. A comparison between the today reachable number of voltage steps with symmetric intermediat circuit and the number of steps with asymmetric intermediate circuit is shown in table I.

TABLE I

COMPARISON OF NUMBER OF VOLTAGE STEPS WITH SYMMETRIC AND ASYMMETRIC INTERMEDIATE CIRCUIT

stage of expansion	today	new topology
0 (two - point)	2	2
1 (three - point)	3	4
2	4	7
3	5	10
⋮	⋮	⋮
7	9	23

For the reverse blocking voltage, there are several cases were switched off valves create an undefined voltage division. Through the tolerances in production, termerature or time depencence the voltage division can't be calculated exactly. Such a case is shown in figure 3, where the valves S_{1n} and S_{2n} from figure 2 are switched on and are therefor represented as a short circuit. The reverse blocking voltages of D_2, S_3 and S_4 are thereby undefined (defined through the parasitic properties of the elements). By using high ohmic resistors like in figure 4 the voltage of all valves can be dimensioned to $\frac{U_d}{3}$ or $\frac{2U_d}{3}$.

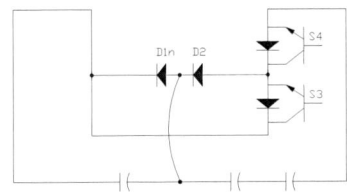

Fig. 3. undefined voltage division

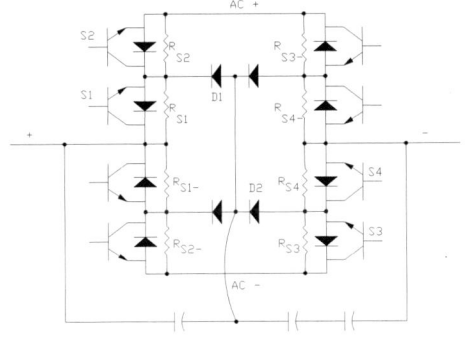

Fig. 4. four - point - converter with ohmic voltage stabilisation

The resistances have to be calculated in a relation of $1 : 2$ (R_{S_4} = $2*R_{S_3}$, for four - point - converter) to achieve the intended voltage distribution of $\left(\frac{1}{3}U_d\right)$ and $\left(\frac{2}{3}U_d\right)$. The blocking voltage of the diode D_2 will be forced to sero. The other resistances

can be calculated in the same way. The resistances can be high ohmic (e.g. >100 kΩ) to reduce or nearly neglect the power losses in the resistances.

The reverse blocking voltage of the valves of a four - point - converter devides in an ration of $\frac{U_d}{3}$ and $\frac{2U_d}{3}$. The ratio of an three - point - converter is always $\frac{U_d}{2}$. If the voltage reserve is large enough the three - point - converter can be easily changed in a four - point - converter by adding a capacitor (or by changing the connection of the existing). If the voltage reserve is not large enough an addional valve has to be installed. If the converter is constructed through serial connections of smaller valves, the change of the distribution can solve the problem of the reverse blocking voltage as well. The sum of the reverse blocking voltages is still the same. The reverse blocking voltages of the different elements are shown in table II.

TABLE II

VOLTAGE OF ELEMENTS, FOUR - POINT - CONVERTER

element	voltage
S_1	$\frac{1}{3}U_d$
S_2	$\frac{2}{3}U_d$
S_3	$\frac{1}{3}U_d$
S_4	$\frac{2}{3}U_d$
D_1	$\frac{1}{3}U_d$
D_2	$\frac{2}{3}U_d$
sum	$3\,U_d$

III. SIMULATION AND MEASUREMENT

The first expansion stage of the circuit is shown in figure 2. The DC - voltage (intermediate circuit) is connected to the + and - clamps. The clamps AC+ and AC- represent the connection point with the load or the net. In table III are the switching configurations for the different potentials presented. The simulation of the circuit results in figure 5.

TABLE III

SWITCHING CONFIGURATION, FOUR - POINT - CONVERTER

U_{AC}	$[S1, S2, S3, S4]$
0	$[1, 1, 0, 0]$
$\frac{1}{3}U_d$	$[1, 1, 1, 0]$
$\frac{2}{3}U_d$	$[0, 1, 1, 1]$
U_d	$[1, 1, 1, 1]$

The experimental solution is shown in figure 6. All expected voltage steps are shown. The THD (5th to 49th harmonic, without harmonics of 3rd order or order 3n) of the measured chart is about 0,13 % which is mainly reasoned by the switching behaviour of the used IGBT - modules (t_r and t_f ≈ 1 μs). An optimised pulse sequence [10] with 18 switching points per quarter periode was used. The theoretical THD is smaller than 0.01 %.

The simulation and measurement of the blocking voltages reaches values like the one shown in table II by using the mentioned parallel resistors. The measured blocking voltages of the elements S_1 and S_2 are shown in figures 8 to 11 as

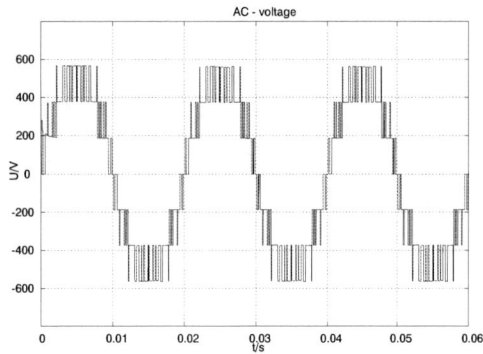

Fig. 5.   AC - voltage, simulation, four - point - converter

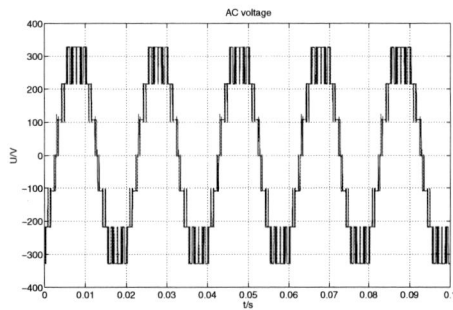

Fig. 6.   AC - voltage, measured, four - point - converter

an example for the change of the blocking voltages by using the parallel resistors. S_2 reaches in both cases a blocking voltage of 222 V ($\frac{2}{3}$ of intermediate circuit voltage). The blocking voltage of S_1 increases without the resistor through the undefined voltage division (like shown in figure 3). The blocking voltage can be reduced to 111 V ($\frac{1}{3}$ of intermediate circuit voltage)) by defining the voltage division. The blocking voltage can therefor be dimensioned through the high ohmic parallel resistors (57 kΩ / 2 * 57 kΩ used) to reduce (neglect) the internal losses.

IV. EXPANSION

An additional branch has to be installed in the circuit to reach the next expansion stage. The number of expansion steps is therby unlimited. A configuration example for an seven - point - converter is shwon in figure 12. The additional branches are marked. Examples for the configuration of the capacitors in the intermediate circuite are shown in table IV. The capacitors should be distributed for example in the fifth expansion stage like $C_1 = 2C_x$, $C_2 = 6C_x$, $C_3 = C_x$, $C_4 = 3C_x$ und $C_5 = C_x$, where C_x is a reference capacity.

The simulation result of the seven - point - converter (figure 12) is shown in figure 13, using the switching configuration like in table V.

The resistors for the voltage distribution (like shown with the four - point - converter) have to be calculated to

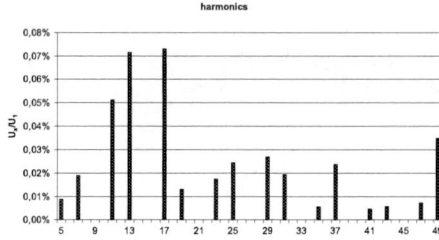

Fig. 7.   amplitude of harmonics (measured)

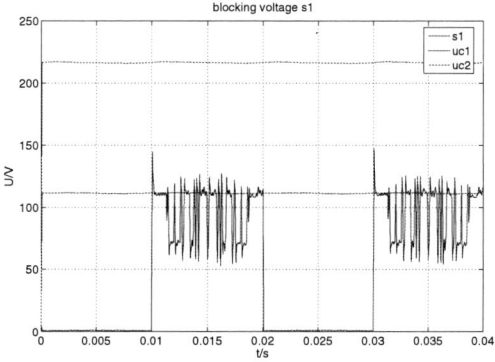

Fig. 8.   blocking voltage S_1 with parallel resistors

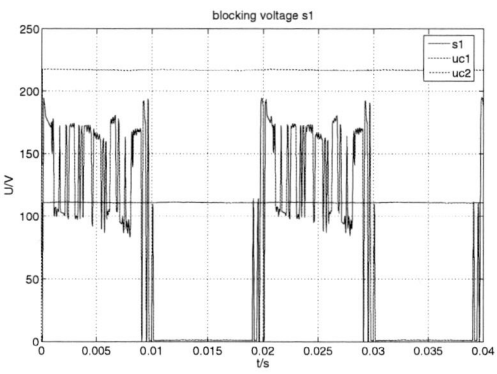

Fig. 9.   blocking voltage S_1 without parallel resistors

TABLE IV
NUMBER OF POTENTIALS (SERO - POTENTIAL IS MENTIONED AS +1)

expansion stage	number of potentials	distribution of capacitors
7	22+1	2,2,2,7,3,5,1
6	17+1	1,4,4,4,2,1
5	13+1	2,6,1,3,1
4	9+1	2,3,3,1
3	6+1	2,3,1
2	3+1	2,1

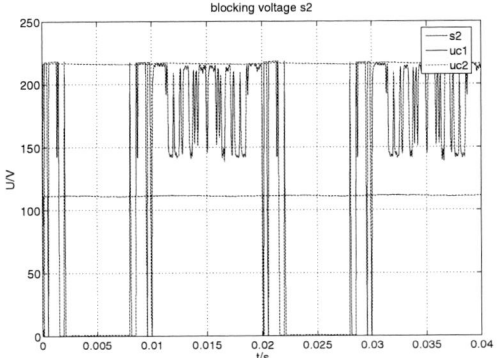

Fig. 10. blocking voltage S_2 with parallel resistors

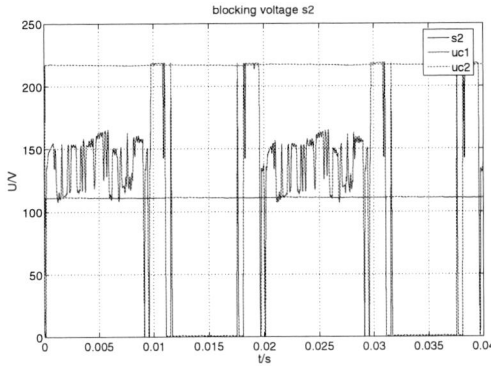

Fig. 11. blocking voltage S_2 without parallel resistors

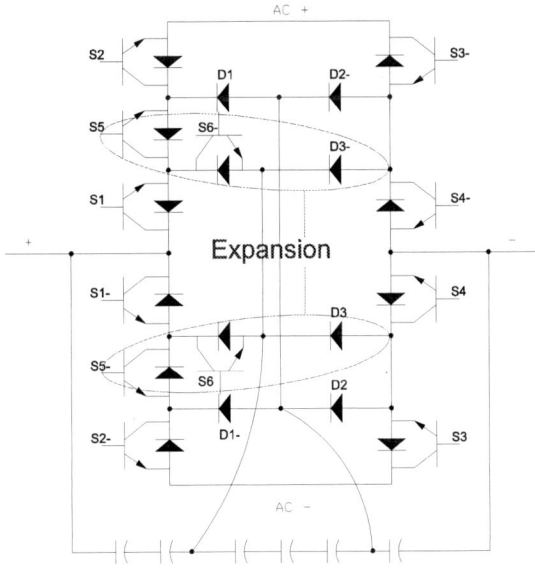

Fig. 12. seven - point - converter

TABLE V
SWITCHING CONFIGURATION, SEVEN - POINT - CONVERTER

U_{AC}	$[S1, S2, S3, S4, S5, S6]$
0	$[1,1,0,0,1,0]$
$\frac{1}{6}U_d$	$[0,1,1,1,0,0]$
$\frac{2}{6}U_d$	$[1,1,0,0,1,1]$
$\frac{3}{6}U_d$	$[0,1,1,0,1,0]$
$\frac{4}{6}U_d$	$[0,1,1,1,1,0]$
$\frac{5}{6}U_d$	$[1,1,1,0,1,0]$
U_d	$[1,1,1,1,1,0]$

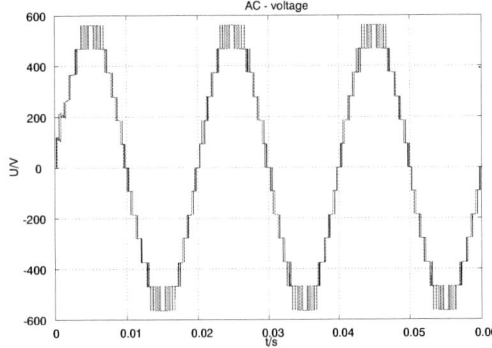

Fig. 13. AC - voltage, simulation, seven - point - converter

$2R_{S1} = 3R_{S5} = R_{S2}$ ($U_{S1} = \frac{2}{6}U_d$, $U_{S5} = \frac{3}{6}U_d$ and $U_{S2} = \frac{1}{6}U_d$). The voltage of the elements S_3 and S_4 can be calculated to $\frac{1}{6}U_d$ and $\frac{5}{6}U_d$, which results in a resistor ratio of 1:5.

The three - phase to three - phase converter has passed the stage of simulation. The commonly known prinziples of voltage- and current - control of inverter and inverse inverter are thereby implemented.

V. SWITCHING FREQUENCY

One of the problems of multi level - converter is the increased switching frequency of some valves. The switching tables are shown in table III and V. To compare the different configurations a defined pulse sequency or scaling has to be used. The comparism will be done by reaching all possible voltage steps one by one. The necessary switching pulses will be counted and scaled to the number of voltage steps.

TABLE VI
SWITCHING FREQUENCY

voltage division	$[S_1, \ S_2, \ S_3, \ S_4, \ S_5, \ S_6]$
$\frac{1}{2}$	$[1, \ 1, \ 1, \ 1]$
$\frac{1}{3}$	$[2, \ 1, \ 1, \ 1]$
$\frac{1}{6}$	$[3, \ 1, \ 2, \ 3, \ 2, \ 1]$

According to table III S_1 has to be switched on two times. The other elements are only switched one time. The notation is $[2, \ 1, \ 1, \ 1]$. Only the scaling is missed which is corrected in table VII. Only S_1 at the four - level - converter reaches a scaled switching frequency which is higher than the one of the three - level - converter. With higher expansion stage only some valves will work at the limit of the switching frequency. Therefor the switching frequency of the whole system can be increased by dimensioning these valves separately (thermal and electrical).

TABLE VII

SWITCHING FREQUENCY

voltage division	$[S_1,$	$S_2,$	$S_3,$	$S_4,$	$S_5,$	$S_6]$
$\frac{1}{2}$		$\frac{1}{2},$	$\frac{1}{2},$	$\frac{1}{2},$	$\frac{1}{2}$	
$\frac{1}{3}$		$\frac{2}{3},$	$\frac{1}{3},$	$\frac{1}{3},$	$\frac{1}{3}$	
$\frac{1}{6}$	$\frac{1}{2},$	$\frac{1}{6},$	$\frac{1}{3},$	$\frac{1}{2},$	$\frac{1}{3},$	$\frac{1}{6}$

The practical pulse sequency is only in some special cases identical with such a standard sequency. In normal working conditions the higher voltage level will be prefered. By switching between level $\frac{5}{6}$ and $\frac{6}{6}$ only S_4 will be switched. A voltage step from $\frac{4}{6}$ and $\frac{5}{6}$ will influence S_1 and S_4. These „standard" switching elements has to be dimensioned as critical elements.

VI. CONCLUSION

A multi - level - converter can be established with the new topology without using an input transformer. The amount of necessary material (switching elements and transformer) can thereby be reduced compared with actual topologies. The increase of levels can be reached by using an asymmetric intermediate circiut. The reverse blocking voltage of the valves can be dimensioned through resistors. The output characteristic can be calculated like the one of any other multi - level - converter.

REFERENCES

[1] J. R. et al., "Multilevel interters: A survey of topologies, controls and applications," *IEEE Transactions on Insutrial Electronics*, vol. 49, no. 4, August 2002.

[2] A. L. R. Marquardt, "A new modular voltage source inverter topology," *EPE 2003*.

[3] F. Peng, "A generalized multilevel inverter topology with self voltage balancing," *IEEE Transactions on industry applications*, vol. 37, no. 2, März 2001.

[4] M. V. A. Rufer, "Control of a hybrid asymmetric multilevel inverter for competitive medium - voltage industrial drives," *IEEE Transactions on Industry Applications*, vol. 41, no. 2, März / April 2005.

[5] M. F. E. et al., "Flying capacitor multileven inverters and dtc motor drive applications," *IEEE Transactions on Industrial Electronics*, vol. 49, no. 4, August 2002.

[6] R. T. et al., "Multilevel inverter by cascading industial vsi," *IEEE Transactions on Insutrial Electronics*, vol. 49, no. 4, August 2002.

[7] T. et al., "Charge balance control scheme for cascade multilevel converter in hybrid electric vehicles," *IEEE Transactions on industrial electronics*, vol. 49, no. 5, Oktober 2002.

[8] K. C. Y. Familiant, "A new cascaded multilevel h-bride drive," *IEEE Transactions on Power Electronics*, vol. 17, no. 1, Januar 2002.

[9] B. P. M. et al., "Multicarrier pwm strategies for multilevel inverters," *IEEE Transactions on Insutrial Electronics*, vol. 49, no. 4, August 2002.

[10] S. S. et al., "Optimum harmonic reduction with a wide range of modulation indexes for multilevel converter," *IEEE Transactions on Industrial Electronics*, vol. 49, no. 4, August 2002.

[11] D. Hasenkopf, "Regelverfahren für einen umrichter zur symmetrierung einphasiger lasten in drehstromnetzen," Ph.D. dissertation, Universität Ulm, 2005.

Frank Grundmann received the M.S. degree in electrical engineering from University of Magdeburg in 2002. He is currently preparing for the Ph.D. degree at the University of Ulm.

Jian Xie received his B.Sc. degree at Jiao - Tong - university Shanghai. He received his M.S. and Ph.D. from University of Darmstadt. He was system engineer for the frequency converter station Jübeck and project engineer at adtranz railway systems. He became a Professor at university of ulm in 1998.

2006 5th International Power Electronics and Motion Control Conference

Model Reference Adaptive Control based on Neural Network for Electrode System in Electric Arc Furnace

ZHANG Shi-feng[*], ZHANG Shao-de[*], LI Kun[*], ZHENG Xiao[**]

[*] School of Electrical Engineering& Information, Anhui University of Technology,
[**] School of Computer Science ,Anhui University of Technology, Ma'anshan , Anhui, China
zhangsf@ahut.edu.cn

Abstract—Control strategy of Model Reference Adaptive Control (MRAC) based on Radial Basis Function Neural Network(RBFNN) online identification is proposed, and a controller is also designed. Which in accordance with the characteristics of the electrode system in electric arc furnace as the high nonlinearity, time-variant, uncertainty and multivariable input and output coupling, The validity of control strategy is verified by result of experimentation.

Keywords- electric arc furnace; electrode control; RBF neural network; online identification; adaptive decouple controller

I. INTRODUCTION

Now multi-variable complex process control system is facing with two main difficulties: first, because of the complexity and serious uncertainty of the system, it is hard to construct on-line by the traditional theories and methods. Second, as the complexity and serious uncertainty of the system, it is hard to depart the coupling by the methods based on departed coupling of parse model. Though lots of researchers have given deep study and statement about the multi-variable process control system, but almost all the theories are based on one precondition: the precise parse model of multi-variable process system must be obtained at first, but it is almost impossible in reality. Especially in such environment conditions: the variant of working condition, serious uncertainty of system structure and parameters, nonlinear, lag, etc, which occurs in the process of control system. According to the fact and background, a new way has been probed in the study of three-phrase AC electric arc furnace electrode which based on neural network real-time recognition, decoupling and model reference adaptive control. This method has following abilities fits rapidly to the change of object and process, identifiable system rapidly and precisely and learns while controlling, all the functions have been proved effective by experiment.

Supported by Anhui Key Scientific Project (No.01012053),Natural Science Foundation of Education Department of Anhui Province P.R.China. (2004KJ059)

II. DESIGN OF MRACS BASED ON NEURAL NETWORK IN ELECTRODE SYSTEM

Discrete time system can be described by the following nonlinear difference equation:

$$y_p(k+1)=f[y_p(k),y_p(k-1),\ldots,y_p(k-n+1)]+\sum_{i=0}^{m-1}\beta u_p(k-i) \quad (1)$$

Because model (1) possesses separate controller, the control signal is separated from the nonlinear object model of the complicated system, participating in the system object to recognize with the control, and need a neural network identifier to recognize the nonlinear part

$$f[u_p(k-1),\ldots,u_p(k),y_p(k),y_p(k-1),\ldots,y_p(k-n+1)]$$
(series-parallel structure).

A strategy of neural network model reference adaptive control is designed, control structure for one phrase system is depicted in Fig.1.

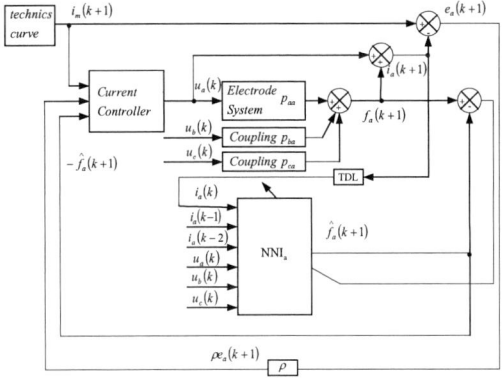

Fig.1 Control structure for one phrase electrode system

The electrode system is droved by a hydraulic pressure device, Which is a complicated system with three- input and three-output, nonlinear, strong coupling (three phrase electrode discharge the steel water in the electric arc furnace, and become the star-type conjunction), variant-time, big delay, many disturbances. System input is control signal $u(k)$, and system output is the electric current of the electrode $i(k+1)$. The MRAC system is set up based on neural network identification, Which

---

1-4244-0448-7/06/$25.00 ©2006 IEEE 1447

treats the electrode system as a generalized object (include a hydraulic pressure device and electrode). In electrode system, three-phrase control signal $u_a(k)$、 $u_b(k)$、 $u_c(k)$ and three phrase output current $i_a(k+1)$、 $i_b(k+1)$、 $i_c(k+1)$ have strong cross coupling, So the output of one phrase current controller can be defined as

$$u_a(k) = -\hat{f}_a(k+1) + i_m(k+1) + \rho e_a(k+1).$$

Fig.1 shows that P_{ba}、 P_{ca} is separately a coupling link between $u_b(k)$、 $u_c(k)$ and $i_a(k+1)$. $i_m(k+1)$ is a desired response for the system ,namely output of the reference model. ρ is feedback coefficient of output error, $-1 < \rho < 0$ .one-phrase output current

$$i_a(k+1) = f_a(k+1) + u_a(k)$$

is in model （1） configuration.

III. ON-LINE DECOUPLING STRATEGY OF THE ELECTRODE SYSTEM

Refer again to Fig.1, the generalized one-phrase object composes of P_{aa} and coupling links P_{ba}、 P_{ca}, nonlinear part $f_a(*)$ includes the effect of $u_b(k)$、 $u_c(k)$,so control signals $u_b(k)$ and $u_c(k)$ are treated as inputs to the one-phrase identifier. This identifier adopts series-parallel configuration, we define input vector, identifier model and output as follows:

$$[u_a(k), i_a(k), i_a(k-1), i_a(k-2), u_b(k), u_c(k)]^T \quad (2)$$

$$\hat{f}_a[u_a(k), i_a(k), i_a(k-1), i_a(k-2), u_b(k), u_c(k)] \quad (3)$$

$$i_a(k+1) = f_a(k+1) + u_a(k)$$

$$= f_a(k+1) - \hat{f}_a(k+1) + i_m(k+1) + \rho e_a(k+1) \quad (4)$$

Based on nearest neighbor clustering algorithm, RBF network real-time identification is adopted, so the identifier learns rapidly, and completely meets on-line requirement. What's more, $\hat{f}_a(k+1)$ approximates $f_a(k+1)$ as closely as possible. Namely

$$\hat{f}_a[u_a(k), i_a(k), i_a(k-1), i_a(k-2), u_b(k), u_c(k)] \approx f_a(k+1) \quad (5)$$

one-phrase current can be described as

$$i_a(k+1) = i_m(k+1) + \rho e_a(k+1) \quad (6)$$

Thus, decoupling control of $i_a(k+1)$ is realized briefly. Both the decoupling methods of $i_b(k+1)$ and that of $i_c(k+1)$ are as well. The equation $|\rho| < 1$ has been proved in reference [5]. $\rho e_a(k+1)$ converges gradually to zero. Namely, system output is able to approximate output of reference model.

Where $i_a(k+1) = i_m(k+1)$.

Samely $i_b(k+1) = i_m(k+1)$, $i_c(k+1) = i_m(k+1)$.

IV. RBFNN INVERSION MODEL LEARNING ALGORITHM BASED ON NEAREST NEIGHBOR CLUSTERING

In Fig.1, the identifier adopts RBF neural network, the activated function of hidden layer is radial basis function, which is given by

$$R_i(x) = \exp\left[-\frac{\|x-c_i\|^2}{2\sigma^2}\right] \quad (7)$$

Where x is n dimension input vectors, c_i is center of the ith RBF, σ is width of RBF.

The nearest neighbor clustering algorithm can be described as follows:

Step 1　Confirm an appropriate clustering radius r, radius correct step h and error threshold value E_m, define A(m) to deposit the sum , B(m) to count the number of sorts of samples, $W(i)$ to deposit weight $(i=1,...,m)$, where m is the number of sort, (c_i, d_i) is the center of ith sort.

Step 2　For the first data (x_1, y_1), let $c_1 = x_1$, $d_1 = y_1$. At the same time, let A(1)=y_1, B(1)=1. The center of this hidden layer is(c_1, d_1), the weight between the hidden layer and output layer is written by $W(1)$=A(1)/B(1).

Step 3　For the second data (x_2, y_2), the distance between x_2 and c_1 is calculated by

$$d = \sqrt{\|x_2 - c_1\|^2 + \|y_2 - d_1\|^2}$$ .If $d \leq r$,then c_1 is the nearest neighbor clustering of x_2, let A(1)=$y_1 + y_2$, B(1)=2， $W(1)$=A(1)/B(1); If $d > r$, then let (x_2, d_2) be a new clustering center, and let $c_2 = x_2$, $d_2 = y_2$, A(2)=y_2, B(2)=1.A neuron is added to the hidden layer of the RBF network, the weight between this neuron and output layer is calculated by $W(2)$=A(2)/B(2).

Step 4　Considering ith sample data(x_i, y_i), $i=3,4,...,N$, we assume that this network has M clustering centers, which is separately (c_1, d_1), (c_2, d_2),..., (c_M, d_M). The distance between (x_i, y_i) and M centers is computed by the following equation

$$H(l) = \sqrt{\|x_i - c_i\|^2 + \|y_i - d_i\|^2}, \qquad l = 1,2,......,M \quad (8)$$

We supposed that $H(k)$ is the minimum distance, that is , (c_k, d_k) is the nearest neighbor clustering of(x_i, y_i).If $H(k) > r$, then let (x_i, y_i) be a new clustering center, $M=M+1$, $c_M = x_i$, $d_M = y_i$, A(M)=y_i, B(M)=1, A(i) and B(i) of the previous M-1 sorts keep constant. The Mth neuron is added to the hidden layer of RBF network. If $H(k) \le r$, then let $A(k)=A(k)+y$, $B(k)=B(k)+1$, and keep the value of $A(i)$ and $B(i)$ $(i = 1, \cdots, M \ and \ i \ne k)$ constant. The weight between hidden layer and output layer is computed by $W(i)=A(i)/B(i)(i=1,\ldots,M)$.

Step 5 After all input samples are considered, the output of RBF network is given by

$$f(x_k) = \sum_{i=1}^{M} W(i) \exp\left(-\frac{\|x_k - c_i\|^2}{r^2}\right) \bigg/ \sum_{i=1}^{M} \exp\left(-\frac{\|x_k - c_i\|^2}{r^2}\right) \tag{9}$$

The error performance target of the entire neural network is calculated by:

$$E = \frac{1}{2} \sum_{i=1}^{N} \left(y_i - f(x_i) \right)^2 \tag{10}$$

where x_i is input samples, y_i is output samples. When E is less than E_m, learning process is finished. otherwise, correct the clustering radius by using $r = r\text{-}h$, and return to step 2.

V. FIELD TEST CONCLUSION

The model reference adaptive decoupling controller for the electrode system of electric arc furnace is implemented by industrial computer. The high-speed data acquisition card- the PXI2010 of LingHua is adopted to acquire three-phrase electric current (two-side),Which carries on data pretreatment and controller design under the VC environment. The system has packed WinAc4.0, in order to make real-time data exchange with the spot of Siemens PLC industry control network .We chose a 50 ton electric arc furnace of a steel factory as experiment object, the electricity parameter for the electric arc furnace is shown in table I.

TABLE I.
ELECTRICAL PARAMETERS FOR THE ELECTRIC ARC FURNACE

Transformer power	Over load ability(Max)	Transformer teams
25 MVA	+ 5%	13 级
Basic arcVoltage Bound	Arc Current (Max)	secondary Current (Max)
110 ~120 V	42.5 kA	44.6 kA

The electrode regulation part in the original system is conventional control which is based on PLC. Neural network identifier firstly self-learn, error threshold value E_m =0.05,while it is in off-line. Error threshold value E_m =0.005, clustering radius correct step h=0.0002,while it is in on-line identification and control. Each initial value of three-phrase clustering radius is 0.01.

On the base of the previous algorithm design, neural network controller is put in practice after system is trained 200 steps off-line. Measured data for the melting and heating stage are shown in table II. The experiment data indicates that the coupling for three-phrase electrode has been decoupled greatly. The electric current fluctuation is obviously smaller than that of the original normal control system. Regulation performance is good, and heating process is more even(4 ℃/min).This strategy provides a possibility for raising the product quality.

TABLE. II
MEASURE DATA FOR THREE-PHRASE CURRENT RMS VALUE

time (s)	A Phase Current(kA)	B Phase Current (kA)	C Phase Current (kA)
5	29.5	28.1	28.7
20	29.2	28.5	28.6
40	28.6	27.9	28.8
60	29.1	28.2	28.4
80	29.2	28.6	27.7
100	29.0	28.5	27.3
120	28.7	29.1	28.5
140	29.2	28.8	28.3
160	28.6	28.9	28.0
180	29.2	28.2	28.4
200	29.0	28.3	28.1
220	29.2	28.9	27.8
240	29.8	29.1	28.3
260	29.3	28.7	28.1
280	29.1	28.5	29.2
300	28.9	28.6	28.8

VI. CONCLUSION

In this paper, the improved nearest neighbor-clustering algorithm is proposed, so neural network is trained quickly. Accordingly on-line model for the electrode system in electric arc furnace is realized. Neural network decoupling is simple and applied, test conclusion illuminates that the strategy is effective.

REFERENCES

[1] E. S. William,and G. B. Norman . "Neural network control system for electric arc furnaces".*MPT*, 18(2),pp.58-64,1995.

[2] A. S. Hauksdóttir and A. Gestsson, "Current control of a three-phase submerged arc ferrosilicon furnace" *Control Engineering Practice*, Vol. 10 Issue 4 ,pp. 457-463,April 2002.

[3] P. C. Morgan. "The continued development of the electric arc furnace,".*Ironmaking & Steelmaking*, Vol. 32 Issue 3,pp.185-192 June 2005.

[4] Yuan wei guo. "review and summarization for computer rise-fall control in steelmaking arc furnaces," *Industry Heating*. 29(1),pp.1-4, Jan 2000.

[5] Chu Yue-zhong,Zhang Shao-de,Zhang She-feng. "Intelligent modeling for the electrode system in ladle furnace based on regular RBF neural network," *Automation and Instrumentation*, 19(5),pp. 5-7, May 2004.

[6] Wang Li-xin. Adaptive fuzzy system and control-design and stability analysis [M].Beijing: National defence industry publish house，1995.

STATCOM ETO Failure Analysis

Zhong Du, Bin Chen, Chong Han, Zhaoning Yang, Wenchao Song,
Subhashish Bhattacharya and Alex Q. Huang
Semiconductor Power Electronics Center
Department of Electrical and Computer Engineering
North Carolina State University
Raleigh, NC 27695-7571, USA
Email: zdu@ncsu.edu

Abstract – **Emitter Turn-Off (ETO) Thyristor based STATic synchronous COMpensator (STATCOM) is one of the promising Flexible AC Transmission System (FACTS) technologies, and it is gaining more popularity in utility grids. A 4.5 MVA ETO based STATCOM failed two ETO devices as over-voltage protection triggered AC connector and disconnected the STATCOM from AC grid supply. This paper presents the investigation of ETO failure reasons. The investigation found that dead time of switching signals is very critical for ETO applications. Protection methods for high power ETO applications are proposed based on failure investigation. Experiments verified the proposed protection method.**

Key words –ETO device; STATCOM; failure analysis

I. INTRODUCTION

Flexible AC transmission systems can increase the capacity of existing transmission networks by allowing energy companies to direct power along specific corridors, aligning the physical flow of power with the commercial transactions. In many instances, power electronics–based controllers can increase power transfer capability by up to 50% while maintaining transmission system security and stability. Because of such promising benefits, an advanced power electronics controller static synchronous compensator (STATCOM) based on the Emitter Turn-Off (ETO) devices and cascaded multilevel converter (CMC) technology is currently being developed. The power rating of the developed STATCOM is 4.5 MVA using Emitter Turn-Off (ETO) devices with 2,500 Amperes peak turn-off current capability and 4,500 Volts voltage blocking capability. The grid-tie operation of the developed 4.5 MVA ETO STATCOM has been successfully achieved in summer 2005 with a TI TMS320C6701 DSP based controller. Because the maximum three phase power

capability of the laboratory is limited, the experiment was conducted up to 480V/500A grid supply [1].

To further improve the STATCOM performance, a new TI TMS320C6713 DSP based controller was developed with human interface to implement new control experiments. During testing experiments in September 2005, two ETO devices and a clamping diode failed. This paper presents the failure analysis to the STATCOM ETO.

II. STATCOM WORKING PRINCIPLE

The single-line diagram of the STATCOM system is illustrated in Fig. 1. In general, a STATCOM system is comprised of three main parts: a voltage source converter, a coupling reactor or a step-up transformer, and a controller. The STATCOM is connected to the power networks at the point of common coupling (PCC). All required voltages and currents are measured and fed into the controller to be compared with the commands. The controller then performs feedback control and outputs a set of switching signals to drive the main semiconductor switches of the power converter accordingly. In a STATCOM, the reactive current IO can be controlled independent of VPCC, this capability is superior compared with conventional Static Var Compensator (SVC) because in a SVC, the output current decreases when VPCC decreases. Due to modularity of H-bridges, it was used to built the 4.5 MVA STATCOM with three output voltage levels which is shown in Fig. 2.

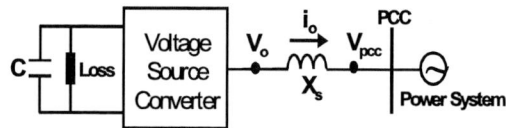

Fig. 1. Single-line Diagram of the Voltage-Source Converter-Based
STATCOM

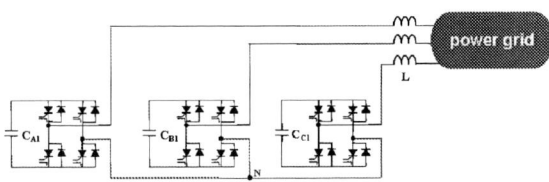

Fig. 2. Three-level 4.5 MVA ETO CMC STATCOM Configuration

The power switches used to build the STATCOM are ETO devices with 2,500 Amperes peak turn-off current capability and 4,500 Volts voltage blocking capability [2][3][4][5][6][7]. The structure and its symbol are shown in Fig. 3. The actual H-bridge converter for the STATCOM is shown in Fig. 4. The STATCOM controller includes a TI TMS320C6713 DSP and a Xilinx XCV300 FPGA. The DSP software is developed using C language. The modulation control method was digital SPWM method. For real-time control, the STATCOM modes can be divided into four modes: stop mode, diode charge mode, PWM boost charge mode and normal operation mode.

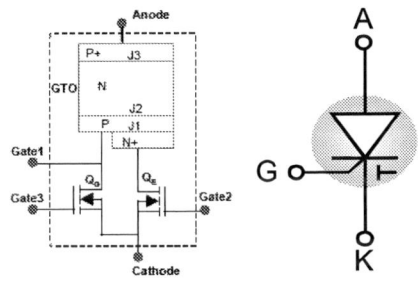

Fig. 3. ETO structure and its symbol

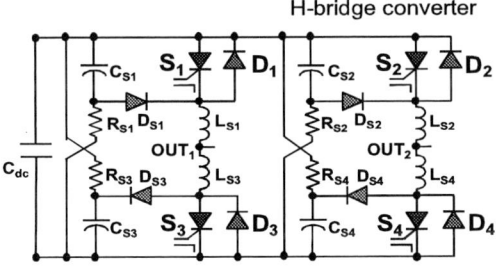

Fig. 4. ETO based H-bridge converter (ETO S_1, S_3 and diode D_{s1} failed)

The FPGA software is developed using VHDL language under Xilinx development environment. The FPGA is used to interface DSP and input/output control signals. It has two major functions. The first function is to generate SPWM switching signals for the ETO switches with 1 kHz triangle carrier signal. The second function is to control A/D converters to collect information of voltages and currents required for the STATCOM control and monitor, and send them back to DSP.

To simplify the DSP control for its real-time performance, DSP only needs to send the duty information to FPGA. A triangle wave generator in FPGA will generate triangle wave carriers, and the duty data from DSP will be used to compare with the carrier to generate the actual switching signals. For shoot through protection, all the switching signals are appended a preset dead time.

III. FAILURE OCCURRENCE

In September of 2005, the new developed DSP and FPGA controller was used on experiments. After setup hardware protection, the operators planed to test the over voltage hardware protection by charging voltages step by step. First, the STATCOM was run in diode charge mode and the capacitors were charged to 340 V; then the STATCOM was run in PWM boost charge mode to charge the capacitors to 1500 V. The first testing is successful and the over voltage protection worked. After the first testing, the capacitors were discharged for the second over voltage protection testing. In the second testing, when the capacitor's voltage reached 1500V, the operators heard a strange sound inside high power lab and watched from the oscilloscope that the capacitor voltage V_{dc} at phase A dropped to zero sharply. The failure event sequence: when dc-bus was charged to 1500V, the hardware protection tripped AC circuit breaker and kept PWM signal sending out. Almost simultaneously, DSP was still running and then detected the "Combined Fault! Check SW signals", which indicated that ETO or DC circuit breaker had already failed and was detected by DSP, and finally DSP blocked PWM output signals 50-60ms later. The water cooling system was not turned on because the charging current was around 20 A which was very low. At the failure point, the output voltage of phase B was on high frequency oscillating which is shown in Figure 5 (Phase A output voltage was not been recorded).

After checking the STATCOM, it was found that two ETO devices in a phase leg A failed short and the peak short current was estimated 10000 A from the recorded data. The failure ETO devices are S_1 and S_3, and the failure diode is D_{s1} in Fig. 4.

1451

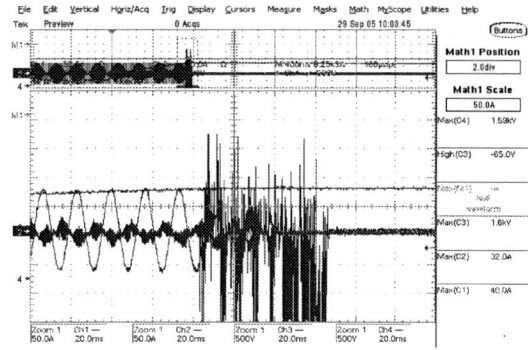

Fig. 5. V_{pcc} voltage (pink), charge current (blue) and capacitor voltage (green)

IV. FAILURE ANALYSIS

From the ETO failure phenomena, it looked that two ETOs in a leg were shoot through failure. The shoot through of the two ETOs caused the clamping diode failure. Based on the failure phenomena, some possible failure reasons could be:

(1) FPGA sent both ON PWM signal commands to top and bottom ETO switches of this leg, and ETO shoot through failure caused large di/dt and/or dv/dt which failed the clamping diode;

(2) FPGA sent out correct switching signals, but the switching signals are absence of dead time, and ETO shoot through failure caused large di/dt and/or dv/dt which failed the clamping diode;

(3) EMI noise on ETO gate drive control circuit caused ETO switches failure, and ETO shoot through failure caused large di/dt and/or dv/dt which failed the clamping diode;

(4) For some reasons, the clamping diode failed first, the following switching on signals failed the two ETOs;

(5) For some reason, one of the ETOs failed short first, the following switching on signal to another ETO failed the second ETO, and ETO shoot through failure caused large di/dt or dv/dt which failed the clamping diode;

(6) Control circuit of gate drive board failure.

Failure reasons (3) (6) were low probability because no large EMI source was nearby at that time and the logic control circuit of the ETO was still working after the ETO failed. Failure reason (4) was low probability because the clamping diode is very high reliability (100 FIT). Failure reason (5) was also low probability because the working current of the GTO devices was very low of 20 A compared its current capability of 2500 A at the failure point.

The investigation to the ETO failure focused on reason (1) and (2). The digital SPWM switching signals are checked one by one using oscilloscope. In the whole modulation range, dead time was observed. But when the modulation index reached 0.95 or above, sometimes the switching signal was very narrow, and the dead time was reduced from normal 25 μs to 6 μs which was shown in Fig. 6. Direct ETO gate signal checking found that if the FPGA switching signals only have 5 μs dead time, then short through occurred between the top and bottom ETOs, which was shown in fig. 7. Further checking found that because the ratio of the carrier and the reference was not an integer, if the reference and the carrier reached their peak values simultaneously. The switching signal will be very narrow and the on-off time of the FPGA control signals reduced to around 5 μs, then the shoot through occurred.

Therefore, the complete failure scenario could be reconstructed: over voltage protection triggered AC connector and disconnected the STATCOM from AC grid supply; AC connector disconnecting caused arc at PCC and caused malfunction of PLL; malfunction of PLL caused current loop PI controller to output modulation index 1 for some time, and the reference and the carrier occasionally (this is because PLL makes the reference variable) reached their peak values simultaneously reduced the FPGA control signal dead time; the wrong dead time of FPGA control signals caused the short-through failure of ETOs.

The investigation showed the ETO failure would occur if two situations were satisfied. (1) The modulation index must be around 1; (2) the reference and the carrier must reach their peak values almost at the same time. The recorded modulation index for Phase A at the failure point supported the investigation findings. In the previous STATCOM experiments, the modulation index was below 0.8. The investigation findings showed the failure was very random with low probability which explained the safe running of STATCOM of previous experiments.

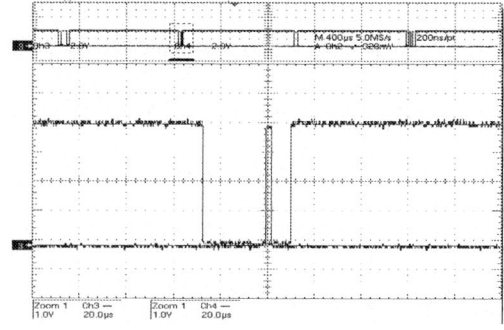

Fig. 6. Switching signals and dead time

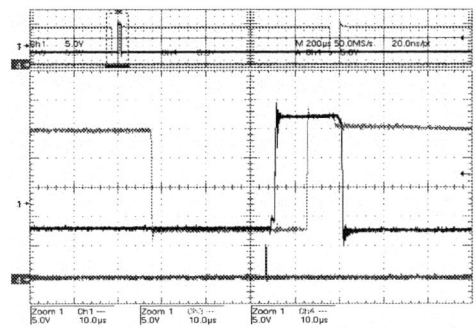

Fig.7. Minmum on/off time setting extend the narrow pulse with non-sufficient deadtime and cause overlap

Based on the failure investigation, FPGA program has been modified to interlock the top and bottom switching signals, and to guarantee 29 μs dead time. Its corresponding ETO gate signals are shown in fig. 8. Fig. 8 also shows the comparison of FPGA coding of dead-time at the ETO drive side. To ensure enough dead-time, the very narrow drive pulse is eliminated. New experiments also verified the failure analysis.

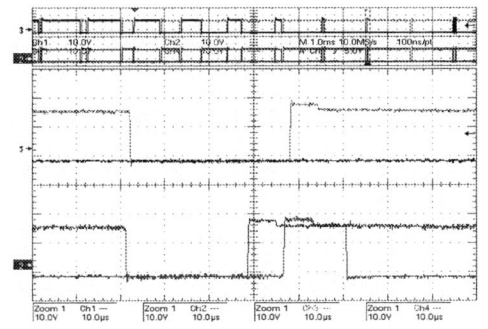

Fig 8. Actual drive signal of ETO devices (top: new FPGA coding, bottom: old FPGA coding)

V. CONCLUSION

The investigation of ETO failure reason of a 4.5 MVA ETO based STATCOM are presented in this paper. The detailed field data and waveforms are used for failure analysis. The most possible failure reason is found and necessary protection method is proposed to avoid failure situations. Failure investigation found that a reliable dead-time function is very critical for high power applications. Therefore it is suggested to apply highly reliable dead time control method to high power ETO applications. The proposed protection method has been implemented using Xilinx XCV300 FPGA. All the experiments showed that the proposed protection method is working and the short-through failure can be avoided.

REFERENCE

[1]   "4.5MVA ETO STATCOM: Design and Field Demonstrations," EPRI annual report, 2005.

[2]   Yuxin Li; Huang, A.Q.; Lee, F.C. "Introducing the emitter turn-off thyristor (ETO)", Industry Applications Conference, 1998. Volume 2, 12-15 Oct. 1998 Page(s):860 - 864 vol.2

[3]   Yuxin Li; Huang, A.Q.; Motto, K.; "Experimental and numerical study of the emitter turn-off thyristor (ETO)", Power Electronics, IEEE Transactions on Volume 15, May 2000 Page(s):561-574

[4]   Bin Zhang; Huang, A.Q.; Xigen Zhou; Yunfeng Liu; Atcitty, S.; "The built-in current sensor and over-current protection of the emitter turn-off (ETO) thyristor", IAS, 2003. Volume 2, 12-16 Oct. 2003 Page(s):1264 - 1269 vol.2

[5]   Zhang, B.; Huang, A.Q.; Chen, B.; Atcitty, S.; Ingram, M.; "SPETO: a superior power switch for high power, high frequency, low cost converters" Industry Applications Conference, 2004, Volume 3, 3-7 Oct. 2004 Page(s):1940 - 1946 vol.3

[6]   B. Chen, N. Zhu, Y. Gao, A. Q. Huang, "Performance of a 4.5 kV, 100A Current-Scalable Emitter Turn-Off (ETO) Thyristor Module", IAS2005, Hong Kong.

[7]   Kevin Motto, Yuxin Li and Alex.Q.Huang, "Comparison of State-of-the-Art High Power IGBTs, IGCTs, and ETO" APEC 2000. Fifteenth Annual IEEE,Volume 2, 6-10 Feb. 2000 Page(s):1129-136 vol.2

Modeling and Control of Three-phase Voltage Source PWM Rectifier

Yao Chen, XinMin Jin

School of Electrical Engineering, Beijing Jiaotong University, Beijing, China

Abstract—**This paper presents the state of the art in the field of three-phase voltage source PWM rectifier with reduced input harmonics and unity power factor. The working principle, modeling procedure, corresponding control strategy as well as typical waveforms are introduced in detail which can provide some guidelines for engineers to analyze, design and implement.**

Keywords-modeling; control strategy; unity power factor; PWM rectifier; power electronics

I. INTRODUCTION

The ac/dc conversion is used increasingly in a wide diversity of applications such as power supplies for microelectronics, battery management, motor drives, etc.

The simplest rectifier uses diodes to transform the electrical energy from ac to dc side. And then thyristors are used to control the energy flow. However these line-commutated rectifiers generate large amount of harmonics and reactive power which can serious pollute the power grid. Although multipulse connections or passive power filters can reduce grid current harmonics to some extent, these methods will greatly increase the volume and the cost which are not desired [1].

Another conceptually different way of harmonics reduction is to adopt new topologies which naturally possess the function of power factor correction. PWM rectifier is one of the most popular topologies which can obtain unity power factor or any active-reactive power combination. PWM rectifiers can be classified as single-phase, three-phase, voltage source and current source rectifier. In recent years, three-phase voltage source PWM rectifier obtained widely application in machine drives and power generation [2]-[4].

This paper is dedicated to this special type of rectifier. Working principle, modeling procedure, corresponding control strategy as well as typical waveforms are introduced in detail which can give some guidelines for engineers to analyze, design and implement three-phase voltage source PWM rectifier.

II. POWER CIRCUIT AND WORKING PRINCIPLE

The power circuit of three-phase voltage-source PWM rectifier (VSR) is showed in Fig. 1 where L is the value of the inductors, R their equivalent resistance and C the value of the dc-side capacitor. Inductances are inserted between

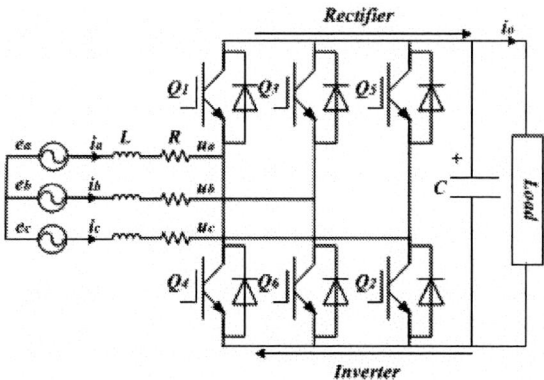

Figure 1. Power circuit of three-phase voltage source PWM rectifier.

grid and converter and insulated gate bipolar transistors (IGBTs) are adopted as the controlled power switches, which are the major differences compared with traditional rectifiers.

The basic operation principle of VSR is to keep the load dc-link voltage V_{dc} at a reference value V_{dc}^* meanwhile obtain desired grid side power factor. This reference dc-link voltage value has to be high enough to keep the diodes of the converter blocked, e.g., in three-phase 380V system, V_{dc}^* must be higher than $\sqrt{2} \times 380\text{V} = 537\text{V}$. Once this condition is satisfied, the dc-link voltage is measured and compared with the reference. The error signal generated from this comparison is used to switch ON and OFF the valves of the VSR. When the dc load current i_o is positive during rectifier operation, the capacitor C is being discharged, and the error signal becomes positive. Under this condition, the control block takes power from the supply by generating the appropriate PWM signals for the six power switches of the VSR. In this way, current flows from the ac to the dc side, and the capacitor voltage is recovered. Inversely, when i_o becomes negative during inverter operation, the capacitor C is overcharged, and the error signal requires the control block to discharge the capacitor by returning power to the ac mains.

To make the rectifier work properly, the frequency of fundamental of converter side voltage u_x must be the same with the power source e_x, $x=a,b,c$. Changing the amplitude of this fundamental, and its phase shift with respect to the mains, the rectifier can be controlled to obtain any active-reactive power combination. There are mainly four kinds of operate modes: unity power factor rectifier, unity power factor inverter, capacity operation at zero power factor and inductor operation at zero power factor [5]. The corresponding waveforms and phasor diagrams are showed in Fig. 2. Obviously, the inductance L plays an extremely importance role in every operation mode. First, the induced voltage it generates allows the VSR working in boost style, which makes dc-link voltage higher than the magnitude of the main so as to block the diodes and working properly. Second, the inductance also works as a filter, which can maintain ac current almost sinusoidal, reducing harmonic contamination to the power supply.

III. SYSTEM MODELING

Considering the displayed variables on the circuit of Fig.1, applying Kirchhoff laws, the following differential equations of the VSR in the three-phase reference frame can be obtained [6]:

$$
\begin{cases}
\dfrac{di_a}{dt} = -\dfrac{R}{L}i_a - \dfrac{1}{L}u_a + \dfrac{1}{L}e_a \\[2mm]
\dfrac{di_b}{dt} = -\dfrac{R}{L}i_b - \dfrac{1}{L}u_b + \dfrac{1}{L}e_b \\[2mm]
\dfrac{di_c}{dt} = -\dfrac{R}{L}i_c - \dfrac{1}{L}u_c + \dfrac{1}{L}e_c
\end{cases}
\tag{1}
$$

By transforming three-phase grid voltage and current of VSR into d-q rotating reference frame synchronized with the main and aligning the d axis on grid voltage vector $\vec{E}$, we can obtain the time-invariant model of VSR as follow,

where $e_d=E$, $e_q=0$ and hence be omitted in the equations.

$$
\begin{cases}
\dfrac{di_d}{dt} = -\dfrac{R}{L}i_d + \omega i_q - \dfrac{1}{L}u_d + \dfrac{1}{L}e_d \\[2mm]
\dfrac{di_q}{dt} = -\dfrac{R}{L}i_q - \omega i_d - \dfrac{1}{L}u_q
\end{cases}
\tag{2}
$$

Considering a symmetrical three-phase system, the transform matrix can be defined as follow where the voltage and current variables are represented by v and $\theta=\omega t$ is the angle between a-axis and d-axis.

$$
\begin{bmatrix} v_d \\ v_q \end{bmatrix} =
\begin{bmatrix} \cos\theta & \sin\theta \\ -\sin\theta & \cos\theta \end{bmatrix}
\begin{bmatrix} 1 & 0 \\ \dfrac{1}{\sqrt{3}} & \dfrac{2}{\sqrt{3}} \end{bmatrix}
\begin{bmatrix} v_a \\ v_b \end{bmatrix}
\tag{3}
$$

The relationship among stationary and rotating reference frames is showed in Fig. 3 where ds-qs represents bi-phase stationary reference frame.

According to (2), we can obtain the equivalent circuits of the VSR in d-q vector space showed respectively in Fig. 4 (a)and (b) where ωLi_d and ωLi_q can be regarded as the additional voltage due to axis transformation.

IV. CONTROL SCHEME AND TYPICAL WAVEFORMS

In the steady state, i_d and i_q are constant components so their derivatives equal to zero. Ignore the equivalent resistance R which is always small, we can obtain the steady state control function according to (2):

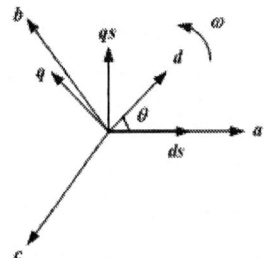

Figure 3. Relationship among stationary and rotating reference frames.

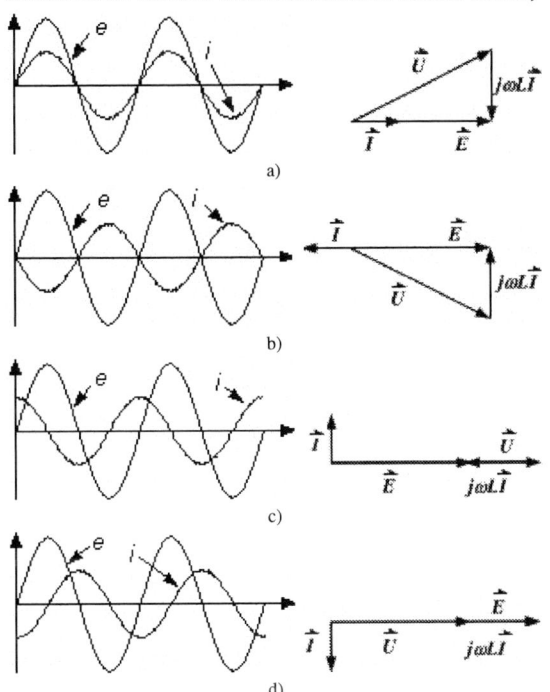

Figure 2. Four modes of operation of the three-phase voltage source PWM rectifier. (a) Unity power factor rectifier. (b) Unity power factor inverter. (c) Capacity operation at zero power factor. (d) Inductor operation at zero power factor.

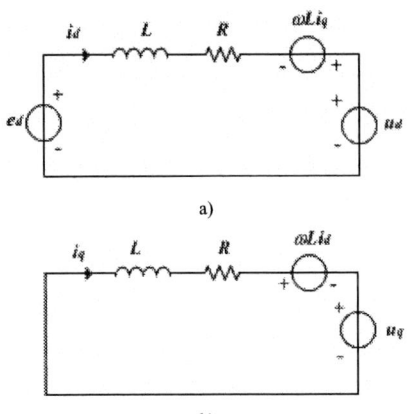

Figure 4. Equivalent circuits in d-q vector space. (a) d-axis equivalent circuit. (b) q-axis equivalent circuit.

$$\begin{cases} u_d = e_d + \omega L i_q \\ u_q = -\omega L i_d \end{cases} \qquad (4)$$

Feedback control components should be added into (4) in order to keep dc-link voltage and ac current at the desired value. Linear control technique such as the PI-based method has been widely applied to both of them.

The PI-based control structure is showed in Fig. 5 where i_d^* and i_q^* are the reference current components in *d-q* frame. This control structure is defined as 'cascade' because the dc voltage controller in outer loop calculates the reference value i_d^* for the *d*-axis current controller and in the inner loop i_d is controlled to perform the dc-link voltage regulation while i_q is controlled to obtain a desired power factor. For unity power factor application i_q^* should be set to zero. SVPWM can be applied directly to generate ON and OFF switching signals right after we obtain the value of u_d and u_q.

Hence the total control equations can be obtained as follow. k_{vp}, k_{vi}, k_{ip} and k_{ii} are proportional and integral gains of voltage and current loop respectively. These parameters can significantly influent the system stability as well as dynamic performance. Generally, the voltage loop parameters are selected according to 'symmetrical optimum' criterion to obtain optimum regulation and stability while the current loop parameters are selected according to 'technical optimum' criterion provides a good control of the overshot to the step change of the reference [7], [8].

$$\begin{cases} i_d^* = k_{vp}(V_{dc}^* - V_{dc}) + k_{vi}\int(V_{dc}^* - V_{dc})dt \\ i_q^* = 0 \\ u_d = e_d + \omega L i_q - k_{ip}(i_d^* - i_d) - k_{ii}\int(i_d^* - i_d)dt \\ u_q = -\omega L i_d - k_{ip}(i_q^* - i_q) - k_{ii}\int(i_q^* - i_q)dt \end{cases} \qquad (5)$$

Some typical waveforms obtained by simulations and experiments are given respectively. The grid-side line voltage rms is 380V, dc-link voltage is 700V and the power stage is 5KW. Fig. 6 represents the grid side voltage and current waveforms when VSR is operating as

a unity power factor rectifier. Fig. 7 represents the grid side voltage and current waveforms when VSR is operating as a unity power factor inverter. Fig. 8 shows the voltage waveforms on grid side inductance and Fig. 9 shows the waveform of converter side line voltage u_{ab}. Waveforms of three-phase upper valves' switching signals generated by SVPWM with lower-pass filter are showed in Fig 10, which can be used to verify the correctness of SVPWM module.

V. CONTROL SCHEME AND TYPICAL WAVEFORMS

This paper has reviewed the state of the art in the field of three-phase voltage source PWM rectifier. The working principle, modeling procedure, control scheme and typical waveforms have been completely introduced to provide some guidelines for the analysis and design. With the sustained theoretical and technological development, three-phase voltage source PWM rectifier will be even widely accepted in industry.

REFERENCES

[1] B. K. Bose, *Modern Power Electronic and AC Drives*. Beijing: Machine Industry Publishing House, 2003.

[2] G. H. Thomas, "A space vector-based rectifier regulator for AC/DC/AC converters", *J. IEEE Trans. Power Electron.*. vol. 8, pp. 30-36, 1993.

[3] T. Shimizu, T. Fujita, G. Kimura, J. Hirose, "A unity power factor PWM rectifier with DC ripple compensation", *J. IEEE Trans. Ind. Electron.*. vol. 44, pp. 447-455, 1997.

[4] X. Zhang, C. W. Zhang, "Study on a new space voltage vector control method about reversible PWM converter", *J. Proceedings of the CSEE*. vol. 21, pp. 102-105, 2001.

[5] J. R. Rodríguez, J. W. Dixon, I. R. Espinoza, J. Pontt, P. Lezana, "PWM regenerative rectifiers: state of the art", *J. IEEE Trans. Ind. Electron.*. vol. 52, pp: 5-22, 2005.

[6] J. F. Silva. "Sliding-mode control of boost-type unity-power-factor PWM rectifiers", *J. IEEE Trans. Ind. Electron.*. vol. 46, pp: 594-603, 1993.

[7] V. Blasko, V. Kaura, "A novel control to actively damp resonance in input *LC* filter of a three-phase voltage source converter", *J. IEEE Trans. Ind. Applicat.*. vol. 33, pp: 542-550, 1997.

[8] M. Liserre, F. Blaabjerg, S. Hansen, "Design and control of an LCL-filter-based three-phase active rectifier", *J. IEEE Trans. Ind. Applicat.*. vol. 41, pp: 1281-1291, 2005.

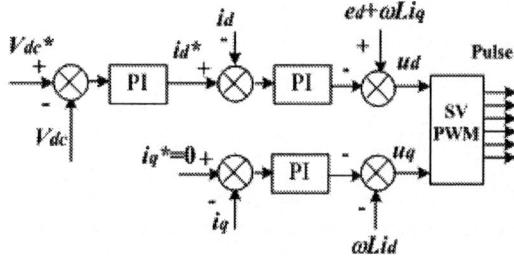

Figure 5. PI-based control structure with *d-q* axis oriented cascade control of three-phase voltage source PWM rectifier

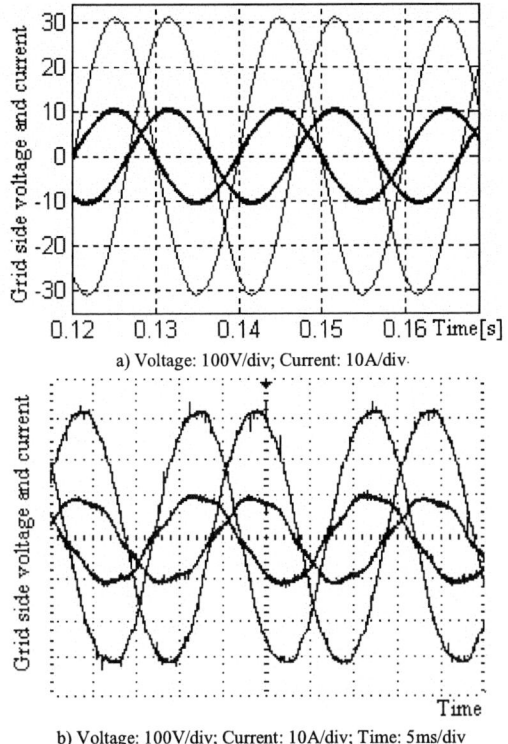

a) Voltage: 100V/div; Current: 10A/div

b) Voltage: 100V/div; Current: 10A/div; Time: 5ms/div

Figure 6.   Grid side voltage and current waveforms at rectification state.
(a) Simulation waveform. (b) Experiment waveform.

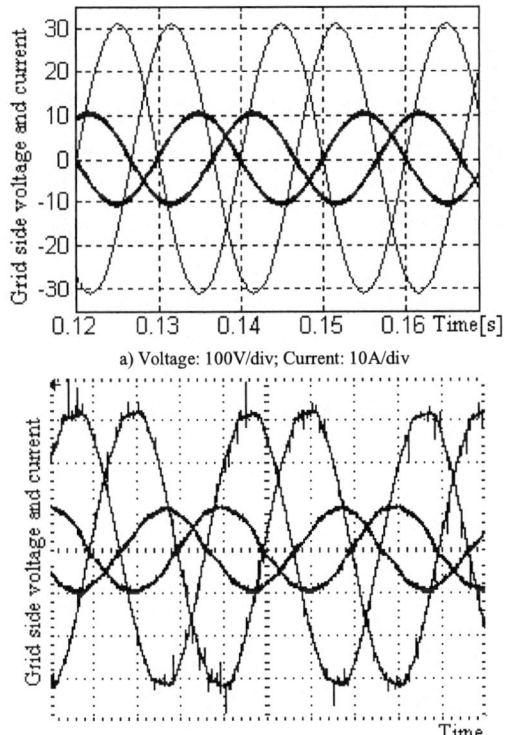

a) Voltage: 100V/div; Current: 10A/div

b) Voltage: 100V/div; Current: 10A/div; Time: 5ms/div

Figure 7.   Grid side voltage and current waveforms at inversion state.
(a) Simulation waveform. (b) Experiment waveform.

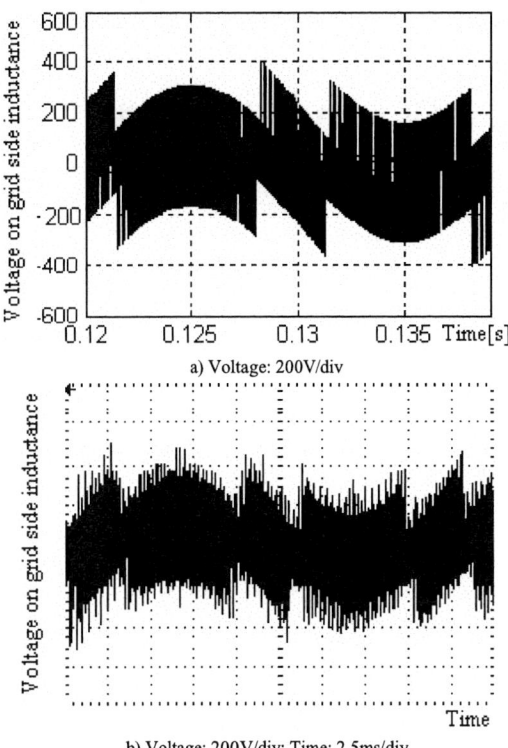

a) Voltage: 200V/div

b) Voltage: 200V/div; Time: 2.5ms/div

Figure 8.   Voltage waveforms on grid side inductance. (a) Simulation
waveform. (b) Experiment waveform.

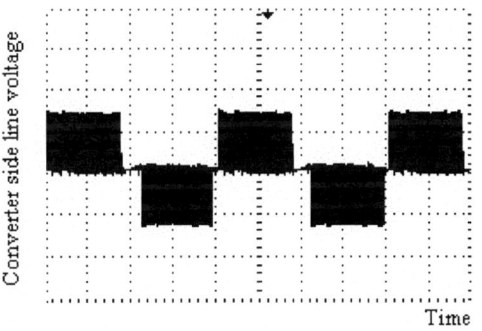

Figure 9.   Waveform of converter side line voltage. Voltage: 500V/div;
Time: 5ms/div

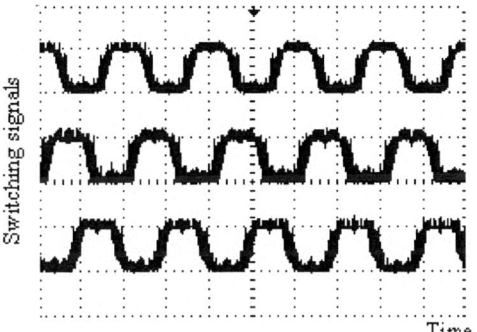

Figure 10. Three-phase upper valves' switching signals generated by
SVPWM with low-pass filter. Time: 10ms/div

Mitigation of Electric Arc Furnace Voltage Flicker Using Static Synchronous Compensator

Y.F. Wang*, J.G. Jiang*, L.S. Ge** and X.J. Yang*

*Dept. of Electrical Eng., Shanghai Jiaotong Univ., Shanghai, China
**Dept. of Electrical Eng., Anhui Uinv.of Technol., Maanshan, China
E-mail: jiang@sjtu.edu.cn

Abstract—Based on the improved model of nonlinear time-varying electric arc resistance, the power supply system model of Electric Arc Furnace (EAF) is established in MATLAB/POWER SYSTEM BLOCKSET, which is used to investigate voltage flicker problem. A flexible power supply strategy is presented, in which static synchronous compensator (STATCOM) is applied to mitigate the voltage flicker produced by EAF. The instantaneous reactive current of EAF impact load is extracted using instantaneous reactive power theory. Then PWM technology is used for current tracing feedback control to make STATCOM produce the required reactive current. The results of simulation indicate that the strategy is valid to mitigate power supply voltage flicker, and the objective of flexible power supply is achieved.

Keywords- EAF; voltage flicker; STATCOM; modeling; simulation

I. INTRODUCTION

Because of the technical and economical superiority of Electric Arc Furnace (EAF) in steel-making, the application of EAFs in metallurgical industry is increasing day by day, the monomer capacity is increasing unceasingly also, and the disturbances produced by EAFs to power system are becoming more and more prominent [1]. The theoretical analysis and the practical experiences show that EAFs are the main impact loads which bring voltage disturbances to power system [2]. The disadvantageous effects produced by EAFs mainly include voltage flicker caused by the large and erratic reactive current swing [3], voltage harmonics caused by the nonlinear characteristic of electric arc resistance [4], and the three-phase imbalance caused by the asymmetric loads [5].

Voltage flicker is the main disturbance and disadvantageous effect caused by EAFs to the power supply system [6]. The large and erratic reactive current swing causes corresponding voltage flicker, and the flicker frequency varies from 0.1Hz to 30Hz.The voltage flicker whose frequency varies from 1Hz to 10Hz can cause the incandescent lamps and the television picture

to glitter, which make human feel agitated. Fierce flicker can even cause electric machinery rotation unstable, the efficiency of electric facilities (include EAF itself) to reduce also [7]. The voltage flicker due to EAFs is a power quality problem that affects our daily lives [8].

The level of voltage flicker is related to short-circuit capacity of the power supply system, the power supply network structure and the load electrical characteristic. Thus, several corresponding measures could be taken to mitigate voltage flicker, such as the improvement of load characteristic, the enhancement of power supply ability and the use of compensators [7]. EAF load creates voltage flicker, whose essence is the impact variation of reactive power [3]. Using the devices which can compensate reactive power, the corresponding voltage flicker can be mitigated, such as using series inductor [9], using series capacitor [10], using static var compensator (SVC) [11], and using static synchronous compensator (STATCOM) [12]. STATCOM is a kind of dynamic reactive power compensator, which has been developed in recent years. As early as in 2000, China has realized independent development and industrial application of STATCOM by Tsinghua University.

In this paper, the power supply system model of EAF is established in MATLAB/POWER SYSTEM BLOCKSET, which is based on the improved model of nonlinear time-varying electric arc resistance. The model does not provide a perfect simulation of EAF behavior, but provides an accurate prediction of the voltage flicker caused by the complex behavior. The model can be used to investigate the effect on flicker compensation using STATCOM. A flexible power supply strategy is presented, in which STATCOM is applied to mitigate the voltage flicker produced by EAF. The instantaneous reactive current of EAF impact load is extracted by using instantaneous reactive power theory. Then PWM technology is used for current tracing feedback control to make STATCOM produce the required reactive current. The results of simulation indicate that the strategy is valid to mitigate the voltage flicker caused by EAF.

---

[a] 973 project *(2005CB221505)*, SRFDP *(20050248058)*.

1-4244-0448-7/06/$25.00 ©2006 IEEE

II. Modeling and Simulation of EAF Power Supply System

Whether or not can we establish the accurate model of EAF is an important prerequisite to investigate the power quality problem. Based on the improved three-phase nonlinear time-varying resistance model which is proposed in [13], the electric arc resistance is defined as:

$$R_a = R_0[1 + \sin(\omega t)] .$$

$$(1)$$

Where $\omega = 2\pi f$, f is the flicker frequency, which often adopts the most sensitive flicker frequency (which is 8.8Hz). R_0 is the constant resistance value which contacts to the running conditions of EAF.

The value of R_0 is determined by the variation range of arc length and the average consumption of active power in EAF load. Here the arc length varies from 0cm to 30cm; the EAF average consumption of active power is 39.7MW, the result of theoretical calculation shows that the value of R_0 is 3.2mΩ.

A 75t steel-making EAF is investigated, the rated capacity of EAF transformer is 50MVA. The high voltage cable connects to the plant headquarters bus directly, which supplies power to the EAF transformer, the secondary voltage of EAF transformer ranges from 375V (tap 1) to 600V (tap 13). The EAF is simulated for melting down at the highest tap setting. The power supply system is shown in Fig. 1.

In Fig. 1, Us is the voltage of infinite power source, PCC is the point of common coupling, $L1$ is the inductance of plant headquarters input cable, $L2$ is the inductance of EAF transformer input cable, T1 is the plant headquarters host transformer, T2 is the EAF transformer, Lc is the inductance in short-net, Rc is the resistance in short-net. Some parameters in EAF power supply system are shown in Table I.

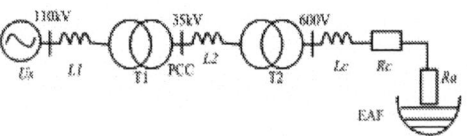

Figure1. EAF power supply system

TABLE I.
SOME PARAMETERS IN EAF POWER SUPPLY SYSTEM

Element	Value(Unit)
Us	110(KV)
$L1$	8.92(mH)
$L2$	0.565(mH)
Lc	0.0127(mH)
Rc	0.15(mΩ)

Supposing the three-phase circuit of EAF is symmetrical. Modeling and simulation of EAF power supply system are achieved in MATLAB/POWER SYSTEM BLOCKSET. The model of total power system includes two transformers and all impedances, which is shown in Fig. 2. A dynamic model of nonlinear time-varying electric arc resistance subsystem is shown in Fig. 3, in which a controlled voltage source is used to simulate EAF. At last, the voltage waveform of phase a in PCC is gained, which is shown in Fig. 4.

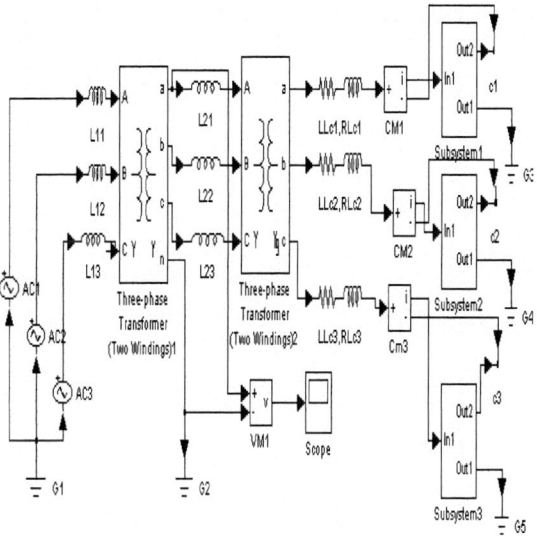

Figure2. The model of EAF power supply system

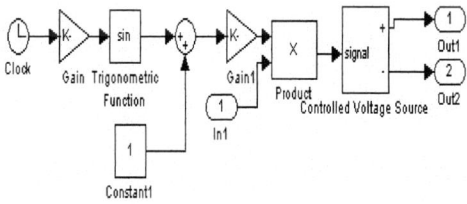

Figure3. The dynamic model of nonlinear time-varying electric arc resistance subsystem

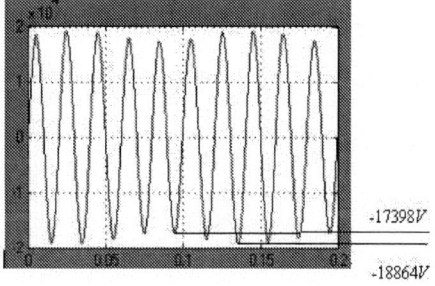

Figure4. The voltage waveform of Phase a in PCC

1459

The voltage fluctuation ratio ($\Delta u/u_n$) in PCC can be calculated using (2).

$$\Delta u/u_n = \frac{U_{max} - U_{min}}{U_n} \times 100\%$$

(2)

Where, U_{max} is the maximum voltage value in PCC, U_{min} is the minimum voltage value in PCC, U_n is the rated voltage value in PCC.

The result of theoretical calculation shows the voltage fluctuation ratio in PCC is 7.25%. According to the Chinese national criterion of voltage fluctuation and flicker (GB12326-2000), when three-phase steel-making EAF runs in short-circuits, the voltage fluctuation ratio in PCC can not surpass 2%. So the corresponding necessary measures should be taken to mitigate the voltage fluctuation and flicker caused by EAF.

III. VOLTAGE FLICKER MITIGATION USING STATCOM

A. Basic principle of STATCOM

The operation of STATCOM is fundamentally different from that of conventional SVC. A SVC operates by selectively connecting passive components (inductors or capacitors) to the power network, a STATCOM is approximately a controlled ac voltage source connected to power network through a suitable tie reactance. By appropriately controlling the STATCOM output voltage, any desired current can be forced to flow through the tie reactance [14]. The basic constitution of STATCOM is shown in Fig. 5, in which self-commutated bridge circuit is paralleled to power network through transformer directly. By adjusting the phase and the amplitude of output voltage in bridge circuit, or by controlling the output current directly, STATCOM can absorb or supply the required reactive current, and compensate the dynamic reactive current instantly. This paper adopts the direct current control strategy.

B. Instantaneous reactive current detection and control

Based on the instantaneous reactive power theory of three-phase circuit, adopting instantaneous active power p and instantaneous reactive power q as the starting condition, the instantaneous reactive current can be obtained by calculation [15], the corresponding detection schematic diagram is shown in Fig. 6.

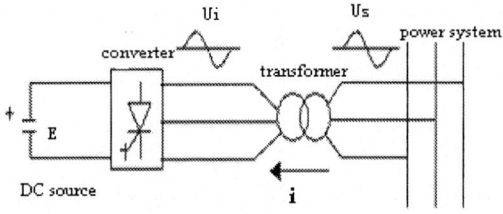

Figure5. The basic circuit of STATCOM

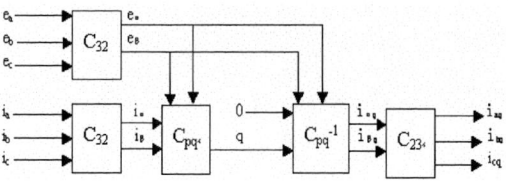

Figure6. Instantaneous reactive current detection schematic diagram

In Fig 6, e_a, e_b and e_c are instantaneous voltage of phase a, b and c,respectively. i_a, i_b and i_c are instantaneous current of phase a, b and c,respectively. The instantaneous voltages and instantaneous currents on the abc coordinates can be transformed into the voltages e_α, e_β and the currents i_α, i_β on the quadrature $\alpha\beta o$ coordinates.

$$\begin{bmatrix} e_\alpha \\ e_\beta \end{bmatrix} = C_{32} \begin{bmatrix} e_a \\ e_b \\ e_c \end{bmatrix}$$

(3)

$$\begin{bmatrix} i_\alpha \\ i_\beta \end{bmatrix} = C_{32} \begin{bmatrix} i_a \\ i_b \\ i_c \end{bmatrix}$$

(4)

Where, $C_{32} = \sqrt{2/3} \begin{bmatrix} 1 & -1/2 & -1/2 \\ 0 & \sqrt{3}/2 & -\sqrt{3}/2 \end{bmatrix}$.

Instantaneous reactive current can be calculated using (5).

$$\begin{bmatrix} i_{aq} \\ i_{bq} \\ i_{cq} \end{bmatrix} = C_{23} C_{pq}^{-1} \begin{bmatrix} 0 \\ q \end{bmatrix}$$

(5)

Where, i_{aq}, i_{bq} and i_{cq} are instantaneous reactive currents of phase a, b and c, respectively.

$$C_{23} = \sqrt{2/3} \begin{bmatrix} 1 & 0 \\ -1/2 & \sqrt{3}/2 \\ -1/2 & -\sqrt{3}/2 \end{bmatrix},$$

$$C_{pq}^{-1} = \begin{bmatrix} \dfrac{e_\alpha}{e_\alpha^2 + e_\beta^2} & \dfrac{e_\beta}{e_\alpha^2 + e_\beta^2} \\ \dfrac{e_\beta}{e_\alpha^2 + e_\beta^2} & -\dfrac{e_\alpha}{e_\alpha^2 + e_\beta^2} \end{bmatrix},$$

$$q=\tfrac{1}{\sqrt{3}}[(e_b-e_c)i_a+(e_c-e_a)i_b+(e_a-e_b)i_c]$$

(6)

When the system voltage keeps constant, controlling the reactive current is same as controlling the reactive power. The direct current control strategy is used, which uses PWM technology to realize the tracking feedback control of instantaneous reactive current. The diagram of direct current control strategy is shown in Fig.7, where the PWM tracking control technology by triangle-wave comparison is used. The instantaneous reactive current reference signal i_{ref} can be extracted from detection circuit, which is shown in Fig. 6.

C. Simulation of flexible power supply system

The detection and control circuit of STATCOM is shown in Fig. 8. In order to analyze simply, the losses of transformer and converter are not considered. The main duty of reference current detection circuit is to extract the instantaneous reactive current i_{ref}, which is corresponded to the required compensatory reactive power. PWM control of converter is realized by the current control circuit and the PWM control circuit.

The model of flexible EAF power supply system is shown in Fig. 9, in which STATCOM is used to mitigate voltage fluctuation and flicker.

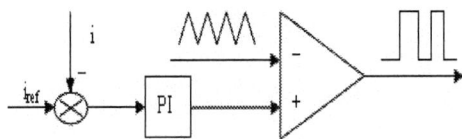

Figure7. PWM technology of direct current control

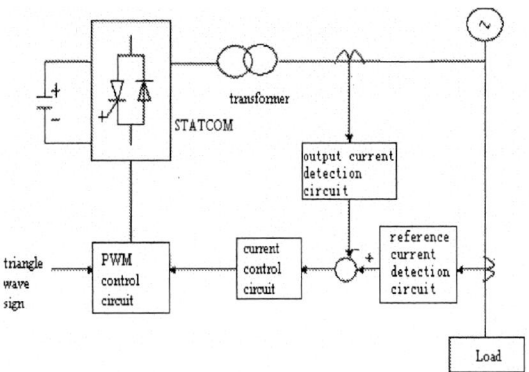

Figure8. The basic detection and control circuit of STATCOM

After repetitive adjustment of *PI* parameters, more ideal voltage waveform is obtained, which is shown in Fig.10. The result shows that the voltage fluctuation ratio in PCC is 1.8% in the novel flexible power supply system. The voltage fluctuation ratio conforms to the Chinese national criterions. Therefore, corresponding voltage flicker is mitigated [16].

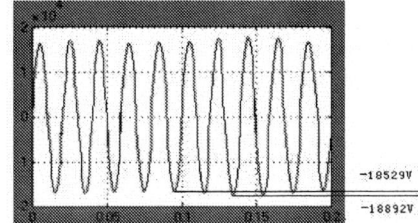

Figure10. Phase *a* voltage waveform of PCC using STATCOM

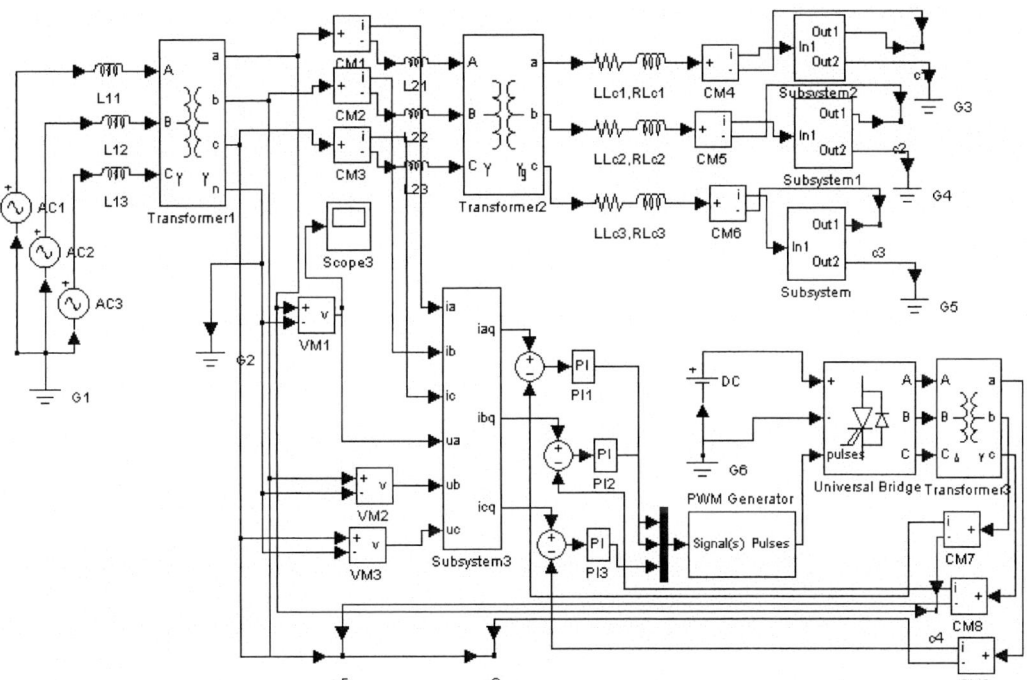

IV. CONCLUSIONS

This paper introduces a flexible power supply strategy for EAF which is realized using STATCOM, in which the main function of STATCOM is to mitigate voltage flicker. Firstly, based on the nonlinear time-varying electric arc resistance model, a power supply system model of EAF is established in MATLAB/POWER SYSTEM BLOCKSET. The model provides an accurate prediction of the voltage flicker caused by EAF, and which can be used to investigate the effect on flicker compensation using STATCOM. Secondly, the flexible power supply strategy is presented, in which the instantaneous reactive current of EAF impact load is extracted using instantaneous reactive power theory. Then PWM technology is used for current tracing feedback control to make STATCOM produce the required reactive current. Finally, the investigation of simulation is accomplished. The results of simulation indicate that the strategy is valid to mitigate the voltage flicker caused by EAF.

REFERENCES

[1] Omer Ozgun and Ali Abur, "Flicker study using a novel arc furnace model," *IEEE Trans. on Power Delivery*, vol.17, no.4, pp.1158-1163, October 2002.

[2] A.Esfandiari and M.Parniani, "Electric arc furnace power quality improvement using shunt active filter and series inductor," *IEEE Region 10 Conference*, vol.4, pp.105-108, November 2004.

[3] Z.Zhang and N.R.Fahmi, "Modeling and analysis of a cascade 11-level inverters-based SVG with control strategies for electric arc furnace (EAF) application," *IEE Proc.-Gener.Transm.Distrib*, vol.150, no.2, pp.217-223, March 2003.

[4] Ying-Pin Chang and Chi-Jui Wu, "Optimal multiobjective Planning of large-scale passive harmonic filters using hybrid differential evolution method considering parameter and loading uncertainty," vol.20, no.1, pp.408-416, January 2005.

[5] D.P.Manjure and E.B.Makram, "Effect of nonlinearity and unbalance on power factor," *IEEE Power Engineering Society Summer Meeting*, vol.2, pp.956-962, July 2000.

[6] A.Hernandez,J.G.Mayordomo,R.Asensi,L.F.Beites, "A method based on interharmonics for flicker propagation applied to arc furnace," *IEEE Trans. on Power Delivery*, vol.20, no.3, pp.2334-2342, July 2005.

[7] C.S.Chen,H.J.Chuang,C.T.Hsu and M.Tseng, "Mitigation of voltage fluctuation for an industrial customer with arc furnace," *IEEE Power Engineering Society Summer Meeting*, vol.3, pp. 1610-1615, July 2001.

[8] M.M.Morcos and J.C.Gomez, "Flicker sources and mitigation," *IEEE Power Engineering Review*, vol.22, no.11, pp.5-10, November 2002.

[9] G.C.Montanari,M.Loggini,L.Pitti,E.Tironi and D.Zaninelli, "The effects of series inductors for flicker reduction in electric power systems supplying arc furnaces," *IEEE Industry Applications Society and Annual Meeting*,pp.1496-1503,1993.

[10] M.W.Marshall,PE, "Using series capacitors to mitigate voltage flicker problems," *41st Annual Rural Electric Power Conference*, pp B3-1-5,1997.

[11] A.Wolf and M.Thamodharan, "Reactive power reduction in three-phase electric arc furnace," *IEEE Trans. on Industrial Electronics*, vol.47, no.4, pp.729-733, August 2000.

[12] J.R.Clouston and J.H.Gurney, "Field demonstration of a distribution static compensator used to mitigate voltage flicker," *Power Engineering Society Winter Meeting*, PP.2568-2576, 2000.

[13] Srinivas Varadan,Elham B.Makram and Adly A.Girgis, "A new time domain voltage source model for an arc furnace using EMTP," *IEEE Trans. on Power Delivery*, vol.11, no.3, pp.1685-1691, July 1996.

[14] Colin Schauder, "STATCOM for compensation of large electric arc furnace installations," *IEEE Power Engineering Society Summer Meeting*, vol.2, pp.1109-1112, July 1999.

[15] Akagi H,Kanazawa Y,Nabae A, "Instantaneous reactive power compensators comprising switching devices without energy storage components," *IEEE Trans. on Industry Applications*, vol.20, no.3, pp.625-630, 1984.

[16] G.C.Montanari,M.Loggini,A.Cavallini,L.Pitti,D.Zanineli, "Arc-furnace model for the study of flicker compensation in electrical networks," *IEEE Trans. on Power Delivery*, vol.9, no.4, pp.2026-2033, October 1994.

2006 5th International Power Electronics and Motion Control Conference

Design of Distributed FACTS Controller and Considerations for Transient Characteristics

Gaidi NING, Shijie HE[*], Yue WANG, Lei YAO, Zhaoan WANG

State Key Laboratory of Electrical Insulation and Power Equipment

Xi'an Jiaotong University, Xi'an, Shaanxi, P.R. China 710049

*Email:hhssdd1@163.com

Abstract—Flexible AC Transmission Systems (FACTS) devices are used to control power flow in the transmission grid to relieve congestion and limit loop flows. However, widespread adoption of this technology has been hampered by high cost and reliability concerns. The concept of Distributed FACTS (D-FACTS) devices, as an alternative approach to realizing cost-effective power flow control, has been recently proposed. This paper discusses design of distributed FACTS controller, and then researches on the transient influence of distributed power control solutions on the power grid, with series active variable inductance (AVI) compensation, changing the impedance of the power line so as to control power flow. Implementation and system effects are presented in the paper, along with experimental results.

Keywords-FACTS; Distributed FACTS; Active power flow control; Active variable inductance

I. INTRODUCTION

The power grid in China and in most of other parts of the world, is aging and under increasing stress. A better method to improve the utilization ratio of the existing electric network is expected. As the awareness of environmental protection is increasing, it becomes critical that existing transmission and distribution resources can be fully utilized.

Flexible AC Transmission Systems devices can be inserted in existing transmission lines to achieve control functions, including enhancement of transient stability, mitigation of system oscillations and so on. Even though FACTS technology is technically proven, it has not seen widespread commercial acceptance due to high voltage (up to 345 kV), high power (multi-hundred MVA), high cost, etc [1][2][3][4]. Wherefore, the concept of distributed FACTS (D-FACTS) devices has recently been proposed as an alternative approach for realizing the functionality of FACTS devices [3][4].

Supported by State Key Laboratory of Electrical Insulation and Power Equipment (Xi'an Jiaotong University)

Each D-FACTS device module, which is rated at about 10kVA and is clamped on the power line as Fig. 1 shown, can be controlled so as to increase or decrease the impedance of the line or to leave it unaltered. With a large number of modules operating together, it is possible to have significant effects on the overall power flow in the line. The low kVA ratings of modules are in line with mass manufactured power electronics components, using mature power conversion techniques to demonstrate the potential for low-cost implementation and high system reliability. D-FACTS is especially suitable for development needs of the electric network in china (weak network, low stability and lack funds).

The structure and the principle of power flow control of DFC based on active variable inductance (AVI) are firstly discussed in this paper. Afterwards, simulation and experimental results of designed DFC system verify the good effectiveness of power flow control. The further simulation on transient state characteristics of DFC is carried on, while DFC is inserted into transmission lines. Conclusions of this paper are presented finally.

II. PRINCIPLE AND EXPERIMENTAL RESEARCH ON DFC

A. Configuration of DFC

The Configuration of DFC is as Fig. 2 shows. Each DFC is basically composed of single turn transformer (STT), single-phase voltage sourced inverter, filter and controlling module. And the self-protection module is essential to apply DFC in practice. The use of the

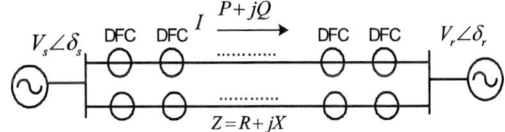

Fig. 1 DFC compensation

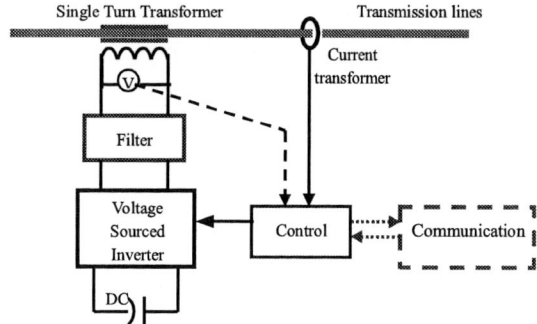

Fig. 2 Configuration of DFC

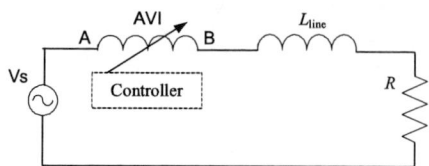

Fig. 4 Basic principles of DFC power flow control

communication module using Power Line Communication (PLC) or wireless brings on controllable power systems.

The main purpose of DFC is to control the effective impedance of transmission lines in a flexible way fast so as to control the total active power flow.

B. Principle of DFC power flow control

DFC changes reactance based on the thought of AVI[5]. The basic principle of the AVI is as Fig. 3 shown. Controller makes AB port network meet the relation between voltage and current of inductance or flux and current of inductance by controlling the inverter. Like this, a port network is equivalent to the inductance, and furthermore it can also be controlled as the negative inductance[6]. The value of AVI can be changed with the definite value of L.

AVI inserted into transmission lines can increase or decrease the totally effective impedance of transmission lines. Fig. 4 provides a basic principle about DFC power flow control based on AVI. By controlling AVI, it is

expected to within the operating constraints of the current-carrying thermal limits of conductors, increase value of transmission assets.

C. Experimental research on DFC

The circuit schematic of AVI is showed in Fig. 5. In the experiment, TI TMS320F2407 DSP is used for digital processing and to generate PWM wave, etc. The (L_f, C_f, R_c) filter is used to eliminate the high-frequency composition outputted by the inverter [7]. The experiment system is as Fig. 6 shows.

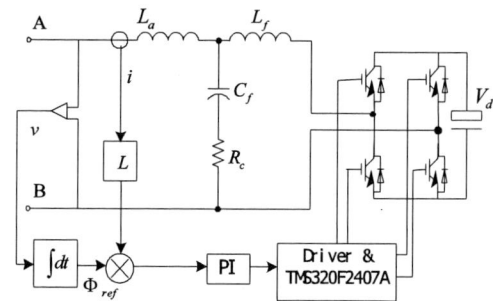

Fig. 5 Circuit schematic of AVI

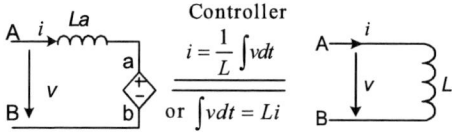

(a)   Principles of AVI

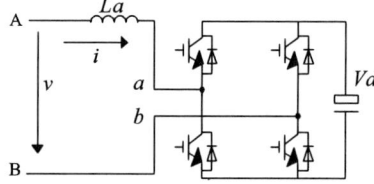

(b)   Practical configuration

Fig. 3 Principles and practical configuration of the AVI

Fig. 6 Experiment system

Experiment parameters are as follows:

V_S=10V; L_{line}＝6mH; R＝5Ω.

Changing the value of AVI, the value of total effective inductance of circuit is reduced from 21mH to 16mH. Experimental waveforms received are showed in Fig. 7.

III. DFC POWER FLOW CONTROL TRANSIENT STATE CHARACTERISTICS

Simulation circuit of DFC power flow control based on AVI is as Fig. 8 shown. Transmission length l=1km, equivalent resistance R=0.027Ω, equivalent reactance X= ωL =0.27Ω, viz. the value of inductance L=0.86mH, the effective value of phase voltage at sending and receiving points V_s=V_r=V=10V, phase angle $\delta = \delta_s - \delta_r = 30°$, the total value of inductance which is inserted by DFC –0.258mH (thirty percent of inductance value in transmission lines). While the value of AVI is reduced from 0H to –0.258mH at 0.08s, it may be gotten that the active power of single phase transient circuit increases from 185.19W to 264.55W according to the theory of circuit. The simulation result is as shown in Fig. 9. The following wave form of Fig. 9 is active power wave form, which agrees well with the calculating value

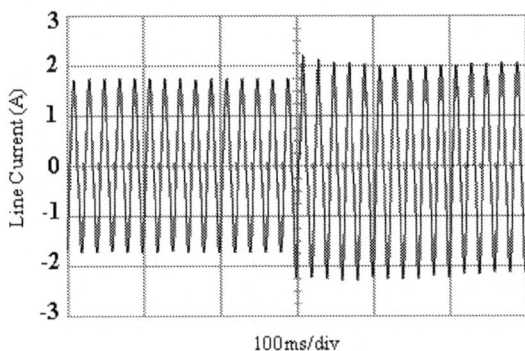

100ms/div

Fig. 7 Experimental Waveform of DFC power flow control

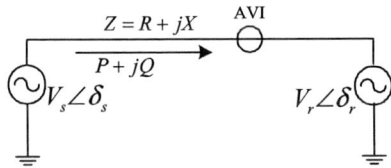

Fig. 8 Simulation circuit of DFC power flow control based on AVI

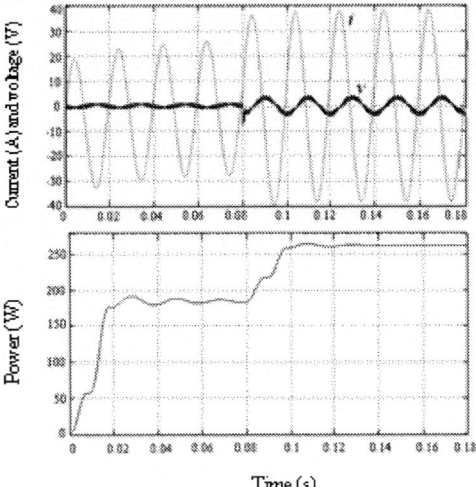

Time (s)

Fig. 9 Simulation waveforms of DFC power flow control

（R=0.027Ω）

at the steady state. The voltage and the current present negative inductance characteristics in the former wave form of Fig. 9. Transformer is not used in the simulation.

The single-phase transient characteristics of system in Fig. 8 is be equivalent to one of the circuit in which one inductance and one resistance are connected in series from one sinusoidal wave voltage. Vs and Vr are the sinusoidal voltage source:

$$u_s = 2\sqrt{2}V\sin\frac{\delta}{2}\sin(\omega t+\frac{180+\delta}{2}*\frac{\pi}{180}) = 5.176\sqrt{2}\sin(\omega t+\frac{105}{180}\pi).$$

Because there are only linear components in the circuit, the electric current can be regarded as superpose for transient state one and steady state one. Zero condition current response of RC in transmission system at 0s:

$$i = \frac{U_m}{z}\sin(\omega t+\psi_u-\varphi) - \frac{U_m}{z}\sin(\psi_u-\varphi)e^{-\frac{t}{\tau}} \quad (1)$$

where ψ_u is initializing angle of source voltage, $z = \sqrt{R^2+(\omega L)^2}$, $\mathrm{tg}\varphi = \frac{\omega L}{R}$ and $\tau = \frac{L}{R}$.

Before L_{AVI} is inserted, the initial value of current of the inductance is calculated through (1):

$$i(0.08_-) = \frac{U_m}{z}\sin(\omega*0.08+\psi_u-\varphi) - \frac{U_m}{z}\sin(\psi_u-\varphi)e^{-\frac{0.08}{\tau}}$$

$$= \frac{U_m}{z}\sin(\psi_u-\varphi)(1-e^{-\frac{0.08}{\tau}}) \approx \frac{U_m}{z}\sin(\psi_u-\varphi).$$

After L_{AVI} ($L_{AVI} = -0.258mH$) is inserted at 0.08s, the current full response in the circuit is:

$$i = [i(0.08_-) - \frac{U_m}{z'}\sin(\psi_u - \phi)]e^{\frac{t-0.08}{t'}} + \frac{U_m}{z'}\sin[\omega(t-0.08) + \psi_u - \phi]$$

$$i = [\frac{U_m}{z}\sin(\psi_u - \phi) - \frac{U_m}{z'}\sin(\psi_u - \phi)]e^{\frac{t-0.08}{t'}} + \frac{U_m}{z'}\sin[\omega(t-0.08) + \psi_u - \phi] \quad (2)$$

where $z' = \sqrt{R^2 + (\omega L')^2}$, $tg\phi = \frac{\omega L'}{R}$, $t' = \frac{L'}{R}$ and $L' = L + L_{AVI}$.

The exponential term in (1) and (2) is the transient component of dynamic course, which decays gradually according to the index law, seen from the wave form of current in Fig. 9. The less phase angle δ and value of equivalent resistance R in transmission Lines, the more mitigative the transient state oscillation is. When $R = 0.0027\Omega$, the simulation wave form is as shown in Fig. 10. The magnified parts of oscillation power in Fig. 9 and Fig. 10 is in Fig. 11. It is easily found that the faster dc component reduces, the shorter course of transient state, which is consistent with the circuit theory. However, the larger value of R, the more oscillation of power as Fig. 11 shown.

IV. CONCLUSIONS

This paper presents a novel DFC——Distributed FACTS Controller based on active variable reactance. The transient state characteristics and power flow control of DFC have carried on the basic research in this paper. Simulations results demonstrate the potential effect of

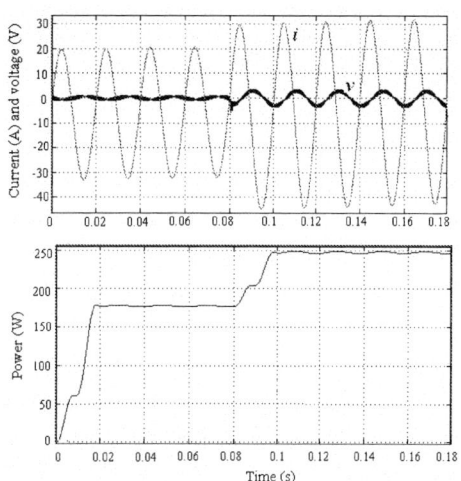

Fig. 10 Influence of transmission lines impedance on DFC

transient characteristics（R=0.0027Ω）

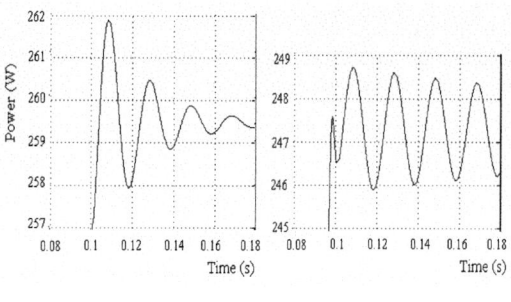

$R=0.027\Omega$ $R=0.0027\Omega$

Fig. 11 Influence of transmission lines impedance on power flow

transient characteristics based on AVI

these devices in terms of better grid utilization and improvement in system reliability.

The following conclusions are also gotten:

1. DFC using commercially available low power devices offers the potential to dramatically reduce the cost of power flow control.

2. The ability to use mature power conversion techniques demonstrates the potential for high reliability and lower-cost implementation.

3. The transient oscillation is more mitigative, as the reactance of transmission lines and phase angle δ descend.

ACKNOWLEDGMENT

The authors would like to thank State Key Laboratory of Electrical Insulation and Power Equipment (Xi'an Jiaotong University) for financial support.

REFERENCES

[1] Song Y. H., Johns A. T. (Ed.). "Flexible AC Transmission Systems (FACTS) [M]," London: IEE Press, IEE Power Engineering Series, 1999. 26~31.

[2] Yinduo Han, Qirong Jiang. "Technical Research Overview to China's Flexible AC Transmission [EB/OL]," (2003 thesis catalogues of thematic seminars on module power and relevant technology and products exhibition) [Online]. Available: http://www.epc.com.cn/proceedings/2003pmt/1/2.htm.

[3] Divan D., Brumsickle W., Schneider R. "A Distributed Static Series Compensator System for Realizing Active Power Flow Control on Existing Power Lines [A]," In: IEEE Proceedings PES[C]. New York, 2004, Vol .2. pp. 654~661.

[4] Divan D., Johal H. "Distributed FACTS – A New Concept for Realizing Grid Power Flow Control [A]," In:Power Electronics

Specialists Conference, 36th Annual IEEE[C]. Recife Brazil, 2005. pp.8~14.

[5] Funato H., Kawamura A. "Proposal of Variable Active-Passive Reactance," 1992 International Conference on Industrial Electronics, Control, Instrumentation, and Automation (IECON'92), PE-10 vol.1,pp.381~388,Nov.1992, in San Diego,U.S.A.

[6] Funato H., Kawamura A., Kamiyama K. "Realization of Negative Inductance using Variable Active-Passive Reactance (VAPAR)," IEEE Transactions on Power Electronics, July.1997, Vol.12, pp.589~596.

[7] Gaidi Ning, Yue Wang, Zhaoan Wang. "Research on Static Series Compensator of Distributed Flexible AC Transmission Systems," Adv. Tech, of Elec. Eng. & Energy, 2005, 24(4): pp.55~58.

2006 5th International Power Electronics and Motion Control Conference

A Wind-Power Generation System Having a Function of Suppressing Line Voltage Deviation

Y. Nakayama[*], S. Fukuda[*], M. Futami[**], M. Ichinose[**], S. Ohara[**], and H. Kita[*]

[*] Graduate School of I.S.T., Hokkaido University, Sapporo, Japan
[**] Hitachi Research Laboratory Co., Ltd., Hitachi, Japan

Abstract— **Wind-power generation (WPG) tends to create voltage deviation in the distribution lines because wind speed always fluctuates. This paper proposes to add a new function to the interface converter of WPG, that is, generation of reactive power to suppress the voltage deviation caused by WPG itself. This paper also proposes a reactive power control strategy for the interface converter to suppress the voltage deviation. The validity is verified by simulation studies.**

Keywords – wind power generation; interface converter; voltage deviation; wind speed fluctuation; reactive power control

I.   INTRODUCTION

Nowadays wind-power generation (WPG) systems tend to employ adjustable speed generators rather than fixed speed ones because of an increase in the generated power due to high efficiency operation and a reduction in the maintenance. Adjustable speed WPG always requires a static interface power converter to connect the variable frequency power to a fixed frequency utility power system. However, when WPG systems are connected to the utility having relatively high line impedances[†] the following problems arise.

1) Fluctuation of wind speed causes voltage deviation at the interconnection point of WPG to the utility.
2) Costly static var controllers or battery storage systems are required to suppress the voltage deviation.

In order to solve these problems simultaneously this paper proposes to add a voltage deviation suppressing function to the utility interface converter. This is because the interface converters can provide reactive power to the utility though their main function is to deliver generated active power by the wind turbine. This paper also proposes a reactive power control strategy for the interface converter taking account of the feature of a WPG system. The validity is verified by simulation-based studies.

II.   INTERFACE CONVERTER AS STATIC VAR CONTROLLER

Generally, WPG systems cause voltage deviation at the point of interconnection (PIC) of WPG to the utility because of the fluctuation of wind speed. To suppress the voltage deviation, usually static var control systems such as a static var compensator (SVC) and a static synchronous compensator (STATCOM) are employed to control reactive power, and battery storage systems are employed to control active power.

A wind turbine does not always provide full power because wind speed always fluctuates. Therefore, there is a margin in the volt-ampere (VA) rating of the interface converter. This paper proposes to utilize the margin to provide reactive power to the utility for suppressing the voltage deviation caused by WPG itself. The interface converter operates as SVC or STATCOM while delivering generated active wind power to the utility. In utility systems, static var controllers are usually used to suppress voltage deviation, but *if the interface converter of WPG had a function of a var controller, an additional var controller such as SVC or STATCOM might be unnecessary.* This is the basic idea of this paper.

Fig. 1 illustrates a wind-power generation system model [1] considered in this paper. A wind-power generator G produces active power P, and it is converted to the utility voltage and frequency by an interface converter. The interface converter delivers the active power P in [pu], and can provide reactive power Q in [pu] as well if it has a margin in the VA rating.

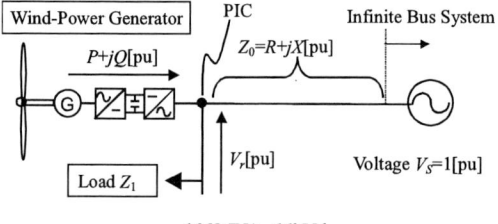

100MVA, 66kV base

Figure 1. Wind-power generation and distribution system model under consideration, P=0.2pu, R=0.12pu, and X=0.2pu.

[†] This sometimes occurs because wind-generation systems are usually located at remote rural areas.

1-4244-0448-7/06/$25.00 ©2006 IEEE

Suppose that WPG and a load system, Z_l, are connected to the utility at the point of interconnection (PIC) as shown in Fig. 1. The power system is based on 100MVA and 66kV, and the system model is assumed as indicated in Fig. 1, where

- impedance of the distribution line: 0.12+0.2j [pu],
- output power of wind-power generation: 0.2 [pu].

If a WPG system delivers active current I_d in [pu] (= active power P) an impedance drop occurs in the distribution line, and the voltage at PIC deviates if I_d deviates. This is the mechanism of voltage deviation caused by WPG systems. If the interface converter provides reactive current I_q in [pu] (= reactive power Q) to the utility, also a voltage drop occurs at PIC. If the interface converter provides an appropriate amount I_q, it compensates for the voltage deviation caused by I_d, and the voltage at PIC can be stabilized. In this case the VA capacity W of the interface converter will be given by

$$W = \sqrt{P^2 + Q^2} \quad [\text{pu}] \tag{1}$$

The VA rating of the interface converter is designed according to maximum active power of wind-turbines. Fig. 2 shows a power curve of a 3MW wind turbine given in [2]. If wind speed is high, the wind turbine is controlled to produce maximum power, 3MW, and the generator produces constant rated active power. Therefore, the voltage deviation at PIC will be almost zero. If wind speed is below the rated speed, the wind turbine generator produces fluctuating active power according to the wind speed fluctuation. As a result, the voltage deviation at PIC will occur. However, the interface converter has a margin in the VA capacity to provide reactive power because WPG is not delivering full power in this case. In other words, large fluctuation of active power occurs when WPG is delivering small output power. Therefore, large reactive power is required to the interface converter for suppressing the voltage deviation. The interface converter can produce larger reactive power Q as it delivers smaller active power P. This feature of WPG is quite suitable for the interface

converter to compensate for the voltage deviation at PIC caused by WPG itself because the addition of a voltage deviation suppressing function does not require an increase in the VA rating of the interface converter.

III. CONTROL SYSTEM OF INTERFACE CONVERTER

A. Control Strategy

Fig. 3 shows a power spectrum [1] of a wind-power generator used for simulation. The active power P contains a wide frequency range due to wind speed fluctuation: low frequency components with large amplitude and high frequency components with some salient amplitude.

If the interface converter attempts to suppress the low frequency deviation components, the required reactive power will be so large and an increase in the converter cost may results because the low frequency deviation includes the voltage deviation caused by the load variations as well. Furthermore, the other economical measures such as a tap-changing transformer are available for suppressing it if the deviation is at quite low frequencies. This paper, therefore, does not attempt to suppress the low frequency voltage deviation, but attempts to suppress the voltage deviation only in a typical frequency range that stems from high frequency active power fluctuation such as an n-component and a $3n$-component [1] depicted in Fig. 3. The n-component corresponds to rotation speed of the wind turbine, and the $3n$-component is known as a result of a tower-shadow effect of WPG systems. The reason of this attempt is that the power fluctuation in that frequency range causes voltage flicker at PIC and, furthermore, the voltage deviation in that frequency range can be suppressed with applying minimum reactive power. Therefore,

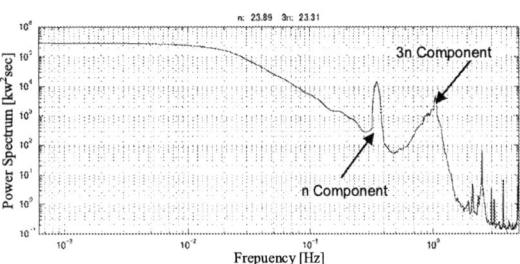

Figure 3. Power spectrum of a wind power generator.

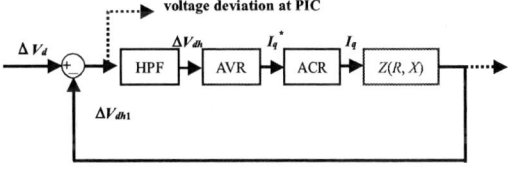

Figure 4. Simplified control system for suppressing voltage deviation.

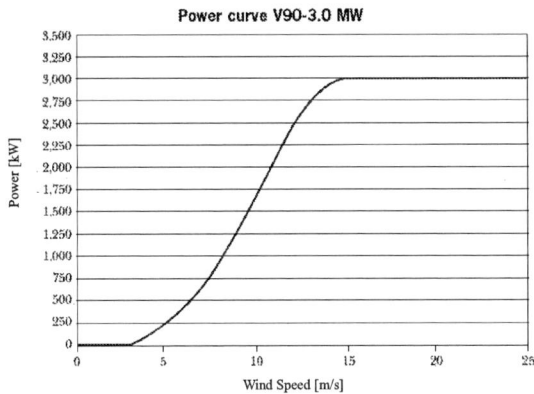

Figure 2. Power curve of a 3MW wind turbine.

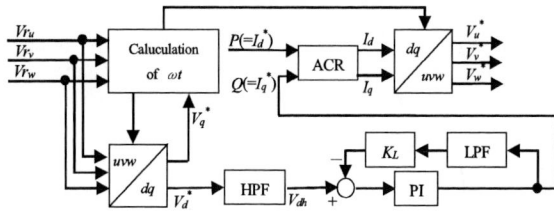

Figure 5. Detailed control system for suppressing voltage deviation.

in most case the suppression can be possible without raising the VA rating of the interphase converter.

B. Control System

Fig. 4 shows the simplified control system for suppressing voltage deviation at PIC. Assume that active current delivered from the wind turbine generator is $I_d + \Delta I_d$, where I_d and ΔI_d represent steady and deviation components, respectively. It is noted that $P = I_d$ and $\Delta P = \Delta I_d$ on the per-unit basis. The ΔI_d acts with the line impedance $Z_0(R, X)$ and causes voltage deviation ΔV_d at PIC. Since the proposed strategy suppresses only high frequency components in ΔV_d, a high-pass filter (HPF) extracts high frequency components ΔV_{dh} from ΔV_d, and it is sent to a voltage regulator (AVR) as a reference voltage.

AVR decides a reference reactive current I_q^* which suppresses ΔV_{dh}, and sends it to a current regulator (ACR). ACR controls the output current of the interface converter and forces the converter to provide a reactive current $I_q = I_q^*$ to the distribution line. The current I_q acts with the line impedance $Z_0(R, X)$ and yields a voltage drop ΔV_{dh1}. Then, the voltage deviation at PIC will be $\Delta V_d - \Delta V_{dh1}$. If $\Delta V_{dh1} = \Delta V_{dh}$, the high frequency voltage deviation is eliminated.

The detailed control system is shown in Fig. 5. The detection of the phase-angle of the voltages at PIC is very important. The amplitude and phase-angle ($\theta = \omega t$) of the infinite bus voltages are base quantities. However, the phase-angle at PIC does not coincide with that of the infinite bus because an impedance drop exists between them if current flows. Thus, the phase-angle at PIC deviates if current deviates. The phase-angle of the interface converter output voltage must follow the phase-angle deviation at PIC to exactly define active and reactive components of the power flow; otherwise control performance of the interface converter concerning active and reactive power will be deteriorated.

The phase-angle detection consists of a PLL (phase locked loop) block, where the voltages at PIC, $V_{ru} - V_{rw}$, are transformed into the d-q rotating coordinate reference frames, V_d and V_q. The d and q axes represent the active and reactive components, respectively. The q-axis voltage V_q is controlled to be zero irrespective of the delivered active and reactive power by adjusting the angular frequency ω through a PI regulator.

In this way the phase-angle $\theta = \omega t$ of the voltages at PIC is obtained, and the phase-angle of the interface converter output voltage can follow the phase-angle deviation at PIC.

In AVR, the d-q transformation block provides the d-axis voltage V_d at PIC. A HPF attenuates the dc and low frequency components in V_d and only the high frequency components are sent to a PI regulator. The PI regulator operates and yields a reactive current reference I_q^* with which the voltage deviation in a specified frequency range can be compensated for. However, since some low frequency components still remain in I_q^*, they are fed back through a low-pass filter (LPF). In this way low frequency components in I_q^* are excluded and only high frequency components are sent to the current regulator (ACR).

In ACR, d-axis steady and deviation currents I_d and ΔI_d (or $I_d^* = I_d + \Delta I_d$) respectively are given. Also a q-axis current reference I_q^* is given from AVR. ACR includes current feedback control loops (not shown in Fig. 5) for I_d and I_q, and it produces the converter output voltages on the d and q axes, V_{d0} and V_{q0} with which the output active and reactive currents coincide with their references $I_d + \Delta I_d$ and I_q^*, respectively. Then, V_{d0} and V_{q0} are transformed into the three-phase system and the actual converter output voltages, $V_u^* - V_w^*$, of the interface converter output terminals are decided. Thus, currents $I_d + \Delta I_d$ and I_q flow into the distribution lines. The voltages at PIC vary depending on the currents, but the q-axis voltage V_q is always kept zero by the PLL block through adjusting the angular frequency ω.

IV. SIMULATION RESULTS

Simulation studies are carried out on a per-unit basis using MATLAB/SIMLINK. Assume that the wind turbine generator is providing active power P (or active current I_d) containing five frequency components, dc, 0.01375Hz, 0.22Hz, 0.66Hz and 2.64Hz with unit amplitude for the dc component and the identical amplitude of 0.01pu for the four deviation components. The load is assumed to be zero ($Z_1 = \infty$). Also assume that the control system is designed to suppress the voltage deviation components only at 0.22Hz and 0.66Hz. The 0.22Hz component corresponds to the deviation caused by a rotation frequency of the wind turbine and is called an n-component. The 0.66Hz component corresponds to the deviation caused by the tower-shadow effect in [1] of the wind turbine having three blades, and is called a $3n$-component. These two frequency components are typical voltage deviation appears in the high frequency range of WPG as illustrated in Fig. 3. The impedance of the distribution line is assumed to be $Z_0 = 0.12 + 0.2j$ [pu].

The Bode plots of the open loop transfer function from ΔV_d to ΔV_{dh1} in Fig. 4 are shown in Fig. 6, where the parameters used are listed in TABLE I. The system characteristic looks like a band-pass-filter (BPF).

1470

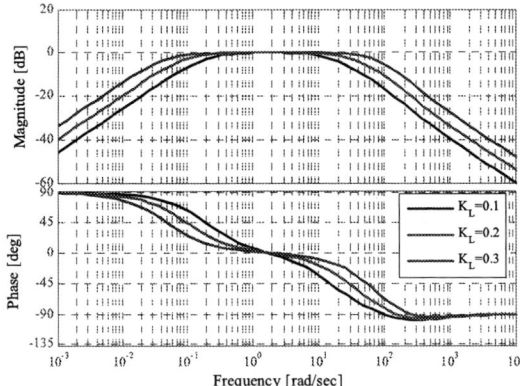

Figure 6. Bode plots of the control system.

TABLE I
PARAMETERS OF CONTROL SYSTEM

Proportional gain of PI regulator	K_P	0.5
Integral gain of PI regulator	K_I	150
Feedback loop gain of AVR	K_L	0.2
Time constant of LPF	T_L	10.0
Time constant of HPF	T_H	10.0

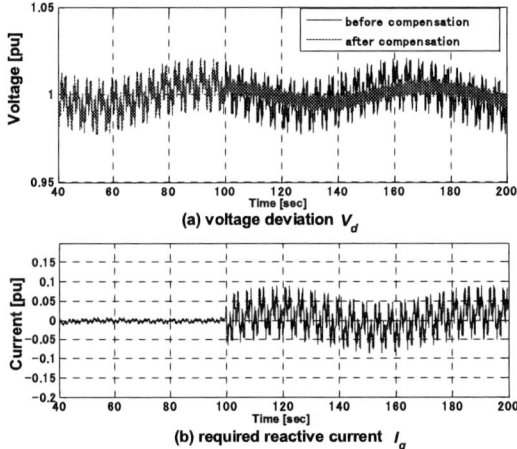

Figure 7. (a) Voltage deviation before and after the compensation, and (b) reactive current required for the compensation.

The gain at the lower and higher frequencies is low, but the gain and phase at the target frequencies is almost unity and zero, respectively. The band width of BPF is adjustable by the gain K_L in the feedback loop of AVR, and the target frequencies of BPF are adjustable by a HPF design in AVR. Thus, any band width and target frequencies are acceptable.

A. Basic Examinations

Fig. 7 shows simulation results. Fig. 7 (a) shows the voltage waveform at PIC. The voltage deviation suppression control was activated at t=100s. One can observe that the designed frequency components at 0.22Hz and 0.66Hz are suppressed quite well and only low frequency 0.01375Hz and high frequency 2.64Hz deviation components are observed.

Fig. 7 (b) shows the reactive current I_q (or reactive power Q) that the interface converter provides to the utility. One can observe that I_q does not include a dc component. This assures that the providing reactive power of the interface converter is minimized.

B. Application of Kaimal Model

In order to treat a nearly actual WPG system, employ a Kaimal turbulence model given in [4] for wind speed. The Kaimal model demonstrates typical wind speed fluctuation and is given in Fig. 8, and a power curve of a 3MW wind-turbine in [2] is already given in Fig. 2. The power output of the wind-turbine generator is calculated using Figs. 2 and 8, and is illustrated in Fig. 9. The output power includes dominant fluctuation components at 0.009Hz, 0.154Hz and 0.5Hz, and these components are used as an input deviation component $\Delta P(=\Delta I_d)$ applied to ACR in Fig. 5.

Fig. 10(a) shows the d-axis voltage V_d at PIC before (in blue ink) and after (in red ink) applying the deviation suppression control. The control was activated at t=100s. One can observe that high frequency components are suppressed quite well.

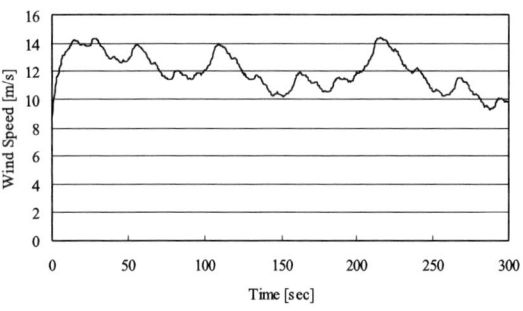

Figure 8. Average wind speed calculated from Kaimal turbulence model.

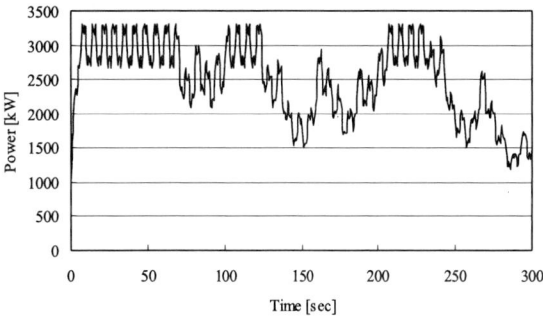

Figure 9. Output active power provided by the wind generator.

1471

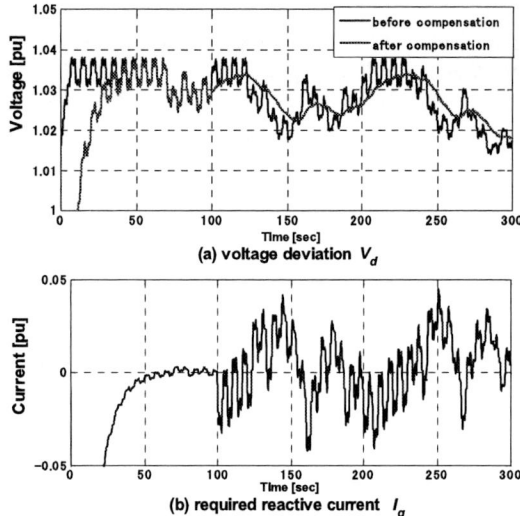

(a) voltage deviation V_d

(b) required reactive current I_q

Figure 10. (a) Voltage deviation before and after the compensation, and (b) reactive current required for the compensation.

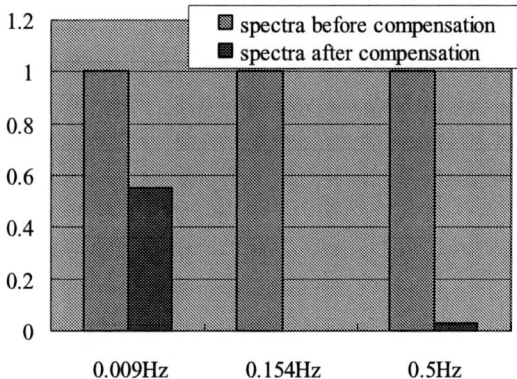

Figure 11. Comparison of voltage deviation before and after the compensation.

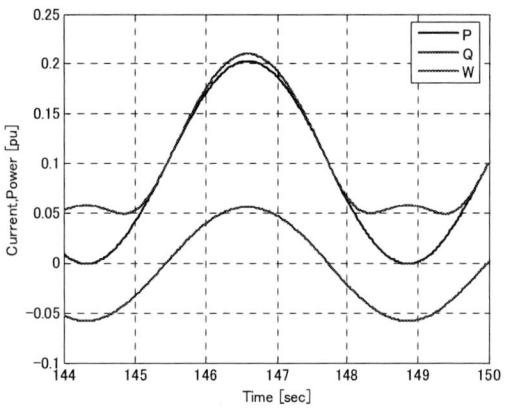

Figure 12. Fluctuation of active power P and required reactive power Q for suppressing the voltage deviation; and VA capacity W of interface converter.

Fig. 10(b) shows the q-axis output current I_q of the interface converter. Since it does not include a dc component, the positive amplitude and negative amplitude are the same. Therefore, the interface converter provides a minimum amount of reactive power. Fig. 11 compares the voltage deviation at PIC before and after applying the suppression control. One can observe that low frequency variation at 0.009Hz remains almost the same, but high frequency deviation components at 0.154Hz and 0.5Hz disappear.

C. VA Capacity of Interface Converter

Fig. 12 illustrates one example of active power P and reactive power Q that the interface converter provides. The active power is assumed

$$P = \overline{P} + \widetilde{P}\sin \omega_n t, \quad f_n = 0.22\text{Hz} .$$

The steady component $\overline{P}$ and fluctuating component $\widetilde{P}$ are assumed to be 0.1pu respectively to maximize $\widetilde{P}$ because the VA rating of the interface converter is 0.2pu. The required reactive power Q to suppress the voltage deviation caused by $\widetilde{P}$ is shown in green ink. Then, the VA capacity W of the interphase converter is given in red ink, and it slightly exceeds the rating of 0.2pu. If $\widetilde{P}$ are slightly less than 0.1, the interface converter will be able to fully suppress the voltage deviation caused by $\widetilde{P}$ by itself.

V. CONCLUSIONS

This paper proposed a new reactive power control strategy suitable for an interface converter of a WPG system. With the proposed strategy, the voltage deviation caused by fluctuation of wind speed can be suppressed by WPG itself. The features of the control strategy are as follows:

(1) Suppression of specified frequency range, which is adjustable, voltage deviation is possible.
(2) Costly reactive power compensator such as SVC and/or STATCOM may be unnecessary.
(3) Required reactive power for interface converter can be minimized. No additional VA capacity is required in the interphase converter.

The validity of the proposed basic idea and control strategy is verified by simulation studies. Experimental verification using a mini model will be the next target to be pursued.

REFERENCES

[1] New Energy and Industrial Technology Development Organization of Japan (NEDO) report 200203 on Stabilization of wind-power generation power systems (2003)

[2] Vestas catalogue on V90-30MW.

[3] Technical committee of IEE Japan report on active and reactive control in power systems, No.743 (1999).

[4] IEC 61400-1 on wind turbine generator systems (1999).

2006 5th International Power Electronics and Motion Control Conference

A Novel Active Islanding Detection Method of Grid-connected Photovoltaic Inverters Based on Current-Disturbing

Zhang Chunjiang   Liu Wei   San Guocheng   Wu Weiyang
College of Electrical Engineering, Yanshan University
Qinhuangdao, Hebei, P. R. China
E-mail: zhangcj@ysu.edu.cn

Abstract—a novel method for islanding detection of a grid-connected photovoltaic system is proposed in this paper. Islanding can be more effectively detected through adding periodical disturbing current to the output current of the inverter. Compared with the active frequency drift method, the method has the many merits that it does no harm to utility, doesn't affect the utility frequency and also it does not generate harmonic current and voltage. In addition to, it has the least NDZ (non detection zone). This method has been proved by simulations and experiments.

Keywords-grid-connected inverter; islanding; active current-disturbing

I. INTRODUCTION

Increasing numbers of photovoltaic arrays are being connected to the power utility through power electronic inverters. This has raised potential problems of network protection. If, due to the action of the inverter or inverters, the local network voltage and frequency remain within regulatory limits when the utility is disconnected, then islanding is said to occur. Islanding phenomenon of grid-connected photovoltaic (PV) inverters refers to their independent operation when the utility is disconnected. The local section energized by self-activated PV inverters becomes an "island" isolated from the remaining power system. When such a phenomenon is raised, it causes danger to uninformed maintenance personnel. Therefore, it is desirable to incorporate detection functions into PV inverters for protection. Islanding detection methods, which have been proposed, are generally classified into passive and active techniques [1-2]. If an inverter has the capability of over voltage protection (OVP), under voltage protection (UVP), over frequency protection (OFP), and under frequency protection (UFP), we say it has the basic islanding detection capability.

Fig.1 shows the schematic diagram of the grid-connected inverter system. Once the voltage level or frequency at the point A exceeds the preset normal range, the situation is regarded as utility malfunction. The inverter should be forced to be shut down to prevent lasting islanding operation.

This paper is supported by the Key Programs of the National Nature Science Foundation of China (No. 50237020) and the Nature Science Foundation of China (50477022)

If the switch S is on, the inverter supplies power:
$$P + jQ,$$
the load obtains power :
$$P_{LOAD} + jQ_{LOAD},$$
the utility supplies power :
$$\Delta P = P_{LOAD} - P \tag{1}$$
$$\Delta Q = Q_{LOAD} - Q \tag{2}$$

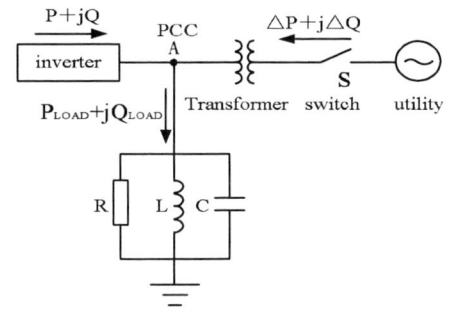

Figure 1. Schematic diagram of the grid-connected inverter system

If $\Delta P \neq 0$, $\Delta Q \neq 0$, when the utility is disconnected, the voltage and frequency at point A changed, so the islanding is detected. However, such detection fails under source-load balanced conditions because the terminal voltage does not change in magnitude or in frequency after the utility is disconnected. In this condition, $\Delta P = 0$, $\Delta Q = 0$, that is , all the power of the load is supplied from the inverter [3-4].

Other methods should be applied to enhance the detection ability. One of them is the voltage harmonics monitoring method. The method is mainly based on the nonlinear characteristics of power transformers in the distribution systems. Without the strong utility voltage source, the current injected from PV inverters into power transformers would cause large voltage harmonics. Continuous monitoring terminal voltages can effectively detect islanding operation when the harmonics level increases. A research has shown that it is hard to select the trip threshold.

Another method called phase jump detection (PJD) method monitors terminal voltages in a different way. PV inverters usually output currents in phase with the utility voltage for unity power factor. When the utility is disconnected, the phase angle between the output current and the terminal voltage of an inverter is determined by the load [5-7]. An instant phase change of the voltage may occur and triggers protection circuits. However, the detection method fails under resistive load conditions.

The detection methods mentioned above all depend on some kinds of monitoring of inverter terminal voltages. They are classified as passive techniques. Their ability of islanding detection is not guaranteed for all load conditions, especially for source-load balanced conditions. As a result, active techniques are invented for improving islanding detection. One simple active technique is to change the frequency of output current of PV inverters periodically. Active frequency drift(AFD) method belongs to this kind. Inverters with this islanding detection method output current to the utility at a frequency slightly lower or higher than the normal frequency. The inverter detects the zero crossing of the network voltage, computes its frequency and introduces a small increase/decrease in the frequency of the output current. This is of no significant consequence during normal operation, when the utility is connected as the inverter current is reset each cycle of the network voltage. However, when the utility is disconnected, the terminal voltage sensed is due to the output current and the load impedance, and thus the frequency, should go up/down to exceed preset shutdown levels [8-9].

This is effective to break source-load balanced conditions. However, the method is impractical because timing synchronization must be made among all of the inverters in a power system or it would generate harmonic wave. To avoid the problem, the paper presents a novel active detection method active current disturbing method. It has the least non detection zone (NDZ), and it does not generate harmonic waves.

II. DESCRIPTIONS OF THE PROPOSED METHOD

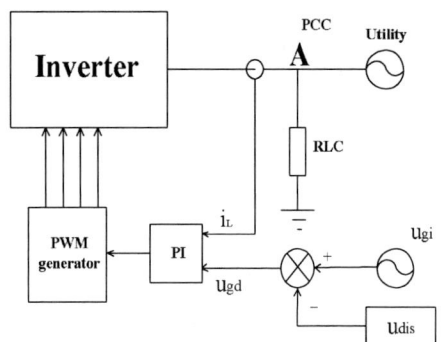

Figure 2. Schematic Diagram of active current-disturbing method system

Fig.2 shows the schematic diagram of the active current-disturbing method for grid-connected inverter system. The proposed system gives the main ideal of the method. The inverter is in current controlled mode. The given signal u_{gd} is the difference between a sinusoidal given signal u_{gi} and a disturbing signal u_{dis}. It minus the inductor current i_L to get the difference which is the input of PI regulator. Thus, the output current of the inverter will be the current with the active disturbing function.

When the inverter is connected to utility, the inverter output voltage is decided by utility. When the utility is disconnected, the output voltage is decided by the output current. Because the magnitude of the current is reduced to half level during the disturbing, the output voltage of the inverter also decreases to exceed the presented threshold, thus, the islanding can be detected.

The detection time of island is according to international standard IEEE Std.2000-929 and UL1741. The main parameters of the standards are listed in TABLE I. We can see that if source-load balanced, the allowed detection time will as long as 2 seconds, because it does less damage to the equipments and load. However, when the source-load unbalanced, the detecting time is no more than 6 cycles.

TABLE I. International standard IEEE Std.2000-929 and UL1741

state	Voltage after ac dump	Frequency after ac dump	The allowed largest detecting time
1	$0.5V_{nom}$	f_{nom}	6 cycles
2	$0.5V_{nom} < V < 0.88V_{nom}$	f_{nom}	2 seconds
3	$0.88V_{nom} \leq V \leq 1.10V_{nom}$	f_{nom}	2 seconds
4	$1.10V_{nom} < V < 1.37V_{nom}$	f_{nom}	2 seconds
5	$1.37V_{nom} \leq V$	f_{nom}	2 cycles
6	V_{nom}	$f < f_{nom} - 0.7Hz$	6 cycles
7	V_{nom}	$f > f_{nom} + 0.5Hz$	6 cycles

III. IMPLEMENTATION OF THE PROPOSED METHOD

A. Simulation

The complete MATLAB circuit is shown in Figure 3. The active current disturbing operates one time every 0.4 second, and maintains two cycles once.

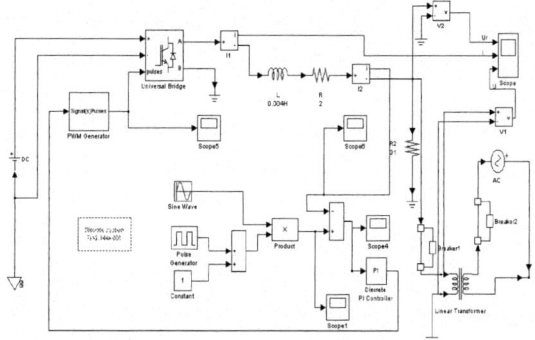

Figure 3. Simulation model of islanding detection with active current-disturbing

1474

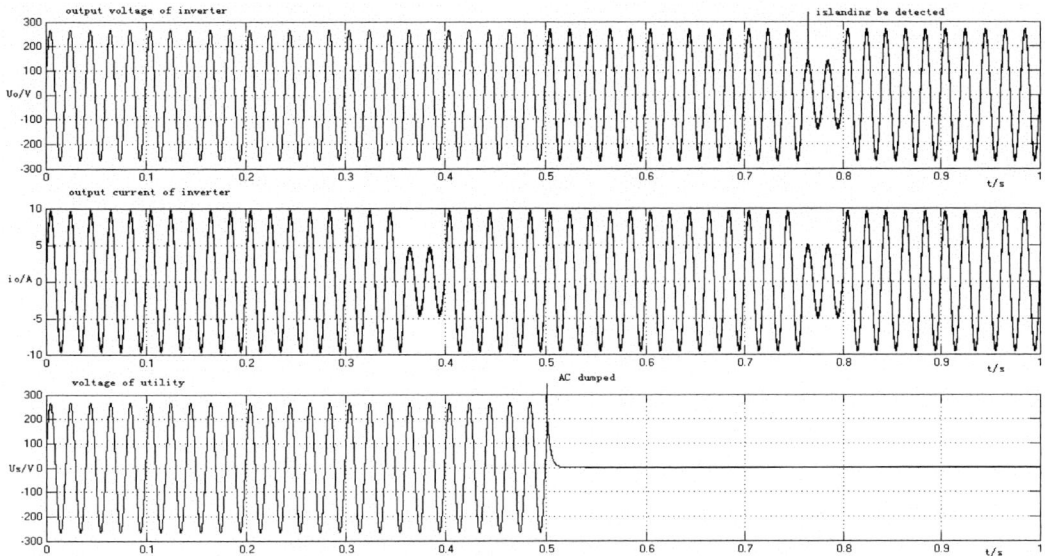

Figure 4. Simulation wave of active current-disturbing method system

Figure 4 shows the Simulation waves of active current-disturbing method under source-load balanced conditions. The first wave is output voltage wave of the inverter, the second is output current wave of the inverter, and the third is utility voltage wave. We can see that the output current of the inverter is disturbed one time every 0.4 second in the second wave, the ac dumped at 0.5 second in the third wave, and at time 0.765 second, the island was detected according to output voltage in the first wave.

B. Experiment

The structure frame diagram of the islanding protection system of the proposed method is shown in Fig.5. In this paper, a 1000W experimental prototype based on the digital unified constant-frequency integration control was constructed. The controller is TI DSP TM320LF2407A. The sample utility voltage signal is sent into DSP. The reference current of the grid-connected inverter is calculated based on the sample signal. The feedback current has two disturbing cycles in one second. During disturbing cycles, the amplitude of current decreases to half level. As soon as AC dumped, the output voltage of the inverter was decided by the output current.

Figure 6 shows the flow chart of active current-disturbing islanding protection program. According to the normal voltage fluctuation of utility, we consider the down

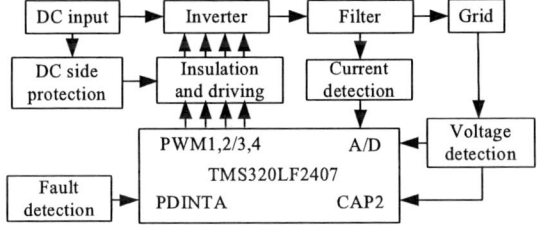

Fig. 5. The structure frame diagram of the system

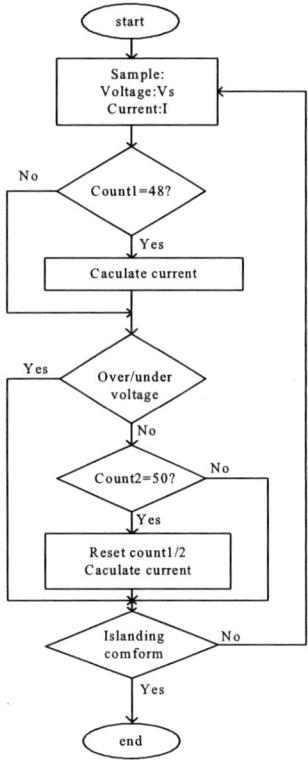

Figure 6. The flow chart of active current-disturb islanding program

threshold voltage as 70% of the normal voltage of utility, and the up threshold as 130% of the normal voltage of utility. To implement the periodic disturbing, we need two counters. When count1 counts to 48, the output current of the inverter is reduced to half, and the current recovers when count2 count to 50. At this time, these two counters are reset.

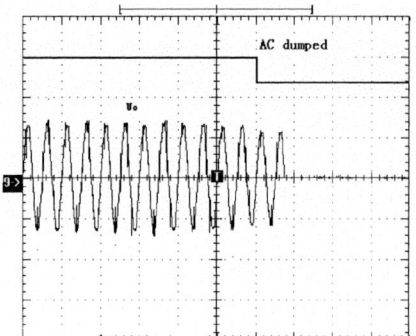

Figure 7. Grid voltage under source-load balanced conditions
(Y-axis: 100V/div, X-axis:40ms/div)

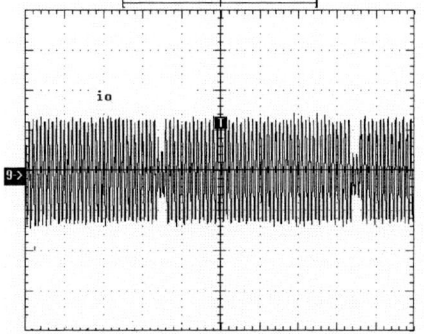

Figure 8 Current wave in the situation of active current disturbing
(Y-axis: 5A/div, X-axis:200ms/div)

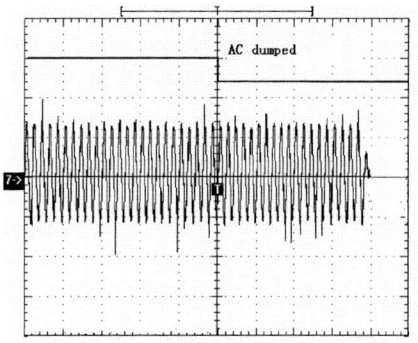

Figure 9. Grid voltage under source-load balanced conditions
islanding protection
(Y-axis:200V/div, X-axis:100ms/div)

Figure 7 shows the output voltage wave of the inverter with the passive islanding protection function under source-load unbalanced conditions. At some moment the AC dumped, the magnitude output voltage of the inverter reduced and after 0.03 second, the islanding protection was implemented.

Figure 8 shows the output current wave of the inverter with the proposed method. We can see that the magnitude of the current is reduced to half for two cycles during the disturbing period.

Figure 9 shows the active islanding protection under source-load balanced condition with the proposed method.

At some moment the AC dumped, the output voltage of the inverter did not distort, and after 0.4 second, the current disturbing occurred, and the output voltage of the inverter also decreases to exceed the presented threshold, thus, the islanding protection was implemented.

IV. CONCLUSION

This paper presents a new islanding detection method. It has the least non detection zone. Compared with the active frequency drift method, it has the greatest merit that it does no harm to utility. Because it does not shift the frequency, at any moment during grid-connection, the utility frequency does not change. In addition to, it does not generate harmonic waves. This detection method has been proved by simulation and experiment.

ACKNOWLEDGMENT

This work was supported by the National Natural Science Foundation of China, NO.50477022.

REFERENCES

[1] A, Kotsopoulos, J.L. Dum, M.A.M. Hendriw, P.J.M. Heskes, "Implementation and testing of anti-islanding algorithms for IEEE 929-2000 compliance of single phase photovoltaic inverters." *Twenty-Ninth* IEEE *Photovoltaic Specialists Conference,* May 2002 pp:1414 – 1419

[2] Woyte, A.; Belmans, R.; Nijs, J.;Energy Conversion," Testing the islanding protection function of photovoltaic inverters," *IEEE Transactions on Volume 18, Issue 1,* March 2003 pp:*157 - 162*

[3] Smith, G.A.; Onions, P.A.; Infield, D.G.;" Predicting islanding operation of grid connected PV inverters," *Electric Power Applications ,* IEE Proceedings-Volume 147, 2000 pp:1 - 6

[4] Woyte, A.; Belmans, R.; Njis, J.;Power "Testing the islanding protection function of photovoltaic inverters," *Engineering Society General Meeting,* 2003, IEEE Volume 4, 13-17 July 2003

[5] Kotsopoulos, A.; Duarte, J.L.; Hendrix, M.A.M.; Heskes, P.J.M.; "Islanding behaviour of grid-connected PV inverters operating under different control schemes," Power Electronics Specialists Conference,. IEEE 33rd Annual Volume 3, 2002 pp:1506 1511

[6] Vachtsevanos, G.J.; Kang,H.; "Simulation studies of islanded behavior of grid-connected photovoltaic systems," *Energy Conversion,* IEEE Transactions on Volume 4, Issue 2, June 1989 pp:177-183

[7] Guo-Kiang Hung; Chih-Chang Chang; Chern-Lin Chen;"Automatic phase-shift method for islanding detection of grid-connected photovoltaic inverters," *Energy Conversion,* IEEE Transactions on Volume 18, Issue 1, March 2003 pp:169 - 173

[8] Jeraputra, C.; Enjeti, P.N, "Development of a robust anti-islanding algorithm for utility interconnection of distributed fuel cell powered generation," *Power Electronics,* IEEE Transactions on Volume 19, Issue5, Sept. 2004 pp: 1163-1170

[9] Huang, S.-J.; Pai, F. S, "Design and operation of grid-connected photovoltaic system with power-factor control and active islanding detection," *Generation, Transmission and Distribution,* IEE Proceedings-Volume 148, Issue 3, May 2001 pp:243 - 250

Grid Connection of Doubly-Fed Induction Generators in Wind Energy Conversion System

Ahmed G. Abo-Khalil [*] , and Dong-Choon Lee [**] , and Se-Hyun Lee [***]

[*],[**] Dept. of Electrical. Eng. Yeungnam Univ., 214-1, Daedong, Gyeongsan, Gyeongbuk, Korea

[***] Electro-Mechanical Research Institute, Hyundai Heavy Industry Co., Ltd, Gyeongki, Korea

E-mail [**] : dclee@yu.ac.kr

Abstract—This paper presents a new synchronization algorithm for grid connection of a doubly fed induction generator (DFIG) in wind generation system. A stator flux-oriented vector control is used for the variable speed DFIG operation. By controlling the generator excitation current the amplitude of the stator EMF is adjusted equal to the amplitude of the grid voltage. To set the generator frequency equal to the grid one, the turbine pitch angle controller accelerates the turbine/generator until it reaches the synchronous speed. A slight difference of stator and grid frequencies may cause the large phase difference between the two voltages. To compensate for this phase difference, a PLL algorithm is used. After the synchronization is achieved, the generator is connected to the grid and is controlled to extract the maximum power. The effectiveness of the proposed synchronization algorithm is verified by simulation results using PSCAD.

Keyword- Wind energy, synchronization, DFIG, PSCAD.

I. INTRODUCTION

Wind energy is one of the most important and promising sources of renewable energy all over the world, mainly because it is considered to be nonpolluting and economically viable. At the same time, there has been a rapid development of related wind turbine technology [1]. A doubly-fed induction generator is based on a wound rotor induction machine. The three-phase rotor windings are supplied with a voltage of controllable amplitude and frequency using an ac/ac converter. Consequently, the speed can be varied while the operating frequency on the stator side remains constant. Depending on the required speed range, the rotor converter rating is usually low compared with the machine rating. Therefore, a DFIG is preferable for variable-speed wind turbine applications [2].

The choice of control strategy incorporated can vary depending on wind turbine generators, but the most popular control scheme for the DFIG of wind turbine generators is a field-oriented control (FOC). This control strategy is well established in the field of variable-speed drives and, when applied to the DFIG control, allows independent control of the electromagnetic torque and stator reactive power [3]. The DFIG using back-to-back PWM converters for the rotor-side control has been well established in wind power system. When used with a wind turbine it offers several advantages over the fixed speed generator systems. These advantages, including speed control and reduced flickers, are primarily achieved by controlling the voltage source converter, with its inherent bi-directional active and reactive power flow [4].

Before connecting the stator of the DFIG to the grid terminals, the stator voltage has to be adjusted to be synchronized with the line voltage. There are only a few papers which handled the DFIG control for the synchronization process. There are some control schemes for DFIG synchronization [5], [6]. In these papers, the transition state for synchronizing duration was not investigated in detail.

This paper describes a smooth and fast synchronization scheme of the DFIG to the grid as well as independent control of active and reactive power of the generator using the stator flux-oriented control at normal operation. During the synchronization process, the blade pitch angle controller adjusts the speed closely to the synchronous speed to make it sure that the stator frequency is the same as that of the grid. The magnitude of stator induced voltage is controlled by adjusting the rotor flux and the phase difference between the stator and grid voltages is compensated by PLL algorithm. The wind turbine control systems are developed using PSCAD software.

II. WIND TURBINE ODEL

A simplified aerodynamic model can be used when the electrical behavior of the wind turbine is the main interest. The relation between the wind speed and aerodynamic torque may be described by the following equation [10]:

$$T_t = \frac{1}{2}\rho\pi R^3 \upsilon^2 \frac{C_p(\lambda,\beta)}{\lambda} \qquad (1)$$

where

T_t : turbine aerodynamic torque [Nm]

ρ : specific density of air [kg/m³];

υ : wind speed [m/s];

R : radius of the turbine blade[m];

C_p : coefficient of power conversion;

β : pitch angle.

The rotational system may therefore be modeled by a equation of motion [11]:

$$J\frac{d\omega_g}{dt} = T_g - T_t - B\omega_g \qquad (2)$$

where J is the system inertia, ω_g is the rotor speed, T_g

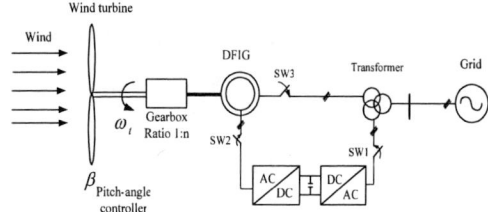

Fig. 1. Basic configuration of DFIG wind turbine

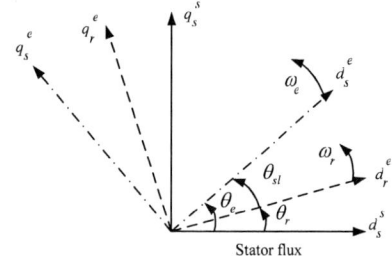

Fig. 2. Vector diagram for stator flux-oriented control

is the generator torque, T_t is the turbine torque and B is the damping coefficient.

III. Control of DFIG

A schematic diagram of the overall system is shown in Fig. 1. Back-to-back PWM converters are connected between the rotor of 2[MW] DFIG and the grid utility. The DFIG is controlled in a rotating d-q reference frame, with the d-axis aligned along the stator-flux vector as shown in Fig. 2. For the stable control of the active and reactive power, it is necessary to independently control them. The stator active and reactive power of the DFIG is controlled by regulating the current and voltage of the rotor windings. Therefore the current and voltage of the rotor windings need to be decomposed into components related to the stator active and reactive power.

A. Stator-Flux Oriented Control of DFIG

For the stator active and reactive power control, a d-q reference frame synchronized with the stator flux is chosen. The stator flux vector is adjusted to be aligned with the d-axis. The flux linkages of the stator and rotor are expressed as [7]:

$$\lambda_s = \lambda_{ds} = L_m i_{ms} = L_s i_{ds} + L_m i_{dr} \qquad (3)$$

$$\lambda_{dr} = \frac{L_m^2}{L_s} i_{ms} + \sigma L_r i_{dr} \qquad (4)$$

$$\lambda_{qr} = \sigma L_r i_{dr} \qquad (5)$$

$$\sigma = 1 - \frac{L_m^2}{L_r L_s} \qquad (6)$$

where

L_m : magnetizing inductance;

L_s : stator self-inductance;

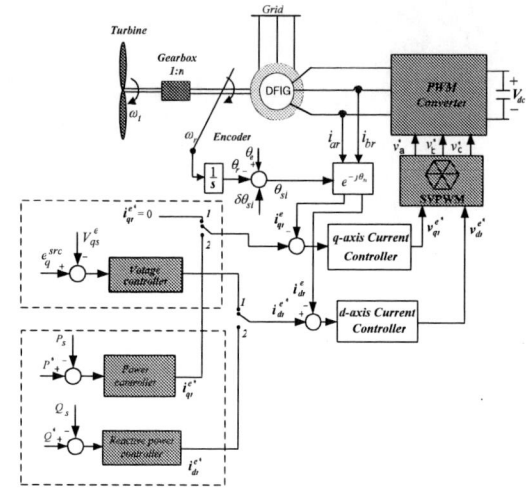

Fig. 3. Control block diagram of DFIG system

L_r : rotor self-inductance;

$\lambda_{ds}, \lambda_{qs}$: stator d-q flux linkage;

$\lambda_{dr}, \lambda_{qr}$: rotor d -q flux linkage;

i_{ms}, i_{ds}, i_{dr} : magnetizing, stator and rotor d-axis currents.

Rotor voltages in d-q reference frame can be expressed as a function of rotor and magnetizing currents

$$v_{dr} = R_r i_{dr} + \sigma L_r \frac{di_{dr}}{dt} - \omega_{sl} \sigma L_r i_{qr} \qquad (7)$$

$$v_{qr} = R_r i_{qr} + \sigma L_r \frac{di_{qr}}{dt} + \omega_{sl}(\sigma L_r i_{dr} + \frac{L_m^2}{L_s} i_{ms}) \qquad (8)$$

where

v_{dr}, v_{qr} : rotor d-q voltages;

R_r : rotor resistance;

ω_{sl} : slip angular frequency.

The stator flux angle is calculated as follows:

$$\lambda_{ds}^s = \int (v_{ds}^s - R_s i_{ds}^s) dt \qquad (9)$$

$$\lambda_{qs}^s = \int (v_{qs}^s - R_s i_{qs}^s) dt \qquad (10)$$

$$\theta_e = \tan^{-1} \frac{\lambda_{qs}^s}{\lambda_{ds}^s} \qquad (11)$$

where a superscript "s" represents quantities in stationary reference frame and

R_s : stator resistance;

θ_e : synchronous frame angle.

B. Power control

Adjustment of the q-axis component of the rotor current controls either the generator developed-torque or the stator-side active power of the DFIG.

$$P_s = \frac{3}{2}(v_{qs} i_{qs} + v_{ds} i_{ds}) = -\frac{3}{2} \cdot \frac{L_m}{L_s} \cdot v_{qs} i_{qr} \qquad (12)$$

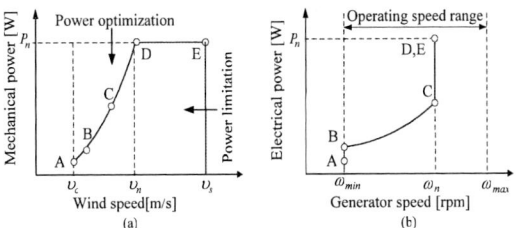

Fig. 4. Wind turbine characteristics

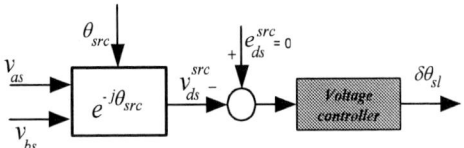

Fig. 5. Phase difference compensation for synchronization

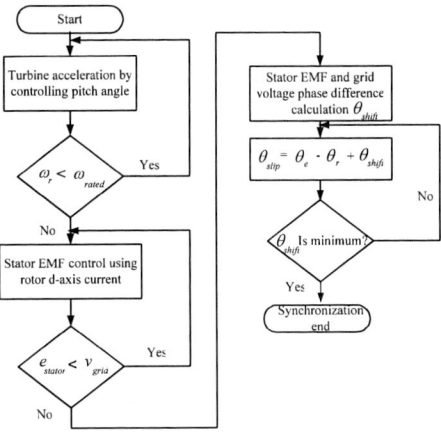

Fig. 6. Sequence of synchronization

On the other hand, regulating the rotor d-axis current component controls directly the stator-side reactive power.

$$Q_s = \frac{3}{2}(v_{qs}i_{ds} - v_{ds}i_{qs}) = \frac{3}{2} \cdot \frac{L_m}{L_s} \cdot v_{qs}(i_{ms} - i_{dr}) \quad (13)$$

It is noticeable that the stator active and reactive power components are proportional to the i_{qr} and i_{dr}, respectively.

Fig. 3 shows the schematic configuration of the DFIG wind turbine system and its simplified control scheme. The stator of the DFIG is connected to the utility grid. The back-to-back PWM converters provide a bidirectional power-flow control thereby enabling the DFIG to operate either in subsynchronous ($\omega_r < \omega_e$) or in supersynchronous modes ($\omega_r > \omega_e$). In both modes the stator active power is generated from the DFIG and delivered to the grid. On the other hand, the rotor active power is either supplied to the machine in the subsynchronous mode or delivered to the grid in the supersynchronous mode. The stator active power is controlled directly assuming that a maximum generator developed power is known from the optimum generator speed value.

The operating curve of the studied wind turbine, which is applied to most modern wind turbines [8], is illustrated in Fig. 4. This curve is characterized by four sections as follows; A~B for minimum rotor speed which is less than that for optimal operation, B~C for an optimal characteristic curve given by $P^* = K_{opt}\upsilon^3$ (where υ is the wind speed) between the cut-in speed and the rated speed, C~D for a constant speed characteristic up to the rated power, and D~ E for a constant power characteristic for higher wind speed than the rated value followed by a blade pitch control.

The optimum power P^* of the DFIG is used as the reference value for the power control loop. In the inner current control loop, the stator-flux vector position is used to establish a reference frame that allows q-axis components of the rotor current to be controlled. As the rotor current reference is expressed in stator-flux

coordinates, these must be transformed into the same reference frame. This is achieved by rotating the rotor current reference vector by an angular position θ_{sl}. Due to the rotor speed variation, θ_{sl} is updated at every sample interval. Once the reference frames for both the reference and measured current vectors are conformed, simple proportional plus integral (PI) regulators can be used to control the d-q components of the rotor current.

C. Synchronization control

The process of connecting the DFIG to the grid consists of two stages, that is, synchronization stage and running stage. At standstill, rotor blades are in a feathering position and the generator is disconnected from the grid. From a complete stop, the first step is to charge the dc link voltage by closing SW1 as shown in Fig. 1. The anemometer measures the wind speed and if the wind speed is higher than the cut-in value, the switch SW2 is closed and the pitch controller changes the blade pitch angle so that the turbine begins to rotate. The controller of the generator rotor side is activated so an excitation current is sent through the rotor.

The excitation current generates the generator flux and then the stator EMF. The turbine accelerates until it reaches near the rated speed. At this point the frequency of the stator EMF is about the same as that of the grid voltage. The amplitude of the stator EMF is about the same as that of the grid. Even slightly different frequencies may cause the phase difference between the two voltages. To compensate for the phase difference between the stator EMF and grid voltage, the phase difference compensation component $\delta\theta_{sl}$ is added to the calculated slip angle as shown in Fig. 5. The compensation component $\delta\theta_{sl}$ is calculated by controlling the stator d-axis voltage component to be zero, equally to the grid d-axis voltage.

The synchronization process is summarized in the flow chart shown in Fig. 6. After the synchronization conditions are achieved, the stator-side contactor is closed, and the generator is connected to the grid.

1479

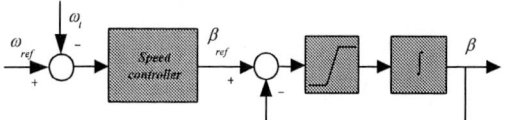

Fig. 7. Block diagram of pitch angle control

The pitch angle controller sets the blade pitch at the optimum point if the blades are not yet at this point. The generator power reference is set to the maximum value which is determined by the wind speed and the pitch angle. Usually the reactive power reference between the grid and the generator is set according to the grid requirements.

D. Pitch angle control

The aerodynamic model of the wind turbine has shown that the aerodynamic efficiency is strongly influenced by variation of the blade pitch angle with respect to the direction of the wind or to the plane of rotation. Small changes in the pitch angle may have a dramatic effect on the power output. In low to moderate wind speeds, the pitch angle should only be at its optimum value to produce the maximum power. In high wind speeds, the pitch angle control provides a very effective means of regulating the aerodynamic power and loads produced by the rotor so that design limits are not exceeded. The relationship between pitch angle and wind speed is shown in Fig. 7.

IV. SIMULATION RESULTS

The proposed model is implemented using PSCAD software and simulated to investigate the DFIG operation during starting and normal running. From a complete stop, the dc link capacitor is connected to the grid utility through the back-to-back PWM converters at t=0[s]. Speed, torque, rotor and stator currents are all zero initially since the rotor and stator are open-circuited.

The pitch angle is controlled from the feathering position to the turbine rated speed position. At t=0.5[s], the rotor terminals are connected to the dc link capacitor through SW2. The stator voltage amplitude increases with the rotor flux current and the phase angle is adjusted using the PLL algorithm. It is noticeable that the synchronization process takes almost two cycles, which means that the synchronization control is fast as shown in Fig. 8. After satisfying the synchronization conditions, the stator contactor is closed and the generator supplies the grid with the power corresponding to the wind speed. From that time, the control algorithm for normal condition replaces the starting algorithm. The rotor d, q-axis currents are adjusted according to the active and reactive power reference. The stator and rotor currents and generator speed are shown in Fig. 9.

During fault condition, the stator terminals are disconnected from the grid while the rotor terminals are kept connected. The pitch angle controller adjusts the pitch angle to the position which reduces the effect of the abnormal condition.

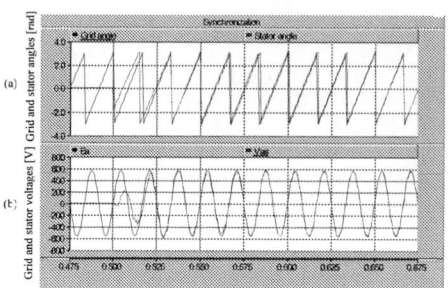

Fig. 8. Stator and grid
(a) Phase angle (b) Voltage

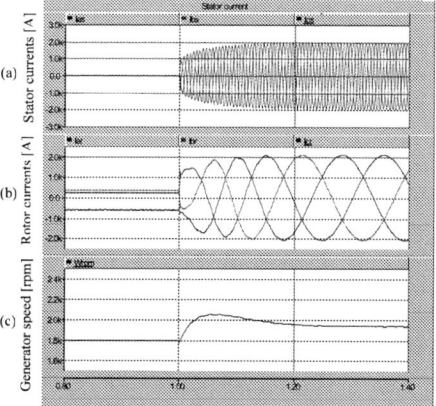

Fig. 9. Generator performance
(a) Stator current (b) Rotor current (c) Speed

After the fault clearing, the synchronization process can be applied again for recovering the generator power. In Fig. 10, the fault occurs after 2 [s], then the aforementioned steps are performed to resynchronize the generator. The stator currents during the disconnection are zeros, while the rotor current is equal to the magnetizing current as shown in Fig. 10(a) and (b). The stator active power is adjusted to the optimum power value during normal operation and is set to zero during faults. Consequently, the generator runs at the synchronous speed during the synchronizing process at starting and re-synchronizing process after fault clearing. After synchronization, the generator runs at a speed corresponding to the optimum power as shown in Fig. 10(c) and (d).

The generator performance for step variations of wind speed is shown in Fig. 11. At low wind speed the controller operates at constant pitch angle and varying rotational speed. The generator active power reference is adjusted to extract the maximum power and the reactive power reference is determined by the grid side requirement. In this study, the reactive power reference is set to zero. The active and reactive power controllers give a fast dynamic response and good steady state performance.

At high wind speeds, the pitch angle controller controls the generator speed at the rated value. It is noticeable that the generator power is limited to the rated value of

2[MW] during the high wind speed operation as shown in Fig. 11(a). It is noticeable that the system is reliable and fast to achieve synchronization as well as excellent control for normal operating condition.

V. CONCLUSIONS

In this paper, a new synchronization scheme of stator flux-oriented DFIG control systems to the utility gird has been proposed. Compared to the existing DFIG synchronization algorithms, the proposed method gives fast starting and can take only 2 cycles to be performed. The stator EMF, frequency and phase angle are adjusted according to the grid values. The pitch angle controller adjusts the turbine speed at the synchronous speed for equal frequencies. The stator EMF is generated then adjusted by controlling the generator d-axis current to be equal to the grid voltage. The voltage phase difference is compensated by comparing the d-axis voltage component of both sides. The proposed synchronization algorithm gives smooth and fast synchronization, which enables the system to be resynchronized quickly after grid fault clearing. The steady state and transient responses of the power, current and pitch angle controllers show excellent performance for the different modes and wind speed. PSCAD simulation has verified that the proposed synchronization and control algorithms are effective and advantageous for 2[MW] DFIG wind power system.

ACKNOWLEDGMENT

This work has been supported by the KEMCO (Korea Energy Management Corporation) under project grant (2004-N-WD12-P-06-3-010-2005).

REFERENCES

[1] A. Tapia, G. Tapia, J. X. Ostolaza, and J. R. Saenz, "Modeling and control of a wind turbine driven doubly-fed induction generator," *IEEE Trans. Energy Conv.*, vol. 18, no. 2, pp. 149–204, June 2003.

[2] Y. Liao, L. Ran, G. A. Putrus, and K. S. Smith, "Evaluation of the effects of rotor harmonics in a doubly-fed induction generator with harmonic induced speed ripple," *IEEE Trans. Energy Conv.*, vol. 18, no. 4, pp. 508–515, Dec. 2003.

[3] A. Mullane and M. O'Malley, "The inertial response of Induction-machine-based wind turbines," *IEEE Trans. Power Syst.*, vol. 20, no. 3, pp. 1496–1503, Aug. 2005.

[4] R. Pena, J. C. Clare, and G. M. Asher, "A doubly fed induction generator using back-to-back PWM converters supplying an isolated load from a variable speed wind turbine," *IEE proc. on Electric Power Appl.*, vol. 143, no. 5, pp. 331-338, Sep. 1996.

[5] S. A. Gomez, and J. L. R. Amenedo, "Grid synchronization of doubly fed induction generators using direct torque control," *IEEE IECON Conf. Proc.*, vol. 4, 2002, pp. 3338-3343.

[6] G. Yuan, J. Chai, and Y. Li "Vector control and synchronization of doubly fed induction wind generator system," *The 4th International PEMC Conf.*, Vol. 2, pp. 886-890, 2004.

[7] L. Zhang, C. Watthansarn and W. Shehered," A matrix converter excited doubly-fed induction machine as a wind power generator," *IEEE IECON Conf. Proc.*, vol. 2, Nov. 2004, pp. 906 - 911.

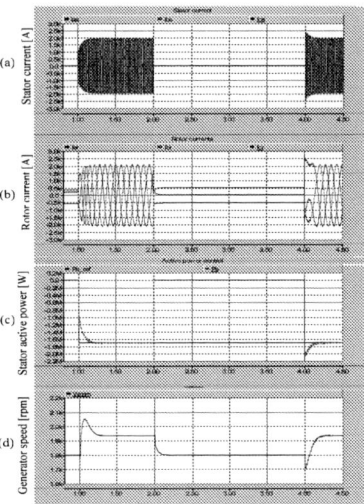

Fig. 10 Generator performance at different control modes

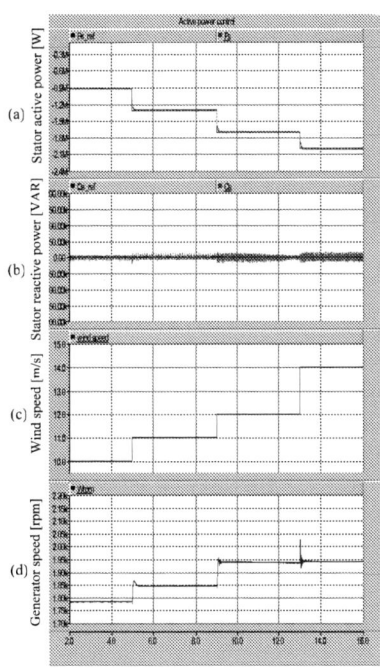

Fig. 11 Generator performance at different wind speeds

[8] T. Sun, Z. Chen, and F. Ballabjerg, "Transient analysis of grid-connected wind turbines with DFIG after an external short- circuit fault," *Nordic Wind Power Conf.*, 1-2 March, 2004, pp. 1-7.

[9] E. Muljadi, and C.P. Butterfield, "Wind farm power system model development," *World Renewable Energy Congress VIII Denver*, August 29 – September 3, 2004, pp. 1-8.

[10] E. Muljadi and C. P. Butterfield, "Pitch-controlled variable speed wind turbine generation," *IEEE Trans. Ind. Appl.*, vol. 37, no. 1, pp. 240-246, Feb. 2001.

[11] M. Chinchilla, S. Arnaltes, and J. L. Rodriguez-Amenedo, "Laboratory set-up for wind turbine emulation," *IEEE ICIT Conf. Proc.*, vol. 1, 2004, pp. 553 - 557.

2006 5th International Power Electronics and Motion Control Conference

Active and Reactive Power Control of DFIG for Wind Energy Conversion under Unbalanced Grid Voltage

Jeong-Ik Jang*, Young-Sin Kim**, Dong-Choon Lee ***

*, *** Dept. of Electrical. Eng. Yeungnam Univ., 214-1, Daedong, Gyeongsan, Gyeongbuk, Korea
** Platform Design 3 Team, LG Philips LCD Co., 642-3, Jinpyung, Kumi, Gyeongbuk, Korea

E-mail***: dclee@yu.ac.kr

Abstract — **In this paper, a novel control method for smoothening the stator active or reactive power ripple components under unbalanced grid voltage for wind power generation using doubly-fed induction generators(DFIG) is proposed. The negative sequence component of the rotor current is used to reduce the active and reactive power ripples, which is extracted by band pass filters. It is shown that reducing the reactive power ripples results in the torque ripple reduction. Simulation results using PSCAD show the proposed control algorithm can reduce the active or reactive power ripple components as well as reducing the torque and speed ripples.**

Keywords – Wind energy conversion, DFIG, unbalanced gird voltage, power control.

I. INTRODUCTION

A doubly fed induction generator is most commonly used in wind power generation. It is a wound rotor induction machine with slip rings attached at the rotor and fed by power converter. With DFIG, generation can be accomplished in variable speed ranging from sub-synchronous speed to super-synchronous speed. The power converters feeding the rotor winding is usually controlled in a current-regulated PWM type, thus the stator current can be adjusted in magnitude and phase angle. The rotor-side converter operates at the slip frequency and the power converter processes only the slip power. Thus if the DFIG is to be varied within ± 30% slip, the rating of the power converter is only about 30% of the rated power of the wind turbine. In this design the net power out of the machine is a summation of the power coming from the stator and the rotor.

The continuity of the power generation in the wind energy system can be affected by the existence of voltage unbalance in utility grid. Disconnecting the generator from the power system during unbalanced grid voltage reduces the utilization of the wind power. Thus, it is desirable to keep the generator connected to the grid as long as possible and to eliminate or reduce the effect of the unbalanced grid voltage [1], [2].

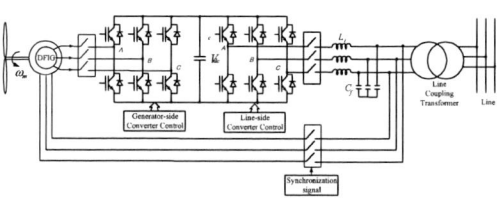

Fig. 1. Configuration of DFIG wind power system.

Due to the direct connection between the stator and the utility grid, the unbalanced grid voltage causes unbalanced stator currents which produce torque pulsations. The torque ripples can be a source of mechanical stress on the drive train and gearbox as well as a source of acoustic noise [3]. A DFIG control algorithm to reduce torque ripple for unbalanced grid voltage was reported in [4], where active and reactive power ripple components were not investigated. On the other hand, Nam et. al. proposed an efficient dual current control algorithm using positive and negative sequence current components in ac/dc PWM converter systems [5]. In DFIG control, however, due to the variation of the rotor frequency the negative sequence component of the rotor current cannot be extracted by symmetric component method.

In this paper, a novel control algorithm to eliminate the torque ripples of the DFIG in case of unbalanced grid voltage is presented. The presented control algorithm aims at eliminating either the double frequency active power or reactive power components to suppress the effect of the unbalanced grid voltage. Simulation results for 2[MW] DFIG system using PSCAD verified that the proposed algorithm is effective.

II. DFIG MODEL AND CONTROL

Configuration of the overall wind generation system is shown in Fig. 1. The stator of DFIG is directly connected to the grid and the rotor is connected through back-to-back PWM converters.

The DFIG is controlled in a rotating d-q reference frame, with the d-axis aligned with the stator flux vector as shown in Fig. 2. The stator active and reactive powers

1-4244-0448-7/06/$25.00 ©2006 IEEE   1482

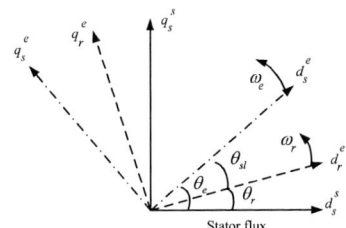

Fig. 2. Vector diagram for stator flux-oriented control.

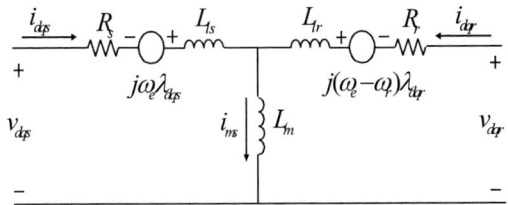

Fig. 3. Equivalent circuit of DFIG

of DFIG are controlled by regulating the current and voltage of the rotor. Therefore the current and voltage of the rotor needs to be decomposed into the components related to stator active and reactive power.

A. DFIG model

Fig. 3 shows the d-q equivalent circuit of DFIG. Under stator flux-oriented control, the fluxes, currents and voltages can be expressed as [6]

$$\lambda_{dqs} = L_s i_{dqs} + L_m i_{dqr} \tag{1}$$

$$\lambda_{dqr} = L_r i_{dqr} + L_m i_{dqs} \tag{2}$$

$$v_{dqs} = R_s i_{dqs} + \frac{d}{dt}\lambda_{dqs} + j\omega_e \lambda_{dqs} \tag{3}$$

$$v_{dqr} = R_r i_{dqr} + \frac{d}{dt}\lambda_{dqr} + j(\omega_e - \omega_r)\lambda_{dqr} \tag{4}$$

where

L_m : Magnetizing inductance;

L_s : Stator self-inductance;

L_r : Rotor self-inductance;

λ_{dqs} : Stator d-q axis flux linkage;

λ_{dqr} : Rotor d-q axis flux linkage;

i_{dqs}, i_{dqr} : Stator and rotor d-q axis currents.

The phase angle of the stator flux vector is calculated as follows;

$$\lambda_{dqs}^s = \int (v_{dqs}^s - R_s i_{dqs}^s)dt \tag{5}$$

$$\theta_e = \tan^{-1}\frac{\lambda_{qs}^s}{\lambda_{ds}^s} \tag{6}$$

where the superscript 's' indicates quantities in the stationary reference frame.

B. Power control

The stator flux vector is adjusted to be aligned with the d-axis. Adjustment of the q-axis component of the rotor current controls either the generator torque or the stator-side active power of the DFIG. On the other hand, regulating the rotor d-axis current component controls directly the stator-side reactive power. Using (1)-(6), the stator active and reactive power can be obtained as

$$P_s = -\frac{3}{2}\frac{L_m}{L_s}v_{qs}i_{qr} \tag{7}$$

$$Q_s = \frac{3}{2}\frac{L_m}{L_s}v_{qs}(i_{ms} - i_{dr}) \tag{8}$$

where i_{ms} is the magnetizing current.

It is noticeable that the stator active power component is proportional to the i_{qr} and the stator reactive power component is proportional to i_{dr}.

III. DFIG CONTROL UNDER UNBALANCED GRID VOLTAGE

A. Generator-side converter control

The stator-side apparent power in unbalanced grid voltage can be expressed in terms of the positive and negative sequence components as [5]

$$S_s = 1.5(V_{dqs}^s I_{dqs}^{s*}) \tag{9}$$

where

$$V_{dqs}^s = e^{j\omega_e t}V_{dqs}^p + e^{j(-\omega_e)t}V_{dqs}^n$$

$$I_{dqs}^s = e^{j\omega_e t}I_{dqs}^p + e^{j(-\omega_e)t}I_{dqs}^n .$$

where ω_e is the stator angular frequency and the superscripts 'p' and 'n' indicate the positive and negative sequence components, respectively. From (9), instantaneous active power $p_s(t)$ and reactive power $q_s(t)$ are obtained as

$$p_s(t) = P_{s0} + P_{sc2}\cos(2\omega_e t) + P_{ss2}\sin(2\omega_e t) \tag{10}$$

$$q_s(t) = Q_{s0} + Q_{sc2}\cos(2\omega_e t) + Q_{ss2}\sin(2\omega_e t) \tag{11}$$

where

$$P_{s0} = 1.5(V_{ds}^p I_{ds}^p + V_{qs}^p I_{qs}^p + V_{ds}^n I_{ds}^n + V_{qs}^n I_{qs}^n)$$

$$P_{sc2} = 1.5(V_{ds}^p I_{ds}^n + V_{qs}^p I_{qs}^n + V_{ds}^n I_{ds}^p + V_{qs}^n I_{qs}^p)$$

$$P_{ss2} = 1.5(V_{qs}^n I_{ds}^p - V_{ds}^n I_{qs}^p - V_{qs}^p I_{ds}^n + V_{ds}^p I_{qs}^n)$$

$$Q_{s0} = 1.5(V_{qs}^p I_{ds}^p - V_{ds}^p I_{qs}^p + V_{qs}^n I_{ds}^n - V_{ds}^n I_{qs}^n)$$

$$Q_{sc2} = 1.5(V_{qs}^p I_{ds}^n - V_{ds}^p I_{qs}^n + V_{qs}^n I_{ds}^p - V_{ds}^n I_{qs}^p)$$

$$Q_{ss2} = 1.5(V_{ds}^p I_{ds}^n + V_{qs}^p I_{qs}^n - V_{ds}^n I_{ds}^p - V_{qs}^n I_{qs}^p)$$

In stator flux-oriented control, the positive and negative sequence components of the stator d-axis voltage are zeros. Hence, the coefficients of the active and reactive power ripple components become

$$P_{sc2} = 1.5(V_{qs}^p I_{qs}^n + V_{qs}^n I_{qs}^p) \tag{12}$$

$$P_{ss2} = 1.5(V_{qs}^n I_{ds}^p - V_{qs}^p I_{ds}^n) \qquad (13)$$

$$Q_{sc2} = 1.5(V_{qs}^p I_{ds}^n + V_{qs}^n I_{ds}^p) \qquad (14)$$

$$Q_{ss2} = 1.5(V_{qs}^p I_{qs}^n - V_{qs}^n I_{qs}^p) \qquad (15)$$

and

$$P_{s2} = \sqrt{(A+B)^2 + (C-D)^2} \qquad (16)$$

$$Q_{s2} = \sqrt{(A-B)^2 + (C+D)^2} \qquad (17)$$

where

$A = V_{qs}^p I_{qs}^n, B = V_{qs}^n I_{qs}^p, C = V_{qs}^n I_{ds}^p, and\ D = V_{qs}^p I_{ds}^n$.

From P_{ss2} and P_{sc2}, it is obvious that the coefficient of the active power ripple P_{s2} is zero when $A=-B$ and $C=D$. Similarly, the coefficient of the reactive power ripple Q_{s2} is zero when $A=B$ and $C=-D$. It means that only either P_{s2} or Q_{s2} can be eliminated, not both.

The power ripple components can be expressed as a function of the rotor and stator currents as below.

$$P_{sc2} = 1.5\{2(R_s + \frac{d}{dt}L_s)I_{qs}^p I_{qs}^n$$
$$+ \frac{d}{dt}L_m(I_{qs}^p I_{qr}^n + I_{qs}^n I_{qr}^p) \qquad (18)$$
$$- \omega_e L_m(I_{qs}^p I_{dr}^n - I_{qs}^n I_{dr}^p)\}$$

$$P_{ss2} = 1.5\{\frac{d}{dt}L_m(I_{qs}^p I_{dr}^n - I_{qs}^n I_{dr}^p)$$
$$- 2\omega_e L_s I_{qs}^p I_{qs}^n \qquad (19)$$
$$- \omega_e L_m(I_{qs}^p I_{qr}^n + I_{qs}^n I_{qr}^p)\}$$

$$Q_{sc2} = -1.5\{\frac{d}{dt}L_m(I_{qs}^p I_{dr}^n + I_{qs}^n I_{dr}^p)$$
$$+ \omega_e L_m(I_{qs}^p I_{qr}^n - I_{qs}^n I_{qr}^p)\} \qquad (20)$$

$$Q_{ss2} = 1.5\{-\frac{d}{dt}L_m(I_{qs}^p I_{qr}^n - I_{qs}^n I_{qr}^p)$$
$$+ \omega_e L_m(I_{qs}^p I_{dr}^n + I_{qs}^n I_{dr}^p)\} \qquad (21)$$

The active power ripples of (18) and (19) cannot be eliminated completely due to the terms of $2(R_s + dL_s/dt)I_{qs}^p I_{qs}^n$ and $2\omega_e L_s I_{qs}^p I_{qs}^n$.

B. Generator torque

For unbalanced grid voltage, (1)-(4) are not sufficient to derive the generator torque equation. Equations (22)-(25) express the negative sequence components for the fluxes and voltages.

$$\lambda_{dqs}^n = L_s I_{dqs}^n + L_m I_{dqr}^n \qquad (22)$$

$$\lambda_{dqr}^n = L_r I_{dqr}^n + L_m I_{dqs}^n \qquad (23)$$

$$V_{dqs}^n = R_s I_{dqs}^n + \frac{d}{dt}\lambda_{dqs}^n + j(-\omega_e)\lambda_{dqs}^n \qquad (24)$$

Fig. 4. Extraction process of rotor negative sequence current.

$$V_{dqr}^n = R_r I_{dqr}^n + \frac{d}{dt}\lambda_{dqr}^n + j(-\omega_e - \omega_r)\lambda_{dqr}^n \qquad (25)$$

The total apparent power of the generator can be expressed as

$$S_T = 1.5(V_{dqs}^s I_{dqs}^{s*} + V_{dqr}^s I_{dqr}^{s*}) \qquad (26)$$

where

$$V_{dqr}^s = e^{j(\omega_e - \omega_r)t} V_{dqr}^p + e^{j(-\omega_e - \omega_r)t} V_{dqr}^n$$

$$I_{dqr}^s = e^{j(\omega_e - \omega_r)t} I_{dqr}^p + e^{j(-\omega_e - \omega_r)t} I_{dqr}^n$$

where ω_r is the rotor speed.

Taking the real part of (26) and dividing it by the mechanical speed, the instantaneous torque is obtained as [7]

$$T_e(t) = T_{e0} + T_{ec2}\cos(2\omega_e t) + T_{es2}\sin(2\omega_e t) \qquad (27)$$

where

$$T_{e0} = 1.5 L_m(I_{qs}^p I_{dr}^p + I_{qs}^n I_{dr}^n)$$
$$T_{ec2} = 1.5 L_m(I_{qs}^p I_{dr}^n + I_{qs}^n I_{dr}^p)$$
$$T_{es2} = 1.5 L_m(I_{qs}^p I_{qr}^n - I_{qs}^n I_{qr}^p) .$$

It is noticeable from (18)-(21) and (27) that T_{ec2} and T_{es2} are related closely to the reactive power ripple component.

Due to the variation of the slip frequency, the negative sequence component of the rotor current cannot be calculated using the symmetric component method. A new method to extract the negative-sequence component is introduced in this paper as shown in Fig. 4. The three-phase rotor currents are transformed into quantities in the synchronous reference frame. Using the band pass filters with a 120[Hz] cut-off frequency, the negative sequence components of the rotor d-q axis currents are extracted.

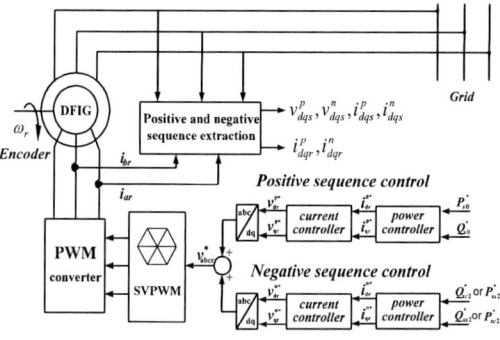

Fig. 5. Power control of DFIG system under unbalance grid voltage.

The negative sequence components are then transformed into the three-phase current. Negative sequence components of the rotor current are used to eliminate the stator active or reactive power ripples.

The complete control block diagram employing two separate current controllers for the positive and negative sequence components is shown in Fig. 5.

IV. SIMULATION RESULTS

To verify the feasibility of the proposed control scheme, computer simulations were performed using PSCAD software. The DFIG used for simulation is rated at 2[MW] and the wind speed is constant at 10[m/s]. The grid voltage is 60[Hz] and 690[V]. For unbalance of the voltage, the magnitude of a-phase voltage is decreased by 20%.

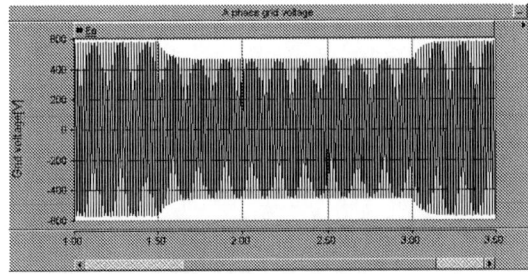

Fig. 6. A-phase gird voltage.

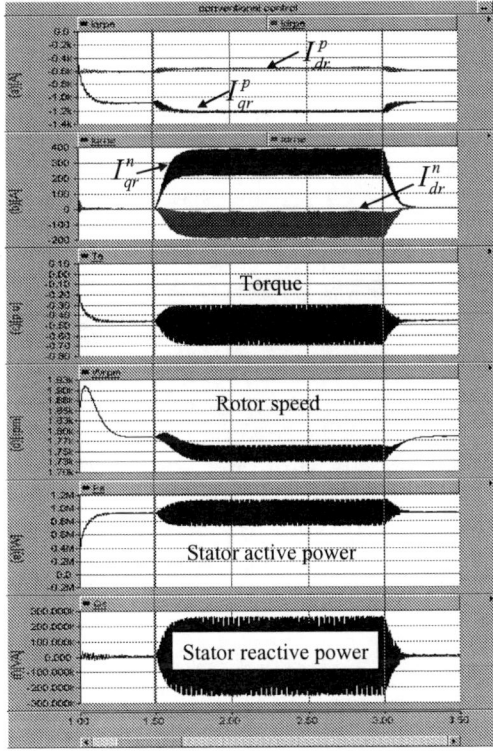

Fig. 7. Output performance under unbalanced grid voltage condition.

Initially, the system runs under balanced grid voltage and then the grid voltage is disturbed after 1.5[s] and then the voltage balance is recovered after 3[s].

Without any compensation for the grid voltage condition shown in Fig. 6, there are significant ripple components in the rotor speed, torque, stator active and reactive power as shown in Fig.7. It is noticeable that the negative sequence components of the rotor d-q axis current contain high ripple components while the positive sequence components have no ripple component as shown in Fig. 7(a) and (b). Controlling the active power ripple component, the rotor d-q axis current ripples are eliminated as shown in Fig. 8 (b) and the active power ripples are reduced as shown in Fig. 8 (e). However, this control method cannot eliminate the torque and rotor speed ripples completely as shown in Fig. 8 (c) and (d). Torque and rotor speed ripples cause a serious mechanical stress on the generator shaft and gearbox as well as acoustic noise. To overcome the aforementioned limitation on the active power ripple control, the reactive power ripple control algorithm can be used. The negative sequence component ripple of the rotor d-q axis current, torque and speed ripples are successfully eliminated as shown in Fig. 9. On the other hand, the active power ripples cannot be eliminated. The proposed reactive power ripple control is as well efficient for variable wind speed. The control algorithm is able to reduce the effect of the high frequency components in the wind speed as shown in

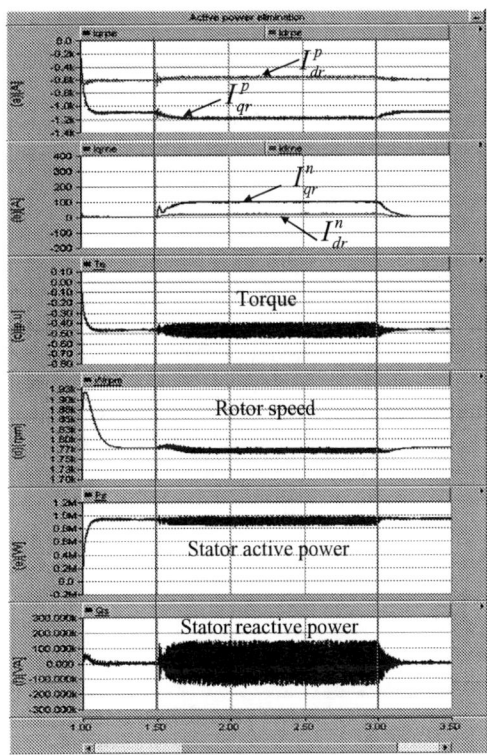

Fig. 8. Power control to eliminate the stator active power ripple.

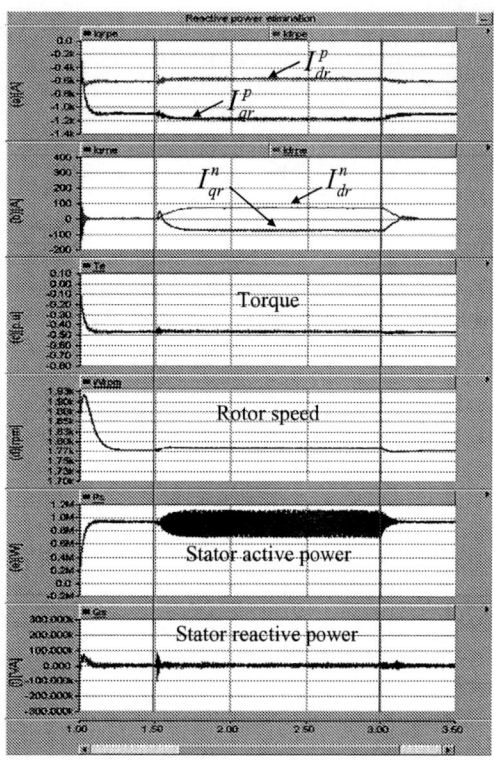

Fig. 9. Power control to eliminate the stator reactive power ripple.

Fig. 10.

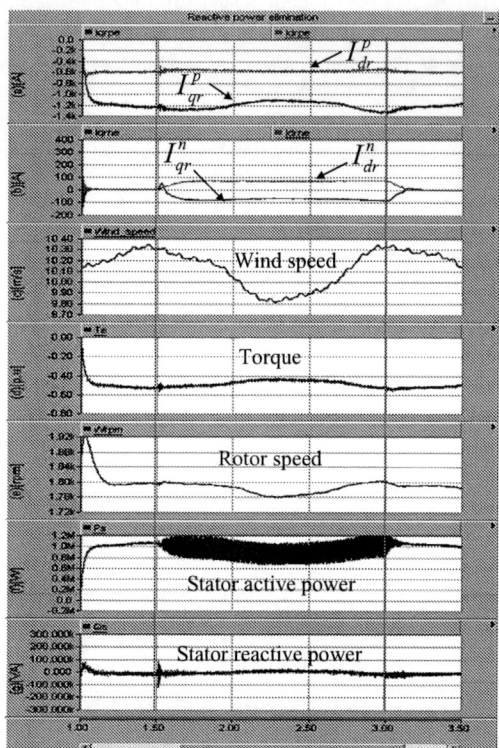

Fig. 10. Power control to eliminate the stator reactive power ripple according to the wind speed variation.

V.   CONCLUSIONS

In this paper, a novel power control scheme of DFIG for wind power generation under unbalanced grid voltage has been proposed and simulated. The proposed control algorithm can reduce the active or reactive power ripples individually. The simulation results show that the reactive power ripple control is more efficient for eliminating the negative sequence component of the rotor current, speed and torque ripples. The proposed algorithm shows a good dynamic performance and ability to reduce the effect of the wind speed variation on the generator toque and speed. It helps the stable operation of the DFIG and power system at unbalanced voltage condition. It can reduce the fatigue of the rotating parts of the system including DFIG and turbine.

REFERENCES

[1]   L. M. Craig, M. Davidson, and N. Jenkins, A. Vaudin, "Integration of wind turbines on weak rural networks,"

Opportunities and Advances in International Power Generation, Conference Publication no. 419, pp 164-167, 1996.

[2]   H. De Battista, P.F. Puleston, R.J. Mantz, and C.F. Christiansen, "Sliding mode control of wind energy systems with DOIG-power efficiency and torsional dynamics optimization," *IEEE Trans. on Power systems*, vol. 15, pp. 728 – 734, May 2000.

[3]   E. Muljadi, T. Batan, D. Yildirim, and C.P. Butterfield, "Understanding the unbalanced-voltage problem in wind turbine generation," *IEEE IAS Conf. Proc.*, pp. 1359-1365, 1999.

[4]   T. Brekken and N. Mohan, "A novel doubly-fed induction wind generator control scheme for reactive power control and torque pulsation compensation under unbalanced grid voltage conditions," *IEEE PESC Conf. Proc.*, vol. 2, pp. 760 - 764, 2003.

[5]   H. S. Song and K. Nam, "Dual current control scheme for PWM converter under unbalanced input voltage conditions," *IEEE Trans. on Ind. Electron.*, vol. 46, no.5, pp. 953-959, Oct., 1999.

[6]   W. Hofman and F. Okafor, "Optimal control doubly fed full controlled induction wind generator with high efficiency," *IEEE IECON Conf. Proc.*, vol. 2, pp.123-1218, 2001.

[7]   C. M. Ong, *Dynamic Simulation of Electric Machinery*, Prentice Hall, Inc., 1998, pp.167–258.

2006 5th International Power Electronics and Motion Control Conference

A BASIC STUDY OF FUZZY-LOGIC-BASED POWER SYSTEM STABILIZATION WITH DOUBLY-FED ASYNCHRONOUS MACHINE

Li Wu, Zhixin Wang
*Shanghai Jiao Tong University, 800 Dongchuan Road, 200240, P.R.China
wuli@sjtu.edu.cn, wangzxin@sjtu.edu.cn

Abstract—**This paper investigates the function of DASM(Doubly-fed ASynchronous Machine) with emphasis placed on its ability to the stabilization of the power system. P(active power) and Q(reactive power) compensation from DASM can be determined by fuzzy logic, and regulated independently through secondary-excitation controlling[1]. Simulation results by EMTP show that such system can restore the power system to a normal operating condition rapidly even following severe transmission-line failures. Comparison studies have also been performed between a conventional controller and proposed fuzzy-logic controller.**

Keywords-**DASM(Doubly–fed ASynchronous Machine)**

I. INTRODUCTION

Nowadays electric power systems are becoming more and more complex, they are often subjected to various stability problems because of the unbalance between the power demand and power supply. Among the different ways being proposed to improve the dynamic performance of the system, a round rotor structure doubly-fed asynchronous machine(DASM) is investigated in the paper to control the active and reactive power of the system.

On the other hand, because of the dynamic and stochastic nature of the operation of the power system, controller parameters that are suitable for one set of operating conditions may not be optimum for another set of operating conditions. Hence instead of the conventional way of calculating the desired power compensation from DASM, we applied fuzzy logic in the control strategy.

The simulation studies on a single machine infinite-bus system show that such proposed fuzzy-logic based power system stabilization way with DASM is effective even when experiencing severe transmission-line failures. Comparison studies have also been performed between a conventional controller and proposed fuzzy logic controller.

II. MODEL SYSTEM CONFIGURATION

The system under consideration is shown in Figure.1, in which a DASM is connected to a single machine infinite-bus system with a double-circuit transmission line[2].

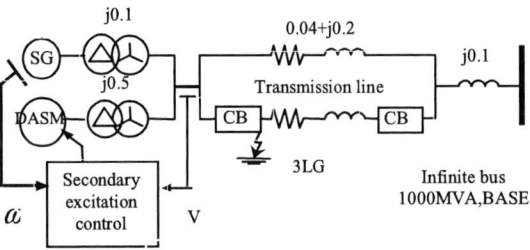

Figure 1. Power system model

The DASM used is based on a wound rotor structure, whose rotor is fed with impressed three phase currents of variable amplitude, frequency and phase so that a constant mmf could be seen from the stator side. Moreover, the rotating reference frame(dq frame) to analyze the DASM is fixed on the space axis of stator voltage.

In the control scheme illustrated in Figure.2, when Δ w (synchronous generator speed deviation) and Δ v(voltage deviation) are detected, the P/Q compensation required from DASM is thus determined through proportional controllers(Kw and Kv). To regulate the error between the desired and detected values of P/Q, a two-step controller is used, the first step of which is APR/AQR, and the second is ACR. Therefore the required field voltage is specified and applied to the rotor side of the DASM.

Controllers Kw, Kv, APR, AQR, ACR and the corresponding optimum gains selected through trial-and-error method are shown in Figure.3.

All the signals used in the control system are per unit values and are transformed to dq frame.

In the inverter side, we simulate the secondary-exciting source with an ideal DC source, and only the fundamental component of inverter output is considered.

With the nominal values mentioned in Table I and II, simulations are carried out. In the simulation, we assume that the mechanical input to DASM is 0, that initially Δ w and Δ v are both 0, and that before fault occurs, active and reactive power from DASM are both 0[pu], while active power and terminal voltage of synchronous machine are 0.95[pu] and 1.0[pu], respectively.

1-4244-0448-7/06/$25.00 ©2006 IEEE

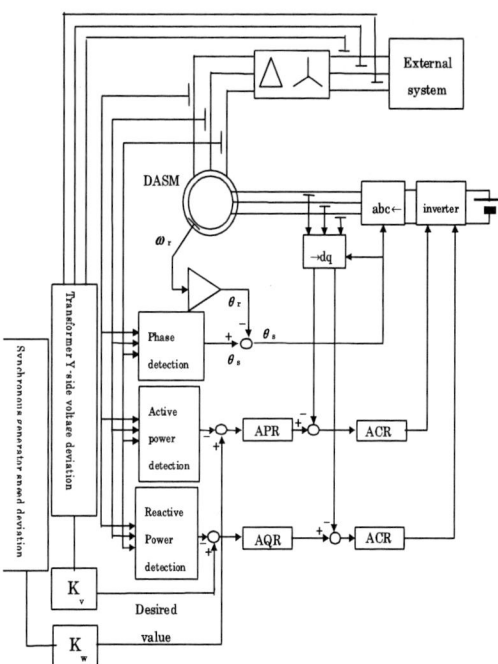

Figure 2. Studied power circuit and regulation block

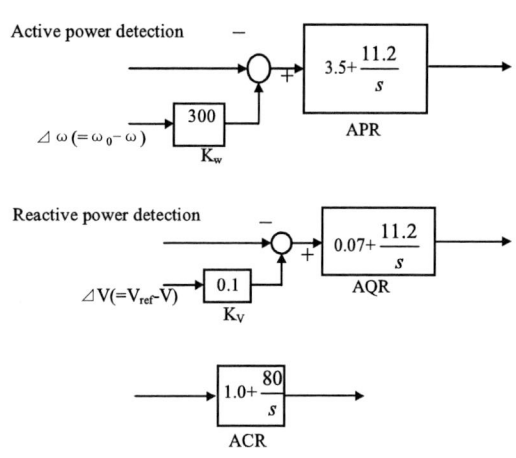

Figure 3. Control signal blocks

TABLE. I SYNCHRONOUS MACHINE NOMINAL VALUES

Rated voltage	20[kV]
Rated output	1000[MVA]
Frequency	50[Hz]
Armature resistance r_a	0.003[pu]
d-axis synchronous reactance X_d	1.790[pu]
q-axis synchronous reactance X_q	1.710[pu]
d-axis transient reactance Xd'	0.169[pu]
q-axis transient reactance Xq'	0.228[pu]
d-axis sub-transient reactance Xd''	0.135[pu]
q-axis sub-transient reactance Xq''	0.200[pu]
d-axis open-circuit transient time constant Tdo'	4.3[sec]
q-axis open-circuit transient time constant Tqo'	0.85[sec]
d-axis open-circuit sub-transient time constant Tdo''	0.032[sec]
q-axis open circuit sub-transient time constant Tqo''	0.05[sec]

TABLE. II DASM NOMINAL

Rated voltage	11 [kV]
Rated output	200 [MVA]
Frequency	50 [Hz]
Magnetizing reactance	2.75 [pu]
Primary leakage reactance	0.142 [pu]
Primary resistance	0.0045 [pu]
Secondary leakage reactance	0.142 [pu]
Secondary resistance	0.0045 [pu]
Inertial moment	$0.18[10^6 kgm^2]$

III. TUNING FOR APPRORIATE PI CONTROLLER GAIN

We first tune for the optimum controller parameters for APR/AQR and ACR without impressing three-line-to-ground fault. Soundness of the PI gains are tested through comparison of the active/reactive responses of 5 times, 1 time, 1/5 time of the selected gain values, which were obtained through trial and error method and shown in Fig.3.

Figures.4(a)~(d) show simulation results when exciting the active power command with unit step function at 1.0 second while maintaining the reactive power command at 0[pu].

Figures.4(e)~(h) show simulation results when exciting the reactive power command with unit step function at 1.0 second while maintaining the active power command at 0[pu].

Figures.4(i)~(l) show simulation results when exciting the active power and reactive power command simultaneously with unit step function at 1.0 second.

From these results, it can be concluded that the selected gain values are effective for the PI controllers, and that any improper setting of the parameters will deteriorate, or sometimes even collapse the system.

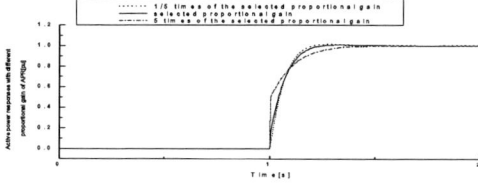

Figure 4(a). Active power response with different proportional gain of APR[pu]

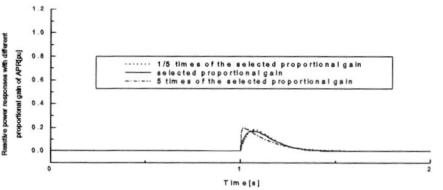

Figure 4(b). Reactive power response with different

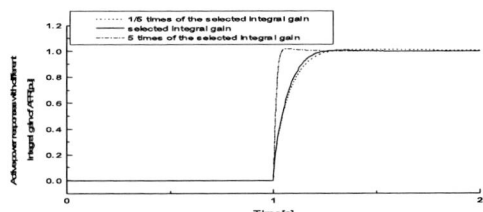

Figure 4(c). Active power response with different integral gain of APR[pu]

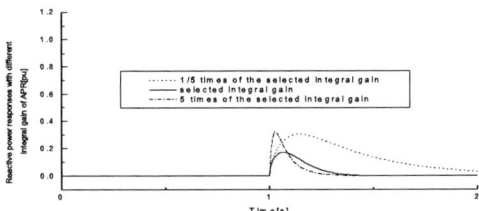

Figure 4(d). Reactive power response with different integral gain of APR[pu]

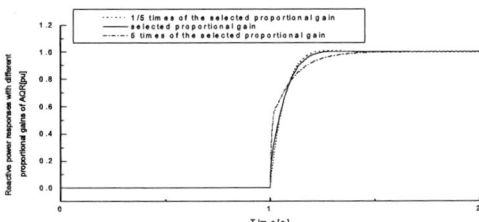

Figure 4(e). Reactive power response with different proportional gain of AQR[pu]

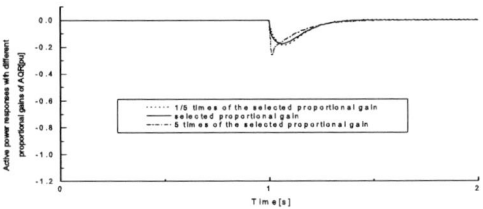

Figure 4(f). Active power response with different proportional gain of AQR[pu]

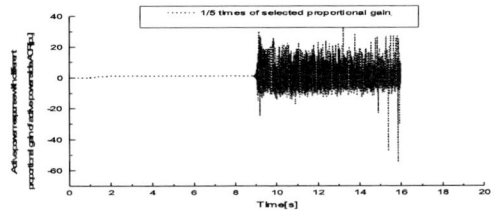

Figure 4(g). Reactive power response with different integral gain of AQR[pu]

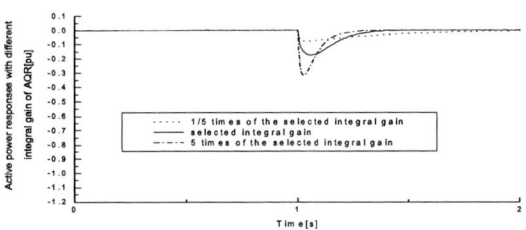

Figure 4(h). Active power response with different integral gain of AQR[pu]

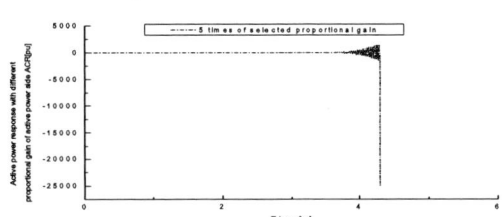

Figure 4(i). Active power response with different proportional gain of ACR[pu]

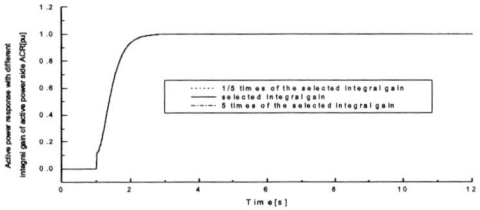

Figure 4(j). Active power response with different integral gain of active power side ACR[pu]

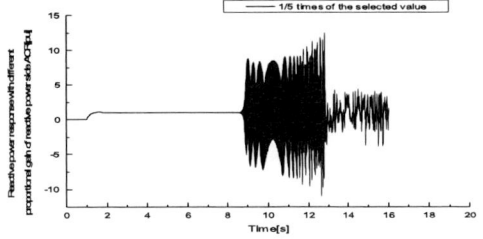

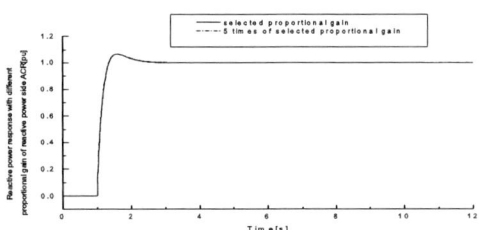

Figure 4(k). Reactive power response with different proportional gain of reactive power side ACR[pu]

1489

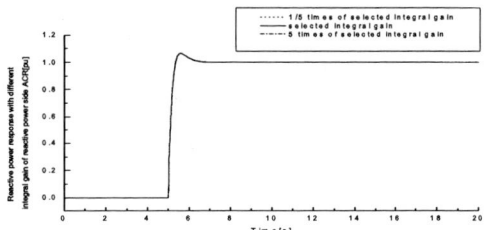

Figure 4(l). Reactive power response with different integral gain of reactive power side ACR[pu]

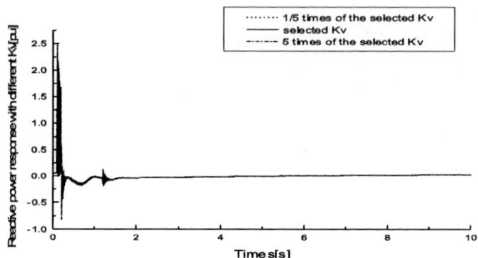

Figure 6. Reactive power response with different Kv[pu]

IV. TUNING FOR THE APPROPRIATE P CONTROLLER GAIN

We applied a proportional controller firstly to calculate the desired active/reactive power compensation from DASM during a fault in the power system. Fixing APR, AQR and ACR on the optimum values described above, soundness of the Kw, Kv gains are tested through comparison of the active/reactive responses of 5 times, 1 time, and 1/5 times of the selected gain values, which were obtained through trial and error method and shown in Figure.3. Conditions of the simulation are that a 3LG fault occurs at 0.1s and cleared at 0.2s. From the results shown in Figures.5 and 6, we note that change of selected Kw gain will deteriorate the response while change of Kv will have little effect on the response. It can also be concluded that selected gain values are effective for the proportional controllers.

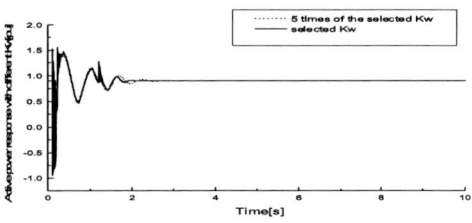

Figure 5(a). Active power response with different Kw[pu]

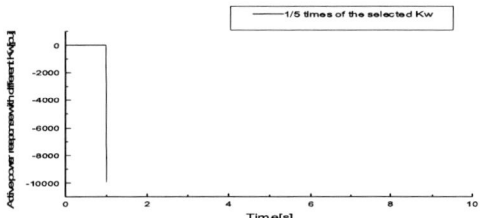

Figure 5(b). Active power response with different Kw[pu]

V. DESIGN OF THE FUZZY CONTROLLER

Instead of the proportional controller shown before, we tried two fuzzy controllers next.

The proposed fuzzy controllers are designed on the previous proportional controllers, and the designing is carried out in the following way:

A. Fuzzification

Synchronous generator speed deviation (Δ w) and the transformer Y-side voltage deviation (Δ v) were selected respectively as inputs to the two controllers. The universe of discourse of each fuzzy variable (Δ w and Δ v) is quantized into three overlapping linguistic NB, ZO, PB, which stand for Negative Big, Zero, Positive Big, respectively. The triangular functions chosen to map the crisp value into fuzzy values are shown in Figures.7 an 8 and can be expressed as

$$A_i(x) = \frac{1}{a_i}(-|x - b_i| + a_i) \vee 0, \qquad a_i > 0 \qquad (1)$$

Where A(x) is the grade of membership values, 'a' is a constant that determines the spread of the i-th membership function, 'b' is the center of the i-th membership function

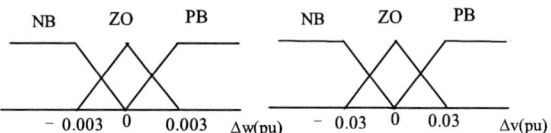

Figure 7. Membership function for Δw(pu)

Figure 8. Membership function for Δv(pu)

B. Rule Creation and Fuzzy Iinference

The fuzzy control rules shown in Table. III and Table. IV have been created with previous knowledge and experience of the controlled system dynamics, and by trial-and-error method. The inference mechanism used is consequence-simplified encoding, whose rule consequence is simplified to a constant crisp value instead of a conventional fuzzy variable. In the table, Kw, Kv represent the optimal proportional gains in the previous proposed proportional controllers.

TABLE.III FUZZY RULE TABLE FOR Δw	
Δw	K′w
NB	1.0Kw
ZO	3.5Kw
PB	1.0Kw

TABLE.IV FUZZY RULE TABLE FOR Δv	
Δv	K′v
NB	0.8Kv
ZO	2.5Kv
PB	0.8Kv

C. Defuzzification

The center-of-area/gravity method is used for the output(Kw, Kv) defuzzification.

In addition, the maximum apparent power command is limited to 1.0 times of the nominal value.

VI. SIMULATION RESULTS

When a three-line-to-ground fault occurs at 0.10s, causing the opening of the CB(circuit breaker) at 0.20s, synchronous generator speed increase while transformer Y-side voltage decreases tremendously, DASM responds to these by consuming/producing adequate amount of active and reactive powers, which, in turn, ask for the corresponding field voltage change and hence the field current tuning. As a result, load angle of the synchronous generator settles down to a normal condition and CB is re-closed at 1.2s. Figures 9(a)~(b) show the various simulation results. Figure9(a) shows the responses of the load angle of the synchronous generator. It is clear that when DASM is not equipped with, after the fault the load angle experiences significant fluctuations. On the other hand, when DASM is equipped with, as both proportional controller and fuzzy controller work well to calculate the desired active/reactive power compensation from DASM, after the fault, load angle of the SG fluctuates for some time, but after a few seconds, it recovers to its initial value. However, the superiority of fuzzy controller in calculating the desired active/reactive power compensation from DASM is apparent, comparing these two cases. Figure 9(b) shows the transformer Δ-side voltage responses of the three cases. It is also clear that the response of applying fuzzy controller is superior to that of applying other two methods. Therefore, it can be concluded from these results that DASM can be properly controlled to decrease the transient of the synchronous generator.

Figure 9(a). Synchronous generator load angle

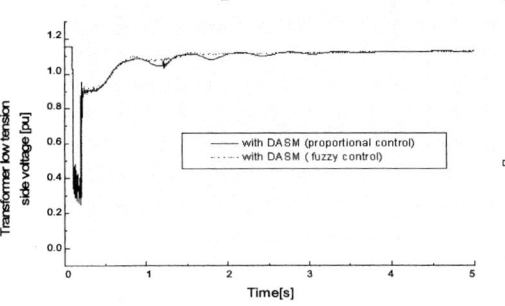

Figure 9(b). Transformer Δ-side voltage

VII. CONCLUSION

The paper proposed a fuzzy-logic-based power system stabilization with DASM. Simulation results show that the proposed fuzzy algorithm, with properly selected parameters, have better operating properties than the simple proportional controller.

ACKNOWLEDGMENT

This Project was granted financial support from China Postdoctoral Science Foundation(No: 2005038435), AND Shanghai Postdoctoral Science Foundation(No: 05R214133).

REFERENCES

[1] Takahashi, Tamura, Tada, Kurita, "Derivation of Model of an Adjustable Speed Hydro Generator and Its Control System", Technical meeting on rotating machinery, RM-00-10,2000.

[2] Nagahama, Wu, Takahashi, Murata, Tamura, "Basic Study of Power System Stabilization by Adjustable Speed Generator", Technical meeting on rotating machinery, RM-00-112,2000.

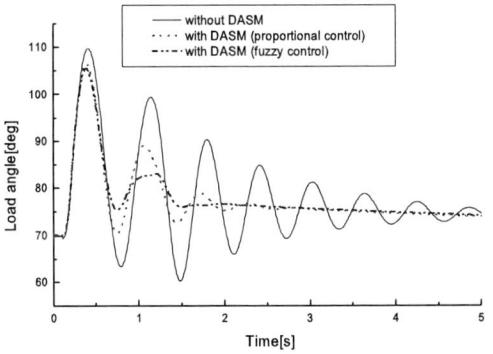

2006 5th International Power Electronics and Motion Control Conference

Quantitative Analysis on Different Modes of Energy Optimal Control for Series Power Quality Controllers

HUANG Xinming, LIU Jinjun, and ZHANG Hui
School of Electrical Engineering, Xi'an Jiaotong University,
28 West Xianning Road, Xi'an 710049, P. R. China
E-mail: xmhuang@stu.xjtu.edu.cn

Abstract—Quantitative analysis on the energy optimal control for series power quality controllers (SPQC) is quite limited so far in previous publications. In this paper, an extensive quantitative analysis is carried out on the energy optimal control of SPQC based on phasor diagram method. The load current phasor is used as the reference phasor and this leads to a simpler and clearer phasor diagram for quantitative relationship. Three different modes for the energy optimal control strategy of SPQC are then proposed according to different relationship between the source voltage magnitude and the load power factor. Subsequently detailed analysis of SPQC under the energy optimal control is provided for different modes in both under voltage/voltage sag and over voltage/voltage swell. In the end, simulation and experimental results are shown to verify the validity of all the analysis and different modes of control.

Keywords- Energy Optimal Control, Series Power Quality Controller, Phasor Diagram, Quantitative Analysis

I. INTRODUCTION

Voltage sag and voltage swell are the most familiar types of voltage disturbances, which can cause bad effects to the quality of products, even makes the equipment can't work properly, consequently results in the financial losses. So the cost-effective solutions, which can make the voltage-sensitive equipment work properly in the time of source voltage disturbance, are required by the industrial field, one such solution is the series power quality controller (SPQC), the system configuration of which is shown in Fig. 1. One of the most important parameters of SPQC is its maximal ride through time in the case of voltage disturbance, which is mainly determined by the relationship between the active power needed in the compensation and the active power that can be supported by the energy storage equipment. In reality, the high rating energy storage equipment is very expensive, so the promising solution to prolong its ride through time is to reduce the active power needed in the compensation itself.

There are several control strategies for SPQC, such as the pre-fault control, the in-phase control and the energy optimal control [1] [2], of which the energy optimal control is the one which aims at reducing active power in the compensation process, and it prolongs the ride through time of SPQC without increasing the rating of the energy

storage equipment, so it is the most promising control strategy for SPQC in the future. But the energy optimal control is much complicated, and there is much confusion and misunderstanding existing even in the mode of this control strategy, not to mention the using in the industrial applications. Most of the current work don't use the source voltage phasor as its reference in the phasor diagram analysis, even don't use the phasor diagram analysis at all, and the results get from the complicated mathematical deduce process are based on the system energy equilibrium, which has several drawbacks such as the complicated deduce process and the abstruse results and having less physical meanings. Besides, most of the previous publications are focused on voltage sag only.

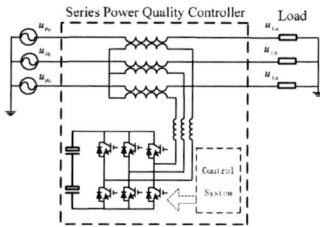

Fig. 1 the system configuration of SPQC

In this paper, the main ideal and its drawback of the energy optimal control is illustrated in part II. The load current phasor is used as the reference phasor for the energy optimal control and three different modes of SPQC are then proposed according to different relationship between the source voltage magnitude and the load power factor, which is covered in part III. The extensive quantitative analysis of SPQC under the energy optimal control is provided for different modes in both voltage sag and voltage swell in part IV. In part V, computer simulation and experimental results are shown to verify the validity of all the analysis and different modes of control.

II. EENERGY OPTIMAL CONTROL

The main aim of the energy optimal control is to compensate the source voltage disturbance with minimal active power. Based on the relationship between the source voltage magnitude and the load power factor in the time of voltage disturbance, there are three modes in this control strategy, which are the zero active power injection

1-4244-0448-7/06/$25.00 ©2006 IEEE

1492

mode, the boundary mode and the minimal active power injection mode. In the previous researcher's publications, this is not a single control strategy but different control strategies [3] [4] [5], like the zero active power injection control and the minimal active power injection control, but according to our analysis, these are not different control strategies, but only one control strategy with different modes.

The main drawbacks of this energy optimal control strategy are having larger compensating voltage magnitude and having phase angle jump problems. But in fact, almost all voltage disturbances are associated with some degree of phase shift problems [6] [7], and we can alleviate its bad effects by taking some actions [8]. The larger compensating voltage magnitude is really a problem in fact, but it is much easier to generate larger compensating voltage magnitude with less active power than to generate smaller compensating voltage magnitude with more active power in fact.

III. PHASOR DIAGRAM ANALYSIS

Different from the previous researchers, the load current phasor is chosen as the reference phasor for the phasor diagram analysis of SPQC in this paper. Based on it, the phasor diagram analysis is done to SPQC using the energy optimal control, from which the boundary working condition for this control strategy in both voltage sag and voltage swell are gotten.

For different types of load, the analysis of the energy optimal control has slightly difference, and the load with inductive power factor is quite common in the industrial applications, so the RL load are chosen in the analysis, and the load voltage phasor is constant comparing with the phasor reference in the phasor diagram. When voltage sag or voltage swell happens, SPQC using the energy optimal control outputs particular compensating voltage to compensate the disturbance.

For different degree of voltage disturbance and the load power factor, SPQC using the energy optimal control works in different modes, and the detailed modes for the energy optimal control are discussed in this section by using the phasor diagram analysis method.

A. Voltage sag situation

When under voltage or voltage sag happens, for different degrees of source voltage disturbance, SPQC using the energy optimal works in the zero active power injection mode, the boundary mode and the minimal active power injection mode respectively.

For the zero active power injection mode, SPQC compensates the voltage disturbance without injecting any active power and the source power factor is not unity, so SPQC only generates reactive power to compensate the under voltage/voltage sag and the compensation voltage phasor is perpendicular to the load current phasor. There are two cases that can fulfill these characteristics and the smaller compensating voltage magnitude case is chosen

for the compensating voltage. In this mode, SPQC doesn't generate any active power but provides some reactive power to the load, which should be provided by the source ordinarily, so the source will provide less reactive power than the load needed, and the source power factor is larger than the load power factor. The phasor diagram for this mode is showed in Fig. 2.

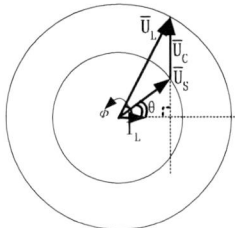

Fig. 2 the phasor diagram of the zero active power injection mode

For the boundary mode, SPQC compensates the voltage disturbance without injecting any active power and the source power factor is unity, so the compensating voltage phasor is also perpendicular to the load current phasor. In this mode, SPQC generates reactive power only, the extent of which is exactly equal to the load needed, and the source only generates active power, which supports all the active power to the load needed. The phasor diagram for this mode is showed in Fig. 3.

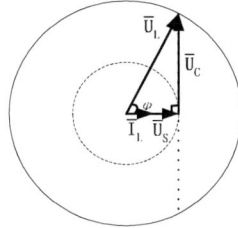

Fig. 3 the phasor diagram of the boundary mode

For the minimal active power injection mode, SPQC generates the minimal active power to compensate the voltage disturbance, and the source voltage phasor is in-phase with the load current phasor, so the source power factor is unity. In this mode, the source provides the maximal active power it can generate, and SPQC supports all the reactive power in addition to the inadequate active power which can not be provided by the source to the load, it can be verified easily that in this case SPQC using the energy optimal control generates the minimal active power, so it is called the minimal active power injection mode, and the phasor diagram for it is showed in Fig. 4.

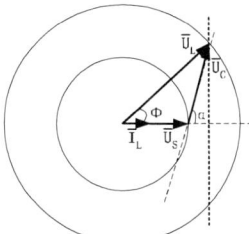

Fig. 4 the phasor diagram for the minimal active power injection mode

B. Voltage swell situation

In the over voltage/voltage swell situation, SPQC using the energy optimal control only works in the zero active power injection mode. Similar to the zero active power injection mode in voltage sag, SPQC compensates the voltage disturbance without injecting any active power, so the compensating voltage phasor is perpendicular to the load current phasor. In this mode, SPQC only absorbs the excessive reactive power generated by the source, and the source has smaller power factor than the load power factor. The phasor diagram for this mode is shown in Fig.5.

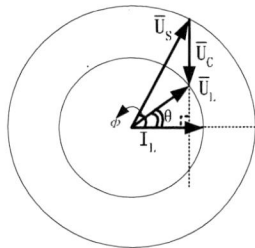

Fig. 5 the phasor diagram for the voltage swell situation

IV. QUANTITATIVE ANALYSIS

If we define the load current as per unit for the current, and the magnitude of the desired load voltage as per unit for the voltage, the power can also be determined quantitatively in per unit. In this section, the detailed quantitative analysis is made to SPQC using the energy optimal control, from which the analytic expressions and curves for the compensation characteristics of related variables are obtained in different kinds of source voltage disturbances.

A. Voltage sag situation

For the zero active power injection mode, the relationship among the source voltage magnitude with the load voltage magnitude and the load power factor is shown in equation (1).

$$U_S > U_L \cdot \cos\phi \tag{1}$$

According to the phasor diagram of this mode shown in Fig. 2, the active power need by this mode is zero, and the compensating voltage magnitude is shown in equation (2).

$$U_C = U_L \cdot \sin\phi - \sqrt{U_S^2 - U_L^2 \cdot \cos^2\phi} \tag{2}$$

Using the per unit definition, the compensating voltage magnitude is shown in equation (3).

$$U_C^* = \sin\phi - \sqrt{\left(U_S^*\right)^2 - \cos^2\phi} \tag{3}$$

From this equation, the quantitative relationship among the compensating voltage magnitude with the source voltage magnitude and the load power factor is shown in Fig. 6. In this figure, x axis is the load power factor, y axis is the source voltage magnitude in per unit, and z axis is the compensating voltage magnitude in per unit.

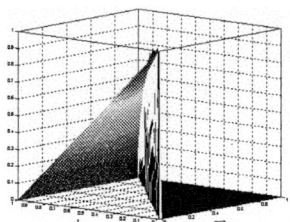

Fig. 6 the compensating voltage magnitude for the zero active power injection mode

For the minimal active power injection mode, the relationship among the source voltage magnitude with the load voltage magnitude and load power factor is shown in equation (4).

$$U_S < U_L \cdot \cos\phi \tag{4}$$

According to the phasor diagram of this mode shown in Fig. 4, the compensating voltage magnitude is shown in equation (5).

$$U_C = \sqrt{U_L^2 + U_S^2 - 2 \cdot U_L \cdot U_S \cdot \cos\phi} \tag{5}$$

Using the per unit definition, the compensating voltage magnitude is shown in equation (6).

$$U_C^* = \sqrt{1 + \left(U_S^*\right)^2 - 2 \cdot U_S^* \cdot \cos\phi} \tag{6}$$

The active power generated by SPQC is shown in equation (7).

$$P_C = U_L \cdot I_L \cdot \cos\phi - U_S \cdot I_L \tag{7}$$

Using the per unit definition, the active power to be generated is shown in equation (8).

$$P_C^* = \cos\phi - U_S^* \tag{8}$$

From equation (6) and (8), the quantitative relationship among the compensating voltage magnitude, the injection active power with the source voltage magnitude and the load power factor are shown in Fig. 7 (a) and (b) respectively. In each figure, x axis is the load power factor, y axis is the source voltage magnitude in per unit, and z axis is the compensating voltage magnitude and the injecting active power in per unit respectively.

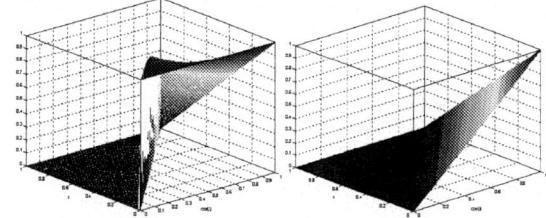

(a) Compensating voltage magnitude   (b) Injecting active power

Fig. 7 quantitative analysis for the minimal active power injection mode

For the boundary mode, the relationship among the source voltage magnitude with the load voltage magnitude and the load power factor is shown in equation (9).

1494

$$U_S = U_L \cdot \cos\phi \qquad (9)$$

According to the phasor diagram of this mode shown in Fig. 3, the active power needed to be generated in the compensation is zero, and the compensating voltage magnitude is shown in equation (10).

$$U_C = U_L \cdot \sin\phi \qquad (10)$$

The compensation voltage magnitude is just the boundary of the zero active power injection mode and the minimal active power injection mode, so it is a space curve with the common boundary of the other two modes.

B. Voltage swell situation

In this situation, SPQC using the energy optimal control only has the zero active power injection mode. According to the phasor diagram of this mode shown in Fig. 5, the active power need is zero, and the compensating voltage magnitude is shown in equation (11).

$$U_C = \sqrt{U_S^2 - U_L^2 \cdot \cos^2\phi} - U_L \cdot \sin\phi \quad (11)$$

Using the per unit definition, the compensating voltage magnitude is shown in equation (12).

$$U_C^* = \sqrt{\left(U_S^*\right)^2 - \cos^2\phi} - \sin\phi \qquad (12)$$

From this equation, the relationship among the source voltage magnitude with the load voltage magnitude and the load power factor is shown in Fig. 8. In this figure, x axis is the load power factor, y axis is the source voltage magnitude in per unit, and z axis is the compensating voltage magnitude in per unit.

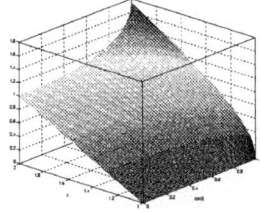

Fig. 8 the compensating voltage magnitude for the zero active power injection mode

V. SIMULATION AND EXPERIMENTAL VERIFICATIONS

From the phasor diagram and the quantitative analysis, this energy optimal control strategy is feasible in compensating voltage sag and voltage swell with zero active power or minimal active power injection. The feasibility and validity of this control strategy is verified by using the simulation and experimental result respectively. In the simulation and the experiment prototype, we choose the RL load with 0.707 lagging power factor, the voltage sag degree is 20%, 30% and 40% respectively, and the voltage swell degree is 40%. Both the simulation results and the experimental results in each figure, four signals are detected from phase C of SPQC system with the source

voltage, the compensating voltage, the load voltage and the load current in descending order.

A. Simulation results

In the simulation, the desired load voltage is 220V in RMS. Voltage sag or voltage swell happens at 0.14S, with 6 cycles of voltage disturbance and it ends at 0.26S. The simulation results of the energy optimal control are shown in Fig. 9, in which (a) is the zero active power injection mode in voltage sag, (b) is the boundary mode in voltage sag, (c) is the minimal active power injection mode in voltage sag, and (d) is the zero active power injection mode in voltage swell.

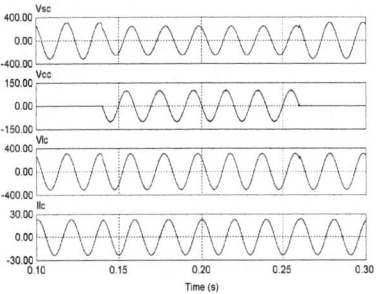

(a) 20% voltage sag

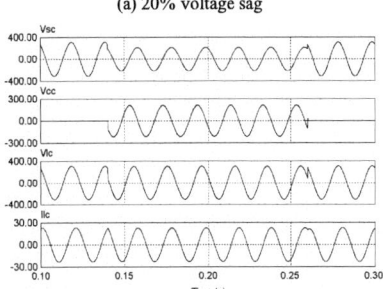

(b) 30% voltage sag

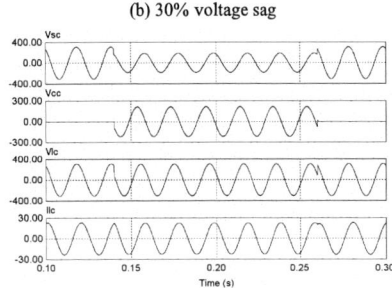

(c) 40% voltage sag

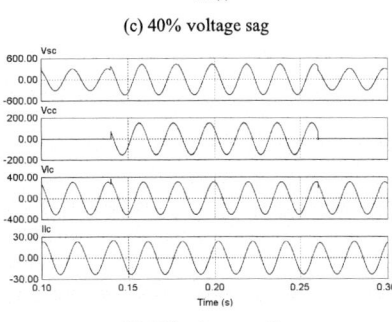

(d) 40%voltage swell

Fig. 9 The simulation results of the energy optimal control

B. Experimental results

With the same parameters as shown in the previous part, except for the desired load voltage, which is 50V in RMS. The experimental results for different cases are shown in Fig. 10, Fig. 11, Fig. 12 and Fig. 13 respectively.

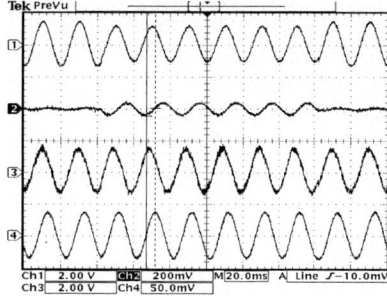

Fig. 10 experimental results with 20% voltage sag

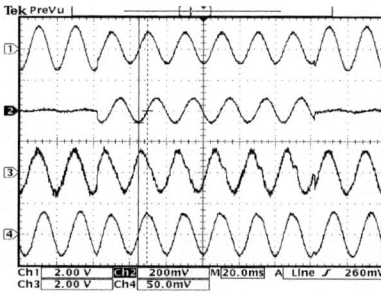

Fig. 11 experimental results with 30% voltage sag

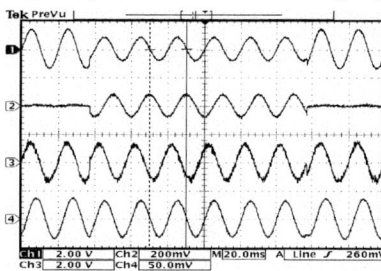

Fig. 12 experimental results with 40% voltage sag

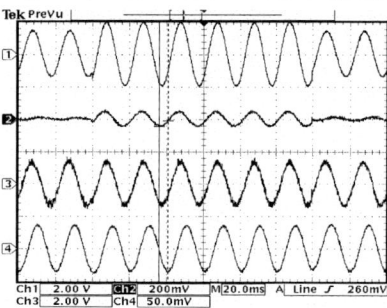

Fig. 13 experimental results with 20% voltage swell

Fig. 10 shows the zero active power injection mode in voltage sag, Fig. 11 shows the boundary mode in voltage sag, Fig. 12 shows the minimal active power injection mode in voltage sag, and Fig. 13 shows the zero active mode in voltage swell. All of these experimental results are consistent with the simulation results, and both of hem

verify the validity of the energy optimal control, not only the choosing of the load current as the phasor reference, the phasor diagram analysis, but also the different modes in the control strategy.

VI. CONCLUSION

Quantitative analysis on energy optimal control for SPQC is quite limited so far in previous publications. In this paper, an extensive quantitative analysis is carried out on the energy optimal control of SPQC based on phasor diagram method. The load current is used as the reference phasor and this leads to a simpler and clearer phasor diagram for quantitative relationship. The zero active power injection mode, the boundary mode and the minimal active power injection mode for the energy optimal control strategy of SPQC are then deduced according to different relationship between the source voltage magnitude and the load power factor. Subsequently detailed analysis of SPQC under energy optimal control is provided for different modes in both voltage sag and voltage swell, from which, the compensating voltage magnitude and the active power needed are gotten quantitatively with relation to the source voltage magnitude and the load power factor. In the end, simulation and experimental results verified the validity of all the analysis and different modes of control.

REFERENCES

[1] D.M. Vilathgamuwa, A.A.D.R. Perera and S.S. Choi. Voltage sag compensation with energy optimized dynamic voltage restorer. IEEE Transactions on Power Delivery, Volume 18, Issue 3, July 2003 Page(s): 928 – 936.

[2] M. Vilathgamuwa, A.A.D. Ranjith Perera, S.S. Choi and K.J. Tseng. Control of energy optimized dynamic voltage restorer. The 25th Annual Conference of the IEEE Industrial Electronics Society, 1999. IECON '99 Proceedings. Vol. 2. 1999 Page(s): 873 – 878.

[3] Hyosung Kim. Minimal energy control for a dynamic voltage restorer. Proceedings of the Power Conversion Conference, 2002. PCC Osaka 2002. vol.2. Page(s): 428 - 433.

[4] Haque, M.H. Compensation of distribution system voltage sag by DVR and D-STATCOM. IEEE Porto Power Tech Proceedings, Sept. 2001 Page(s):5 pp. vol.1

[5] S.H. Hosseini, M.R. Banael. A new minimal energy control of the DC link energy in four-wire dynamic voltage restorer. The 30th Annual Conference of the IEEE Industrial Electronics Society, 2004. IECON 2004. Volume: 3. Page(s): 3048 – 3053.

[6] S.W. Middlekauff and E.R. Collins Jr. System and customer impact: considerations for series custom power devices. IEEE Transactions on Power Delivery. Vol. 13, Issue 1. Jan. 1998. Page(s): 278 – 282

[7] M. H. J. Bollen, Understanding Power Quality Problems: Voltage Sags and Interruptions. Piscataway, NJ: IEEE Press, 2000.

[8] S.S. Choi, B.H. Li, D.M. Vilathgamuwa. Dynamic voltage restoration with minimum energy injection. IEEE Transactions on Power Systems. Feb. 2000. Vol.15, Issue 1. Page(s): 51 – 57.

2006 5th International Power Electronics and Motion Control Conference

Resonance inverter power system for improving plasma sterilization effect

Y.M Kim*, J.Y Kim, M. C Jo, S.H Lee, S.P Mun, H.W Lee, S.K Kwon, K.Y Suh

Department of Electrical Engineering, Kyung-nam University, Masan, Korea
* Department of Computer & Electrical Engineering, Masan College, Masan, Korea
e-mail : skiyoung@kyungnam.ac.kr

Abstract— This paper presents a novel prototype of a current source resonant inverter using insulated gate bipolar transistors for driving a streamer reactor, streamer generation technology has been recognized as one of the best methods for water treatment, disinfection, industrial wastes utilization, and so on. However, some technological difficulties related to efficient streamer production have been significant problems restricting streamer usage in the industrial plants. Introduced in this paper is a pulse density modulated high frequency inverter for a plasma generate, which is developed with the aim to improve power conversion and control characteristics of the streamer reactor by using advances in power electronic technology. The developed system implements the feed forward control-based pulse density modulation control scheme with pulse width modulation feedback control strategy to compensate temperature and other environmental influences on streamer discharge

Keywords- sterilization, phase-shift full-bridge Inverter

I. INTRODUCTION

In recent years, the plasma has been widely utilized for chemical processing of water treatment and exhausted smoke treatment, depolarization, color removal and disinfection in industrial systems and public pipeline facilities. It is also particularly recognized that in semiconductor manufacturing industry fields, the broad applications of streamer reactor is hindered primarily because of low efficiency for the plasm generation. To meet this requirement, much effort for improving streamer generation and the switching mode power supplies using power transistors to drive the streamer generation has been directed to raise the streamer generation efficiency. However, there are only a few studies on miniaturization in physical size and weight, high efficiency, high performance, control stabilization of the power supply system from a power electronics point of view which drives nonlinear silent discharge based capacitive load with an active DC voltage source corresponding to the discharge sustaining voltage. Compared to conventional power conversion circuit, this study uses the power converter for plasma sterilizer as proposed in this study in order to stabilize the supply of power source by reducing switching losses and harmonics through resonant soft switching method, while enhancing bactericidal activity and energy efficiency by inducing balanced plasma, and attempts to increase power factor to higher level by making input current work in discontinuous ways to transforming input current to sine wave. Furthermore, this study also intends to remove any harmonics by changing the waveforms of output line voltage into sine waves by means of output filter.

In this paper, a new unique power regulation scheme of a single phase current source type parallel inductor compensated parallel load resonance high frequency inverter using a single two terminal switched capacitor type edge resonant DC link snubber is developed for driving the streamer discharge type plasma rector. The steady state power regulation characteristics of this active resonant DC link snubber assisted current source type high frequency soft switching parallel load resonant inverter which works under the principle of PDM control strategy or PDM and PWM hybrid control strategy are illustrated and evaluated as the next generation high - performance streamer generator.

In addition, this study analyzes hard and soft switching losses upon zero voltage and current transition via auxiliary partial resonance, as well as increasing output voltage and performance of eliminating harmonics.

All these considerations will be validated through experimental investigations.

II. HIGH-FREQUENCY INVERTER POWER SUPPLY SYSTEM FOR PLASMA GENERATOR

The streamer discharge principle in the discharge space is more widely used for the ozone gas generation, CO_2 laser generation, excimer lamp based streamer generation, plasma display panels and electric dust collector in the fields of a variety of industrial and consumer applications. In particular, the main operating principle of the streamer discharge type plasma generation driven by the current source type high frequency inverter using Mos-gate

1-4244-0448-7/06/$25.00 ©2006 IEEE

controlled power semiconductor switching devices is schematically illustrated in Fig.1. The internal structure of a newly developed high concentration and high efficient streamer generation is depicted in Fig. 1(a). This has a cylindrical structure with a stainless steel ground electrode in its outside and a high voltage stainless steel electrode in its inside frame. By supplying gas toward the inlet of the streamer generation tube across two electrodes with a stable streamer discharge area. The streamer generation produced here is composed of the discharge gap between two electrodes and the dielectric material substrate of glass spacer as a dielectric barrier inserted into two high voltage AC electrodes. The equivalent electrical circuit model considering discharge and non-discharge operation modes of the streamer discharge type generation driven and controlled by the current source type parallel load resonant high-frequency inverter using IGBTs is illustrated in Fig. 1(b). In a non-discharge period of the streamer generation, the electric circuit model of the streamer generation tube represented by the capasitor with the capacitance C_a relating to the discharge gap in series with the capacitor with the capacitance C_g relating to glass dielectric substrate as a dielectric barrier make the capacitive load with the resultant capacitance of C_a and C_g.

On the other hand ,in the discharge period, the average value of the voltage across the discharge gap is approximately kept constant as the discharge property of the discharge gap, which is termed as the discharge sustaining voltage V_z represented by the AC voltage source ,the positive voltage $+V_z$ and the negative voltage $-V_z$. In the electric circuit model of the streamer generation, the nonlinear capacitive circuit is depicted with DC voltage source V_z via the full bridge diode rectifier connected in parallel with the discharge gap capacitor with the C_a is connected to a series capacitance C_g of glass dielectric barrier. Fig. 1(c) shows the operating voltage and current waveforms of the streamer generation driven by the AC current source provided by the current source type inverter. Fig. 2 shows a schematic configuration of power conversion circuit for a next generation ozonizer driven by a current source type high-frequency parallel load resonant inverter using two terminal switched capacitor type active quasi resonant DC link snubber, which can operate under a principle of zero current soft switching side, high frequency step up transformer, parallel compensation resonance reactor L_p streamer generation represented by a nonlinear capacitive load, The high voltage AC has to be applied for the streamer generation tube Especially, the peak voltage of streamer generation tube is necessary for sustaining a stable silent discharge.

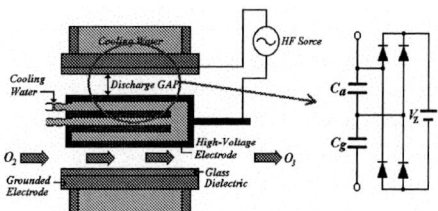

a) Schematic structure of streamer generation

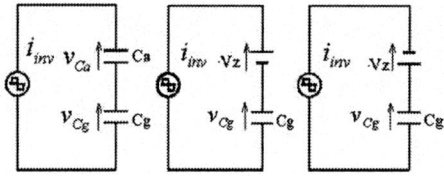

(b) Equivalent circuit in case of non-discharging

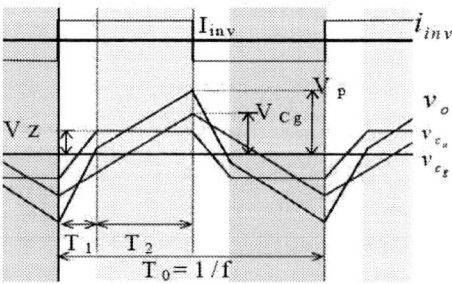

(c) Operation waveforms in discharge and non-discharge of streamer generation

Figure 1. Magnetization as a function of applied field. Note how the caption is centered in the column

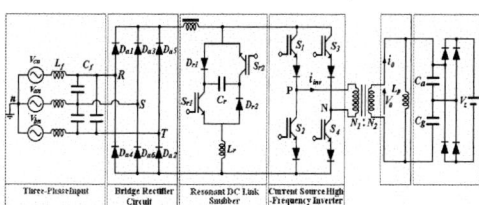

a) Experimental circuit of power converter for plasma sterilizer.

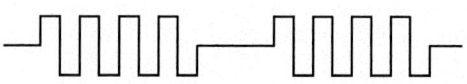

(b) Equivalent circuit in case of non-discharging

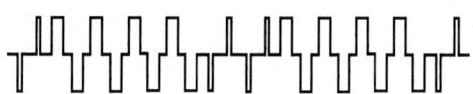

(c) The principle of PDM and PWM control

Figure 2. Current source high frequency inverter

A PDM or PDM and PWM hybrid-based power regulation methods proposed newly is effectively introduced for the 60[Hz] streamer generation load which is implemented into the current source type parallel load resonant inverter using IGBTs. The advanced new conceptual of PDM and PWM hybrid based power regulation strategy for the current source high frequency load resonant soft switching inverter with a single active resonant DC link snubber is illustrated in Fig.2(c). The effective power delivered into the streamer generation can be continuously regulated by means of high frequency AC current based PDM and the PWM hybrid control scheme. On the other hand, the PWM strategy is effectively implemented during the discharge condition fluctuation. This PDM control procedure is based upon a time ratio control or duty cycle control of the high frequency AC current pulse number modulation, which is produced by the current source type load resonant inverter employing the magnetizing inductor and the additional inductor as one of the parasitic circuit components of the high-frequency high voltage transformer.

The auxiliary active DC resonant link commutation switched capacitor circuit connected in current-source DC busline is composed of the resonant capacitor C_r, each IGBT (S_{r1} or S_{r2}) in series with each fast recovery diode (D_{r1} or D_{r2}) and resonant inductor L_r. Current commutation of a single power switch in the bridge leg side of the full bridge inverter circuit to another active power switch is to be performed on the basis of zero current soft switching transition PDM, or PDM and PWM scheme. In this case, the DC bus-line current of this inverter has to be commutated from the full bridge resonant inverter power circuit to the auxiliary active resonant snubber circuit. This allows all the active power switch of the full bridge inverter to turn on and off under zero current switching conditions in spite of PWM regulation in addition to some load parameter disturbances.

Fig. 3 illustrates the soft commutation operating principle based on the equivalent circuits and the steady state voltage and current operating waveforms of the resonant inductor current i_{Lr} and resonant capacitor voltage V_{cr} in the current source type switched capacitor type quasi resonant DC link snubber circuit. The steady state circuit operation of this current type parallel load resonant soft switching inverter incorporating the current source type switched capacitor resonant DC link snubber is illustrated as follows.

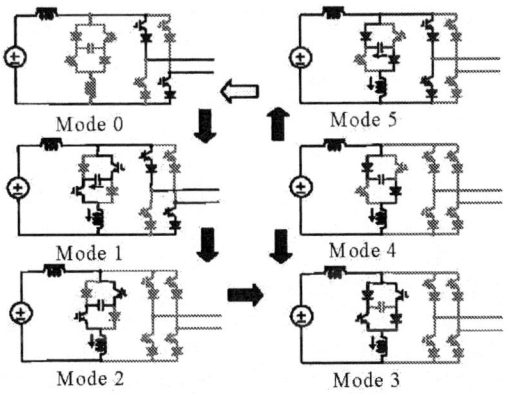

a) Current source high frequency inverter

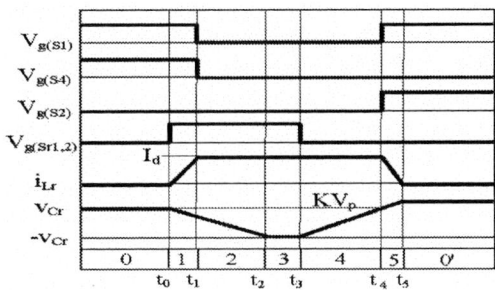

(b) Timing pulse segucuces of gate voltage and operating waveforms of resonant

Figure 3. Current source high frequency inverter

III. REVIEW OF EXPERIMENT RESULTS

In this study, commercial AC power at 380[V], 60[Hz] is boosted up to 12[kV], 60[Hz] by means of voltage type inverter and neon transformer under the influence of DC voltage without any ripple as induced by diode rectifier and partial resonant converter. Then the voltage required for discharge becomes generated by serial resonant circuit Fig. 4 shows the apparatus of power converter for plasma sterilizer as embodied in practice.

Figure 4. The apparatus of power converter for plasma sterilizer

The voltage and current waveforms of this parallel load resonant inverter operating under a new Pulse Density Modulation (PDM) and Pulse Width Modulation (PWM) hybrid regulation scheme are shown in Fig. 5(a). The stable inverter operation in repetitive steady state condition includes a positive polarity based charging and discharging modes in streamer generation tube as well as a negative polarity based charging and discharging modes in the streamer generation. The voltage across the discharge gap during a discharge period maintains a certain high voltage Vz when this gap voltage exceeds a specified voltage enough to start a streamer discharge. The streamer discharge type generation driven by a current source type high frequency parallel load resonant soft switching inverter using IGBTs has non-linear characteristics

Accordingly, the high voltage AC across the streamer generation has to be kept a specified peak value (7[kV]-

12[kV]). In this case, its output power is delivered from the current source type high-frequency parallel load-resonant soft switching inverter. If the high-frequency AC voltage across two high voltage AC electrodes is to be much lower, the discharge phenomena does not occur at all and comes to a partial discharge phenomena. In case of a rated voltage across two high voltage AC electrodes, the electrical insulation is destructed on an edge of an streamer generation and a harmful discharge is possible to generate around two high voltage AC electrodes.

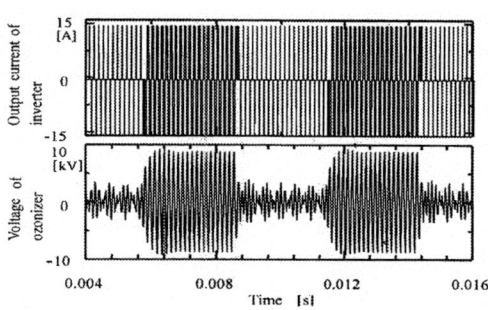

(a) Voltage and current waveforms

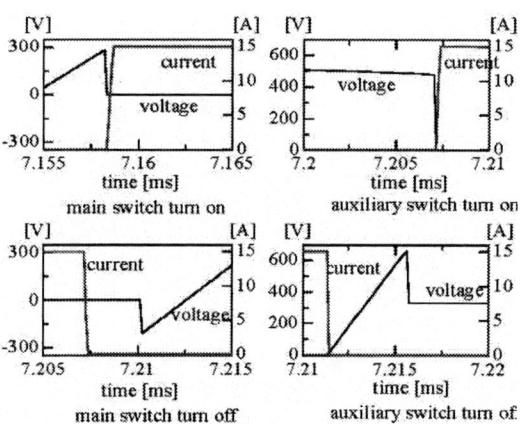

(b) Voltage and current switching wafeworms

Figure 5. Characteristic of current-type ZCS-PDM&PWM inverter

Fig. 5(b) shows the magnified operating voltage and current switching waveforms in the case of the turn on and off soft-commutation of the main active power switch of the bridge arm of the full bridge inverter. It is proven that the switching mode transition is basically performed under a condition of no switching power losses. In addition, this figure illustrates the turning on and turning off voltage and current waveforms of the auxiliary active power switches in two terminal capacitor type active resonant DC link sunbber. It is understood that all the auxiliary active power switches in two terminal switched capacitor type resonant DC link snubber circuit do not essentially generate the switching power losses of IGBTs because of zero voltage soft switching as well as electromagnetic interference noise.

In order to control the output power of the inverter, it is necessary to regulate the pulse width in accordance with the input side voltage source E of the inverter for changing and discharging sustaining voltage V_0. The relationship between pulse width and the input side voltage source E of the inverter is shown in Fig. 6. In addition, the relationship between pulse width and the discharging sustaining voltage V_0 is estimated graphically in Fig. 7.

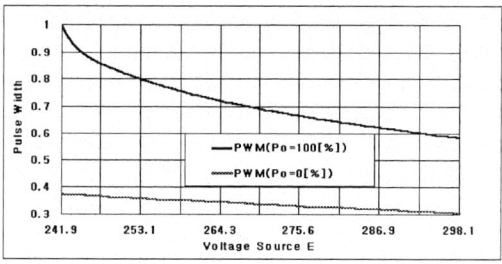

Figure 6. The relationship between pulse width and voltage source variable E

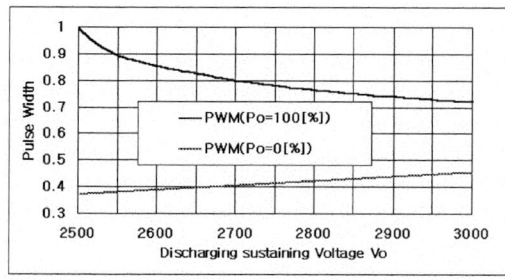

Figure 7. The relationship between pulse width and discharging sustaining Voltage V_0

Figure 8. Plasma reactor

Fig. 8 indicates the plasma reactor used in the experiments of this study. The reactor is flat-bed reactor that can generate the most typical unequal electric field. This reactor allows us to clearly identify the growth of plasma depending on the boost of input voltage. Therefore, the flat-bed reactor is used for this experiment, because it is considered most fit for shape analysis to determine sterilization effects depending on input voltage.

Fig. 9 shows the picture of moulds cultured according to input voltage and time when sterilizing real moulds. The shapes of specimen used for this experiment belong to fine powder type(at the size of several micron[u]) like wheat flour. Fig. 10 shows the variation in the number of moulds sterilized depending on input voltage level and time upon sterilization of real moulds. Conclusively, as the input voltage was boosted up and processing time went by, sterilization rate reached 98[%] at 10[kV], 10[min] and even saturation curves were shown under circumstance beyond 10[kV], 10[min]. Notably, it was found that sterilization efficiency could reach the best at 10[kV], 10[min] and 11[kV], 5[min]. The reason why sterilization rate was a little more decreased at 11[kV] than 10[kV] is possibly attributed to error resulting from failure to irradiate plasma energy within reactor at constantly uniform density

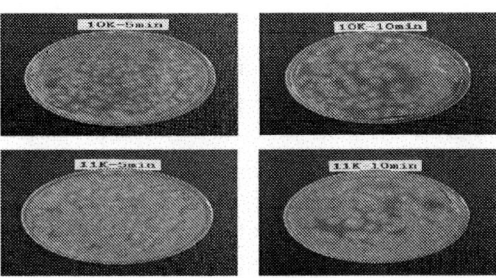

Figure 9. The picture of mould cultivated after sterilization depending on input voltage and time

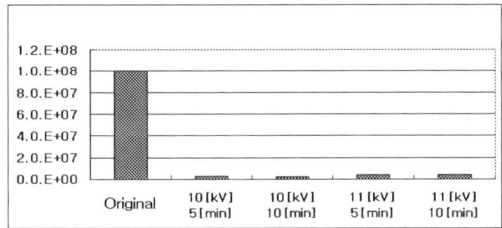

Figure 10. The relationship between mould sterilized, input voltage and time

IV. CONCLUSION

In this study, we devised and designed a power converter for plasma sterilizer to determine sterilization rate by applying powder including moulds to said experimental apparatus As a result of experiment; we came to the following conclusions:

1) The resonant converter system as proposed in this study may facilitate electric protection and control of entire system, higher efficiency, higher power factor and lower loss of electronic noise switching.

2) In the future generation the effective output power of this current type soft switching inverter could be linearly regulated over a wide PDM duty factor range from 10[%] to 100[%] of the rated power.

3) Linear control of inverter output power across broader range helps keep and supply recovery current pulse to

enhance the responsiveness of ozone generator cells to increasing voltage.

4) In conventional power supplies, arc pulse flows into the input end to induce imbalanced input voltage, while the power supply proposed herein has advantage of constant input voltage.

5) The novel current source inverter based PDM control method was more cost effective for efficiency improvement, performance enhancement of this power inverter and much more improvement of the power regulation characteristics on the nonlinear discharge and non discharge capacitance load as the streamer generation load.

6) The rate of sterilization applied to power moulds reached 97-98[%], so this power converter is useful as plasma power supply for powder mould sterilization with low power consumption.

Summing up, all these considerations were validated through appropriate experiments in this study. Hopefully, if the power converter circuit as proposed herein is employed to power supply for plasma generation, it will be properly used as a controller with prompt responsiveness.

ACKNOWLEDGMENT

This work was financially supported by MOCIE through IERC program

REFERENCES

[1] H.Fujita and H.Akagi: rmxulse Density Modulation power control of a 4kW, 450kHz voltage-source inverter for induction melting applications, *IEEE Transactions on Industry Applications*, Vol.32, No.2, pp.279-286 ,1996.

[2] J.Ferreira, A.Ross and J.Wyk: 'A hybrid phase arm power module with nonlinear resonant tank , *Records of IEEE-IAS Annual Meeting*, Vol 3 pp.1679-1685, 1990.

[3] Y.Konishi, S.P.Wang, S.Shirakawa and M.Nakaoka: 己 ulse density modulated high-frequency load resonant inverter for ozonizer and its feasible performance evaluations , *Records of IEEE-IAS Annual Meeting*, Vol 2 pp.1313-1319 ,1998.

[4] A. Tsul, Commutating SOA Capability of Power MOSFET, *IEEE APEC*, pp . 481～485, 1990

[5] Shengpei Wang, Yoshihiro Konishi, Voltage-Fed Pulse Density and Pulse Width Modulation Resonant Inverter for silent Discharge Type Ozonizer, *IEE Japan Industry Applications Society*, 1966

[6] H.J. Song, K.S.Lee, D.I.Lee, A study on the high voltage nozzle type ozonizer, *11th International Conference on Gaseous Discharge and Their Applications*. Vol. 2, pp. 320～323, 1995

[7] K.Kit sum and Bruse W. Carsten, Trends in High Frequency Power conversion ,*HFPC*, pp.198～204 , May, 1998

2006 5th International Power Electronics and Motion Control Conference

Generic optimization for SMPS design with Smart Scan and Genetic Algorithm

Heidi H.T. Yeung*, N. K. Poon* and Stephen L. Lai*
* PowerELab Limited, Hong Kong, HKSAR
heidi@powerelab.com
nkpoon@powerelab.com
stephen.lai@ieee.org

Abstract—the paper presents a new approach for generating optimized solutions of a Switched-Mode Power Supply with the higher efficiency. At the very beginning, we initialize a preliminary power supply design with a known topology (e.g. Fly-back). Then, we choose a set of alternative parts for some critical components such as the MOSFET and transformer. It is quite a complicated and time consuming task to obtain a design with the highest efficiency and lowest cost from numerous combinations. Multi-objective Genetic Algorithm is one generic solution to solve such optimization problems. In order to encode the electrical parts as the basic units of GA, the chromosomes, and evaluate them with a numerical function, we have to model an entire power supply circuits into a Component-based System and simulate the electrical reactions by the numerical characteristics of components. This approach not only reduces the time in the design stage but provide a more convincing design before the production. The experiment results are presented to show the robustness and the effectiveness of this approach.

Keywords- Genetic Algorithm; Power supply optimization; Component based system;

I. INTRODUCTION

In the design stage of Switched-Mode Power Supply (SMPS), the engineers use their expert knowledge and experience to draft the power supply prototype, fabricate a sample and examine its efficiency on different aspects such as power dissipation. To refine the performance and adjust the cost, they usually change the parts at some critical nodes. These steps are reiterated until the product meets certain criteria and it may consume weeks or even months. It is no doubt that these steps are unavoidable to ensure the higher quality and the lower cost of the power supply. Indeed, the procedures themselves may not be definitely completed by manipulation. We generally classify this kind of problem searching for the best one from a set of combinations as a Discrete Optimization Problem.

A. Techniques for solving Discrete Optimization Problem

The following is a typical Discrete Optimization Problem (DOP). Some boxes are put in the fixed size container. The size of boxes is varying and there is a set of positioning and placement combinations for those boxes. There must be at least one solution that most boxes fit in the containers.

There are some well-known techniques to solve the DOP. Full Search, or try-an-error, is the most reliable method to obtain the solution. All combinations are evaluated with a fitness function and the one with the highest score is the best solution. Most industrial processes will apply this technique if the optimization task is too complicated to be analyzed. However, this method is very expensive and time-consuming if the number of combinations is too large. There are some other methods to tackle DOP likes Binary Search and Nearest Neighbor Search. Despite the exponential reduction in the number of evaluation for combinations by these methods, it is still difficult to appraise the resource for optimization.

With limited resource, some evolutionary computing techniques such as Genetic Algorithm (GA) [4] and Particle Swarm Optimization (PSO) [6] are proposed. They are based on the evolutional algorithm to find the solution approximating to the best one at assigned resources. Even though the solution is not always the best as Full Search does, it is not far away from that. Most important thing is that we can control the resources (e.g. the size of pool and the number of generations) to achieve different effectiveness of optimization. There are some previous works applying GA for solving optimization problems in SMPS. Reference [1] employs GA for synthesizing low power circuits. GA is applied to search the optimal commitment of thermal units in power generation in [3]. A research conducted in [5] generates the pattern of high power supply noise to estimate the maximum power supply noise of chip. Moreover, there are some works on finding the best circuit configuration in power supply controller evaluated with transfer function, as in [7].

---

[a] Infineon provides the Evaluation Design circuit in our experiment

B. Outlines

In this paper, we propose using a Component-based System (CBS) accompanied with Genetic Algorithm (GA) to optimize a power supply from different combinations of the real components for higher efficiency. In the section II, the idea of GA solving Discrete Optimization Problem is overviewed. Then, the CBS of power supply in online power supply design software, "PowerESim", is introduced for preparing candidates in the optimization in the section III. In section IV, the fitness function to evaluate the efficiency of the modeled power supply is presented. Finally, the successful rate to obtain the efficient power supply is demonstrated in the experiment results in section V.

II. GENETIC ALGORITHM

A. Operations of GA

Assume an optimization problem contains at least one solution and the searching space is finite. Then, Genetic Algorithm (GA) [4] is able to locate the optimal solutions within the searching space.

The basic unit of GA is chromosome or candidate. They are parameterized as a list of numbers which are the features representing the chromosome. In the optimization problems, there are some parameters to be optimized and a set of parameters forms one candidate. A few candidates are generated randomly and put in a pool. They are evaluated with the fitness function, or cost function. The fittest group of candidates always survives in the pool and they are mated as the parents for next generation. These parents form some pairs and born the offspring by the crossover and mutation operation [4]. The better candidates, supposedly, are evolved from the competitions among the candidates in last generation. After several generations, the combinations remained are the elites and the final solution is the best inside the pool.

B. Multi-objective GA

For multi-objective optimization problems, we need multi-objective GA to solve it. For instance, we have to evaluate the efficiency, the unit price of product and the stress at extreme conditions while designing a SMPS. These evaluations are conflict to each other and no solution with the highest scores at all criteria can be achieved. In order to fit all the criteria, we can record the more than one high score candidates for each evaluations and the one with the highest average score for those criteria. User can select one of them to be the final solution. In this paper, we are going to evaluate the efficiency of SMPS ONLY such that it does not complicate the idea we proposed.

III. COMPONENT BASED SYSTEM OF SMPS

A. Definition of Component-based system

Component-based system (CBS) [2] is a widely used approach for computerizing the industrial processes into the software. In general, any process with the descriptive participants, the procedural actions and the measurable values can be implemented as a CBS. SMPS is obvious a CBS. The components are the physical parts like resistors, capacitors and transformers. Their connections lead the electrical response from the input source and each component participates in its position. The current and voltage across the components are measurable by the meters. In our experiment, a well-developed software, called "PowerESim", which is a power supply simulator built as a CBS is applied as the testing platform. Initially, user selects a converter topology and provides the specification (e.g. the range of input voltage, expected output voltage and current output current) from the interface. Then, the software provides the basic design which just fulfilled the specification and simulates the entire power dissipation at one operational cycle *1/fs* where *fs* is the switching frequency of power supply in several milliseconds.

B. Modeling component from real parts

To model a real part to be a component in the software, the essential characteristics of a particular component should be defined. Use MOSFET as an example. The characteristics of the MOSFET are modeled in the software according to the specifications are shown as

TABLE I.
MODELING CHARACTERISTICS OF MOSFET

Characteristics	Values
Max. Drain to Source Voltage V_{DSS}	730V
Max. Gate to Source Voltage V_{GS}	20V
Max. Continuous Drain Current I_D	7.3A
Max. Pulsed Drain Current I_{DM}	21.9A
Max. Pulsed Avalanche Rating E_{AS}	230mJ
Max. Power Dissipation P_D	83W
Max. Operating Temperature T_J	150°C
Typical Gate to Source Threshold Voltage V_{GS}	3V
Max. Total Gate Charge $Q_{g(TOT)}$	21nC
Max. Body Diode Reverse Recovery Timet_{rr}	400ns

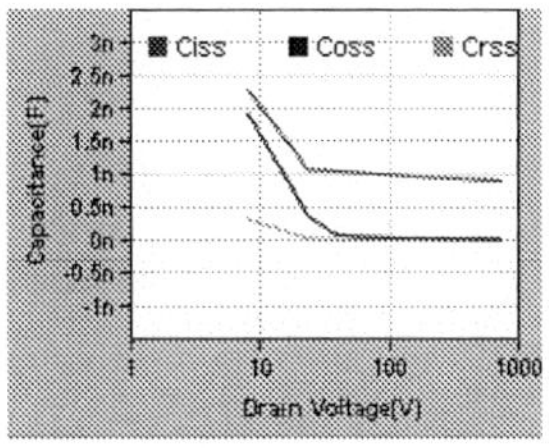

Figure 2. Typical Capacitance Vs. Drain to Source Voltage caption is centered in the column

following.

Thousands of modeled components are stored as table entries in the database. User can modify the components inside the converter, controller and feedback circuits. With a known converter topology, the power dissipation of every individual part can be estimated according to their modeled characteristics. Power dissipation of the main primary MOSFET in the Fly-back converter is defined as the integration of *voltage × current* within duration of *1/fs* (1) in the software. The software provides a multiple selection interface of components whereas users can select the alternatives as the parameters for optimization.

$$Loss_{M1} = \int_0^{1/fs} V \cdot I dt \qquad (1)$$

C. Modeling the transformer

Transformer takes a very important role in a SMPS. The main difference between transformer and other parts is that transformer constituted by the sub-parts such as magnetic core, the magnetic wires, the bobbin etc. Different arrangements (e.g. number of turns and number of parallel wires) of those sub-parts directly affect its behavior and performance. Formerly, it is not easy to model and simulate such complicated structure in component-based system. The proposed software has developed a subtle tool called Magnetic Builder providing a construction interface and simulator of the magnetic characteristics and power dissipation in a well-formed transformer. To find the best construction of transformer, we need to select multiple cores, wires and number of turns at particular winding. The number of turns in winding is quantized as a set of step values. For optimization, there are several combinations of transformer are generated by the crossover of all those alternative parameters.

D. The effective index of power supply

Efficiency is always the most important measurement for a power supply. In the software, the efficiency simulated is the Effective index defined as (2) where Po_{total} is the total output power and Pd_{total} is the total power dissipation. This becomes the fitness function for the optimization where the higher value induces higher efficiency and the maximum value is 1.

$$Effective\_index = \frac{Po_{total}}{Pd_{total} + Po_{total}} \qquad (2)$$

IV. EXPERIMENT RESULTS

A. Using concurrent converter topology

We choose an Evaluation Design of a power supply using Fly-back topology in the software. The input voltage range is 85 – 264V. The expected one output voltage is 16 V and current output current is 3.75 A. The circuit diagram is shown as Figure 3. The initial Effective Index of the design is 0.79. There are some critical nodes in the

converter have been selected as the parameters for optimization in Table II.

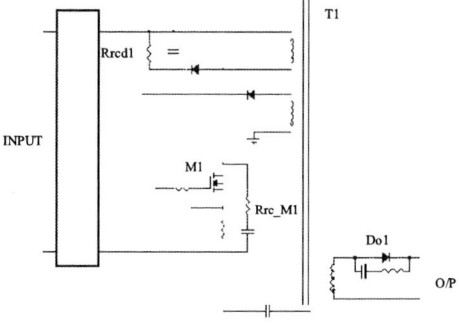

TABLE II.
THE COMPONENTS FOR OPTIMIZATION

Index (i)	Part no.	Description	Number of alternatives (*ksel_i*)
0	Rrcd1	RCD Clamper resistor	4
1	M1	Primary Main MOSFET	10
2	Rrc_M1	Slobber resistor	2
3	Do2	Output Diode	10
4	T1	Main Transformer	12 (4 cores and 3 number of turns at primary windings)

B. Prepare the first pool by Smart Scan Algorithm

Any parts containing only one alternative will not be considered as the parameters for optimization so to reduce the complexity. The chromosome is designed as Figure 4.

The chromosomes in the first pool are generally

Rrcd1$_0$	M1$_0$	Rrc_M1$_0$	Do2$_0$	T1$_0$

Figure 4. The first chromosome selecting the first alternative of each parameter

selecting from the combination of alternatives randomly. If the maximum number of executed evaluation function, *Iteration_allowed*, is small, some alternatives might be eliminated and never participate in the optimizing space. Hence, the best combination is not guaranteed in the pool.

In the power supply design, some dominating alternatives lead to high efficiency no matter what other parameters are. With this feature, an election method is proposed called Smart Scan to form the first pool intellectually as the procedures below.

1504

i. One of the parameters is randomly selected for the dominating scan, e.g. Rrcd1. Other parameters are restricted to be the one randomly selected from their list of alternatives. One chromosome is formed certainly by the first alterative of Rrcd1 and the fixed alternatives of other parameters.

ii. The chromosome is evaluated by the fitness function and both the score and the chromosome are pushed into the empty and fixed size pool.

iii. Another alternative for Rrcd1 is chosen to form the second chromosome. It is evaluated again and pushed into the pool. When all alternatives of Rrcd1 are examined, one round of scan is finished.

iv. The alternative of Rrcd1 in the chromosome with highest score is the dominated alternative. The second round of scan starts for other parameter, e.g. M1. The alternative of Rrcd1 in the chromosome is always set to the dominating one and others remain unchanged.

v. If the pool is getting full, the chromosome with lowest score is popped up to reserve space for the better one. Finally, a pool is filled by the strong chromosomes.

vi. Repeat the scan until the dominating alternatives of all the parameters are found.

This election method ensures the chromosomes are strong enough to generate better offspring. However, it requires some of evaluating iterations during the election. The number of fitness function executed is calculated by (3) where N is number of parameters. If the $Iteration_{scan}$ is more or equal to the upper limit of iterations $Iteration_{allowed}$ set by user, some of alternatives may be eliminated until it is within the limitation.

$$Iteration_{scan} = \left(\sum_{i=0}^{N-1} ksel_i - (N-1) \right) \times N \qquad (3)$$

C. Genetic Algorithm

After electing the pool, the chromosomes are automatically copied to the pool for next generation. The pairs of chromosomes are randomly selected from the first pool for Crossover and Mutation to generate the children. Crossover operation of the chromosomes exchanges the combination of alternatives from one to another and Mutation modifies one alternative of one parameter in the chromosome randomly. The probability of the mutation occurs for each crossover operation is 0.2.

If the child is better than one in the second pool, the worst one is popped up and the new one is pushed to the pool. Otherwise, it is eliminated. This method guarantees the best throughout the generation must survive in the pool. Then the third pool is duplicated from the second pool. The optimization is terminated when the maximum number of generation is finished.

$$N_{generation} = trunc \left(\frac{\left(Iteration_{allowed} - Iteration_{scan} \right)}{2 \cdot N_{pair}} \right) \qquad (4)$$

$$Iteration_{total} = Iteration_{Scan} + 2 \cdot N_{generation} \cdot N_{pair} \qquad (5)$$

The number of generations in GA allowed is estimated by (4) given that $Iteration_{allowed}$ is set by user. The pool size (N_{pool}) is 50 and the number of parent pairs (no duplication) selected from the pool (N_{pair}) is 50. Then, the total number of iterations actually is $Iteration_{total}$ found by (5). In the case $N_{generation}$ is 0, this means the limited iterations does not allow the GA operations and the best in the first pool becomes the final solution.

Figure 5 depicts the flowchart of the proposed optimization process.

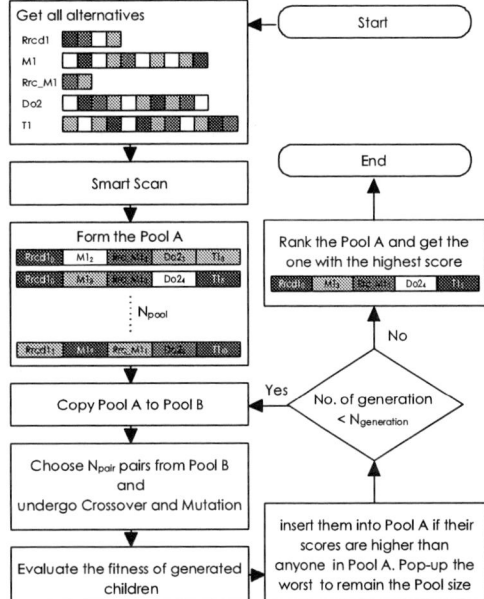

Figure 5. The flowchart of GA optimization

D. Summaries

The experimental result of Full search is the baseline of the optimization performance. It finds the highest and the lowest Effective Index from all combinations shown as Table III. Indeed, the total number of evaluations required in Full search is the total combination of all parameters calculated by (6) where $ksel_i$ is number of alternatives at i^{th} parameter defined in Table II.

TABLE III.
THE EXPERIMENT RESULT FROM FULL SEARCH

The no. of iterations run	Highest Effective Index	Lowest Effective Index
9600	0.817321	0.725137

TABLE IV.
THE EXPERIMENT RESULT FROM GA OPTIMIZATION AND SMART SCAN

$Iteration_{allowed}$	The no. of iterations run	Average Highest Effective index	$N_{generation}$	Number of alternatives are eliminated	Successful rate to find the best combination
400	370	0.817321	2	0	100%
300	270	0.817321	1	0	100%
200	170	0.817315	0	0	95%
150	150	0.815618	0	4	85%
100	100	0.815403	0	14	70%

$$Total\ combination = \prod_{i=0}^{Ncomp-1} ksel_i \qquad (6)$$

We have repeated the Smart Scan and GA optimization with different $Iteration_{allowed}$ from 100 to 400 and each test is repeated for 20 times. The results are summarized at the Table IV.

The tests with 300 and 400 $Iteration_{allowed}$ complete at least one GA round. Both of them find the best combination perfectly in all tests. When $Iteration_{allowed}$ is 200, it is resulted from Smart Scan only. The successful rate to get the best solution is 95% and the average highest effective index is 6×10^{-6} less that the highest Effective index. It induces that one of test cannot achieve the highest score but it is very close to the best one.

After reducing $Iteration_{allowed}$ to 150 and 100, some of alternatives are eliminated in the pool. 4 of alternatives are removed randomly for 150 iterations and 14 of that are removed for 100 iterations. Only 85% and 70% of tests can found the best combination in these two cases respectively. Nonetheless, the average best score results are still approximate to the highest score of all combination.

To conclude, the number of the iterations of Full search is about 26 times more than that of the guarantee optimization with GA and Smart Scan proposed.

V. CONCLUSIONS

Achieving the high efficient converter is considered in every SMPS design. Engineers conventionally choose the alternatives and put it in a real power supply. Its efficiency is estimated by the thermal analysis in an enclosed environment. These steps are repeated until the expected efficiency is obtained. However, these procedures are expensive and time-consuming.

With using the proposed method, the simulated power supply optimization is optimized by the Smart Scan of the real components modeled in the software, which parameterizes the entire power supply into a CBS and estimates the efficiency in seconds, and Genetic Algorithm.

There are several extensions to this work. First, the power losses, thermal effects and stress are evaluated at the same time by a multi-objective Smart Scan and GA optimizer. Second, this approach is also applicable in design the controller and feedback compensator like [7]. The differences are the real components are selected to form the circuit and it is estimated by the expected cut-off frequency and expected phase margin behaved in the power supply.

ACKNOWLEDGMENT

This project has been implemented as the "Smart optimizer" in "www.PowerESIM.com" developed by PowerELab Limited.

REFERENCES

[1] T. Arslan, E. Ozdemir, M. S. Bridge, and D. H. Horrocks, "Generic Synthesis Techniques for Low-Power Digital Signal Processing Circuits," *Proc. Of the IEE Colloquium On Digital Synthesis*, pp 7/1 – 7/5, February 1996.

[2] I. Crnkovic, J. A. Stafford, and H. W. Schmidt, *Component-based Software Engineering*, Springer-Verlag Berlin Heidelberg, 2004.

[3] D. Dasgupta, and D. R. McGregor, "Short Term Unit Commitment Using Genetic Algorithms," *Technical Report*, IKBS-16-93, August 1993.

[4] D. E. Goldberg, *Genetic Algorithms*, Addison Wesley, 1988.

[5] Y. M. Jiang, and K. T. Cheng, "Vector Generation for Power Supply Noise Estimation and Verification of Deep Submicron Designs," *IEEE Trans. VLSI Syst.*, 9(2), pp 329-340, April 2001.

[6] J. Kennedy, and R. C. Eberhart, "Particle swarm optimization," *Proceedings of the 1995 IEEE International Conference on Neural Networks (Perth, Australia)*, pp. 1942-1948, 1995.

[7] A. Maiden, A. Purvis, and M. Kinghorn, "The Development of a Digital Switched-Mode Power Supply Controller and Controller Design Tool," *Proc. International Signal Processing Conference*, Dallas, Texas, March 2003.

[8] D. Whitley, "A genetic algorithm tutorial," *Statistics and Computing*, vol. 4, pp. 65–85, 1994.

2006 5th International Power Electronics and Motion Control Conference

Novel Single-Stage Isolated Buck-Boost Inverter Based on Improved SPWM Control Method

Guang-Hui Tan, Fanpeng Zeng, Yanchao Ji, Xi Chen and Hua Wang
Department of Electrical Engineering, Harbin Institute of Technology, Harbin, China
hearpc@163.com

Abstract—**A novel single-stage isolated buck-boost inverter is proposed in this paper, which can generate an ac output voltage larger or lower than the input one in a single power stage. A high-frequency transformer operating as energy storage element is used to perform electrical isolation between the dc input and the ac output. The operation principle is analyzed in detail and the mathematics model is founded based on the state-space average method. An improved SPWM control method employing dc bias sine modulation wave is presented to optimize the inverter performance. The proposed control method eliminates the waveform distortion on output voltage zero-crossing area as caused by linear approximation in traditional SPWM method. The new inverter circuit topology provides the main switches for turn-on at zero current switching (ZCS) to achieve a low switching loss by a series-resonant tank. Laboratory experimental results from a 500W prototype are given to verify the validity and effectiveness of the proposed inverter.**

Keywords—*inverter; buck-boost; SPWM; series-resonant; state-space average*

I. INTRODUCTION

Power inverters have been used in a wide variety of industrial applications, such as uninterruptible power supplies (UPS), adjustable speed drives (ASD) and distributed power generators (DG) due to their capability in allowing continuous and linear control of the frequency and fundamental component of the output voltage. Generally, the major structure of the dc-ac inversion includes the inversion topology and waveshaping circuit. Buck inverter, which is probably the most popular and important power converter topology, is employed in the conventional inversion circuit [1]. However, one of the characteristics of the buck inverter is that the instantaneous average output voltage is always lower than the input dc voltage. It will result in the application range decrease. As a consequence, when an output voltage larger than the input one is needed, a boost converter should be used between the dc source and the inverter [2]. This solution will result in some problems, such as high cost, complicated structure and low efficiency.

In order to minimize power components and step up voltage, single-stage boost or buck-boost inverters were

proposed in [3]-[9], which implement boosting and inverting functions in a single power stage. However, some of them still have high component count and complicated operations [3], [4], thus compromising their benefits. Others limit their applications by either imposing on high switching frequency [5], [6], or presenting low power ratings [5].

A large body of voltage or current-mode strategies has devoted to the control of inverters in power electronics. Of those voltage control strategies, traditional SPWM method is used to control many single-stage buck-boost inverters [7], [8]. However, it results in output voltage or current waveforms distortion. Closed SPWM method and nonlinear control strategy are employed to optimize inverter output performance, but requiring complicated and inconvenient control circuits [8], [9].

In this paper, a novel single-stage series-resonant buck-boost inverter is proposed, which can generate an ac output voltage larger or lower than the input dc voltage. The electrical isolation between the dc input and the ac output is performed by a high-frequency transformer operating as energy storage element. The power inverter is realized by combining two sets of high frequency buck-boost choppers in discontinuous conduction mode (DCM). Improved SPWM control method, which employs the dc bias sine wave instead of the conventional sine modulation wave, is designed to eliminate the waveform distortion on output voltage zero-crossing area. The operation principle, theoretical analysis, and experimental results of the new single-stage inverter, rated 500W and operating at 10kHz, are provided in this paper to verify the performance of the proposed inverter.

II. OPERATION ORINCIPLE

Fig. 1 shows the basic configuration of the proposed single-stage buck-boost inverter, in which two sets of buck-boost choppers are combined in a power circuit and IGBTs are used as the main switching devices. It is composed of a series-resonant power stage and an output filter. The power stage is built by two main power switches S_1 and S_2, two complementary power switches S_3 and S_4 with two diodes D_1 and D_2, two resonant inductors L_1 and L_2, and a resonant capacitor C. The energy-storage components, L_1 and L_2, are the primary and secondary windings of a high-frequency transformer

1-4244-0448-7/06/$25.00 ©2006 IEEE

and have the identical turns and inductances as L. The ac voltage waveform is synthesized alternately by each half period of the desired output. In other words, the positive half (negative half) waveform of the output voltage is synthesized with the power switches S_1 (S_2) and S_3 (S_4) and the diode D_1 (D_2).

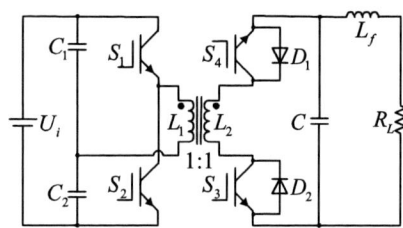

Figure 1.   The proposed single-stage isolated buck-boost inverter

There are six basic operational states as shown in Fig. 2 as:

1) Positive half cycle (PHC) charging state as shown in Fig. 2(a),

2) PHC resonant state in Fig. 2(b),

3) PHC discharging state in Fig. 2(c),

4) Negative half cycle (NHC) charging state in Fig. 2(d),

5) NHC resonant state in Fig. 2(e),

6) NHC discharging state in Fig. 2(f).

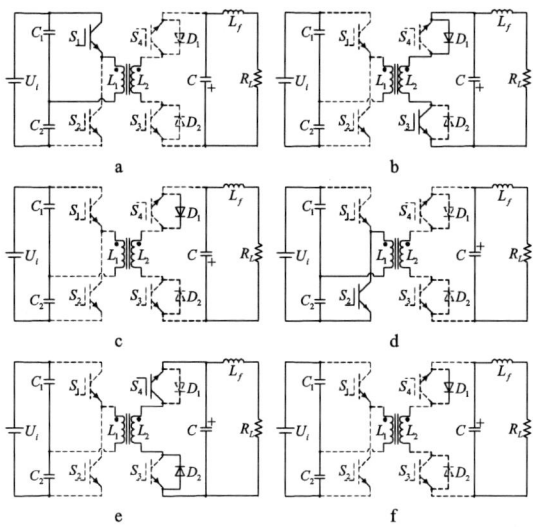

Figure 2.   The six basic operational states

III.   SYSTEM ANALYSIS

For convenience in analysis, only the positive half cycle of the output voltage is interpreted. The power switches are prescribed to operate with DCM in the series-resonant dynamics. In a switching cycle, there are three dynamic states including charging, resonant and discharging states and the corresponding waveforms are shown in Fig. 3. In order to simplify the mathematical analysis of the module,

the assumptions are supposed as: all the devices are ideal, the filter inductor L_f is large enough to present nearly constant dc current I_{Lf} during the entire switching cycle, and the stray losses of L_1, L_2, L_f, and C are neglected. The three dynamic states in the positive-half output are respectively described as follows.

1) Charging state, t in $[t_0, t_1]$: In this state, switch S_1 is on and switches S_2, S_3, and S_4 are off. This state begins as the power switch S_1 turns on with ZCS at $t = t_0$, the resonant inductor L_1 charges linearly energy from input DC voltage source until $t = t_1$. And the resonant capacitor C discharges its energy to the load until $t = t_1$. With the initial conditions $i_L(t_0) = 0$, and $u_C(t_0)$, the portraits $i_L(t)$ and $u_C(t)$ are represented by

$$i_L(t) = \frac{U_i}{2L}(t - t_0) \tag{1}$$

$$u_C(t) = u_C(t_0) - \frac{I_{Lf}}{C}(t - t_0) \tag{2}$$

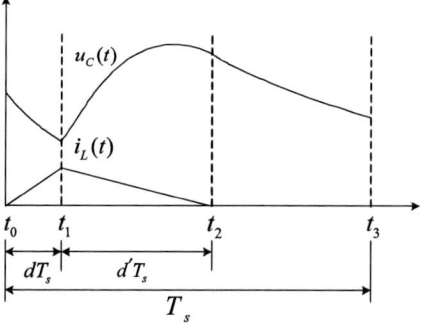

Figure 3.   Waveforms of the series resonant capacitor voltage and series resonant inductor current in a switching cycle

2) Resonant state, t in $[t_1, t_2]$: In this state, switch S_3 is on and switches S_1, S_2, and S_4 are off. This state begins when the power switch S_1 turns off at $t = t_1$, the diode D_1 promptly turns on, the series-resonant loop is then formed by L_2, S_3, D_1, and C. The energy stored in L_2 is transferred to C until null at $t = t_2$. The resonant portraits $i_L(t)$ and $u_C(t)$ can be given by

$$i_L(t) = I_{Lf} + [(i_L(t_1) - I_{Lf})\cos\omega(t - t_1)]$$
$$- \frac{u_C(t_1)}{Z}\sin\omega(t - t_1) \tag{3}$$

$$u_C(t) = u_C(t_1)\cos\omega(t - t_1)$$
$$+ Z[i_L(t_1) - I_{Lf}]\sin\omega(t - t_1) \tag{4}$$

where $\omega = 1/\sqrt{LC}$ is the resonant angular frequency, and $Z = \sqrt{L/C}$ is the characteristic impedance.

3) Discharging state, t in $[t_2, t_3]$: In this state, switches S_1, S_2, S_3, and S_4 are all off. This state begins when the energy stored in L_2 discharges to null at $t = t_2$, and the diode D_1 promptly ceases conduction, the resonant capacitor C then discharges its energy to the load until the subsequent driving pulse arrives at $t = t_3$. The discharging portraits of $i_L(t)$ and $u_C(t)$ can be given by

1508

$$i_L(t) = 0 \qquad (5)$$

$$u_C(t) = u_C(t_2) - \frac{I_{Lf}}{C}(t - t_2) \qquad (6)$$

Combination of (1), (2), (3), (4), (5), and (6) yields the average inductor current as following

$$\overline{i_L} = \frac{1}{T_s}\int_{t_0}^{t_3} i_L(t)dt = \frac{1}{T_s}(\frac{U_i}{4L}d^2T_s^2 + I_{Lf}T_s) \qquad (7)$$

In the proposed buck-boost inverter, the main switches are prescribed to operate with DCM and the steady-state description in a switching cycle can be determined by the state-space average method. For each of the three states in Fig. 3, there exists a corresponding state-space equation as follows.

State 1), for the interval dT_s, t in $[t_0, t_1]$

$$\frac{d}{dt}\begin{bmatrix} i_L(t) \\ u_C(t) \\ i_{Lf}(t) \\ u_o(t) \end{bmatrix} = \begin{bmatrix} 0 & 0 & 0 & 0 \\ 0 & 0 & -\frac{1}{C} & 0 \\ 0 & \frac{1}{L_f} & 0 & -\frac{1}{L_f} \\ 0 & 0 & -\frac{1}{C} & \frac{1}{R_LC} \end{bmatrix}\begin{bmatrix} i_L(t) \\ u_C(t) \\ i_{Lf}(t) \\ u_o(t) \end{bmatrix} + \begin{bmatrix} \frac{1}{L} \\ 0 \\ 0 \\ 0 \end{bmatrix}\frac{U_i}{2}$$

$$(8)$$

State 2), for the interval $d'T_s$, t in $[t_1, t_2]$

$$\frac{d}{dt}\begin{bmatrix} i_L(t) \\ u_C(t) \\ i_{Lf}(t) \\ u_o(t) \end{bmatrix} = \begin{bmatrix} 0 & -\frac{1}{L} & 0 & 0 \\ \frac{1}{C} & 0 & -\frac{1}{C} & 0 \\ 0 & \frac{1}{L_f} & 0 & -\frac{1}{L_f} \\ 0 & 0 & -\frac{1}{C} & -\frac{1}{R_LC} \end{bmatrix}\begin{bmatrix} i_L(t) \\ u_C(t) \\ i_{Lf}(t) \\ u_o(t) \end{bmatrix} \qquad (9)$$

State 3), for the interval $(1-d-d')T_s$, t in $[t_2, t_3]$

$$\frac{d}{dt}\begin{bmatrix} i_L(t) \\ u_C(t) \\ i_{Lf}(t) \\ u_o(t) \end{bmatrix} = \begin{bmatrix} 0 & 0 & 0 & 0 \\ 0 & 0 & -\frac{1}{C} & 0 \\ 0 & \frac{1}{L_f} & 0 & \frac{1}{L_f} \\ 0 & 0 & -\frac{1}{C} & -\frac{1}{R_LC} \end{bmatrix}\begin{bmatrix} i_L(t) \\ u_C(t) \\ i_{Lf}(t) \\ u_o(t) \end{bmatrix} \qquad (10)$$

Base on the state-space average method, the above equations corresponding to the three states in a switching cycle are time averaged, and the following expression can be given:

$$\frac{d}{dt}\begin{bmatrix} i_L \\ u_C \\ i_{Lf} \\ u_o \end{bmatrix} = \begin{bmatrix} 0 & -\frac{d'}{L} & 0 & 0 \\ \frac{d'}{C} & 0 & -\frac{1}{C} & 0 \\ 0 & \frac{1}{L_f} & 0 & -\frac{1}{L_f} \\ 0 & 0 & \frac{1}{C} & -\frac{1}{R_LC} \end{bmatrix}\begin{bmatrix} i_L \\ u_C \\ i_{Lf} \\ u_o \end{bmatrix} + \begin{bmatrix} \frac{d}{L} \\ 0 \\ 0 \\ 0 \end{bmatrix}\frac{U_i}{2}$$

$$(11)$$

IV. CONTROL METHOD

According to the mathematics model given by (11), and the inverter is assumed to operate around the steady-state points (I_L, U_C, I_{Lf}, U_o, D, D'), we can obtain the desired steady-state model as following

$$\begin{bmatrix} 0 & -\frac{D'}{L} & 0 & 0 \\ \frac{D'}{C} & 0 & -\frac{1}{C} & 0 \\ 0 & \frac{1}{L_f} & 0 & -\frac{1}{L_f} \\ 0 & 0 & \frac{1}{C} & -\frac{1}{R_LC} \end{bmatrix}\begin{bmatrix} I_L \\ U_C \\ I_{Lf} \\ U_o \end{bmatrix} + \begin{bmatrix} \frac{D}{L} \\ 0 \\ 0 \\ 0 \end{bmatrix}\frac{U_i}{2} = 0 \quad (12)$$

This leads to

$$I_L = \frac{U_o}{D'R_L} \qquad (13)$$

$$I_{Lf} = \frac{U_o}{R_L} \qquad (14)$$

$$D' = \frac{U_i}{2U_C}d \qquad (15)$$

$$U_C = U_o \qquad (16)$$

And $I_L = \overline{i_L}$, so from (7), the average current across the resonant inductor can be written as

$$I_L = \frac{1}{T_s}(\frac{U_i}{4L}D^2T_s^2 + I_{Lf}T_s) \qquad (17)$$

Therefore, the inherent relationship between the output voltage U_o and the duty cycle of the main switch D is finally obtained as

$$M = \frac{R_LT_s}{2LM}D^3 + D \qquad (18)$$

where

$$M = \frac{2U_o}{U_i} \qquad (19)$$

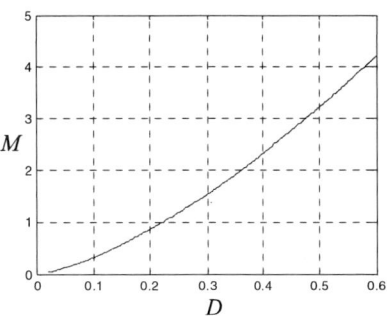

Figure 4. The curve of the output voltage versus the main switch duty cycle under $R_L = 10\Omega$

The curve of the output voltage versus the main switch duty cycle under $R_L = 10\Omega$ is shown in Fig. 4, from which, we can see that it is not a purely straight line. That is to say, when the input voltage U_i keeps constant, the output voltage U_o is not linear with the main switch duty cycle D. And so, traditional SPWM control method, which is based on simple linear approximation, is not appropriate to be used directly in the DCM inverter.

Basically, a sine reference wave, serving as modulating signal, is compared with a triangular carrier wave, and the intersection points determine the switching angles and pulse widths in traditional SPWM method. In this paper, a novel improved SPWM control method is deployed to optimize the single-stage isolated inverter performance. A dc bias sine modulation wave is employed to compare with the sawtooth carrier wave, and the new improved SPWM gating signal pattern is realized as shown in Fig. 5.

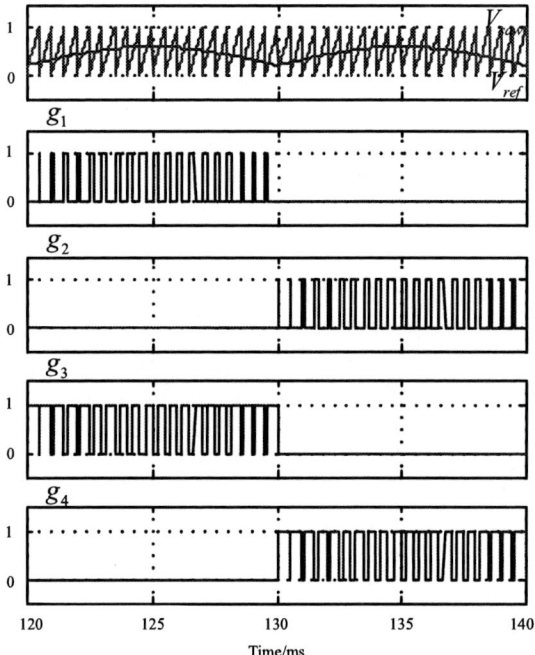

Figure 5. The improved SPWM gating signal pattern

In Fig. 5, V_{ref} is the modulating signal, V_{saw} is the sawtooth carrier wave, g_1, g_2, g_3 and g_4 are the drive signals for the power switches S_1, S_2, S_3 and S_4, respectively. The difference in the two SPWM control methods is that there is dc bias value in the modulating signal of the improved SPWM method but not of the traditional SPWM method. And the modulating signal of the improved SPWM method can be expressed as

$$V_{ref} = V_{dc} + \alpha \mid \sin(2\pi f_{out} t) \mid \qquad (20)$$

where V_{dc} is the dc bias value of the modulating signal, α is the peak value of the sine component in modulating signal, and f_{out} is the frequency of ac output voltage.

The proposed inverter with improved SPWM controller is designed to work under fixed switching frequency and discontinuous current mode. Furthermore, an output voltage cycle is divided into two half-cycles, only two

switches are operating at switching frequency and the rests are stationary off. So the switching loss is reduced and the efficiency is improved.

V. EXPERIMENTAL RESUTLS

To verify the validity of theoretical analysis and feasibility of the proposed single-stage isolated buck-boost inverter based on the improved SPWM scheme, an experimental unit with maximum power capacity of 500W is built. A single-chip digital signal processor (TMS320F2812) is adopted as the kernel in the implementation of a digital controller. The dc input is obtained after a rectifier from an adjustable three-phase autotransformer. The main parameters used were: $L_1 = L_2 = L = 10\mu H$, $C = 10\mu F$, $L_f = 2mH$, $R_L = 10\Omega$, $U_i = 80V$, $f_{out} = 50Hz$, and $U_{om} = 100V$. The sawtooth carrier frequency is 10kHz.

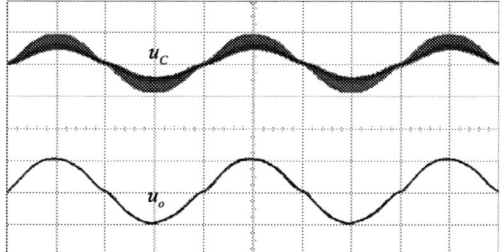

Figure 6. Experimental results with traditional SPWM method (u_C, u_o 100V/div; time 5ms/div)

Fig. 6 shows the experimental results with traditional SPWM method ($V_{dc} = 0$, $\alpha = 0.45$). In the figure, the upper waveform represents the voltage of resonant capacitor u_C and the below waveform is the ac output voltage u_o. Due to the linear approximation in traditional SPWM method, it is founded that there is serious waveform distortion on output voltage zero-crossing area.

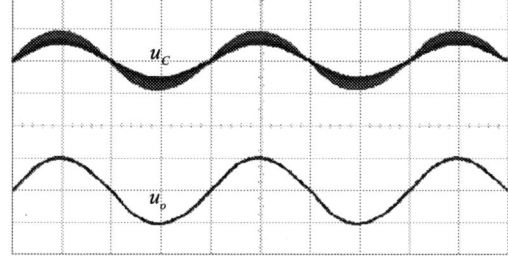

Figure 7. Experimental results with the improved SPWM method (u_C, u_o 100V/div; time 5ms/div)

In contrast with the traditional SPWM method, Fig. 7 shows the experimental results of the proposed improved SPWM control method ($V_{dc} = 0.05$, $\alpha = 0.40$). In the figure, the upper waveform and the below waveform represent the voltage of resonant capacitor u_C and the ac output voltage u_o, respectively. The experimental results show that the new PWM technique eliminates the waveform distortion resulted from the linear approximation method and exhibits high quality output voltage waveform.

VI. CONCLUSIONS

A novel single-phase isolated buck-boost inverter with compact circuit configuration has been presented to generate an ac output voltage larger or lower than the input one in a single power stage. Only four switches are used in the proposed inverter. The electrical isolation between the dc input and ac output is performed by a high-frequency transformer operating as energy storage element. A resonant cell is built into the power stage to realize ZCS for turning on the main switches to achieve a low switching loss. The state-space average approach is used to estimate the system performance. Two control methods, the traditional SPWM method and the improved SPWM method, are implemented on a 500W inverter prototype. Experimental results show that the improved SPWM technique eliminates the output voltage distortion resulted from the linear approximation of traditional SPWM method and exhibits good ac output voltage waveform. Therefore, it is very suitable for the inverter with low input voltage such as a fuel cell, battery and solar cell input in stand-alone or grid-connected system.

REFERENCES

[1] S. Y. R. Hui, S. Gogani, and J. Zhang, "Analysis of a quasiresonant circuit for soft-switched inverters," *IEEE Trans. Power Electron.*, vol. 11, pp. 106-114, Jan. 1996.

[2] T. Boutot and L. Chang, "Development of a single-phase inverter for small wind turbines," *Proc. IEEE Elect. Comput. Eng. Can. Conf. (CCECE'98)*, May 24-28, 1998, Waterloo, Canada, pp. 305-308.

[3] M. Nagao and K. Harada, "Power flow of photovoltaic system using buck-boost PWM power inverter," *Proc. IEEE-PEDS'97*, May 26-29, 1997, Singapore, pp. 144-149.

[4] S. B. Kjaer and F. Blaabjerg, "Design optimization of a single phase inverter for photovoltaic applications," *Proc. IEEE-PESC'03*, Jun. 15-19, 2003, Acapulco, Mexico, pp.1183-1190.

[5] M. Kusakawa, H. Nagayoshi, K. Kamisako, and K. kurokawa, "Further improvement of a transformerless, voltage-boosting inverter for ac modules," *Solar Energy Material and Solar Cells*, vol. 67, pp. 379-387, Mar. 2001.

[6] T. Shimizu, K. Wada, and N. Nakamura, "A flyback-type single phase utility interactive inverter with low-frequency ripple current reduction on the dc input for an ac photovoltaic module system," *Proc. IEEE-PESC'02*, Jun. 23-27, 2002, Cairns, Australia, pp. 1483-1488.

[7] C. -M. Wang, "A novel single-stage series-resonant buck-boost inverter," *IEEE Tran. Ind. Electron.*, vol. 52, pp. 1099-1108, Aug. 2005.

[8] Y. S. Xue and L. C. Chang, "Closed-loop SPWM control for grid-connected buck-boost inverters," *Proc. IEEE-PESC'04*, Jun. 20-25, 2004, Aachen, Germany, pp. 3366-3371.

[9] C. -M. Wang, "A novel single-stage full-bridge buck-boost inverter," *IEEE Trans. Power Electron.*, vol. 19, pp. 150-159, Jan. 2004.

2006 5th International Power Electronics and Motion Control Conference

On the Effects of Voltage Loop in Paralleled Converters Under Master-Slave Current Sharing

Yuehui Huang and Chi K. Tse

Department of Electronic and Information Engineering

The Hong Kong Polytechnic University, Kowloon, Hong Kong

Email: yuehui.huang@polyu.edu.hk, encktse@polyu.edu.hk

Abstract— This paper studies the effects of the presence of voltage loops in parallel connected buck switching converters under master-slave current sharing scheme. The system employs a typical proportional-integral (PI) controller for regulation. Comparisons are made for the cases where the slave modules are controlled with and without a voltage loop. Generally, we find that the voltage loop in the slave is helpful in widening the stability range though it is theoretically redundant for the purpose of controlling the output voltage in the small-signal sense. Such a loop provides stable current reference for the slave modules. Effectively, each slave module is under current-mode control by virtue of the current sharing loop, making it a current source. Simulation results under different control configurations are presented to demonstrate the phenomenon.

I. INTRODUCTION

Power supplies based on paralleling switching converters offer a few advantages over a single, high-power, centralized power supply. They enjoy low component stresses, increased reliability, ease of maintenance and repair, improved thermal management, etc. [1], [2]. Paralleling of standardized converters is an approach used widely in distributed power systems for both front-end and load converters. Since current sharing has to be maintained among the paralleled converters, some form of control has to be used to equalize the individual currents in the converters. One widely used method for balancing currents is the *master-slave current sharing* method [3], [4].

For paralleled converters, we have to control the current distribution as well as the output voltage. Typically, it contains a main voltage loop and a current sharing loop in voltage mode control; alternatively there may be a main voltage loop, a current loop and a current sharing loop in current mode control. The dynamic behavior becomes complex in N-paralleled converters because of the interaction between these loops. Intuitively, the main voltage loop is necessary for regulating the output voltage as in a stand-alone converter. The current sharing loop helps to regulate the reference voltage to get the expected output [4], [5], [6], [7]. However, all outputs of the converters are connected to one node (the load side). From circuit theory, paralleled branches should behave like current sources with large output impedance in order to ensure stable operation [8], [9]. Consequently, one voltage loop is enough to control the output voltage for the paralleled system. In the master-slave current sharing system, the master will control the output voltage and the slaves are required to follow the current of the master.

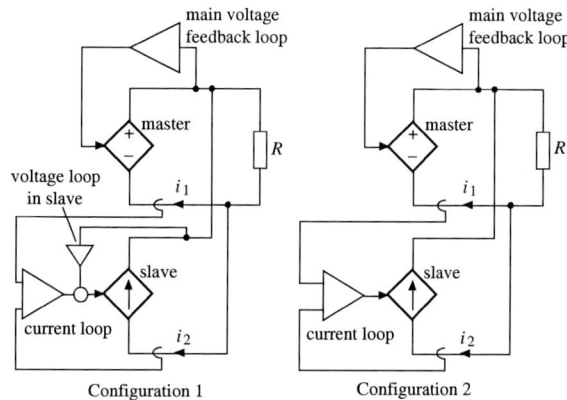

Fig. 1. Master-slave parallel system. Configuration 1: voltage loop in slave. Configuration 2: no voltage loop in slave

For simplicity, the system under study in this paper is a parallel connected system of two buck converters. Under the master-slave scheme, one of the converters is the master and the other is the slave. The master has a main feedback loop consisting of a typical proportional-integral (PI) control, to regulate the output voltage. The slave basically sets its current to equal that of the master via an active loop involving comparison of the currents of the two converters.

In this paper, we will study the effects of the slave's voltage loop and compare the stability boundaries of feedback parameters with and without the voltage loop in the slave converter. We find that the stable region in the parameter space can be larger when the slave converter contains a voltage loop. The voltage loop in the slave converter is therefore useful. The role of it is not to control the output voltage directly, but to provide a stable current reference for the slave. Furthermore, we confirm that as the slave converter is under current mode control with current sharing feedback, it behaves effectively as a current source.

II. SYSTEM DESCRIPTION AND OPERATION

Figure 1 shows the circuit model of two paralleled buck converters under master-slave control. Configuration 1 incorporates an additional voltage loop in the slave, whereas Configuration has no such a loop. Figure 2 (a) is the converter circuit, Figs 2 (b) and (c) form the control circuit for

1-4244-0448-7/06/$25.00 ©2006 IEEE

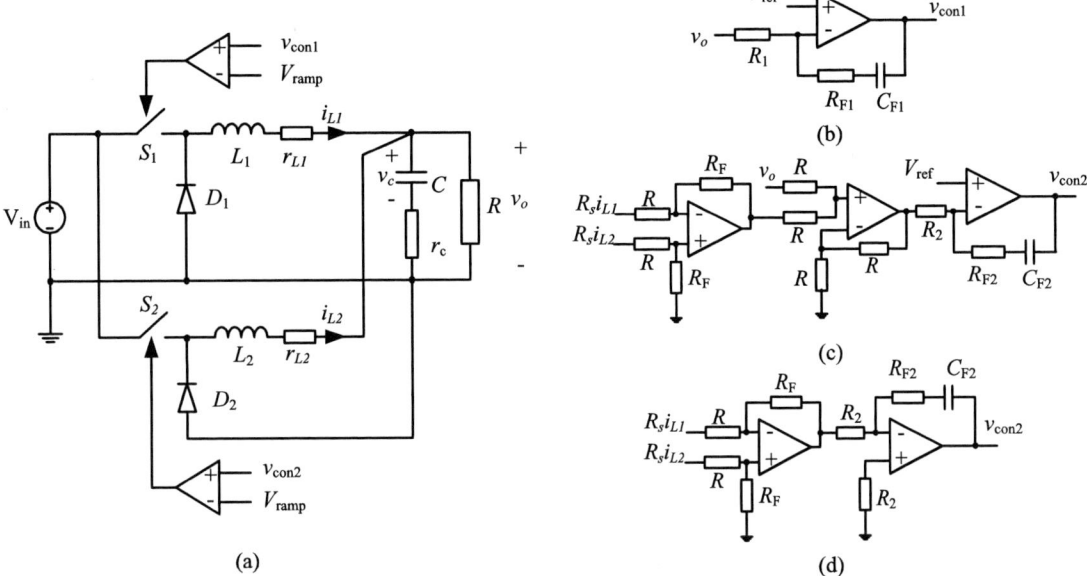

Fig. 2. Paralleled buck converters under master-slave current sharing and PI control. (a) Power stage; (b) controller for the master; (c) controller for the slave with voltage loop; (d) controller for the slave without voltage loop.

Configuration 1, and Figs. 2 (b) and (d) form the control circuit for Configuration 2.

In this circuit, S_1 and S_2 are switches, which are controlled by a standard pulse-width modulator consisting of a comparator comparing a control signal and a ramp signal. The ramp signal is given by [10]

$$V_{\mathrm{ramp}} = V_L + (V_U - V_L)\left(\frac{t}{T_s} \bmod 1\right) \quad (1)$$

where V_L and V_U are the lower and upper thresholds of the ramp, respectively, and T_s is the switching period. Basically, switch S_i ($i = 1, 2$) is closed if $v_{\mathrm{con}i} > V_{\mathrm{ramp}}$ and is open otherwise.

The control signals $v_{\mathrm{con}1}$ and $v_{\mathrm{con}2}$ are derived from the feedback compensator, as shown in Figs. 2 (b) and (c). Here the compensator is a PI controller, i.e.,

$$\frac{V_{\mathrm{con}1}(s)}{E(s)} = -K_p\left(1 + \frac{1}{T_i s}\right) \quad (2)$$

where $V_{\mathrm{con}1}(s)$ and $E(s)$ are the Laplace transforms of $v_{\mathrm{con}1}(t)$ and $e(t)$; $e(t)$ is the error between reference and output; K_p and T_i are the control parameters. With respect to the slave, an extra current sharing signal is included. We can likewise derive the equation.

We assume that the converter operates in continuous conduction mode (CCM) and diodes D_1 and D_2 are always in complementary state to S_1 and S_2. Consequently, the state

equations of the converter stage of Fig. 2 are

$$\begin{cases} \dot{x}_1 = -\frac{1}{L1}\left[(r_{L1} + \frac{Rr_c}{R+r_c})x_1 + \frac{Rr_c}{R+r_c}x_2 + \frac{R}{R+r_c}x_3 - q_1 V_{in}\right] \\ \dot{x}_2 = -\frac{1}{L2}\left[\frac{Rr_c}{R+r_c}x_1 + (r_{L2} + \frac{Rr_c}{R+r_c})x_2 + \frac{R}{R+r_c}x_3 - q_2 V_{in}\right] \\ \dot{x}_3 = \frac{1}{C(R+r_c)}(Rx_1 + Rx_2 - x_3) \end{cases} \quad (3)$$

where x_1, x_2, x_3 are the converter state variables defined as

$$[x_1 \quad x_2 \quad x_3] = [i_{L1} \quad i_{L2} \quad v_c] \quad (4)$$

and q_1 and q_2 are the switching function decided by the controllers. They are time varying functions given by

$$q_i(t) = \begin{cases} 1, & \text{if } v_{\mathrm{con}i} \geq V_{\mathrm{ramp}}, \\ 0, & \text{if } v_{\mathrm{con}i} < V_{\mathrm{ramp}}. \end{cases} \quad (5)$$

According to the feedback circuit in Figs. 2 (b), (c) and (d), we can derive the control equations. For Configuration 1, we have

$$\frac{dv_{\mathrm{con}1}}{dt} = -K_1 \frac{dv_o}{dt} - \frac{K_1}{\tau_{F1}} v_o + \frac{K_1}{\tau_{F1}} V_{\mathrm{ref}} \quad (6)$$

$$\frac{dv_{\mathrm{con}2}}{dt} = -K_2 \frac{dv_o}{dt} - \frac{K_2}{\tau_{F2}} v_o + K_2 K_i\left(\frac{di_{L1}}{dt} - \frac{di_{L2}}{dt}\right)$$
$$+ \frac{K_2 K_i}{\tau_{F2}}(i_{L1} - i_{L2}) + \frac{K_2}{\tau_{F2}} V_{\mathrm{ref}} \quad (7)$$

and for Configuration 2, we have

$$\frac{dv_{\mathrm{con}1}}{dt} = -K_1 \frac{dv_o}{dt} - \frac{K_1}{\tau_{F1}} v_o + \frac{K_1}{\tau_{F1}} V_{\mathrm{ref}} \quad (8)$$

$$\frac{dv_{\mathrm{con}2}}{dt} = K_2 K_i\left(\frac{di_{L1}}{dt} - \frac{di_{L2}}{dt}\right) + \frac{K_2 K_i}{\tau_{F2}}(i_{L1} - i_{L2}). \quad (9)$$

1513

Also, v_o can be written as

$$v_o = v_c + r_c i_c = v_c + r_c(i_{L1} + i_{L2} - \frac{v_o}{R}) \qquad (10)$$

where K_1 and K_2 are the proportional gains, τ_{F1} and τ_{F2} are the integral coefficients, K_i is the current sharing gain, and V_{ref} is the reference voltage (expected output voltage). In circuit terms, $K_1 = R_{F1}/R_1$, $\tau_{F1} = R_{F1}C_{F1}$, $K_2 = R_{F2}/R_2$, $\tau_{F2} = R_{F2}C_{F2}$, $K_i = R_F R_s/R$, where R_s is the current sensing resistance. Equations (6) and (7), together with (3), form the complete set of state equations for Configuration 1 and equations (8), (9) and (3) for Configuration 2.

III. SIMULATION RESULTS

The simulations performed using the state equations derived in the foregoing section and hence are exact cycle-by-cycle simulations. We are primarily concerned with the system stability in relation to the feedback parameters of the PI controller, i.e., K_1, K_2, τ_{F1}, τ_{F2}. We assume that the inductances in the converters are generally different and fix the current sharing parameter at $K_i = 1$. The circuit parameters and component values are listed in Table I.

TABLE I
COMPONENT VALUES USED IN SIMULATIONS

Circuit Components	Values
Switching Period T_s	10 μs
Input Voltage V_{in}	12 V
Reference Voltage V_{ref}	5 V
Ramp Voltage V_L,V_U	0 V, 2 V
Inductance L_1, ESR r_{L1}	55 μH, 0.01 Ω
Inductance L_2, ESR r_{L2}	110 μH, 0.05 Ω
Capacitance C, ESR r_c	126 μF, 0.01 Ω
Load Resistance R	0.5 Ω
Current sensing Resistance R_s	0.01 Ω

Firstly, we fix the control parameters of the master, K_1 and $1/\tau_{F1}$, and identify the stable region in the space of K_2 and $1/\tau_{F2}$. Figure 3 shows the stability boundary in the space of K_2 and $1/\tau_{F2}$ for different values of K_1 and $1/\tau_{F1}$. Then, we fix the control parameters of the slave, K_2 and $1/\tau_{F2}$, and identify the stability boundary in the space of K_1 and $1/\tau_{F1}$. Figure 4 shows the stability boundary in the space of K_1 and $1/\tau_{F1}$.

From Fig. 3, we may conclude that when the master is fixed, the maximum stable value of K_2 is more or less unchanged regardless of variation of $1/\tau_{F2}$ for both Configurations 1 and 2. Moreover, the maximum values of K_2 are unaffected by the parameters of the master when it operates in a stable region. In a previous publication [11], it has been shown that for stand-alone converters, the stable range of proportional gain K diminishes rapidly as the integral time constant parameter $1/\tau_F$ increases in the voltage mode control, which is clearly different from the results shown in Fig. 3. We can explain this phenomenon in terms of the characteristic of current-mode control. When under average current-mode control, the buck converter shows a single-pole

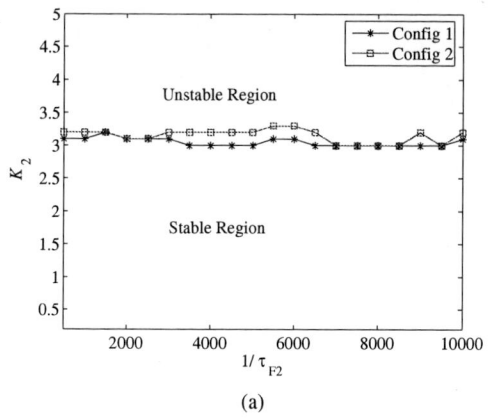

(a)

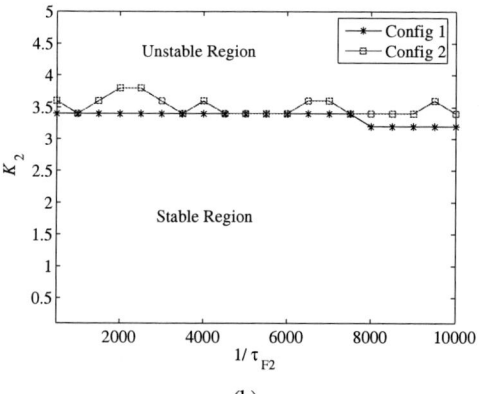

(b)

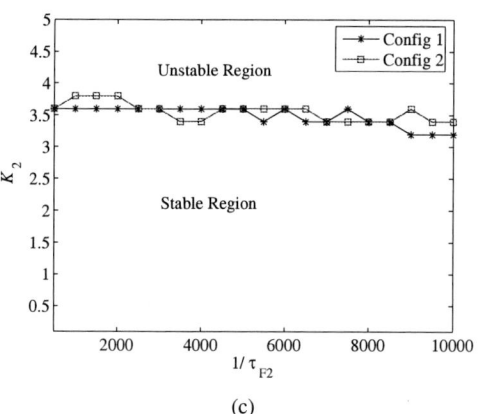

(c)

Fig. 3. Stability boundaries of feedback parameters for Configuration 1 and Configuration 2. (a) $K_1 = 1, 1/\tau_{F1} = 1000$; (b) $K_1 = 0.2, 1/\tau_{F1} = 1000$; (c) $K_1 = 0.2, 1/\tau_{F1} = 10000$.

behavior since the inductor has been controlled by the current feedback. The unstable behavior is caused by saturation, i.e., the control signal overruns the range of the ramp signal. In the paralleled system, since the slave is effectively under current-mode control, the stable range of proportional control gain K_2 will not be affected by the variation of $1/\tau_{F2}$. The current sharing feedback makes the slave a current source in both

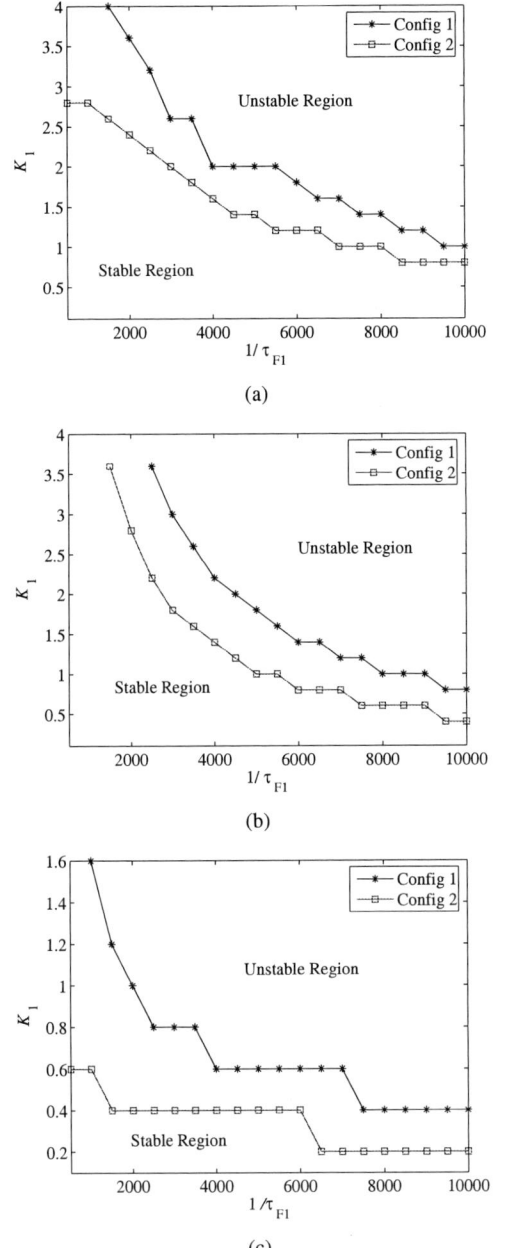

(a)

(b)

(c)

Fig. 4. Stability boundaries of feedback parameters for Configuration 1 and Configuration 2. (a) $K_2 = 3, 1/\tau_{F2} = 5000$; (b) $K_2 = 0.3, 1/\tau_{F2} = 5000$; (c) $K_2 = 0.3, 1/\tau_{F2} = 20000$.

Configurations 1 and 2 when the master is fixed.

From Fig. 4, we observe that the stable range of K_1 decreases greatly as the value of $1/\tau_{F1}$ increases, which is consistent with the characteristic of voltage-mode controlled buck converters, for both Configurations 1 and 2 when the slave is fixed. And the stability boundary in Configuration 1 is always larger than that in Configuration 2. Moreover, K_2

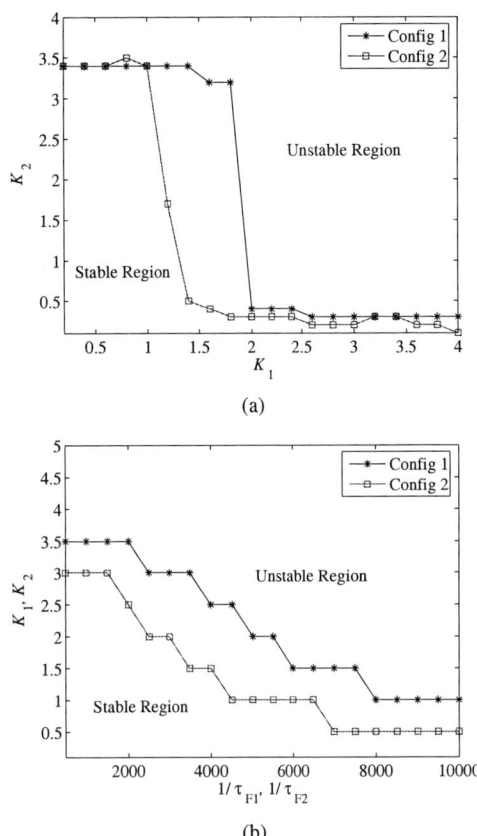

(a)

(b)

Fig. 5. Stability boundaries of feedback parameters for Configuration 1 and Configuration 2. (a) $1/\tau_{F1} = 1/\tau_{F2} = 2000$; (b) $K_1 = K_2, 1/\tau_{F1} = 1/\tau_{F2}$.

and $1/\tau_{F2}$ affect the stable region in the plane of K_1–$1/\tau_{F1}$, as shown in Figs. 4 (a), (b) and (c). The larger K_2 and $1/\tau_{F2}$ are, the smaller the stable region.

Figure 5 (a) shows the stability boundary of K_1 versus K_2 for $1/\tau_{F1} = 1/\tau_{F2} = 2000$. When K_1 is small, K_2 has a large stable range, but when it becomes large, the stable range of K_2 falls rapidly. Figure 5 (b) shows the stability boundary when the feedback parameters of the master and slave are changed simultaneously. Again, the stable region for Configuration 2 is smaller than that for Configuration 1.

Based on the stability boundaries of the feedback parameters, we can identify the parameter range for stable operation. Also, we want to know the relationship between the input voltage and feedback parameters. Figure 6 shows the input voltage range for different values of K_1 and K_2. We observe that when K_1 is small (e.g. $K_1 = 0.2$), the stability boundaries are almost identical in the plane of K_2–V_{in} for Configurations 1 and 2, as shown in Fig. 6 (b) and (c). Also, the stable region diminishes greatly in the plane of K_1–V_{in} as K_2 increases from Fig. 6 (d) and (e). Furthermore, if K_1 and K_2 change simultaneously, the system can be operated under

1515

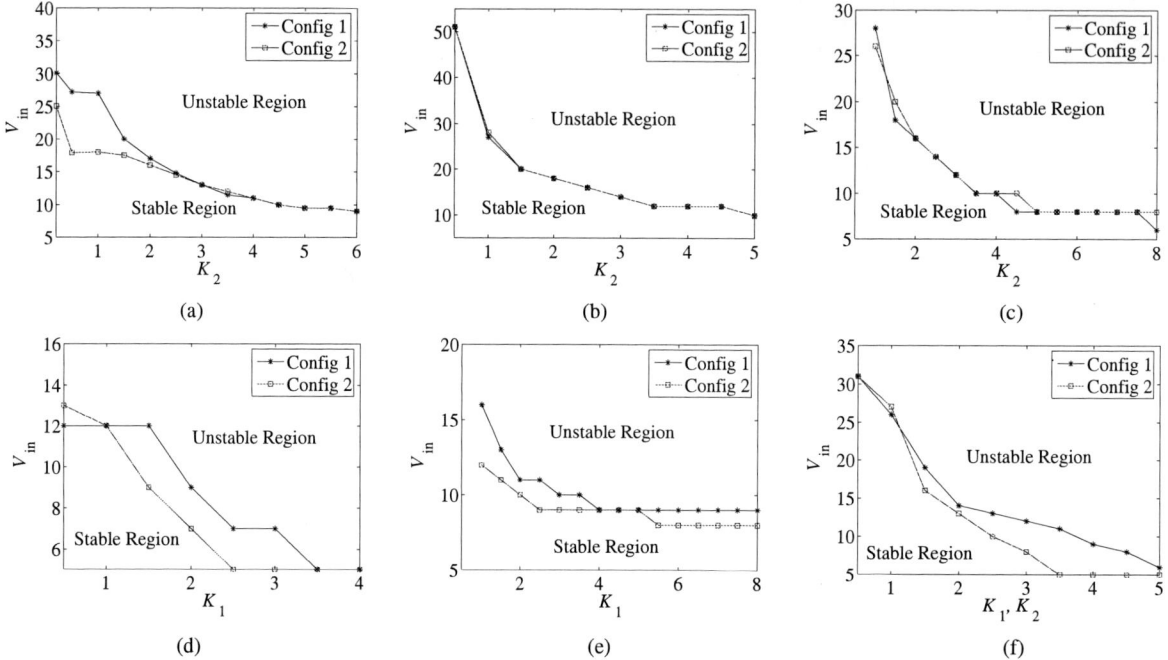

Fig. 6. Stability boundaries of V_{in} versus feedback parameters for Configuration 1 and Configuration 2. (a) $K_1 = 2, 1/\tau_{F1} = 1/\tau_{F2} = 1000$; (b) $K_1 = 0.2, 1/\tau_{F1} = 1000, 1/\tau_{F2} = 2000$; (c) $K_1 = 0.2, 1/\tau_{F1} = 1000, 1/\tau_{F2} = 10000$; (d) $K_2 = 3, 1/\tau_{F1} = 1/\tau_{F2} = 5000$; (e) $K_2 = 0.3, 1/\tau_{F1} = 1/\tau_{F2} = 5000$; (f) $1/\tau_{F1} = 1/\tau_{F2} = 2000$.

a wide range of input voltage when K_1 and K_2 are small. Comparing the stability boundaries in Fig. 6, we know that the stable input voltage range for Configuration 2 is smaller than that for Configuration 1.

IV. CONCLUSION

In the paper, we study the effects of feedback parameters in paralleled buck converters under a master-slave current sharing control. In particular we consider the effects of the inclusion of a voltage feedback loop in the control of the slave in addition to the current sharing loop. In general, we find that the system's stability range is wider with a voltage loop in the slave. Therefore, we may conclude that the voltage feedback in the slave converter is useful in enlarging the stable operation range. However, the role of this voltage loop is not to control the output voltage directly, but to provide a better regulated current reference for the slave. In brief, the slave gets a more stable current reference under a wider parameter range. Furthermore, we note that the slave converter is under current-mode control with the current sharing feedback, behaving effectively as a current source which should be expected in paralleled converters.

ACKNOWLEDGMENT

This work was supported by Hong Kong Research Grants Council under a CERG project (Ref. PolyU 5237/04E).

REFERENCES

[1] V. J. Thottuvelil and G. C. Verghese, "Analysis and control of paralleled dc/dc converters with current sharing," *IEEE Trans. Power Electron.,* vol. 13, no. 4, pp. 635–644, July 1998.

[2] J. Rajagopalan, K. Xing, Y. Guo, F. C. Lee, and B. Manners, "Configurationing and dynamic analysis of paralleled DC/DC converters with master-slave current sharing control," *Proc. IEEE APEC'96,* pp. 678–684, Feb 1996.

[3] Y. Panov, J. Rajagopalan, and F. C. Lee, "Analysis and design of N paralleled DC-DC converters with master-slave current-sharing control," *Proc. IEEE APEC'97,* pp. 436–442, Feb 1997.

[4] K. Siri, C. Q. Lee, and T. F. Wu, "Current distribution control for parallel connected converters: Part I and Part II," *IEEE Trans. Aerospace Electron. Syst.,* vol. 28, no. 3, pp. 829–851, July 1992.

[5] S. Luo, Z. Ye, R.-L. Lin, and F. C. Lee, "A classification and evaluation of paralleling methods for power supply modules," *Proc. IEEE PESC'99,* pp. 901–908, June 1999.

[6] J. Liu, W. Xu, Y. Qiu, and J.-H. Park, "A comparative elaluation of current-sharing methods for paralleled power modules," *2001 VPEC Seminar Record,* pp. 361–366, 2001.

[7] J. Sun, Y. Qiu, B. Lu, M. Xu, F. C. Lee, and W. C. Tipton, "Dynamic performance analysis of outer-loop current sharing control for paralleled DC-DC converters," *Proc. IEEE APEC'05,* pp. 1346–1352, Feb 2005.

[8] Y. Panov and M. M. Jovanovic, "Stability and dynamic performance of current-sharing control for paralleled voltage regulator modules," *IEEE Trans. Power Electron.,* vol. 17, no. 2, pp. 172–179, March 2002.

[9] J. Rajagopalan, K. Xing, Y. Guo, F. C. Lee, and B. Manners, "Configurationing and dynamic analysis of paralleled DC/DC converters with master-slave current sharing control," *Proc. IEEE APEC'96,* pp. 678–684, Feb 1996.

[10] H. H. C. Iu and C. K. Tse, "Bifurcation behavior of parallel-connected buck converters," *IEEE Trans. Circ. Syst. I,* vol. 48, no. 2, pp. 233–240, Feb 2001.

[11] Y. Huang and C. K. Tse, "On the basins of attraction of parallel connected buck switching converters," *Proc. IEEE ISCAS'06,* May 2006.

Improved Control for Parallel Inverter with Current-Sharing Control Scheme

Zhao Qinglin, Chen Zhongying and Wu Weiyang

College of Electrical Engineering, Yanshan University, Qinhuangdao, Hebei, China

Abstract—A novel current-sharing control scheme is presented for redundant parallel inverters. The control scheme employs the share bus which interconnects all the parallel inverters to transfer synchronous signal and average current signal. Current-sharing of all the parallel modules is realized through distributed average circuit. Average current is subtracted from the output current of each parallel inverter module and the subtraction result provides the instantaneous current deviation. The average current deviation of half line cycle is used to revise the voltage reference of next line cycle, which can achieve current-sharing and improve overall reliability. The instantaneous current deviation of switching period is fed back to the current-loop to improve the system dynamic response. Theoretical analysis and experimental results indicate that the control scheme performs a precise current sharing in dynamic state and steady state.

Keywords-parallel inverter; current-sharing; average current deviation; instantaneous current deviation

I. INTRODUCTION

SPWM inverter has been widely used for critical loads such as computer systems, instrumentation plants, communication systems and hospital equipment to supply constant voltage and constant frequency power [1]. To improve the power capacity and stability, module power supply has been one of the development trends of SPWM inverter. Parallel operation of inverters is a promising way to expand power capacity and achieve N+1 redundancy in power electronics. Each parallel module takes its share of load, so the current stress on power switch is reduced greatly [2]. Higher reliability, larger power capacity, lower cost and the shorter development cycle can be achieved and system configuration becomes flexible by using parallel inverter modules. However, the circulation current between parallel modules results in damage of power semiconductors in the parallel inverter. In order to reduce the effect of circulation current, the output voltage of all paralleled inverters must be strictly consistent in frequency, phase and amplitude to guarantee the output power sharing.

In this paper, a novel current-sharing control scheme is introduced into redundant parallel inverters. The proposed control scheme employs the current-sharing bus, current-sharing controller, the instantaneous current controller and the instantaneous voltage controller. The current-sharing extracts the output current of each inverter module and the subtraction provides the current deviation which reflects circulation current. The average current deviation of half line cycle is used to revise the voltage reference of next line cycle to realize current-sharing control of parallel inverters. The average current deviation of every line cycle is used to compensate AD offset to restrain the circulation current caused by the DC offset of output voltage. The instantaneous current deviation of switching period is fed back to the current-loop to improve the dynamic current-sharing.

II. SYSTEM STRUCTURE

The structure of parallel inverter system is shown in Fig.1. As an example, this system only consists of two inverter units connected to a common load. In order to simplify analysis, it is assumed that $L_1 = L_2 = L$, $C_1 = C_2 = C$ and serial equivalent resistance is r. When output voltages of parallel inverters are different, circulation current will arise and flow between two inverters. Circulation current is defined as

$$\vec{I}_H = \frac{\vec{I}_{L1} - \vec{I}_{L2}}{2} = \frac{\vec{U}_{m1} - \vec{U}_{m2}}{2(r + j\omega L)} \tag{1}$$

Equation (1) indicates that even a small difference between two parallel inverters output voltage will lead to large circulation current. So the output voltage of the parallel inverters must be strictly consistent in frequency, phase and amplitude to guarantee the equally output power sharing. Therefore, a proper control scheme is required to share load equally.

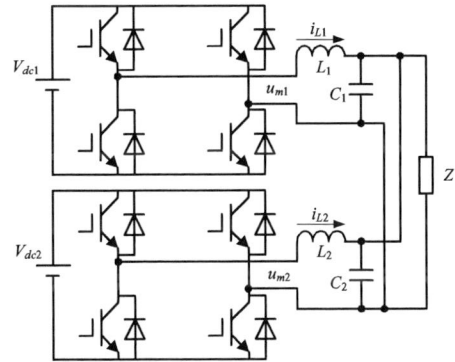

Figure1. The structure of the parallel inverter system

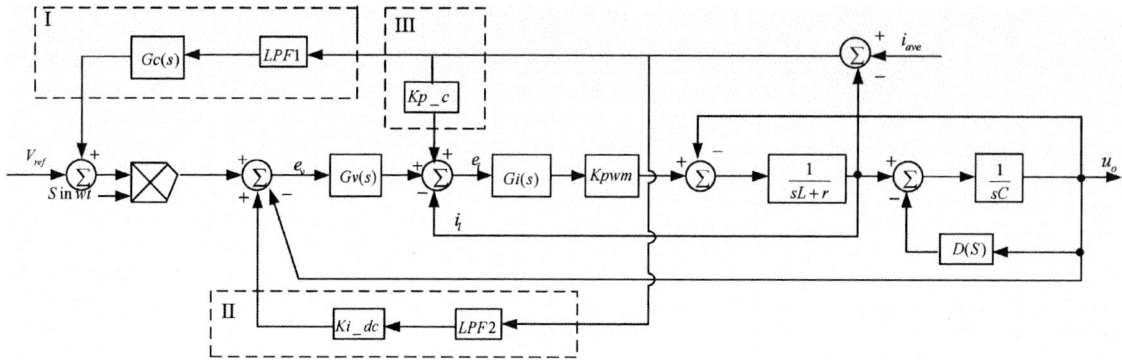

Figure2. Block diagram of the proposed control system

The reference voltage of each inverter module is synchronized by synchronization circuit. Distributed averaging circuit can calculate the precise current of the redundancy paralleled inverters. In this paper, an instantaneous voltage, instantaneous current and current-sharing controllers are proposed to implement control of the parallel inverter system. The block diagram of the proposed control system is shown in Fig.2.

III. PARALLEL INVERTER CONTROL SCHEME

A. Synchronization circuit

In this paper taking n-module inverter parallel system for example, the synchronization control principle is demonstrated in Fig.3. The reverse signal of the synchronization signal Syn_i of each inverter is transferred to the common synchronization Syn bus line. The fastest synchronization signal in the parallel operation mode is automatically tracked by other inverters as the synchronization signal. The principle of synchronization is shown in Fig.4. It can be expressed as follows:

$$Syn = \overline{(Syn_1) + (Syn_2) + \cdots\cdots + (Syn_n)} \quad (2)$$

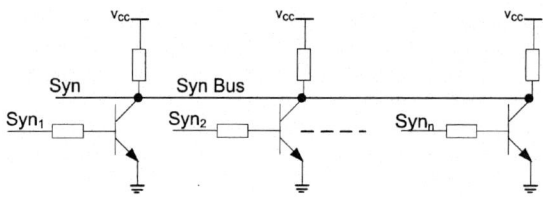

Figure3. Synchronization circuit diagram

In the parallel inverter system, each inverter adjusts its phase and frequency according to the synchronization signal Syn. If the inverter of the parallel system lags behind the synchronization signal less than one switch period, the phase and frequency won't be adjusted. Otherwise, its phase will be adjudsted to keep itself synchronized with the common synchronization signal Syn.

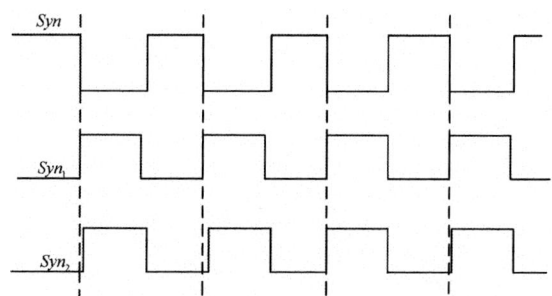

Figure4. The principle of synchronization signal

B. Current-sharing control

The current-sharing control of the parallel inverter system is based on the average current to realize instantaneous current sharing, the averaging circuit is shown in Fig.5. i_{li} is the sensed inductor current and i_{ave} means the average current every inverter shared. Suppose an n-module in parallel, average current i_{ave} is expressed as:

$$\left(i_{l1} - i_{ave}\right)/R + \left(i_{l2} - i_{ave}\right)/R + \cdots\cdots + \left(i_{ln} - i_{ave}\right)/R = 0 \quad (3)$$

then

$$i_{ave} = \frac{i_{l1} + i_{l2} + \dots i_{ln}}{n} \quad (4)$$

The averaging circuit unit of each module gets the sampling current of all the paralleled inverters and the output current i_{ave} is the average current among modules in the parallel system. Compared with its sensed inductor current, the average current deviation is used to compensate the voltage reference and AD offset. The instantaneous current deviation is used to adjust the inductor current to be as equal as possible to the average current i_{ave}.

1518

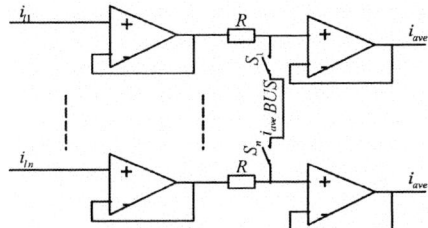

Figure5. Averaging circuit diagram

C. Analysis of circulation current

1. Feedback coefficient effect on circulation current

Synchronization signal, the voltage reference and transfer function of all paralleled inverter are realized by DSP which serves as a core controller. As premise of synchronization theoretically the parameters of the paralleled inverters are identical except for the sampling coefficient f_v. Take the parallel system which consists of two inverters for example. The control block diagram is shown in Fig.6, where $G_v(s)$ is the transfer function of the voltage regulator, $G(s)$ is the transfer function of the inner current loop and f_{vi} is the corresponding voltage feedback coefficient, $i = 1, 2$.

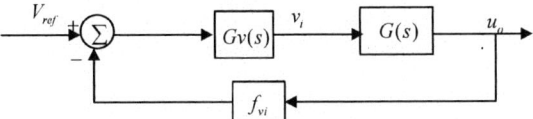

Figure 6. Control diagram of single inverter

According to Fig.6, the output voltage u_{oi} in stand-alone operation mode is obtained:

$$u_{o1} = \frac{V_{ref} G_v(s) G(s)}{1 + G_v(s) G(s) f_{v1}} \qquad (5)$$

$$u_{o2} = \frac{V_{ref} G_v(s) G(s)}{1 + G_v(s) G(s) f_{v2}} \qquad (6)$$

From (5) and (6), the difference of output voltage against voltage feedback coefficient in stand-alone operation mode is shown as follow:

$$|u_{o1} - u_{o2}| = \left| \frac{V_{ref} G_v^2(s) G^2(s)(f_{v2} - f_{v1})}{(1 + G_v(s) G(s) f_{v1})(1 + G_v(s) G(s) f_{v2})} \right| \qquad (7)$$

According to the simplified parallel operation mode shown in Fig.7:

$$U_O = [1 - r_1 /(r_1 + R_L // r_2)]U_{m1} + [1 - r_2 /(r_2 + R_L // r_1)]U_{m2} \qquad (8)$$

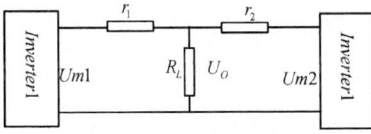

Figure7. Simplified parallel system block diagram

As premise of $r_1 = r_2 = r$ and $r \ll R_L$:

$$U_O = [1 - r /(r + R_L // r)](U_{m1} + U_{m2}) \approx (U_{m1} + U_{m2})/2 \qquad (9)$$

In parallel operation mode, u'_{oi} is the corresponding voltage of the paralleled inverters, $i = 1, 2$. According to Fig.6, (9) can be re-written as:

$$U_O \approx (U_{m1} + U_{m2})/2 = (u'_{o1} + u'_{o2})/2 \qquad (10)$$

$$u'_{o1} = G_v(s) G(s)(V_{ref} - U_O f_{v1}) \qquad (11)$$

$$u'_{o2} = G_v(s) G(s)(V_{ref} - U_O f_{v2}) \qquad (12)$$

According to (10), (11) and (12)

$$u'_{o1} = \frac{V_{ref} G_v(s) G(s)(2 + G_v(s) G(s)(f_{v1} - f_{v2}))}{2 + G_v(s) G(s)(f_{v1} + f_{v2})} \qquad (13)$$

$$u'_{o2} = \frac{V_{ref} G_v(s) G(s)(2 + G_v(s) G(s)(f_{v2} - f_{v1}))}{2 + G_v(s) G(s)(f_{v1} + f_{v2})} \qquad (14)$$

From (13) and (14), the difference of output voltage against voltage feedback coefficient in parallel operation mode is shown as follow:

$$|u'_{o1} - u'_{o2}| = \left| \frac{V_{ref} G_v^2(s) G^2(s)(f_{v2} - f_{v1})}{1 + G_v(s) G(s)(f_{v1} + f_{v2})/2} \right| \qquad (15)$$

Comparing (4) and (15), it is clear:

$$| u'_{o1} - u'_{o2} | > |u_{o1} - u_{o2}| \qquad (16)$$

The conclusion can be drawn that the difference of the voltage or current feedback coefficient will lead to larger difference of output voltage in parallel operation mode than in stand-alone operation mode. Furthermore the difference between the amplitude of sinusoidal reference voltage signal will enlarge the difference of output voltage in parallel operation mode[3], so the reference voltage should be revised in order to restrain the circulation current.

2. Amplitude difference effect on circulation current

In the case of synchronization control and no AD offset, there is only ac amplitude difference between their output voltages, as is shown in Fig.8. The output voltage of the parallel inverter is individually expressed as: $u_{oi} = V_i \sin \omega t$, where V_i is the amplitude of sinusoidal wave, u_{oi} is the output voltage and $i=1,2$.

$$\begin{cases} u_{o1} = V_1 \sin \omega t \\ u_{o2} = V_2 \sin \omega t \end{cases} \qquad (17)$$

According to (17) and Fig.8, the difference is:

$$\Delta u_o = u_{o1} - u_{o2} = (V_1 - V_2) \sin \omega t \qquad (18)$$

Because of the serial equivalent impedance is very small between parallel inverters, a little difference on amplitude of output voltage can result in large circulation current. It can be restrained by introducing the circulation current compensation into reference voltage generated by DSP.

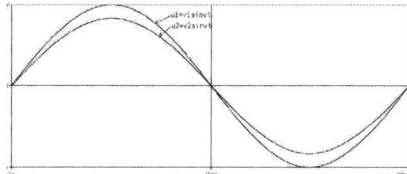

Figure8. Output voltages with different amplitude

3. DC offset effect on circulation current

It is hard to avoid DC offset of inverter output voltage without any compensation measures. The output voltage waveforms with DC offset are shown in Fig.9. The output voltages can be expressed as:

$$\begin{cases} u_{o1} = V\sin\omega t + a_1 \\ u_{o2} = V\sin\omega t + a_2 \end{cases} \quad (19)$$

Where V is the amplitude, a_1 and a_2 are DC offset. According to (19) and Fig.9, the difference is:

$$\Delta u_o = u_{o1} - u_{o2} = a_1 - a_2 \quad (20)$$

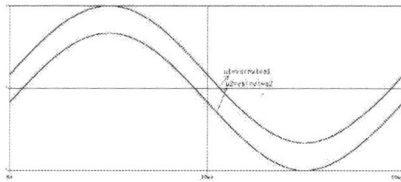

Figure9. Analysis of DC offset

Usually the DC impedance of output filter inductor can be neglected, the equivalent output resistor r against DC voltage offset is very small and the DC circulation current caused by DC offset may be infinite, so some measures must be implemented to reduce DC offset. It can be restrained by introducing the DC circulation current compensation into the error signal of voltage regulator.

D. System control strategies

Many control strategies of the parallel inverter have been proposed in recent researches, such as instantaneous current control and average current control [4][5]. While the instantaneous current control is simple, implemented easily and has good dynamic performance, it has vital drawbacks in the reactive condition when the load current lags behind reference voltage. When the corresponding instantaneous circulation current is introduced into the error signal, it means that the reference voltage phase is changed, and thus the synchronization of parallel inverters may be failed so that the parallel system can't operate. As for average current control, it depends on the average circulation current to revise the reference voltage amplitude for current sharing. Although high stability and

good performance in current sharing, it has disadvantage such as bad dynamic performance.

A novel current-sharing multi-loop control scheme is proposed in the paper. The reference signals of the parallel inverters are in phase with each other via synchronous buses, which enable their output voltages are in phase likewise. The AC steady-state circulation current can be suppressed by regulating reference voltage amplitude according to average circulation current in half line cycle. The instantaneous circulation current is introduced to current loop to improve dynamic circulation current rejecting capability. The DC circulation current caused by output voltage DC offset is suppressed by introducing DC offset compensation of reference signal according to circulation current DC offset.

Fig.2 shows the block diagram of the parallel inverter system. Area I means that the current deviation of half line cycle passing *LPF1* and $G_c(s)$ is used to revise the amplitude and reduce the influence of the feedback coefficient. Area II means that the current deviation of every line cycle passing *LPF2* and a proportional controller is used to revise AD offset. Thus the key factor which causes circulation current is solved and the circulation current is reduced clearly. Area III means the instantaneous current deviation of switching period is added to current-loop for improving the dynamic response.

IV. EXPERIMENTAL RESULTS

In order to verify the proposed control scheme, a parallel system with two inverter units has been built. The control unit is implemented based on TMS320LF2407 and EPM7128.The parameters of the unit listed as Table 1.

Table 1 THE PARAMETERS OF INVERTER UNIT

Utility voltage	220V rms, 50Hz
DC bus voltage	400V
Switching frequency	20KHz
Filter inductor	1.4mH
Filter capacitor	8.4uF
Deadbeat time	1.4us

Fig.10 shows the experimental waveforms of the two-module parallel inverter system with the control strategy of adjusting the reference voltage amplitude by the average current deviation of half line cycle only, where i_{L1} and i_{L2} are sampling inductor current respectively, with the sampling coefficient of 5A/div. It is obvious that circulation current i_H keep negative all the time and DC offset is the key factor. In order to reduce the influence of DC offset, the controller of compensating AD offset should be required.

Fig.11shows the experimental waveforms of the parallel operation mode with the proposed control strategy of both adjusting the reference voltage and compensating AD offset by the average current deviation every line cycle, where i_{L1} and i_{L2} have the same definition as in

Fig.9. In view of introducing the controller of compensating AD offset, the circulation current is an AC offset and it is restrained availably by the filter inductor. Compared Fig.11 with Fig.10, the circulation current i_H in Fig.11 is smaller than in Fig.9, which means that the proposed control strategies can restraint circulation current successfully and overcome the influence of the DC offset in parallel inverters system in some sense.

The experimental results of the parallel inverter system under a resistive step load are shown in Fig.12, where CH1 is the sampling output voltage with the sampling coefficient of 500V/div, CH2 is the sampling output current with the sampling coefficient of 10A/div, CH3 and CH4 are the sampling inductor current with the sampling coefficient of 5A/div, CH5 is the circulation between the parallel inverters. The results prove that the circulation current induced by DC offset is reduced largely and the dynamic response of the parallel system is improved.

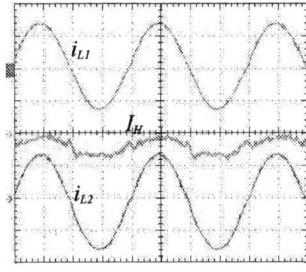

Figure10. The waveforms of the parallel system with adjusting the reference voltage

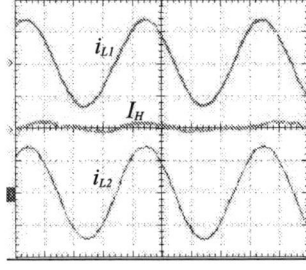

Figure11. The waveforms of the parallel system with the proposed control scheme

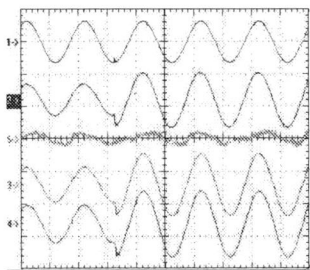

Figure12. Experimental waveforms of output voltage and current under load variation

V. CONCLUSION

In this paper, a novel current-sharing control scheme is proposed. A detailed theoretical analysis has been performed to enable the proposed the parallel inverter system to achieve high performance in both dynamic response and current sharing. Especially the proposed analysis shows that DC offset greatly enlarges circulation current and how to reduce DC offset. Experiment results validate that the compensation of AD offset availably restraint the circulation current by DC offset. Meanwhile, revising the reference voltage every line cycle reduces the circulation current in view of the different sampling feedback coefficient in the parallel system. The instantaneous current deviation of switching period is added to current-loop for improving the dynamic response.

ACKNOWLEDGMENT

This work was supported by the National Natural Science Foundation of China, NO. 50237020.

REFERENCES

[1] Jiann-Fuh Chen, Ching-Lung Chu, "Combination voltage-controlled and current-controlled PWM inverters for UPS parallel operation," *IEEE Trans.Pow.Electron,* vol.10, pp. 547-558,Sept.1995

[2] Lee, C.S., Kim, C.B.,Hong, S.C.; Yoo, J.S., Kim, S.W., Kim, C.H., Woo, S.H., Sun, S.Y., " A novel instantaneous current sharing control for parallel connected UPS," *Telecommunication Energy Conference,1998.INTELEC.Twentieth* pp. 513-519, Oct.1998.

[3] Zhongyi He, Yan Xing, Yuwen Hu, "Low cost compound current sharing control for inverters in parallel operation," *Power Electronics Specialists Conference 2004,* vol. 1, pp. 20 –25,June 2004.

[4] Xiao Sun,Yim-Shu Lee, Dehong Xu, "Modeling analysis and implementation of parallel multi-inverter systems with instantaneous average-current-sharing scheme," *Power Electronics, IEEE Transactionl,* vol. 18, pp. 844– 856, May 2003.

[5] Yan Xing, Lipei Huang, Sun. S,Yangguang Yan, "Novel control for redundant parallel UPS with instantaneous current sharing ," *Power Conversion Conference,2002 PCC Osaka,* vol. 3, pp. 2-5, April 2002.

2006 5th International Power Electronics and Motion Control Conference

A Novel Digital Controlled battery charger for High power UPS application

Fang. Luo   , Yong.Kang   , Shan.Xu.Duan   , Xueliang.Wei
College of electrical & electronic engineering
Huazhong University of Science and technology   Wuhan   Hubei   P.R.China
email: lfxa@263.net

Abstract-In this paper, a digital controlled battery charger based on microprocessor for high power UPS application is presented. A two-stage charging strategy along with the control method is analyzed. A new DC current measuring method using current transformer is put forward in this design, which can both reducing the cost of current sensor component and fulfilling the requirement of current detecting, simply adding a diode and a resister .

Keywords- Charger; current transformer; magnetic reset

I.   Introduction

High power Un-interrupt Power Supplies are widely used in telecommunication systems and computer systems which require stable power supplies. Constantly high power UPS system without output transformer needs a high DC bus voltage up to 600V~700V. Usually a group of series connected batteries are used as back up power. Charging management for those batteries has a strong infection on the reliabilities of UPS systems. Properly design of battery chargers can extend the battery life while reducing the charging time.

Charging strategies for high capability batteries always divided into following ways: constant voltage, constant voltage with charging current limitation, two-stage, multi-stage. The first two charging strategies involve an individual charging voltage control, which is usually a settled voltage equals to the charging termination voltage. Because the charging current is not properly controlled while using the constant voltage charging strategy, and the battery internal impedance is very low, normally at the number of milliohm even to micro-ohm, the charging current will be very high especially at the beginning of charging operation. Constant voltage with charging current limitation can limit the peak value of charging current, but the current is varying in a wide range along with the process. Those two charging strategies may cause over charge which does harms to battery packs.

According to the characteristics of the batteries, two-stage and multi-stage charger are considered to give

most benefits to the batteries, charging with less time while producing more smooth charging current. In practical industrial application, two-stage charging method is considered to be more simple while brings the same charging effect as multi-stage method.

Approaching to the two-stage charging contains output current control and output voltage control. At the initial state of charging, the batteries are controlled in constant current (CC) charging mode. Output current of charge was extremely controlled at 0.1C(C is defined as a basic unit of charging current according to the capability of batteries, for example, 100Ah battery means C is 100A).

While constant current charging, the terminal voltage of batteries is gradually increasing. When it's reaching the trickle voltage, the charger switched from constant current control to constant voltage (CV) control. During constant voltage charging operation, the charging current is not under strict control and the terminal voltage of the battery pack is clamped to the trickle voltage. The charging current will drop to nearly zero as long as the CV charging operation behaves. Charging current appears at times in a low magnitude which can make up the battery energy loss in backup state. The relationship between charger output current and voltage is shown is fig.1.

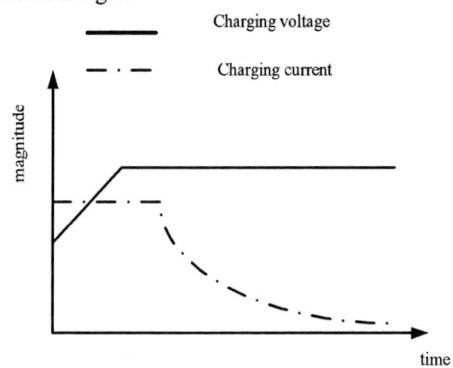

Fig.1 two-stage charging

II. Description of the charger design

A.   System Overview

The charger is designed for a 32 cell series battery

1-4244-0448-7/06/$25.00 ©2006 IEEE   1522

pack, output voltage ranging from 360Vdc to 432Vdc, while Input voltage is DC voltage from 3 phase diode rectifier, ranging from 400V to 640V, with programmable output current control, and the maxim output current is up to 10Adc. A series boost-buck structure is provided for this high power charger as shown in fig.2 .

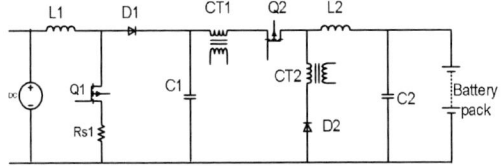

Fig.2 main circuit of the battery charger

The boost part gives a voltage higher than 615V as an input for buck circuit. It works whenever the voltage of capacitor C is lower than 615V. This part ensures the buck convertor working under a proper pulse wide not too large. This is important for the buck control using current transformers. Here a peak-current control pulse wide modulator chip UC3843 is used to control the output of boost convertor. Microprocessor simply gives on/off signal to this part.

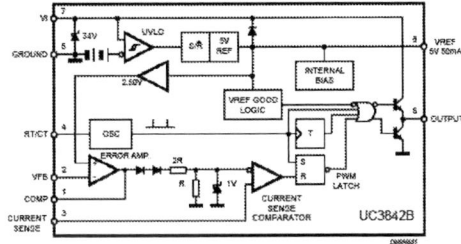

Fig.3 block diagram of UC3843

All charger functions were implemented in proper control of buck convertor. To meet the goal of different control aim in two stage of charging process, feedback control is essentially needed. An output voltage control loop and a charging current loop is parallel connected, switching under the control of battery pack voltage. When battery pack voltage is lower than trickle voltage, the charger is controlled in constant current mode, otherwise, the charger is in constant voltage mode. UC3843 is also used in buck convertor, giving the function of peak inductor current limitation.

B. Inductor Current detecting using current transformer on buck circuit

In considerations of system cost, sensing components play an important part of whole system. Hall sensors can give an isolated solution for voltage and current measuring with a fast response in the cost of money; Sample resistances can involve perfect linearity of measurement, but can not provide isolated measuring;

current transformers can not only bring the benefit of good linearity and fast transit response, but also give a low cost isolated current sensing solution. But transformers need symmetrical bipolar primary current without DC current offset in order to keep magnetic flux balance, which limits the usage of current transformer in single-polarity current applications such as boost and buck convertor. Generally current transformer is considered not to be proper for DC detecting or the magnetic core will be saturated.

According to the characters of magnetic component, there are two ways used in magnetic resetting: self-reset and forced reset. The only goal of reset is to make positive and negative volt-second of transformer wining to be equal as shown in following formula:

$$V_{on} \times T_{on} = V_{off} \times T_{off} \qquad (1)$$

Self-reset can generate reset voltage depending on the exciting current. When the measured current disappears, the exciting current in the transformer winding will continue according to the character of inductors. Thus the exciting current will go through the sample resister and produce a inverse voltage over the secondary winding. In this way, the magnetic core of transformer can be reset. The value of sample resistor and reset time are so small that can not generate enough reset volt-second. This will cause magnetic core saturation. This solution usually used when pulse wide D is far less than 0.5.

(a) current measuring circuit using current transformer

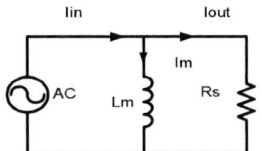

(b) equalized circuit of current transformer when current appears in primary winding

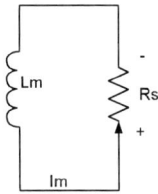

(c)equalized circuit of magnetic core self reset without forced reset circuit

Fig.4 equalized circuits of current transformer

Forced reset needs an inverse polarity reset voltage on the primary winding during the eliminate time of the measured current,   producing an inverse reset exciting current for the magnetic core. The value of the reset voltage is related to the impedance. Forced reset required an extra voltage generator working synchronously with the measured current.   The equalized circuit of forced reset is shown in following figure.

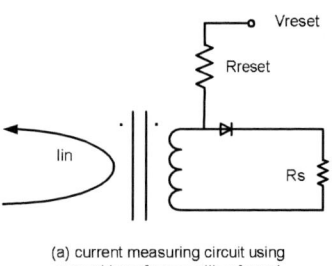

(a) current measuring circuit using current transformer with a forced reset circuit

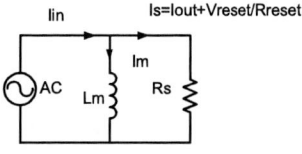

(b) equalized circuit of current transformer when current appears in primary winding

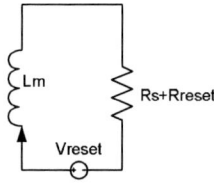

(c)equalized circuit of magnetic core reset with forced reset circuit

Fig.5 equalized circuits of forced reset current transformer

In order to obtain the benefit of transformer in the single polarity application, we brought forward an advanced self-reset circuit by adding a reset resistor and a diode shown as below.

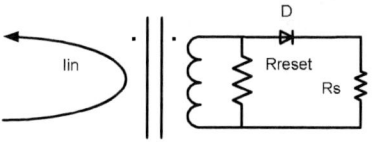

(a) current measuring circuit using current transformer with an advanced self reset circuit

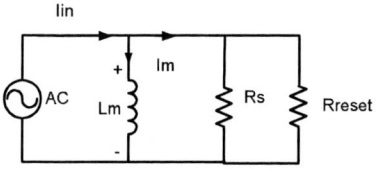

(b) equalized circuit of current transformer when current appears in primary winding

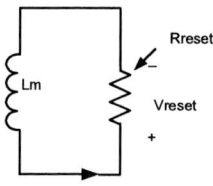

(c)equalized circuit of magnetic core reset with advanced self reset circuit

Fig.6 equalized circuits of current transformer forced reset

As show above, when the current in primary winding rises, the diode is turned on and a induced current in secondary winding and produce a sample voltage on the sample resistor R_s ; else when the current in primary winding disappears, exciting current I_m continues in its own direction, going through the reset resistor. Because the diode was shut down by the inverse current, no current pass the branch of sample resistor. Resistor used for reset is much larger than sample resistors in order to generate enough reset voltage. Equalized circuit changed form fig (b) to fig (c). In this way, the single polarity current I can be detected by the current transformer. Here $R_{reset} >> R_s$, thus the reset resistor won't cause much error on current measuring.

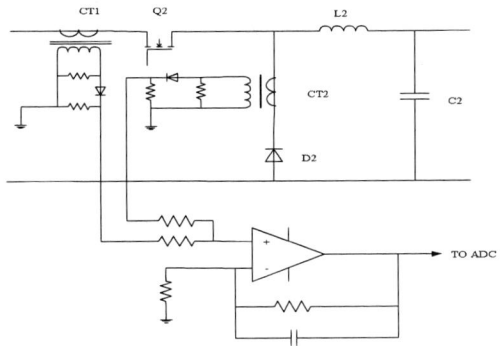

Fig.7 block scheme of buck control using transformer

Using two transformers we can construct the current waveform provide by buck circuit. As shown in fig.7, CT1 detects the turning on current of switch Q2, and CT2 detects the continued current going through D2, the two parts synthesized together equals the average current of the buck output current. Fig shows an analog adder made by error amplifier, this combined turned on current and continued current to form the average output current. The waveforms of the transformers and average current is shown in fig.8.(CH1 shows the IGBT gate driving pulse while CH2 shows voltage waveforms on reset resister, which is the same as voltage waveform on secondary winding of the transformer)

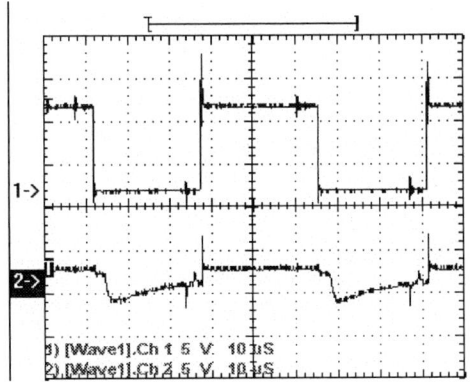

Fig.8 reset voltage waveform at D=0.5

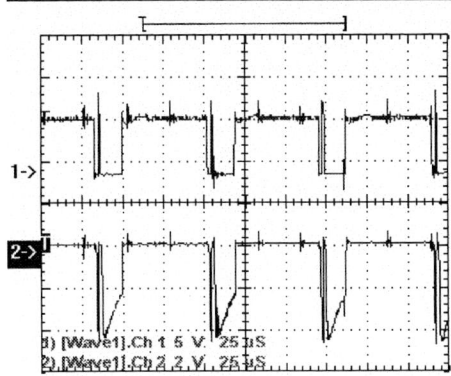

Fig.9 reset voltage waveform at D=0.8

Here for the purpose of ensure that the magnet core is reset while brings least measuring error, Rreset is properly 10~100 times higher than Rs.

C. *Control Strategies*

The whole control block diagram is shown as fig.10

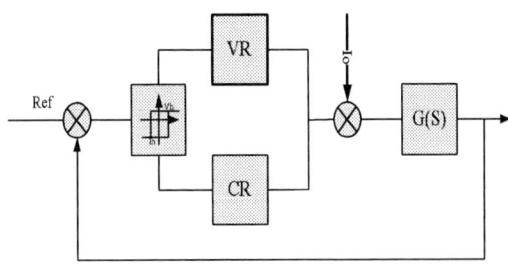

Fig.10 control diagram of the charger

As shown above, a voltage control loop and a current control loop is activated by the microprocessor. UC3843 used as an inner current loop to limits the peak value of the output current, performing the function of over loading protection. The switch point of the two control loop is controlled by microprocessor, depending on whether the battery pack voltage reaches the trickle voltage.

To avoid frequent switch between two control loops, we set a dead-band on the switch point. If the voltage is higher than V_h, controller begins CV control; else if the load current suddenly increases to more than I_h, controller turns to CC control.

The inner current loop works under high frequency up to 30kHz, and outer control loop calculates every 2ms, the stability is ensured by the inner current loop, so the UC3843 along with the inner loop and the switch can be modeled like following form.

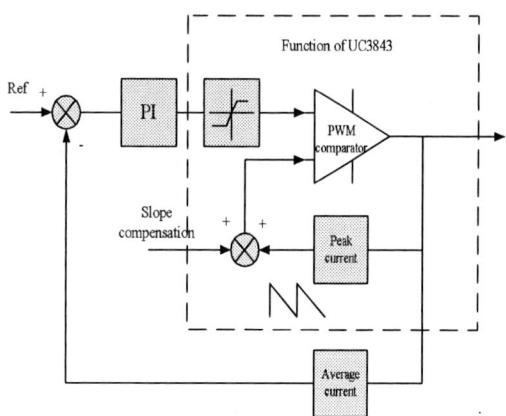

Fig.11 equalized diagram of UC3843

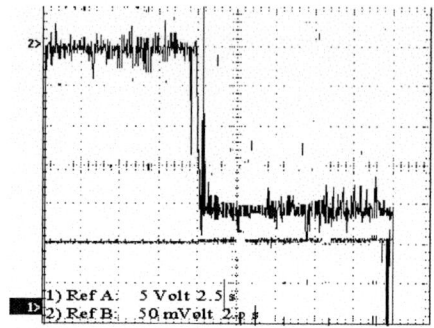

Fig.12 charging current at sudden load

Using the microcontroller up78f9116, we can implement a digital control using PI regulator. The discrete form of PI algorithm is shown like this.

$$R(k) = Kp \times E(k) + Ki \times \sum_{i=0}^{k} E(i) \quad (2)$$

For the conveniences of realizing on 8 bit micro-processor, usually we use increment instead of absolutely integration. From (2) we can get the expression of R(k-1):

$$\triangle R = (Kp + Ki) \times E(k) - Ki \times E(k-1)$$

$$R(k) = R(k-1) + \triangle R$$

$$R(k-1) = Kp \times E(k-1) + Ki \times \sum_{i=0}^{k-1} E(i) \quad (3)$$

(2)-(3):

$$\triangle R = (Kp + Ki) \times E(k) - Ki \times E(k-1) \quad (4)$$

$$R(k) = R(k-1) + \triangle R \quad (5)$$

Hence we can simply calculate the increment of every step, no need to calculate the whole accumulation of error.

III.Experimental Results and Conclusions

The design was realized and test as a full function charger. Using two current transformers (ratio: 1:200, secondary winding inductor value 384uH) to obtain the average output current. Scope waveforms are obtain from a external hall sensor .Experimental results are shown as below.

Tab. I   Test Result in Constant Current Mode

Given value	Output value
10A	9.9A
5A	4.8A
4A	3.9A
2A	2.0A

From the result we can get a conclusion that using transformers can precisely measure the single polarity current by simply adding a diode and a resister. It's a good choice for low cost products.

References

[1] Hongshan.Tao, Xiehua.Wu, "Application of Current Transformer In SMPS ", Power supply technologies and application, Vol.6 :pp29-32 Aug.2003

[2] Mohamad A. S. Masoum, Seyed Mahdi Mousavi Badejani, and Ewald F. Fuchs, "Microprocessor-Controlled New Class of Optimal Battery Chargers for Photovoltaic Applications", IEEE transactions on energy conversion , Vol.19, No.3, Sept.2004,pp.599-606

[3] W.J.Ho , J.B.Lio , W.S.Feng, "Economic UPS structure with phase-controlled battery charger and input-power-factor improvement", IEE Proc -Electr. Power Appl., Vol. 144, No.4, July 1997.pp.221-226

[4] Ma Xuejun, Kang Yong , Chen Jian, "Design of the Novel Charge and Discharge Circuit in Online UPS" , Power electronics, Vol.38 No.1, Feb.2004, pp.75-77

[5] Chen Jing-jin, Yu Ning-mei, "Study on the multi-stage constant current charging method for VRLA battery system", Chinese Journal of Power Sources, Vol.28,No.1,Jan 2004,pp.32-33

2006 5th International Power Electronics and Motion Control Conference

A Novel High Input Power Factor Single-Stage Single-Phase AC/AC Converter

Chien-Ming Wang[*], *member, IEEE,* Chien-Yeh Ho[**] and Maoh-Chin Jiag[*]

[*] National Ilan University /Department of Electrical Engineering, Taiwan, China
[**] Lunghwa University of Science and Technology /Department of Electronic Engineering, Taiwan, China

Abstract—This paper proposes novel high input power factor single-stage single-phase ac/ac converter to give high input power factor and low current distortion on the rectifier side and provide clean and stable ac voltage on the inverter side. The proposed ac/ac converter can not only provide regulation of the output ac voltage magnitude but provide variable output ac voltage frequency. A significant reduction in the conduction losses is achieved, since the rectifier in the proposed converter uses a single converter instead of the conventional configuration composed of a four-diode front-end rectifier followed by a boost converter. An average-current-mode control is employed in the rectifier side of proposed converter to detect the transition time and synthesize a suitable low harmonics sinusoidal waveform for the input current. The sinusoidal pulse-width modulation (SPWM) control strategy is employed in the inverter of proposed converter to achieve well dynamic regulation. A design example of 1000W soft-switching single-stage single-phase ac/ac converter is examined to assess the converter performance.

Keywords-ac/ac converter; SPWM; average-current-mode

I. INTRODUCTION

Matrix converters have received considerable attention [1]-[3] due to their potentiality to provide direct AC/AC conversion without energy storage. However, they turned into wide application due to severe requirements: four-quadrant switches, critical timing, sensing of switch voltage and current, snubber circuits needed to absorb overvoltages coming from the inductive commutation. As a result, circuit efficiency and reliability are affected. More popular is the indirect ac/dc/ac conversion by means of PWM rectifier-inverter systems with dc link. As compared to matrix converters, these systems show improved reliability and allow a greater output voltage. In these system, a big tank capacitor in the dc link provides decoupling between the rectifier and the inverter, so that the two converters can be driven independently according to usual PWM techniques [4]-[5], providing excellent input and output performances. In fact, this system is the combination of the boost rectifier and the buck inverter. The boost rectifier performs the functions of power factor correction and boost ac/dc conversion. The buck inverter performs the function of buck dc/ac conversion with output voltage of variable amplitude and frequency. Therefore, these ac/dc/ac systems have been widely used in industrial application such as uninterruptible power supplies (UPS), static frequency changes and variable speed drives. By similar

concept, a novel high input power factor single-stage single-phase ac/ac converter is proposed in this paper. The proposed ac/ac converter not only can provide regulation of the output ac voltage magnitude but provide variable output ac voltage frequency. A significant reduction in the conduction losses is achieved, since the rectifier in the proposed converter uses a single converter instead of the conventional configuration composed of a four-diode front-end rectifier followed by a boost converter. An average-current-mode control is employed in the rectifier side of proposed converter to detect the transition time and synthesize a suitable low harmonics sinusoidal pulse-width modulation (SPWM) control strategy is employed in the inverter of proposed converter to achieve well dynamic regulation. System analysis for predicting and evaluating the ac/ac converter performance are conducted.

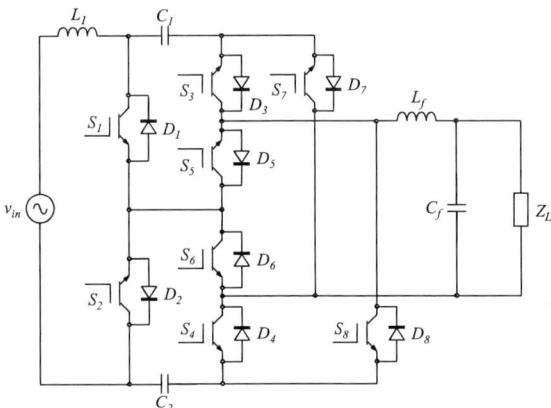

Fig. 1 Proposed high input power factor single-stage single-phase ac/ac converter.

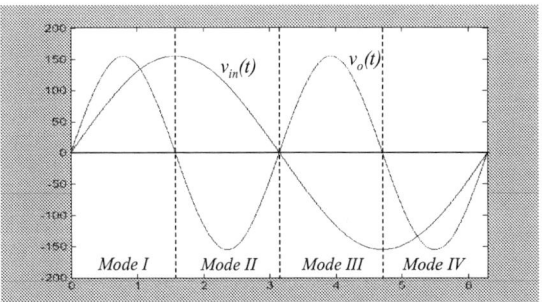

Fig. 2 The definition of circuit operation mode

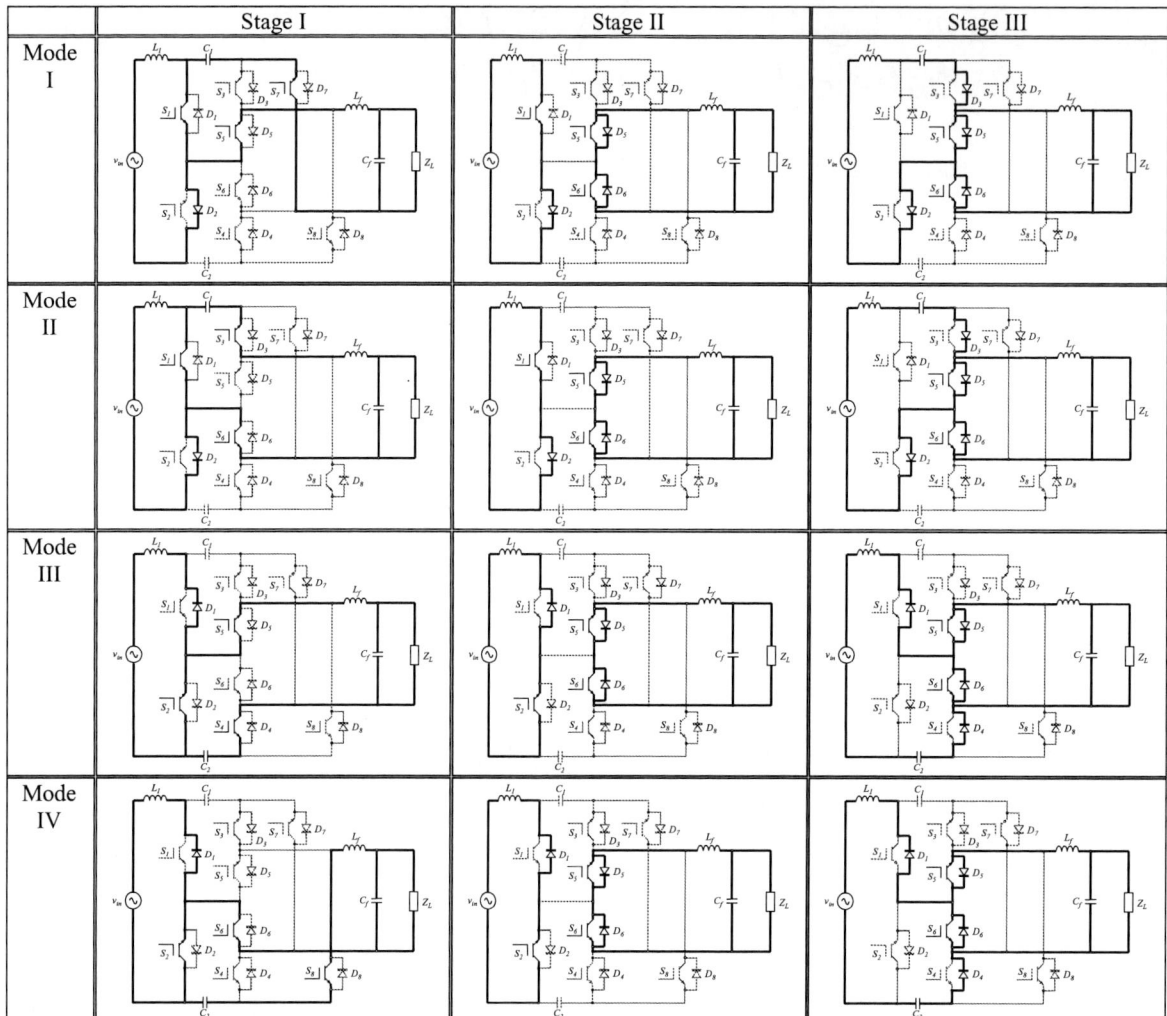

Fig. 3 The topology stages of the proposed high input power factor single-stage single-phase ac/ac

II. OPERATION PRINCIPLE

A. Main Circuit and Operation

The power stage diagram of the novel high input power factor single-stage single-phase ac/ac converter is shown in Fig. 1. The circuit can be divided in two sections. The first section is a PWM buck inverter with unipolar voltage switching. It is composed by the switches S_3, S_4, S_5, S_6, S_7, S_8, D_3, D_4, D_5, D_6, D_7, D_8, and output filter L_f, C_f. This section performs the function of buck dc/ac conversion with output voltage of variable amplitude and frequency. The second section is the pulse-width-modulation continuous conduction mode step-up ac/dc converter, composed by L_1, S_1, S_2, D_1, D_2, D_3, D_4, D_5, D_6, C_1, and C_2. This section performs the functions of power factor correction and boost ac/dc conversion at fixed frequency. The switches operate on a PWM pattern to shape both the input current and output voltage to follow the reference commands. The inductor L_1 provides voltage boost operation, and L_f and C_f also provide filter

operation of the output voltage. The dc-link capacitor C_1 and C_2 acts as a dc voltage source and provides filtering operation. For satisfying the step up/down function in the proposed converter, the dc-link capacitor voltage must be sufficiently higher than the peak voltage of the ac main source. Because the frequencies of input voltage and the desired output voltage are not the same, we can divide the circuit operation into four modes during one line voltage period. The definition of circuit operation mode is shown in Fig. 2 and the dynamic equivalent circuits during one switching period for each mode are shown in Fig. 3. To simplify the analysis, it is assumed that the proposed single-stage single-phase ac/ac converter is operating in steady-state, all devices are ideal and the losses in L_1, L_f, C_1, C_2, and C_f are all neglected. The operational principle of the proposed converter can be described as follow.

Mode I: $v_{in}(t) > 0$ and $v_o(t) > 0$
Stage I: Before this stage, the switches S_1, S_2, S_3, S_4, S_7, and S_8 maintain turn-off state, the switches S_5 and S_6

1528

maintain turn-on state. The energy stored in inductor L_1 is delivered to capacitor C_1 through D_3, D_5, and D_2 while the output loop of the inverter is in a freewheeling state and the freewheeling loop is formed by S_5, S_6, D_5, D_6, and output filter loop. This stage begins when S_1 and S_7 turn on and S_6 turns off. The input inductor L_1 is charged from input voltage v_{in} through the switch S_1 and the body diode of switch S_2. The energy stored in capacitor C_1 supplies inverter stage through S_3 and S_7.

Stage II: During this stage, the input inductor L_1 is continuously charged from input voltage v_{in} through the switch S_1 and the body diode of switch S_2. The energy stored in capacitor C_1 does not supply inverter stage and the output loop of the inverter returns a freewheeling state.

Stage III: In this stage, the output loop of the inverter is still in a freewheeling state. The energy stored in inductor L_1 is delivered to the filter capacitor C_1 through D_3, D_5, and D_2.

Mode II: $v_{in}(t)>0$ and $v_o(t)<0$

Stage I: Before this stage, the switches S_1, S_2, S_3, S_4, S_7, and S_8 maintain turn-off state, the switches S_5 and S_6 maintain turn-on state. The energy stored in inductor L_1 is delivered to capacitor C_1 through D_3, D_5, and D_2 while the output loop of the inverter is in a freewheeling state and the freewheeling loop is formed by S_5, S_6, D_5, D_6, and output filter loop. This stage begins when S_1 and S_3 turn on and S_5 turns off. The input inductor L_1 is charged from input voltage v_{in} through the switch S_1 and the body diode of switch S_2. The energy stored in capacitor C_1 supplies inverter stage through S_3 and S_6.

Stage II: During this stage, the input inductor L_1 is continuously charged from input voltage v_{in} through the switch S_1 and the body diode of switch S_2. The energy stored in capacitor C_1 does not supply inverter stage and the output loop of the inverter returns a freewheeling state.

Stage III: In this stage, the output loop of the inverter is still in a freewheeling state. The energy stored in inductor L_1 is delivered to the filter capacitor C_1 through D_3, D_5, and D_2.

Mode III: $v_{in}(t)<0$ and $v_o(t)>0$

Stage I: Before this stage, the switches S_1, S_2, S_3, S_4, S_7, and S_8 maintain turn-off state, the switches S_5 and S_6 maintain turn-on state. The energy stored in inductor L_1 is delivered to capacitor C_2 through D_4, D_6, and D_1 while the output loop of the inverter is in a freewheeling state and the freewheeling loop is formed by S_5, S_6, D_5, D_6, and output filter loop. This stage begins when S_2 and S_4 turn on and S_6 turns off. The input inductor L_1 is charged from input voltage v_{in} through the switch S_2 and the body diode of switch S_1. The energy stored in capacitor C_2 supplies inverter stage through S_4 and S_5.

Stage II: During this stage, the input inductor L_1 is continuously charged from input voltage v_{in} through the switch S_2 and the body diode of switch S_1. The energy stored in capacitor C_2 does not supply inverter stage and the output loop of the inverter returns a freewheeling state.

Stage III: In this stage, the output loop of the inverter is still in a freewheeling state. The energy stored in inductor

L_1 is delivered to the filter capacitor C_2 through D_4, D_6, and D_1.

Mode IV: $v_{in}(t)<0$ and $v_o(t<0$

Stage I: Before this stage, the switches S_1, S_2, S_3, S_4, S_7, and S_8 maintain turn-off state, the switches S_5 and S_6 maintain turn-on state. The energy stored in inductor L_1 is delivered to capacitor C_2 through D_4, D_6, and D_1 while the output loop of the inverter is in a freewheeling state and the freewheeling loop is formed by S_5, S_6, D_5, D_6, and output filter loop. This stage begins when S_2 and S_8 turn on and S_5 turns off. The input inductor L_1 is charged from input voltage v_{in} through the switch S_2 and the body diode of switch S_1. The energy stored in capacitor C_2 supplies inverter stage through S_6 and S_8.

Stage II: During this stage, the input inductor L_1 is continuously charged from input voltage v_{in} through the switch S_2 and the body diode of switch S_1. The energy stored in capacitor C_2 does not supply inverter stage and the output loop of the inverter returns a freewheeling state.

Stage III: In this stage, the output loop of the inverter is still in a freewheeling state. The energy stored in inductor L_1 is delivered to the filter capacitor C_2 through D_4, D_6, and D_1.

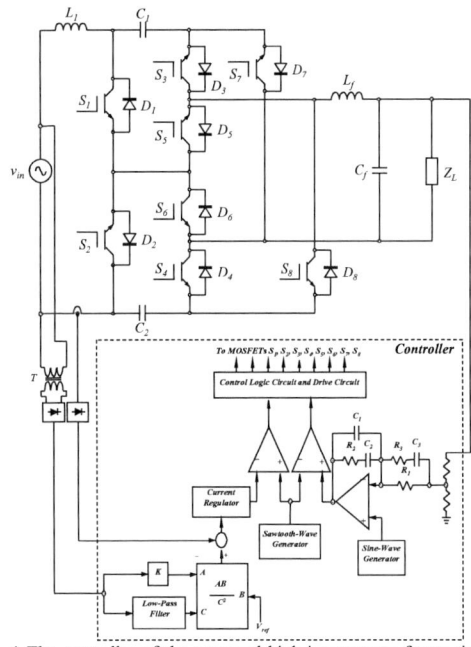

Fig. 4 The controller of the proposed high input power factor single-stage single-phase ac/ac converter.

B. Power Factor Correction

Traditionally, conversion of the ac line voltage from the utilities has been dominated by phase-controlled or diode rectifiers. The nonideal character of the input current drawn by these rectifiers creates a number of problems for the power distribution network and for other electrical systems in the vicinity of the rectifier, such as high input current harmonics, low input power factor, lower rectifier efficiency, ac source voltage distortion, and high reactive-component size. Therefore, the optimal rectifier would be one in which the input would draw a pure sinusoidal current at unity power factor. The topology usually employed in power factor correction single-phase power supplies in composed by a front end rectifier followed by a boost converter. In this topology, the boost converter in continuous conduction mode (CCM) with the average current control and pulse-width modulation (PWM) technique has been the most popular circuit [1]-[3]. But the significant conduction loss in the PFC circuit always includes two diode losses from the front-end bridge rectifier and one (or two) power switch loss, the conduction losses is larger. For improve this problem, the rectifier circuit has been revised. The presented rectifier circuit has only two power semiconductor conduction drops in the power flow path. Therefore, the conduction loss is considerably reduced.

C. Inverter

The PWM inverter is required to synthesize a sinusoidal waveform at its output port under different types of loads. Since the PWM inverter plays such an important role in converting a dc voltage to an ac voltage, the performance of an ac power conditioning system in highly dependent on the built-in controller of the PWM inverter. To minimize the harmonic distortion of the output waveform of a PWM inverter, many methods based on modulation strategies have been proposed. Any PWM technique can be used with this inverter. A general sinusoidal PWM technique is employed in this paper for simplicity. In this case, because D_{Ik} is directly proportional to the amplitude of output voltage in the kth switching period, it can be designed with ease according to the desired load voltage.

$$|V_{ok}| = D_{Ik} \bullet V_C \tag{1}$$

Therefore, the magnitude of the output voltage can be simply obtained using (1). Because the frequencies of ac line voltage and desired output voltage are not the same, the frequency of the desired output voltage can be decided by the reference signal of controller.

D. Control Strategy

The controller of the proposed high input power factor single-stage single-phase ac/ac converter is shown in Fig. 4. In this controller, the average current mode is used as the control reference in the boost power factor pre-regulator. The boost power factor pre-regulator is designed to operate in continuous-conduction mode (CCM). This average current controller can prescribe the shape and the frequency of the input current due to its inherently synchronous feedback loop. In order to obtain almost unity power factor, the synchronous signal is sensed from a rectified sinusoidal waveform. Thus, the signal of bridge rectifier is necessary to obtain the desired synchronous signal and the *rms* input voltage for the control IC. Hall Effect sensor for detecting the input current is installed for the average current mode control. The reference current is then generated by a multiplier/divider combination of the synchronous feedback loop and input voltage feed-forward loop. In the

buck inverter, a sinusoidal PWM (SPWM) technique is used to regulate the system dynamics. The feedback circuit includes sinusoidal generator, error amplifier and compensator network. The sawtooth-wave generator is common for eliminating noise interference each other.

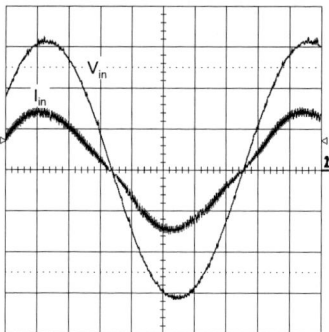

Fig. 5 Input voltage (V_{in}) and input current (I_{in}). V_{in}: 100V/div; I_{in}:5A/div, time:2ms.

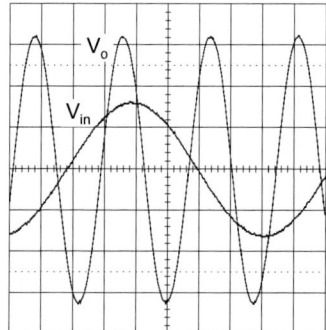

Fig. 6 Input and output voltage: V_{in}:100V/div, V_o:100V/div, time:2ms/div.

III. EXPERIMENTAL RESULTS

An example of a 1kW single-stage single-phase ac/ac converter is designed and realized. The implemented power stage circuit is shown in Fig. 1. The boost inductance value L_1 and the filter capacitances C_1, and C_2 to minimize the ripple voltage of voltage V_C are calculated as $L_1=650\,\mu H$, $C_1=C_2=940\,\mu F$. The output filter inductor L_f and capacitor C_f to minimize the undesired harmonics of the output ac voltage are specified as $L_f=1mH$, $C_f=4.7\,\mu F$. In hardware realization, we use IGBT's IRG4PC50UD as the power switches and diodes. The waveforms of the input voltage and current of the proposed high input power factor single-stage single-phase ac/ac converter at the related 1000W are shown in Fig. 5, in which the waveforms of the input voltage and current are almost in phase and the measured power factor is over 0.99. The input and output voltage is also measured in Fig. 6. In order to assess the dynamic performance of the presented inverter, three kinds of loads (resistive R, inductive RL, and rectifier with RLC) are examined in Fig. 7, in which the measured total harmonic distortions (THDs) for the mentioned loads are given as 1.91% for the R load, 2.63% for the RL load, and 5.25% for the rectifier with the RLC load.

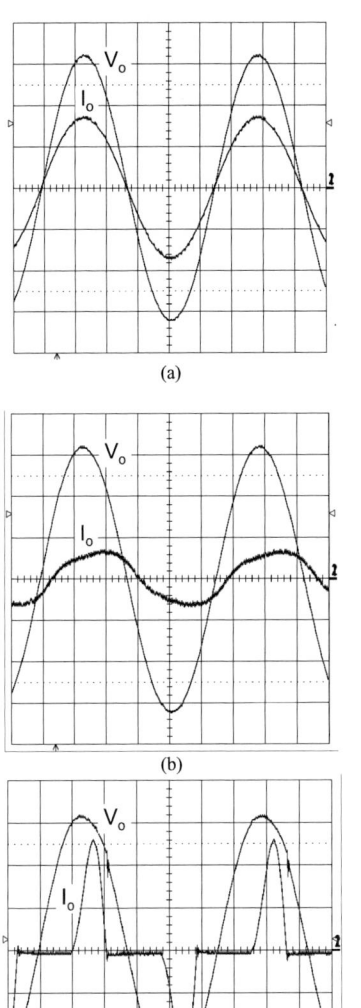

(a)

(b)

(c)

Fig. 7 Experimental results of $v_o(t)$ and $i_o(t)$ and their spectra with (a) resistive load (R_L=48Ω), (b) inductive load (R_L=`100Ω, L_L=0.26H) and (c) rectifier with RLC load (R_L=48Ω, L_L=1mH, and C_L=1740μF), where f_o=180Hz. (V_o: 100V/div; I_o:5A/div, time:1ms/div)

IV. CONCLUSION

A novel single-stage single-phase ac/ac converter with high input power factor and clean ac output voltage is presented. Its configuration is simple and compact. Thus, the proposed ac/ac converter is applicable in UPS and ac source design. An average-current-mode control is employed in the rectifier side of proposed converter to detect the transition time and synthesize a suitable low harmonics sinusoidal waveform for the input current. The sinusoidal pulse-width modulation (SPWM) control strategy is employed in the inverter of proposed converter to achieve well dynamic regulation. The circuit operation has been described and discussed. The design procedure and example of the novel high input power factor single-stage single-phase ac/ac converter is described. Some experiment results prove the truth of the theoretical prediction.

ACKNOWLEDGMENT

The authors gratefully acknowledge the National Science Council Sponsored the work, Project no. NSC94-2213 - E-197-005.

REFERENCES

[1] M. Venturini, "A new sine wave in, sine wave out, conversion technique eliminates reactive elements," in *Powercon* 7, San Diego, CA, 1980, pp. E3-El5.

[2] J. Oyama and T. Higuchi *er al.*, "Novel control strategy for matrix converter," pp. 360-367.

[3] P. Tenti, L. Malesani, and L. Rossetto, "Optimum control of N-input K-output matrix converters," in *IEEE Trans. Power Electron.* vol. 7, no. 4, pp. 707-713, Oct. 1992.

[4] D. Divan, T. Habetler, and T. Lipo, "PWM techniques for voltage source inverters," in *IEEE PESC Conf. Rec.* 1990, Tutorial Notes.

[5] R. Wu, S. Dewan, and G. Slemon, "Analysis of an ac to dc voltage source converter using PWM with phase and amplitude control," in *IEEE Ind. Applicat. Soc. Annu. Meefing,* San Diego, CA, Oct. 1989, pp. 1156-1163.

[6] H. W. Park, S. J. Park, J. G. Park, and C. U. Kim, "A novel high-performance voltage regulator for single-phase ac sources," *IEEE Trans. Ind. Electron.,* vol. 48. pp. 554-562, June 2001.

[7] B. H. Kwon, G.Y. Jeong, S.H. Han, and D. H. Lee, "Novel line conditioner with voltage up/down capability," *IEEE Trans. Ind. Electron.,* vol. 49. pp. 1110-1119, Oct. 2002.

[8] J. Y. Lee, Y. M. Chang, and F. U. Liu, "A new UPS topology employing a PFC boost rectifier cascaded high-frequency tri-port converter," *IEEE Trans. Ind. Electron.,* vol. 46. pp. 803-813, Aug. 1999.

Author Index

A

Abbasian, M.A. 1043
Abdelhamid, T. H. 411
Abedini, A. 224
Abjadi, N. R. 1917
Abo-Khalil, Ahmed G. 1477
Abramovitz, A. 1412
Agarwal, Anant K. 157
Agarwal, Vivek 281
Ahmed, Nabil A. 242
Ahn, C. H. 1198
Aide, Xu 1162
Ai-Juan, Jin 1421, 1426
Ait-Amirat, Y. 1882
Ajjarapu, Venkataramana 505
Akagi, Hirofumi 23
Akimasa, Koji 1613
Andersen, Henrik Rosendal 1032
Ando, Tatsuo 947
Arpilliere, M. 249
Ashida, M. 2000
Askari, J. 1917

B

Badica, M. 1751
Bai, Haijun 219, 826
Bai, Zhifeng 1581
Baihua, ... 161
Balda, J.C. 1353
Banaei, M. R. 759, 764
Bao, G.Q. 813
Baocheng, Wang 569, 1991
Baoming, Ge 918
Barsoum, ... 1148
Bendjedia, M. 1882
Berthon, A. 515, 1882, 2005
Bhattacharya, Subhashish 1450
Bin, Su 406
bin, Wu 1368
Binder, A. 842
Bing, Chen 1401
Bisogno, F.E. 1117
Blaaberg, F. 46, 1107, 2029
Bo, Chen 122
Böcker, Joachim 1112
Bodson, M. 1912
Bojoi, R. 1651
Boroyevich, D. 249
Boroyevich, Dushan 1836
Bréhaut, Stéphane 92
Brouji, H. El 1663

C

C, Sreekumar 281
Cailin, Wang 1167
Calderon-Lopez, G. 1328
Calverley, S.D. 977
Camara, M.B. 515
Câmpeanu, A. 1751
Cao, Binggang. 1581
Cao, R. X. 510
Cao, Yanjie 2015
Carazo, A. V. 1117
Cartes, David 774
Cen, Yuwan 1986
Chan, C.C. 57
Chang, Chung-Hsing 291
Chang, Duan Qi 1551
Chang, H.-H. 1343
Chang, Jie (Jay) 102
Chang, Lon-Kou 417
Chang, Yuan 1722
Changhong, Wang 1793
Changzheng, Zhang 489, 739
Chau, K. T. 1788
Chen, Bin 1450
Chen, C.-C. 117
Chen, Cheng-Hu 913
Chen, Chern-Lin 332
Chen, Chien-An 967
Chen, Guiyou 1202
Chen, Guocheng 1560
Chen, Guozhu 794
Chen, H. 933
Chen, H. G. 1129
Chen, J. 194
Chen, Jian 1218
Chen, Jiann-Fuh 361, 1178
Chen, Jiaxin 346, 831
Chen, Jie 113
Chen, Jun-Ning 199, 286, 1283, 1392
Chen, Min 442
Chen, Qiaoliang 433, 642
Chen, Rui 1171
Chen, Ruijuan 1253, 1877
Chen, Tso-Min 332
Chen, Wei 171, 1081
Chen, Xi 1507
Chen, Xiangjun 607
Chen, XuWu 438
Chen, Y.-M. 108, 117
Chen, Yao 1454
Chen, Yen-Ming 1763
Chen, Yuan-rui 1397
Chen, Yunpeng 236
Chen, Z. 49, 499, 1773, 2029
Chen, Zongxiang 142
Chenchen 386
Cheng, Chun-An 1178
Cheng, Ming 1746, 1815
Cheng, Ming-Yang 913

A-1

Author Index

Cheng-ning, Zhang 1027
Chengsheng, Wang 589
Cherifi, A. 574
Chi, Song 890, 1825
Chiang, Huann-Keng 967
Chiasson, J. N. 1703, 1912
Chiu, Huang-Jen 291
Cho, Yun-hyun 1238, 1784
Choi, E.S. 1382
Chongjian, Li 589
Chun, Dong 1623
Chun, YonDo 1784
Chung, Jung Kee 1736
Chunjiang, Zhang 554, 559, 1473, 1618
Corzine, Keith A. 637
Costa, François 92
Crausaz, A. 2005
Cui, Bo 798
Cui, Jiefan 657
Cui, Junwei 1436
Cvetkovski, G. 254

D

Dai, Ke 789
Dai, Renchang 1122
Dai, Yue-Hua 199
Da-ming, Liu 1674
Danhe, Li 1991
De Doncker, R. W. 31
Deng, Jianming 923
Deng, Yan 1931
Dianguo, Xu 301, 1713
Ding, Xiaoyu 1560
Ding, Ye 1223
Divan, D. 16
Divan, Deepak M. 2010
Doi, Toshimitsu 356, 1302, 1307
Dong, Jiang 880
Donghua, Luo 1634
Dongsheng, Zuo 468
Dong-Shoutian, ... 484
DongYu, ... 1623
Dou, Sen 537
Du, Guiping 316
Du, Zhong 1450
Duan, Baoxing 70
Duan, Huijuan 798
Duan, Shan Xu 1522
Duan, Shanxu 1218
Duarte, Jorge L. 784

E

Ebrahimi, Yousef 779
Eiuo, Bin 1358
El Din, Ashraf Salah El Din Zein 847
Elbanhawy, Alan 342. 1967

Endo, Tsunehiro 947
Ertugrul, Nesimi 147, 962

F

Fa, Naiguang 1253, 1877
Fang, Liang 817
Fang, Xin 789
Fang, Xu-Peng 166
Fang, Yu 1406
Fang, Zhuo 122, 1542
Fathy, Khairy 356, 1302, 1307, 1358, 1363
Fei, Wanmin 1138
Fei-peng, Xu 647
Feng, D. 499
Feng, Huang 1401
Feng, L. 842
Feng, Zhao 622, 679
Feng, Zheng 585
Fengxiang, Wang 449, 903
Feyzi, M. Reza 204, 1228
Forrest, S. J. 977
Forsyth, A. J. 1323, 1328
Francis, Jerry 1836
Friedrichs, Peter 132
Fröhleke, Norbert 1112
Fuchs, F.-W. 325
Fujita, Kouetsu 1971
Fukuda, S. 1468
Fukushima, Kentaro 1333
Funian, Hua 449
Furuya, Atsushi 1598
Fu-sheng, Wang 463
Futami, M. 1468

G

Gallay, R. 2005
Gang, Ma 378
Gao, F. 1107
Gao, Yan 157
Gao, Yang 113, 1159
Gao, Yong 1198
Gao, Z. Y. 1071
Garinto, Dodi 306
Ge, L.S. 1458
Ge, Lu-sheng 1368, 1576
Ge, Qiongxuan 1171
Geng, Pan 789
Goharrizi, A. Yazdanpanah 1697
Gong, Yu 1223, 1788
Grabner, C. 999
Graczkowski, J. J. 1096
Grantham, Colin 1207, 1858
Gruenberger, Hans Pert 219, 826
Grundmann, Frank 870, 1442
Gu, G. 175
Gu, Herong 473, 1585

Author Index

Gu, Yilei 171, 276
Gualous, H. 515, 2005
Guan, Xiaohan 688
Guang, Zeng 734, 1669
Guangzheng, NI 1091
Guenther, D. 842
Guilan, Chen 554, 1006, 1049
Gui-xin, Shao 1027
Guiyou, Chen 1630, 1634
Guo, Hongche 612, 853
Guo, Qingding 612, 853, 896
Guo, Wei 952
Guo, Xin 337
Guo, Youguang 346, 831
Guobiao, Gu 808
Guocheng, San 1473
Guojun, Lu 802
Guoxin, Zhu 1802
Gustin, F. 515

H

Habetler, Thomas G. 836
Haibing, Hu 937
Haijie, Xu 937
Haiping, Xu 684, 1298
Haitao, Zhang 161
Halász, S. 693
Han, B.D. 1143
Han, Chong 1450
Han, Chong Zhao 652
Han, F. T. 1071
Han, S.K. 1382
Hang-Tian, Li 1421, 1426
Harley, Ronald G. 836, 2010
Hartavi, A.E. 2018
Hashimoto, Takayoshi 1333
He, Guofeng 657
He, Junping 1081
He, Shijie 1463
He, Xiangning 83, 1931
He, Zhongyi 1537
Hemin, Wang 1849
Heming, Li 458
Hendrix, Marcel A. M. 784
Herong, Gu 559, 1618
Hirao, Mitsuhiro 1768
Ho, Chien-Yeh 1527, 1995
Ho, S. L. 1901
Hong, Peng 899
Hong, Shen 559, 1273, 1618
Hong-mei, LI 463
Hongren, Yin 401
Hori, Yoichi 1797
Hosseini, S. H. 759, 764, 1697
Hosseini, Seyyed Hossein 753, 779, 1679
Howe, D. 908. 928, 1841

Hsu, Kai-Sheng 967
Hsu, Ken-Chuan 1957
Hsu, W.P. 718
Hu, D.Q. 1143
Hu, Haibing 127, 1183
Hu, Jiangang 703
Hu, Qing 1806
Hu, Qingbo 526
Hu, Songqin 351
Hu, Weihao 321, 585
Hu, Wenhua 397
Hu, Xuezhi 438
Hu, Y. 2029
Hu, Z. L. 1708
Hu, Zongbo 316
Hua, Li 729, 1571
Hua, Wei 1746, 1758
Huade, Li 401
Huang, Zhenyue 1213
Huang, Alex Q. 113, 157, 1159, 1450
Huang, Chien-Lan 748, 1278
Huang, Congsheng 1288
Huang, Jin 1017
Huang, Xuwen 1288
Huang, Yafeng 542
Huang, Yi 1076
Huang, Yuehui 580, 1512
Hui, Li 729
Hui, Wu 862
Hui, Zhang 1032, 1492
Hui-jie, Xiang 1942

I

Ichinose, M. 1468
Inoue, Kaoru 1233, 1613
Inoue, Shigenori 23
Iov, F. 46
Iwanski, G. 494

J

Jang, Jeong-Ik 1482
Jangwanitlert, A. 1353
Járdán, R.K. 1338
Jeon, K. S. 1198
Jewell, G. W. 977
Ji, Yanchao 627, 1507
Jia, C. 1323
Jia, Y. 2000
Jia, Y.P. 1143
Jiag, Maoh-Chin 1527
Jian, Chen 74, 769
Jian, Cui 1431
Jian, Liu 272
Jian, Wu 1713
Jiang , Chang 1213
Jiang-Hui, Chen 1401

A-3

Author Index

Jiang, J. Z. 1788
Jiang, J.G. 1458, 1952
Jiang, J.J. 152
Jiang, J.Z. 813
Jiang, Jianguo 1081
Jiang, Jianzhong 1223
Jiang, Xianglong 1608
Jiang, Xiaochun 1896
Jianguo, Jiang 468
Jianlin, Zhu 1557
Jianru, Wan 1273
Jian-Ru, Wan 1431
Jian-wen, Zhang 1657
Jianze, Wang 1693
Jiarong, Kan 1532
JiaYi, Yuan 808
Jie, Shuo 1806
Jie, Wang 569
Jiefan, Cui 862, 1849
Jin, Jianxun 831
Jin, Mengjia 885, 1872
Jin, Shun 617
Jin, Tianjun 1183
Jin, Wenxi 127
Jin, Xin Min 1454
Jing, Liu 88
Jin-gang, Li 549
Jing-Gang, Zhang 1669
Jinjun, Liu 1061, 1492, 1722
Jinlong, Zhang 401
Jinupun, P. 1887
Jiqiang, Wang 903
Jiuhe, Wang 401
Jo, WonYoung 1784
Johal, H. 16
Johnson, C. M. 977
Joseph, Alan 1076
Jou, H.L 718
Jun, Liu 1012
Jun, Wang 899
Jung, Kun-seok 1238
Junjuan, Sun Xiaofeng Wu 569
Junmin, Zhang 1684
Junzhu, Wan 1849
Jwo, Ko-Wen 1590

K

Kaijie, Feng 1741
Kaipei, Liu 209, 1684
Kang, B.W. 1143
Kang, Ju-Sung 1358
Kang, Y. 194
Kang, Yong 97, 564, 789, 1218, 1522, 1981
Karimi, E. 1697
Kato, Tomohiko 1971
Kato, Toshiji 1233, 1613

Ke, Dao-Ming 199
Ke, Fu-Jing 1154
Ke, Yi-Jing 1392
Kerkman, Russel J. 1054
Kesong, Ye 699, 1832
Khaehintung, Noppadol 137, 368
Khajee, M. Darkalee 759
Khan, Mahamnad Mansoor 386
Kim, E. D. 1198
Kim, Jang-Hwan 662
Kim, Joo Han 1736
Kim, Young-Sin 1482
Kimura, Noriyuki 1768
Kiranon, Wiwat 137
Kita, H. 1468
Koczara, W. 494
Konghirun, M. 972
Koo, DaeHyun 1784
Kou, X. 1096
Krishnaswami, Sumi 157
Ku, Chung-Ping 1590
Kumar, Pavan 537
Kun, Li 1447
Kunakorn, Anantawat 368
Kuo, J. -S. 1343

L

L., M. 1148
Lai, Ching-Ming 1590
Lai, Stephen L. 1502
Lai, Y. M. 296
Lang, Yongqiang 708
Lee, Chi-Yang 1192
Lee, Dong-Choon 1477, 1482
Lee, Fred C. 1
Lee, Hyun Woo 392, 1302, 1307, 1358, 1363, 1372, 1377
Lee, Se-Hyun 1477
Lemberg, Nicholas 989
Li, Chongjian 995
Li, Dong 1202
Li, Dongsheng 947
Li, F. 152
Li, H. 1773
Li, Han 1006, 1049
Li, Hongtao 688, 1248
Li, M. 1703, 1912
Li, Ma 88, 209
Li, Mingzhu 1537
Li, Min-zu 97
Li, Qi 79
Li, Qunzhan 423
Li, Rongyuan 1112
Li, Shijie 1171
Li, Tianbo 1947
Li, Wen 1797

Author Index

Li, Wenguang 1815
Li, Xia 1674
Li, Y.W. 1101
Li, Yaohua 995
Li, Yong 428
Li, Yongbin 942
Li, Yongdong 1892
Li, Zhanlong 1416
Li, Zhaoji 70, 79
Li, Zheng-Ping 286
Li, Zhou 1630, 1634
Liang, L. 1129
Liang, Tsorng-Juu 361, 1178
Liang, Zhonghua 607
Liao, Changming 102
Li-Jiahui, ... 484
Lijie, Chen 1595
Li-jun, Hang 406
Lijun, Zhao 862, 1849
Lili, Jiang 1849
Liming, Liu 74
Lin, Bor-Ren 748, 967, 1278
Lin, Chang-Hua 1957
Lin, Fei 184, 1976
Lin, Liangrui 1243
Lin, Li-Wei 291
Lin, Ray-Lee 361
Lin, Ruan 808
Lin, Ruiguang 885, 1872
Lin, W.-C. 108
Lin, Yang-Sheng 1178
Lin, Ying-De 1154
Lin, Yu-Tzung 1192
Ling, Xia 899
Lingjie, Meng 674
Lipo, T.A. 989
Liqiang, Yuan 161
Liu, Cheng-Tsung 1763
Liu, Ching-Hsiung 361
Liu, Guiqiu 896
Liu, Hongwei 798
Liu, Hsing-Fu 417
Liu, Jian 1267
Liu, Jianqiang 184
Liu, Jingbo 703
Liu, Jinjun 713, 1726
Liu, K. 1071
Liu, Kaipei 453
Liu, Shu-Lin 1267
Liu, Tien-Shuo 1957
Liu, Wei-Shih 361, 1178
Liu, Wenhua 542
Liu, Wenji 1248
Liu, Xiang 229
Liu, Xiaodong 351
Liu, Xinhua 1223

Liu, Yuanchao 236, 1248
Liu, Zhengang 688
Liu-Xueli, ... 484
Liwei, Zhang 1012
Loh, P. C. 1107
Lorenz, L. 39
Lou, Z. L. 373
Lu, Bin 836
Lu, Bing 1
Lu, Cheng 489, 739
Lu, Haihui 1054
Lu, P.-C. 117
Lu, Shuai 637
Lu, Xiaodong 83
Lu, Zhengyu 127, 171, 276, 526, 1183
Luk, P.C.K 478, 1872, 1887
Luo, Fang 789, 1522
Lyons, James 1122

M

Ma, Hao 1312, 1637
Ma, Hongfei 708
Ma, Wenchuan 627
Ma, Xiangfei 1836
Ma, Xuejun 438, 1288
Maeda, Toshihiro 1971
Manmek, Thip 1207
Mansouri, O. 574
Mao, Hong 1267
Mathew, Anu 442
Matsumoto, Shuji 1598
Matsuse, Kouki 1598
Mayor, J. Rhett 2010
Meghriche, K. 574
Member, Student 1825
Meng, Zheng 236
Mi, Chris 942
Miao, Guan 744
Miao, Zhao 549
Miller, Nicholas 1122
Ming, Cheng 1758
Ming, Zhou 1431
Ming, Zong 449, 903
Ming-fu, Zhao 1623
Mingli, Ding 1793
Min-qian, Ke 734
Miyatake, Masafumi 242
Moghbelli, H. 597
Mohr, M. 325
Molinas, Marta 63
Moon, G.W. 1382
Morimoto, Keiki 356, 1302, 1307
Morizane, Toshimitsu 1768
Mou, Shann-Chyi 291
Mu, Gang 542
Mudannayake, Chathura P. 1207

Author Index

N

N., N. 1148
Na, He 1713
Nagy, I. 1338
Naidu, S. R. 1731
Nakaoka, Mutsuo 356, 392, ...1302, 1307, 1358, 1363, 1372, 1377
Nakayama, Y. 1468
Nan, C. H. 1708
Nan, Liu 214, 1942
Nan, Zhao 918
Nasiri, A. 224
Neff, K. L. 1096
Niasar, A. Halvaei 597
Ning, Gaidi 1463
Ninomiya, Tamotsu 1333
Nishimae, Kazuya 1233
Nittayarumphong, S. 1117
Niu, Shuangxia 1788
Nolle, Eugen 219, 826
Nondahl, Thomas A. 1054
Notohara, Yasuo 947
Nozawa, Yusuke 1598
Nuttall, D. R. 1328

O

Ogiwara, Hiroyuki 1307, 1358
Ohara, S. 1468
Oka, Kazuo 1598
Okude, Takaaki 1363
Oleschuk, V. 1651
Omata, Ryuji 1598
Omori, Hideki 392, 1358, 1363, 1372, 1377
Ou, Chung-Lun 1957
Ouyang, Wen 989

P

Pan, Junmin 142, 267, 1348
Pan, Ming-Ho 1590
Pan, Sanbo 1348
Pang, Da-Chen 1763
Paponpen, K. 972
Park, J. D. 1198
Paweletz, A. 842
Payam, A. Farrokh 1906
Pedersen, John K. 1773
Pei, Yunqing 321
Peng, Fang Z. 1076
Pengcheng, Zhu 74
Petchjatuporn, Panom 137
Petkovska, L. 254
Piwko, Richard 1122
Poon, N. K. 1502
Poure, P. 1663
Prado, R. N. do 1117

Pratt, Annabelle 537
Profumo, F. 1651

Q

Qi, Feng 1637
Qi, Wang 1793
Qian, Lewei 774
Qian, Zhaoming 127, 171, 276, 1076, 1183
Qiang, Li 1853
Qiang, Mei 1926
Qiao, Ermin 337
Qiao, Wei 836
Qiaofu, Chen 489, 739
Qi-gang, Fu 734
Qing, Sun 862
Qingding, Guo 1802, 1846
Qingdong, Zhou 1793
Qingfan, Zhang 1634
Qinglin, Zhao 532, 1387, 1517
Qingyu, Yang 699, 1832
Qinmu, Wu 1820
Qiu, Dongyuan 316, 1293
Qiu, Jianqi 885, 1872
Qiu, Zhiling 794
Qizhi, Zhan 1542

R

Radecker, M. 1117
Rafik, F. 2005
Ragon, S. 249
Rahman, M. F. 983, 1646, 1867, 2023
Rahman, M. Faz 1858
Rahnavard, Reza 779
Rajagopalan, Satish 2010
Ren, Hai Peng 652
Ren, Shi 699, 1832
Rentschler, A. 842
Rhyu, Se Hyun 1736
Rosado, Sebastian 1836
Rosario, L.C. 478
Ruan, L. 175
Ruan, Xinbo 1936
Ruixia, Wang 1853
Ruliang, Zhang 1167
Ruxi, Wang 1061

S

Saadate, S. 1663
Sabahi, Mehran 753, 1228, 1679
Sabzali, A. 411
Saha, Bishwajit 392, 1372
Sahinkaya, M.N. 2018
Sakamoto, Kiyoshi 947
Sanchez-Gasca, Juan 1122
Scozzie, Charles 157
Segawa, Takeshi 1333

Author Index

Shancheng, Xing 1023
Shanxu, Duan 769
Shao, Changhong 607
Shao-De, Zhang 1447
Shaojun, Xie 1532
Shao-Long, Li 1421, 1426
Sharifian, M. B. B. 204
Shen, Guoqiao 1566
Shen, Hong 1603
Shen, J.X. 908, 928
Shen, Miaosen 1076
Shen, W. 249
Sheng, K. 1188
Sheng, Weihui 995
Shergin, V. V. 1133
Shi, Cenwei 885, 1872
Shi, Y. F. 1841
Shiang, J. -Z. 1086
Shibata, R. 2000
Shi-feng, Zhang 1447
Shiri, A. 821
Shoulaie, A. 821
Shu, Mantang 1560
Shu, Zhibing 1811
Shun, Jin ....
Shutong, Qiao 468
Shyu, Kuo-Kai 1590
Sibo, Ge 699, 1832
Sirisuk, Phaophak 137, 368
Skorokhod, Y. Y. 1133
Sneineh, Anees Abu 1318
Soltani, J. 1038, 1043, 1906, 1917
Song, Wenchao 1450
Song, Wenxiang 1560
Songboonkaew, J. 1353
Songhua, Shen 744
Soong, Wen Liang 962
Souza, E. V. N. 1731
Stankovic, A.M. 1651
Stefanovic, V. 249
Su, Hongsheng 423
Sugimoto, Hidehiko 1258
Sugimura, Hisayuki 392, 1372, 1377
Sul, Seung-Ki 662
Sun, Chin 417
Sun, Jia-E 199
Sun, Jian 442
Sun, Sizhou 351
Sun, Wei 866
Sun, Xiaofeng 674
Sun, Yuxin 179
Sunat, Khamron 137
Sung, Ha Kyeong 1736
Suul, Jon Are 63
Suzuki, Takahiro 947

T

Tai, Wei-Chih 913
Takahashi, Toshio 428
Tan, Guang-Hui 627, 1507
Tan, Ruimin 1032
Tan, Siew-Chong 296
Tanaka, Chikara 947
Tang, Yan 866
Tang, Yupeng 1416
Tang, Yu-peng 669
Taniguchi, Katsunori 1768
Tao, Haimin 784
Tao, Liu 122
Tenconi, A. 1651
Teodorescu, R. 46
Tezcan, Ibrahim 1546
Thammasiriroj, W. 1353
Tian, Kai 1318
Toba, Akio 1971
Tolbert, L. M. 1703, 1912
Tongjing, Sun 1630
Tsai, C.-T. 108
Tsai, Ming-Fa 1154
Tsay, Shuh-Chuan 1278
Tse, Chi K. 296, 580, 1512
Tseng, S. -Y. 1086, 1343
Tseng, S.-H. 1086
Tuncay, R.N. 2018
Tzou, Ying-Yu 1192

U

Undeland, Tore 63

V

Vahedi, A. 597, 821
van der Broeck, Heinz 1546
Vansencc, Flalph 875
Varjasi, I. 693
Venkataramanan, Giri 259
Volskiy, S. I. 1133

W

Walther, B. 1882
Wan, Deyu 1585
Wan, Shuyun 1608
Wang, Bin 674
Wang, Bingsen 259
Wang, Changkun 1258
Wang, Chengxue 2015
Wang, Chien-Ming 1527, 1995
Wang, Deyu 473
Wang, F. 249
Wang, Fred 1836
Wang, Gang 1986
Wang, Hua 1507

Author Index

Wang, J. 977
Wang, J.K. 813
Wang, Jianhui 1436
Wang, Jian-quan 229
Wang, Juan 798
Wang, Jui-Kum 1154
Wang, Li-Li 1283
Wang, Linbing 311
Wang, Liqiao 632, 724
Wang, Ming-Yan 1318
Wang, Pei-zhen 1576
Wang, Qingyi 1608
Wang, Qun-jing 857
Wang, Shuo 1
Wang, Xiaofeng 1931
Wang, Xiaoyu 713, 1726
Wang, Y.F. 1458
Wang, Yaonan 866
Wang, Yue 1463
Wang, Yunfei 1896
Wang, Z. A. 1708
Wang, Z. S. 373, 1901
Wang, Zhaoan 321, 433, 642, 713, 1463, 1726
Wang, Zhixin 520, 1487
Watkins, S. J. 1551
Wei, Dong 189
Wei, Guo 880
Wei, Liu 1473
Wei, Shi 1061
Wei, Wen 684, 1298
Wei, Xueliang 789, 1522
Weibin, Cheng 602
Weiguo, Liu 679
Wei-ping, Zhou 1674
Weiyang, Wu 189, 532, 554, 559, 569,
1273, 1387, 1473, 1517, 1618, 1991
Wei-Yang, Wu 1926
Wen , Z. 175
Wen, H.-T. 1343
Wen, Xuhui 337, 622
Wenjuan, Dong 1542
Wenlang, Deng 1557
Wenlong, Qu 378
Wenqing, Shi 684, 1298
Wenxi, Yao 937
Wetzel, Hermann 1112
Wiseman, J. 1101
Wu, B. 1101
Wu, Bin 397
Wu, C.-Y. 117
Wu, Chih-Yu 417
Wu, Hongxia 438, 1288
Wu, J.C 718
Wu, Jiaju 1258
Wu, Jiande 83
Wu, Li 1487

Wu, Li 520
Wu, Q. P. 1071
Wu, Shanshan 1892
Wu, T.-F. 108, 117
Wu, Tao 1936
Wu, Wei-yang 1603
Wu, Weiyang 473, 632, 674, 724, 1585
Wu, Wilson 537
Wu, Yong 1608

X

Xi, Zhai 122
Xia, Kun 857
Xiangjun, Zhang 301
Xiangrong, Li 301
Xiangyun, Fu 1693
Xianmin, Ma 1922
Xianmin, Mu 1693
Xiao, D. 983, 1867, 2023
Xiao, G. C. 1708
Xiao, Lei 209
Xiao, Wenxun 1293
Xiao, Zheng 1447
Xiaobo, Yang 1273
Xiaofeng, Sun 189, 1991
Xiaofeng, Zhang 958, 1626
Xiaohuan, Wang 1273
Xiaojie, Wu 468
Xiao-ping, Yang 1718
Xiaoqiang, Guo 532
Xiaotan, Zhao 589
Xiaoxia, Wei 1693
Xiaoyi, Jin 189
Xiaoyu, Wang 1722
Xie, Jian 870, 1442
Xie, S.S. 1143
Xie, Yong 1406
Ximei, Zhao 1802, 1846
Xindong, Tian 808
Xing, Yan 1406, 1537
Xinming, Huang 1492
Xinxin, Wang 1162
Xu, Cai 1657
Xu, D. 1101
Xu, Dehong 1566
Xu, Dianguo 708
Xu, Jianping 1066
Xu, Jia-peng 669
Xu, Jinbang 1608
Xu, Longya 703, 890, 1779
Xu, Ming 1
Xu, Wancai 627
Xu, Y. N. 194
Xu, Yanping 1863
Xu, Longya 1825
Xuan-fang, Yang 1674

A-8

Author Index

Xue, H. 1952
Xue, Shan 622
Xuhui, Wen 684, 1006, 1012, 1049, 1298
Xun, Li 769

Y

Yabin, LI 458
Yamamura, N. 2000
Yan, Caizhong 1811
Yan, Chen 1623
Yan, Gangui 542
Yan, Wang 1718
Yanchao, Ji 1693
Yanfeng, Wu 729
Yang, Bo 83, 311
Yang, Chun-Sheng 1278
Yang, Geng 1896
Yang, Hui 1863
Yang, J.J. 718
Yang, Jia-qiang 1017
Yang, Junyou 657, 1253, 1877
Yang, R. 152
Yang, S. Y. 510
Yang, Sheng 505
Yang, X.J. 1458
Yang, Xiao-bo 1603
Yang, Xi-jun 229
Yang, Xing-hua 229
Yang, Xu 433, 642
Yang, Zhaoning 1450
Yang, Zhongping 184
Yang, Zilong 473
Yanhui, He 1061
Yanliang, Xu 1741
Yan-min, Su 734, 1669
Yan-ru, Zhong 382, 549, 602, 1718
Yansong, Hou 1571
Yao, Duan 880
Yao, Lei 1463
Yao, Ruoping 989
Yao, Tianjun 127
Yao, Yue-feng 1397
Yaogang, ...386
Yaohua, Li 589
Yao-Xin, ...484
Yazdanpanah, R. 1038
Ye, Min 1581
Ye, Pengsheng 142
Ye, Peng-sheng 229
Yeic, Zhuliang 875
Yesong, Li 1820
Yeung, Heidi H.T. 1502
Yi, Qin 1820
Yi, Wen 1387
Yi, Zhang 862
Yidan, Sun 449, 903

YII, ...1148
Yin, Qiang 1054
Ying, Jiang 1942
Ying, Li 147
Yinhai, Fan 1162
Yong, Gao 88, 1167, 1595
Yong, Kang 74, 769
Yong, Wang 744
Yongchang, Zhang 161
Yonglong, Peng 458
Yoon, H.K. 1382
You, Keping 1646
You, Xiaojie 1976
Youbin, Zhao 739
Yougui, Guo 1557
Youn, M.J. 1382
Yu, Dongmei 1806
Yu, Hongxiang 627
Yu, L.C. 1188
Yu, Y. H. 1129
Yu, Zhang 489
Yuan, Chang 713, 1726
Yuan, Xiaoming 593, 1122, 1566
Yuan, Yang 1595
Yuan, Zhou 647
Yuanbin, Wang 272
Yuanfang, Wen 802
Yuanyuan, Liu 378, 1162
Yuda, Chen 739
Yue, Wang 1061
Yue-feng, Yang 406
Yu-fan, Xi 1669
Yu-gang, Yang 1942
Yugang, Yang 214
Yun-qing, Pei 585
Yun-Xiang, Xie 1401

Z

Zargari, N. 1101
Zeng, Fanpeng 1507
Zeng, Guohong 1689
Zhai, Xiaohua 866
Zhang , B. 152
Zhang, Bo 70, 316, 505, 1293
Zhang, C. L. 1198
Zhang, C. W. 510
Zhang, D. 813
Zhang, Dong 1788
Zhang, Dongyan 236, 1243
Zhang, Fengge 219, 826
Zhang, Hairong 1811
Zhang, Handong 1986
Zhang, Hongyan 794
Zhang, Hui 453
Zhang, Jia 1689
Zhang, Jianzhong 1746

Author Index

Zhang, Jun 1858
Zhang, Kai 564, 1981
Zhang, L. 1551
Zhang, Luan-guo 229
Zhang, Qian 1976
Zhang, Qiang 774
Zhang, Shifu 219, 826
Zhang, Tengchao 179
Zhang, Weiping 236, 688, 1243, 1248
Zhang, X. 510
Zhang, Xi 267
Zhang, Xianmiao 1183
Zhang, Xiaoqiang 1248
Zhang, Xueguang 708
Zhang, Yanli 1138
Zhang, Yingchao 952
Zhang, Yingqi 593
Zhang, Yonggao 564, 1981
Zhang, Yongping 316
Zhang, Yu 1218
Zhang, Yuan 1779
Zhao, Wenxiang 1746, 1815
Zhao, Xiaotan 995
Zhao, Xusen 688, 1243
Zhao, Y. 194
Zhao, Zhengming 952
Zhaoan, Wang 122, 1061, 1542, 1722
Zhaomin, Fanyinhai 1962
Zhao-ming, Qian 406
Zhao-yong, Zhou 647
Zhao-Yulin, ... 484
Zhe, Chen 802, 1387
Zhe, Zhang 559, 1618
Zheng- guo, Wu 1674
Zheng, Shi-cheng 1576
Zheng, Trillion Q. 184, 1012, 1976
Zheng, Zedong 1892
Zhengfeng, Ming 382, 1091
Zhengguo, Wu 729, 1023
Zhengming, Zhao 161, 880
Zheng-Na, ... 484
Zhengyu, Lu 406, 937, 958, 1626
Zhen-lin, Xu 1926
Zhi, Na 1198
Zhili, Tan 769
Zhi-Qiang, Wei 1431
Zhi-yuan, Zhang 1623
Zhong, Yanru 1863
Zhong, Yan-ru 617
Zhongmin, Wang 1630
Zhongnan, Guo 554
Zhongying, Chen 1517
Zhou, Qian-zhi 1368
Zhou, Qianzhi 397
Zhou, Tao 1066
Zhou, Wenqi 1312, 1637

Zhou, Y. M. 1129
Zhou, Yang 923, 1947
Zhou, Yu-Fei 286, 1283, 1392
Zhou, Yunbin 564, 1981
Zhu, Guo-rong 97
Zhu, Huangqiu 179, 817, 923, 1213, 1947
Zhu, Jianguo 346, 831
Zhu, Jingwei 962
Zhu, Xiaoyong 1746, 1815
Zhu, Yuran 798
Zhu, Yu-wu 1238
Zhu, Z.Q. 908, 928, 1841
Zou, X. D. 194

CURRAN ASSOCIATES INC.
proceedings
.com

9781424404483